FOR STUDENTS

- Career Opportunities
- Career Fitness Program
- Becoming an Electronics Technician
- Free On-Line Study Guides (companion web sites)

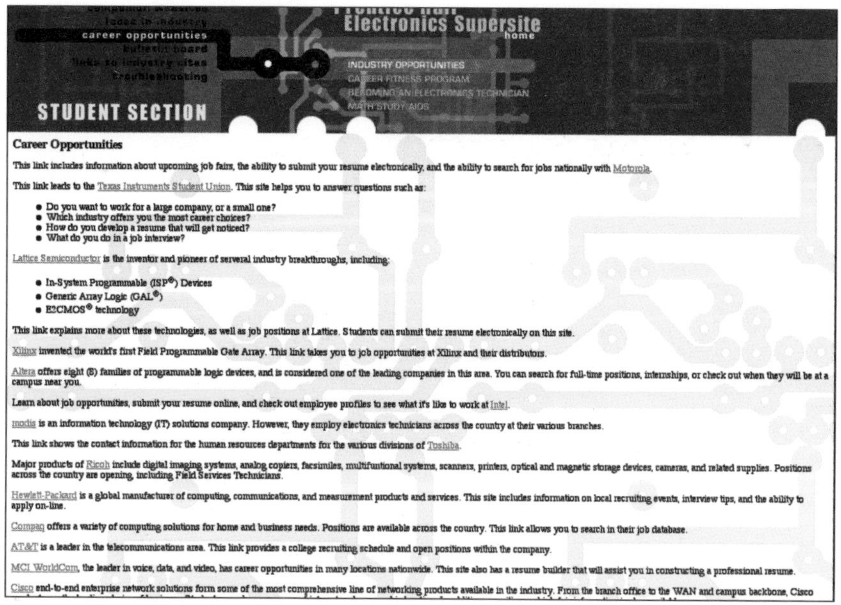

FOR FACULTY*

- Supplements
- On-Line Product Catalog
- Electronics Technology Journal
- Prentice Hall Book Advisor

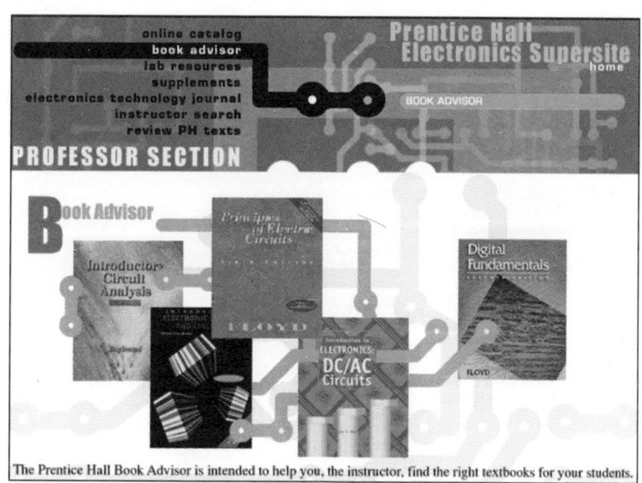

* *Please contact your Prentice Hall representative for passcode*

Electronics Technology Fundamentals

Robert T. Paynter

B. J. Toby Boydell

Upper Saddle River, New Jersey
Columbus, Ohio

Library of Congress Cataloging-in-Publication Data
Paynter, Robert T.
Electronics technology fundamentals / Robert T. Paynter, B.J. Toby Boydell.--1st ed.
p. cm.
Includes index.
ISBN 0-13-032340-3
1. Electronics. I. Boydell, B.J. Toby. II. Title.
TK7816 .P39 2001
621.381—dc21

Editor-in-Chief: Stephen Helba
Product Manager: Scott J. Sambucci
Associate Editor: Kate Linsner
Production Manager: Pat Tonneman
Production Editor: Rex Davidson
Production Coordination: TECHBOOKS
Text Designer: Jill Little
Design Coordinator: Karrie Converse-Jones
Cover Designer: Jason Moore

This book was set in Times Roman by **TECH**BOOKS and was printed and bound by Courier Kendallville, Inc. The cover was printed by Phoenix Color Corp.

Pearson Education Ltd., *London*
Pearson Education Australia Pty. Limited, *Sydney*
Pearson Education Singapore Pte. Ltd.
Pearson Education North Asia Ltd., *Hong Kong*
Pearson Education Canada, Ltd., *Toronto*
Pearson Educación de Mexico, S.A. de C.V.
Pearson Education—Japan, *Tokyo*
Pearson Education Malaysia Pte. Ltd.
Pearson Education, *Upper Saddle River, New Jersey*

10 9 8 7 6 5 4 3 2 1
ISBN 0-13-032340-3

This book is dedicated to our wives, families, and friends for their endless support and patience

Bob Paynter
Toby Boydell

Preface

The fundamentals of electronics (dc circuits, ac circuits, and devices) have traditionally been taught over the course of four semesters, trimesters, or quarters. With the constant development of new applications courses, however, many educators have expressed a need for a single text that presents these fundamentals in a more condensed format, allowing them to be covered in a shorter period of time. *Electronics Technology Fundamentals* was written to fulfill this need. The first 8 chapters cover dc circuit fundamentals, the second 8 chapters cover ac circuit fundamentals, and the final 10 chapters cover discrete devices and circuits, op-amps, and op-amp circuits.

Those who have taught the fundamentals of circuits and devices may wonder how dc circuits, ac circuits, and devices can be adequately introduced in a single volume. If you compare this text to the more "traditional" books, you'll see that we have included all the information your students need to continue with more advanced courses. At the same time, we have eliminated the equation derivations and redundant examples that take up so much space in traditional books. (Both, however, are available in supplements for those who wish to use them.) We also forego many sections on circuit applications and troubleshooting for several reasons: First, we discovered (through contact with a variety of instructors) that troubleshooting sections generally are reserved for assigned reading (to save precious class time). In addition, this book was designed to allow more time for future applications courses, so we felt that including the common applications for every principle was unnecessary. What remains is a text that is complete and concise.

Learning Aids

As always, our primary goal has been to produce an introductory book that students can really *use* in their studies. To this end, the most useful learning aids from Paynter's *Introductory Electric Circuits* and *Introductory Electronic Devices and Circuits* have been incorporated into this text:

- **Performance-based objectives** provide a handy overview of the chapter organization and a map to student learning.
- **Objective identifiers** in the margins cross-reference the objectives with the chapter material. This helps students to locate the material needed to fulfill a given objective.
- **Margin notes** highlight the differences between theory and practice and provide brief reminders of principles covered in earlier sections and chapters.
- **In-chapter practice problems** included in the examples provide students with an immediate opportunity to apply the principles and procedures being demonstrated.
- **Summary illustrations** provide a convenient description of circuit operating principles and applications. Many also provide comparisons between related circuits.

The following have also been incorporated into *Electronics Technology Fundamentals* to help reinforce student learning:

- **Section Review Questions.** Each section of the text ends with a series of questions that students can use to check their learning.
- **Critical Thinking Questions.** Most section reviews include critical thinking questions designed to encourage students to think beyond the scope of the topic discussions.
- **Practice Problems.** An extensive set of practice problems appears at the end of each chapter. In addition to standard practice problems, most of the problem sets include problems relating to previous chapters **(Looking Back)** and challengers **(Pushing the Envelope)**

Multisim Applications Problems

Multisim has been incorporated in a manner that allows instructors to choose (on an individual basis) whether to include it in their curriculum. The CD-ROM included with the text contains exercises that were developed and written by **George Shaiffer** (Pike's Peak Community College, Colorado Springs, CO).

Various figures throughout the text are marked with a Multisim icon. A list of **Multisim Applications Problems** at the end of each chapter provides the file numbers for the appropriate figures. (The directions for accessing the individual files are included with the disk.)

The CD-ROM contains over 120 applications files that are tied-in directly to figures in the text. While these files are provided by Prentice Hall at no cost, Multisim® software is not. The applications files can be accessed using your **Multisim Version 6.21** software. They can be downloaded in **Electronics Workbench® (EWB) Version 5** format from the companion website for this text (www.prenhall.com/paynter).

Supplements

A variety of supplements have been developed to help you and your students. The following are available at no cost:

- **PH Test Manager.** This resource contains a variety of text questions that can be used to develop your tests or as additional review materials for your students.
- **PowerPoint® Presentation Files.**
- **Instructor's Resource Manual.** This manual includes the equation derivations that have been omitted from the text, the results for the lab manual exercises, and a Test Item File.

The following supplements are available for purchase and may be ordered through your Prentice Hall representative:

- **Lab Manual to accompany *Electronics Technology Fundamentals.*** This manual includes more than 40 exercises to supplement the text material. Many labs include Multisim troubleshooting simulation exercises.
- **Practice Problems Manual.** This supplement includes a worked-out example and five (or more) practice problems for most of the numbered equations in the text. The answers to the practice problems are included in the supplement.

Acknowledgments

A project of this size could not have been completed without help from a variety of capable and concerned individuals. A special thanks goes out to the following professionals for their quality reviews of selected chapters:

Leonard Coates, Heald College
David G. Delker, Kansas State University–Salina
Norman Grossman, DeVry Institute of Technology
Philip Regalbuto, Trident Technical College
Michael Rodriguez, DeVry Institute of Technology
Suga Suganthan, DeVry Institute of Technology
Asad Yousuf, Savannah State University

We would also like to thank the staff at Prentice Hall for their "behind-the-scenes" efforts on this text. The following people deserve special recognition:

Kate Linsner, Associate Editor, for overseeing the review process and the initial stages of production.
Rex Davidson, Production Editor, for overseeing the production process and helping to keep things on track.
Scott Sambucci, Product Manager, for having enough faith in the project to go out on many limbs.
Steve Helba, Editor-in-Chief, for believing in the project and providing the support needed to make it a reality.
and

Lara Dugan, Editorial Assistant, for finding the answers to all our inquiries and handling a million details.

We would also like to thank the following individuals for their efforts, which went well beyond the call of duty:

Carolyn Greene, Production Coordinator, **TECH**BOOKS Production Services
Denise Keller, Art Coordinator, **TECH**BOOKS Production Services

and

Wesley Morrison, Copy Editor, Morrison Communications

Their combined efforts transformed a rough manuscript into a quality book in near-record time.

Finally, a special thanks goes out to our families and friends for their constant support and patience.

Bob Paynter
Toby Boydell

To The Student

Why Am I Learning This?

Have you ever found yourself asking this question? If you have, then take a moment to read on.

Any subject is easier to learn if you understand *why* it is being taught and how it relates to your long-term goals. For this reason, we're going to take a moment to discuss a few things:

- How the material in this book relates to a career in electronics technology
- How you can get the most out of this course

Few developments have affected out lives during the past 30 years as profoundly as those in electronics technology. Most of the electronic "gizmos" we take for granted, such as cellular phones, laptop computers, home theaters, pagers, and personal audio systems, have been developed during this time. These items, and many others, have been made possible by advances in production technology. As a result of these advances, many electronic systems that once filled an entire room can now be held in the palm of your hand. Even so, these systems are extremely complex devices that contain a wide variety of components, and each of these components operates according to one or more fundamental principles. These components and their operating principles are the subjects of this book.

Learning how to work on various electronic systems begins with learning the components and principles that are common to all of them. These principles may not always have a direct bearing on *how* to repair a specific electronic system, but they must be learned if you are to understand *why* things work the way they do. Learning why things work allows you to grow beyond the scope of any book (or course).

The material in this book forms a foundation for the courses that are to follow. This means that learning this material is critical if your knowledge is to advance beyond the point where it is now.

How Can I Get the Most from This Course?

There are several steps you can take to ensure that you will successfully complete this course and advance to the next. The first is to accept the fact that *learning electronics requires active participation on your part.* If you are going to learn the material in this book, you must take an active role in your education. It's like learning to play a musical instrument. You need to practice on a regular basis. You can't learn how to play simply by "reading the book." The same can be said about learning electronics. You must be actively involved in the learning process.

How do you get involved in the learning process? Here are some habits that will take you a long way toward successfully completing your course of study:

- *Attend class on a regular basis.*
- *Take part in classroom problem-solving sessions.* This means getting out your calculator and solving problems along with the rest of the class.
- *Do all the assigned homework.* Circuit analysis is a skill. As with any skill, you gain competency only through practice.

- *Take part in classroom discussions.* More often that not, classroom discussions can clarify points that may be confusing otherwise.
- *Read the material before it is discussed in class.* When you know what is going to be discussed in class, read the related material *before* the discussion. That way, you'll know which parts (if any) are causing you problems before the class begins.
- *Become an active reader.*

Being an *active reader* means that you must do more than simply "read the book." When you are studying new material, there are several things that you should do:

1. *Learn the terminology.* You are taught new terms because you need to know what they mean and both how and when to use them. When you come across a new term, take time to commit that term to memory. How do you know when a new term is being introduced? Throughout this text, new terms are identified using **bold** text. When you see a new term, stop and check its meaning before going on to the next section. (If you are unsure of its meaning, look it up in the glossary.)
2. *Use your calculator to work through the examples.* When you come across an example, get out your calculator, and try the example for yourself. When you do this, you develop the skill necessary to solve problems on your own.
3. *Solve the example practice problems.* Most examples in this book end with a practice problem that is identical in nature to the example. When you see these problems, solve them. Then, you can check your answer(s) by looking them up at the end of the chapter.
4. *Use the chapter objectives to measure your learning.* Each chapter begins with an extensive list of performance-based objectives. These objectives tell you what you should be able to do as a result of learning the material.

Throughout this text, *objective identifiers* are included in the margins. For example, if you look on page 47, you'll see "Objective 4" printed in the margin. This identifier tells you that this is the point where you are taught the skill mentioned in Objective 4 at the opening page of the chapter (page 41). These identifiers can be used to help you with your studies. If you don't know how to perform the action called for in a specific objective, just flip through the chapter until you see the appropriate identifier. At that point, you'll find the information you need to successfully meet the objective.

One Final Note

Being an active learner involves a lot of work. However, the extra effort will pay off in the end. Your understanding of electronics will be better as a result of your efforts. We wish you the best of success.

Bob Paynter
Toby Boydell

Contents

INTRODUCTION 1

I.1 Electronics as a Career 2
I.2 Electronic Systems 3
I.3 Electronic Components and Units 6
I.4 Scientific Notation 11
I.5 Engineering Notation 17
Key Terms 20
Practice Problems 21
Answers to the Example Practice Problems 22
Multisim File References 22

Part I BASIC DC PRINCIPLES AND CIRCUITS 23

1 PRINCIPLES OF ELECTRICITY 23

1.1 The Starting Point: Elements, Atoms, and Charge 24
1.2 Current 28
1.3 Voltage 31
1.4 Resistance and Conductance 32
1.5 Conductors, Insulators, and Semiconductors 34
Key Terms 38
Practice Problems 38
Answers to the Example Practice Problems 40
Multisim File References 40

2 COMPONENTS AND CIRCUIT MEASUREMENTS 41

2.1 Conductors and Insulators 42
2.2 Resistors 45
2.3 Resistor Color Codes and Related Topics 49
2.4 Potentiometers 56
2.5 Batteries 58
2.6 DC Power Supplies 61
2.7 Switches and Circuit Protectors 64
2.8 Measuring Current, Voltage, and Resistance 68
Key Terms 71
Practice Problems 72
Answers to the Example Practice Problems 73
Multisim File References 73

3 OHM'S LAW AND POWER 74

3.1 Ohm's Law 75
3.2 Basic Circuit Calculations 77
3.3 Using Ohm's Law to Diagnose a Circuit Failure 80
3.4 Power 83
3.5 Related Topics 87
Key Terms 91
Practice Problems 91

Answers to the Example Practice Problems 94
Multisim File References 95

4 SERIES CIRCUITS 96

4.1 Series Circuit Characteristics 97
4.2 Voltage Relationships: Kirchhoff's Voltage Law, Voltage Sources, and the Voltage Divider 102
4.3 Related Topics 107
Key Terms 114
Practice Problems 114
Answers to the Example Practice Problems 122
Multisim File References 122

5 PARALLEL CIRCUITS 123

5.1 Parallel Circuit Characteristics 124
5.2 Parallel Resistance Relationships 128
5.3 Current Relationships: Kirchhoff's Current Law, Current Sources, and the Current Divider 131
5.4 Related Topics 136
Key Terms 139
Practice Problems 139
Answers to the Example Practice Problems 144
Multisim File References 144

6 SERIES-PARALLEL CIRCUITS 145

6.1 An Introduction to Series-Parallel Circuits 146
6.2 Analyzing Series-Parallel Circuits 152
6.3 Circuit Loading 158
6.4 Bridge, Delta, and Wye Circuits 165
Key Terms 172
Practice Problems 172
Answers to the Example Practice Problems 180
Multisim File References 180

7 CIRCUIT ANALYSIS TECHNIQUES 181

7.1 Superposition 182
7.2 Voltage and Current Sources 186
7.3 Thevenin's Theorem 191
7.4 Applications of Thevenin's Theorem 198
7.5 Norton's Theorem 203
Key Terms 207
Practice Problems 207
Answers to the Example Practice Problems 214
Multisim File References 214

8 MAGNETISM 215

8.1 Magnetism: An Overview 216
8.2 Magnetic Characteristics of Materials 219
8.3 Electromagnetism 223
8.4 Related Topics 229
Key Terms 231

Practice Problems 231
Answers to the Example Practice Problems 233

PART II BASIC AC PRINCIPLES AND CIRCUITS 234

9 ALTERNATING CURRENT AND AC MEASUREMENTS 234

9.1 Alternating Current (AC): Overview and Time Measurements 235
9.2 Magnitude Values and Measurements 242
9.3 Sine Waves: Phase Measurements and Instantaneous Values 248
9.4 Related Topics 261
Key Terms 269
Practice Problems 270
Answers to the Example Practice Problems 279
Multisim File References 279

10 INDUCTORS 280

10.1 Inductance 281
10.2 The Phase Relationship Between Inductor Current and Voltage 287
10.3 Connecting Inductors in Series and Parallel 290
10.4 Inductive Reactance (X_L) 293
10.5 Transformers 297
10.6 Related Topics 304
Key Terms 309
Practice Problems 309
Answers to the Example Practice Problems 316
Multisim File References 316

11 RESISTIVE-INDUCTIVE (*RL*) CIRCUITS 317

11.1 Series *RL* Circuits 318
11.2 Power Characteristics and Calculations 329
11.3 Parallel *RL* Circuits 332
11.4 Series-Parallel Circuit Analysis 339
Key Terms 345
Practice Problems 345
Answers to the Example Practice Problems 350
Multisim File References 351

12 CAPACITORS 352

12.1 Capacitors and Capacitance: An Overview 353
12.2 Alternating Voltage and Current Characteristics 357
12.3 Series and Parallel Capacitors 361
12.4 Capacitive Reactance (X_C) 364
12.5 Related Topics 367
Key Terms 372
Practice Problems 372
Answers to the Example Practice Problems 376
Multisim File References 376

13 *RC* CIRCUITS 377

13.1 Series *RC* Circuits 378
13.2 Power Characteristics and Calculations 388
13.3 Parallel *RC* Circuits 392
13.4 Series-Parallel *RC* Circuit Analysis 399
Key Terms 404
Practice Problems 404
Answers to the Example Practice Problems 408
Multisim File References 408

14 *RLC* CIRCUITS 409

14.1 Series *LC* Circuits 410
14.2 Parallel *LC* Circuits 414
14.3 Resonance 419
14.4 Series and Parallel *RLC* Circuits 424
14.5 Series-Parallel *RLC* Circuit Analysis 435
Key Terms 441
Practice Problems 441
Answers to the Example Practice Problems 446
Multisim File References 446

15 FREQUENCY RESPONSE AND PASSIVE FILTERS 447

15.1 Frequency Response: Curves and Measurements 448
15.2 Amplitude Measurements: dB Power Gain 458
15.3 Amplitude Measurements: dB Voltage and Current Gain 463
15.4 *RC* and *RL* Low-Pass Filters 466
15.5 *RC* and *RL* High-Pass Filters 472
15.6 Bandpass and Notch Filters 476
Key Terms 486
Practice Problems 486
Answers to the Example Practice Problems 492
Multisim File References 492

16 *RL* AND *RC* CIRCUIT PULSE RESPONSE 493

16.1 *RL* Circuit Pulse Response 494
16.2 *RC* Circuit Pulse Response 505
Key Terms 513
Practice Problems 513
Answers to the Example Practice Problems 515
Multisim File References 515

PART III ELECTRONIC DEVICES AND CIRCUITS 516

17 INTRODUCTION TO SOLID-STATE COMPONENTS: DIODES 516

17.1 Semiconductors 517
17.2 Doping 519
17.3 The *PN* Junction 521
17.4 Bias 523

17.5 *PN* Junction Diodes 526
17.6 Diode Ratings 529
17.7 Other Diode Characteristics 531
17.8 Diode Specifications 534
17.9 Zener Diodes 537
17.10 Zener Diode Ratings 539
17.11 Light-Emitting Diodes 542
17.12 Diodes: A Comparison 545
Key Terms 545
Practice Problems 546
Answers to the Example Practice Problems 553
Multisim File References 553

18 BASIC DIODE CIRCUITS 554

18.1 Half-Wave Rectifiers 555
18.2 Full-Wave Rectifiers 560
18.3 Full-Wave Bridge Rectifiers 563
18.4 Filtered Rectifiers 567
18.5 Clippers (Limiters) 573
18.6 Clampers (DC Restorers) 578
18.7 Voltage Multipliers 582
Key Terms 585
Practice Problems 586
Answers to the Example Practice Problems 593
Multisim File References 593

19 BIPOLAR JUNCTION TRANSISTOR OPERATION AND BIASING 594

19.1 Introduction to Bipolar Junction Transistors 595
19.2 Transistor Construction and Operation 597
19.3 Transistor Currents 602
19.4 Introduction to Transistor Biasing 607
19.5 Base Bias 610
19.6 Voltage-Divider Bias 613
19.7 Other Transistor Biasing Circuits 618
19.8 Related Topics 625
Key Terms 631
Practice Problems 631
Answers to the Example Practice Problems 637
Multisim File References 637

20 BJT AMPLIFIERS 638

20.1 Amplifier Properties 639
20.2 BJT Amplifier Configurations 645
20.3 Common-Emitter Amplifiers: Fundamental Concepts 648
20.4 Common-Emitter Amplifiers: Gain and Impedance Characteristics 652
20.5 The Emitter Follower (Common-Collector Amplifier) 656
20.6 The Common-Base Amplifier 661
20.7 Class B Amplifiers 663
20.8 Related Topics 669
Key Terms 671
Practice Problems 671
Answers to the Example Practice Problems 678
Multisim File References 678

21 FIELD-EFFECT TRANSISTORS AND CIRCUITS 679

21.1 Introduction to JFETs 680
21.2 JFET Biasing Circuits 686
21.3 Common-Source Amplifiers 695
21.4 Common-Drain and Common-Gate Amplifiers 701
21.5 Introduction to MOSFETs: D-MOSFET Operation and Biasing 706
21.6 E-MOSFETs 712
21.7 Complementary MOSFETs (CMOS): A MOSFET Application 715
21.8 Related Topics 716
Key Terms 718
Practice Problems 719
Answers to the Example Practice Problems 725
Multisim File References 726

22 OPERATIONAL AMPLIFIERS 727

22.1 Operational Amplifiers: An Overview 728
22.2 Operation Overview 730
22.3 Differential Amplifiers and Op-Amp Specifications 735
22.4 Inverting Amplifiers 741
22.5 Noninverting Amplifiers 744
22.6 Op-Amp Frequency Response 747
22.7 Op-Amp Circuits 750
Key Terms 760
Practice Problems 760
Answers to the Example Practice Problems 770
Multisim File References 770

23 ACTIVE FILTERS AND OSCILLATORS 771

23.1 Tuned Amplifier Characteristics 772
23.2 Active Filters: An Overview 774
23.3 Low-Pass and High-Pass Filters 777
23.4 Bandpass and Notch Filters 781
23.5 Introduction to Oscillators 788
23.6 Phase-Shift Oscillators 793
23.7 The Wien-Bridge Oscillator 794
23.8 Discrete *LC* Oscillators: The Colpitts Oscillator 797
23.9 Other *LC* Oscillators 799
23.10 Crystal-Controlled Oscillators 802
Key Terms 804
Practice Problems 804
Answers to the Example Practice Problems 810
Multisim File References 811

24 SWITCHING CIRCUITS 812

24.1 Introductory Concepts 813
24.2 Basic Switching Circuits: Practical Considerations 819
24.3 Schmitt Triggers 826
24.4 Multivibrators: The 555 Timer 832
Key Terms 839
Practice Problems 840
Answers to the Example Practice Problems 847
Multisim File References 847

25 DISCRETE AND INTEGRATED VOLTAGE REGULATORS 848

25.1 Voltage Regulation: An Overview 849
25.2 Series Voltage Regulators 852
25.3 Shunt Voltage Regulators 856
25.4 Linear IC Voltage Regulators 857
25.5 Switching Regulators 862
Key Terms 868
Practice Problems 868
Answers to the Example Practice Problems 870
Multisim File References 870

26 THYRISTORS AND OTHER DEVICES 871

26.1 Introduction to Thyristors: The Silicon Unilateral Switch 872
26.2 Silicon-Controlled Rectifiers 875
26.3 Diacs and Triacs 882
26.4 Unijunction Transistors 885
26.5 Discrete Photodetectors 888
26.6 Optoisolators and Optointerrupters 893
26.7 Varactor Diodes 894
26.8 Transient Suppressors 896
26.9 Tunnel Diodes 898
Key Terms 900
Practice Problems 901
Answers to the Example Practice Problems 902
Multisim File References 902

Appendix A Conversions and Units 903
Appendix B Component Standard Values and Color Codes 908
Appendix C Specification Sheets 910
Appendix D Selected Equation Derivations 917
Appendix E H-Parameters 934
Appendix F Polar and Rectangular Notations 938
Appendix G Glossary 947
Appendix H Answers to Selected Odd-Numbered Problems 961
Index 967

Introduction

Objectives

After studying the material in this chapter, you should be able to:

1. Compare and contrast the duties and working environments of *electronics engineers, electronics technicians,* and *electronics technologists.*
2. Describe the purposes served by *communications, computer, industrial,* and *biomedical* systems.
3. Identify and describe the most commonly used pieces of electronics test equipment.
4. Describe the function, unit of measure, and schematic symbols for *resistors.*
5. Describe the function, unit of measure, and schematic symbols for *capacitors.*
6. Describe the function, unit of measure, and schematic symbols for *inductors.*
7. Compare and contrast *vacuum tubes* and *semiconductors.*
8. Work with values that are in *scientific notation.*
9. Convert numbers back and forth between standard form and scientific notation, with or without a calculator.
10. Identify commonly used *engineering notation* prefixes, along with their symbols and ranges.
11. Work with values that are written using engineering notation.

The fact that you are reading this book says something about you. It says that you have chosen a career in electronics technology or a related field. Good choice! The Bureau of Labor Statistics (a division of the U.S. Department of Labor) predicts that the need for highly trained professionals will continue to grow during the foreseeable future. As a result of industry growth and personnel changes, thousands of new positions continue to open every year. By enrolling in this course, you have taken the first step toward filling one of those positions.

I.1 Electronics as a Career

Electronics is the field that deals with *the design, manufacturing, installation, maintenance, and repair of electronic circuits and systems.* Television receivers, audio systems, word processors, radar systems, and industrial control systems are just a few examples of electronic systems.

While a variety of trained professionals (such as sales persons, technical writers, equipment assemblers, and customer service representatives) are employed in the electronics industry, the primary technical duties are performed by *electronics engineers, electronics technicians,* and *electronics technologists.*

Electronics Engineers

OBJECTIVE 1 ➤ The *Dictionary of Occupational Titles* (a publication of the U.S. Department of Labor) describes an **electronics engineer** as someone who:

- Designs electronic components, circuits, and systems.
- Develops new applications for existing electronics technologies.
- Assists field technicians in solving unusual problems.
- Supervises the construction and initial testing of system **prototypes** (initial working models).

In most cases, engineers are graduates of four-year college or university degree programs. In some cases, they are former electronics engineering technologists who have demonstrated an in-depth understanding of circuit and system design principles. Electronics engineers are also referred to as *design engineers* (among other titles).

Electronics Technicians

The *Dictionary of Occupational Titles* describes an **electronics technician** as someone who:

- Locates and repairs faults in electronic systems and circuits using electronic test equipment.
- Performs periodic maintenance (such as aligning, calibrating, etc.) on electronic systems.
- Maintains records of system faults and maintenance.
- Installs systems that are purchased by a customer.

In contrast to the engineer, the electronics technician generally deals with the maintenance and repair aspects of electronics. Electronics technicians are also referred to as *service representatives* and *field-service engineers* (among other titles).

Electronics Technologists

An **electronics technologist** performs some of the duties of both the engineer and the technician. Basically, a technologist is someone who:

- Performs the duties of a technician.
- Builds and tests prototype systems as directed by an engineer.
- Maintains records on the performance of prototype systems.
- Recommends circuit and system modifications based on the testing of prototypes.
- Takes part in developing installation, maintenance, and testing procedures for newly designed systems.

As you can see, the technologist is often involved in the design and manufacturing aspects of electronics.

Most companies require a technician to have significant field experience before he or she can be considered for a position as a technologist in a research and development group. However, graduates of accredited programs have been known to go straight from school into this type of working environment. Electronics technologists are also referred to as *electronics engineering technologists* and *electronics engineering technicians* (among other titles).

Section Review

1. What is *electronics*?
2. What are the job functions of an *electronics engineer*?
3. What are the job functions of an *electronics technician*?
4. What are the job functions of an *electronics technologist*?

1.2 Electronic Systems and Test Equipment

In the broadest sense, most common electronic systems can be classified as *communications, computer, industrial,* or *biomedical* systems. In this section, we're going to look at each of these classifications. We'll also take a brief look at some common types of test equipment.

◄ *OBJECTIVE 2*

Communications Systems

Communications systems are *those designed to transmit and/or receive information.* For example, radio systems transmit audio information from the radio station to your AM or FM receiver. Television systems transmit both audio and video information from the television station to your TV set.

A Practical Consideration: The classifications listed here are extremely broad. Each can be broken down into several areas of specialization.

Telecommunications is *the area of communications that deals with the transmission of data between two or more locations.* The Internet is perhaps the best-known example of a telecommunications system. Telecommunications terminals usually communicate with each other via telephone lines, satellite links, or cable (wire or optical).

Computer Systems

Computer systems are *those designed to store and process information.* Personal computers, word processors, printers, and mainframes are a few examples of computer systems.

Many electronic systems can be classified in more than one category. For example, a *facsimile* (fax) machine transmits documents from one point to another over telephone lines. The fax itself is a computer system (it processes and stores the information in the document), but it is also classified as a telecommunications system. Thus, the line between one type of system and another is not always clearly defined.

Industrial Systems

Industrial systems are *those designed for use in industrial (manufacturing) environments.* Control circuits for industrial welders, robotics systems, power distribution systems, and motor control systems (such as automotive electronic systems) are some examples of industrial systems.

Biomedical Systems

Biomedical systems are *those designed for use in diagnosing, monitoring, and treating medical problems.* Electrocardiographs, x-ray machines, diagnostic treadmills, and surgical laser systems are all examples of biomedical systems.

Because biomedical systems are used in the diagnosis and treatment of medical problems, biomedical technicians are generally required to have some practical knowledge of anatomy and physiology.

An Important Point

You may be wondering what you will need to learn to work on a specific type of system. Certain electronic principles do apply more to some types of systems than to others, but you will not encounter any of these principles until later in your education.

In this course, you will be taught the principles of electronics that apply equally to all the electronic systems mentioned in this section. As you study, keep in mind that the material you are learning *does* apply to your area of interest, even though the connection is not always easy to see at first.

Electronic Test Equipment

OBJECTIVE 3 ➤ Many electronic system failures are not visible to the naked eye. For this reason, engineers and technicians must use a wide variety of test equipment to measure circuit values and, thus, analyze and diagnose circuit failures.

Many pieces of electronic test equipment are almost universal. That is, they are used by most engineers and technicians to take the most commonly needed measurements.

A **digital multimeter (DMM)** is *a meter that measures three of the most basic electrical properties: voltage, current, and resistance.* Many DMMs are capable of making additional measurements, depending on the model. A DMM is shown in Figure I.1.

The DMM is a descendant of the **volt-ohm-milliammeter (VOM).** The VOM (Figure I.2) is *a meter that uses an analog scale (rather than a digital scale) to display readings.*

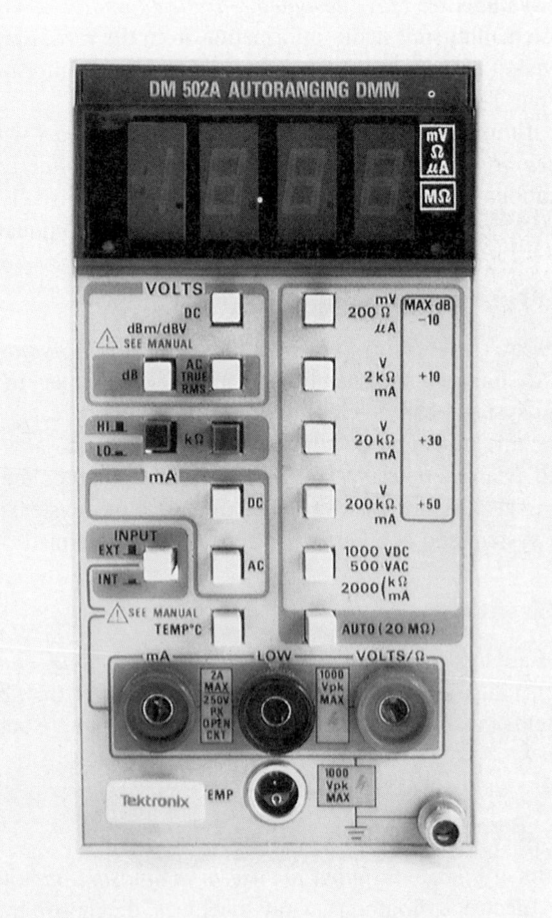

Figure I.1 A digital multimeter (DMM).

Figure I.2 An analog volt-ohm-milliammeter (VOM).

The DMM is preferred over the VOM, because it is easier to read and usually provides more accurate readings. However, some circuit tests are easier to perform using a VOM.

The **frequency counter** (Figure I.3a) is *a piece of test equipment that measures the number of times certain events occur every second.* For example, Figure I.3b shows a **function generator,** which is *a piece of test equipment used to generate a variety of*

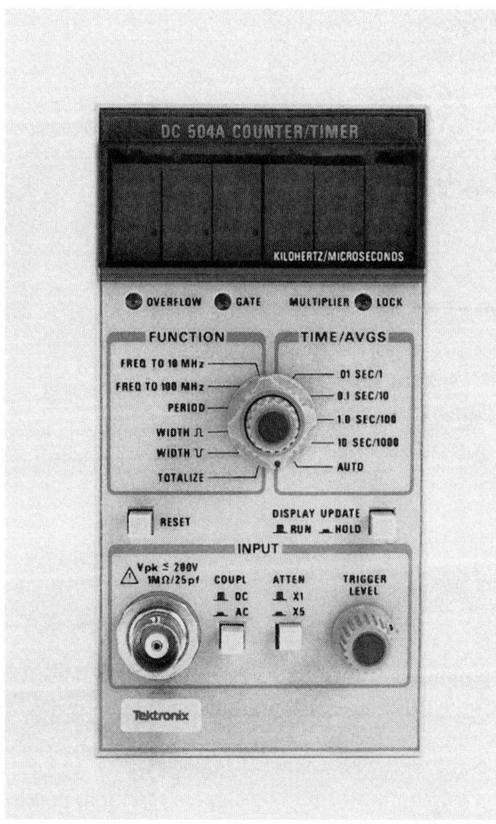

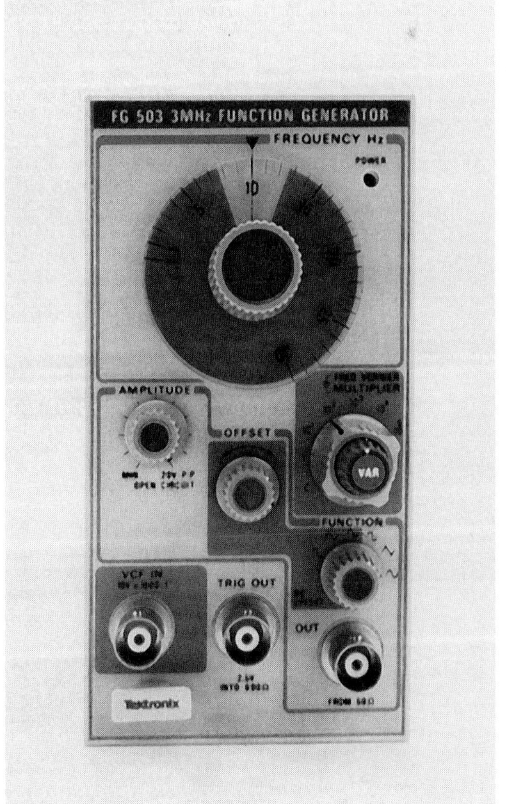

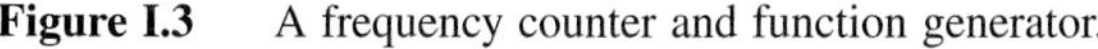

Figure I.3 A frequency counter and function generator.

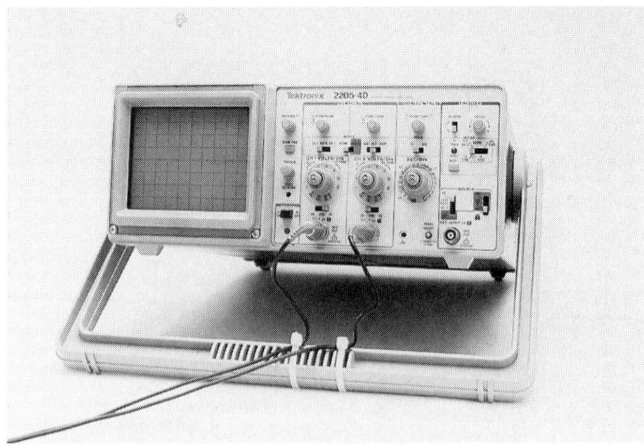

Figure I.4 An oscilloscope.

waveforms. A frequency counter can be used to determine how many times the function generator produces a complete waveform each second.

The **oscilloscope** (Figure I.4) is *a piece of test equipment that provides a visual display for a variety of voltage and time measurements.* The oscilloscope may look a bit intimidating, but it is relatively easy to use with proper training, and is one of the most versatile pieces of test equipment.

An Important Point

When used and maintained properly, a given piece of test equipment will provide accurate readings even after many years of service. However, if a piece of test equipment is not used and maintained properly, it will provide faulty readings and may even be damaged or destroyed. For this reason, you must make sure that you understand *how* to use a given piece of test equipment before attempting to use it. The rule of thumb is simple: *If you aren't sure . . . ask before acting!* Learn the proper technique, then use your test equipment to take the measurements you need. In the long run, you'll be glad you did.

Section Review

1. What are *communications systems*?
2. What are *computer systems*?
3. What are *industrial systems*?
4. What are *biomedical systems*?
5. What is a *DMM*? What is it generally used to measure?
6. Why are DMMs generally preferred over VOMs?
7. What is a *frequency counter*?
8. What is an *oscilloscope*?
9. What precaution should be taken before using any piece of test equipment?

I.3 Electronic Components and Units

All electronic systems contain *circuits.* Loosely defined, a **circuit** is *a group of components that performs a specific function.* The function performed by a given circuit depends on the components used and how they are connected together.

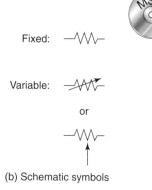

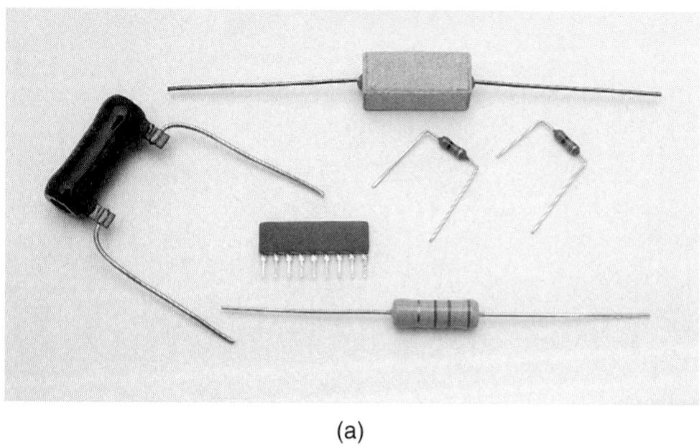

(a)

Fixed:

Variable:

or

(b) Schematic symbols

MSM

Figure I.5 Resistors.

It may surprise you to learn that almost all circuits, no matter how complex, are made up of the same basic types of components. The number of components used and their arrangement vary from one circuit to another, but the *types* of components used are almost universal. In this section, we will take a brief look at the basic types of components and the units that are used to express their values.

Resistors

◄ *OBJECTIVE 4*

We have all heard the term *current* as it applies to a river or stream: It is the flow of water from one point to another. In electronics, the term *current* is used to describe *the flow of charge from one point in a circuit to another.* The **resistor** is *a component that is used to limit the current in a circuit,* just as a valve can be used to limit the flow of water through a pipe.

Resistance is represented using the letter *R*. The unit of measure for resistance is the **ohm,** which is represented using the Greek letter *omega* (Ω). Note that the higher the ohmic value of a resistor, the more it limits the current in a given circuit.

Resistance (R) is measured in *ohms (Ω).*

Several *fixed resistors* are shown in Figure I.5a. The value of each resistor is a function of its physical construction and cannot be altered by the user. In contrast, the **potentiometer** is *a variable resistor whose value can be adjusted (within a fixed range) by the user.*

The symbols shown in Figure I.5b are used in **schematic diagrams,** which are *the electronics equivalent of blueprints.* The schematic diagram of a circuit shows you how the components are interconnected. Note that the schematic symbol for a resistor indicates whether it is fixed or variable in nature.

Capacitors

◄ *OBJECTIVE 5*

A **capacitor** is *a component that opposes a change in voltage.* The applications for such a component are too complex to discuss at this time, but capacitors are used for a variety of purposes in all kinds of circuits. Several types of capacitors and the most common capacitor schematic symbols are shown in Figure I.6.

Capacitance is represented using the letter *C*. The unit of measure for capacitance is the **farad (F).** Note that capacitors are often referred to as **condensers** in older electronics manuals.

Capacitance (C) is measured in *farads (F).*

Inductors

◄ *OBJECTIVE 6*

An **inductor** is *a component that opposes a change in current.* This may seem similar (in purpose) to a capacitor, but inductors and capacitors are nearly opposites in terms of their electrical characteristics. Several types of inductors and the basic inductor schematic symbols are shown in Figure I.7.

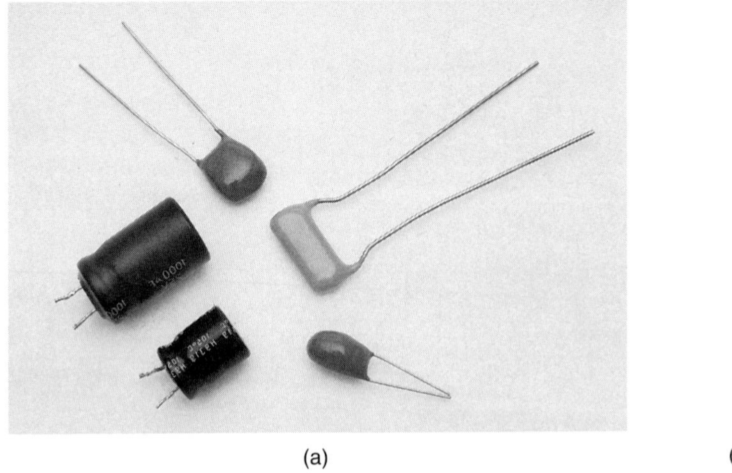

(a)

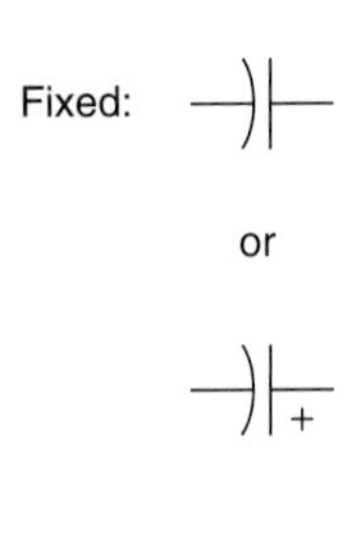

(b) Schematic symbols

Figure I.6 Capacitors.

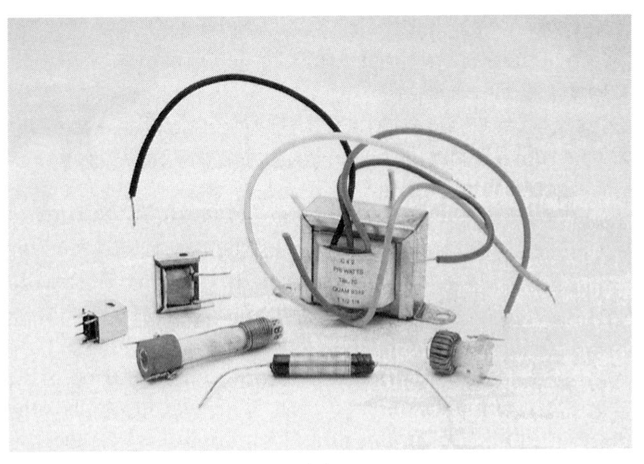

(a)

Fixed:

Variable:

Transformer:

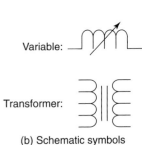

(b) Schematic symbols

Figure I.7 Inductors.

Inductance (L) is measured in *henries (H).*

Inductance is represented using the letter *L.* The unit of measure for inductance is the **henry (H).** It should be noted that inductors are often referred to as **coils** or **chokes.**

OBJECTIVE 7 ➤

Vacuum Tubes

The term **vacuum tube** is used to describe *any component whose operation is based on the flow of charge through a vacuum.* Several vacuum tubes and vacuum tube schematic symbols are shown in Figure I.8.

Unlike resistors, capacitors, and inductors, there is no single unit of measure for vacuum tubes. Note that vacuum tubes like the ones shown in Figure I.8 are used primarily in high-power systems (such as radio and television transmitters). They are also used in many high-fidelity audio applications (such as musical instrument amplifiers).

The **cathode-ray tube (CRT)** is *a vacuum tube that is used as the video display for many televisions, video games, and oscilloscopes.* A CRT is shown in Figure I.9.

Semiconductors

At one time, vacuum tubes were used in almost every type of electronic system. However, a group of components called **semiconductors** started to emerge during the early

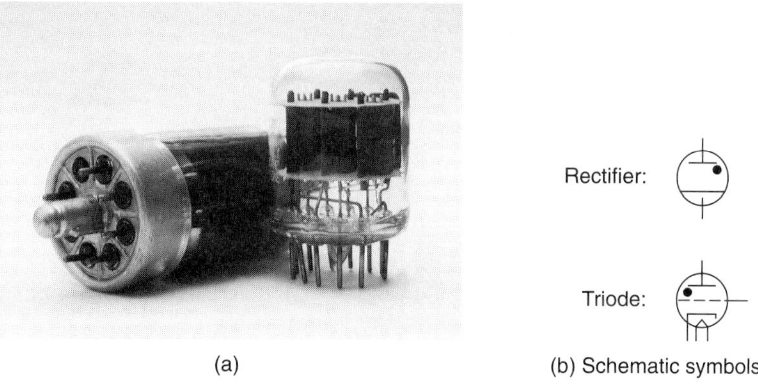

(a) (b) Schematic symbols

Figure I.8 Vacuum tubes.

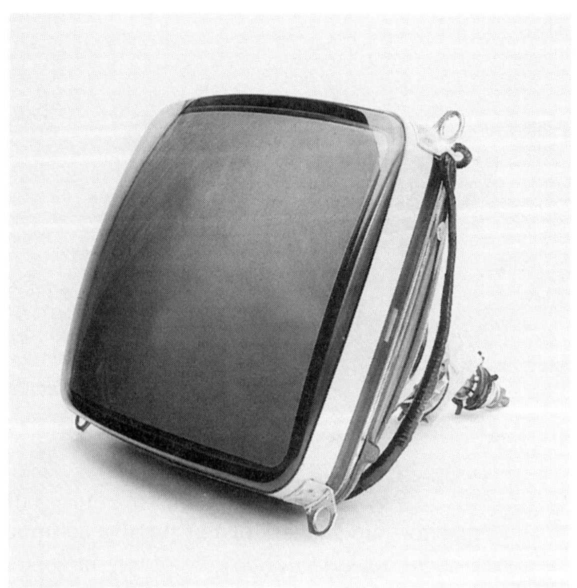

Figure I.9 A cathode-ray tube (CRT).

1950s. These components are capable of performing most of the functions of vacuum tubes but are smaller, cheaper, far more efficient, and more rugged. Some common semiconductors and semiconductor schematic symbols are shown in Figure I.10.

Because of the advantages that semiconductors offer, they have replaced vacuum tubes in the vast majority of applications. However, semiconductors do not (yet) have the power-handling capabilities of vacuum tubes. This is why vacuum tubes are still used in some high-power applications.

Integrated Circuits (ICs)

Integrated circuits (ICs) are *semiconductor components that contain entire groups of components or circuits housed in a single package.* Several types of ICs are shown in Figure I.11.

The development of ICs led to the availability of home computers and many other consumer electronic systems that we now take for granted. Before the 1970s, the cost of producing a computer (as well as the size required) kept computers out of the home market. However, when ICs made it possible to construct entire circuits in a relatively small space, home computers (and many other consumer systems) became a reality.

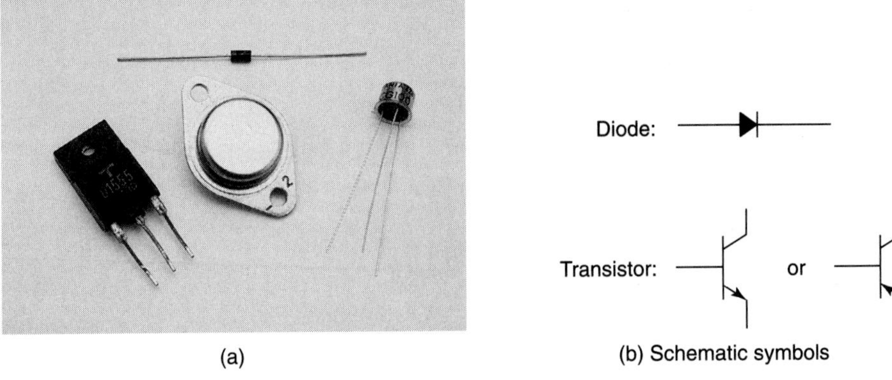

Figure I.10 Semiconductors.

Figure I.11 Integrated circuits (ICs).

Active Versus Passive Components

Resistors, inductors, and capacitors are all classified as **passive components.** This means that *their values are not set using an external power source.* In contrast, vacuum tubes and most semiconductors are classified as **active components.** This means that *their values are set using an external power source.* That is, their operating characteristics can be varied (within limits) using an external power source.

Most electronic circuits contain both active and passive components. In many cases, the passive components are used to help control the operation of the active components in one fashion or another. This point will become clear as your study of electronics advances.

Other Electrical Quantities and Units

You have been shown that *resistance (R)* is measured in *ohms (Ω), capacitance (C)* in *farads (F),* and *inductance (L)* in *henries (H).* Other electrical quantities also have their own units of measure. Some of the more common quantities and their units of measure are listed in Table I.1.

One Final Note

Some of the quantities listed in this section tend to have extremely high values (such as a frequency of 10,000,000 Hz), while others tend to have extremely low values (such as a capacitance of 0.000000001 F). Because many electronic values tend to be extremely high or low in value, they are usually represented using either *scientific notation* or *engineering notation.* Scientific and engineering notations allow us to easily represent very large and small numbers, as you will see in the following sections.

TABLE I.1 Common Quantities and Their Units of Measure

Quantity	*. . . is measured in . . .*	*Unit*
Charge (Q)		coulombs (C)
Conductance (G)		siemens (S)[a]
Current (I)		amperes (A)
Frequency (f)		hertz (Hz)
Impedance (Z)		ohms (Ω)
Power (P)		watts (W)
Reactance (X)		ohms (Ω)
Time (t)		seconds (s)
Voltage (V)		volts (V)

[a] *Conductance is sometimes given in* mhos, *represented by an upside-down omega* (℧).

Section Review

1. What is a resistor? What is a potentiometer?
2. What is a schematic diagram?
3. What is the unit of measure of resistance? What symbol is used to represent this unit of measure?
4. What is a capacitor? What is the unit of measure for capacitance?
5. What is an inductor? What is the unit of measure of inductance?
6. What are vacuum tubes? How do they differ from semiconductors?
7. What distinguishes *passive* and *active* components?
8. Describe the relationship between *quantities* and *units.*

I.4 Scientific Notation

◄ *OBJECTIVE 8*

One of the most basic principles of electronics, called *Ohm's law,* can be stated mathematically as

$$V = IR$$

which reads, "voltage equals current times resistance." Now, consider the following values of current (I) and resistance (R):

$$I = 0.000045 \text{ A} \qquad R = 1{,}500{,}000\ \Omega$$

The odds of making a mistake when entering these values into your calculator are better than average. However, the odds of making a mistake are greatly reduced when we use *scientific notation.*

Scientific notation is *a method of expressing numbers where a value is represented as the product of a number and a whole-number power of ten.* For example, look at the following values and their scientific notation equivalents:

Value	*Scientific Notation Equivalent*
2,300	$\mathbf{2.3} \times 10^{2}$
47,500	$\mathbf{4.75} \times 10^{4}$
0.0683	$\mathbf{6.83} \times 10^{-2}$
0.000000355	$\mathbf{3.55} \times 10^{-7}$

TABLE I.2 Powers of Ten

$10{,}000{,}000 = 1 \times 10^7$	
$1{,}000{,}000 = 1 \times 10^6$	
$100{,}000 = 1 \times 10^5$	Positive powers of ten
$10{,}000 = 1 \times 10^4$	
$1{,}000 = 1 \times 10^3$	
$100 = 1 \times 10^2$	
$10 = 1 \times 10^1$	
$1 = 1 \times 10^0$	
$0.1 = 1 \times 10^{-1}$	
$0.01 = 1 \times 10^{-2}$	
$0.001 = 1 \times 10^{-3}$	
$0.0001 = 1 \times 10^{-4}$	Negative powers of ten
$0.00001 = 1 \times 10^{-5}$	
$0.000001 = 1 \times 10^{-6}$	
$0.0000001 = 1 \times 10^{-7}$	

In each case, the value is represented as a number (shown in **bold**) multiplied by a whole-number power of ten. Note that the number is written as a value between 1 and $9.\overline{999}$. *This is always the case when a given number is written in scientific notation.*

Table I.2 lists the most commonly used positive and negative powers of ten. Several observations can be made with the help of this list. First, note that all values greater than (or equal to) ten have a *positive* power of ten, while all the values less than one have a *negative* power of ten. The power of ten is *zero* for all values between 1 and 10 (zero is not considered to be either positive or negative). Second, note that *the power of ten in each case equals the number of places that the decimal point must be shifted to end up on the immediate right of the one (1).* For example, the first line in the table reads

A Practical Consideration: When a number does not contain a decimal point, its assumed position is at the end of the number (just as a period falls at the end of a sentence).

$$10{,}000{,}000 = 1 \times 10^7$$

Imagine moving the decimal point from its implied position (at the end of the number) so that it falls to the immediate *right* of the one (1). It would have to be moved *seven* places, and the power of ten is seven. A close look at Table I.2 shows that this principle holds true for the all the numbers shown.

At this point, we can establish the rules for converting a number from standard form to scientific notation:

OBJECTIVE 9 ➤

1. Count the number of positions that the decimal point must be shifted to fall to the immediate right of the most significant digit in the number.
2. If the decimal point must be shifted to the *left,* the power of ten is *positive.* If the decimal point must be shifted to the *right,* the power of ten is *negative.* If the decimal point does not need to be shifted (i.e., it already falls to the immediate right of the most significant digit), the power of ten is *zero.*

The following series of examples illustrates the process for converting a given number from standard form to scientific notation.

Example I.1

Convert the number 465,000 to scientific notation.

Solution: To place the decimal point to the immediate right of the most significant digit, it must be shifted *five places to the left* (as shown below).

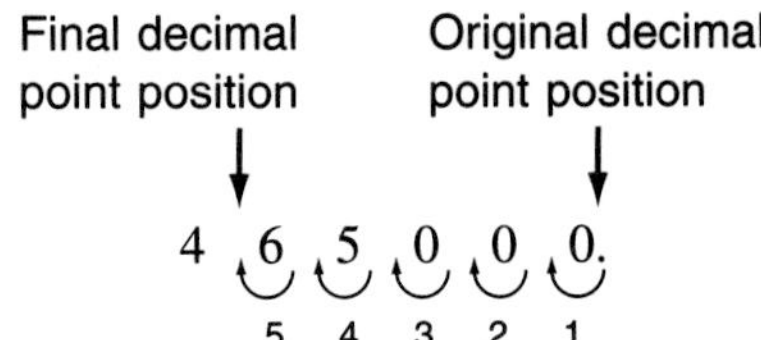

Thus, the power of ten is positive five, and the number is written as

$$465{,}000 = 4.65 \times 10^5$$

Practice Problem I.1

Convert the number 1,238,000 to scientific notation.

Example I.2

Convert the number 0.00794 to scientific notation.

Solution: To place the decimal point to the immediate right of the most significant digit, it must be shifted *three places to the right* (as shown below).

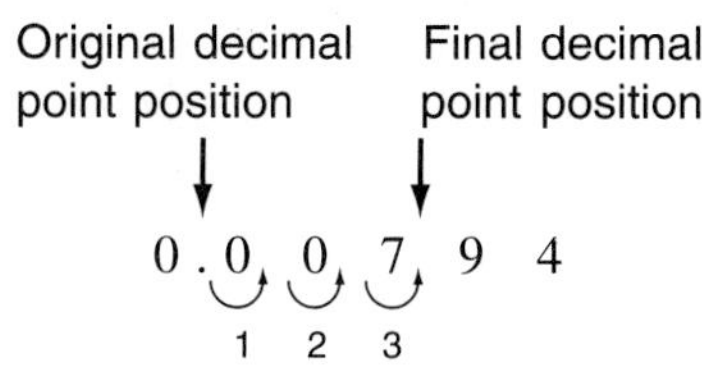

Thus, the power of ten is negative three, and the number is written as

$$7.94 \times 10^{-3}$$

Practice Problem I.2

Convert the number 0.000000000993 to scientific notation.

Example I.3

Convert the number 5.98 to scientific notation.

Solution: Because the decimal point already falls to the immediate right of the most significant digit (the five), it does not need to be shifted, and the power of ten is *zero*. Thus, the number is written as

$$5.98 \times 10^0$$

Practice Problem I.3

Convert the following numbers from standard form to scientific notation:

6,800 0.00358 9.5 10 0.02

Using Your Calculator

A value in scientific notation is entered into your calculator using the *exponent* key, which is most commonly labeled as **EE** or **EXP.** When the exponent key is pressed, it tells the calculator that *the next value entered is a power of ten.* For example, the number 5.28×10^2 would be entered into your calculator as

5.28 **EE** 2

A Practical Consideration: Some calculators use labels other than EE or EXP to denote the exponent key. If you cannot locate the exponent key on your calculator, refer to its instruction booklet.

Once the value is entered into your calculator, you should get a display that is similar to the one shown in Figure I.12. Note that the value (5.28) is shown on the left of the display while the power of ten (02) is shown on the right.

Several important points should be made at this time. First, *the exponent key takes the place of the "$\times$ 10" portion of a value that is written in scientific notation.* In other words, you would not enter 3.4×10^8 using

3.4 **×** 10 **EE** 8 (incorrect)

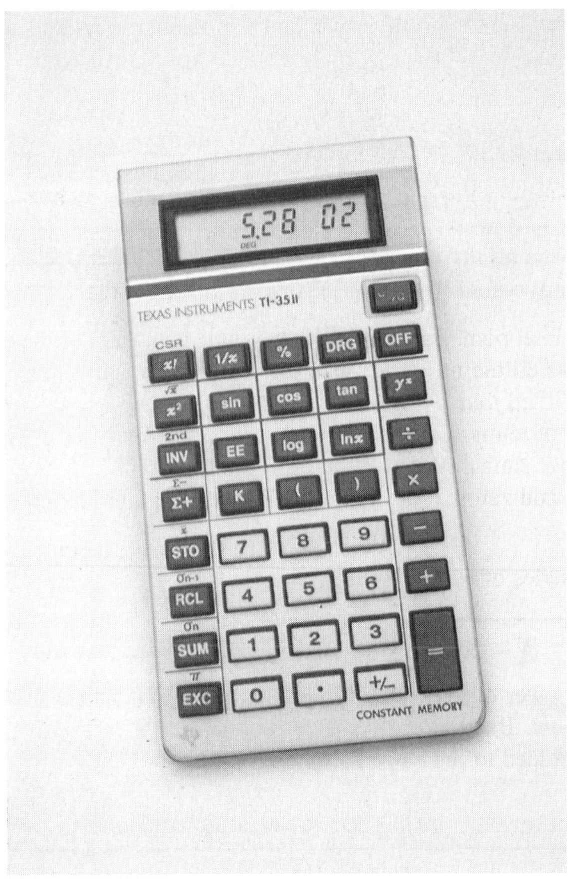

Figure I.12 A calculator display.

because the calculator assumes the "× 10" portion of the value when the EE key is pressed. Thus, the value would simply be entered as

3.4 **EE** 8 (correct)

Note that using the incorrect key sequence causes an incorrect value to be entered into your calculator.

Second, *negative* powers of ten are entered into the calculator using the *sign* (+/−) key. When the power of ten is negative, the value is entered as before. Then, *after pressing the EE key,* the sign key is pressed. For example, the number 5.5×10^{-2} would be entered into the calculator as

5.5 **EE** +/− 2

Note that the calculator will allow you to press the sign key either before or after entering the number 2. However, for the *exponent* to be negative (rather than the value of 5.5), the sign key must be pressed *after* the EE key.

Calculator Modes

Scientific calculators are capable of operating in a *scientific notation mode.* When this mode is used, all values entered in standard form are automatically converted to scientific notation, and all values displayed are given in proper scientific notation. Your calculator instruction booklet describes the procedure for putting your calculator into this operating mode.

Converting from Scientific Notation to Standard Form

Even though we use scientific notation, most of us don't *think* in terms of scientific notation. For example, most people don't think of a year as containing 3.65×10^2 days.

After you have worked with scientific notation for awhile, you will find yourself being able to form a mental image of the actual value of any number written in this form. In the meantime, you'll need to know how to convert a number from scientific notation to standard form if the results of a problem are going to have any meaning to you. A number written in scientific notation is converted to standard form as follows:

1. Shift the decimal point by a number of positions equal to the power of ten. For example, to convert the number 1.48×10^4 to standard form, the decimal point will need to be shifted *four* places.
2. If the power of ten is *positive,* shift the decimal point to the *right.* If the power of ten is *negative,* shift the decimal point to the *left.*
3. If necessary, add zeros to the value to allow the required number of decimal point shifts.

The following series of examples illustrates the use of this conversion technique.

Example I.4

Convert 2.34×10^3 to standard form.

Solution: The power of ten is *positive* three, so the decimal point is shifted *three places to the right.* Because there are only *two* digits to the right of the decimal point, a zero is added to provide the required number of digits. Thus,

$$2.34 \times 10^3 = 2{,}340.0$$

As you can see, the decimal point has been shifted three places to the right.

Practice Problem I.4

Convert 6.22×10^5 to standard form.

Example I.5

Convert 7.92×10^{-5} to standard form.

Solution: Since the power of ten is *negative* five, the decimal point must be shifted *five places to the left.* There is only one digit to the left of the decimal point, so *four zeros must be added to the front of the number* to allow the required number of shifts. Thus,

$$7.92 \times 10^{-5} = .0000792$$

As you can see, the decimal point has been shifted five places to the left.

Practice Problem I.5

Convert the number 6.17×10^{-2} to standard form.

When dealing with standard form values that are less than one, many technicians and engineers tend to add another zero *to the left of the decimal point* to firmly establish its position. Thus, the value of .0000792 from Example I.5 may be written as 0.0000792.

Solving Scientific Notation Problems

When using your calculator to solve problems written in scientific notation, the values are entered using the key sequence given earlier in this section. For example, to solve the problem

$$(2.34 \times 10^4) + (6.2 \times 10^3) = \_\_\_\_\_$$

you would use the following key sequence:

2.34 **EE** 4 + 6.2 **EE** 3 =

As you can see, the numbers and operations are entered in the same order as they are shown in the problem. The following example illustrates this point further.

Example I.6

Determine the key sequence needed to solve the following problem:

$$\frac{3.45 \times 10^{-4}}{2.22 \times 10^{2}} = \_\_\_\_\_$$

Solution: The problem is entered exactly as it is shown. The needed key sequence is

3.45 **EE** +/− 4 ÷ 2.22 **EE** 2 =

When this key sequence is used, your calculator gives you a result of 1.55×10^{-6}.

Practice Problem I.6

Use your calculator to solve the following problem:

$$\frac{5.6 \times 10^{4}}{3.3 \times 10^{-2}} = \_\_\_\_\_$$

Order of Operations

Calculators are designed to solve mathematical operations in a predictable order. This order of operations is as follows:

1. *Single-variable operations,* such as *radicals, square roots,* and *reciprocals.*
2. *Multiplication/division* (in the order entered).
3. *Addition/subtraction* (in the order entered).

It is important to know the order of operations for your calculator, because it affects the way that you enter a given equation. For example, let's say that you want your calculator to solve

$$\frac{10}{2 + 3} = \_\_\_\_\_$$

If you enter this problem into your calculator using

10 ÷ 2 + 3 =

the result given by your calculator will be **8,** which is clearly wrong. The problem in this case is the order of operations: The calculator performs the division operation *before* it performs the addition operation. To correctly solve the equation, *you must indicate that the values in the denominator (2 and 3) are to be combined first.* This is accomplished by using the parentheses keys, as follows:

10 ÷ (2 + 3) =

When entered as shown above, the calculator provides the correct result: **2.**

Avoiding order-of-operations errors.

There is no fail-safe rule that prevents order-of-operations errors. However, you can reduce your chances of making a mistake by using the following guidelines:

1. *When an equation contains any parentheses, enter them into your calculator as shown.* This ensures that any groupings given in the equation are followed by the calculator.
2. *When dealing with fractions, solve any unknowns in the numerator and the denominator before solving the fraction.* For example, in the problem:

$$\frac{24 + 62}{98} = \underline{\qquad}$$

the addition problem in the numerator must be solved before the division. This can be accomplished *by adding your own parentheses,* as follows:

$$(\quad 24 \quad + \quad 62 \quad) \quad \div \quad 98 \quad =$$

When the parentheses are used, the calculator solves the addition problem before the division problem is entered. You could also use another equals sign (rather than the parentheses), as follows:

$$24 \quad + \quad 62 \quad = \quad \div \quad 98 \quad =$$

Again, the addition problem is solved before the division problem is entered into the calculator.

You will find that most order-of-operations errors can be avoided when you follow these simple guidelines.

One Final Note

In the next section, you will be introduced to *engineering notation,* another method that is commonly used to represent extremely large and small values. As you will see, engineering notation is nothing more than a slightly altered form of scientific notation.

Section Review

1. Describe the scientific notation method of representing values.
2. List the steps for converting a number from standard form to scientific notation.
3. Describe the function of the EE button on your calculator.
4. List the steps for converting a number from scientific notation to standard form.

I.5 Engineering Notation

◄ OBJECTIVE 10

You have probably heard people refer to 1000 meters as 1 kilometer, and you have probably heard people use the term *megabucks* to refer to millions of dollars. *Kilo-* and *mega-* are only two of the prefixes that are used in **engineering notation,** which is *a form of scientific notation that uses standard prefixes to designate specific ranges of values.* Most of the prefixes used and the ranges they represent are shown in Table I.3.

When a number is written in engineering notation, the appropriate prefix is used *in place of the power of ten.* For example, here are some values and their engineering notation equivalents:

Value	*Engineering Notation Equivalent*
12.3×10^3 ohms	12.3 kΩ (kilo-ohms)
284×10^6 hertz	284 MHz (megahertz)
4.7×10^{-3} seconds	4.7 ms (milliseconds)
51×10^{-12} farads	51 pF (picofarads)

TABLE 1.3 Engineering Notation Ranges, Prefixes, and Symbols

Range	Prefix	Symbol
(1 to 999) $\times 10^{12}$	tera-	T
(1 to 999) $\times 10^{9}$	giga-	G
(1 to 999) $\times 10^{6}$	mega-	M
(1 to 999) $\times 10^{3}$	kilo-	k
(1 to 999) $\times 10^{0}$	(none)	(none)
(1 to 999) $\times 10^{-3}$	milli-	m
(1 to 999) $\times 10^{-6}$	micro-	μ
(1 to 999) $\times 10^{-9}$	nano-	n
(1 to 999) $\times 10^{-12}$	pico-	p

In the first value, the $\times 10^3$ has been replaced by the prefix *kilo- (k).* In the second, the $\times 10^6$ has been replaced by the prefix *mega- (M),* and so on.

Why engineering notation prefixes are used.

The advantage of using engineering notation is that it makes it much easier to convey large and small values to others. For example, it is much easier to say, "I need a 510 k-ohm resistor," than it is to say, "I need a 5.1×10^5 ohm resistor." Writing values in engineering notation is also much faster than writing them in standard form or scientific notation.

Converting Numbers to Engineering Notation

OBJECTIVE 11 ➤

Converting values from standard form to engineering notation is simple when you remember that *engineering notation prefixes group digits in the same fashion that you do.* For example, consider the number 1,000,000,000.000,000,000. The commas separate the digits into groups of three. The engineering notation prefixes are used to identify each group of three digits, as shown in Figure I.13. As you can see, *kilo-* is just another way of saying *thousands, giga-* is just another way of saying *billions,* and so on.

A Practical Consideration: Most calculators will perform this conversion automatically. However, learning to do it mentally decreases your dependence on your calculator.

To convert a number from standard form to engineering notation:

1. Determine the appropriate prefix by noting *the position of the most significant digit* in the number.
2. Reading left to right, replace the first comma with a decimal point.

The following series of examples illustrates the use of this procedure.

Example I.7

Convert 3,420 Hz to engineering notation.

Solution: Since the most significant digit (the three) is in the *thousands,* the prefix we will use is *kilo- (k).* Replacing the comma with a decimal point, we get

$$3{,}420 \text{ Hz} = 3.42 \text{ kHz}$$

Practice Problem I.7

Convert 357,800 Hz to engineering notation.

Example I.8

Convert 54,340,000,000 W to engineering notation.

Solution: Since the most significant digit (the five) is in the *billions,* the prefix we will use is *giga- (G).* Replacing the first comma with a decimal point, we get

$$54{,}340{,}000{,}000 \text{ W} = 54.34 \text{ GW}$$

Note that it is unnecessary to keep the remaining zeros in the original number, because the prefix *giga-* indicates that the value is in billions.

Practice Problem I.8

Convert 82,982,000 W to engineering notation.

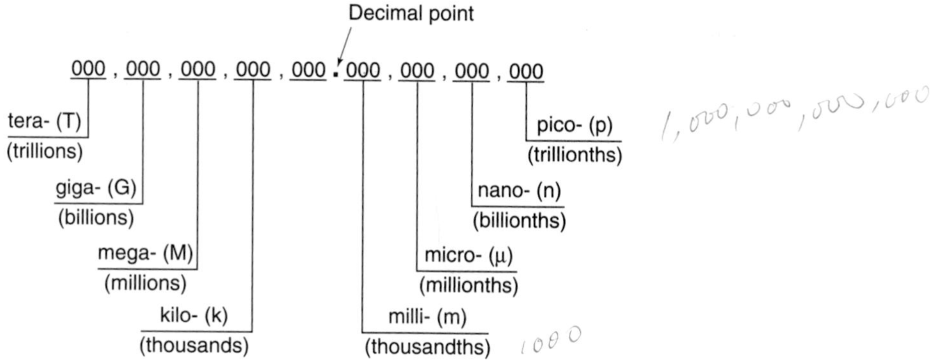

Figure I.13 Engineering notation groups.

Example I.9

Convert 0.00346 A to engineering notation.

Solution: Since the most significant digit (the three) is in the *thousandths,* the prefix we will use is *milli- (m).* If we were to use a comma to form groups of three digits (starting at the decimal point), the number would be written as 0.003,46. Replacing this comma with a decimal point, we get

$$0.00346 \text{ A} = 3.46 \text{ mA}$$

Practice Problem I.9

Convert 0.0000492 A to engineering notation.

Converting from scientific notation to engineering notation is easy if you start by solving a simple problem:

1. Determine the highest *whole-number multiple of three that is less than (or equal to) the power of ten in the number to be converted.*
2. Determine the difference between the value found in step one and the power of ten in the number.

For example, let's say that you want to convert 6.8×10^7 to engineering notation:

1. *Six* is the highest whole-number multiple of three that is less than seven.
2. The difference between six and seven is *one.*

The *six* indicates *the power of ten for the prefix,* and the *one* indicates *the number of places that the decimal point must be shifted to the right.* Therefore,

$$6.8 \times 10^7 = 68 \times 10^6 = 68 \text{ M}$$

Note that *mega-* corresponds to 10^6, and that the decimal point in the number has been shifted *one* place to the right. The following series of examples further illustrates the use of this simple conversion technique.

Example I.10

Convert 3.82×10^{11} Hz to engineering notation.

Solution: The highest whole-number multiple of 3 that is less than 11 is *nine,* and the difference between 9 and 11 is *two.* Since 10^9 corresponds to the prefix *giga-,* this is the prefix we use. Shifting the decimal point *two* places to the right, we get

$$3.82 \times 10^{11} \text{ Hz} = 382 \times 10^9 \text{ Hz} = 382 \text{ GHz}$$

Practice Problem I.10

Convert 4.55×10^4 Hz to engineering notation.

Example I.11

Convert 3.22×10^{-8} A to engineering notation.

Solution: The highest whole-number multiple of 3 that is less than -8 is -9, which corresponds to the engineering prefix *nano- (n)*. The difference between -8 and -9 is 1. Shifting the decimal point one place to the right, we get

$$3.22 \times 10^{-8}\text{ A} = 32.2 \times 10^{-9}\text{ A} = 32.2\text{ nA}$$

Practice Problem I.11

Convert 7.62×10^{-1} A to engineering notation.

As you can see, the conversion from scientific notation to engineering notation is relatively simple. In fact, it won't be long before you'll be doing it in your head (if you aren't already).

Other Engineering Notation Prefixes

In this section, you have been introduced to the most commonly used engineering notation prefixes and their value ranges. Some less frequently used prefixes and their value ranges are as follows:

exa- (E) is used for (1 to 999) $\times 10^{18}$
peta- (P) is used for (1 to 999) $\times 10^{15}$
fempto- (f) is used for (1 to 999) $\times 10^{-15}$
atto- (a) is used for (1 to 999) $\times 10^{-18}$

One Final Note

The prefixes listed above (and in Table I.3) all show a value range of (1 to 999). There are cases, however, when values outside of this range are used *to keep the notation constant.* A classic example of using constant notation can be found on the *AM dial* of any radio, which uses the following sequence:

530 kHz 600 kHz 700 kHz 800 kHz 1000 kHz 1200 kHz 1400 kHz 1700 kHz

The last four values shown could be written as 1 MHz, 1.2 MHz, 1.4 MHz, and 1.7 MHz. However, they are written as shown so that the kHz notation remains constant.

Section Review

1. List the commonly used engineering prefixes, along with their symbols and ranges.
2. What purpose is served by using engineering notation?
3. List the steps for converting a number from standard form to engineering notation.
4. How do you convert a number from scientific notation to engineering notation?

Key Terms

The following terms were introduced in this chapter on the pages indicated:

Active component, 10
Biomedical systems, 3
Capacitor, 7
Cathode-ray tube (CRT), 8
Choke, 8
Circuit, 6
Coil, 8
Communications systems, 3
Computer systems, 3
Condenser, 7
Digital multimeter (DMM), 4
Electronics, 2
Electronics engineer, 2
Electronics technician, 2
Electronics technologist, 2
Engineering notation, 17
Farad (F), 7

Frequency counter, 5
Function generator, 5
Henry (H), 8
Inductor, 7
Industrial systems, 3
Integrated circuit (IC), 9
Ohm, 7
Oscilloscope, 6
Passive component, 10
Potentiometer, 7
Prototype, 2
Resistor, 7
Schematic diagram, 7
Scientific notation, 11
Semiconductor, 8
Telecommunications, 3
Vacuum tube, 8
Volt-ohm-milliammeter (VOM), 4

Practice Problems

1. Convert the following numbers to scientific notation without the use of your calculator.
   a. 3,492
   b. 922
   c. 23,800,000
   d. −476,000
   e. 22,900
   f. 1,220,000
   g. −82,000
   h. 1
   i. 3,970,000,000
   j. −3,800
2. Convert the following numbers to scientific notation without the use of your calculator.
   a. 0.143
   b. 0.000427
   c. −0.0000401
   d. 0.000000023
   e. −0.0381
   f. 0.0039
   g. 0.07018
   h. −0.629
   i. 0.0000097
   j. 0.0000000000702
3. Write the key sequence you would use to enter each of the following values into your calculator.
   a. 1.83×10^{23}
   b. -5.6×10^{4}
   c. 4.2×10^{-9}
   d. -9.22×10^{-7}
   e. 3.3×10^{-12}
   f. 10^{6}
4. Convert each of the following values to standard form without the use of your calculator.
   a. 3.2×10^{4}
   b. 6.6×10^{8}
   c. 9.24×10^{0}
   d. 5.7×10^{12}
   e. -1.1×10^{5}
   f. -4.98×10^{2}
   g. 8.77×10^{6}
   h. -1.26×10^{1}
   i. 2.84×10^{3}
   j. 4.45×10^{7}
5. Convert each of the following values to standard form without the use of your calculator.
   a. 4.43×10^{-2}
   b. -2.8×10^{-5}
   c. 7.62×10^{-4}
   d. 3.38×10^{-10}
   e. -5.8×10^{-8}
   f. 6.34×10^{-1}
   g. 1.11×10^{-9}
   h. 9.02×10^{-3}
   i. -8.3×10^{-6}
   j. 4.44×10^{-7}
6. Write the key sequence you would use to solve each of the following problems with your calculator.
   a. $(2.43 \times 10^{4})(-3.9 \times 10^{-4}) =$ ______
   b. $\dfrac{-4 \times 10^{6}}{2.8 \times 10^{-2}} + 30 =$ ______
   c. $(4.5 \times 10^{-5}) + (-1.1 \times 10^{-4}) =$ ______
   d. $(8 \times 10^{-2})(5.1 \times 10^{3}) - (9 \times 10^{2}) =$ ______
7. Use your calculator to solve each of the following problems.
   a. $(1 \times 10^{4})(2.2 \times 10^{-2}) + 96 =$ ______
   b. $\dfrac{2.4 \times 10^{6}}{3.3 \times 10^{5}} - (3 \times 10^{2}) =$ ______

c. $(9 \times 10^0) + (9 \times 10^{-1}) - (-4 \times 10^{-2}) =$ ______

d. $\dfrac{(-6.2 \times 10^{-4})(-3 \times 10^3)}{6.8 \times 10^{-3}} =$ ______

8. Use your calculator to solve each of the following problems. Write your answers in standard form.
   a. $(8.3 \times 10^{-3})(3.3 \times 10^6) - (1.1 \times 10^2) =$ ______
   b. $(4.22 \times 10^0) - (1.2 \times 10^1)(3.3 \times 10^{-2}) =$ ______
   c. $(-5 \times 10^3) + \dfrac{4 \times 10^2}{6.28 \times 10^{-1}} =$ ______
   d. $\dfrac{4.82 \times 10^4}{(2.28 \times 10^{-2})(1.8 \times 10^{-1})} =$ ______

9. Write each of the following values in proper engineering notation.
   a. 38,400 m
   b. 234,000 W
   c. 44,320,000 Hz
   d. 175,000 V
   e. 60 W
   f. 1,800 Ω
   g. 3.2 A
   h. 2,500,000 Ω
   i. 4,870,000,000 Hz
   j. 22,000 Ω

10. Write each of the following values in proper engineering notation.
   a. 0.022 H
   b. 0.00047 F
   c. 0.00566 m
   d. 0.000055 A
   e. 0.935 s
   f. 0.288 A
   g. 0.00000000244 s
   h. 0.000033 H
   i. 0.000000000051 F
   j. 0.00355 C

11. Convert each of the following values to engineering notation form.
   a. 2.2×10^7 Ω
   b. 3.3×10^3 W
   c. 7×10^{11} Hz
   d. 5.1×10^5 Ω
   e. 8.5×10^4 W
   f. 4.7×10^{-5} F
   g. 1×10^{-1} H
   h. 6.6×10^{-4} s
   i. 4×10^{-8} S
   j. 2.2×10^{-10} F

Answers to the Example Practice Problems

I.1. 1.238×10^6

I.2. 9.93×10^{-10}

I.3. 6.8×10^3
3.58×10^{-3}
9.5×10^0
1×10^1
2×10^{-2}

I.4. 622,000

I.5. 0.0617

I.6. 1.7×10^6

I.7. 357.8 kHz

I.8. 82.982 MW

I.9. 49.2 μA

I.10. 45.5 kHz

I.11. 762 mA

Multisim File References

Figure	File Reference
I.1	ETFi_1.msm
I.3	ETFi_3.msm
I.5	ETFi_5.msm
I.7	ETFi_7.msm

Part I
Basic DC Principles and Circuits

chapter 1

Principles of Electricity

Objectives

After studying the material in this chapter, you should be able to:

1. Describe the relationship among matter, elements, and atoms.
2. Describe the structure of the atom.
3. Explain the concept of *charge* as it relates to the atom.
4. State the relationship between *current, charge,* and *time*.
5. Discuss the *electron flow* and *conventional current* approaches to describing current.
6. Contrast *direct current (dc)* and *alternating current (ac)*.
7. Define *voltage,* and describe its relationship to energy and charge.
8. Define *resistance* and *conductance*, and describe the relationship between the two.
9. Compare and contrast *conductors, insulators,* and *semiconductors*.
10. Discuss the relationship between resistance and each of the following: *resistivity, material length, material cross-sectional area,* and *temperature.*

When you mention the term *electricity,* most people form a mental image of lightning, a power transmission station, or some other evidence of the *presence* of electricity. However, few people actually understand what electricity *is.*

In this chapter, we're going to discuss electricity and its properties. As you will see, the study of electricity starts with a discussion of a basic building block of matter, the atom.

1.1 The Starting Point: Elements, Atoms, and Charge

A battery has two terminals: a negative terminal and a positive terminal. While the terms *negative* and *positive* are commonly associated with electricity, most people don't know they actually describe charges that are found in the atom. In this section, we're going to take a look at matter, the structure of the atom, and the nature of charge.

OBJECTIVE 1 ➤

Matter

Matter can be defined as *anything that has weight and occupies space.* All matter is made of various combinations of elements. An **element** is *a substance that cannot be broken down into a combination of simpler substances.* For example, copper (an element) cannot be broken down into a combination of other elements. While elements each have their own properties, they are all made up of *atoms.*

OBJECTIVE 2 ➤

Atoms

The **atom** is *the smallest particle of matter that retains the physical characteristics of an element.* The simplest model of the atom is called the **Bohr model**. The Bohr model of the helium atom is shown in Figure 1.1. It contains a central core (called the **nucleus**) that is orbited by two particles (called **electrons**). The electrons revolve around the nucleus in the same way that planets orbit the sun. The nucleus itself contains two other types of particles: **protons** and **neutrons**.

The number of electrons, protons, and neutrons contained in the atom varies from one element to another. This point is illustrated in Figure 1.2. As you can see, the hydrogen atom has one proton and one electron, the helium atom has two of each, and the lithium atom has three of each. Note that *the number of protons in each atom* equals the **atomic number** of that element. Thus, the elements represented in Figure 1.2 have the atomic numbers shown. (Elements are commonly identified by their atomic numbers.)

Figure 1.2 also illustrates a very important point: *In its natural state, a given atom contains an equal number of protons and electrons.* The significance of this relationship will be discussed in the next section.

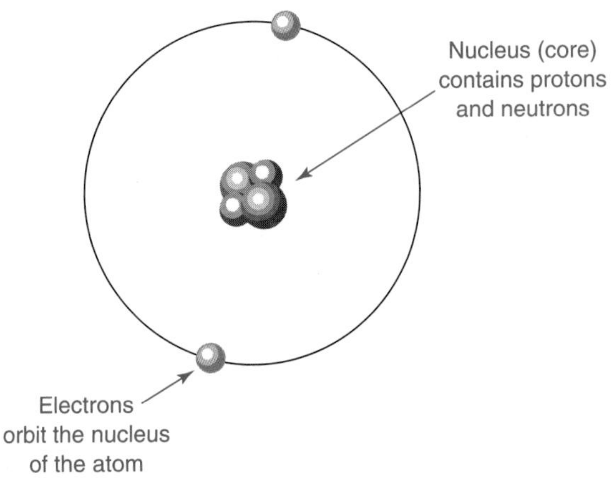

Figure 1.1 The helium atom.

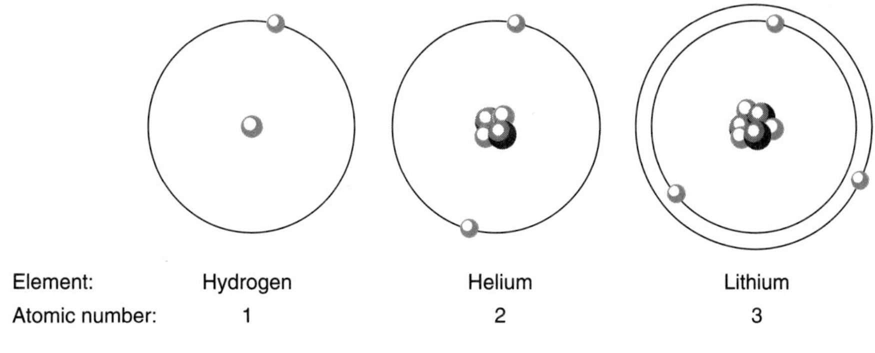
Element:
Hydrogen
Helium
Lithium
Atomic number:
1
2
3
Note: Hydrogen is the only atom that has no neutrons

Figure 1.2 The hydrogen, helium, and lithium atoms.

Electrons travel around the nucleus of the atom in *orbital paths* that are referred to as **shells**. For example, the copper atom in Figure 1.3 contains 29 electrons that travel around the nucleus in four different orbital shells. The maximum number of electrons that each shell can contain is found as $2n^2$, where n is the number of the shell. For example, the third orbital shell (or M shell) can hold up to $2(3^2) = 18$ electrons.

The *outermost shell* of the atom is called the **valence shell**. Several points should be made regarding this shell:

- The electrons that occupy the valence shell are referred to as **valence electrons**.
- The valence shell of a given atom cannot hold more than eight electrons. When the valence shell of an atom contains eight electrons, that shell is described as being *complete.*

Later in this section, you will see how the valence shell of an atom determines the electrical characteristics of the element.

Charge

◀ *OBJECTIVE 3*

Charge is *a force that causes two particles to be attracted to, or repelled from, each other.* There are two types of charge: positive charge and negative charge. Note that the terms *positive* and *negative* are used to signify that these charges are *opposites.*

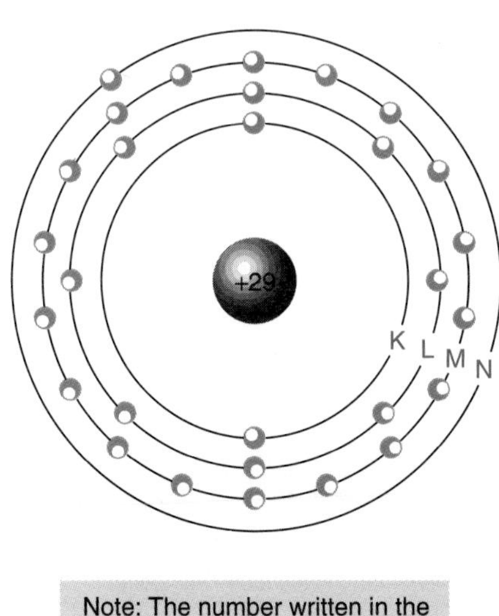
+29
K
L
M
N
Note: The number written in the nucleus represents the number of protons in the atom.

Figure 1.3 The copper atom.

Figure 1.4
Atomic charges.

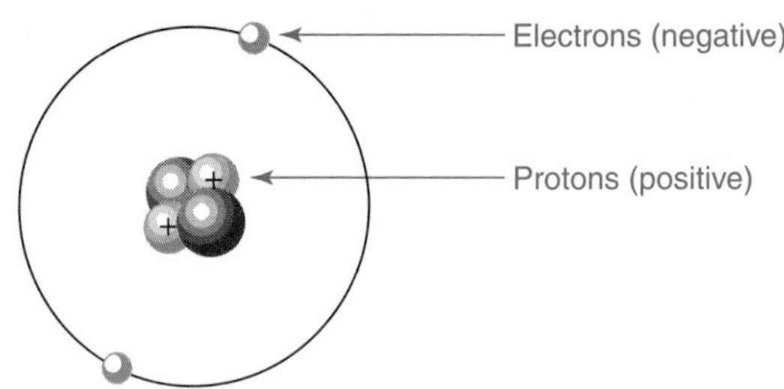

Because they are opposites, positive and negative charges offset each other. For example, an atom that contains two positive charges and two negative charges has a net charge of *zero*.

Protons → positive.
Electrons → negative.
Neutrons → neutral.

Protons and electrons are the sources of charge in the atom. Each proton has a *positive* charge, and each electron has a *negative* charge, as illustrated in Figure 1.4. The atom shown has a balance of positive and negative charges, and thus, a net charge of *zero*. Note that neutrons are electrically *neutral* (which is how they got their name).

Attraction and Repulsion

The fundamental law of charges is that *like charges repel each other and opposite charges attract each other.* This law is illustrated in Figure 1.5. As you can see, two positive charges repel each other, as do two negative charges. However, a negative charge is attracted to a positive charge, and vice versa.

Figure 1.5
Attraction and repulsion of charges.

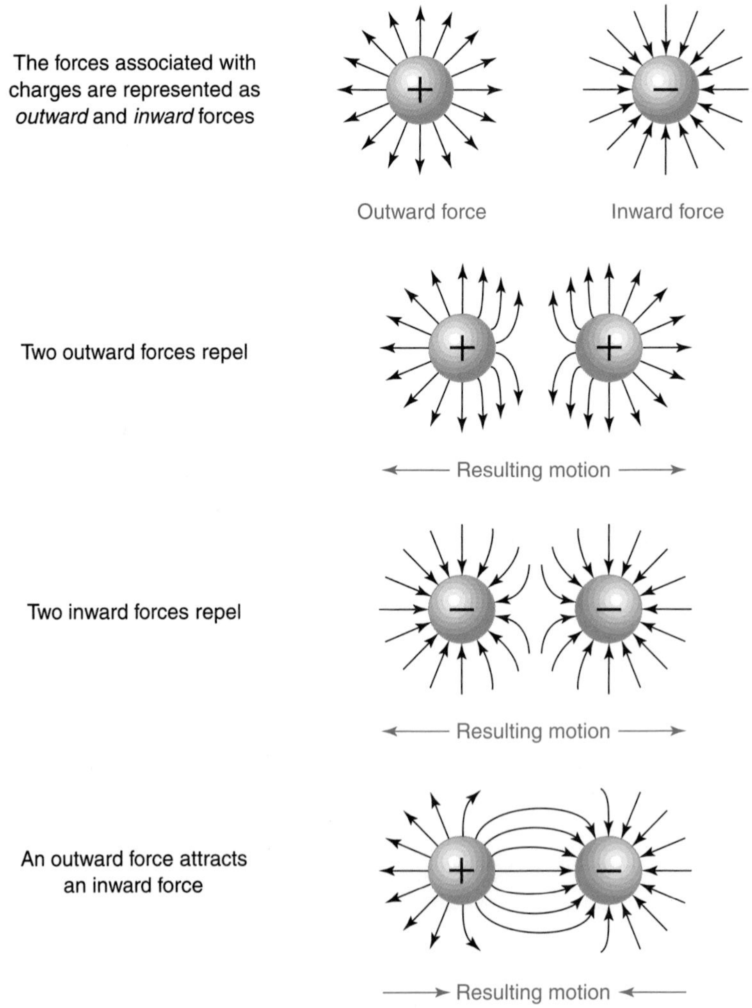

The forces associated with charges are represented as *outward* and *inward* forces

Outward force

Inward force

Two outward forces repel

Resulting motion

Two inward forces repel

Resulting motion

An outward force attracts an inward force

Resulting motion

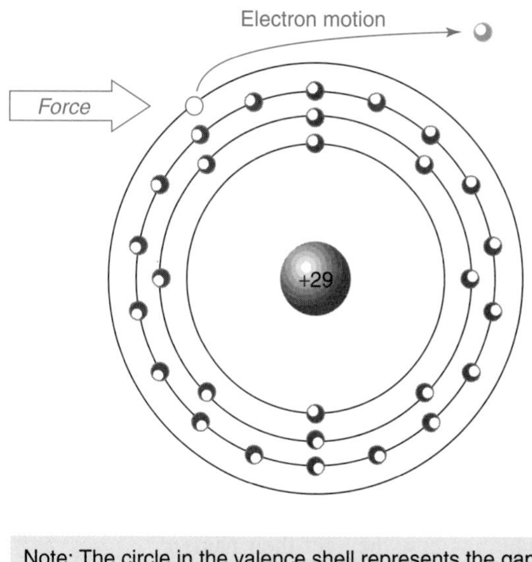

Figure 1.6 Generating a positive ion and a free electron.

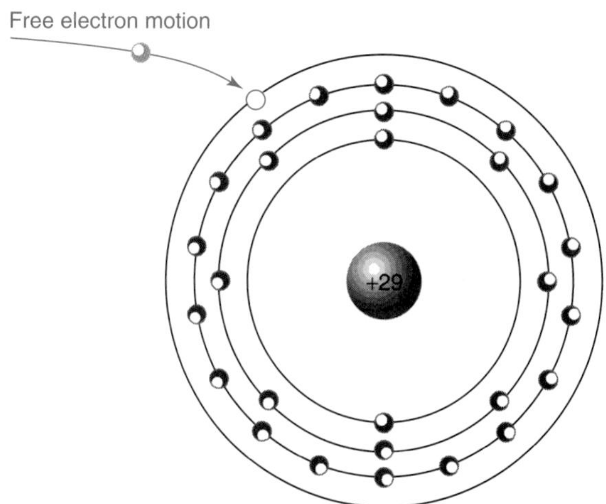

Figure 1.7 Neutralizing a positive ion.

Ions

The force of attraction between the negatively charged electron and the positively charged proton is the *inward force* that offsets the centrifugal force of an orbiting electron. When these forces are balanced, the electron remains in its orbit.

When an outside force is applied to an atom, the balance between the forces acting on a given electron can be disturbed, forcing the electron to leave its orbit. This point is illustrated in Figure 1.6.

When the electron leaves its orbit, the atom is left with fewer electrons than protons, and thus, has a net charge that is positive. An atom with a positive net charge is referred to as a **positive ion**.

Just as an outside force can cause an atom to lose an electron, it can also cause an atom to effectively *gain* an electron. When this occurs, the atom has more electrons than protons and, thus, a net charge that is *negative.* In this case, the atom is referred to as a **negative ion**.

Free Electrons

When an outside force causes an electron to break free from its parent atom, that electron is referred to as a *free electron.* A **free electron** is *an electron that is not bound to any particular atom*, and thus, is free to drift from one atom to another.

If a free electron drifts into the vicinity of a positive ion, that electron neutralizes the ion. For example, consider the free electron and the atom shown in Figure 1.7. When the electron nears the atom, it is attracted by the overall positive charge on the ion and is pulled into the gap in the valence shell. As a result, the atom once again has an equal number of electrons and protons and a net charge of *zero.*

Figure 1.7 and its related discussion illustrate an important point: *Positive and negative charges always seek to neutralize each other.* As you will see in the next section, this is the basis of one of the primary electrical properties, *current.*

Section Review

1. What is *matter*? What is an *element*?
2. What is an atom? Describe its physical makeup.
3. What is an *orbital shell*? What is the *valence shell*?

4. What is a *complete* valence shell?

5. What are the two types of charges? How do they relate to each other?

6. What is the net charge on an atom in its natural state? Explain.

7. What is the fundamental law of charges?

8. What are *positive ions* and *negative ions*? How is each created?

9. What is a *free electron*?

10. Describe what happens when a free electron drifts into the vicinity of a positive ion.

1.2 Current

Current, voltage, and *resistance* are three of the fundamental electrical properties. Stated simply:

- **Current** is *the directed flow of charge through a conductor.*
- **Voltage** is *the force that causes the directed flow of charge* (current).
- **Resistance** is *the opposition to current that is provided by the conductor.*

In this section, we're going to take a close look at the first of these properties, *current.*

As you know, an outside force can cause an electron to break free from its parent atom. In copper (and many other elements), very little external force is needed to generate free electrons. In fact, the thermal energy (heat) present at room temperature (25°C) is sufficient to cause free electrons to be generated. The number of electrons generated varies directly with temperature.

The motion of the free electrons in copper is random when no directing force is applied. That is, the free electrons move in every direction, as shown in Figure 1.8. Because the free electrons are moving in every direction, the *net flow of electrons in any one direction is zero.*

Figure 1.9 illustrates what happens if we apply an external force to the copper that causes all the electrons to move in the same direction. In this case, a *negative potential* is applied to one end of the copper and a *positive potential* is applied to the other. As a result, the free electrons all move from negative to positive, and we can say that we have a *directed flow of charge* (electrons). This directed flow of charge is referred to as *current.*

OBJECTIVE 4 ➤ Current is represented by the letter I (for *intensity*). The intensity of current is the rate at which charge flows, measured in *charge per unit time.* Stated mathematically,

$$I = \frac{Q}{t} \tag{1.1}$$

where

I = the intensity of the current

Q = the amount of charge

t = the time (in seconds) required for the charge (Q) to pass

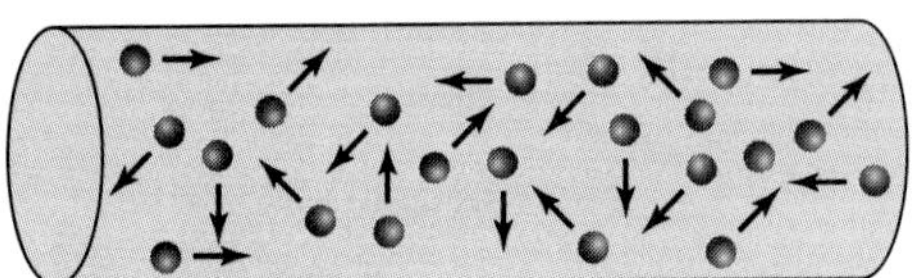

Figure 1.8 Random electron motion in copper.

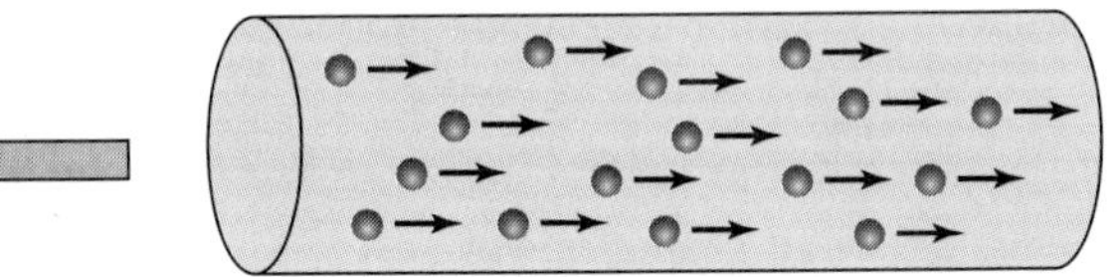

Figure 1.9 Directed electron motion in copper.

The charge of a single electron is too small to provide a practical unit of charge. Therefore, the basic unit of charge is the **coulomb (C),** which represents *the total charge of approximately* 6.25×10^{18} *electrons.* When one coulomb of charge passes a point in one second, we have one **ampere (A)** of current. Note that the ampere is the basic unit of measure for current (I), given as

$$1 \text{ ampere} = 1 \text{ coulomb/second}$$

The total current passing a point (in amperes) can be found by dividing the total charge (in coulombs) by the time (in seconds).

Example 1.1

Three coulombs of charge pass through a point in a copper wire every two seconds. Calculate the total current through the wire.

Solution: Using equation (1.1), the current is found as

$$I = \frac{Q}{t} = \frac{3 \text{ C}}{2 \text{ s}} = 1.5 \text{ C/s} = 1.5 \text{ A}$$

Practice Problem 1.1

Ten coulombs of charge pass through a point in a copper wire every four seconds. Calculate the total current through the wire.

It should be noted that current is not commonly calculated using equation (1.1), because it is impractical to directly measure coulombs of charge. As you will be shown later, there are far more practical ways of calculating current.

Electron Flow Versus Conventional Current

◄ *OBJECTIVE 5*

There are actually two ways to describe current, and both are used in practice. The **electron flow** approach describes current as *the flow of charge from negative to positive.* The **conventional flow** approach describes current as *the flow of charge from positive to negative.* Which definition is correct? Both are.

If we assume that "the flow of charge" refers to *electrons* (as shown in Figure 1.9), then we are taking the *electron flow* approach to describing current. If we assume that it refers to *positive ions*, then we are taking the *conventional flow* approach to describing current. Both approaches are illustrated in Figure 1.10. Each time an electron moves to

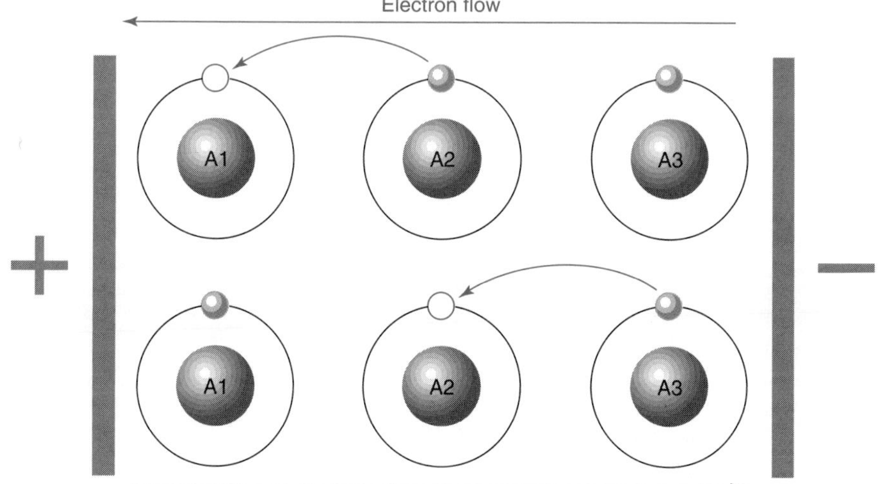

Figure 1.10
Conventional flow and electron flow.

the left, a positive ion is left behind. The result is that electrons are flowing from negative to positive, and that a positive ion is *effectively* flowing from positive to negative.

Conventional current is sometimes referred to as *hole flow.*

The *electron flow* view of current describes the action in Figure 1.10 as *the flow of electrons from negative to positive.* The *conventional current* approach describes the same action as *the flow of positive ions from positive to negative.* As you can see, both approaches are valid.

How we describe current does not affect circuit calculations or measurements. Even so, you should take the time to get comfortable with both descriptions, because both are used by many engineers, technicians, and technical publications. In this text, we will take the *electron flow* approach to current. That is, we will treat current as *the flow of electrons from negative to positive.*

OBJECTIVE 6 ➤

Direct Current Versus Alternating Current

Current is generally classified as being either *direct* or *alternating.* The differences between direct and alternating current are illustrated in Figure 1.11.

Direct current (dc) is *unidirectional,* meaning that *the flow of charge is always in the same direction.* For example, consider the diagram and graph in Figure 1.11a. The diagram represents a 1 A current through a resistor. As shown in the graph, this current is constant in both magnitude and direction.

Alternating current (ac) is *bidirectional,* meaning that *the flow of charge changes direction periodically.* For example, consider the graph and diagrams in Figure 1.11b. The two diagrams represent the current through a single resistor *during two different time periods.* During the portion of the graph that falls between time intervals t_0 and t_1, the resistor current is in the direction indicated. During the portion of the graph that falls between t_1 and t_2, the resistor current changes direction (as indicated in the diagram).

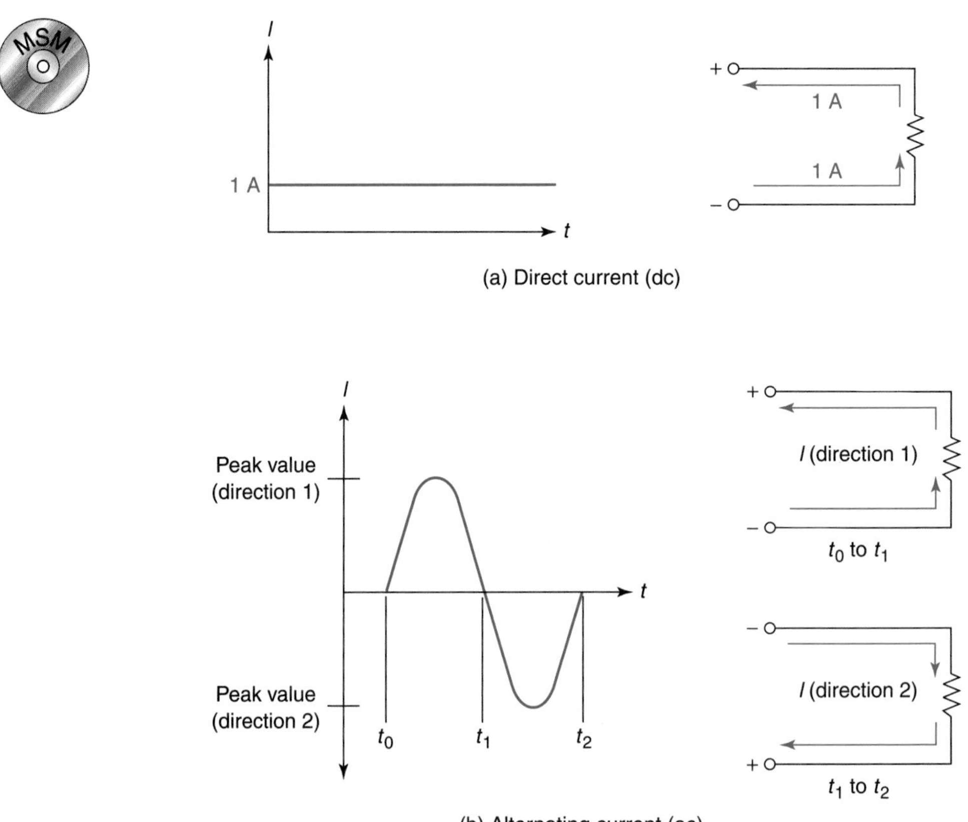

Figure 1.11 Direct current (dc) versus alternating current (ac).

Initially, we will deal exclusively with *direct current* and *dc circuits.* Then, our emphasis will shift to *alternating current* and *ac circuits.* As your electronics education continues, you will find that most electronic systems contain both dc and ac circuits.

Section Review

1. How are free electrons generated in a conductor at room temperature?
2. When undirected, the net flow of free electrons in any single direction is zero. Why?
3. What is *current*? What factors affect the *intensity* of current?
4. What is a *coulomb*?
5. What is the basic unit of current? How is it defined?
6. Contrast the *electron flow* and *conventional flow* descriptions of current.
7. Describe the difference between *direct current (dc)* and *alternating current (ac).*

1.3 Voltage

Voltage can be described as *a "difference of potential" that can generate a directed flow of charge (current) through a circuit.* In this section, we'll take a closer look at voltage and the means by which it produces current. ◄ *OBJECTIVE 7*

The battery represented in Figure 1.12a has two terminals. One terminal is shown to have an excess of positive ions and is described as having a *positive potential.* The other terminal is shown to have an excess of negative ions and is described as having a *negative potential.* Thus, there is a *difference of potential,* or voltage (*V*), between the two terminals. If we connect the two terminals of the battery with a copper wire (as shown in Figure 1.12b), a current is produced in the wire as the negative and positive charges seek to neutralize each other. In other words, there is a directed flow of charge between the terminals of the battery.

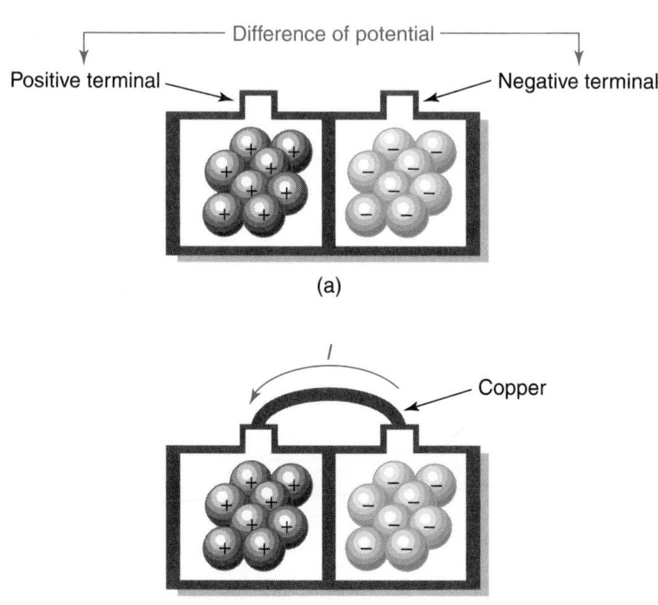

Figure 1.12

A Practical Consideration: The terminal connection shown in Figure 1.12b is used to illustrate a point. In practice, wire is never connected across the terminals of any voltage source. The reason for this is made clear in Chapter 2.

Several important points need to be made at this time:

- Voltage is a force that moves electrons. For this reason, it is often referred to as **electromotive force (emf)**.
- Current and voltage are not the same thing. *Current* is the flow of charge between the two terminals. *Voltage* is the difference of potential between the terminals that generates the current. In other words, *current occurs as a result of an applied voltage* (difference of potential).

The means by which a battery produces a difference of potential is discussed briefly in Chapter 2.

Note that the *volt* (V) is the basic unit of voltage. Technically defined, one **volt** is *the difference of potential that uses one joule of energy to move one coulomb of charge.* That is,

$$1 \text{ volt} = 1 \text{ joule/coulomb}$$

Section Review

1. What is *voltage*?
2. How does voltage generate a current through a wire?
3. What is the unit of measure for voltage? How is it defined?

Critical Thinking

4. How would you define a *coulomb* in terms of voltage and energy?
5. How would you define a *joule* in terms of voltage and charge?

1.4 Resistance and Conductance

OBJECTIVE 8 ➤ *Resistance* and *conductance* are basically opposites. While resistance is the opposition to current in a circuit, conductance is a measure of the ability of a circuit to conduct (pass current).

Resistance

All elements provide some amount of opposition to current. This opposition to current is called resistance. The higher the resistance of an element, component, or circuit, the lower the current produced by a given voltage.

Resistance (*R*) is measured in *ohms.* Ohms are represented using the Greek letter *omega* (Ω). Technically defined, one **ohm** is *the amount of resistance that limits current to one ampere when a one volt difference of potential is applied.* This definition is illustrated in Figure 1.13.

The schematic diagram shown in Figure 1.13 contains a *battery* (on the left) and a *resistor.* A **resistor** is *a component that is designed to provide a specific amount of resistance.* Note that the long end-bar on the battery schematic symbol represents the *positive* terminal, and that the short end-bar represents the *negative* terminal.

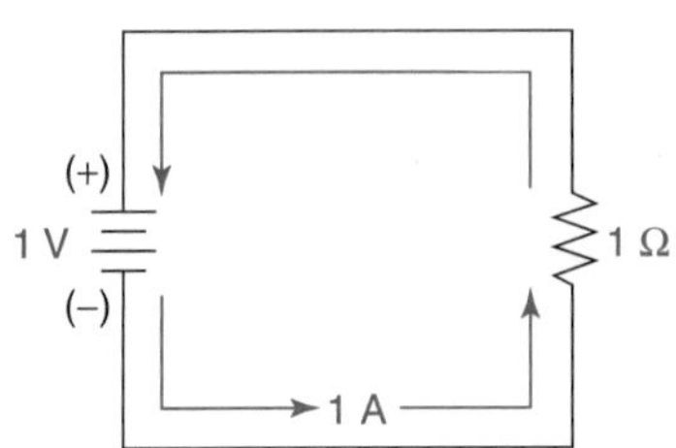

Figure 1.13

Conductance

Conductance is *a measure of the ease with which current will pass through a component or circuit*, found as the reciprocal of resistance. By formula,

$$G = \frac{1}{R} \tag{1.2}$$

where

G = the conductance of the circuit or component

R = the resistance of the circuit or component

Note: At one time, conductance was measured in *mhos*, symbolized by an upside-down omega (℧). Some older publications still use this unit of measure.

Conductance is most commonly measured in *siemens (S)*. The following example illustrates the calculation of conductance when resistance is known.

Example 1.2

Calculate the conductance of a 10 kΩ resistor.

Solution: Using equation (1.2), the conductance of the resistor is found as

$$G = \frac{1}{R} = \frac{1}{10\ \text{k}\Omega} = 1 \times 10^{-4}\ \text{S} = 100\ \mu\text{S}$$

Practice Problem 1.2

Calculate the conductance of a 33 kΩ resistor.

Just as resistance can be used to calculate conductance, conductance (when known) can be used to calculate resistance, as follows:

$$R = \frac{1}{G} \tag{1.3}$$

Example 1.3

A circuit has a conductance of 25 mS. Calculate the resistance of the circuit.

Solution: Using equation (1.3), the circuit resistance is found as

$$R = \frac{1}{G} = \frac{1}{25\ \text{mS}} = 40\ \Omega$$

Practice Problem 1.3

A circuit has a conductance of 10 mS. Calculate the resistance of the circuit.

Putting It All Together

We have now defined charge, current, voltage, resistance, and conductance. For convenience, these electrical properties are summarized in Table 1.1.

Many of the properties listed in Table 1.1 can be defined in terms of the others. For example, in our discussion on resistance, we said that one *ohm* is the amount of resistance that limits current to 1 A when a 1 V difference of potential is applied. By the same token, we can redefine the *ampere* and the *volt* as follows:

- One *ampere* is the amount of current that is generated when a 1 V difference of potential is applied across 1 Ω of resistance.
- One *volt* is the difference of potential required to generate 1 A of current through 1 Ω of resistance.

As you will see, these definitions relate closely to the most basic law of electronics, called *Ohm's law.*

TABLE 1.1 Basic Electrical Properties

Property	*Unit*	*Description*
Charge (Q)	Coulomb (C)	The total charge of 6.25×10^{18} electrons
Current (I)	Ampere (A)	The flow of charge; equal to 1 coulomb per second (C/s)
Voltage (V)	Volt (V)	Difference of potential; the force required to move a given amount of charge; equal to 1 joule per coulomb
Resistance (R)	Ohm (Ω)	Opposition to current; 1 Ω of resistance limits current to 1 A when 1 V is applied.
Conductance (G)	Siemens (S)	The reciprocal of resistance; a measure of the relative ease with which current passes through a component or circuit

Section Review

1. What is *resistance*?
2. What is the basic unit of resistance? How is it defined?
3. What is *conductance*? What is its unit of measure?

1.5 Conductors, Insulators, and Semiconductors

All materials are classified according to their ability (or inability) to conduct, as shown in Table 1.2.

A good example of a *conductor* is copper. Copper wire (which is used to interconnect components in a circuit) passes current with little opposition. A good example of an *insulator* is rubber. Rubber is used to coat the handles of many tools that are used in electrical work (such as pliers, screwdrivers, etc.) It takes an extremely high voltage to force rubber to conduct. A good example of a *semiconductor* is graphite (a form of carbon), which is used to make many common resistors. Graphite limits the amount of current that can be generated by a given voltage.

In this section, we will take a look at some of the characteristics of conductors, insulators, and semiconductors.

Conductors

Conductors are *materials that provide little opposition to the flow of charge* (current). A low opposition to current means that little energy is required to generate current. The materials with the lowest resistance are the best conductors.

TABLE 1.2 Material Classifications

OBJECTIVE 9 ➤

Material	*Characteristics*
Conductor	Extremely *low* resistance (high conductance). Conducts with very little voltage applied.
Insulator	Extremely *high* resistance (low conductance). Conducts only when an extremely high voltage is applied.
Semiconductor	Resistance that falls approximately midway between that of a conductor and an insulator. Limits the current at a given voltage.

In Section 1.1, you were told that the valence shell of an atom determines its electrical characteristics. The *conductivity* of an element is determined by:

- *The number of valence shell electrons.* The more valence shell electrons that an atom contains, the more difficult it is to force the atom to give up (or to accept) free electrons. *The best conductors contain one valence electron per atom.*
- *The number of atoms per unit volume.* With more atoms per unit volume, a given voltage can generate more free electrons. *The best conductors contain a high number of atoms per unit volume.*

A Practical Consideration: The concept of "atoms per unit volume" applies only to gases.

Insulators

Insulators are *materials that normally block current.* For example, the insulation that surrounds a power cord prevents the current through the cord from reaching you when it is touched. Some elements (like neon) are natural insulators. Most insulators, however, are compounds such as rubber, Teflon, and mica (among others).

As you have probably guessed, conductors and insulators have opposite characteristics. In general, insulators are materials that have:

- Complete valence shells.
- Very few atoms per unit volume.

With these characteristics, an extremely high voltage is required to force an insulator into conduction.

Semiconductors

Semiconductors are *materials that are neither good conductors nor good insulators.* For example, graphite (a form of carbon) does not conduct well enough to be considered a conductor. At the same time, it does not block current well enough to be considered an insulator. Some other examples of semiconductors are *silicon, germanium,* and *gallium arsenide.*

In general, semiconductors have the following characteristics:

- Valence shells that are half complete (i.e., contain four valence electrons).
- A relatively high number of atoms per unit volume.

Other Factors that Affect Resistance

◄ *OBJECTIVE 10*

The resistance of a conductor, insulator, or semiconductor at a given temperature depends on three factors:

- The *resistivity* of the material used.
- The *length* of the material.
- The *cross-sectional area* of the material.

Resistivity is *the resistance of a specified volume of an element or compound.* Resistivity is commonly rated using one of the following units of measure:

- Circular-mil ohms per foot (CM-Ω/ft)
- Ohm-centimeters (Ω-cm)

These two units of measure are described in the paragraphs that follow.

The **mil** is *a unit of measure equal to one-thousandth of an inch (0.001 in.).* The **circular-mil (CM)** is *a unit of area found by squaring the diameter (in mils) of the conductor.* One **circular-mill foot** of material has *a diameter of 1 mil and a length of 1 ft,* as shown in Figure 1.14. When this volume of material is used to measure the resistivity of a given material, the rating is given in **circular-mil ohms per foot.**

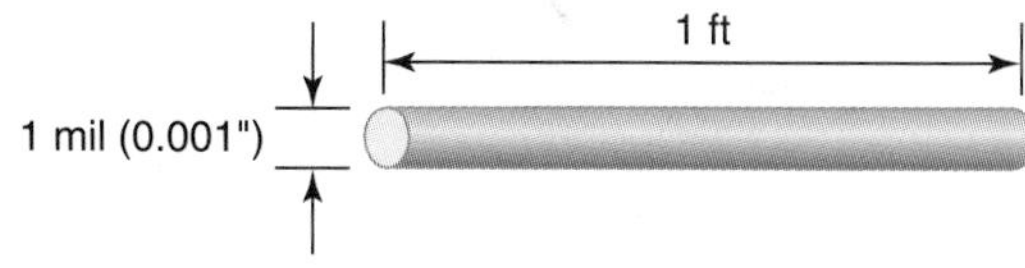

Figure 1.14 One circular-mill foot.

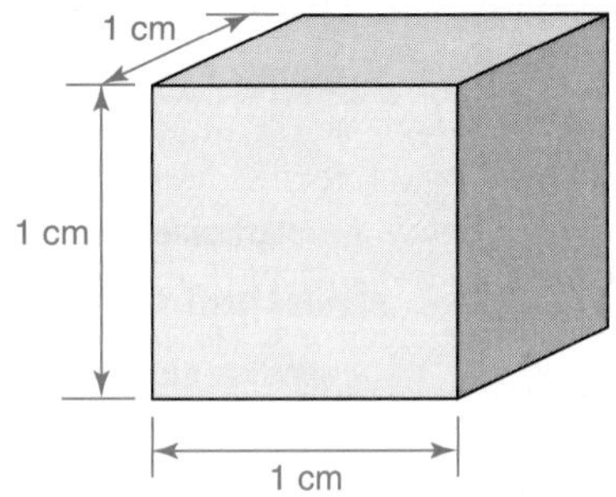

Figure 1.15. One cubic centimeter.

An important point: When the resistance of one circular-mil foot is used to determine the resistivity of a material, it would seem that the proper unit of measure would be *ohms per circular-mill foot* (Ω/CM-ft). However, in any calculation, the units of measure must combine in such a way that the result has the proper unit of measure. Later in this section, you will see that the units of measure combine properly when resistivity (in CM-Ω/ft) is used to calculate the resistance of a material.

The second unit of volume used for rating resistivity is the *cubic centimeter,* which is represented in Figure 1.15. When this volume is used to measure the resistivity of a given material, the rating is given in **ohm-centimeters (Ω-cm).** Again, the unit of measure may appear awkward, but it combines properly with other units of measure when calculating the resistance of a material (as will be demonstrated later in this section).

Table 1.3 shows the resistivity ratings of some elements that are commonly used in electronics.

Calculating the Resistance of a Conductor

When the length and cross-sectional area of a conductor are known, the resistance of the conductor can be found as

$$R = \rho \frac{\ell}{A} \tag{1.4}$$

where

ρ = the resistivity of the material
ℓ = the length of the conductor
A = the cross-sectional area of the conductor

Note that the Greek letter *rho* (ρ) is commonly used to represent resistivity. The following examples illustrate the calculation of the resistance of a conductor.

A Practical Consideration: Most resistivity ratings are given in ohm-centimeters. The circular-mil ohms per foot rating is used almost exclusively for lengths of cable.

TABLE 1.3 The Resistivity Ratings of Some Common Elements

	Resistivity	
Element	*CM-Ω/ft*	*Ω-cm*
Silver	9.9	1.645×10^{-6}
Copper	10.37	1.723×10^{-6}
Gold	14.7	2.443×10^{-6}
Aluminum	17.0	2.825×10^{-6}
Iron	74.0	12.299×10^{-6}
Carbon	2.1×10^{4}	3.50×10^{-3}

Calculate the resistance of a six inch length of copper that has a diameter of 0.005 in.

Example 1.4

Solution: Copper has a resistivity rating of 10.37 CM-Ω/ft (as shown in Table 1.3). Six inches equals 0.5 ft, and 0.005 in. equals 5 mils. With a diameter of 5 mils, the cross-sectional area is found as

$$\text{Area} = (5 \text{ mils})^2 = 25 \text{ CM}$$

Now, using equation (1.4), the resistance of the copper is found as

$$R = \rho\frac{\ell}{A} = \left(10.37\frac{\text{CM-}\Omega}{\text{ft}}\right)\left(\frac{0.5 \text{ ft}}{25 \text{ CM}}\right) = 0.207\ \Omega = 207\ \text{m}\Omega$$

With this kind of resistance, it is easy to see that copper provides very little opposition to current.

Practice Problem 1.4

Calculate the resistance of an 18 inch length of copper that has a diameter of 0.005 in.

Note how the units of measure for resistivity, length, and area combine, leaving ohms (Ω) as the unit of measure for the result. This is why resistivity is rated in CM-Ω/ft (rather than ohms per circular-mil foot). By the same token, Example 1.5 demonstrates why resistivity is rated in Ω-cm (rather than ohms per cubic centimeter).

Calculate the resistance of a 25 cm length of copper that has a cross-sectional area of 0.04 cm^2.

Example 1.5

Solution: Copper has a resistivity rating of 1.723×10^{-6} Ω-cm (as shown in Table 1.3). Using this value and equation (1.4), the resistance of the copper is found as

$$R = \rho\frac{\ell}{A} = (1.723 \times 10^{-6}\ \Omega\text{-cm})\left(\frac{25 \text{ cm}}{0.04 \text{ cm}^2}\right) = 1.08 \times 10^{-3}\ \Omega = 1.08\ \text{m}\Omega$$

Practice Problem 1.5

Calculate the resistance of a 25 cm length of copper that has a cross-sectional area of 0.1 cm^2.

What Equation (1.4) Shows Us

Every equation serves two purposes. First, it shows us how to calculate the value of interest. Second (and probably most important), it shows us the relationship between the value of interest and its variables. For example, equation (1.4) states that

$$R = \rho\frac{\ell}{A}$$

This relationship shows us that:

- *The resistance of a material is directly proportional to its resistivity.* That is, increases (or decreases) in resistivity cause a similar change in resistance.
- *The resistance of a conductor is directly proportional to its length.*
- *The resistance of a conductor is inversely proportional to its cross-sectional area.* That is, changes in cross-sectional area and resistance are opposites. If one increases, the other decreases, and vice versa.

The Effects of Temperature on Resistance

The calculations in Examples 1.4 and 1.5 assume an operating temperature of 20°C. The resistivity of a given material is affected by changes in operating temperature. The effect of temperature on the resistivity of a material depends on whether the material has a *positive temperature coefficient* or a *negative temperature coefficient.*

A **positive temperature coefficient** is *a rating that indicates the resistance of a given material increases when temperature increases.* A **negative temperature coefficient** is *a rating that indicates the resistance of a given material decreases when temperature increases,* and vice versa.

Conductors have *positive* temperature coefficients. Semiconductors and insulators generally have *negative* temperature coefficients. Thus, the resistance of a conductor increases when temperature increases. At the same time, the resistance of a semiconductor or insulator generally decreases when temperature increases.

Section Review

1. What is a *conductor*? An *insulator*? A *semiconductor*?
2. List the factors that determine the conductivity of an element.
3. What is *resistivity*?
4. Describe the units of measure that are commonly used to express resistivity.
5. What is a *positive temperature coefficient*?
6. What is a *negative temperature coefficient*?
7. Identify the temperature coefficient for each of the following: a conductor, an insulator, and a semiconductor.

Key Terms

The following terms were introduced in this chapter on the pages indicated:

Alternating current (ac), 30
Ampere (A), 29
Atom, 24
Atomic number, 24
Bohr model, 24
Charge, 25
Circular-mil (CM), 35
Circular-mil foot, 35
Circular-mil ohms per foot (CM-Ω/ft), 35
Conductance, 33
Conductor, 34
Conventional flow, 29
Coulomb, 29
Current, 28
Direct current (dc), 30
Electromotive force (emf), 32
Element, 24
Electron, 24
Electron flow, 29
Free electron, 27
Insulator, 35
Matter, 24
Mil, 35
Negative ion, 27
Negative temperature coefficient, 38
Neutron, 24
Nucleus, 24
Ohm (Ω), 32
Ohm-centimeter (Ω-cm), 36
Positive ion, 27
Positive temperature coefficient, 38
Proton, 24
Resistance, 28
Resistivity, 35
Resistor, 32
Semiconductor, 35
Shell, 25
Valence electron, 25
Valence shell, 25
Volt, 32
Voltage, 28

Practice Problems

1. One coulomb of charge passes a point every 20 seconds. Calculate the value of the current through the point.
2. A total charge of 2.5×10^{-3} C passes a point every 40 seconds. Calculate the value of the current through the point.

3. A total charge of 4.0×10^{-6} C passes a point every second. Calculate the value of the current through the point.

4. A total charge of 50×10^{-3} C passes a point every 2.5 seconds. Calculate the value of the current through the point.

5. Calculate the conductance of a 4.7 Ω resistor.

6. Calculate the conductance of a 1.5 MΩ resistor.

7. Calculate the conductance of a 510 kΩ resistor.

8. Calculate the conductance of a 22 kΩ resistor.

9. A circuit has 50 μS of conductance. Calculate the circuit resistance.

10. A circuit has 4.25 mS of conductance. Calculate the circuit resistance.

11. A circuit has 250 nS of conductance. Calculate the circuit resistance.

12. A circuit has 375 μS of conductance. Calculate the circuit resistance.

13. Calculate the resistance of a 4 inch length of copper that has a cross-sectional area of 50 CM.

14. Calculate the resistance of a 12 inch length of aluminum that has a diameter of 0.003 in.

15. Calculate the resistance of a 30 cm length of copper that has a cross-sectional area of 0.08 cm^2.

16. Calculate the resistance of a 10 cm length of iron that has a cross-sectional area of 0.05 cm^2.

Pushing the Envelope

17. A 10 V source uses two joules of energy per second. Calculate the current being generated by this voltage source.

18. The current through a wire is equal to 100 mA. The energy being used by the voltage source is 2.5×10^{-3} j. What is the value of the voltage source?

19. If one value increases at the same rate that another decreases (or vice versa), the two values are said to be *inversely proportional.* Prove that resistance and conductance are inversely proportional.

20. If two values increase (or decrease) at the same rate, the two values are said to be *directly proportional.* Prove that current is directly proportional to charge and inversely proportional to time.

Looking Back

These problems relate to material presented in earlier chapters. In this case, the problems are drawn from the Introduction.

21. Convert the following numbers to scientific notation without the use of your calculator.
    a. 0.0006835
    b. 42,977

22. Convert each of the following values to standard form without the use of your calculator.
    a. 5.92×10^{-4}
    b. 9.47×10^{7}

23. Convert each of the following to engineering notation.
    a. 1.11×10^{8} Hz
    b. 0.022×10^{-6} F

24. Convert each of the following to engineering notation.
    a. 0.00128 A
    b. 0.000047 H

Answers to the Example Practice Problems

1.1. 2.5 A
1.2. 30.3 μS
1.3. 100 Ω
1.4. 622 mΩ
1.5. 431 μΩ

Multisim File References

Figure	File Reference
1.11	ETF1_11.msm
1.13	ETF1_13.msm

chapter 2

Components and Circuit Measurements

Objectives

After studying the material in this chapter, you should be able to:

1. Describe the commonly used types of wire and the American Wire Gauge (AWG) system of wire sizing.
2. Describe the ratings commonly used for conductors and insulators.
3. List and describe the various types of resistors.
4. Determine the range of values for a given resistor.
5. Describe and use the standard (non-precision) resistor color code.
6. Determine the value of a resistor coded with an alphanumeric code.
7. List the guidelines for replacing resistors.
8. Describe the operation of commonly used potentiometers.
9. Describe the construction and characteristics of commonly used batteries.
10. Discuss the operation and use of a basic dc power supply.
11. List and describe the various types of switches.
12. Describe the construction, operation, ratings, and replacement procedures for fuses.
13. Compare and contrast fuses with circuit breakers.
14. Describe the procedures for measuring current, voltage, and resistance.

There are two aspects to every science: *theory* and *practice.* Theory deals with the concepts of the science, while practice deals with using those concepts to accomplish something. In Chapter 1, you took the first step toward understanding the theory of electricity and electronics. In this chapter, we will take the first step toward understanding its practical side.

2.1 Conductors and Insulators

In Chapter 1, we discussed the characteristics of conductors and insulators. In this section, you will be introduced to conductor and insulator ratings and applications.

OBJECTIVE 1 ➤

Wires

All wires are either *solid* or *stranded,* as shown in Figure 2.1a. **Solid wire** contains a solid-core conductor and is easier to manufacture than stranded wire. **Stranded wire** contains a group of very thin wires that are wrapped together to form a larger diameter wire and is more *flexible* than solid wire. Both solid and stranded wires are usually coated with an insulating layer. Wires are produced in a wide range of diameters, as shown in Figure 2.1b.

Why Copper Is Used

Copper is the most commonly used conductor, because *it provides the best balance between resistance and cost per unit length.* Silver has a lower resistance (per unit length) than copper but is much more expensive. Aluminum is lower in cost, but it has almost twice the resistance (per unit length) of copper and very low elasticity. Its low elasticity can cause problems when aluminum is connected to a copper connector (which has high elasticity).

Note: Each wire gauge has approximately:
a. *one-half* the resistance (per unit length) of the wire that is three gauges higher.
b. *one-tenth* the resistance (per unit length) of the wire that is ten gauges higher.

Wire Sizes

Wires are produced in standard sizes. The size of a wire is determined by its *cross-sectional area.* The **American Wire Gauge (AWG)** system uses numbers to identify these standard wire sizes, as shown in Table 2.1. For example, 24 gauge wire has a cross-sectional area of 404.01 circular mils (CM) and a resistance of 25.67 Ω per 1000 ft.

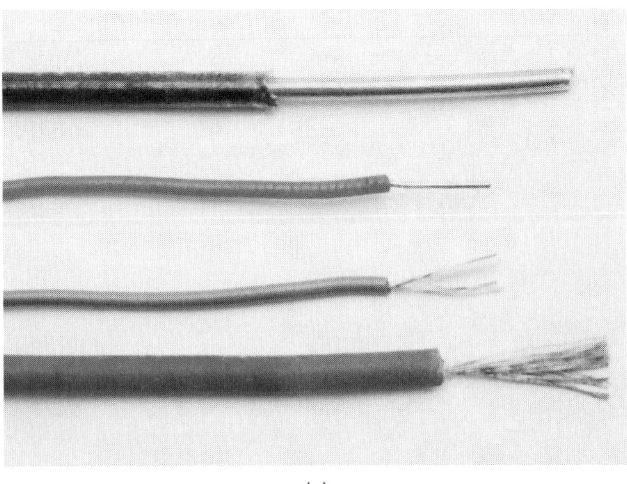

(a)

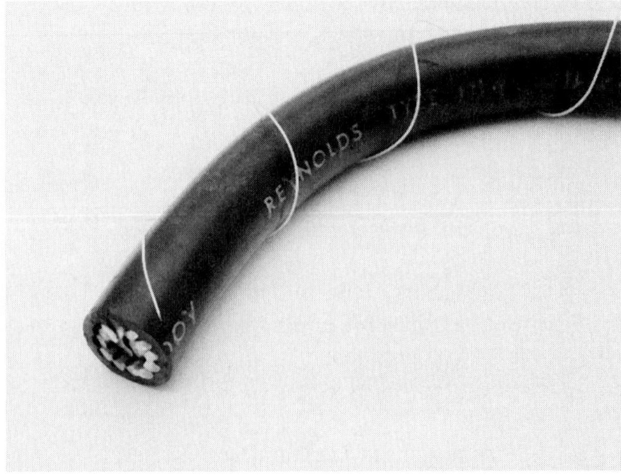

(b)

Figure 2.1 Several sizes of solid and stranded wire.

TABLE 2.1 American Wire Gauge (AWG) System

AWG Number	Area (CM)		Ω/1000 ft	Ampacity (A)	Applications/Examples
0000	211,600	(Wire size increases as gauge number decreases)	0.0490	312	
000	167,810		0.0618	262	Commercial power distribution
00	133,080		0.0780	220	
0	105,530		0.0983	185	
1	83,694		0.1240	156	
2	66,373		0.1563	131	
3	52,634		0.1970	110	
4	41,742		0.2485	92	Commercial/residential wiring and other electrical applications
5	33,102		0.3133	78	(battery cables, house wiring, lighting, etc.)
6	26,250		0.3951	65	
7	20,816		0.4982	55	
8	16,509		0.6282	46	
9	13,094		0.7921	39	
10	10,381		0.9989	33	
11	8,234.0		1.260	27	
12	6,529.0		1.588	23	
13	5,178.4		2.003	19	
14	4,106.8		2.525	16	
15	3,256.7		3.184	14	
16	2,582.9		4.016	12	
17	2,048.2		5.064	10	
18	1,624.3		6.358	8	
19	1,228.1		8.051		
20	1,021.5		10.15		
21	810.10		12.80		
22	642.40		16.14		
23	509.45		20.36		
24	404.01		25.67		General electronic circuit applications (radio, video, etc.)
25	320.40		32.37		
26	254.10		40.81		
27	201.50		51.47		
28	159.79		64.90		
29	126.72		81.83		
30	100.50		103.2		
31	79.70		130.1		
32	63.21		164.1		
33	50.13		206.9		
34	39.75		260.9		
35	31.52		329.0		Low-current electronic circuit applications (hand-held
36	25.00		414.8		devices, etc.)
37	19.83		523.1		
38	15.72		659.6		
39	12.47		831.8		
40	9.89		1,049.0		

Current Capacity

◄ *OBJECTIVE 2*

Some wire gauges are rated according to their *ampere capacity,* or *ampacity.* **Ampacity** is *the maximum allowable current that can be safely carried by a given wire gauge,* measured in amperes. Note that:

- Ampacity increases as wire size increases. The lower the wire gauge, the higher the ampacity.
- Only wire gauges 0000 through 18 have ampacity ratings. (Higher-gauge wires are generally used in relatively low-current applications, making ampacity ratings unnecessary.)

The ampacity ratings for the lower-gauge wires are provided in Table 2.1.

(a)

(b)

Figure 2.2 A printed circuit (PC) board.

PC Board Traces

Most electronic circuits are built on **printed circuit boards,** or *PC boards.* A PC board is shown in Figure 2.2. Usually, components are mounted on one side of the board, though *multisided* PC boards have components mounted on both sides. The conductors that connect the components are referred to as copper **traces.** Like wires, traces have limits on the amount of current they can carry.

Insulator Ratings

Any insulator can be forced to conduct if a sufficient voltage is applied. The **average breakdown voltage** rating of an insulator is *the voltage (per unit thickness) that will force the insulator to conduct,* measured in *kilovolts per centimeter (kV/cm).* Some average breakdown voltage ratings are listed in Table 2.2.

It should be noted that there are several types of ratings. One type, called a **parameter,** is *a limit.* When any component parameter is exceeded, that component will not operate properly and may be damaged or destroyed. The values given in Table 2.2 are all parameters. (Other types of ratings are introduced—as needed—later in the text.)

TABLE 2.2 Typical Insulator Breakdown Voltage Ratings

Material	*Average Breakdown Voltage (kV/cm)*
Air	30
Rubber	270
Paper	500
Teflon	600
Glass	900
Mica	2000

Section Review

1. Why is copper the preferred conductor in most applications?
2. What is the relationship between a wire and one that is three wire gauges higher?

3. What is the relationship between a wire and one that is ten wire gauges higher?
4. What is *ampacity*? What is the relationship between ampacity and wire gauge?
5. Define *average breakdown voltage.*
6. What is a *parameter*? Why are parameters important?

2.2 Resistors

A **resistor** is *a component that is designed to provide a specific amount of resistance.* Resistors are classified as either fixed or variable. A **fixed resistor** is *one that has a specific value that cannot be adjusted.* A **variable resistor** is *one that has a value that can be adjusted within a specified range.* ◄ *OBJECTIVE 3*

Carbon-Composition Resistors

One common type of fixed resistor is the **carbon-composition resistor** (illustrated in Figure 2.3). As you can see, the resistor has two metal leads (conductors) that are separated by carbon. The relatively high resistivity of the carbon is the source of the resistor's opposition to current.

The value of a given carbon-composition resistor is determined by two main factors:

- *The purity of the carbon.* By adding impurities (other elements) to carbon, the resistivity of the carbon can be increased or decreased.
- *The spacing between its leads.* This point is illustrated in Figure 2.4. The current in Figure 2.4b must pass through more carbon than the current in Figure 2.4a. Because the resistance of a material is directly proportional to its length, the current in Figure 2.4b encounters more resistance.

Carbon-composition resistors are produced in a wide variety of values. A list of standard resistor values is provided later in this section.

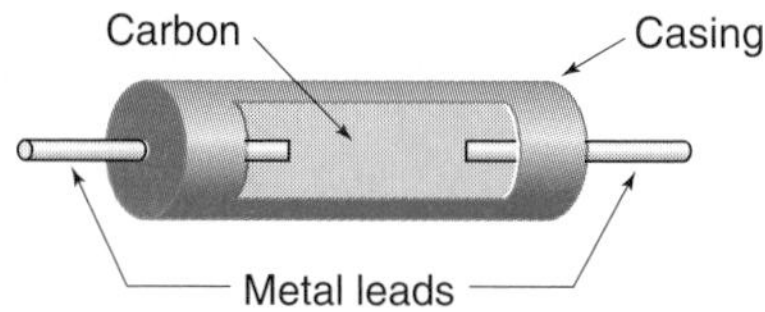

Figure 2.3 A carbon-composition resistor.

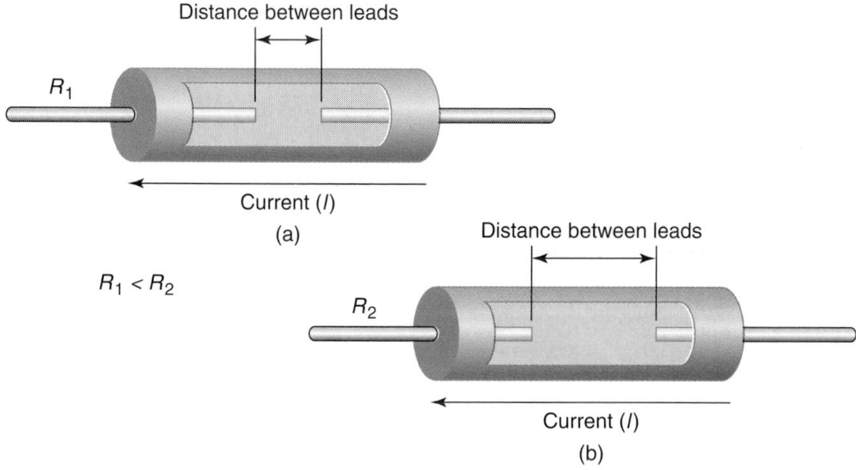

Figure 2.4 Lead spacing.

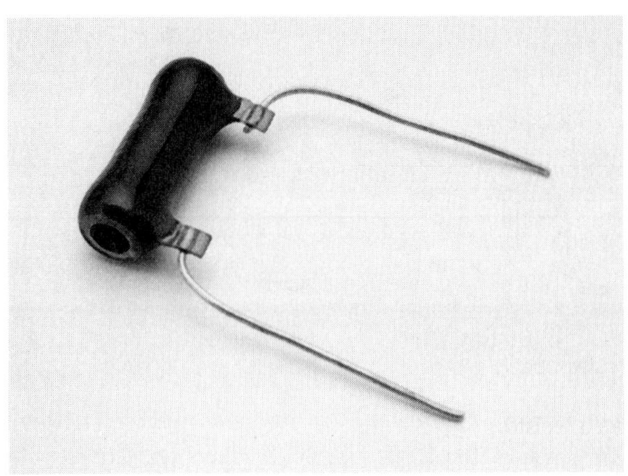

Figure 2.5 A wire-wound resistor.

Other Types of Resistors

A **wire-wound resistor** is *a component that uses the resistivity of a length of wire to produce a desired value of resistance.* A wire-wound resistor is shown in Figure 2.5. The component contains a length of wire that is wrapped in a series of loops around a hollow ceramic base. The length of the wire determines the value of the component. Any current through the component passes through the entire length of the wire.

Wire-wound resistors are used primarily in applications where the components must be able to *dissipate* (throw off) a relatively high amount of heat. A wire-wound resistor can tolerate more heat than a carbon-composition resistor for several reasons:

- Copper can withstand much more heat than the carbon-based compound that is used to make carbon-composition resistors.
- Since a wire-wound resistor is constructed on a hollow ceramic base, its surface area is much greater than that of a comparable carbon-composition resistor. The greater surface area of the wire-wound resistor allows the component to dissipate more heat.

The **metal-film resistor** is *a resistor designed to have a low temperature coefficient.* As a result, the resistance of a metal-film resistor is relatively independent of temperature. For example, a temperature coefficient of 0.0005 Ω/°C would be typical for a 10 Ω metal-film resistor. (The temperature coefficient for a comparable carbon-composition resistor would be approximately 100 times as high.) Several metal-film resistors are shown in Figure 2.6a.

The internal construction of a metal-film resistor is illustrated in Figure 2.6b. The *metal film* is the source of the component's resistance. It is molded in a spiral shape, with the spirals separated by a ceramic layer. Any current through the component passes through the metal-film spiral.

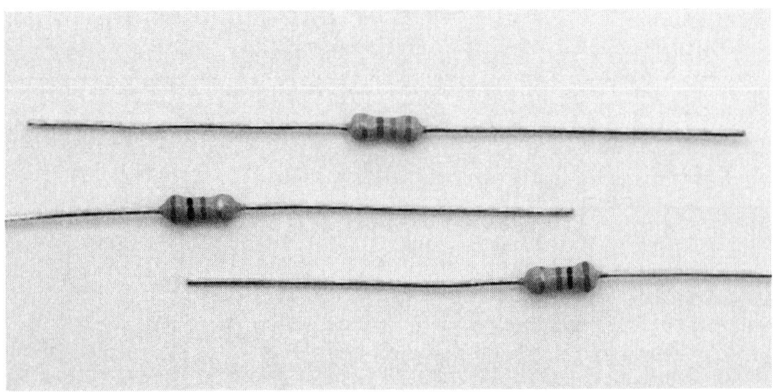

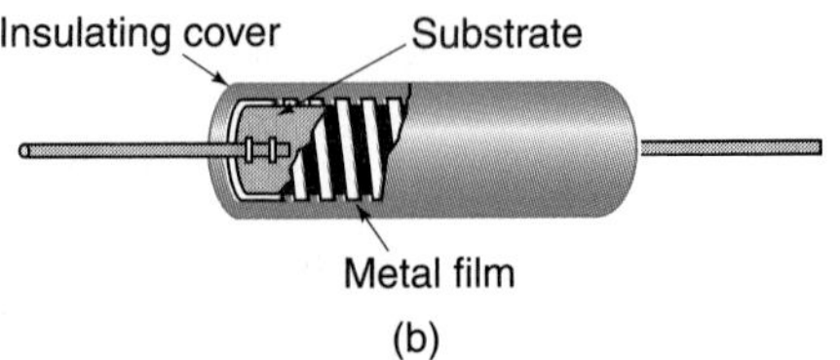

Figure 2.6 A metal-film resistor.

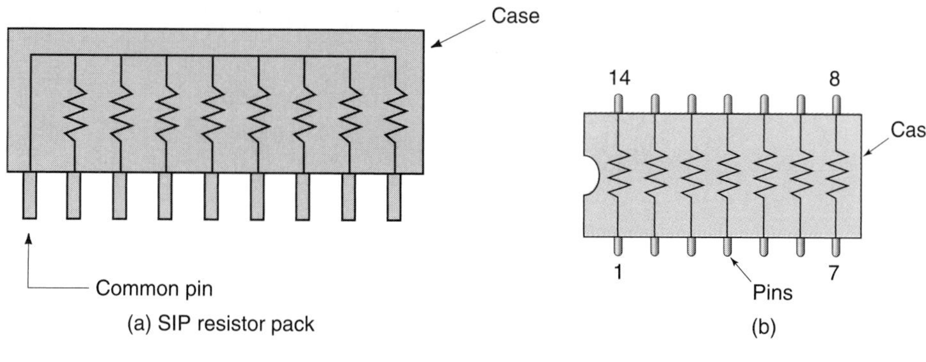

Figure 2.7 Integrated resistors.

Note: The resistor packs in Figure 2.7 are drawn larger than actual size.

The **carbon-film resistor** is similar (in structure) to the metal-film resistor. In a way, it is like a cross between the metal-film resistor and the carbon-composition resistor. A carbon-based compound is formed in the shape of a spiral (like the metal-film spiral in Figure 2.6b). The makeup of the carbon-based compound and the length of material between the component leads determine the resistance of the component.

Integrated resistors are *micro-miniature resistors made using semiconductors other than carbon.* Because they are so small, many can be housed in a single casing. For example, an 8-input SIP (*single in-line package*) is illustrated in Figure 2.7a. This package contains eight resistors with individual input pins and one common output pin. The resistor pack illustrated in Figure 2.7b is a DIP (*dual in-line package*) that contains all the components shown in the diagram. Note that the resistor packs shown have the advantage of small size but are limited to low-power applications; that is, they are limited to applications where the components are not required to dissipate any significant amount of heat.

Surface-Mount Packages

Many of the resistors discussed in this section can be manufactured in *surface-mount packages (SMPs).* A **surface-mount package** is *one that is designed for mounting directly on the surface of a printed circuit board.* SMPs typically measure 20 mm × 7 mm × 7 mm or smaller. Carbon-composition, wire-wound, and metal-film resistors can all be produced in SMPs.

Standard Resistor Values

◄ *OBJECTIVE 4*

Resistors are commercially produced in a variety of values. The standard resistor values are listed in Table 2.3.

If you look at any row of numbers in Table 2.3, you'll see that it contains a series of values that all start with the same two digits. The only difference between the values is a *power-of-ten multiplier.* For example, the values shown in the bottom row could be written as follows:

$$91 \times 10^{-2}$$
$$91 \times 10^{-1}$$
$$91 \times 10^{0}$$
$$91 \times 10^{1}$$
$$91 \times 10^{2}$$
$$91 \times 10^{3}$$
$$91 \times 10^{4}$$
$$91 \times 10^{5}$$

As these values indicate, we can designate any resistor value using only the first two digits and a power-of-ten multiplier. As you will see later, this is the basis of the **resistor color code** (*a method used to designate the value of a given resistor*).

TABLE 2.3 Standard Resistor Values (2% Tolerance and Higher)

Ω				kΩ			MΩ	
0.10	1.0	10	100	1.0	10	100	1.0	10
0.11	1.1	11	110	1.1	11	110	1.1	11
0.12	1.2	12	120	1.2	12	120	1.2	12
0.13	1.3	13	130	1.3	13	130	1.3	13
0.15	1.5	15	150	1.5	15	150	1.5	15
0.16	1.6	16	160	1.6	16	160	1.6	16
0.18	1.8	18	180	1.8	18	180	1.8	18
0.20	2.0	20	200	2.0	20	200	2.0	20
0.22	2.2	22	220	2.2	22	220	2.2	22
0.24	2.4	24	240	2.4	24	240	2.4	
0.27	2.7	27	270	2.7	27	270	2.7	
0.30	3.0	30	300	3.0	30	300	3.0	
0.33	3.3	33	330	3.3	33	330	3.3	
0.36	3.6	36	360	3.6	36	360	3.6	
0.39	3.9	39	390	3.9	39	390	3.9	
0.43	4.3	43	430	4.3	43	430	4.3	
0.47	4.7	47	470	4.7	47	470	4.7	
0.51	5.1	51	510	5.1	51	510	5.1	
0.56	5.6	56	560	5.6	56	560	5.6	
0.62	6.2	62	620	6.2	62	620	6.2	
0.68	6.8	68	680	6.8	68	680	6.8	
0.75	7.5	75	750	7.5	75	750	7.5	
0.82	8.2	82	820	8.2	82	820	8.2	
0.91	9.1	91	910	9.1	91	910	9.1	

Reference: Precision resistors (1% tolerance and lower) are available in values not listed here. A list of precision resistor standard values is provided in Appendix B.

Resistor Tolerance

Every commercially produced resistor is guaranteed to have a measured value that falls within a given percentage of its rated value. The **tolerance** of a resistor is *a rating that indicates the limits on the resistance of the component, given as a percentage of its nominal (rated) value.*

Most common resistors have 2% or 5% tolerance ratings, which means that their measured values are guaranteed to fall within ±2% or ±5% of their rated values. Some older resistors have 10% and 20% tolerance ratings, but they are rarely encountered in modern electronic circuits.

Integrated resistors like those shown in Figure 2.7 often have very poor tolerance ratings—as high as 30% in some cases. As a result, their use is limited to applications where high tolerance will have no effect on the circuit operation.

Advances in resistor manufacturing technology have led to the development of **precision resistors.** These are *components with tolerance ratings below 2%*. Some commercially available precision resistors have tolerances as low as 0.001%. As a result of their low tolerance ratings, precision resistors have a unique list of standard values. This list is provided in Appendix B.

To determine the guaranteed range of values for a resistor:

1. Multiply the rated value of the resistor by its tolerance to get the *maximum variation* in resistance.
2. Add the maximum variation to the rated value of the component to find the *upper limit.*
3. Subtract the maximum variation from the rated value of the component to find the *lower limit.*

The use of this procedure is demonstrated in the following example.

Example 2.1

Determine the guaranteed range of measured values for a 47 kΩ resistor with a 2% tolerance.

Solution: First, the maximum variation in resistance is found as

$$(47\text{ k}\Omega) \times 2\% = 940\ \Omega \qquad (0.94\text{ k}\Omega)$$

The maximum and minimum component values are found as

$$47\text{ k}\Omega + 0.94\text{ k}\Omega = 47.94\text{ k}\Omega \qquad \text{(maximum)}$$

and

$$47\text{ k}\Omega - 0.94\text{ k}\Omega = 46.06\text{ k}\Omega \qquad \text{(minimum)}$$

Practice Problem 2.1

Determine the guaranteed range of values for a 68 kΩ resistor with a 2% tolerance.

Two points should be made regarding resistor tolerance:

- The ideal (perfect) resistor would have a 0% tolerance, meaning that its measured value would always equal its rated value.
- Even though they are no longer manufactured, you may still see 10% and 20% tolerance resistors in some older electronic systems. Lower tolerance resistors can be used to replace these components (when necessary).

In the next section, you will be shown how the tolerance of a resistor is indicated on the body of the component.

Section Review

1. What physical factors determine the value of a *carbon-composition resistor*?
2. What is a *wire-wound resistor*?
3. Why are *wire-wound resistors* more tolerant of heat than comparable *carbon-composition resistors*?
4. Describe the construction of the *metal-film resistor.*
5. Compare and contrast *carbon-composition resistors* and *carbon-film resistors.*
6. What are the primary restrictions on the use of *integrated resistors*?
7. How do you determine the guaranteed range of values for a resistor?
8. When replacing a resistor, it is acceptable to use one that has a *lower* tolerance rating than the original component. Why is this true? (Critical Thinking)

2.3 Resistor Color Codes and Related Topics

Several methods are used to indicate the value of a given resistor. Many use a series of color bands on the component while others use numeric codes. In this section, we will take a look at several resistor-marking schemes.

The Color Code for Standard (Nonprecision) Resistors

◀ *OBJECTIVE 5*

In most cases, the value of a resistor is indicated by a series of *color bands* on the component. For example, look at the resistors shown in Figure 2.8. Each of these resistors has four color bands that are numbered in the order shown. Note that the band closest to one end of the component is the first color band.

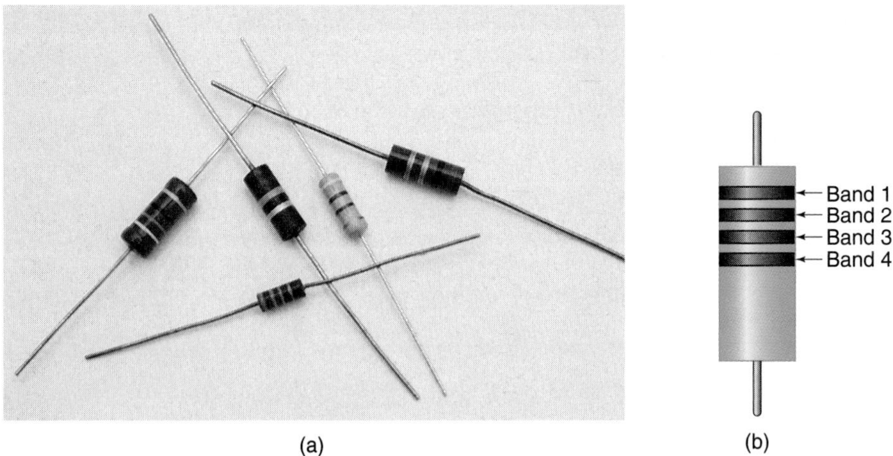

Figure 2.8 Resistor color bands.

TABLE 2.4 Resistor Color Band Code

Color	*Value*	*Color*	*Value*
Black	0	Green	5
Brown	1	Blue	6
Red	2	Violet	7
Orange	3	Gray	8
Yellow	4	White	9

Earlier, you were shown that the value of a resistor could be designated using the first two digits of the value and a power-of-ten multiplier. The first three bands on a resistor designate its rated value in this fashion as follows:

Reference: The band designations for precision resistors (1% tolerance and lower) are provided in Appendix B.

Band 1: The color of this band designates the first digit in the resistor value.
Band 2: The color of this band designates the second digit in the resistor value.
Band 3: The color of this band designates the power-of-ten multiplier for the first two digits. (In most cases, this is simply the number of zeros that follow the first two digits.)

The colors most often used in these three bands are coded as shown in Table 2.4.

The following example shows how the first three color bands are used to determine the value of a given resistor.

Example 2.2

Determine the rated value of the resistor shown in Figure 2.9.

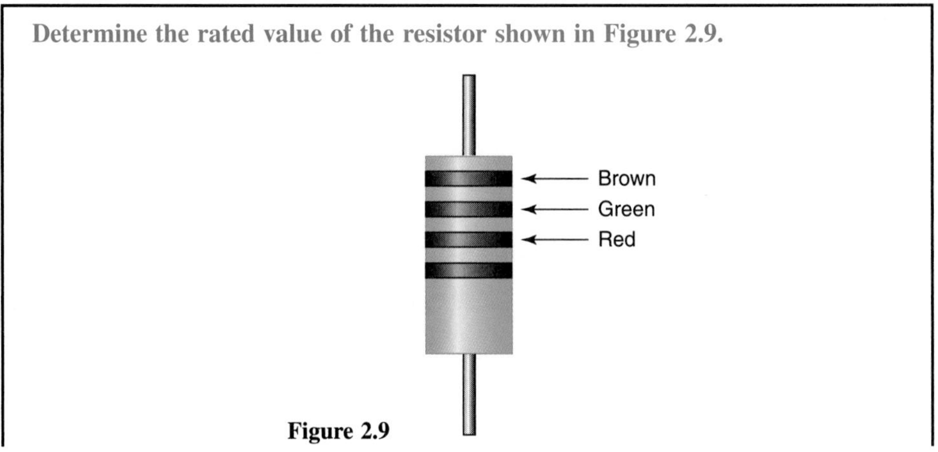

Figure 2.9

Solution: Brown = 1, so the first digit in the resistor value is 1.
Green = 5, so the second digit in the resistor value is 5.
Red = 2, so the power-of-ten multiplier is $10^2 = 100$.

Therefore, the rated value of the resistor is found as

$$(15 \times 100)\ \Omega = 1500\ \Omega = 1.5\ \text{k}\Omega$$

Note that the multiplier band value (2) equals the number of zeros in the value of the resistor.

Practice Problem 2.2

The first three bands on a resistor are as follows: yellow—violet—orange. Determine the rated value of the component.

Black multiplier bands. When the multiplier band on a resistor is black, you have to be careful not to make a common mistake. Because black corresponds to zero, you might initially assume that a black multiplier band indicates a zero after the first two digits. Actually, a black multiplier band indicates that *no zeros* follow the first two digits, as illustrated in the following example.

Example 2.3

Determine the rated value of the resistor shown in Figure 2.10.

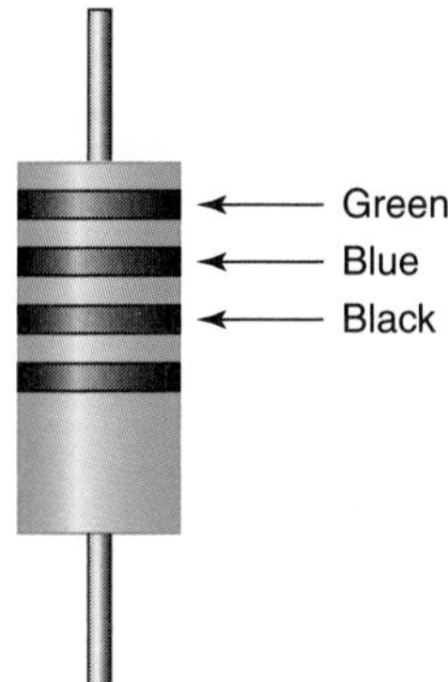

Figure 2.10

Solution: Green = 5, so the first digit in the value is 5.
Blue = 6, so the second digit in the value is 6.
Black = 0, so the multiplier is $10^0 = 1$.

Therefore, the rated value of the component is found as

$$(56 \times 1)\ \Omega = 56\ \Omega$$

Practice Problem 2.3

The first three bands on a resistor are as follows: blue—gray—black. Determine the rated value of the component.

Gold and silver multiplier bands. Two other colors may appear in the multiplier band. These colors and their power-of-ten values are as follows:

$$\text{Gold} = -1 \qquad \text{Silver} = -2$$

These colors are decoded as illustrated in the following examples.

Example 2.4

Determine the rated value of the resistor shown in Figure 2.11.

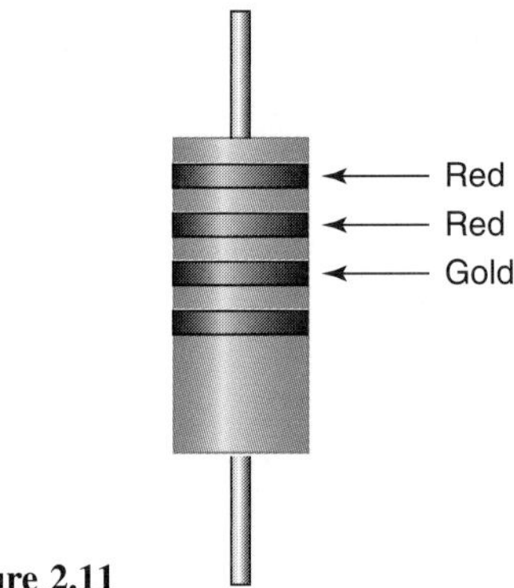

Figure 2.11

Solution: Red = 2, so the first digit in the component value is 2.
Red = 2, so the second digit in the component value is (also) 2.
Gold = −1, so the multiplier is $10^{-1} = 0.1$.

Therefore, the rated value of the component is found as

$$(22 \times 0.1)\ \Omega = 2.2\ \Omega$$

Practice Problem 2.4

The first three bands on a resistor are as follows: violet—green—gold. Determine the rated value of the component.

Example 2.5

Determine the rated value of the resistor shown in Figure 2.12.

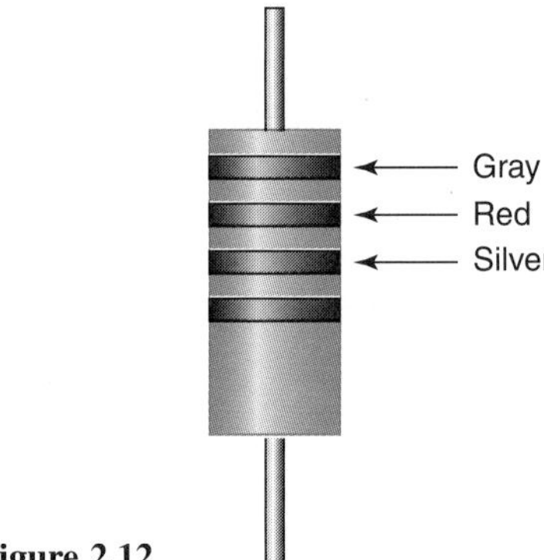

Figure 2.12

Solution: Gray = 8, so the first digit in the component value is 8.
Red = 2, so the second digit in the component value is 2.
Silver = −2, so the multiplier is $10^{-2} = 0.01$.

Therefore, the rated value of the component is found as

$$(82 \times 0.01)\ \Omega = 0.82\ \Omega$$

Practice Problem 2.5

The first three bands on a resistor are as follows: brown—gray—silver. Determine the rated value of the component.

As often as not, you will need to determine the color code for a specified resistance. When this is the case, take a moment to write the component value in standard form and determine the color bands as demonstrated in the following example.

Example 2.6

You need to locate a 360 kΩ resistor. Determine the colors of the first three bands on the component.

Solution: Written in standard form, 360 kΩ = 360,000 Ω. The first three bands are coded for 3, 6, and 4 (which is the number of zeros in the value). The colors that correspond to these numbers are:

3 = orange
6 = blue
4 = yellow

These are the colors (in order) of the first three bands.

Practice Problem 2.6

Determine the colors of the first three bands of a 24 kΩ resistor.

Resistor tolerance. The fourth band on a standard (nonprecision) resistor designates its tolerance. The colors used in the tolerance band are as follows:

Red = 2% Gold = 5% Silver = 10%

When there is no fourth band, the resistor has a 20% tolerance. The following example shows how the first four bands on a resistor are used to determine the guaranteed range of values for the component.

A Practical Consideration: 10% and 20% tolerance resistors are rarely encountered, but their designations are provided so you'll be able to recognize them in older electronic systems.

Example 2.7

Determine the guaranteed range of values for the resistor in Figure 2.13.

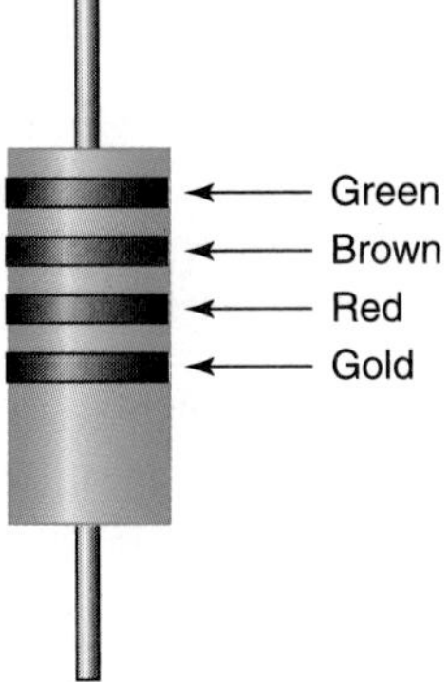

Figure 2.13

Solution: Green = 5, so the first digit in the component value is 5.
Brown = 1, so the second digit in the component value is 1.
Red = 2, so the multiplier is $10^2 = 100$.
Gold = 5% (the tolerance of the component).

The rated value of the component is 5100 Ω, or 5.1 kΩ, and

$$(5.1\text{ k}\Omega) \times 5\% = 255\ \Omega$$
$$5.1\text{ k}\Omega + 255\ \Omega = 5.355\text{ k}\Omega \qquad \text{(maximum)}$$
$$5.1\text{ k}\Omega - 255\ \Omega = 4.845\text{ k}\Omega \qquad \text{(minimum)}$$

> ***Practice Problem 2.7***
>
> The first four bands on a resistor are as follows: brown—blue—green—silver. Determine the guaranteed range of measured values for the component.

A Practical Consideration: Reliability bands are rarely seen on the resistors used in commercial systems. (They were developed primarily for use in military equipment and applications.)

When a fifth band is present on a 5% or 10% tolerance resistor, this band designates the **reliability** of the component. For example, if the reliability rating of a resistor is 1%, it means that no more than 1% (1 out of every 100) resistors will fall out of tolerance after 1000 hours of use. The colors that are used to indicate reliability are coded as follows:

Brown = 1% Red = 0.1% Orange = 0.01% Yellow = 0.001%

If a fifth band is not present (which is usually the case), the component does not have a reliability rating.

Color Code Summary

The color code used for standard resistors is summarized in Table 2.5.

Precision Resistors

Modern resistor technology has led to the production of *precision resistors* with tolerances as low as 0.001%. Precision resistors have a unique color code and list of standard values. A list of these values and a diagram of the color code can be found in Appendix B.

OBJECTIVE 6 ➤

Alphanumeric Codes

An **alphanumeric code** is *one that uses both letters and numbers to express the value of a component.* When the resistor color code cannot be used (because of component size or shape), an alphanumeric code is used. One common resistor alphanumeric code is illustrated in Figure 2.14 and works as described in Table 2.6.

TABLE 2.5 Summary of the Standard (Nonprecision) Resistor Color Code

Band	*Designates*	*Color Values*	
1	1st value digit	Black = 0	Green = 5
2	2nd value digit	Brown = 1	Blue = 6
3	Multiplier (number of zeros after the first two digits)	Red = 2	Violet = 7
		Orange = 3	Gray = 8
		Yellow = 4	White = 9
	Multiplier (values less than zero)	Gold = −1	(10^{-1} = 0.1)
		Silver = −2	(10^{-2} = 0.01)
4	Tolerance	Red = 2%	Gold = 5%
		Silver = 10%	No band = 20%
5	Reliability	Brown = 1%	Red = 0.1%
		Orange = 0.01%	Yellow = 0.001%

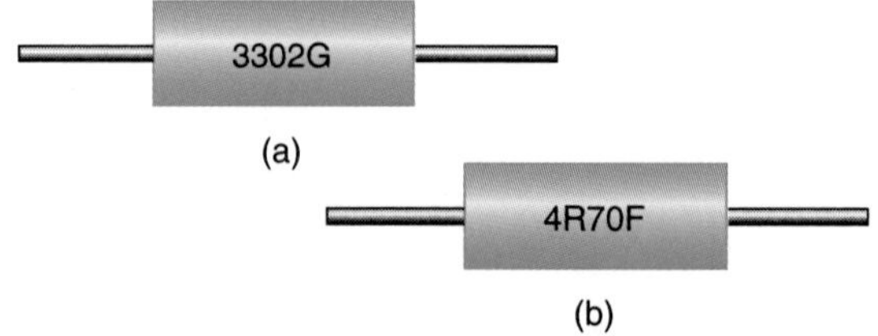

Figure 2.14 An alphanumeric resistor code.

TABLE 2.6 How the Alphanumeric Code in Figure 2.14 Works

When R ≥100 Ω	*When R <100 Ω*
The first three digits in the code are the digits in the resistor value. The fourth digit is the *multiplier.* The letter indicates the *tolerance* of the component and is decoded as follows: F = ±1% G = ±2% K = ±10% M = ±20%	The first three digits in the code are the digits in the resistor value. The letter "R" is used to designate the position of a decimal point within the component value. The final letter indicates the *tolerance* of the component and is decoded in the same manner as shown for higher-value resistors.

According to Table 2.6, the resistor in Figure 2.14a has a value of

$$3302G = 330 \times 10^2 = 33{,}000\ \Omega = 33\ k\Omega\ (\pm 2\%)$$

and the resistor in Figure 2.14b has a value of

$$4R70F = 4.70\ \Omega\ (\pm 1\%)$$

Note that the decimal point has been placed at the point where the "R" is positioned in the value.

Resistor Size

The size of a resistor is usually a good indicator of the amount of heat that it can dissipate. For example, consider the carbon-composition resistors illustrated in Figure 2.15. Each resistor size has an associated *power rating.* In Chapter 3, we will discuss the concept of power in depth. For now, we will simply say that the **power rating** of a component is *a measure of its ability to dissipate heat, measured in watts (W).* The power rating of a carbon-composition resistor (and some other types as well) can be determined by the size of the component. Some other types of resistors (such as wire-wound resistors) have their power ratings printed on the body of the component.

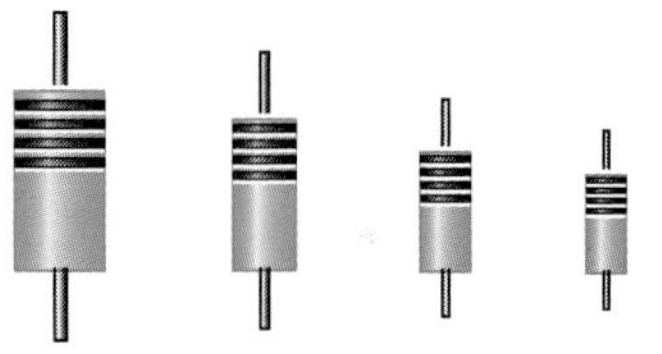

1 W $\frac{1}{2}$ W $\frac{1}{4}$ W $\frac{1}{8}$ W

Figure 2.15
Carbon-composition resistor sizes and power ratings.

Resistor Substitution

OBJECTIVE 7

When substituting one resistor for another, there are three rules to follow:

1. Make sure that the substitute component has the same rated value as the original component.
2. Make sure that the tolerance of the substitute component is equal to (or *lower than*) that of the original component.
3. Make sure that the power rating of the substitute component is equal to (or *greater than*) that of the original component.

As long as these guidelines are followed, the new component will have the proper rated value, be within circuit tolerance, and not be damaged from excessive heat.

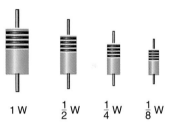

A Practical Consideration: The higher the power rating of a resistor, the greater the diameter of its leads. This may limit your ability to substitute a higher-power resistor for a low-power resistor on a PC board.

Section Review

1. What is indicated by each color band in the standard resistor color code?
2. What does a *black* multiplier band indicate?
3. What does a *gold* multiplier band indicate?
4. What does a fifth color band on a standard resistor indicate?
5. What is an *alphanumeric* code? When is one used to designate the value of a resistor?
6. What is the relationship between the size of a resistor and its power rating?

7. What are the guidelines for substituting one resistor for another?

Critical Thinking

8. Table 2.6 shows how the alphanumeric code in Figure 2.14 is decoded. If you were using this code, how would you know which decoding scheme to use?

2.4 Potentiometers

OBJECTIVE 8 ➤

The **potentiometer** is *a three-terminal resistor whose value can be adjusted (within set limits) by the user.* Throughout our discussion on potentiometer operation, we will use the labels shown in Figure 2.16 to identify the component terminals and resistance values. The potentiometer (or "pot") is designed so that the resistance between the middle terminal (*b*) and each of the outer terminals (*a* and *c*) changes when the control shaft is turned.

(a)

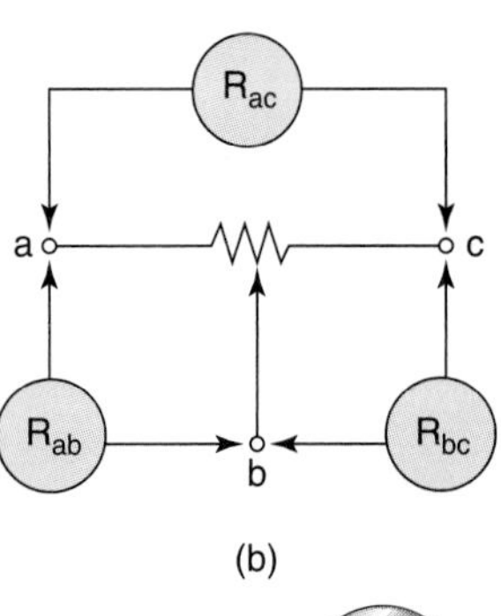

(b)

Figure 2.16 The potentiometer.

Potentiometer Construction and Operation

The construction of a typical pot is illustrated in Figure 2.17. The *a* and *c* terminals are connected to the ends of a length of carbon. A *sliding contact* is moved along the surface of the carbon by rotating the *control shaft.* The **wiper arm** is *a conductor that connects the sliding contact to the control shaft and the* b *terminal.* (Note: The *b* terminal is commonly referred to as the wiper arm, even though the term actually describes the entire mechanism that connects the terminal to the sliding contact.)

Figure 2.18 shows the effect of rotating the control shaft on the component resistance values. When the sliding contact is in the center position (Figure 2.18a), the lengths of carbon between the contact and the outer terminals are equal, and $R_{ab} = R_{bc}$. If we assume

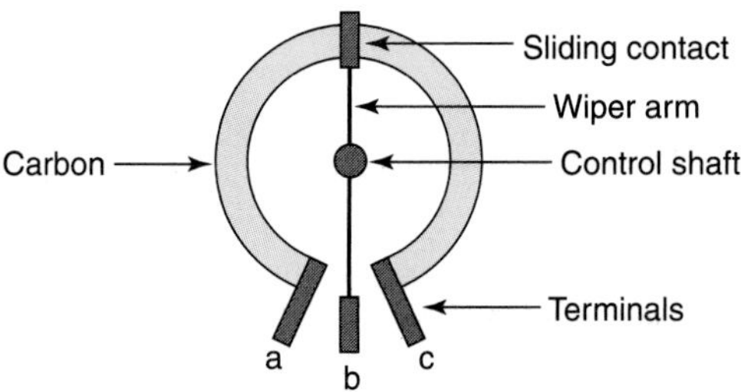

Note: Most potentiometers are constructed so that the moving contact can rotate approximately 350°.

Figure 2.17 Potentiometer construction.

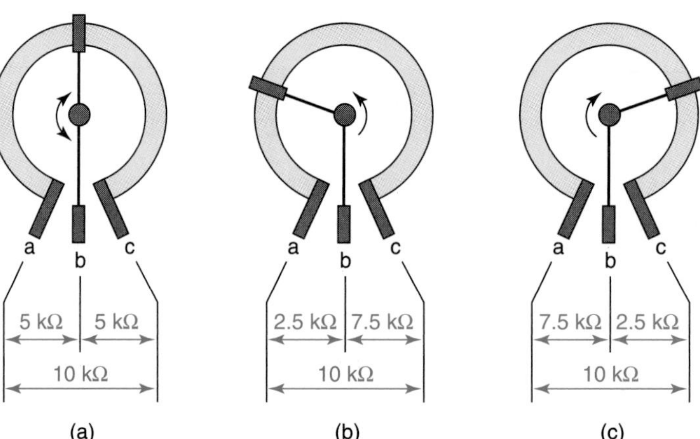

Figure 2.18 The effect of turning the control shaft on the component resistances.

that the potentiometer has a value of $R_{ac} = 10\ \text{k}\Omega$, then R_{ab} and R_{bc} must each be half this value, or $5\ \text{k}\Omega$.

The relationship between the values of R_{ab}, R_{bc}, and R_{ac} in Figure 2.18a can be stated mathematically as

$$R_{ac} = R_{ab} + R_{bc} \tag{2.1}$$

This relationship holds true regardless of the position of the sliding contact.

If the control shaft is turned counterclockwise, the sliding contact moves as shown in Figure 2.18b. The amount of carbon between the wiper arm and the *a* terminal decreases, so R_{ab} decreases. At the same time, R_{bc} increases. Note that the sum of the two resistances is still equal to R_{ac}. Turning the control shaft clockwise (Figure 2.18c) causes the sliding contact to move toward the *c* terminal. As a result, R_{bc} decreases and R_{ab} increases.

Potentiometer Ratings

A potentiometer is rated in terms of its resistance and power dissipation capability, just like a resistor. Specifically:

- The **resistance rating** of a pot is *the resistance between its outer terminals.*
- The **power dissipation** rating is *a measure of the maximum amount of heat that the component can dissipate (in watts).*

The resistance rating of a pot is normally printed on the component. The resistance may be written as a straightforward value (such as $5\ \text{k}\Omega$, $10\ \text{k}\Omega$, etc.) or in an alphanumeric code similar to the one discussed in Section 2.3. For example, a pot may be marked as shown in Figure 2.19. This simple code works as follows:

- The first two digits are the digits in the value of the potentiometer.
- The third digit is the *multiplier.*

Thus, the maximum value of the pot in Figure 2.19 is $50 \times 10^2 = 5000\ \Omega = 5\ \text{k}\Omega$.

Figure 2.19 Potentiometer value code.

The power rating of a pot is not normally printed on the device. Generally, you must consult the manufacturer's catalog to determine the device's maximum power rating.

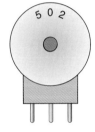

Trimmer Potentiometers

The **trimmer potentiometer** is *a potentiometer that is used in low-power applications.* Several trimmer pots are shown in Figure 2.20a. The control shaft on most potentiometers can be turned one full rotation (360°) or less. However, **multi-turn potentiometers** (like the *ten-turn* pot in Figure 2.20b) are *potentiometers that have control shafts geared to allow rotations that are greater than 360°.*

The advantage of multi-turn potentiometers is that they have better *resolution* than single-turn pots. The **resolution** of a potentiometer is *the change in resistance per degree of control shaft rotation.* Multi-turn potentiometers have higher resolution than single-turn pots, meaning that one degree of rotation causes a much smaller change in resistance. This means it is easier to adjust a multi-turn pot to a very specific value.

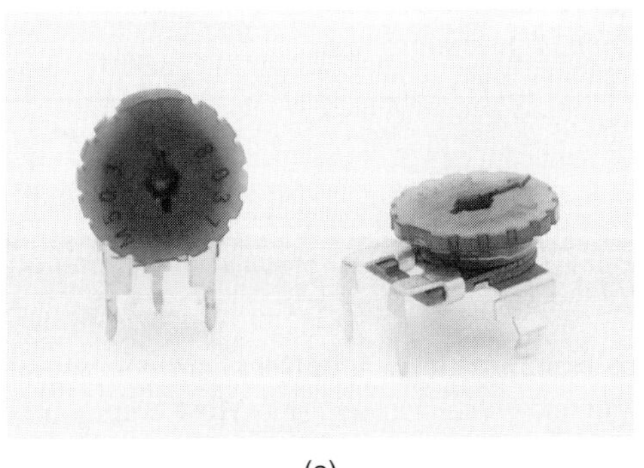

(a)

(b)

Figure 2.20 Trimmer potentiometers.

Figure 2.21 Gang-mounted potentiometers.

Gang-Mounted Potentiometers

In some cases, potentiometers are **gang-mounted** as shown in Figure 2.21. The pots on the left share a common control shaft. When the value of one pot is varied, the value of the other is varied by the same amount. The gang-mounted pots on the right actually have separate control shafts. The outer shaft controls one of the pots, while the inner controls the other.

Section Review

1. What is a *potentiometer*?
2. Briefly describe the operation of a potentiometer.
3. Describe the construction of a typical potentiometer.
4. Describe the relationship between the three interterminal resistances of a potentiometer.
5. List and describe the various potentiometer ratings.
6. What is a *trimmer* potentiometer?
7. What is a *multi-turn* potentiometer? What advantage does a multi-turn pot have over a standard potentiometer?
8. Describe the two types of *gang-mounted* potentiometers.

2.5 Batteries

OBJECTIVE 9 ➤ In simpler dc circuits, voltage is obtained from a battery. A **battery** is *a component that converts chemical, thermal, or light energy into electrical energy.* Most common batteries produce a difference of potential across their terminals as a result of a continuous chemical reaction.

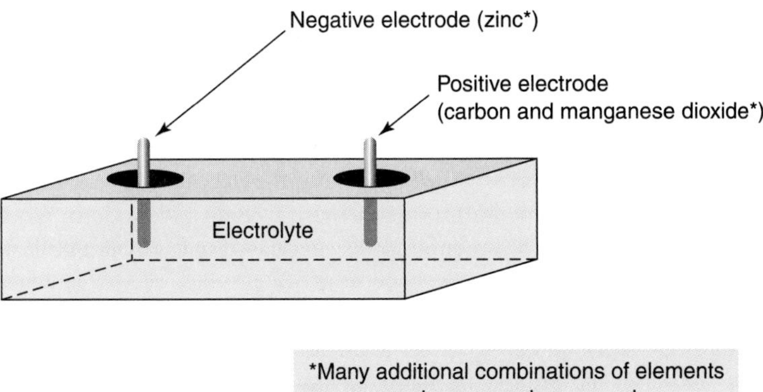

Figure 2.22 A chemical cell.

Cells

A battery is made up of one or more cells. A **cell** is *a unit designed to produce electrical energy through thermal (heat), chemical, or optical (light) means.* A typical cell produces a voltage through chemical means. A basic cell can be represented as shown in Figure 2.22.

The cell shown in Figure 2.22 has two *electrodes* and a chemical called the *electrolyte.* The **electrodes** are *the battery terminals.* These terminals are made using various elements or compounds. The electrodes are in constant contact with an **electrolyte,** which is *a chemical that interacts with the electrodes and serves as a conductor between them.* As a result of the chemical interaction between the electrolyte and the electrodes, the negative terminal generates electrons, and the positive terminal is depleted of electrons. Thus, a difference of potential (voltage) exists between the electrodes.

The schematic symbol for a single cell is shown in Figure 2.23a. A battery may contain a single cell or a series of cells, as shown in Figure 2.23b. Note that the voltage between the terminals of the battery is equal to the sum of the cell voltages.

Battery Capacity

Batteries are generally rated in terms of their output voltage and capacity. The **capacity** rating of a battery is *a measure of how long the battery will last at a given output current, measured in ampere-hours (Ah).*

The **ampere-hour (Ah)** is *a unit of measure that equals the product of discharge current* (the current supplied by the battery) *and time.* For example, a battery with a capacity of 1 Ah will last for 1 hour at a discharge rate of 1 A, or for 10 hours at a discharge rate of 100 mA.

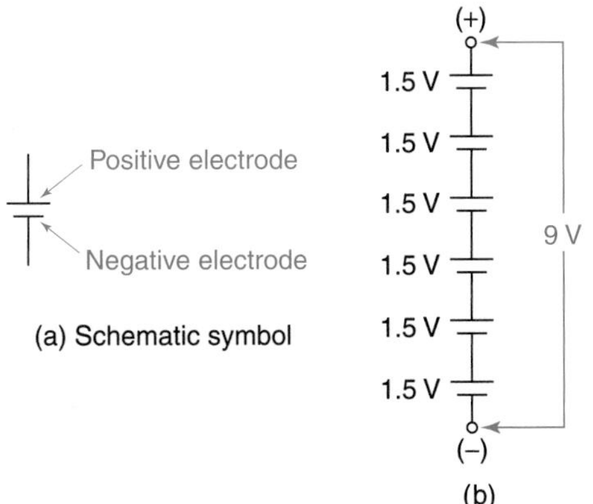

Figure 2.23

Common Batteries (Primary Cells)

In some cells, the chemical action between the electrolyte and the electrodes causes permanent changes in the chemical structure of the device. This type of cell is called a *primary cell.* A **primary cell** is *a type of cell that cannot be recharged.* Primary cells are often referred to as *voltaic cells* or **dry cells** (because they contain dry electrolytes).

In general, batteries are named after one or more of their electrode or electrolyte elements (or compounds). Some commonly used batteries are described as follows:

- *Carbon-zinc* batteries contain one or more 1.5 V primary cells. They are extremely common and inexpensive but have relatively low capacity (Ah) ratings. As a result, they drain relatively quickly in most applications.
- *Alkaline* batteries are similar in many ways to carbon-zinc batteries. Alkaline batteries contain one or more 1.55 V primary cells. The electrolyte used is an *alkali* (high pH) compound. These batteries have higher capacity (Ah) ratings than carbon-zinc batteries, so they last longer in most applications.
- *Mercury* batteries are relatively small components that contain 1.35 V primary cells. The positive electrode is made of mercuric oxide, which is where the component gets its name. These batteries have high capacity (Ah) ratings and a long shelf life, and are typically used in consumer goods such as digital watches, calculators, and cameras.
- *Lithium-iodine* batteries are relatively small components that contain 2.88 V primary cells. Like mercury batteries, they have high capacity (Ah) ratings and a long shelf life, but they are also very expensive. Like mercury batteries, these batteries are typically used in consumer goods such as digital watches, personal computers, calculators, and cameras.

Common Batteries (Secondary Cells)

A **secondary cell** is *a type of cell that can be recharged.* By generating a current through a secondary cell (in the opposite direction of the normal cell current), the chemical structure of the cell is restored, and the battery is recharged. Secondary cells are commonly referred to as *rechargeable cells* or **wet cells** (because they contain liquid electrolytes). The most commonly used rechargeable batteries are described as follows:

- *Lead-acid* batteries contain 2 V secondary cells. Lead-acid cells have relatively high capacity ratings (up to 20 Ah) and can last for hundreds of charge/discharge cycles. Typically, lead-acid batteries can be used for up to four years before having to be replaced.
- *Nickel-cadmium (Ni-Cd)* batteries contain 1.15 V secondary cells. These cells have capacity ratings up to 5 Ah and can last for hundreds of charge/discharge cycles. At equal capacity ratings, Ni-Cd cells are capable of higher output currents than lead-acid batteries, and they can be used for up to 10 years before having to be replaced.

Connecting Batteries

In many applications, more than one battery is needed to supply a required voltage or current. Figure 2.24 shows one of the two methods that are used to connect several batteries.

When two (or more) batteries are connected as shown in Figure 2.24, the batteries are said to be connected in *series.* Note that the negative terminal of one battery is connected to the positive terminal of the other. When two or more batteries are connected in this fashion, the results are as follows:

- The total voltage provided by the batteries is equal to the *sum* of their individual voltages.
- The maximum possible output current for the two is equal to the *maximum current* for either battery alone.

The other method of connecting two (or more) batteries is shown in Figure 2.25. When two or more batteries are connected in this fashion, the batteries are said to be connected in *parallel.* Note that the two batteries are connected negative-to-negative and positive-to-positive. The results of connecting two or more batteries in this fashion are as follows:

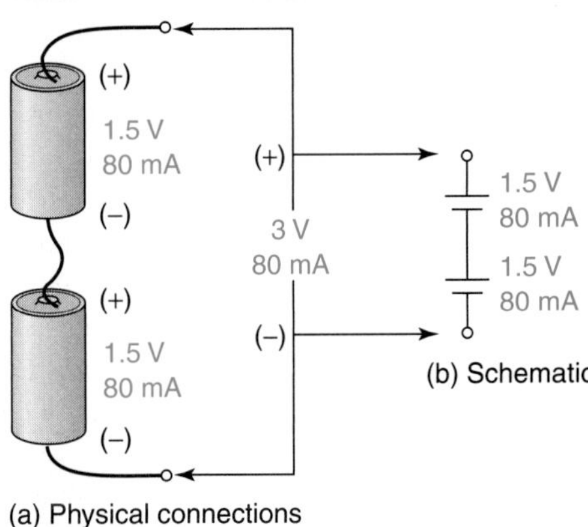

Figure 2.24 Series-connected batteries.

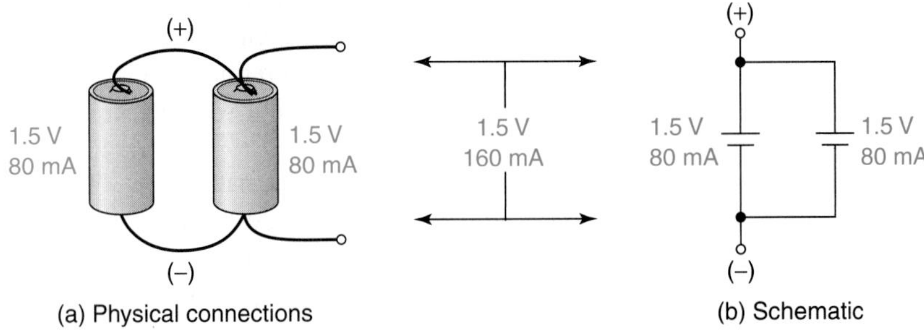

Figure 2.25 Parallel-connected batteries.

- The total voltage is equal to the *individual* battery voltages.
- The maximum possible output current is equal to the *sum* of the battery currents.

In summary, batteries can be connected in series to increase the total *voltage*, or they can be connected in parallel to increase the maximum possible *current*.

Section Review

1. What is a *battery*? What is a *cell*?
2. Briefly describe the operation of a cell.
3. How are cells connected to form a battery?
4. What are the differences between *primary cells* and *secondary cells*?
5. What is *battery capacity*? What is its unit of measure?
6. List and describe the commonly used primary and secondary cells.
7. What is the result of connecting batteries in *series*? In *parallel*?

2.6 DC Power Supplies

◄ *OBJECTIVE 10*

One method of obtaining a variety of dc voltages is to use a *dc power supply.* A **dc power supply** is *a piece of equipment with dc outputs that can be adjusted to provide any voltage within its supply limits.* The front panel of a typical dc power supply is represented in Figure 2.26. It should be noted that dc power supplies are produced by a variety of manufacturers, and not all of them have the same features. However, the controls shown in Figure 2.26 are typical of those found on most dc power supplies.

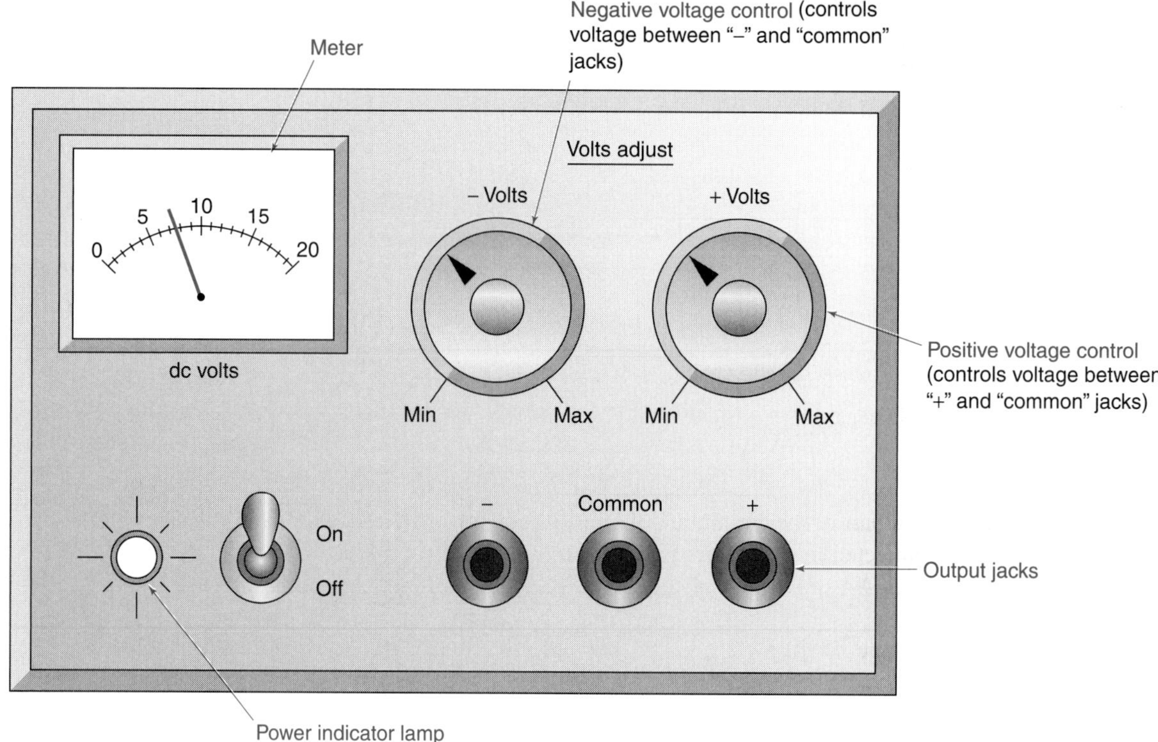

Figure 2.26 A dc power supply panel.

DC Outputs

As shown in Figure 2.26, there are three dc output jacks on the dc power supply control panel. A simplified representation of these outputs is shown in Figure 2.27. The power supply contains two internal voltage sources with a common connection point.

Normally, the dc power supply is connected to a circuit in one of three fashions. When connected as shown in Figure 2.28a, current takes the path indicated by the arrows.

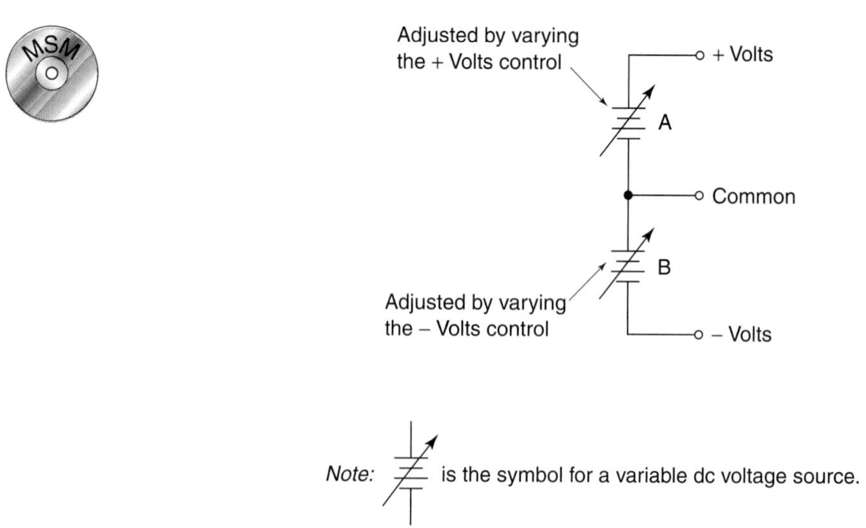

Figure 2.27 A simplified representation of a dc power supply.

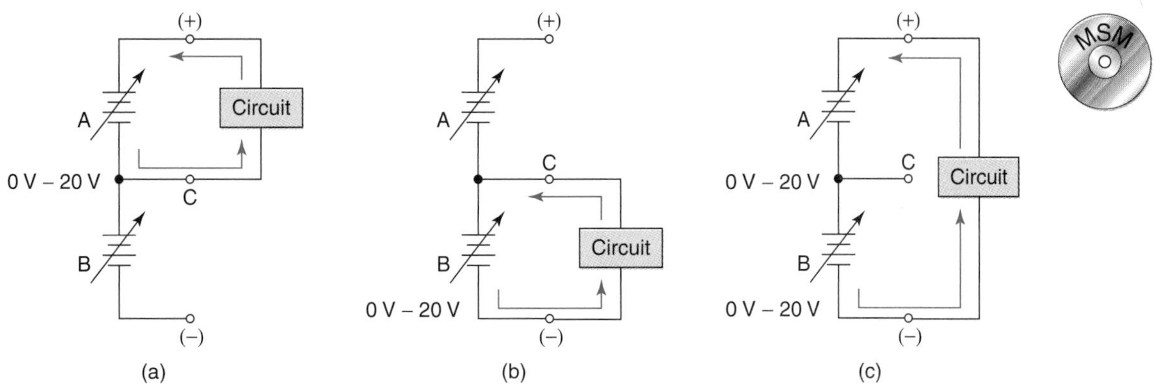

Figure 2.28 Power supply connections.

Voltage source B is not a part of the circuit, and the maximum power supply output voltage is 20 V. When connected as shown in Figure 2.28b, current takes the path indicated, and voltage source A is not a part of the circuit. Once again, the maximum output voltage is 20 V. When connected as shown in Figure 2.28c, current takes the path indicated, and both voltage sources are a part of the circuit. The voltage sources are connected in *series,* so the maximum output voltage is equal to the sum of the source voltages, or 40 V.

Positive Voltage Versus Negative Voltage

Voltages are normally identified as having a specific **polarity;** that is, the voltage is defined as either *positive* or *negative.* Consider the positive and negative terminals of the voltage source shown in Figure 2.29. We could describe this source in either of two ways:

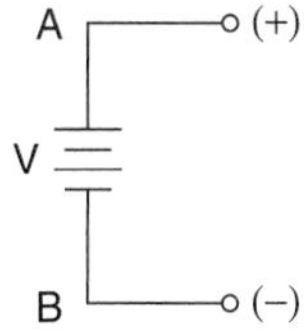

Figure 2.29

1. Side A is positive with respect to side B.
2. Side B is negative with respect to side A.

While both statements are correct, they do not assign a specific polarity to the source. To assign a polarity to any voltage, we must agree on a *reference point.*

The **common terminal** of the dc power supply shown in Figure 2.28 is *the terminal that is common to both voltage sources, and therefore, is used as the reference point*; that is, the output polarity of either source is described in terms of its relationship to this point. For example, the *negative* terminal of source A is conntected to the common point. Since the top side of source A is positive with respect to this point, we say that the output from source A is a *positive dc voltage.* By the same token, the *positive* terminal of source B is connected to the common point, so we say that the output from source B is a *negative dc voltage.* Note that the common point is considered to be at 0 V.

Using a DC Power Supply

When you are in the process of constructing a circuit, the circuit is built and inspected for errors *before the dc power supply is turned on.* Once you are sure that you have built the circuit correctly:

1. Adjust the voltage controls on the dc power supply to their 0 V settings.
2. Connect the dc power supply to the circuit.
3. Turn on the dc power supply.
4. Slowly increase the output from the power supply to the desired value.

One Final Note

We have touched on some of the basics of working with dc power supplies. However, as with all electronic equipment, you should take the time to get thoroughly familiar with *your* dc power supply before attempting to use it.

Section Review

1. What is a *dc power supply*?
2. What determines the *polarity* of a voltage?
3. What is the *common terminal* of a power supply?
4. List the steps that you should take when connecting a dc power supply to a circuit.

2.7 Switches and Circuit Protectors

In any circuit, there must be a complete path between the terminals of the voltage source. A complete path is required to have a flow of charge (current) between the terminals. Switches and circuit protectors (such as fuses and circuit breakers) are devices that make or break the current path in a circuit under certain circumstances.

OBJECTIVE 11 ➤

Switches

A **switch** is *a device that allows you to make or break the connection between two or more points in a circuit.* A switch is connected in a circuit as shown in Figure 2.30. Note that a path for current is made when the switch is closed, and that the current path is broken when the switch is open.

The moving contact in a switch is called a **pole.** A single-pole switch (like the one represented in Figure 2.30) has only one moving contact. A double-pole switch (like the one represented in Figure 2.31) has two moving contacts.

The switches represented in Figures 2.30 and 2.31 can be closed in only one position; that is, they can make or break a connection with only one nonmoving contact. *The nonmoving contact in a switch* is referred to as a **throw,** and the switches shown in these figures are classified as single-throw switches. In contrast, several double-throw switches are shown in Figure 2.32.

Switches are generally described in terms of the number of poles and throws they contain. For example, the switch represented in Figure 2.30 is described as a *single-pole, single-throw* (SPST) switch, meaning that it has a single moving contact that can be made

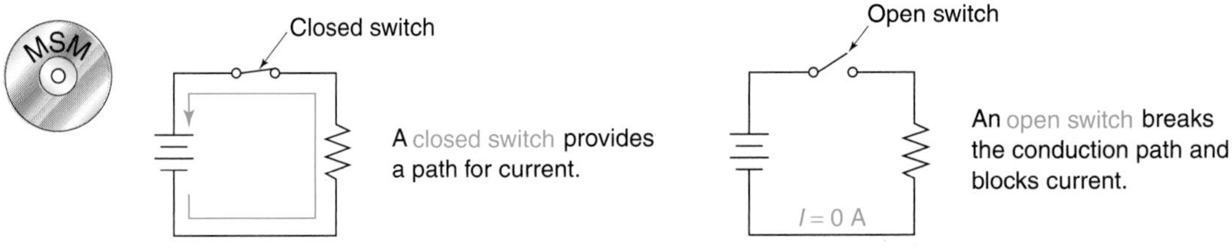
MSM
Closed switch
A closed switch provides a path for current.
Open switch
I = 0 A
An open switch breaks the conduction path and blocks current.

Figure 2.30 Closed and open switches.

MSM
Note: The dashed line in the symbol indicates that the poles are not independent. Both are open, or both are closed.

Figure 2.31 Double-pole switches.

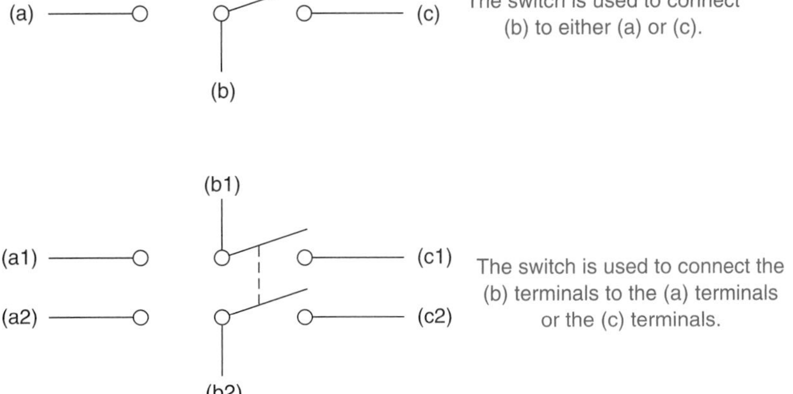
(a)
(b)
(c)
The switch is used to connect (b) to either (a) or (c).
(b1)
(a1)
(c1)
(a2)
(c2)
(b2)
The switch is used to connect the (b) terminals to the (a) terminals or the (c) terminals.

Figure 2.32 Double-throw switches.

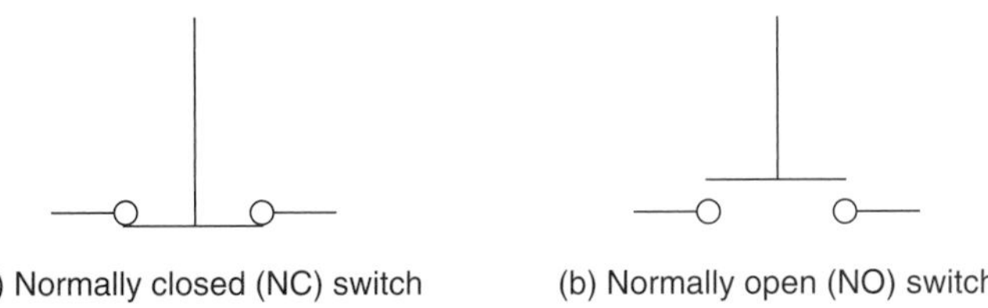
(a) Normally closed (NC) switch
(b) Normally open (NO) switch

Figure 2.33 Normally closed (NC) and normally open (NO) switches.

to close in only one position. The switch in Figure 2.31 is described as a *double-pole, single-throw* (DPST) switch, meaning that it has two moving contacts that can be made to close in only one position. Figure 2.32 shows a *single-pole, double-throw* (SPDT) switch and a *double-pole, double-throw* (DPDT) switch.

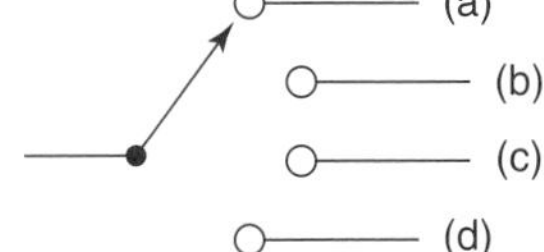
(a)
(b)
(c)
(d)

Figure 2.34 A rotary switch.

Normally Open and Normally Closed Switches

A **normally closed (NC)** switch is *a push-button switch that must be activated (pushed) to break the connection between its terminals.* When the button is *not* being pushed, the switch makes a connection between its terminals. The schematic symbol for an NC switch is shown in Figure 2.33a.

A **normally open (NO) switch** is *a push-button switch that must be activated (pushed) to make a connection between its terminals.* When the button is *not* being pushed, the connection between its terminals is broken. The schematic symbol for this type of switch is shown in Figure 2.33b. Most NC and NO switches are **momentary switches,** meaning that a *connection is made or broken only as long as the button is pushed.*

Rotary Switches

The simplest explanation of a **rotary switch** is that it is *a switch with one or more poles and any number of throws.* The schematic symbol for a simple rotary switch is shown in Figure 2.34. By turning the control shaft, a connection is made between the pole and one of the throws.

DIP Switches

DIP (dual in-line package) switches are *SPST switches that are grouped in a single case.* A group of DIP switches and their schematic symbol are shown in Figure 2.35. DIP switches are the most commonly used mechanical switches in modern electronic circuits and systems.

(a)

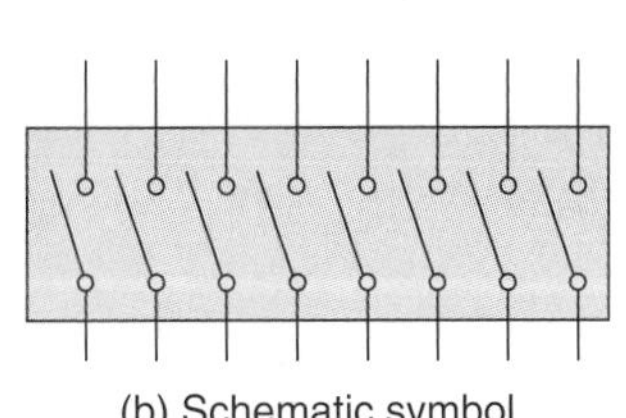

(b) Schematic symbol

Figure 2.35 DIP switches.

OBJECTIVE 12 ➤

Fuses

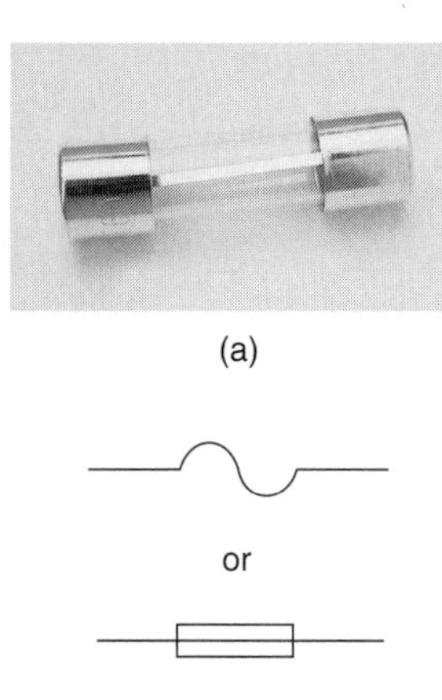
(a)

or

(b) Schematic symbols

Figure 2.36
A fuse and its schematic symbols.

A **fuse** is *a device that is designed to open automatically if its current exceeds a specified value.* In some ways, a fuse is like a *normally closed* switch. Under normal circumstances, a fuse provides a path for the flow of charge (current). However, unlike an NC switch, it must be replaced once it has opened. The simplest type of fuse and its schematic symbols are shown in Figure 2.36.

As current passes through a conductor, heat is produced. The amount of heat varies directly with the amount of current. The fuse in Figure 2.36 contains a thin conductor that is designed to melt at a given temperature. If the current through the fuse reaches a specified level, the conductor heats to its melting point. When the conductor melts, the connection between the ends of the fuse is broken, acting as an open switch.

The purpose of a fuse is to protect a circuit or system from excessive current. For this reason, the fuse is normally placed between the power supply and the circuit, as shown in Figure 2.37. As you can see, the total circuit current in Figure 2.37 passes through the fuse. If the fuse opens, the current path is broken. This protects the circuit.

Fuse Ratings

Fuses have a *current rating* and a *voltage rating.* The **current rating** of a fuse is *the maximum allowable fuse current.* If the current rating of a fuse is exceeded, the fuse will **blow** (open). The **voltage rating** of a fuse is *the maximum voltage that an open fuse can block.* If the voltage across an open fuse exceeds its voltage rating, the air in the fuse may **ionize** (charge), causing the open fuse to conduct again. The result is that the circuit is no longer protected.

Replacing a Fuse

When a fuse is blown, it must be replaced. When replacing a fuse, these guidelines should be followed:

- Make sure that all power is removed from the system. For example, when replacing the fuse in a stereo system, make sure that the system is unplugged.
- Use an *exact* replacement. That is, use a replacement fuse that is the same size and has the same current and voltage ratings as the original fuse.
- *Never* replace a fuse with one that has a higher current rating!

Once a fuse has been replaced, one of two things will happen:

- The circuit operation will return to normal.
- The new fuse will blow when power is restored.

If the circuit operation returns to normal, then the problem was probably nothing more than an old or defective fuse. If the new fuse blows, then there is a problem that is causing excessive current to be drawn from the power supply. In this case, the problem must be diagnosed and repaired before attempting to replace the fuse again.

Types of fuses There are basically three types of fuses: *high-speed instantaneous, normal instantaneous,* and *time delay.* As you have probably guessed, these categories rate fuses in terms of the time required for them to blow.

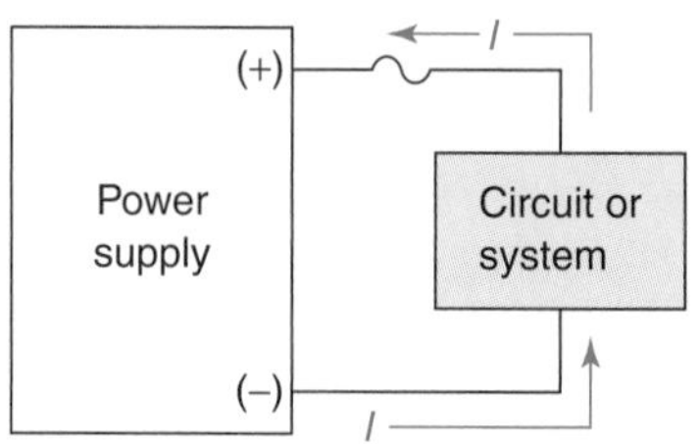

Figure 2.37 A typical fuse application.

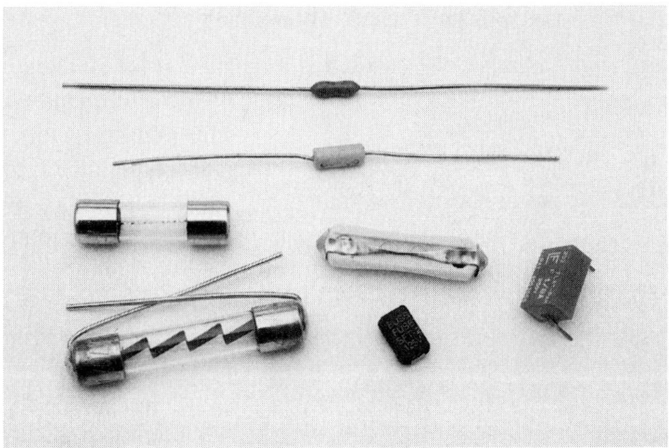

Figure 2.38 Fuses of various shapes and sizes.

High-speed instantaneous fuses react faster to an overload (excessive current) condition than any other type of fuse. High-speed instantaneous fuses are also known as *fast-acting* or *quick-acting fuses.*

Normal instantaneous fuses are somewhat slower to respond to an overload condition. Normal instantaneous fuses are also known as *instantaneous* or *normal-blow fuses.*

Time-delay (slow-blow) fuses are similar to normal-blow fuses, but they can handle short-duration overloads without blowing. Many appliances (such as clothes dryers) produce a current surge when they are turned on. If a slow-blow fuse is in the circuit of such an appliance, the current surge will not blow the fuse.

Fuses come in a variety of shapes and sizes, as shown in Figure 2.38. The current and voltage ratings for a given fuse are printed on the component. The fuse type can be determined by looking up the part number in the manufacturer's catalog.

Circuit Breakers

◀ *OBJECTIVE 13*

Circuit breakers are *devices that are designed to protect circuits from overload conditions,* just like fuses. However, unlike fuses, circuit breakers are not destroyed when they are activated. A typical circuit breaker and its schematic symbol are shown in Figure 2.39.

In essence, a circuit breaker acts as a normally-closed switch. When an overload condition develops, the circuit breaker opens, and the circuit is protected. The reset switch on the breaker returns the device to its NC position, and the breaker can then be used again.

Section Review

1. In terms of their operating characteristics, what is the primary difference between a *switch* and a *circuit protector*?
2. What is a *pole*? What is a *throw*?

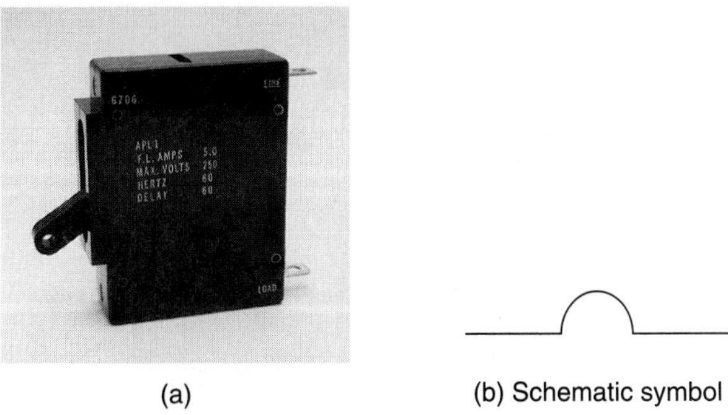

Figure 2.39 A circuit breaker and its schematic symbol.

3. List and describe the four basic types of switches.
4. What is a *normally open switch*? What is a *normally closed switch*?
5. What is a *momentary switch*?
6. Describe the operation of a basic fuse.
7. List and describe the common fuse ratings.
8. List the guidelines for replacing a fuse.
9. What is the primary difference between a circuit breaker and a fuse?

2.8 Measuring Current, Voltage, and Resistance

In the lab manual that accompanies this text, there are exercises that deal with measuring current, voltage, and resistance. This section serves as an introduction to these exercises.

Multimeters

Many types of meters are used to measure various electrical properties. For example, the **ammeter** is *a meter that is used to measure current,* the **voltmeter** is *a meter that is used to measure voltage,* and the **ohmmeter** is *a meter that is used to measure resistance.* A **multimeter** is *a single piece of test equipment that can operate as any of these three meters.* That is, it can be used to measure current, voltage, or resistance. A typical *digital multimeter (DMM)* is shown in Figure 2.40.

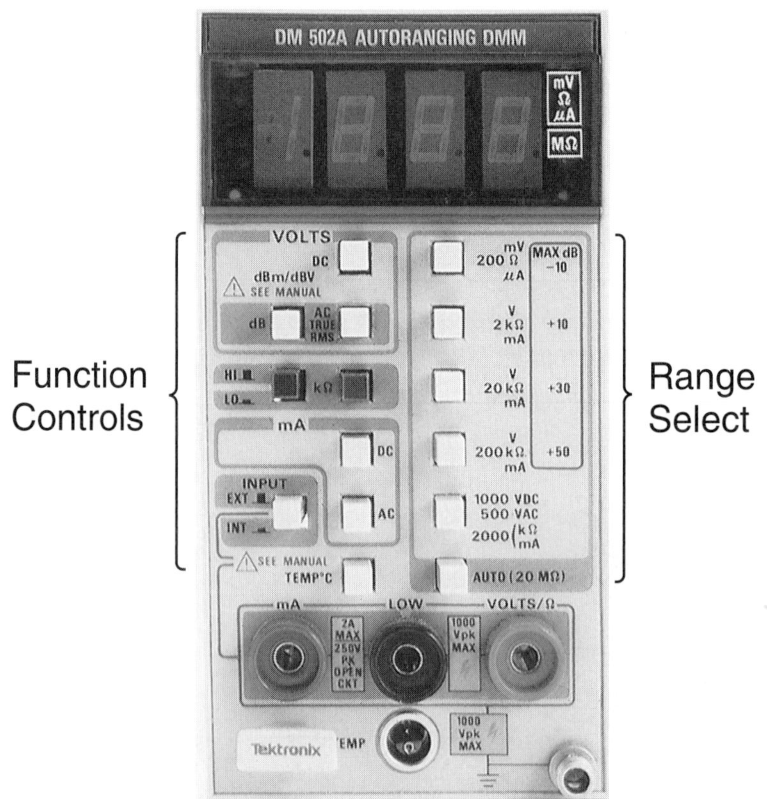

Figure 2.40 A typical digital multimeter (DMM).

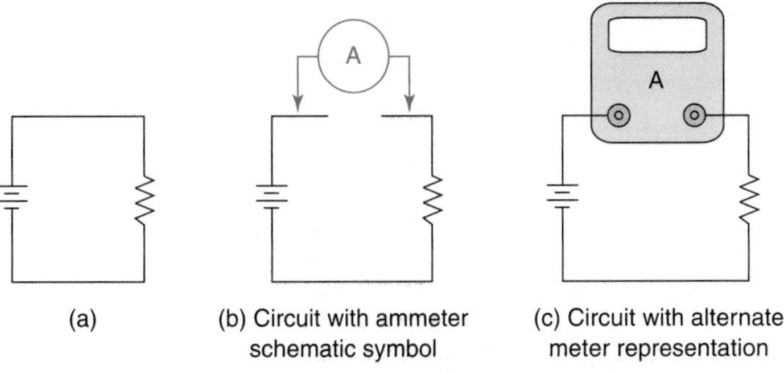

Figure 2.41 Ammeter connection.

In lab, you will be shown how to set the controls on your particular DMM. In this section, you will be shown how to connect the meter to a circuit or component to take the desired measurement.

Measuring Current

◄ *OBJECTIVE 14*

Current is measured by inserting the meter *in the current path.* Let's say that we want to measure the current in the circuit shown in Figure 2.41a. You must insert the meter so that the current passes through the meter. This is accomplished by *breaking the current path* and then inserting the meter in its place, as shown in Figure 2.41b. Figure 2.42 shows the actual connection of a DMM for measuring current.

Measuring Voltage

Voltage is measured by connecting the voltmeter *across* a component. This means that one of the voltmeter leads is connected to one side of the component while the other lead is connected to the other side of the component, as shown in Figure 2.43.

As you can see, the circuit current path is not broken when you are measuring voltage. The actual connection of a DMM for measuring voltage is shown in Figure 2.44.

Measuring Resistance

Measuring resistance can get a little tricky, because resistance cannot be measured with power applied to the circuit. In most cases, the safest and most reliable method of measuring resistance is to simply remove the resistor from the circuit. Once the resistor is removed, the meter is connected *across* the component just like a voltmeter, as shown in Figure 2.45.

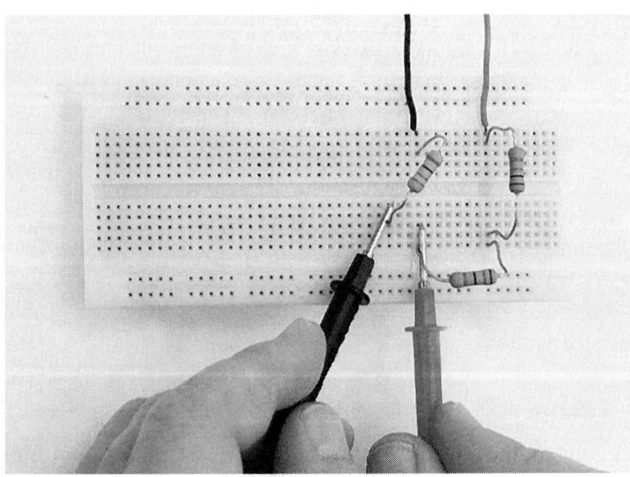

Figure 2.42 Proper DMM connections for measuring current.

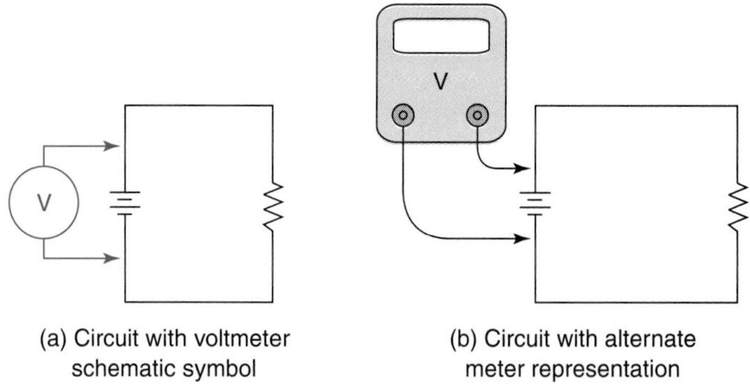

(a) Circuit with voltmeter schematic symbol

(b) Circuit with alternate meter representation

Figure 2.43 Voltmeter connection.

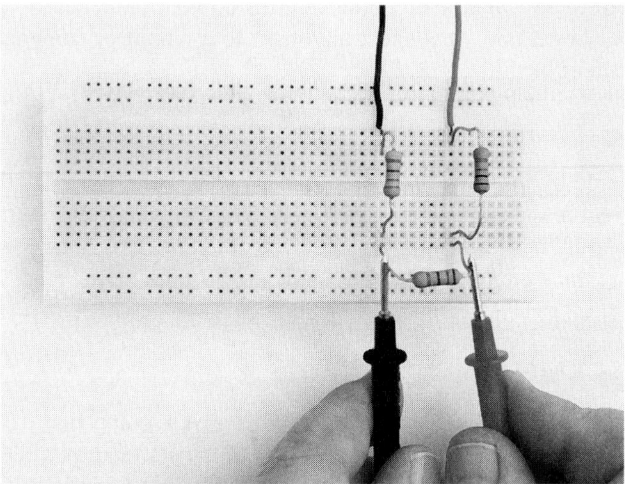

Figure 2.44 Proper DMM connection for measuring voltage.

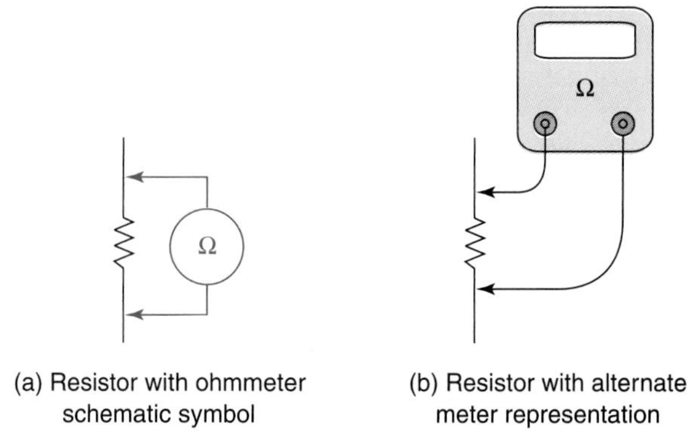

(a) Resistor with ohmmeter schematic symbol

(b) Resistor with alternate meter representation

Figure 2.45 Ohmmeter connection.

An important point needs to be made at this time: *If you hold both leads of a resistor while measuring its resistance, you'll get a faulty reading.* When you hold both ends of a resistor, the resistance of your skin affects the reading. To avoid this problem, you should hold only one resistor lead, as shown in Figure 2.46.

One Final Note

There is a lot to know about the proper use of test equipment. When you are in lab, make sure that you take the time to get thoroughly familiar with the meters before attempting

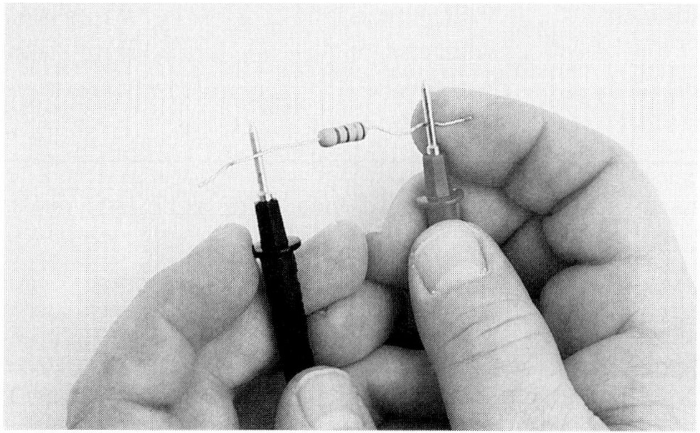

Figure 2.46 Proper DMM connection for measuring resistance.

to use them. This will help you to avoid faulty readings and possibly damaging the meters as well.

Section Review

1. What does an *ammeter* measure? How is it connected in a circuit?
2. Why is an ammeter connected as described in Question 1?
3. How is a *voltmeter* connected in a circuit?
4. How is an *ohmmeter* connected to a component?
5. What precautions must be taken when measuring resistance?

Key Terms

The following terms were introduced in this chapter on the pages indicated:

Alphanumeric code, 54
American Wire Gauge (AWG), 42
Ammeter, 68
Ampacity, 43
Ampere-hour (Ah), 59
Average breakdown voltage, 44
Battery, 58
Blow, 66
Capacity, 59
Carbon-composition resistor, 45
Carbon-film resistor, 47
Cell, 59
Circuit breaker, 67
Common terminal, 63
Current rating, 66
DC power supply, 61
Dry cell, 60
Dual in-line package (DIP) switch, 65
Electrode, 59
Electrolyte, 59
Fixed resistor, 45
Fuse, 66
Gang-mounted potentiometer, 58
High-speed instantaneous fuse, 67
Integrated resistor, 47
Ionize, 66
Metal-film resistor, 46
Momentary switch, 65
Multimeter, 68
Multi-turn potentiometer, 57
Negative dc voltage, 63
Normal instantaneous fuse, 67
Normally closed (NC) switch, 65
Normally open (NO) switch, 65
Ohmmeter, 68
Parameter, 44
Polarity, 63
Pole, 64
Positive dc voltage, 63
Potentiometer, 56
Power dissipation, 57
Power rating, 55
Precision resistor, 48
Primary cell, 60
Printed circuit board, 44
Reliability, 54
Resistance rating, 57
Resistor, 45
Resistor color code, 47
Resolution, 57
Rotary switch, 65
Secondary cell, 60
Solid wire, 42
Stranded wire, 42
Surface-mount package (SMP), 47
Switch, 64
Throw, 64
Time delay (slow-blow) fuse, 67
Tolerance, 48
Trace, 44

Trimmer potentiometer, 57
Variable resistor, 45
Voltage rating, 66
Voltmeter, 68
Wet cell, 60
Wiper arm, 56
Wire-wound resistor, 46

Practice Problems

1. Determine the guaranteed range of values for a 110 kΩ resistor with a 10% tolerance.
2. Determine the guaranteed range of values for a 1.5 MΩ resistor with a 2% tolerance.
3. Determine the guaranteed range of values for a 360 kΩ resistor with a 5% tolerance.
4. Determine the guaranteed range of values for a 6.8 Ω resistor with a 10% tolerance.
5. For each of the color codes listed, determine the rated value of the component.

Band 1	*Band 2*	*Band 3*	*Value*
a. Brown	Blue	Brown	_____
b. Red	Violet	Green	_____
c. Orange	White	Yellow	_____
d. Green	Brown	Black	_____
e. Yellow	Violet	Gold	_____
f. Green	Blue	Orange	_____

6. For each of the color codes listed, determine the rated value of the component.

Band 1	*Band 2*	*Band 3*	*Value*
a. Red	Black	Brown	_____
b. Orange	Black	Silver	_____
c. Green	Brown	Black	_____
d. White	Brown	Green	_____
e. Brown	Blue	Blue	_____
f. Blue	Gray	Silver	_____

7. Determine the colors of the first three bands for each of the following resistor values.
   a. 33 kΩ
   b. 15 Ω
   c. 910 kΩ
   d. 2.2 kΩ
   e. 120 kΩ
   f. 10 MΩ
   g. 1.8 Ω
   h. 0.24 Ω
8. Determine the colors of the first three bands for each of the following resistor values.
   a. 11 Ω
   b. 20 MΩ
   c. 680 Ω
   d. 3.3 Ω
   e. 160 kΩ
   f. 0.13 Ω
   g. 9.1 MΩ
   h. 36 Ω
9. For each of the color codes listed, determine the rated component value and the guaranteed range of measured values.
   a. Red—Yellow—Yellow—Gold
   b. Orange—Blue—Gold—Red
   c. Gray—Red—Black
   d. Blue—Gray—Silver—Silver
10. For each of the color codes listed, determine the rated component value and the guaranteed range of measured values.
   a. Yellow—Orange—Silver—Gold
   b. Brown—Gray—Yellow—Red
   c. Yellow—Violet—Orange—Silver
   d. Brown—Green—Silver—Gold

11. Determine the value and tolerance indicated by each of the following resistor codes.
  a. 2203K   c. R360G
  b. 51R0F   d. 4703M

Looking Back

These problems relate to material presented in earlier chapters. The chapters are identified in brackets.

12. Convert the following numbers to engineering notation form without the use of your calculator. [Introduction]
  a. $4.83 \times 10^4\ \Omega$
  b. -7.21×10^{-1} V
13. Calculate the conductance of a 5.1 MΩ resistor. [Chapter 1]
14. Three coulombs of charge pass a point every 150 ms. Calculate the value of the circuit current. [Chapter 1]
15. Calculate the resistance of an 18 in. length of copper that has a cross-sectional area of 0.008 in. [Chapter 1]

Pushing the Envelope

16. What is the resistance of a 12.8 cm length of 40 gauge wire?
17. What length of 14 gauge wire (in centimeters) has a resistance of 2 mΩ?

Answers to the Example Practice Problems

2.1. 66.64 to 69.36 kΩ
2.2. 47 kΩ
2.3. 68 Ω
2.4. 7.5 Ω
2.5. 0.18 Ω
2.6. Red—Yellow—Orange
2.7. 1.44 to 1.76 MΩ

Multisim File References

Figure	File Reference
2.7	ETF2_7.msm
2.16	ETF2_16.msm
2.27	ETF2_27.msm
2.28	ETF2_28.msm
2.30	ETF2_30.msm
2.31	ETF2_31.msm
2.32	ETF2_32.msm
2.41	ETF2_41.msm
2.42	ETF2_42.msm

chapter 3

Ohm's Law and Power

Objectives

After studying the material in this chapter, you should be able to:

1. Describe the relationship between *voltage, current,* and *resistance.*
2. Predict how a change in either voltage or resistance will affect circuit current.
3. Use *Ohm's law* to calculate any one of the following values, given the other two: current, voltage, and resistance.
4. Describe how Ohm's law can be used in circuit troubleshooting.
5. List the characteristics of an *open circuit.*
6. List the characteristics of a *short circuit.*
7. Discuss *power* and its relationship to current and voltage.
8. Calculate power, given any two of the following: current, voltage, and resistance.
9. Describe the relationship between power and heat.
10. Determine the minimum acceptable power rating for a resistor in any circuit.
11. Define and calculate *efficiency* and the *kilowatt-hour* (*kWh*) unit of energy.
12. Calculate any two of the following values, given the other two: current, voltage, resistance, and power.
13. Calculate the range of current values for a given circuit.
14. Define the terms *load* and *full load.*

In Chapter 1, you were shown that *current* is the *flow of charge* in a circuit, *voltage* is the *difference of potential* that causes the current, and *resistance* is the *opposition to current* that is provided by the circuit. You were also shown that the units of current, voltage, and resistance can be defined in terms of each other, as follows:

- One *ampere* is the amount of current that is generated when a 1 V difference of potential is applied across 1 Ω of resistance.
- One *volt* is the difference of potential required to generate 1 A of current through 1 Ω of resistance.
- One *ohm* is the amount of resistance that limits current to 1 A when a 1 V difference of potential is applied.

These statements clearly indicate that there is a relationship between current, voltage, and resistance. This relationship, called *Ohm's law,* is a powerful tool that can be used to:

- Predict how circuit current will respond to a change in voltage or resistance.
- Calculate the value of any one basic circuit property (current, voltage, or resistance) when the other two values are known.
- Determine the possible causes of a fault in a circuit.

In this chapter, we will discuss Ohm's law and its applications. We will also discuss *power* (another basic circuit property) and its relationship to current, voltage, and resistance. Finally, we will take a look at some miscellaneous topics that apply to basic circuits.

3.1 Ohm's Law

In the early nineteenth century, a German physicist named Georg Simon Ohm found that *current is directly proportional to voltage and inversely proportional to resistance.* This statement has come to be known as **Ohm's law.** Ohm's law tells us that:

◄ *OBJECTIVE 1*

1. The current through a fixed resistance:
   a. *Increases* when the applied voltage *increases.*
   b. *Decreases* when the applied voltage *decreases.*
2. The current generated by a fixed voltage:
   a. *Decreases* when the resistance *increases.*
   b. *Increases* when the resistance *decreases.*

We'll start our discussion by taking a closer look at these relationships.

The Relationship Between Current and Voltage

According to Ohm's law, the greater the difference of potential across a circuit, the greater the resulting current. This relationship is illustrated in Figure 3.1.

In Figure 3.1a, the meter shows that the source is set to 10 V, generating 100 mA of current. When the applied voltage is doubled (to 20 V), as shown in Figure 3.1b, the current also doubles (to 200 mA). In other words, *the current increases proportionally to the*

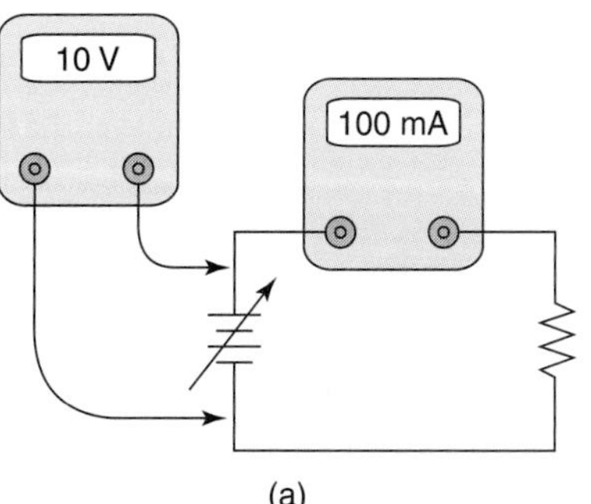

(a)

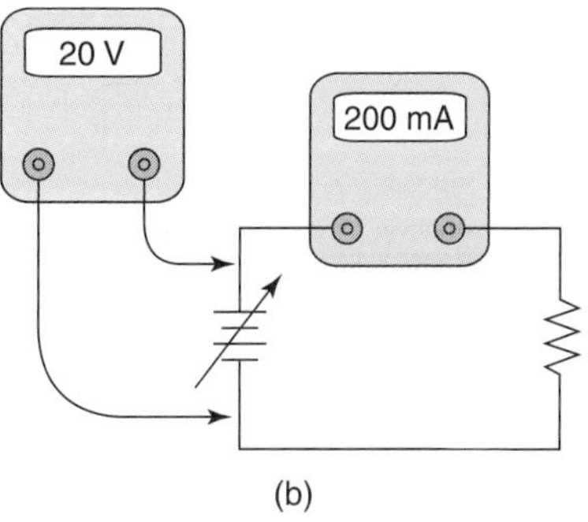

(b)

Figure 3.1 The relationship between current and voltage.

A Practical Consideration: Current varies in direct proportion to voltage only if resistance is fixed (does not change). If resistance also changes, current will not vary in direct proportion to voltage.

A Practical Consideration: Current varies in inverse proportion to resistance only if voltage is fixed (does not change). If voltage also changes, current will not vary in inverse proportion to resistance.

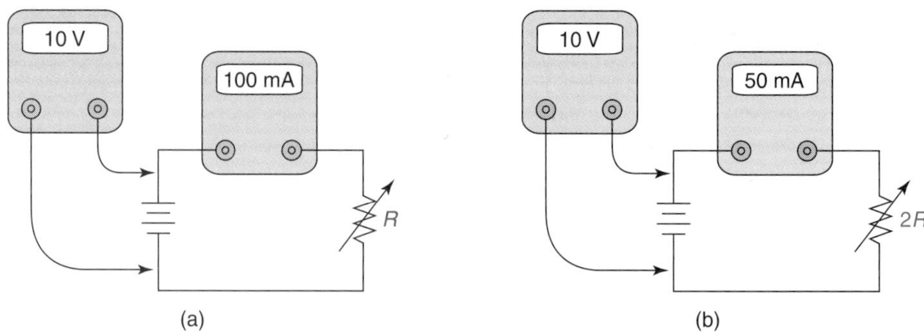

Figure 3.2 The relationship between current and resistance.

increase in voltage. By the same token, a *decrease* in voltage causes a proportional *decrease* in current. In either case, the current changes *as a result of the change in voltage.*

The Relationship Between Current and Resistance

According to Ohm's law, the greater the resistance, the lower the current at a given voltage. This relationship is illustrated in Figure 3.2.

In Figure 3.2a, the applied voltage is fixed at 10 V, generating 100 mA of current. When the circuit resistance is *doubled,* as shown in Figure 3.2b, the circuit current drops to half its original value (50 mA), even though the applied voltage has not changed. In other words, *the current decreases proportionally to the increase in resistance.* By the same token, a *decrease* in resistance causes a proportional *increase* in current. In either case, the current changes *as a result of the change in resistance.*

Predicting Circuit Behavior

OBJECTIVE 2 ➤ Ohm's law indicates the current changes as a result of a change in voltage or a change in resistance. The relationships illustrated in Figures 3.1 and 3.2 can be summarized as shown in Table 3.1.

As you can see, Ohm's law can be used to predict how a change in circuit voltage or resistance will affect circuit current. At the same time, there is a subtle message here: *When you adjust any circuit component or setting, there will always be some type of response to that change.*

TABLE 3.1 A Summary of Ohm's Law Relationships

Cause	*. . . Results in . . .*	*Effect*
Increased voltage		Increased current[a]
Decreased voltage		Decreased current[a]
Increased resistance		Decreased current[b]
Decreased resistance		Increased current[b]

[a]Assumes that resistance is fixed (i.e., does not change).

[b]Assumes that voltage is fixed (i.e., does not change).

Section Review

1. What is Ohm's law?
2. Describe the relationship between *voltage* and *current.*
3. Describe the relationship between *resistance* and *current.*
4. What happens to the current through a variable resistor when you adjust its value? Explain your answer.

5. The resistance in a circuit is increasing. If it continues to increase, what will eventually happen to the circuit current? Critical Thinking

6. If circuit voltage and resistance *both* double in value, what happens to the value of circuit current? Explain your answer.

3.2 Basic Circuit Calculations

You have seen how Ohm's law can be used to predict the response of a circuit to a change in voltage or current. In this section, you will be shown how it is used to calculate circuit current, voltage, or resistance when the other two values are known.

Using Ohm's Law to Calculate Current

◄ *OBJECTIVE 3*

Ohm's law also provides a specific *mathematical* relationship between the values of current, voltage, and resistance in a circuit. This relationship is given as

$$I = \frac{V}{R} \tag{3.1}$$

where

I = the circuit current
V = the applied voltage
R = the circuit resistance

Equation (3.1) allows us to calculate the value of current in a circuit when the applied voltage and circuit resistance are known, as demonstrated in the following example.

Example 3.1

Calculate the value of current for the circuit in Figure 3.3.

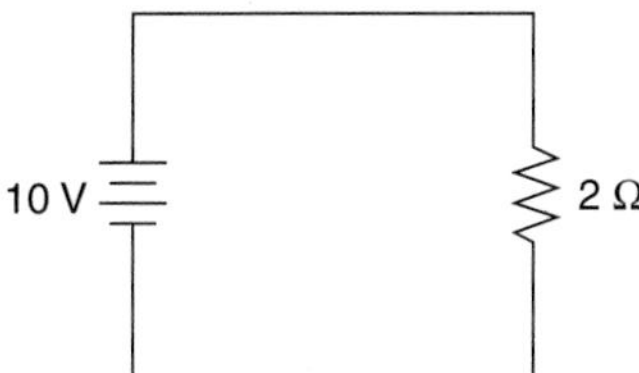

Figure 3.3

Solution: Using the values of V and R given in Figure 3.3, the value of the circuit current is found as

$$I = \frac{V}{R} = \frac{10\text{ V}}{2\ \Omega} = 5\text{ A}$$

Practice Problem 3.1

A circuit like the one in Figure 3.3 has values of $V = 18$ V and $R = 100\ \Omega$. Calculate the value of the current in the circuit.

Using Ohm's Law to Calculate Voltage

If we multiply both sides of equation (3.1) by R, we get the following useful equation:

$$V = IR \tag{3.2}$$

Equation (3.2) allows us to calculate the applied voltage when the circuit current and resistance values are known, as demonstrated in the following example.

Example 3.2

Calculate the value of the applied voltage in the circuit shown in Figure 3.4.

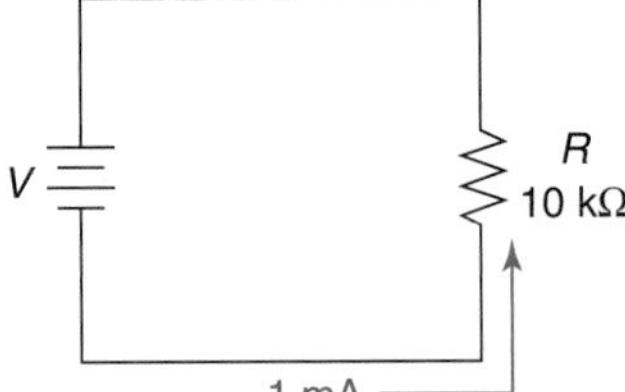

Figure 3.4

Solution: Using the values of I and R given in Figure 3.4, the value of the applied voltage is found as

$$V = IR = (1\text{ mA})(10\text{ k}\Omega) = 10\text{ V}$$

Practice Problem 3.2

A circuit like the one in Figure 3.4 has values of $I = 500\ \mu\text{A}$ and $R = 3.3\text{ k}\Omega$. Calculate the value of the applied voltage for the circuit.

Using Ohm's Law to Calculate Resistance

If we divide both sides of equation (3.2) by I, we get the another useful equation:

$$R = \frac{V}{I} \tag{3.3}$$

This equation can be used to calculate the circuit resistance when the current and voltage values are known, as demonstrated in the following example.

Example 3.3

Calculate the value of R for the circuit in Figure 3.5.

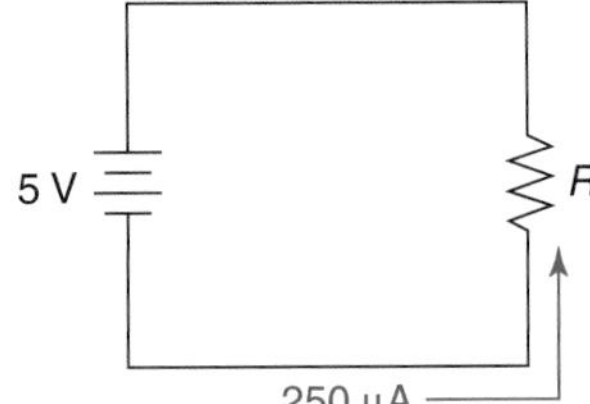

Figure 3.5

Solution: Using the values of current and voltage given in Figure 3.5, the value of R is found as

$$R = \frac{V}{I} = \frac{5\text{ V}}{250\ \mu\text{A}} = 20\text{ k}\Omega$$

Practice Problem 3.3

A circuit like the one shown in Figure 3.5 has values of $V = 12$ V and $I = 40$ mA. Calculate the value of the circuit resistance.

Verifying the Effects of Changes in Voltage and Resistance on Current

It was stated earlier that the current through a fixed resistance is *directly proportional* to the applied voltage. This relationship is verified in the following example.

Example 3.4

The circuit in Figure 3.6 has the values shown. If the applied voltage doubles, how does the circuit current respond?

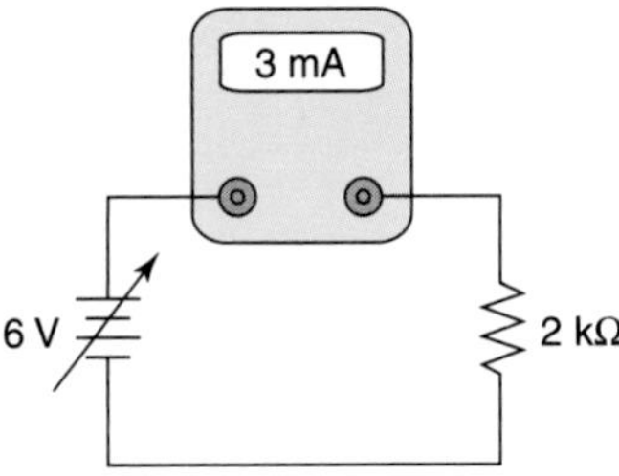

Figure 3.6

Solution: Using $V = 12$ V, the circuit current is found as

$$I = \frac{V}{R} = \frac{12 \text{ V}}{2 \text{ k}\Omega} = 6 \text{ mA}$$

As you can see, the circuit current doubles when the applied voltage doubles. Therefore, the two are directly proportional.

Practice Problem 3.4

Verify that the circuit current in Figure 3.6 will drop to half its value if the applied voltage is cut in half.

You have been told that the current generated by a fixed voltage is *inversely proportional* to the circuit resistance. This relationship is verified in the following example.

Example 3.5

The circuit shown in Figure 3.7 has the values shown. If the circuit resistance doubles, how does the circuit current respond?

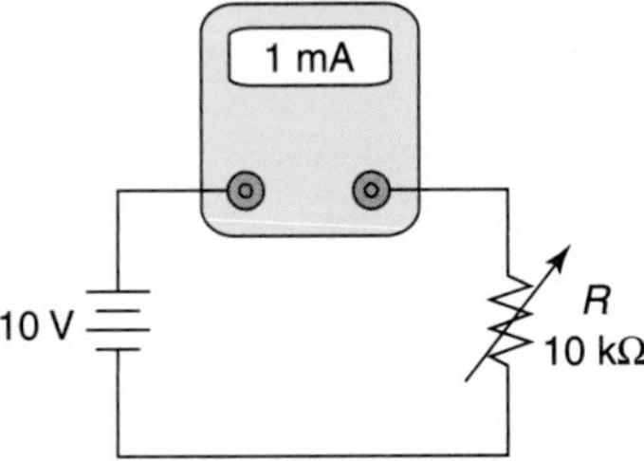

Figure 3.7

Solution: Using $R = 20$ kΩ, the circuit current is found as

$$I = \frac{V}{R} = \frac{10 \text{ V}}{20 \text{ k}\Omega} = 500 \ \mu\text{A}$$

As you can see, doubling the circuit resistance causes the current to drop to half its original value. Therefore, the two are inversely proportional.

Practice Problem 3.5

Verify that the current in Figure 3.7 will double if the circuit resistance is cut in half.

Summary

In this section, you have been shown how Ohm's law can be used to calculate any of the basic circuit properties (current, voltage, or resistance) when the values of the other two are known. The mathematical forms of Ohm's law are

$$I = \frac{V}{R} \qquad V = IR \qquad R = \frac{V}{I}$$

Section Review

1. What are the three forms of Ohm's law?
2. How does current respond to a change in *voltage*?
3. How does current respond to a change in *resistance*?

Critical Thinking

4. Using the circuit in Figure 3.6, verify the validity of the following statement: *A graph of voltage versus current for the circuit is a straight line.*
5. A portable cassette player uses two AA batteries to generate a 3 V dc supply voltage. How could you determine the resistance of the cassette player without the use of an ohmmeter?

3.3 Using Ohm's Law to Diagnose a Circuit Failure

OBJECTIVE 4 ➤ You have seen how Ohm's law is used to predict circuit behavior and to calculate circuit values of voltage, current, and resistance. In this section, you will see how Ohm's law can be used to help you diagnose the cause of a circuit failure.

Troubleshooting

Troubleshooting is *the process of locating faults in a circuit or system.* When a circuit doesn't operate as expected, Ohm's law is one of the best tools you have to help diagnose the problem. For example, Ohm's law tells you that:

1. Current *increases* as a result of:
   a. An increase in the applied voltage.
   b. A decrease in the circuit resistance.
2. Current *decreases* as a result of:
   a. A decrease in the applied voltage.
   b. An increase in the circuit resistance.

The first of these statements can be used to determine the possible causes of the problem shown in Figure 3.8a. According to Ohm's law, the measured current in the circuit should

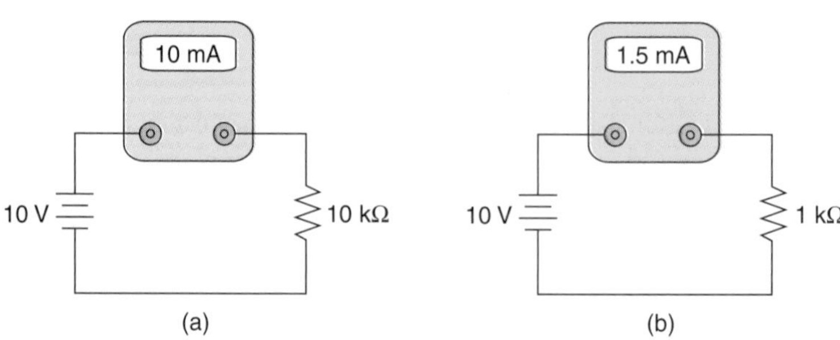

Figure 3.8 Current readings indicating circuit faults.

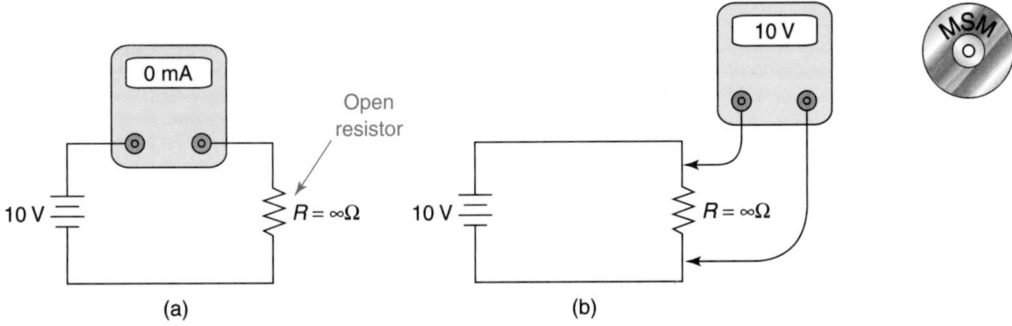

Figure 3.9 Open-circuit characteristics.

be 1 mA. However, the ammeter shows a reading of 10 mA. This high current reading indicates that the applied voltage is too high or the circuit resistance is too low.

The second statement can be used to determine the possible causes of the problem shown in Figure 3.8b. Ohm's law tells us that the measured current in the circuit should be 10 mA, but the ammeter reads 1.5 mA. This low current reading indicates that the applied voltage is too low or the circuit resistance is too high.

In each case, Ohm's law told you the types of problems that you were looking for. Once you know the type of problem, a few simple measurements will tell you which problem is actually present.

Circuit Faults: The Open Circuit

◄ *OBJECTIVE 5*

Two common types of circuit faults are referred to as the *open circuit* and the *short circuit.* An **open circuit** is *a physical break in a conduction path.* Figure 3.9a illustrates the current and resistance characteristics of an open circuit. As you can see, the path for conduction is broken, and the current through that component drops to zero. If you attempt to measure the resistance of an open component, its resistance will be too high for the ohmmeter to measure.

When an open develops in a basic circuit, the full applied voltage is felt across the open. This point is illustrated in Figure 3.9b. When the voltmeter is placed across the open, it is effectively connected across the open terminals of the battery. Thus, we measure the full applied voltage across the open. As a summary, the characteristics of an open circuit are:

- The circuit current drops to zero.
- The resistance of the component is too high to measure.
- The full applied voltage is measured across the open.

Circuit Faults: The Short Circuit

◄ *OBJECTIVE 6*

A *short circuit* is the opposite of an open circuit. A **short circuit** is *a low resistance path between two points that does not normally exist.* For example, the resistor in Figure 3.10 has been "shorted" with a wire that has a resistance (R_W) of 0.5 Ω. With a total short-circuit resistance of 0.5 Ω, the current in Figure 3.10 is found as

$$I = \frac{V}{R_W} = \frac{10\text{ V}}{0.5\ \Omega} = 20\text{ A}$$

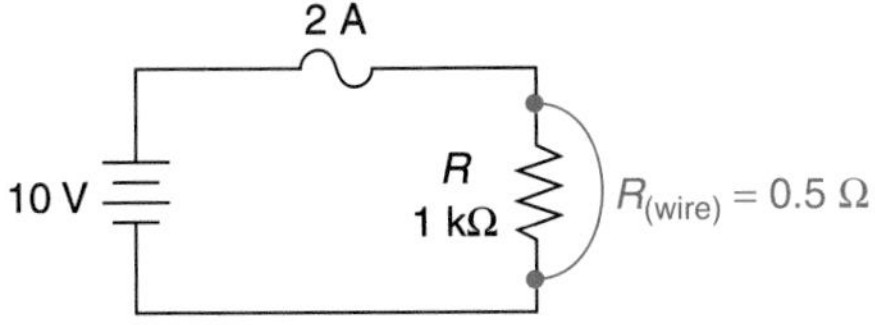

Figure 3.10 A short circuit.

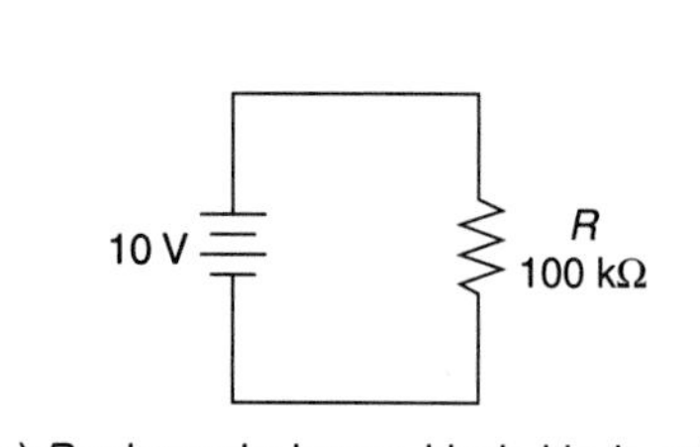

(a) *R* color code: brown, black, black, gold

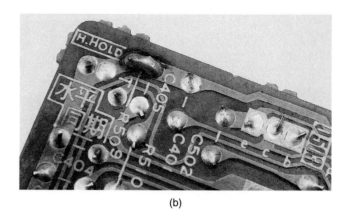

(b)

Figure 3.11

In this case, the circuit current exceeds the rating of the fuse, and the fuse opens. If the circuit had no fuse, the excessive current would eventually cause the short-circuit current path (or the power supply) to burn open. Thus, the current in a short circuit is extremely high until the fuse (or some other component) burns open as a result of the high current. As a summary, the characteristics of a short circuit are:

- The resistance of the shorted component is extremely low.
- The current through the short-circuit current path is extremely high. This high current may cause a circuit fuse (or some other component) to open.
- The voltage across the shorted component is typically extremely low.

Causes of apparently shorted resistors.

Note that carbon-composition resistors do not internally short circuit. When a carbon-composition resistor appears to be shorted, the short is usually caused by one of the conditions shown in Figure 3.11. In Figure 3.11a, the schematic calls for a 100 kΩ resistor. However, the color code of the component indicates that its value is 10 Ω. In this case, the resistor has accidentally been replaced with one that has a much lower value. This low resistance appears as a shorted component to the voltage source.

Note: The solder bridge in Figure 3.11b is exaggerated. In practice, a solder bridge is usually more difficult to see.

Figure 3.11b shows another possible cause of a shorted resistor. Components are connected to a PC board using a compound called **solder,** which is *a high-conductivity compound with a low melting point* (temperature). When a drop of melted solder is applied to the connection point between the resistor and the PC board, it cools and forms a solid electrical connection between the two points. *If the melted solder accidentally makes a connection between two points that shouldn't be connected,* we have what is called a **solder bridge.** The solder bridge in Figure 3.11b forms a low resistance path between the two sides of a component, effectively shorting that component. In this case, the short is repaired by removing the solder bridge.

It should be noted that the other types of resistors (wire-wound, metal-film, integrated, etc.) can and do form internal short circuits. When one of these components shorts internally, it must be replaced.

Summary

When a fault develops in a circuit, Ohm's law can help you to identify the nature of that fault. An *open* results in a reduction of circuit current, while a *short* causes it to increase significantly. As you will see in future chapters, these symptoms make themselves apparent in a variety of ways.

Section Review

1. What is *troubleshooting*?
2. What purpose does Ohm's law serve when troubleshooting? Give specific examples.
3. What is an *open circuit*?
4. What are the primary symptoms of an open circuit?
5. What is a *short circuit*?

6. What are the primary symptoms of a short circuit?
7. What can cause a carbon-composition resistor to appear to be shorted?
8. When a soldering iron is plugged into a known-good receptacle, the circuit breaker trips. What is the most likely cause of the problem? Explain your answer.

Critical Thinking

9. The coil in a space heater is essentially a length of wire that heats when a current passes through it. If a space heater blows only cold air, what is the most likely cause of the problem? Explain your answer.

3.4 Power

◀ *OBJECTIVE 7*

In Chapter 2, you were told that the power rating of a component is a measure of its ability to dissipate (throw off) heat. In this section, we will take a closer look at power and how its value is calculated.

Power

Technically defined, **power (*P*)** is *the amount of energy used per unit time.* When *a voltage source uses one joule of energy per second to generate current,* we say that it has used one **watt (W)** of power. By formula,

$$1 \text{ watt} = 1 \text{ joule/second}$$

This relationship is illustrated in Figure 3.12. The power supply in the figure is using one joule of energy to move one coulomb of charge per second. Therefore, it is using one joule of energy per second. This rate of energy usage is equal to 1 watt (W). Note that the voltage and current relationships shown in the figure were introduced in Chapter 1.

Calculating Power

◀ *OBJECTIVE 8*

Figure 3.12 implies that power is related to both voltage and current. The mathematical relationship among the three values is given as

$$P = VI \tag{3.4}$$

where

P = the power used, in watts (W)
V = the applied voltage, in volts (V)
I = the generated current, in amperes (A)

The validity of this relationship can be demonstrated using the circuit in Figure 3.13. In the circuit shown, 1 V is used to generate 2 A of current. For this circuit,

$$P = VI = (1 \text{ V})(2 \text{ A}) = \left(1 \frac{\text{joule}}{\cancel{\text{coulomb}}}\right)\left(2 \frac{\cancel{\text{coulombs}}}{\text{second}}\right) = 2 \frac{\text{joules}}{\text{second}} = 2 \text{ W}$$

As you can see, the product of voltage and current yields a result in watts (W). The use of equation (3.4) is demonstrated further in Example 3.6.

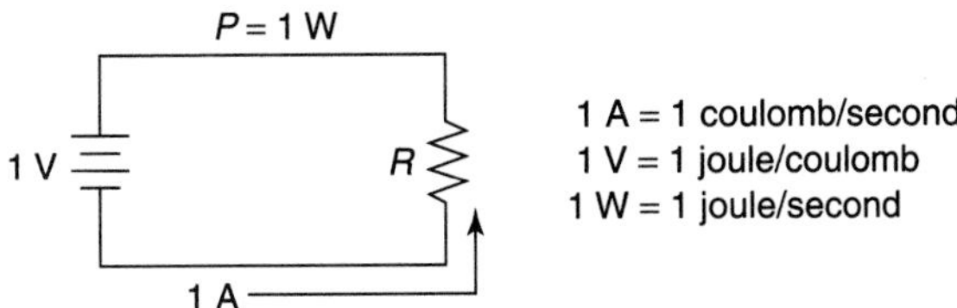

Figure 3.12

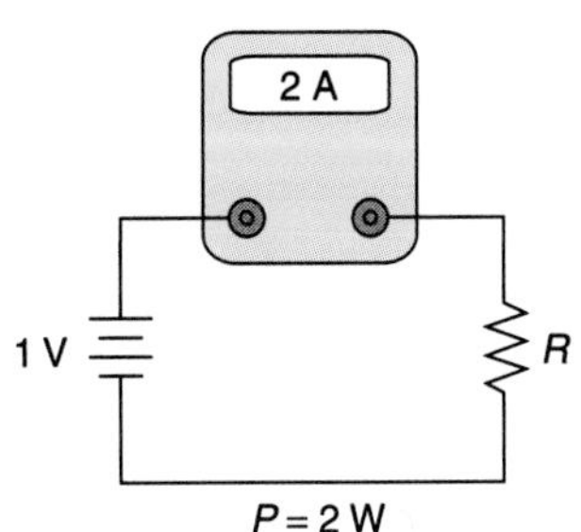

Figure 3.13

Example 3.6

Calculate the value of source power for the circuit in Figure 3.14.

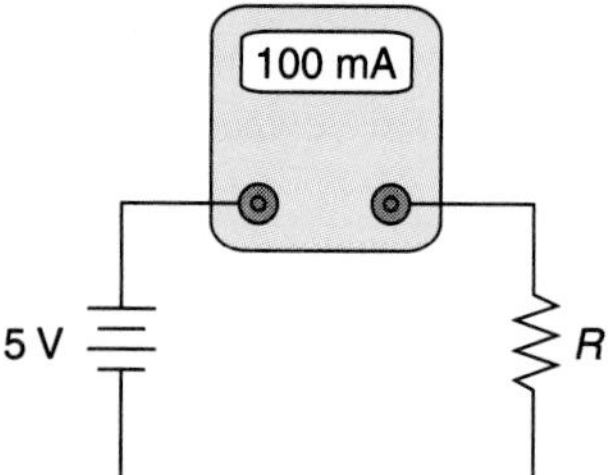

Figure 3.14

Solution: Using the values of voltage and current shown in Figure 3.14, the power being used by the source is found as

$$P = VI = (5 \text{ V})(100 \text{ mA}) = 500 \text{ mW}$$

Practice Problem 3.6

A 12 V power supply is generating 80 mA of current. What is the value of the source power?

Other Power Equations

We can use Ohm's law to derive two other useful power equations. By using *IR* in place of *V* in equation (3.4), we obtain an equation that defines power in terms of current and resistance, as follows:

$$P = I^2R \tag{3.5}$$

The following example demonstrates the use of this equation.

Example 3.7

Calculate the value of source power for the cicuit in Figure 3.15.

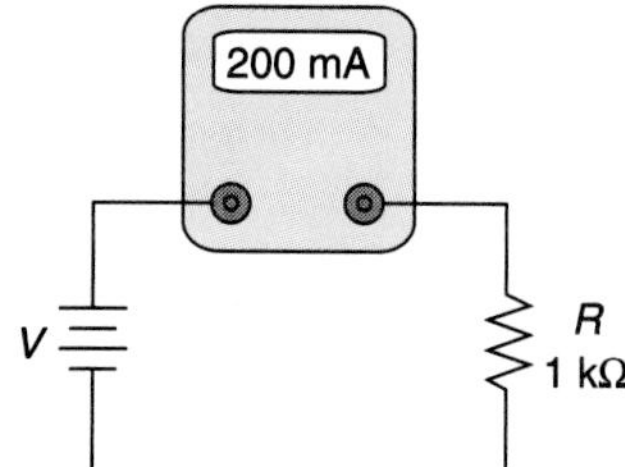

Figure 3.15

Solution: Using the values of current and resistance shown in Figure 3.15, the power being used by the voltage source is found as

$$P = I^2R = (200 \text{ mA})^2(1 \text{ k}\Omega) = 40 \text{ W}$$

Practice Problem 3.7

A circuit like the one in Figure 3.15 has values of $I = 80$ mA and $R = 1.8$ kΩ. What is the value of source power?

By substituting $\frac{V}{R}$ for the value of *I* in equation (3.4), we obtain an equation that defines power in terms of voltage and resistance, as follows:

$$P = \frac{V^2}{R} \tag{3.6}$$

The following example demonstrates the use of this equation.

Example 3.8

Calculate the value of source power for the circuit in Figure 3.16.

Figure 3.16

Solution: Using the values shown in Figure 3.16, the power being used by the voltage source is found as

$$P = \frac{V^2}{R} = \frac{(12\text{ V})^2}{10\text{ k}\Omega} = 14.4\text{ mW}$$

Practice Problem 3.8

A circuit like the one in Figure 3.16 has values of $V = 10$ V and $R = 4.7$ kΩ. What is the value of source power?

As you can see, power can be calculated using any *two* basic circuit properties and the appropriate power equation. The basic power relationships are

$$P = VI \qquad P = I^2R \qquad P = \frac{V^2}{R}$$

Power and Heat

◄ *OBJECTIVE 9*

You have been told that the *power rating* of a component is *a measure of its ability to dissipate heat.* In our basic science classes, we were taught that matter cannot be created or destroyed. The same principle applies to energy. When energy is transferred to a resistor, the resistor absorbs that energy. Since this energy cannot be destroyed, the resistor must convert it into another form—in this case, heat. Some other examples of energy conversion are as follows:

- A toaster converts electrical energy into heat.
- A light bulb converts electrical energy into light and heat.

Note that *any device designed to convert energy from one form to another* is referred to as a **transducer**.

As you know, *power* is defined as *energy used per unit time.* If a component is absorbing a given amount of energy per unit time, it must be capable of converting that energy into heat at the same rate. This is why the power rating of a component is critical. If you exceed the power rating of a component, that component will keep getting hotter, and will eventually be destroyed by the excessive heat.

◄ *OBJECTIVE 10*

You were told in Chapter 2 that the power rating of a resistor is a parameter that indicates its maximum power dissipation capability. When choosing a resistor for a particular application, the power rating of the component should be much greater than the calculated value of component power. This ensures that the resistor can easily absorb the applied power. A practical application of this principle is demonstrated in the following example.

Example 3.9

You want to build the circuit in Figure 3.17. Determine the minimum acceptable power rating for the resistor.

Figure 3.17

Solution: The circuit shown in Figure 3.17 has values of $V = 15$ V and $R = 1$ kΩ. Using these values,

$$P = \frac{V^2}{R} = \frac{(15\text{ V})^2}{1\text{ k}\Omega} = 225\text{ mW}$$

A Practical Consideration: A common guideline for selecting a resistor is to use one with twice the required power-dissipation capability. As such, you would probably want to use a $\frac{1}{2}$ W resistor for the application in Example 3.9.

Since this power will be absorbed by the resistor, its power rating must be greater than 225 mW. Thus, the resistor must have a power rating of at least $\frac{1}{4}$ W (250 mW).

Practice Problem 3.9

A circuit like the one in Figure 3.17 has values of $V = 18$ V and $R = 910\ \Omega$. Determine the minimum acceptable power rating for the resistor.

OBJECTIVE 11 ➤

Efficiency

The dc power supply in any circuit contains a variety of components that each use some amount of power. Therefore, the output power from the voltage source must be less than its input power.

The **efficiency** (η) of a circuit or component is *the ratio of its output power to its input power, given as a percentage.* By formula,

$$\eta = \frac{P_o}{P_i} \times 100 \tag{3.7}$$

where

η = the efficiency, as a percentage

P_o = the output power

P_i = the input power

Note that efficiency is represented using the Greek letter *eta* (η). Example 3.10 shows how the efficiency of a dc power supply is calculated.

Example 3.10

The dc power supply represented in Figure 3.18 has the following ratings: $P_o = 5$ W (maximum) and $P_i = 12$ W (maximum). Calculate the efficiency of the power supply.

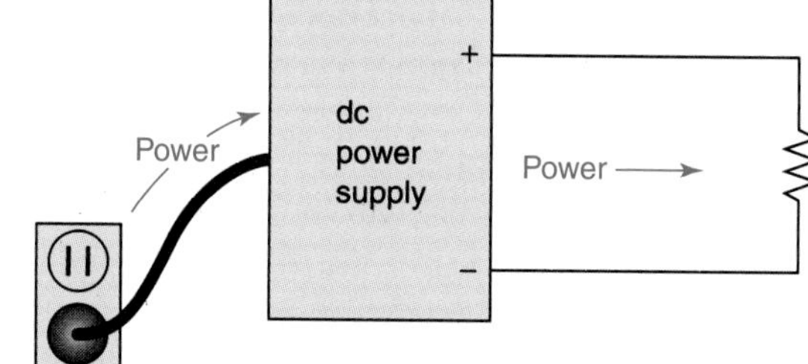

Figure 3.18

Solution: The efficiency of the power supply is found as

$$\eta = \frac{P_o}{P_i} \times 100 = \frac{5\text{ W}}{12\text{ W}} \times 100 = 41.7\%$$

This means that 41.7% of the power supply's input power is converted to output power. The other 58.3% of the input power is used by the power supply's internal circuitry.

Practice Problem 3.10

A dc power supply has the following maximum ratings: $P_o = 20$ W and $P_i = 140$ W. Calculate the efficiency of the power supply.

A few points should be made at this time. First, it is not possible to have an efficiency that is greater than 100%. An efficiency of greater than 100% would imply that the output power is greater than the input power, which is impossible. (Remember, energy cannot be created.) Also, because all electronic circuits absorb some amount of power, every practical efficiency rating is less than 100%.

Energy Measurement

Every appliance in your home uses energy. The practical measurement for the energy used by a utility customer is the **kilowatt-hour (kWh).** *One kilowatt-hour* is *the amount of energy used by a 1000 W device that is run for one hour.* The energy you use, in kilowatt-hours, is found as

$$E_{kWh} = \frac{P \cdot t}{1000} \tag{3.8}$$

where

P = the power used, in watts

t = the time, in hours

Note that the 1000 in the denominator of the equation converts watts into kilowatts. The following example demonstrates the use of equation (3.8).

Example 3.11

Calculate the power used to run twenty 60 W light bulbs for six hours.

Solution: The combined power being used by the light bulbs is found as

$$20 \times 60\ \text{W} = 1200\ \text{W}$$

Using this combined power, the total energy used is found as

$$E_{kWh} = \frac{P \cdot t}{1000} = \frac{(1200\ \text{W})(6\ \text{h})}{1000} = 7.2\ \text{kWh}$$

Practice Problem 3.11

Calculate the total energy used by a 1500 W dishwasher, a 3600 W clothes dryer, and a 750 W air conditioner that are all being used for 2 hours.

Section Review

1. What is *power*?
2. Define the *watt* unit of power.
3. Describe the relationship between *power* and *current.*
4. Describe the relationship between *power* and *voltage.*
5. Discuss the relationship between *power* and *heat.*
6. What is a *transducer*?
7. What is *efficiency?* How is it calculated?
8. Why can't a circuit or device have an efficiency that is greater than 100%?
9. What unit is used by power companies to measure the amount of energy you use? What does this unit equal?

3.5 Related Topics

In this section, we're going to discuss several topics that relate to basic circuits. These include circuit problem solving, the effects of resistor tolerances, and some frequently used terms.

OBJECTIVE 12 ➤ **Solving Basic Circuit Problems**

In any circuit problem, you have two types of variables: those with *known values* (given in the problem), and those with *unknown values.* Generally, you must define one or more of the unknown variables in terms of the known variables. For example, consider the simple problem shown in Figure 3.19. To solve for the circuit current (the unknown value), we must define it in terms of voltage and resistance (the known values). In this case, the solution is simple. We just use Ohm's law in the form of

$$I = \frac{V}{R}$$

This equation defines our unknown value (current) in terms of our known values (voltage and resistance). Solving the equation solves the problem.

In many cases, the solution is not as obvious as it is in Figure 3.19. For example, consider the problem in Figure 3.20. In this case, we need to solve for the value of the resistor, using the given values of applied voltage and power. Now, consider the equations we have used so far:

$$I = \frac{V}{R} \qquad V = IR \qquad R = \frac{V}{I}$$

$$P = \frac{V^2}{R} \qquad P = I^2R \qquad P = VI$$

None of these equations directly defines resistance in terms of voltage and power. However, if you look closely, you will see that one equation contains all three of our variables. That one is

$$P = \frac{V^2}{R}$$

If we transpose this equation, we get

$$R = \frac{V^2}{P}$$

Now we have an equation that defines our unknown value (resistance) in terms of our known values (voltage and power), and we can solve the problem.

When you encounter any basic circuit problem, you should:

1. Find an equation that contains all the variables that are directly involved in the problem.
2. If necessary, transpose the equation to obtain one that defines the unknown value in terms of the known values.

The use of this procedure is demonstrated further in the following example.

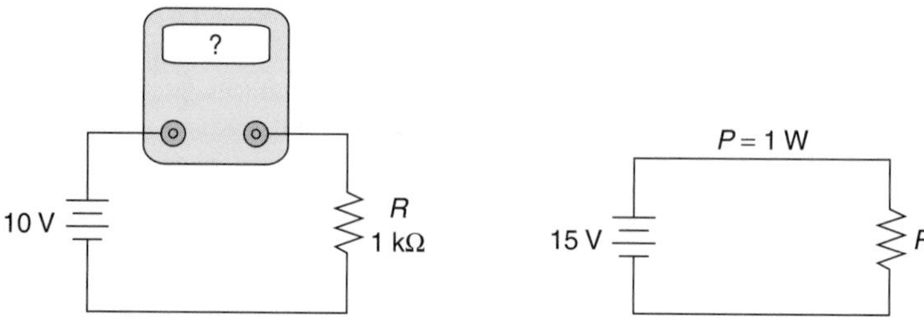

Figure 3.19 **Figure 3.20**

Example 3.12

Calculate the value of current in the circuit shown in Figure 3.21.

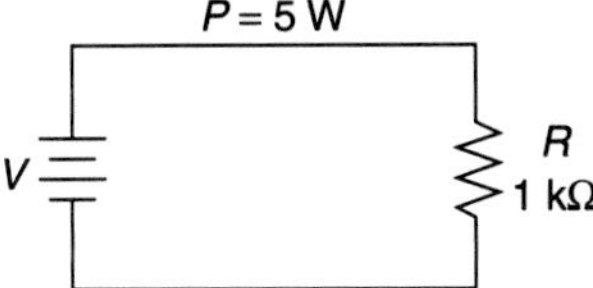

Figure 3.21

Solution: We are trying to calculate current using known values of power and resistance. The basic equation that contains all three of these properties is

$$P = I^2R$$

Transposing this equation to solve for current, we get

$$I = \sqrt{\frac{P}{R}}$$

In this case,

$$I = \sqrt{\frac{P}{R}} = \sqrt{\frac{5\text{ W}}{1\text{ k}\Omega}} = 70.7\text{ mA}$$

Practice Problem 3.12

A circuit like the one shown in Figure 3.21 has the following given values: $P = 2$ W and $R = 1.8$ kΩ. Calculate the value of the applied voltage.

By following the procedure demonstrated in the example, you can solve for any one of the basic circuit properties (current, voltage, resistance, or power) when any two values are known.

Resistor Tolerance

◄ *OBJECTIVE 13*

As you learned in Chapter 2, the value of a given resistor can fall within a range of values that is determined by the component tolerance. For example, a 1 kΩ resistor with a 5% tolerance rating can have a value that falls anywhere between 950 and 1050 Ω.

Resistor tolerance is often ignored when solving basic circuit problems. As a result, it is easy to forget that circuit current actually falls within a range of values. The limits of this range are found using

$$I_{min} = \frac{V}{R_{max}} \tag{3.9}$$

and

$$I_{max} = \frac{V}{R_{min}} \tag{3.10}$$

Example 3.13 shows how this range of current values is determined for a basic circuit.

Example 3.13

Calculate the possible range of current values for the circuit in Figure 3.22.

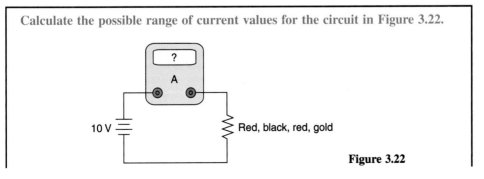

?
A
10 V
Red, black, red, gold

Figure 3.22

Solution: The color code of the resistor in Figure 3.22 indicates that it is a 2 kΩ resistor with a 5% tolerance. The range of possible values for the component is found as

$$(2\ \text{k}\Omega) \times 5\% = 100\ \Omega$$

$$2\ \text{k}\Omega - 100\ \Omega = 1.9\ \text{k}\Omega \qquad \text{(minimum)}$$

$$2\ \text{k}\Omega + 100\ \Omega = 2.1\ \text{k}\Omega \qquad \text{(maximum)}$$

The *maximum* circuit current is now found as

$$I_{max} = \frac{V}{R_{min}} = \frac{10\ \text{V}}{1.9\ \text{k}\Omega} = 5.26\ \text{mA}$$

and the *minimum* circuit current is found as

$$I_{min} = \frac{V}{R_{max}} = \frac{10\ \text{V}}{2.1\ \text{k}\Omega} = 4.76\ \text{mA}$$

Thus, the actual circuit current may fall anywhere between 4.76 mA and 5.26 mA.

Practice Problem 3.13

A circuit like the one in Figure 3.22 has a 12 V source and a resistor that is color coded as follows: blue—gray—red—silver. Determine the range of current values for the circuit.

Had we considered only the rated value of the resistor in Figure 3.22, we would have predicted a circuit current value of 5 mA. As demonstrated in the example, the actual circuit current could fall anywhere between 4.76 mA and 5.26 mA. When working in lab, keep in mind that the tolerance of a resistor may cause a measured current to vary slightly from its predicted value.

OBJECTIVE 14 ➤

Circuit Loads

The simplest circuit contains two components: one that supplies power, and one that absorbs it. As illustrated in Figure 3.23, we generally refer to the components as the *source* and the *load.* The **source** *supplies the power,* and the **load** *absorbs* (uses) *the power.*

Every circuit or system, no matter how simple or complex, is designed to deliver power (in one form or another) to one or more loads. In many cases, the load on a circuit is represented as a single load resistor (R_L), as shown in Figure 3.23. The current through the load is called the *load current* (I_L), and the voltage across the load is referred to as the *load voltage* (V_L). The power that is being absorbed by the load is referred to as the *load power* (P_L).

When the resistance of a load is variable (as shown in Figure 3.24), the minimum load resistance is referred to as a *full load.* A **full load** is *one that draws maximum current from the source.* As Ohm's law tells us, maximum current is produced when the load resistance is at its minimum value.

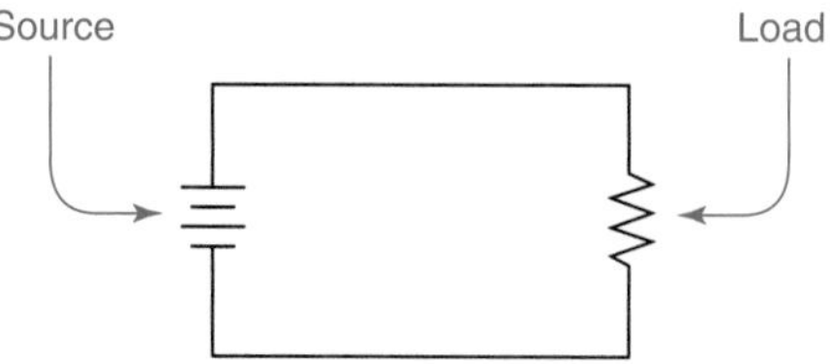

Figure 3.23

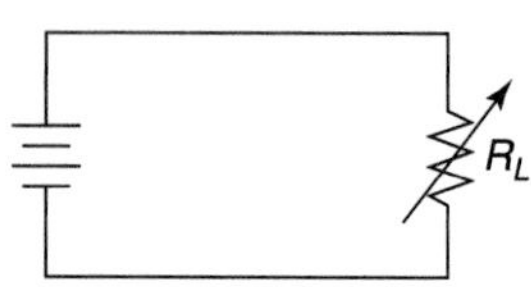

Figure 3.24

Section Review

1. Describe the procedure for selecting the proper equation to solve a basic circuit problem.
2. What effect does *resistor tolerance* have on a measured value of circuit current?
3. What is a *source*? What is a *load*?
4. What is a *full load*?
5. Does resistor tolerance have an impact on circuit power values? Explain your answer. Critical Thinking

Key Terms

The following terms were introduced in this chapter on the pages indicated:

Efficiency (η), 86
Full load, 90
Kilowatt-hour (kWh), 87
Load, 90
Ohm's law, 75
Open circuit, 81
Power (*P*), 83
Short circuit, 81
Solder, 82
Solder bridge, 82
Source, 90
Transducer, 85
Troubleshooting, 80
Watt (W), 83

Practice Problems

1. For each combination of voltage and resistance, calculate the resulting current.
   a. $V = 12$ V, $R = 2.2$ kΩ
   b. $V = 80$ mV, $R = 1.8$ kΩ
   c. $V = 8$ V, $R = 470$ Ω
   d. $V = 16$ V, $R = 330$ kΩ
   e. $V = 120$ μV, $R = 6.8$ Ω
2. For each combination of voltage and resistance, calculate the resulting current.
   a. $V = 15$ V, $R = 4.7$ kΩ
   b. $V = 8.6$ V, $R = 510$ Ω
   c. $V = 22.8$ V, $R = 82$ kΩ
   d. $V = 10$ V, $R = 1.5$ MΩ
   e. $V = 2$ V, $R = 12$ Ω
3. For each combination of current and resistance, calculate the voltage required to generate the current.
   a. $I = 10$ mA, $R = 820$ Ω
   b. $I = 65$ mA, $R = 100$ Ω
   c. $I = 130$ μA, $R = 2.7$ kΩ
   d. $I = 24$ mA, $R = 1.1$ kΩ
   e. $I = 800$ mA, $R = 3.6$ Ω
4. For each combination of current and resistance, calculate the voltage required to generate the current.
   a. $I = 20$ mA, $R = 910$ Ω
   b. $I = 33$ mA, $R = 51$ Ω
   c. $I = 14$ mA, $R = 180$ Ω
   d. $I = 60$ μA, $R = 24$ kΩ
   e. $I = 100$ μA, $R = 16$ kΩ
5. For each combination of voltage and current, determine the resistance needed to limit the current to the given value.
   a. $V = 6$ V, $I = 4$ mA
   b. $V = 94$ V, $I = 20$ mA
   c. $V = 11$ V, $I = 500$ μA
   d. $V = 9$ V, $I = 50$ mA
   e. $V = 33$ V, $I = 330$ μA
6. For each combination of voltage and current, calculate the resistance needed to limit the current to the given value.
   a. $V = 8.8$ V, $I = 24$ mA
   b. $V = 12$ V, $I = 16$ mA
   c. $V = 28$ V, $I = 8$ mA
   d. $V = 180$ mV, $I = 260$ μA
   e. $V = 1$ V, $I = 250$ nA
7. For each combination of voltage and current, calculate the power that is being supplied by the voltage source.
   a. $V = 14$ V, $I = 160$ mA
   b. $V = 120$ V, $I = 40$ mA
   c. $V = 5$ V, $I = 2.2$ A
   d. $V = 24$ V, $I = 600$ μA
   e. $V = 15$ V, $I = 850$ mA

8. For each combination of voltage and current given in Problem 5, calculate the power that is being supplied by the voltage source.

9. For each combination of current and resistance, calculate the power that is being supplied by the voltage source.
   a. $I = 25$ mA, $R = 3.3$ kΩ
   b. $I = 16$ mA, $R = 12$ kΩ
   c. $I = 4.6$ mA, $R = 9.1$ kΩ
   d. $I = 7.5$ mA, $R = 11$ kΩ
   e. $I = 420$ μA, $R = 150$ kΩ

10. For each combination of current and resistance given in Problem 3, calculate the power that is being supplied by the voltage source.

11. For each combination of voltage and resistance, calculate the power that is being supplied by the voltage source.
    a. $V = 15$ V, $R = 4.7$ kΩ
    b. $V = 8$ V, $R = 47$ kΩ
    c. $V = 120$ mV, $R = 14$ Ω
    d. $V = 3.8$ V, $R = 510$ Ω
    e. $V = 28$ V, $R = 120$ kΩ

12. For each combination of voltage and resistance given in problem 1, calculate the power that is being supplied by the voltage source.

13. Calculate the minimum acceptable power rating for the resistor in Figure 3.25a.

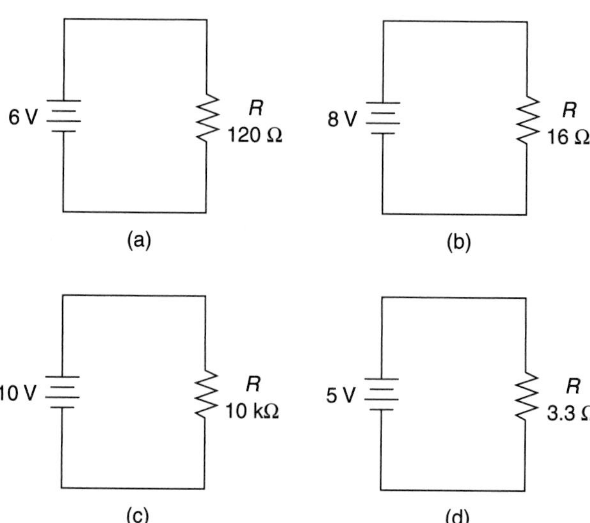

Figure 3.25

14. Calculate the minimum acceptable power rating for the resistor in Figure 3.25b.

15. Calculate the minimum acceptable power rating for the resistor in Figure 3.25c.

16. Calculate the minimum acceptable power rating for the resistor in Figure 3.25d.

17. For each combination of input and output power, calculate the efficiency of the circuit.
    a. $P_i = 2$ W, $P_o = 36$ mW
    b. $P_i = 6$ W, $P_o = 445$ mW
    c. $P_i = 8.5$ W, $P_o = 6.7$ W
    d. $P_i = 100$ W, $P_o = 34$ W
    e. $P_i = 15$ kW, $P_o = 600$ mW

18. Calculate the energy used (in kWh) to run twelve 150 W light bulbs for eight hours.

19. Calculate the energy used (in kWh) to run a 500 W stereo system and a 2400 W air conditioner for three hours.

20. Calculate the energy used to run the following combination of appliances for the times given: a 700 W microwave oven for 20 minutes, a 3800 W dishwasher for 30 minutes, a 2200 W air conditioner for 12 hours, and an 1800 W clothes dryer for one hour.

21. Complete the chart below.

Current	*Voltage*	*Resistance*	*Power*
a. 10 mA	_____	_____	4 W
b. _____	32 V	_____	16 mW
c. _____	_____	3.3 kΩ	231 mW
d. 15 mA	45 V	_____	_____
e. 24 mA	_____	1.2 kΩ	_____

22. Complete the chart below.

Current	*Voltage*	*Resistance*	*Power*
a. _____	_____	1 kΩ	240 mW
b. 50 mA	2 V	_____	_____
c. _____	8 V	200 Ω	_____
d. _____	16 V	_____	800 mW
e. 35 mA	_____	_____	1.4 W

23. Calculate the resistance of a 110 V, 60 W light bulb.

24. A television has a 110 V input and an average power rating of 180 W. Calculate the television's average input current.

25. The power supply for a backyard lighting system can provide a maximum output current of 5 A. The lights designed to be used with the system contain 24 V, 5 W lamps. How many lights can be connected to the power supply without exceeding 80% of its maximum output current rating?

26. A variable dc power supply is connected to a 250 Ω, $\frac{1}{2}$ W resistor. What is the maximum allowable setting for the power supply output voltage?

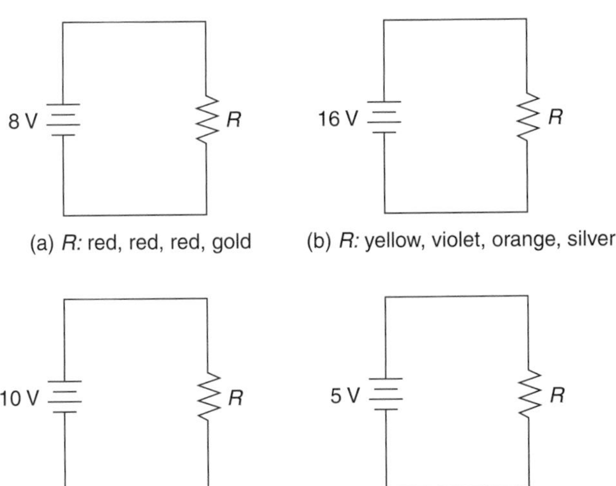

(a) *R:* red, red, red, gold (b) *R:* yellow, violet, orange, silver

(c) *R:* orange, orange, red, gold (d) *R:* brown, green, green, silver

Figure 3.26

27. Calculate the range of possible current values for the circuit in Figure 3.26a.

28. Calculate the range of possible current values for the circuit in Figure 3.26b.

29. Calculate the range of possible current values for the circuit in Figure 3.26c.

30. Calculate the range of possible current values for the circuit in Figure 3.26d.

Looking Back

These problems relate to material presented in earlier chapters. The chapters are identified in brackets.

31. Convert the following numbers to engineering notation without the use of your calculator. [Introduction]

a. 0.000376 A
b. −3,842 V
c. 472,980 Hz
d. 0.00000788 F

32. Five coulombs of charge pass through a circuit every minute. What is the value of the circuit current? [Chapter 1]

33. Determine the guaranteed range of values for a resistor with the following color code: red—yellow—black—red. [Chapter 2]

Pushing the Envelope

34. Calculate the minimum allowable setting for the potentiometer in Figure 3.27.

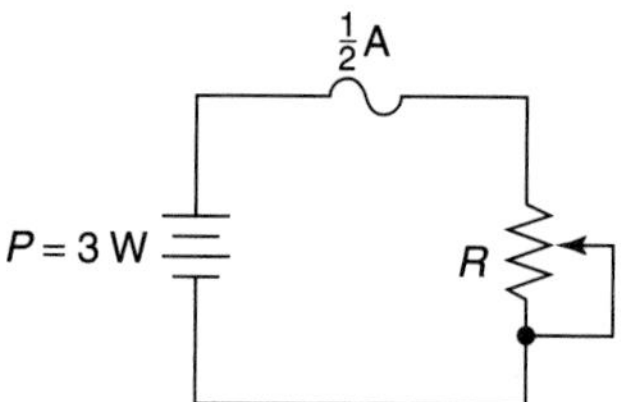

Figure 3.27

35. Calculate the maximum allowable setting for the voltage source in Figure 3.28.

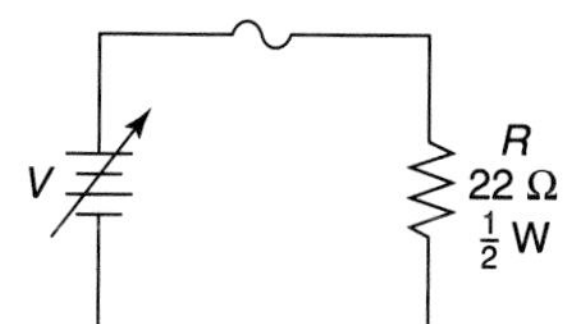

Figure 3.28

36. Ohm's law states that:
   a. The current through a fixed resistance is directly proportional to voltage.
   b. The current generated by a fixed voltage is inversely proportional to resistance.
   Using the circuit in Figure 3.29 as a model, show that Ohm's law doesn't necessarily predict the change in current when both voltage and resistance change.

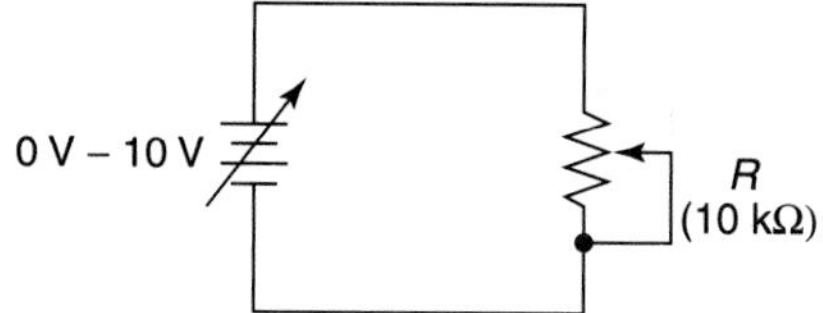

Figure 3.29

37. Without substituting power (P) for either expression, prove the following:

$$\frac{V^2}{R} = I^2R$$

Answers to the Example Practice Problems

3.1. 180 mA

3.2 1.65 V

3.3 300 Ω

3.4 When $V = 3$ V, $I = 1.5$ mA (half the original value).

3.5. When $R = 5$ kΩ, $I = 2$ mA (twice the original value).

3.6. 960 mW

3.7. 11.52 W

3.8. 21.3 mW

3.9. $P = 356$ mW, so the resistor must have a minimum rating of $\frac{1}{2}$ W (500 mW).

3.10. 14.3%

3.11. 11.7 kWh

3.12. 60 V

3.13. 1.60 to 1.96 mA

Multisim File References

Figure	File Reference
3.8	ETF3_8.msm
3.9	ETF3_9.msm
3.10	ETF3_10.msm
3.25	ETF3_25.msm

chapter 4

Series Circuits

Objectives

After studying the material in this chapter, you should be able to:

1. State the current characteristic that distinguishes *series circuits* from *parallel circuits.*
2. Calculate any *resistance value* in a series circuit, given the other resistances.
3. Determine the *potentiometer setting* required to provide a specific amount of resistance in a series circuit.
4. Determine the *total current* in any series circuit.
5. Describe the *voltage* and *power relationships* in a series circuit.
6. State and explain *Kirchhoff's voltage law.*
7. Calculate the voltage from any point in a series circuit to ground.
8. Calculate the voltage across any resistor (or group of resistors) in a series circuit using the voltage-divider equation.
9. Describe the use of a potentiometer as a *variable voltage divider.*
10. Describe the *ideal voltage source.*
11. Discuss the effects of *source resistance* on circuit voltage values.
12. Discuss the *maximum power transfer theorem,* and calculate the *maximum load power* for a given series circuit.
13. Calculate the *total voltage* and *current values* for a series circuit containing two *series-aiding* or *series-opposing* voltage sources.
14. Compare and contrast *earth ground* with *chassis ground.*

Two types of circuits that are found in virtually every type of electronic system are the *series circuit* and the *parallel circuit.* These two types of circuits are shown in Figure 4.1. As you can see, the two circuits look very different from each other. In fact, they not only look different, they are nearly opposites in terms of their operating characteristics. Most electronic circuits, no matter how complex, are series circuits, parallel circuits, or some combination of the two.

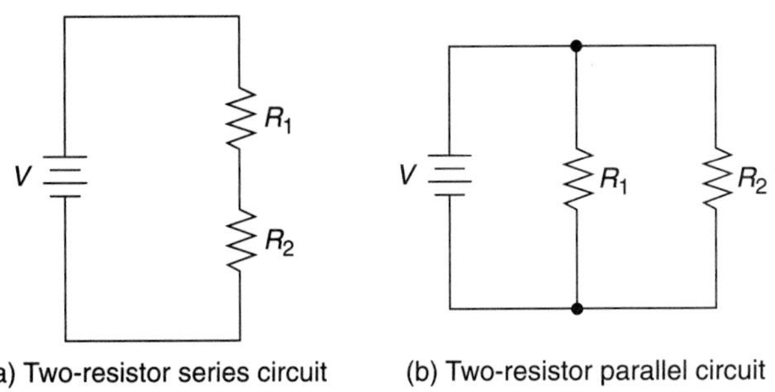

Figure 4.1 The simplest series and parallel circuits.

4.1 Series Circuit Characteristics

Simply defined, a **series circuit** is *a circuit that contains only one current path.* For example, consider the circuits shown in Figure 4.2. In each case, the current generated by the voltage source has only one path, and that path contains all of the components in the circuit. (In contrast, the parallel circuit in Figure 4.1b contains two current paths between the terminals of the voltage source; one through R_1 and one through R_2.) ◀ *OBJECTIVE 1*

Total Series Resistance

As shown in Figure 4.2, the current in a series circuit passes through all of the resistors in the circuit. Therefore, the total opposition to current (R_T) is equal to the sum of the individual resistor values. By formula, ◀ *OBJECTIVE 2*

$$R_T = R_1 + R_2 + \ldots + R_n \tag{4.1}$$

where

R_T = the total circuit resistance

R_n = the highest numbered resistor in the circuit

The total resistance in a series circuit is calculated as shown in the following example.

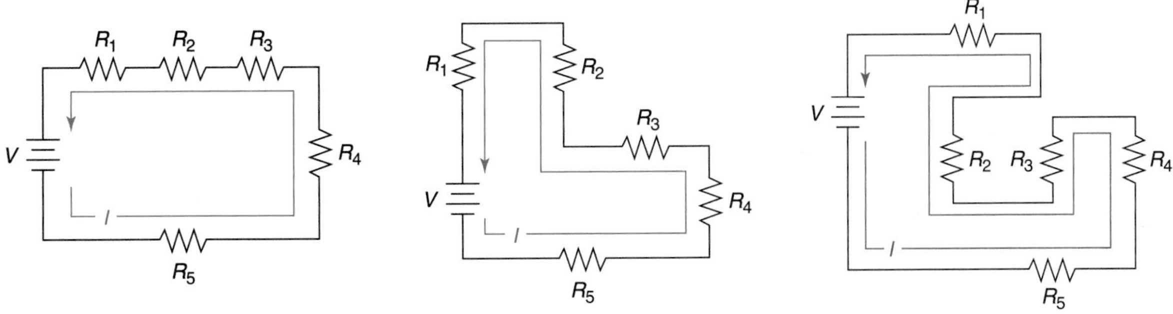

Figure 4.2

Example 4.1

Calculate the total resistance in the circuit shown in Figure 4.3.

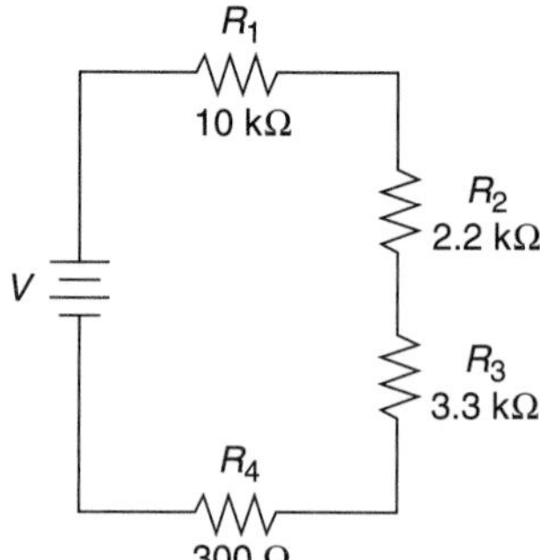

Figure 4.3

Solution: Using equation (4.1), the total resistance is found as

$$R_T = R_1 + R_2 + R_3 + R_4$$
$$= 10\text{ k}\Omega + 2.2\text{ k}\Omega + 3.3\text{ k}\Omega + 300\ \Omega = 15.8\text{ k}\Omega$$

Practice Problem 4.1

A circuit like the one shown in Figure 4.3 has the following values: $R_1 = 680\ \Omega$, $R_2 = 1.5\text{ k}\Omega$, $R_3 = 470\text{ k}\Omega$, and $R_4 = 3.6\text{ k}\Omega$. Calculate the total resistance in the circuit.

OBJECTIVE 3 ➤

When you need to find the value of an *unknown* resistance in a series circuit, you can calculate it by subtracting the sum of the known resistances from the total circuit resistance. This technique is demonstrated in the following example.

Example 4.2

Determine the potentiometer setting required in Figure 4.4 to provide a total circuit resistance of 12 kΩ.

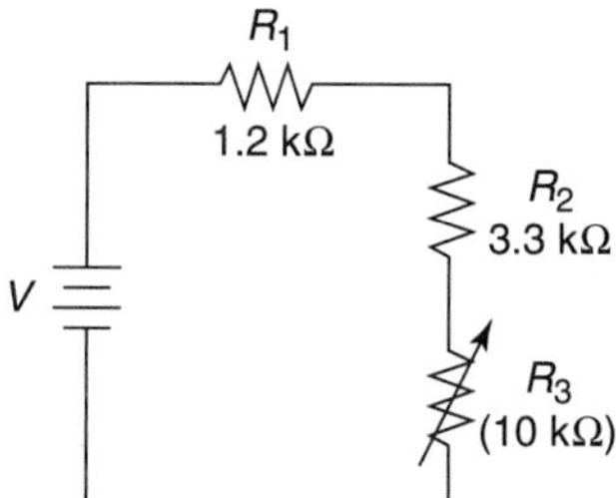

Figure 4.4

Solution: The combined resistance of R_1 and R_2 is found as

$$R_1 + R_2 = 1.2\text{ k}\Omega + 3.3\text{ k}\Omega = 4.5\text{ k}\Omega$$

R_3 must account for the difference between 4.5 kΩ and the desired total of 12 kΩ. Therefore, its adjusted value is found as

$$R_3 = R_T - (R_1 + R_2) = 12\text{ k}\Omega - 4.5\text{ k}\Omega = 7.5\text{ k}\Omega$$

Practice Problem 4.2

A circuit like the one shown in Figure 4.4 has values of $R_1 = 47\text{ k}\Omega$ and $R_2 = 110\text{ k}\Omega$. Determine the potentiometer setting that will provide a total circuit resistance of 159 kΩ.

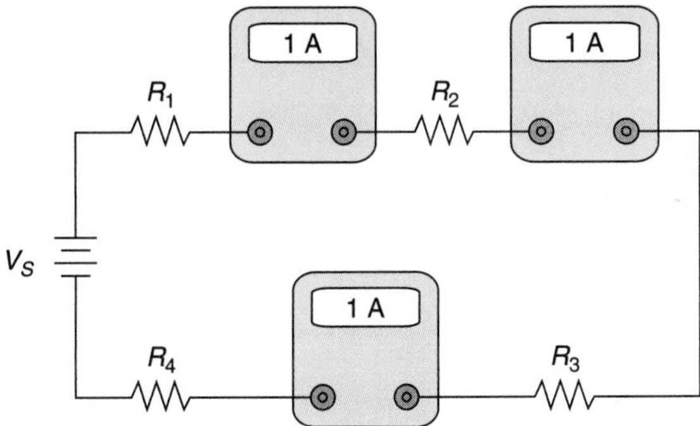

Figure 4.5

Current Characteristics

Since a series circuit contains only one current path, we can state that the current at any point in a series circuit must equal the current at every other point in the circuit. This principle is illustrated in Figure 4.5. Each of the ammeters shows a reading of 1 A. Since charge is flowing at a rate of one coulomb per second (1 A), it must be flowing at the same rate at all points in the circuit. Figure 4.5 also illustrates the fact that you can measure the total current at any point in a series circuit. ◄ *OBJECTIVE 4*

The actual value of current in a series circuit depends on the source voltage (V_S) and the total circuit resistance (R_T), as demonstrated in the following example.

Example 4.3

Calculate the total current (I_T) through the circuit shown in Figure 4.6.

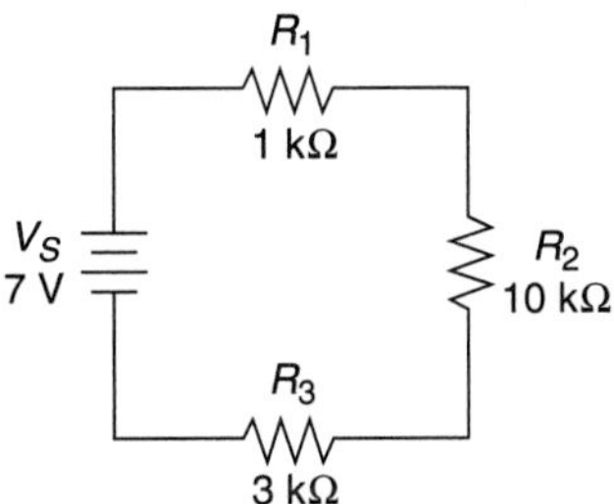

Figure 4.6

Solution: First, the total resistance in the circuit is found as

$$R_T = R_1 + R_2 + R_3 = 1\text{ k}\Omega + 10\text{k}\Omega + 3\text{ k}\Omega = 14\text{ k}\Omega$$

Now, using $R_T = 14\text{ k}\Omega$ and $V_S = 7$ V, the total circuit current is found as

$$I_T = \frac{V_S}{R_T} = \frac{7\text{ V}}{14\text{ k}\Omega} = 500\ \mu\text{A}$$

Practice Problem 4.3

A series circuit like the one shown in Figure 4.6 has the following values: $R_1 = 12\text{ k}\Omega$, $R_2 = 1.5\text{ k}\Omega$, $R_3 = 7.5\text{ k}\Omega$, and $V_S = 15$ V. Calculate the total circuit current.

In summary, the current through a series circuit is equal at all points in the circuit, and thus, can be measured at any point in the circuit. The actual value of current in a series circuit is determined by the source voltage and the total circuit resistance.

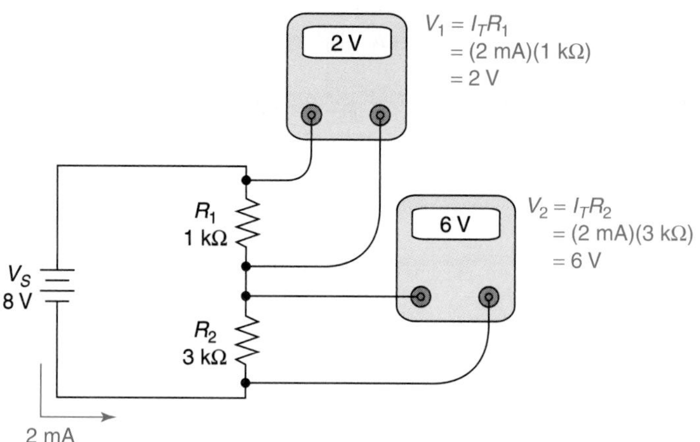

Figure 4.7

Voltage Characteristics

OBJECTIVE 5 ➤ Whenever current passes through a resistance, a difference of potential (voltage) is developed across that resistance, as given by the relationship

$$V = IR$$

Since the current in a series circuit passes through all of the resistors, a voltage is developed across each resistor. This principle is demonstrated in Figure 4.7. As shown in the figure, the voltage across each resistor is determined by the circuit current and the value of the resistor.

If you look closely at Figure 4.7, you'll see that the sum of the component voltages is equal to the source (or total) voltage. This relationship, which holds true for all series circuits, is given as

$$V_S = V_1 + V_2 + \ldots + V_n \tag{4.2}$$

where

V_S = the source (or total) voltage

V_n = the voltage across the highest numbered resistor in the circuit

Equation (4.2) is one way of expressing a relationship called *Kirchhoff's voltage law,* which is discussed in more detail later in this chapter. Example 4.4 shows how the relationship can be used to determine the total voltage applied to a series circuit.

Example 4.4

Determine the value of the source voltage in Figure 4.8.

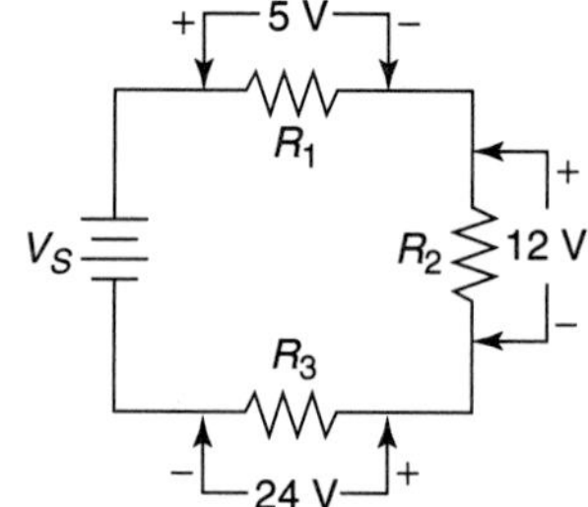

Figure 4.8

Solution: Using the values given in Figure 4.8 and equation (4.2), the value of the source voltage is found as

$$V_S = V_1 + V_2 + V_3 = 5\text{ V} + 12\text{ V} + 24\text{ V} = 41\text{ V}$$

Practice Problem 4.4

A series circuit like the one shown in Figure 4.8 has the following values: $V_1 = 2$ V, $V_2 = 16$ V, and $V_3 = 8.8$ V. Determine the value of the source voltage.

Power Characteristics

Whenever current passes through a resistance, some amount of power is dissipated by the component. For example, the circuit in Figure 4.9 has values of

$$\begin{aligned} P_1 &= I^2R_1 \\ &= (2\text{ mA})^2(1\text{ k}\Omega) \\ &= 4\text{ mW} \end{aligned} \quad \text{and} \quad \begin{aligned} P_2 &= I^2R_2 \\ &= (2\text{ mA})^2(3\text{ k}\Omega) \\ &= 12\text{ mW} \end{aligned}$$

The total power dissipated by the resistors in a series circuit is equal to the total power being supplied by the source. By formula,

$$P_S = P_1 + P_2 + \ldots + P_n \tag{4.3}$$

where

P_S = the total power being supplied by the source

P_n = the power that is dissipated by the highest numbered resistor in the circuit.

Remember: The law of conservation of energy states that energy cannot be created or destroyed. Therefore, the components in a circuit must use all the power provided by the source.

For the circuit in Figure 4.9,

$$P_S = P_1 + P_2 = 4\text{ mW} + 12\text{ mW} = 16\text{ mW}$$

The value of P_S can also be found as

$$P_S = V_S I_T = (8\text{ V})(2\text{ mA}) = 16\text{ mW}$$

which validates the relationship given in equation (4.3).

Summary

A series circuit is one that contains a single path for current. The basic series circuit characteristics are summarized in Figure 4.10.

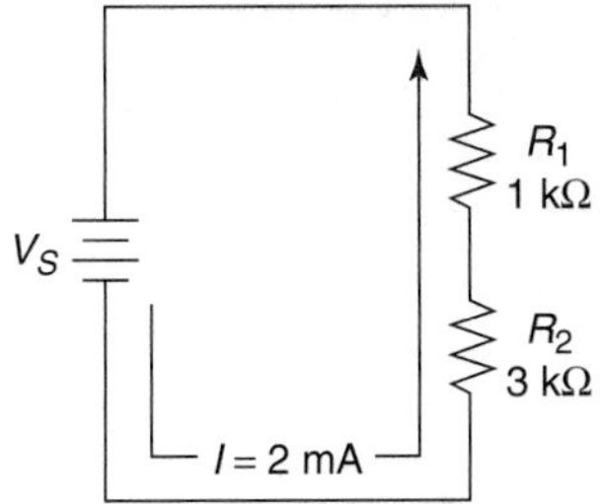

Figure 4.9

Section Review

1. What is a *series circuit*? How does it differ from a *parallel circuit*?
2. Describe the relationship between the total resistance in a series circuit and the values of the individual resistors.

SUMMARY ILLUSTRATION

Series Circuit Characteristics

Sample schematic:

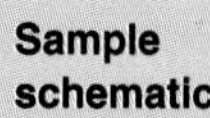

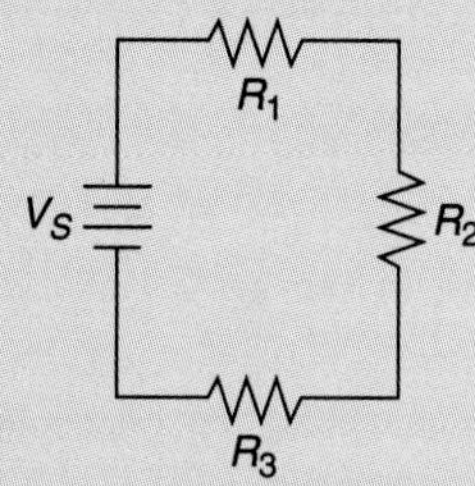

Resistance: The total resistance is equal to the sum of the individual resistances. By formula:

$$R_T = R_1 + R_2 + \ldots + R_n$$

Current: The current at any point in the circuit equals the current at all other points. The value of current depends on the source voltage and the total circuit resistance. By formula:

$$I_T = \frac{V_S}{R_T}$$

Voltage: The sum of the component voltages must equal the source voltage. By formula:

$$V_S = V_1 + V_2 + \ldots + V_n$$

Power: The total power used by the circuit must equal the power delivered by the source. By formula:

$$P_S = P_1 + P_2 + \ldots + P_n$$

Figure 4.10

3. What value is constant throughout any series circuit?
4. Describe the procedure for setting the current in a series circuit to a specific value using a potentiometer.
5. Which series circuit values are additive?
6. Every resistor in a series circuit dissipates some amount of power. Why?

Critical Thinking

7. Refer to the circuit in Figure 4.10. If the value of R_3 doubles, what happens to the voltages across the other two resistors (V_1 and V_2)? Explain your answer.
8. Refer to the circuit in Figure 4.10. If we short out R_3, what happens to the voltages across the other two resistors (V_1 and V_2)? Explain your answer.

4.2 Voltage Relationships: Kirchhoff's Voltage Law, Voltage References, and the Voltage Divider

In this section, we are going to discuss some topics that relate to the source and component voltages in a series circuit.

Kirchhoff's Voltage Law

OBJECTIVE 6 ➤ As you know, *the sum of the component voltages in a series circuit must equal the source voltage.* This relationship, which was first described in the 1840s by German physicist

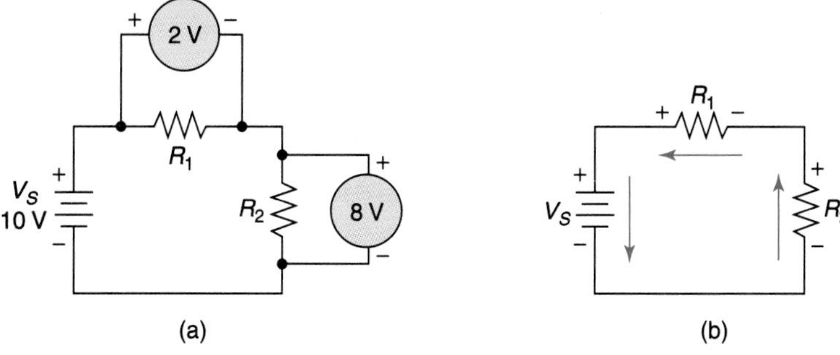

Figure 4.11

Gustav Kirchhoff, has come to be known as **Kirchhoff's voltage law.** Kirchhoff's voltage law is commonly expressed as

$$V_S = V_1 + V_2 + \ldots + V_n$$

Kirchhoff's voltage law is actually worded as follows: *The algebraic sum of the voltages around a closed loop is zero.* The meaning of Kirchhoff's voltage law may not seem clear at first, but it can easily be explained using the circuit in Figure 4.11a. According to Kirchhoff's voltage law,

A Practical Consideration: Kirchhoff used the term *closed loop* to describe a *series circuit.*

$$V_S + V_1 + V_2 = 0 \text{ V}$$

In Figure 4.11a, $V_S = 10$ V, $V_1 = 2$, and $V_2 = 8$ V. It would seem that the sum of these voltages is 20 V. However, Kirchhoff's voltage law takes into account the *polarity* of each voltage.

The polarities of the voltages in a series circuit can be established as shown in Figure 4.11b. The current through R_1 is going from one potential to a more positive potential. Therefore, we represent the voltage across the component as a *positive* value. The same holds true for the voltage across R_2. At the same time, the current through the voltage source is going from one potential to a more-negative potential, so we represent the source voltage as a *negative* value. We now have the following values for the circuit:

$$V_S = -10 \text{ V} \qquad V_1 = +2 \text{ V} \qquad V_2 = +8 \text{ V}$$

If we put these values into our Kirchhoff's loop equation, we get

$$V_S + V_1 + V_2 = -10 \text{ V} + 2 \text{ V} + 8 \text{ V} = 0 \text{ V}$$

which agrees with the actual wording of Kirchhoff's voltage law.

If we had taken the voltage polarities into account from the start, the loop equation for the circuit in Figure 4.11 would have been written as

$$-V_S + V_1 + V_2 = 0 \text{ V}$$

or

$$V_1 + V_2 = V_S$$

Thus, Kirchhoff's voltage law is just another way of saying that the sum of the component voltages in a **closed loop** (*series circuit*) must equal the source voltage.

Voltage References

In any circuit, there is *a point that serves as the 0 V reference.* This point is referred to as **ground** (or *common*) and is designated by the symbol shown in Figure 4.12a. When the ground symbol appears in a schematic, the source and other circuit voltages are referenced

◄ *OBJECTIVE 7*

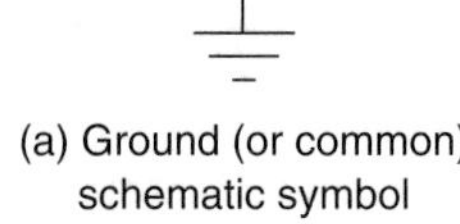

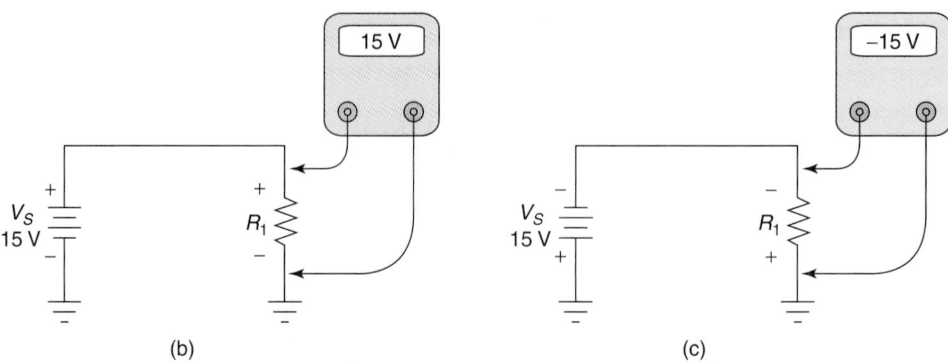

Figure 4.12

to that point. For example, the negative sides of the source and the resistor in Figure 4.12b are both returned to ground. Thus, V_S and V_1 are both *positive with respect to ground* (the reference). The positive sides of the source and the resistor in Figure 4.12c are returned to ground, so V_S and V_1 are both *negative with respect to ground.*

When the ground symbol appears more than once in a circuit, it indicates that the various points are physically connected, as shown in Figure 4.13. In the circuit shown, the voltage from point A to ground equals the source voltage, 20 V. The voltage from point B to ground is equal to 14 V. Generally, we say that the 6 V difference between the voltages at points A and B is "dropped across" R_1. (The term **voltage drop** is commonly used to describe *a change from one potential to a lower potential.*)

The Voltage Divider Relationship

The term **voltage divider** is often used to describe *a series circuit,* because the source voltage is divided across the components in the circuit. For example, consider the circuits shown in Figure 4.14. In each case, the source voltage is *divided* across the resistors. As always, the sum of the resistor voltages in each series circuit equals the source voltage.

If you look closely at the circuits in Figure 4.14, you'll see that *the ratio of any resistor voltage to the source voltage equals the ratio of the resistor value to the total resistance.*

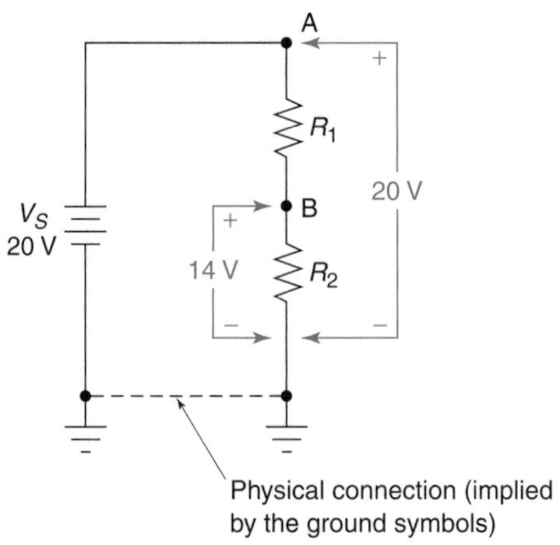

Figure 4.13

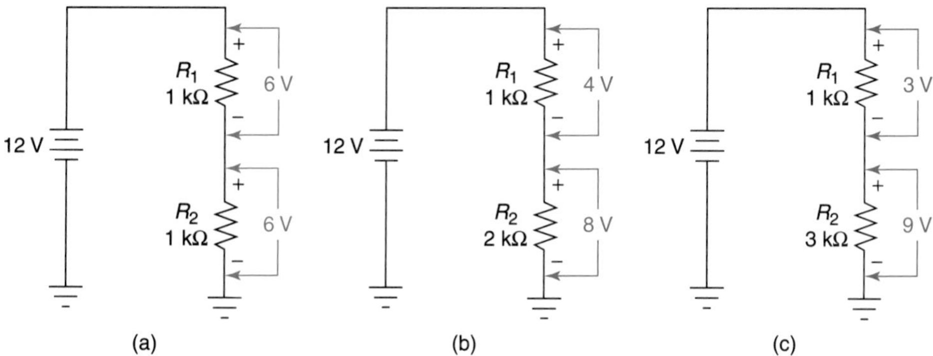

Figure 4.14

For example, in Figure 4.14a, R_1 accounts for one-half the total circuit resistance (2 kΩ) and is dropping one-half the source voltage (12 V). By the same token:

- In Figure 4.14b, R_1 equals one-third of R_T and V_1 equals one-third of V_S.
- In Figure 4.14c, R_1 equals one-fourth of R_T, and V_1 equals one-fourth of V_S.

In each case, the ratio of R_1 to R_T equals the ratio of V_1 to V_S. This relationship, which holds true for all series circuits, is stated mathematically as

$$\frac{V_n}{V_S} = \frac{R_n}{R_T} \tag{4.4}$$

where

R_n = the resistor of interest

V_n = the voltage drop across R_n (where n is the component number)

If we multiply both sides of equation (4.4) by V_S, we get: ◄ *OBJECTIVE 8*

$$V_n = V_S \frac{R_n}{R_T} \tag{4.5}$$

Equation (4.5) is commonly referred to as the *voltage-divider equation.* It allows us to calculate the voltage across a resistor without first calculating the value of the circuit current, as demonstrated in the following example.

Example 4.5

Determine the voltage across R_3 in Figure 4.15.

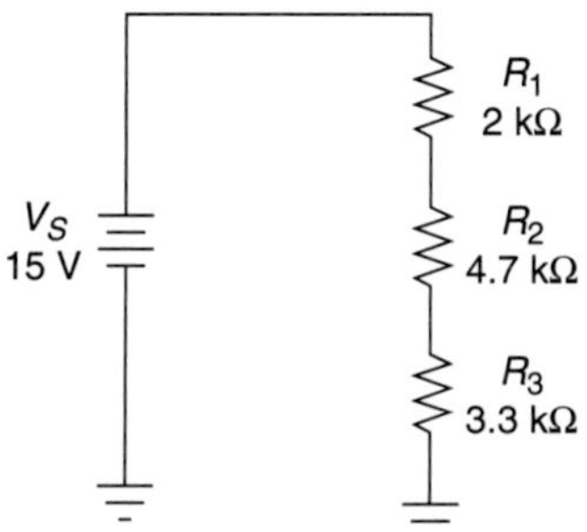

Figure 4.15

Solution: Since we are trying to find the voltage across R_3, we will use V_3 and R_3 in place of V_n and R_n in the voltage-divider equation, as follows:

$$V_3 = V_S \frac{R_3}{R_T} = (15\text{ V})\frac{3.3\text{ k}\Omega}{10\text{ k}\Omega} = (15\text{ V})(0.33) = 4.95\text{ V}$$

Practice Problem 4.5

A circuit like the one shown in Figure 4.15 has values of $V_S = 12$ V, $R_1 = 1$ kΩ, $R_2 = 2.2$ kΩ, and $R_3 = 3.3$ kΩ. Using the voltage-divider equation, calculate the value of V_2.

Example 4.6 demonstrates how the voltage-divider equation can be used with Kirchhoff's voltage law to calculate the values of the component voltages in a circuit.

Example 4.6

Determine the values of V_1 and V_2 for the circuit in Figure 4.16.

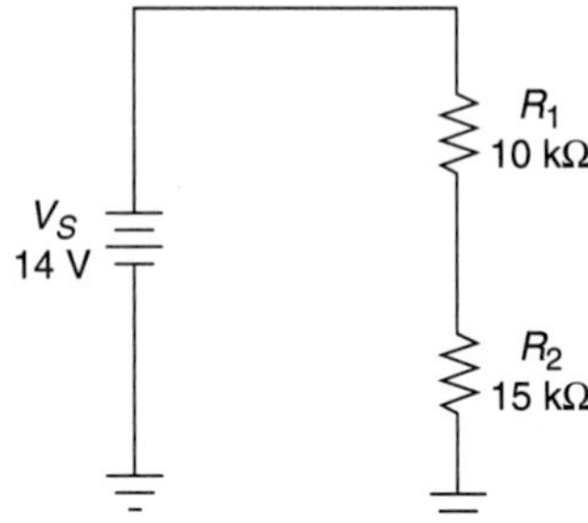

Figure 4.16

Solution: The value of V_1 is found as

$$V_1 = V_S \frac{R_1}{R_T} = (14\text{ V})\frac{10\text{ k}\Omega}{25\text{ k}\Omega} = (14\text{ V})(0.4) = 5.6\text{ V}$$

V_2 can now be found as the difference between V_S and V_1, as follows:

$$V_2 = V_S - V_1 = 14\text{ V} - 5.6\text{ V} = 8.4\text{ V}$$

Practice Problem 4.6

A circuit like the one shown in Figure 4.16 has values of $V_S = 20$ V, $R_1 = 33$ kΩ, and $R_2 = 47$ kΩ. Determine the values of V_1 and V_2 for the circuit using the voltage-divider equation and Kirchhoff's voltage law.

Had we used the previously established method of finding V_1 and V_2, we would have calculated the values of R_T and I_T and then used I_T to calculate the values of V_1 and V_2 found in the example. The combination of the voltage-divider equation and Kirchhoff's voltage law has allowed us to determine the resistor voltages with less effort.

The voltage-divider equation can also be used to find the voltage across a group of resistors, as demonstrated in the following example.

Example 4.7

For the circuit shown in Figure 4.17, determine the value of the voltage from point A to ground (V_A).

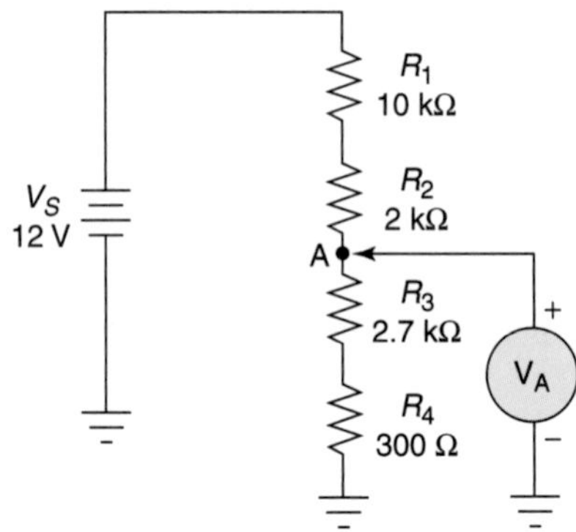

Figure 4.17

Solution: According to Kirchhoff's voltage law, the voltage from point A to ground must equal the sum of V_3 and V_4. Therefore,

$$R_n = R_3 + R_4 = 3 \text{ k}\Omega$$

and

$$V_A = V_S \frac{R_n}{R_T} = (12 \text{ V})\frac{3 \text{ k}\Omega}{15 \text{ k}\Omega} = 2.4 \text{ V}$$

Practice Problem 4.7

A circuit like the one shown in Figure 4.17 has the following values: $V_S = 18$ V, $R_1 = 2.2$ kΩ, $R_2 = 4.7$ kΩ, $R_3 = 3.3$ kΩ, and $R_4 = 10$ kΩ. Determine the voltage from point A to ground.

Avoiding a Common Mistake: In Figure 4.17, V_A equals the sum of the voltages between point A and ground ($V_3 + V_4$). However, it does not equal the sum of ($V_1 + V_2$). Make sure you correctly identify the voltages of interest when using the voltage-divider equation.

Once again, the voltage-divider equation has provided an efficient method of solving the problem.

One Final Note

In this section, we have established two extremely important relationships that apply to virtually every series circuit: *Kirchhoff's voltage law* and the *voltage-divider equation.* As you will see, these relationships are used almost as commonly as Ohm's law.

Section Review

1. What does *Kirchhoff's voltage law* state?
2. When developing the Kirchhoff's loop equation for a series circuit, how do you assign the polarities of the resistor voltages?
3. What is *ground*? By what other name is it known?
4. What is indicated by the appearance of the ground symbol at two or more points in a circuit?
5. Explain the *voltage-divider* relationship.
6. Using your knowledge of power relationships and series circuit characteristics, show that the following *power-divider relationship* is valid for a series circuit: Critical Thinking

$$P_n = P_T \frac{R_n}{R_T}$$

4.3 Related Topics

In this section, we're going to complete the chapter by discussing some miscellaneous topics that relate to series circuits.

The Potentiometer as a Voltage Divider

A single potentiometer can be used as a variable voltage divider when connected as shown in Figure 4.18a. The voltage from the wiper arm to ground (V_b) is determined by the setting of the potentiometer. For example, the potentiometer in Figure 4.18a is shown to be set to $R_{bc} = 2$ kΩ. Since the outer terminal resistance of the potentiometer is 10 kΩ, the value of V_b is found as ◄ *OBJECTIVE 9*

$$V_b = V_S \frac{R_{bc}}{R_{ac}} = (10 \text{ V})\frac{2 \text{ k}\Omega}{10 \text{ k}\Omega} = 2 \text{ V}$$

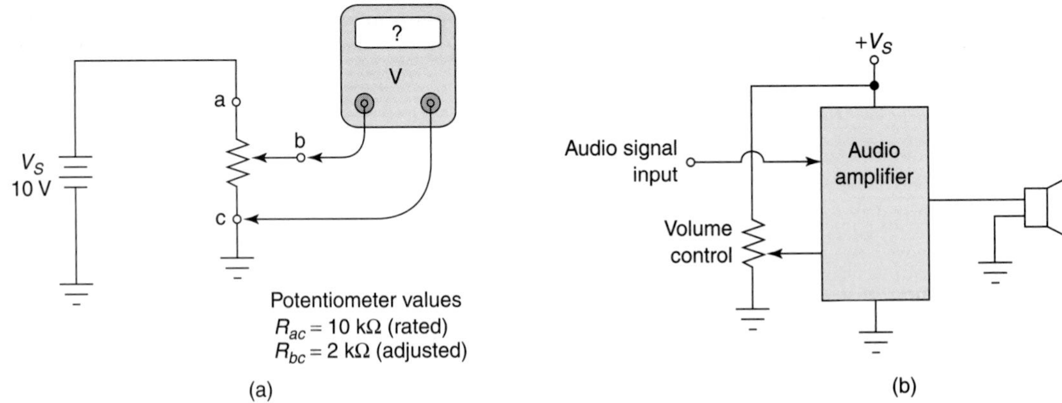

Figure 4.18 The potentiometer as a voltage divider.

If the potentiometer is adjusted to $R_{bc} = 6\text{ k}\Omega$, V_b changes to:

$$V_b = V_S \frac{R_{bc}}{R_{ac}} = (10\text{ V})\frac{6\text{ k}\Omega}{10\text{ k}\Omega} = 6\text{ V}$$

One application of a variable voltage divider is shown in Figure 4.18b. The block in the figure represents an *audio amplifier.* The "volume control" on the audio amplifier is a potentiometer. When the potentiometer setting is varied, the wiper arm voltage to the audio amplifier increases (or decreases). By design, this voltage affects the output power from the audio amplifier.

Source Resistance: A Practical Consideration

OBJECTIVE 10 ➤ The **ideal voltage source** is *one that maintains a constant output voltage regardless of the resistance of its load.* For example, let's say that the voltage source in Figure 4.19a is an ideal voltage source. As shown, the voltage across the open terminals of the source is

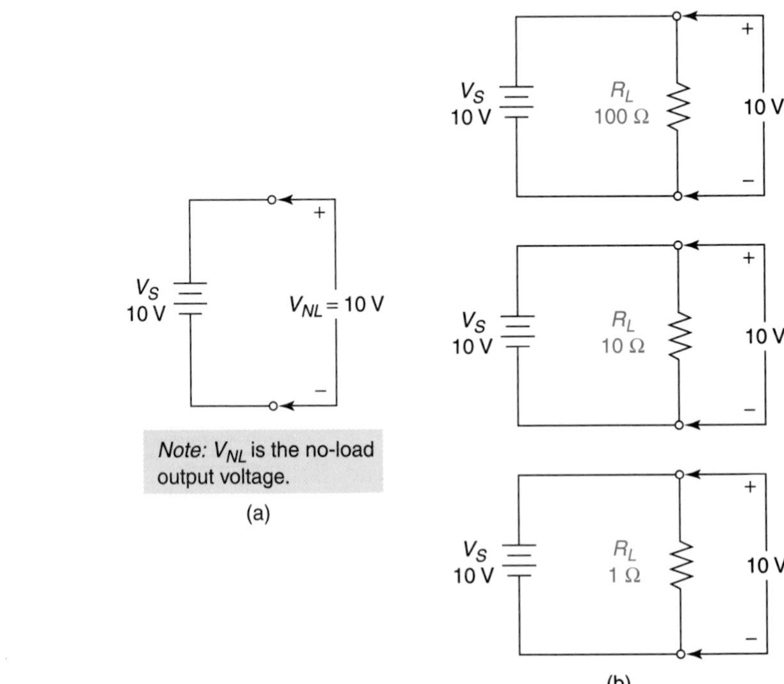

Figure 4.19

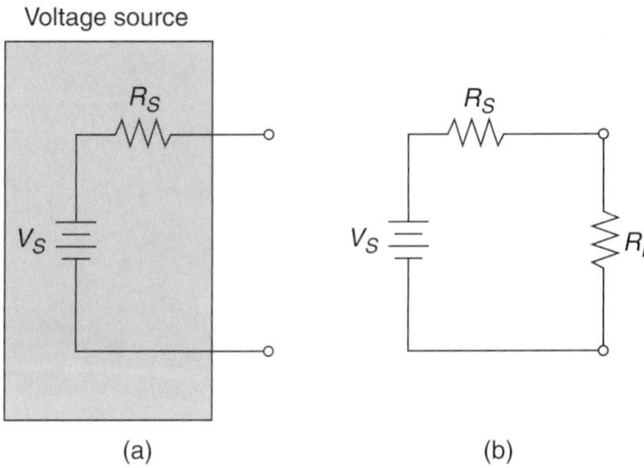

Figure 4.20
Voltage source internal resistance.

10 V. *The output from a voltage source when its terminals are open* is referred to as the **no-load output voltage (V_{NL}).** Note that V_{NL} is the maximum possible output from a given voltage source.

When the various load resistances shown in Figure 4.19b are connected to the source, it maintains the same 10 V output. Thus, for the ideal voltage source, $V_L = V_{NL}$ regardless of the value of R_L.

◄ *OBJECTIVE 11*

Unfortunately, the ideal voltage source does not exist at this time. For any practical voltage source, a decrease in load resistance results in a decrease in the source voltage. This is because every voltage source has some amount of *internal resistance* (R_S), as shown in Figure 4.20a. When a load is connected to the source (as shown in Figure 4.20b), it forms a voltage divider with the internal resistance of the source. This causes V_L to be lower than V_{NL}, as demonstrated in the following example.

Example 4.8

The no-load output voltage of the source in Figure 4.21 is 12 V. Calculate the values of V_L for $R_L = 1\text{ k}\Omega$ and $R_L = 100\ \Omega$.

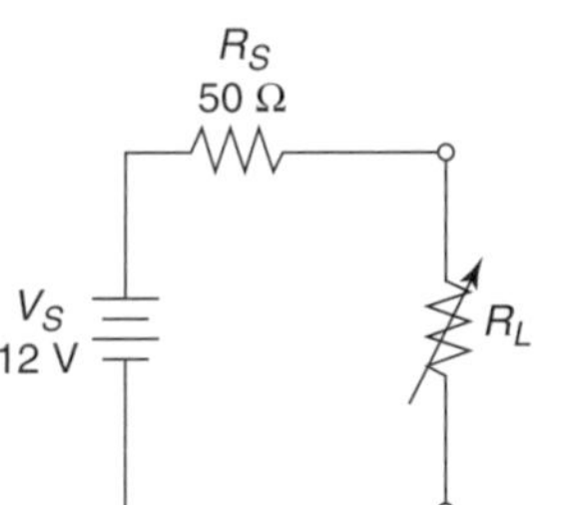

Figure 4.21

Solution: The load resistance forms a voltage divider with the internal resistance of the source (R_S). When $R_L = 1\text{ k}\Omega$, V_L is found as

$$V_L = V_S \frac{R_L}{R_T} = (12\text{ V})\frac{1\text{ k}\Omega}{1.05\text{ k}\Omega} = 11.43\text{ V}$$

Where $R_T = R_S + R_L$. When $R_L = 100\ \Omega$, V_L is found as

$$V_L = V_S \frac{R_L}{R_T} = (12\text{ V})\frac{100\ \Omega}{150\ \Omega} = 8\text{ V}$$

As you can see, the output from the practical voltage source decreases as the load resistance decreases.

Practice Problem 4.8

A voltage source with 40 Ω of internal resistance has a no-load output of 14 V. Determine the values of V_L when $R_L = 1\text{ k}\Omega$ and $R_L = 120\ \Omega$.

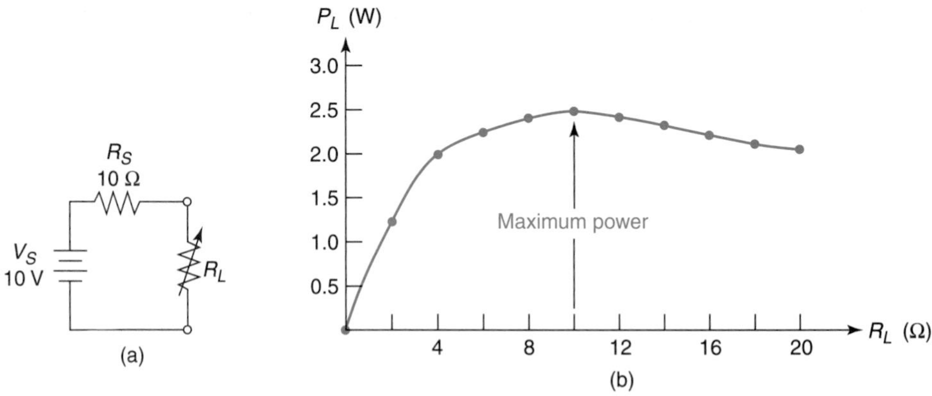

Figure 4.22 Load power as a function of load resistance.

You have seen that the internal resistance of a voltage source causes its output voltage to decrease when load resistance decreases. The ideal voltage source would not experience this decrease in output voltage, because *its internal resistance would be 0 Ω.*

The internal resistance of most dc voltage sources is typically 50 Ω or less. This doesn't pose a problem for loads in the low kΩ range or higher. However, source resistance can impact circuit operation under several circumstances:

- The internal resistance of the source can cause a significant drop in output voltage for a low resistance load.
- The internal resistance of the source determines (in part) the maximum amount of power that can be transferred from the source to its load.

At this point, we'll take a look at the relationship between source resistance, load resistance, and maximum load power.

Maximum Power Transfer

OBJECTIVE 12 ➤

In Chapter 3, you were told that electronic systems are designed to deliver power (in one form or another) to a given load. So, it follows that we would be interested in knowing the maximum possible output power for a given source.

TABLE 4.1
Example Load Values for Figure 4.22

R_L	V_L	P_L
0 Ω	0 V	0 W
2 Ω	1.66 V	1.38 W
4 Ω	2.86 V	2.04 W
6 Ω	3.75 V	2.34 W
8 Ω	4.44 V	2.47 W
10 Ω	**5.00 V**	**2.50 W**
12 Ω	5.45 V	2.48 W
14 Ω	5.83 V	2.43 W
16 Ω	6.15 V	2.37 W
18 Ω	6.43 V	2.30 W
20 Ω	6.67 V	2.23 W

The **maximum power transfer theorem** establishes a relationship between source resistance, load resistance, and maximum load power. According to this theorem, *maximum power transfer from a voltage source to its load occurs when the load resistance is equal to the source resistance.* This means that load power reaches its maximum possible value when $R_L = R_S$. Consider the circuit shown in Figure 4.22a. According to the maximum power transfer theorem, the voltage source will transfer maximum power to its load when the potentiometer is set to 10 Ω (the value of R_S).

In Table 4.1, we see the values of load voltage and power that occur when the load in Figure 4.22a is varied from 0 to 20 Ω (in 2 Ω increments). If you look at the load power (P_L) column of values, you'll see that the maximum value of load power (2.50 W) occurs when R_L is set to 10 Ω, the value of the source resistance.

If we plot a graph of P_L versus R_L, we get the curve shown in Figure 4.22b. Again, you can see that maximum load power in this circuit occurs when $R_L = R_S$.

The maximum power transfer theorem allows us to easily calculate the maximum possible load power for a circuit when the source resistance is known. This is demonstrated in the following example.

Example 4.9

Calculate the maximum possible load power for the circuit in Figure 4.23.

R_S
25 Ω
V_S
10 V
R_L
(100 Ω)

Figure 4.23

Solution: Maximum power is delivered to the load when $R_L = R_S$. When the potentiometer is set to this value (25 Ω), the load voltage is one-half the no-load source voltage, and

$$P_L = \frac{V_L^2}{R_L} = \frac{(5\text{ V})^2}{25\ \Omega} = 1\text{ W} \qquad \text{(maximum)}$$

Practice Problem 4.9

A circuit like the one in Figure 4.23 has values of $V_{NL} = 32$ V and $R_S = 40\ \Omega$. Calculate the maximum possible load power for the circuit.

An important point needs to be made: The maximum power transfer theorem holds true when source resistance is fixed and load resistance is variable. However, *it fails to hold true when load resistance is fixed and source resistance is variable.* For example, consider the circuit shown in Figure 4.24. In this circuit, the load resistance is fixed at 10 Ω, and the source resistance is variable. You would think that this load would receive maximum power when the source resistance is set to 10 Ω. However, this is not true. If R_S is set to 10 Ω, the load voltage is 5 V (according to the voltage-divider equation), and the load power is found as

The maximum power transfer theorem fails to hold true when load resistance is fixed and source resistance is variable.

$$P_L = \frac{V_L^2}{R_L} = \frac{(5\text{ V})^2}{10\ \Omega} = 2.5\text{ W}$$

If R_S is set to 0.1 Ω, the voltage divider provides a load voltage of 9.9 V, and the load power is found as

$$P_L = \frac{V_L^2}{R_L} = \frac{(9.9\text{ V})^2}{10\ \Omega} = 9.8\text{ W}$$

As you can see, maximum power transfer to this *specific* load resistance occurs when $R_S << R_L$. For a specific source resistance, maximum power transfer occurs when $R_L = R_S$. However, for a specific load resistance, maximum power transfer occurs when R_S is *set to its lowest possible value.*

A couple of points should be made before we move on to the next topic:

- Maximum load power does not necessarily correspond to maximum load voltage. This point can be verified using the values in Table 4.1.
- Maximum power transfer does not mean that the power supply is 100% efficient. In fact, efficiency can be considerably less than 100% when maximum power transfer occurs. For example, a circuit like the one shown in Figure 4.23 experiences maximum power transfer when $R_L = R_S$. In this case, the circuit efficiency is only 50%, because the power is divided equally between R_L and R_S.

Series-Connected Voltage Sources

In later chapters, we will encounter situations where two (or more) voltage sources are connected in series. These voltage sources are connected either as *series-aiding* or *series-opposing* sources, as shown in Figure 4.25.

◀ *OBJECTIVE 13*

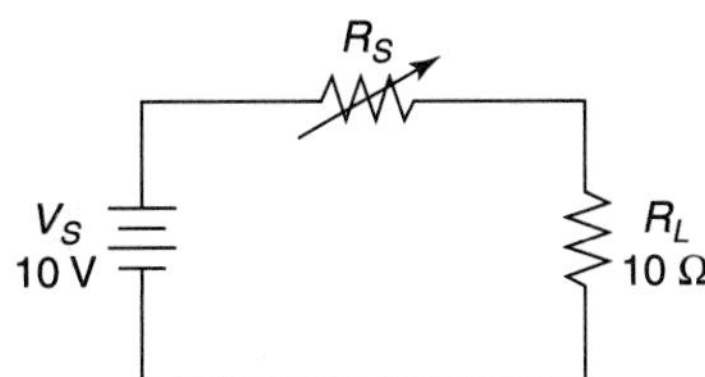

Figure 4.24

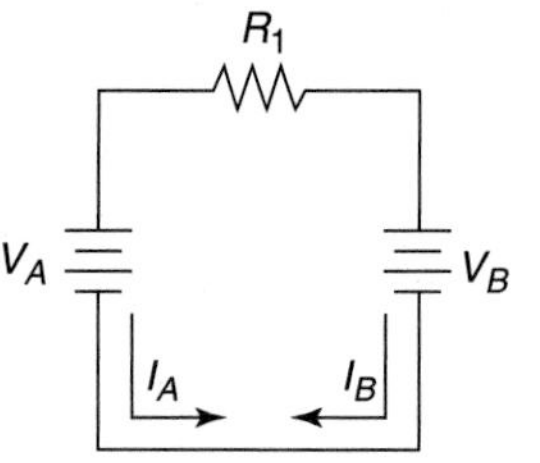

(a) Series-aiding voltage sources (b) Series-opposing voltage sources

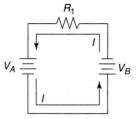

Note: The current directions shown indicate the direction each current would take without the presence of the other source.

Figure 4.25

Series-aiding voltage sources (like those shown in Figure 4.25a) are *those connected so that the currents generated by the individual sources are in the same direction.* As a result, each source is *aiding* the other. In contrast, **series-opposing voltage sources** (like those shown in Figure 4.25b) are *those connected so that the currents generated by the individual sources are in opposition to each other.*

Series-aiding voltage sources When two voltage sources are connected as shown in Figure 4.25a, the total circuit voltage equals the sum of the individual source voltages. This point is illustrated in the following example.

Example 4.10

Calculate the total current value for the circuit in Figure 4.26a.

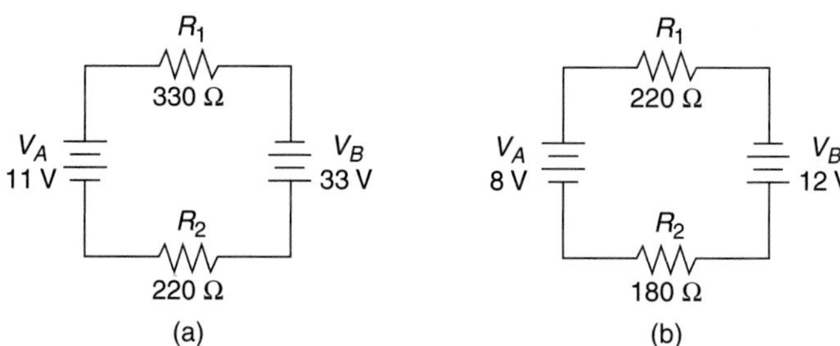

Figure 4.26

Solution: Since the sources are connected in series-aiding fashion, the total voltage can be found as

$$V_T = V_A + V_B = 11\text{ V} + 33\text{ V} = 44\text{ V}$$

Now, the total current can be found as

$$I_T = \frac{V_T}{R_T} = \frac{44\text{ V}}{550\ \Omega} = 80\text{ mA}$$

Practice Problem 4.10

Using the approach demonstrated in this example, calculate the value of I_T for the circuit in Figure 4.26b.

Series-opposing voltage sources When two voltage sources are connected as shown in Figure 4.25b, the total circuit voltage equals the difference between the individual source voltages.

The analysis of a circuit with series-opposing sources is usually performed by finding the difference between the source voltages and using Ohm's law to determine the circuit current, as demonstrated in the following example.

Example 4.11

Calculate the total current for the circuit in Figure 4.27a.

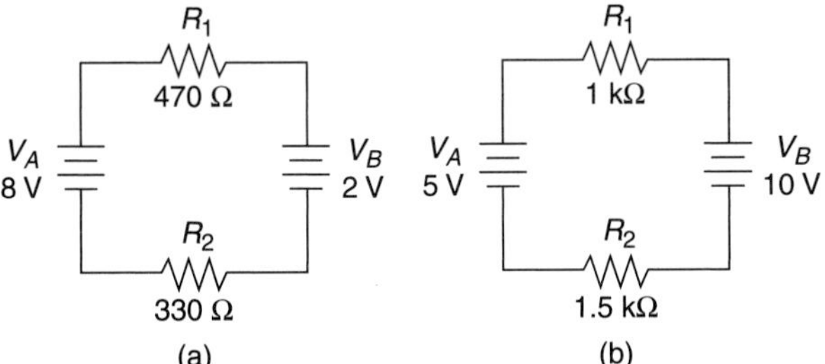

Figure 4.27

Solution: The total voltage in the circuit is found as

$$V_T = V_A - V_B = 8 \text{ V} - 2 \text{ V} = 6 \text{ V}$$

Now, the total circuit current can be found as

$$I_T = \frac{V_T}{R_T} = \frac{6 \text{ V}}{800 \ \Omega} = 7.5 \text{ mA}$$

Practice Problem 4.11

Calculate the value of I_T for the circuit in Figure 4.27b.

Determining the direction of circuit current For any pair of series-opposing sources, the total current is in the direction of the source with the greater magnitude. In the circuit shown in Figure 4.27a, the magnitude of V_A (8 V) is greater than that of V_B (2 V). Therefore, the circuit current is in the direction indicated by source A.

Earth Ground Versus Chassis Ground

◀ *OBJECTIVE 14*

Earlier in the chapter, you were introduced to ground as a 0 V reference and common connection point in a circuit. There are actually two common types of ground connections, called *earth ground* and *chassis ground.* The symbols for each are shown in Figure 4.28.

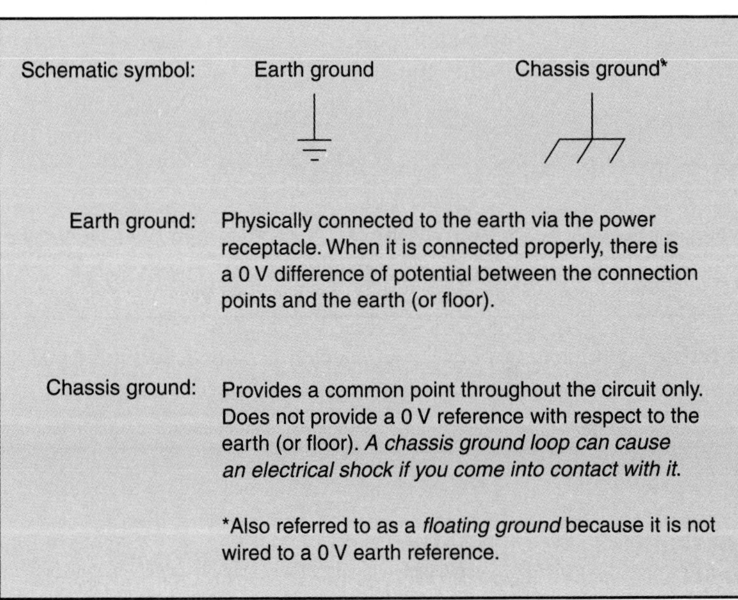

Figure 4.28 Earth ground versus chassis ground.

As its name implies, an **earth ground** *provides a physical connection to the earth (via the wall outlet).* This ensures that the ground in the circuit is at a 0 V potential with respect to the earth. Chassis ground, on the other hand, is not returned to earth. In other words, a **chassis ground** *provides a 0 V reference for the components in the circuit, but not necessarily between the circuit and the earth.* As a result, there may be a difference of potential between the circuit ground and the earth. This difference of potential could be felt by you, so always be careful when dealing with a chassis ground loop—it can hurt you if you happen to come into contact with it!

Section Review

1. Explain the use of a potentiometer as a variable voltage divider.
2. What happens to the output of an *ideal* voltage source when load resistance changes?
3. What happens to the output of a *practical* voltage source when load resistance changes?
4. In terms of source resistance, how does the ideal voltage source differ from a practical voltage source?
5. When is maximum power transferred from a voltage source to its load?
6. How do *series-aiding* voltage sources differ from *series-opposing* voltage sources?
7. Why must you be careful when working with a chassis-grounded circuit?

Critical Thinking

8. One voltage source has a value of $R_S = 100\ \Omega$, and another has a value of $R_S = 25\ \Omega$. Which would be considered the higher-quality source? Explain your answer.
9. You were told that the maximum output from a voltage source is called the *no-load output voltage.* The opposite of a *no-load* condition is called a *full load.* Using the words *minimum* and *maximum,* answer each of the following questions. Then, explain the reason for your choice.
   a. A full load has ________ resistance.
   b. A full load draws ________ current from the source.

Key Terms

The following terms were introduced in this chapter on the pages indicated:

Chassis ground, 114
Closed loop, 103
Earth ground, 114
Ground, 103
Ideal voltage source, 108
Kirchhoff's voltage law, 103
Maximum power transfer theorem, 110
No-load output voltage, 109
Series-aiding voltage source, 112
Series circuit, 97
Series-opposing voltage sources, 112
Voltage divider, 104
Voltage drop, 104

Practice Problems

1. Each resistor combination below is connected as shown in Figure 4.29. For each combination, calculate the total circuit resistance.

	R_1	R_2	R_3	R_4
a.	1 kΩ	220 Ω	330 Ω	1.1 kΩ
b.	10 Ω	18 Ω	47 Ω	200 Ω
c.	150 Ω	220 Ω	820 Ω	51 Ω
d.	10 kΩ	91 kΩ	5.1 kΩ	300 Ω

2. Each resistor combination below is connected as shown in Figure 4.29. For each combination, calculate the total circuit resistance.

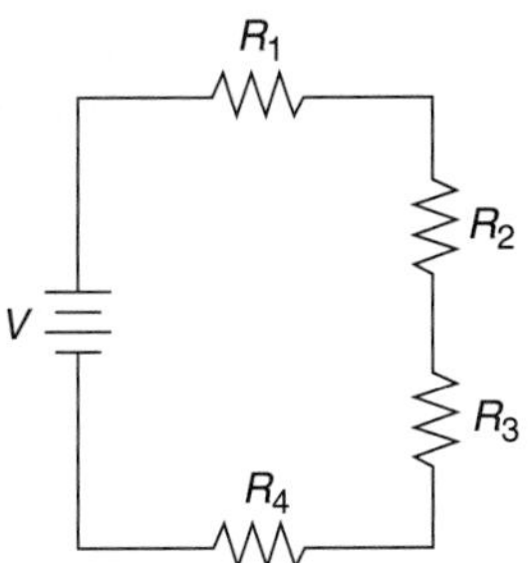

Figure 4.29

	R_1	R_2	R_3	R_4
a.	1 MΩ	470 kΩ	270 kΩ	51 kΩ
b.	22 kΩ	39 kΩ	12 kΩ	75 kΩ
c.	360 Ω	1.1 kΩ	68 Ω	2.2 kΩ
d.	8.2 kΩ	3.3 kΩ	9.1 kΩ	5.1 kΩ

3. The resistor combinations below are for a circuit like the one shown in Figure 4.30. In each case, determine the unknown resistor value.

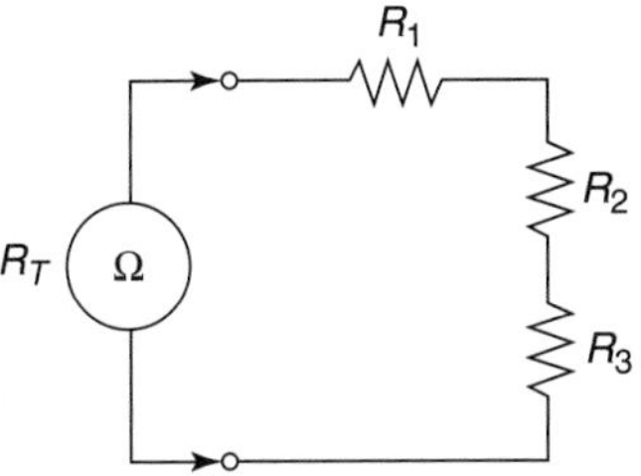

Figure 4.30

	R_1	R_2	R_3	R_T
a.	1.1 kΩ	330 Ω	_____	1.9 kΩ
b.	_____	47 kΩ	91 kΩ	165 kΩ
c.	33 kΩ	_____	6.2 kΩ	44.3 kΩ
d.	27 Ω	39 Ω	82 Ω	_____

4. The resistor combinations below are for a circuit like the one shown in Figure 4.30. In each case, determine the unknown resistor value.

	R_1	R_2	R_3	R_T
a.	200 kΩ	33 kΩ	_____	308 kΩ
b.	_____	82 kΩ	75 kΩ	204 kΩ
c.	510 Ω	_____	68 Ω	611 Ω
d.	1.5 MΩ	2.2 MΩ	10 MΩ	_____

5. Determine the potentiometer setting required in Figure 4.31a to provide a total resistance of 52 kΩ.

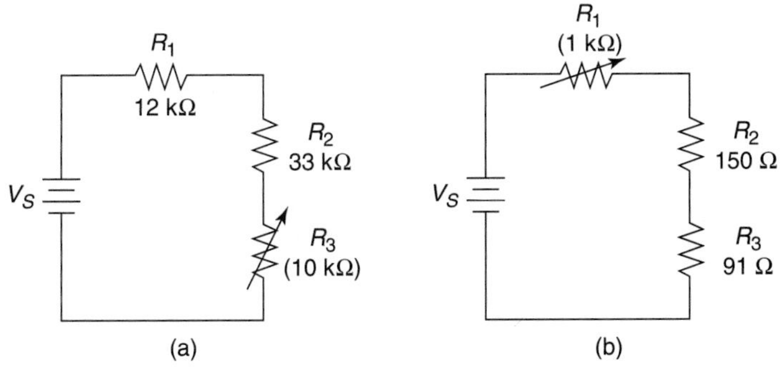

Figure 4.31

6. Determine the potentiometer setting required in Figure 4.31b to provide a total resistance of 872 Ω.

7. Calculate the total current for the circuit shown in Figure 4.32a.

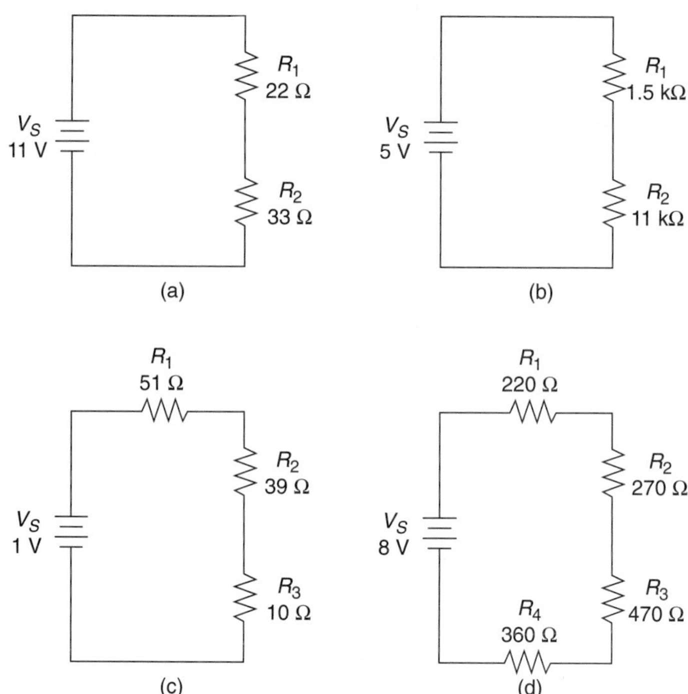

Figure 4.32

8. Calculate the total current for the circuit shown in Figure 4.32b.

9. Calculate the total current for the circuit shown in Figure 4.32c.

10. Calculate the total current for the circuit shown in Figure 4.32d.

11. Determine the potentiometer setting needed to set the current in Figure 4.33a to 18 mA.

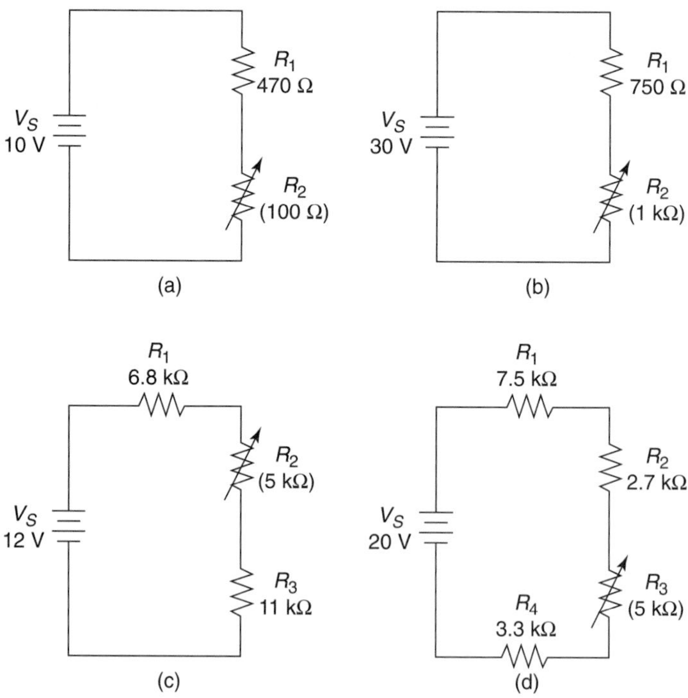

Figure 4.33

12. Determine the potentiometer setting needed to set the current in Figure 4.33b to 27 mA.

13. Determine the potentiometer setting needed to set the current in Figure 4.33c to 560 μA.

14. Determine the potentiometer setting needed to set the current in Figure 4.33d to 1.22 mA.

15. Calculate the value of the source voltage in Figure 4.34a.

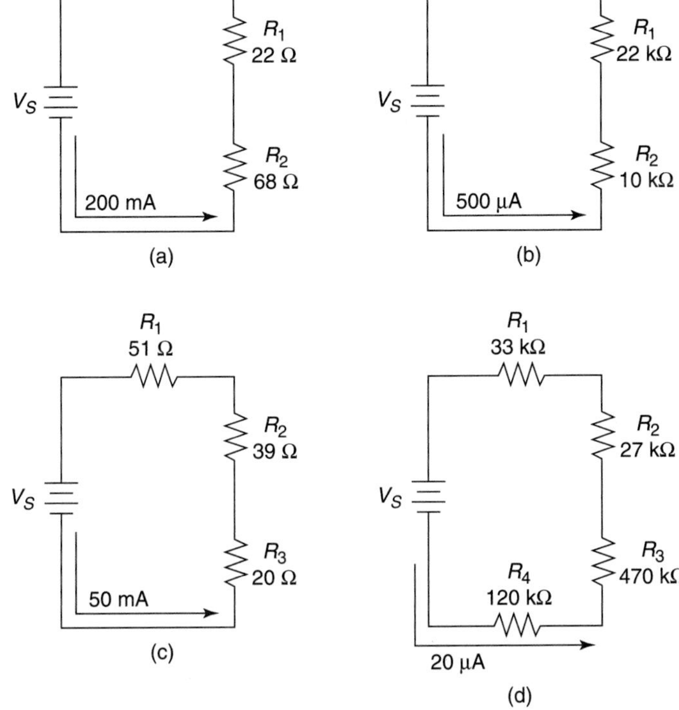

Figure 4.34

16. Calculate the value of the source voltage in Figure 4.34b.

17. Calculate the value of the source voltage in Figure 4.34c.

18. Calculate the value of the source voltage in Figure 4.34d.

19. For the circuit shown in Figure 4.35a, verify that the total power dissipated by the resistors is equal to the power being supplied by the source.

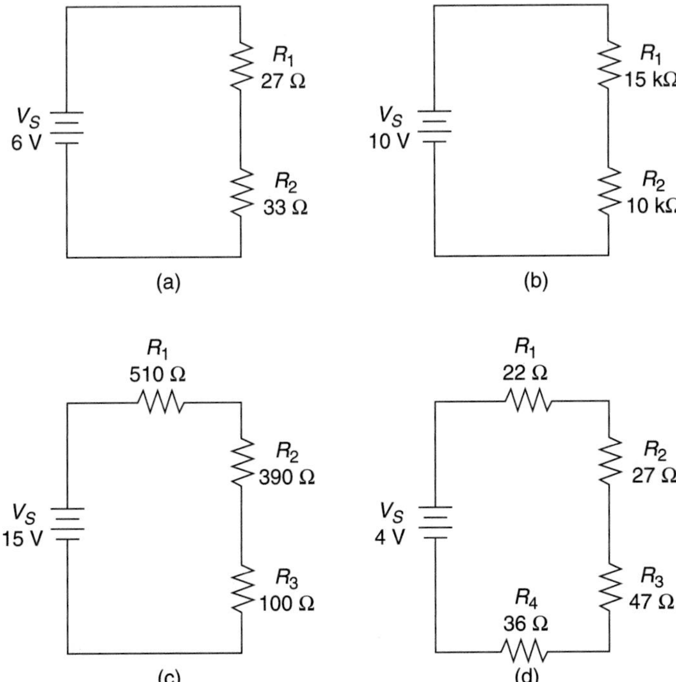

Figure 4.35

20. Repeat problem 19 for the circuit shown in Figure 4.35b.

21. Repeat problem 19 for the circuit shown in Figure 4.35c.

22. Repeat problem 19 for the circuit shown in Figure 4.35d.

23. Write the Kirchhoff's loop equation for the circuit shown in Figure 4.36a, and verify that the voltages add up to 0 V.

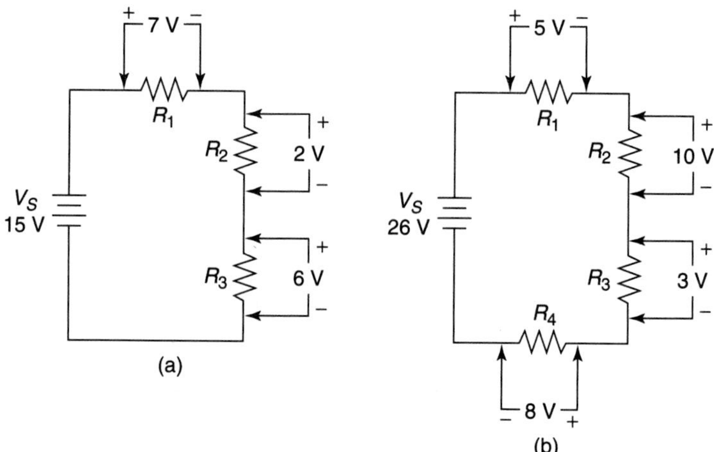

Figure 4.36

24. Repeat problem 23 for the circuit shown in Figure 4.36b.

25. Calculate the value of the current in Figure 4.37a.

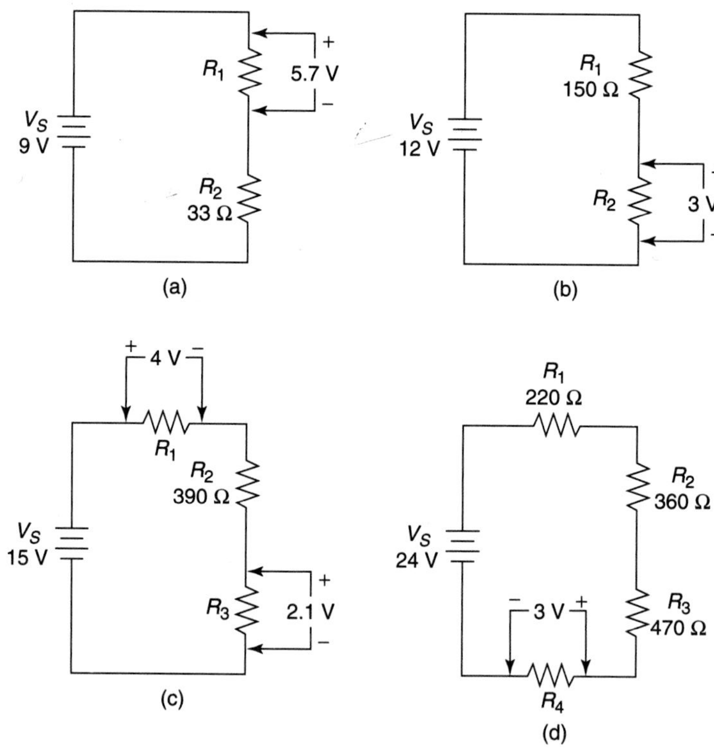

Figure 4.37

26. Calculate the value of the current in Figure 4.37b.

27. Calculate the value of the current in Figure 4.37c.

28. Calculate the value of the current in Figure 4.37d.

29. For the circuit shown in Figure 4.38a, calculate the voltage from point A to ground.

30. For the circuit shown in Figure 4.38b, calculate the voltage from point A to ground.

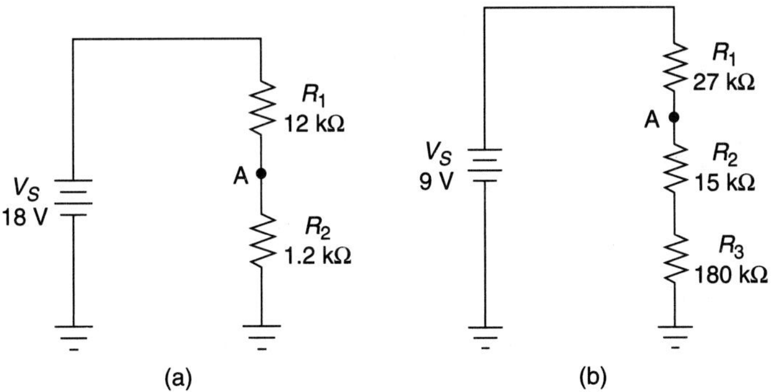

Figure 4.38

31. Using the voltage-divider equation, calculate the voltage across each of the resistors in Figure 4.39a.

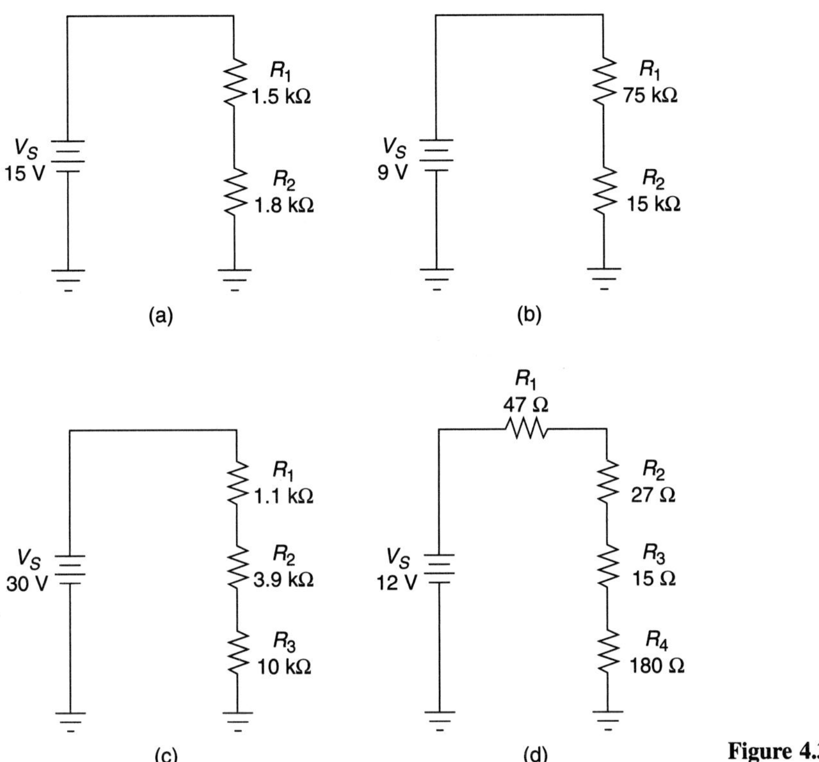

Figure 4.39

32. Repeat problem 31 for the circuit in Figure 4.39b.
33. Repeat problem 31 for the circuit in Figure 4.39c.
34. Repeat problem 31 for the circuit in Figure 4.39d.
35. Using the voltage-divider equation, determine the voltage from point A to ground for the circuit shown in Figure 4.40a.

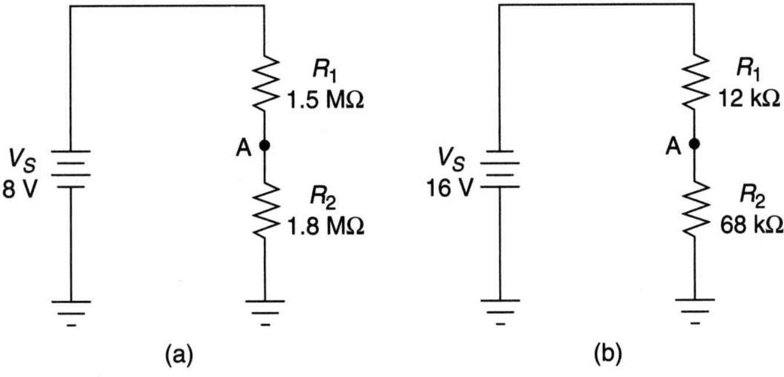

Figure 4.40

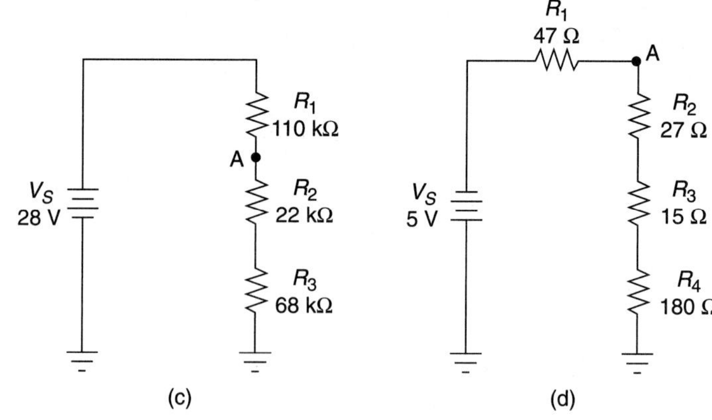

Figure 4.40

36. Repeat problem 35 for the circuit shown in Figure 4.40b.

37. Repeat problem 35 for the circuit shown in Figure 4.40c.

38. Repeat problem 35 for the circuit shown in Figure 4.40d.

39. Calculate the component voltage and power values for the circuit shown in Figure 4.41a, along with the circuit values of R_T, I_T, and P_T.

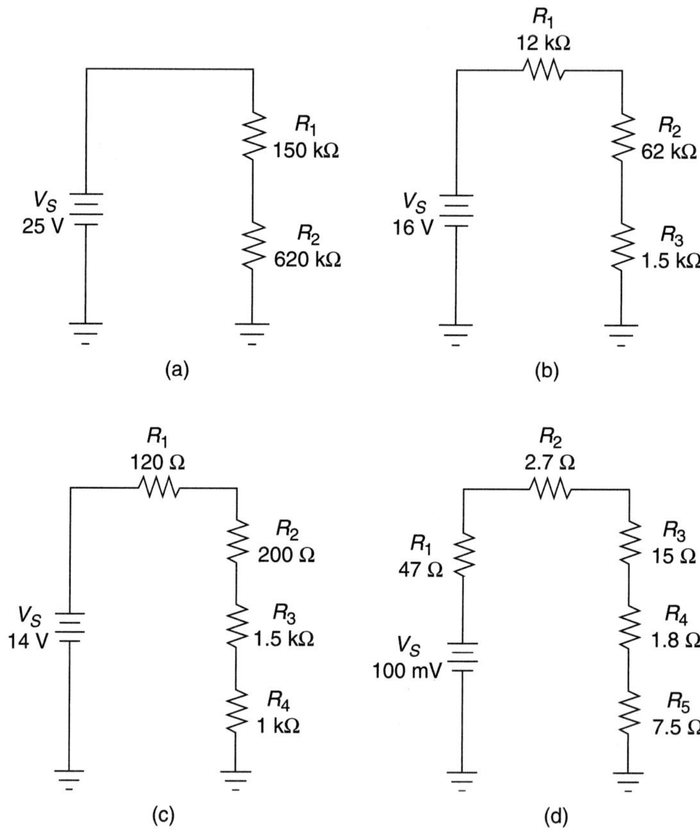

Figure 4.41

40. Calculate the component voltage and power values for the circuit shown in Figure 4.41b, along with the circuit values of R_T, I_T, and P_T.

41. Calculate the component voltage and power values for the circuit shown in Figure 4.41c, along with the circuit values of R_T, I_T, and P_T.

42. Calculate the component voltage and power values for the circuit shown in Figure 4.41d, along with the circuit values of R_T, I_T, and P_T.

43. Calculate the maximum possible value of load power for the circuit shown in Figure 4.42a.

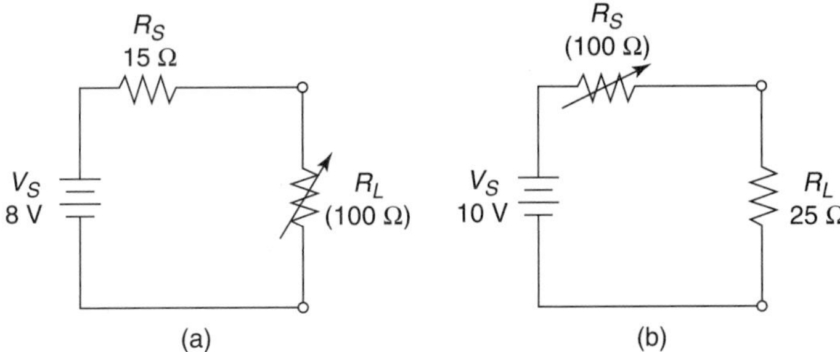

Figure 4.42

44. Calculate the maximum possible value of load power for the circuit shown in Figure 4.42b.

Looking Back

These problems relate to material presented in earlier chapters. The chapters are identified in brackets.

45. What is the resistance of a 100 V, 5 W soldering iron? [Chapter 3]

46. A 15 V source is connected to a 910 Ω resistor with a 5% tolerance. Calculate the maximum value of circuit current. [Chapter 2]

47. Convert the following values to engineering notation form without using your calculator. [Introduction]

a. 0.000047 F
b. 12,560 Ω
c. 0.000000010 m
d. 3,500,000,000 Hz

48. A circuit contains a 12 V source and a resistor. The circuit current always falls within ±2% of 6 mA. What is the color code of the resistor? [Chapters 2 and 3]

Pushing the Envelope

49. Each of the resistors shown in Figure 4.43 has a tolerance of 5%. Determine the minimum and maximum values of I_T for the circuit (assuming that all resistors are within tolerance).

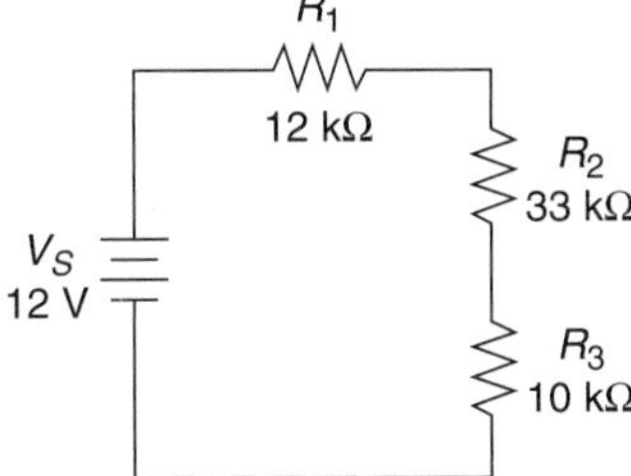

Figure 4.43

50. For the circuit shown in Figure 4.44:

a. Calculate the values of V_L for R_L settings of 0 to 1 kΩ (at 100 Ω increments).
b. Using your calculated values, plot a curve of V_L versus R_L.
c. Determine (from your curve) the minimum allowable value of R_L, assuming that V_L can drop as far as 10% below its maximum value.

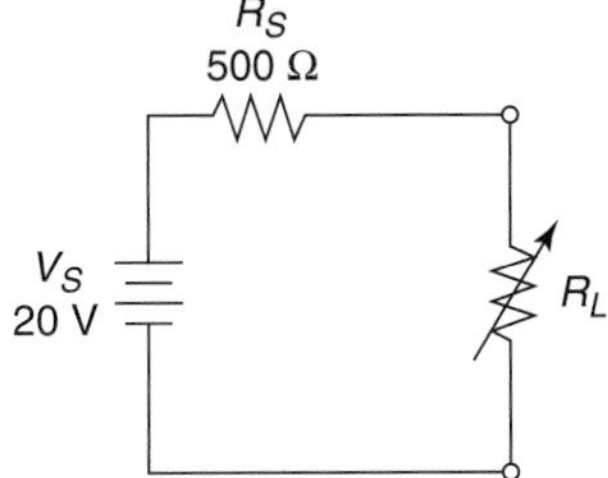

Figure 4.44

Answers to the Example Practice Problems

4.1. 6.25 kΩ

4.2. 2 kΩ

4.3. 714 μA

4.4. 26.8 V

4.5. 4.06 V

4.6. $V_1 = 8.25$ V, $V_2 = 11.75$ V

4.7. 11.85 V

4.8. When $R_L = 1$ kΩ, $V_L = 13.46$ V
When $R_L = 120$ Ω, $V_L = 10.5$ V

4.9. 6.4 W (maximum)

4.10. $I_T = 50$ mA

4.11 $I_T = 2$ mA

Multisim File References

Figure	File Reference
4.3	ETF4_3.msm
4.12	ETF4_12.msm
4.13	ETF4_13.msm
4.18	ETF4_18.msm
4.19	ETF4_19.msm
4.25	ETF4_25.msm

chapter 5

Parallel Circuits

Objectives

After studying the material in this chapter, you should be able to:

1. Identify a parallel circuit.
2. Describe the current and voltage characteristics of parallel circuits.
3. Calculate any current value in a parallel circuit, given the source voltage and branch resistance values.
4. State the relationship between the branch resistances and the total resistance of any parallel circuit.
5. Compare and contrast the power characteristics of series and parallel circuits.
6. Calculate the total resistance in any parallel circuit, given the branch resistance values.
7. Describe *Kirchhoff's current law,* and solve the Kirchhoff's current equation for any point within a parallel circuit.
8. Describe the characteristics of a *current source* and calculate its output current.
9. Use the *current-divider* equations to solve for either branch current in a two-branch circuit.
10. Compare and contrast the characteristics of a *practical* current source to those of the *ideal* current source.
11. Discuss the *maximum power transfer theorem* as it relates to parallel circuits.
12. Calculate the maximum load power that can be provided by a current source.

In this chapter, we continue our coverage of basic circuits by discussing the operating characteristics of parallel circuits. As you will see, these circuits are nearly the opposite of series circuits in many ways.

5.1 Parallel Circuit Characteristics

OBJECTIVE 1 ➤ A **parallel circuit** is *a circuit that provides more than one current path between any two points.* Several parallel circuits are shown in Figure 5.1. As you can see, each circuit contains two or more paths for current. Note that *each current path in a parallel circuit* is referred to as a **branch.** For example, the circuit in Figure 5.1a contains two branches, the circuit in Figure 5.1b contains three branches, and so on.

Current Characteristics

OBJECTIVE 2 ➤ Using the circuit in Figure 5.1a, let's take a look at the current action in a typical parallel circuit: The current leaves the negative terminal of the source and splits when it reaches the first connection point in the circuit (A). Part of the current then passes through R_1, and the remainder passes through R_2. When these currents reach the second connection point (B), they recombine into a single current and return to the source.

The total current in a parallel circuit (I_T) equals the sum of the individual branch current values. By formula,

$$I_T = I_1 + I_2 + \ldots + I_n \tag{5.1}$$

where

I_n = the current through the highest-numbered branch in the circuit.

Equation (5.1) is one way of expressing a relationship called *Kirchhoff's current law,* which is discussed in detail in Section 5.2. Example 5.1 shows how it is used to determine the total current in a parallel circuit.

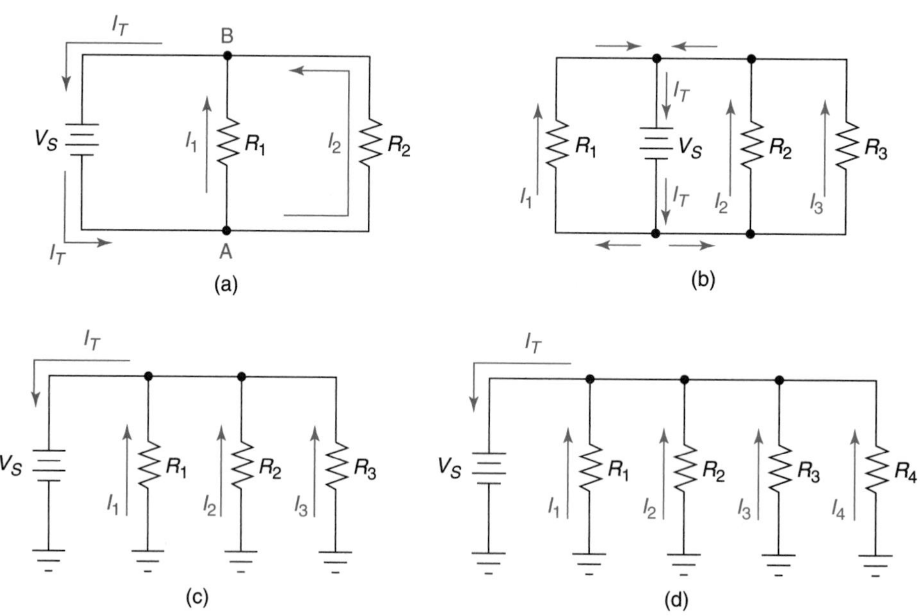

Figure 5.1 Parallel circuits.

Example 5.1

Calculate the total current through the circuit in Figure 5.2.

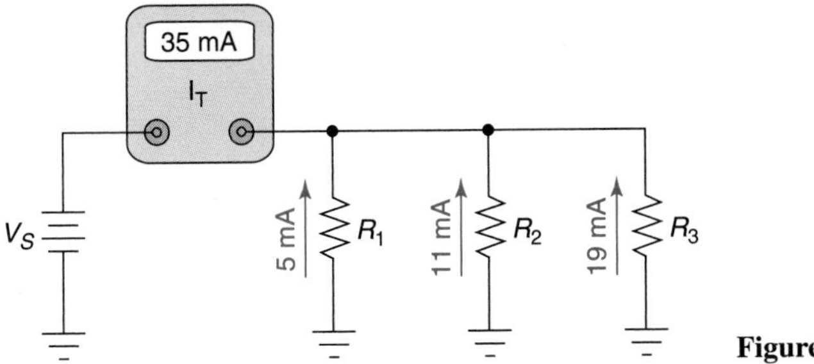

Figure 5.2

Solution: Using equation (5.1), the total circuit current is found as

$$I_T = I_1 + I_2 + I_3 = 5\text{ mA} + 11\text{ mA} + 19\text{ mA} = 35\text{ mA}$$

Practice Problem 5.1

A circuit like the one shown in Figure 5.2 has the following values: $I_1 = 6.4$ mA, $I_2 = 3.3$ mA, and $I_3 = 14.8$ mA. Calculate the total current in the circuit.

When you think about it, the result in Example 5.1 makes sense. The branch current values given must come from the circuit's voltage source, so I_T must equal the sum of the branch currents.

Voltage and Current Values

When two or more components are connected in parallel, the voltage across each component is equal to the voltage across all of the others. This characteristic is illustrated in Figure 5.3.

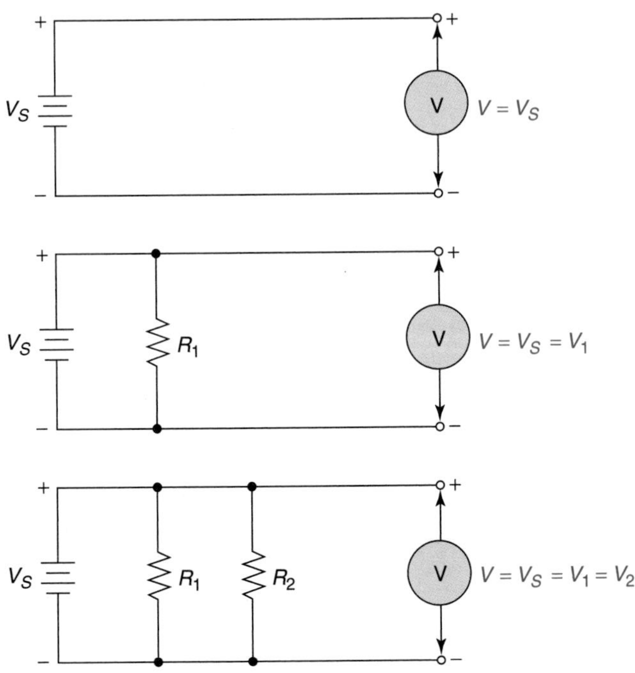

Figure 5.3 The voltages across parallel components are equal.

As the circuits in Figure 5.3 show, the source voltage produces a difference of potential between the (+) and (−) conductors. Since the resistors are connected between these conductors, the voltage across each resistor is equal to the source voltage. By formula,

$$V_S = V_1 = V_2 = \ldots = V_n \tag{5.2}$$

where

V_n = the voltage across the highest-numbered branch

OBJECTIVE 3 ➤ The current through each branch of a parallel circuit is determined by the source voltage and the resistance of the branch. By formula,

$$I_n = \frac{V_S}{R_n}$$

Once we have calculated the value of each branch current, we simply add those values to determine the total circuit current, as demonstrated in the following example.

Example 5.2

Calculate the total current in the circuit shown in Figure 5.4.

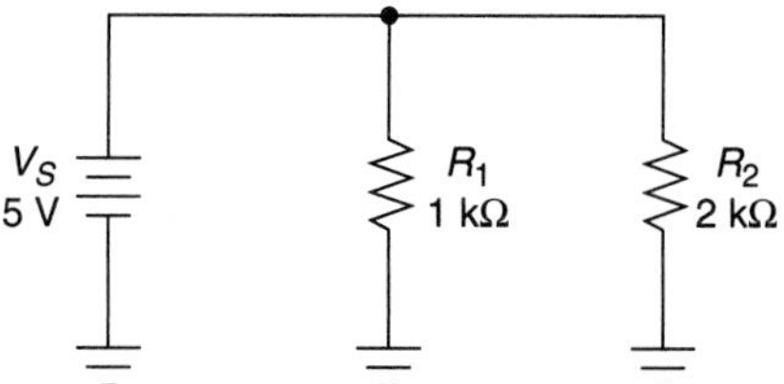

Figure 5.4

Solution: First, we calculate the individual branch currents, as follows:

$$I_1 = \frac{V_S}{R_1} = \frac{5\text{ V}}{1\text{ k}\Omega} = 5\text{ mA} \quad \text{and} \quad I_2 = \frac{V_S}{R_2} = \frac{5\text{ V}}{2\text{ k}\Omega} = 2.5\text{ mA}$$

Now, the total circuit current is found as

$$I_T = I_1 + I_2 = 5\text{ mA} + 2.5\text{ mA} = 7.5\text{ mA}$$

Practice Problem 5.2

A circuit like the one shown in Figure 5.4 has the following values: $V_S = 12$ V, $R_1 = 240\ \Omega$, and $R_2 = 360\ \Omega$. Calculate the total current in the circuit.

Before we move on, let's compare the voltage and current characteristics of parallel circuits with those of series circuits. These characteristics are summarized in Table 5.1. As you can see, the voltage and current characteristics of the parallel circuit are the opposite of those for the series circuit.

You may recall that a *series circuit* is often referred to as a *voltage divider,* because it divides the source voltage among the circuit components. By the same token, the term **current divider** is often used to describe a parallel circuit because *the source current is divided among the circuit branches.* This topic is discussed in greater detail later in this chapter.

TABLE 5.1 Comparison of Series and Parallel Circuit Characteristics

Circuit Type	*Voltage Characteristics*	*Current Characteristics*
Series	$V_T = V_1 + V_2 + \ldots + V_n$	Equal at all points
Parallel	Equal across all branches	$I_T = I_1 + I_2 + \ldots + I_n$

Resistance Characteristics

In Example 5.2, the total current supplied by a 5 V source was found to be 7.5 mA. If we apply Ohm's law to these values, the total circuit resistance is found as ◄ *OBJECTIVE 4*

$$R_T = \frac{V_S}{I_T} = \frac{5\text{ V}}{7.5\text{ mA}} = 667\ \Omega$$

As you can see, this value is lower than the value of either resistor in the circuit.

In any parallel circuit, *the total circuit resistance is always lower than any of the branch resistance values.* This relationship is not as difficult to understand as it may seem. Consider the circuit shown in Figure 5.5. According to Ohm's law, current varies inversely with resistance (when voltage is fixed). Since the total current in Figure 5.5 is *greater than* any individual branch current, the total resistance must be *less than* any individual branch resistance.

In Section 5.2, you will be shown how to calculate the value of R_T using the branch resistance values. At this point, we are interested only in establishing that R_T is lower than any branch resistance value in the circuit.

Power Characteristics

The power characteristics of parallel circuits are nearly identical to those of series circuits. For example, the total power in a parallel circuit is found as the sum of the power dissipation values for the individual components in that circuit. Also, component power dissipation values in any parallel circuit are calculated using the same standard power relationships that we used in series circuits. ◄ *OBJECTIVE 5*

In terms of power dissipation, there is one major difference between series and parallel circuits. In a series circuit, the higher the value of a resistor, the higher the portion of the total power it dissipates. In a parallel circuit, the opposite is true; that is, the *lower* the value of branch resistance, the higher the portion of the total power it dissipates. For example, the circuit in Figure 5.4 (Example 5.2) has the following power values:

$$P_1 = \frac{V_S^2}{R_1} = \frac{(5\text{ V})^2}{1\text{ k}\Omega} = 25\text{ mW} \qquad \text{and} \qquad P_2 = \frac{V_S^2}{R_2} = \frac{(5\text{ V})^2}{2\text{ k}\Omega} = 12.5\text{ mW}$$

As you can see, the smaller resistor is dissipating more power. This relationship is due to the fact that power is inversely proportional to resistance when voltage is fixed.

One Final Note

As you have been told, series and parallel circuits are opposites in many respects. Figure 5.6 provides a comparison of a basic parallel circuit and a basic series circuit. As you can see, the current and voltage characteristics are exact opposites. The power characteristics for the two circuits appear to be the same, but as you were shown in this section, the *lowest* resistance in a parallel circuit dissipates the greatest amount of power. The opposite

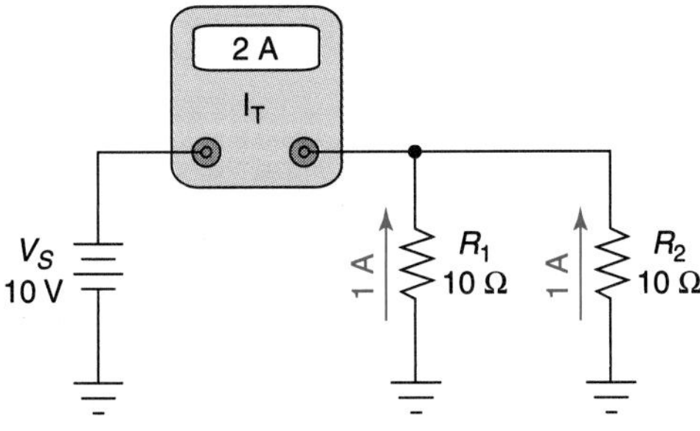

Figure 5.5

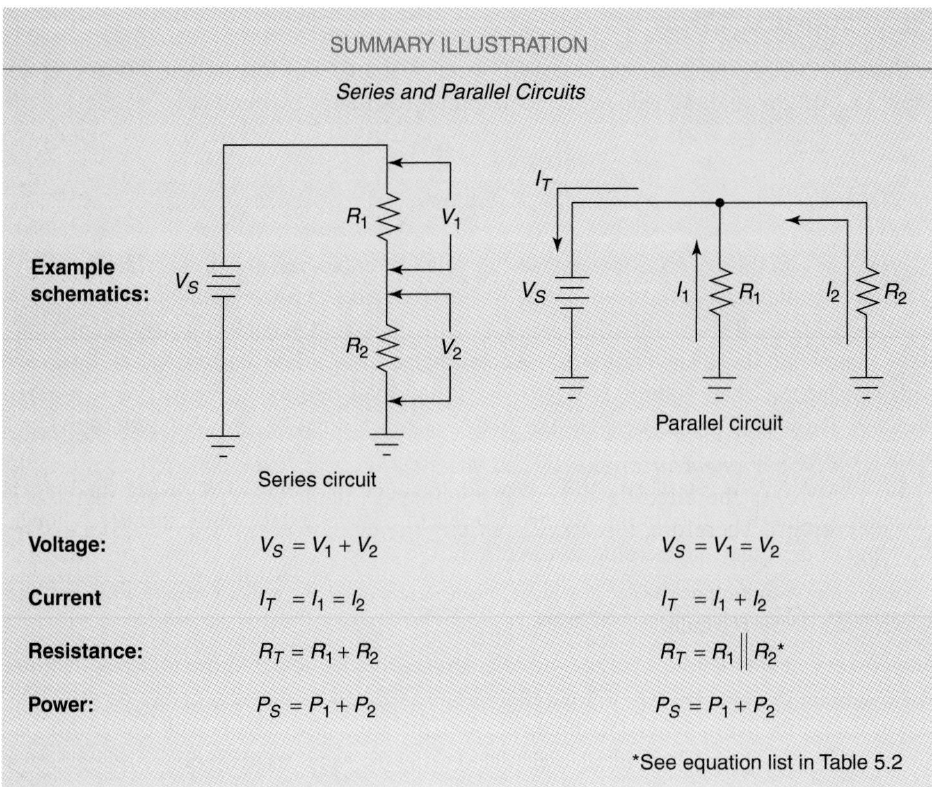

Figure 5.6

holds true for a series circuit, where the *highest* resistance dissipates the greatest amount of power.

Section Review

1. In terms of current, how does a *parallel circuit* differ from a *series circuit*?
2. What is a *branch*?
3. Describe the voltage characteristics of a parallel circuit.
4. Why are parallel circuits referred to as *current dividers*?
5. Why is the *total resistance* of a parallel circuit lower than any of the branch values?
6. Compare and contrast the power characteristics of series and parallel circuits.

Critical Thinking

7. Refer to the parallel circuit in Figure 5.5. What would happen to the current through R_2 if R_1 was to *open*? Explain your answer.

5.2 Parallel Resistance Relationships

In the last section, you were shown that the total resistance in a parallel circuit is lower than the resistance in any individual branch. In this section, we will take a closer look at the relationship between R_T and the branch resistance values.

Conductance

To establish a mathematical relationship between the branch resistances and the total resistance in a parallel circuit, we need to take a brief look at the *conductance* characteristics of parallel circuits. You may recall that conductance (G) is *a measure of the*

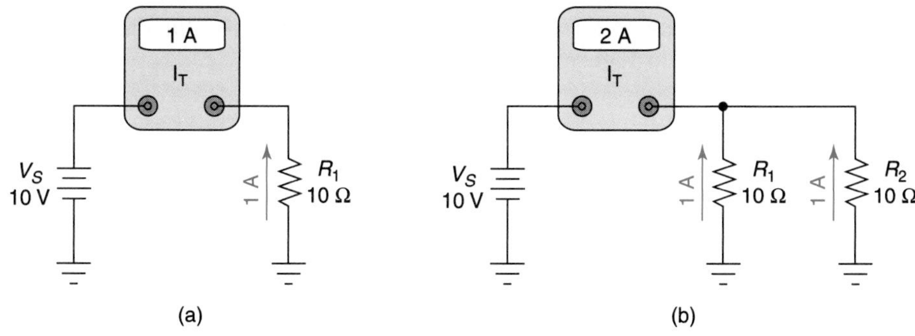

Figure 5.7

ability of a component or circuit to conduct. In Figure 5.7, the addition of a parallel branch increases the circuit's ability to conduct (as demonstrated by the increase in circuit current). Therefore, *the addition of the branch increases the conductance of the circuit.*

The total conductance (G_T) in a parallel circuit equals the sum of the branch conductance values. By formula,

$$G_T = G_1 + G_2 + \ldots + G_n \tag{5.3}$$

where

G_n = the conductance of the highest-numbered branch in the circuit

Calculating Total Resistance: The Reciprocal Method

We can use the relationship between conductance and resistance to develop an equation for the total resistance in a parallel circuit. Resistance and conductance are reciprocal values, so the total resistance of a parallel circuit can be found as ◄ *OBJECTIVE 6*

$$R_T = \frac{1}{G_T}$$

Since the total conductance (G_T) in a parallel circuit is equal to the sum of the branch conductance values, the above equation can be rewritten as

$$R_T = \frac{1}{G_1 + G_2 + \ldots + G_n} \tag{5.4}$$

Finally, since $G = \frac{1}{R}$, the relationship given in equation (5.4) can be rewritten as

$$R_T = \frac{1}{\frac{1}{R_1} + \frac{1}{R_2} + \ldots + \frac{1}{R_n}} \tag{5.5}$$

Equation (5.5) may look intimidating at first, but it's actually very easy to solve on any standard calculator. Just enter the values into your calculator as you would for any addition problem. However,

1. You press the reciprocal [1/x] key after entering each resistor value.
2. You press the [=] key, followed by the [1/x] key, after all the values have been entered.

The use of this key sequence is demonstrated in the following example.

Example 5.3

Determine the calculator key sequence you would use to solve for the value of R_T for the circuit in Figure 5.8.

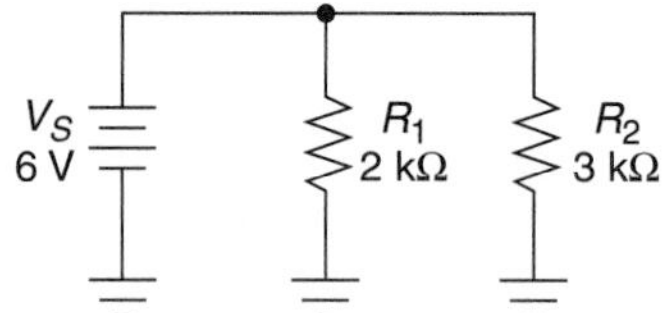

Figure 5.8

Solution: The problem is solved using the following key sequence:

2 [EE] 3 $\left[{}^{1}\!/\!_{x}\right]$ [+] **3 [EE] 3** $\left[{}^{1}\!/\!_{x}\right]$ [=] $\left[{}^{1}\!/\!_{x}\right]$

This key sequence yields a result of 1.2 kΩ.

Practice Problem 5.3

A circuit like the one shown in Figure 5.8 has the following values: $R_1 = 240\ \Omega$ and $R_2 = 360\ \Omega$. Solve for the value of R_T for the circuit.

Calculating Total Resistance: The Product-Over-Sum Method

Another commonly used method to solve for the total resistance of a parallel circuit is the *product-over-sum method.* This method allows you to solve for the total resistance of a *two*-branch circuit as follows:

Note: Equation (5.6) is derived as shown in Appendix D.

$$R_T = \frac{R_1 R_2}{R_1 + R_2} \tag{5.6}$$

The use of this equation is demonstrated in the following example.

Example 5.4

Solve for the total resistance of the circuit in Figure 5.8 in Example 5.3.

Solution: The value of R_T for the circuit can be found as

$$R_T = \frac{R_1 R_2}{R_1 + R_2} = \frac{(2\text{ k}\Omega)(3\text{ k}\Omega)}{2\text{ k}\Omega + 3\text{ k}\Omega} = 1.2\text{ k}\Omega$$

Practice Problem 5.4

Solve Practice Problem 5.3 using the product-over-sum method.

Calculating Total Resistance When Branch Resistances Are Equal

There is another "special-case" method for finding the value of R_T. When a given number of *equal resistances* are connected in parallel, the total resistance is equal to the value of one resistor divided by the number of branches. For example, consider the circuit shown in Figure 5.9. Here we have three 5.1 kΩ resistors that are connected in parallel. The total resistance of this circuit can be found as

$$R_T = \frac{5.1\text{ k}\Omega}{3} = 1.7\text{ k}\Omega$$

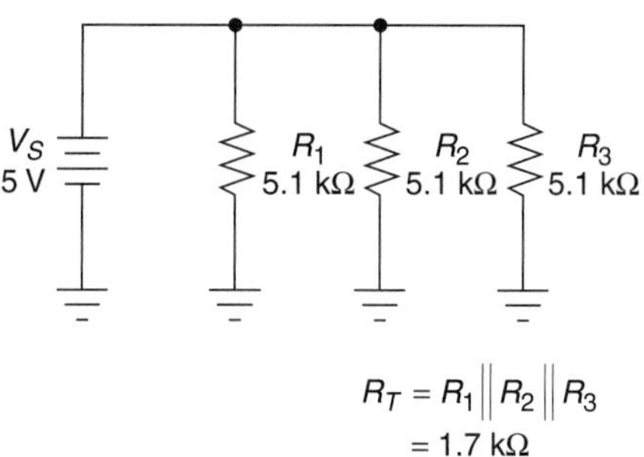

Figure 5.9

Note that the equation $R_T = R_1 \| R_2 \| R_3$ shown in Figure 5.9 indicates that the total resistance is equal to the parallel combination of R_1, R_2, and R_3. We will use this short-hand notation from now on.

TABLE 5.2 Methods Used to Solve for Total Parallel Resistance

Equation	*Applications*
$R_T = \dfrac{1}{\dfrac{1}{R_1} + \dfrac{1}{R_2} + \ldots + \dfrac{1}{R_n}}$	All circuits
$R_T = \dfrac{R_1 R_2}{R_1 + R_2}$	Circuits containing only two branches
$R_T = \dfrac{\text{the value of one resistor}}{\text{the number of branches}}$	Circuits containing branches with equal resistance values

Remember: When the source voltage and total current are known, R_T can also be found using Ohm's law.

Putting It All Together

There are three methods that are commonly used to calculate the total resistance of a parallel circuit. These methods and their applications are summarized in Table 5.2.

Section Review

1. Which method(s) can be used to solve $R_T = R_1 \| R_2 \| R_3$ when the resistor values are unequal?
2. Which method(s) can be used to solve $R_T = R_1 \| R_2 \| R_3$ when the resistor values are equal?
3. Which method(s) can be used to solve $R_T = R_1 \| R_2$ when the resistor values are unequal?
4. A parallel circuit has a value of $R_T = R_1 \| R_2 \| R_3$. What happens to the value of R_T if any one resistor opens? Explain your answer. Critical Thinking

5.3 Current Relationships: Kirchhoff's Current Law, Current Sources, and Current Dividers

You were shown earlier that the sum of the branch currents in a parallel circuit equals the total circuit current. In this section, we're going to take a closer look at the relationships between the various currents in a parallel circuit.

Note: The currents *entering* a point are represented as *positive* values in Kirchhoff's current equations and the currents *leaving* a point are represented as *negative* values. For example, the current equation for Figure 5.10a is

$I_T + I_1 + I_2$
$= 10 \text{ mA} + (-2 \text{ mA}) + (-8 \text{ mA})$
$= 0 \text{ mA}$

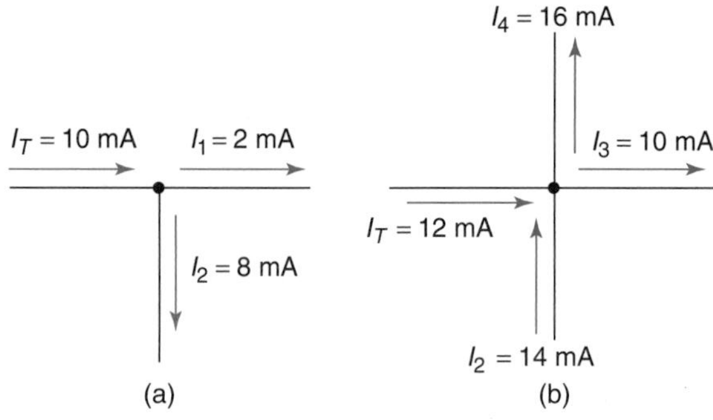

Figure 5.10

Kirchhoff's Current Law

OBJECTIVE 7 ➤

As you know, *Kirchhoff's voltage law* describes the relationship between the various voltages in a series circuit. In a similar fashion, *Kirchhoff's current law* describes the relationship between the various currents in a parallel circuit.

According to **Kirchhoff's current law,** the algebraic sum of the currents entering and leaving a point must equal zero. In other words, *the total current leaving a point must equal the total current entering that point.* This relationship is illustrated in Figure 5.10. In Figure 5.10a, the total current shown entering the connection point (I_T) is 10 mA. The sum of the currents leaving the point ($I_1 + I_2$) is also 10 mA. In Figure 5.10b, $(I_1 + I_2) = (I_3 + I_4) = 26$ mA. In each circuit, the total current leaving the connection point is equal to the total current entering the point. Note that *points connecting two or more current paths* (like those in Figure 5.10) are referred to as **nodes.**

There are a variety of practical applications for Kirchhoff's current law. One such application is demonstrated in the following example.

Example 5.5

Note: In Figure 5.11, the potentiometer value shown in parentheses is the *rated* value of the component.

Determine the potentiometer setting required to adjust the total circuit current in Figure 5.11 to 28 mA.

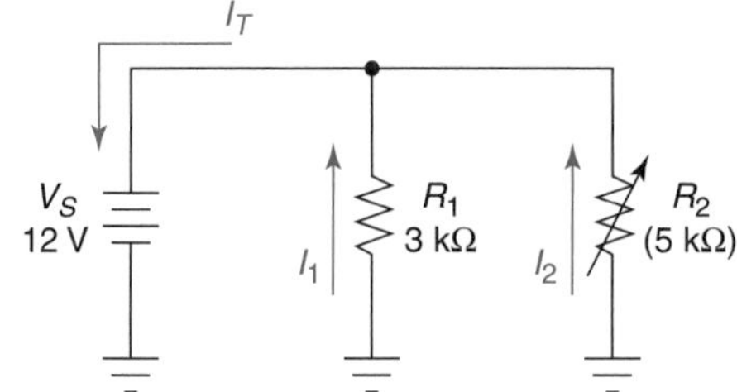

Figure 5.11

Solution: First, the value of I_1 is found using Ohm's law, as follows:

$$I_1 = \frac{V_S}{R_1} = \frac{12 \text{ V}}{3 \text{ k}\Omega} = 4 \text{ mA}$$

The value of I_2 required to set I_T to the desired value of 28 mA is found as

$$I_2 = I_T - I_1 = 28 \text{ mA} - 4 \text{ mA} = 24 \text{ mA}$$

Note that the above equation is simply a form of Kirchhoff's current law. Finally, Ohm's law is used to determine the potentiometer setting, as follows:

$$R_2 = \frac{V_S}{I_2} = \frac{12 \text{ V}}{24 \text{ mA}} = 500 \ \Omega$$

This potentiometer setting will adjust the total circuit current to 28 mA.

Practice Problem 5.5

A circuit like the one shown in Figure 5.11 has the following values: $V_S = 18$ V, and $R_1 = 2$ kΩ. Determine the potentiometer setting required to adjust the total source current to 15 mA.

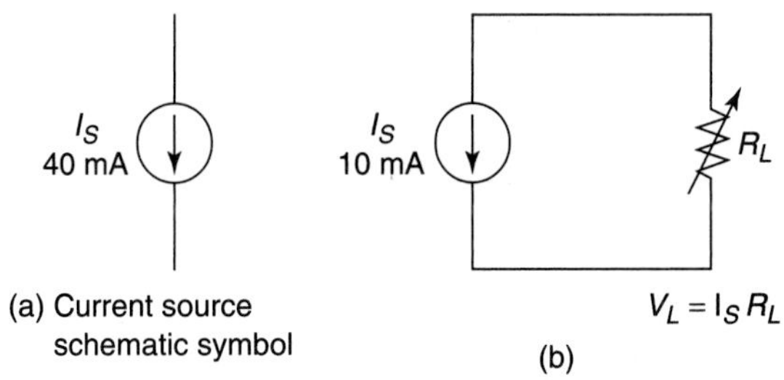

Figure 5.12

Current Sources

A **current source** is *a source that is designed to provide an output current value that remains relatively constant over a wide range of load resistance values.* The schematic symbol for a dc current source is shown in Figure 5.12a. Note that the output current value of the source is given, and that the arrow in the symbol indicates the direction of the current. ◄ *OBJECTIVE 8*

Figure 5.12b contains a 10 mA current source and a variable load resistance. When the load resistance is changed, the circuit current remains at the rated value of the source. Thus, *a change in the load resistance must result in a change in load voltage* (in keeping with Ohm's law). This point is demonstrated in the following example.

Example 5.6

Calculate the change in V_L that occurs in Figure 5.13 if R_L is adjusted from 1 kΩ to 4 kΩ.

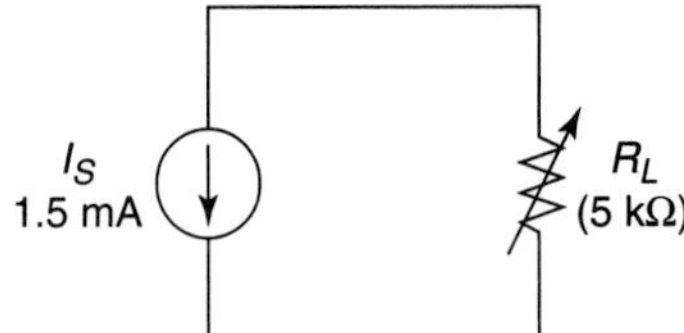

Figure 5.13

Solution: When $R_L = 1$ kΩ, the value of V_L is found as

$$V_L = I_S R_L = (1.5 \text{ mA})(1 \text{ k}\Omega) = 1.5 \text{ V}$$

The value of the circuit current is fixed at 1.5 mA. Therefore, when R_L changes to 4 kΩ, V_L changes as follows:

$$V_L = I_S R_L = (1.5 \text{ mA})(4 \text{ k}\Omega) = 6 \text{ V}$$

Therefore, the change in load resistance causes V_L to change from 1.5 to 6 V.

Practice Problem 5.6

A circuit like the one shown in Figure 5.13 has a 100 μA current source. Calculate the change in V_L that occurs if R_L is changed from 12 kΩ to 32 kΩ.

An important point needs to be made: The relationship between source current and load resistance demonstrated in Example 5.6 is *ideal.* The ideal current source has *infinite* internal resistance, which keeps the circuit current constant when the load resistance changes. (Since source resistance is infinite, a change in load resistance has no effect on the total circuit resistance.) In practice, however, the internal resistance of a current source is less than infinite, so its output current *does* vary slightly when the load resistance changes. This point is discussed further in Section 5.4.

When a current source is used in a parallel circuit, the values of the branch currents can be calculated as demonstrated in the following example.

Example 5.7

Calculate the values of I_1 and I_2 for the circuit in Figure 5.14a.

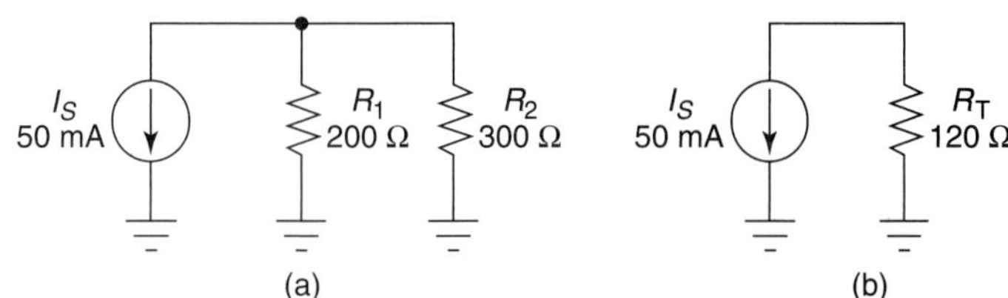

Figure 5.14

Solution: Before we can find the values of I_1 and I_2, we need to calculate the source voltage. We start by finding the value of R_T for the circuit, as follows:

$$R_T = R_1 \| R_2 = (200\ \Omega) \| (300\ \Omega) = 120\ \Omega$$

Once we have calculated the total circuit resistance, we can use it to draw an *equivalent* of the original circuit. This equivalent circuit is shown in Figure 5.14b. Applying Ohm's law to this circuit, we can calculate the value of the source voltage as follows:

$$V_S = I_S R_T = (50\ \text{mA})(120\ \Omega) = 6\ \text{V}$$

Using $V_S = 6$ V, the values of the branch currents are found as

$$I_1 = \frac{V_S}{R_1} = \frac{6\ \text{V}}{200\ \Omega} = 30\ \text{mA}$$

and

$$I_2 = \frac{V_S}{R_2} = \frac{6\ \text{V}}{300\ \Omega} = 20\ \text{mA}$$

Note that the sum of I_1 and I_2 is 50 mA, the rated value of the current source.

Practice Problem 5.7

A circuit like the one in Figure 5.14a has the following values: $I_S = 25$ mA, $R_1 = 100\ \Omega$, and $R_2 = 150\ \Omega$. Calculate the values of I_1 and I_2 for the circuit.

Note: The solution for R_T in Example 5.7 is written using the "shorthand" notation described earlier. The actual value is calculated using equation (5.5) or equation (5.6).

There is a simpler way to solve a problem like the one in Example 5.7. When a current source is used in a parallel circuit, we can treat the circuit as a current divider.

Current Dividers

OBJECTIVE 9 ➤ You may recall that a series circuit can be viewed as a voltage divider, because the source voltage is divided among the resistors in the circuit. In a similar fashion, a parallel circuit can be viewed as a *current divider,* because the source current is divided among the branches in the circuit.

A comparison of voltage dividers and current dividers is provided in Figure 5.15. Note the difference in the resistance ratios used in the divider equations. For the voltage

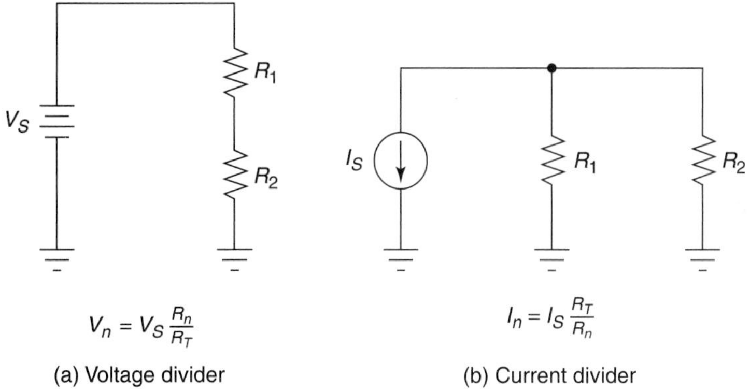

Figure 5.15

divider, the source voltage is multiplied by the ratio of the component resistance to the total circuit resistance. For the current divider, the source current is multiplied by the ratio of the total circuit resistance to the component (or branch) resistance, as follows:

$$I_n = I_S \frac{R_T}{R_n} \tag{5.7}$$

Example 5.8 demonstrates the validity of this relationship.

Example 5.8

Calculate the branch currents for the circuit in Figure 5.16. (This is the same circuit we analyzed in Example 5.7).

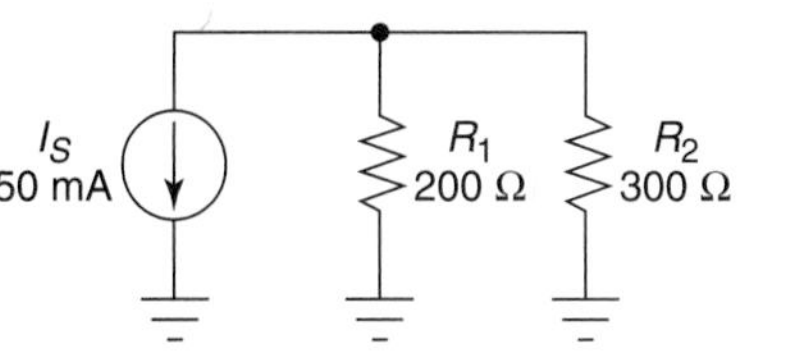

Figure 5.16

Solution: First, the total resistance in the circuit is found as

$$R_T = R_1 \| R_2 = 200\ \Omega \| 300\ \Omega = 120\ \Omega$$

Now, the branch current values can be found as

$$I_1 = I_S \frac{R_T}{R_n} = (50\text{ mA})\frac{120\ \Omega}{200\ \Omega} = 30\text{ mA}$$

and

$$I_2 = I_S \frac{R_T}{R_n} = (50\text{ mA})\frac{120\ \Omega}{300\ \Omega} = 20\text{ mA}$$

Of course, once we determined the value of I_1, we could have simply used Kirchhoff's current law to find the value of I_2.

Practice Problem 5.8

Solve Practice Problem 5.7 using the current-divider equation.

When a current source is used in a parallel circuit, the current-divider equations can be used to calculate the values of the branch currents.

Section Review

1. What does *Kirchhoff's current law* state?
2. When solving a Kirchhoff's current equation, how are the currents entering a point distinguished from those leaving the point?
3. What is a *node*?
4. What is a *current source*?
5. How does a circuit with a current source respond to a change in load resistance?

Critical Thinking

6. Refer to Figure 5.15. What change (if any) occurs in the voltage across R_1 if R_2 opens? Explain your answer.
7. Refer to Figure 5.15. What change (if any) in source current occurs if R_2 shorts? Explain your answer.

5.4 Related Topics

In this section, we will close out the chapter by discussing some miscellaneous topics that relate to parallel circuits.

OBJECTIVE 10 ➤

Practical Current Sources: The Effects of Source Resistance

(a)

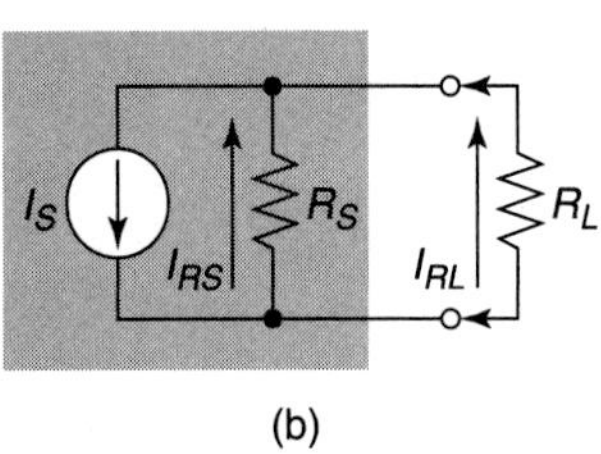

(b)

Figure 5.17
Practical current source.

In Section 5.3, you were told that an *ideal* current source is one that:

- Maintains a constant output current over a wide range of load resistance values.
- Has infinite internal resistance.

In practice, the current generated by a current source *does* vary slightly when there is a change in load resistance, because the internal resistance of the current source (R_S) is *not* infinite. The internal resistance of a current source, which is typically in the high-kΩ range, is represented as shown in Figure 5.17a.

When a load is connected to the practical current source, it forms a current divider with the source resistance, as shown in Figure 5.17b. As a result, there is a slight change in the output from the current source when the load resistance changes. This concept is demonstrated in Example 5.9.

Example 5.9

The current source in Figure 5.18 has a variable load resistance. Calculate the change in load current that occurs if the load resistance is changed from 1 kΩ to 2 kΩ.

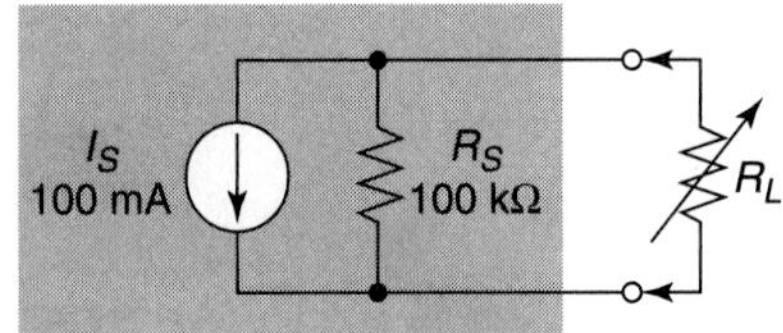

Figure 5.18

Solution: When $R_L = 1\ \text{k}\Omega$,

$$R_T = R_S \| R_L = 100\ \text{k}\Omega \| 1\ \text{k}\Omega = 990\ \Omega$$

and

$$I_L = I_S \frac{R_T}{R_L} = (100\ \text{mA}) \frac{990\ \Omega}{1\ \text{k}\Omega} = 99\ \text{mA}$$

When $R_L = 2\ \text{k}\Omega$,

$$R_T = R_S \| R_L = 100\ \text{k}\Omega \| 2\ \text{k}\Omega = 1.96\ \text{k}\Omega$$

and

$$I_L = I_S \frac{R_T}{R_L} = (100 \text{ mA}) \frac{1.96 \text{ k}\Omega}{2 \text{ k}\Omega} = 98 \text{ mA}$$

The change in current is, therefore, 1 mA.

Practice Problem 5.9

A circuit like the one shown in Figure 5.18 has a value of $R_S = 200 \text{ k}\Omega$ and $I_S = 50$ mA. Calculate the change in load current that occurs if the load resistance changes from 5 kΩ to 15 kΩ.

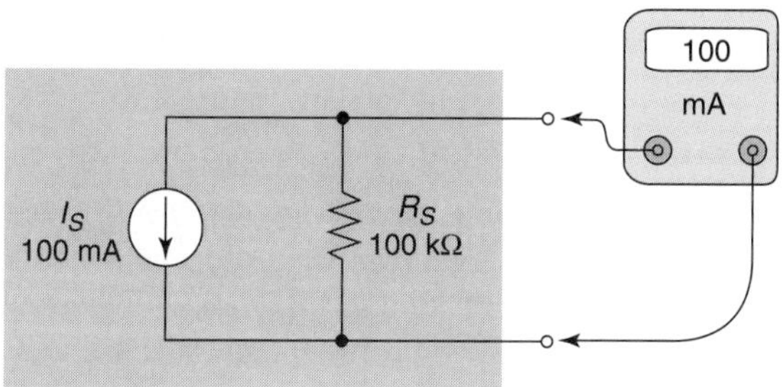

Figure 5.19
Measuring the shorted-load output current.

The change in load current calculated in Example 5.9 is very slight when you consider that the load resistance *doubled.* Had the same change in load resistance occurred in a circuit with a voltage source, the load current would have changed by approximately 50%.

The *maximum output from a current source* is referred to as the **shorted-load output current (I_{SL}).** The value of I_{SL} can be measured as shown in Figure 5.19. Assuming that the internal resistance of the current meter is 0 Ω, all of the source current passes through the meter when it is connected as shown, because the meter shorts out R_S. Thus, I_{SL} equals the rated value of the source current (I_S).

It should be noted that $I_L < I_{SL}$ when $R_L > 0\ \Omega$. This point is demonstrated in the following example.

Example 5.10

Calculate the values of I_{SL} and I_L for the circuit shown in Figure 5.20.

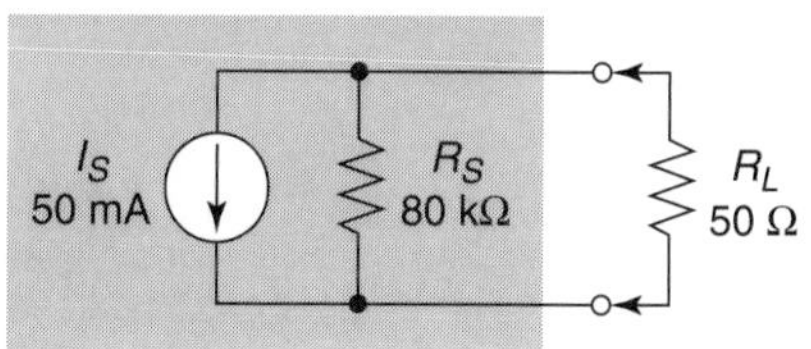

Figure 5.20

Solution: When the load resistance is shorted, the current through the short equals the rated source current, as follows:

$$I_{SL} = I_S = 50 \text{ mA}$$

The value of I_L can be found as

$$I_L = I_S \frac{R_T}{R_L} = (50 \text{ mA}) \frac{80 \text{ k}\Omega \| 50\ \Omega}{50\ \Omega} = (50 \text{ mA}) \frac{49.97\ \Omega}{50\ \Omega} = 49.97 \text{ mA}$$

Practice Problem 5.10

A circuit like the one shown in Figure 5.20 has values of $I_S = 25$ mA, $R_S = 50 \text{ k}\Omega$, and $R_L = 1.5 \text{ k}\Omega$. Calculate the values of I_{SL} and I_L for the circuit.

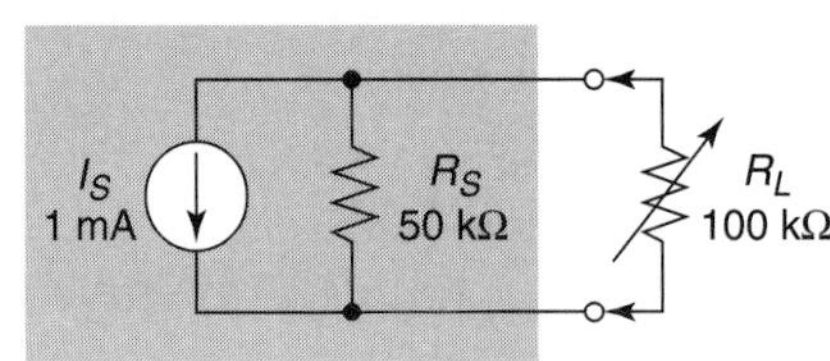

R_L	I_L	P_L
10 kΩ	833 μA	6.94 mW
20 kΩ	714 μA	10.20 mW
30 kΩ	625 μA	11.72 mW
40 kΩ	556 μA	12.37 mW
50 kΩ	**500 μA**	**12.50 mW**
60 kΩ	455 μA	12.42 mW
70 kΩ	417 μA	12.17 mW
80 kΩ	385 μA	11.86 mW
90 kΩ	357 μA	11.47 mW
100 kΩ	333 μA	11.09 mW

Figure 5.21 Maximum power transfer from a current source.

The internal resistance of a current source is typically much greater than the resistance of its load. When this is the case, R_S has little effect on the source output current (as demonstrated in Example 5.10).

Maximum Power Transfer

OBJECTIVE 11 ➤ In Chapter 4, you were introduced to the *maximum power transfer theorem,* which states that maximum power transfer from a source to its load occurs when load resistance (R_L) equals the source resistance (R_S). You were also told that:

- The theorem holds true when source resistance is *fixed* and load resistance is *variable.*
- The theorem *fails* to hold true when source resistance is *variable* and load resistance is *fixed.*

At this point, we are going to see how this theorem applies to a current source.

OBJECTIVE 12 ➤ If a current source has a fixed value of R_S and a variable value of R_L, maximum power transfer occurs when $R_L = R_S$. This concept is illustrated in Figure 5.21. The table in that figure shows the changes in P_L that occur as R_L increases from 10 kΩ to 100 kΩ (at 10 kΩ increments). As you can see, P_L reaches its maximum value (12.5 mW) when $R_L = R_S = 50$ kΩ.

As with voltage sources, the maximum power transfer theorem fails to hold true for a current source when source resistance is *variable* and load resistance is *fixed.* For a current source with a fixed load resistance, *maximum power transfer occurs when $R_S >> R_L$.* This point is verified by the calculations shown in Figure 5.22.

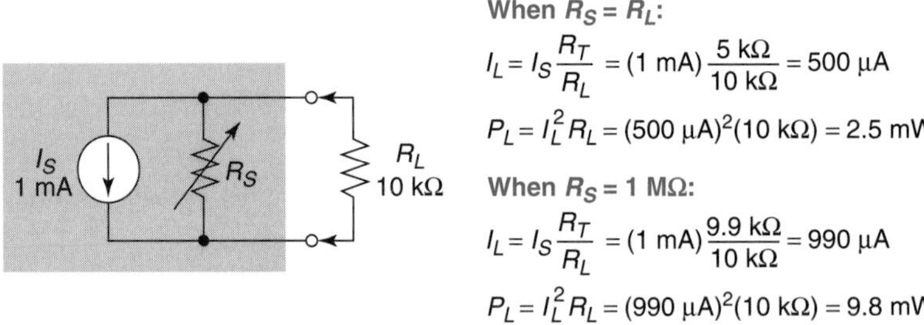

I_S
1 mA
R_S
R_L
10 kΩ
When $R_S = R_L$:
$I_L = I_S \frac{R_T}{R_L} = (1\text{ mA})\frac{5\text{ k}\Omega}{10\text{ k}\Omega} = 500\ \mu\text{A}$
$P_L = I_L^2 R_L = (500\ \mu\text{A})^2(10\text{ k}\Omega) = 2.5\text{ mW}$
When $R_S = 1$ MΩ:
$I_L = I_S \frac{R_T}{R_L} = (1\text{ mA})\frac{9.9\text{ k}\Omega}{10\text{ k}\Omega} = 990\ \mu\text{A}$
$P_L = I_L^2 R_L = (990\ \mu\text{A})^2(10\text{ k}\Omega) = 9.8\text{ mW}$

Figure 5.22 Maximum power transfer to a fixed load resistance.

Section Review

1. How do the characteristics of a *practical* current source differ from those of the *ideal* current source?
2. What type of circuit is formed when a load is connected to a current source?
3. How does the output from a practical current source respond to a change in load resistance?
4. What is the relationship between the *shorted-load output current* (I_{SL}) and the I_S rating of a current source?

5. When does maximum power transfer occur from a current source to its load?

6. The equivalent circuit for a practical current source contains an *ideal* current source and an internal resistance (R_S). Why is the resistor represented as being in *parallel* with the source? Critical Thinking

Key Terms

The following terms were introduced in this chapter on the pages indicated:

Branch, 124
Current divider, 126
Current source, 133
Kirchhoff's current law, 132
Node, 132
Parallel circuit, 124
Shorted-load output current (I_{SL}), 137

Practice Problems

1. The table below refers to Figure 5.23. For each combination of branch currents, find the total circuit current.

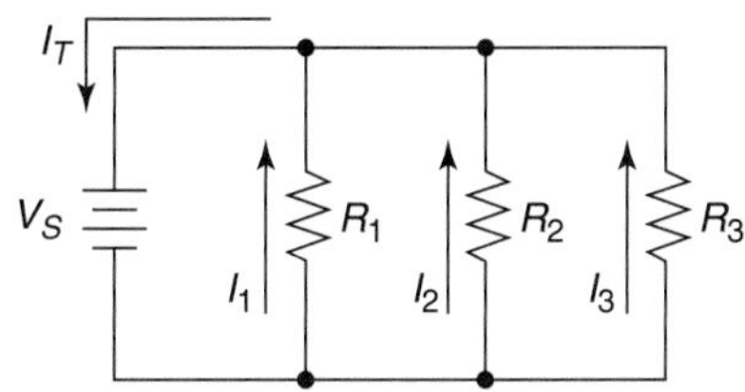

Figure 5.23

	I_1	I_2	I_3	I_T
a.	12 mA	2.8 mA	6.6 mA	_______
b.	5.5 mA	360 μA	1.04 mA	_______
c.	43 mA	25 mA	3.3 mA	_______
d.	500 μA	226 μA	1.08 mA	_______

2. The table below refers to Figure 5.23. For each combination of branch currents, find the total circuit current.

	I_1	I_2	I_3	I_T
a.	1.5 mA	3.3 mA	17 mA	_______
b.	480 μA	20 mA	5.1 mA	_______
c.	52 mA	28 mA	800 μA	_______
d.	10 mA	112 mA	34 mA	_______

3. Determine the values of the branch currents and the total current for the circuit shown in Figure 5.24a.

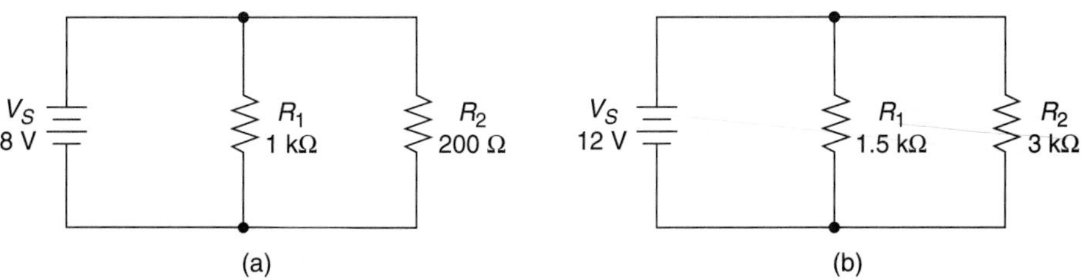

Figure 5.24

4. Determine the values of the branch currents and the total current for the circuit shown in Figure 5.24b.

5. Determine the values of the branch currents and the total current for the circuit shown in Figure 5.25a.

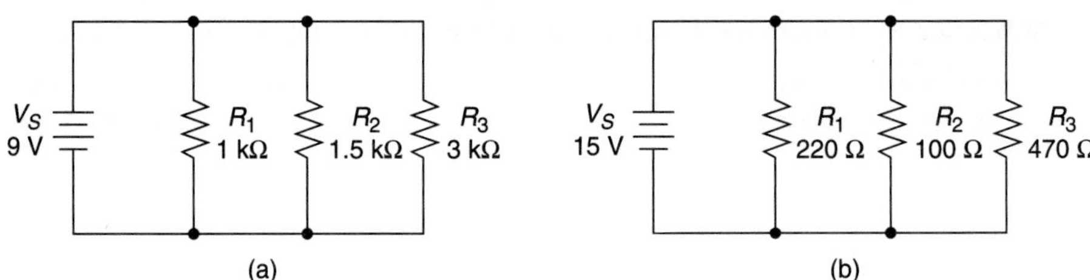

Figure 5.25

6. Determine the values of the branch currents and the total current for the circuit shown in Figure 5.25b.
7. Calculate the value of R_T for the circuit shown in Figure 5.24a.
8. Calculate the value of R_T for the circuit shown in Figure 5.24b.
9. Calculate the value of R_T for the circuit shown in Figure 5.25a.
10. Calculate the value of R_T for the circuit shown in Figure 5.25b.
11. Calculate the value of R_T for the circuit shown in Figure 5.26a.
12. Calculate the value of R_T for the circuit shown in Figure 5.26b.

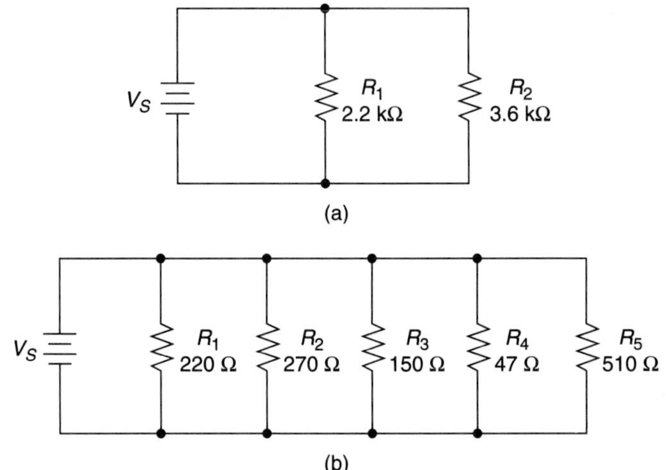

Figure 5.26

13. Calculate the value of R_T for the circuit shown in Figure 5.27a.
14. Calculate the value of R_T for the circuit shown in Figure 5.27b.

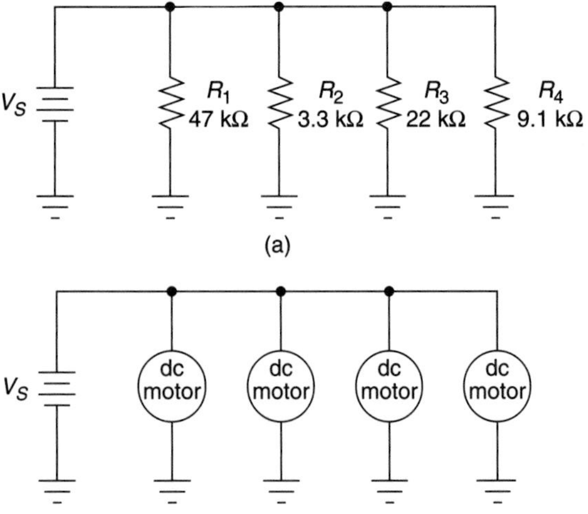

Figure 5.27

15. Determine the potentiometer setting required to set the total circuit current in Figure 5.28a to 50 mA.

16. Determine the potentiometer setting required to set the total circuit current in Figure 5.28b to 36 mA.

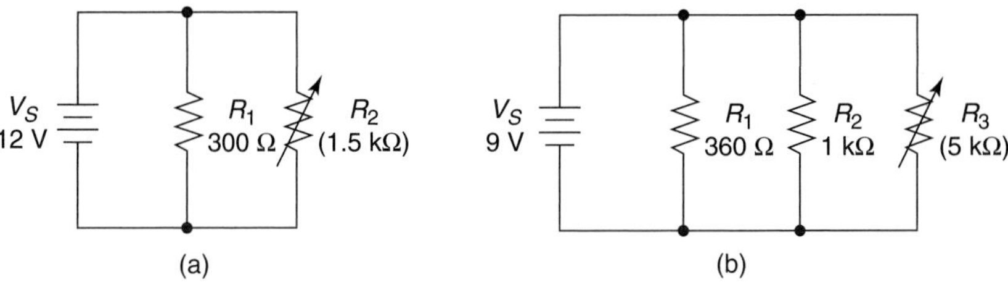

Figure 5.28

17. For the circuit shown in Figure 5.29a, determine the change in V_L that occurs when R_L is changed from 1.2 kΩ to 3.3 kΩ.

18. For the circuit shown in Figure 5.29b, determine the change in V_L that occurs when R_L is changed from 470 Ω to 220 Ω.

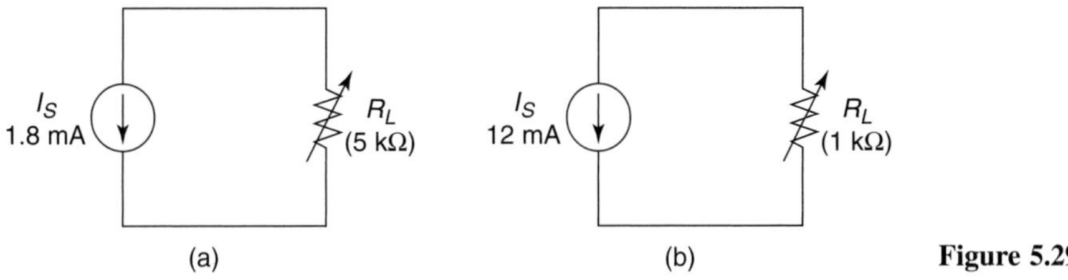

Figure 5.29

19. Determine the branch current values for the circuit shown in Figure 5.30a.

20. Determine the branch current values for the circuit shown in Figure 5.30b.

21. Determine the branch current values for the circuit shown in Figure 5.30c.

22. Determine the branch current values for the circuit shown in Figure 5.30d.

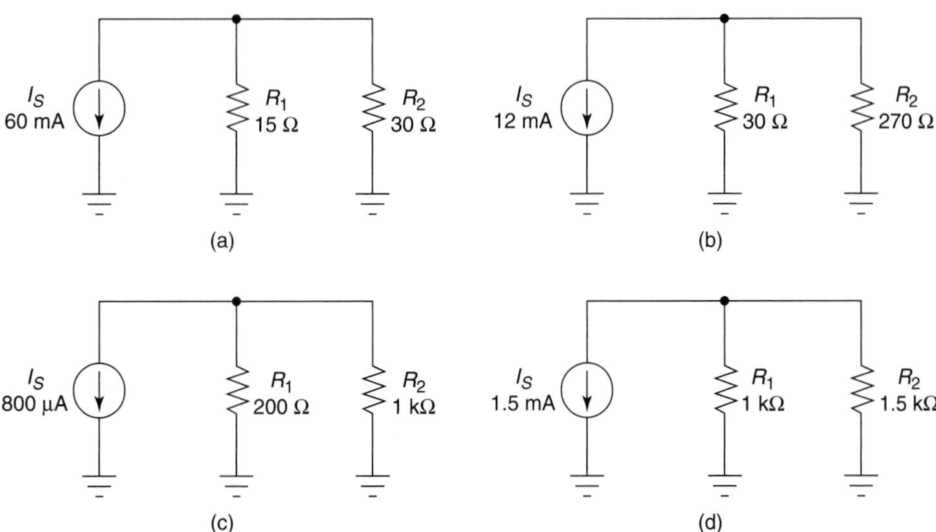

Figure 5.30

23. The circuit in Figure 5.31a has the values shown. Calculate the change in load current that occurs when R_L changes from 2 kΩ to 10 kΩ.

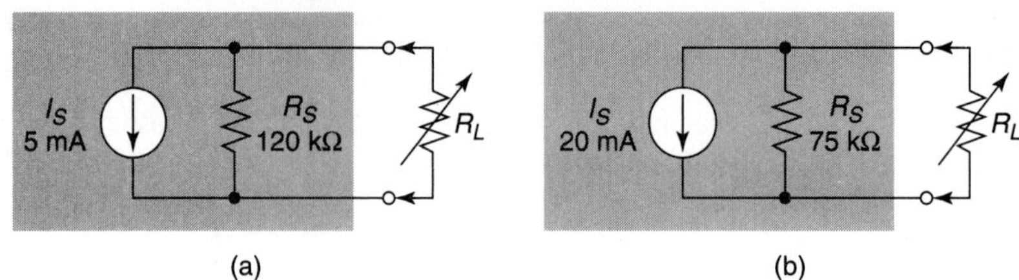

Figure 5.31

24. The circuit in Figure 5.31b has the values shown. Calculate the change in load current that occurs when R_L changes from 500 Ω to 1.5 kΩ.
25. Calculate the values of I_{SL} and I_L for the circuit shown in Figure 5.32a.
26. Calculate the values of I_{SL} and I_L for the circuit shown in Figure 5.32b.

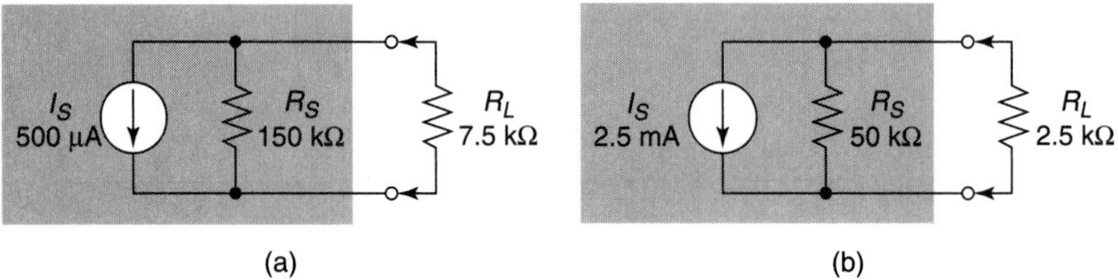

Figure 5.32

27. Determine the maximum possible load power for the circuit shown in Figure 5.33a.
28. Determine the maximum possible load power for the circuit shown in Figure 5.33b.

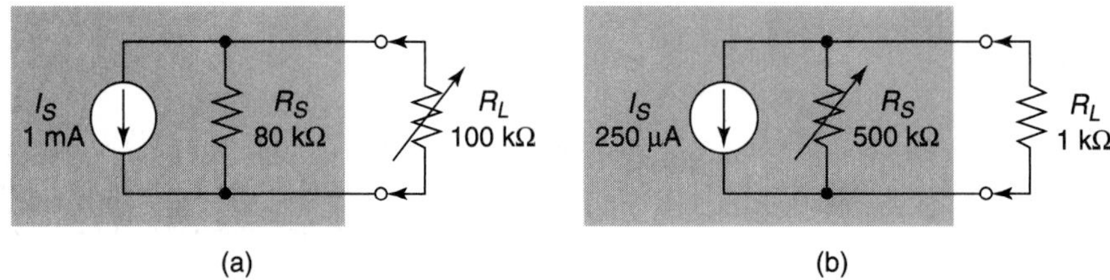

Figure 5.33

29. Calculate the branch values of current and power for the circuit shown in Figure 5.34. Also, calculate the circuit values of I_T, R_T, and P_T.

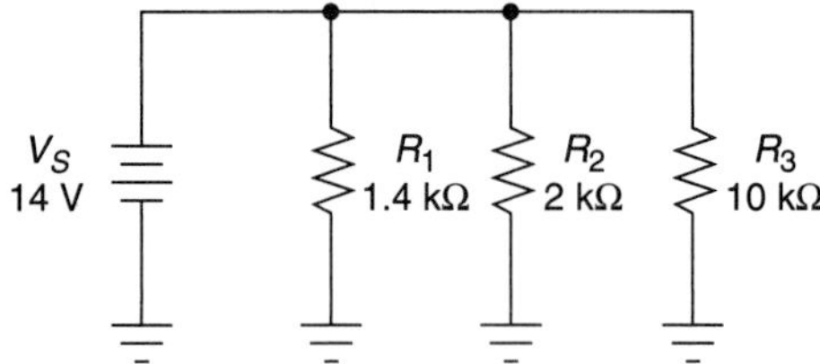

Figure 5.34

30. Calculate the branch values of current and power for the circuit shown in Figure 5.35. Also, calculate the circuit values of I_T, R_T, and P_T.

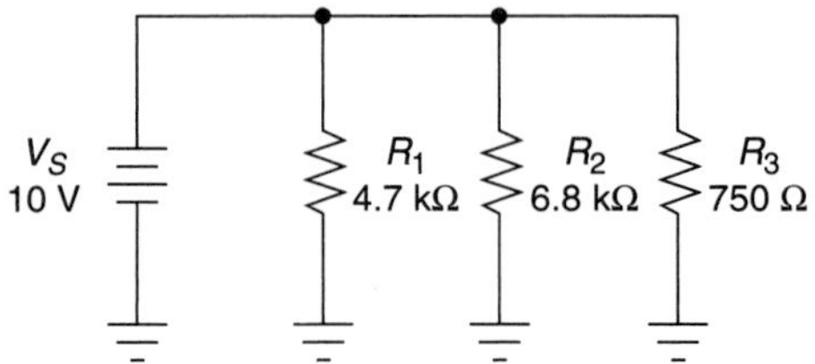

Figure 5.35

31. Calculate the branch values of current and power for the circuit shown in Figure 5.36. Also, calculate the circuit values of I_T, R_T, and P_T.

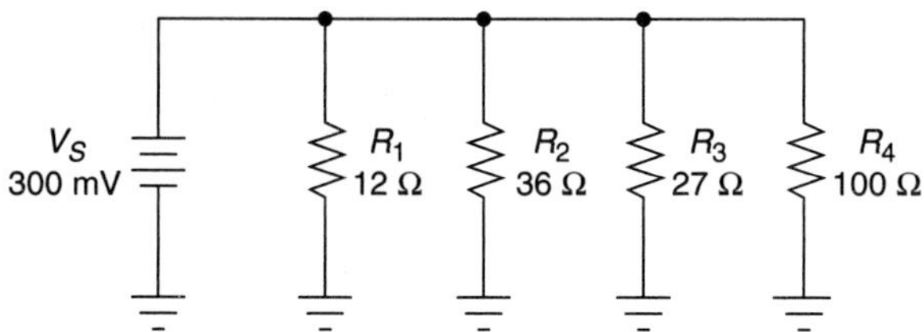

Figure 5.36

32. Calculate the branch values of current and power for the circuit shown in Figure 5.37. Also, calculate the circuit values of I_T, R_T, and P_T.

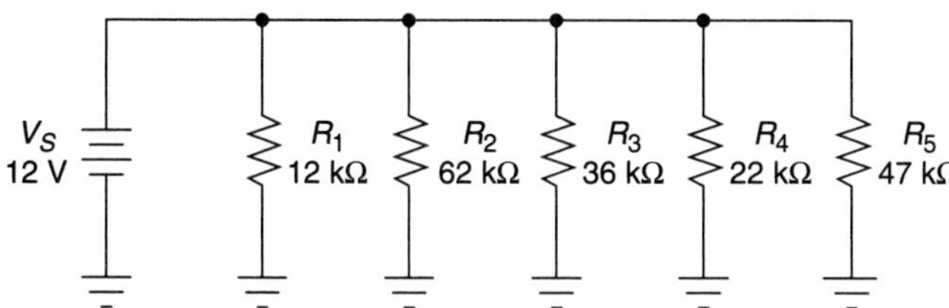

Figure 5.37

Looking Back

These problems relate to material presented in earlier chapters. The chapters are identified in brackets.

33. A dc power supply uses 5 joules of energy to move 2 coulombs of charge per second. How much power is the circuit is expending? [Chapter 1]

34. The output from a battery is rated at 3 Ah. How long will the battery last with a continuous output of 2 mA? [Chapter 2]

35. A series circuit has the following values: $V_S = 12$ V, $R_1 = 1$ kΩ, and $R_2 = 2$ kΩ. Calculate the change in V_1 that occurs if R_2 doubles in value. [Chapter 4]

36. The conductivity of aluminum is 2.825×10^{-6} Ω-cm. Calculate the resistance of a 50 cm length of aluminum with a cross-sectional area of 0.05 cm^2. [Chapter 1]

Pushing the Envelope

37. For each of the circuits shown in Figure 5.38, find the missing value(s).

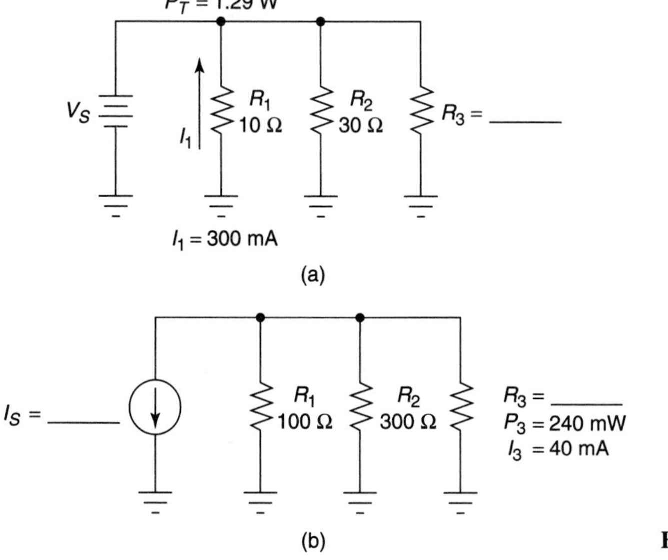

Figure 5.38

38. Figure 5.39 shows a potentiometer wired as a current divider. Determine the values of R_{ab} and R_{bc} that will provide values of $I_1 = 4$ mA and $I_2 = 6$ mA.

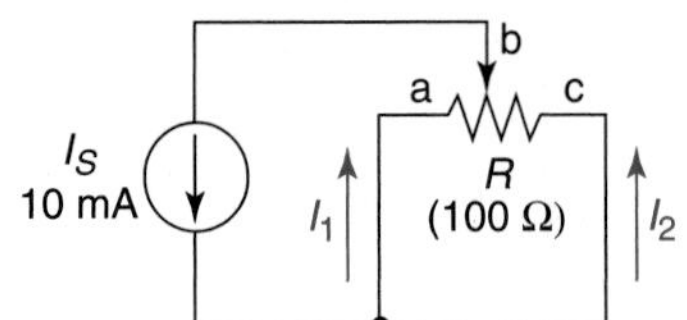

Figure 5.39

Answers to the Example Practice Problems

5.1. 24.5 mA

5.2. 83.3 mA

5.3. 144 Ω

5.4. 144 Ω

5.5. 3 kΩ

5.6. V_L changes from 1.2 V to 3.2 V.

5.7. $I_1 = 15$ mA, $I_2 = 10$ mA

5.8. $I_1 = 15$ mA, $I_2 = 10$ mA

5.9. The current changes by 2.27 mA (from 48.78 mA to 46.51 mA).

5.10. $I_{SL} = 25$ mA, $I_L = 24.27$ mA

Multisim File References

Figure	File Reference
5.4	ETF5_4.msm
5.13	ETF5_13.msm
5.19	ETF5_19.msm
5.25	ETF5_25.msm

chapter 

Series-Parallel Circuits

Objectives

After studying the material in this chapter, you should be able to:

1. Describe and analyze the operation of series circuits that are connected in parallel.
2. Describe and analyze the operation of parallel circuits that are connected in series.
3. Derive the *series-equivalent* or *parallel-equivalent circuit* for a series-parallel circuit.
4. Solve a given series-parallel circuit for any current, voltage, or power value.
5. List the steps used to solve for any value in a series-parallel circuit.
6. Discuss the effects of load resistance on the operation of a loaded voltage divider.
7. Discuss the concept of *voltage-divider stability.*
8. Describe the construction and operation of a *variable voltage divider.*
9. Discuss *voltmeter loading* and its effects on voltage measurements.
10. Describe the construction and operation of the *Wheatstone bridge.*
11. Describe the use of a Wheatstone bridge as a *resistance-measuring circuit* and as a sensing circuit.
12. Describe the *delta* (Δ) and *wye* (Y) circuit configurations.
13. Perform *Δ-to-Y* and *Y-to-Δ* conversions.

In practice, most circuits contain a combination of both *series* and *parallel* connections. Several examples of *series-parallel circuits* are shown in Figure 6.1. As you can see, some of the components in each circuit are clearly in series or parallel, while other connections are not so easily identified.

In this chapter, you will be shown how to analyze series-parallel circuits. You will also be introduced to some specific circuits that are commonly used in a variety of applications.

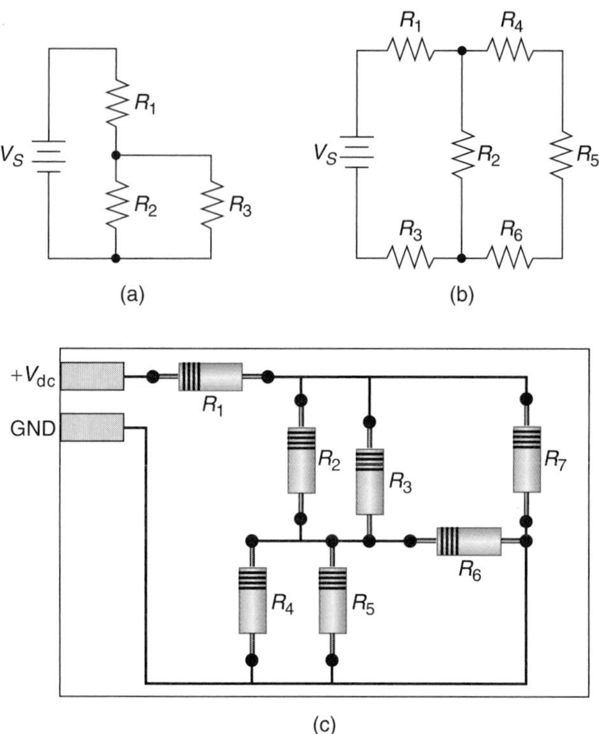

Figure 6.1 Examples of series-parallel circuits.

6.1 Introduction to Series-Parallel Circuits

Series-parallel circuits vary widely in their applications and complexity. For example, a stereo system can be viewed (loosely) as one extremely complex series-parallel circuit. At the same time, series-parallel circuits can be as simple as the ones shown in Figure 6.2.

The circuit in Figure 6.2a contains two series circuits that are connected in parallel. One series circuit consists of R_1 and R_2, while the other consists of R_3 and R_4. Figure 6.2b shows two parallel circuits that have been connected in series. R_1 and R_2 make up one of the parallel circuits, while R_3 and R_4 make up the other.

Connecting Series Circuits in Parallel

OBJECTIVE 1 ➤ Let's take a closer look at the circuit in Figure 6.2a. For the sake of discussion, the branches in the circuit are identified as shown in Figure 6.3. Since R_1 and R_2 form a series circuit, branch A retains all the characteristics of any series circuit. By the same token, R_3 and R_4 form a series circuit, so branch B retains all the characteristics of any series circuit. These branch characteristics are summarized in Table 6.1.

Since the branches in Figure 6.3 form a parallel circuit, the combination of branch A and branch B retains the characteristics of any parallel circuit. Therefore:

- The voltage across each branch is equal to the source voltage [$V_A = V_B = V_S$].
- The sum of the branch currents is equal to the source current [$I_A + I_B = I_t$].

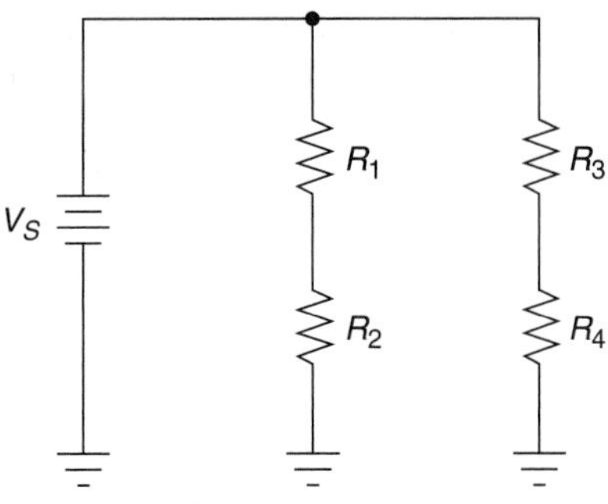

(a) Series circuits connected in parallel

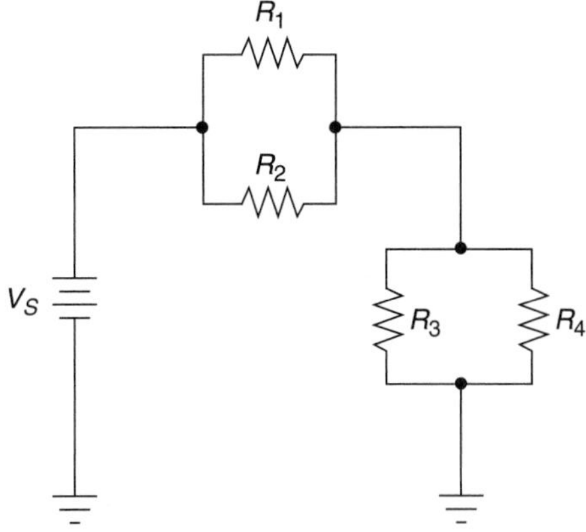

(b) Parallel circuits connected in series

Figure 6.2 Basic series-parallel configurations.

- The total circuit resistance is equal to the parallel combination of R_A and R_B $[R_T = R_A \| R_B]$.

When analyzing a circuit like the one in Figure 6.3, each branch is treated as an individual series circuit. Once the branch values are calculated, they are used (as they would be in any parallel circuit) to determine the overall values for the circuit. This approach to analyzing the circuit is demonstrated in Example 6.1.

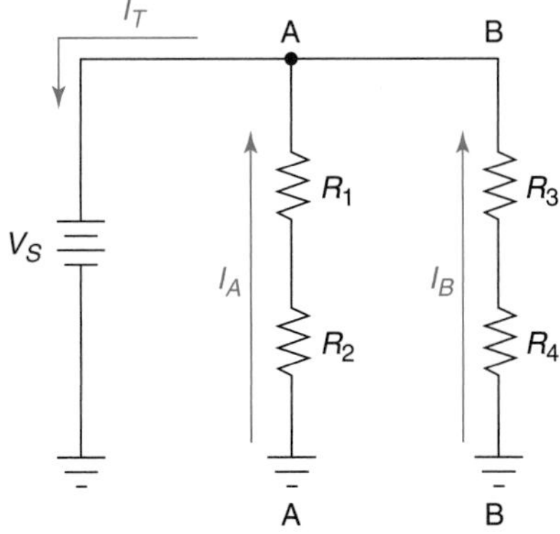

Figure 6.3

TABLE 6.1 Series Circuit Characteristics for Figure 6.3

Characteristic	*Branch A (R_1 and R_2)*	*Branch B (R_3 and R_4)*
Current	$I_1 = I_1$	$I_3 = I_4$
Voltage	$V_1 + V_2 = V_S$	$V_3 + V_4 = V_S$
Resistance	$R_A = R_1 + R_2$	$R_B = R_3 + R_4$

Example 6.1

Calculate the values of the branch currents, component voltages, total current, and total resistance for the circuit in Figure 6.4a.

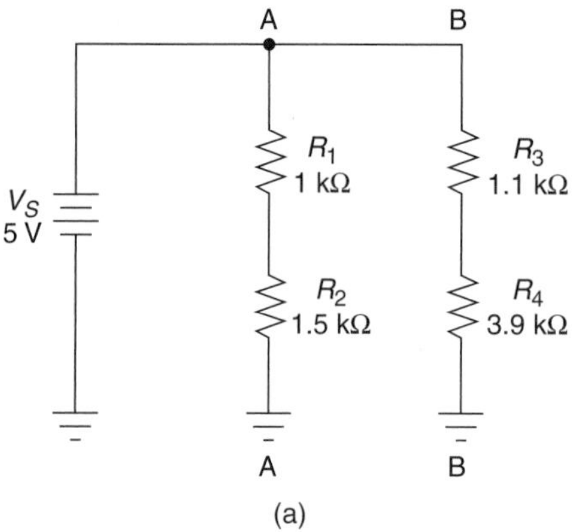

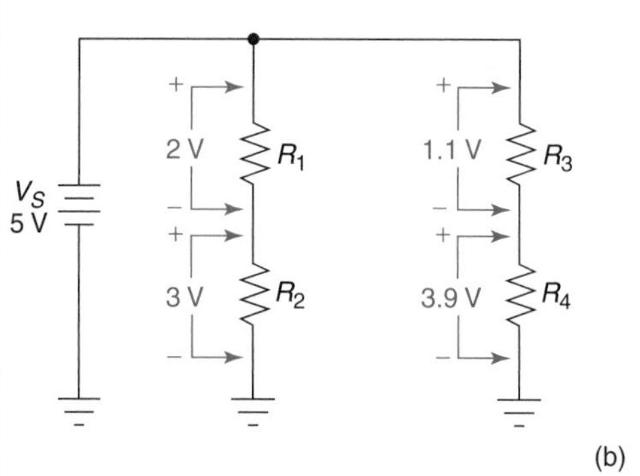

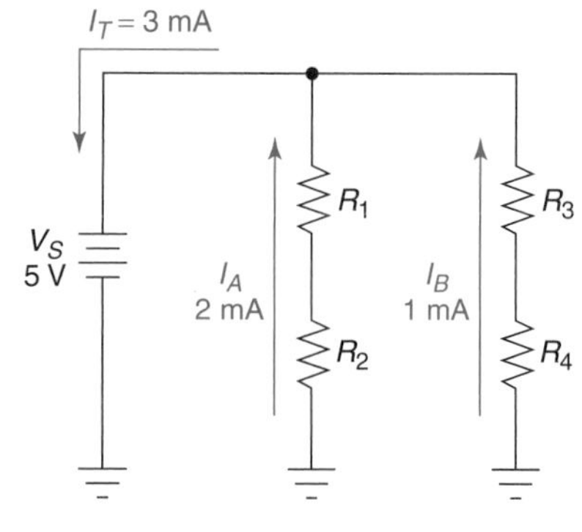

(b)

Figure 6.4

Solution: We'll start with the analysis of branch A. The total resistance of this branch is found as

$$R_A = R_1 + R_2 = 1\text{ k}\Omega + 1.5\text{ k}\Omega = 2.5\text{ k}\Omega$$

The current through the branch can now be found as

$$I_A = \frac{V_S}{R_A} = \frac{5\text{ V}}{2.5\text{ k}\Omega} = 2\text{ mA}$$

Since I_A is the current through R_1 and R_2, the values of V_1 and V_2 can be found as

$$V_1 = I_A R_1 = (2\text{ mA})(1\text{ k}\Omega) = 2\text{ V}$$

and

$$V_2 = I_A R_2 = (2\text{ mA})(1.5\text{ k}\Omega) = 3\text{ V}$$

Now, the same sequence of calculations is repeated for branch B, as follows:

$$R_B = R_3 + R_4 = 1.1\text{ k}\Omega + 3.9\text{ k}\Omega = 5\text{ k}\Omega$$

$$I_B = \frac{V_S}{R_B} = \frac{5\text{ V}}{5\text{ k}\Omega} = 1\text{ mA}$$

$$V_3 = I_B R_3 = (1\text{ mA})(1.1\text{ k}\Omega) = 1.1\text{ V}$$

and

$$V_4 = I_B R_4 = (1\text{ mA})(3.9\text{ k}\Omega) = 3.9\text{ V}$$

Since the branches are in parallel, the total circuit current is found as

$$I_T = I_A + I_B = 2 \text{ mA} + 1 \text{ mA} = 3 \text{ mA}$$

and the total circuit resistance is found as

$$R_T = R_A \| R_B = 2.5 \text{ k}\Omega \| 5 \text{ k}\Omega = 1.67 \text{ k}\Omega$$

Note that solving $R_T = \frac{V_S}{I_T}$ would have yielded the same result for R_T.

Practice Problem 6.1

A circuit like the one shown in Figure 6.4 has the following values: $R_1 = 220\ \Omega$, $R_2 = 330\ \Omega$, $R_3 = 120\ \Omega$, $R_4 = 100\ \Omega$, and $V_S = 11$ V. Calculate the values of the branch currents, component voltages, total current, and total resistance for the circuit.

The voltages and currents calculated in Example 6.1 are shown in Figure 6.4b. Note that the branch voltages conform to Kirchhoff's voltage law and the branch currents to Kirchhoff's current law.

Connecting Parallel Circuits in Series

Figure 6.5 shows two parallel circuits (identified as *loop A* and *loop B*) that are connected in series with a single voltage source. Since R_1 and R_2 form a parallel circuit, loop A retains all the characteristics of any parallel circuit. By the same token, R_3 and R_4 form a parallel circuit, so loop B retains all the characteristics of any parallel circuit. These characteristics are summarized in Table 6.2. ◄ *OBJECTIVE 2*

Since the loops in Figure 6.5 form a *series circuit,* the combination of loop A and loop B retains the characteristics of any series circuit. Therefore:

- The sum of V_A and V_B equals the source voltage $[V_A + V_B = V_S]$.

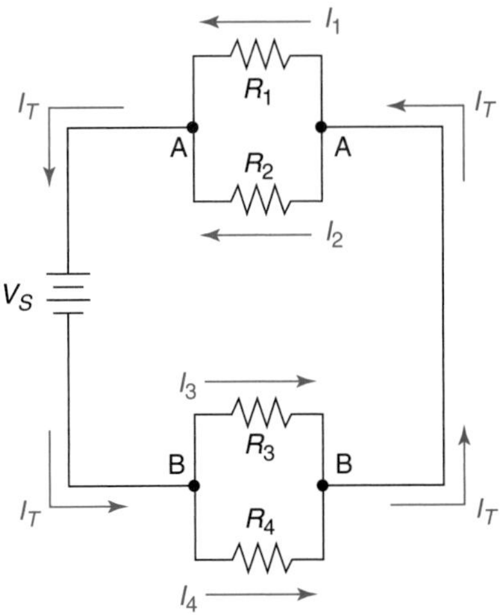

Figure 6.5

TABLE 6.2 Parallel Circuit Characteristics for Figure 6.5

Characteristic	*Loop A (R_1 and R_2)*	*Loop B (R_3 and R_4)*
Current	$I_t = I_1 + I_2$	$I_t = I_3 + I_4$
Voltage	$V_A = V_1 = V_2$	$V_B = V_3 = V_4$
Resistance	$R_A = R_1 \| R_2$	$R_B = R_3 \| R_4$

- The currents through the two loops must equal the source current $[(I_1 + I_2) = (I_3 + I_4) = I_T]$.
- The total circuit resistance equals the sum of R_A and R_B $[R_T = R_A + R_B]$.

When analyzing a circuit like the one shown in Figure 6.5, each loop is reduced to a single equivalent resistance, and the voltage across each resistance is determined. Once the loop voltages are calculated, they are used (as they would be in any parallel circuit) to determine the component current values. This approach to analyzing the circuit is demonstrated in the following example.

Example 6.2

Calculate the values of the branch currents, component voltages, total current, and total resistance for the circuit in Figure 6.6a.

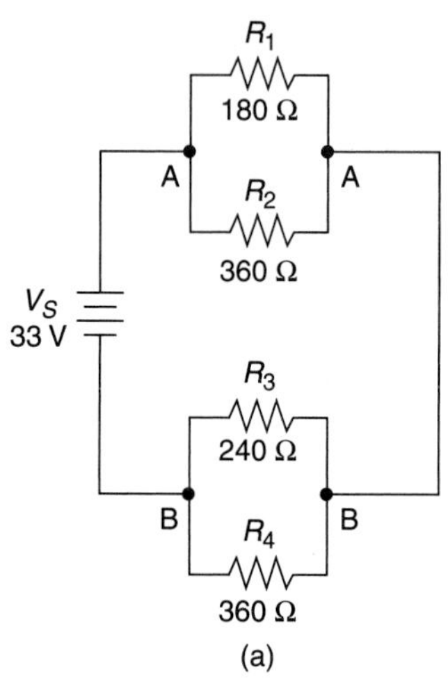

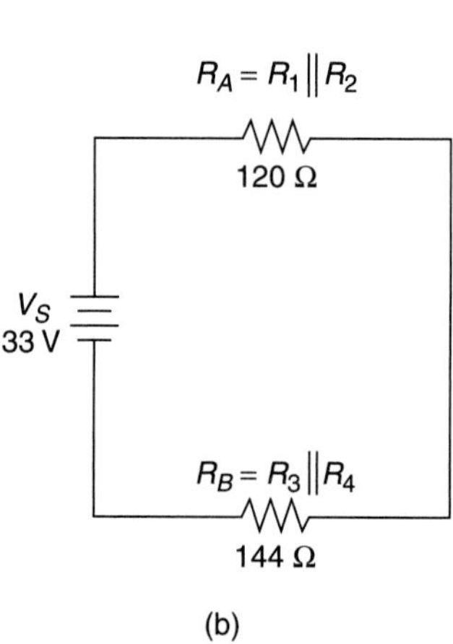

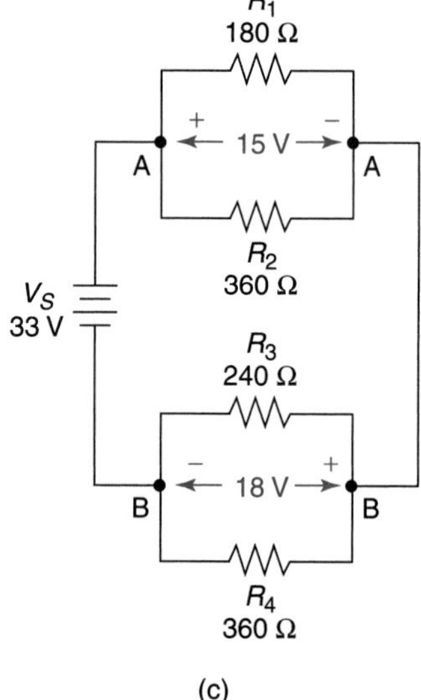

Figure 6.6

Solution: We'll start by calculating the total resistance of each loop. For loop A,

$$R_A = R_1 \| R_2 = 180\ \Omega \| 360\ \Omega = 120\ \Omega$$

For loop B,

$$R_B = R_3 \| R_4 = 240\ \Omega \| 360\ \Omega = 144\ \Omega$$

These values are used to redraw the circuit, as shown in Figure 6.6b. Using this circuit, we can calculate the voltage across each loop as

$$V_A = V_S \frac{R_A}{R_A + R_B} = (33\text{ V})\frac{120\Omega}{264\ \Omega} = 15\text{ V}$$

and

$$V_B = V_S \frac{R_B}{R_A + R_B} = (33\text{ V})\frac{144\Omega}{264\ \Omega} = 18\text{ V}$$

These calculations show that we have 15 V across loop A and 18 V across loop B, as shown in Figure 6.6c. Using these values and the branch resistance values, the resistor currents can be found as

$$I_1 = \frac{V_A}{R_1} = \frac{15\text{ V}}{180\ \Omega} = 83.33\text{ mA}$$

$$I_2 = \frac{V_A}{R_2} = \frac{15\text{ V}}{360\ \Omega} = 41.67\text{ mA}$$

$$I_3 = \frac{V_B}{R_3} = \frac{18\text{ V}}{240\ \Omega} = 75\text{ mA}$$

and

$$I_4 = \frac{V_B}{R_4} = \frac{18\text{ V}}{360\ \Omega} = 50\text{ mA}$$

The total circuit resistance can then be found as

$$R_T = R_A + R_B = 120\ \Omega + 144\ \Omega = 264\ \Omega$$

and the circuit current can be found as

$$I_T = \frac{V_S}{R_T} = \frac{33\text{ V}}{264\ \Omega} = 125\text{ mA}$$

Practice Problem 6.2

A circuit like the one shown in Figure 6.6a has the following values: $R_1 = 100\ \Omega$, $R_2 = 150\ \Omega$, $R_3 = 180\ \Omega$, $R_4 = 120\ \Omega$, and $V_S = 330$ mV. Calculate the values of the branch currents, component voltages, total current, and total resistance for the circuit.

In this section, we have analyzed only two of the most basic series-parallel circuits. As you will see, more complex series-parallel circuits can be dealt with effectively when you:

1. Use the proper approach.
2. Remember the basic principles of series and parallel circuits.

Section Review

1. List the current, voltage, and resistance characteristics of series circuits.
2. List the current, voltage, and resistance characteristics of parallel circuits.
3. Using component numbers, describe the path that the circuit current takes in Figure 6.1c.
4. Refer to Figure 6.7a. Determine which (if any) of the following statements are true for the circuit shown.* *(Critical Thinking)*
   a. Under normal operating conditions, $V_3 = V_4$.
   b. R_2 is in series with R_3.
   c. The value of V_S can be found by adding the values of V_1, V_2, and V_4.

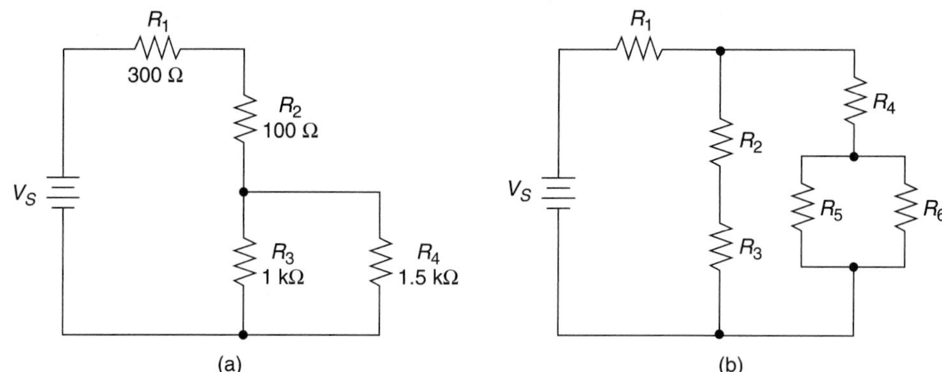

Figure 6.7

5. Answer each of the following questions for the circuit shown in Figure 6.7b.*
   a. Which two current values add up to equal the value of I_4?
   b. Which two current values add up to equal the value of I_1?
   c. Is R_1 in series with R_2? Explain your answer.
   d. What three component voltage values could be added to determine the value of V_S?
6. Refer back to Example 6.2. Describe an alternate approach to solving for the circuit voltage and current values.

**The answers to these questions are included with the Practice Problem answers at the end of the chapter.*

6.2 Analyzing Series-Parallel Circuits

In most cases, the goal of any series-parallel circuit analysis is to determine the value of voltage, current, and/or power for one or more components. Finding a single component value in a series or a parallel circuit is usually a simple, straightforward problem. However, this is not the case with series-parallel circuits. To calculate a single voltage, current, or power value in a series-parallel circuit, we usually find that we must calculate several other values first. This point is demonstrated in Example 6.3.

Example 6.3

Calculate the value of load power for the circuit in Figure 6.8a.

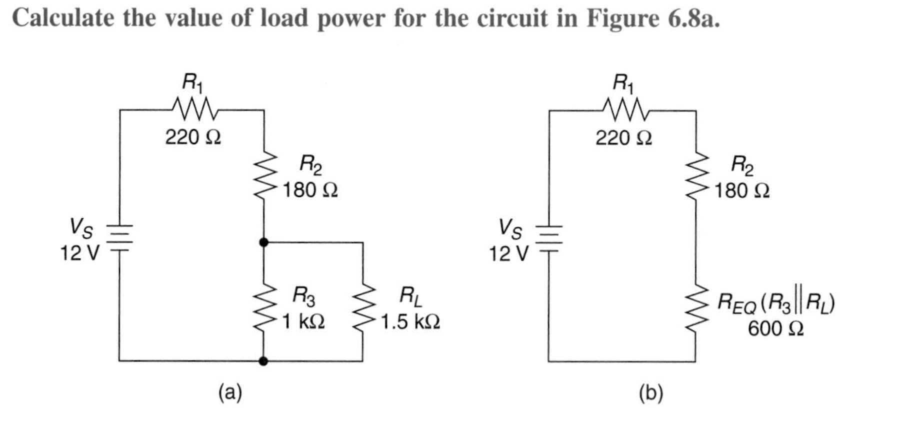

Figure 6.8

Solution: First, R_3 and R_L are combined into a single equivalent resistance (R_{EQ}), as follows:

$$R_{EQ} = R_3 \| R_L = 1\text{ k}\Omega \| 1.5\text{ k}\Omega = 600\ \Omega$$

When we replace the parallel circuit with its equivalent resistance, we get the circuit shown in Figure 6.8b. The voltage across the equivalent resistance (V_{EQ}) can now be found as

$$V_{EQ} = V_S \frac{R_{EQ}}{R_1 + R_2 + R_{EQ}} = (12\text{ V})\frac{600\ \Omega}{1\text{ k}\Omega} = 7.2\text{ V}$$

The voltage across the equivalent resistance (V_{EQ}) is equal to the voltage across the original parallel circuit. Therefore, we can find the value of load power as

$$P_L = \frac{V_L^2}{R_L} = \frac{(7.2\text{ V})^2}{1.5\text{ k}\Omega} = 34.56\text{ mW}$$

Practice Problem 6.3

A circuit like the one shown in Figure 6.8a has the following values: $R_1 = 120\ \Omega$, $R_2 = 150\ \Omega$, $R_3 = 300\ \Omega$, $R_L = 100\ \Omega$, and $V_S = 6$ V. Calculate the value of load power for the circuit.

Equivalent Circuits

◄ *OBJECTIVE 3*

Equivalent circuits are derived by combining groups of parallel and/or series components to obtain an equivalent—but simpler—circuit. For example, consider the circuits shown in Figure 6.9. Here we have a series-parallel circuit and its series equivalent. The equivalent circuit shown was derived as follows:

1. R_2 and R_3 were combined into a single equivalent resistance (R_{EQ1}) as follows:

$$R_{EQ1} = R_2 \| R_3 = 2\text{ k}\Omega \| 3\text{ k}\Omega = 1.2\text{ k}\Omega$$

2. R_5 and R_L were combined into another equivalent resistance (R_{EQ2}) as follows:

$$R_{EQ2} = R_5 \| R_L = 2\text{ k}\Omega \| 18\text{ k}\Omega = 1.8\text{ k}\Omega$$

Using these values, the equivalent circuit was drawn as shown in the figure.

The process of deriving an equivalent circuit ends when the original series-parallel circuit has been simplified to a *series-equivalent* or a *parallel-equivalent* circuit. A **series-equivalent circuit** is *an equivalent circuit made up entirely of components that are connected in series.* A **parallel-equivalent circuit** is *an equivalent circuit made up entirely*

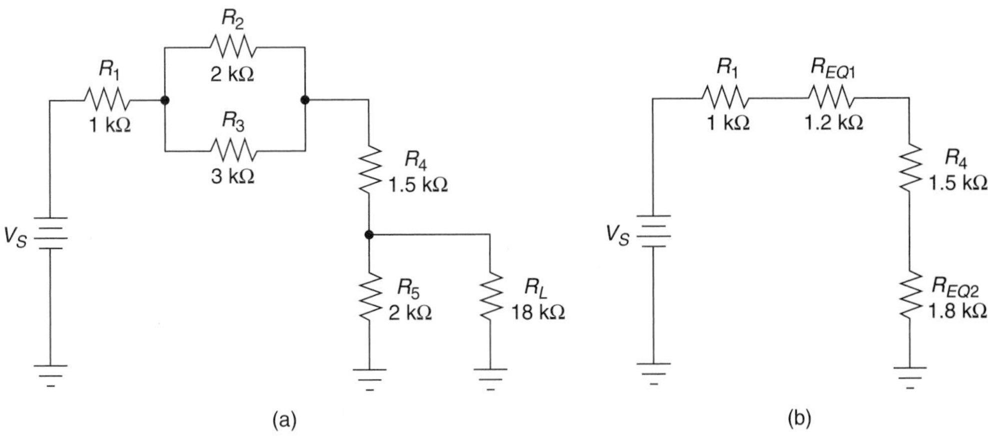

Figure 6.9

of components that are connected in parallel. An equivalent circuit is derived as demonstrated in Example 6.4.

Example 6.4

Derive an equivalent for the series-parallel circuit in Figure 6.10a.

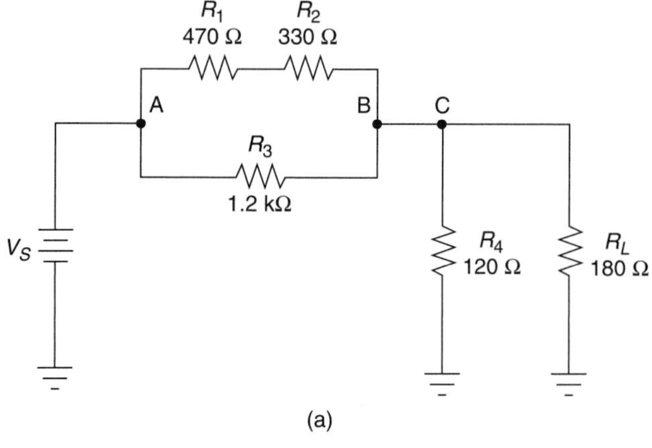

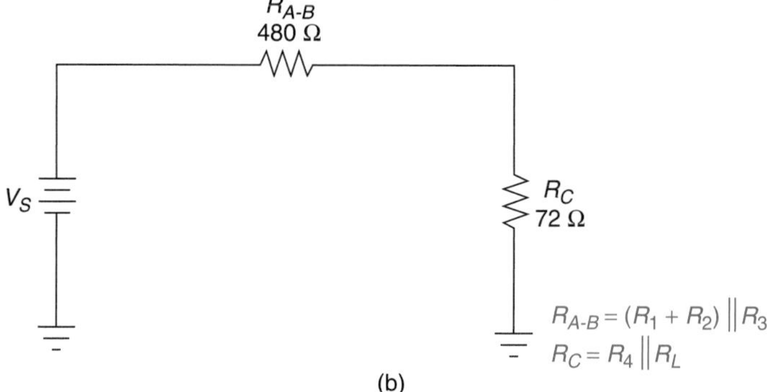

Figure 6.10

Solution: The series-parallel circuit shown in Figure 6.10a can be reduced to a simple series-equivalent circuit by:

1. Combining the resistors between points A and B into a single equivalent resistance ($R_{A\text{-}B}$).
2. Combining the resistors between point C and ground into a single equivalent resistance (R_C).

Between points A and B, R_1 and R_2 form a series circuit in parallel with R_3. Therefore;

$$\begin{aligned} R_{A\text{-}B} &= (R_1 + R_2)\|R_3 = (470\ \Omega + 330\ \Omega)\|1.2\ \text{k}\Omega \\ &= 800\ \Omega\|1.2\ \text{k}\Omega = 480\ \Omega \end{aligned}$$

Between point C and ground, we have the parallel combination of R_4 and the load. Therefore,

$$R_C = R_4\|R_L = 120\ \Omega\|180\ \Omega = 72\ \Omega$$

Using the calculated values of $R_{A\text{-}B}$ and R_C, the series-equivalent of the original circuit is drawn as shown in Figure 6.10b.

Practice Problem 6.4

Derive the series equivalent for the circuit shown in Figure 6.11.

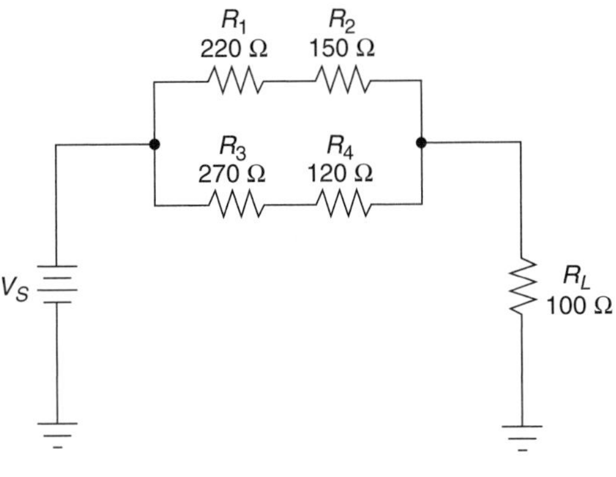

Figure 6.11

Series-Parallel Circuit Analysis

◀ *OBJECTIVE 4*

In some circuit analysis problems, you must reduce an entire circuit to its series or parallel equivalent to determine a given current, voltage, and/or power value. In others, a given circuit value can be determined by reducing only a portion of the original circuit. This point is demonstrated in the following example.

Example 6.5

Determine the value of load power (P_L) for the circuit in Figure 6.12a.

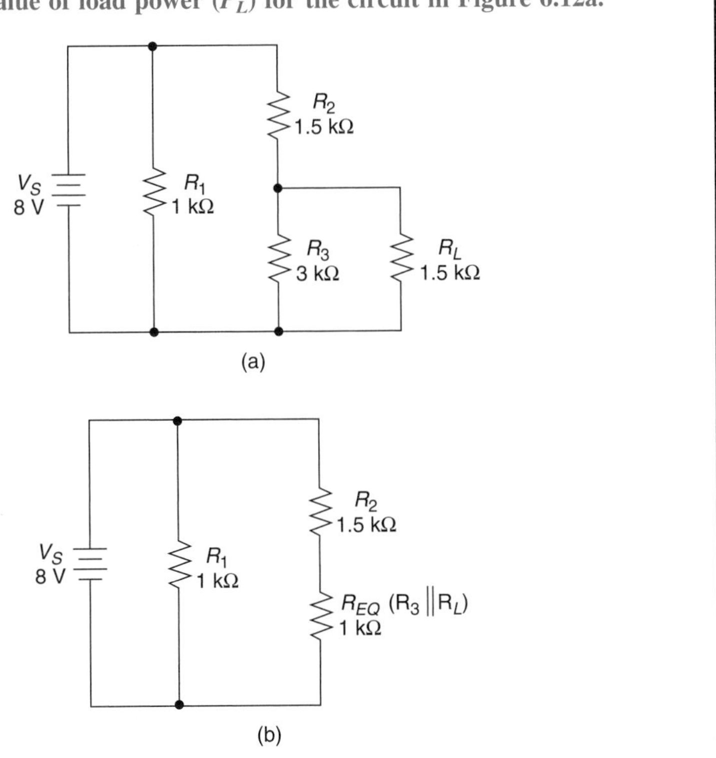

Figure 6.12

Solution: As shown in Figure 6.12b,

$$R_{EQ} = R_3 \| R_L = 3\text{ k}\Omega \| 1.5\text{ k}\Omega = 1\text{ k}\Omega$$

The source voltage is applied to the series combination of R_2 and R_{EQ}. Therefore, the voltage across R_{EQ} can be found as

$$V_{EQ} = V_S \frac{R_{EQ}}{R_2 + R_{EQ}} = (8 \text{ V}) \frac{1 \text{ k}\Omega}{2.5 \text{ k}\Omega} = 3.2 \text{ V}$$

As you know, $V_{EQ} = V_3 = V_L$. Therefore, the load power can be found as

$$P_L = \frac{V_L^2}{R_L} = \frac{(3.2 \text{ V})^2}{1.5 \text{ k}\Omega} = 6.83 \text{ mW}$$

Practice Problem 6.5

A circuit like the one shown in Figure 6.12a has the following values: $V_S = 12$ V, $R_1 = 2$ kΩ, $R_2 = 3$ kΩ, $R_3 = 12$ kΩ, and $R_L = 18$ kΩ. Calculate the value of load power for the circuit.

In Example 6.5, we didn't need to reduce the entire circuit to solve the problem. At the same time, calculating values such as total resistance (R_T), current (I_T), or power (P_T) requires a complete reduction of the circuit to its simplest parallel equivalent.

Example Analysis Problems

As you can see, the key to solving for any given value in a series-parallel circuit is knowing how far the circuit needs to be reduced in order to solve the problem. At this point, we will work through two analysis problems to further demonstrate this principle.

Example 6.6

Determine the value of load current (I_L) for the circuit in Figure 6.13a.

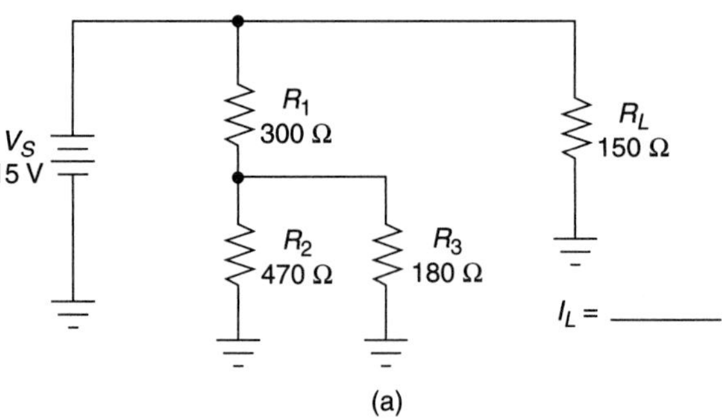

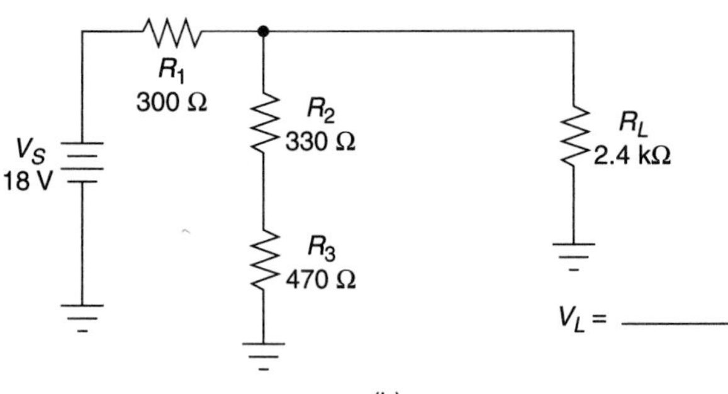

Figure 6.13

Solution: This problem is much easier to solve than it may seem at first. Since the load is in parallel with the rest of the circuit, $V_L = V_S$, and

$$I_L = \frac{V_L}{R_L} = \frac{15 \text{ V}}{150 \text{ }\Omega} = 100 \text{ mA}$$

As you can see, no circuit reduction was needed to solve for I_L in Figure 6.13a. Remember, even though they are more complex, series-parallel circuits still operate by the same rules as other circuits. If you keep the basics in mind, many series-parallel circuit problems end up being simpler to solve than they first appear to be.

Example 6.7

Determine the value of load voltage (V_L) for the circuit in Figure 6.13b.

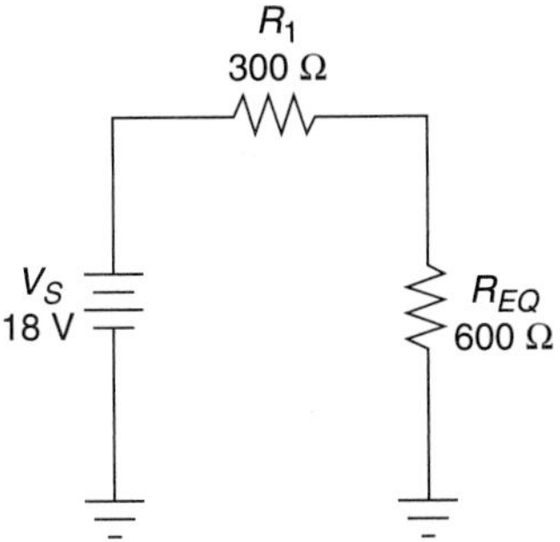

Figure 6.14

Solution: To solve this problem, we must reduce the circuit to the series equivalent shown in Figure 6.14. The value of R_{EQ} (which represents the total resistance of the parallel branches) was found as

$$R_{EQ} = (R_2 + R_3)\|R_L = (330\ \Omega + 470\ \Omega)\|2.4\ \text{k}\Omega = 600\ \Omega$$

The voltage across this equivalent resistance (which we will designate as V_{EQ}) can be found using the voltage-divider equation, as follows:

$$V_{EQ} = V_S \frac{R_{EQ}}{R_1 + R_{EQ}} = (18\ \text{V})\frac{600\ \Omega}{900\ \Omega} = 12\ \text{V}$$

V_{EQ} is the voltage across the parallel branches. Therefore,

$$V_L = V_{EQ} = 12\ \text{V}$$

Practice Problem 6.7

A circuit like the one shown in Figure 6.13b has the following values: $V_S = 12$ V, $R_1 = 200\ \Omega$, $R_2 = 100\ \Omega$, $R_3 = 300\ \Omega$, and $R_L = 1.2\ \text{k}\Omega$. Determine the value of load voltage for the circuit.

Again, the problem was solved using an equivalent circuit and basic circuit relationships—in this case, the voltage-divider equation and the relationship between parallel voltages.

Summary

◀ *OBJECTIVE 5*

The analysis of most series-parallel circuits involves three steps:

1. Reduce the circuit to a series or a parallel equivalent, depending on the type of circuit.
2. Apply the basic series and parallel circuit principles and relationships to the equivalent circuit.
3. Use the values obtained from the equivalent circuit to solve the original for the desired value(s).

When you use this three-step approach, you will rarely have a problem solving for any single value in a series-parallel circuit.

Section Review

1. What is a *series-equivalent circuit*?
2. What is a *parallel-equivalent circuit*?
3. How are equivalent circuits derived?
4. What are the steps involved in solving most practical series-parallel circuit problems?

Critical Thinking

5. When looking at a series-parallel circuit schematic, how can you determine whether it reduces to a *series* equivalent or to a *parallel* equivalent?

6.3 Circuit Loading

Most electronic circuits are designed to supply power (in one form or another) to one or more loads. The effect that a load has on the operation of its source circuit is a concept that applies to virtually every area of electronics. In this section, we will discuss the effects that loads can have on the operation of a voltage divider.

Loaded Voltage Dividers

OBJECTIVE 6 ➤ Voltage dividers are commonly used to provide a specific voltage for one or more loads. For example, consider the circuit shown in Figure 6.15. Here, we have a *loaded voltage divider* that is being used to supply the voltages required for the operation of two loads (designated as *Load A* and *Load B*). The voltage and current ratings for each load are given in the illustration. Assuming that the circuit is operating properly, the voltage divider is providing the values of $V_A = 5$ V and $V_B = 8$ V required by the loads.

No-Load Output Voltage (V_{NL})

The **no-load output voltage (V_{NL})** from a voltage divider is *the voltage measured across its load terminals with the load removed.* The measurement of V_{NL} is illustrated in Figure 6.16. As you can see, the load is removed from the circuit, and the voltmeter is connected directly to the open load terminals.

The no-load output voltage for a voltage divider is always greater than the value of V_L for the circuit. For example, in Figure 6.16a, the load voltage is found as

$$V_L = V_S \frac{R_{EQ}}{R_1 + R_{EQ}}$$

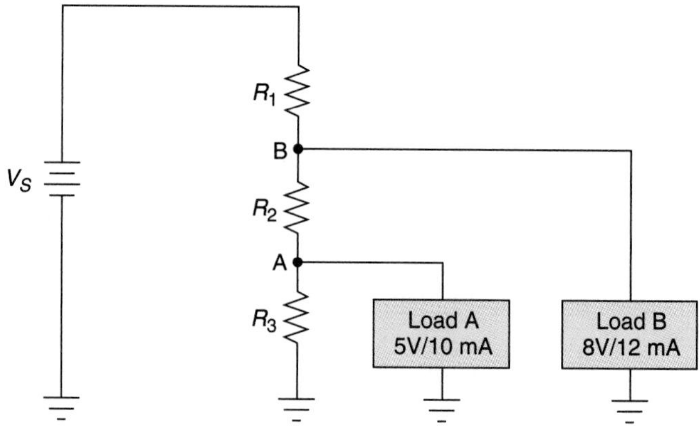

Figure 6.15 A loaded voltage divider.

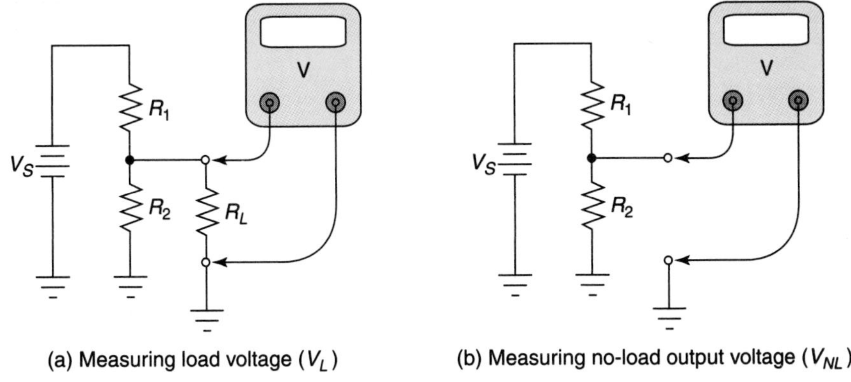

Figure 6.16 Measuring no-load output voltage.

where $R_{EQ} = R_2 \| R_L$. On the other hand, the load voltage in Figure 6.16b is found as

$$V_{NL} = V_S \frac{R_2}{R_1 + R_2}$$

Since R_2 must be greater than the value of R_{EQ}, the value of V_{NL} must be greater than the value of V_L. This point is demonstrated further in the following example.

Example 6.8

Determine the values of V_L and for V_{NL} the circuit in Figure 6.17a.

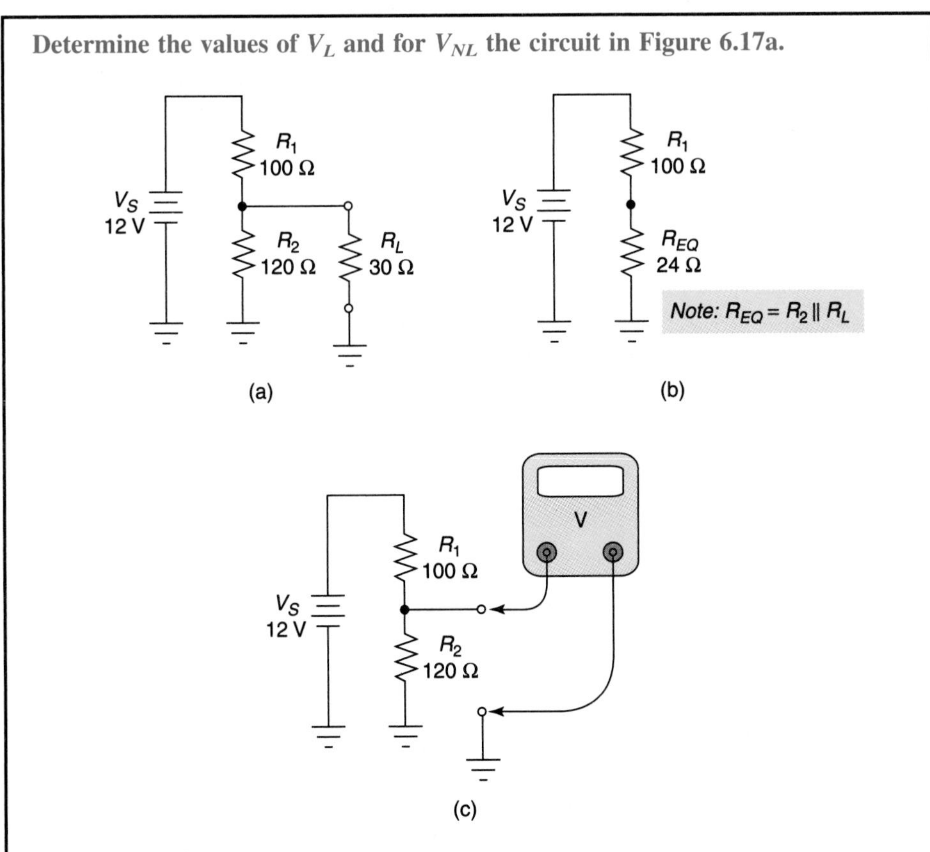

Figure 6.17

Solution: We start by combining R_2 and R_L into a single equivalent resistance, as follows:

$$R_{EQ} = R_2 \| R_L = 120\ \Omega \| 30\ \Omega = 24\ \Omega$$

Using this value, the circuit can now be redrawn as shown in Figure 6.17b. The voltage across R_{EQ} can now be found as

$$V_{EQ} = V_S \frac{R_{EQ}}{R_1 + R_{EQ}} = (12\text{ V})\frac{24\ \Omega}{124\ \Omega} = 2.32\text{ V}$$

Since V_{EQ} is across the parallel combination of R_2 and R_L, $V_L = V_{EQ} = 2.32$ V. As shown in Figure 6.17c, V_{NL} is the voltage across R_2 with the load removed. Therefore,

$$V_{NL} = V_S \frac{R_2}{R_1 + R_2} = (12\text{ V})\frac{120\ \Omega}{220\ \Omega} = 6.55\text{ V}$$

Practice Problem 6.8

A circuit like the one in Figure 6.17a has the following values: $V_S = 15$ V, $R_1 = 1$ kΩ, $R_2 = 1.5$ kΩ, and $R_L = 3$ kΩ. Calculate the values of V_L and V_{NL} for the circuit.

As you can see, the value of V_{NL} is significantly greater than the value of V_L. This illustrates an important relationship between load resistance and load voltage: *Load voltage varies with changes in load resistance.* This concept is illustrated further in Example 6.9.

Example 6.9

Determine the range of output voltages for the circuit in Figure 6.18a.

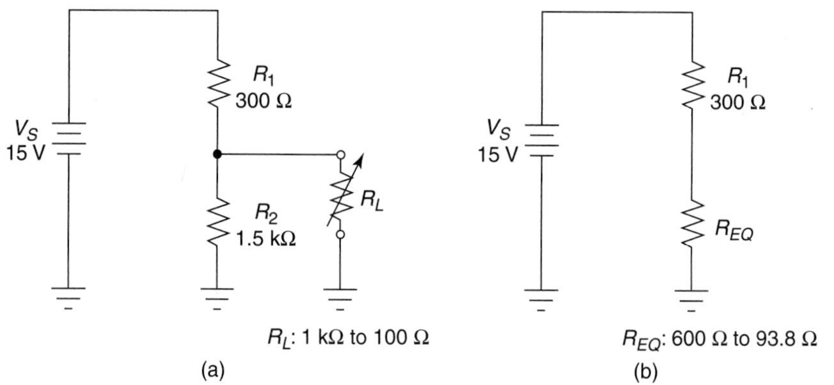

Figure 6.18

Solution: The load is shown as being variable over a range of 1 kΩ to 100 Ω. When $R_L = 1$ kΩ,

$$R_{EQ} = R_2 \| R_L = 1.5\text{ k}\Omega \| 1\text{ k}\Omega = 600\ \Omega$$

and

$$V_L = V_S \frac{R_{EQ}}{R_1 + R_{EQ}} = (15\text{ V})\frac{600\ \Omega}{900\ \Omega} = 10\text{ V}$$

When $R_L = 100\ \Omega$,

$$R_{EQ} = R_2 \| R_L = 1.5\text{ k}\Omega \| 100\ \Omega = 93.8\ \Omega$$

and

$$V_L = V_S \frac{R_{EQ}}{R_1 + R_{EQ}} = (15\text{ V})\frac{93.8\ \Omega}{393.8\ \Omega} = 3.57\text{ V}$$

Practice Problem 6.9

A circuit like the one shown in Figure 6.18a has the following values: $V_S = 12$ V, $R_1 = 220\ \Omega$, $R_2 = 330\ \Omega$, and $R_L = 50\ \Omega$ to $500\ \Omega$. Determine the range of output voltages for the circuit.

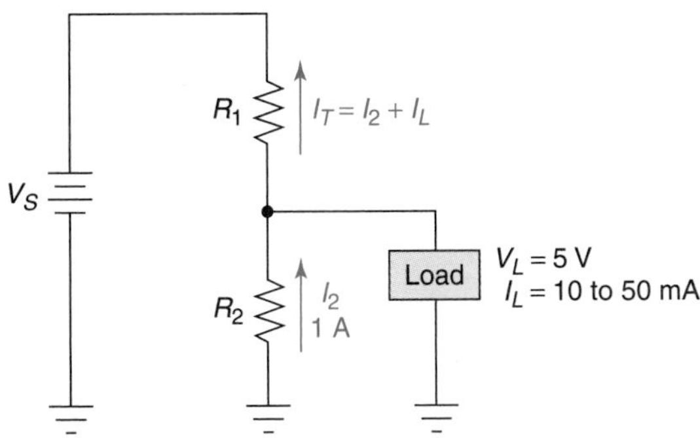

Figure 6.19

◀ *OBJECTIVE 7*

In practice, voltage dividers are usually designed so that the output voltage is relatively stable. That is, they are designed so that any variations in output voltage caused by a change in load resistance are held to a minimum.

How voltage-divider stability is achieved.

Voltage-divider stability is accomplished by designing the circuit so that *the load current makes up a relatively small percentage of the total circuit current.* For example, the circuit shown in Figure 6.19 is designed to have a value of $I_2 = 1$ A. Since the total current in the circuit equals the sum of I_2 and I_L, it has a range of

$$I_T = 1\text{ A} + 10\text{ mA} = 1.01\text{ A} \quad \text{to} \quad I_T = 1\text{ A} + 50\text{ mA} = 1.05\text{ A}$$

As you can see, the change in load current has no *practical* effect on the value of I_T. Since

$$V_1 = I_T R_1$$

the value of V_1 is also relatively stable against changes in I_L. This means that the load voltage must also be relatively stable, as given in the relationship

$$V_L = V_S - V_1$$

The concept of voltage-divider stability is demonstrated further in the following example.

Example 6.10

Calculate the normal range of V_L for the loaded voltage divider in Figure 6.20a.

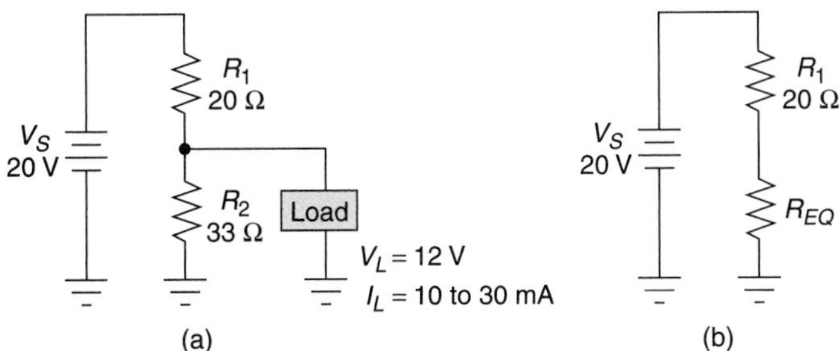

Figure 6.20

Solution: The load is shown to have the following ratings: $V_L = 12$ V and $I_L =$ 10 mA to 30 mA. Using these values,

$$R_{L(\min)} = \frac{12\text{ V}}{30\text{ mA}} = 400\ \Omega \quad \text{and} \quad R_{L(\max)} = \frac{12\text{ V}}{10\text{ mA}} = 1.2\text{ k}\Omega$$

For reference, the series equivalent of the loaded voltage divider is shown in Figure 6.20b. When $R_L = 400\ \Omega$,

$$R_{EQ} = R_2 \| R_L = 33\ \Omega \| 400\ \Omega = 30.5\ \Omega$$

and

$$V_L = V_S \frac{R_{EQ}}{R_1 + R_{EQ}} = (20\ \text{V}) \frac{30.5\ \Omega}{50.5\ \Omega} = 12.08\ \text{V}$$

When $R_L = 1.2\ \text{k}\Omega$,

$$R_{EQ} = R_2 \| R_L = 33\ \Omega \| 1.2\ \text{k}\Omega = 32.1\ \Omega$$

and

$$V_L = V_S \frac{R_{EQ}}{R_1 + R_{EQ}} = (20\ \text{V}) \frac{32.1\ \Omega}{52.1\ \Omega} = 12.32\ \text{V}$$

Practice Problem 6.10

A circuit like the one shown in Figure 6.20a has the following values: $V_S = 18$ V, $R_1 = 68\ \Omega$, and $R_2 = 75\ \Omega$. The load on the circuit is rated at $V_L = 9$ V and $I_L = 2$ mA to 12 mA. Determine the normal range of V_L values for the circuit.

As you have seen, voltage divider stability is achieved by designing the circuit so that the total circuit current is much greater than the load current. In a loaded voltage divider, $I_T >> I_L$ when the current through the component that is parallel to the load (R_2) is much greater than the load current. Note that *the current through the component that is parallel to the load* is referred to as the **bleeder current (I_B).** The relationship among I_T, I_B, and I_L can be stated mathematically as follows:

$$I_T >> I_L \text{ when } I_B >> I_L \qquad (6.1)$$

A Practical Consideration: When $I_B >> I_L$, the power dissipated by the bleeder resistor is much greater than the power dissipated by the load. Therefore, the bleeder resistor must have a much higher power-dissipation rating than the load.

The validity of this relationship can be demonstrated using the results from Example 6.10. In that example, I_2 is the bleeder current. When V_L is at its minimum value, the bleeder current can be found as

$$I_B = I_2 \cong \frac{V_L}{R_2} = \frac{12.08\ \text{V}}{33\ \Omega} = 366\ \text{mA} \qquad \text{(minimum)}$$

and the value of I_L can be found as

$$I_L = \frac{V_L}{R_{L(\text{min})}} = \frac{12.08\ \text{V}}{400\ \Omega} = 30.2\ \text{mA} \qquad \text{(maximum)}$$

As you can see, $I_B >> I_L$. The circuit fulfills the relationship given in equation (6.1).

OBJECTIVE 8 ➤

Variable Voltage Dividers

A potentiometer can be (and often is) used as a *variable voltage divider.* A **variable voltage divider** is *one that can be varied to provide a specific load voltage.* A variable voltage divider is shown in Figure 6.21. When the potentiometer setting is varied, it changes the values of R_{ab} and R_{bc}. This causes the voltages between the wiper arm and the outer potentiometer terminals to change. Since the load is in parallel with R_{bc}, the load voltage also changes.

A range of load voltages is provided by varying the potentiometer from one extreme to the other. When set to one extreme, the value of R_{bc} is approximately 0 Ω. In this case, the load voltage is approximately 0 V. When set to the other extreme, R_{ab} is approximately 0 Ω. Because of this, all of the source voltage is dropped across the parallel combination

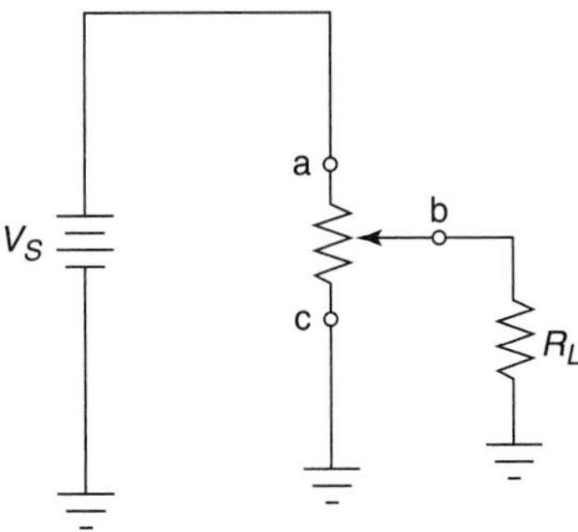

Figure 6.21 A potentiometer as a variable voltage divider.

of R_{bc} and the load. In this case, the load voltage is approximately equal to the source voltage.

Potentiometer resistance In Chapter 2, you were told that the value of R_{ac} for a given potentiometer is constant. However, this does not hold true when a potentiometer is connected as a variable voltage divider. As the potentiometer setting is varied, the *effective* value of R_{ac} also changes, as illustrated in Figure 6.22. When the potentiometer is set to $R_{bc} = 0\ \Omega$, the value of R_{ac} is equal to the value of R_{ab}, or 10 kΩ. However, when set to the other extreme,

$$R_{ac} = R_{bc} \| R_L = 10\text{ k}\Omega \| 10\text{ k}\Omega = 5\text{ k}\Omega$$

Therefore, changing the setting on a loaded voltage divider causes the effective value of R_{ac} to change.

Voltmeter Loading

When used to measure any voltage in a circuit, *a voltmeter may cause the reading to be significantly lower than the actual component voltage.* This **voltmeter loading** is due to the effect of the meter's *internal resistance,* as shown in Figure 6.23. ◀ *OBJECTIVE 9*

When the meter is connected as shown in the figure, its internal resistance (R_M) is connected in parallel with R_2. This essentially converts the circuit into a loaded voltage divider, with the meter's internal resistance acting as the load. If the internal resistance of the meter is low enough, the impact on the voltage being measured can be fairly severe, as demonstrated in Example 6.11.

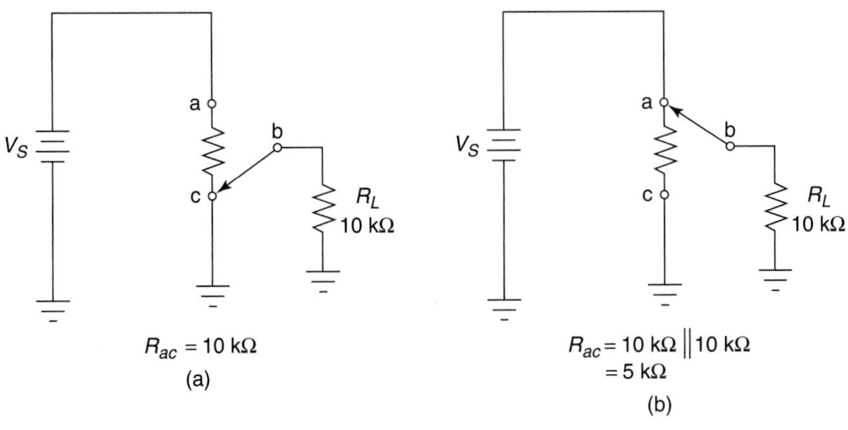

Note: The potentiometer symbols have been altered in this figure to illustrate a point. They are not normally drawn as shown here.

Figure 6.22

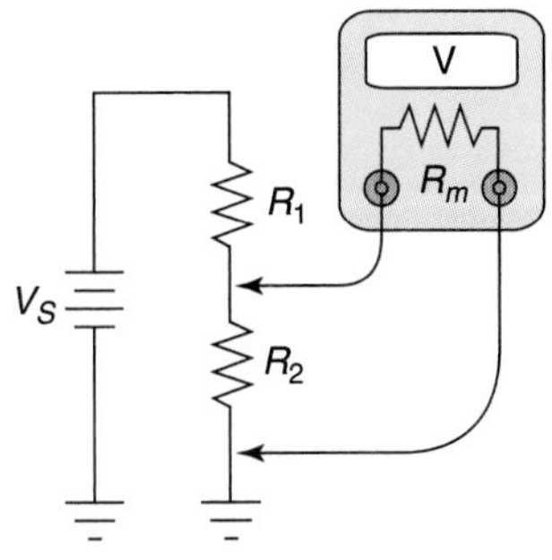

Figure 6.23

Example 6.11

Determine the value of V_2 for the circuit in Figure 6.24a. Then, determine the value of V_2 when the voltmeter is connected as shown in Figure 6.24b.

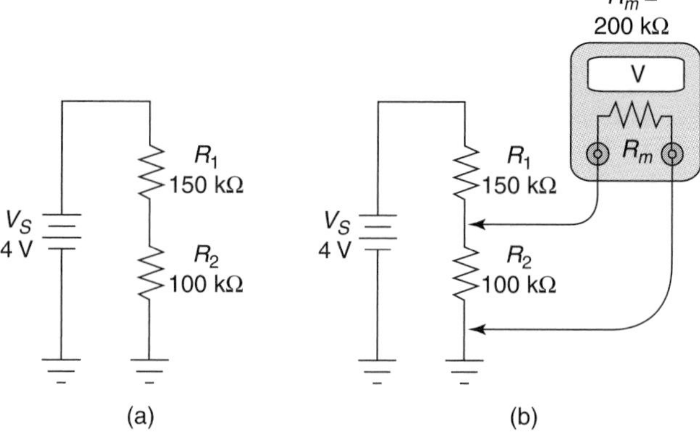

Figure 6.24

Solution: The value of V_2 for the series circuit is found as

$$V_2 = V_S \frac{R_2}{R_1 + R_2} = (4 \text{ V}) \frac{100 \text{ k}\Omega}{250 \text{ k}\Omega} = 1.6 \text{ V}$$

When the meter is connected as shown in Figure 6.24b, R_2 and the internal resistance of the meter (R_m) form a parallel circuit. The resistance of this circuit is found as

$$R_{EQ} = R_2 \| R_m = 100 \text{ k}\Omega \| 200 \text{ k}\Omega = 66.7 \text{ k}\Omega$$

and

$$V_2 = V_S \frac{R_{EQ}}{R_1 + R_{EQ}} = (4 \text{ V}) \frac{66.7 \text{ k}\Omega}{216.7 \text{ k}\Omega} = 1.23 \text{ V}$$

Practice Problem 6.11

The value of V_2 for the circuit in Figure 6.24a is measured using a DMM with an internal resistance of 10 MΩ. Calculate the value of V_2 with the meter connected. Compare your result to those obtained in this example.

As you can see, connecting a voltmeter to a circuit can have a significant effect on the accuracy of the reading. However, the impact of the connection can be reduced by using a voltmeter with *high internal resistance.* For example, consider the circuits shown in Figure 6.25. The internal resistance of the VOM in Figure 6.25a produces a parallel-

Figure 6.25

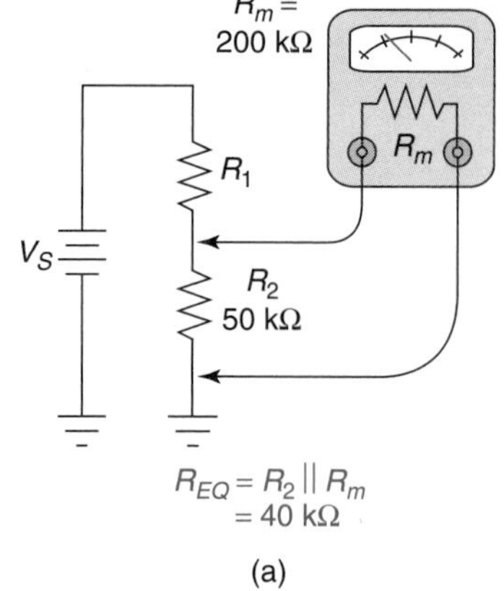

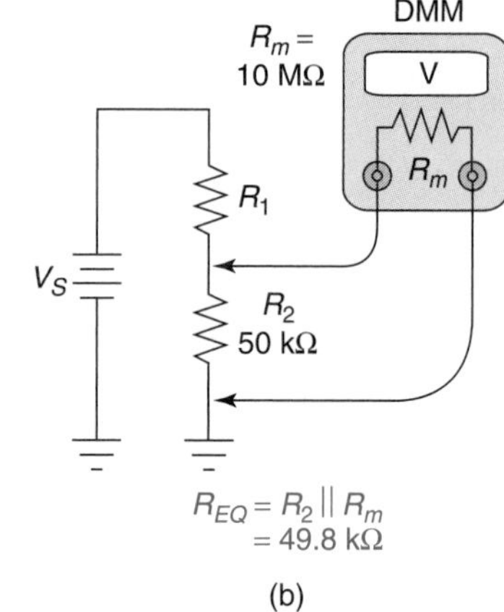

equivalent resistance of 40 kΩ. At the same time, the DMM in Figure 6.25b produces a parallel-equivalent resistance of 49.5 kΩ.

DMMs typically have much higher input resistance than VOMs. When you consider the impact of circuit loading, it is easy to understand why DMMs are preferred over VOMs for most voltage measurements.

Why DMMs are preferred over VOMs for voltage measurements.

Section Review

1. What is a loaded voltage divider?
2. What is the *no-load output voltage* (V_{NL}) of a voltage divider?
3. For a loaded voltage divider, what is the relationship between
   a. V_L and V_{NL}?
   b. R_L and V_L?
4. What is meant by the term *voltage-divider stability*?
5. What design requirement must be met to produce a *stable* voltage divider?
6. Describe the use of a potentiometer as a *variable* voltage divider.
7. How does varying the setting of a variable voltage divider affect the effective value of R_{ac} for the potentiometer?
8. Describe the effect of *voltmeter loading* on series circuit voltage measurements.
9. When $I_B = 10I_{L(\max)}$, any anticipated change in load resistance causes the output from a loaded voltage divider to change by no more than 10%. What relationship would be required between the values of I_B and I_L to limit any change in output voltage to 1%? Explain your answer.

Critical Thinking

6.4 Bridge, Delta, and Wye Circuits

Series-parallel circuits come in a near-infinite variety of configurations. In this section, we will focus on several circuits, starting with the *Wheatstone bridge.* The **Wheatstone bridge** is *a circuit containing a meter "bridge" that provides extremely accurate resistive measurements and responds to any changes in resistance.*

Bridge Construction

◀ OBJECTIVE 10

A basic Wheatstone bridge consists of a dc voltage source, four resistors, and a meter, as shown in Figure 6.26. The resistors form two series circuits that are connected in parallel. The meter is connected as a "bridge" between the series circuits (which is how the circuit got its name).

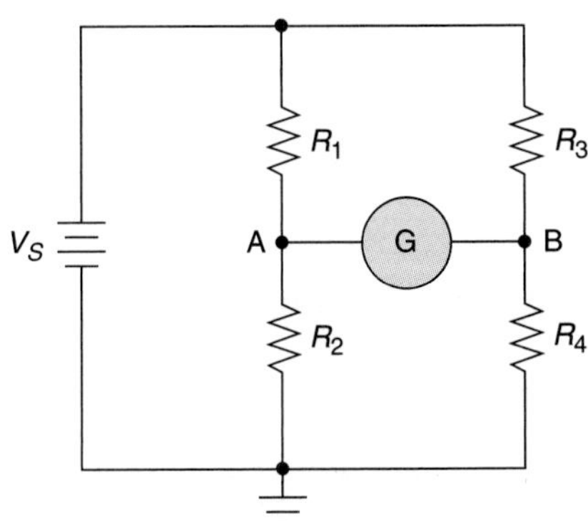

Figure 6.26 The Wheatstone bridge.

Note that the bridge between the resistive branches forms a unique connection; that is, it is connected in such a way that it is not in series, parallel, or series-parallel with any specific component. Even though it is approached as a series-parallel circuit, the bridge is an electrical anomaly. It is the only connection that cannot be classified specifically as series, parallel, or series-parallel.

The meter shown in Figure 6.26 is a **galvanometer.** A galvanometer is *a current meter that indicates both the magnitude and the direction of a low-value current.* As you will see, the galvanometer can indicate *the relationship between the voltages at points A and B in the circuit.*

Bridge Operation

Each branch in the Wheatstone bridge is a voltage divider. If we remove the galvanometer (as shown in Figure 6.27), the values of V_A and V_B can be found as

A Practical Consideration: Equations (6.2) and (6.3) hold true only when the galvanometer is not connected (or not conducting). When the galvanometer is connected, another approach must be used to analyze the circuit.

$$V_A = V_S \frac{R_2}{R_1 + R_2} \tag{6.2}$$

and

$$V_B = V_S \frac{R_4}{R_3 + R_4} \tag{6.3}$$

The operation of the Wheatstone bridge is based on the fact that it has *three possible operating states;* that is, three possible combinations of V_A and V_B. These operating states are as follows:

$$V_A = V_B \qquad V_A > V_B \qquad V_A < V_B$$

Each of these operating states is illustrated in Figure 6.28.

In Figure 6.28, R_4 is a potentiometer. By adjusting the value of this component, we can set V_A to any value (within limits). In Figure 6.28a, R_4 is adjusted so that $V_A = V_B$. When these voltages are equal, there is no difference of potential across the galvanometer, and the galvanometer current (I_G) equals 0 A. When in this operating state, the bridge is said to be *balanced.*

In Figure 6.28b, R_4 is adjusted so that $V_B < V_A$. Since the voltages are no longer equal, there is a difference of potential (voltage) across the galvanometer. As a result, a measurable value of I_G is generated. The galvanometer reading indicates the magnitude and direction of I_G.

In Figure 6.28c, R_4 has been adjusted so that $V_B > V_A$. Again, the voltages are unequal, and there is a difference of potential across the galvanometer. Again, a measurable value of I_G is generated, and the galvanometer indicates its magnitude and direction.

Resistance Ratios

By definition, a Wheatstone bridge is balanced when $V_A = V_B$. For these voltages to be equal, the resistance ratios in equations (6.2) and (6.3) must be equal. In other words, a bridge is balanced only when

$$\frac{R_2}{R_1} = \frac{R_4}{R_3} \tag{6.4}$$

Figure 6.27

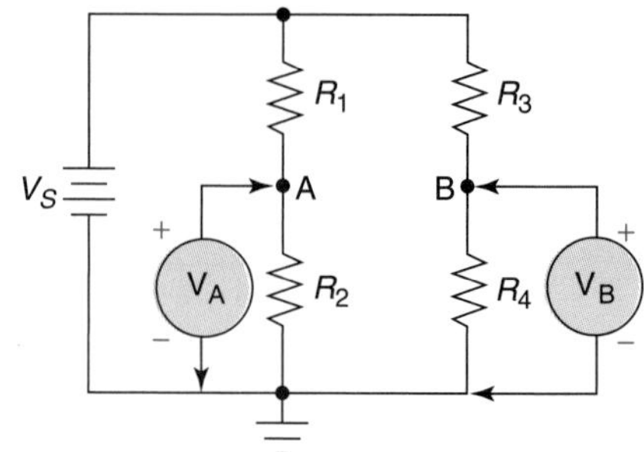

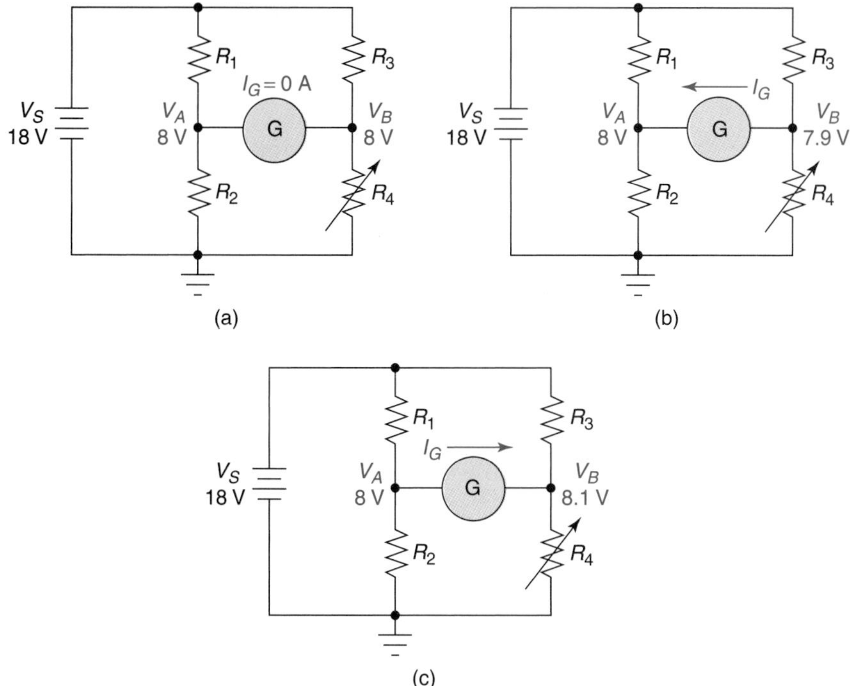

Figure 6.28 Wheatstone bridge operation.

TABLE 6.3
Bridge Resistance Ratios and Operating States

Resistance Ratio	*Resulting Operating State*
$\frac{R_2}{R_1} = \frac{R_4}{R_3}$	$V_A = V_B$
$\frac{R_2}{R_1} > \frac{R_4}{R_3}$	$V_A > V_B$
$\frac{R_2}{R_1} < \frac{R_4}{R_3}$	$V_A < V_B$

Note that this relationship is derived by setting equations (6.2) and (6.3) equal to each other and simplifying the result. As Table 6.3 shows, each of the operating states is produced by a specific relationship between the resistance ratios.

Measuring Resistance with a Wheatstone Bridge

◄ *OBJECTIVE 11*

A Wheatstone bridge can be used as an effective resistance measuring circuit when set up as shown in Figure 6.29. In the bridge circuit shown:

- The unknown resistance (R_x) has been placed in the R_3 position.
- R_4 has been replaced by a *calibrated decade box.*

A **calibrated decade box** is *a device that contains several series-connected potentiometers.* By adjusting the individual potentiometers, the overall resistance of the decade box (R_{db}) can be set to any value within its limits, usually between 1 Ω and 9.999999 MΩ. (The value of R_{db} for any setting is indicated by the calibrated readouts on the decade box.)

When connected as shown in Figure 6.29, the value of the unknown resistance is found by balancing the bridge; that is, by adjusting the value of R_{db} so that the galvanometer reads 0 A. When the bridge is balanced, the resistance ratios of the series circuits must be equal. Therefore, the value of R_x can be found using a modified form of equation (6.4), as follows:

$$R_x = R_{db}\frac{R_1}{R_2} \tag{6.5}$$

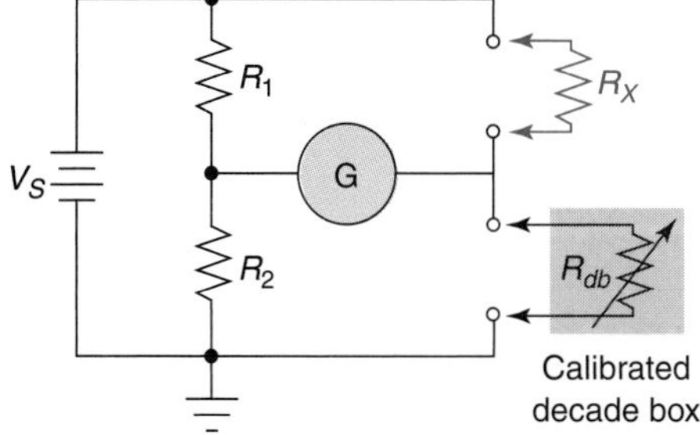

Figure 6.29 Using a Wheatstone bridge to measure resistance.

Note that this equation was derived by taking the reciprocals of the resistance ratios in equation (6.4), substituting R_x and R_{db} for R_3 and R_4, and then solving for R_x.

Measuring the value of R_x is easier when a Wheatstone bridge is constructed using equal values of R_1 and R_2. When R_1 and R_2 are equal, the fraction in equation (6.5) equals 1, and the value of R_x can be found as

$$R_x = R_{db} \qquad (\text{when } R_1 = R_2) \tag{6.6}$$

This means that a bridge with equal values of R_1 and R_2 will balance when R_{db} equals the value of R_x. This simplifies the overall procedure for measuring the value of an unknown resistance.

A Wheatstone Bridge Application

While the Wheatstone bridge can be used to determine the value of an unknown resistance, it also has more common applications. One such application is illustrated in Figure 6.30, where a Wheatstone bridge is used in a smoke detector.

The Wheatstone bridge in Figure 6.30a contains two *photoconductive cells.* A **photoconductive cell** is *a solid-state component that acts as a light-controlled resistance.* When the light applied to a photoconductive cell *increases,* the resistance of the component *decreases.* The photoconductive cells are positioned within the smoke detector as shown in Figure 6.30b.

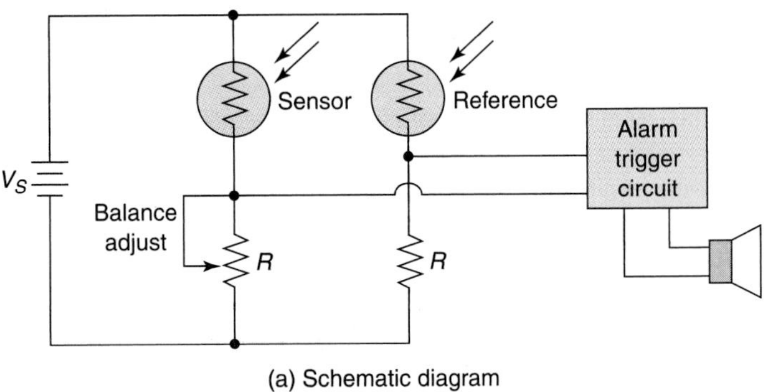

(a) Schematic diagram

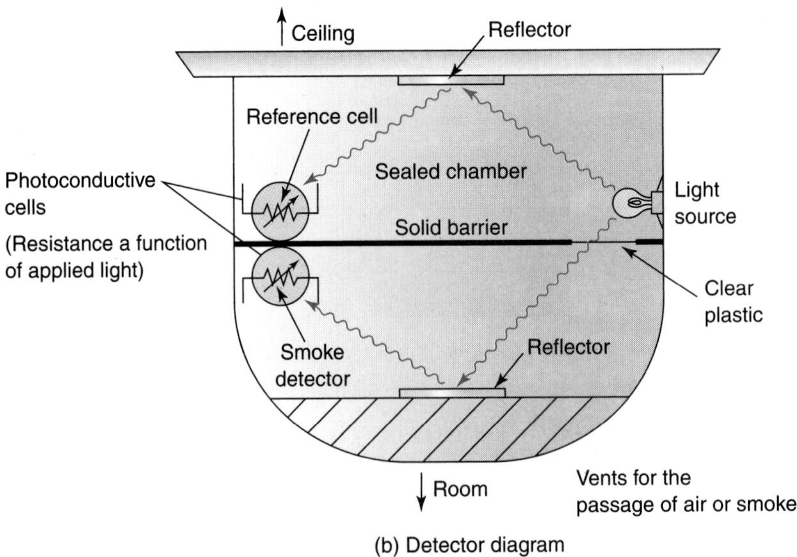

(b) Detector diagram

Figure 6.30 A smoke detector. (Figure 6.30b reprinted from Boylestad, *Introductory Circuit Analysis,* 9th ed., Figure 8.85c. Copyright 1999 by Pearson Education. Used by permission.)

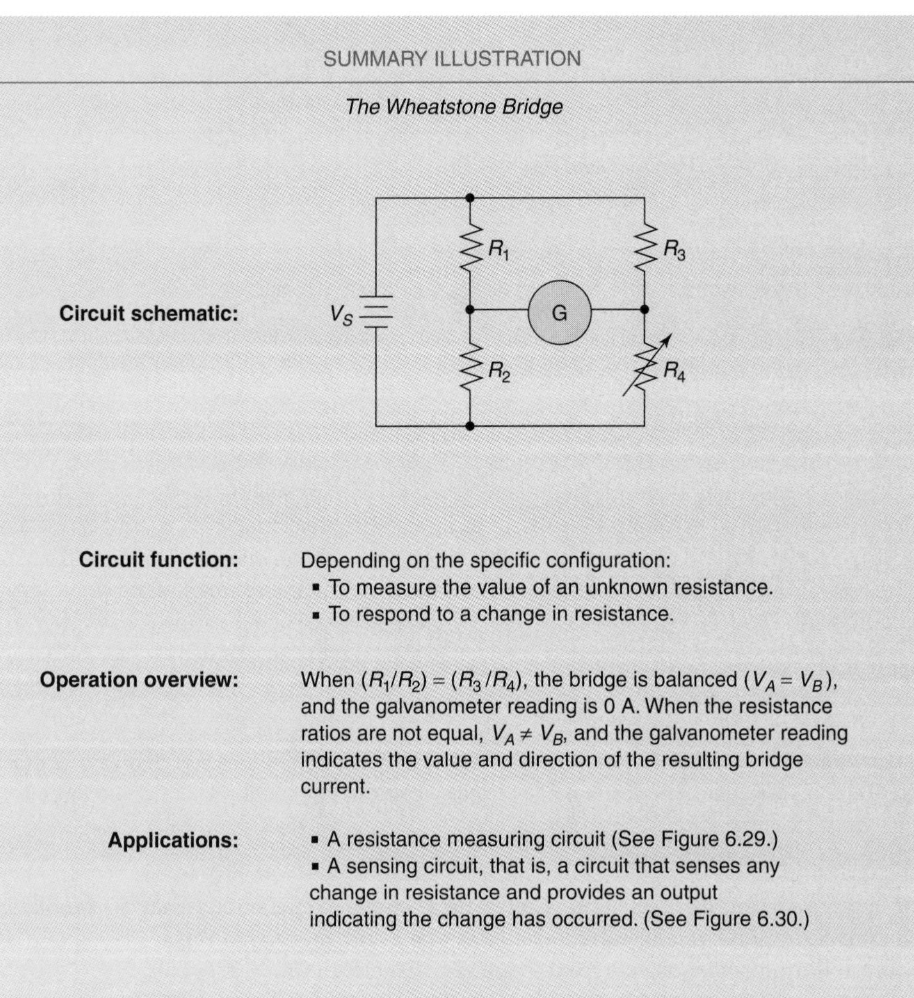

SUMMARY ILLUSTRATION
The Wheatstone Bridge
Circuit schematic:
V_S
R_1
R_3
G
R_2
R_4
Circuit function:
Depending on the specific configuration:
▪ To measure the value of an unknown resistance.
▪ To respond to a change in resistance.
Operation overview:
When $(R_1/R_2) = (R_3/R_4)$, the bridge is balanced ($V_A = V_B$), and the galvanometer reading is 0 A. When the resistance ratios are not equal, $V_A \neq V_B$, and the galvanometer reading indicates the value and direction of the resulting bridge current.
Applications:
▪ A resistance measuring circuit (See Figure 6.29.)
▪ A sensing circuit, that is, a circuit that senses any change in resistance and provides an output indicating the change has occurred. (See Figure 6.30.)

Figure 6.31

When the air is clear, the bridge is balanced, and the input to the alarm trigger circuit is 0 V. In this case, the alarm is inactive. If smoke enters the open chamber of the smoke detector, the resistance of the sensing cell increases, causing an imbalance in the bridge. In this case, a difference of potential is applied to the alarm trigger circuit, and the alarm is activated. Note that the *reference cell* is kept in a sealed chamber. This prevents the alarm from sounding if the light source should dim, since a problem with the light source affects *both* conductive cells equally. Only a difference in the resistance values of the cells will cause the alarm to activate.

When used as shown in Figure 6.30, the Wheatstone bridge is acting as a *sensor.* In this case, the circuit is "sensing" whether smoke is present in the room. Using different resistive elements, a basic Wheatstone bridge can be modified to sense changes in pressure, temperature, or magnetic fields. Even when special-purpose resistive elements are used (such as the photoconductive cells shown in Figure 6.30), the circuit maintains the characteristics discussed in this section. As a reference, these characteristics are summarized in Figure 6.31.

Delta and Wye Circuits

Two common series-parallel configurations that we have yet to discuss are the **delta** and **wye circuits.** These circuits, which are shown in Figure 6.32a, are named for their resemblance to the Greek letter *delta* (Δ) and the letter *Y.* Sometimes, the circuits are drawn ◀ *OBJECTIVE 12*

A Practical Consideration: Delta (Δ) and wye (Y) circuit configurations are used extensively in ac power distribution applications.

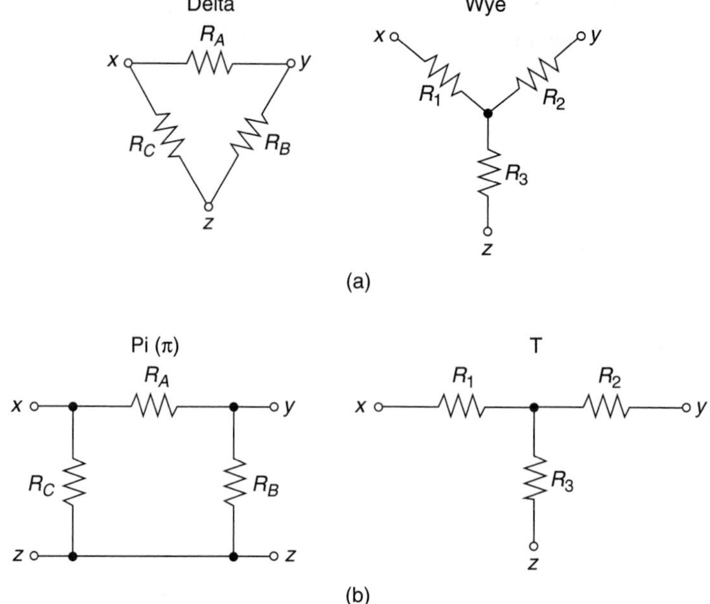

Figure 6.32 Delta and wye circuits (with π and T representations).

as shown in Figure 6.32b. When drawn as such, they are sometimes referred to as *Pi* (π) and *T* circuits. (Regardless of the names used, we're talking about the same two circuits.)

Circuit Conversions

OBJECTIVE 13 ➤ For every Δ circuit, there exists an equivalent Y circuit. Some circuit analysis problems can be made easier by converting a Δ circuit to a Y circuit, or vice versa.

The following equations are used to convert a Δ circuit into a Y circuit.

$$R_1 = \frac{R_A R_C}{R_A + R_B + R_C} \tag{6.7}$$

$$R_2 = \frac{R_A R_B}{R_A + R_B + R_C} \tag{6.8}$$

and

$$R_3 = \frac{R_B R_C}{R_A + R_B + R_C} \tag{6.9}$$

The relationships given in equations (6.7) through (6.9) would be difficult to commit to memory, but there is a pattern that makes them relatively easy to remember. This

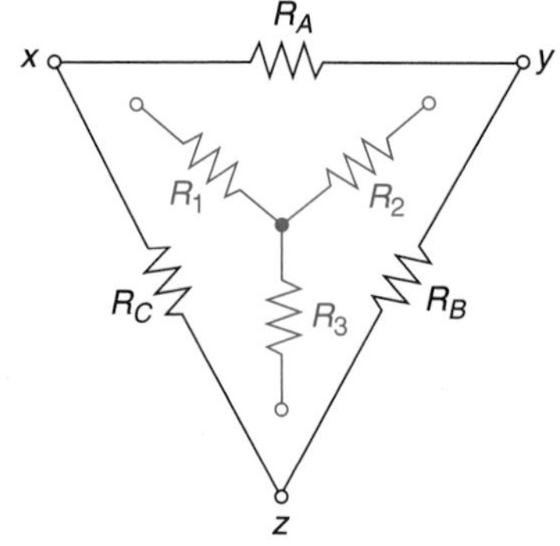

Figure 6.33

pattern can be seen in Figure 6.33, which shows a Y circuit positioned within a Δ circuit. First, note that each equation is in *product-over-sum* form. The pattern used in the numerator of each equation is as follows: Each numerator contains the product of the two Δ resistors that are adjacent to the Y resistor. For example, R_1 is adjacent to R_A and R_C, which are the two terms that appear in the numerator of the R_1 equation. Finally, each denominator contains the sum of the three resistors in the Δ circuit.

The following equations are used to convert a Y circuit into a Δ circuit:

$$R_A = \frac{R_1R_2 + R_2R_3 + R_1R_3}{R_3} \tag{6.10}$$

$$R_B = \frac{R_1R_2 + R_2R_3 + R_1R_3}{R_1} \tag{6.11}$$

and

$$R_C = \frac{R_1R_2 + R_2R_3 + R_1R_3}{R_2} \tag{6.12}$$

As before, there is a pattern in the conversion formulas that can be seen with the aid of Figure 6.33. The numerator of each equation contains the sum of the three possible two-resistor products. The denominator of each equation contains the resistance in the Y circuit that is *opposite* the resistor in the Δ circuit. For example, R_3 is opposite R_A in Figure 6.33, and R_3 appears in the denominator for the R_A equation.

Δ-to-Y Conversion: An Application

It is difficult to calculate the values of the component voltages and currents in a Wheatstone bridge because of the unique connection formed by the bridge component (or meter). For example, the total resistance of the circuit in Figure 6.34a cannot be calculated, because R_L cannot be assumed to be in series, parallel, or series-parallel with any specific component. If you cannot determine the total circuit resistance, you cannot determine the component currents or voltages.

The analysis of a bridge circuit is made possible by using a Δ-to-Y conversion, as shown in Figure 6.34b. The combination of R_2, R_L, and R_4 in Figure 6.34a forms a Δ circuit (drawn in the π form). Performing a Δ-to-Y conversion, we get the circuit shown in Figure 6.34b. As you can see, we now have two parallel branches that are in series with a single resistor. This circuit can be analyzed easily using the techniques established earlier in this chapter.

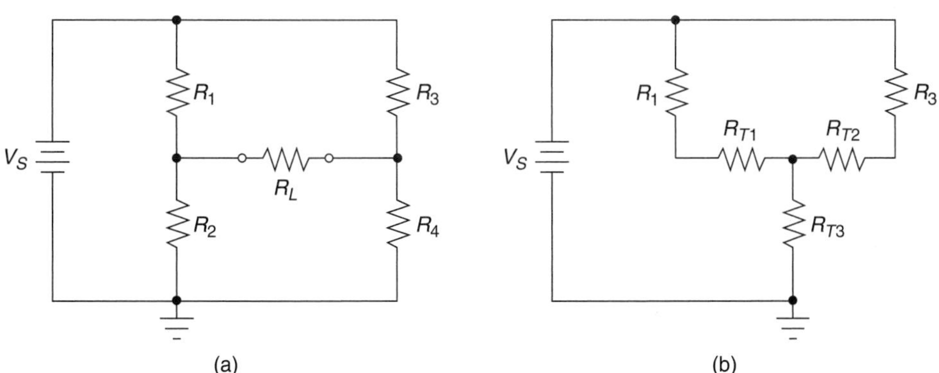

Figure 6.34 Using Δ-to-Y conversion on a bridge circuit.

Section Review

1. What is a *galvanometer*?
2. What are the three operating states of a *Wheatstone bridge*?
3. Describe a *balanced* bridge.

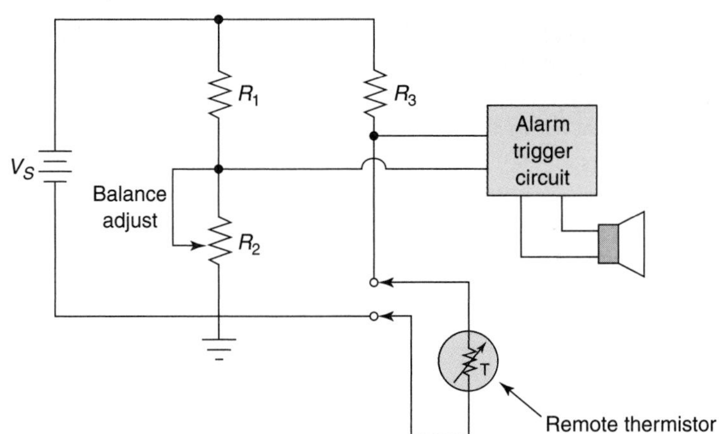

Figure 6.35

4. Describe the operating states of a Wheatstone bridge in terms of *resistance ratios.*
5. Describe the procedure for *measuring resistance* with a Wheatstone bridge.
6. Describe the relationships used to:
   a. Derive the *Y equivalent* of a Δ circuit.
   b. Derive the Δ *equivalent* of a *Y* circuit.

Critical Thinking

7. A *thermal resistor,* or *thermistor,* is a solid-state component whose resistance drops significantly when temperature increases over a rated range of values. The schematic in Figure 6.35 contains a thermistor. Explain how the circuit would respond to a change in temperature.

Key Terms

The following terms were introduced in this chapter on the pages indicated:

Bleeder current, 162
Calibrated decade box, 167
Delta (Δ) circuit, 169
Galvanometer, 166
No-load output voltage (V_{NL}), 158
Parallel-equivalent circuit, 153
Photoconductive cell, 168
Series-equivalent circuit, 153
Variable voltage divider, 162
Voltage-divider stability, 161
Voltmeter loading, 163
Wheatstone bridge, 165
Wye (*y*) circuit, 169

Practice Problems

1. Calculate the component voltages and branch currents for the circuit shown in Figure 6.36, along with the values of I_T and R_T.

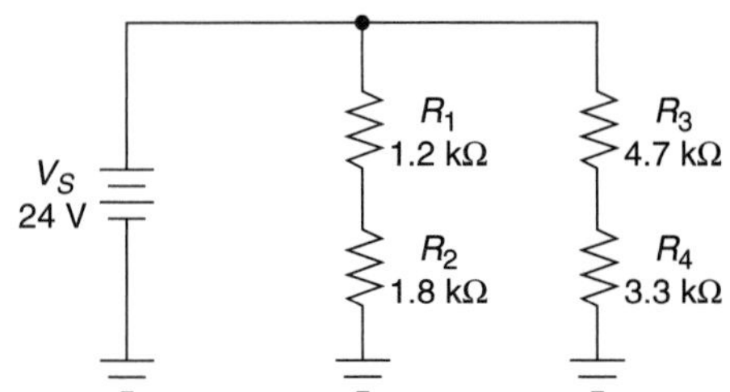

Figure 6.36

2. Calculate the component voltages and branch currents for the circuit shown in Figure 6.37, along with the values of I_T and R_T.

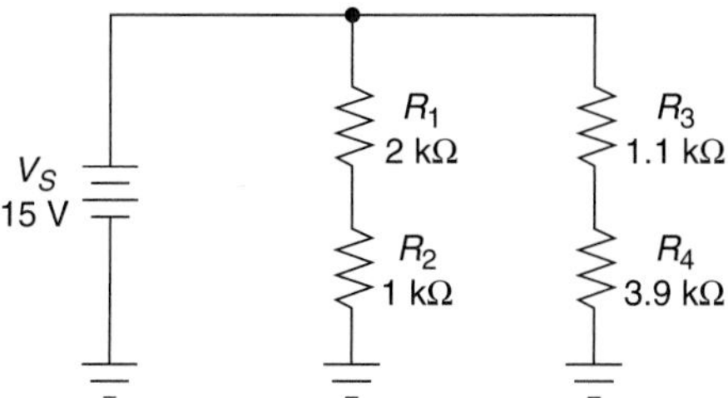

Figure 6.37

3. Calculate the component voltages and branch currents for the circuit shown in Figure 6.38, along with the values of I_T and R_T.

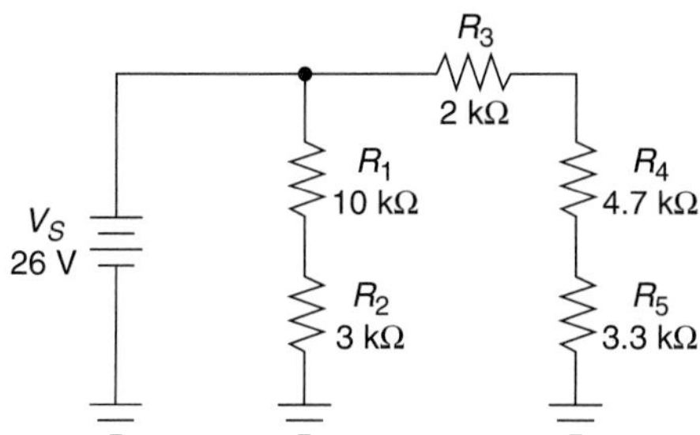

Figure 6.38

4. Calculate the component voltages and branch currents for the circuit shown in Figure 6.39, along with the values of I_T and R_T.

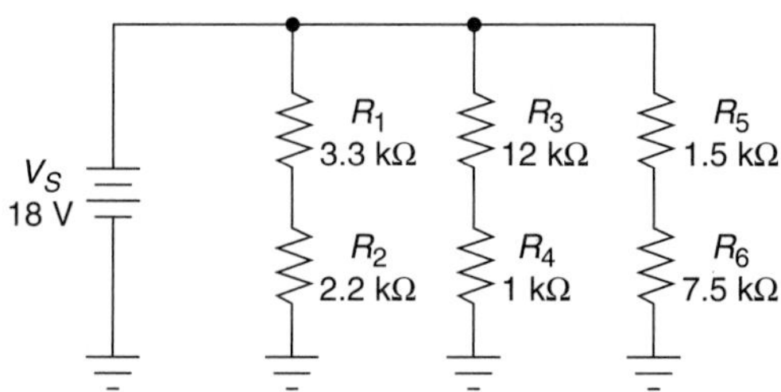

Figure 6.39

5. Calculate the component currents and loop voltages for the circuit shown in Figure 6.40, along with the values of I_T and R_T.

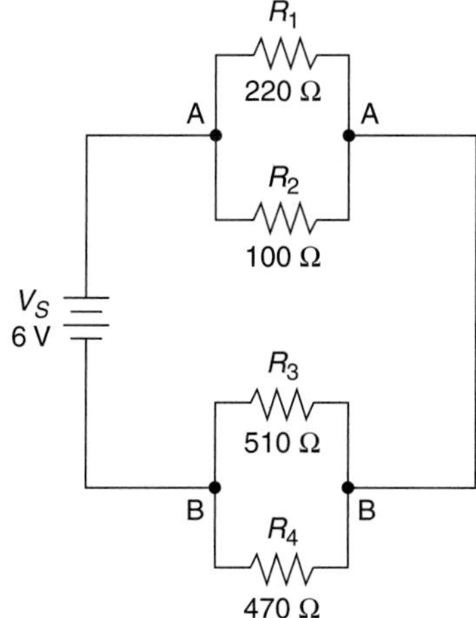

Figure 6.40

6. Calculate the component currents and loop voltages for the circuit shown in Figure 6.41, along with the values of I_T and R_T.

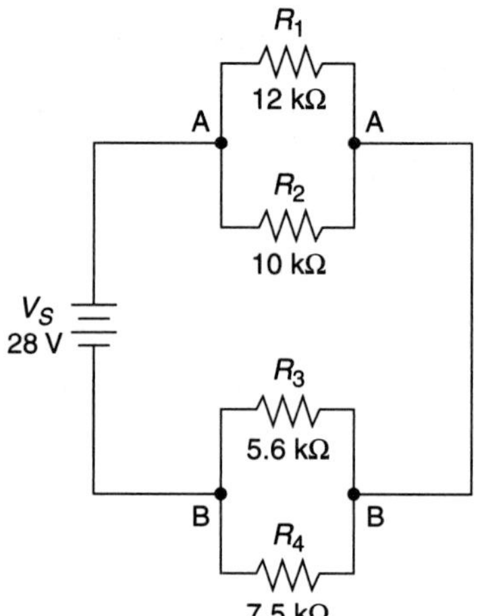

Figure 6.41

7. Calculate the component currents and loop voltages for the circuit shown in Figure 6.42, along with the values of I_T and R_T.

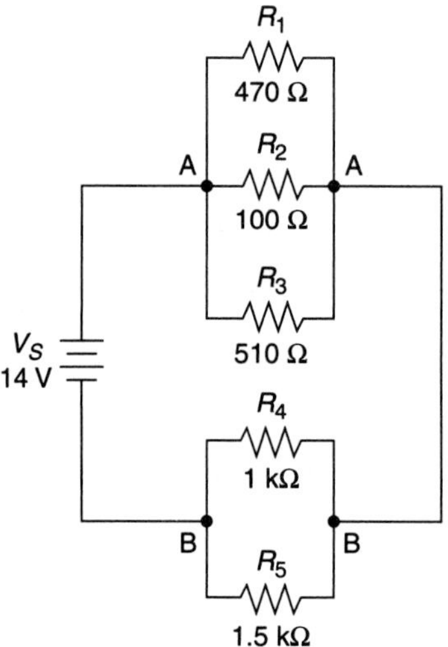

Figure 6.42

8. Calculate the component currents and loop voltages for the circuit shown in Figure 6.43, along with the values of I_T and R_T.

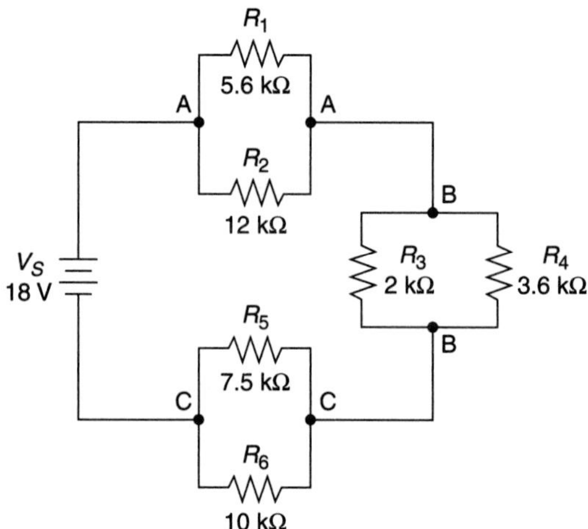

Figure 6.43

9. Derive the series equivalent of the circuit shown in Figure 6.44a.

10. Derive the series equivalent of the circuit shown in Figure 6.44b.

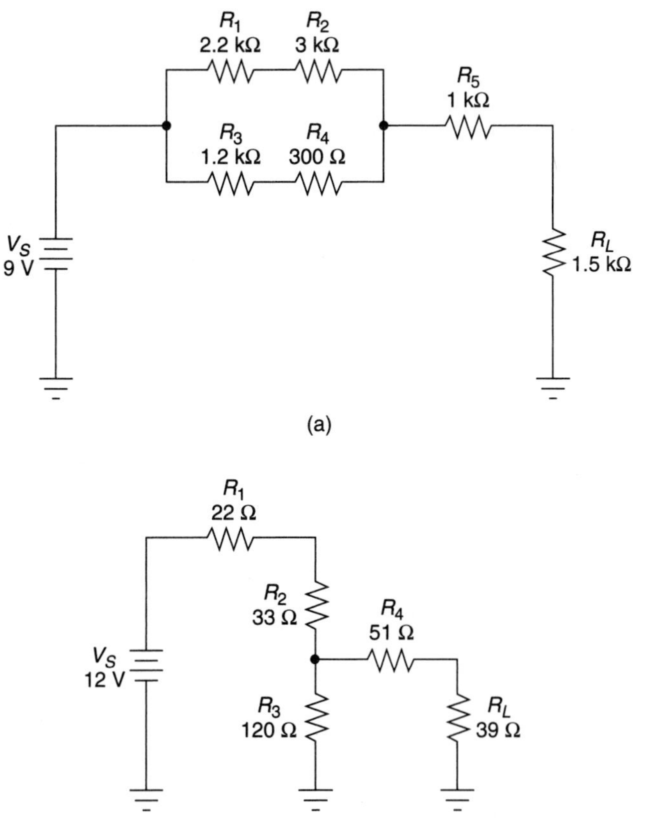

Figure 6.44

11. Derive the parallel equivalent of the circuit shown in Figure 6.45a.

12. Derive the parallel equivalent of the circuit in Figure 6.45b.

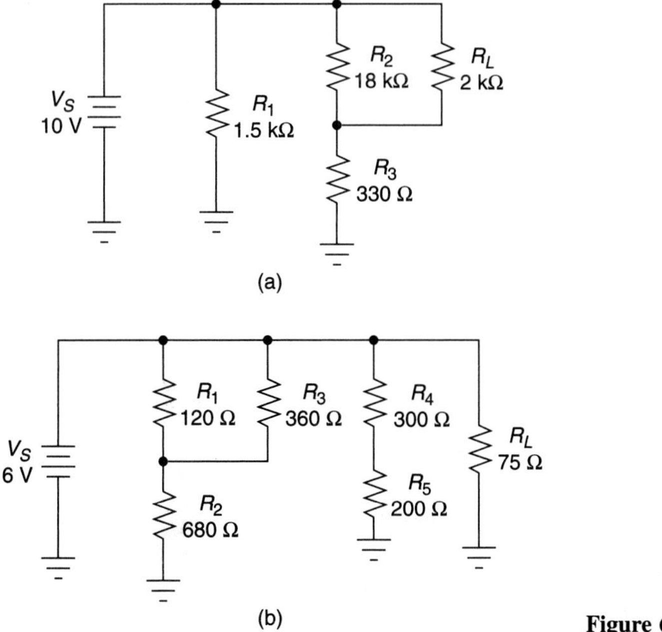

Figure 6.45

13. Determine the value of load voltage (V_L) for the circuit shown in Figure 6.44a.

14. Determine the value of load voltage (V_L) for the circuit shown in Figure 6.44b.

15. Determine the value of load current (I_L) for the circuit shown in Figure 6.45a.

16. Determine the value of load current (I_L) for the circuit shown in Figure 6.45b.

17. Determine the value of load power (P_L) for the circuit shown in Figure 6.46a.

18. Determine the value of load power (P_L) for the circuit shown in Figure 6.46b.

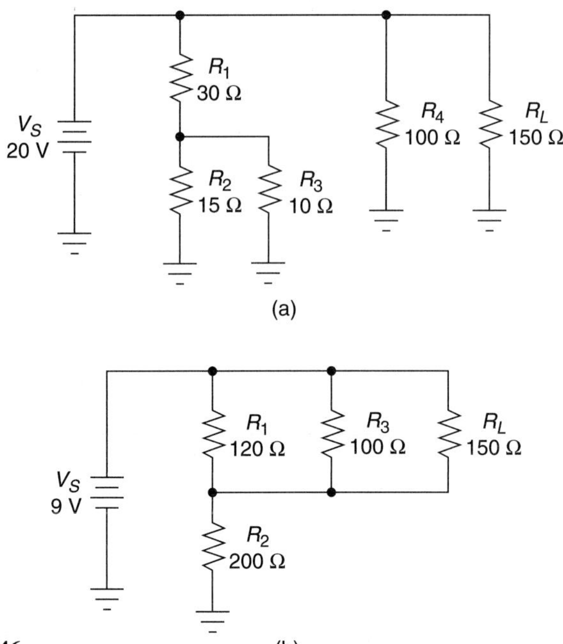

Figure 6.46

19. Determine the values of V_L and V_{NL} for the circuit shown in Figure 6.47a.

20. Determine the values of V_L and V_{NL} for the circuit shown in Figure 6.47b.

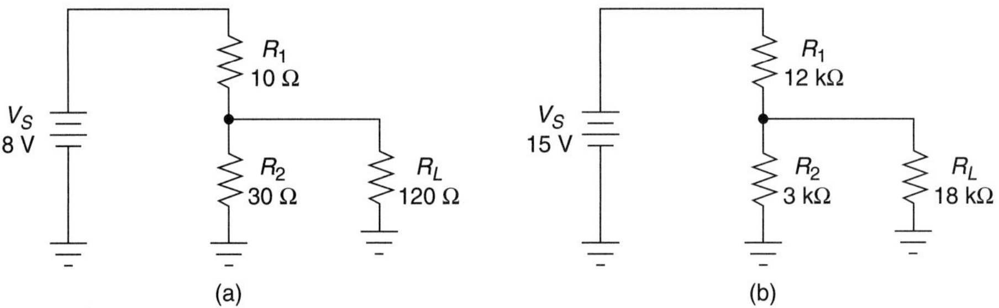

Figure 6.47

21. Determine the normal range of output voltages for the circuit shown in Figure 6.48a.

22. Determine the normal range of output voltages for the circuit shown in Figure 6.48b.

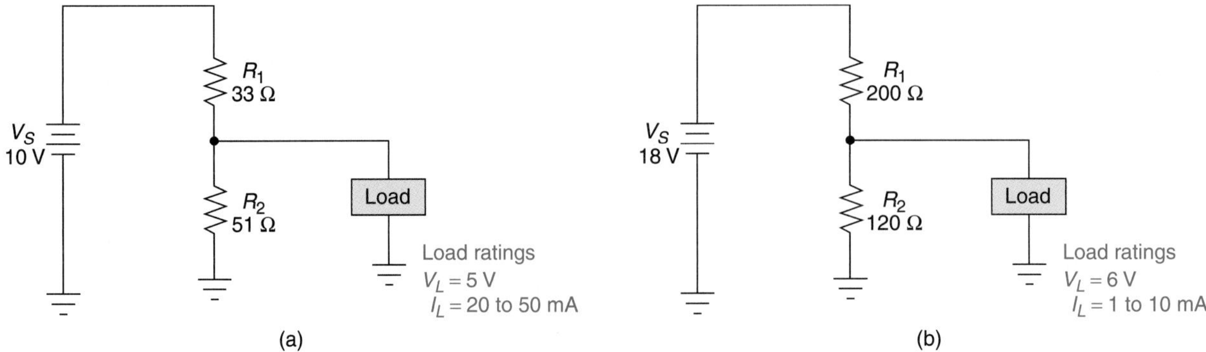

Figure 6.48

23. Determine the normal range of output voltages for the circuit shown in Figure 6.49a.

24. Determine the normal range of output voltages for the circuit shown in Figure 6.49b.

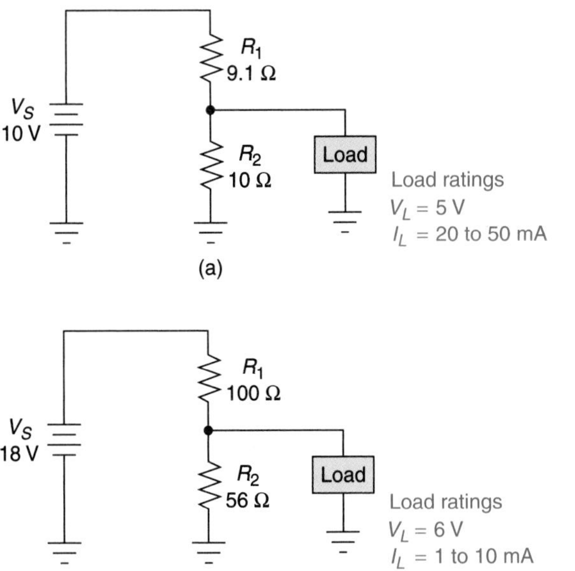

Figure 6.49

25. Determine the value of V_2 for the circuit shown in Figure 6.50a. Then, determine the reading that will be produced when the voltmeter is connected as shown in Figure 6.50b.

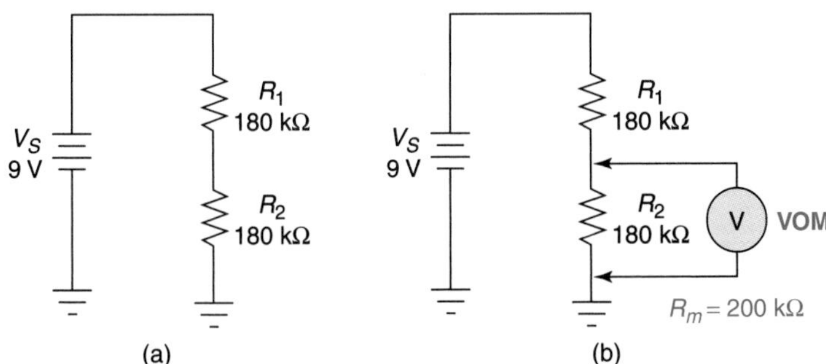

Figure 6.50

26. Repeat problem 25 for the circuit shown in Figure 6.51.

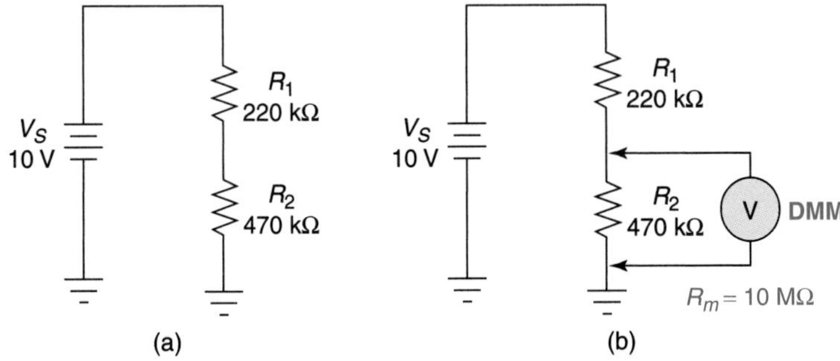

Figure 6.51

27. Perform a Δ-to-Y conversion on the circuit in Figure 6.52a.

28. Perform a Δ-to-Y conversion on the circuit in Figure 6.52b.

29. Perform a Y-to-Δ conversion on the circuit in Figure 6.52c.

30. Perform a Y-to-Δ conversion on the circuit in Figure 6.52d.

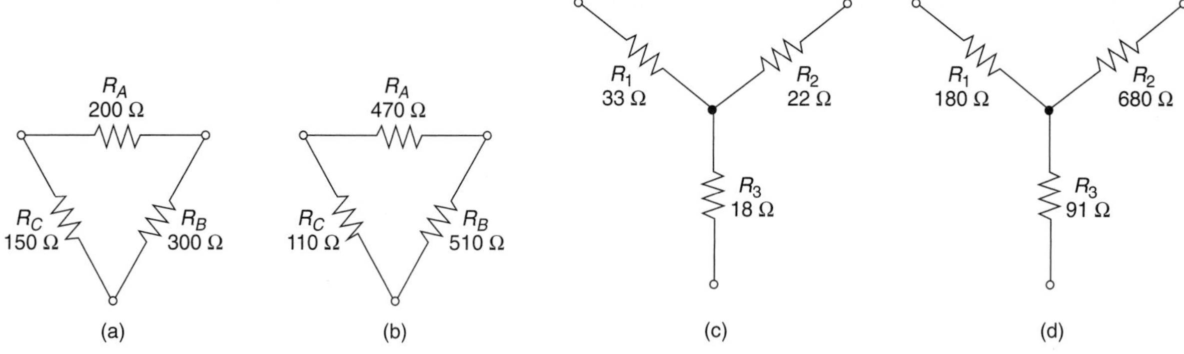

Figure 6.52

Looking Back

These problems relate to material presented in earlier chapters. The chapters are identified in brackets.

31. Calculate the voltage required to generate 200 mA through a 7.5 kΩ resistor. [Chapter 3]

32. The resistivity of copper is 10.37 CM-Ω/ft. What is the resistance of a 1000 ft long wire that has a diameter of 0.005 in.? [Chapter 1]

33. What is the AWG rating of the wire described in problem 32? [Chapter 2]

34. A battery is rated for a continuous output of 1.5 V at 10 mA. If three of these batteries are connected in series, what are the current and voltage capabilities of the resulting voltage source? [Chapter 2]

35. A series circuit has the following values: $V_S = 10$ V, $R_1 = 2.2$ kΩ, and $R_2 = 1$ kΩ. Calculate the value of source power for the circuit. [Chapter 5]

36. A parallel circuit has the following values: $V_S = 4$ V, $R_1 = 180$ Ω, and $R_2 = 330$ Ω. Calculate the value of source power for the circuit. [Chapter 5]

Pushing the Envelope

37. For the circuit shown in Figure 6.53, determine the adjusted value of R_2 that will cause Load 1 to run at its rated values.

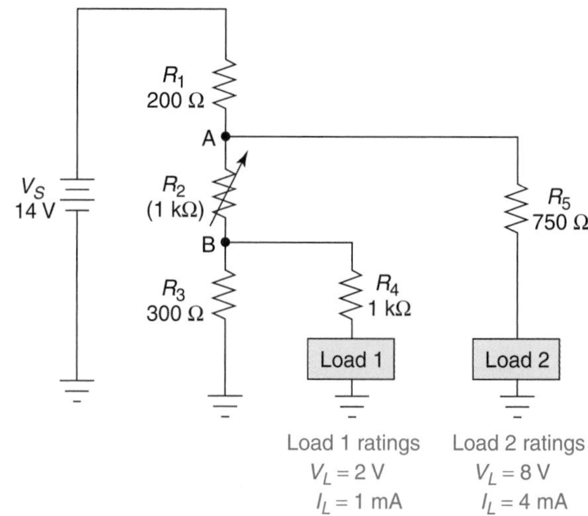

Figure 6.53

38. Determine the value of V_6 for the circuit shown in Figure 6.54.

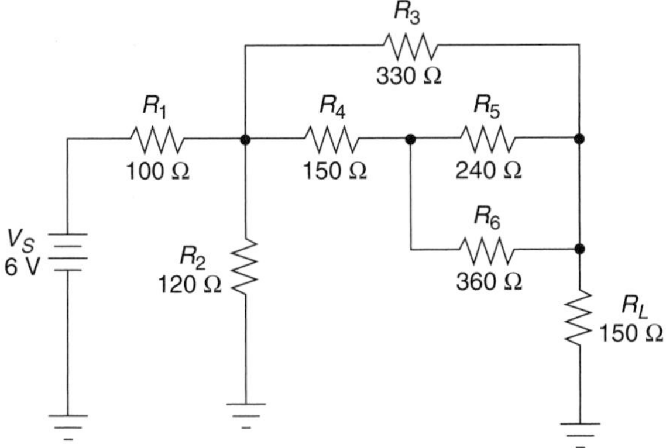

Figure 6.54

39. Figure 6.55 shows a simplified block diagram for a basic AM receiver. The power supply is a circuit that supplies the dc voltages required for each block to function. Each block performs a specific operation on the input signal received from the antenna. If you treat each block as a single component, is the receiver a series, parallel, or series-parallel circuit? Explain your answer.

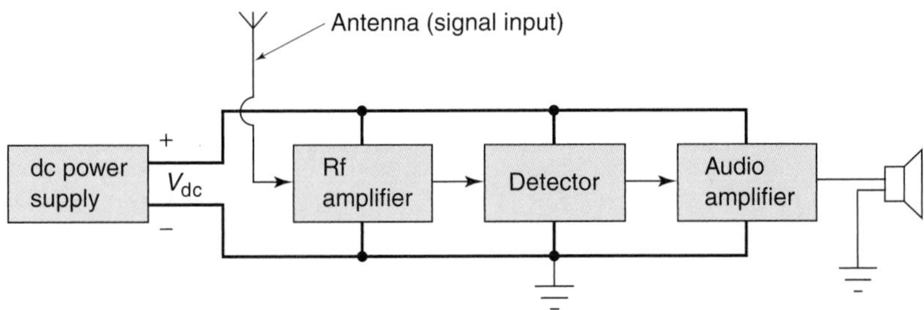

Figure 6.55

Answers to the Example Practice Problems

6.1 $I_A = 20$ mA, $V_1 = 4.4$ V, $V_2 = 6.6$ V, $P_1 = 88$ mW, $P_2 = 132$ mW, $I_B = 50$ mA, $V_3 = 6$ V, $V_4 = 5$ V, $P_3 = 300$ mW, $P_4 = 250$ mW, $I_T = 70$ mA, $R_T = 157\ \Omega$, $P_T = 770$ mW

6.2 $R_A = 60\ \Omega$, $R_B = 72\ \Omega$, $V_A = 150$ mV, $V_B = 180$ mV, $I_1 = 1.5$ mA, $I_2 = 1$ mA, $I_3 = 1$ mA, $I_4 = 1.5$ mA, $P_1 = 225\ \mu$W, $P_2 = 150\ \mu$W, $P_3 = 180\ \mu$W, $P_4 = 270\ \mu$W, $I_T = 2.5$ mA, $R_T = 132\ \Omega$, $P_T = 825\ \mu$W

6.3 17 mW

6.4 See Figure 6.56

6.5 3.99 mW

6.7 7.2 V

6.8 $V_L = 7.5$ V, $V_{NL} = 9$ V

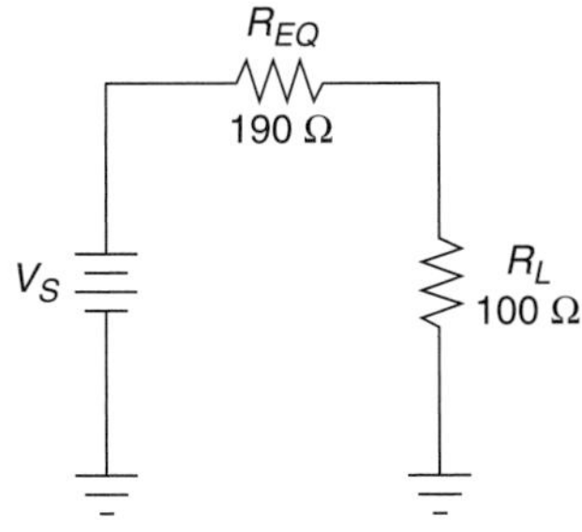

$R_{EQ} = (R_1 + R_2) \| (R_3 + R_4)$ **Figure 6.56**

6.9 1.98 V to 5.7 V

6.10 9.01 V to 9.37 V

6.11 The DMM will provide a reading of 1.59 V. This is very close to the calculated value of V_2.

Answers to the Section 6.1 Review Problems

4. a. True
   b. False
   c. True

5. a. $I_4 = I_5 + I_6$
   b. $I_1 = I_2 + I_4$ or $I_1 = I_3 + I_4$
   c. No. It is in series with the parallel combination of the two branches.
   d. Any of the following combinations could be added to determine the value of V_S: $(V_1 + V_2 + V_3)$, $(V_1 + V_4 + V_5)$, or $(V_1 + V_4 + V_6)$.

Multisim File References

Figure	File Reference
6.4	ETF6_4.msm
6.13	ETF6_13.msm
6.24	ETF6_24.msm
6.32	ETF6_32.msm

chapter 7

Circuit Analysis Techniques

Objectives

After studying the material in this chapter, you should be able to:

1. Explain how the *superposition theorem* is used to analyze multisource circuits.
2. Perform the analysis of a two-source circuit using the superposition theorem.
3. List and contrast the characteristics of *ideal* and *practical* voltage and current sources.
4. Convert a voltage source to a current source, and vice versa.
5. Describe the purposes served by the load analysis of a circuit.
6. Discuss *Thevenin's theorem* and the purpose it serves.
7. Determine the values of *Thevenin voltage* (V_{TH}) and *Thevenin resistance* (R_{TH}) for a series-parallel circuit.
8. Derive the *Thevenin equivalent* of a given series-parallel circuit.
9. Use Thevenin's theorem to determine the range of output voltages for a series-parallel circuit with a variable load.
10. Use Thevenin's theorem to determine the maximum possible value of load power for a series-parallel circuit with a variable load.
11. Use Thevenin's theorem to analyze the operation of multiload circuits.
12. Use Thevenin's theorem to analyze the operation of bridge circuits.
13. Discuss *Norton's theorem* and the purpose it serves.
14. Determine the values of *Norton current* (I_N) and *Norton resistance* (R_N) for a series-parallel circuit.
15. Analyze the operation of a series-parallel circuit using a Norton equivalent circuit.
16. Describe and perform *Thevenin-to-Norton* and *Norton-to-Thevenin* conversions.

In Chapter 6, you were introduced to the fundamentals of series-parallel circuit operation and analysis. In this chapter, we're going to look at some **network theorems.** These theorems are most commonly used to:

- Analyze circuits that cannot be solved easily using the techniques established in Chapter 6.
- Analyze series-parallel circuits that contain more than one voltage (or current) source.
- Predict how a series-parallel circuit will respond to changes in load demand.

7.1 Superposition

OBJECTIVE 1 ➤ All of the series-parallel circuits we've discussed to this point have contained a single voltage source. A **multisource circuit** is *a circuit that contains more than one voltage and/or current source.* Several examples of multisource circuits are shown in Figure 7.1.

The **superposition theorem** states that *the response of a circuit to more than one source can be determined by analyzing the circuit's response to each source (alone) and then combining the results.* For example, consider the circuit shown in Figure 7.2. According to the superposition theorem, we can determine the component voltages by:

1. Analyzing the circuit as if V_A were the only voltage source.
2. Analyzing the circuit as if V_B were the only voltage source.
3. Combining the results from steps 1 and 2.

When using the superposition theorem to solve a multisource circuit, it is very helpful to draw the original circuit as several single-source circuits. For example, the circuit in Figure 7.2 can be redrawn as shown in Figure 7.3. Note that:

- Each circuit is identical to the original, except for the fact that one source has been removed.
- When a voltage source is removed, it is replaced with a wire. This simulates the ideal value of $R_S = 0\ \Omega$ for the missing source.

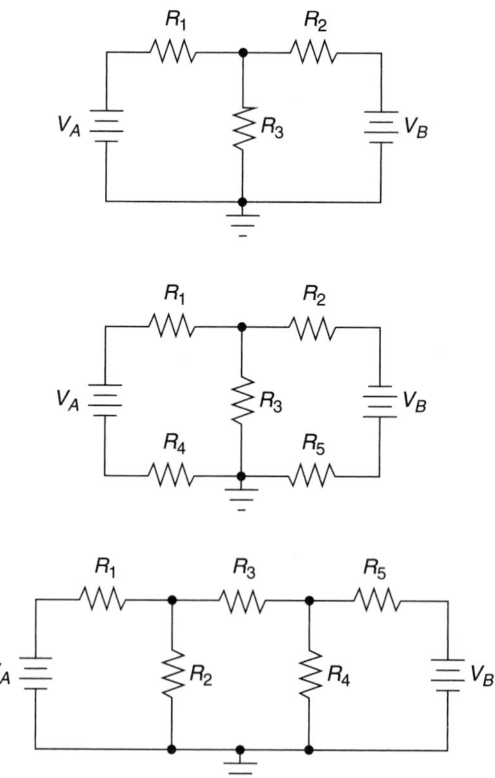

Figure 7.1 Some basic multisource circuits.

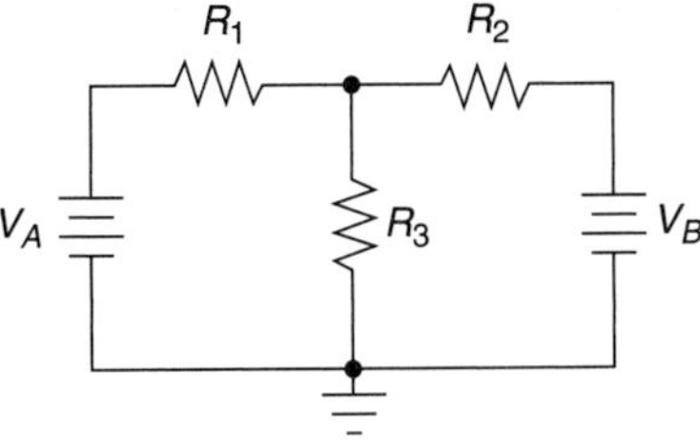

Figure 7.2

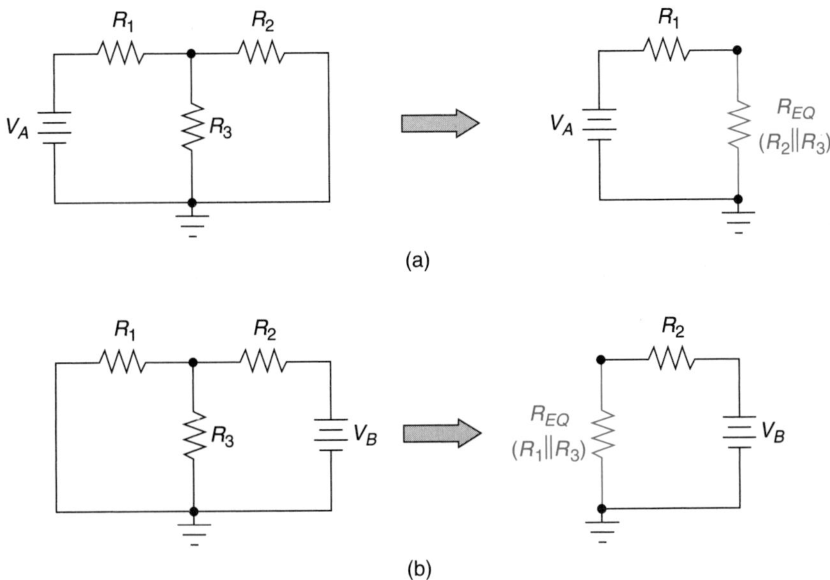

Figure 7.3

When drawn as shown, the analysis of the circuit for each voltage source becomes a problem like those we covered in Chapter 6. For example, the circuit in Figure 7.3a can be simplified to a series-equivalent by combining R_2 and R_3 into a parallel-equivalent resistance. The circuit in Figure 7.3b is simplified to a series-equivalent in the same fashion: by combining R_2 and R_3 into a parallel-equivalent resistance.

After calculating the voltage values for the circuits shown in Figure 7.3, we combine the values to obtain the actual circuit voltages. However, care must be taken to ensure that the voltages are combined properly. For example, Figures 7.4a and 7.4b show the polarities of the voltages that are produced by their respective sources. If we combine these voltages as shown in Figure 7.4c, you can see that:

1. V_{1A} and V_{1B} are opposing voltages (as indicated by the mismatch in the polarity signs), and the actual component voltage equals *the difference between the two.*

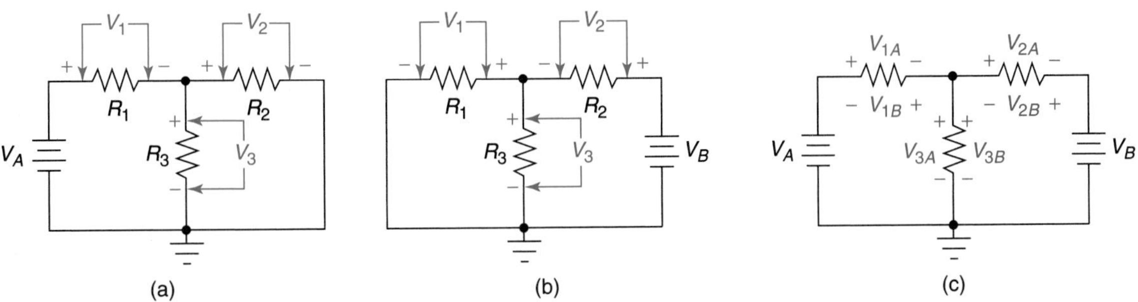

Figure 7.4

2. V_{2A} and V_{2B} are also opposing voltages. Again, the actual component voltage equals the difference between the two.
3. V_{3A} and V_{3B} are aiding voltages (as indicated by the matching polarity signs), and the actual component voltage equals *the sum of the two.*

OBJECTIVE 2 ➤ The following example demonstrates the superposition analysis of a multisource circuit.

Example 7.1

Determine the values of V_1, V_2, and V_3 for the circuit shown in Figure 7.5a.

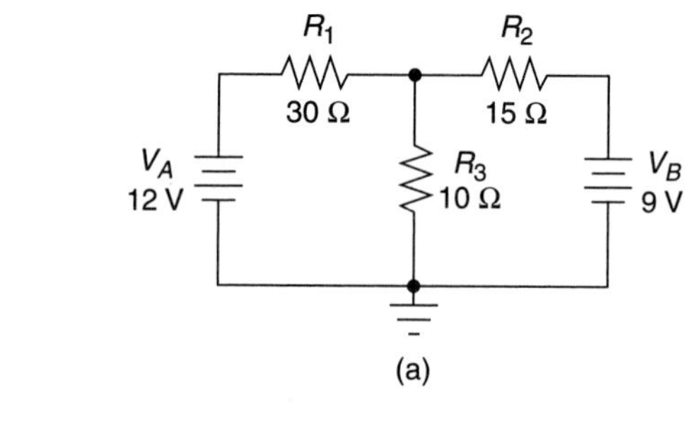

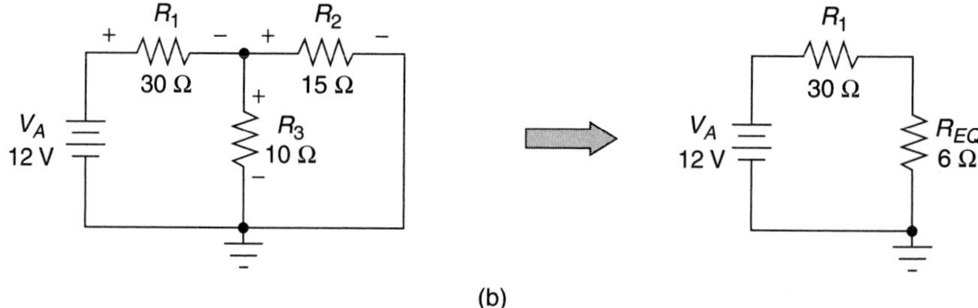

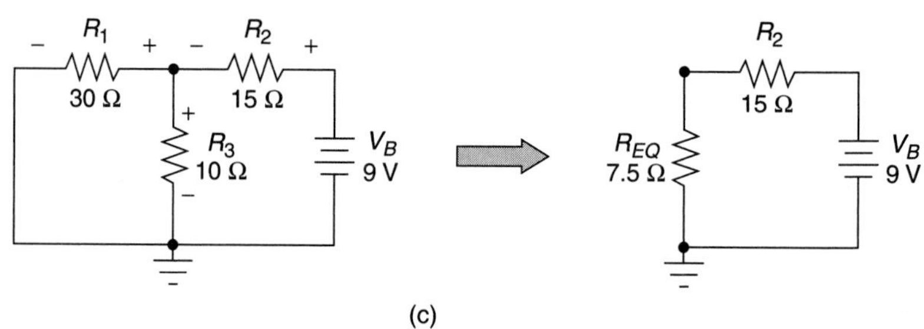

Figure 7.5

Solution: The first step is to split the circuit into two single-source circuits. These circuits (along with their series equivalents) are shown in Figures 7.5b and 7.5c. For Figure 7.5b,

$$R_{EQ} = R_2 \| R_3 = 15\ \Omega \| 10\ \Omega = 6\ \Omega$$

and

$$V_{EQ} = V_A \frac{R_{EQ}}{R_{EQ} + R_1} = (12\ \text{V}) \frac{6\ \Omega}{36\ \Omega} = 2\ \text{V}$$

Since this voltage is across the parallel combination of R_2 and R_3,

$$V_{2A} = V_{3A} = 2\ \text{V}$$

and

$$V_1 = V_A - V_{EQ} = 12\ \text{V} - 2\ \text{V} = 10\ \text{V}$$

At this point, we have calculated the following values:

$$V_{1A} = 10\ \text{V} \qquad V_{2A} = 2\ \text{V} \qquad V_{3A} = 2\ \text{V}$$

Using the same approach on the circuit in Figure 7.5c, we get

$$R_{EQ} = R_1 \| R_3 = 30\ \Omega \| 10\ \Omega = 7.5\ \Omega$$

$$V_{EQ} = V_B \frac{R_{EQ}}{R_{EQ} + R_1} = (9\text{ V}) \frac{7.5\ \Omega}{22.5\ \Omega} = 3\text{ V}$$

Since this voltage is across the parallel combination of R_1 and R_3,

$$V_{1B} = V_{3B} = 3\text{ V}$$

and

$$V_2 = V_B - V_{EQ} = 9\text{ V} - 3\text{ V} = 6\text{ V}$$

For the circuit in Figure 7.5c, we have calculated the following values:

$$V_{1B} = 3\text{ V} \qquad V_{2B} = 6\text{ V} \qquad V_{3B} = 3\text{ V}$$

Finally, we can determine the actual component voltages by combining the A and B results. If you compare the polarities of V_{1A} and V_{1B} in Figure 7.5, you can see that they are opposing. Therefore,

$$V_1 = V_{1A} - V_{1B} = 10\text{ V} - 3\text{ V} = 7\text{ V}$$

Figure 7.5 also shows V_{2A} and V_{2B} to be opposing. Therefore,

$$V_2 = V_{2B} - V_{2A} = 6\text{ V} - 2\text{ V} = 4\text{ V}$$

Since V_{3A} and V_{3B} are aiding voltages, the actual value of V_3 is found as

$$V_3 = V_{3A} - V_{3B} = 2\text{ V} + 3\text{ V} = 5\text{ V}$$

These are the component voltages for the circuit in Figure 7.5a.

Practice Problem 7.1

A circuit like the one shown in Figure 7.5a has the following values: $V_A = 14$ V, $V_B = 7$ V, $R_1 = 100\ \Omega$, $R_2 = 150\ \Omega$, and $R_3 = 150\ \Omega$. Calculate the values of V_1, V_2, and V_3 for the circuit.

The results from Example 7.1 can be used to demonstrate several important points. Figure 7.6a shows the original circuit with the component voltages calculated in the example. Note the polarities of V_1 and V_2. In both cases, the assumed polarity matches that of the greater voltage, as follows:

- Since $V_{1A} > V_{1B}$, the polarity of V_1 matches that of V_{1A}.
- Since $V_{2B} > V_{2A}$, the polarity of V_2 matches that of V_{2B}.

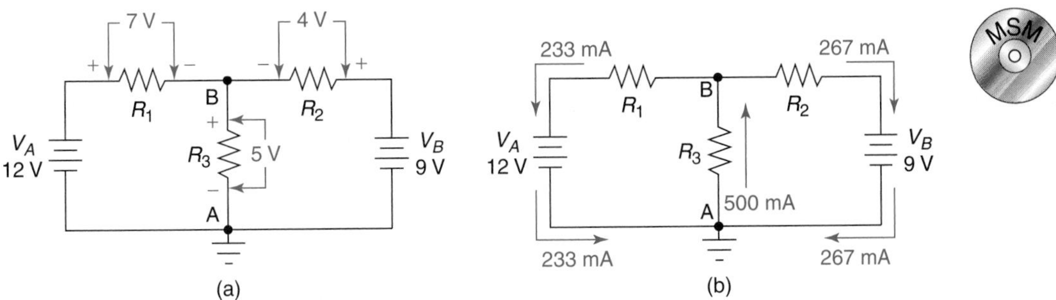

Figure 7.6 The results from Example 7.1.

Figure 7.6a also demonstrates that *the voltages in each loop fulfill Kirchhoff's voltage law,* as follows:

$$V_A = V_1 + V_3 = 7\text{ V} + 5\text{ V} = 12\text{ V}$$

and

$$V_B = V_2 + V_3 = 4\text{ V} + 5\text{ V} = 9\text{ V}$$

The value in solving the Kirchhoff's loop equations is that *they verify that all our circuit calculations were correct.* Had we made any mistakes in solving for the component voltages, at least one of the loop equations would not have worked.

Figure 7.6b shows the circuit currents and their values. Note that the value of each current was found using Ohm's law and the values shown in Figure 7.6a. Using the current values shown, the overall operation of the circuit can be described as follows:

- V_A generates a total current of 233 mA.
- V_B generates a total current of 267 mA.
- At point A, the source currents combine. The sum of these currents (500 mA) passes through R_3.
- At point B, the currents split. Each current returns to its source via the series resistor.

Figure 7.6b also demonstrates that the circuit operates according to Kirchhoff's current law. As you can see, the total current entering a given point or component equals the total current leaving that point or component.

Even in a multisource circuit, the fundamental laws of circuit operation hold true. Using these laws provides you with a means of quickly checking the results of any circuit analysis problem.

Section Review

1. What is the *superposition theorem*? To what type of circuit does it apply?
2. Why is it helpful to draw a multisource circuit as several single-source circuits when using the superposition theorem?
3. When calculating the component voltages in a multisource circuit, how can you perform a quick check of your results?

Critical Thinking

4. Using the values shown in Figure 7.6a:
   a. Show that the voltages around the outer loop (V_A, V_B, V_1, and V_2) conform to Kirchhoff's voltage law.
   b. Show that the sum of the resistor power-dissipation values equals the sum of the source power values (P_A and P_B).

7.2 Voltage and Current Sources

Some circuit analysis problems require the ability to convert a voltage source to an equivalent current source, or vice versa. As you will see, these conversions play a role in some problems presented later in this chapter.

Voltage Sources

OBJECTIVE 3 ➤ In Chapter 4, you were introduced to the operating characteristics of voltage sources. These characteristics are summarized in Figure 7.7. As shown in the figure, the primary difference between ideal and practical voltage sources is the effect that *source resistance* has on load voltage. (A more detailed discussion on voltage sources can be found in Section 4.2.)

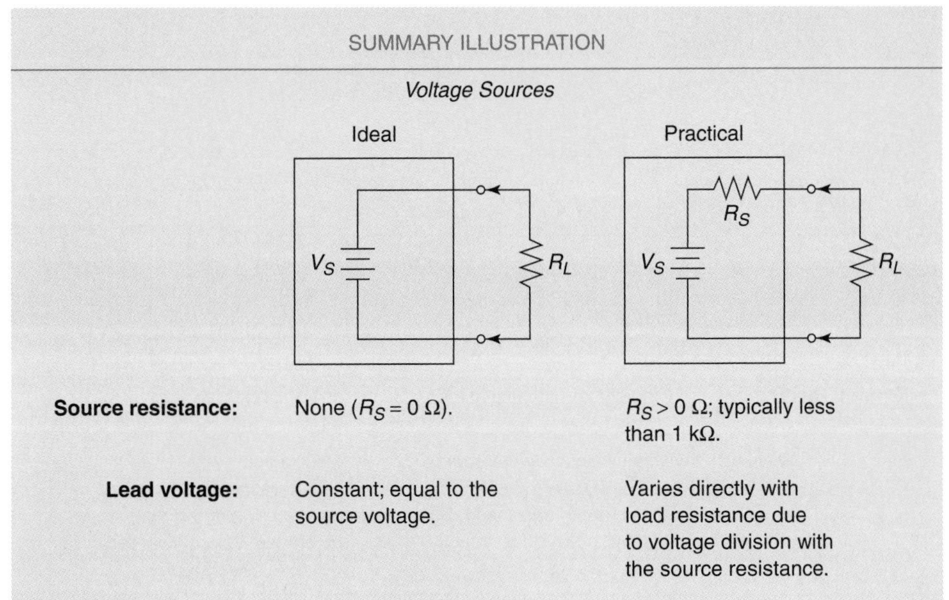

SUMMARY ILLUSTRATION
Voltage Sources
Ideal
Practical
V_S
R_L
R_S
Source resistance:
None ($R_S = 0\ \Omega$).
$R_S > 0\ \Omega$; typically less than 1 kΩ.
Lead voltage:
Constant; equal to the source voltage.
Varies directly with load resistance due to voltage division with the source resistance.

Figure 7.7

Current Sources

In Chapter 5, you were introduced to the operating characteristics of current sources. These characteristics are summarized in Figure 7.8. As shown in the figure, the primary difference between ideal and practical current sources is the effect that *source resistance* has on load current. (A more detailed discussion on current sources can be found in Section 5.2.)

Voltage and Current Sources: A Comparison

The characteristics of voltage and current sources are compared in Figure 7.9. As shown, the internal resistance of the voltage source causes the circuit to act as a *voltage divider.*

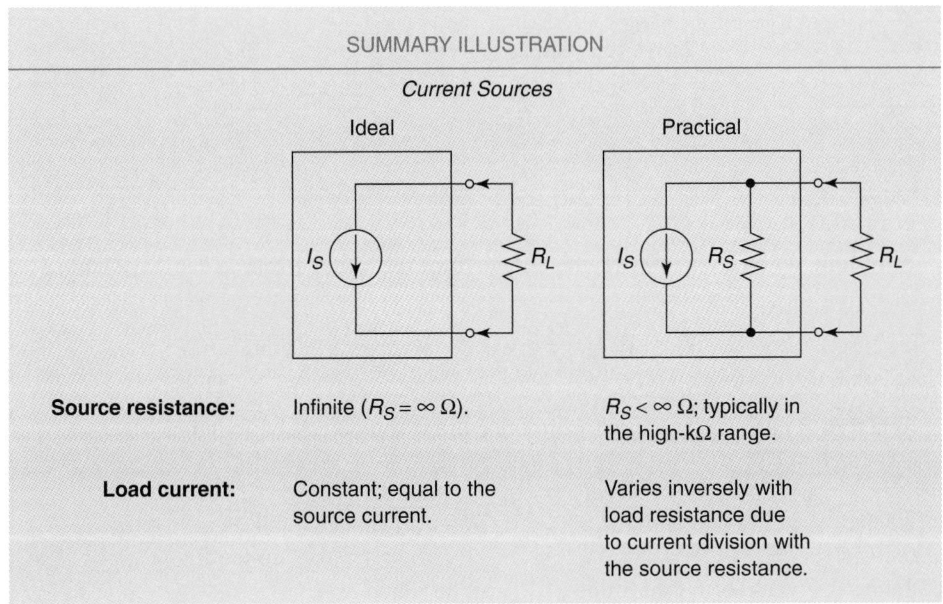

SUMMARY ILLUSTRATION
Current Sources
Ideal
Practical
I_S
R_L
R_S
Source resistance:
Infinite ($R_S = \infty\ \Omega$).
$R_S < \infty\ \Omega$; typically in the high-kΩ range.
Load current:
Constant; equal to the source current.
Varies inversely with load resistance due to current division with the source resistance.

Figure 7.8

SUMMARY ILLUSTRATION

Voltage and Current Sources: A Comparison

	Voltage Source	Current Source
	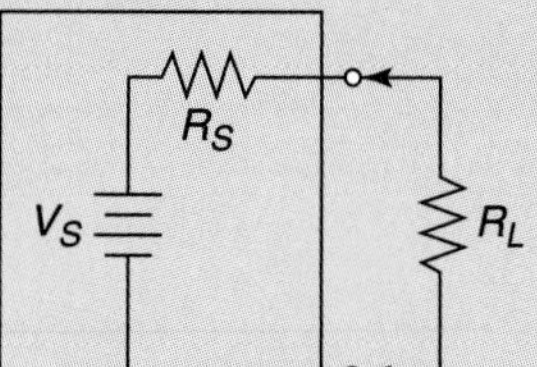	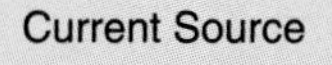 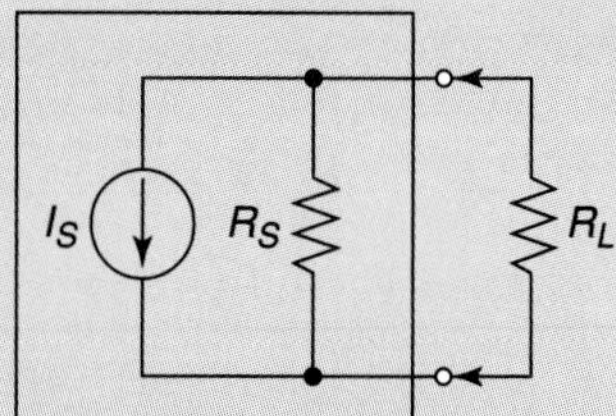
Source resistance:	Ideally zero; typically less than 1 kΩ.	Ideally infinite; typically in the high-kΩ range.
Output:	Determined by the source voltage rating, the source resistance, and the load resistance. Varies directly with load resistance.	Determined by the source current rating, the source resistance, and the load resistance. Varies inversely with load resistance.

Figure 7.9

The internal resistance of the current source causes that circuit to act as a *current divider.* In each case, the appropriate "divider" relationship is used to determine the output value when a load is connected to the source.

Equivalent Voltage and Current Sources

For every voltage source, there exists an equivalent current source, and vice versa. For any value of load resistance, equivalent voltage and current sources provide the same output values, as demonstrated in Example 7.2.

Example 7.2

Figure 7.10 shows a voltage source and its current source equivalent. Determine the value of V_L provided by each of the circuits when $R_L = 400\ \Omega$.

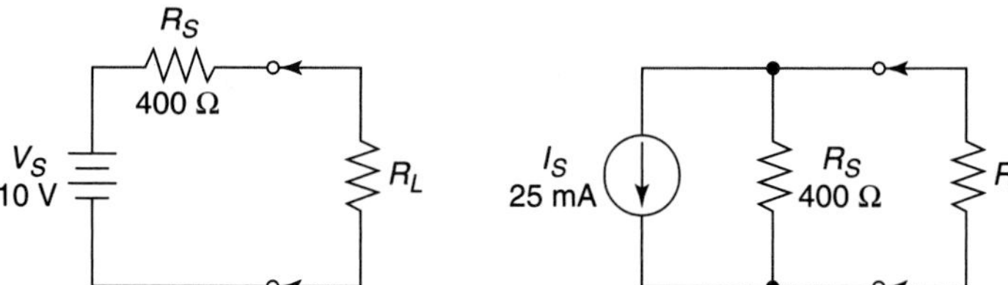

Figure 7.10

Solution: When $R_L = 400\ \Omega$, the output from the voltage source can be found as

$$V_L = V_S \frac{R_L}{R_L + R_S} = (10\text{ V})\frac{400\ \Omega}{800\ \Omega} = 5\text{ V}$$

When $R_L = 400\ \Omega$, the output from the current source can be found as

$$I_L = I_S \frac{R_L \| R_S}{R_L} = (25\text{ mA})\frac{200\ \Omega}{400\ \Omega} = 12.5\text{ mA}$$

Using this value of I_L, the value of V_L can be found as

$$V_L = I_L R_L = (12.5\text{ mA})(400\ \Omega) = 5\text{ V}$$

Practice Problem 7.2

Show that the circuits in Figure 7.10 will produce the same load voltage for values of $R_L = 100\ \Omega$.

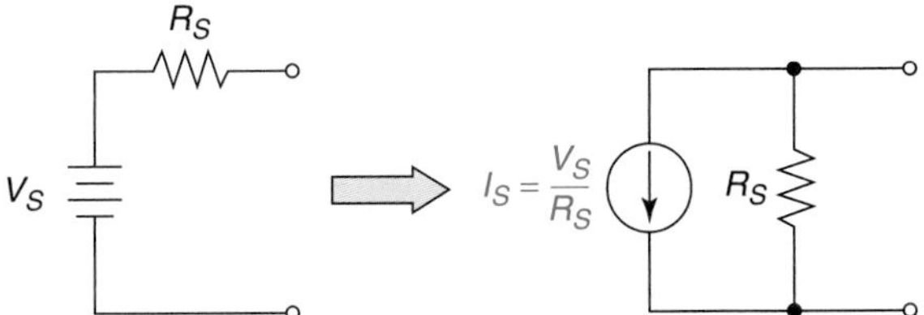

(a) Voltage source to current source

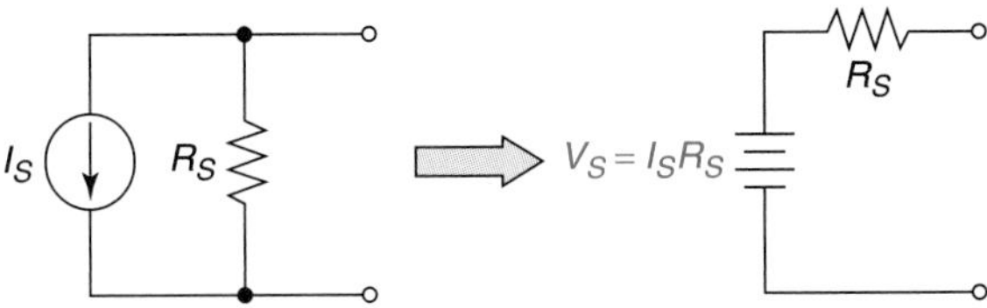

(b) Current source to voltage source

Figure 7.11 Source conversions.

Source Conversions

◀ *OBJECTIVE 4*

Source conversions are actually simple to perform. The relationships used to derive the equivalent of each type of source are shown in Figure 7.11. As you can see, the value of R_S does not change from one circuit to the other. When converting a voltage source to a current source (as shown in Figure 7.11a), the value of I_S is found as

$$I_S = \frac{V_S}{R_S} \tag{7.1}$$

Once the value of I_S has been calculated, the current source is drawn as shown in Figure 7.11a. Note that the direction of the arrow in the current source matches the direction of the current from the voltage source. The process for converting a voltage source to a current source is demonstrated further in Example 7.3.

Example 7.3

Convert the voltage source shown in Figure 7.12a to an equivalent current source.

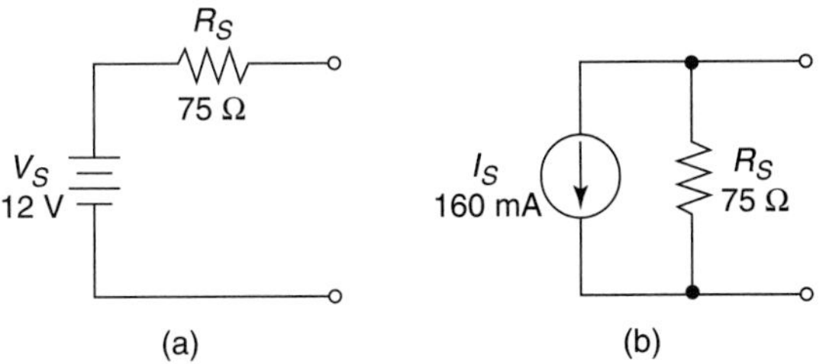

Figure 7.12

Solution: The value of source resistance remains the same as shown in Figure 7.12a, 75 Ω. The value of I_S is found using equation (7.1), as follows:

$$I_S = \frac{V_S}{R_S} = \frac{12\text{ V}}{75\ \Omega} = 160\text{ mA}$$

Using the calculated values, the current source is drawn as shown in Figure 7.12b.

Practice Problem 7.3

A voltage source like the one shown in Figure 7.12a has values of $V_S = 8$ V and $R_S = 16\ \Omega$. Convert the voltage source to an equivalent current source.

When converting a current source to a voltage source (as shown in Figure 7.11b), the value of source voltage is found as

$$V_S = I_S R_S \tag{7.2}$$

Once the value of V_S is known, the voltage source is drawn as shown in Figure 7.11b. Again, care must be taken to ensure that the polarity of the voltage source matches the direction of current indicated in the current source. The process for converting a current source to a voltage source is demonstrated further in Example 7.4.

Example 7.4

Convert the current source shown in Figure 7.13a to an equivalent voltage source.

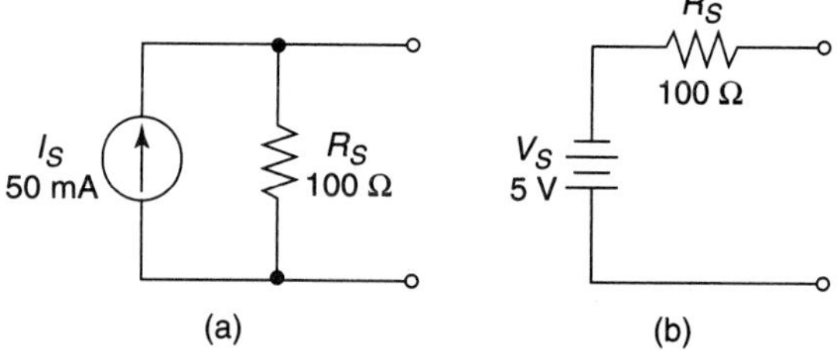

Figure 7.13

Solution: The value of the source resistance remains the same as shown in Figure 7.13a, 100 Ω. The value of V_S is found using equation (7.2), as follows:

$$V_S = I_S R_S = (50\ \text{mA})(100\ \Omega) = 5\ \text{V}$$

Using the calculated values, the voltage source is drawn as shown in Figure 7.13b.

Practice Problem 7.4

A current source like the one in Figure 7.13 has values of $I_S = 40$ mA and $R_S = 150\ \Omega$. Convert the current source to an equivalent voltage source.

Section Review

1. In terms of source resistance, how does the *ideal* voltage source differ from the *practical* voltage source?
2. In terms of load voltage, how does the *ideal* voltage source differ from the *practical* voltage source?
3. In terms of source resistance, how does the *ideal* current source differ from the *practical* current source?
4. In terms of load current, how does the *ideal* current source differ from the *practical* current source?
5. Compare and contrast voltage sources and current sources.
6. Describe the process for converting a voltage source to a current source.
7. Describe the process for converting a current source to a voltage source.

7.3 Thevenin's Theorem

The **load analysis** of a circuit is *an analysis that provides a means of predicting how the circuit will respond to a change in load resistance.* Using the analysis techniques we have established so far, the load analysis of a series-parallel circuit could be long and tedious. For example, consider the circuit shown in Figure 7.14. Solving this problem for each value of load resistance could take quite some time. Fortunately, the type of problem represented in Figure 7.14 can be simplified by using *Thevenin's theorem.* ◄ *OBJECTIVE 5*

Thevenin's theorem states that *any resistive circuit or network, no matter how complex, can be represented as a voltage source in series with a source resistance.* For example, the series-parallel circuit shown in Figure 7.15a can be represented by the *Thevenin equivalent circuit* shown in Figure 7.15b. Note that the values in the Thevenin equivalent circuit are referred to as the *Thevenin voltage* (V_{TH}) and *Thevenin resistance* (R_{TH}). ◄ *OBJECTIVE 6*

In this section, you will learn how to derive the Thevenin equivalent of a given series-parallel circuit. First, let's take a look at how this theorem can be used to simplify the load analysis of series-parallel circuits.

The Purpose Served by Thevenin's Theorem

The value of the **Thevenin equivalent circuit** is that it *produces the same output values as the original circuit for any given value of load resistance,* as demonstrated in Example 7.5.

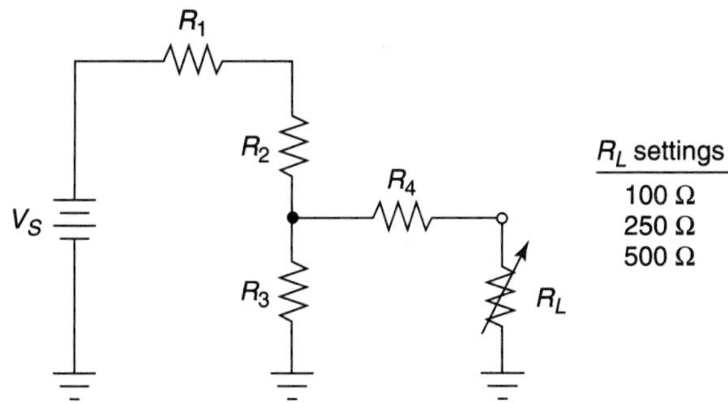

Figure 7.14

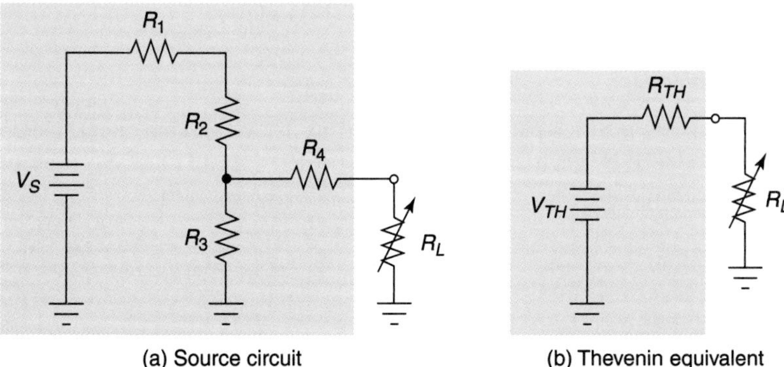

Figure 7.15 A series-parallel circuit and its Thevenin equivalent.

Example 7.5

Figure 7.16 shows a loaded voltage divider and its Thevenin equivalent circuit. Determine the load voltage produced by each circuit when $R_L = 100\ \Omega$ and when $R_L = 1\ k\Omega$.

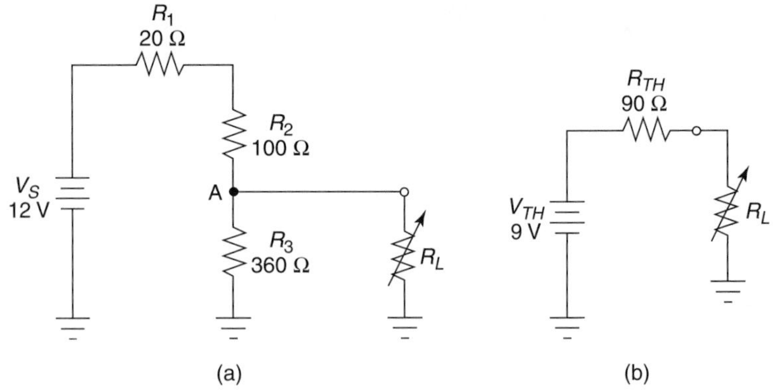

Figure 7.16

Solution: When $R_L = 100\ \Omega$, the loaded voltage divider in Figure 7.16a has the following values:

$$R_{EQ} = R_3 \| R_L = 360\ \Omega \| 100\ \Omega = 78.3\ \Omega$$

and

$$V_{EQ} = V_S \frac{R_{EQ}}{R_T} = (12\ \text{V}) \frac{78.3\ \Omega}{198.3\ \Omega} = 4.74\ \text{V}$$

When $R_L = 1\ \text{k}\Omega$, the loaded voltage divider has the following values:

$$R_{EQ} = R_3 \| R_L = 360\ \Omega \| 1\ \text{k}\Omega = 264.7\ \Omega$$

and

$$V_{EQ} = V_S \frac{R_{EQ}}{R_T} = (12\ \text{V}) \frac{264.7\ \Omega}{384.7\ \Omega} = 8.26\ \text{V}$$

Using the Thevenin equivalent circuit, we could solve the same problem as follows: When $R_L = 100\ \Omega$,

$$V_L = V_{TH} \frac{R_L}{R_L + R_{TH}} = (9\ \text{V}) \frac{100\ \Omega}{190\ \Omega} = 4.74\ \text{V}$$

When $R_L = 1\ \text{k}\Omega$,

$$V_L = V_{TH} \frac{R_L}{R_L + R_{TH}} = (9\ \text{V}) \frac{1\ \text{k}\Omega}{1.09\ \text{k}\Omega} = 8.26\ \text{V}$$

Practice Problem 7.5

Figure 7.17 shows a series-parallel circuit and its Thevenin equivalent. Determine the output produced by each circuit for values of $R_L = 150\ \Omega$ and $R_L = 30\ \Omega$.

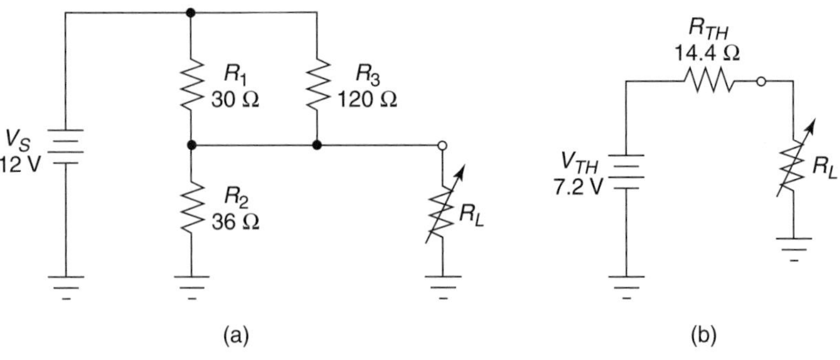

Figure 7.17

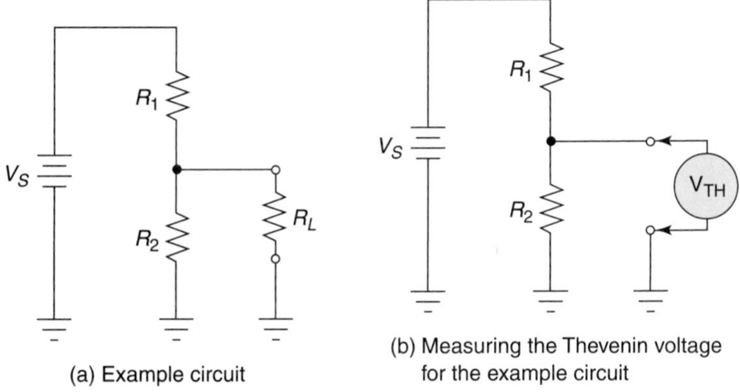

Figure 7.18 Measuring Thevenin voltage.

As you can see, the Thevenin equivalent for a series-parallel circuit allows us to determine the output voltage for any given value of load resistance both quickly and easily.

It should be stressed that the Thevenin equivalent of a series-parallel circuit is used to predict *only* the change in load values that result from a change in load resistance. If you want to know how any of the other circuit values change (such as V_1, V_2, etc.), you must use the standard analysis techniques discussed in Chapter 6.

Thevenin Voltage (V_{TH})

◄ *OBJECTIVE 7*

Deriving the Thevenin equivalent of a series-parallel circuit begins with determining the value of *Thevenin voltage* (V_{TH}). The **Thevenin voltage** for a series-parallel circuit is *the voltage present at the output terminals of the circuit when the load is removed.* Although we've given it a new name, V_{TH} is nothing more than the *no-load output voltage* (V_{NL}) for the circuit. The value of V_{TH} is measured as shown in Figure 7.18. The process for calculating the value of V_{TH} is demonstrated in Example 7.6.

Example 7.6

Determine the value of V_{TH} for the series-parallel circuit shown in Figure 7.19a.

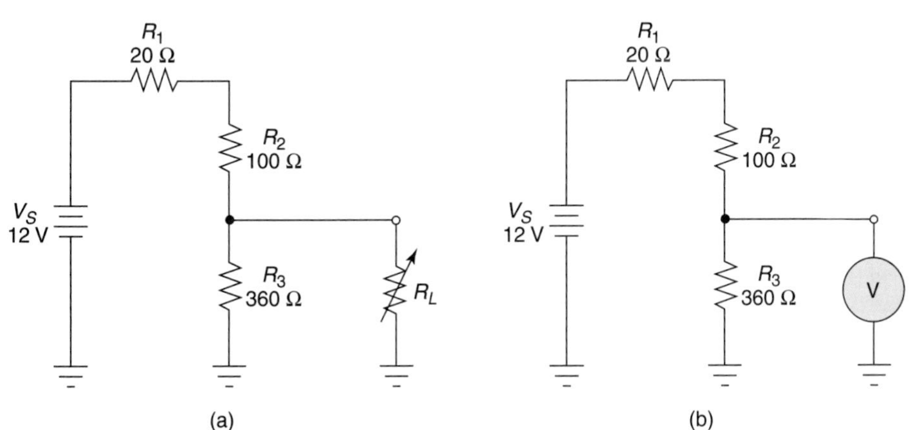

Figure 7.19

Solution: Since V_{TH} is equal to the no-load output voltage, the first step is to remove the load. This gives us the circuit shown in Figure 7.19b. For this circuit, the voltage across the open load terminals is equal to the voltage across R_3. Therefore, V_{TH} can be found as

$$V_{TH} = V_S \frac{R_3}{R_T} = (12 \text{ V}) \frac{360 \ \Omega}{480 \ \Omega} = 9 \text{ V}$$

Practice Problem 7.6

Determine the value of V_{TH} for the circuit shown in Figure 7.20.

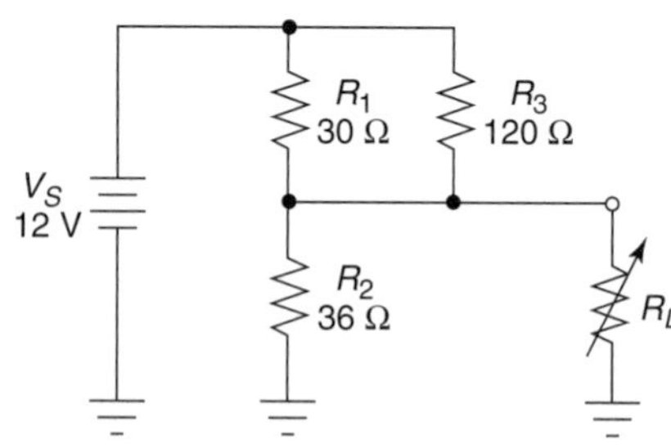

Figure 7.20

Thevenin Resistance (R_{TH})

The second step in deriving a Thevenin equivalent circuit is determining its value of *Thevenin resistance* (R_{TH}). The **Thevenin resistance** of a circuit is *the resistance measured across the output terminals with the load removed.* In essence, R_{TH} can be thought of as the *no-load output resistance* (R_{NL}) of a circuit. The value of R_{TH} is measured as shown in Figure 7.21. Note that:

- The load has been removed.
- The voltage source has been replaced by a wire.

Why replace the voltage source with a wire? For the resistance measurement to be accurate, we must replace the source with an *equivalent resistance.* If we assume that the source has an ideal value of $R_S = 0\ \Omega$, then it must be replaced with a $0\ \Omega$ resistance. The wire represents this ideal value of source resistance.

The currents shown in Figure 7.22 represent the circuit currents that are generated by the meter. These currents indicate that (from the meter's perspective):

- R_1 is in parallel with R_2.
- R_3 is in series with the other two resistors.

A Practical Consideration: The value of $R_S = 0\ \Omega$ represented by the wire in Figure 7.21b is an ideal value. Because every voltage source has *some* amount of internal resistance, the value of R_{TH} calculated (or measured) with a shorted source is always slightly lower than the actual value.

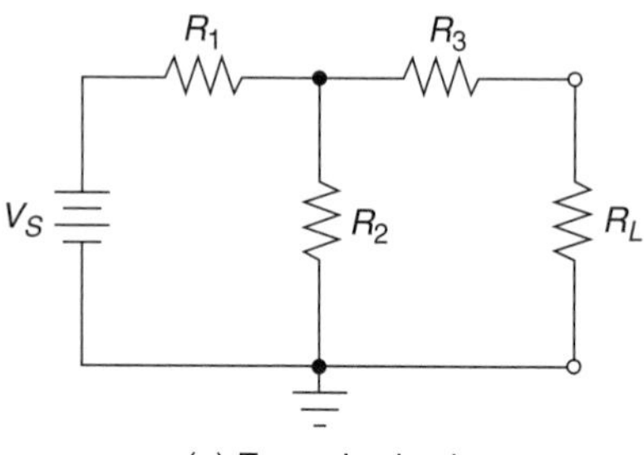

(a) Example circuit

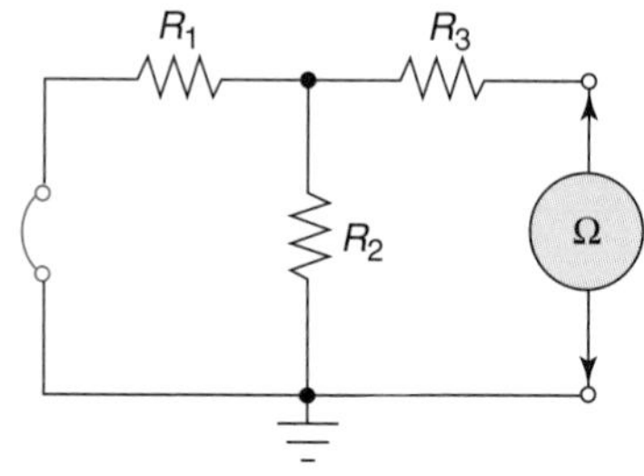

(b) Measuring the Thevenin resistance for the example circuit. (*Note:* that the source has been replaced by a jumper wire.)

Figure 7.21 Measuring Thevenin resistance.

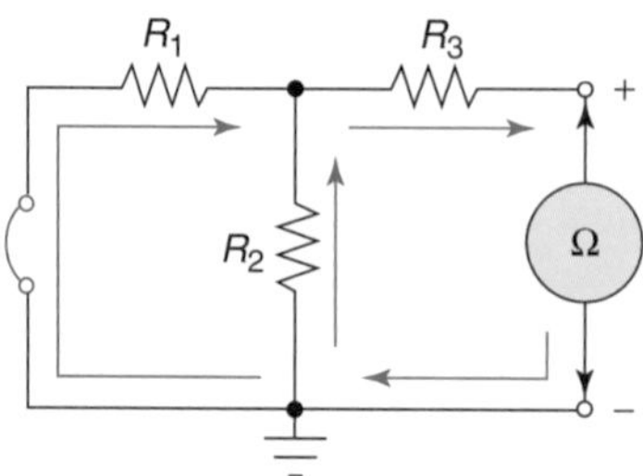

Figure 7.22 Currents produced by the ohmmeter.

Therefore, the Thevenin resistance of the circuit is found as

$$R_{TH} = (R_1 \| R_2) + R_3$$

Example 7.7 demonstrates the complete procedure for calculating the Thevenin resistance of a series-parallel circuit.

Example 7.7

Determine the value of Thevenin resistance for the circuit shown in Figure 7.23a.

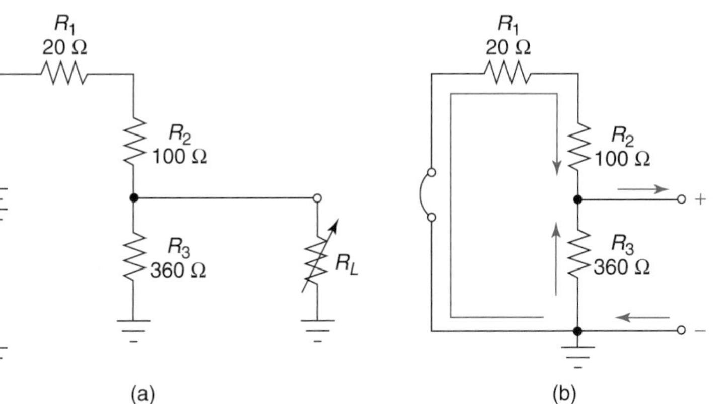

Figure 7.23

Solution: First, the circuit is drawn as shown in Figure 7.23b. Note that the load has been removed, and that the source has been replaced by a wire. The current arrows shown represent the current that would be produced by an ohmmeter connected to the load terminals. These currents indicate that:

- R_1 is in series with R_2.
- R_3 is in parallel with the other two resistors.

Therefore, the Thevenin resistance for the circuit can be found as

$$R_{TH} = (R_1 + R_2) \| R_3 = 120\ \Omega \| 360\ \Omega = 90\ \Omega$$

Practice Problem 7.7

Determine the Thevenin resistance for the circuit shown in Figure 7.20 (see Example 7.6).

The results from the last two examples are illustrated in Figure 7.24. As you can see, the values of V_{TH} and R_{TH} were combined into a single Thevenin equivalent circuit.

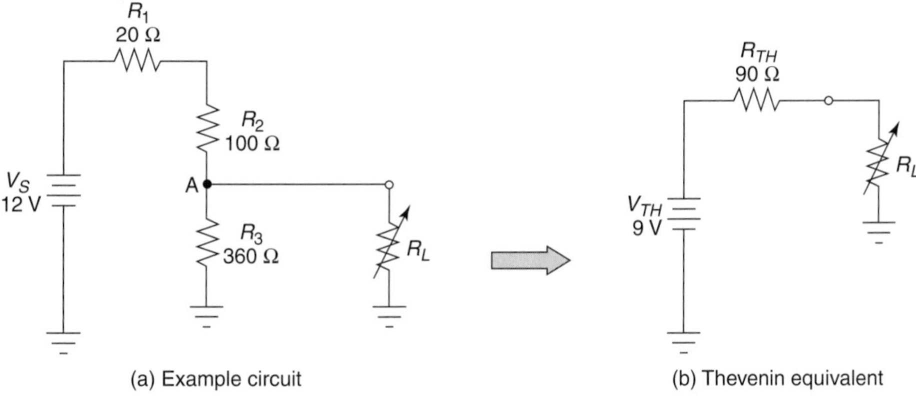

Figure 7.24

Deriving Thevenin Equivalent Circuits

OBJECTIVE 8 ➤ You have been shown how to determine the Thevenin equivalent voltage and resistance for a series-parallel circuit. At this point, we'll put it all together by deriving the Thevenin equivalents for two example circuits.

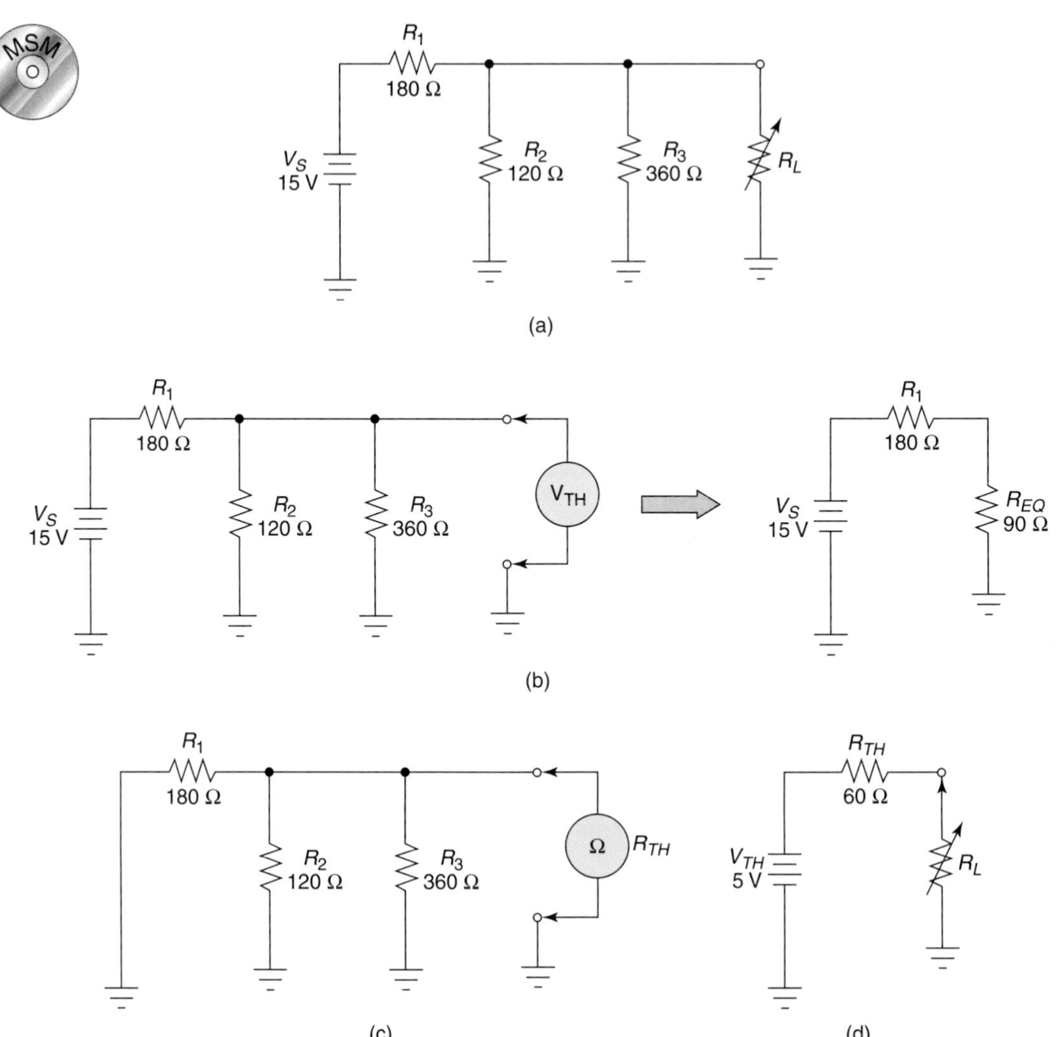

Figure 7.25

Example 7.8

Derive the Thevenin equivalent of the circuit shown in Figure 7.25a.

Solution: As shown in Figure 7.25b, the voltage across the open load terminals (V_{TH}) equals the voltage across the parallel combination of R_2 and R_3. Combining these two resistors into a single equivalent resistance, we get

$$R_{EQ} = R_2 \| R_3 = 120\ \Omega \| 360\ \Omega = 90\ \Omega$$

Since V_{TH} is measured across R_{EQ}, its value can be found as

$$V_{TH} = V_S \frac{R_{EQ}}{R_1 + R_{EQ}} = (15\text{ V})\frac{90\ \Omega}{270\ \Omega} = 5\text{ V}$$

The value of R_{TH} is measured as shown in Figure 7.25c. From the perspective of the meter, the three resistors are in parallel. Therefore,

$$R_{TH} = R_1 \| R_2 \| R_3 = 180\ \Omega \| 120\ \Omega \| 360\ \Omega = 60\ \Omega$$

Using our calculated values, the Thevenin equivalent of the original circuit is drawn as shown in Figure 7.25d.

Practice Problem 7.8

Derive the Thevenin equivalent of the circuit shown in Figure 7.26.

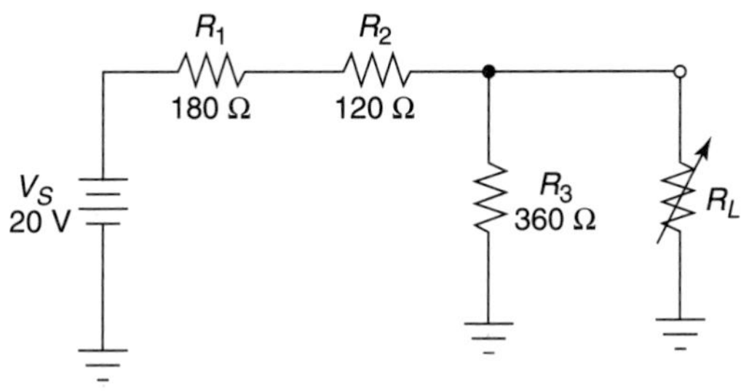

Figure 7.26

When a circuit contains branches in parallel with the source, one or more resistors in the circuit may have no effect on the values of V_{TH} and R_{TH}, as demonstrated in Example 7.9.

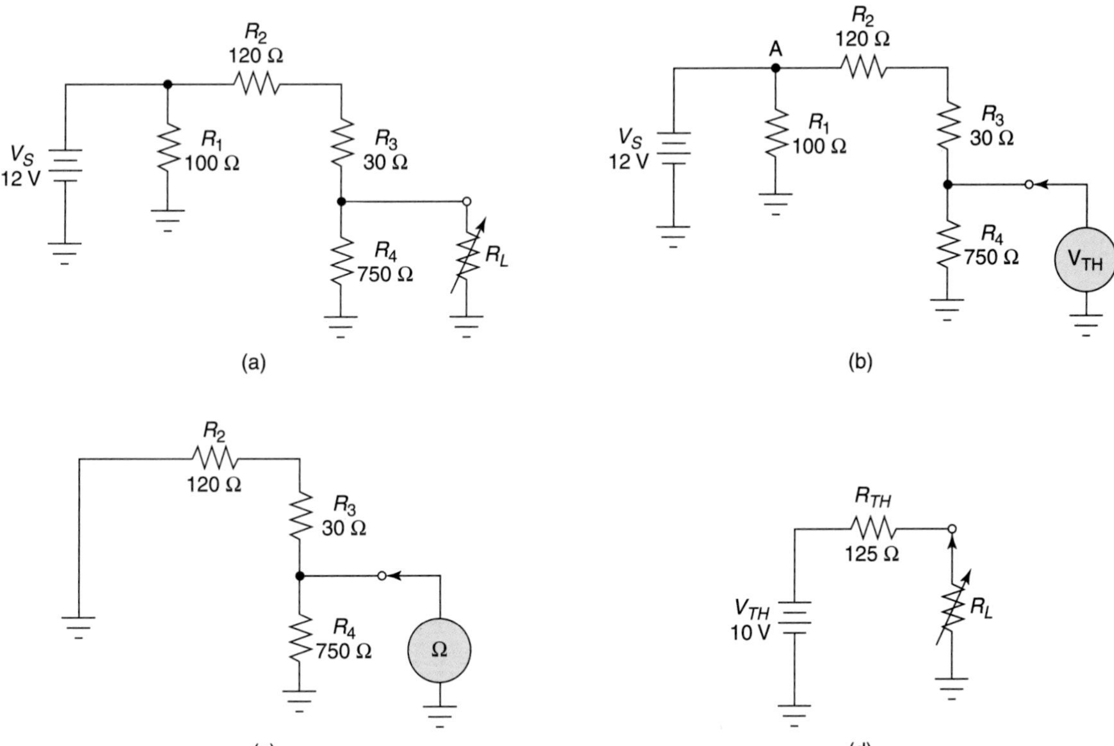

Figure 7.27

Example 7.9

Derive the Thevenin equivalent of the circuit shown in Figure 7.27a.

Solution: V_{TH} is measured as shown in Figure 7.27b. In this circuit, R_1 is in parallel with the branch containing the other three resistors. Since $V_{TH} = V_4$, only R_2, R_3, and R_4 are involved in the calculation of V_{TH}, and

$$V_{TH} = V_S \frac{R_4}{R_2 + R_3 + R_4} = (12 \text{ V})\frac{750\ \Omega}{900\ \Omega} = 10 \text{ V}$$

The value of R_{TH} is measured as shown in Figure 7.27c. R_1 is shorted when the voltage source is shorted, because the two are in parallel. (This is why R_1 is not shown in the resistance diagram.) For the circuit shown, the series combination of R_2 and R_3 is in parallel with R_4. Therefore,

$$R_{TH} = (R_2 + R_3)\|R_4 = 150\ \Omega\|750\ \Omega = 125\ \Omega$$

Using the calculated values, the Thevenin equivalent circuit is drawn as shown in Figure 7.27d.

Practice Problem 7.9

Derive the Thevenin equivalent for the circuit shown in Figure 7.28.

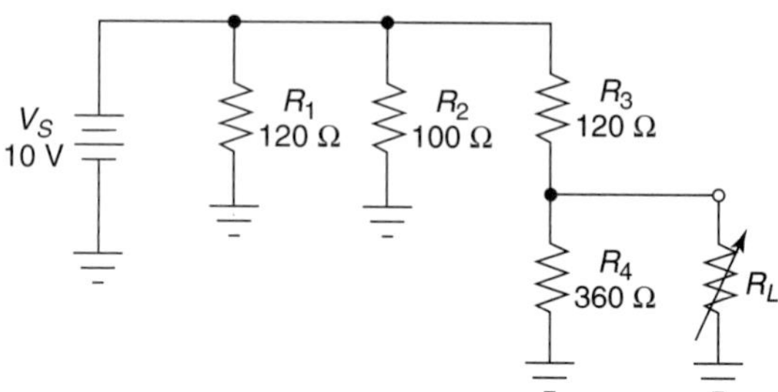

Figure 7.28

One Final Note

Though it may seem complicated at first, deriving the Thevenin equivalents of series-parallel circuits becomes relatively simple with practice. Once you become comfortable with using it, Thevenin's theorem can be used to solve circuit problems that would be extremely difficult to solve otherwise.

You were told at the beginning of this section that the strength of Thevenin's theorem can be seen in its circuit applications. In the next section, we'll take a look at several of these applications.

Section Review

1. What is *load analysis*?
2. What is *Thevenin's theorem*? What purpose does it serve?
3. What is the limitation on Thevenin's theorem in circuit analysis problems?
4. What is V_{TH}? How is it measured?
5. What is R_{TH}? How is it measured?

Critical Thinking

6. Derive V_{TH} and R_{TH} equations that will work for any series circuit.*
7. Derive V_{TH} and R_{TH} equations that will work for any parallel circuit.*

**Hint: Draw a circuit, and derive its Thevenin equivalent.*

7.4 Applications of Thevenin's Theorem

In the last section, you were shown how to derive the Thevenin equivalent of a given series-parallel circuit. In this section, you will be shown how Thevenin equivalent circuits can be applied to several circuit analysis problems.

Load Voltage Ranges

OBJECTIVE 9 ➤ Thevenin's theorem is most commonly used to predict the change in load voltage that will result from a change in load resistance. Since many loads are variable, it is important to be able to predict how a change in load will affect the output from the source circuit.

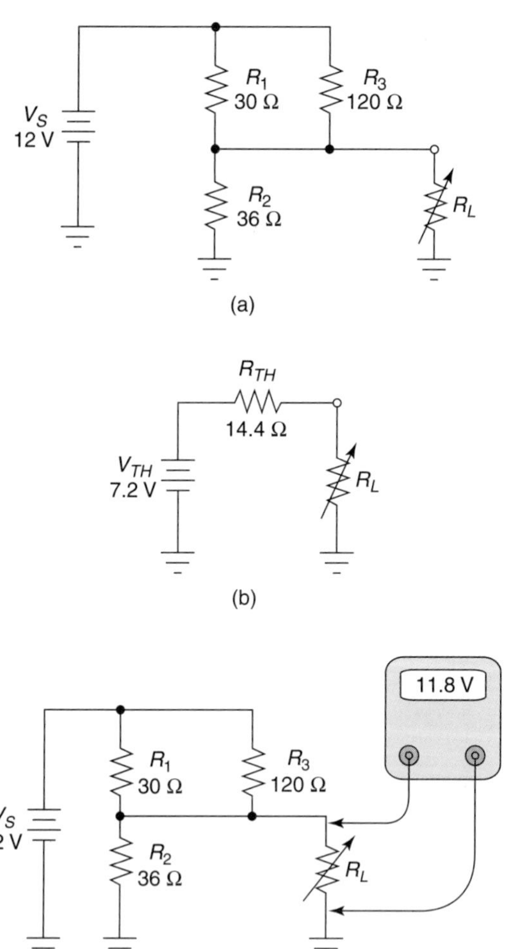

Figure 7.29

The Thevenin equivalent of a circuit can be used to determine the normal range of V_L for that circuit. For example, assume that the load shown in Figure 7.29a has a range of 10 Ω to 100 Ω. Using the Thevenin equivalent circuit shown in Figure 7.29b, the limits on the value of V_L can be found as follows:

$$V_L = V_{TH}\frac{R_L}{R_L + R_{TH}} = (7.2\text{ V})\frac{10\ \Omega}{24.4\ \Omega} = 2.95\text{ V} \qquad (\text{when } R_L = 10\ \Omega)$$

and

$$V_L = V_{TH}\frac{R_L}{R_L + R_{TH}} = (7.2\text{ V})\frac{100\ \Omega}{114.4\ \Omega} = 6.29\text{ V} \quad (\text{when } R_L = 100\ \Omega)$$

If we know the normal range of V_L, it is easy to determine whether a particular load voltage reading indicates a problem in the circuit. For example, let's say the output from the circuit is measured as shown in Figure 7.29c. We determined that the output voltage normally falls within the range of 2.95 V to 6.29 V. Since the measured voltage is well outside this range, we know that a fault exists somewhere in the circuit.

Maximum Power Transfer

In Chapter 4, you were introduced to the maximum power transfer theorem. As you may recall, *maximum power transfer* from a circuit to a variable load occurs *when the load resistance equals the source resistance.* ◄ *OBJECTIVE 10*

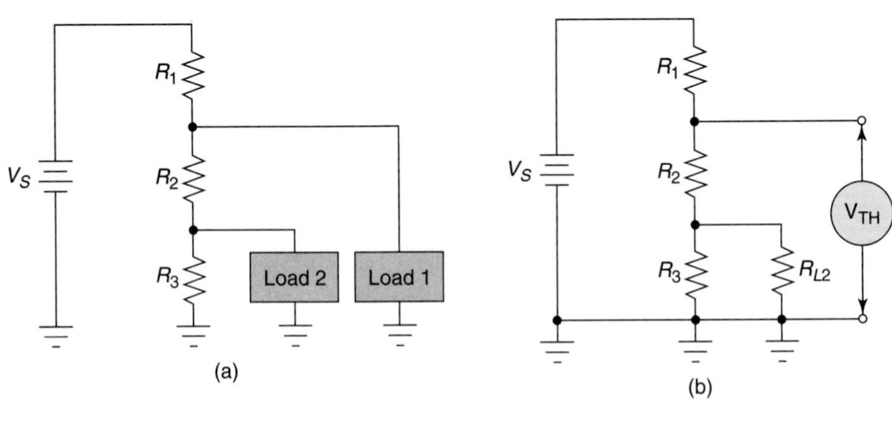

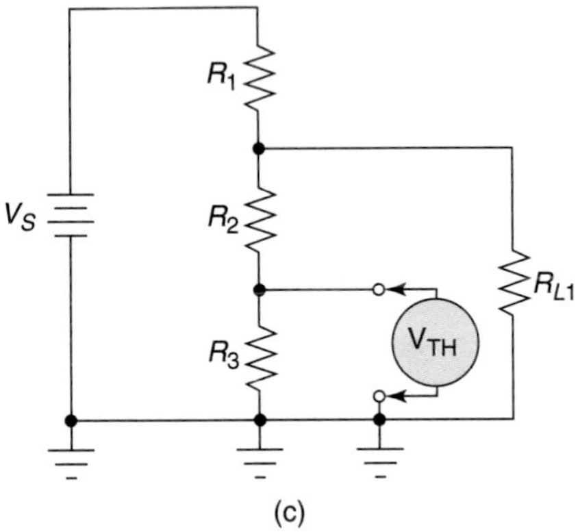

Figure 7.30

When maximum power transfer occurs in a series-parallel circuit.

For a series-parallel circuit with a variable load, maximum power transfer occurs when the load resistance equals the Thevenin resistance of the circuit. By formula,

$$P_L = P_{L(\max)} \text{ when } R_L = R_{TH} \tag{7.3}$$

Thus, by calculating the value of R_{TH} for a circuit, we determine the value of R_L that results in maximum power transfer from the circuit to the load.

Multiload Circuits

OBJECTIVE 11 ➤

Loaded voltage dividers may have more than one load, as illustrated in Figure 7.30. When using Thevenin's theorem to analyze a circuit like the one shown in Figure 7.30, there are several points to keep in mind:

- Each load has its own Thevenin equivalent circuit. Since they are connected to the circuit at different points, each load has its unique Thevenin equivalent of the original circuit.
- When deriving the Thevenin equivalent for either load, the resistance of the other load must be taken into account.

These points are illustrated in Figure 7.30. Figure 7.30b shows how V_{TH} is measured for Load 1; Figure 7.30c shows how V_{TH} is measured for Load 2. Comparing the schematics, you can see that the source circuit for Load 1 is different than the one for Load 2. The analysis of one load in a multiload circuit is demonstrated in Example 7.10.

Example 7.10

Determine the value of $P_{L(\max)}$ for Load 1 in Figure 7.31a. Assume that R_{L2} is fixed at 100 Ω.

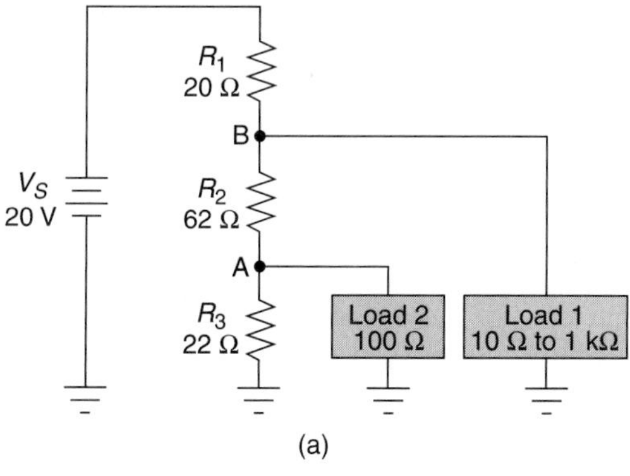

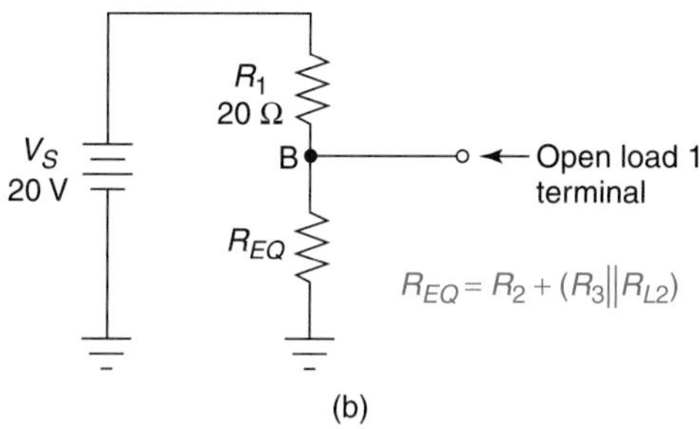

Figure 7.31

Solution: In Figure 7.31b, R_2, R_3, and R_{L2} are combined into a single equivalent resistance. The value of this resistance is found as

$$R_{EQ} = R_2 + (R_3 \| R_{L2}) = 62\ \Omega + (22\ \Omega \| 100\ \Omega) = 62\ \Omega + 18\ \Omega = 80\ \Omega$$

Using this value of R_{EQ}, the voltage at the open load terminal can be found as

$$V_{TH1} = V_S \frac{R_{EQ}}{R_1 + R_{EQ}} = (20\text{ V})\frac{80\ \Omega}{100\ \Omega} = 16\text{ V}$$

The value of R_{TH} in Figure 7.31b is found as

$$R_{TH1} = R_1 \| R_{EQ} = 20\ \Omega \| 80\ \Omega = 16\ \Omega$$

Maximum load power will be generated when $R_{L1} = R_{TH1}$. Using $R_{L1} = 16\ \Omega$,

$$V_L = V_{TH} \frac{R_L}{R_L + R_{TH}} = (16\text{ V})\frac{16\ \Omega}{32\ \Omega} = 8\text{ V}$$

and

$$P_{L(\max)} = \frac{V_L^2}{R_L} = \frac{(8\text{ V})^2}{16\ \Omega} = 4\text{ W}$$

Practice Problem 7.10

Assume that the positions of Load 1 and Load 2 in Figure 7.31a are reversed. Calculate the value of $P_{L(\max)}$ for Load 1.

The analysis technique demonstrated in Example 7.10 can be used on any number of loads. However, there is one restriction: *Each load (other than the load of interest) must be assumed to have a fixed resistance.* If you were to assume a range of values for more than one load, you wouldn't be able to determine set values of V_{TH} and R_{TH} for the load of interest. Without set values of V_{TH} and R_{TH}, attempting to solve the problem would be pointless.

Bridge Circuits

OBJECTIVE 12 ➤

In Chapter 6, you were introduced to the *Wheatstone bridge.* In general, the term "bridge" is used to describe any circuit that is constructed as shown in Figure 7.32. Note that the load (R_L) forms the "bridge" between the two series circuits.

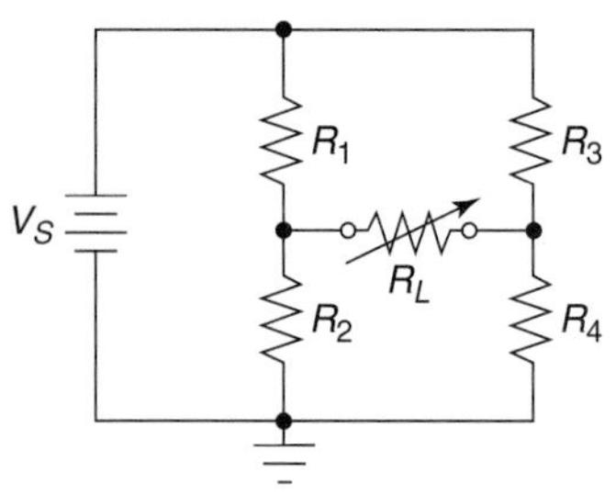

Figure 7.32
Bridge configuration.

Thevenin's theorem can be used to predict the magnitude of the load voltage in a bridge circuit like the one shown in Figure 7.32. When the load is removed from a bridge circuit (as shown in Figure 7.33a), the voltage across the open load terminals is equal to the difference between V_A and V_B. Therefore, the magnitude of V_{TH} for the circuit is found as

$$V_{TH} = |V_A - V_B| \tag{7.4}$$

The Thevenin resistance of a bridge circuit is measured as shown in Figure 7.33b. As usual, the ohmmeter is connected across the open load terminals. Note that the line around the outside of the circuit connects the bottom sides of R_2 and R_4 with the top sides of R_1 and R_3. Therefore, the circuit can be redrawn as shown in Figure 7.33c. As the currents in this circuit indicate,

$$R_{TH} = (R_1 \| R_2) + (R_3 \| R_4) \tag{7.5}$$

Once the values of V_{TH} and R_{TH} are known, the Thevenin equivalent circuit can be used to make the same types of predictions as those we made for the other types of circuits.

One Final Note

In this section, you were shown that Thevenin's theorem can be a powerful load analysis tool. In many cases, it can be used to solve problems that would be extremely difficult to solve otherwise. Thevenin's theorem is used more often than all the rest of the network theorems combined.

Section Review

1. When does maximum power transfer occur in a series-parallel circuit with a variable load?
2. Describe the process used to predict the value of maximum load power for a series-parallel circuit with a variable load.
3. Describe how Thevenin's theorem can be used to perform the load analyses of a multiload circuit.
4. What restriction applies to the Thevenin analysis of a multiload circuit?
5. Describe the process you would use to predict the value of maximum load power for a bridge circuit with a variable load.

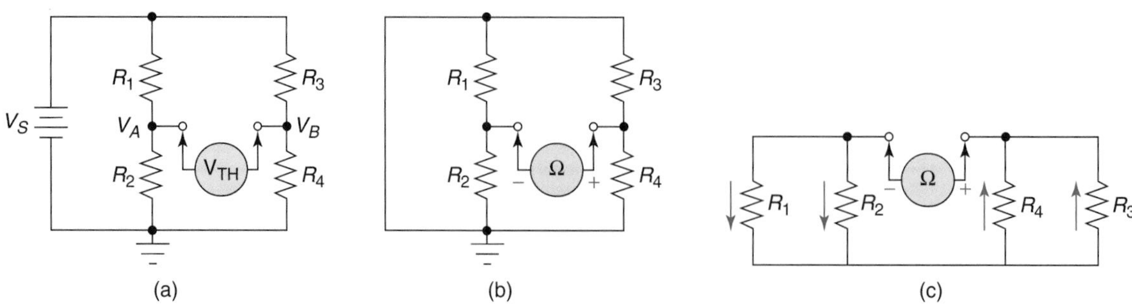

Figure 7.33 Thevenin voltage and resistance.

7.5 Norton's Theorem

In this section, we'll discuss another network theorem that can be used in the load analysis of a given series-parallel (or bridge) circuit. As you will see, *Norton's theorem* is very closely related to Thevenin's theorem.

Norton's Theorem: An Overview

Norton's theorem states that *any resistive circuit or network, no matter how complex, can be represented as a current source in parallel with a source resistance.* A series-parallel circuit is shown with its Norton equivalent circuit in Figure 7.34. Note that the source current and resistance are referred to as the *Norton current* (I_N) and the *Norton resistance* (R_N). ◄ *OBJECTIVE 13*

Obviously, Norton's theorem is very similar to Thevenin's theorem. The primary difference is that Norton represents a complex circuit as a practical *current source* while Thevenin represents that same circuit as a practical *voltage source.*

Norton Current (I_N)

By definition, **Norton current (I_N)** is *the current through the shorted load terminals.* The value of I_N for the circuit in Figure 7.34 is measured as illustrated in Figure 7.35. As you can see, the load has been replaced with a current meter. Since the internal resistance of a current meter is (ideally) 0 Ω, this is the equivalent of shorting the load. I_N is the current through the shorted load terminals. Note that I_N is simply the *shorted-load output current* (I_{SL}) of a circuit. ◄ *OBJECTIVE 14*

Norton Resistance (R_N)

Norton resistance (R_N) is *the resistance measured across the open load terminals of a circuit.* The value of R_N for a series-parallel circuit is measured and calculated exactly like R_{TH}. That is:

1. The load is removed.
2. The source is replaced by its resistive equivalent.
3. The resistance (as measured at the open load terminals) is determined.

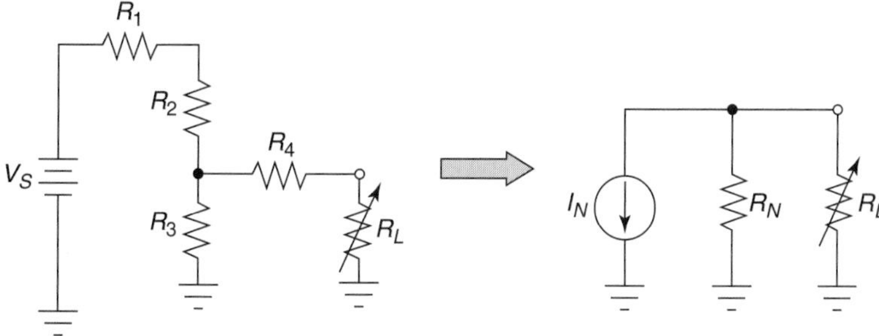

Figure 7.34 A series-parallel circuit and its Norton equivalent.

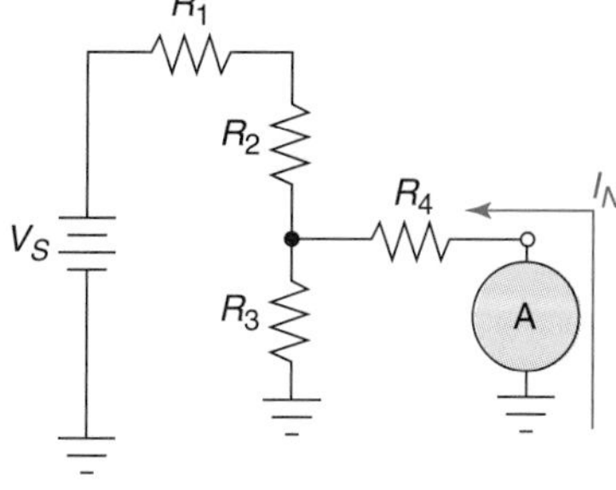

Figure 7.35 Measuring Norton current in a series-parallel circuit.

Load Analysis Using Norton Equivalent Circuits

OBJECTIVE 15 ➤ Norton equivalent circuits are used in the same fashion as Thevenin equivalent circuits. Once the Norton equivalent of a circuit is derived, the load is added, and the current divider relationship is used to determine the value of load current (I_L). The following example demonstrates how a load analysis problem can be solved using Norton equivalent circuits.

Example 7.11

Predict the values of I_L that will be produced by the circuit in Figure 7.36a for values of $R_L = 25\ \Omega$ and $R_L = 75\ \Omega$.

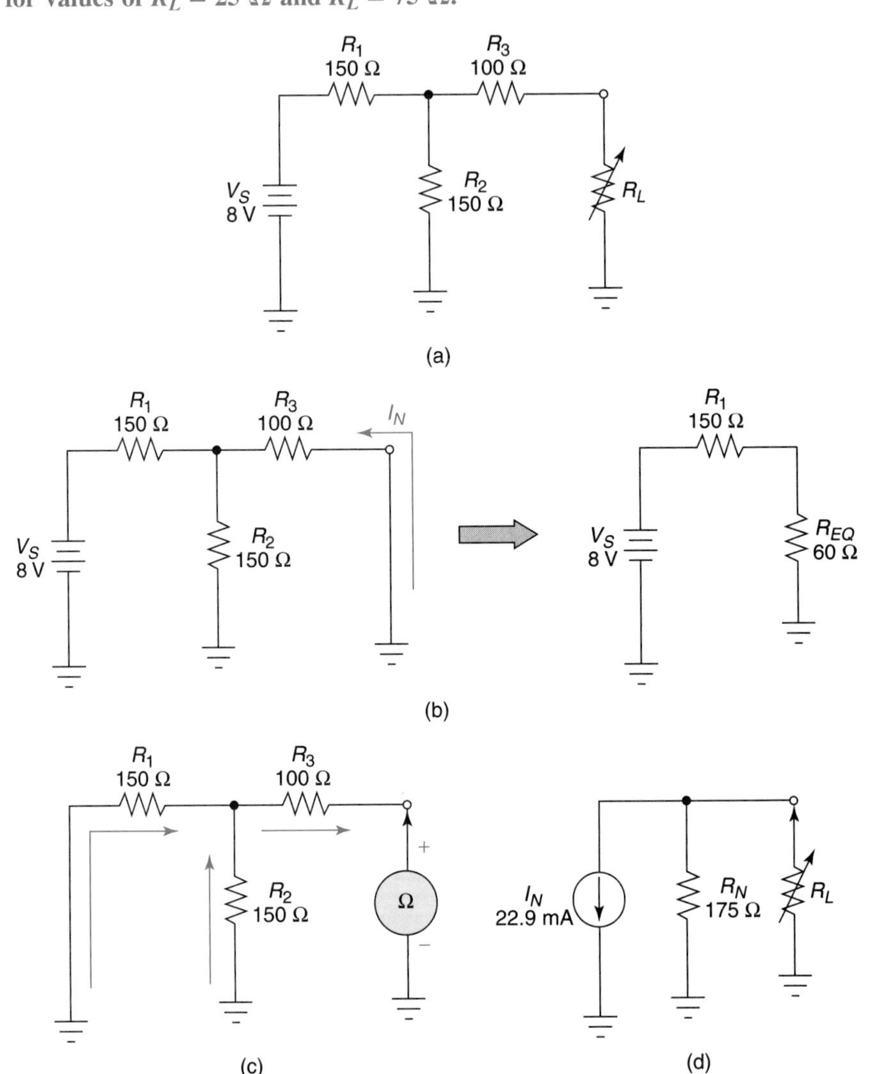

Figure 7.36

Solution: With the load shorted as shown in Figure 7.36b, I_N equals the current through R_3. Using the series equivalent circuit shown, the value of I_3 is found using

$$R_{EQ} = R_2 \| R_3 = 150\ \Omega \| 100\ \Omega = 60\ \Omega$$

$$V_{EQ} = V_S \frac{R_{EQ}}{R_1 + R_{EQ}} = (8\text{ V})\frac{60\ \Omega}{210\ \Omega} = 2.29\text{ V}$$

and

$$I_N = I_3 = \frac{V_{EQ}}{R_3} = \frac{2.29\text{ V}}{100\ \Omega} = 22.9\text{ mA}$$

The value of R_N is measured as shown in Figure 7.36c. The currents shown are those that would be produced by the ohmmeter. As these arrows indicate, R_N can be found as

$$R_N = (R_1 \| R_2) + R_3 = (150\ \Omega \| 150\ \Omega) + 100\ \Omega = 175\ \Omega$$

Using the calculated values, the Norton equivalent circuit is drawn as shown in Figure 7.36d. When $R_L = 25\ \Omega$,

$$I_L = I_N \frac{R_L \| R_N}{R_L} = (22.9\text{ mA})\frac{25\ \Omega \| 175\ \Omega}{25\ \Omega} = (22.9\text{ mA})\frac{21.88\ \Omega}{25\ \Omega} = 20\text{ mA}$$

When $R_L = 75\ \Omega$,

$$I_L = I_N \frac{R_L \| R_N}{R_L} = (22.9\text{ mA})\frac{75\ \Omega \| 175\ \Omega}{75\ \Omega} = (22.9\text{ mA})\frac{52.5\ \Omega}{75\ \Omega} = 16\text{ mA}$$

Practice Problem 7.11

Predict the values of I_L that will be produced by the circuit shown in Figure 7.37 for values of $R_L = 100\ \Omega$ and $R_L = 150\ \Omega$.

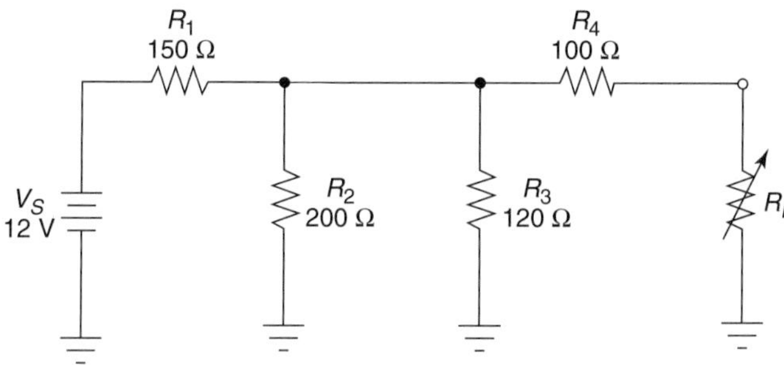

Figure 7.37

Norton-to-Thevenin and Thevenin-to-Norton Conversions

Every Thevenin equivalent circuit has a matching Norton equivalent circuit, and vice versa. For example, Figure 7.38 contains a series-parallel circuit along with its Norton and Thevenin equivalents. The load calculations shown for each equivalent circuit were made using an assumed value of $R_L = 60\ \Omega$ (the value of the source resistance). As you can see, the equivalent circuits provide identical values of load voltage, current, and power. ◄ *OBJECTIVE 16*

Using the techniques already discussed for source conversions, you can convert either type of equivalent circuit to the other. Both conversions are shown in Figure 7.39. As you can see, the conversion formulas used are the same as those you learned in Section 7.2.

If we apply the Thevenin-to-Norton relationships to the Thevenin circuit in Figure 7.38, we get

$$R_N = R_{TH} = 60\ \Omega \qquad \text{and} \qquad I_N = \frac{V_{TH}}{R_{TH}} = \frac{5\text{ V}}{60\ \Omega} = 83.3\text{ mA}$$

These values match those in the Norton equivalent circuit in Figure 7.37.

One Final Note

As your study of electronics continues, you'll find that some analysis situations lend themselves more toward one theorem than to the other. That is, some circuits are better represented as voltage sources, while others are better represented as current sources.

In most cases, the choice of which theorem to use for load analysis is up to you. Many technicians find it easier to become very proficient at using one type of theorem and then

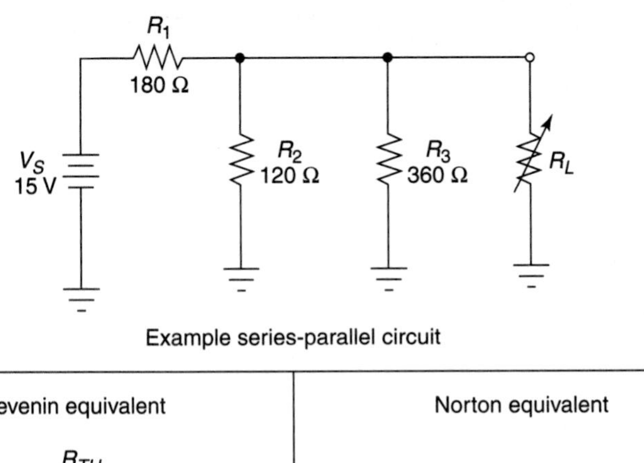

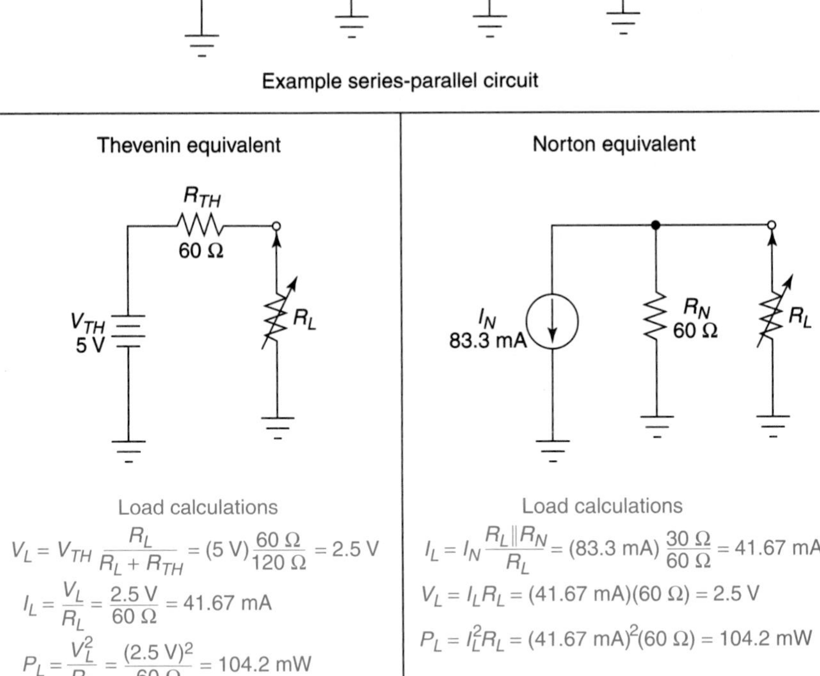

Thevenin equivalent
R_{TH}
60 Ω
V_{TH}
5 V
R_L
Load calculations
$V_L = V_{TH}\frac{R_L}{R_L + R_{TH}} = (5\text{ V})\frac{60\ \Omega}{120\ \Omega} = 2.5\text{ V}$
$I_L = \frac{V_L}{R_L} = \frac{2.5\text{ V}}{60\ \Omega} = 41.67\text{ mA}$
$P_L = \frac{V_L^2}{R_L} = \frac{(2.5\text{ V})^2}{60\ \Omega} = 104.2\text{ mW}$
Norton equivalent
I_N
83.3 mA
R_N
60 Ω
R_L
Load calculations
$I_L = I_N\frac{R_L \| R_N}{R_L} = (83.3\text{ mA})\frac{30\ \Omega}{60\ \Omega} = 41.67\text{ mA}$
$V_L = I_L R_L = (41.67\text{ mA})(60\ \Omega) = 2.5\text{ V}$
$P_L = I_L^2 R_L = (41.67\text{ mA})^2(60\ \Omega) = 104.2\text{ mW}$

Figure 7.38 Comparing Norton and Thevenin equivalent circuits.

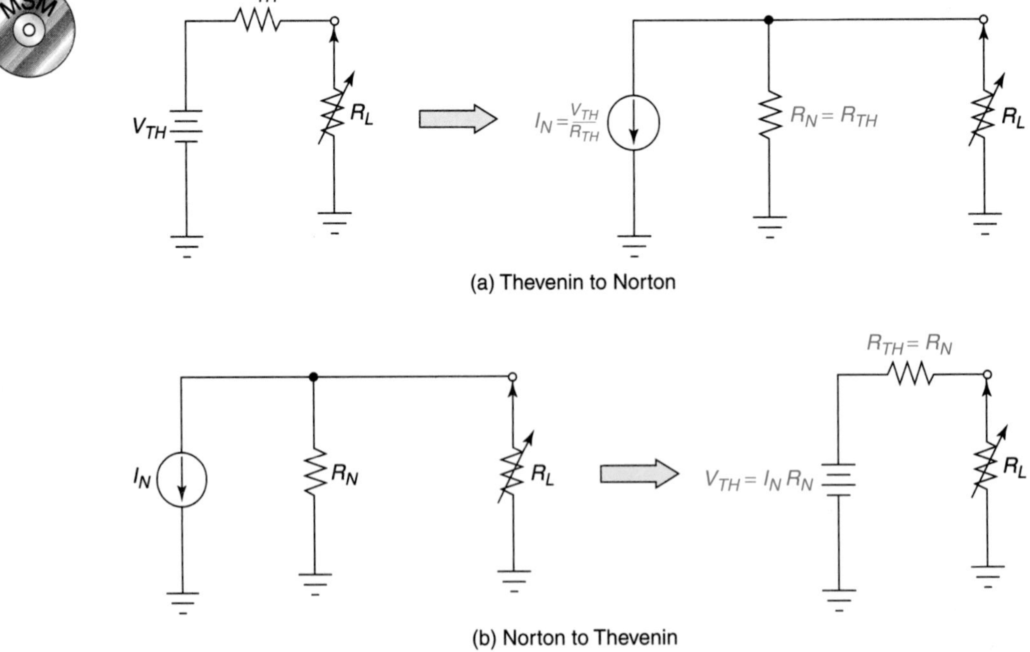

Figure 7.39 Thevenin-to-Norton and Norton-to-Thevenin conversions.

using source conversions to derive the other (when the need arises). Others find it beneficial to become proficient at both Norton's and Thevenin's theorems.

Section Review

1. Describe the process for determining the value of Norton current for a series-parallel circuit.
2. Describe the process for determining the value of Norton resistance for a series-parallel circuit.
3. In terms of load analysis, how does Norton's theorem differ from Thevenin's theorem?
4. Describe the relationship between the Norton and Thevenin equivalents of a given circuit.
5. How do you convert from one equivalent circuit (Norton or Thevenin) to the other?

Key Terms

The following terms were introduced in this chapter on the pages indicated:

Load analysis, 191
Multisource circuit, 182
Network theorem, 182
Norton current (I_N), 203
Norton resistance (R_N), 203
Norton's theorem, 203
Superposition theorem, 182
Thevenin equivalent circuit, 191
Thevenin resistance (R_{TH}), 194
Thevenin's theorem, 191
Thevenin voltage (V_{TH}), 193

Practice Problems

1. Determine the values of V_1, V_2, and V_3 for the circuit shown in Figure 7.40a.
2. Determine the values of V_1, V_2, and V_3 for the circuit shown in Figure 7.40b.

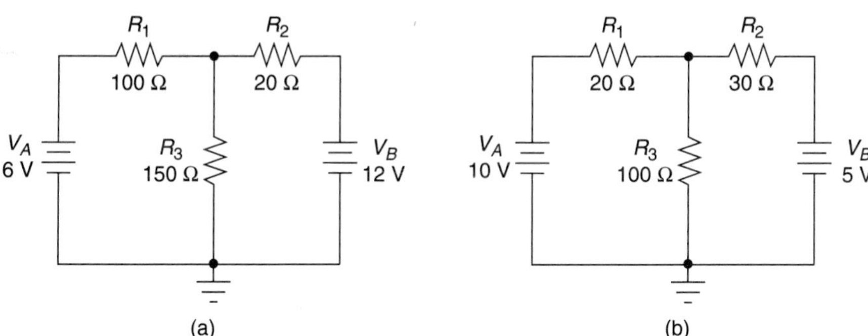

Figure 7.40

3. Determine the values of V_1, V_2, and V_3 for the circuit shown in Figure 7.41a.

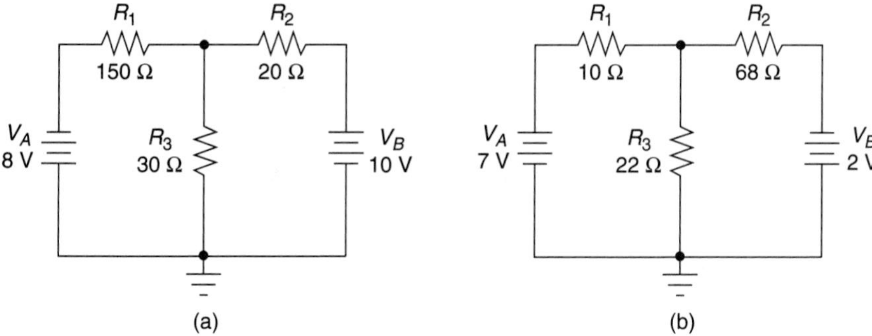

Figure 7.41

4. Determine the values of V_1, V_2, and V_3 for the circuit shown in Figure 7.41b.

5. Determine the values of V_1 through V_4 for the circuit shown in Figure 7.42a.

6. Determine the values of V_1 through V_4 for the circuit shown in Figure 7.42b.

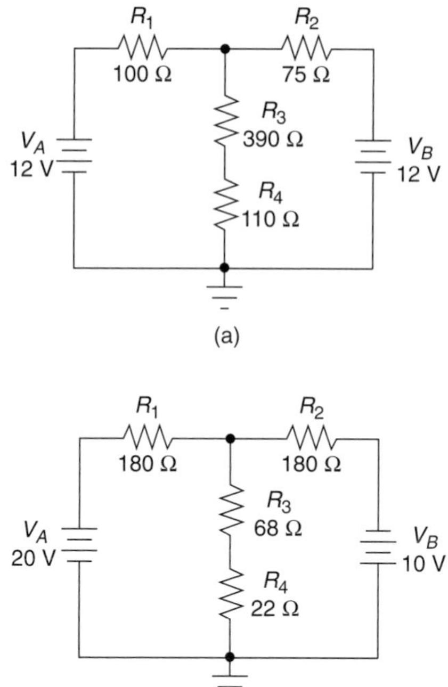

Figure 7.42

7. Using a value of $R_L = 100\ \Omega$, determine whether or not the circuits shown in Figure 7.43a are equivalent circuits.

8. Using a value of $R_L = 2.2\ \text{k}\Omega$, determine whether or not the circuits shown in Figure 7.43b are equivalent circuits.

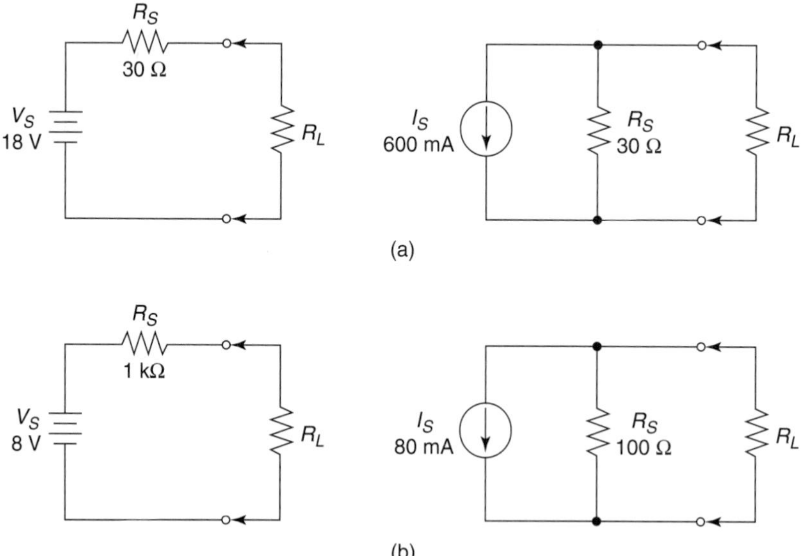

Figure 7.43

9. Convert the voltage source shown in Figure 7.44a to an equivalent current source.

10. Convert the voltage source shown in Figure 7.44b to an equivalent current source.

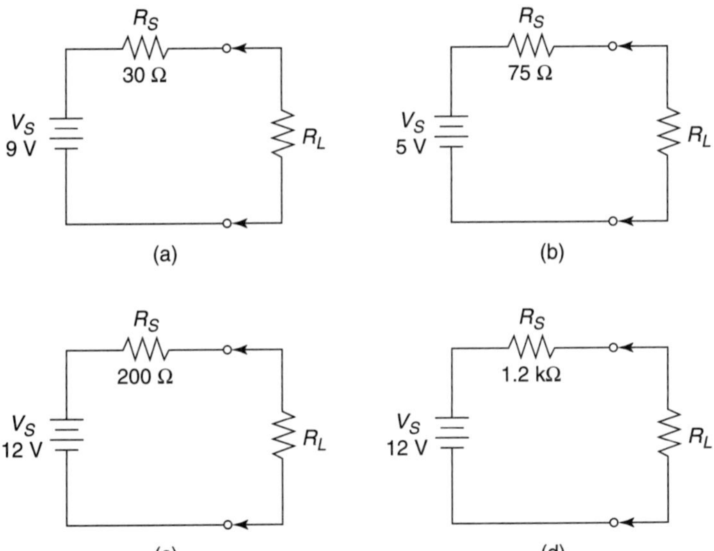

Figure 7.44

11. Derive the current source equivalent of the voltage source shown in Figure 7.44c. Then, calculate the values of V_L produced by each circuit for values of R_L = 100 Ω and R_L = 330 Ω.

12. Derive the current source equivalent of the voltage source shown in Figure 7.44d. Then, calculate the values of I_L produced by each circuit for values of R_L = 180 Ω and R_L = 1 kΩ.

13. Convert the current source shown in Figure 7.45a to an equivalent voltage source.

14. Derive the voltage source equivalent of the current source shown in Figure 7.45b. Then, calculate the values of V_L produced by each circuit for values of R_L = 75 Ω and R_L = 150 Ω.

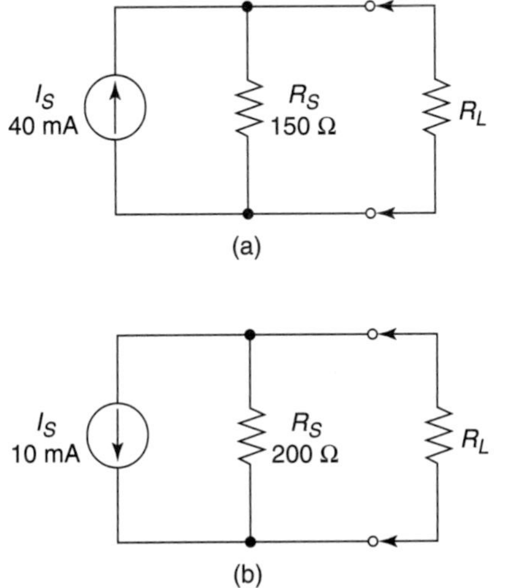

Figure 7.45

15. Derive the Thevenin equivalent of the circuit shown in Figure 7.46a.

16. Derive the Thevenin equivalent of the circuit shown in Figure 7.46b.

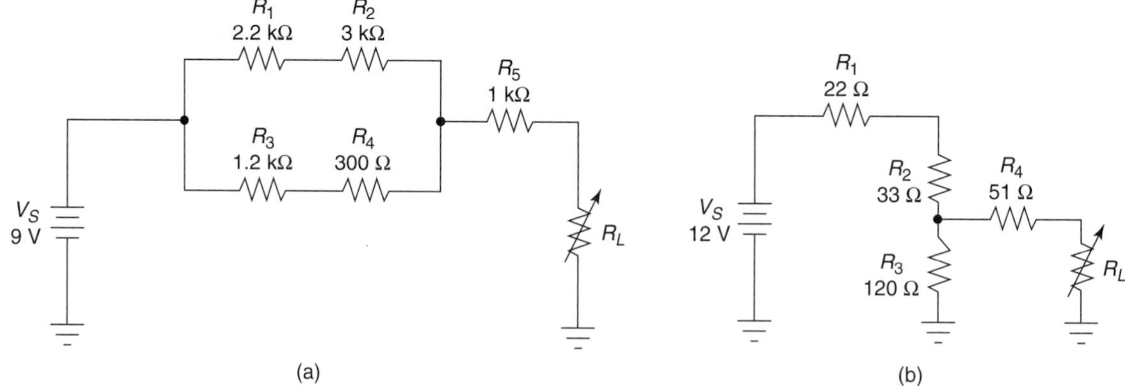

Figure 7.46

17. Derive the Thevenin equivalent of the circuit shown in Figure 7.47a.
18. Derive the Thevenin equivalent of the circuit shown in Figure 7.47b.

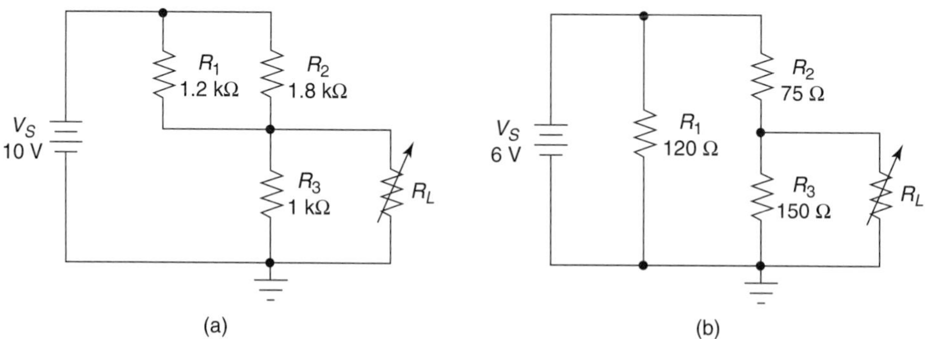

Figure 7.47

19. The circuit shown in Figure 7.48a has a load with a range of $R_L = 100\ \Omega$ to $1\ \text{k}\Omega$. Determine normal range of V_L for the circuit using its Thevenin equivalent.
20. The circuit shown in Figure 7.48b has a load with a range of $R_L = 10\ \Omega$ to $100\ \Omega$. Determine normal range of V_L for the circuit using its Thevenin equivalent.

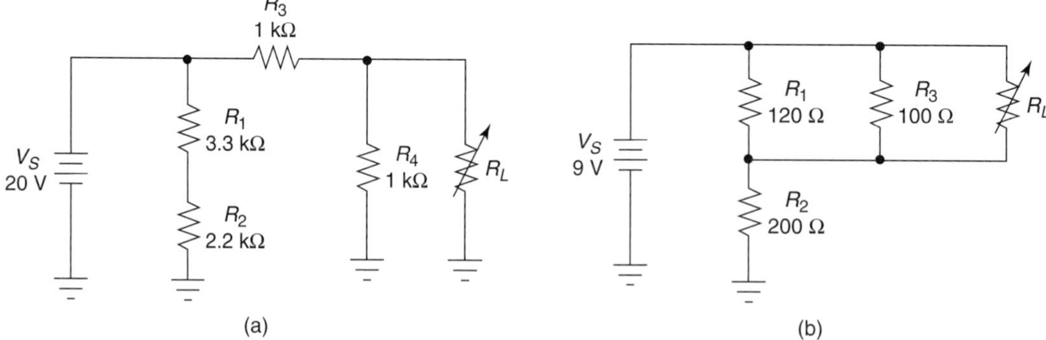

Figure 7.48

21. Calculate the maximum possible load power for the circuit shown in Figure 7.49a.
22. Calculate the maximum possible load power for the circuit shown in Figure 7.49b.

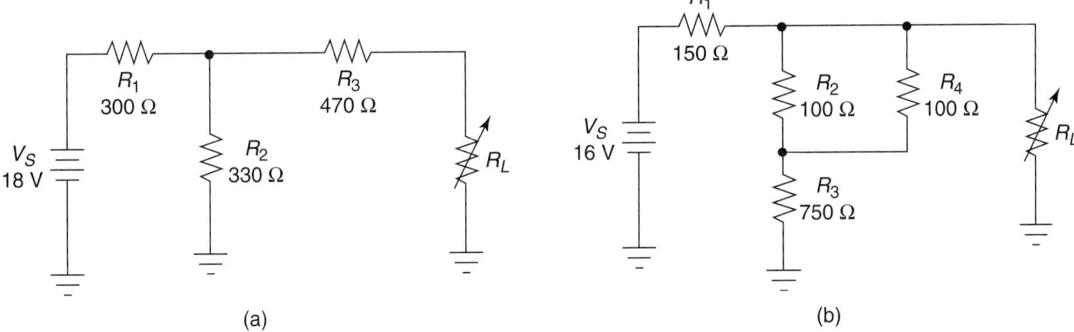

Figure 7.49

23. Calculate the normal range of output voltages for the circuit shown in Figure 7.50a.
24. Calculate the normal range of output voltage values for the circuit shown in Figure 7.50b.

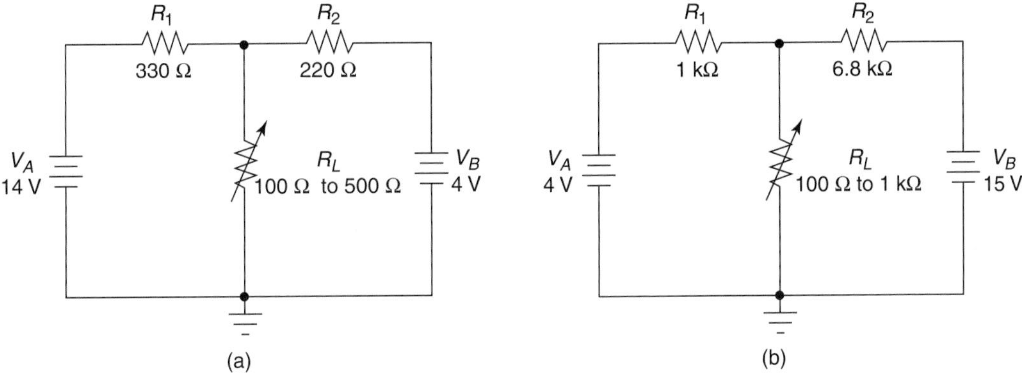

Figure 7.50

25. Calculate the maximum possible load power for the circuit shown in Figure 7.51a.
26. Calculate the maximum possible load power for the circuit shown in Figure 7.51b.

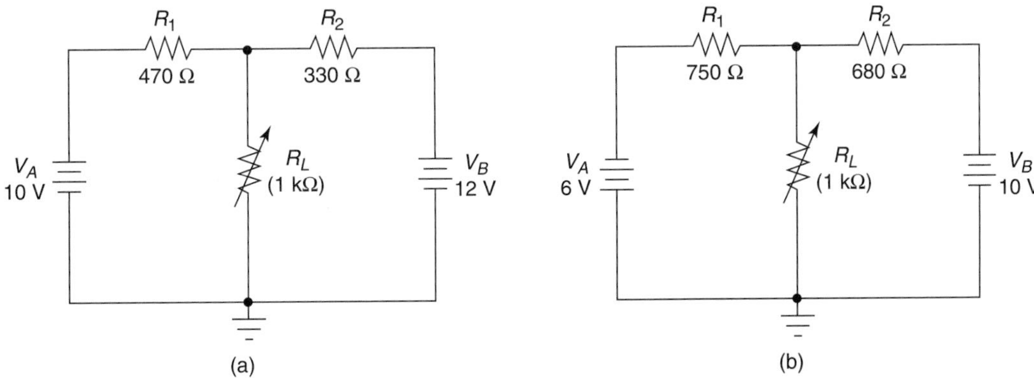

Figure 7.51

27. Derive the Thevenin equivalent circuit for Load 1 in Figure 7.52.
28. Derive the Thevenin equivalent circuit for Load 2 in Figure 7.52.

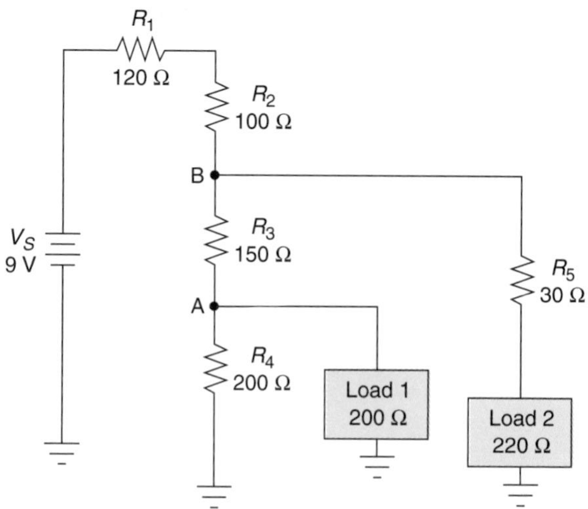

Figure 7.52

29. Determine the maximum possible load power for the bridge circuit shown in Figure 7.53a.

30. Determine the maximum possible load power for the bridge circuit shown in Figure 7.53b.

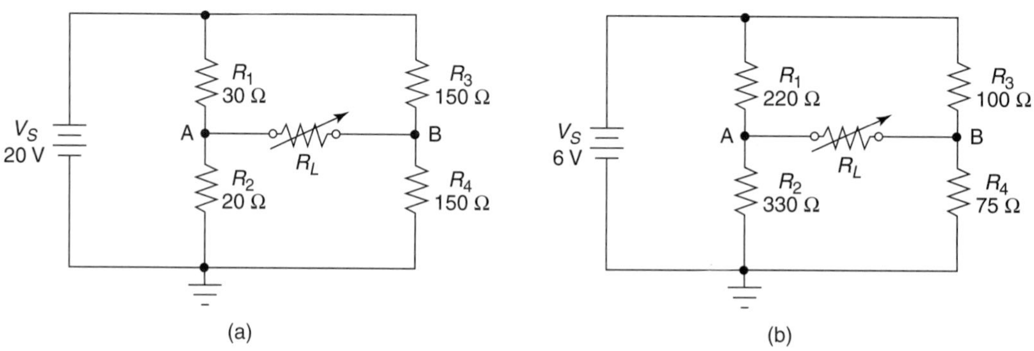

Figure 7.53

31. Derive the Norton equivalent for the circuit shown in Figure 7.46a.

32. Derive the Norton equivalent for the circuit shown in Figure 7.46b.

33. Derive the Norton equivalent for the circuit shown in Figure 7.47a.

34. Derive the Norton equivalent for the circuit shown in Figure 7.47b.

35. Using the Norton equivalent for the circuit shown in Figure 7.54a, determine its normal range of load current values.

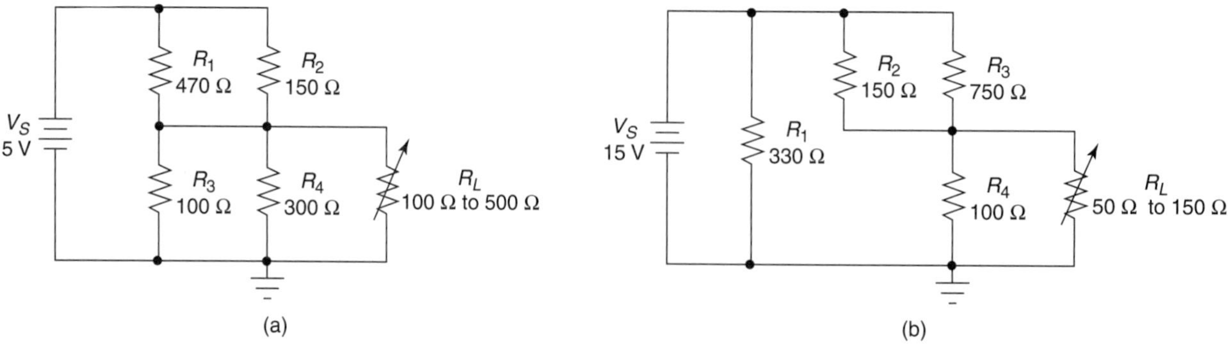

Figure 7.54

36. Using the Norton equivalent for the circuit shown in Figure 7.54b, determine its normal range of load current values.

Looking Back

These problems relate to material presented in earlier chapters. The chapters are identified in brackets.

37. Calculate the value of current that will cause a 150 Ω resistor to dissipate 600 mW. [Chapter 3]

38. Solve each of the following: [Introduction]
   a. 3.87 mW = ______ μW
   b. 400 kΩ = ______ Ω
   c. 54 MHz = ______ GHz

39. A resistor has the following alphanumeric code: **2203G.** Determine the color code for a resistor having the same value. [Chapter 2]

40. A potentiometer is connected in series with a 220 Ω resistor, a 1 kΩ resistor, and a 15 V source. What is the potentiometer setting when $P_S = 150$ mW? [Chapter 5]

Pushing the Envelope

41. Calculate the range of V_L for each load shown in Figure 7.55.

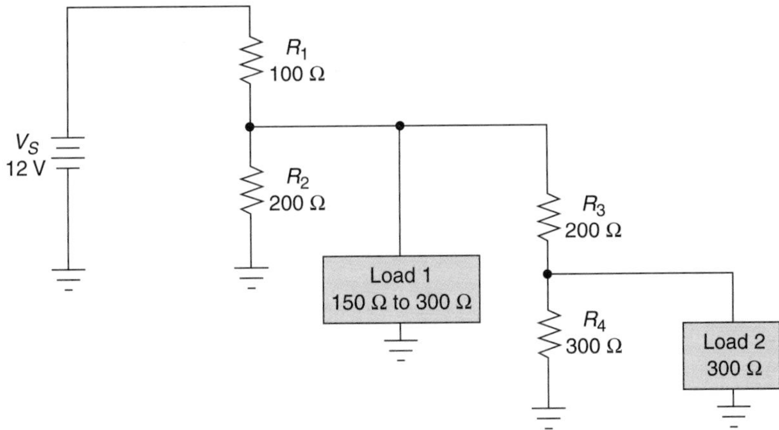

Figure 7.55

42. Calculate the range of V_L for the circuit shown in Figure 7.56.

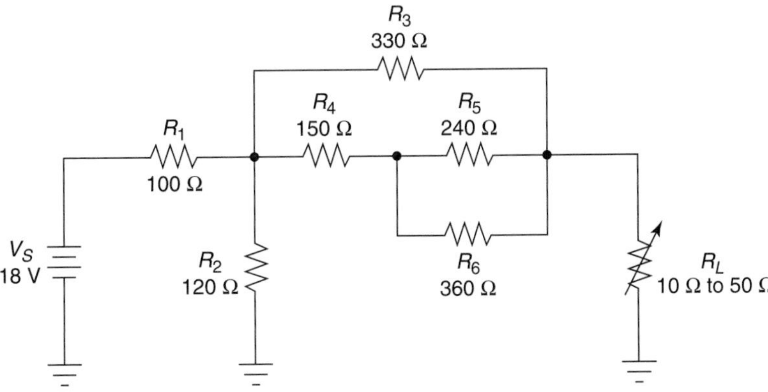

Figure 7.56

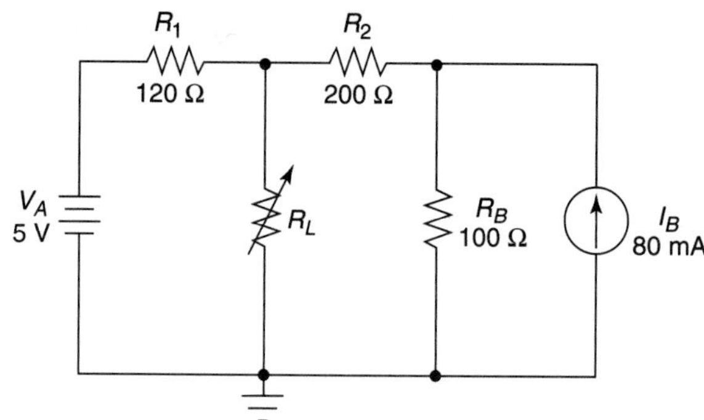

Figure 7.57

43. Determine the maximum possible load power for the circuit shown in Figure 7.57.

Answers to the Example Practice Problems

7.1. $V_1 = 6$ V, $V_2 = 1$ V, $V_3 = 8$ V

7.2. For both circuits: $V_L = 2$ V, $I_L = 20$ mA

7.3. $I_S = 500$ mA, $R_S = 16\ \Omega$

7.4. $V_S = 6$ V, $R_S = 150\ \Omega$

7.5. For $R_L = 150\ \Omega$: $V_L = 6.57$ V. For $R_L = 30\ \Omega$: $V_L = 4.86$ V

7.6. 7.2 V

7.7. 14.4 Ω

7.8. $V_{TH} = 10.9$ V, $R_{TH} = 164\ \Omega$

7.9. $V_{TH} = 7.5$ V, $R_{TH} = 90\ \Omega$

7.10. 191 mW

7.11. When $R_L = 100\ \Omega$: $I_L = 16$ mA. When $R_L = 150\ \Omega$: $I_L = 13.33$ mA

Multisim File References

Figure	File Reference
7.6	ETF7_6.msm
7.25	ETF7_25.msm
7.39	ETF7_39.msm

chapter 

Magnetism

Objectives

After studying the material in this chapter, you should be able to:

1. Describe *magnetic force.*
2. Identify the poles of a magnet and state the relationships between them.
3. List and define the common units of magnetic flux.
4. Contrast the following: *magnetic fields, magnetic flux,* and *flux density.*
5. List and describe the units of flux density.
6. Calculate the flux density at any point in a magnetic field, given the cross-sectional area and the total flux.
7. Compare and contrast *permeability* with *relative permeability.*
8. Given the permeability of a material, calculate its relative permeability.
9. Discuss the *domain theory* of the source of magnetism.
10. Describe the *magnetic induction* methods of producing a magnet.
11. Define *retentivity* and discuss its relationship to permeability.
12. Define *reluctance* and discuss its relationship to permeability.
13. List and describe the magnetic classifications of materials.
14. Describe the relationship between current and a resulting magnetic field.
15. Compare the basic magnetic and electric quantities.
16. Discuss the relationships that exist among the following: *magnetomotive force, ampere-turns, coil current, core permeability,* and *coil flux density.*
17. Define and discuss *hysteresis.*
18. Describe *magnetic shielding.*
19. Contrast *ring magnets* with other types of magnets.
20. Describe the proper care and handling of magnets.

Magnetism plays an important role in the operation of many electronic components and systems. For example, audio and video systems, electric motors, and many analog meters utilize the properties of magnetism (to one extent or another).

In this chapter, we're going to discuss only a few of the basic properties of magnetism. A complete study of the subject would require several volumes, so our discussions will be limited to those principles that relate directly to the study of electricity and electronics.

8.1 Magnetism: An Overview

OBJECTIVE 1 ➤ As you know, a magnet can attract (or repel) another magnet or any piece of iron. *The force that a magnet exerts on the objects around it* is referred to as **magnetic force.**

When exposed to magnetic force, iron filings line up as shown in Figure 8.1a. For this reason, magnetic force is commonly represented as a series of lines. These lines form closed loops around (and through) the magnet, as illustrated in Figure 8.1b.

Magnetic Poles

OBJECTIVE 2 ➤ *The points where magnetic lines of force leave (and return to) a magnet* are referred to as the **poles.** By convention, the lines of force are assumed to emanate from the **north-seeking pole (N)** and to return to the magnet via the **south-seeking pole (S).** Within the magnet, the lines of force continue from (S) to (N).

Like and Unlike Poles

You may recall that like charges repel (and unlike charges attract) each other. Magnetic poles have similar effects on each other. That is:

- Like poles repel each other.
- Unlike poles attract each other.

These interactions between the poles are illustrated in Figure 8.2.

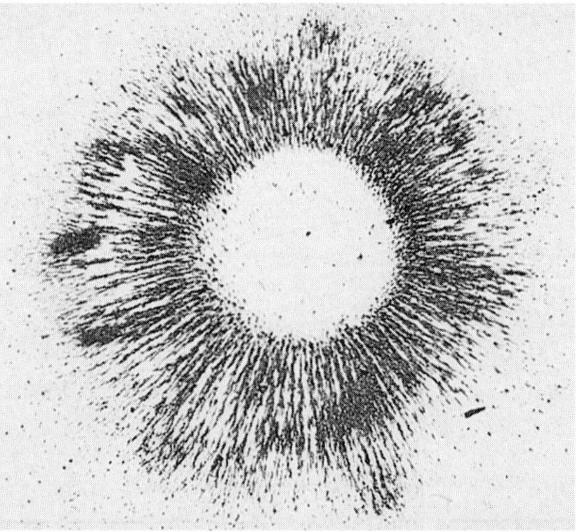

(a) Iron filings lining up along lines of magnetic force

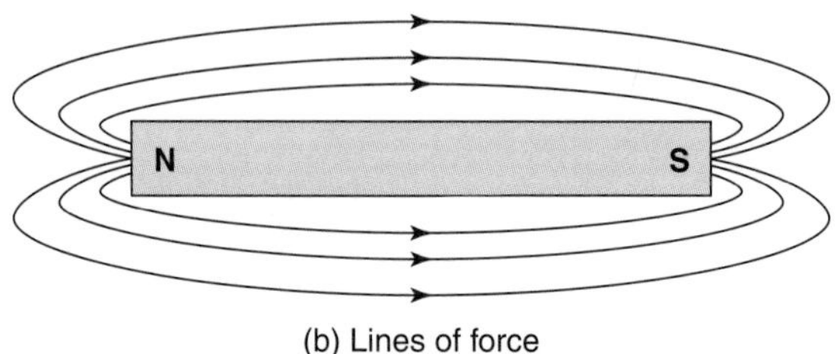

(b) Lines of force

Figure 8.1

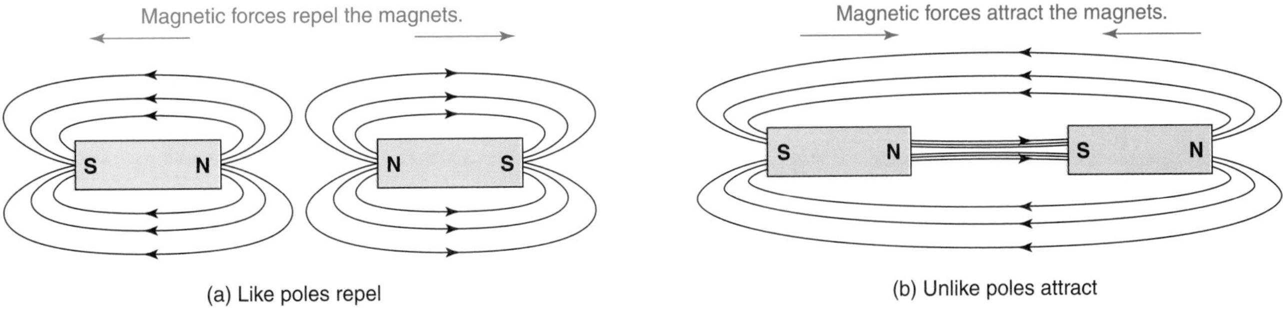

Figure 8.2 Pole interactions.

Magnetic Flux

The lines of force produced by a magnet are collectively referred to as **magnetic flux.** There are two fundamental units of measure for magnetic flux. These units are identified and defined as follows:

◄ *OBJECTIVE 3*

Unit of Measure	*Unit System*	*Value (Defined)*
maxwell (Mx)	cgs	1 Mx = 1 line of force
weber (Wb)	SI	$1\text{ Wb} = 1 \times 10^8\text{ Mx}$

The *maxwell (Mx)* is easy to visualize. The magnet represented in Figure 8.1b has six lines of force. Therefore, the magnetic flux (in maxwells) is given as

$$\phi = 6\text{ Mx}$$

where ϕ (the Greek letter *Phi*) is used to represent magnetic flux.

The maxwell is not a practical unit of measure, because magnets produce flux in the thousands of maxwells (and higher). On the other hand, the **weber (Wb)** is often too large a unit of measure to be useful. In such cases, the **microweber (μWb)** is the preferred unit of measure. By formula,

$$1\ \mu\text{Wb} = 100\text{ Mx} \tag{8.1}$$

That is, 1 μWb equals 100 lines of magnetic flux.

Magnetic characteristics are typically measured in both *cgs* (centimeter-gram-second) units and *SI* (Systeme International) units. Throughout the remainder of this chapter, we will focus on the SI units of measure for magnetic quantities, because they are used more often than cgs units. The cgs units, along with any applicable conversions, will be identifed in the margins. (You need to be familiar with both, as both are used in practice.)

Flux Density

The area of space surrounding a magnet that contains magnetic flux is referred to as a **magnetic field.** A two-dimensional representation of a magnetic field is shown in Figure 8.3.

◄ *OBJECTIVE 4*

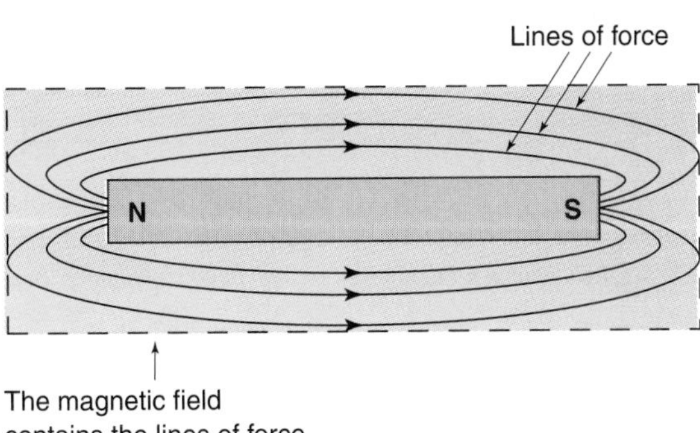

Figure 8.3 Magnetic field.

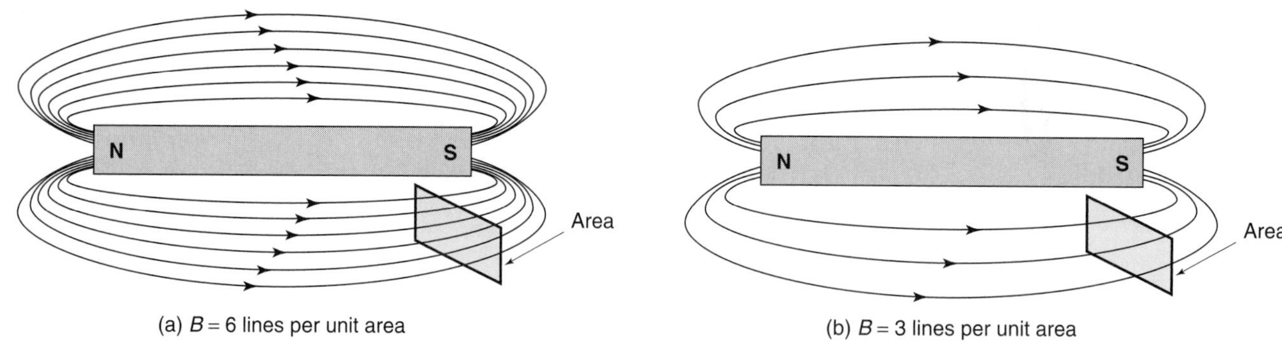

Figure 8.4 Flux density.

Note that the field is actually a three-dimensional space (similar to a cylinder) that surrounds the magnet.

The "strength" of the force produced by a magnet is a function of the field size and the amount of flux that it contains. The greater the amount of *flux per unit area,* the stronger the magnetic force. **Flux density** is *a measure of flux per unit area,* and therefore, an indicator of magnetic strength. (In fact, *flux density* is often referred to as *field strength.*) By formula, flux density is found as

$$B = \frac{\phi}{A} \tag{8.2}$$

where

B = the flux density
ϕ = the amount of flux
A = the cross-sectional area containing the flux

OBJECTIVE 5 ➤

Note: The cgs unit of measure for flux density is the *gauss (G).* One gauss equals 1 maxwell per square centimeter (Mx/cm^2).

The concept of flux density is illustrated in Figure 8.4. As you can see, the shaded area in Figure 8.4a has six lines of flux passing through it. The same area in Figure 8.4b has only three lines of flux. Since flux density is a measure of flux per unit area, the flux density in Figure 8.4a is twice that in Figure 8.4b.

The SI unit of measure for flux density is the **tesla (T),** defined as *one weber of flux per square meter (Wb/m^2).* The calculation of flux density (in teslas) is demonstrated in the following example.

Example 8.1

Calculate the flux density for the shaded area in Figure 8.5. Assume that each line of force represents 40 μWb of flux.

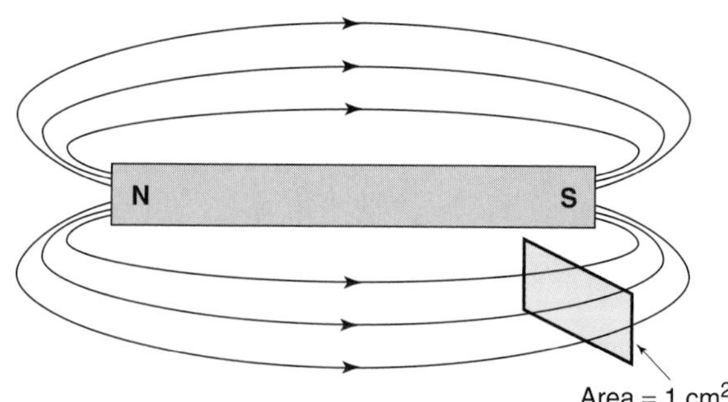

Figure 8.5

OBJECTIVE 6 ➤

Solution: The shaded area is shown to be one square centimeter (1 cm^2), which is equal to 1×10^{-4} square meters. Therefore, the flux density represented in Figure 8.5 is found as

$$B = \frac{\phi}{A} = \frac{(3)(40 \times 10^{-6}\ \text{Wb})}{1 \times 10^{-4}\text{m}^2} = 1.2\ \text{Wb/m}^2 = 1.2\text{T}$$

Practice Problem 8.1

Refer to Figure 8.6. If each line in Area A represents 75 μWb of flux, what is the flux density at the pole of the magnet? (Assume Area A = 3 cm^2).

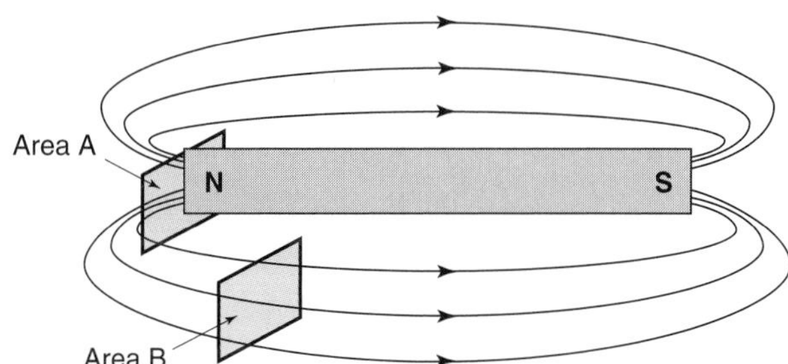

Figure 8.6 Flux density is greatest at the poles of a magnet.

For any given magnet, the maximum flux density is measured at the poles, as illustrated in Figure 8.6. As you can see, there is twice as much flux in Area A as in Area B. Since all lines of flux emanate from (and return to) the magnetic poles, the flux density must be greater at the poles than anywhere else in the magnetic field.

Section Review

1. What is *magnetic force*?
2. How is magnetic force commonly represented?
3. What are the *poles* of a magnet? What are they called?
4. How do *like* and *unlike* poles affect each other?
5. What is *magnetic flux*?
6. List and define the common units of measure for magnetic flux.
7. What is a *magnetic field*?
8. What is *flux density*? What does it indicate?
9. List and define the units of measure for flux density.
10. Demonstrate that 1 μWb equals 100 lines of magnetic flux. Critical Thinking

8.2 Magnetic Characteristics of Materials

Some materials respond to the presence of a magnetic force, while others do not. In this section, we will discuss the basic magnetic characteristics and classifications of materials.

Permeability

◄ OBJECTIVE 7

Permeability (μ) is *a measure of the ease with which lines of magnetic force are established within a material.* Materials with high permeability have the ability to concentrate magnetic lines of force, as illustrated in Figure 8.7. The bar shown is made of soft iron, a material with a permeability that is much higher than that of the surrounding air. Since the iron bar more readily accepts the lines of force, they align themselves in the iron bar as shown.

Since the lines of force shown in Figure 8.7 are drawn to the iron bar, the flux density at any point in the bar is greater than it is in the surrounding air. Thus, high-permeability materials can be used to increase flux density.

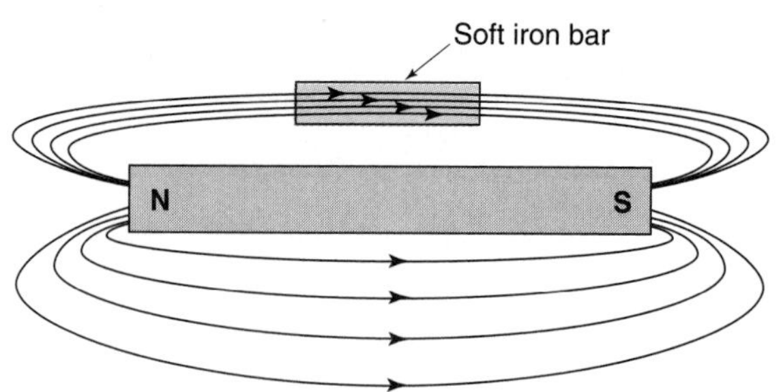

Figure 8.7 The effect of a high-permeability material on magnetic flux.

Relative Permeability

In general, materials are rated according to their **relative permeability (μ_r)**, which is *the ratio of the material's permeability to that of free space.* In SI units, the permeability of free space is known to have a value of

Note: The value of $4\pi \times 10^{-7}$ given in the permeability equation is often written as 1.26×10^{-6}. The two values are the same.

$$\mu_0 = 4\pi \times 10^{-7} \frac{\text{Wb}}{\text{A} \cdot \text{m}}$$

where $\frac{\text{Wb}}{\text{A} \cdot \text{m}}$ (webers per ampere–meter) is the unit of measure for permeability. (The *ampere-meter* is discussed later in this chapter.)

OBJECTIVE 8 ➤ When the permeability of a material is known, its relative permeability is found as

$$\mu_r = \frac{\mu_m}{\mu_0} \tag{8.3}$$

where

μ_r = the relative permeability of the material

μ_m = the permeability of the material, in $\frac{\text{Wb}}{\text{A} \cdot \text{m}}$

μ_0 = the permeability of free space

Since relative permeability is a ratio of one permeability value to another, it has no unit of measure.

Why relative permeability is important.

Relative permeability is important, because it equals the ratio of the flux density within a material to the flux density in free space. That is, it tells us how much flux density is increased by a given material, relative to the space surrounding it. For example, let's assume that an iron bar has a value of $\mu_r = 200$. If the flux density in the space surrounding the bar equals 2×10^{-3} T, the flux density within the iron bar is 200 times as great, or 4×10^{-1} T.

The Source of Magnetism: Domain Theory

OBJECTIVE 9 ➤ As you know, all *atoms* contain *electrons* that orbit a *nucleus.* It is believed that *every electron in a given atom has both an electric charge and a magnetic field.* The electric charge is always *negative* and is offset by a matching *positive* charge in the nucleus. The polarity of the magnetic field, however, depends on *the direction in which the electron spins on its axis.*

Each electron in an atom spins in a clockwise or a counterclockwise direction. For example, let's assume that each electron in Figure 8.8 is spinning in the direction indicated. When an atom contains an equal number of electrons spinning in each direction, their magnetic fields cancel each other. As a result, the atom does not generate magnetic force. However, when the electrons spinning in one direction outnumber the ones spinning in

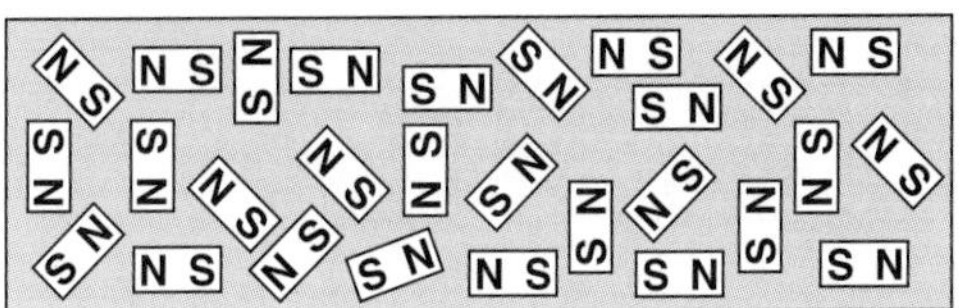

(a) Random poles

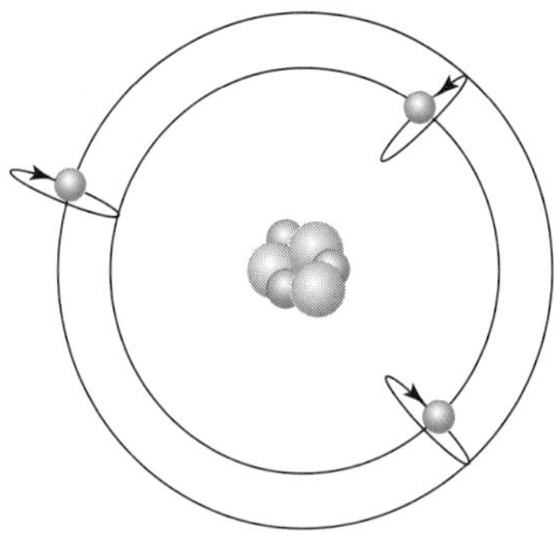

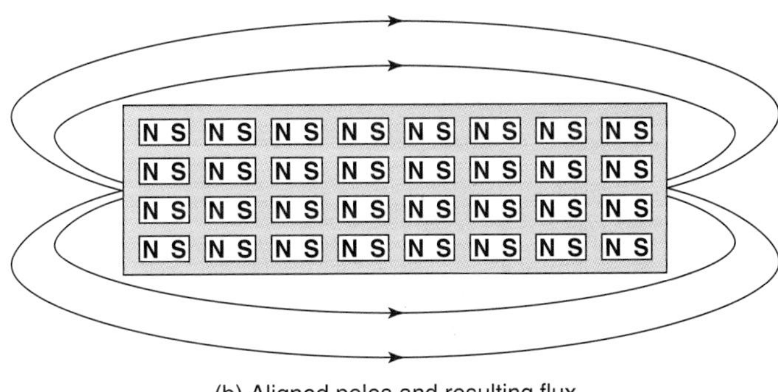

(b) Aligned poles and resulting flux

Figure 8.8 Electron spin.

Figure 8.9

the other direction, the atom acts as a source of magnetic force. The polarity of the magnetic force depends on the net spin direction of the electrons.

Domain theory states that *atoms with like magnetic fields join together to form magnetic domains.* Each domain acts as a magnet, with two poles and a magnetic field. In a nonmagnetized material, the domains are randomly positioned, as shown in Figure 8.9a. As a result, the magnetic fields produced by the domains cancel each other, and the material has no net magnetic force. However, if the domains can be aligned (as shown in Figure 8.9b), the material generates an external magnetic force.

Magnetic Induction

Most magnets are artificially produced by an external force. In Figure 8.10a, a bar of steel is stroked with a magnet. The flux generated by the magnet magnetizes the steel. It is also possible to magnetize an object simply by placing it within the field of a magnet, as illustrated in Figure 8.10b. Note that the poles in the iron bar shown are opposite those in the magnet. ◀ *OBJECTIVE 10*

The process of producing an artificial magnet is referred to as **induction.** The methods described above are referred to as **magnetic induction,** because *an external magnetic force is used to magnetize the material.* An iron bar can also be magnetized using a strong dc current, as shown in 8.10c. We will discuss the relationship between electric currents and magnetic fields later in this chapter. For now, we will simply state that a strong dc current can be used to magnetize a material.

High-permeability materials concentrate magnetic flux. As a result, they are easily magnetized. On the other hand, low-permeability materials do not concentrate magnetic flux and are not as easily magnetized.

Retentivity

High-permeability materials are easily magnetized, but quickly lose most of their magnetic strength. On the other hand, low-permeability materials are difficult to magnetize, but they tend to retain most of their magnetic strength for a long period of time. ◀ *OBJECTIVE 11*

Retentivity is *the ability of a material to retain its magnetic characteristics after a magnetizing force has been removed.* High-retentivity materials, such as hardened steel, are used to produce *permanent* magnets. Once magnetized, they remain magnetic. Low-retentivity materials, such as soft iron, act as *temporary magnets.* That is, they lose most

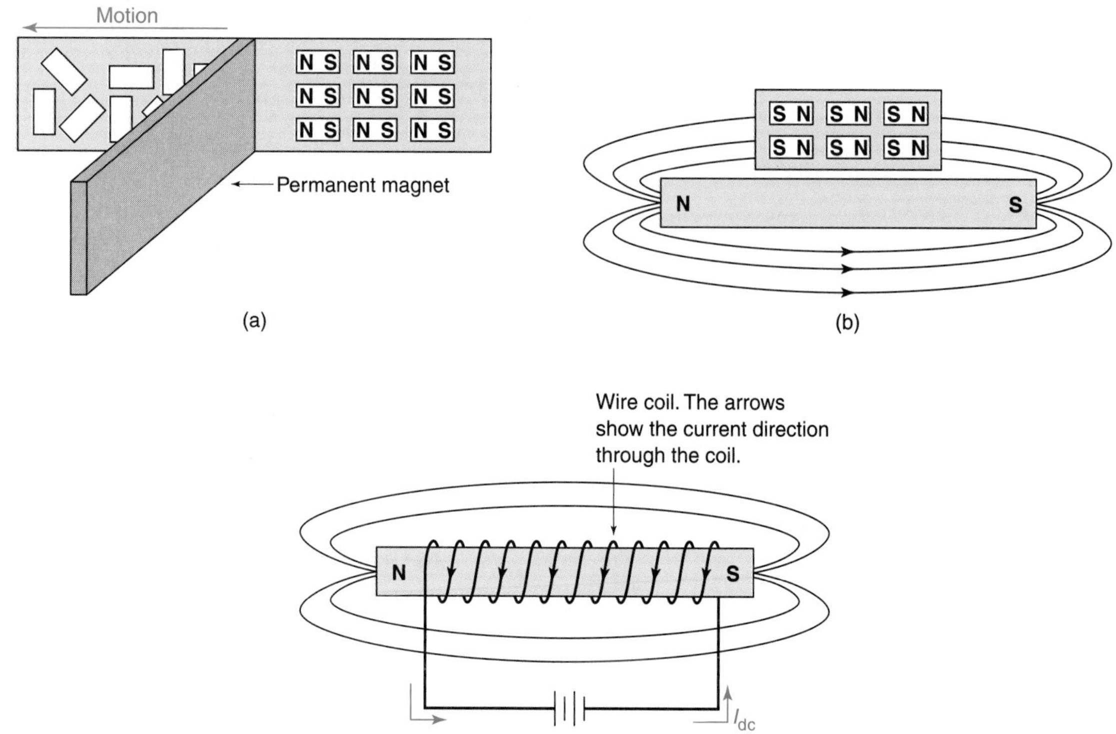

Figure 8.10 Aligning magnetic domains.

of their magnetic strength soon after being isolated from an external source of magnetic force.

Reluctance

OBJECTIVE 12 ➤ **Reluctance** is *the opposition that a material presents to magnetic lines of force.* In essence, reluctance can be viewed as the magnetic resistance of a given material.

The reluctance of a material can be found as

$$\mathcal{R} = \frac{\ell}{\mu A} \tag{8.4}$$

where

$\mathcal{R}$ = the reluctance of the material
ℓ = the length of the material
μ = the permeability of the material
A = the cross-sectional area of the material

Note: As discussed in the next section, current can be used to generate flux through a material. The A/Wb unit of measure for reluctance indicates that the amount of current required to generate flux is directly proportional to the reluctance of the material.

Equation (8.4) demonstrates that *reluctance varies inversely with permeability.* That is, high-permeability materials have low reluctance values, and vice versa. This makes sense when you consider that:

- Permeability is a measure of how easily a material passes magnetic flux.
- Reluctance is a measure of opposition to magnetic flux.

Reluctance is typically given as a *ratio.* For example, the SI unit of measure for reluctance is *amperes per weber* (A/Wb). In the cgs system, reluctance is measured in *gilberts per maxwell* (Gb/Mx).

TABLE 8.1 Magnetic Classifications of Materials

	Ferromagnetic	*Paramagnetic*	*Diamagnetic*
Relative permeability (μ_r)	Much greater than 1	Slightly greater than 1	Less than 1
Retentivity	Low	Moderate	High
Reluctance	Low	Moderate	High
Considered to be	Magnetic	Slightly magnetic	Nonmagnetic
Examples	Iron	Aluminum	Copper
	Steel	Air	Silver
	Nickel	Platinum	Gold

Material Classifications

◄ *OBJECTIVE 13*

All materials can be classified according to their magnetic characteristics. The magnetic classifications of materials are identified in Table 8.1, along with their characteristics. High-permeability (*ferromagnetic*) materials are considered to be magnetic, while all others are generally considered to be nonmagnetic.

Section Review

1. What is *permeability*?
2. What effect does a high-permeability material have on magnetic flux?
3. What is *relative permeability*?
4. What is a *magnetic domain*?
5. List and describe the methods of *magnetic induction.*
6. What is *retentivity*?
7. What types of materials are used to produce permanent magnets? Temporary magnets?
8. What is *reluctance*? By what other name is it known?
9. What is the relationship between reluctance and permeability? Explain your answer.
10. Using the information contained in Table 8.1, compare and contrast the magnetic characteristics of aluminum and iron.
11. Using equations (8.4) and (1.4), describe the similarities between *reluctance* and *resistance.* Critical Thinking

8.3 Electromagnetism

◄ *OBJECTIVE 14*

In 1820, H.C. Oersted (a Danish physicist) discovered that an electric current produces a magnetic field. Using a structure similar to the one shown in Figure 8.11, he found that the compass always aligned itself at 90° angles to a current-carrying wire. When the direction of the current reversed, so did the compass alignment (as shown in the figure).

Oersted's work showed that a current-carrying wire generates lines of magnetic force that circle the wire as shown in Figure 8.12. As illustrated, the polarity of the flux depends on the direction of the current.

The Left-Hand Rule

The **left-hand rule** is *a memory aid that helps you to determine the polarity of the magnetic field that results from the current through a wire.* The left-hand rule is illustrated in Figure 8.13. As you can see, the fingers indicate the polarity of the flux when the thumb points in the direction of the current. If the current is reversed, the left hand is

Figure 8.11

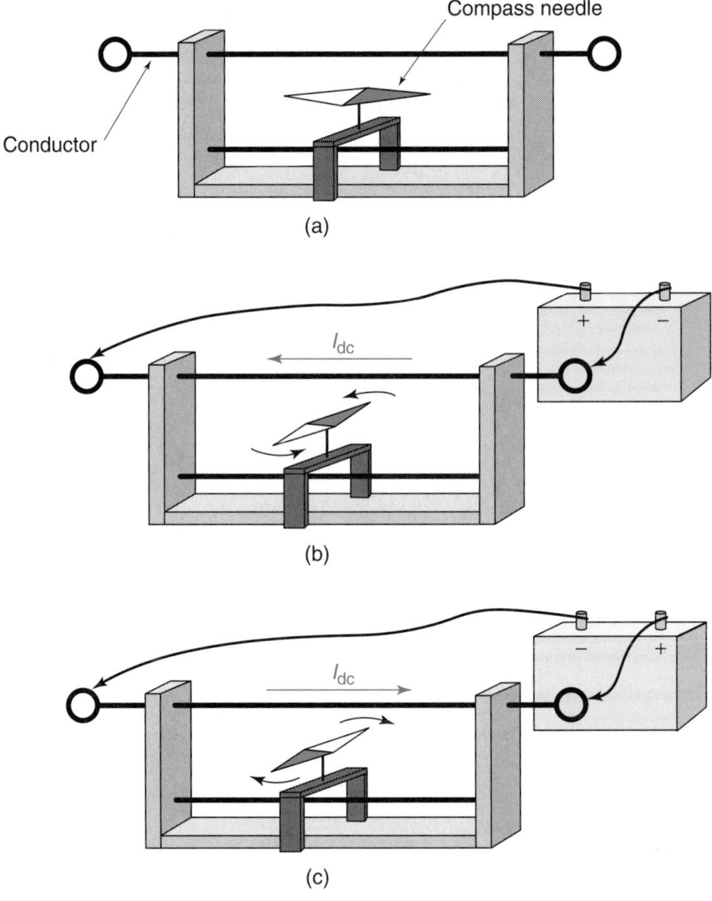

Figure 8.12
Current generating magnetic lines of force.

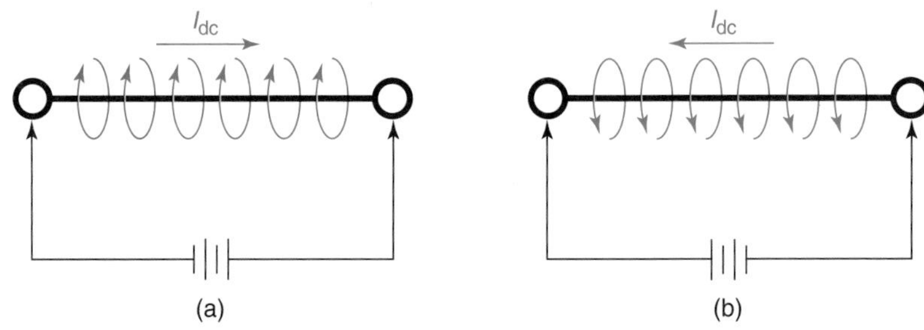

Figure 8.13
The left-hand rule.

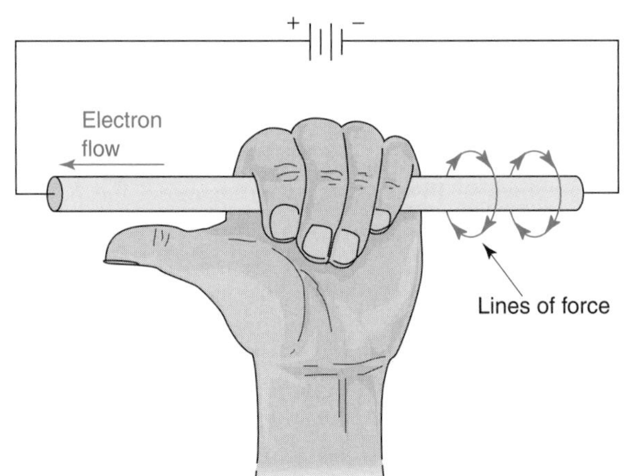

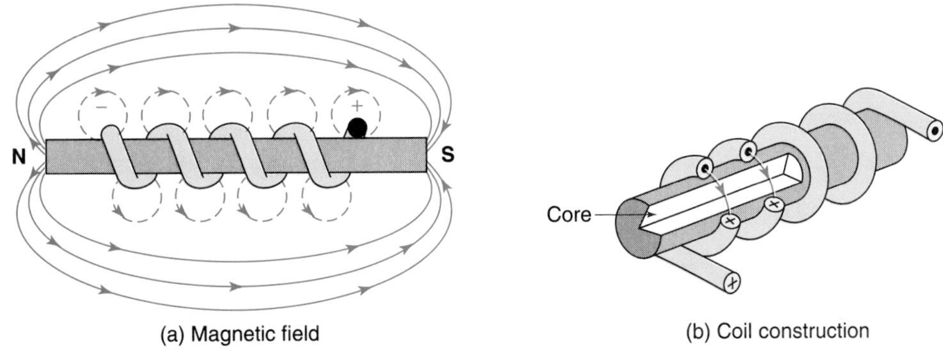

Figure 8.14 The magnetic field generated by the current through a coil.

turned the other way. When repositioned, the fingers indicate that the flux polarity has also reversed.

The Coil

A **coil** is *a wire wrapped into a series of loops for the purpose of concentrating magnetic lines of force.* When a current passes through a coil, the lines of force that are generated by the current add together to form a magnetic field, as shown in Figure 8.14. Note that *the space inside of the coil* is called the **core.** If the core is made using an iron bar, the lines of flux are concentrated in the bar. As described in the last section, the iron bar becomes magnetic by induction. (However, the magnetic characteristics quickly fade after power is removed from the coil because of iron's low retentivity.)

The left-hand rule can be applied to a coil, as shown in Figure 8.15. In this case, the thumb points to the north-seeking pole (N) when the fingers point in the direction of the coil current. As illustrated, the magnetic poles reverse when the current direction through the coil reverses.

Magnetomotive Force

In Figure 8.15, the current through the coil is generating magnetic flux through the reluctance of the core. Generating magnetic flux through any value of reluctance requires

◀ *OBJECTIVE 15*

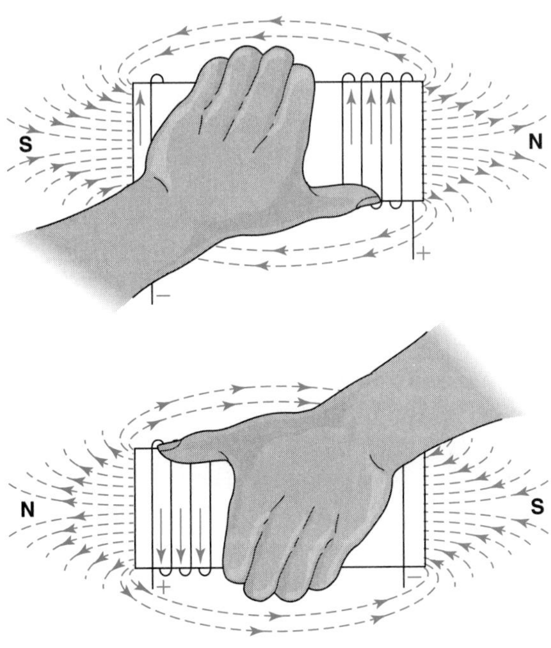

Figure 8.15 The left-hand rule for a coil.

Note: In cgs units, mmf is measured in *gilberts (Gb)*, flux is measured in *maxwells (Mx)*, and reluctance is measured in *gilberts per maxwell (Gb/Mx)*.

TABLE 8.2 Comparison of Basic Electric and Magnetic Quantities

	Electric Circuit	*Magnetic Circuit*
Force	Electromotive force (emf) (measured in volts)	Magnetomotive force (mmf) (measured in ampere-turns)
Result	Current (I) (measured in amperes)	Flux (ϕ) (measured in webers)
Opposition	Resistance (R) (measured in ohms)	Reluctance ($\mathcal{R}$) (measured in ampere-turns/weber)

a certain amount of **magnetomotive force (mmf).** As Table 8.2 indicates, mmf is to magnetic circuits what electromotive force (emf) is to electric circuits.

OBJECTIVE 16 ➤ Just as Ohm's law defines the relationship between emf, current, and resistance, *Rowland's law* defines the relationship between mmf, flux, and reluctance. **Rowland's law** states that *magnetic flux is directly proportional to magnetomotive force and inversely proportional to reluctance.* By formula,

$$\phi = \frac{F}{\mathcal{R}} \tag{8.5}$$

where

ϕ = the magnetic flux, in webers (Wb)

F = the magnetomotive force, in *ampere-turns* (A · t)

$\mathcal{R}$ = the reluctance, in ampere-turns per weber $\left(\frac{\text{A} \cdot \text{t}}{\text{Wb}}\right)$

Equation (8.5) shows that there are two ways to increase the amount of flux produced by a magnet like the one illustrated in Figure 8.15. The amount of flux produced by the magnet can be increased by either:

- Increasing the amount of magnetomotive force.
- Using a lower-reluctance material as the core.

For example, iron-core magnets have more flux than equivalent air-core magnets.

Ampere-Turns

The magnetomotive force produced by a coil is proportional to the coil current and the number of turns. These two factors are combined into a single value called the **ampere-turn (A · t).** The ampere-turn rating of a coil is the product of the number of turns and the coil current, as illustrated in Figure 8.16. Note that ampere-turns are commonly represented using *NI* (*N* for number of turns, and *I* for current).

The magnetomotive force produced by a coil equals the ampere-turns value of the coil. By formula,

Note: In the cgs system, mmf (in Gb) is found as

$mmf = 1.2566NI$

$$mmf = NI \tag{8.6}$$

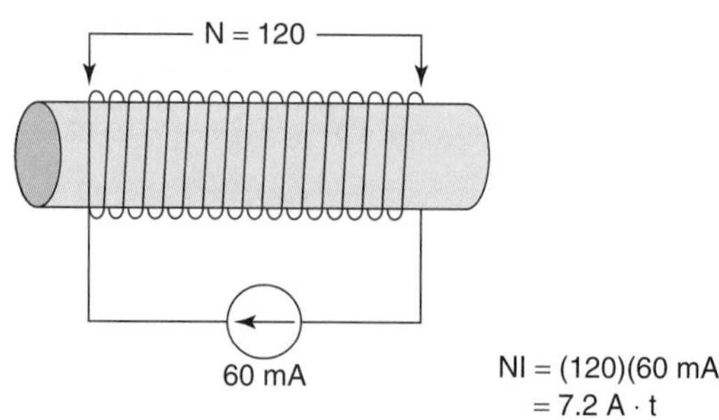

Figure 8.16 The ampere-turns value for a coil.

The flux density produced by a coil depends on three factors: the permeability of the core, the ampere-turns of the coil, and the length of the coil. When these values are known, the flux density produced by a coil (in teslas) is found as

$$B = \frac{\mu_m NI}{\ell} \tag{8.7}$$

where

B = the flux density, in teslas (T)

μ_m = the permeability of the core material, in $\frac{\text{Wb}}{\text{A} \cdot \text{m}}$

NI = the ampere-turns value of the coil

ℓ = the length of the coil, in meters (m)

If you take a look at equation (8.6), you'll see that coil current (I) is the only value that can be easily adjusted to vary flux density. All other factors are physical attributes of the coil.

Hysteresis

◄ *OBJECTIVE 17*

As you know, most materials experience a drop in magnetic strength (flux density) when a magnetizing force is removed. For example, the iron core shown in Figure 8.17 experiences a significant drop in flux density when current is removed from the coil. The time required for the flux density to drop after the magnetizing force is removed depends on the *retentivity* of the material. *The time lag between the removal of a magnetizing force and the drop in flux density* is referred to as **hysteresis.**

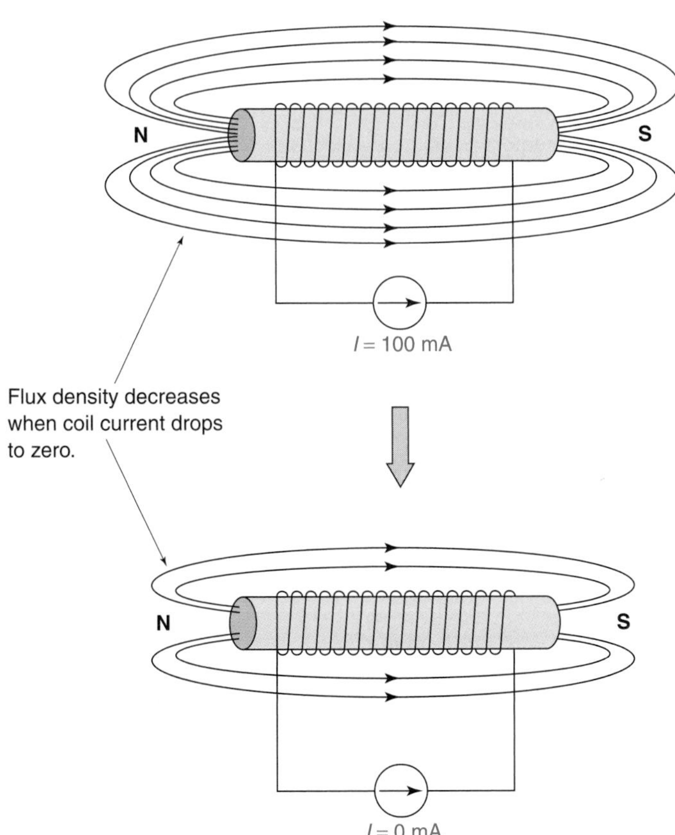

Figure 8.17 Hysteresis.

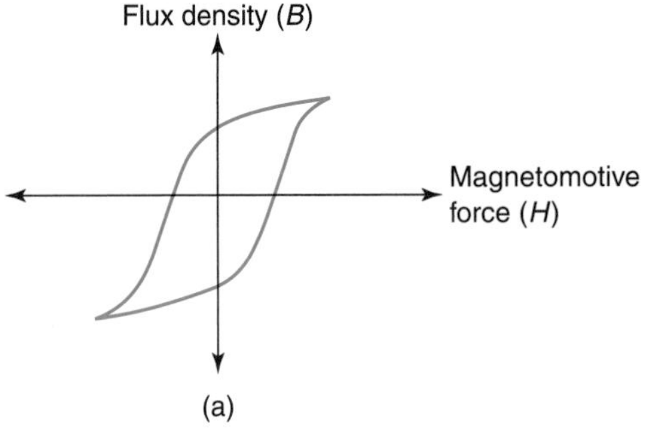

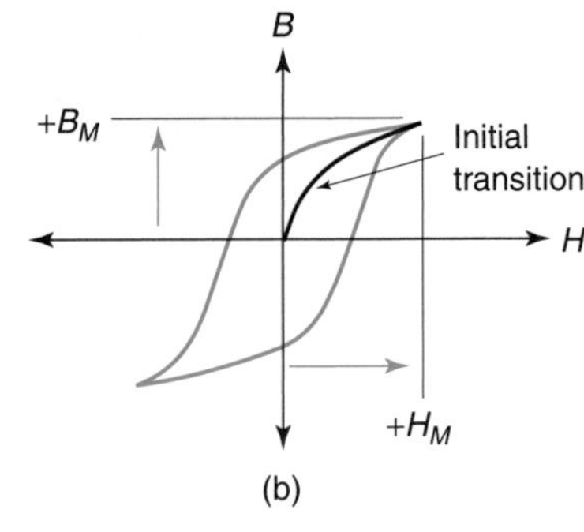

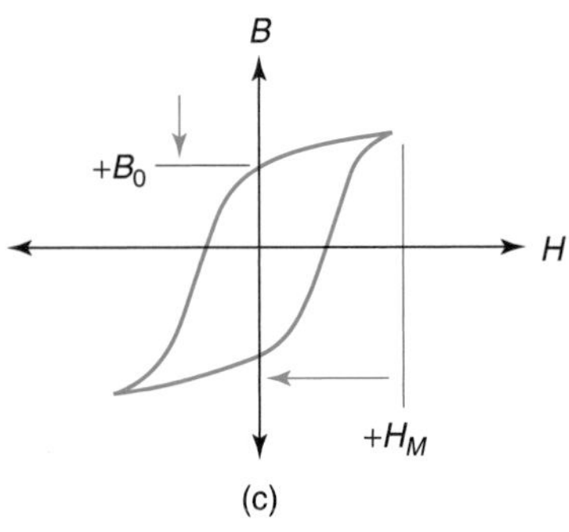

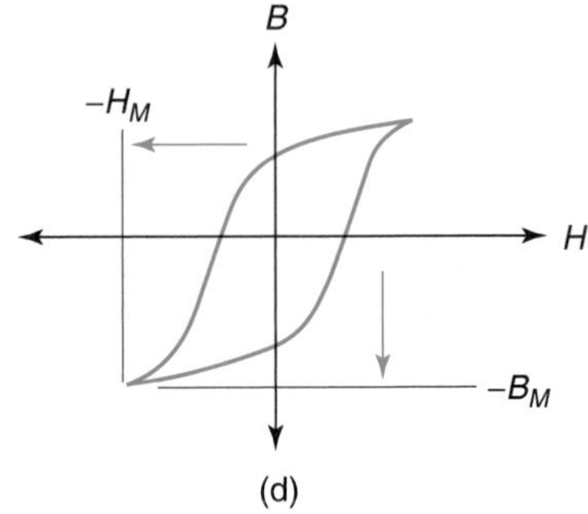

Figure 8.18 Hysteresis curves.

The retentivity of a material is commonly represented using the **hysteresis curve** shown in Figure 8.18a. This curve represents *flux density (B) as a function of the field intensity (H)* of the magnetizing force. Figure 8.18b shows what happens when a current is applied to a coil like the one shown in Figure 8.17. The field intensity (H) of the magnetizing force reaches its maximum value ($+H_M$), as does the flux density through the core. When power is removed from the coil, the value of H drops to zero. However, some **residual flux density** remains in the core, as represented by the positive value of $+B_0$ in the curve (Figure 8.18c). The *time* required for the flux density to drop as well as the *amount* of flux density depend on the *retentivity* of the material. The higher the retentivity of the material:

- The longer it takes for B to drop to $+B_0$.
- The greater the value of $+B_0$.

Note: Energy must be used to overcome any residual magnetism in a material when switching from one magnetic polarity to the other. This loss of energy is referred to as *hysteresis loss.*

If the current through the coil is reversed, the polarity of the magnetomotive force and the resulting flux density also reverse. This condition is represented in Figure 8.18d. Again, if power is removed, magnetomotive force drops to zero, and the residual flux density drops to the value represented on the curve. Note that *the magnetomotive force required to return the value of flux density (B) to zero* is referred to as the **coercive force.**

Section Review

1. What relationship did Orstead discover between magnetism and current?
2. What happens to the polarity of a magnetic field when the current is reversed?
3. What is a *coil*? What is the *core* of a coil?
4. What is *magnetomotive force (mmf)*?
5. What is *Rowland's law*? Which law of electrical circuits does it resemble?
6. What is an *ampere-turn*?
7. What is the relationship between ampere-turns and flux density?
8. In a practical sense, how is the amount of magnetomotive force produced by a given coil usually varied?
9. What is *hysteresis*?

10. What is *residual flux density*?
11. Using the hysteresis curve, describe what happens when power is applied to and removed from a coil.

8.4 Related Topics

In this section, we're going to discuss briefly several topics that relate to magnetism. Among the topics we will cover are magnetic shielding, common shapes of magnets, and their proper care and storage.

Magnetic Shielding

◄ *OBJECTIVE 18*

Some instruments (like analog meter movements) are sensitive to magnetic flux and may operate improperly when placed in a magnetic field. To overcome this problem, **magnetic shielding** is *used to insulate the instrument from the effects of the magnetic flux.*

Note that *there is no known magnetic insulating material.* Flux passes through any material, regardless of its reluctance. Because of this, *magnetic shielding is accomplished by diverting lines of flux around the object to be shielded.* The means by which this is accomplished is illustrated in Figure 8.19.

Figure 8.19 shows the lines of force between two magnetic poles. As you can see, a *soft-iron ring* has been placed between the poles. Rather than follow a straight-line path between the poles, the lines of force bend to pass through the high-permeability material. If a magnetically sensitive device is placed in the middle of the iron ring, the lines of force are diverted around the device. As a result, the device is shielded from the flux.

Ring Magnets

◄ *OBJECTIVE 19*

The space directly between two magnetic poles is called an **air gap.** Figure 8.20 shows the air gap for several types of magnets.

Ring magnets are *magnets without any identifiable poles or air gaps that do not generate an external magnetic field.* Two ring magnets are shown in Figure 8.21. Figure 8.21a shows a ring magnet that is actually made up of *two horseshoe magnets.* When connected as shown, the lines of flux remain in the ring magnet.

When constructed as shown in Figure 8.21b, a ring magnet has no identifiable poles. Rather, a continuous flow of magnetic flux circles within the magnet. Ring magnets are most commonly used in *analog meter movements.*

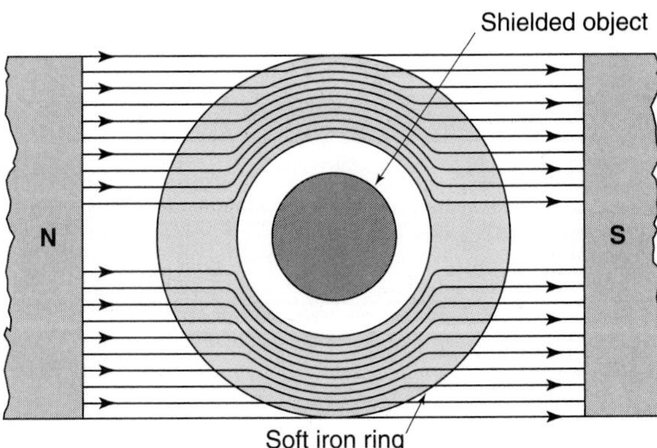

Figure 8.19 Magnetic shielding.

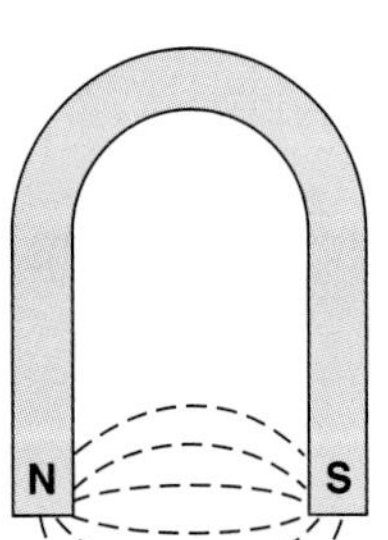

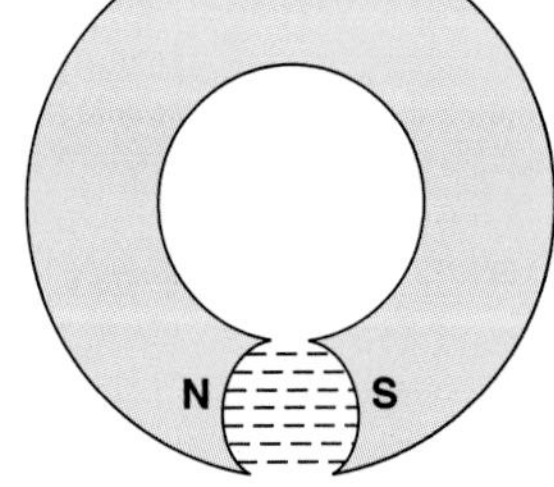

Figure 8.20 Horseshoe and round magnet air gaps.

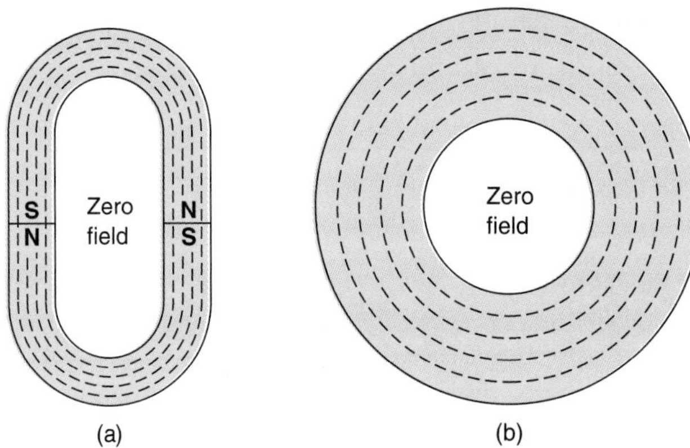

Figure 8.21 Ring magnets.

Care and Storage

OBJECTIVE 20 ➤ Over time, even a "permanent" magnet can lose most of its magnetic strength if not properly cared for. The most common causes of loss of magnetic strength for a permanent magnet are:

- Allowing the magnet to be jarred or dropped.
- Exposing the magnet to sufficiently high temperatures.
- Improperly storing a magnet.

When a magnet is jarred or dropped, the impact can knock the magnetic domains out of alignment. As a result, magnetic strength is lost. Excessive heat can also cause a misalignment of the domains, resulting in a loss of magnetic strength.

When storing magnets, care must be taken to ensure that *the flux produced by the magnet is not lost externally.* This is accomplished by storing one or more magnets in one of the fashions shown in Figure 8.22. Each of the methods shown has the same effect as constructing a ring magnet. That is, the flux generated by the magnet(s) stays within a loop rather than being lost externally.

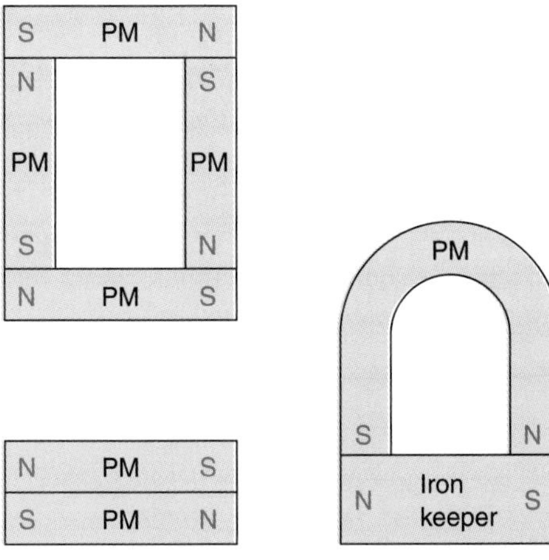

Figure 8.22 Storing magnets.

Section Review

1. What is *magnetic shielding*? How is it accomplished?
2. What is an *air gap*?
3. What is the relationship between air gap width and magnetic strength?
4. What characteristics distinguish a *ring magnet* from other types of magnets?

5. What is the most common application for ring magnets?
6. List the most common causes of loss of magnetic strength for a permanent magnet.
7. Describe the proper storage of a magnet.

Key Terms

The following terms were introduced in this chapter on the pages indicated:

Air gap, 229
Ampere-turns (A · t), 226
Coercive force, 228
Coil, 225
Core, 225
Domain theory, 221
Flux density, 218
Hysteresis, 227
Hysteresis curve, 228
Induction, 221
Left-hand rule, 223
Magnetic field, 217
Magnetic flux, 217
Magnetic force, 216
Magnetic induction, 221
Magnetic shielding, 229
Magnetomotive force (mmf), 226
Microweber (μWb), 217
North-seeking pole (N), 216
Permeability (μ), 219
Poles, 216
Relative permeability (μ_r), 220
Reluctance, 222
Residual flux density, 228
Retentivity, 221
Ring magnet, 229
Rowland's law, 226
South-seeking pole (S), 216
Tesla (T), 218
Weber (Wb), 217

Practice Problems

1. Calculate the flux density for the shaded area in Figure 8.23a. Assume that each line of force represents 120 μWb of flux.
2. Calculate the flux density for the shaded area in Figure 8.23b. Assume that each line of force represents 200 μWb of flux.

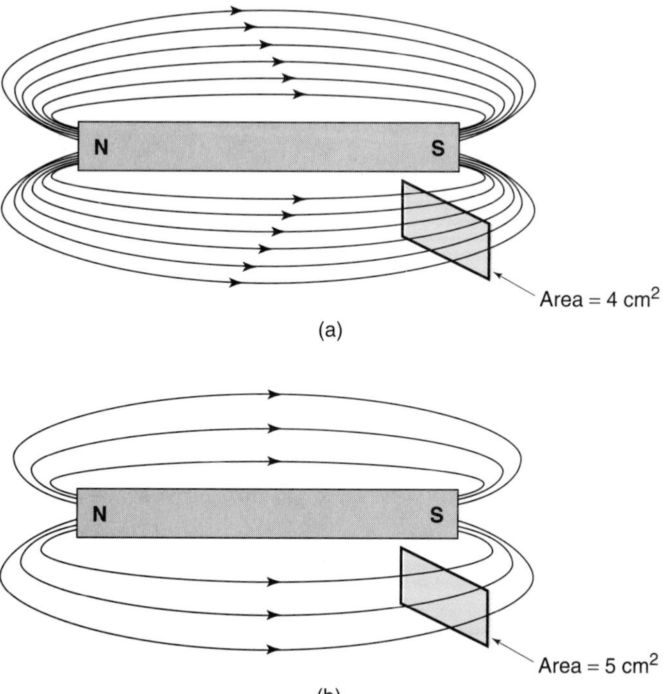

Figure 8.23

3. A magnet with a cross-sectional area of 1.5 cm^2 generates 300 μWb of flux at its poles. Calculate the flux density at the poles of this magnet.
4. A magnet with a cross-sectional area of 3 cm^2 generates 1200 μWb of flux at its poles. Calculate the flux density at the poles of this magnet.

5. The permeability of a piece of low-carbon steel is determined to be 7.54×10^{-4} Wb/A · m. Calculate the relative permeability of this material.

6. Calculate the relative permeability of free space.

7. The permeability of a piece of permalloy (an alloy made primarily of iron and nickel) is determined to be 0.1 Wb/A · m. Calculate the relative permeability of this material.

8. The permeability of a piece of iron ingot is determined to be 3.77×10^{-3} Wb/A · m. Calculate the relative permeability of this material.

9. Calculate the flux density produced by the coil shown in Figure 8.24a.

10. Calculate the flux density produced by the coil shown in Figure 8.24b.

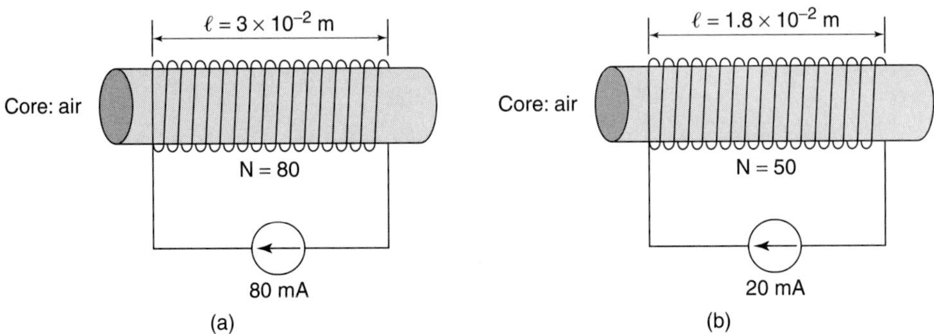

Figure 8.24

11. The permeability of iron is determined to be 2.5×10^{-3} Wb/A · m. Calculate the flux density produced by the coil shown in Figure 8.25a.

12. The permeability of iron is determined to be 2.5×10^{-3} Wb/A · m. Calculate the flux density produced by the coil shown in Figure 8.25b.

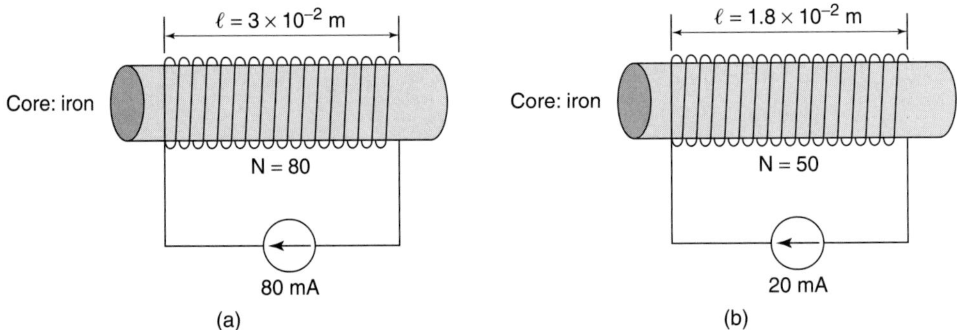

Figure 8.25

Looking Back

These problems relate to material presented in earlier chapters. The chapters are identified in brackets.

13. Calculate the value of R_4 that will balance the bridge shown in Figure 8.26.

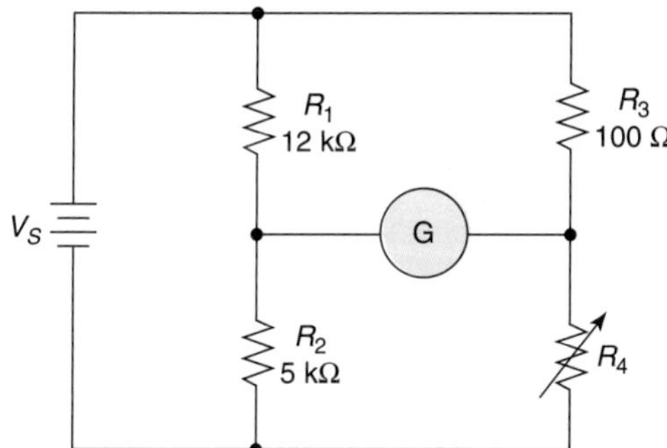

Figure 8.26

14. The following components are connected in parallel with a 15 V source: a potentiometer, a 220 Ω resistor, and a 1 kΩ resistor. What is the potentiometer setting when $P_S = 3W$? [Chapter 5]

15. Derive the Thevenin equivalent for the circuit shown in Figure 8.27 from the perspective of Load 3. Assume that $R_{L1} = R_{L2} = 300\ \Omega$. [Chapter 7]

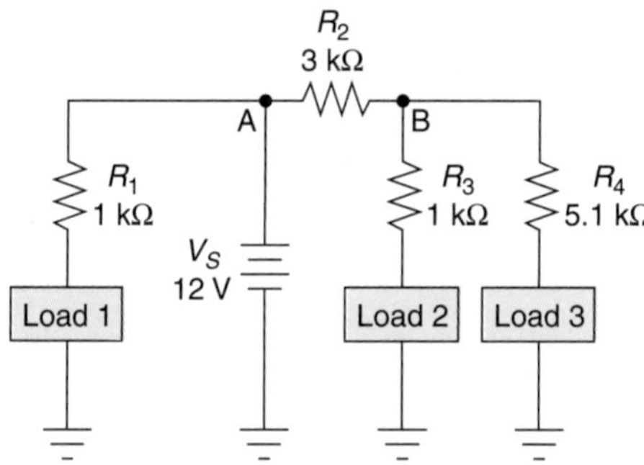

Figure 8.27

16. Determine the value of P_{L3} for the circuit shown in Figure 8.27. Assume that $R_{L1} = R_{L2} = R_{L3} = 300\ \Omega$. [Chapter 6]

Answer to the Example Practice Problem

8.1. 4.5 T

Part II
Basic AC Principles and Circuits

chapter 

Alternating Current and AC Measurements

Objectives

After studying the material in this chapter, you should be able to:

1. Describe *alternating current (ac).*
2. Describe the makeup of a *sinusoidal waveform.*
3. Describe the relationship between *cycle time* and *frequency.*
4. Determine the cycle time and frequency of a waveform that is displayed on an oscilloscope.
5. Define and identify each of the following magnitude-related values: *peak, peak-to-peak, instantaneous, full-cycle average,* and *half-cycle average.*
6. Describe the relationship between *average ac power* and *rms values.*
7. Describe the relationship between *peak* and *rms values.*
8. Describe how *effective* (rms) values are measured.
9. Describe the magnetic induction of current.
10. Compare and contrast phase and time measurements.
11. Describe the *phase angle* between two waveforms.
12. Calculate any instantaneous sine wave value using the *degree approach.*
13. Describe a sine wave in terms of *radians* and *angular velocity.*
14. Calculate any instantaneous sine wave value using the *radian approach.*
15. Compare and contrast the degree and radian approaches to measuring *phase.*
16. Compare and contrast *static* and *dynamic* values.
17. Discuss the effects of a *dc offset* on sine-wave measurements.
18. Describe *wavelength* and calculate its value for any waveform.
19. Discuss the relationship between a waveform and its harmonics.
20. Describe and analyze *rectangular waveforms.*
21. Describe *sawtooth waveforms.*

In Chapter 1, you were told that there are two types of current: *direct current (dc)* and *alternating current (ac)*. In dc circuits, charge flows in only one direction.

The receptacles (wall outlets) in your home provide power in the form of alternating current and voltage. Everything you connect to the receptacles relies on the alternating current and voltage for their operation. In this chapter, we will focus on ac measurements, sources, and characteristics. As your studies continue, you will see that most systems contain both direct and alternating currents and voltages.

9.1 Alternating Current: Overview and Time Measurements

Generally, the term **alternating current (ac)** is used to describe *any current that periodically changes direction.* For example, Figure 9.1 shows how a current changes as a function of time (t). The vertical axis of the graph represents the **magnitude** (*value*) and the direction of the current (I), while the horizontal axis represents time (t). The current builds to a peak value in one direction and returns to zero (t_0 to t_1). It then builds to a peak value in the other direction and returns to zero (t_1 to t_2). Note that the current not only changes direction but also changes in magnitude. ◄ *OBJECTIVE 1*

A graph of the relationship between magnitude and time is referred to as a **waveform.** For reasons we will discuss later, the waveform shown in Figure 9.1 is called a **sinusoidal waveform** or **sine wave.**

The sine wave is not the only type of ac waveform. We will look at some nonsinusoidal waveforms later in this chapter.

Basic AC Operation

The operation of a simple AC circuit is illustrated in Figure 9.2. Using the time periods shown in the figure, the operation of the circuit can be described as follows:

Time Period	*Circuit Operation*
t_0 to t_1	As the output from the voltage source increases from 0 V to its positive peak value (+V peak), the current also increases from 0 A to its positive peak value. Note the voltage polarity and current direction.
t_1 to t_2	The output from the voltage source returns to 0 V, and the circuit current returns to 0 A.
t_2 to t_3	The output from the voltage source increases to its negative peak value (−V peak), as does the circuit current. Note that both voltage polarity and current direction have changed.
t_3 to t_4	The output from the voltage source returns to 0 V, and the circuit current drops to 0 A.

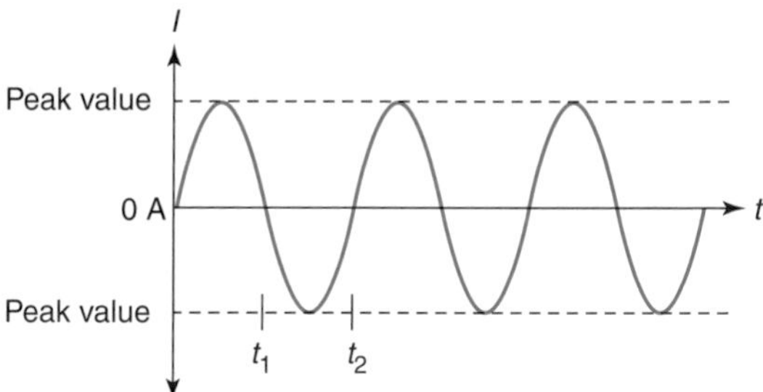

Figure 9.1 Alternating current (ac).

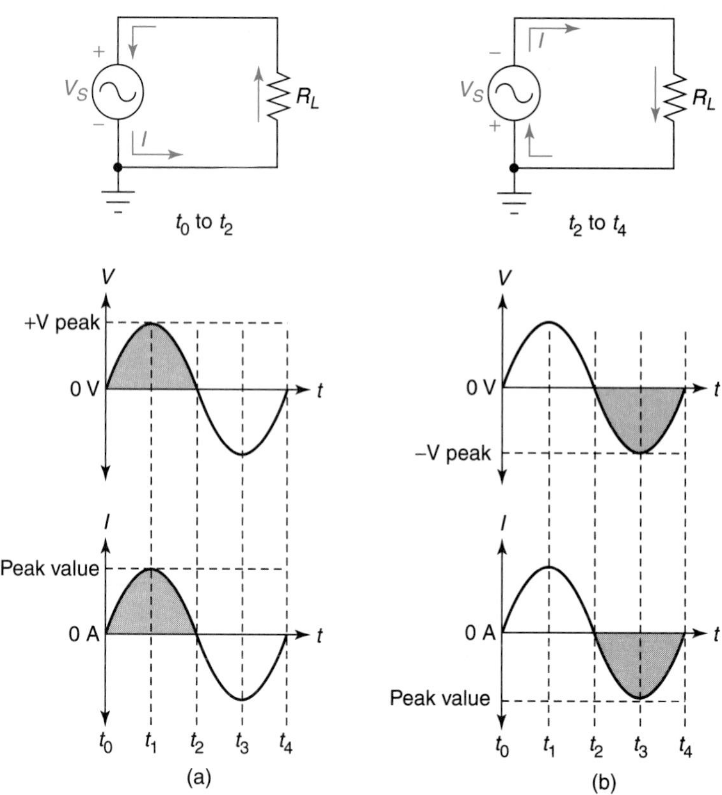

Figure 9.2 Basic ac operation.

Alternations and Cycles

OBJECTIVE 2 ➤ A sine wave is normally described using the labels shown in Figure 9.3a. *The positive and negative transitions* are referred to as **alternations.** *The complete transition through one positive alternation and one negative alternation* is referred to as a **cycle.** For this reason, *each alternation* is often referred to as a **half-cycle.** As shown in Figure 9.3b, a sine wave is made of a continuous stream of cycles.

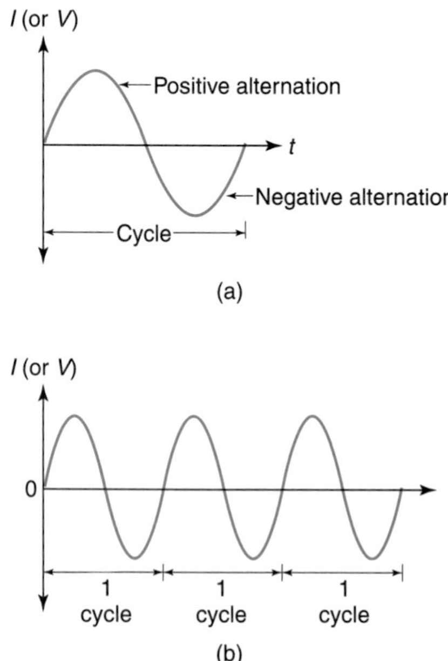

Figure 9.3 Waveform makeup.

Cycle Time (Period) and Frequency

It takes some measurable period of time for an ac *waveform* (or **signal**) to complete each cycle. *The time required to complete one cycle of a signal* is referred to as its **cycle time** or **period.** Example 9.1 demonstrates the concept of cycle time. ◄ *OBJECTIVE 3*

Example 9.1

Determine the cycle time of the waveform in Figure 9.4.

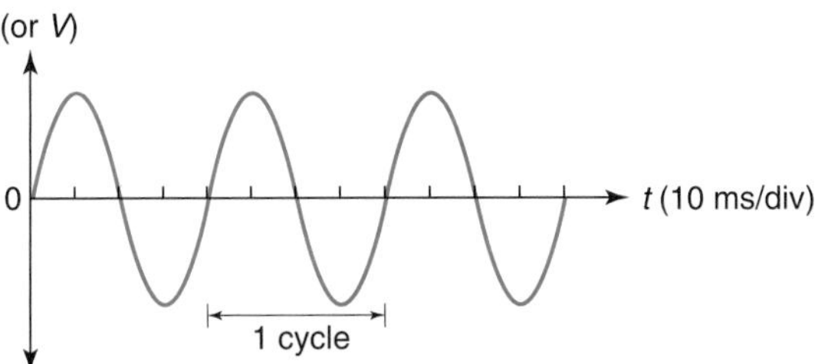

Figure 9.4

Solution: Each division on the horizontal axis of the graph represents a time period of 10 ms. One cycle of the waveform is four divisions in length. Therefore,

$$T_C = (4 \text{ div}) \times 10 \frac{\text{ms}}{\text{div}} = 40 \text{ ms}$$

Practice Problem 9.1

Calculate the cycle time of the waveform shown in Figure 9.5.

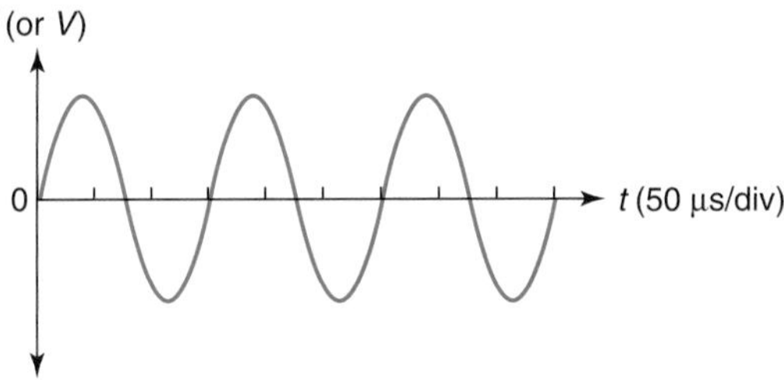

Figure 9.5

The period of a waveform can be measured from any given point on the waveform to the identical point in the next cycle, as illustrated in Figure 9.6. With a horizontal scale of 5 ms/div, the total cycle time (regardless of where it is measured) works out to be

$$T_C = (4 \text{ div}) \times 5 \frac{\text{ms}}{\text{div}} = 20 \text{ ms}$$

The **frequency** of a waveform is *the rate at which the cycles repeat themselves, in cycles per second.* For example, the waveform shown in Figure 9.7 has a cycle time of

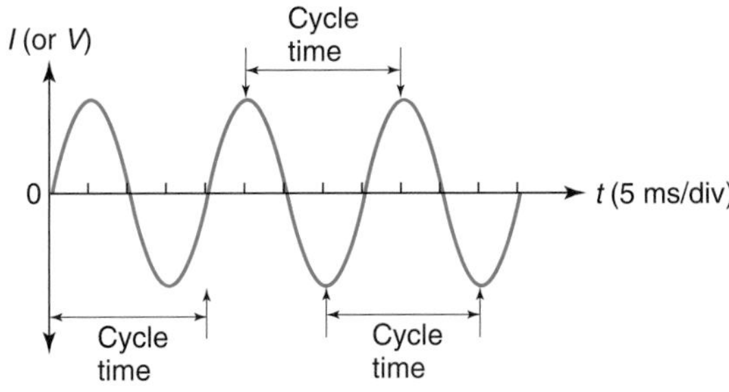

Figure 9.6

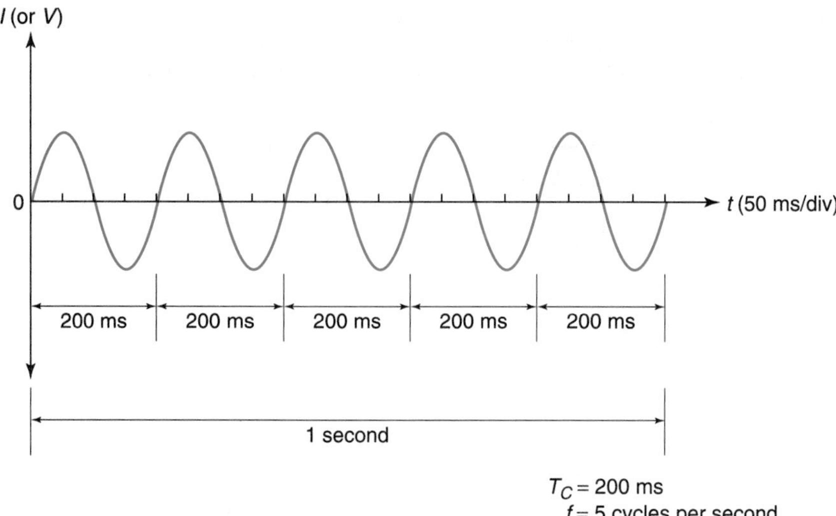

Figure 9.7 Cycle time and frequency.

200 ms. Therefore, the cycle repeats itself five times every second, and the frequency (f) of the waveform is given as

$$f = 5 \text{ cycles per second}$$

The unit of measure for frequency is **hertz (Hz).** One hertz is equal to one cycle per second. Therefore, the frequency of the waveform shown in Figure 9.7 is written as

$$f = 5 \text{ Hz}$$

When the cycle time of a waveform is known, the frequency of the waveform can be found as

$$f = \frac{1}{T_C} \tag{9.1}$$

where

T_C = the cycle time of the waveform in seconds.

The use of this equation is demonstrated in Example 9.2.

Example 9.2

Calculate the frequency of a waveform with a 100 ms cycle time.

Solution: With a cycle time of 100 ms,

$$f = \frac{1}{T_C} = \frac{1}{100 \text{ ms}} = \frac{1}{100 \times 10^{-3} \text{ s}} = 10 \text{ Hz}$$

This result indicates that a waveform with a 100 ms cycle time repeats itself ten times each second.

Practice Problem 9.2

A sine wave has a cycle time of 40 ms. Calculate the frequency of the waveform.

When the frequency of a waveform is known, the cycle time can be found as

$$T_C = \frac{1}{f} \tag{9.2}$$

The use of this relationship is demonstrated in Example 9.3.

Example 9.3

Calculate the cycle time of a sine wave with a frequency of 400 Hz.

Solution: With a frequency of 400 Hz,

$$T_C = \frac{1}{f} = \frac{1}{400\text{ Hz}} = 2.5\text{ ms}$$

This result means that a 400 Hz waveform completes one cycle every 2.5 ms.

Practice Problem 9.3

Calculate the cycle time of a 1 kHz sine wave.

Equations (9.1) and (9.2) indicate that frequency and cycle time are inversely proportional.

Oscilloscope Time and Frequency Measurements

◄ *OBJECTIVE 4*

An **oscilloscope** is a *piece of equipment that provides a visual representation of a voltage waveform.* This visual display can be used to make a variety of magnitude- and time-related measurements.

An oscilloscope display screen is represented in Figure 9.8. As you can see, the screen is divided into a series of major and minor divisions. The divisions along the x-axis are used for time measurements, while those along the y-axis are used for voltage measurements.

The **time base control** on the oscilloscope *determines the amount of time represented by the major divisions along the x-axis.* The value of any time interval equals the product of the time base setting and the number of major divisions, as illustrated in Figure 9.9a.

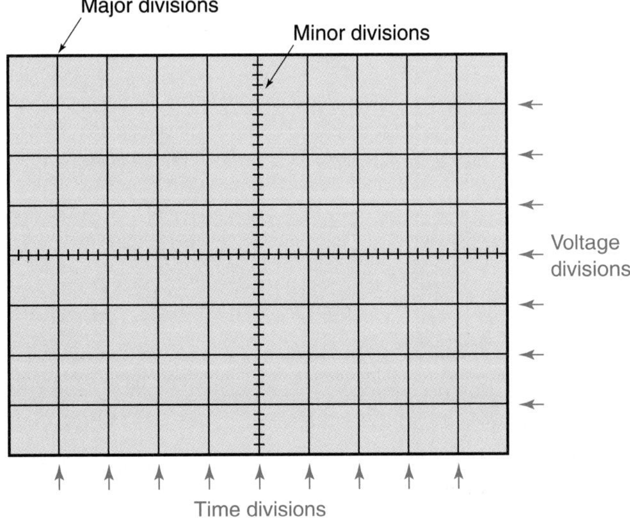

Figure 9.8 Oscilloscope display screen.

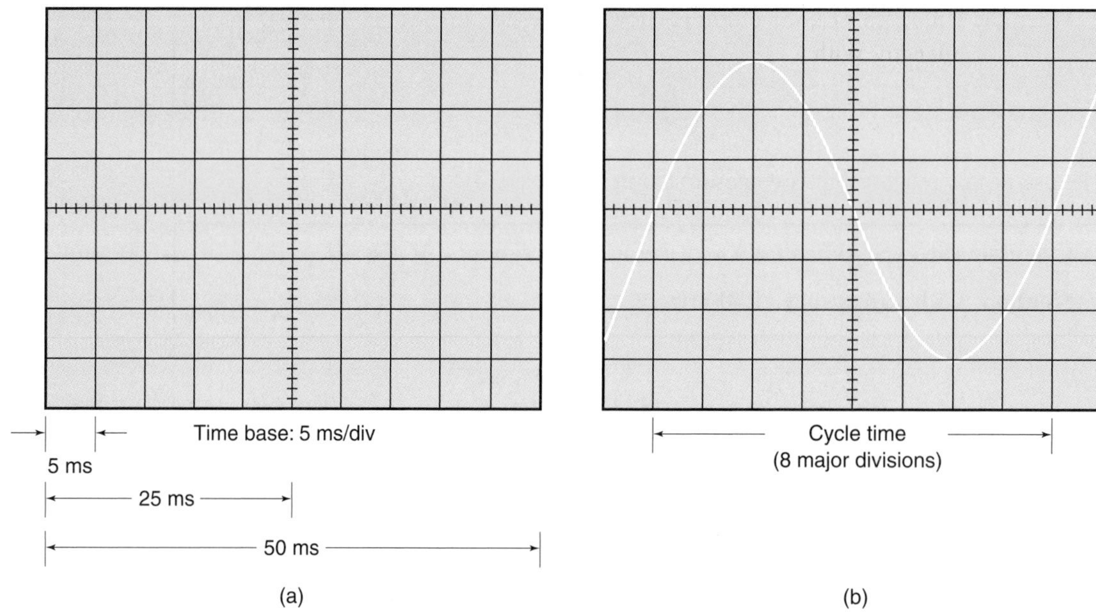

Figure 9.9 Measuring cycle time.

Calculating cycle time.

Figure 9.9b shows how cycle time is measured with an oscilloscope. One cycle of the waveform is eight major divisions in length. The cycle time equals the product of the number of divisions and the setting on the time base control. For example, if we assume that the time base control is set to 100 μs/div, the cycle time of the waveform shown in Figure 9.9b is found as

$$T_C = (8 \text{ div}) \times \frac{100 \ \mu\text{s}}{\text{div}} = 800 \ \mu\text{s}$$

The frequency of a waveform can be determined by measuring its cycle time and then calculating its frequency, as demonstrated in Example 9.4.

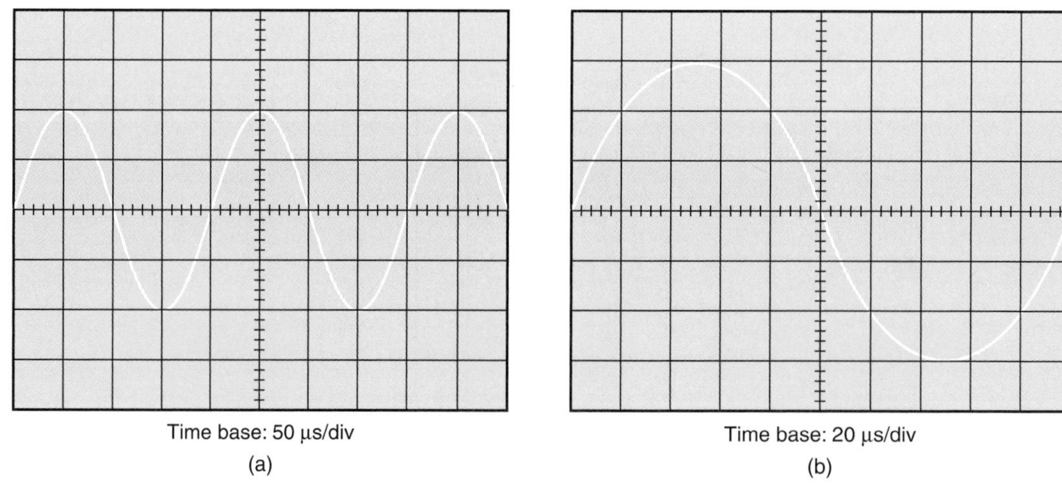

Figure 9.10

Example 9.4

Determine the frequency of the waveform shown in Figure 9.10a.

Solution: The display shows 2.5 cycles of the waveform, with each cycle being four divisions in length. With a time base of 50 μs/div,

$$T_C = (4 \text{ div}) \times \frac{50 \ \mu\text{s}}{\text{div}} = 200 \ \mu\text{s}$$

and

$$f = \frac{1}{T_C} = \frac{1}{200 \ \mu\text{s}} = 5 \text{ kHz}$$

Practice Problem 9.4

Calculate the frequency of the waveform shown in Figure 9.10b.

When you solve Practice Problem 9.4, you find that the waveforms shown in Figure 9.10 have the same frequency—even though they look very different on the oscilloscope display. This demonstrates that *changing the time base control on the oscilloscope does not change the frequency of the waveform.* It only changes the time represented by each division on the horizontal axis, and therefore, the display.

The time-related characteristics of ac waveforms are summarized in Figure 9.11. In the next section, we'll look at several magnitude-related characteristics.

An Analogy: Changing the time base setting is like changing the power setting on a telescope. Changing a telescope power setting makes the object appear closer (or further away); it doesn't change the actual distance to the object. By the same token, changing the time base setting changes only the display, not the actual waveform values.

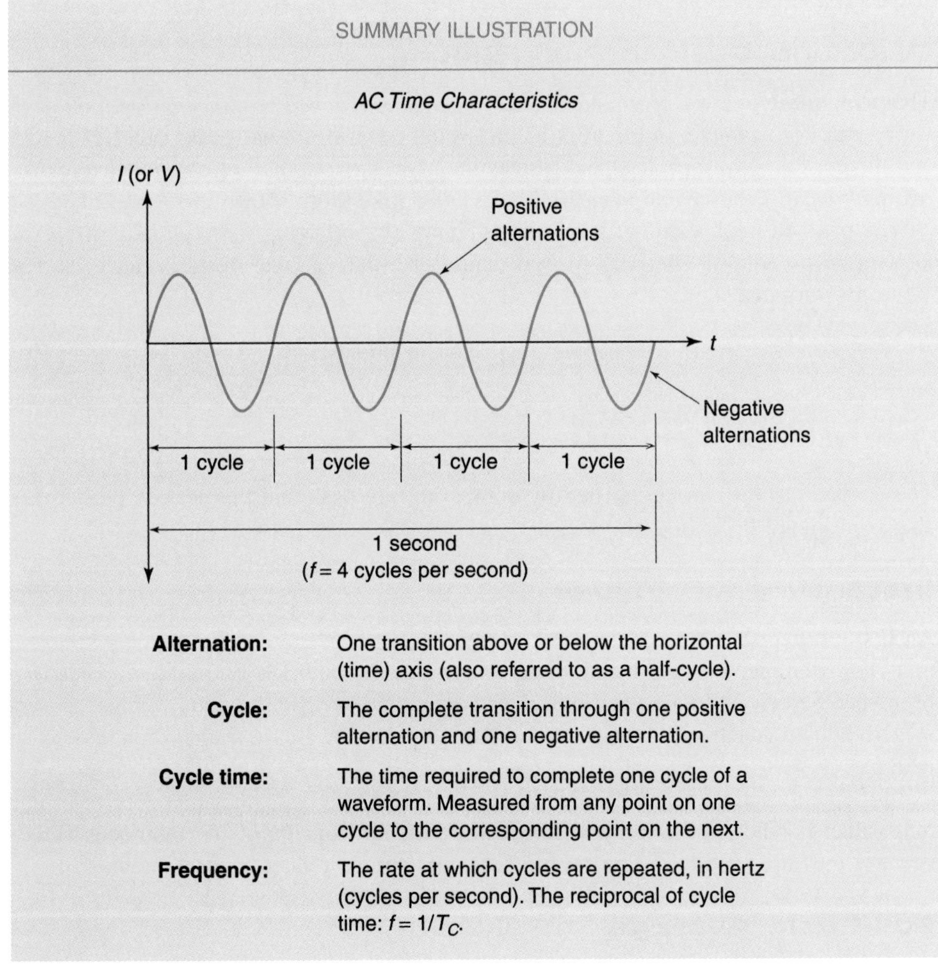

SUMMARY ILLUSTRATION
AC Time Characteristics
I (or V)
Positive alternations
t
Negative alternations
1 cycle
1 cycle
1 cycle
1 cycle
1 second
(f = 4 cycles per second)
Alternation:
One transition above or below the horizontal (time) axis (also referred to as a half-cycle).
Cycle:
The complete transition through one positive alternation and one negative alternation.
Cycle time:
The time required to complete one cycle of a waveform. Measured from any point on one cycle to the corresponding point on the next.
Frequency:
The rate at which cycles are repeated, in hertz (cycles per second). The reciprocal of cycle time: f = 1/T_C.

Figure 9.11

Section Review

1. What is a *waveform*?
2. What is the relationship between *alternations* and *cycles*?
3. What is *cycle time*? By what other name is it known?
4. Where on a sine wave can cycle time be measured?
5. What is *frequency*? What is its unit of measure?
6. Describe the relationship between cycle time and frequency.
7. What is an *oscilloscope*?
8. What is adjusted using the *time base control* on an oscilloscope?
9. How do you measure cycle time with an oscilloscope? Frequency?

Critical Thinking

10. Assume you double the frequency of a waveform but do not adjust any of the controls on an oscilloscope. What happens to the number of cycles on the display?

9.2 Magnitude Values and Measurements

OBJECTIVE 5 ➤ There are many different ways to describe and measure the magnitude of an ac waveform. In this section, we will discuss the common methods used to describe and measure waveform magnitude.

Peak and Peak-to-Peak Values

The **peak value** of a waveform is *the maximum value reached by either alternation of the waveform.* For example, in Figure 9.12, the voltage waveform has peak values of ± 10 V, and the current waveform has peak values of ± 1 mA. Note that a **pure ac waveform** is *one with equal positive and negative peak values* (like those shown in Figure 9.12).

The **peak-to-peak value** of a waveform equals *the difference between its positive and negative peak values.* The peak-to-peak value of a pure ac waveform is twice its peak value. By formula,

$$V_{PP} = 2\ V_{pk} \qquad \text{(for pure ac)} \tag{9.3}$$

and

$$I_{PP} = 2\ I_{pk} \qquad \text{(for pure ac)} \tag{9.4}$$

These relationships can be verified by comparing the peak and the peak-to-peak values shown in Figure 9.12. In each case, the peak-to-peak value is twice the peak value.

Instantaneous Values

An **instantaneous value** is *the magnitude of a voltage or current at a specified point in time.* For example, consider the sine wave shown in Figure 9.13. Two time intervals have been specified (t_1 and t_2), along with their corresponding instantaneous voltages (v_1 and v_2). Note that instantaneous values are identified using lowercase (rather than uppercase) letters.

In the next section, you will learn how to calculate any instantaneous voltage or current value. For now, we simply want to introduce the concept of instantaneous values, because they are used to define several other magnitude-related measurements.

Full-Cycle Average

The **full-cycle average** of a waveform is *the average of all the instantaneous values of voltage (or current) throughout one complete cycle.* It is calculated as shown in Figure 9.14.

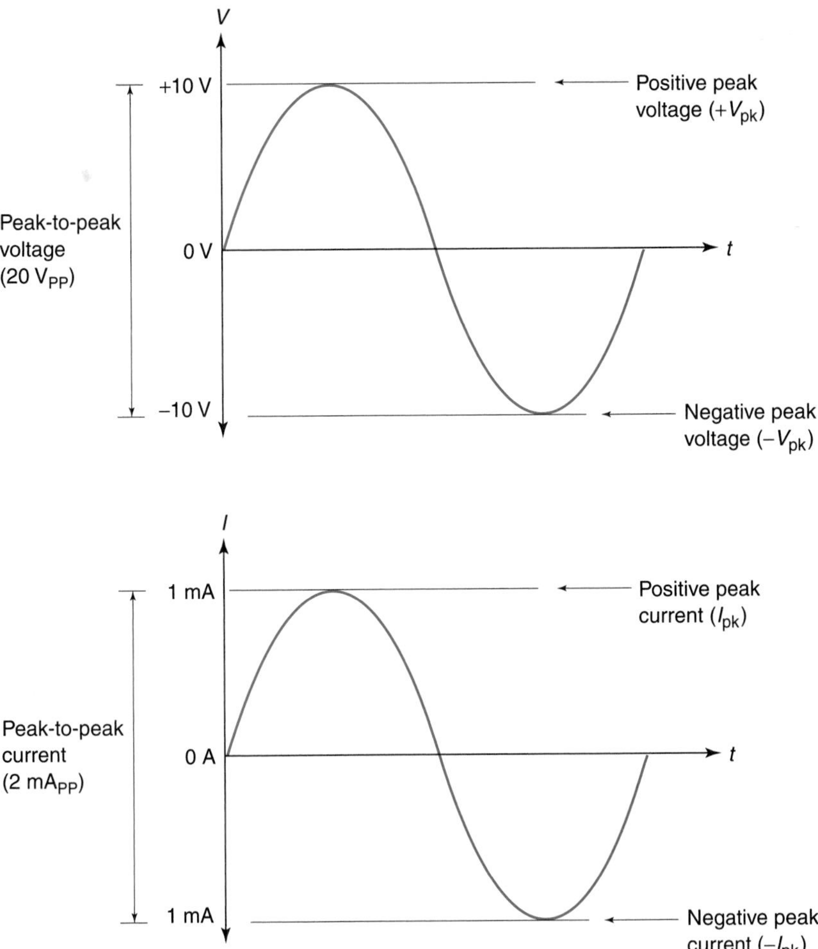

Figure 9.12 Peak and peak-to-peak values.

For a pure ac waveform, the full-cycle average is always 0 V (and 0 A). This is because every positive instantaneous value has a negative equivalent, and when these two values are added, the result is always zero. The concept is important, however, because many waveforms are not pure ac. When that is the case, the full-cycle average can have a value other than zero. (This point is demonstrated later in this chapter.)

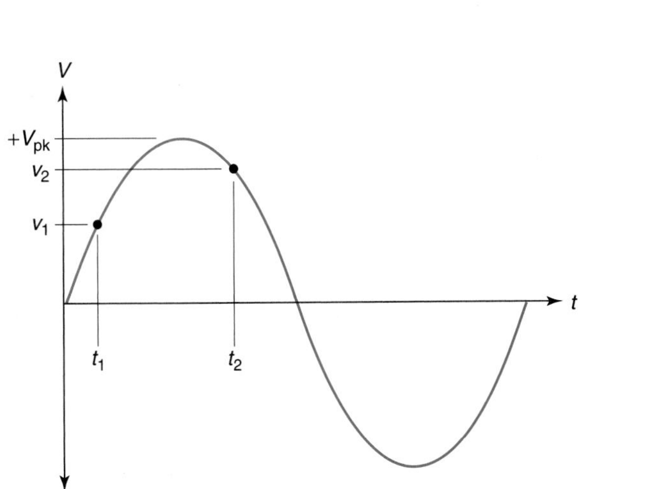

Figure 9.13 Instantaneous voltages.

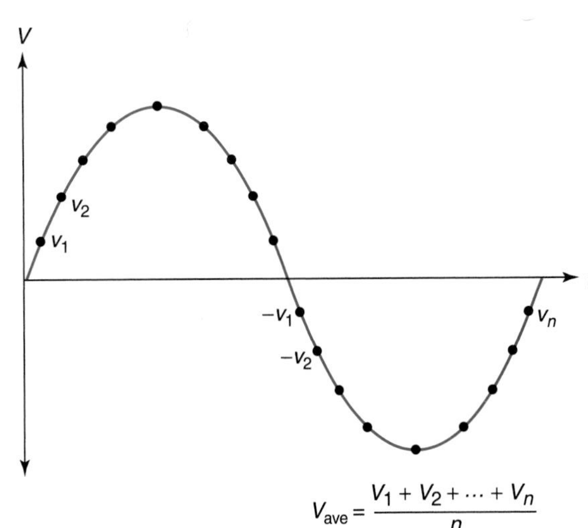

Figure 9.14 Determining the full-cycle average of a sine wave.

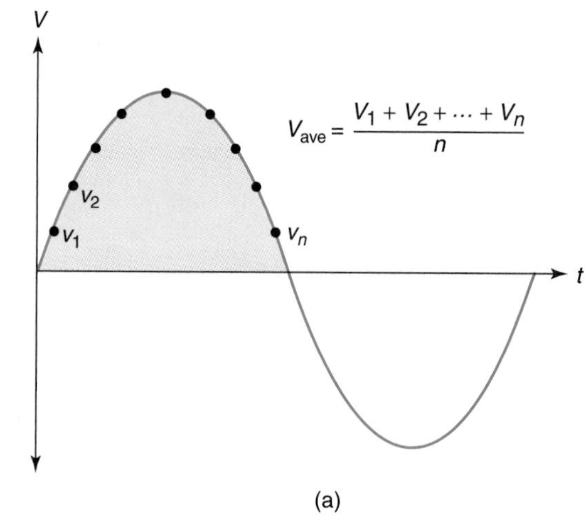

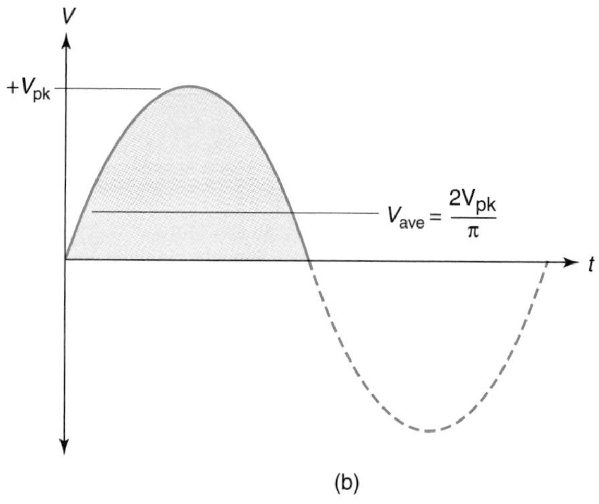

Figure 9.15 Determining the half-cycle average of a sine wave.

Half-Cycle Average

In some applications, it is helpful to know the *half-cycle average* of a waveform. The **half-cycle average** of a waveform is *the average of all its instantaneous values of voltage (or current) through either of its alternations.* The half-cycle average of a waveform can be determined as shown in Figure 9.15.

Note: Since $2/\pi \cong 0.637$, the value of V_{ave} can be found using

$$V_{ave} \cong 0.637V_{pk}$$

You will see the above form of equation (9.5) in later chapters.

It is not necessary to measure and average a series of instantaneous values to determine the value of V_{ave} for a given sine wave. The half-cycle average for a pure ac sine wave can be approximated as

$$V_{ave} \cong \frac{2V_{pk}}{\pi} \tag{9.5}$$

where

V_{pk} = the peak value of the waveform

The value of V_{ave} can be calculated as shown in Example 9.5.

Example 9.5

A sine wave has peak values of ±15 V. Calculate the half-cycle average voltage (V_{ave}) for the waveform.

Solution: Using the peak value given, V_{ave} is found as

$$V_{ave} = \frac{2V_{pk}}{\pi} = \frac{(2)(15\ \text{V})}{\pi} = 9.55\ \text{V}$$

> ***Practice Problem 9.5***
>
> A sine wave has peak values of ±24 V. Calculate the half-cycle average voltage (V_{ave}) for the waveform.

Average AC Power and RMS Values

In Chapter 3, you were shown that load power can be found as ◀ *OBJECTIVE 6*

$$P_L = \frac{V^2}{R_L}$$

Calculating load power for a dc circuit is relatively simple: We just use the values of V_L and R_L in the power equation. However, calculating load power for an ac circuit gets a bit more complicated because the source voltage is constantly changing. Therefore, the value of load power must also be changing (according to the power equation above).

When dealing with ac circuits, we are interested in *the average ac power generated throughout each cycle.* The concept of average ac power is illustrated in Figure 9.16a. If we assume that the curve shown represents the variations in load power over one cycle of the load voltage, the **average ac power (P_{ac})** is *the value that falls midway between 0 W and the peak value.* (The basis for positioning the curve as shown will become clear as this discussion continues.)

As you know, power is proportional to the square of voltage. If we assume that R_L in Figure 9.16a has a value of 1 Ω, then the load power for the circuit can be found as

$$P_L = \frac{V_L^2}{1\ \Omega}$$

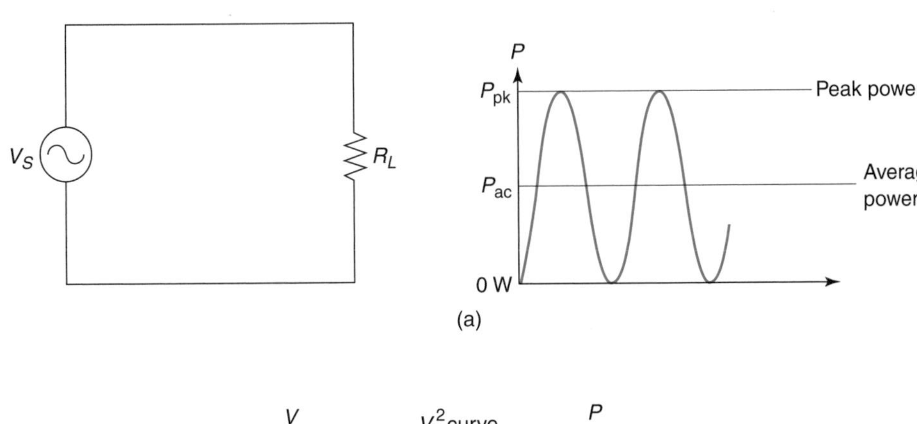

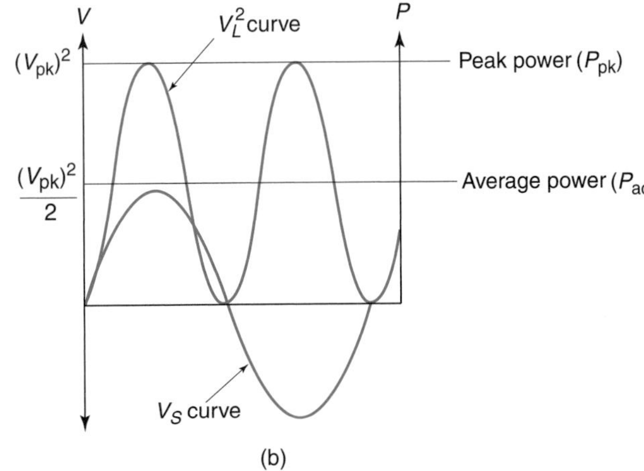

Figure 9.16 Sine wave voltage and power curves.

or

$$P_L = V_L^2 \qquad \text{(when } R_L = 1\ \Omega\text{)}$$

This relationship indicates that the P_L and V_L^2 curves for this circuit are identical. Keeping this relationship in mind, look at the waveforms shown in Figure 9.16b. The V_L^2 curve is derived by *squaring the values in the source voltage* (V_S) *curve.* According to the curve, $P_L = P_{ac}$ when

$$V_L^2 = \frac{V_{pk}^2}{2}$$

which is the average value of the V_L^2 waveform. Another way of expressing this relationship is

$$P_L = P_{ac} \quad \text{when} \quad V_L = \sqrt{\frac{V_{pk}^2}{2}} \tag{9.6}$$

OBJECTIVE 7 ➤ Equation (9.6) defines what is called the **root-mean-square (rms)** value of voltage for the circuit in Figure 9.16. The *rms value* of a voltage (or current) waveform is *the value that, when used in the appropriate power equation, will give you the average ac power of the waveform.* By formula,

$$V_{rms} = \sqrt{\frac{V_{pk}^2}{2}} \tag{9.7}$$

Equation (9.7) establishes the relationship between *peak* and *rms* voltage. However, it can be greatly simplified, as follows:

$$V_{rms} = 0.707V_{pk} \tag{9.8}$$

By the same token, the rms value of a *current* sine wave can be found as

$$I_{rms} = 0.707I_{pk} \tag{9.9}$$

Example 9.6 demonstrates the use of these equations for determining rms values.

Example 9.6

Calculate the rms values of voltage and current for the circuit shown in Figure 9.17a.

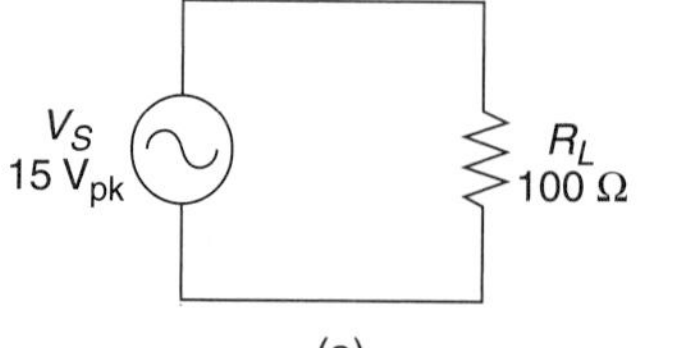

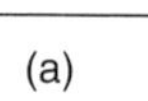

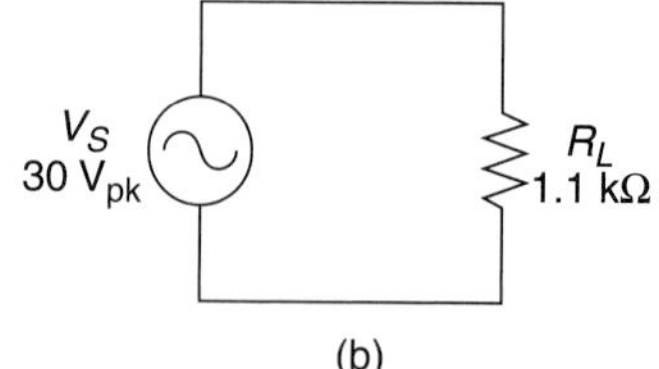

Figure 9.17 (a) (b)

Solution: Using the peak value shown, the rms voltage is found as

$$V_{rms} = 0.707V_{pk} = (0.707)(15\text{ V}) = 10.6\text{ V}$$

The peak current through the circuit can be found as

$$I_{pk} = \frac{V_{pk}}{R_L} = \frac{15\text{ V}}{100\ \Omega} = 150\text{ mA}$$

The rms current can now be found as

$$I_{rms} = 0.707I_{pk} = (0.707)(150\text{ mA}) = 106\text{ mA}$$

(Of course, we could have simply divided the rms voltage by the value of R_L to find the value of rms current. The result would have been the same.)

Practice Problem 9.6

Calculate the rms values of voltage and current for the circuit shown in Figure 9.17b.

Example 9.7 demonstrates the process for calculating average load power when the peak voltage is known.

Example 9.7

A circuit like those shown in Figure 9.17 has values of $V_{pk} = 12$ V and $R_L = 330\ \Omega$. Calculate the load power for the circuit.

Solution: The load power is found as follows:

$$V_{rms} = 0.707V_{pk} = (0.707)(12\text{ V}) = 8.48\text{ V}$$

and

$$P_L = \frac{V_{rms}^2}{R_L} = \frac{(8.48\text{ V})^2}{330\ \Omega} = 218\text{ mW}$$

Practice Problem 9.7

A circuit like those in Figure 9.17 has values of $V_{pk} = 10$ V and $R_L = 1.5$ kΩ. Calculate the average load power for the circuit.

Several more points need to be made regarding rms values and their applications:

- Average ac power is the equivalent of dc power. That is, a 20 V_{rms} source delivers the same average power to a load as a 20 V_{dc} source.
- Since power is a primary concern with any electrical or electronic system, rms values are used more often than any other magnitude-related measurement (such as peak, average, and so on).
- When a magnitude-related measurement is not specifically identified (as peak, peak-to-peak, or average), it is assumed to be an rms value.

Measuring Effective Values

◄ *OBJECTIVE 8*

The rms values of an ac waveform are commonly referred to as *effective* values. The name stems from the fact that rms values produce the same heating effect as their equivalent dc values.

Effective (rms) voltage is measured using an ac voltmeter, while effective current is measured using an ac ammeter. On most DMMs, switching from a dc scale to an ac scale is simply a matter of using the selector switch to choose the type of measurement desired. On others, you may also need to switch your probe to a specially designated input.

Summary

In this section, you have been introduced to peak, peak-to-peak, average, and effective (rms) values. The various sine-wave voltage values are summarized in Figure 9.18. Even though voltages are emphasized in the figure, sine-wave currents are labeled and defined in a similar fashion.

Section Review

1. What is the relationship between the two peak values of a *pure* ac waveform?
2. How do you calculate the peak-to-peak value of a waveform?
3. What are *instantaneous values*? How are they designated?
4. What is the *full-cycle average* of a waveform?

5. What is the *half-cycle average* of a sine wave?
6. What is the *root-mean-square (rms)* value of a sine wave?
7. What is the relationship between *rms* voltage (and current) and *dc equivalent power*?
8. In terms of voltage (or current) designations, how do you know when a given value is an *rms* value?
9. What is used to measure effective (rms) voltage? Effective (rms) current?

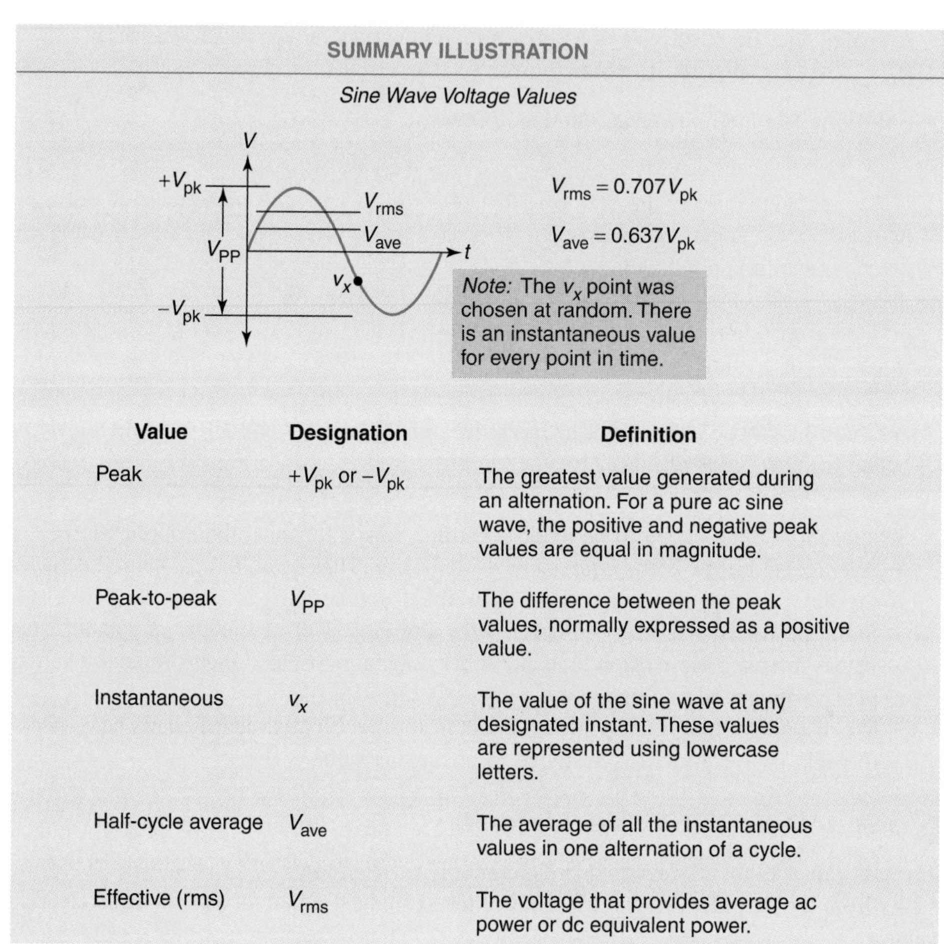

Value	Designation	Definition
Peak	$+V_{pk}$ or $-V_{pk}$	The greatest value generated during an alternation. For a pure ac sine wave, the positive and negative peak values are equal in magnitude.
Peak-to-peak	V_{PP}	The difference between the peak values, normally expressed as a positive value.
Instantaneous	v_x	The value of the sine wave at any designated instant. These values are represented using lowercase letters.
Half-cycle average	V_{ave}	The average of all the instantaneous values in one alternation of a cycle.
Effective (rms)	V_{rms}	The voltage that provides average ac power or dc equivalent power.

Figure 9.18

9.3 Sine Waves: Phase Measurements and Instantaneous Values

In Chapter 8, you were shown that the current through a coil can be used to generate a magnetic field. As a review, this principle of magnetic induction is illustrated in Figure 9.19. Just as current can be used to produce a magnetic field, the field produced by a magnet can be used to generate current through a conductor. In this section, you will be shown how a magnetic field can be used to generate a sinusoidal alternating current.

Magnetic Induction of Current

OBJECTIVE 9 ➤ When perpendicular lines of magnetic force cut a conductor, maximum current is induced in that conductor. This principle is illustrated in Figure 9.20. In Figure 9.20a, the

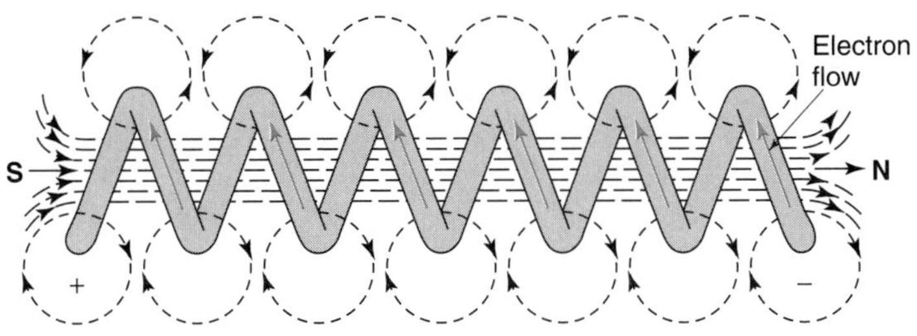

Figure 9.19 Magnetic induction.

conductor is at a 90° angle to the lines of force. When the conductor is moved downward through the lines of force, a current is generated through the conductor in the direction shown. In Figure 9.20b, the conductor direction is reversed, which reverses conductor current as well.

In Figures 9.20a and b, the conductor, lines of force, and direction of conductor motion are all at 90° angles to each other. This relationship is illustrated further in Figure 9.20c. As shown, the conductor and the lines of force are at a 90° angle on the horizontal plane. The directions of conductor motion are at 90° angles to the plane and, therefore, at 90° angles to the conductor and the lines of force. As long as the angles are maintained at 90°, maximum current is generated in the wire. However, if any one of the angles changes, the amount of current through the conductor decreases. For example, consider the conductor motion illustrated in Figure 9.20d. When the conductor motion is in the direction of the flux, the current generated in the conductor drops to 0 A because the conductor is no longer cutting the lines of force.

The Left-Hand Rule for Generators

You may recall that the left-hand rule is a memory aid designed to help you remember the relationship between a current and its associated magnetic field. We can use the left-hand rule to help determine the direction of the current that is generated in a conductor

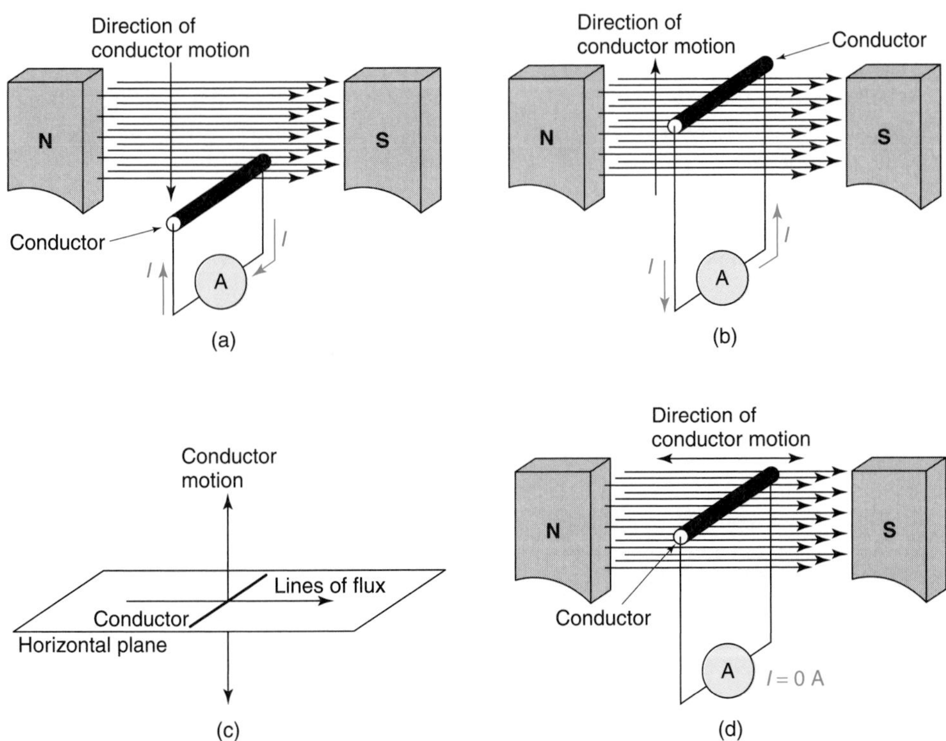

Figure 9.20 Current generated by magnetic induction.

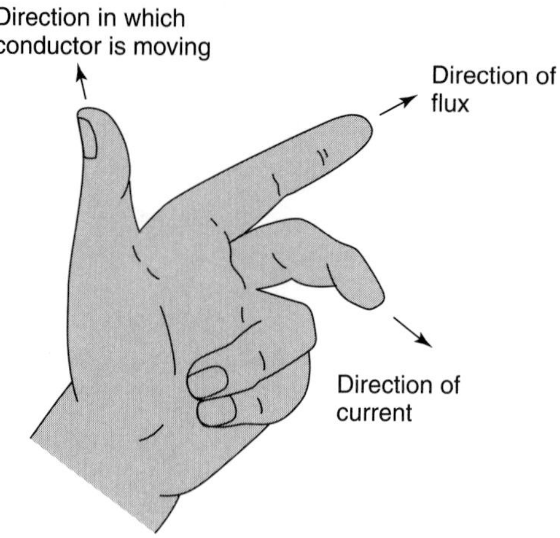

Figure 9.21 The left-hand rule for generators.

passing through a magnetic field. This application of the left-hand rule is illustrated in Figure 9.21.

Generating a Sine Wave

A sine wave can be generated by rotating a "loop conductor" through a stationary magnetic field. A simplified sine-wave generator is illustrated in Figure 9.22. As you can see, a *loop conductor* (or **rotor**) is positioned in the magnetic field. Each end of the rotor is connected to a slip ring. When a load is connected across the slip rings, the circuit is completed, providing a path for any current produced by the generator.

As it rotates through the magnetic field, a current is induced through the rotor. Both the magnitude and direction of the current produced are determined by the angle of conductor movement, relative to the magnetic lines of force. The generator operation is illustrated in Figure 9.23. The voltage curves shown represent the load voltage that would be produced by any generated current. (To simplify this illustration, the slip rings and load have been omitted.)

When side A of the conductor is at 0°, its motion is *parallel* to the lines of force. Since it is not cutting the lines of force, the current through the conductor is 0 A, and there is no voltage across the load. When side A has rotated to the 90° point, it is:

- Perpendicular to the lines of force.
- Traveling in a downward direction.

Since the angle of the conductor motion (relative to the lines of force) is 90°, maximum current is induced through the conductor. For the sake of discussion, this current is assumed to generate a positive load voltage, as shown in the voltage curve.

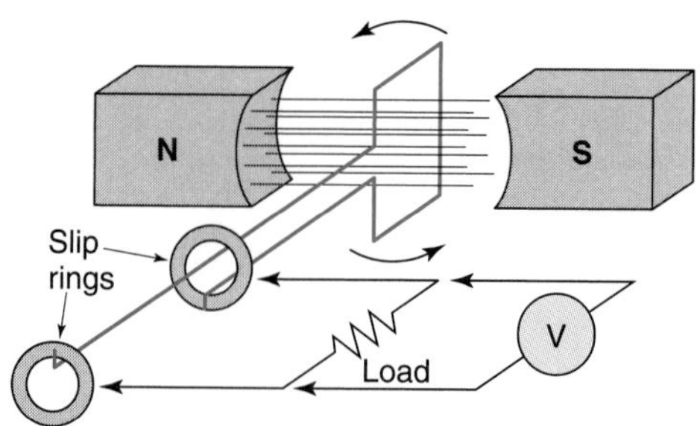

Figure 9.22 A simplified sine-wave generator.

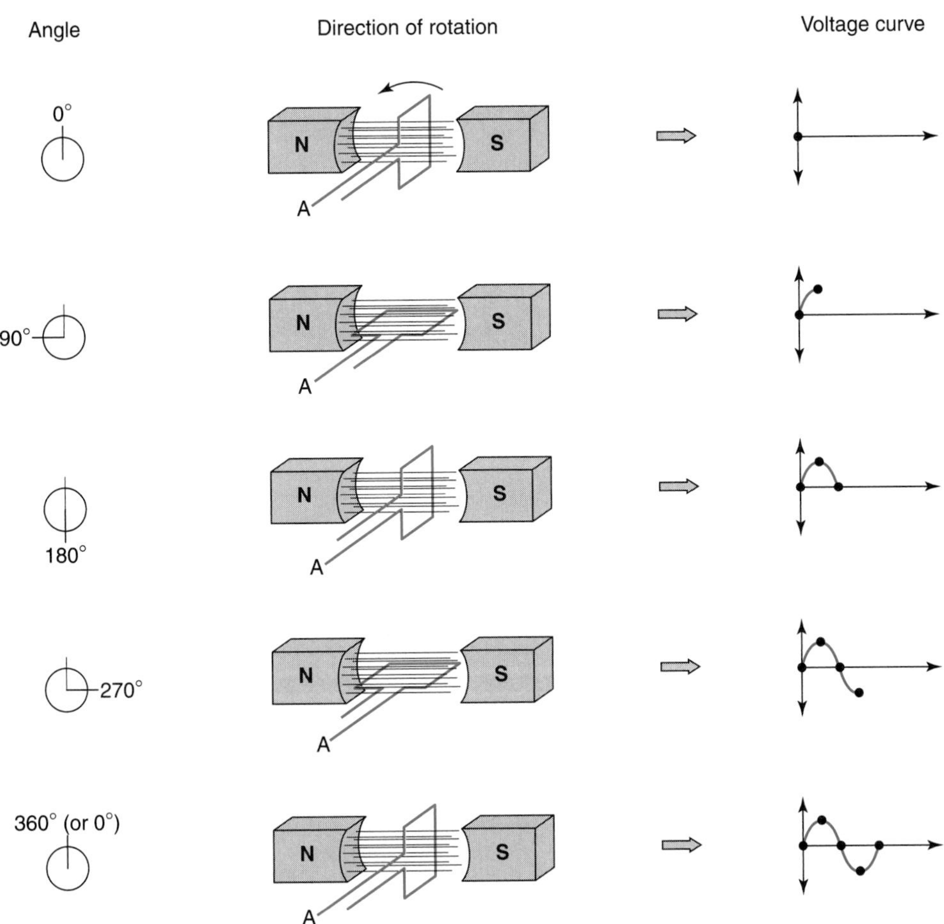

Figure 9.23 Generating a sine wave.

When side A of the conductor reaches 180°, it is parallel once again to the lines of force, and both the current and voltage return to zero. When side A of the conductor has rotated to the 270° point, it is:

- Perpendicular (again) to the lines of force.
- Traveling in an upward direction.

Once again, current through the conductor reaches a maximum value. However, because the conductor motion has changed (from downward to upward), direction of load current and load voltage polarity have reversed. When side A of the conductor reaches 360° (its starting position), the load current and voltage both return to zero, and the cycle then repeats itself.

Phase

If you compare each curve in Figure 9.23 with its corresponding angle of rotation, you'll see that: ◄ *OBJECTIVE 10*

- The positive peak occurs when the angle of rotation reaches 90°.
- The waveform returns to 0 V when the angle of rotation reaches 180°.
- The negative peak occurs when the angle of rotation reaches 270°.
- The waveform returns to 0 V when the angle of rotation reaches 360° (or 0°).

These angles are all relative to the starting point of the waveform, which is illustrated further in Figure 9.24. Note that the horizontal axis of the graph is used to represent the angles of rotation, designated by the Greek letter *theta* (θ).

Any point on a waveform can be identified by its *phase*. The **phase** of a given point is *its position relative to the start of the waveform, expressed in degrees.* Several points are identified by phase in Figure 9.25.

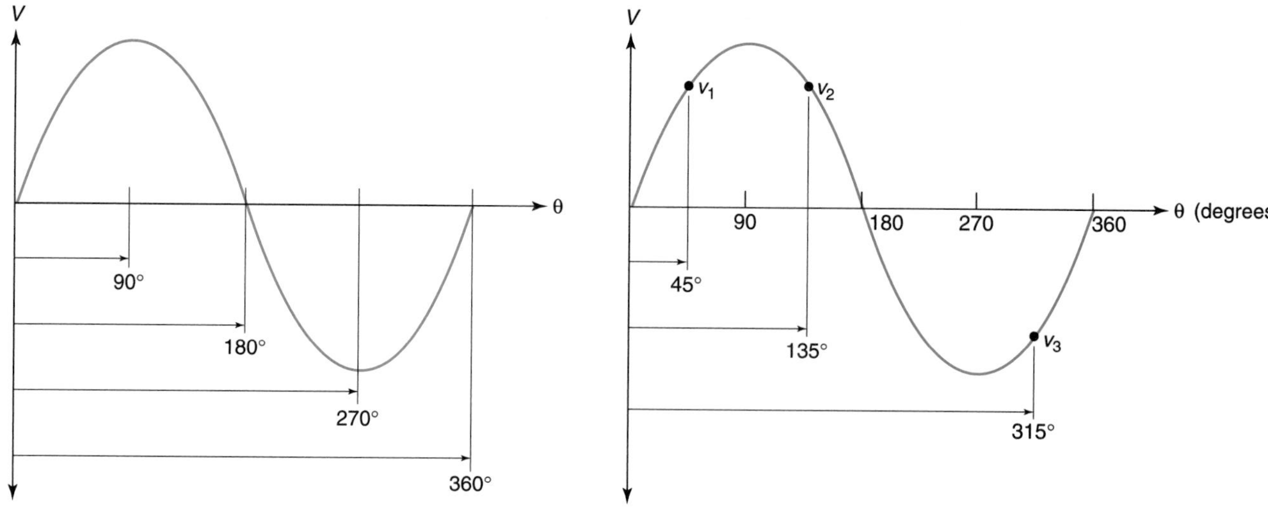

Figure 9.24 Degree measurements.

Figure 9.25 Example phase measurements.

Phase and Time Measurements

Phase and time measurements are related as shown in Figure 9.26. As the figure implies, the ratio of phase (θ) to 360° equals the ratio of instantaneous time (t) to cycle time (T_C). By formula,

$$\frac{\theta}{360^\circ} = \frac{t}{T_C} \tag{9.10}$$

where

θ = the phase of the point (v_x)

t = the time from the start of the cycle to v_x

T_C = the time required for one complete cycle

For example, assume that each time increment in Figure 9.26 represents 1 ms. If we apply equation (9.10) to the 90° point on the waveform, we get

$$\frac{90^\circ}{360^\circ} = \frac{1 \text{ ms}}{4 \text{ ms}}$$

which is true.

Figure 9.26
Phase/time measurements.

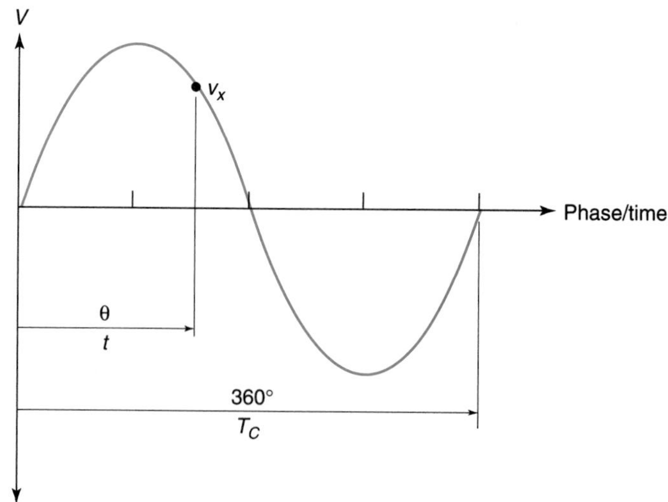

When the cycle time of a waveform is known, the time from the start of a cycle to a given phase (θ) can be found as

$$t = T_C \frac{\theta}{360°} \tag{9.11}$$

Example 9.8 demonstrates the use of this relationship.

Example 9.8

Assume that the waveform shown in Figure 9.26 has a value of $f = 15$ kHz. Determine the time from the start of the cycle until the $\theta = 120°$ point on the waveform is reached.

Solution: First, the cycle time of the waveform is found as follows:

$$T_C = \frac{1}{f} = \frac{1}{15 \text{ kHz}} = 66.67\ \mu\text{s}$$

The time required to reach an angle of 120° can now be found as

$$t = T_C \frac{\theta}{360°} = (66.67 \text{ ms}) \frac{120°}{360°} \cong 22.2\ \mu\text{s}$$

Practice Problem 9.8

Assume that the waveform shown in Figure 9.26 has a value of $f = 250$ Hz. Determine the time from the start of the cycle until the $\theta = 135°$ point on the waveform is reached.

Equation (9.11) can also be modified to provide an equation for θ, as follows:

$$\theta = (360°) \frac{t}{T_C} \tag{9.12}$$

This relationship allows us to determine the phase (in degrees) for any point on an oscilloscope display, as demonstrated in Example 9.9.

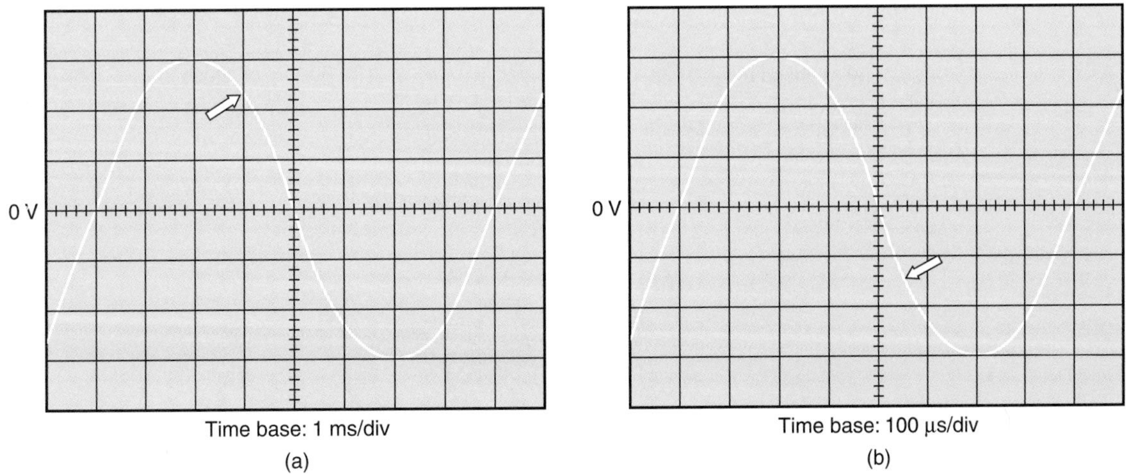

Figure 9.27

Example 9.9

Determine the value of θ for the point highlighted in Figure 9.27a.

Solution: With a time base of 1 ms/div, the cycle time of the waveform is found as

$$T_C = (8 \text{ div})\frac{1 \text{ ms}}{\text{div}} = 8 \text{ ms}$$

The highlighted point occurs three divisions into the waveform. Therefore, the value of t for the point is found as

$$t = (3 \text{ div})\frac{1 \text{ ms}}{\text{div}} = 3 \text{ ms}$$

Finally, the phase measured at the point is found as

$$\theta = (360°)\frac{t}{T_C} = (360°)\frac{3 \text{ ms}}{8 \text{ ms}} = 135°$$

Practice Problem 9.9

Determine the value of θ for the point highlighted in Figure 9.27b.

Phase Angles

OBJECTIVE 11 ➤ Two (or more) sine waves of equal frequency that reach their peaks at the same time are said to be *in phase.* Two (or more) sine waves of equal frequency that do *not* reach their peaks at the same time are said to be *out of phase.* Both waveform relationships are illustrated in Figure 9.28. Note that the peaks and zero-crossings of the two waveforms in Figure 9.28a occur at the same time.

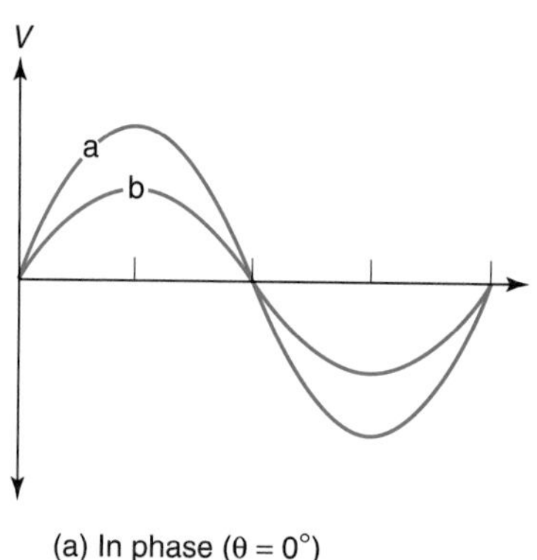

(a) In phase (θ = 0°)

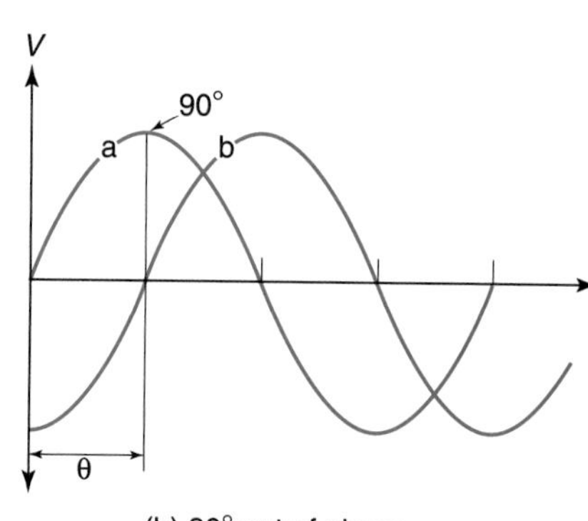

(b) 90° out of phase

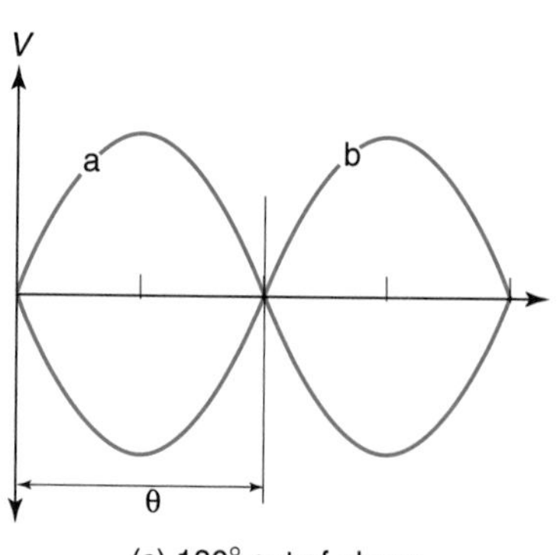

(c) 180° out of phase

Figure 9.28

The two waveforms in Figure 9.28b have the same cycle time, but they do not reach their peak values and zero-crossings at the same time. The **phase angle** (θ) is *the phase difference between two waveforms, measured in degrees.*

Sometimes it is easy to determine the phase relationship between a pair of waveforms. The waveforms shown in Figure 9.28b are 90° out of phase. This is easy to see because the 0° point of one corresponds (in time) to the 90° point of the other. The waveforms shown in Figure 9.28c are 180° out of phase. This is easy to see because one waveform reaches its positive peak at the same time that the other reaches its negative peak.

The waveforms shown in Figure 9.28a demonstrate another important point: *The phase angle between two (or more) waveforms is strictly a function of time.* The peak voltage (or current) values have nothing to do with the phase relationship between the waveforms.

Instantaneous Values

Earlier in this section, you were told that an instantaneous value is the magnitude of a voltage (or current) at a specified point on the waveform. For example, v_1 in Figure 9.29a is an instantaneous value. Now that we have discussed phase relationships, we are ready to look at the means that are used to determine the magnitude of any instantaneous value. ◄ *OBJECTIVE 12*

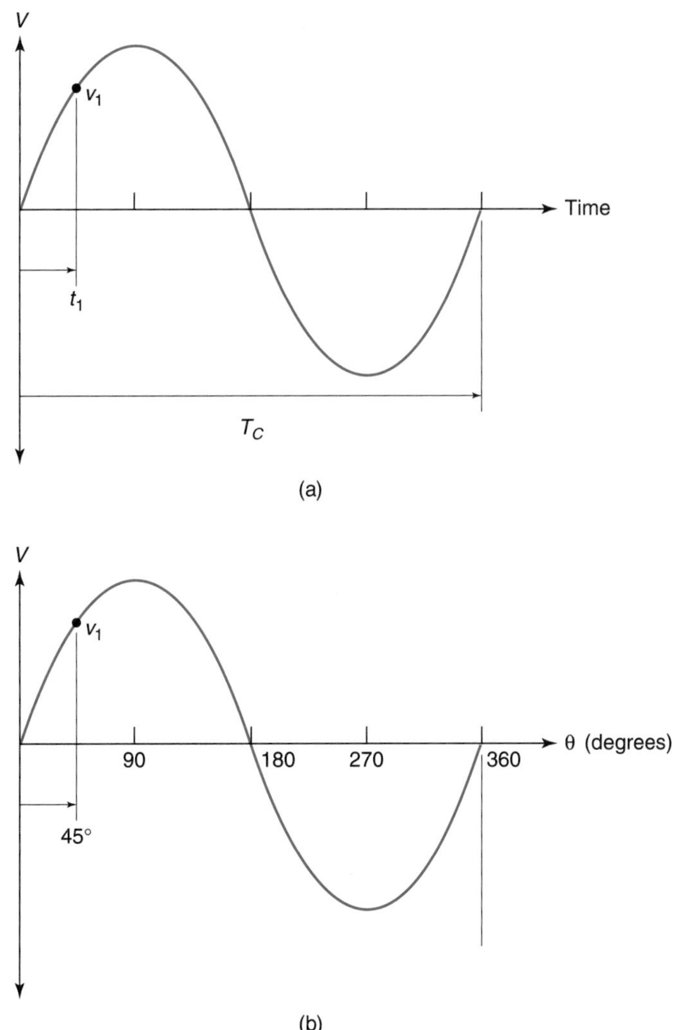

Figure 9.29

Note: The sine values shown at each increment in Figure 9.30a can be verified by entering the angle into your calculator and pressing the sin key.

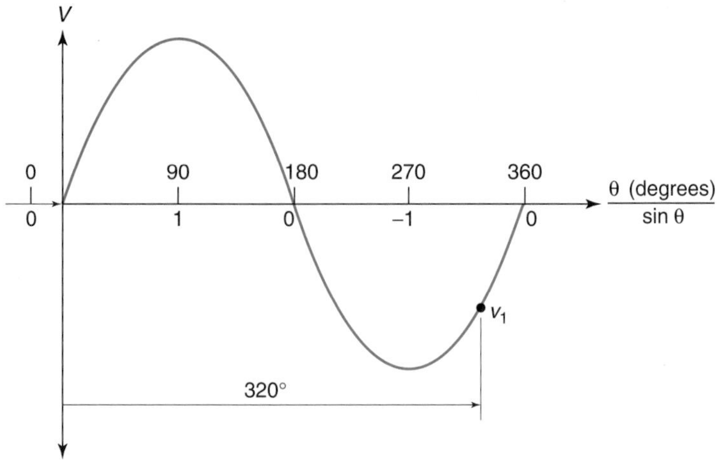

Figure 9.30

The instantaneous value of a sine-wave voltage (or current) is proportional to the peak value of the waveform and the sine of the phase angle. This relationship is given as

$$v = V_{pk} \sin \theta \tag{9.13}$$

Example 9.10 demonstrates the use of this equation.

Example 9.10

Determine the value of v_1 in Figure 9.29b. Assume that the waveform has a peak value of 10 V.

Solution: As shown in the figure, v_1 occurs when $\theta = 45°$. With a peak value of 10 V, the value of v_1 is found as

$$v_1 = V_{pk} \sin \theta = (10 \text{ V})(\sin 45°) = (10 \text{ V})(0.707) = 7.07 \text{ V}$$

Practice Problem 9.10

A sine wave has a peak value of 12 V. Determine the instantaneous value of the waveform when $\theta = 60°$.

The basis for the term *sine wave.*

The peak value of a sine wave is a constant, but the phase angle changes from one instant to the next. Since the instantaneous values are proportional to the sine of the phase angle ($\sin \theta$), the waveform is said to vary at a "sinusoidal" rate. (This is where the name *sine wave* comes from.)

Figure 9.30 shows a sine wave divided into 90° increments. At each increment, the value of $\sin \theta$ is shown. From the $\sin \theta$ values shown, we can draw the following conclusions:

- During the positive alternation of the waveform, $\sin \theta$ is positive. Therefore, ($V_{pk} \sin \theta$) is also positive.
- During the negative alternation of the waveform, $\sin \theta$ is negative. Therefore, ($V_{pk} \sin \theta$) is also negative.

This means that equation (9.13) yields the correct polarity for any instantaneous value in either alternation, as demonstrated further in Example 9.11.

Example 9.11

Determine the value of v_1 in Figure 9.30. Assume the waveform has a peak value of 15 V.

Solution: Using equation (9.13), the value of v_1 is found as

$$v_1 = V_{pk} \sin \theta = (15 \text{ V})(\sin 320°) = (15 \text{ V}_{pk})(-0.643) \cong -9.65 \text{ V}$$

Practice Problem 9.11

A sine wave has a peak value of 30 V. Determine the instantaneous value of the waveform when $\theta = 200°$.

To determine the instantaneous value of a waveform at a specified time interval, we must determine the phase angle at that time interval, as demonstrated in Example 9.12.

Example 9.12

Determine the value of v_1 in Figure 9.31a.

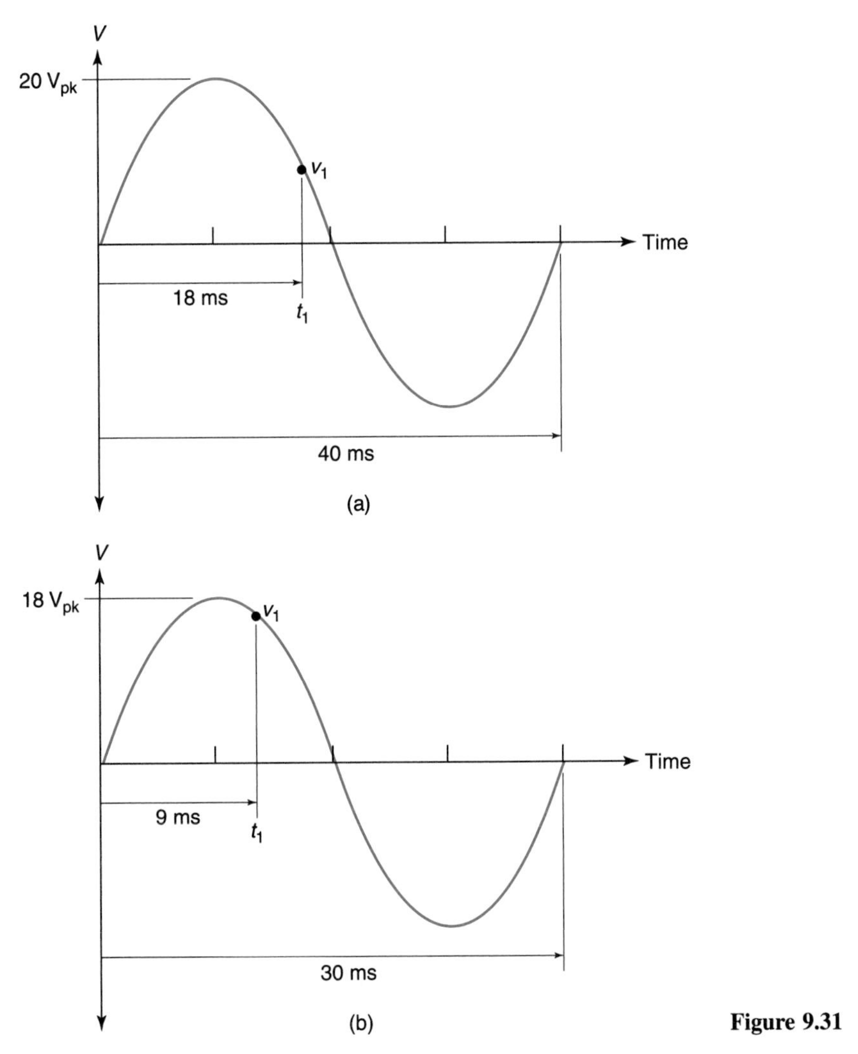

Figure 9.31

Solution: First, we determine the phase angle of the waveform at t_1. Using equation (9.12), the value of θ is found as

$$\theta = (360°)\frac{t_1}{T_C} = (360°)\frac{18 \text{ ms}}{40 \text{ ms}} = 162°$$

Now, the value of v_1 can be found as

$$v_1 = V_{pk} \sin \theta = (20 \text{ V})(\sin 162°) = (20 \text{ V})(0.309) = 6.18 \text{ V}$$

Practice Problem 9.12

Determine the value of v_1 in Figure 9.31b.

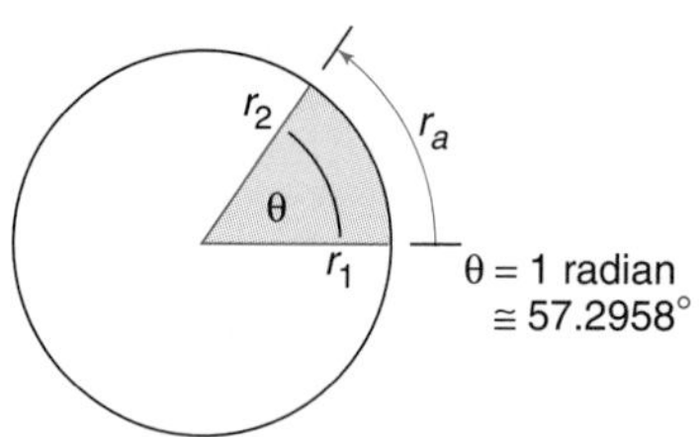

Figure 9.32 A vector rotation equal to one radian.

Phase Measurements: The Radian Method

OBJECTIVE 13 ➤ So far, we have used *degrees* to represent phase. Another way of representing phase involves the use of *radians.* Technically defined, a **radian** is *the angle formed at the center of a circle by two radii separated by an arc of equal length.* A radian is formed as shown in Figure 9.32.

The two radii (r_1 and r_2) are separated by an arc (r_a). The lengths of r_1, r_2, and r_a are all equal. The angle formed by r_1 and r_2 (which is labeled θ in the figure) is one radian.

The radian is independent of the size of the circle. For example, if we were to double the size of the circle shown in Figure 9.32, the angle designated by θ would not change, because r_1, r_2, and r_a would all increase by the same factor. Thus, *the number of degrees contained in a radian is the same for every circle.*

So, how many degrees are in one radian? There are 2π radians in a circle. Stated mathematically,

$$2\pi(\text{rad}) = 360°$$

Therefore,

$$1 \text{ rad} = \frac{360°}{2\pi} \cong 57.296°$$

Angular velocity Earlier in this section, you were shown how a sine wave can be generated by rotating a *conducting loop* (or *rotor*) through a permanent magnetic field. *The speed of rotation for the rotor* is referred to as **angular velocity.** The angular velocity (in radians per second) can be found as

$$\omega = 2\pi f \tag{9.14}$$

where

ω = the angular velocity, in radians per second
2π = the number of radians in one cycle
f = the number of cycles per second (frequency)

Angular velocity is of interest, because it indicates *the rate at which the phase and instantaneous values of a sine wave change.* This can be seen by referring back to Figure 9.23. If the rotor is spinning at a set rate, then the phase of the sine wave must be changing at the same rate. Since the instantaneous values of the sine wave are proportional to the phase, they must also be changing at the same rate. As you will learn in upcoming chapters, this rate of change is used in many ac circuit calculations.

OBJECTIVE 14 ➤ **Instantaneous values** Earlier, you were shown that an instantaneous value of sine wave voltage can be found as

$$v = V_{pk} \sin \theta$$

When you know the frequency of the waveform, the instantaneous value at a designated time interval (t) can be found using

$$v = V_{pk} \sin \omega t \tag{9.15}$$

where

$\omega = 2\pi f$ (the angular velocity, in radians per second)

t = the designated time interval from the start of the cycle

Equation (9.15) always provides the same result as ($V_{pk} \sin \theta$) *provided that your calculator is set for the "radian" mode of operation.* This calculator mode automatically converts your radian input ($2\pi f$) to degrees. As a result, the *sine* function provides the correct result. If the calculator is left in the "degree" mode of operation, your result will be incorrect. This point is demonstrated in Example 9.13.

The waveform shown in Figure 9.33a has the values shown. Calculate the value of v_1 using both the *degree* and *radian* methods.

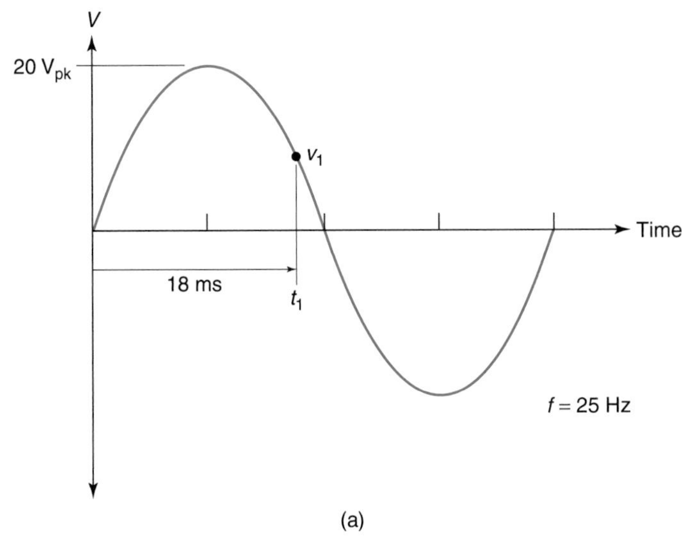

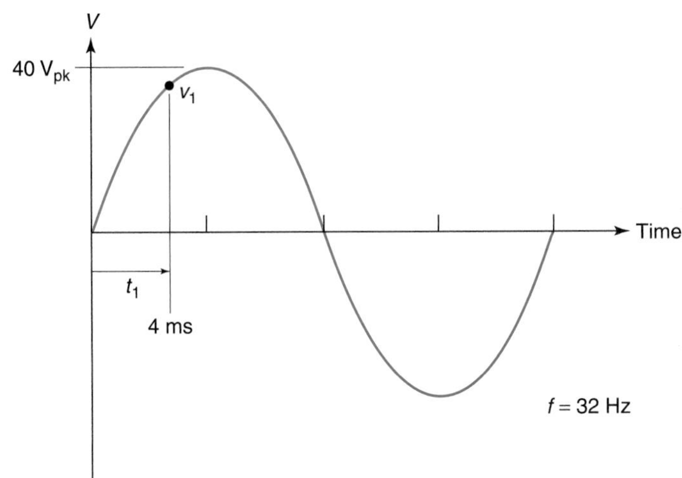

Figure 9.33

Solution: The waveform shown is the same one that we analyzed in Example 9.12. In that example, we calculated the following values:

$$\theta = 162^\circ \qquad \text{and} \qquad v_1 = 6.18 \text{ V}$$

After switching your calculator to the *radian* mode of operation, the value of v_1 can be found as

$$v_1 = V_{pk} \sin \omega t = (20 \text{ V}) \sin (2\pi \times 25\text{Hz} \times 18 \text{ ms})$$
$$= (20 \text{ V}) \sin (28.274) = (20 \text{ V})(0.3090) = 6.18 \text{ V}$$

This is the same value of v_1 we calculated in Example 9.12. Had we attempted to use the *degree* mode of operation, we would have obtained

$$v_1 = (20 \text{ V}) \sin (28.274) = (20 \text{ V})(0.4737) = 9.474 \text{ V}$$

which does not agree with the result obtained in Example 9.12.

Practice Problem 9.13

Calculate the value of v_1 for the waveform shown in Figure 9.33b using both the degree and radian approaches. (A model of the degree approach can be found in Example 9.12.)

Putting It All Together: Phase Measurements and Calculations

OBJECTIVE 15 ➤ A sine wave contains an infinite number of instantaneous values. Certain circuit analysis problems require you to calculate an instantaneous value of sine wave voltage (or current). To calculate one of these values, you must be able to determine:

- The peak value of the waveform.
- The position of the point relative to the start of the waveform cycle, that is, its *phase,* which is measured either in *degrees* or in *radians.*

The degree and radian approaches to solving for an instantaneous voltage are compared in Figure 9.34. As you can see, the *degree approach* is preferable when a phase value (in degrees) is known. However, if the waveform frequency and the instant of time (t) are known, you may find the *radian approach* to be easier.

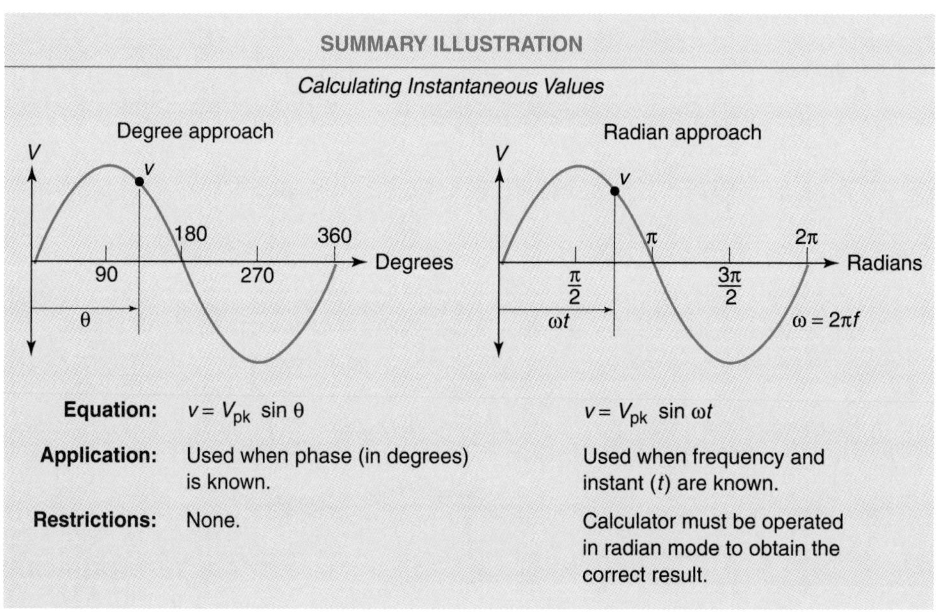

Figure 9.34

Section Review

1. Briefly describe how current can be generated by magnetic induction.
2. Briefly describe the operation of the simple ac generator shown in Figure 9.23.

3. What is the *phase* of a point on a waveform?
4. Why are sine waves commonly described in terms of phase?
5. How can *phase* and *time* measurements be related?
6. What is meant by the term *phase angle*?
7. Describe the relationship between *phase angle* and each of the following: *frequency, cycle time, peak voltage* (or current).
8. What is a *radian*?
9. How many radians are contained in one alternation of a sine wave? In one complete cycle?
10. What is *angular velocity*? Why is it significant?
11. Compare the *degree* and *radian* approaches to measuring phase.
12. When using the *radian approach* to calculating instantaneous values, what adjustment must be made to your calculator?
13. When would you use the *degree approach* to calculating an instantaneous value? When would you use the *radian approach*?
14. Velocity is defined as distance per unit time. Using $2\pi =$ *radians per cycle* and $f =$ *cycles per second,* show that $2\pi f$ is a measure of velocity. Critical Thinking

9.4 Related Topics

In this section, we're going to complete the chapter by discussing a variety of topics that relate to ac waveforms and their measurements.

Static and Dynamic Values

Most circuit values are classified as being either *static* or *dynamic.* **Static values** are *those that do not change during the normal operation of a circuit.* For example, the resistor shown in Figure 9.35a has a value that remains constant (static) at $R_L = 1\text{k}\Omega$. On the other hand, **dynamic values** are *those that do change during the normal operation of the circuit.* For example, the instantaneous value of a sine wave is dynamic, because it changes from one instant to the next. ◄ *OBJECTIVE 16*

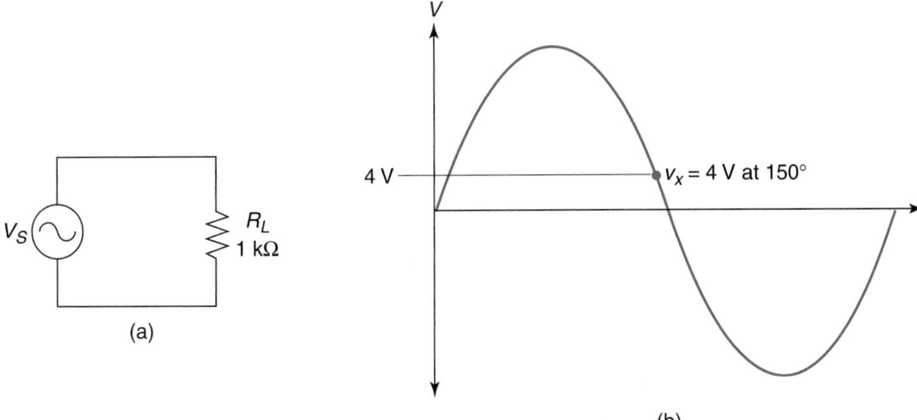

Figure 9.35

A dynamic value always has one or more specified conditions; that is, some conditions under which the value is valid. For example, look at the waveform shown in Figure 9.35b. The value of v_x must include both the value and the phase angle, as follows:

$$v_x = 4\ \text{V}\angle 150^\circ$$

where

4 V is the value and $\angle 150^\circ$ indicates the phase angle.

The phase angle is required because the waveform has two 4 V points in the positive half-cycle. Therefore, we must indicate which 4 V point we are referring to. (Incidentally, the equation above is written in a form that is called *polar notation.* This notation, and others, are discussed in detail in Appendix F.)

DC Offsets

OBJECTIVE 17 ➤ Earlier in the chapter, you were told that a pure ac waveform is one that has:

- Equal positive and negative peak values.
- An average value of 0 V (or 0 A).

In some cases, an ac waveform contains a **dc offset.** A sine wave with a dc offset is shown in Figure 9.36a. As the figure implies, a sine wave with a dc offset has:

- Unequal positive and negative peak values.
- An average value that is equal to the value of the dc offset (rather than zero).

Offsets can be generated by a variety of sources. For example, an amplifier can introduce a dc offset, as shown in Figure 9.36b. An **amplifier** is *a circuit that is used to increase the power level of the sine wave.* It does so by transferring power from its dc source ($+V_{dc}$) to the waveform. In the process, it also introduces a dc offset into the sine wave. Depending on the design of the amplifier, the dc offset will fall somewhere within the range shown in the figure.

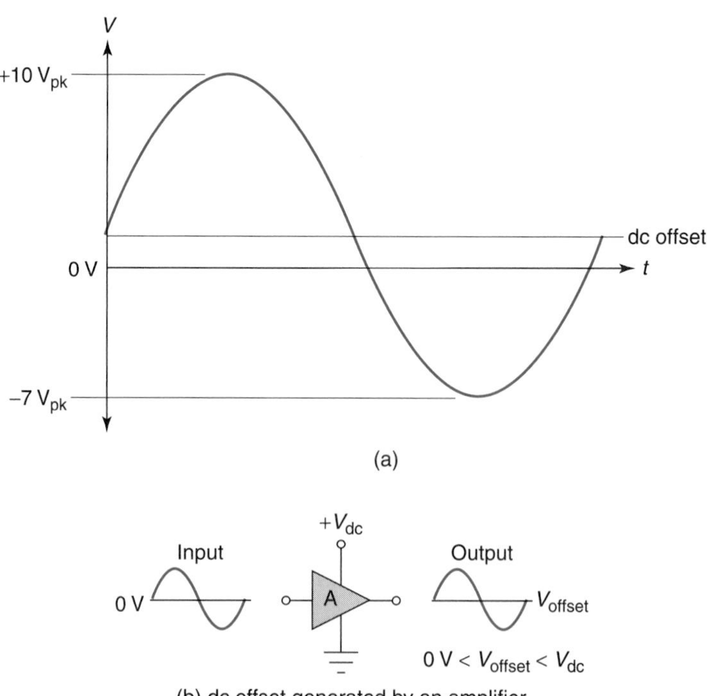

Figure 9.36 DC offsets.

Wavelength

A sine wave can be transmitted through space as a series of electromagnetic waves. An **electromagnetic wave** is *a waveform that consists of perpendicular electric and magnetic fields,* as illustrated in Figure 9.37. ◀ *OBJECTIVE 18*

The **wavelength** of a transmitted waveform is *the physical length of one cycle of the waveform.* The wavelength of a waveform is determined by:

- The speed at which it travels through space.
- The cycle time of the waveform.

In a vacuum, all electromagnetic waves travel at the velocity of electromagnetic radiation (which is commonly referred to as *the speed of light*). Some commonly used speed of light measurements are:

- 1.86×10^5 m/s (miles per second)
- 3×10^8 m/s (meters per second)
- 300 m/μs (meters per microsecond)
- 984 ft/μs (feet per microsecond)

The concept of wavelength is illustrated in Figure 9.38. The waveform shown is being transmitted into space from an antenna. In free space, the waveform travels at a speed of 3×10^8 meters per second. With a cycle time of one second, point A on the waveform will have traveled 3×10^8 meters by the time point B leaves the source. Thus, the waveform shown would have a wavelength of 3×10^8 m.

The wavelength of a transmitted waveform is represented using the Greek letter *lambda* (λ), and can be found as the product of velocity and cycle time. By formula,

$$\lambda = cT_C \tag{9.16}$$

where

c = the speed of light

T_C = the cycle time of the waveform

For the waveform shown in Figure 9.38,

$$\lambda = cT_C = (3 \times 10^8 \text{ m/s})(1 \text{ s}) = 3 \times 10^8 \text{ m}$$

which agrees with the description of the waveform given earlier.

There is another (and more commonly used) method of calculating wavelength. Since $T = \frac{1}{f}$, wavelength can be found as

$$\lambda = \frac{c}{f} \tag{9.17}$$

where

f = the frequency of the waveform being transmitted.

The use of this equation is demonstrated in Example 9.14.

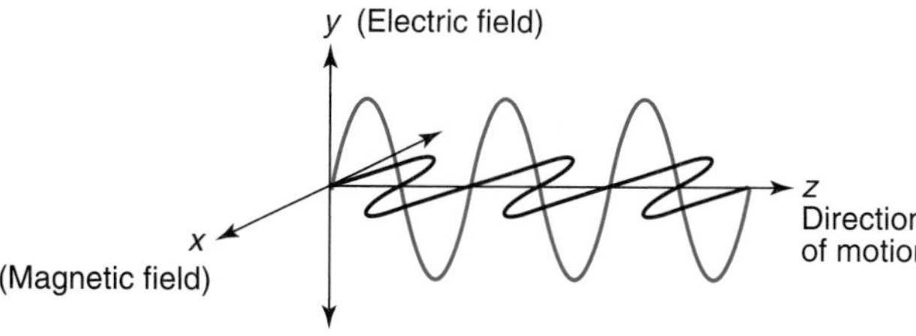

Figure 9.37 An electromagetic wave.

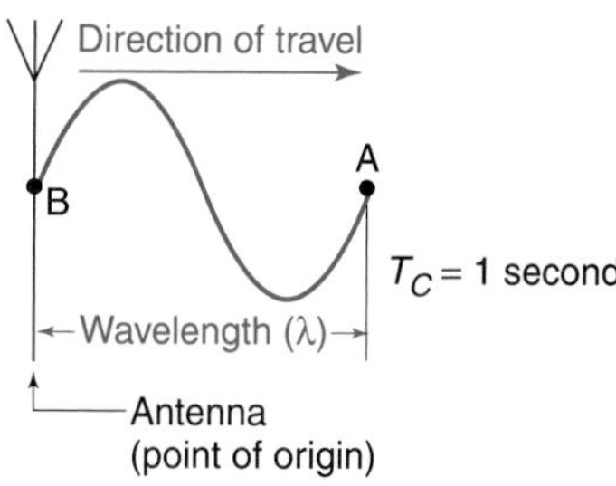

Figure 9.38 Wavelength.

Example 9.14

Calculate the wavelength (in meters) for a 150 MHz sine wave.

Solution: Using $c = 3 \times 10^8$ m/s and the signal frequency,

$$\lambda = \frac{c}{f} = \frac{3 \times 10^8 \text{ m/s}}{150 \times 10^6 \text{ Hz}} = 2 \text{ m}$$

Practice Problem 9.14

Calculate the wavelength (in meters) of a 250 MHz sine wave.

As equation (9.17) indicates, wavelength and frequency are inversely proportional. For example, if the operating frequency given in Example 9.14 *doubles,* the wavelength decreases to

$$\lambda = \frac{c}{f} = \frac{3 \times 10^8 \text{ m/s}}{300 \times 10^6 \text{ Hz}} = 1 \text{ m}$$

As your studies continue, you'll see that wavelength is an important consideration in a variety of communications applications. Some examples of wavelength-sensitive devices are antennas and fiber-optic cables.

Harmonics

OBJECTIVE 19 ➤ A **harmonic** is *a whole-number multiple of a given frequency.* For example, a 2 kHz sine wave has harmonics of:

$$2 \text{ kHz} \times 2 = 4 \text{ kHz}$$
$$2 \text{ kHz} \times 3 = 6 \text{ kHz}$$
$$2 \text{ kHz} \times 4 = 8 \text{ kHz}$$
$$2 \text{ kHz} \times 5 = 10 \text{ kHz}$$

and so on. Note that *a group of related frequencies* like the one shown above is referred to as a **harmonic series.** The *reference frequency in a harmonic series* (2 kHz in this case) is referred to as the **fundamental frequency.**

The fundamental frequency is the lowest frequency in any harmonic series. The other frequencies are generally identified as ***n*-order harmonics,** as illustrated in Table 9.1.

Two of the frequency relationships in Table 9.1 are often identified using other names. *When frequency changes by a factor of two,* the change is referred to as one **octave.** Thus, the second-order harmonic is said to be one octave above the fundamental frequency. By the same token, the fourth-order harmonic is said to be two octaves above the fundamental frequency (100 Hz × 2 × 2 = 400 Hz). The frequency that is n octaves above a given fundamental frequency (f_0) can be found using

$$f = 2^n f_0 \tag{9.18}$$

TABLE 9.1 Harmonic Identifiers

Fundamental Frequency	*Multiplier*	*Harmonic Frequency*	*Identifier*
100 Hz	2	200 Hz	2nd-order harmonic
100 Hz	3	300 Hz	3rd-order harmonic
100 Hz	4	400 Hz	4th-order harmonic
⋮	⋮	⋮	⋮
100 Hz	10	1 kHz	10th-order harmonic

For example, the frequency that is five octaves above 100 Hz can be found as

$$f = 2^n f_0 = (2^5)(100 \text{ Hz}) = (32)(100 \text{ Hz}) = 3.2 \text{ kHz}$$

When frequency changes by a factor of ten, the change is referred to as one **decade.** Thus, the tenth-order harmonic (ten times the fundamental frequency) is said to be one decade above the fundamental frequency. The frequency that is n decades above a given fundamental frequency can be found using

$$f = 10^n f_0 \tag{9.19}$$

For example, the frequency that is three decades above 100 Hz can be found as

$$f = 10^n f_0 = (10^3)(100 \text{ Hz}) = (1000)(100 \text{ Hz}) = 100 \text{ kHz}$$

In Chapter 11, we will begin to discuss circuits whose outputs are affected directly by the frequency of operation. At that time, you will be shown how *octave* and *decade* frequency intervals are often used to describe circuit frequency response.

Nonsinusoidal Waveforms: Rectangular Waves

Rectangular waves are *waveforms that alternate between two dc levels,* as shown in Figure 9.39a. The waveform shown alternates between the $-V_{dc}$ and $+V_{dc}$ levels.

◄ *OBJECTIVE 20*

Figure 9.39b shows the response of a circuit to a positive rectangular-wave input. From t_1 to t_2, the source has the polarity and the current direction shown. From t_2 to t_3, the source polarity and current direction have reversed, as shown in Figure 9.39c. Even though the rectangular wave is made of alternating dc levels, it produces current that alternates directions. Therefore, it is correctly classified as an ac waveform.

Terminology and time measurements The parts of a rectangular waveform are identified in Figure 9.40. The *positive half-cycle* is referred to as the **pulse width.** The *negative half-cycle* is referred to as the **space width.** The sum of the pulse width and the space width equals the cycle time of the waveform.

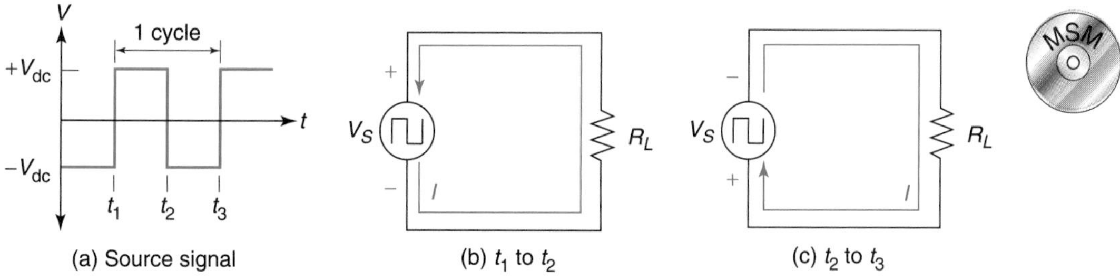

Figure 9.39 Circuit response to a rectangular input.

MSM

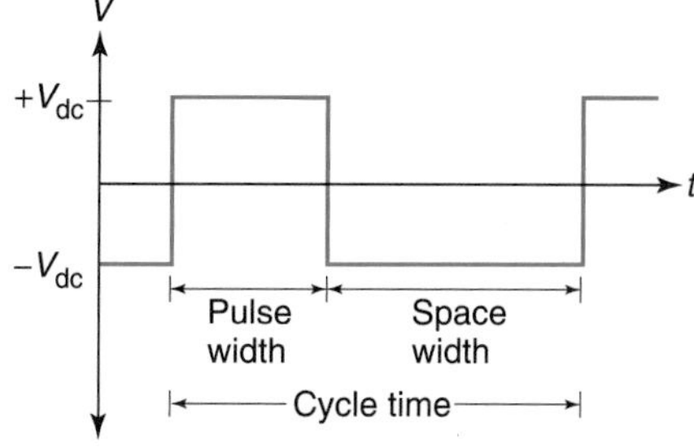

Figure 9.40 Pulse width, space width, and cycle time.

A Practical Consideration: The exact values of $-V_{dc}$ and $+V_{dc}$ aren't important at this point. As long as one voltage is positive and the other is negative, the waveform qualifies as an ac waveform.

Duty cycle Circuits that are designed to respond to rectangular waveforms are generally referred to as *pulse* and *switching* circuits, or *digital* circuits. Because of certain design factors, many digital circuits have a limit on *the ratio of pulse width to cycle time.* This ratio, which is commonly given as a percentage, is referred to as **duty cycle.** By formula,

$$\text{duty cycle}(\%) = \frac{PW}{T_C} \times 100 \tag{9.20}$$

where

PW = the pulse width of the circuit input
T_C = the cycle time of the circuit input

Example 9.15 demonstrates the calculation of duty cycle for a rectangular waveform.

Example 9.15

Determine the duty cycle of the waveform shown in Figure 9.41a.

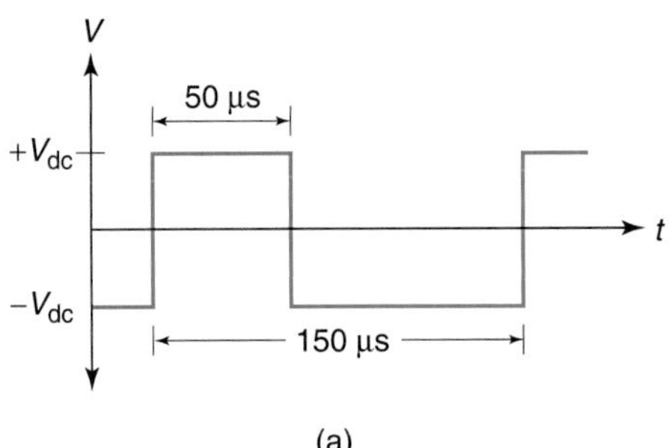

(a)

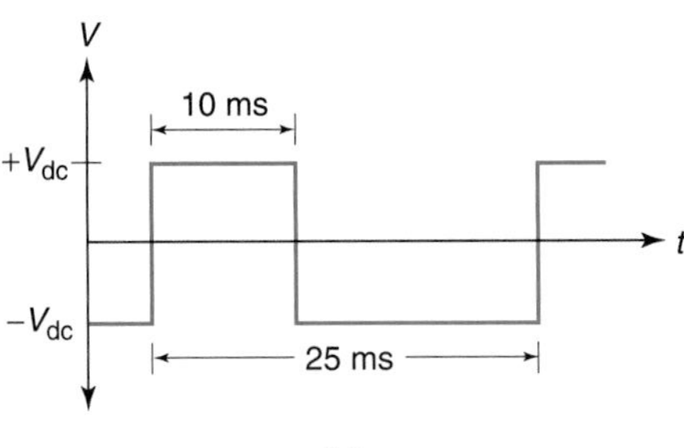

Figure 9.41 (b)

Solution: Using the values shown, the duty cycle is found as

$$\text{duty cycle} = \frac{PW}{T_C} \times 100 = \frac{50\ \mu\text{s}}{150\ \mu\text{s}} \times 100 = 33.3\%$$

This value indicates that the pulse width makes up 33.3% of the total cycle time.

Practice Problem 9.15

Calculate the duty cycle of the waveform shown in Figure 9.41b.

When you study digital electronics, you will see that the duty cycle rating of a component is a limit. If the duty cycle limit of a component is exceeded, the component will respond erratically.

Square Waves

The **square wave** is *a special-case rectangular waveform that has equal pulse width and space width values,* as shown in Figure 9.42. Since the values of PW and SW are equal, the duty cycle of a square wave is always 50%.

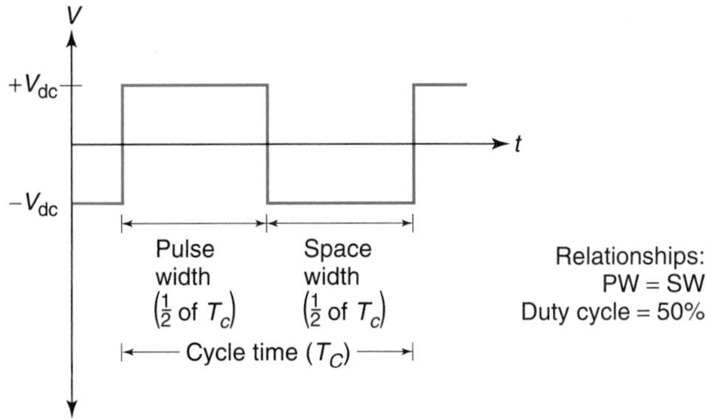

Figure 9.42 Square wave measurements.

Symmetrical and asymmetrical waveforms A **symmetrical waveform** is *one with cycles that are made up of identical halves.* For example, look at the square wave in Figure 9.43a. In this case:

- PW = SW, so the waveform is symmetrical in time.
- The waveform varies equally above and below 0 V, so it is symmetrical in amplitude.

By definition, a square wave is symmetrical in time. However, a square wave may be *asymmetrical* in amplitude. That is, it may vary unequally above and below 0 V. For example, consider the waveform shown in Figure 9.43b. This waveform has peak values of +5 V and −15 V, so the waveform is considered to be asymmetrical in amplitude.

A square wave with equal positive and negative peak values has an average value of 0 V. For a square wave with asymmetrical peak values, the average voltage can be found as

$$V_{\text{ave}} = \frac{+V_{\text{pk}} + (-V_{\text{pk}})}{2} \quad (9.21)$$

Note: Equation (9.21) applies only to square waves. Equation (9.22) is a more *general-purpose* relationship, meaning that it can be used for any rectangular waveform.

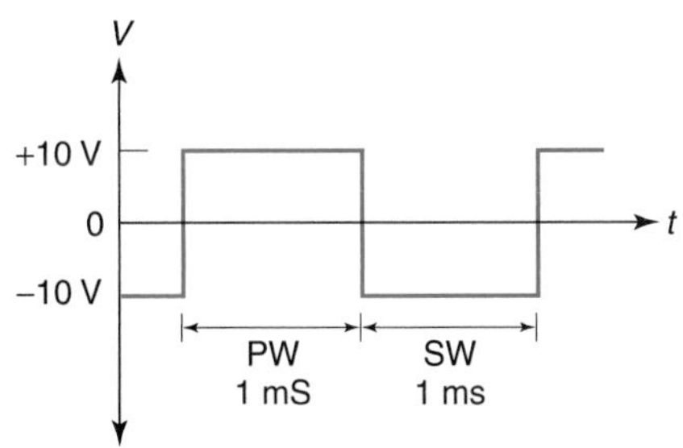

(a) Symmetrical time and amplitude

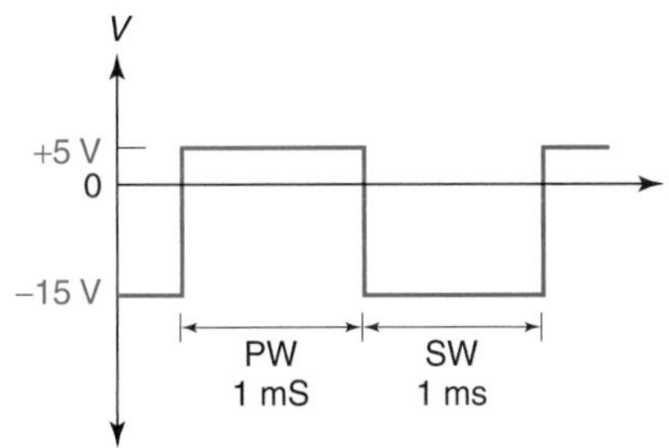

(b) Symmetrical time, asymmetrical amplitude

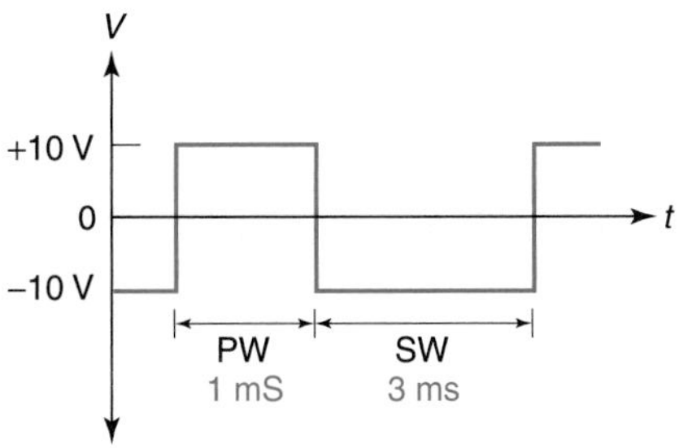

(c) Asymmetrical time, symmetrical amplitude

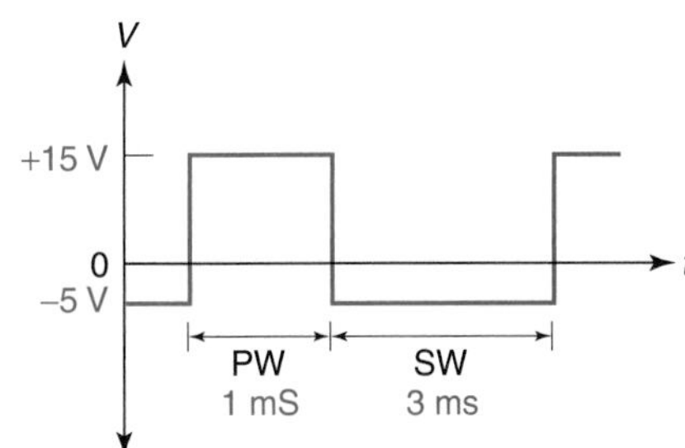

(d) Asymmetrical time and amplitude

Figure 9.43 Symmetrical and asymmetrical waveforms.

For the waveform in Figure 9.43b, the average value can be found as

$$V_{\text{ave}} = \frac{+V_{\text{pk}} + (-V_{\text{pk}})}{2} = \frac{+5\text{V} + (-15\text{ V})}{2} = \frac{-10\text{ V}}{2} = -5\text{ V}$$

Figure 9.43c shows a waveform that has symmetrical peak values, but is asymmetrical in time. For this type of waveform, the average value is found as

$$V_{\text{ave}} = \frac{(PW)(+V_{\text{pk}}) + (SW)(-V_{\text{pk}})}{T_C} \tag{9.22}$$

This equation weighs in the difference in time spent at each peak value. For the waveform shown in Figure 9.43c,

$$V_{\text{ave}} = \frac{(PW)(+V_{\text{pk}}) + (SW)(-V_{\text{pk}})}{T_C} = \frac{(1\text{ ms})(+10\text{ V}) + (3\text{ ms})(-10\text{ V})}{4\text{ ms}}$$

$$= \frac{(+10\text{ V}\cdot\text{ms}) - (30\text{ V}\cdot\text{ms})}{4\text{ ms}} = \frac{-20\text{ V}\cdot\text{ms}}{4\text{ ms}} = -5\text{ V}$$

Equation (9.22) can also be used to find the average value of a rectangular waveform that is asymmetrical in both time and amplitude. For example, look at the waveform shown in Figure 9.43d. The average value of this waveform can be found as

$$V_{\text{ave}} = \frac{(PW)(+V_{\text{pk}}) + (SW)(-V_{\text{pk}})}{T_C}$$

$$= \frac{(1\text{ mS})(+15\text{ V}) + (3\text{ ms})(-5\text{ V})}{4\text{ ms}} = \frac{(15\text{ V}\cdot\text{ms}) - (15\text{ V}\cdot\text{ms})}{4\text{ ms}} = 0\text{ V}$$

As this result indicates, a rectangular waveform can have an average of 0 V and still be considered asymmetrical in amplitude. Even though the peaks vary unequally above and below 0 V, the average (in this case) still works out to 0 V.

Nonsinusoidal Waveforms: Sawtooth and Triangular Waves

OBJECTIVE 21 ➤ The term **sawtooth** is used to describe *a waveform that changes constantly at a linear rate.* That is, it is made up of straight-line changes back and forth between its peak values, as shown in Figure 9.44a. Because of its shape, the *sawtooth waveform* is sometimes referred to as a **ramp.**

The **slope** of a sawtooth waveform is *the rate at which its value changes, measured in volts per unit time.* Slope is measured as shown in Figure 9.44b. Note that a 6 V change over a period of 4 ms translates into a slope of 1.5 V/ms.

The slopes of the positive-going and negative-going transitions of a sawtooth are not necessarily equal, as shown in Figure 9.44a. For this waveform, the slope of the negative-going transition is much greater than that of the positive-going transition. The **triangular waveform** is *a symmetrical sawtooth waveform.* That is, the positive-going and negative-going transitions are mirror images of each other, as shown in Figure 9.44c.

Section Review

1. What is a *static value*? What is a *dynamic value*?
2. How can you distinguish a dynamic value from a static value?
3. What is a *dc offset*? What effect does it have on the average value of a sine wave?
4. What is an *electromagnetic wave*?
5. What is *wavelength*?

6. What factors determine the wavelength of a given waveform?
7. When calculating wavelength, what determines the unit of measure for the result?
8. What is a *harmonic*?
9. In a string of harmonic frequencies, which is the *fundamental*?
10. What is an *octave*? What is a *decade*?
11. What is a *rectangular* waveform?
12. Draw a rectangular waveform, and identify each of the following: *pulse width, space width,* and *cycle time.*
13. What is *duty cycle*?
14. What is a *square wave*?
15. What is a *symmetrical* waveform?
16. What is a *sawtooth* waveform? By what other name is it known?
17. What is the slope of a ramp?
18. What is a *triangular* waveform?

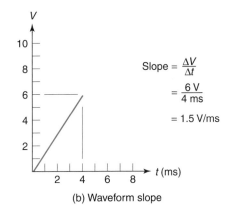

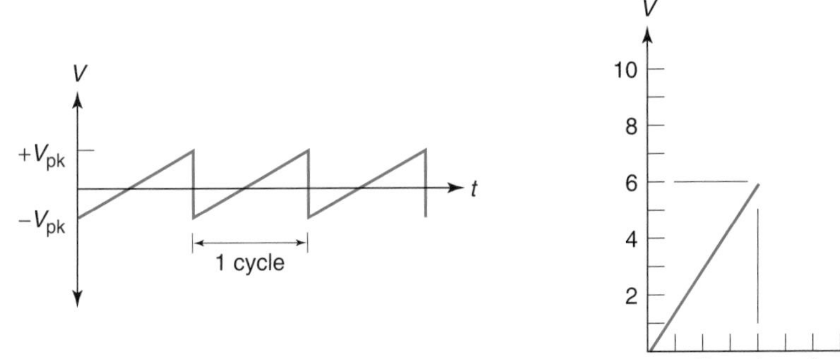

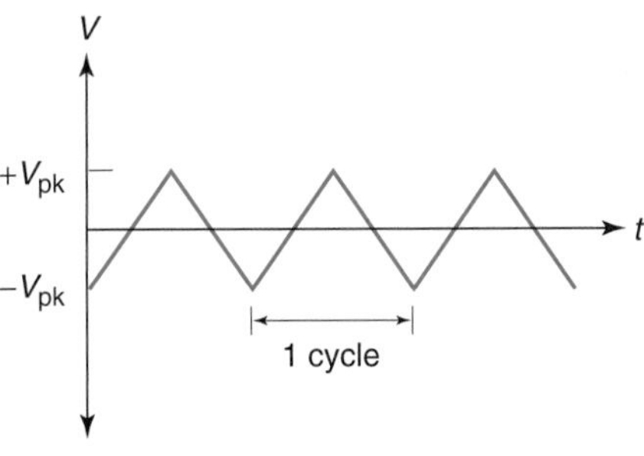

Figure 9.44 Ramp waveforms.

Key Terms

The following terms were introduced in this chapter on the pages indicated:

Alternating current (ac), 235
Alternations, 236
Angular velocity, 258
Amplifier, 262
Average ac power (P_{ac}), 245
Cycle, 236
Cycle time, 237
DC offset, 262
Decade, 264
Duty cycle, 266
Dynamic value, 261
Electromagnetic wave, 263

Frequency, 237
Full-cycle average, 242
Fundamental frequency, 264
Half-cycle, 236
Half-cycle average, 244
Harmonic, 264
Harmonic series, 264
Hertz (Hz), 238
Instantaneous value, 242
Magnitude, 235
n-Order harmonic, 264
Octave, 264
Oscilloscope, 239
Peak value, 242
Peak-to-peak value, 242
Period, 237
Phase, 251
Phase angle (θ), 255
Pulse width, 265
Pure ac waveform, 242
Radian, 258
Ramp, 268
Rectangular wave, 265
Root-mean-square (rms), 246
Rotor, 250
Sawtooth wave, 268
Signal, 237
Sine wave, 235
Sinusoidal waveform, 235
Slope, 268
Space width, 265
Square wave, 266
Static value, 261
Symmetrical waveform, 267
Time base control, 239
Triangular waveform, 268
Waveform, 235
Wavelength, 263

Practice Problems

1. Calculate the cycle time of the waveform shown in Figure 9.45.

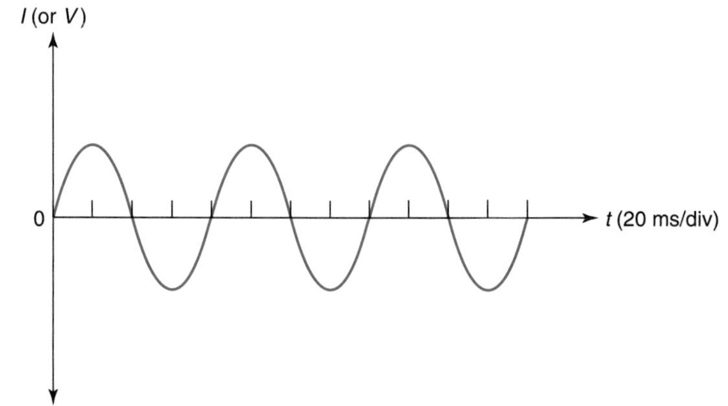

Figure 9.45

2. Calculate the cycle time of the waveform shown in Figure 9.46.

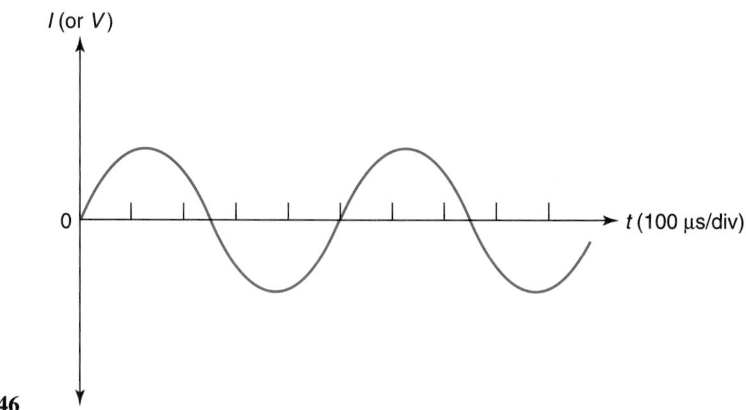

Figure 9.46

3. Calculate the cycle time of the waveform shown in Figure 9.47a.
4. Calculate the cycle time of the waveform shown in Figure 9.47b.
5. Calculate the frequency of the waveform shown in Figure 9.45.
6. Calculate the frequency of the waveform shown in Figure 9.46.

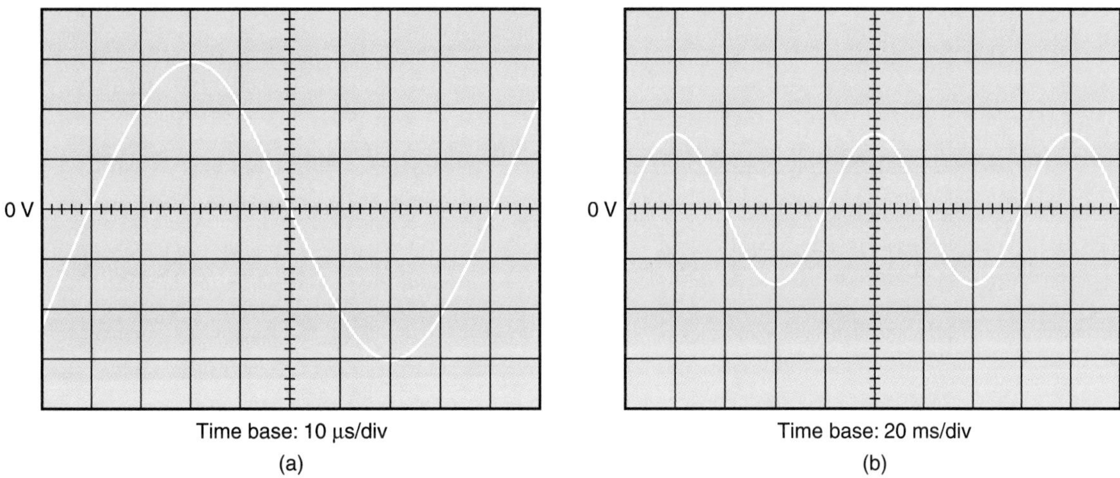

Figure 9.47

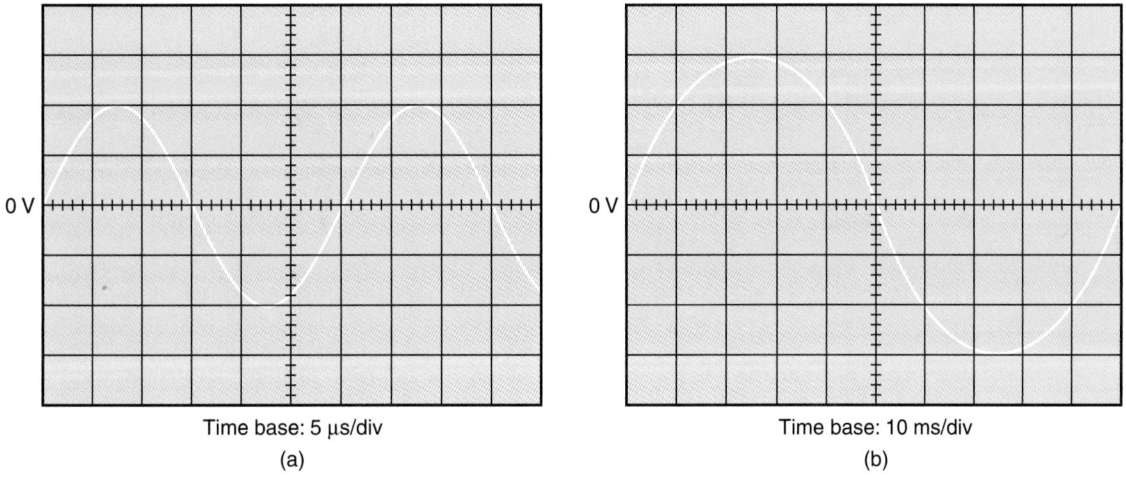

Figure 9.48

7. Calculate the frequency of the waveform shown in Figure 9.48a.
8. Calculate the frequency of the waveform shown in Figure 9.48b.
9. Calculate the cycle time that corresponds to each of the following:
   a. 500 Hz
   b. 2.5 kHz
   c. 1 MHz
   d. 75 kHz
10. Calculate the range of cycle time values for each of the following:
    a. 100 Hz to 300 Hz
    b. 1 kHz to 5 kHz
    c. 20 Hz to 20 kHz
    d. 1.5 MHz to 10 MHz
11. A sine wave has peak values of ± 12 V. Calculate the waveform's half-cycle average voltage.
12. A sine wave has peak values of $\pm$ 20 V. Calculate the waveform's half-cycle average voltage.
13. Determine the peak-to-peak and half-cycle average voltages for the waveform shown in Figure 9.49a.
14. Determine the peak-to-peak and half-cycle average voltages for the waveform shown in Figure 9.49b.

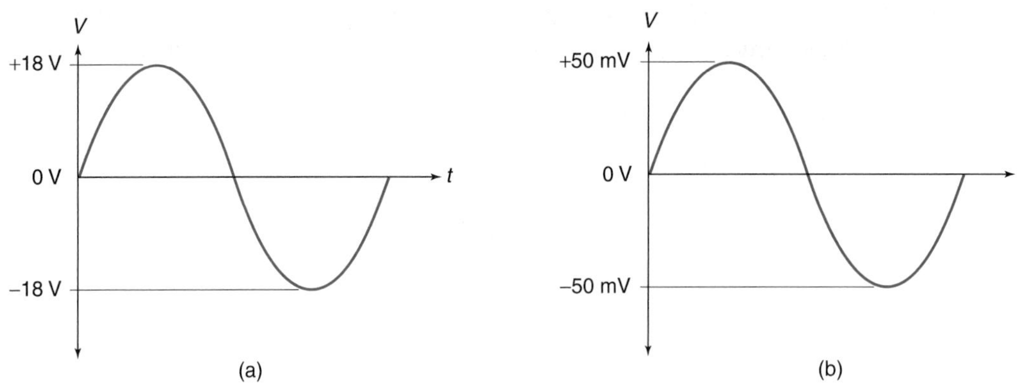

Figure 9.49

15. Calculate the rms values of load voltage and current for the circuit shown in Figure 9.50a.

16. Calculate the rms values of load voltage and current for the circuit shown in Figure 9.50b.

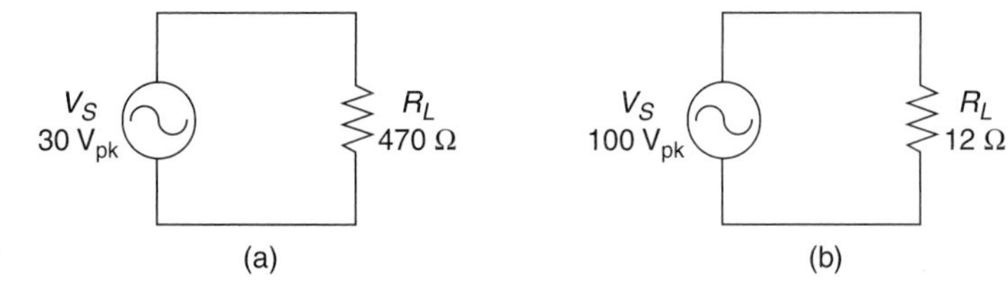

Figure 9.50

17. Calculate the rms values of load voltage and current for the circuit shown in Figure 9.51a.

18. Calculate the rms values of load voltage and current for the circuit shown in Figure 9.51b.

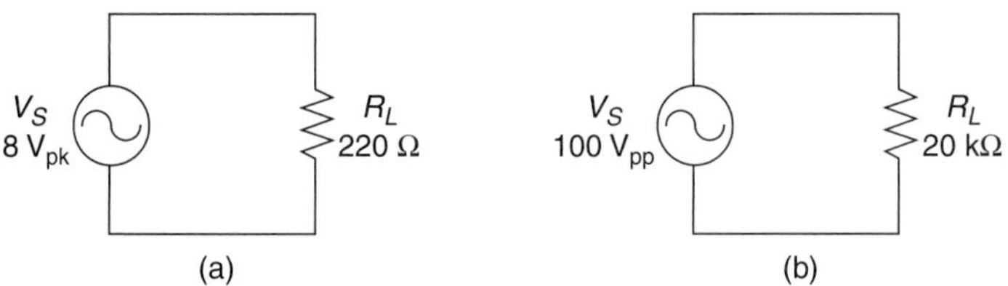

Figure 9.51

19. Calculate the load power for the circuit shown in Figure 9.52a.

20. Calculate the load power for the circuit shown in Figure 9.52b.

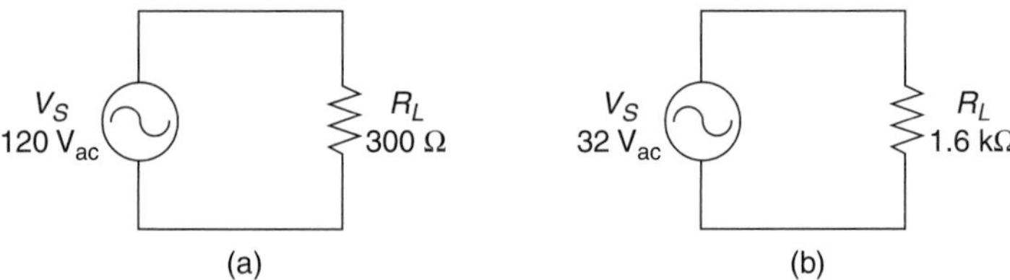

Figure 9.52

21. Determine the peak and peak-to-peak values of the waveform shown in Figure 9.47a. Assume that the vertical sensitivity setting of the oscilloscope is 50 mV/div.

22. Determine the peak and peak-to-peak values of the waveform shown in Figure 9.47b. Assume that the vertical sensitivity setting of the oscilloscope is 100 μV/div.

23. Determine the peak and peak-to-peak values of the waveform shown in Figure 9.48a. Assume that the vertical sensitivity setting of the oscilloscope is 10 V/div.

24. Determine the peak and peak-to-peak values of the waveform shown in Figure 9.48b. Assume that the vertical sensitivity setting of the oscilloscope is 200 mV/div.

25. For a 20 kHz sine wave:
   a. $\theta = 30°$ when $t =$ ______.
   b. $\theta = 150°$ when $t =$ ______.
   c. $\theta =$ ______ when $t = 25\ \mu s$.
   d. $\theta =$ ______ when $t = 10\ \mu s$.

26. For a 1.8 kHz sine wave:
   a. $\theta = 200°$ when $t =$ ______.
   b. $\theta = 15°$ when $t =$ ______.
   c. $\theta =$ ______ when $t = 300\ \mu s$.
   d. $\theta =$ ______ when $t = 55\ \mu s$.

27. Complete the table below.

Frequency	*Cycle Time*	*Instantaneous Time (t)*	*Phase (θ)*
______	10 ms	2 ms	______
120 kHz	______	______	60°
______	45 ms	______	100°
______	______	10 ms	90°

28. Complete the table below.

Frequency	*Cycle Time*	*Instantaneous Time (t)*	*Phase (θ)*
______	100 μs	40 μs	______
2 MHz	______	______	200°
______	50 ms	______	300°
______	______	1.25 ms	60°

29. Determine the value of θ for the highlighted point in Figure 9.53a.

30. Determine the value of θ for the highlighted point in Figure 9.53b.

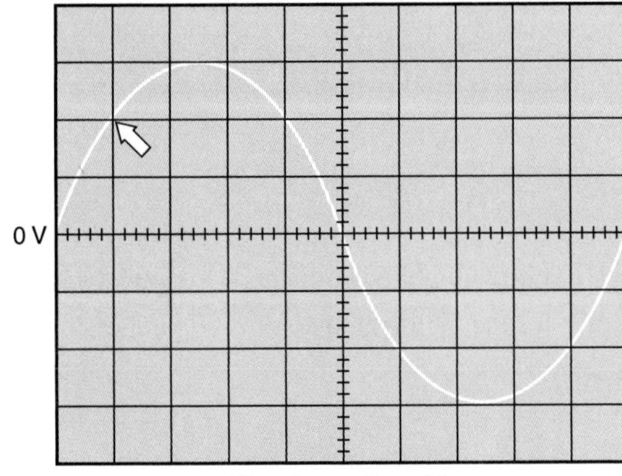

Time base: 20 ms/div
(a)

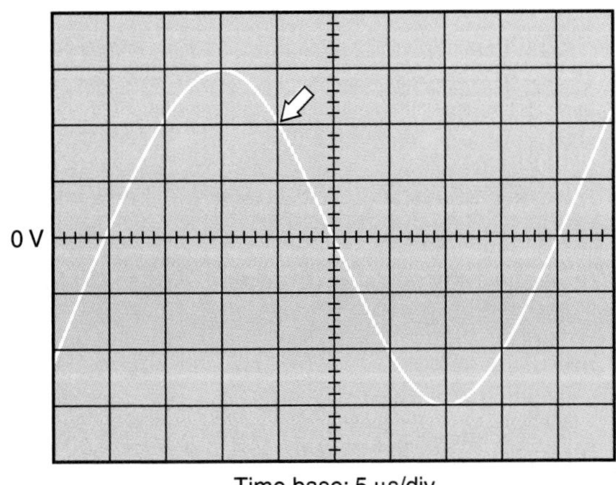

Time base: 5 μs/div
(b)

Figure 9.53

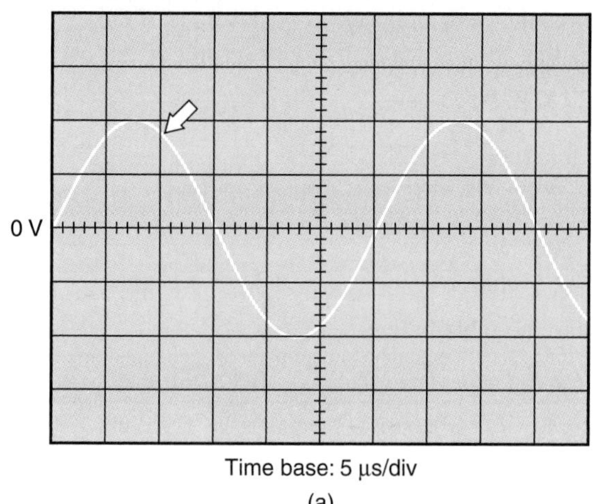

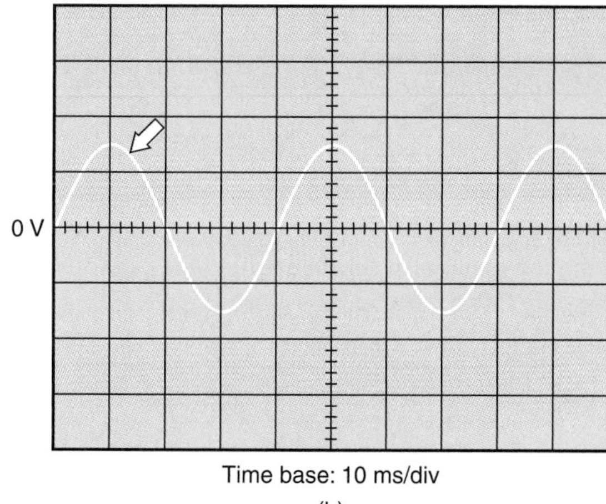

Figure 9.54

31. Determine the value of θ for the highlighted point in Figure 9.54a.

32. Determine the value of θ for the highlighted point in Figure 9.54b.

33. For a $\pm 15V_{pk}$ sine wave:
   a. $v =$ ______ when $\theta = 60°$.
   b. $v =$ ______ when $\theta = 150°$.
   c. $v =$ ______ when $\theta = 240°$.
   d. $v =$ ______ when $\theta = 350°$.

34. For a $\pm 170V_{pk}$ sine wave:
   a. $v =$ ______ when $\theta = 45°$.
   b. $v =$ ______ when $\theta = 190°$.
   c. $v =$ ______ when $\theta = 10°$.
   d. $v =$ ______ when $\theta = 180°$.

35. A sine wave has the following values: $T_C = 200\ \mu s$ and $V_{pk} = 15$ V. Determine the instantaneous voltage when $t = 120\ \mu s$.

36. A sine wave has the following values: $T_C = 16$ ms and $V_{pk} = 170$ V. Determine the instantaneous voltage when $t = 4$ ms.

37. A sine wave has the following values: $f = 1.5$ MHz and $V_{pk} = 50$ mV. Determine the instantaneous voltage when $t = 67$ ns. (Use the degree method.)

38. A sine wave has the following values: $f = 200$ kHz and $V_{pk} = 5$ V. Determine the instantaneous voltage when $t = 2.5\ \mu s$. (Use the degree method.)

39. Solve Problem 37 using the radian method.

40. Solve problem 38 using the radian method.

41. The waveform shown in Figure 9.55a has the values shown. Calculate the value of v_1 using both the degree and radian methods.

42. The waveform shown in Figure 9.55b has the values shown. Calculate the value of v_1 using both the degree and radian methods.

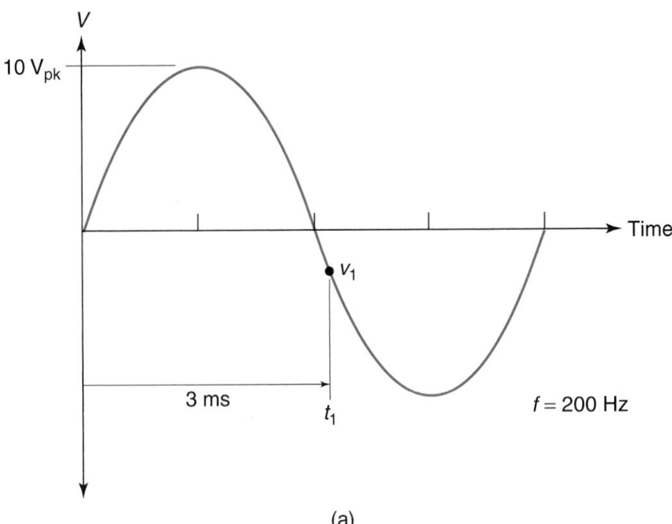

(a)

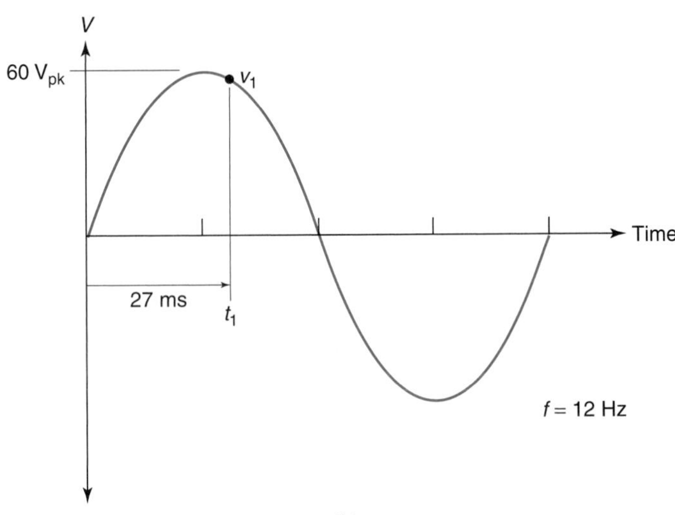

(b)

Figure 9.55

43. The waveforms shown in Figure 9.56a are 45° out of phase. Determine the instantaneous value of waveform A when waveform B is at 0°.

44. The waveforms shown in Figure 9.56b are 60° out of phase. Determine the instantaneous value of waveform A when waveform B is at 20°.

45. Calculate the wavelength (in centimeters) for a 300 MHz sine wave.

46. Calculate the wavelength (in meters) for a 220 kHz sine wave.

47. Calculate the wavelength (in miles) for a 60 Hz sine wave.

48. Calculate the wavelength (in meters) for a 500 kHz sine wave.

49. A sine wave has a frequency range of 15 kHz to 55 kHz. Calculate the range of wavelength values (in miles) for the waveform.

50. A sine wave has a frequency range of 30 MHz to 100 MHz. Calculate the range of wavelength values (in meters) for the waveform.

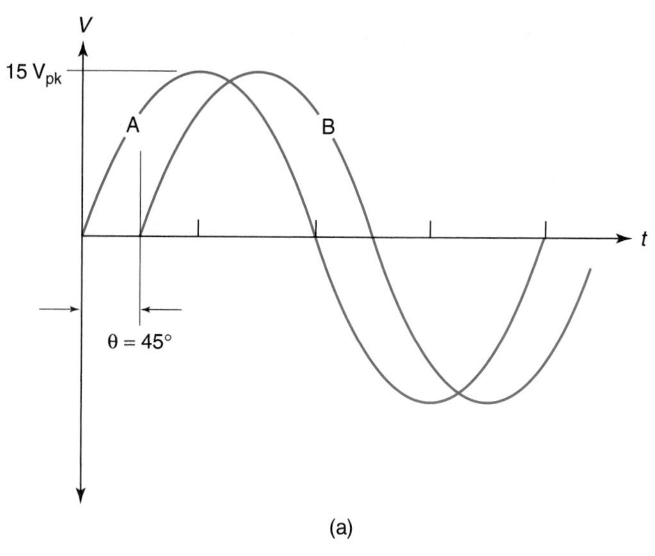

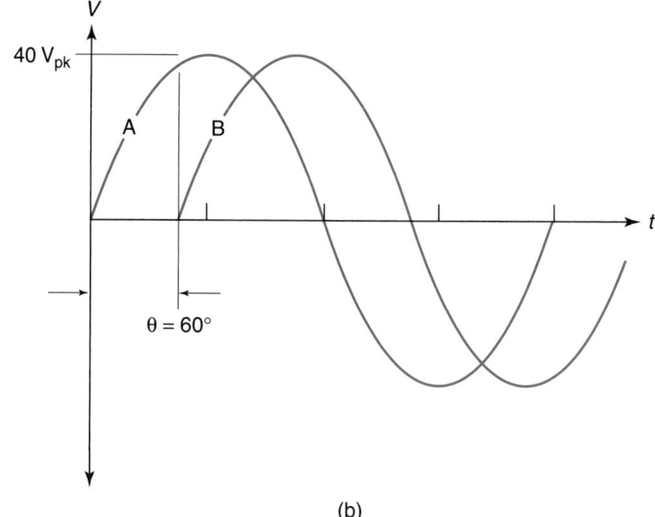

Figure 9.56

51. Determine the frequency that is four octaves above 1 kHz.
52. Determine the frequency that is 12 octaves above 300 Hz.
53. Determine the frequency that is 1.5 decades above 12 kHz.
54. Determine the frequency that is three decades above 10 Hz.
55. Determine the duty cycle of the waveform shown in Figure 9.57a.

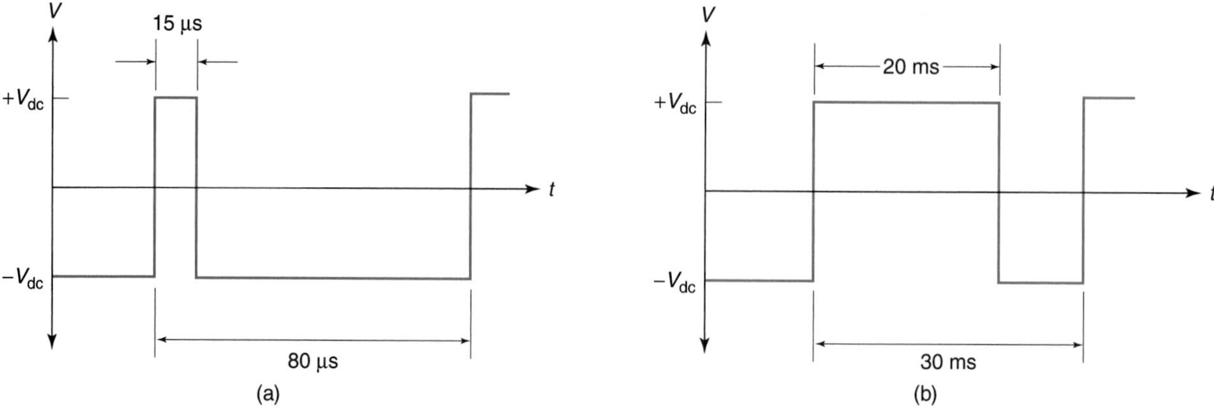

Figure 9.57

56. Determine the duty cycle of the waveform shown in Figure 9.57b.

57. Determine the duty cycle of the waveform shown in Figure 9.58a.

58. Determine the duty cycle of the waveform shown in Figure 9.58b.

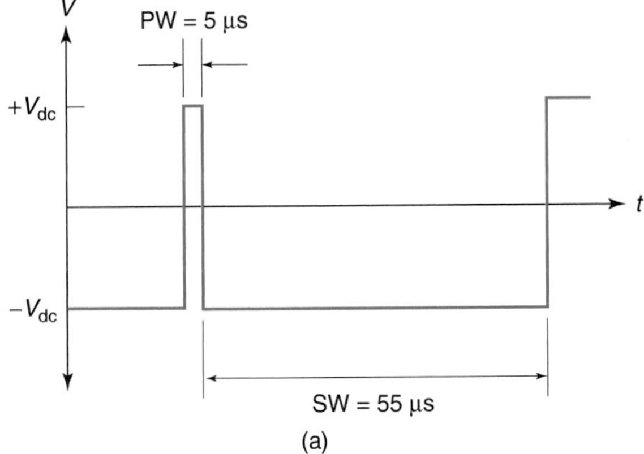

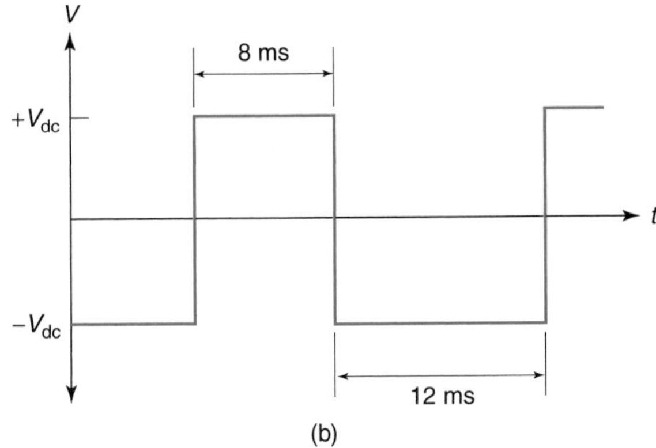

Figure 9.58

59. Determine the average value of the waveform shown in Figure 9.59a.

60. Determine the average value of the waveform shown in Figure 9.59b.

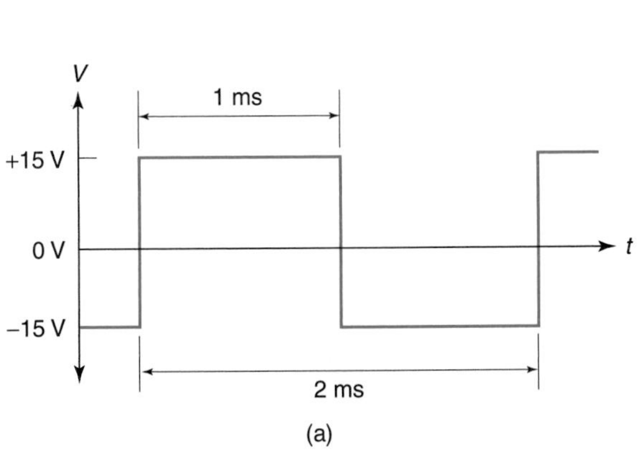

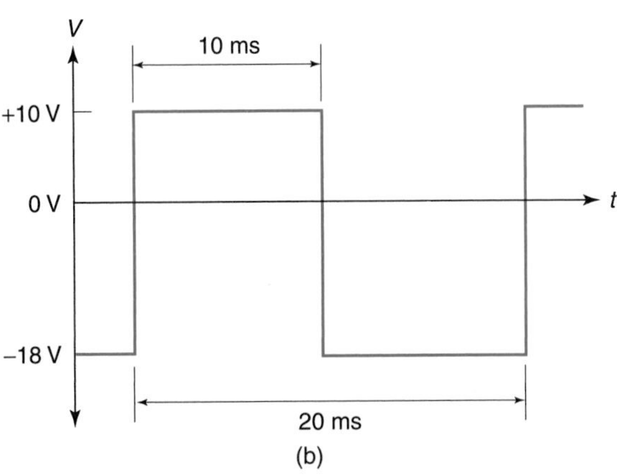

Figure 9.59

61. Determine the average value of the waveform shown in Figure 9.60a.

62. Determine the average value of the waveform shown in Figure 9.60b.

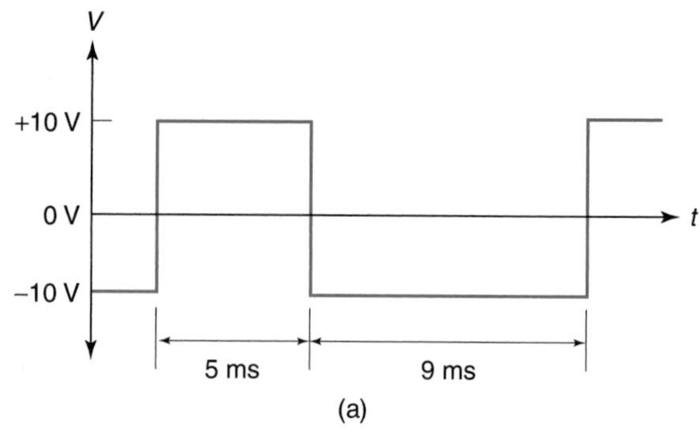

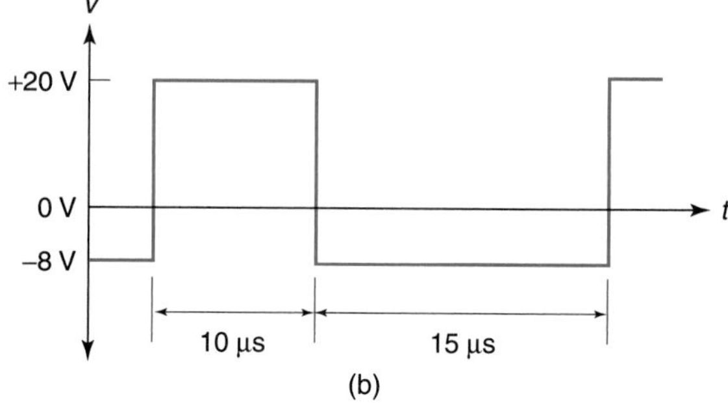

Figure 9.60

Looking Back

These problems relate to material presented in earlier chapters. The chapters are identified in brackets.

63. A series circuit has the following values: $P_S = 200$ mW, $R_T = 500\ \Omega$, and $R_1 = 100\ \Omega$. Determine the value of P_1 for the circuit. [Chapter 4]

64. Determine the change in load voltage that occurs in Figure 9.61 when the load current varies over its rated range. [Chapter 6]

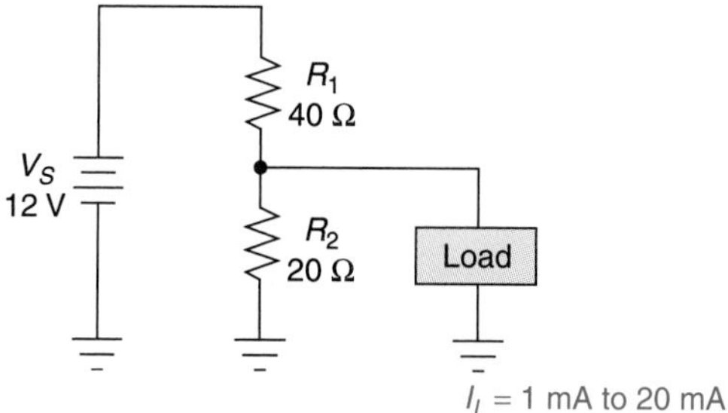

Figure 9.61

65. Calculate the maximum load power for the circuit shown in Figure 9.62. [Chapter 4]

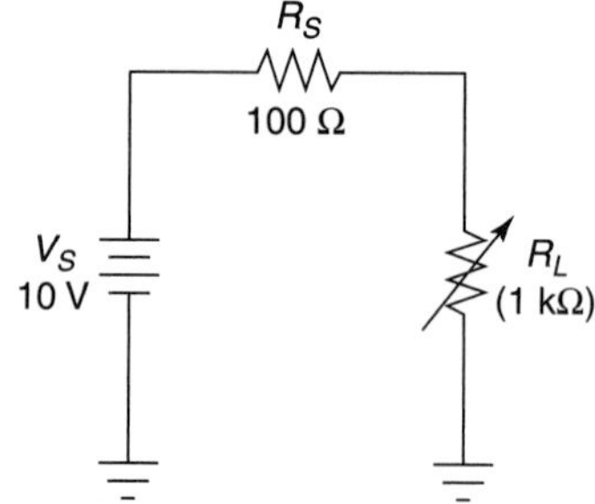

Figure 9.62

66. A magnet with a cross-sectional area of 6 cm^2 generates approximately 1800 μWb of flux at its poles. Calculate the flux density at the poles of the magnet. [Chapter 8]

67. A sine wave has the following values: $V_{pk} = +12$ V and $v_1 = +4.8$ V when $t = 50\ \mu s$. Determine the two lowest possible frequencies of the waveform. [*Hint:* $\sin\theta = \sin(180° - \theta)$.]

68. A sine wave takes 50 μs to reach a phase of 60°. Calculate the wavelength (in meters) of the waveform.

69. A sine wave has a value of $T_C = 25\ \mu s$. Determine the wavelength (in centimeters) of its fifth-order harmonic.

70. A rectangular waveform has a 40% duty cycle and a space width of 60 μs. Determine its wavelength (in centimeters).

71. Determine the negative peak value of the waveform shown in Figure 9.63.

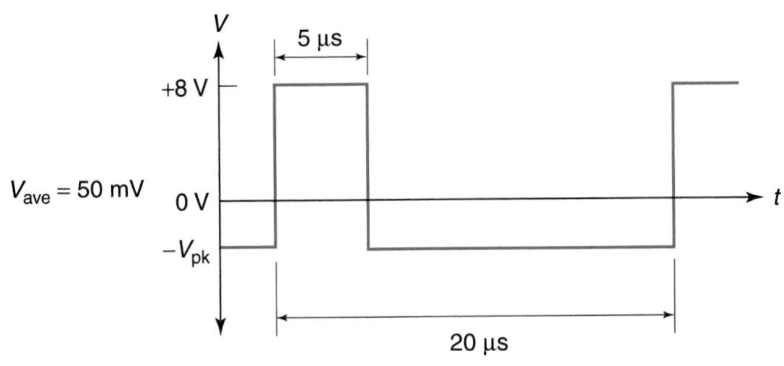

Figure 9.63

Answers to the Example Practice Problems

9.1 150 μs

9.2 25 Hz

9.3 1 ms

9.4 5 kHz

9.5 15.28 V

9.6 $V_{rms} = 21.21$ V, $I_{rms} = 19.28$ mA

9.7 33.3 mW

9.8 1.5 ms

9.9 Approximately 203°

9.10 10.4 V

9.11 −10.26 V

9.12 17.1 V

9.13 For both methods, $v_1 = 28.8$ V

9.14 1.2 m

9.15 40%

Multisim File References

Figure	File Reference
9.17	ETF9_17.msm
9.28	ETF9_28.msm
9.36	ETF9_36.msm
9.39	ETF9_39.msm
Table 9.1	ETFtb9_1.msn

chapter 10

Inductors

Objectives

After studying the material in this chapter, you should be able to:

1. Describe the effect of varying current on a magnetic field.
2. Describe the concept of *self-induction.*
3. Explain the concept of *counter emf.*
4. Define the *henry (H)* unit of inductance.
5. Explain the phase relationship between inductor current and voltage.
6. Calculate the total inductance for any group of inductors connected in series.
7. Calculate the total inductance for any group of inductors connected in parallel.
8. Discuss *mutual inductance,* the factors that affect its value, and its effect on series- and parallel-connected inductors.
9. Discuss *inductive reactance* (X_L) and its source.
10. Calculate the value of X_L for any inductor operated at a specified frequency.
11. Perform basic circuit calculations involving inductive reactance.
12. Describe the construction of a typical transformer.
13. List the three basic types of transformers, and state the voltage input/output relationship of each.
14. Briefly describe transformer operation in terms of electromagnetic induction.
15. Perform voltage calculations for a transformer using its turns ratio.
16. List and describe the power losses that occur in a typical transformer.
17. Perform current calculations for a transformer using its turns ratio.
18. Perform impedance calculations for a transformer using its turns ratio.
19. Describe, compare, and contrast the following: *apparent power, true power,* and *reactive power.*
20. Discuss the concept of inductor *quality* (Q) and calculate its value for a given inductor operating at a specified frequency.
21. Compare and contrast *iron-core* and *air-core* inductors.
22. List and describe the various types of inductors.

You have seen how magnetic flux can be used to induce a current through a conductor, and vice versa. In this chapter, we're going to take a look at a property that is based on the relationship between magnetism and current called *inductance.*

Technically defined, **inductance** is *the ability of a component with a changing current to induce a voltage across itself or a nearby circuit by generating a changing magnetic field.* Based on its overall effect, inductance is often described as:

- The ability of a component to oppose a change in current.
- The ability of a component to store energy in an electromagnetic field.

An **inductor** is *a component designed to provide a specific measure of inductance.* An inductor opposes any change in current and can store energy in an electromagnetic field. These characteristics make it useful in a variety of applications.

10.1 Inductance

In Chapter 9, you learned that a current-carrying coil generates magnetic flux, as shown in Figure 10.1. As a review, here are some of the principles we established regarding the relationship between current and a resulting flux:

- The current through a wire produces magnetic lines of force.
- When a wire is wrapped in a series of loops, called a *coil,* the lines of force generated by the current add together to form a magnetic field.
- The strength of the magnetic field is determined by the amount of *magnetomotive force (mmf)* produced by the current through the coil.
- The amount of mmf is determined by the current (I) and the number of turns (N) in the coil. The product of these factors is called the *ampere-turn* ($A \cdot t$).
- The polarity of the flux generated by a current is determined by the direction of that current.

As you will see, all these factors play a role in the principle of inductance.

The Effect of Varying Current on a Magnetic Field

A coil like the one shown in Figure 10.1 is referred to as an *inductor.* When current passes through an inductor, magnetic flux is generated as shown in the figure. The flux density is found as ◄ *OBJECTIVE 1*

$$B = \frac{\mu_m NI}{\ell} \quad (10.1)$$

Reference: Equation (10.1) was first introduced in Chapter 8.

where

B = the flux density, in webers per square meter $\left(\frac{Wb}{m^2}\right)$

μ_m = the permeability of the core material

NI = the ampere-turns product

ℓ = the length of the coil, in meters

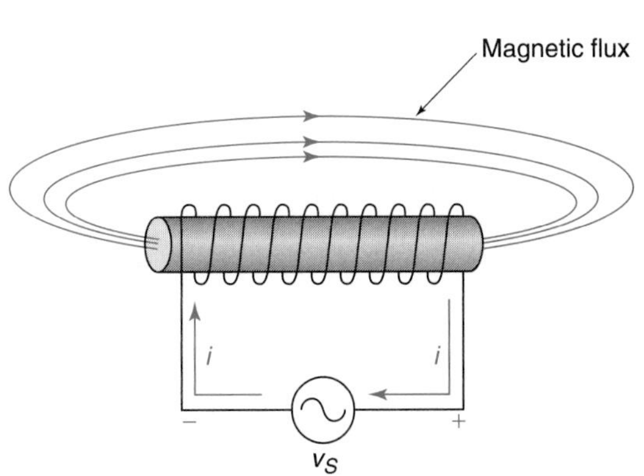

Figure 10.1 Current-carrying coil and the resulting magnetic flux.

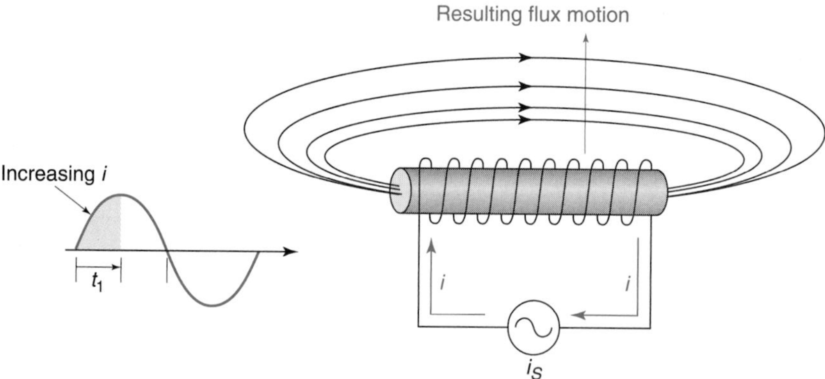

(a) An increase in current causes the magnetic field to expand outward.

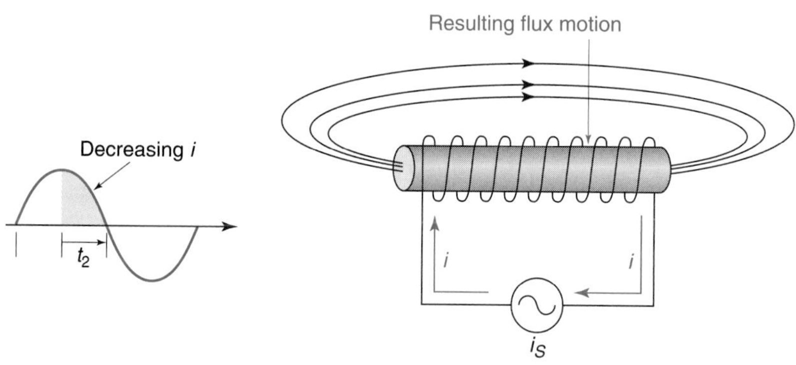

Figure 10.2 Effects of changing current on magnetic flux.

According to equation (10.1), *flux density varies in direct proportion to inductor current.* The effect of varying inductor current on the resulting magnetic field is illustrated in Figure 10.2. As the inductor current increases, flux density increases, and the magnetic field expands outward from the component. As the inductor current decreases, flux density decreases, and the magnetic field collapses back into the component. The impact of this changing magnetic field is described by several of *Faraday's laws of induction.*

Faraday's Laws of Induction

OBJECTIVE 2 ➤ In 1831, an English scientist named Michael Faraday discovered that *a magnetic field could be used to generate (or induce) a voltage across a coil* through a process called *electromagnetic induction.* Faraday's observations on the results of his experiments have come to be known as *Faraday's laws of induction.* The first three of these laws can be paraphrased as follows:

Law 1: To induce a voltage across a wire, there must be a relative motion between the wire and the magnetic field.

Law 2: The voltage induced is proportional to the rate of change in magnetic flux encountered by the wire.

Law 3: When a wire is cut by 10^8 perpendicular lines of force per second, 1 V is induced across that wire.

Faraday's first law states that induction takes place when there is relative motion between the conductor and the magnetic field. In the case of the inductor, its own expanding and collapsing magnetic field produces the necessary relative motion between the conductor (coil) and the magnetic field.

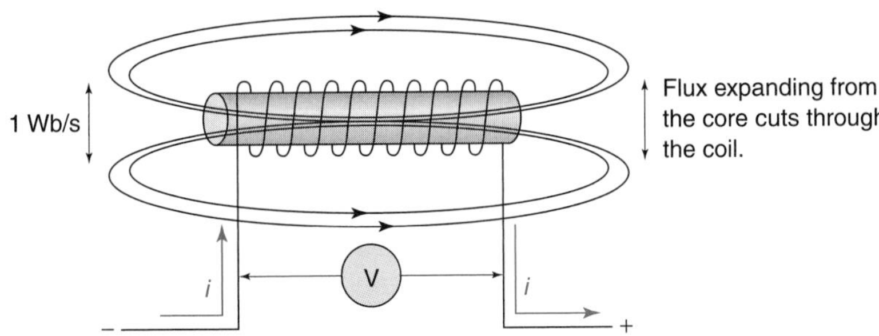
1 Wb/s
Flux expanding from
the core cuts through
the coil.
V
i
i
-
+

Figure 10.3 One volt is induced across each turn when flux cuts the coil at a rate of 1 Wb/s.

Faraday's second law states that induced voltage is proportional to the rate of change in the magnetic flux. This law can be stated mathematically as

$$V = N\frac{d\phi}{dt} \tag{10.2}$$

where

N = the number of turns in the coil

$\frac{d\phi}{dt}$ = the instantaneous rate of change in the flux (in webers per second)

The faster that the field changes in intensity (expands and collapses), the greater the value of $\frac{d\phi}{dt}$, and the greater the induced voltage. For example, Figure 10.3 shows a magnetic field expanding outward from the core of an inductor. When *1 weber* (10^8 lines) of magnetic flux pass through the inductor per second, a 1 V difference of potential is induced *across each turn* in the component. If 1 Wb of flux cuts through the inductor every 100 ms, the rate of change in flux (per turn) is found as

Note: Based on Faraday's third law, we can now redefine the *weber (Wb)* as the amount of flux that, when passing through a perpendicular coil in one second, induces 1 V across that coil.

$$\frac{d\phi}{dt} = \frac{1\text{ Wb}}{100\text{ ms}} = 10\text{ Wb/s}$$

Therefore, 10 V is induced across each turn in the component (1 V for each Wb/s).

We have now established the following:

- The current through an inductor generates magnetic flux.
- As the inductor current increases, the magnetic field expands. As the inductor current decreases, the magnetic field collapses.
- As the magnetic field expands and collapses, it cuts through the stationary inductor. This produces a relative motion between the two.
- The relative motion caused by the changing magnetic field induces a voltage across the inductor.

The process described above is referred to a **self-inductance.** To understand the impact of self-inductance, we have to take a look at *Lenz's law.*

Lenz's Law

◄ *OBJECTIVE 3*

In 1834, Heinrich Lenz derived the relationship between a magnetic field and the voltage it induces. This relationship, known as **Lenz's law,** can be paraphrased as follows: *An induced voltage always opposes its source.* That is, the polarity of the voltage induced by a magnetic field opposes the change in current that produced the magnetic field, as illustrated in Figure 10.4.

In Figure 10.4a:

- An increase in the inductor current causes the magnetic field to expand.
- As the magnetic field expands, it cuts through the coil, inducing a voltage.
- The polarity of the induced voltage (v_L) opposes the increase in current.

Figure 10.4
An inductor opposes a change in current.

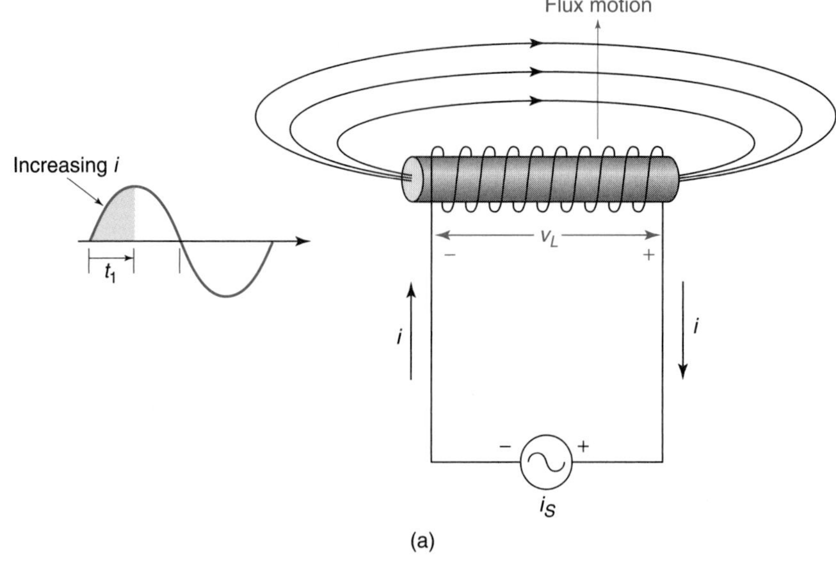

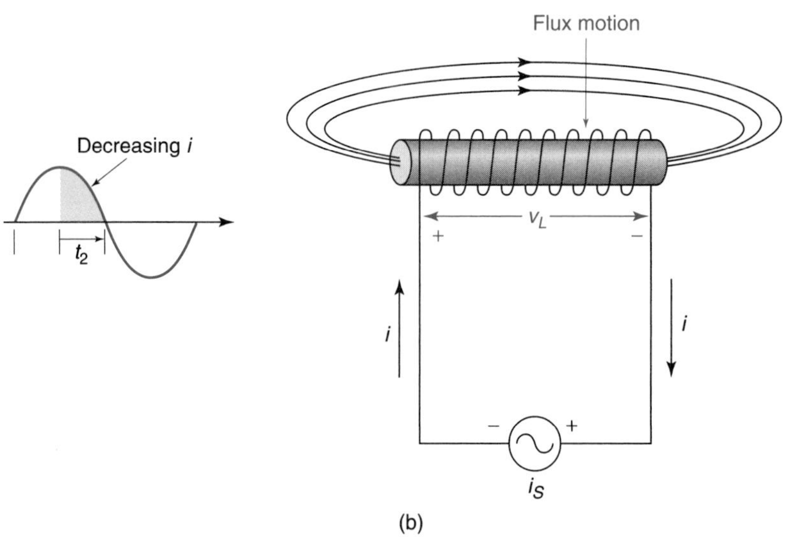

In Figure 10.4b:

- A decrease in the inductor current causes the magnetic field to collapse.
- As the magnetic field collapses, it cuts through the coil, inducing a voltage across the component. (Note the polarity of the voltage has reversed.)
- The polarity of the induced voltage opposes the decrease in current.

Because it always opposes the change in coil current, the *induced voltage* is referred to as **counter emf.**

Induced Voltage

The voltage that is induced across an inductor when the current through the component changes can be found as

$$-v_L = L\frac{di}{dt} \tag{10.3}$$

where

v_L = the instantaneous value of induced voltage

L = the inductance of the coil, measured in henries (H)

$\frac{di}{dt}$ = the instantaneous rate of change in inductor current (in amperes per second)

Note: *di/dt* is simply a way of expressing an instantaneous rate of change in current. For example, if current is changing at a rate of 1 A/s at an instant in time, then

$$\frac{di}{dt} = 1 \text{ A/s}$$

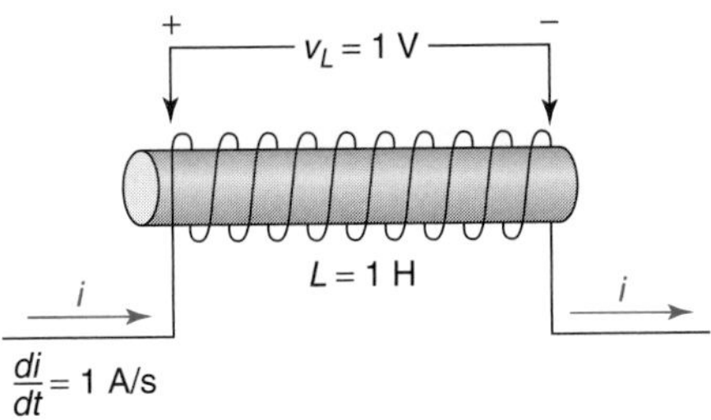

Figure 10.5 The henry (H) unit of inductance.

Note that the minus sign in equation (10.3) is in keeping with Lenz's law, which defines induced voltage as a *counter* emf.

The **henry (H),** which is *the unit of measure of inductance,* is easy to define when equation (10.3) is transposed as follows:

◄ *OBJECTIVE 4*

$$L = \frac{v_L}{\left(\frac{di}{dt}\right)} \tag{10.4}$$

Note: Equation (10.4) is being used to define the henry (H). Since a unit of measure is neither positive nor negative, the minus sign from equation (10.3) has been dropped.

This equation indicates that inductance is measured in *volts per rate of change in current.* The current through the inductor shown in Figure 10.5 is changing at a rate of one ampere per second (A/s). This rate of change is inducing 1 V across the inductor. When a change of 1 A/s induces 1 V across an inductor, the amount of inductance is said to be 1 H.

The relationship between inductance and inductor current and voltage is demonstrated further in Example 10.1.

Example 10.1

The schematic symbol for an air-core inductor is shown in Figure 10.6. For the component shown, calculate the value of v_L when the current changes at a rate of 100 mA/s.

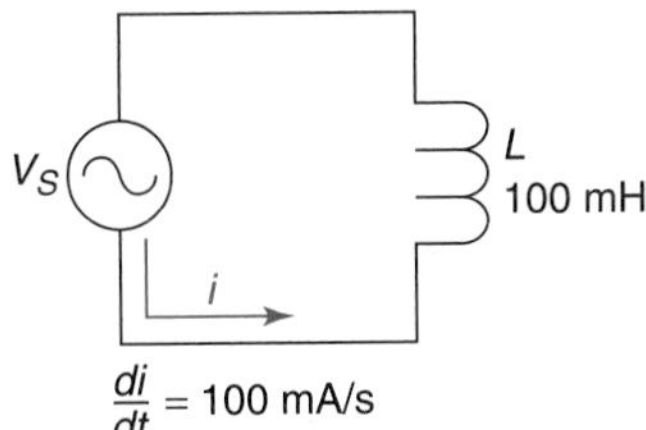

Figure 10.6

Solution: Using equation (10.3), the value of the induced voltage can be found as

$$-v_L = L\frac{di}{dt} = (100 \text{ mH})(100 \text{ mA/s}) = 10 \text{ mV}$$

Practice Problem 10.1

The current through a 33 mH inductor changes at a rate of 200 mA/s. Calculate the value of the induced voltage.

Equation (10.4) defines inductance in terms of induced voltage and rate of change in current. It is also possible to approximate the value of an inductor based on its physical characteristics, as follows:

$$L \cong \frac{\mu_m N^2 A}{\ell} \tag{10.5}$$

Note: Equation (10.5) holds true only if the length of the coil is at least ten times greater than the diameter of the coil.

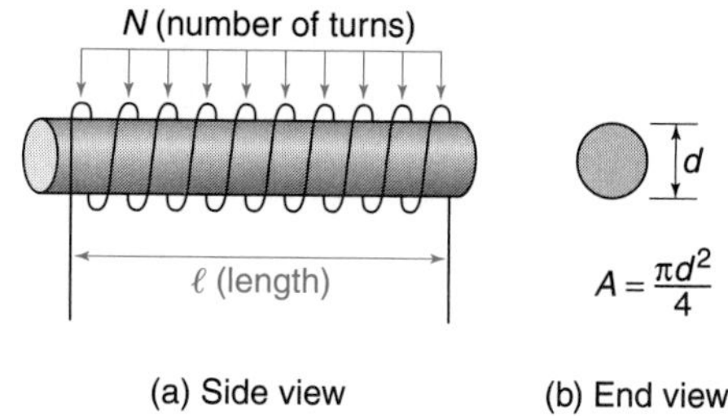

Figure 10.7 Inductor physical dimensions.

where

μ_m = the permeability of the inductor core

N^2 = the square of the number of turns

A = the cross-sectional area of the inductor core, in square centimeters (cm^2)

ℓ = the length of the coil, in centimeters (cm)

The physical characteristics of an inductor are illustrated in Figure 10.7. Note that the area of the core is calculated as shown in the figure. According to equation (10.5), inductance is:

- Directly proportional to the permeability of the inductor core, the square of the number of turns, and the cross-sectional area of the core.
- Inversely proportional to the length of the component.

Putting It All Together

An *inductor* is a component designed to provide a specific amount of inductance. Inductance is measured in *volts per rate of change in current.* The unit of measure of inductance is the *henry (H).* When the current through 1 H of inductance changes at a rate of 1 A/s, the difference of potential induced across the coil is 1 V.

The value of an inductor (in henries) is determined by its physical dimensions. Many of the properties of inductance are summarized in Figure 10.8.

Section Review

1. List the three commonly accepted definitions of *inductance.*
2. What is an *inductor?*
3. What is the relationship between flux density and:
   a. core permeability?
   b. the ampere-turns of a coil?
   c. coil length?
4. How does a change in coil current affect the magnetic field generated by the component?
5. How is Faraday's first law of induction fulfilled by a coil and its magnetic field?
6. What is *Lenz's law?*
7. What is the unit of measure for inductance? In terms of volts per change in current, what does this unit equal?
8. Describe the relationship between the physical dimensions of an *inductor* and its *inductance.*

Critical Thinking

9. Two inductors have identical physical dimensions. The only difference between the two is that one has an air core and the other has an iron core. Which has the higher value of inductance? Explain your answer.

10. Equation (10.3) states that $-v_L = L\dfrac{di}{dt}$. Using the definition of the henry (H), show how the units of L and $\dfrac{di}{dt}$ resolve to give an answer in volts (V).

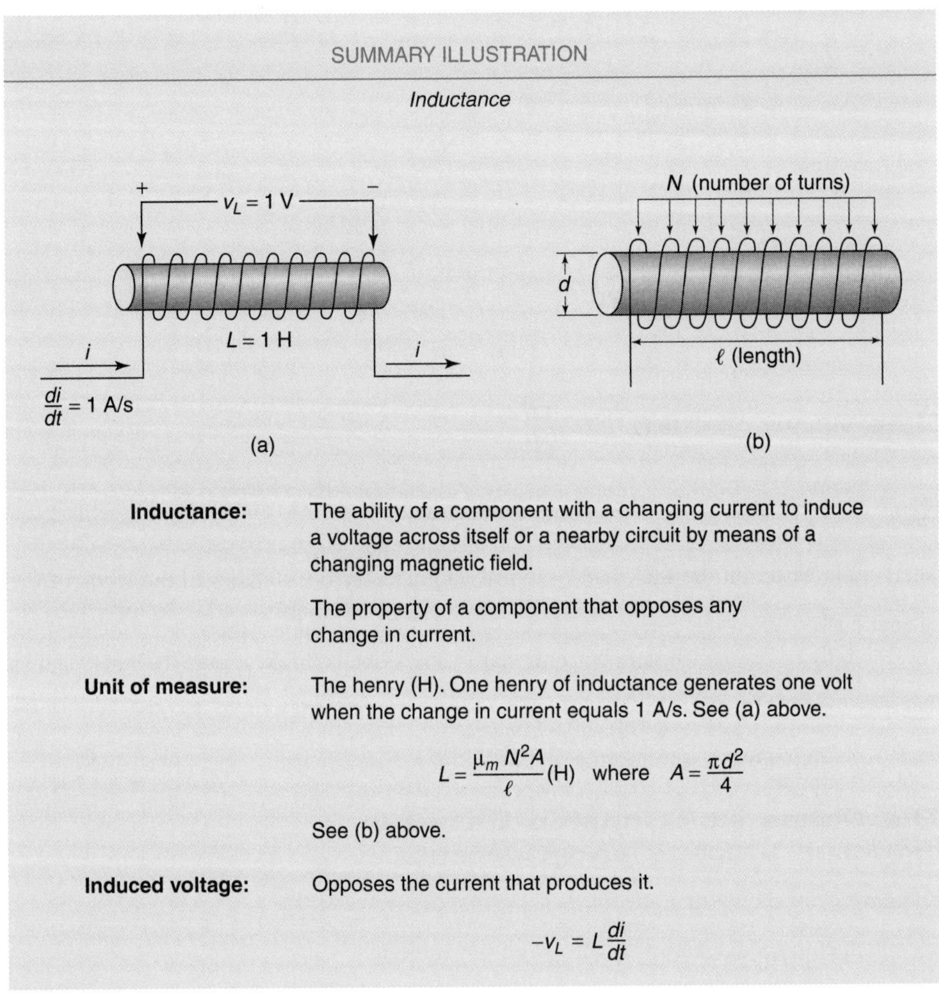

SUMMARY ILLUSTRATION
Inductance
+
−
$v_L = 1$ V
$L = 1$ H
i
i
$\frac{di}{dt} = 1$ A/s
(a)
N (number of turns)
d
ℓ (length)
(b)
Inductance:
The ability of a component with a changing current to induce a voltage across itself or a nearby circuit by means of a changing magnetic field.
The property of a component that opposes any change in current.
Unit of measure:
The henry (H). One henry of inductance generates one volt when the change in current equals 1 A/s. See (a) above.
$L = \frac{\mu_m N^2 A}{\ell}$ (H) where $A = \frac{\pi d^2}{4}$
See (b) above.
Induced voltage:
Opposes the current that produces it.
$-v_L = L\frac{di}{dt}$

Figure 10.8

10.2 The Phase Relationship Between Inductor Current and Voltage

In a purely resistive circuit, current and voltage are directly proportional. In a purely inductive circuit, the relationship between current and voltage gets more complicated. The relationship between inductor voltage and circuit current was given earlier in the chapter as ◄ *OBJECTIVE 5*

$$-v_L = L\frac{di}{dt}$$

Assuming that the value of L is constant, the v_L equation indicates that induced voltage varies directly with the rate of change in current $\left(\dfrac{di}{dt}\right)$.

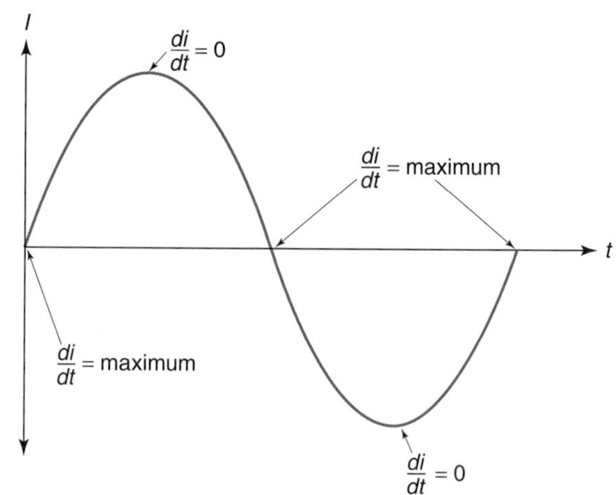

Figure 10.9 Sine-wave values of $\frac{di}{dt}$.

Sine-Wave Values of $\frac{di}{dt}$

Figure 10.9 shows a current sine wave. As illustrated, $\frac{di}{dt} \cong 0$ A/s when $i = I_{pk}$. This characteristic is based on the fact that $\frac{di}{dt}$ is just another way of referring to the *slope* of the current curve. When the current reaches either of its peak values, there is an instant during which there is no change in current; that is, the slope is 0 A/s. Figure 10.9 also shows that the slope of the curve $\left(\frac{di}{dt}\right)$ is greatest at the zero-crossing points. In other words, $\frac{di}{dt}$ reaches its maximum value when $i = 0$.

An Analogy: This analogy may help you to understand why *di/dt* is zero at the peaks of a sine wave: Picture a ball thrown straight up into the air. At the moment it reaches its maximum height, it must stop for an instant before it begins to fall. At that instant, its rate of change in position is zero.

The Phase Relationship Between Inductor Voltage and Current

You have been shown that:

- Inductor voltage varies directly with the value of $\frac{di}{dt}$.
- The value of $\frac{di}{dt}$ varies inversely with inductor current.

These relationships are combined in Table 10.1.

The voltage and current values in Table 10.1 indicate that inductor voltage and current are *90° out of phase,* as illustrated in Figure 10.10. Table 10.2 describes the relationship between the inductor current and voltage waveforms.

As shown in Figure 10.10, there is a 90° phase shift between the voltage and current waveforms. If we use the voltage waveform as a reference, the inductor current begins its positive half-cycle 90° after the inductor voltage. This relationship is normally described in one of two ways:

- Voltage **leads** current by 90°.
- Current **lags** voltage by 90°.

TABLE 10.1 The Relationships Among Inductor Instantaneous Values

Current (i)	*di/dt*	*Induced Voltage (v_L)*
Minimum ($i = 0\ A$)	Maximum	Maximum ($v_L = V_{pk}$)
Maximum ($i = I_{pk}$)	Minimum	Minimum ($v_L = 0\ V$)

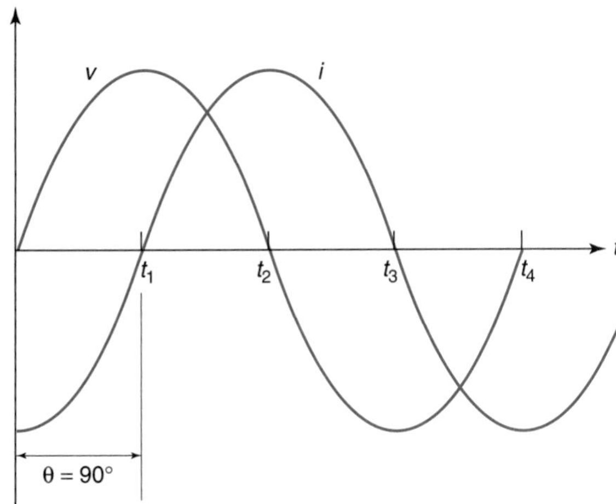

Figure 10.10 The phase relationship between inductor voltage and current.

TABLE 10.2 The Current and Voltage Relationship Shown in Figure 10.10

Time	*Description*
t_1	Current is at the zero crossing, $\frac{di}{dt}$ is at its maximum value, and inductor voltage is at its maximum value ($v_L = V_{pk}$).
t_2	Current reaches its positive peak, $\frac{di}{dt} = 0$ A/s, and $v_L = 0$ V.
t_3	Current crosses zero again, so $\frac{di}{dt}$ and v_L reach their peak (maximum) values.
t_4	Current reaches its negative peak, $\frac{di}{dt} = 0$ A/s, and $v_L = 0$ V.

You will see that the terms *lead* and *lag* are commonly used when describing the phase relationships in ac circuits.

It is important to note that the current versus voltage relationship described in this section applies only to a purely inductive circuit. If we add one or more resistors to an inductive circuit, we have what is called a *resistive-inductive (RL) circuit.* As you will see in Chapter 11, the phase relationship between circuit current and source voltage changes when any significant amount of resistance is added to an inductive circuit.

Section Review

1. What is the relationship between the voltage induced across a coil and the rate of change in current $\left(\frac{di}{dt}\right)$?
2. What is the relationship between $\frac{di}{dt}$ and the instantaneous value of inductor current?
3. What is the relationship between the voltage induced across a coil and the instantaneous value of inductor current?
4. Using the above relationships, describe the phase relationship shown in Figure 10.10.
5. What is meant by the terms *lead* and *lag?*
6. Under what condition does the phase relationship described in this section no longer apply?

10.3 Connecting Inductors in Series and Parallel

OBJECTIVE 6 ➤ When inductors are connected in series, the total inductance of the circuit can be found as

$$L_T = L_1 + L_2 + \cdots + L_n \tag{10.6}$$

where

L_n = the highest numbered inductor in the circuit

As you can see, total inductance (in this case) is found in the same fashion as total resistance in a series circuit. The use of this equation is demonstrated in Example 10.2.

Example 10.2

Calculate the total inductance of the coils in Figure 10.11.

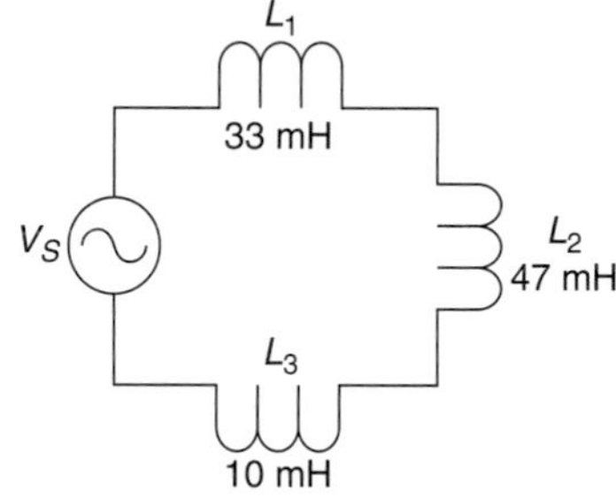

Figure 10.11

Solution: The component values are simply added, as follows:

$$L_T = L_1 + L_2 + L_3 = 33\text{ mH} + 47\text{ mH} + 10\text{ mH} = 90\text{ mH}$$

Practice Problem 10.2

A circuit like the one shown in Figure 10.11 has the following values: $L_1 = 100\ \mu\text{H}$, $L_2 = 3.3$ mH, and $L_2 = 330\ \mu\text{H}$. Calculate the value of L_T.

Parallel-Connected Coils

OBJECTIVE 7 ➤ When coils are connected in parallel, the total inductance of the circuit can be found as

$$L_T = \frac{1}{\dfrac{1}{L_1} + \dfrac{1}{L_2} + \cdots + \dfrac{1}{L_n}} \tag{10.7}$$

where

L_n = the value of the highest-numbered inductor in the circuit

The following example demonstrates the use of this relationship.

Example 10.3

Calculate the total inductance for the circuit in Figure 10.12.

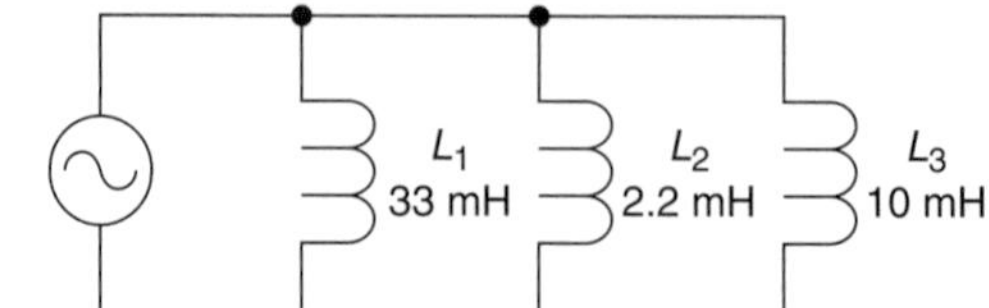

Figure 10.12

Solution: Using equation (10.7), the total inductance in the circuit can be found as

$$L_T = \frac{1}{\dfrac{1}{L_1} + \dfrac{1}{L_2} + \dfrac{1}{L_3}}$$

$$= \frac{1}{\dfrac{1}{33 \text{ mH}} + \dfrac{1}{2.2 \text{ mH}} + \dfrac{1}{10 \text{ mH}}} = \frac{1}{0.5848} \text{ mH} = 1.71 \text{ mH}$$

Practice Problem 10.3

A circuit like the one shown in Figure 10.12 has the following values: $L_1 = 330\ \mu\text{H}$, $L_2 = 1$ mH, and $L_3 = 470\ \mu\text{H}$. Calculate the total inductance in the circuit.

As you can see, total inductance is calculated in the same fashion as total resistance in a parallel circuit. This means that we can also calculate total parallel inductance using

$$L_T = \frac{L_1 L_2}{L_1 + L_2} \tag{10.8}$$

for an inductive circuit containing only two branches. For a circuit containing any number of branches with equal inductances, the total inductance can be found using

$$L_T = \frac{L}{n} \tag{10.9}$$

where

n = the number of branches

Since you have dealt extensively with the resistive counterparts of equations (10.8) and (10.9), we do not need to elaborate on them further.

Mutual Inductance

When one inductor is placed in close proximity to another, the flux produced by each coil can induce a voltage across the other. This phenomenon, called **mutual inductance,** is illustrated in Figure 10.13. When positioned as shown in Figure 10.13a, the flux produced ◄ *OBJECTIVE 8*

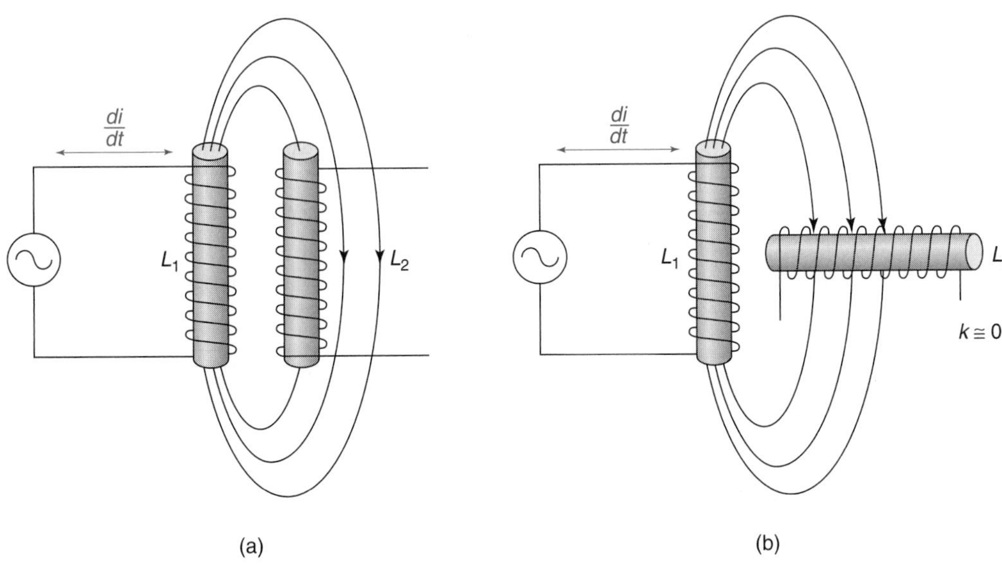

Figure 10.13 Mutual inductance.

by the current through L_1 cuts through the turns of L_2. As a result, a voltage is induced across L_2. Note that two or more components (or circuits) are said to be *coupled* when *energy is transferred from one to the other.* In Figure 10.13a, energy is transferred between the inductors via the lines of force that pass between them.

The amount of mutual inductance that exists between two coils can be found as

$$L_M = k\sqrt{L_1 L_2} \tag{10.10}$$

where k represents the *coefficient of coupling.* The **coefficient of coupling (k)** between two or more coils is *a measure of the degree of coupling that takes place between the coils.* Expressed mathematically,

Note: Since it is a ratio, k has no unit of measure.

$$k = \frac{\phi_2}{\phi_1} \tag{10.11}$$

where

ϕ_1 = the amount of flux generated by L_1

ϕ_2 = the amount of ϕ_1 that passes through L_2 at a 90° angle to the turns of the coil

A Practical Consideration: The *ideal* condition of $k = 1$ is called **unity coupling.** Unity coupling cannot be achieved in practice. However, by wrapping two coils around the same form, values of $k > 0.9$ can be reached.

The higher the coefficient of coupling (k), the greater the mutual inductance between the coils, and the greater the energy transfer between the components. Note that the value of k is always less than one, because the flux generated by one coil *expands in all directions.* Therefore, it is impossible for *all the flux* generated around L_1 to pass through L_2, and $\phi_2 < \phi_1$.

The coefficient of coupling between two coils depends on several factors, including:

- The physical distance between the components
- The relative angle between them.

The first of these factors is easy to visualize. To have mutual inductance, the coils must be near enough to each other for the flux from one to pass through the turns of the other. If the coils are far enough apart, then $k = 0$, and there is no mutual inductance.

The second factor is illustrated in Figure 10.13b. In this case, the coils are positioned so that the flux from L_1 passes through L_2. However, the coils are perpendicular, so the lines of flux are *parallel to the turns of* L_2. Therefore, no voltage is induced in the second coil, $k = 0$, and there is no mutual inductance between the coils.

The Effects of Mutual Inductance on L_T

When $k \neq 0$ for series-connected inductors, their mutual inductance becomes a factor in the total circuit inductance. The total value of two coils with some mutual inductance between them is found as

$$L_T = L_1 + L_2 \pm 2L_M \tag{10.12}$$

Whether the value of $2L_M$ is added or subtracted depends on how the components are connected.

When the inductors are connected as shown in Figure 10.14a, *the magnetic poles are such that the flux produced by the coils add together.* This increases the overall flux, which has the same effect as *increasing* the inductance. In this case, the inductors are in a *series-aiding configuration,* and

$$L_T = L_1 + L_2 + 2L_M$$

When connected as shown in Figure 10.14b, *the flux generated by each coil opposes the flux generated by the other.* This reduces the overall flux, which has the same effect as *decreasing* the total inductance. In this case, the inductors are in a *series-opposing configuration,* and

$$L_T = L_1 + L_2 - 2L_M$$

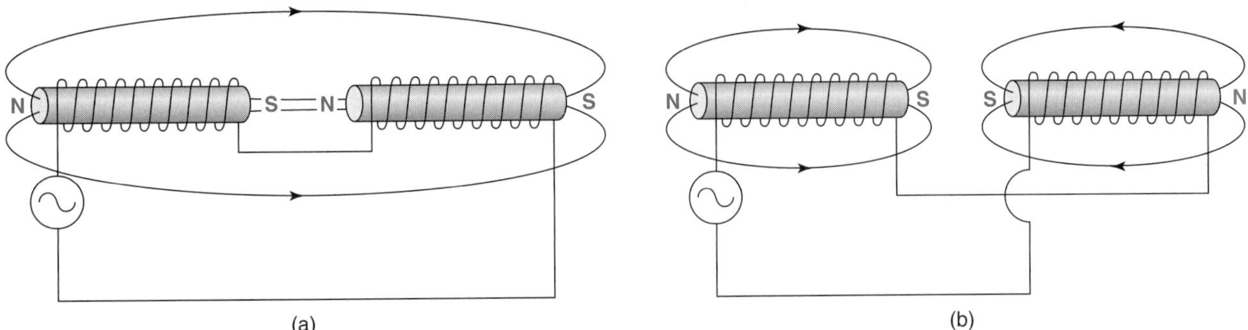

Figure 10.14 Series-aiding and series-opposing connections.

If two parallel inductors are wired in close proximity to each other, there may be some measurable amount of mutual inductance between them. In this case, the total inductance in the circuit can be found using a modified version of equation (10.8), as follows:

$$L_T = \frac{(L_1 \pm L_M)(L_2 \pm L_M)}{(L_1 \pm L_M) + (L_2 \pm L_M)} \tag{10.13}$$

A Practical Consideration

In the course of circuit design and layout, the potential effects of mutual inductance are usually taken into account and avoided. That is, coils are normally positioned to reduce (or eliminate) any mutual inductance. However, you may occasionally find a discrepancy between measured component values and total inductance. When this is the case, the probable cause is some amount of mutual inductance that is increasing (or reducing) the total inductance in the circuit.

Section Review

1. What is *mutual inductance*?
2. List the factors that determine the amount of mutual inductance between two (or more) coils.
3. What do we mean when we say that two (or more) components are electrically coupled?
4. What is indicated by the *coefficient of coupling* (k) between two coils?
5. How do you determine the total inductance in a series circuit when $k \neq 0$?
6. How do you determine the total inductance in a parallel circuit when $k \neq 0$?

10.4 Inductive Reactance (X_L)

◄ *OBJECTIVE 9*

Any component that has a voltage across it and a current through it is providing some measurable *opposition* to that current. For example, consider the inductive waveforms and circuit shown in Figure 10.15. As the waveforms indicate, there is a 90° phase difference between the inductor current and voltage. At the same time, each waveform has a measurable rms (or effective) value. Using the rms values shown, we can calculate this opposition to current that the inductor presents as

$$Opposition = \frac{V_{\text{rms}}}{I_{\text{rms}}} = \frac{10\text{ V}}{1\text{ mA}} = 10\text{ k}\Omega$$

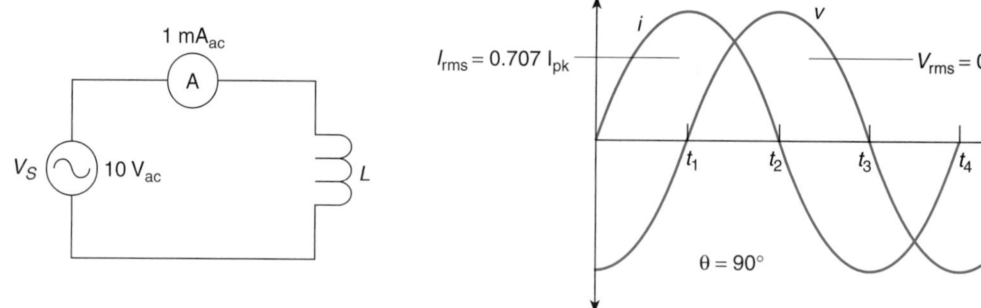

Figure 10.15

As you know, an inductor is basically a wire that is wrapped into a series of loops. If we were to measure the resistance of a coil, we would obtain an extremely low reading (because wire has little resistance). For example, if the coil shown in Figure 10.15 was made up of 2 ft of 30 gauge wire, its resistance would be approximately 0.21 Ω. This is significantly lower than the 10 kΩ of opposition that we calculated using Ohm's law. Assuming that these values are correct, there is only one way to account for the difference between our calculated values of 0.21 Ω and 10 kΩ: *The inductor is providing some type of opposition to current other than resistance.* This "other" opposition to current is known as *inductive reactance* (X_L).

Reactance (X)

Earlier in this chapter, you were told that inductance can be defined as the ability of a component to oppose a change in current. In an inductive circuit with a sinusoidal input (like the one in Figure 10.15), the current provided by the source is constantly changing. *The opposition (in ohms) that an inductor presents to this changing current* is referred to as **inductive reactance (X_L).** Note that the letter X is used to designate *reactance,* and the subscript L is used to identify the reactance as being *inductive.* (In Chapter 12, you will be introduced to another type of reactance.)

Calculating the Value of X_L

OBJECTIVE 10 ➤

You have already been shown one way to calculate the value of X_L for a given inductor. When the rms values of inductor current and voltage are known, the reactance of the component can be found as

$$X_L = \frac{V_{rms}}{I_{rms}} \tag{10.14}$$

A Practical Consideration: Nearly all electronics certification exams (like the FCC, CET, and NARTE exams) require that you know the relationship in equation (10.15). For this reason, you may want to commit the equation (and its meaning) to memory.

There is another (and more commonly used) method for calculating the reactance of a given inductor. When the value of an inductor and the frequency of operation are known, the reactance of the inductor can be determined using

$$X_L = 2\pi fL \tag{10.15}$$

The use of this relationship is demonstrated in Example 10.4.

Example 10.4

Calculate the value of X_L for the inductor shown in Figure 10.16.

Figure 10.16

Solution: Using the values of $L = 1$ mH and $f = 50$ kHz,

$$X_L = 2\pi fL = 2\pi(50 \text{ kHz})(1 \text{ mH}) \cong 314\ \Omega$$

Practice Problem 10.4

A 470 mH inductor is operated at a frequency of 20 kHz. Calculate the reactance provided by the component.

X_L and Ohm's Law

Inductive reactance (X_L) can be used with Ohm's law to determine the total circuit current (just like resistance). The analysis of a simple inductive circuit is demonstrated in Example 10.5. ◀ *OBJECTIVE 11*

Example 10.5

Calculate the total current for the circuit shown in Figure 10.17.

Figure 10.17

Solution: Using the values shown, the circuit current is found as

$$I = \frac{V_S}{X_L} = \frac{12 \text{ V}}{1 \text{ k}\Omega} = 12 \text{ mA}$$

Practice Problem 10.5

A circuit like the one shown in Figure 10.17 has the following values: $V_S = 18$ V, and $X_L = 5$ kΩ. Calculate the total circuit current.

The circuit current can be calculated as demonstrated in Example 10.6 when the values of inductance, source voltage, and operating frequency for a circuit are known.

Example 10.6

Calculate the total current for the circuit shown in Figure 10.18a.

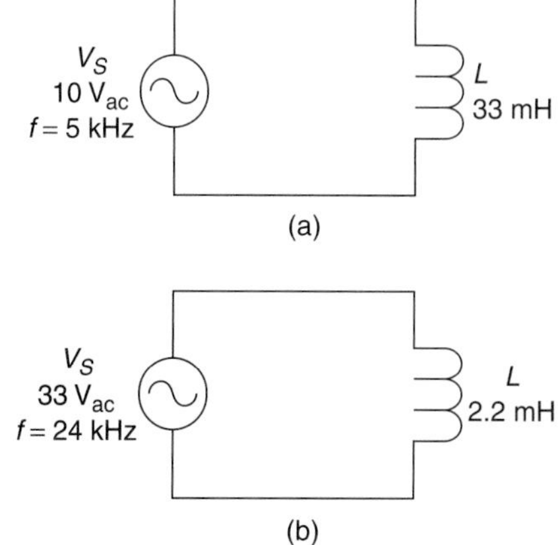

Figure 10.18

Solution: First, the reactance of the coil is found as

$$X_L = 2\pi fL = (5 \text{ kHz})(33 \text{ mH}) = 1.04 \text{ k}\Omega$$

Now the circuit current can be found as

$$I = \frac{V_S}{X_L} = \frac{10 \text{ V}}{1.04 \text{ k}\Omega} = 9.62 \text{ mA}$$

Practice Problem 10.6

Calculate the value of current for the circuit shown in Figure 10.18b.

Series and Parallel Reactances

When any number of reactive components are connected in series or parallel, the total reactance is found in the same fashion as total resistance. This point is illustrated in Figure 10.19. (Since we have used similar equations for resistance and inductance calculations, we do not need to rework them here.)

Resistance, Reactance, and Impedance (Z)

Resistance is usually a static value, meaning that it usually does not vary when the circuit conditions change. Reactance, on the other hand, is a dynamic value. As you know, the reactance of an inductor (X_L) changes when the circuit operating frequency changes. Although they are both oppositions to current (measured in ohms), resistance and reactance cannot be added algebraically to determine the total opposition to current in a circuit.

In Chapter 11, you will learn how to combine resistance and reactance into a single value, called *impedance.* Technically defined, **impedance (Z)** is *the total opposition to current in an ac circuit, consisting of resistance and/or reactance.* As you will see, impedance is found as the *geometric* (vector) sum of resistance and reactance.

Section Review

1. How would you measure the opposition to current provided by an inductor?
2. Why is the resistance of an inductor extremely low?
3. What is *inductive reactance* (X_L)?
4. What two relationships are commonly used to calculate the value of X_L?
5. How is total current determined in a purely inductive circuit?
6. How is total reactance calculated in a series circuit? A parallel circuit?
7. What is *impedance* (Z)? How is it calculated?

Critical Thinking

8. In what type of circuit would inductive reactance (X_L) be a *static value*? Explain your answer.
9. Show how the units of measure in $2\pi fL$ resolve themselves to yield a result measured in ohms.

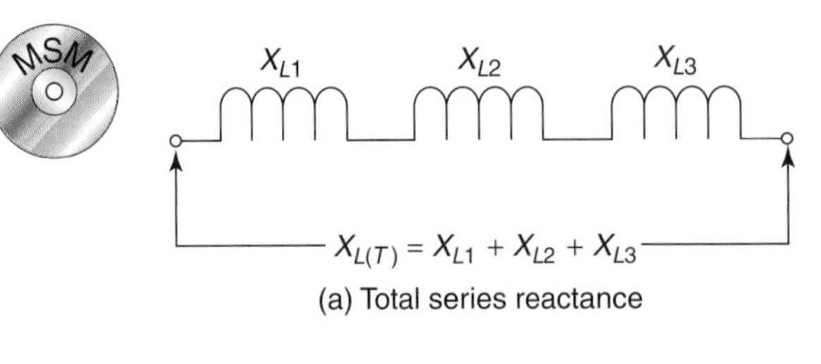

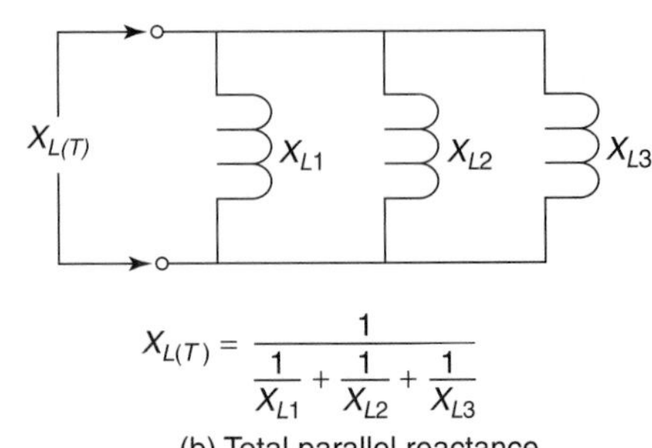

Figure 10.19 Total series and parallel reactance.

10.5 Transformers

A **transformer** is *a two-coil component that uses electromagnetic induction to pass an ac signal from its input to its output while providing dc isolation between the two.* The ac and dc input/output characteristics of a transformer are illustrated in Figure 10.20. ◄ *OBJECTIVE 12*

When a component (or circuit) allows any signal or potential to pass from one point to another, it is said to provide **coupling** between the points. *If it prevents a signal or potential from passing between points,* it is said to provide **isolation** between the two. Therefore, a transformer provides both ac coupling and dc isolation between its input and output terminals.

Transformer Construction and Symbols

A transformer is made up of two coils, called the **primary** and the **secondary.** Two types of transformers are illustrated in Figure 10.21. Note that the *primary* serves as the transformer *input* and the *secondary* serves as its *output.*

Figure 10.22 shows two commonly used transformer schematic symbols. In each case, the coils are identified by the labels on the input and output leads. P_1 and P_2 are the primary terminals. S_1 and S_2 are the secondary terminals.

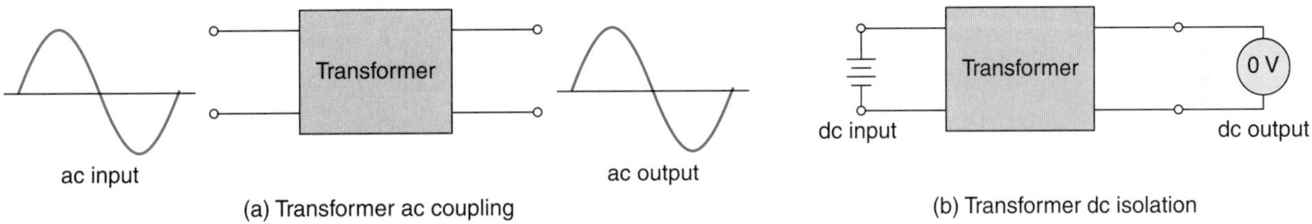

Figure 10.20 Transformer input/output functions.

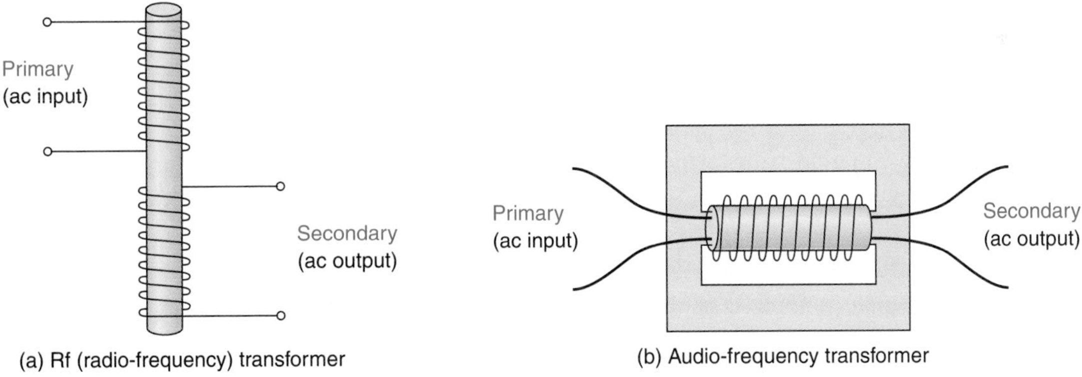

Figure 10.21 Basic transformers.

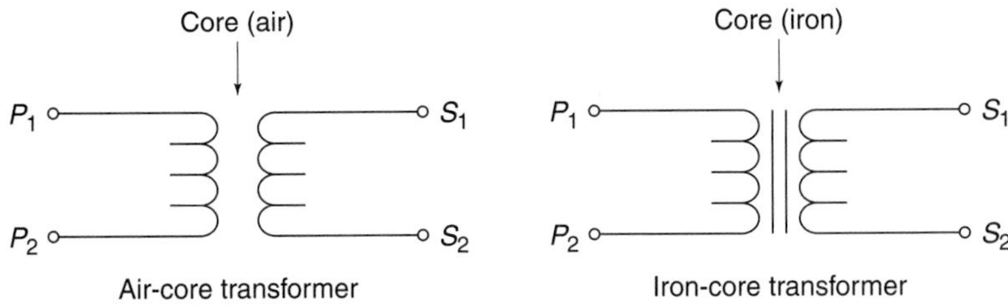

Figure 10.22 Transformer schematic symbols.

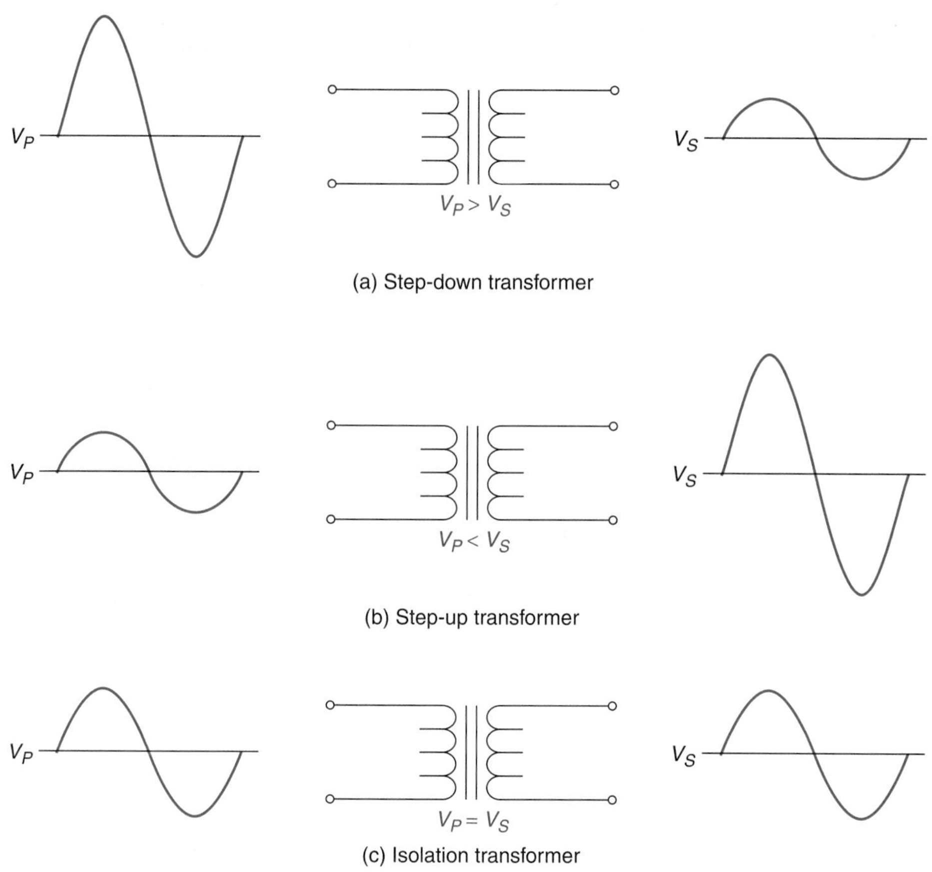

Figure 10.23

Transformer Classifications

OBJECTIVE 13 ➤ A given transformer is commonly described in terms of the relationship between its input and output voltages. For example, Figure 10.23 shows three types of transformers. The **step-down transformer** shown in Figure 10.23a *provides a secondary (output) voltage that is less than the primary (input) voltage.* In contrast, the **step-up transformer** shown in Figure 10.23b *provides an output voltage that is greater than its input voltage.* The **isolation transformer** shown in Figure 10.23c is *one with equal input and output voltages.*

Transformer Operation

OBJECTIVE 14 ➤ When an ac input is applied to the primary of a transformer, the changing primary current produces a magnetic field, as shown in Figure 10.24a. This changing field cuts through

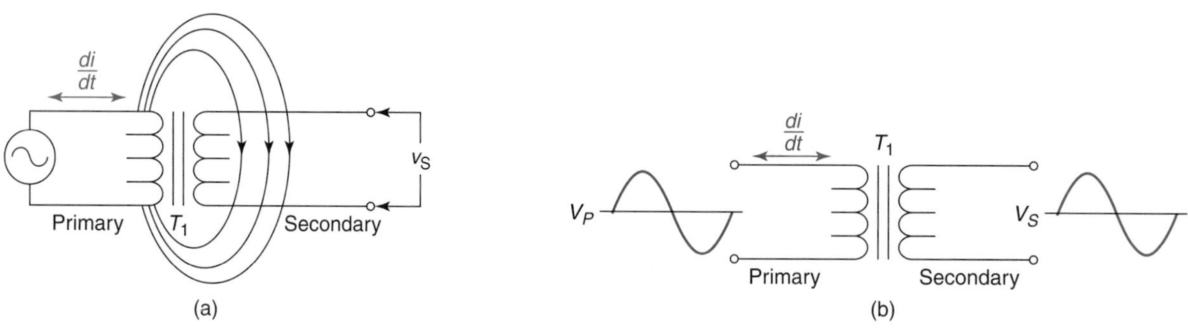

Figure 10.24

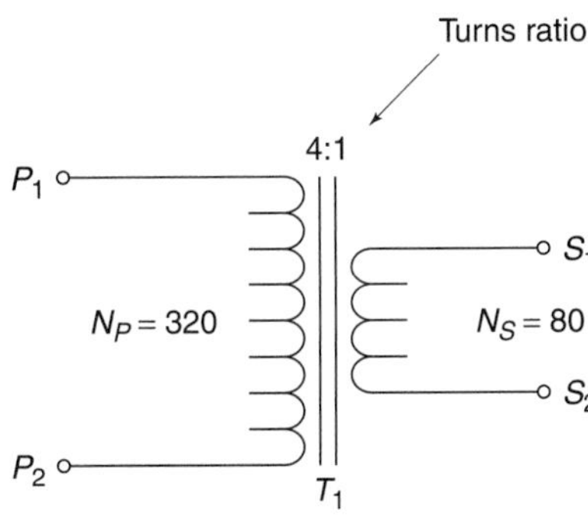

Figure 10.25 Transformer turns ratio.

A Practical Consideration: Transformers are often described in terms of their turns ratios. For example, the transformer in Figure 10.25 would be referred to as a *four-to-one* transformer.

the secondary coil, inducing a voltage across the secondary. Assuming that the input to the transformer is a sine wave, the output is also a sine wave (as shown in Figure 10.24b).

Note that the primary and secondary inductors are not physically connected. Therefore, any dc voltage applied to L_1 cannot be coupled to L_2. If a dc voltage is applied to the primary, a changing current is not generated and no coupling occurs.

Turns Ratio

The **turns ratio** of a transformer is *the ratio of primary turns to secondary turns.* For example, the transformer shown in Figure 10.25 is shown to have 320 turns in the primary and 80 turns in the secondary. Using these values, the turns ratio of the component is found as

$$\frac{N_P}{N_S} = \frac{320 \text{ turns}}{80 \text{ turns}} = \frac{4}{1}$$

Note that the turns ratio of a given transformer is normally written in the form shown in Figure 10.25.

The turns ratio of a transformer is important because it determines the ratio of input voltage to output voltage. By formula, ◄ *OBJECTIVE 15*

$$\frac{N_P}{N_S} = \frac{V_P}{V_S} \tag{10.16}$$

For the transformer shown in Figure 10.25, this equation indicates that the voltage across the primary is four times the voltage across the secondary when the component has an ac input. As such, the transformer shown in the figure is a *step-down transformer.*

Secondary Voltage (V_S)

If we solve equation (10.16) for secondary voltage (V_S), we get

$$V_S = V_P \frac{N_S}{N_P} \tag{10.17}$$

Equation (10.17) allows us to determine the secondary voltage when the primary voltage and turns ratio of a transformer are known, as demonstrated in Example 10.7.

Example 10.7

Determine the secondary voltage (V_S) for the transformer shown in Figure 10.26.

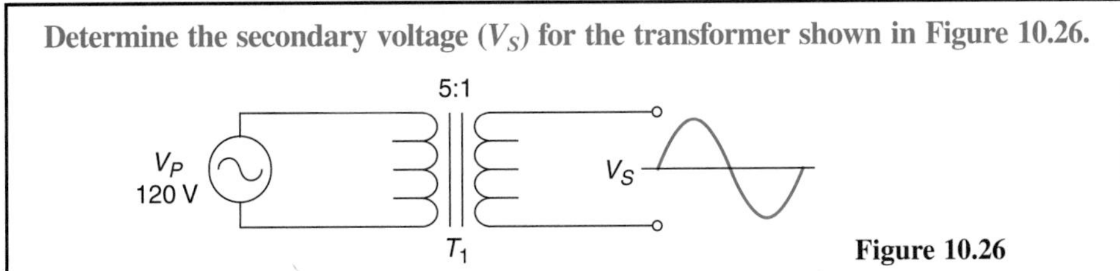

Figure 10.26

Solution: The secondary voltage is found as

$$V_S = V_P\frac{N_S}{N_P} = (120\text{ V})\frac{1}{5} = 24\text{ V}$$

Practice Problem 10.7

A transformer like the one shown in Figure 10.26 has a 100 V input and a 10:1 turns ratio. Calculate the value of V_S for the component.

Power Transfer

OBJECTIVE 16 ➤ Under ideal conditions, all the power applied to the primary (input) of a transformer is transferred to the secondary (output). By formula,

$$P_P = P_S \qquad \text{(ideal)}$$

or

$$I_P V_P = I_S V_S \qquad \text{(ideal)} \tag{10.18}$$

As you know, there are no ideal components. In practice, the value of secondary power is always slightly lower than the value of primary power, because a number of power losses occur in any practical transformer. At this point, we will take a brief look at these losses.

Copper loss All wire has some measurable amount of resistance. Even though this resistance is very low (less than 100 Ω), it still causes some amount of power to be dissipated in both the primary and secondary coils. Note that copper loss is often referred to as *I^2R loss.*

Loss due to eddy currents This type of loss is unique to iron-core transformers. The transformer's magnetic flux generates *a current within the core,* called an **eddy current.** As shown in Figure 10.27, this current travels in a circular motion through the core. As eddy currents pass through an iron core, the resistance of the core causes some amount of power dissipation in the form of heat.

Hysteresis loss In Chapter 9, we defined *retentivity* as *the ability of a material to retain its magnetic characteristics after a magnetizing force has been removed.* Simply put, the term **hysteresis loss** refers to *the energy expended to overcome the core's retentivity.* Each time the input current reverses polarity, the polarity of the magnetic field must also reverse. To do this, it must overcome the retentivity of the core material.

Even though the losses just described are very real, each results in relatively little power loss. For this reason, upcoming discussions are based on the ideal transformer.

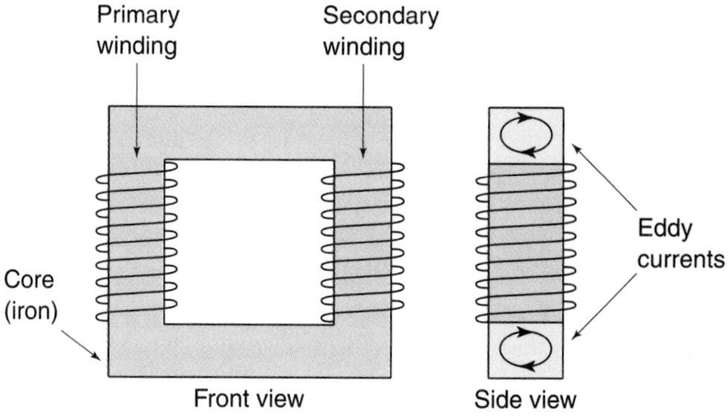

Figure 10.27 Eddy currents in an iron core.

Transformer Input and Output Current

As you know, the ideal transformer has equal input and output power values. This relationship was described earlier using ◄ *OBJECTIVE 17*

$$I_P V_P = I_S V_S$$

This relationship can be rewritten as

$$I_S = I_P \frac{V_P}{V_S} \tag{10.19}$$

Equation (10.19) is used to determine the value of I_S in Example 10.8.

Example 10.8

Determine the secondary current (I_S) for the transformer shown in Figure 10.28.

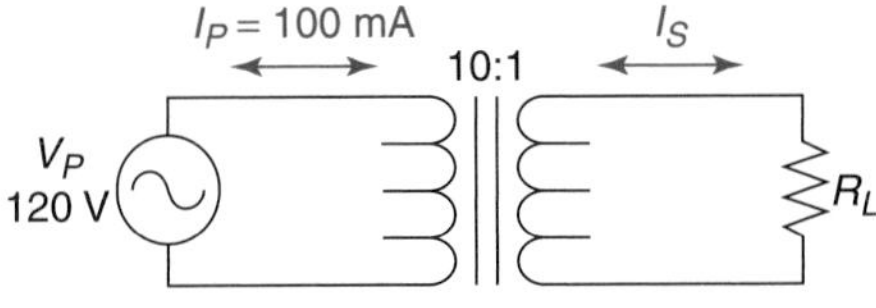

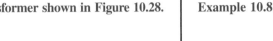

Figure 10.28

Solution: The value of secondary voltage for the transformer is found as

$$V_S = V_P \frac{N_S}{N_P} = (120 \text{ V}) \frac{1}{10} = 12 \text{ V}$$

Now the value of I_S can be found as

$$I_S = I_P \frac{V_P}{V_S} = (100 \text{ mA}) \frac{120 \text{ V}}{12 \text{ V}} = 1 \text{ A}$$

Practice Problem 10.8

A transformer like the one shown in Figure 10.28 has a 40 V input, a turns ratio of 2:1, and a value of $I_P = 1.5$ A. Calculate I_S for the transformer.

Earlier, you were told that voltage varies directly with the turns ratio of a transformer. Since current and voltage vary inversely, current varies inversely with the turns ratio. By formula,

$$\frac{I_S}{I_P} = \frac{N_P}{N_S} \tag{10.20}$$

If we solve this relationship for secondary current (I_S), we get

$$I_S = I_P \frac{N_P}{N_S} \tag{10.21}$$

This equation can be used to demonstrate the relationship between the values of voltage and current for a *step-up transformer,* as shown in Example 10.9.

Example 10.9

Calculate the output voltage and current values for the step-up transformer shown in Figure 10.29.

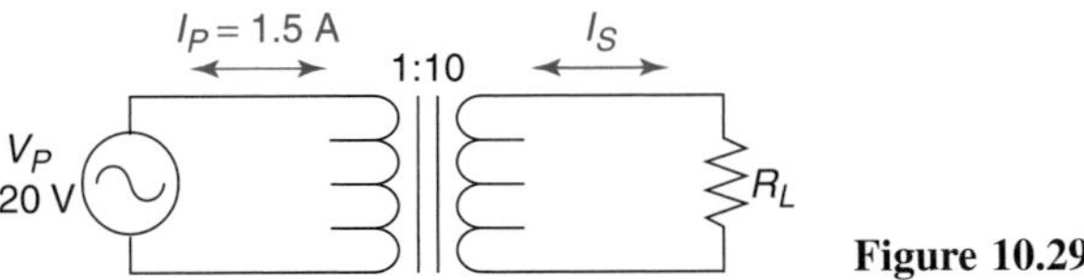

Figure 10.29

Solution: The secondary voltage is found as

$$V_S = V_P \frac{N_S}{N_P} = (120 \text{ V}) \frac{10}{1} = 1.2 \text{ kV}$$

Now the value of the secondary current can be found as

$$I_S = I_P \frac{N_P}{N_S} = (1.5 \text{ A}) \frac{1}{10} = 150 \text{ mA}$$

Practice Problem 10.9

A transformer like the one shown in Figure 10.29 has a 240 V input, a 1:5 turns ratio, and a value of $I_P = 500$ mA. Calculate the secondary voltage and current values for the transformer.

If we compare the results from Examples 10.8 and 10.9, we get a clearer picture of the relationships among the turns ratio, voltage values, and current values for step-up and step-down transformers. For convenience, the results from the two examples are listed in Table 10.3. As shown, the step-down transformer:

- Has a turns ratio of 10 : 1.
- Decreases voltage by a factor of ten.
- Increases current by a factor of ten.

On the other hand, the step-up transformer:

- Has a turns ratio of 1 : 10.
- Increases voltage by a factor of ten.
- Decreases current by a factor of ten.

These results agree with the statements made earlier regarding the effect of a transformer's turns ratio on the input and output voltage and current values.

For future reference, the input/output relationships established so far are summarized in Figure 10.30. Note that the input and output values for the isolation transformer are equal in every category. As a result, the isolation transformer is used strictly to provide dc isolation between two ac circuits, without affecting the ac coupling from one to the other.

Primary Impedance (Z_P)

OBJECTIVE 18 ➤ The primary impedance of a given transformer is directly proportional to the square of the turns ratio and the secondary impedance (Z_S), as follows:

$$Z_P = Z_S \left(\frac{N_P}{N_S} \right)^2 \tag{10.22}$$

where

Z_S = the total opposition to current in the secondary (generally assumed to equal the opposition provided by the load)

An application of this relationship is provided in Example 10.10.

TABLE 10.3 Results from Examples 10.8 and 10.9

Transformer	*Circuit*	*Turns*[a]	*Voltage*	*Current*	*Power*
Step-down	Primary	10	120 V	100 mA	12 W
	Secondary	1	12 V	1 A	12 W
Step-up	Primary	1	120 V	1.5 A	180 W
	Secondary	10	1.2 kV	150 mA	180 W

[a]These are the numbers given in the turns ratio. They do not equal the actual number of turns in the primary and second windings.

SUMMARY ILLUSTRATION

Transformer Input/Output Relationships

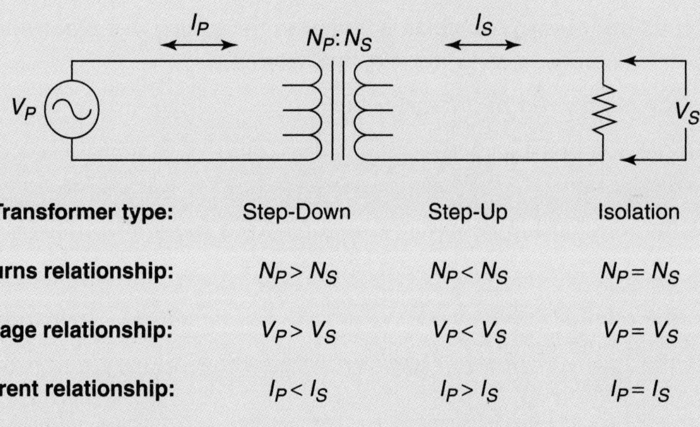

Transformer type:	Step-Down	Step-Up	Isolation
Turns relationship:	$N_P > N_S$	$N_P < N_S$	$N_P = N_S$
Voltage relationship:	$V_P > V_S$	$V_P < V_S$	$V_P = V_S$
Current relationship:	$I_P < I_S$	$I_P > I_S$	$I_P = I_S$
Power relationship:	$P_P \cong P_S$	$P_P \cong P_S$	$P_P \cong P_S$

Figure 10.30

Example 10.10

A transformer has a 5:1 turns ratio and a 100 Ω load. Calculate the primary impedance of the component.

Solution: With a turns ratio of 5:1 and a 100 Ω load, the value of the primary impedance can be found as

$$Z_P = Z_S\left(\frac{N_P}{N_S}\right)^2 = (100\ \Omega)\left(\frac{5}{1}\right)^2 = (100\ \Omega)(25) = 2.5\ \text{k}\Omega$$

Practice Problem 10.10

A transformer has a 12:1 turns ratio and a 50 Ω load. Calculate the primary impedance of the component.

Putting It All Together

You have been shown that the voltage, current, and impedance ratios of a given transformer relate to its turns ratio as follows:

$$\frac{N_P}{N_S} = \frac{V_P}{V_S} \qquad \frac{N_P}{N_S} = \frac{I_S}{I_P} \qquad \left(\frac{N_P}{N_S}\right)^2 = \frac{Z_P}{Z_S}$$

Based on these relationships, we can summarize the voltage, current, and impedance characteristics for every *step-down transformer* as follows:

$$N_P > N_S \qquad V_P > V_S \qquad I_P < I_S \qquad Z_P > Z_S$$

For the *step-up transformer,* the voltage, current, and impedance characteristics can be summarized as follows:

$$N_P < N_S \qquad V_P < V_S \qquad I_P > I_S \qquad Z_P < Z_S$$

Section Review

1. What is a *transformer*?
2. Describe the ac and dc input/output relationships for a transformer.

3. Describe the construction of a typical transformer.
4. Which coil serves as a transformer's *input*? Which serves as its *output*?
5. List and describe the three basic types of transformers.
6. Describe the process by which an ac input to the primary of a transformer causes a voltage to be induced across the secondary circuit.
7. What is the *turns ratio* of a transformer?
8. What is the relationship between the turns ratio of a transformer and its input/output voltage ratio?
9. What is the relationship between N_P and N_S for a *step-down* transformer?
10. What is the relationship between N_P and N_S for a *step-up* transformer?
11. What is the relationship between N_P and N_S for an *isolation* transformer?
12. What is the *ideal* relationship between transformer primary power and secondary power?
13. Describe each of the following:
    a. *Copper loss.*
    b. *Eddy current loss.*
    c. *Hysteresis loss.*
14. Describe the relationship between a transformer's voltage ratio and its current ratio.
15. Describe the relationship between a transformer's turns ratio and its current ratio.
16. Describe the relationship between a transformer's turns ratio and its impedance ratio.

10.6 Related Topics

In this section, we will discuss some miscellaneous topics that relate to inductors, their applications, and ratings.

Apparent Power (P_{APP})

OBJECTIVE 19 ➤ When used in an ac circuit, any inductor has measurable values of voltage and current. If these values are plugged into the $P = VI$ equation, they provide a numeric result (as does any combination of voltage and current). For the circuit shown in Figure 10.31, this power equation yields a result of

$$P = V_S I_L = (10\text{ V})(12\text{ mA}) = 120\text{ mW}$$

The only problem is that the result of this calculation is not a true indicator of energy used. As such, it is referred to as **apparent power (P_{APP}).** The term *apparent power* is used because most of the energy is actually stored in the electromagnetic field generated by the inductor. Only a small portion of the value obtained is actually dissipated as heat. This point can be explained with the help of Figure 10.32.

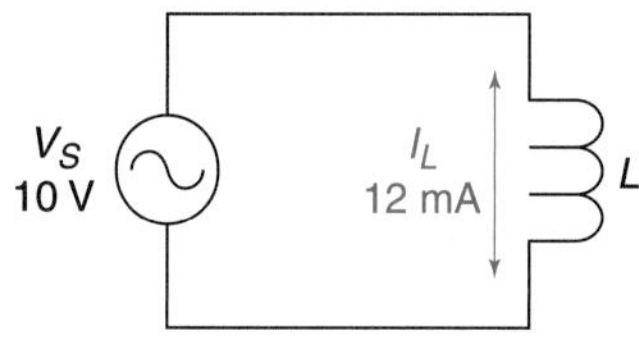

Figure 10.31

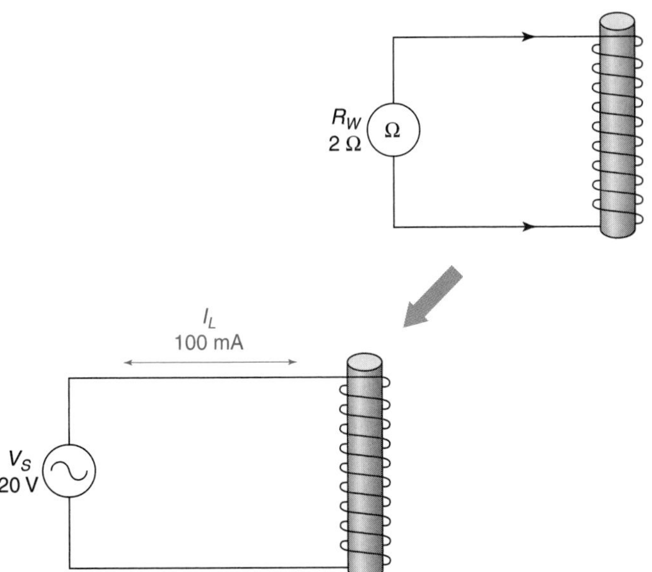

Figure 10.32

Power is dissipated whenever current passes through any measurable amount of resistance. If we assume that the **winding resistance (R_W)** of the coil shown in Figure 10.32 has a value of 2 Ω, then the power dissipated by the component can be found as

$$P = I_L^2 R_W = (100 \text{ mA})^2\ (2\ \Omega) = 20 \text{ mW}$$

At the same time, the inductor voltage and current values shown in the figure indicate that the component is dissipating

$$P = I_L V_L = (100 \text{ mA})(20 \text{ V}) = 2 \text{ W}$$

Obviously, there is a discrepancy between these two calculations.

The discrepancy lies not in our calculations, but in our interpretation of the results. The inductor only converts 20 mW into heat. This power is referred to as **resistive power,** or **true power,** and it is measured in watts (W). The remaining energy is transferred from the ac source to the coil's magnetic field.

A Practical Consideration: In most cases, the power dissipated by the winding resistance of an inductor is a very small portion of the apparent power. This is why inductors normally remain relatively cool, even when used in high-current applications.

The field generated by the inductor contains magnetic energy. Since energy cannot be created (or destroyed), this magnetic energy had to come from the ac source. In a sense, the inductor can be viewed as an energy converter, because it converts electrical energy into magnetic energy.

The energy (per second) that is transferred to and from the inductor's magnetic field is referred to as **reactive power.** Its unit of measure is the **volt-amperes-reactive (VAR).** Because it is not actually dissipated, *reactive power* is often referred to as **imaginary power.** *The combination of resistive (true) and reactive (imaginary) power* is *apparent power,* and is measured in **volt-amperes (VA).**

Three important points need to be made at this time:

- Since neither reactive power nor apparent power represent a *true* power value, they have their own unit of measure.
- When the magnetic field surrounding an inductor collapses (due to a polarity change in the ac source), the energy that it contains is returned to the circuit, and thus, to the ac source. The only power loss experienced by the source is the power used by any resistance in the circuit. That is why this power loss is referred to as true power.
- In practice, every inductor has some measurable amount of winding resistance, and therefore, dissipates some amount of power. The amount of winding resistance determines (in part) the *quality* of an inductor.

Note: Resistive and reactive power values cannot be added together algebraically. The method used to combine resistive and reactive power values (to calculate apparent power) is discussed in Chapter 11.

Inductor Quality (Q)

The ideal inductor (if it could be produced) would have no winding resistance. As a result, all the energy drawn from the ac source would be transferred to the inductor's magnetic field.

OBJECTIVE 20 ➤ The **quality (Q)** rating of an inductor is *a numeric value that indicates how close the inductor comes to the power characteristics of the ideal component.* The Q of a given inductor can be found as the ratio of reactive power to true power for the component. By formula,

$$Q = \frac{P_X}{P_{RW}} \tag{10.23}$$

where

P_X = the reactive power of the component, measured in VARs

P_{RW} = the true power dissipation of the component, measured in watts

The higher the Q for a given inductor, the closer it comes to having the power characteristics of the ideal component. Note that Q is a ratio, so it has no unit of measure.

We can modify equation (10.23) to develop a simpler means of calculating the value of Q for a given inductor. Based on the fundamental power relationships, equation (10.23) can be written as

$$Q = \frac{P_X}{P_{RW}} = \frac{I_L^2 X_L}{I_L^2 R_W}$$

or

$$Q = \frac{X_L}{R_W} \tag{10.24}$$

The use of this equation to solve for inductor Q is demonstrated in Example 10.11.

Example 10.11

Calculate the Q of the inductor shown in Figure 10.33.

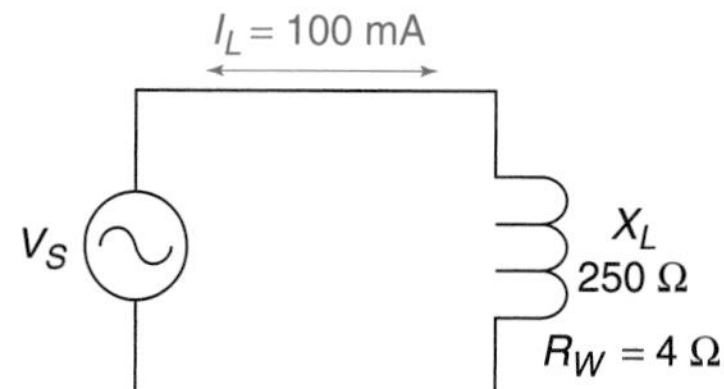

Figure 10.33

Solution: The Q of the inductor can be found as

$$Q = \frac{X_L}{R_W} = \frac{250\ \Omega}{4\ \Omega} = 62.5$$

Practice Problem 10.11

An inductor like the one shown in Figure 10.33 has values of $X_L = 7.14\ \text{k}\Omega$ and $R_W = 6.8\ \Omega$. Calculate the Q of the component.

Equation (10.24) demonstrates a very important relationship. Since, $X_L = 2\pi fL$, the Q of a given inductor varies directly with the operating frequency of the component. This relationship is demonstrated in the following example.

The ac source shown in Figure 10.34a is variable over a frequency range of 20 Hz to 20 kHz. Calculate the range of Q for the inductor. **Example 10.12**

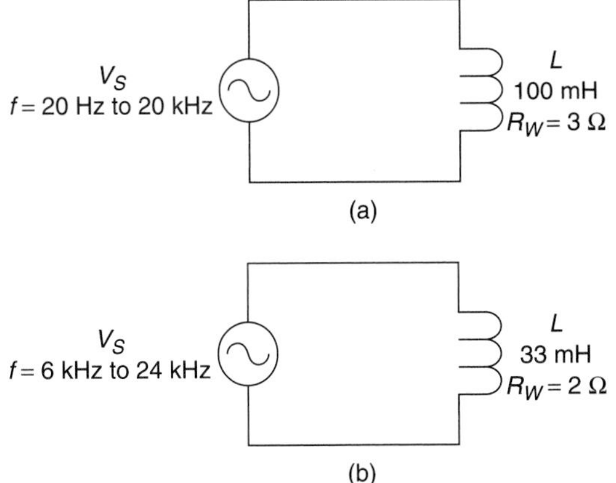

Figure 10.34

Solution: When $f = 20$ Hz,

$$X_L = 2\pi fL = 2\pi(20\text{ Hz})(100\text{ mH}) = 12.6\ \Omega$$

and

$$Q = \frac{X_L}{R_W} = \frac{12.6\ \Omega}{3\ \Omega} = 4.2$$

When $f = 20$ kHz,

$$X_L = 2\pi fL = 2\pi(20\text{ kHz})(100\text{ mH}) = 12.6\text{ k}\Omega$$

and

$$Q = \frac{X_L}{R_W} = \frac{12.6\text{ k}\Omega}{3\ \Omega} = 4200 \qquad (4.2 \times 10^3)$$

Thus, when the ac source is varied across its frequency range, the Q of the inductor varies between 4.2 and 4200.

Practice Problem 10.12

The ac source shown in Figure 10.34b is variable over a frequency range of 6 kHz to 24 kHz. Calculate the range of Q for the inductor.

In Chapter 15, you will be shown that the Q of an inductor plays a role in the *frequency response* of any circuit containing the component. Inductor Q also helps to explain why certain types of inductors are preferred over others for use in circuits that operate within specified frequency limits.

Types of Inductors

Many types of inductors are designed for specific applications. Even though these inductors each have unique characteristics, they all work according to the principles covered in this chapter.

Iron-core versus air-core inductors Most inductors have either an iron core or an air core. Iron- and air-core inductor schematic symbols are shown in ◄ *OBJECTIVE 21*

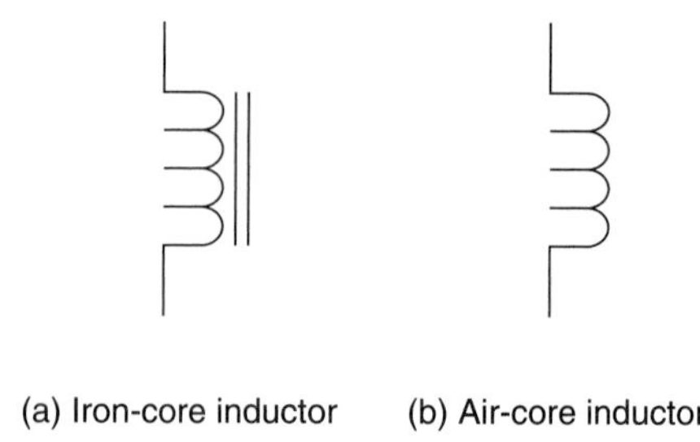

Figure 10.35 Iron-core and air-core inductor symbols.

Note: An iron-core inductor is shown (on the left) in Figure 10.37. An air-core inductor is shown (on the right) in the same figure.

Figure 10.35. Iron-core inductors are better suited for use in low-frequency applications, such as audio and dc power supply circuits, because:

- The dc winding resistance of an iron-core inductor is much lower than that of an equal-value air-core inductor.
- The Q of an iron-core inductor is much higher that of an equal-value air-core inductor.

These characteristics are based on the relative permeability of iron. The permeability of iron is approximately 200 times that of air, so it takes far fewer turns (less wire) to produce a given value of inductance. As a result, the winding resistance (R_W) of an iron-core inductor is always significantly less than that of an equal-value air-core inductor. With a lower R_W, the iron-core inductor:

- Has a much higher Q and tends to be much smaller in size than an equal-value air-core inductor.
- Is better suited for use in high-current applications, such as dc power supply filter circuits.

At the same time, iron cores can experience significant power losses when operated at higher frequencies, making air-core inductors more suitable for such applications.

Air cores The turns of an air-core inductor are either self-supporting or are physically supported by a nonmagnetic form, such as a ceramic tube.

OBJECTIVE 22 ➤

Toroids A **toroid** is *a coil that is wrapped around a doughnut-shaped magnetic core,* as shown in Figure 10.36. Because of its shape, nearly all the flux produced by the coil remains in the core. As a result, toroids have:

- Inductance values that are much greater than their physical size would indicate.
- Extremely high quality (Q) ratings.
- Extremely accurate rated values.

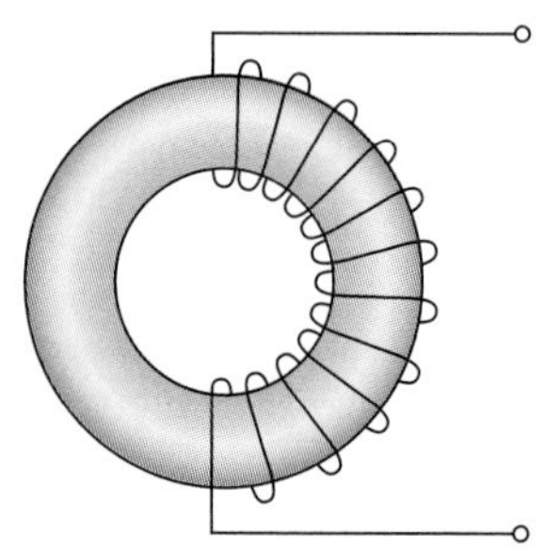

Figure 10.36
A toroid.

Chokes A **choke** is *a low-resistance inductor that is designed to provide a high opposition to current (reactance) within a designated frequency range.* The two most common types of chokes are **filter chokes** and **radio-frequency (RF) chokes.** Typical filter and RF chokes are shown in Figure 10.37.

The filter choke is designed for operation in low-frequency circuits, typically operating at a frequency lower than 1 kHz. The RF choke is designed for operation within the

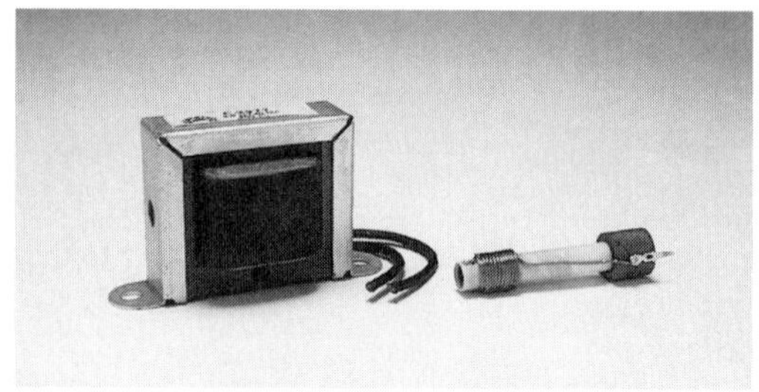

Figure 10.37. A filter choke (left) and an RF choke (right).

radio-frequency range, or approximately 30 kHz to 30 GHz. In Figure 10.37, the filter choke has an iron core, and the RF choke has an air core. This is consistent with our previous discussion on core materials and frequency.

Section Review

1. Why is the term *apparent power* used to describe the product of inductor current and voltage?
2. Describe the inductor as an *energy converter.*
3. What terms are commonly used to describe *resistive power* and *reactive power*?
4. What are the *units of measure* for reactive power and apparent power?
5. List two characteristics of the *ideal inductor.*
6. What is the *quality* (Q) rating of an inductor?
7. What is the relationship between inductor Q and:
   a. Winding resistance?
   b. Operating frequency?
8. Why are iron-core inductors better suited for low-frequency applications than air-core inductors?
9. Why are air-core inductors better suited for high-frequency applications than iron-core inductors?
10. What is a *toroid*? What are the significant characteristics of this component?
11. What is a *choke*?

Key Terms

The following terms were introduced in this chapter on the pages indicated:

Apparent power (P_{APP}), 304
Choke, 308
Coefficient of coupling (k), 292
Counter emf, 284
Coupling, 297
Eddy current, 300
Filter choke, 308
Henry (H), 285
Hysteresis loss, 300
Imaginary power, 305
Impedance (Z), 296
Inductance, 281
Inductive reactance (X_L), 294
Inductor, 281
Isolation, 297
Isolation transformer, 298
Lag, 288
Lead, 288
Lenz's law, 283
Mutual inductance, 291
Primary, 297
Quality (Q), 306
Radio-frequency (RF) choke, 308
Reactive power, 305
Resistive power, 305
Secondary, 297
Self-inductance, 283
Step-down transformer, 298
Step-up transformer, 298
Toroid, 308
Transformer, 297
True power, 305
Turns ratio, 299
Unity coupling, 292
Volt-amperes (VA), 305
Volt-amperes-reactive (VARs), 305
Winding resistance (R_W), 305

Practice Problems

1. Calculate the value of v_L developed across a 33 mH coil when the instantaneous rate of change in component current equals 200 mA/s.
2. Calculate the value of v_L developed across a 4.7 H coil when the instantaneous rate of change in component current equals 1.5 A/s.
3. Calculate the value of v_L developed across a 100 mH coil when the instantaneous rate of change in component current equals 25 mA/ms. (*Hint:* Convert the current change into A/s.)

4. Calculate the value of v_L developed across a 470 μH coil when the instantaneous rate of change in component current equals 2 mA/μs.

5. Calculate the value of the air-core inductor shown in Figure 10.38a.

6. Calculate the value of the air-core inductor shown in Figure 10.38b.

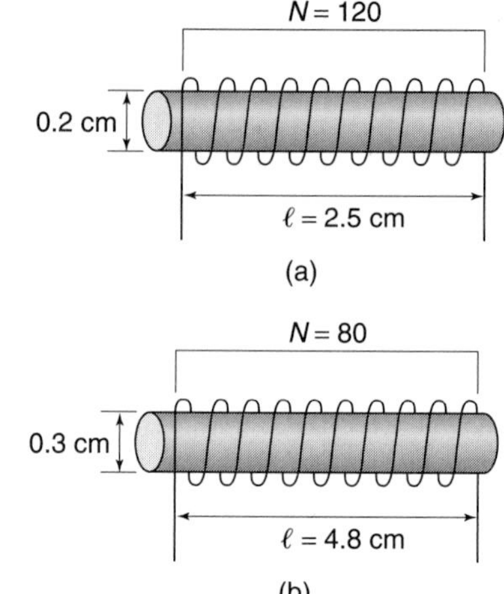

Figure 10.38

7. Calculate the value of the iron-core inductor shown in Figure 10.39a.

8. Calculate the value of the iron-core inductor shown in Figure 10.39b.

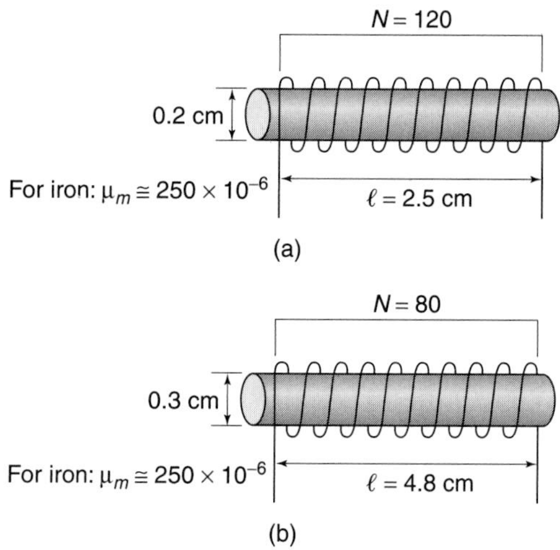

Figure 10.39

9. The coils shown in Figure 10.40a have a value of $k = 0$. Calculate the total inductance in the circuit.

10. The coils shown in Figure 10.40b have a value of $k = 0$. Calculate the total inductance in the circuit.

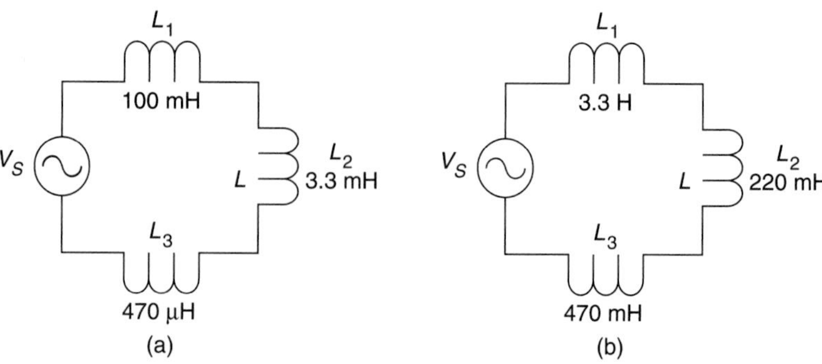

Figure 10.40

11. The inductors shown in Figure 10.41a are connected in series-aiding fashion. Assuming that the coefficient of coupling between the two is 0.25, calculate the total inductance in the circuit.

12. The inductors shown in Figure 10.41b are connected in series-opposing fashion. Assuming that the coefficient of coupling between the two is 0.33, calculate the total inductance in the circuit.

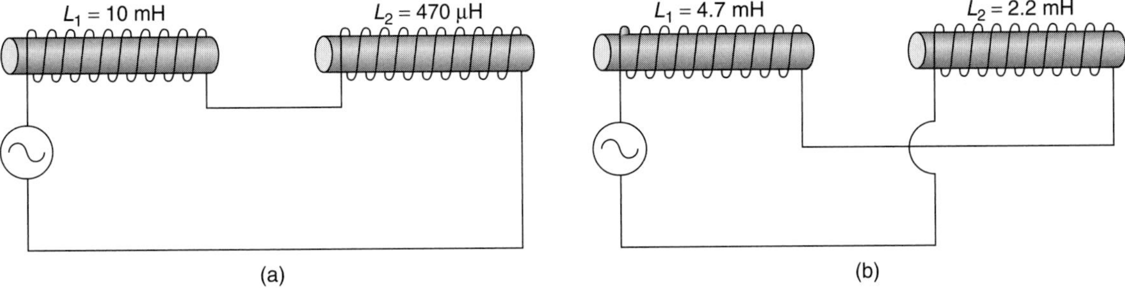

Figure 10.41

13. Determine the coefficient of coupling for the circuit shown in Figure 10.42a.

14. Determine the coefficient of coupling for the circuit shown in Figure 10.42b.

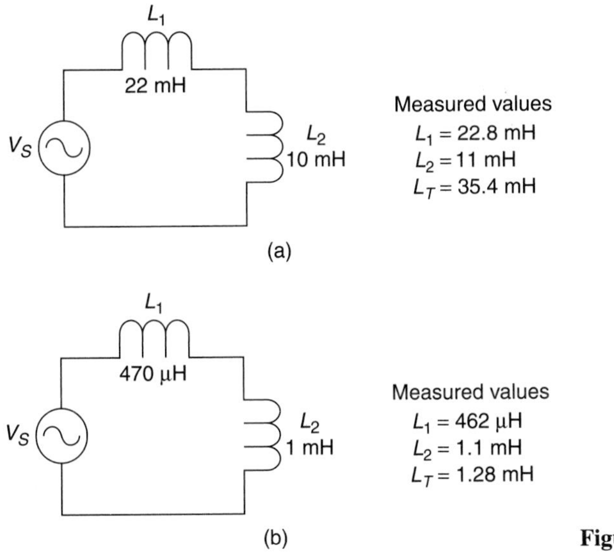

Figure 10.42

15. The circuit shown in Figure 10.43a has a value of $k = 0$. Calculate the total inductance in the circuit.

16. The inductors shown in Figure 10.43b are wired in parallel-aiding fashion with a value of $k = 0.7$. Calculate the total inductance in the circuit.

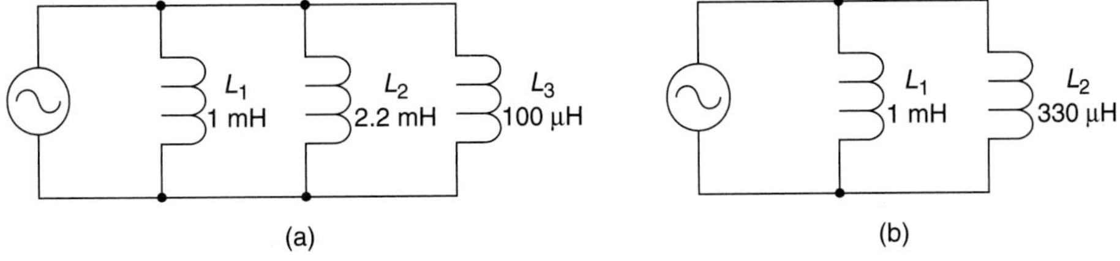

Figure 10.43

17. Calculate the value of X_L for the circuit shown in Figure 10.44a.

18. Calculate the value of X_L for the circuit shown in Figure 10.44b.

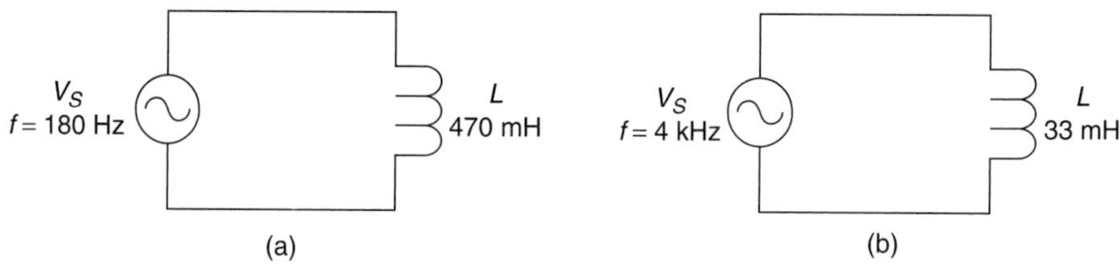

Figure 10.44

19. The following table lists values for an inductive circuit. Complete the table.

V_{rms}	I_{rms}	X_L
12 V	______	2 kΩ
______	30 mA	180 Ω
20 V	480 μA	______

20. The following table lists values for an inductive circuit. Complete the table.

V_{rms}	I_{rms}	X_L
180 mV	______	200 Ω
______	40 μA	18 MΩ
2 V	100 μA	______

21. Determine the operating frequency of the circuit shown in Figure 10.45a.

22. Determine the operating frequency of the circuit shown in Figure 10.45b.

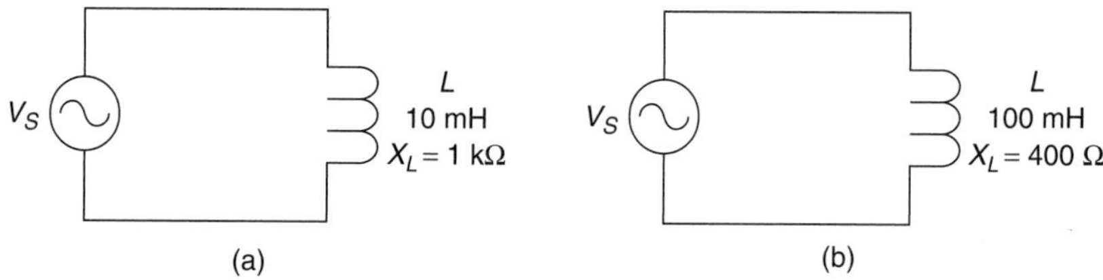

Figure 10.45

23. Calculate the Q of the inductor shown in Figure 10.46a.

24. Calculate the Q of the inductor shown in Figure 10.46b.

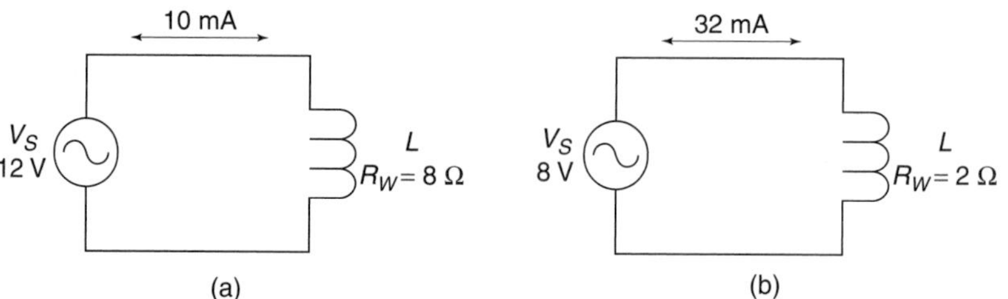

Figure 10.46

25. Calculate the Q of the inductor shown in Figure 10.47a.

26. Calculate the Q of the inductor shown in Figure 10.47b.

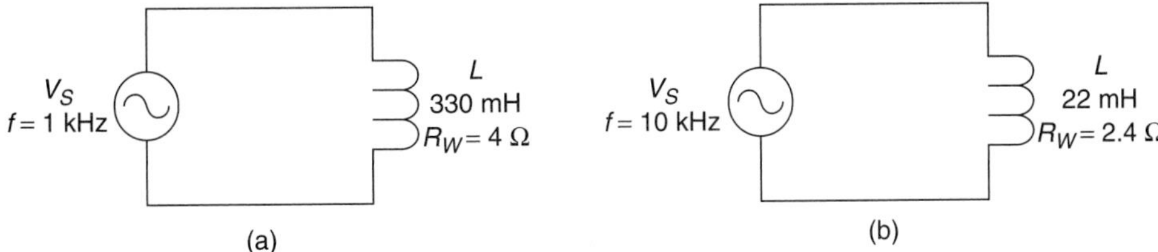

Figure 10.47

27. The source shown in Figure 10.48a is variable over a frequency range of 10 to 50 kHz. Calculate the range of Q values for the inductor.
28. The source shown in Figure 10.48b is variable over a frequency range of 100 Hz to 8 kHz. Calculate the range of Q values for the inductor.

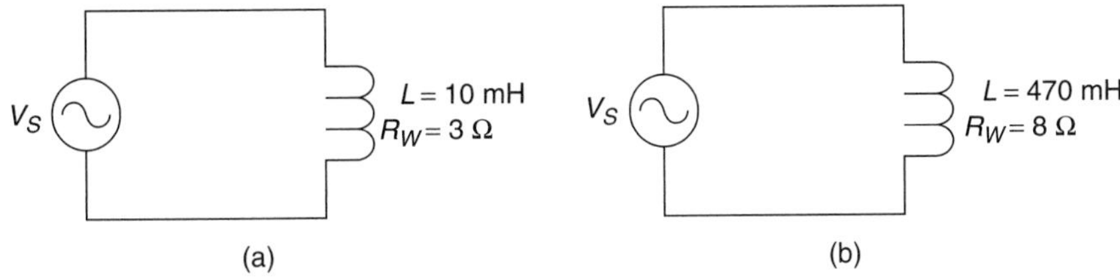

Figure 10.48

29. A transformer has a 30 V input and a 10:1 turns ratio. Determine the secondary voltage.
30. A transformer has a 75 V input and a 3:1 turns ratio. Determine the secondary voltage.
31. A transformer has a 120 V input and a 1:8 turns ratio. Determine the secondary voltage.
32. A transformer has a 250 mV input and a 1:100 turns ratio. Determine the secondary voltage.
33. Determine the secondary current (I_S) for the transformer shown in Figure 10.49a.
34. Determine the secondary current (I_S) for the transformer shown in Figure 10.49b.

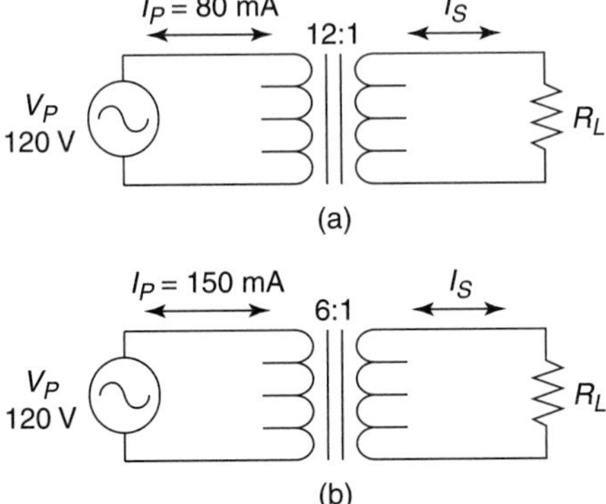

Figure 10.49

35. Calculate the secondary voltage and current for the transformer shown in Figure 10.50a.
36. Calculate the secondary voltage and current for the transformer shown in Figure 10.50b.

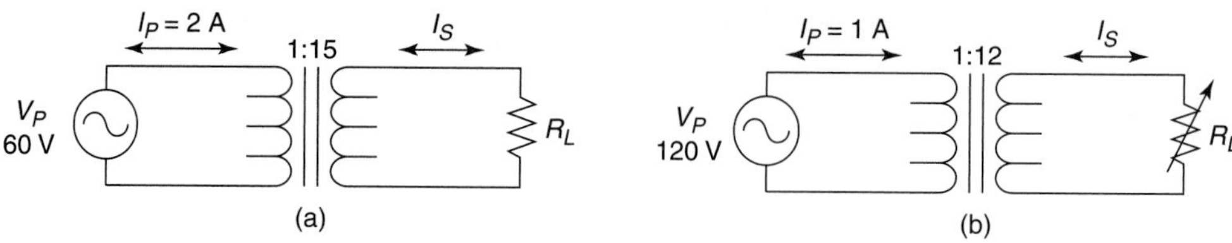

Figure 10.50

37. Determine the primary current (I_P) for the circuit shown in Figure 10.51a.
38. Determine the primary current (I_P) for the circuit shown in Figure 10.51b.

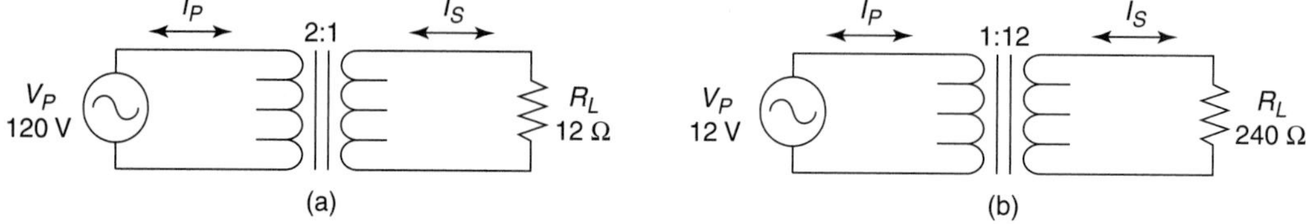

Figure 10.51

39. The peak primary current in Figure 10.52a must be limited to 80% of the primary fuse rating. Determine the minimum allowable setting for R_L.
40. The peak primary current in Figure 10.52b must be limited to 80% of the primary fuse rating. Determine the minimum allowable setting for R_L.

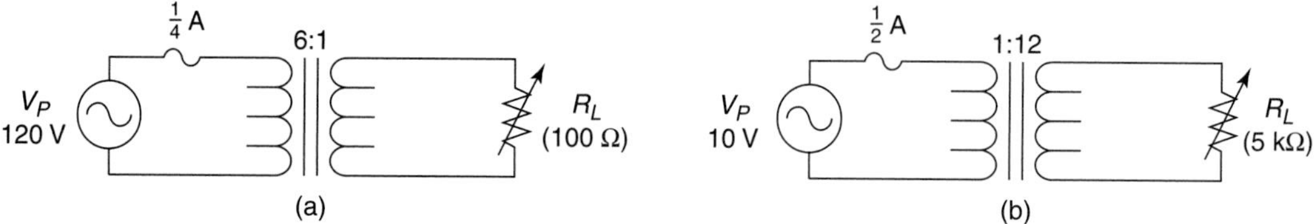

Figure 10.52

41. Refer to Figure 10.51a. Calculate the primary impedance (Z_P) for the circuit shown.
42. Refer to Figure 10.51b. Calculate the primary impedance (Z_P) for the circuit shown.

Looking Back

These problems relate to material presented in earlier chapters. The chapters are identified in brackets.

43. Determine the adjusted value of R_2 in Figure 10.53 that will result in a value of $P_T = 50$ mW. [Chapter 5]

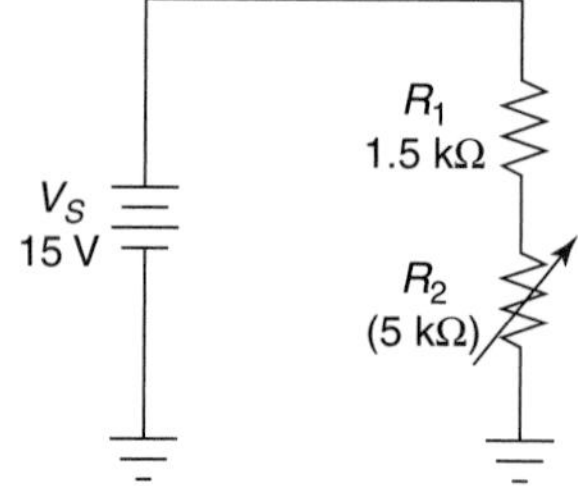

Figure 10.53

44. Determine the minimum allowable setting for R_3 in Figure 10.54. Assume that the circuit current must be limited to 80% of the rated value of the fuse. [Chapter 6]

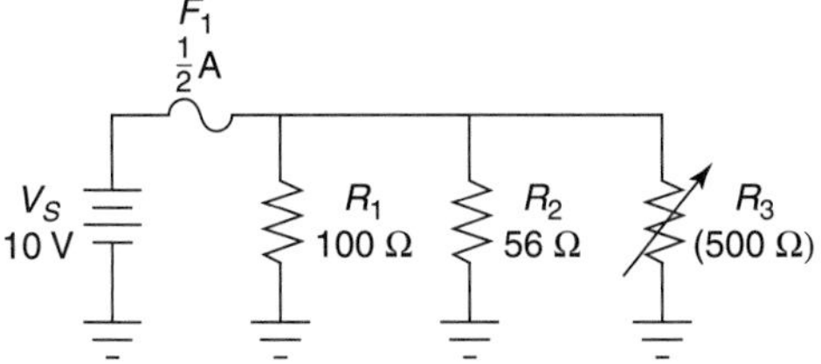

Figure 10.54

45. A wire has a value of $R = 1.5\ \Omega$. What is the resistance of a second wire that has five times the length and twice the cross-sectional area of the first? [Chapter 1]

46. A series circuit has the following values: $V_S = 10$ V, $R_1 = 51$ kΩ, and $R_2 = 91$ kΩ. V_2 is measured using a voltmeter with an internal resistance of 200 kΩ. How far off is the meter reading from the actual value of V_2? [Chapter 6]

47. How much energy is used by twelve 100 W light bulbs over the course of 24 hours? [Chapter 1]

48. A sine wave has an effective value of 10 V. What is the value of the waveform at the 60° point? [Chapter 9]

Pushing the Envelope

49. Determine the frequency of operation that will provide a current of 100 mA through the circuit shown in Figure 10.55.

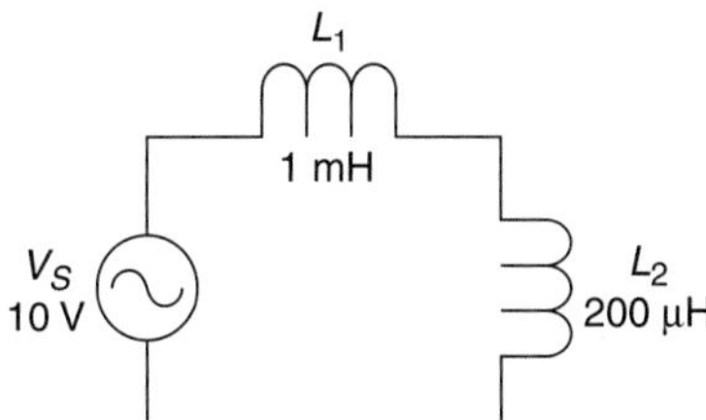

Figure 10.55

50. The circuit current shown in Figure 10.56 has a measured value of 100 mA. Assuming that the circuit has a value of $k = 0$, determine the value of L_2.

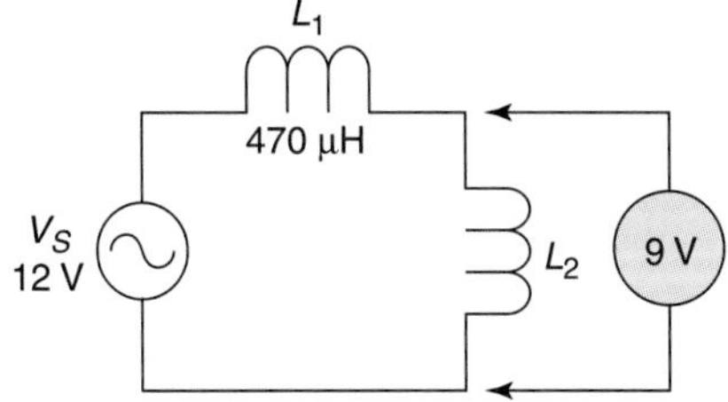

Figure 10.56

51. The operating frequency of the circuit shown in Figure 10.57 is variable. Determine the values of I_{rms} and f when the inductor has a value of $Q = 25$.

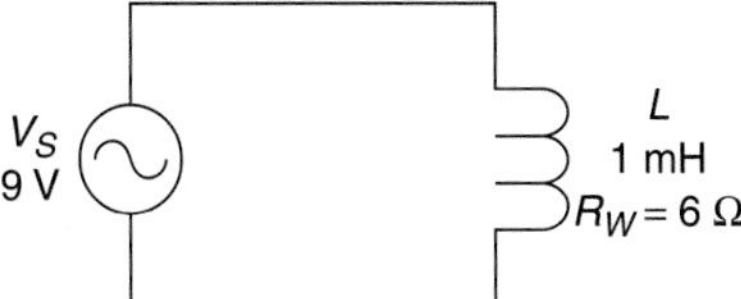

Figure 10.57

52. You have an unlimited length of 20 gauge wire, and you need a 5 H inductor. If you had to construct the inductor yourself (using air as a core):
   a. What physical dimensions would it have?
   b. What value of Q would it have at an operating frequency of 1 kHz?
   (*Note:* The resistance of 20 gauge wire is provided in Chapter 2.)

Answers to the Example Practice Problems

10.1	6.6 mV	**10.7**	10 V
10.2	3.73 mH	**10.8**	3 A
10.3	162.4 μH	**10.9**	1.2 kV, 100 mA
10.4	59.1 kΩ	**10.10**	7.2 kΩ
10.5	3.6 mA	**10.11**	1050
10.6	99.5 mA	**10.12**	622 to 2488

Multisim File References

Figure	File Reference
10.6	ETF10_6.msm
10.10	ETF10_10.msm
10.16	ETF10_16.msm
10.19	ETF10_19.msm

chapter 11

Resistive-Inductive (*RL*) Circuits

Objectives

After studying the material in this chapter, you should be able to:

1. Compare and contrast series resistive, inductive, and resistive-inductive (*RL*) circuits.
2. Describe the voltage and phase angle characteristics for a series *RL* circuit.
3. Describe the impedance and current characteristics of a series *RL* circuit.
4. Calculate the impedance, voltage, and current values for any series *RL* circuit.
5. Describe and analyze the frequency response of a series *RL* circuit.
6. Perform all relevant power calculations for a series *RL* circuit.
7. Compare and contrast parallel resistive, inductive, and resistive-inductive (*RL*) circuits.
8. Calculate any current or impedance value for a parallel *RL* circuit.
9. Describe and analyze the frequency response of a parallel *RL* circuit.
10. Calculate the impedance, voltage, and current values for any series-parallel *RL* circuit.

Up to this point, we have limited our discussions to circuits that contain only resistance or inductive reactance. In practice, most circuits contain both resistive and reactive components. In this chapter, we are going to begin our analysis of such circuits by studying the combined effects of resistance and inductive reactance.

11.1 Series *RL* Circuits

A *resistive-inductive* (*RL*) *circuit* is *one that contains any combination of resistors and inductors.* Several examples of basic *RL* circuits are shown in Figure 11.1.

Series Circuits: A Comparison

OBJECTIVE 1 ➤ As you know, series resistive and inductive circuits have the following characteristics:

- Current is the same at all points throughout the circuit.
- The applied voltage equals the sum of the component voltages.
- The total opposition to current (resistance or reactance) equals the sum of the component values.

The current, voltage, and impedance (*Z*) characteristics of series *RL* circuits are listed in Figure 11.2. As you can see, current is the same at all points throughout the circuit (as is the case for the resistive and inductive circuits). However:

- The applied voltage (V_S) equals the *geometric sum* of V_L and V_R.
- The total opposition to current (Z_T) equals the *geometric sum* of X_L and *R*.

We will now discuss the V_S and Z_T relationships shown in Figure 11.2.

Series Voltages

OBJECTIVE 2 ➤ In Chapter 10, you were told that *voltage leads current by 90°* in an inductive circuit. In a resistive circuit, *voltage and current are always in phase.* When we connect an inductor and resistor in series with a voltage source, we get the circuit and waveforms shown

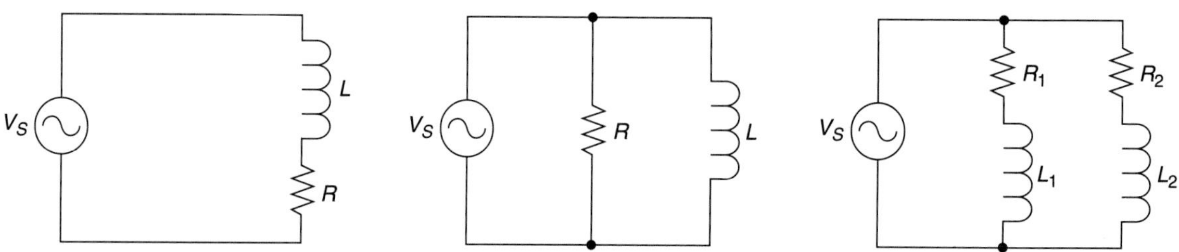

Figure 11.1 Examples of basic *RL* circuits.

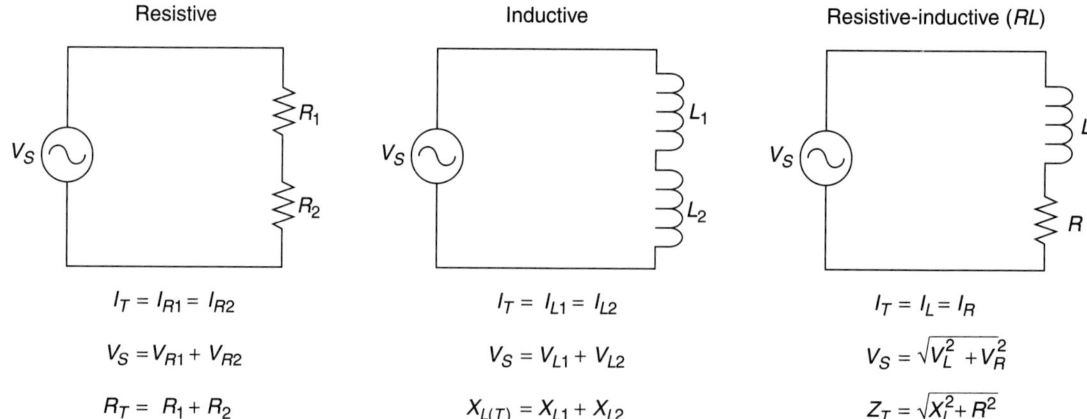

Figure 11.2 Series circuit characteristics.

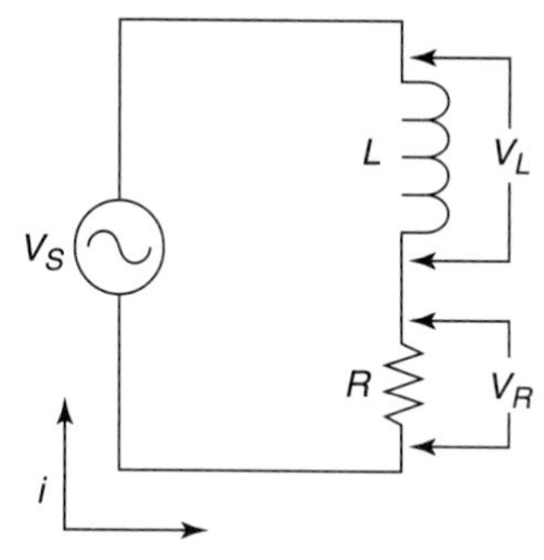

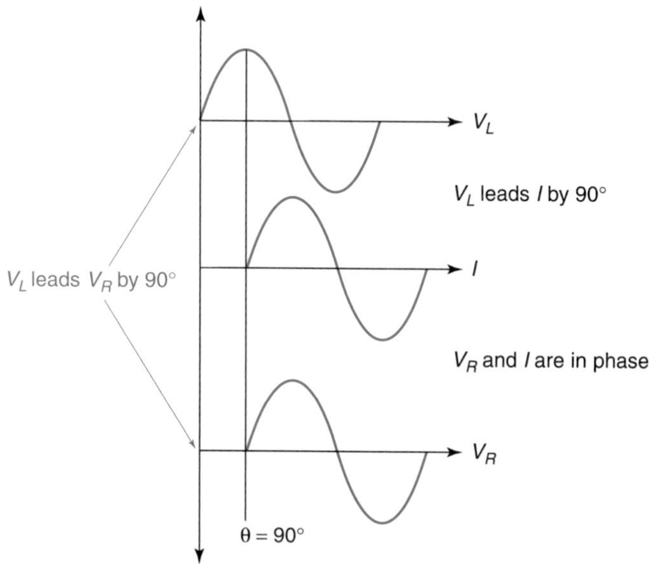

Figure 11.3 Voltage phase relationships in a series *RL* circuit.

in Figure 11.3. Since current is constant throughout a series circuit, the inductor and resistor voltage waveforms are referenced to a single current waveform. As you can see:

- The inductor voltage leads the circuit current by 90°.
- The resistor voltage is in phase with the circuit current.

Therefore, *inductor voltage leads resistor voltage by 90°.*

Since V_L leads V_R by 90°, the values of the component voltages are usually represented using *phasors.* A **phasor** is *a vector used to represent a value that contantly changes phase* (such as the instantaneous value of a sine wave). A phasor representation of V_L and V_R is shown in Figure 11.4a. By plotting the value of V_R on the x-axis (0°V) and the value of V_L on the y-axis (90°), the phase relationship between the two is represented in the graph.

Note: The distinction between phasors and vectors is made here for technical accuracy. Most technicians refer to both types of lines as vectors.

Like any other series circuit, the source voltage in an *RL* circuit must equal the sum of the component voltages. Since the component voltages in an *RL* circuit are at 90° angles,

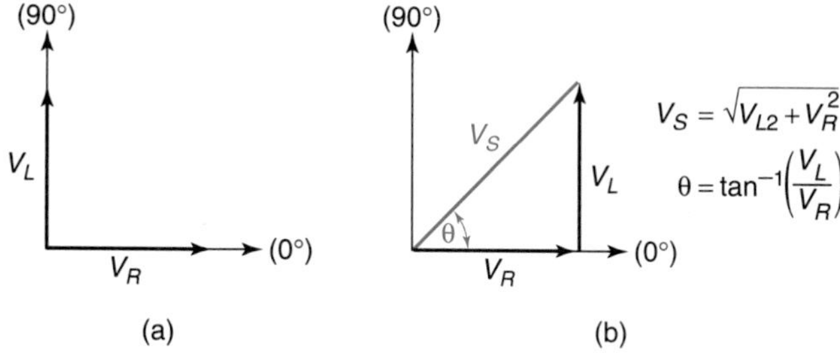

Figure 11.4 Geometric addition of V_L and V_R.

they must be added *geometrically.* As described in Appendix F, the sum of two phasors (vectors) at a 90° angle is found as

$$z = \sqrt{x^2 + y^2}$$

where

z = the sum of the two phasor magnitudes
x and y = the magnitudes of the phasors to be added

Relating this equation to Figure 11.4b, the sum of V_L and V_R can be found as

$$V_S = \sqrt{V_L^2 + V_R^2} \tag{11.1}$$

Note that the V_S phasor represents the geometric sum of V_R and V_L.

The *phase angle* (θ) identified in Figure 11.4b represents the phase difference between V_S and V_R. Since V_R is in phase with the circuit current, *θ represent the phase difference between the source voltage and the circuit current.* As shown in Figure 11.4b, θ is one of three angles in a right triangle. In Appendix F, this angle is shown to have a value of

$$\theta = \tan^{-1}\left(\frac{y}{x}\right)$$

where

$\tan^{-1}$ = the *inverse tangent* of the fraction
y = the magnitude of the y-axis value
x = the magnitude of the x-axis value

Relating this equation to Figure 11.4b, the phase angle of the source voltage can be found as

$$\theta = \tan^{-1}\left(\frac{V_L}{V_R}\right) \tag{11.2}$$

Example 11.1 demonstrates the calculation of source voltage for a series RL circuit.

Example 11.1

The component voltages shown in Figure 11.5a were measured using an ac voltmeter. Calculate the source voltage for the circuit.

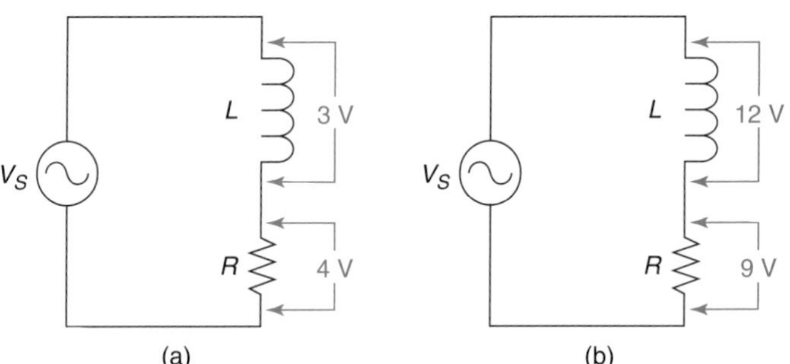

Figure 11.5

Solution: Using equation (11.1), the *magnitude* of V_S is found as

$$V_S = \sqrt{V_L^2 + V_R^2} = \sqrt{(3\text{ V})^2 + (4\text{ V})^2} = \sqrt{25\text{ V}^2} = 5\text{ V}$$

The *phase angle* of V_S (relative to the circuit current) can be found as

$$\theta = \tan^{-1}\left(\frac{V_L}{V_R}\right) = \tan^{-1}\left(\frac{3\text{ V}}{4\text{ V}}\right) = \tan^{-1}(0.75) \cong 36.9°$$

Therefore, the source voltage for the circuit is $5\text{ V}\angle 36.9°$.

Note: Here's a quick check for the result of any geometric addition: The result should be greater than either value and less than the algebraic sum of the two. For example, the result in Example 11.1 is 5 V, which is greater than either component voltage and less than the algebraic sum of the two (7 V).

Practice Problem 11.1

Calculate the value of V_S for the circuit in Figure 11.5b.

The result in Example 11.1 indicates that the source voltage:

- Has a magnitude of 5 V.
- Leads the circuit current by 36.9°.

For any *RL* circuit, the limits on the phase angle between the source voltage and circuit current are given as follows:

$$0° < \theta < 90° \tag{11.3}$$

This range of values for θ makes sense when you consider that:

- The phase angle in a *purely resistive* circuit is 0°.
- The phase angle in a *purely inductive* circuit is 90°.

A circuit containing both resistance and inductance is neither *purely resistive* nor *purely inductive.* Therefore, it must have a phase angle that is greater than 0° but less than 90°. As you will see, the value of θ for a given *RL* circuit depends on *the ratio of inductive reactance to resistance.*

Series Impedance

◄ *OBJECTIVE 3*

In an *RL* circuit, the total impedance (opposition to alternating current) is made up of resistance and inductive reactance. Like series voltages, the values of R and X_L for a series circuit are phasor quantities that must be added geometrically. The reason for this can be explained with the help of Figure 11.6.

In Chapter 10, you learned that inductive reactance can be found using

$$X_L = \frac{V_L}{I_L}$$

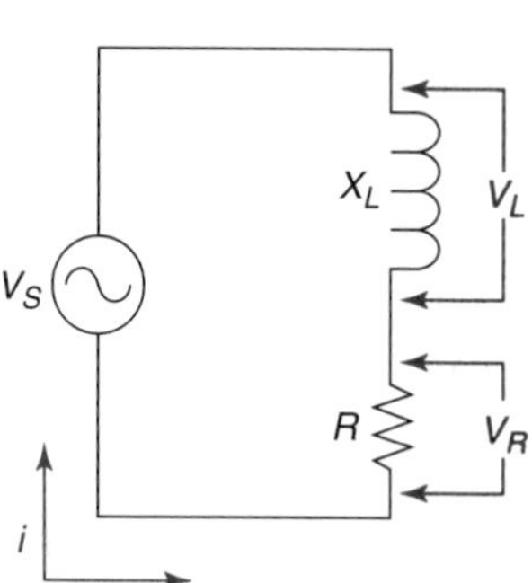

Figure 11.6
Series *RL* voltages and circuit current.

Since Chapter 10 dealt with purely inductive circuits, it was not necessary to consider any phase angle associated with X_L. Now that we are dealing with *RL* circuits, we need to establish the fact that X_L *does* have a phase angle. For the circuit shown in Figure 11.6, the phase angle of X_L can be found as

$$X_L = \frac{V_L\angle 90°}{I\angle 0°} = \frac{V_L}{I}\angle 90°$$

While this result does not include a magnitude, it does indicate that inductive reactance leads circuit current by 90°. At the same time, the phase angle for the circuit resistance can be found as

$$R = \frac{V_R\angle 0°}{I\angle 0°} = \frac{V_R}{I}\angle 0°$$

These calculations indicate that X_L leads R by 90°. As a result, the total circuit impedance (Z_T) is found as the *geometric sum* of X_L and R. By formula,

$$Z_T = \sqrt{X_L^2 + R^2} \tag{11.4}$$

The phase angle of the circuit impedance (relative to the circuit current) is found in the same fashion as the phase angle of the source voltage. By formula,

$$\theta = \tan^{-1}\left(\frac{X_L}{R}\right) \tag{11.5}$$

Example 11.2 demonstrates the procedure for determining the total impedance in a series *RL* circuit.

Example 11.2

Calculate the total impedance for the circuit shown in Figure 11.7a.

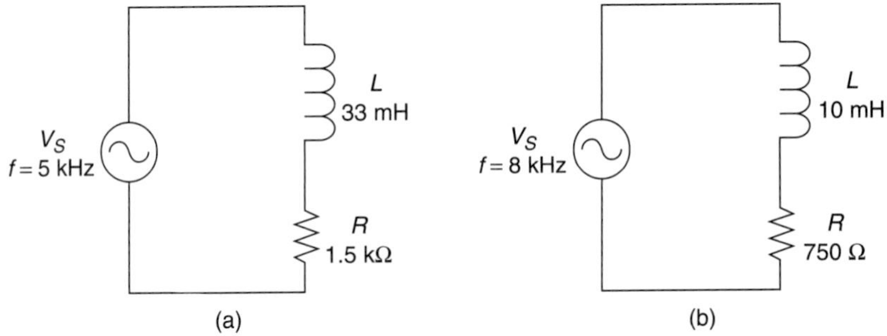

Figure 11.7

Solution: First, the value of X_L is found as

$$X_L = 2\pi f L = 2\pi(5 \text{ kHz})(33 \text{ mH}) \cong 1.04 \text{ k}\Omega$$

Now the magnitude of Z_T is found as

$$Z_T = \sqrt{X_L^2 + R^2} = \sqrt{(1.04 \text{ k}\Omega)^2 + (1.5 \text{ k}\Omega)^2} = 1.83 \text{ k}\Omega$$

and its phase angle is found as

$$\theta = \tan^{-1}\left(\frac{X_L}{R}\right) = \tan^{-1}\left(\frac{1.04 \text{ k}\Omega}{1.5 \text{ k}\Omega}\right) = 34.7°$$

Therefore, the total circuit impedance equals $1.83 \text{ k}\Omega\angle 34.7°$

Practice Problem 11.2

Calculate the total impedance for the circuit in Figure 11.7b.

Resolving Series Circuit Phase Angles

Like standard numeric values, those written in polar notation conform to the basic laws of circuit operation (such as Ohm's law and Kirchhoff's laws). For example, Figure 11.8 contains the circuit we analyzed in Example 11.2. The voltages were calculated using an assumed current of 1 mA. According to equations (11.1) and (11.2),

$$V_S = \sqrt{V_L^2 + V_R^2} = \sqrt{(1.04 \text{ V})^2 + (1.5 \text{ V})^2} = 1.83 \text{ V}$$

and

$$\theta = \tan^{-1}\left(\frac{V_L}{V_R}\right) = \tan^{-1}\left(\frac{1.04 \text{ V}}{1.5 \text{ V}}\right) = \tan^{-1}(0.693) \cong 34.7°$$

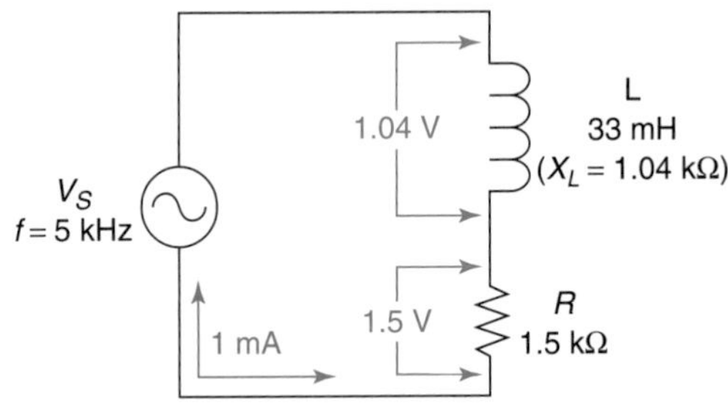

Figure 11.8

In Example 11.2, we calculated a total value of $Z_T = 1.83\ \text{k}\Omega\angle 34.7°$. If we use these two values, Ohm's law provides a circuit current value of

$$I = \frac{V_S}{Z_T} = \frac{1.83\ \text{V}\angle 34.7°}{1.83\ \text{k}\Omega\angle 34.7°} = 1\ \text{mA}\angle 0°$$

which agrees with our assumed value of circuit current.

As you can see, the phase angles for voltage, current, and impedance values all conform to basic circuit principles. This will always be the case when a circuit is properly analyzed.

Calculating Series Circuit Current

When you use Ohm's law to calculate the total current in a series *RL* circuit, your result will indicate that the circuit current has a *negative* phase angle. For example, consider the circuit shown in Figure 11.9. The circuit impedance has a magnitude of

$$Z_T = \sqrt{X_L^2 + R^2} = \sqrt{(12.6\ \text{k}\Omega)^2 + (15\ \text{k}\Omega)^2} = 19.6\ \text{k}\Omega$$

at a phase angle of

$$\theta = \tan^{-1}\left(\frac{X_L}{R}\right) = \tan^{-1}\left(\frac{12.6\ \text{k}\Omega}{15\ \text{k}\Omega}\right) = \tan^{-1}(0.84) = 40°$$

Using these values, Ohm's law provides us with a circuit current of

$$I = \frac{V_S}{Z_T} = \frac{20\ \text{V}\angle 0°}{19.6\ \text{k}\Omega\angle 40°} = 1.02\ \text{mA}\angle -40°$$

As you can see, the calculated circuit current includes a *negative* phase angle.

The meaning of a negative current phase angle.

The negative current phase angle may seem to be incorrect—but it isn't. It simply indicates that *I_T lags V_S by 40°*, which is the same as saying that *V_S leads I_T by 40°*. The phase relationship between V_S and I_T can be expressed as

$$V_S = 20\ \text{V}\angle 0° \qquad\qquad V_S = 20\ \text{V}\angle 40°$$
$$I = 1.02\ \text{mA}\angle -40° \quad \text{or} \quad I = 1.02\ \text{mA}\angle 0°$$

The only difference is that the first set of values references the phase angle to the source voltage, and the second set references the phase angle to the circuit current. Since current is constant throughout a series circuit, using current as the 0° reference is the preferred method.

Series Circuit Analysis

◀ *OBJECTIVE 4*

The basic analysis of a simple series *RL* circuit is performed as demonstrated in Example 11.3.

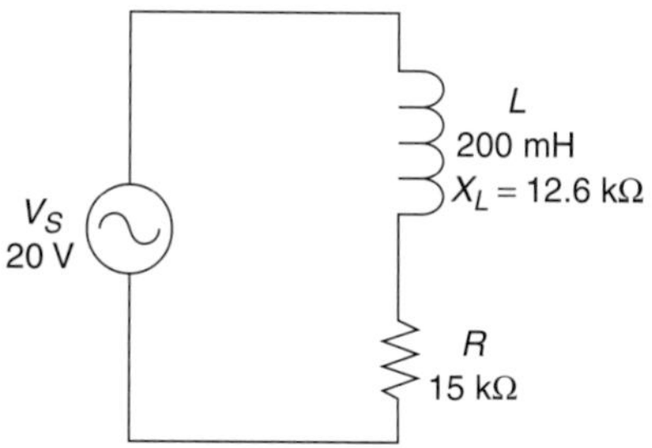

Figure 11.9

Example 11.3

Determine the voltage, current, and impedance values for the circuit shown in Figure 11.10a.

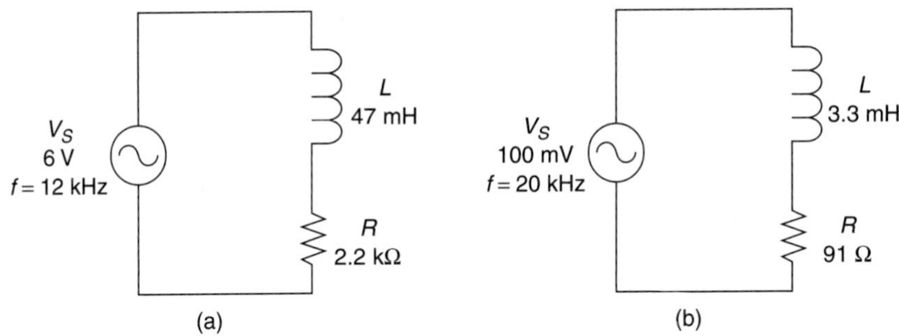

Figure 11.10

Solution: First, the value of X_L is found as

$$X_L = 2\pi fL = 2\pi(12\text{ kHz})(47\text{ mH}) = 3.54\text{ k}\Omega$$

The circuit impedance has a magnitude of

$$Z_T = \sqrt{X_L^2 + R^2} = \sqrt{(3.54\text{ k}\Omega)^2 + (2.2\text{ k}\Omega)^2} = 4.17\text{ k}\Omega$$

at a phase angle of

$$\theta = \tan^{-1}\left(\frac{X_L}{R}\right) = \tan^{-1}\left(\frac{3.54\text{ k}\Omega}{2.2\text{ k}\Omega}\right) = 58.1^\circ$$

Using the values of Z_T and V_S, the circuit current is found as

$$I = \frac{V_S}{Z_T} = \frac{6\text{ V}\angle 0^\circ}{4.17\text{ k}\Omega\angle 58.1^\circ} = 1.44\text{ mA}\angle -58.1^\circ$$

Since current is the 0° reference in a series circuit, the values of V_S and I are rewritten as $V_S = 6\text{ V}\angle 58.1^\circ$ and $I = 1.44\text{ mA}\angle 0^\circ$. Now, the values of V_L and V_R are found as

$$V_L = I \cdot X_L = (1.44\text{ mA}\angle 0^\circ)(3.54\text{ k}\Omega\angle 90^\circ) = 5.19\text{ V}\angle 90^\circ$$

and

$$V_R = I \cdot R = (1.44\text{ mA})(2.2\text{ k}\Omega) = 3.17\text{ V}$$

Note: Phase angles have been omitted from the V_R calculation because the current and resistance both have a phase angle of 0°.

Practice Problem 11.3

Determine the voltage, current, and impedance values for the circuit shown in Figure 11.10b.

The solution to Example 11.3 demonstrates an important point: *In a series RL circuit where I_T is used as the 0° phase reference, V_S and Z_T always have the same phase angle.*

RL Voltage Dividers

In a series *RL* circuit, we can calculate the component voltages using a voltage-divider relationship, just as we did in dc circuits. In this case,

$$V_n = V_S \frac{Z_n}{Z_T} \qquad (11.6)$$

where

Z_n = the magnitude of R or X_L

V_n = voltage across that component

Example 11.4 demonstrates the use of this voltage-divider relationship.

Example 11.4

Calculate the component voltages for the circuit shown in Figure 11.11a.

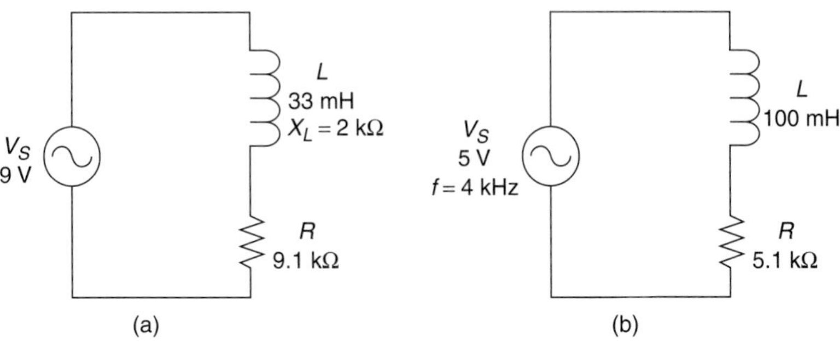

Figure 11.11

Solution: The circuit impedance has a magnitude of

$$Z_T = \sqrt{X_L^2 + R^2} = \sqrt{(2\text{ k}\Omega)^2 + (9.1\text{ k}\Omega)^2} = 9.32\text{ k}\Omega$$

at an angle of

$$\theta = \tan^{-1}\left(\frac{X_L}{R}\right) = \tan^{-1}\left(\frac{2\text{ k}\Omega}{9.1\text{ k}\Omega}\right) = 12.4°$$

Now the inductor voltage can be found as

$$V_L = V_S\frac{X_L}{Z_T} = (9\text{ V}\angle 12.4°)\frac{2.07\text{ k}\Omega\angle 90°}{9.32\text{ k}\Omega\angle 12.4°} = 2.00\text{ V}\angle -90°$$

Finally, the resistor voltage can be found as

$$V_R = V_S\frac{R}{Z_T} = (9\text{ V}\angle 12.4°)\frac{9.1\text{ k}\Omega}{9.32\text{ k}\Omega\angle 12.4°} = 8.79\text{ V}$$

Practice Problem 11.4

Calculate the component voltages for the circuit in Figure 11.11b.

Note: The phase angles of V_S and Z_T cancel each other out in the V_R and V_L calculations. For this reason, they can be ignored in the voltage divider equation without losing any accuracy in the result.

Two important points should be emphasized at this time:

- The component voltages calculated in Example 11.4 might appear to be incorrect. Remember, however, that V_S equals the *geometric* sum (not the algebraic sum) of the component voltages.
- As mentioned earlier, the phase angles of Z_T and V_S are always equal. This means that we can ignore these phase angles when using the voltage-divider equation, as they always cancel each other out. (This is illustrated in Example 11.4.)

When a series *RL* circuit contains more than one inductor and/or resistor, the analysis of the circuit takes a few more steps, as demonstrated in Example 11.5.

Example 11.5

Calculate the values of Z_T, V_{L1}, V_{L2}, and V_R for the circuit shown in Figure 11.12a.

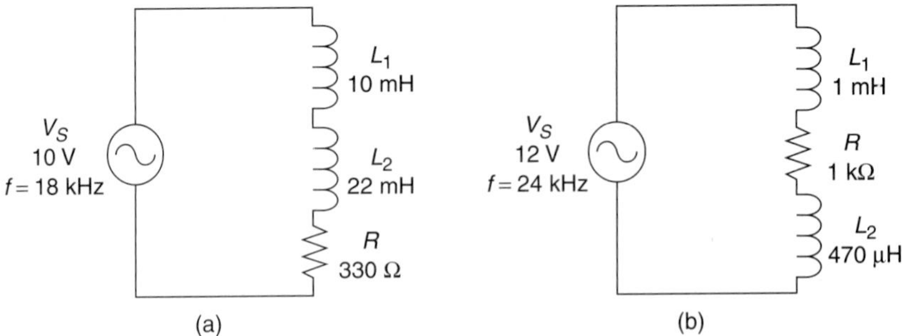

Figure 11.12

Solution: The inductor reactance values are found as

$$X_{L1} = 2\pi f L_1 = 2\pi(18 \text{ kHz})(10 \text{ mH}) = 1.13 \text{ k}\Omega$$

and

$$X_{L2} = 2\pi f L_2 = 2\pi(18 \text{ kHz})(22 \text{ mH}) = 2.49 \text{ k}\Omega$$

Now the total circuit reactance is found as

$$X_{L(T)} = X_{L1} + X_{L2} = 1.13 \text{ k}\Omega + 2.49 \text{ k}\Omega = 3.62 \text{ k}\Omega$$

The circuit impedance has a magnitude of

$$Z_T = \sqrt{X_{LT}^2 + R^2} = \sqrt{(3.62 \text{ k}\Omega)^2 + (330 \ \Omega)^2} = 3.64 \text{ k}\Omega$$

at an angle of

$$\theta = \tan^{-1}\left(\frac{X_L}{R}\right) = \tan^{-1}\left(\frac{3.62 \text{ k}\Omega}{330 \ \Omega}\right) = 84.8°$$

Now we can use the voltage divider equation to calculate the magnitudes of the component voltages:

$$V_{L1} = V_S\frac{X_{L1}}{Z_T} = (10 \text{ V})\frac{1.13 \text{ k}\Omega}{3.64 \text{ k}\Omega} = 3.10 \text{ V}$$

$$V_{L2} = V_S\frac{X_{L2}}{Z_T} = (10 \text{ V})\frac{2.49 \text{ k}\Omega}{3.64 \text{ k}\Omega} = 6.84 \text{ V}$$

and

$$V_R = V_S\frac{R}{Z_T} = (10 \text{ V})\frac{330 \ \Omega}{3.64 \text{ k}\Omega} = 907 \text{ mV}$$

Practice Problem 11.5

Calculate the values of Z_T, V_{L1}, V_{L2}, and V_R for the circuit shown in Figure 11.12b. (*Note:* The placement of the components does not affect the circuit analysis.)

Once again, we can verify our results. Since V_{L1} and V_{L2} are in phase, the total inductor voltage is found as

$$V_{L(T)} = V_{L1} + V_{L2} = 3.10 \text{ V} + 6.84 \text{ V} = 9.94 \text{ V}$$

The source voltage has a magnitude of

$$V_S = \sqrt{V_{LT}^2 + V_R^2} = \sqrt{(9.94\text{ V})^2 + (904\text{ mV})^2} = 9.98\text{ V}$$

Since our result is approximately equal to 10 V (the given magnitude of the source), we can safely assume that all our calculations are correct.

Series *RL* Circuit Frequency Response

The term **frequency response** is used to describe *any changes that occur in a circuit as a result of a change in operating frequency.* Example 11.6 demonstrates the effects of a change in operating frequency on series *RL* circuit values. ◀ *OBJECTIVE 5*

Example 11.6

The source in Figure 11.13a has the frequency limits shown. Calculate the circuit voltage, current, and impedance values at the frequency limits of the source.

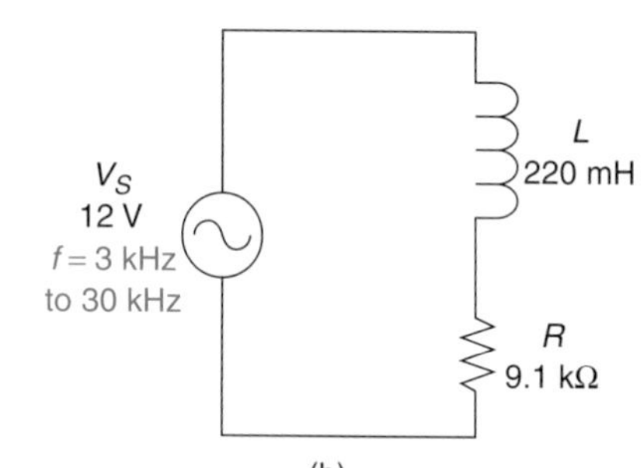

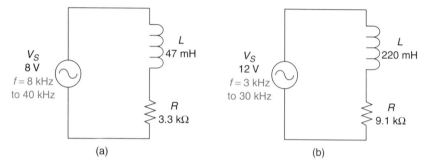

Figure 11.13

Solution: When the operating frequency of the circuit is 8 kHz, the reactance of the inductor is found as

$$X_L = 2\pi fL = 2\pi(8\text{ kHz})(47\text{ mH}) = 2.36\text{ k}\Omega$$

The circuit impedance has a magnitude of

$$Z_T = \sqrt{X_L^2 + R^2} = \sqrt{(2.36\text{ k}\Omega)^2 + (3.3\text{ k}\Omega)^2} = 4.06\text{ k}\Omega$$

at a phase angle of

$$\theta = \tan^{-1}\left(\frac{X_L}{R}\right) = \tan^{-1}\left(\frac{2.36\text{ k}\Omega}{3.3\text{ k}\Omega}\right) = \tan^{-1}(0.7152) = 35.6^\circ$$

Now the circuit current is found as

$$I_T = \frac{V_S}{Z_T} = \frac{8\text{ V}\angle 35.6^\circ}{4.06\text{ k}\Omega\angle 35.6^\circ} = 1.97\text{ mA}$$

The magnitudes of the component voltages can be found as

$$V_L = I_T X_L = (1.97\text{ mA})(2.36\text{ k}\Omega) = 4.65\text{ V}$$

and

$$V_R = I_T R = (1.97\text{ mA})(3.3\text{ k}\Omega) = 6.50\text{ V}$$

When the operating frequency increases to 40 kHz, the reactance of the inductor increases to

$$X_L = 2\pi fL = 2\pi(40\text{ kHz})(47\text{ mH}) = 11.8\text{ k}\Omega$$

This increase in X_L gives us a circuit impedance of

$$Z_T = \sqrt{X_L^2 + R^2} = \sqrt{(11.8\ \text{k}\Omega)^2 + (3.3\ \text{k}\Omega)^2} = 12.3\ \text{k}\Omega$$

at a phase angle of

$$\theta = \tan^{-1}\left(\frac{X_L}{R}\right) = \tan^{-1}\left(\frac{11.8\ \text{k}\Omega}{3.3\ \text{k}\Omega}\right) = \tan^{-1}(3.576) = 74.4^\circ$$

The increased circuit impedance reduces circuit current to

$$I_T = \frac{V_S}{Z_T} = \frac{8\ \text{V}\angle 74.4^\circ}{12.3\ \text{k}\Omega\angle 74.4^\circ} = 650\ \mu\text{A}$$

Finally, the magnitudes of the component voltages can be found as

$$V_L = I_T X_L = (650\ \mu\text{A})(11.8\ \text{k}\Omega) = 7.67\ \text{V}$$

and

$$V_R = I_T R = (650\ \mu\text{A})(3.3\ \text{k}\Omega) = 2.15\ \text{V}$$

Practice Problem 11.6

The voltage source in Figure 11.13b has the frequency limits shown. Calculate the circuit voltage, current, and impedance values at the frequency limits of the source.

The values obtained in Example 11.6 are listed in Table 11.1. We can summarize these results using a series of cause-and-effect relationships, as follows:

- The increase in frequency causes X_L to increase.
- The increase in X_L causes Z_T and θ to increase.
- The increase in Z_T causes I_T to decrease.
- The increase in X_L causes V_L to increase.
- The increase in V_L causes V_R to decrease.

Throughout the remainder of this text, we will touch on the concept of circuit frequency response. As you will learn, frequency response is an integral part of any discussion regarding ac circuits.

TABLE 11.1 Comparison of Results in Example 11.6

Variable	*Value at f = 8 kHz*	*Value at f = 40 kHz*	*Effect*
X_L	2.36 kΩ	11.8 kΩ	Increased
Z_T	4.06 kΩ	12.3 kΩ	Increased
θ	35.6°	74.4°	Increased
I_T	1.97 mA	650 μA	Decreased
V_L	4.65 V	7.67 V	Increased
V_R	6.50 V	2.15 V	Decreased

Section Review

1. What are the three fundamental characteristics of every series circuit?
2. What is the *geometric* sum of two variables? What does it represent?
3. In a series *RL* circuit, what is the phase relationship between V_L and V_R?
4. In a series *RL* circuit, what does the phase angle of the source voltage represent?
5. In a series *RL* circuit, what is the phase relationship between X_L and R?
6. In a series *RL* circuit, what does the phase angle of the circuit impedance represent?

7. When using Ohm's law to calculate the circuit current in a series RL circuit, what is indicated by a *negative* phase angle in the result? How is this phase angle normally resolved?

8. Describe the effect that a change in operating frequency has on the voltage, current, and impedance values in a series RL circuit.

9. The cause-and-effect statements on page 328 describe the response of a series RL circuit to an *increase* in operating frequency. Rewrite the statements to describe the response of the circuit to a *decrease* in operating frequency. Critical Thinking

11.2 Power Characteristics and Calculations

In Chapter 10, we discussed *apparent power, true power,* and *reactive power* in terms of inductor operation. Each of these power values is described (in terms of RL circuit values) in Table 11.2. ◄ *OBJECTIVE 6*

TABLE 11.2 Power Values in Resistive-Reactive Circuits

Value	*Definition*
Resistive power (P_R)	The power dissipated by the resistance in an RL circuit. Also known as *true power.*
Reactive power (P_X)	The value found using $P = I^2X_L$. Also known as *imaginary power.* The energy stored by the inductor in its electromagnetic field. P_X is measured in volt-amperes-reactive (VARs) to distinguish it from true power.
Apparent power (P_{APP})	The combination of resistive (true) power and reactive (imaginary) power. Measured in volt-amperes (VAs).

The values of resistive and reactive power for a series RL circuit are calculated as shown in Example 11.7.

Example 11.7

The circuit shown in Figure 11.14a has the values shown. Assuming that the inductor is an ideal component, calculate the values of resistive and reactive power for the circuit.

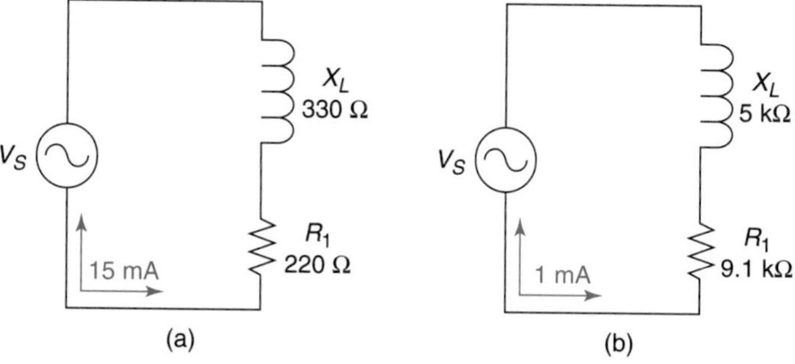

Figure 11.14

Solution: Using the values shown, the reactive power is found as

$$P_X = I^2X_L = (15\text{ mA})^2(330\ \Omega\angle 90°) = 74.3\text{ mVAR}\angle 90°$$

and the resistive power is found as

$$P_R = I^2R_1 = (15\text{ mA})^2(220\ \Omega) = 49.5\text{ mW}$$

> ***Practice Problem 11.7***
>
> Calculate the values of P_X and P_R for the circuit shown in Figure 11.14b. Assume that the inductor is an ideal component.

Note that the winding resistance of the coil (R_W) typically does affect the calculated value of resistive power. It is usually very low compared to total circuit resistance and thus can normally be ignored.

Calculating Apparent Power (P_{APP})

Apparent power represents the total power supplied by the source in a resistive-reactive circuit. The value of P_{APP} can be found as

$$P_{APP} = (V_S\angle\theta)(I_T) \tag{11.7}$$

Apparent power in a series *RL* circuit can also be found as *the geometric sum of resistive power and reactive power.* The basis for this relationship is illustrated in Figure 11.15.

The graph shown in Figure 11.15a shows the phase relationship between X_L and R. As indicated, X_L has a 90° phase angle relative to R. If we consider this phase angle in our P_X equation, we get

$$P_X = I^2X_L = (I^2\angle 0°)(X_L\angle 90°) = (I^2X_L)\angle 90°$$

While the result does not commit to a magnitude, it does indicate that *reactive power has a phase angle of 90°* (relative to the circuit current). If we perform the same type of calculation for P_R, we get

$$P_R = I^2R = (I^2\angle 0°)(R\angle 0°) = (I^2R)\angle 0°$$

Again, the result does not commit to a magnitude. However, it does indicate that *resistive power is in phase with the circuit current.* Since reactive power leads circuit current by 90°, it also leads resistive power by the same angle. Therefore, apparent power equals *the geometric sum of P_X and P_R*, as shown in Figure 11.15b. By formula,

$$P_{APP} = \sqrt{P_X^2 + P_R^2} \tag{11.8}$$

where

P_{APP} = the apparent power, measured in volt-amperes (VAs)

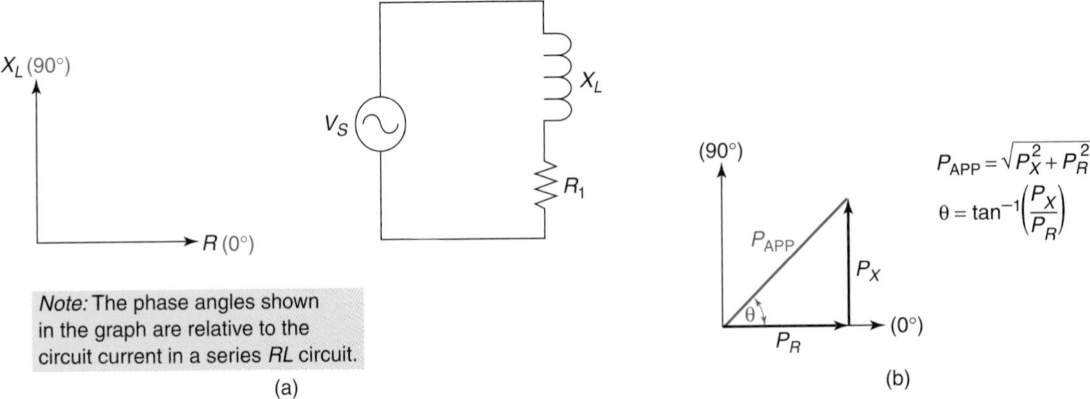

Figure 11.15 Power phase relationships in a series *RL* circuit.

The phase angle of apparent power (relative to the circuit current) is found in the same fashion as the phase angle for impedance and source voltage. By formula,

$$\theta = \tan^{-1} \frac{P_X}{P_R} \tag{11.9}$$

Example 11.8 demonstrates the use of these equations.

Example 11.8

Calculate the value of apparent power (including its phase angle) for the circuit in Example 11.7.

Solution: In Example 11.7, we calculated values of $P_X = 74.3$ mVAR and $P_R = 49.5$ mW. Using these values, apparent power is found as

$$P_{APP} = \sqrt{P_X^2 + P_R^2} = \sqrt{(74.3 \text{ mVAR})^2 + (49.5 \text{ mW})^2} = 89.3 \text{ mVA}$$

at a phase angle of

$$\theta = \tan^{-1} \frac{P_X}{P_R} = \tan^{-1} \left(\frac{74.3 \text{ mVAR}}{49.5 \text{ mW}} \right) = 56.3°$$

Practice Problem 11.8

Calculate the value of P_{APP} for the circuit in Practice Problem 11.7.

Power Factor

The **power factor (PF)** for an *RL* circuit is *the ratio of resistive power to apparent power.* It indicates the portion of the apparent power that is actually being dissipated in the circuit. By formula,

$$\text{PF} = \frac{P_R}{P_{APP}} \tag{11.10}$$

The cosine of the phase angle (cos θ) in Figure 11.16 equals the ratio of the length of the adjacent (adj) side of the triangle to the length of the hypotenuse (hyp). By formula,

$$\cos \theta = \frac{\text{adj}}{\text{hyp}}$$

Applying this relationship to the power triangle in Figure 11.16, we get

$$\cos \theta = \frac{P_R}{P_{APP}} \tag{11.11}$$

which, by definition, is the power factor for the circuit.

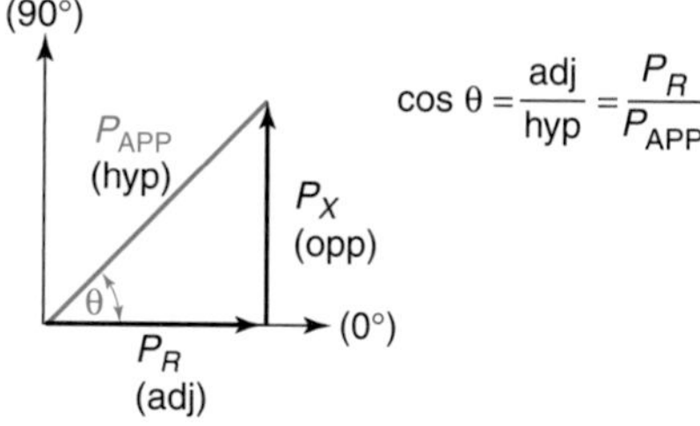

Figure 11.16 Power triangle.

One Final Note

We have now completed the basic analysis of series *RL* circuits. At this point, we're going to move on to parallel *RL* circuits.

Section Review

1. Compare and contrast the following: *apparent power, resistive power,* and *reactive power.*
2. Why is the winding resistance (R_W) of a coil normally ignored in the resistive power calculations for an *RL* circuit?
3. State and explain the phase relationship between reactive power (P_X) and resistive power (P_R).
4. What is the *power factor* for an *RL* circuit? What does it tell you?
5. How is the power factor of an *RL* circuit determined?

Critical Thinking

6. Derive a series of calculations that will determine the power dissipation in a series *RL* circuit using the values of V_S, I_T, and θ.

11.3 Parallel *RL* Circuits

A *parallel RL circuit* contains *one or more resistors in parallel with one or more inductors.* Several parallel *RL* circuits are shown in Figure 11.17. As you can see, none of the circuit branches contain more than one component. (If they did, we'd be dealing with a series-parallel *RL* circuit.)

Parallel Circuits: A Comparison

OBJECTIVE 7 ➤ As we did with series circuits, we're going to start by comparing a parallel *RL* circuit to purely resistive and purely inductive parallel circuits. A comparison between these circuits is provided in Figure 11.18.

In Chapters 5 and 10, we established the following fundamental characteristics of resistive and inductive parallel circuits:

- All branch voltages equal the source voltage.
- The circuit current equals the sum of the branch currents.
- The total impedance (resistance or reactance) is always lower than the lowest branch value.

The current and impedance relationships shown for the *RL* circuit in Figure 11.18 are explained later in this section. Even so, you can see that the total current in a parallel *RL* circuit equals the *geometric* sum of the branch currents. By the same token, the calculation of total impedance involves the geometric sum of inductive reactance and resistance (in the denominator of the equation). As you may have guessed, these geometric relationships exist because of phase angles within the circuit.

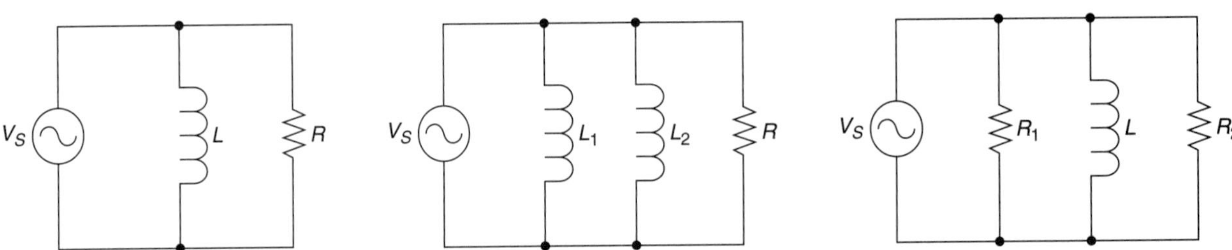

Figure 11.17 Some basic parallel *RL* circuits.

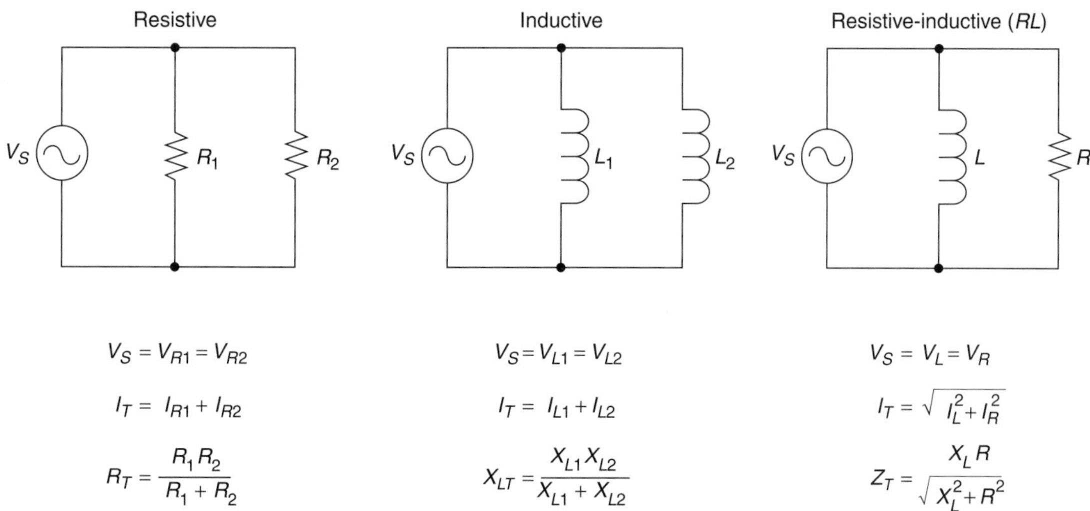

Figure 11.18 Parallel circuit characteristics.

Branch Currents

When connected in parallel, a resistor and an inductor produce the waveforms shown in Figure 11.19. As you know: ◄ *OBJECTIVE 8*

- Current lags voltage by 90° in an inductive circuit.
- Current and voltage in a resistive circuit are always in phase.

Since voltage is equal across parallel branches, inductor current lags both V_S and I_R by 90°. This phase relationship holds true for every parallel *RL* circuit with an ac voltage source.

Since I_L lags I_R by 90°, the values of the component currents are represented using phasors, as shown in Figure 11.20a. By plotting the value of I_R on the *x*-axis (0°) and the value of I_L on the negative *y*-axis (−90°), the phase relationship between the two is represented in the graph.

As in any parallel circuit, the sum of I_L and I_R must equal the total circuit current. In this case, the phase relationship between the two values requires that they be added geometrically, as follows:

$$I_T = \sqrt{I_L^2 + I_R^2} \tag{11.12}$$

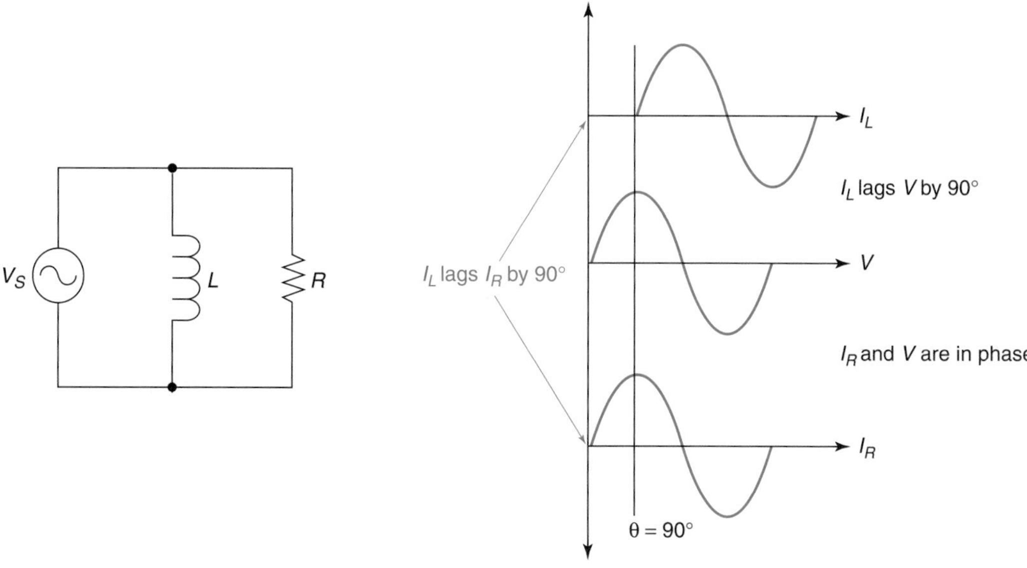

Figure 11.19 Current phase relationships in a parallel *RL* circuit.

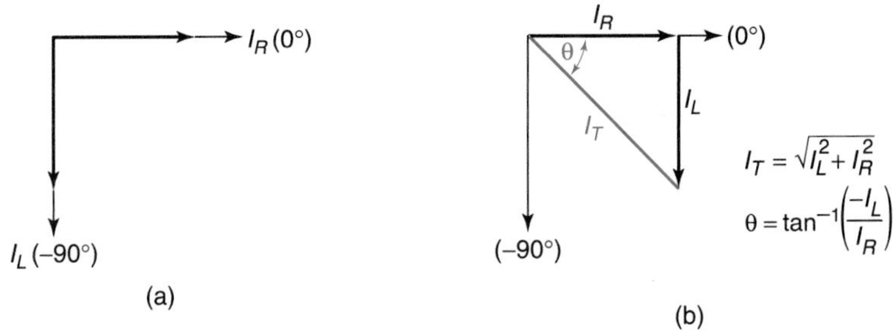

Figure 11.20 Geometric addition of I_L and I_R.

The *phase angle* of I_T in a parallel *RL* circuit is *the phase difference between the circuit current and the source voltage.* By formula,

$$\theta = \tan^{-1}\left(\frac{-I_L}{I_R}\right) \tag{11.13}$$

Why I_L is negative in equation (11.13).

If you refer back to Figure 11.20b, you'll see that the phasor for I_T lies below the *x*-axis of the graph. This means that *it must have a negative phase angle* (relative to the resistor current). Note that the negative value of I_L in equation (11.13) results in a *negative* value of θ, which indicates that the inductor current *lags* the source voltage. Example 11.9 demonstrates the calculation of total current for a parallel *RL* circuit.

Example 11.9

The current values in Figure 11.21a were measured as shown. Calculate the total current for the circuit.

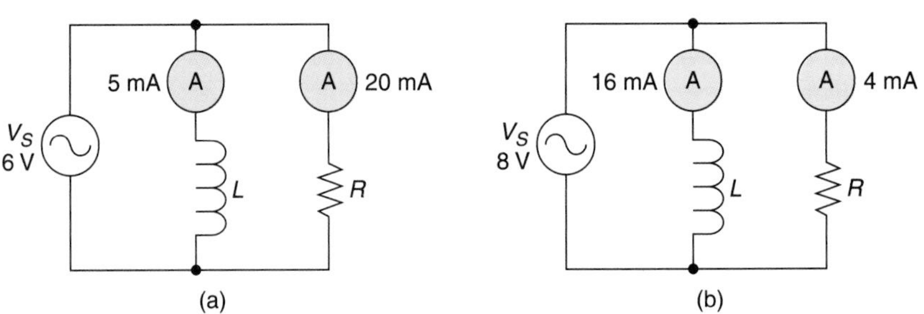

Figure 11.21

Solution: The magnitude of the circuit current is found as

$$I_T = \sqrt{I_L^2 + I_R^2} = \sqrt{(5 \text{ mA})^2 + (20 \text{ mA})^2} = 20.6 \text{ mA}$$

The phase angle of the circuit current (relative to the source voltage) is found as

$$\theta = \tan^{-1}\left(\frac{-I_L}{I_R}\right) = \tan^{-1}\left(\frac{-5 \text{ mA}}{20 \text{ mA}}\right) = \tan^{-1}(-0.25) = -14°$$

Practice Problem 11.9

Calculate the value of I_T for the circuit in Figure 11.21b.

Remember: In a mathematical relationship involving two negative numbers, the less-negative number is considered to be the greater of the two. This is why θ is written as being greater than −90° in equation (11.14).

For any parallel *RL* circuit, the limits on the phase angle between circuit current and source voltage are given as follows:

$$-90° < \theta < 0° \tag{11.14}$$

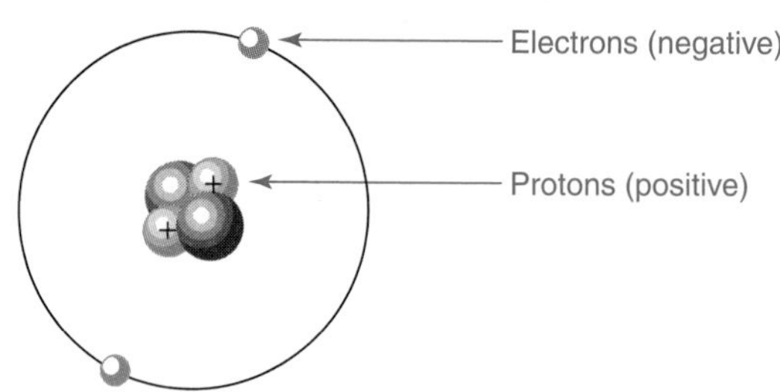

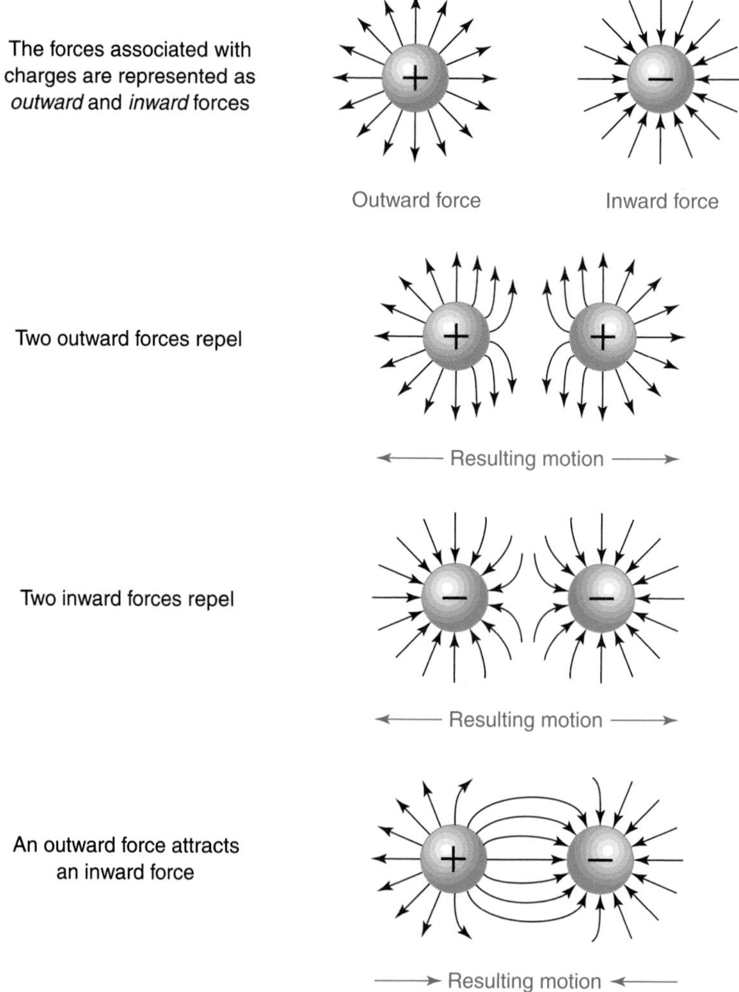

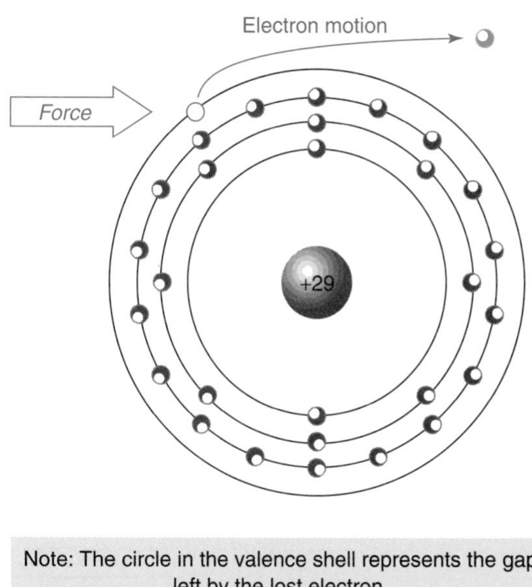

Note: The circle in the valence shell represents the gap left by the lost electron.

Figure 1.6 Generating a positive ion and a free electron.

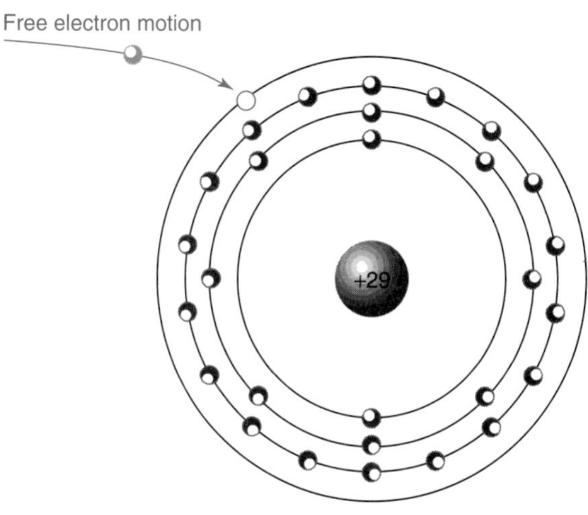

Note: The circle in the valence shell represents the gap left by a lost electron.

Figure 1.7 Neutralizing a positive ion.

Ions

The force of attraction between the negatively charged electron and the positively charged proton is the *inward force* that offsets the centrifugal force of an orbiting electron. When these forces are balanced, the electron remains in its orbit.

When an outside force is applied to an atom, the balance between the forces acting on a given electron can be disturbed, forcing the electron to leave its orbit. This point is illustrated in Figure 1.6.

When the electron leaves its orbit, the atom is left with fewer electrons than protons, and thus, has a net charge that is positive. An atom with a positive net charge is referred to as a **positive ion**.

Just as an outside force can cause an atom to lose an electron, it can also cause an atom to effectively *gain* an electron. When this occurs, the atom has more electrons than protons and, thus, a net charge that is *negative*. In this case, the atom is referred to as a **negative ion**.

Free Electrons

When an outside force causes an electron to break free from its parent atom, that electron is referred to as a *free electron.* A **free electron** is *an electron that is not bound to any particular atom*, and thus, is free to drift from one atom to another.

If a free electron drifts into the vicinity of a positive ion, that electron neutralizes the ion. For example, consider the free electron and the atom shown in Figure 1.7. When the electron nears the atom, it is attracted by the overall positive charge on the ion and is pulled into the gap in the valence shell. As a result, the atom once again has an equal number of electrons and protons and a net charge of *zero.*

Figure 1.7 and its related discussion illustrate an important point: *Positive and negative charges always seek to neutralize each other.* As you will see in the next section, this is the basis of one of the primary electrical properties, *current*.

Section Review

1. What is *matter*? What is an *element*?
2. What is an atom? Describe its physical makeup.
3. What is an *orbital shell*? What is the *valence shell*?

Example 1.4

Calculate the resistance of a six inch length of copper that has a diameter of 0.005 in.

Solution: Copper has a resistivity rating of 10.37 CM-Ω/ft (as shown in Table 1.3). Six inches equals 0.5 ft, and 0.005 in. equals 5 mils. With a diameter of 5 mils, the cross-sectional area is found as

$$\text{Area} = (5 \text{ mils})^2 = 25 \text{ CM}$$

Now, using equation (1.4), the resistance of the copper is found as

$$R = \rho \frac{\ell}{A} = \left(10.37 \frac{\cancel{\text{CM}}\text{-}\Omega}{\cancel{\text{ft}}}\right)\left(\frac{0.5 \cancel{\text{ft}}}{25 \cancel{\text{CM}}}\right) = 0.207\ \Omega = 207\ \text{m}\Omega$$

With this kind of resistance, it is easy to see that copper provides very little opposition to current.

Practice Problem 1.4

Calculate the resistance of an 18 inch length of copper that has a diameter of 0.005 in.

Note how the units of measure for resistivity, length, and area combine, leaving ohms (Ω) as the unit of measure for the result. This is why resistivity is rated in CM-Ω/ft (rather than ohms per circular-mil foot). By the same token, Example 1.5 demonstrates why resistivity is rated in Ω-cm (rather than ohms per cubic centimeter).

Example 1.5

Calculate the resistance of a 25 cm length of copper that has a cross-sectional area of 0.04 cm^2.

Solution: Copper has a resistivity rating of 1.723×10^{-6} Ω-cm (as shown in Table 1.3). Using this value and equation (1.4), the resistance of the copper is found as

$$R = \rho \frac{\ell}{A} = (1.723 \times 10^{-6}\ \Omega\text{-}\cancel{\text{cm}})\left(\frac{25 \cancel{\text{cm}}}{0.04 \cancel{\text{cm}^2}}\right) = 1.08 \times 10^{-3}\ \Omega = 1.08\ \text{m}\Omega$$

Practice Problem 1.5

Calculate the resistance of a 25 cm length of copper that has a cross-sectional area of 0.1 cm^2.

What Equation (1.4) Shows Us

Every equation serves two purposes. First, it shows us how to calculate the value of interest. Second (and probably most important), it shows us the relationship between the value of interest and its variables. For example, equation (1.4) states that

$$R = \rho \frac{\ell}{A}$$

This relationship shows us that:

- *The resistance of a material is directly proportional to its resistivity.* That is, increases (or decreases) in resistivity cause a similar change in resistance.
- *The resistance of a conductor is directly proportional to its length.*
- *The resistance of a conductor is inversely proportional to its cross-sectional area.* That is, changes in cross-sectional area and resistance are opposites. If one increases, the other decreases, and vice versa.

4. What relationships are commonly used to calculate the value of X_C?

5. Which of the relationships that you listed to answer question 4 is most often used? Why?

6. How is the magnitude of current in a capacitive circuit calculated?

7. How is the total capacitive reactance calculated in a series circuit? In a parallel circuit?

8. If an ohmmeter is connected to an *uncharged* capacitor, the meter reading is initially very low. Over the course of several seconds, the meter reading climbs to *infinite* ohms. Why? Critical Thinking

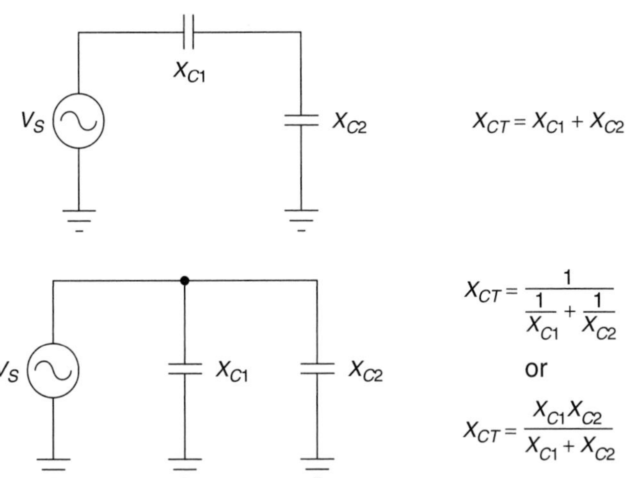

$$X_{CT} = X_{C1} + X_{C2}$$

$$X_{CT} = \frac{1}{\frac{1}{X_{C1}} + \frac{1}{X_{C2}}}$$

or

$$X_{CT} = \frac{X_{C1}X_{C2}}{X_{C1} + X_{C2}}$$

Figure 12.18

12.5 Related Topics

In this section, we will discuss some miscellaneous topics that relate to capacitors, their applications, and ratings.

Capacitor Voltage Ratings

◄ OBJECTIVE 12

As you know, the plates of a capacitor are separated by a dielectric (insulator). Every insulator, no matter how strong, will break down and conduct if the voltage across the component is sufficient. The **breakdown voltage rating** of a capacitor indicates *the voltage that will cause the dielectric of a capacitor to break down and conduct.* If this rating is exceeded, the capacitor is usually destroyed.

Leakage Current

◄ OBJECTIVE 13

Ideally, there is no current through a working capacitor. In practice, however, this is not the case. There is always some **leakage current** because the dielectric resistance of a capacitor is not infinite. Typically, dielectric resistance is in the mega-ohm range or higher. The fact that capacitors have finite and measurable values of dielectric resistance is important, because it affects their *quality* (Q) rating.

Capacitor Quality (Q)

◄ OBJECTIVE 14

In Chapter 10, you were told that the quality (Q) rating of an inductor is a figure of merit that indicates how close the component comes to having the characteristics of an *ideal* component. Capacitors also have quality ratings. The higher the Q of a capacitor, the closer it comes to having the characteristics of an ideal component. An ideal capacitor:

- Has infinite dielectric resistance and, therefore, no leakage current.
- Does not dissipate any power (because it prevents any leakage current).
- Has an infinite breakdown voltage rating (because of its infinite dielectric resistance).

Calculating the quality (Q) of a capacitor In Chapter 10, you learned that Q equals the ratio of reactive power to resistive power. For an inductor, this relationship simplifies to

$$Q = \frac{X_L}{R_W}$$

where

R_W = the winding resistance of the coil

The Q of a capacitor equals the ratio of dielectric resistance to capacitive reactance. By formula,

Remember: A high value of Q means that most of the energy delivered to the component is stored and later returned to the circuit. In the case of a capacitor, the energy is stored in its electrostatic field. Very little is used by the component itself.

$$Q = \frac{R_D}{X_C} \qquad (12.12)$$

where

X_C = the reactance of the capacitor at the frequency of operation

R_D = the dielectric resistance of the component

Earlier, we stated that the value of R_D is typically in the mega-ohm range or higher, which results in a very high Q for the typical capacitor. Capacitor Q is typically in the thousands and higher. For example, a 1 μF capacitor operated at a frequency of 1 kHz has a value of $X_C \cong 159\ \Omega$. If we assume that the component has a dielectric resistance of only 2 MΩ, the Q of the component (at this operating frequency) is

$$Q = \frac{R_D}{X_C} = \frac{2\ \text{M}\Omega}{159\ \Omega} = 12.58 \times 10^3$$

Fixed-Value Capacitors

Most capacitors are named for the material used as the dielectric. Table 12.2 summarizes some common types of capacitors and their characteristics. The values and applications given in the table are typical; that is, you may come across uses and component values that are not listed here.

OBJECTIVE 15 ➤ **Dielectric absorption** Table 12.2 makes reference to a property called *dielectric absorption.* **Dielectric absorption** can be defined as *the tendency of a dielectric to absorb charge.* This is important in some applications, because it takes time for a capacitor to release any charge in its dielectric. For this reason, capacitors with high dielectric absorption (like electrolytic capacitors) have limited high-frequency capabilities.

OBJECTIVE 16 ➤ **Electrolytic capacitors** Electrolytic capacitors are extremely common components that are used in low-frequency circuits, such as dc power supplies. These capacitors are housed in metal cans, which makes them relatively easy to identify. An electrolytic capacitor is shown in Figure 12.19b.

Electrolytic capacitors are *capacitors which contain an electrolyte that makes it possible to produce relatively small, high-capacity components.* However, the high capacity of the electrolytic capacitor is offset by its low Q, which limits its use to frequencies below 1 kHz.

TABLE 12.2 Common Capacitors[a]

Type	Value Range	Voltage Range	Tolerance	Applications/Comments	Frequency Range
Ceramic	10 pF–1 μF	50 V–30 kV	± (20+)%	Inexpensive, very common	500 Hz–500 MHz
Double layer	0.1–10 F	1.5–6 V	± 20%	Memory backup	1 kHz and below
Electrolytic	0.1 μF–1.6 F	3–600 V	−20 to +100%	Power supply filters (usually polarized, very short life)	1 kHz and below
Glass	10–1000 pF	100–600 V	± 10%	Long-term stability	1 kHz–5 GHz
Mica	1 pF–0.01 μF	100–600 V	± 10%	Radio-frequency (RF) circuits (excellent overall)	1 kHz–5 GHz
Mylar	1 nF–50 μF	50–600 V	± 20%	Inexpensive, very common	500 Hz–500 MHz
Oil	0.1–20 μF	200 V–10 kV	± 10%	High-voltage filters	1 kHz and below
Polycarbonate[b]	100 pF–30 μF	50–800 V	± 10%	High quality, small size	dc to 1000 MHz
Polypropylene[b]	100 pF–50 μF	100–800 V	± 5%	High quality, low dielectric absorption	dc to 1000 MHz
Polystyrene[b]	10 pF–2.7 μF	100–600 V	± 10%	Signal filters High quality, large size	dc to 1000 MHz
Porcelain	100 pF–0.1 μF	50–400 V	± 10%	Good long-term stability	500 Hz–500 MHz
Tantalum	0.1–500 μF	6–100 V	± 20%	High capacity, polarized	1 kHz and below
Teflon	1 nF–2 μF	50–200 V	± 5%	Highest quality, lowest dielectric absorption	dc to 1000 MHz
Vacuum	1–5000 pF	2–36 kV	± 10%	Transmitters	500 Hz–500 MHz

[a]*All values and ranges are typical.*

[b]*These are collectively referred to as* plastic film capacitors.

Polarized electrolytic capacitors Most electrolytic capacitors are *polarized.* The schematic symbol for a polarized electrolytic capacitor is shown in Figure 12.19a. The negative (or positive) terminal of a polarized electrolytic capacitor is usually identified as shown in Figure 12.19b.

When replacing a polarized capacitor, care must be taken to match the polarity of the capacitor to that of any dc voltage in the circuit. When inserted correctly (as shown in Figure 12.20a), the polarization of the capacitor matches the polarity of the voltage source. If a polarized electrolytic capacitor is inserted incorrectly (as shown in Figure 12.20b),

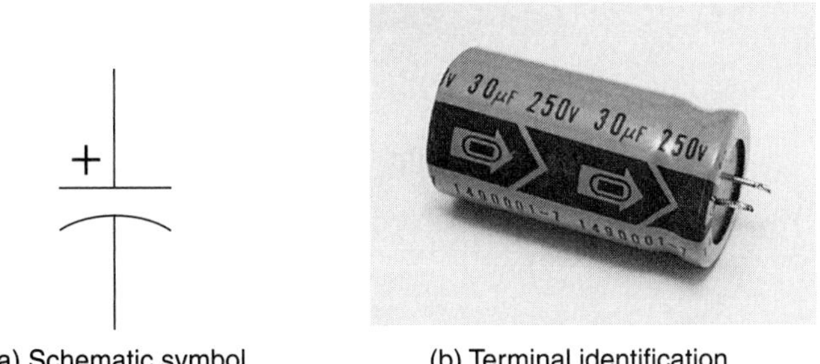

Figure 12.19 Polarized electrolytic capacitors.

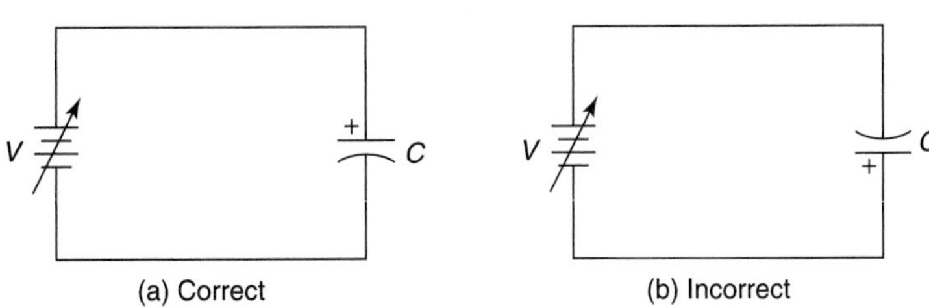

Figure 12.20 Connecting polarized capacitors.

the component will become extremely hot. This condition can be dangerous because the heat generated can cause the electrolyte of the capacitor to break down and burn. The resulting gasses trapped within the metal can eventually cause it to explode.

As a polarized electrolytic heats up, it produces a strong burning odor. If you smell such an odor, immediately remove power from the circuit. Be sure to allow several minutes for the component to cool before attempting to handle it.

Variable Capacitors

A variety of variable capacitors allow the user to adjust for specific values of capacitance. At this point, we're going to briefly discuss several types of variable capacitors and their applications.

Interleaved-plate capacitors *Interleaved-plate capacitors* are found primarily in high-power systems and older consumer electronics systems, such as stereos and televisions. An **interleaved-plate capacitor** *consists of two groups of plates and an air dielectric,* as shown in Figure 12.21a.

The capacitor is divided into two parts: the *stator* and the *rotor.* When the rotor is adjusted, the rotor plates cut into the spaces between the stator plates. The air between the plates is the dielectric. The more interleaved the plates, the greater the effective plate area and, thus, the greater the capacitance.

Ganged capacitors Interleaved-plate capacitors are often **ganged,** meaning that *two or more capacitors share a common rotor shaft.* Two ganged interleaved-plate capacitors are shown in Figure 12.22a. When the rotor shaft is turned, the two capacitors experience the same change in capacitance. In a schematic, ganged capacitors are identified as shown in Figure 12.22b.

Transmitter-type interleaved capacitors These components are designed specifically for high-power applications. The *vanes* (metal plates) are spaced

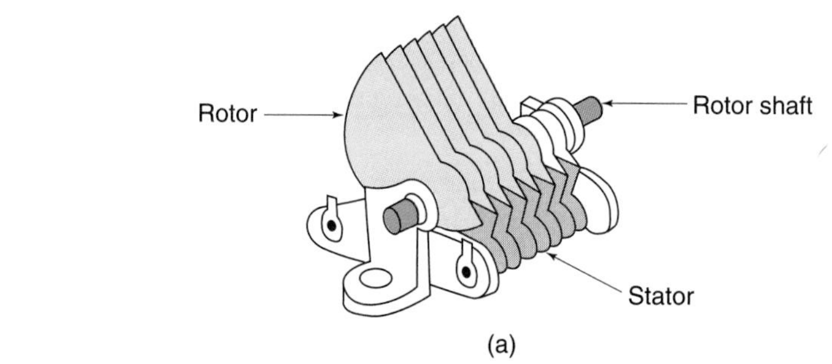

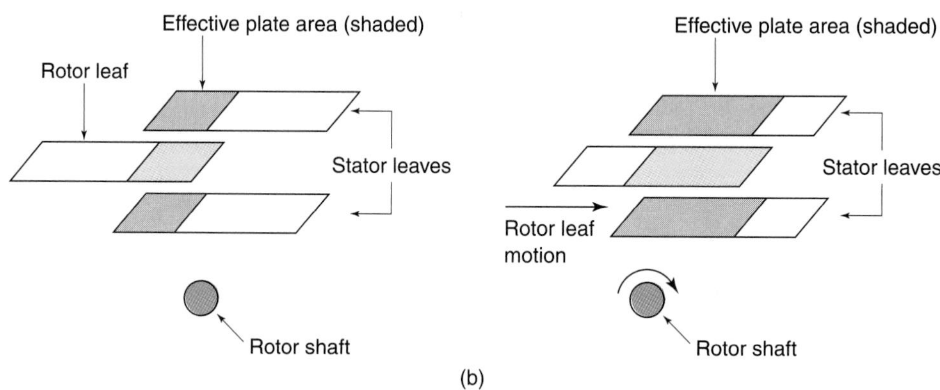

Figure 12.21 An interleaved plate capacitor.

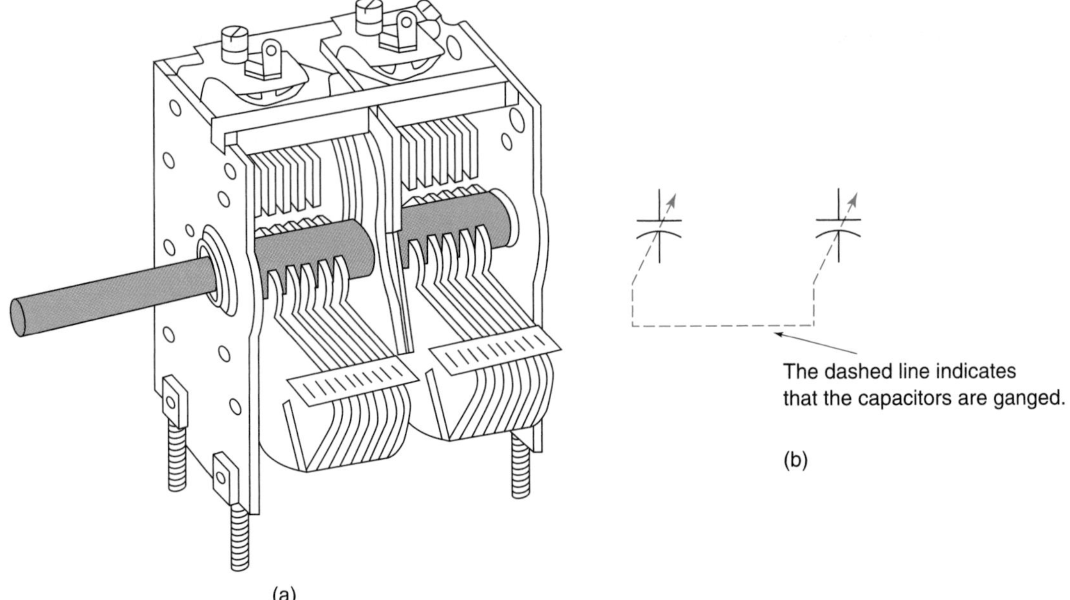

Figure 12.22 Ganged capacitors.

further apart than they are for a common interleaved-plate capacitor. Oil dielectrics may be used to provide higher capacity and breakdown voltage ratings.

Variable precision capacitors These capacitors are used primarily as lab standard components and in measurement instruments. Normally, they have a maximum capacity of 5000 pF or less. The capacity ratings have tolerances as low as 0.02%.

Trimmer capacitors **Trimmer capacitors** are *low-value components that are used to make fine adjustments to the total capacitance in a circuit.* For example, a trimmer capacitor may be connected as shown in Figure 12.23. When connected as shown, the total capacitance between points A and B equals the sum of the capacitor values. By adjusting the trimmer capacitor, the total capacitance can be varied to any value between 100.5 pF and 110 pF.

Trimmer capacitors normally have values in the low picofarad range. Most often, their dielectrics are air, mica, or ceramic. Typically, trimmer capacitors are used in circuits that operate at frequencies up to (and including) the radio-frequency range.

A Practical Consideration: Trimmer capacitors are normally adjusted using a plastic alignment tool in place of a screwdriver, because a metal screwdriver can actually affect the value of a low-capacity (pF) component.

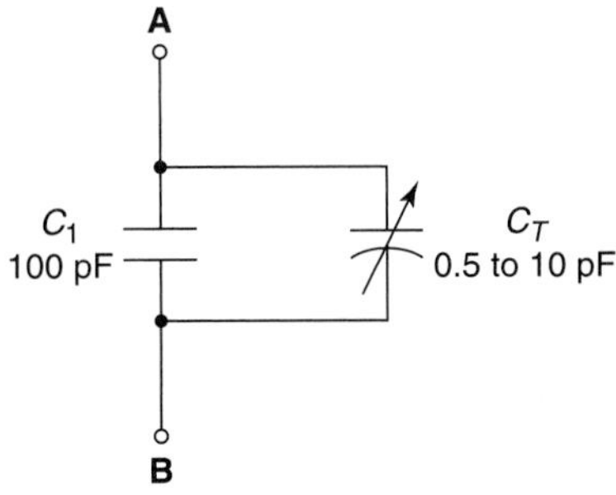

Figure 12.23 A parallel-connected trimmer capacitor.

Section Review

1. What does the *breakdown voltage rating* of a capacitor indicate?
2. What is *leakage current*?

3. List the characteristics of the *ideal* capacitor.

4. How is the *quality* (Q) of a capacitor calculated?

5. What is the relationship between capacitor Q and operating frequency?

6. What is *dielectric absorption*?

7. Explain how dielectric absorption affects the discharge time of a capacitor?

8. Why are electrolytic capacitors easy to identify?

9. What is the advantage that electrolytic capacitors have over other types of capacitors?

10. List the drawbacks of using electrolytic capacitors.

11. What precautions must be taken when working with *polarized* electrolytic capacitors?

12. What are *interleaved-plate capacitors*?

13. What are *ganged capacitors*?

14. What are *trimmer capacitors* used for?

Critical Thinking

15. What is the relationship between capacitor Q and operating frequency? [*Hint:* Use equation (12.12) as a starting point.]

Key Terms

The following terms were introduced in this chapter on the pages indicated:

Breakdown voltage rating, 367
Capacitance, 353
Capacitive reactance (X_C), 365
Capacitor, 353
Capacity, 355
Coupling, 357
Dielectric, 353
Dielectric absorption, 368
Electrolytic capacitor, 368
Farad (F), 355
Ganged, 370
Interleaved-plate capacitor, 370
Isolation, 357
Leakage current, 367
Permittivity, 356
Plates, 353
Relative permeability, 357
Trimmer capacitor, 371

Practice Problems

1. A capacitor is storing 200 mC of charge when the difference of potential across its plates is 25 V. Calculate the value of the component.

2. A capacitor is storing 100 μC of charge when the difference of potential across its plates is 40 V. Calculate the value of the component.

3. Determine the charge that a 22 μF capacitor stores when the difference of potential across its plates is 18 V.

4. Determine the charge that a 100 pF capacitor stores when the difference of potential across its plates is 25 V.

5. Calculate the total capacitance in the series circuit shown in Figure 12.24a.

Figure 12.24

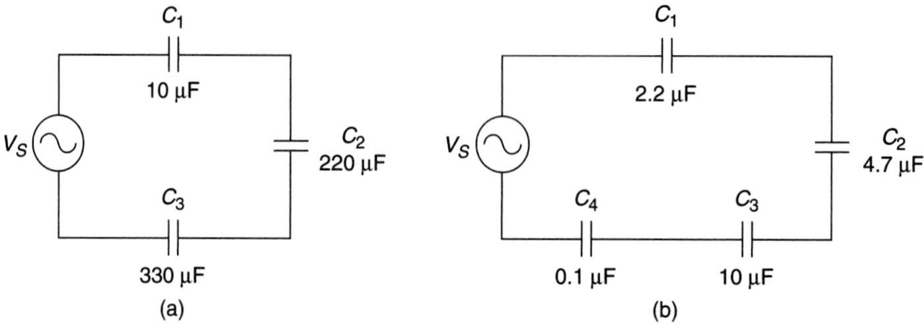

6. Calculate the total capacitance in the series circuit shown in Figure 12.24b.
7. Calculate the total capacitance in the series circuit shown in Figure 12.25a.
8. Calculate the total capacitance in the series circuit shown in Figure 12.25b.

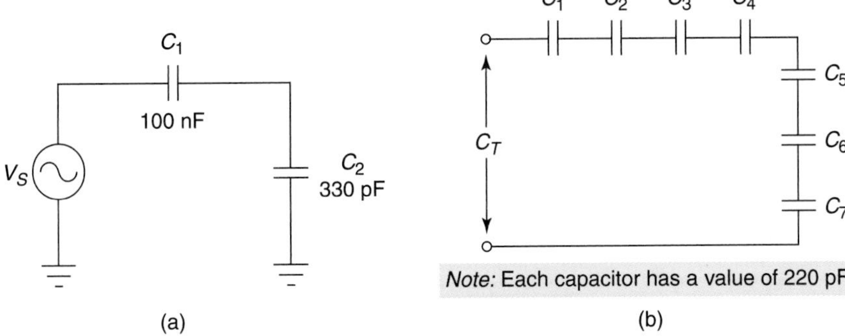

Figure 12.25

9. Calculate the total capacitance in the parallel circuit shown in Figure 12.26a.
10. Calculate the total capacitance in the parallel circuit shown in Figure 12.26b.

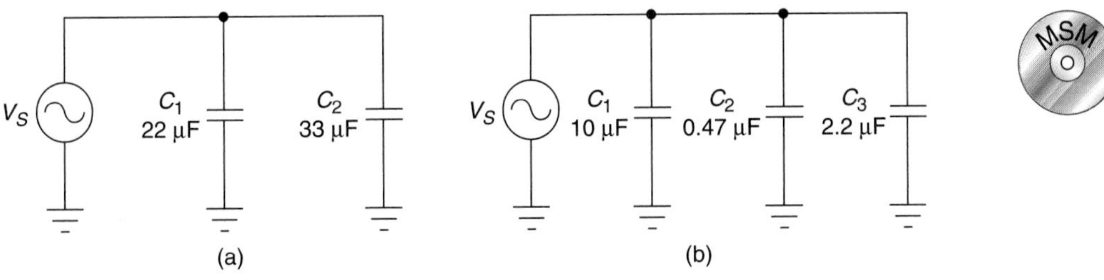

Figure 12.26

11. Calculate the value of X_C for the circuit shown in Figure 12.27a.
12. Calculate the value of X_C for the circuit shown in Figure 12.27b.

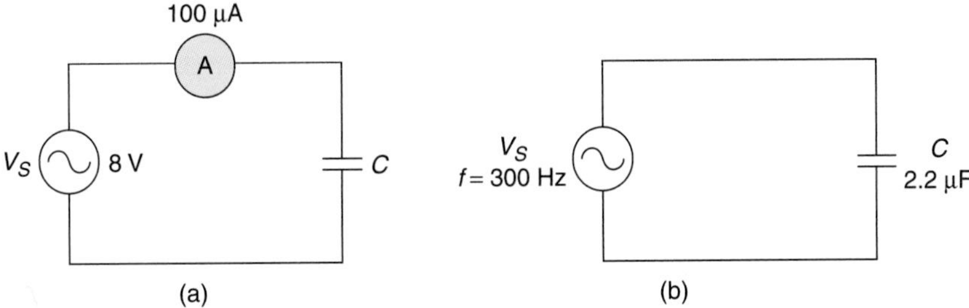

Figure 12.27

13. Calculate the value of X_C for the circuit shown in Figure 12.28a.
14. Calculate the value of X_C for the circuit shown in Figure 12.28b.

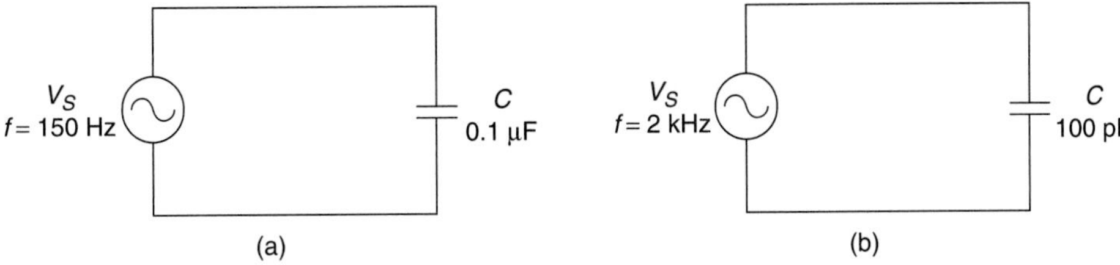

Figure 12.28

15. Calculate the value of total current for the circuit shown in Figure 12.29a.

16. Calculate the value of total current for the circuit shown in Figure 12.29b.

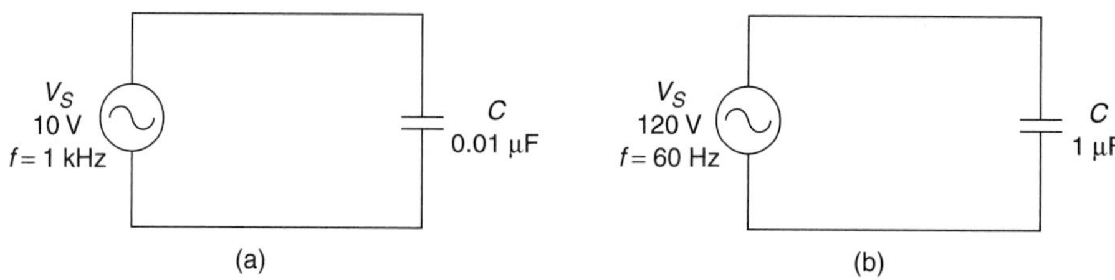

Figure 12.29

17. Calculate the total reactance in the circuit shown in Figure 12.30a.

18. Calculate the total reactance in the circuit shown in Figure 12.30b.

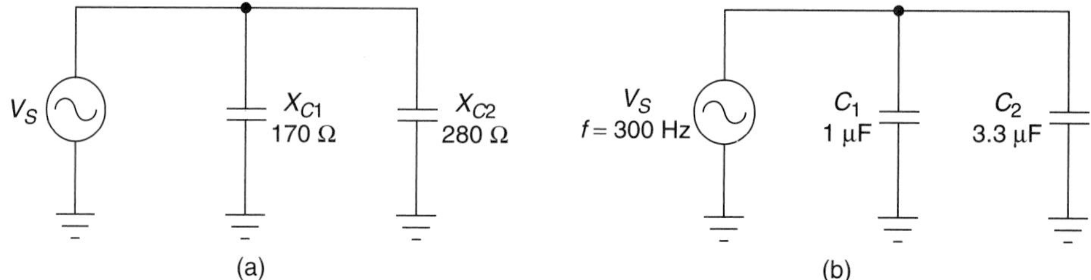

Figure 12.30

19. Determine whether or not a 50 V capacitor can be used in the circuit shown in Figure 12.31a.

20. Determine whether or not a 100 V capacitor can be used in the circuit shown in Figure 12.31b.

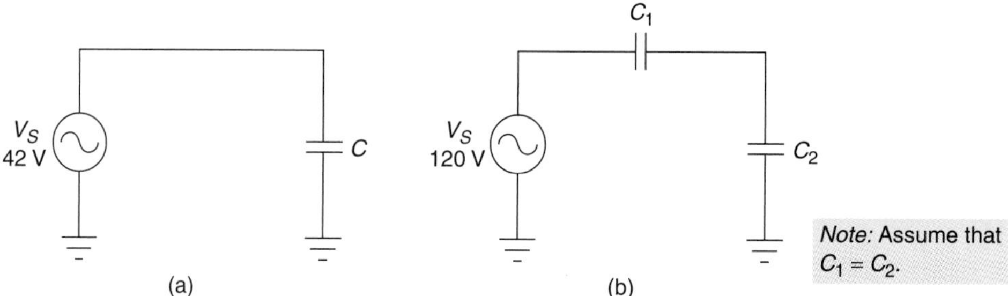

Figure 12.31

21. The capacitor shown in Figure 12.32a has a dielectric resistance of 500 MΩ. Calculate the Q of the component.

22. The capacitor shown in Figure 12.32b has a dielectric resistance of 180 MΩ. Calculate the Q of the component.

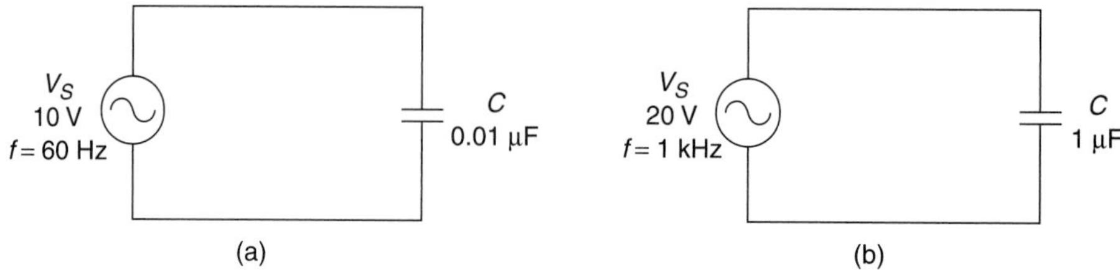

Figure 12.32

23. The source in Figure 12.33a has the frequency range shown. Calculate the range of Q for the circuit, assuming that the capacitor has a dielectric resistance of 1000 MΩ.

24. The source in Figure 12.33b has the frequency range shown. Calculate the range of Q for the circuit, assuming that the capacitor has a dielectric resistance of 200 MΩ.

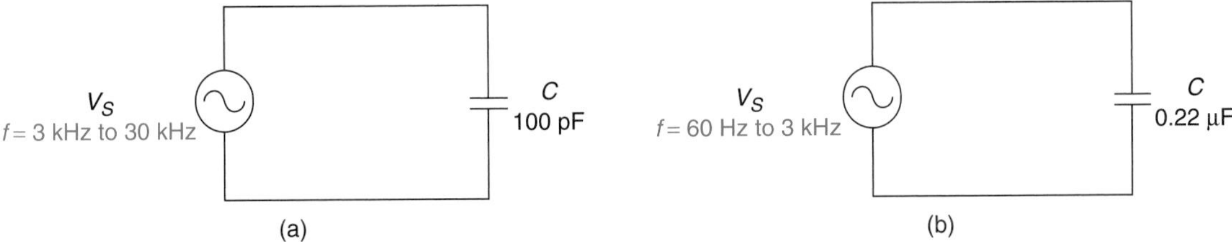

Figure 12.33

Looking Back

These problems relate to material presented in earlier chapters. The chapters are identified in brackets.

25. Convert each of the following values to engineering notation form. [Introduction]
   a. 1.32×10^5 Hz = ______
   b. 3.5×10^7 Ω = ______
   c. 8.2×10^{-8} H = ______
   d. 6.45×10^{-4} s = ______

26. A sine wave has the following values: T_C = 200 μs, and V_{pk} = 15 V. Determine the instantaneous voltage when t = 120 μs. [Chapter 9]

27. Calculate the energy used (in kWh) to run a 500 W stereo and a 2400 W air conditioner for 3 hours. [Chapter 3]

28. A 0.33 mH inductor is operated at a frequency of 15 kHz. Calculate the current through the component when V_L = 5 V. [Chapter 10]

29. A 5 W resistor has a value of 150 Ω. Assuming that the power cannot exceed 80% of the rated limit, determine the maximum allowable value of resistor voltage. [Chapter 3]

30. A magnet with a cross-sectional area of 3 cm^2 generates 1200 μWb of flux at its poles. Calculate the flux density at the poles of the magnet. [Chapter 8]

Pushing the Envelope

31. The relative permittivity of a material is the ratio of its permittivity to that of free space. The relative permittivity ratings of several common dielectrics are as follows:

Dielectric	*Relative Permittivity*
Air	1.001
Glass	7.500
Mica	5.000
Porcelain	6.000
Teflon	2.000

Determine which of these dielectrics is used in the capacitor shown in Figure 12.34.

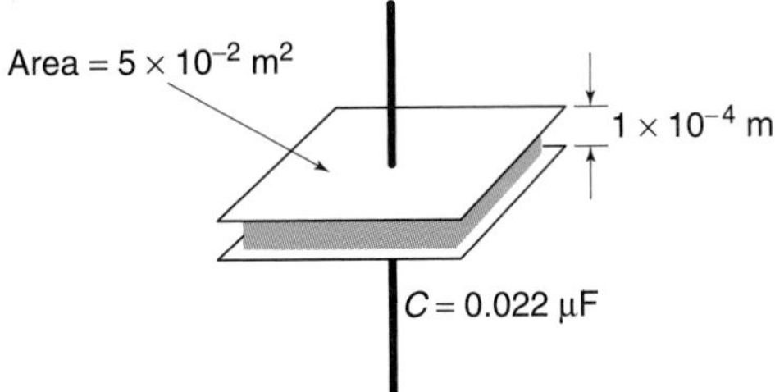

Figure 12.34

Answers to the Example Practice Problems

12.1	673.5 nF	12.3	603 Ω
12.2	3.3 μF	12.4	56.5 mA

Multisim File References

Figure	File Reference
12.9	ETF12_9.msm
12.15	ETF12_15.msm
12.18	ETF12_18.msm
12.26	ETF12_26.msm

chapter 13

Resistive-Capacitive (*RC*) Circuits

Objectives

After studying the material in this chapter, you should be able to:

1. Compare and contrast series *resistive, capacitive,* and *resistive-capacitive* (*RC*) *circuits.*
2. Describe the voltage and phase angle characteristics for a series *RC* circuit.
3. Describe the impedance and current characteristics of a series *RC* circuit.
4. Calculate the impedance, voltage, and current values for any series *RC* circuit.
5. Describe and analyze the frequency response of a series *RC* circuit.
6. Perform all relevant power calculations for a series *RC* circuit.
7. Compare and contrast parallel *resistive, capacitive,* and *resistive-capacitive* (*RC*) *circuits.*
8. Calculate any current or impedance value for a parallel *RC* circuit.
9. Describe and analyze the frequency response of a parallel *RC* circuit.
10. Calculate the impedance, voltage, and current values for any series-parallel *RC* circuit.

Capacitors (like inductors) are typically used in conjunction with resistors. When a circuit contains both capacitors and resistors, it is referred to as a *resistive-capacitive (RC) circuit.* As you will see, the principles of *RC* circuits are very similar to those of the *RL* circuits that we discussed in Chapter 11.

13.1 Series RC Circuits

A **resistive-capacitive (RC) circuit** is *one that contains any combination of resistors and capacitors.* Several examples of basic RC circuits are shown in Figure 13.1.

Series Circuits: A Comparison

OBJECTIVE 1 ➤ We have already established the fundamental characteristics of resistive and capacitive series circuits. These characteristics are as follows:

- Current is the same at all points throughout the circuit.
- The applied voltage equals the sum of the individual component voltages.
- The total impedance (resistance or reactance) equals the sum of the individual component *values.*

These relationships are provided (as equations) in Figure 13.2.

The current, voltage, and impedance characteristics of series *RC* circuits are listed in Figure 13.2. As you can see, current is the same at all points throughout the circuit (as is the case for the resistive and capactivie circuits). However:

- The applied voltage (V_S) equals the *geometric sum* of V_C and V_R.
- The total opposition to current (Z_T) equals the *geometric sum* of X_C and R.

These series voltage and impedance relationships are nearly identical to those for the series *RL* circuit. The voltage and impedance relationships for series *RC* and series *RL* circuits are compared in Figure 13.3.

Series Voltages

OBJECTIVE 2 ➤ As you learned in Chapter 12, current leads voltage by 90° in a capacitive circuit. In a resistive circuit, voltage and current are always in phase. When we connect a capacitor

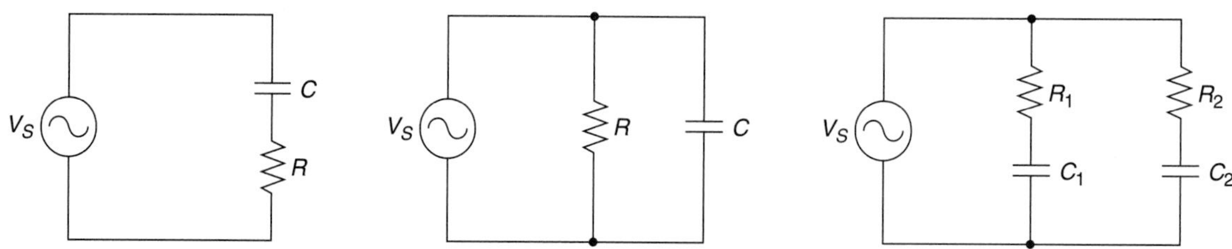

Figure 13.1

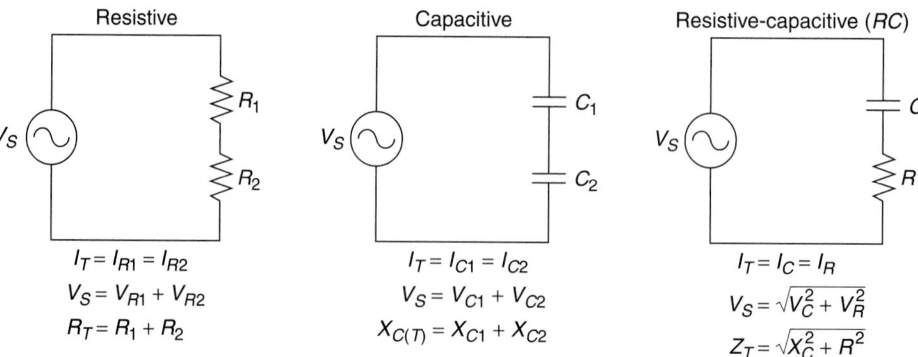

Figure 13.2

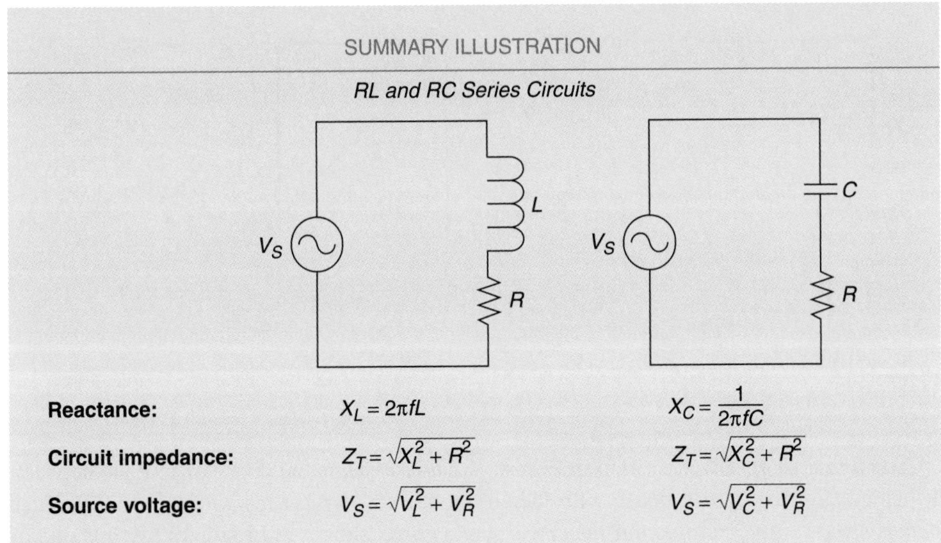

Figure 13.3

and a resistor in series with a voltage source, we get the circuit and waveforms shown in Figure 13.4. Since current is constant throughout a series circuit, the capacitor and resistor voltage waveforms are referenced to a single current waveform. As you can see:

- The capacitor voltage *lags* the circuit current by 90°.
- The resistor voltage is *in phase* with the circuit current.

Therefore, *capacitor voltage lags resistor voltage by 90°*. By plotting the value of V_R on the x-axis (0°) and the value of V_C on the negative y-axis (−90°), the phase relationship between the two is represented as shown in Figure 13.5a.

As in any series circuit, the source voltage in a series *RC* circuit must equal the sum of V_C and V_R. However, as shown in Figure 13.5, these voltages must be added geometrically. By formula,

$$V_S = \sqrt{V_C^2 + V_R^2} \tag{13.1}$$

Note that the V_S phasor represents the geometric sum of V_R and V_C.

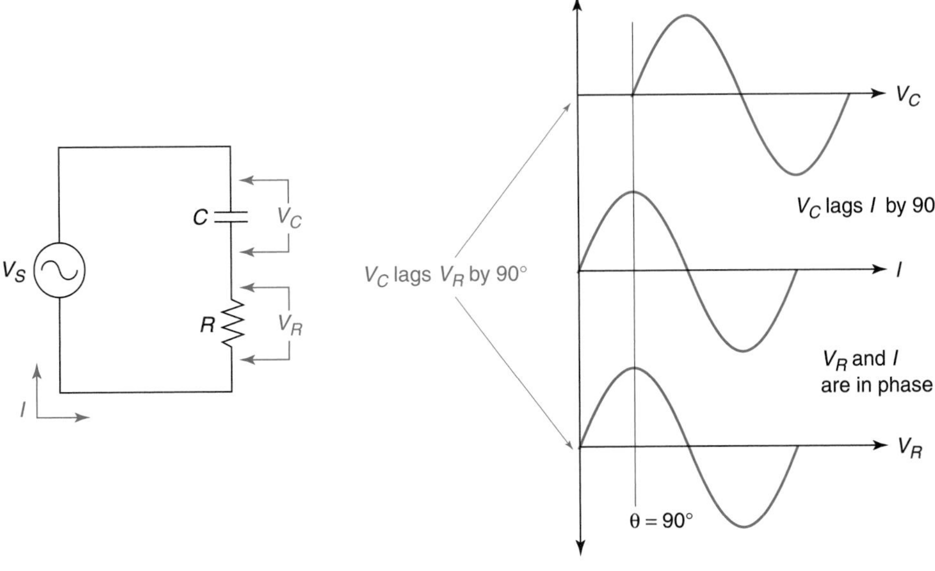

Figure 13.4

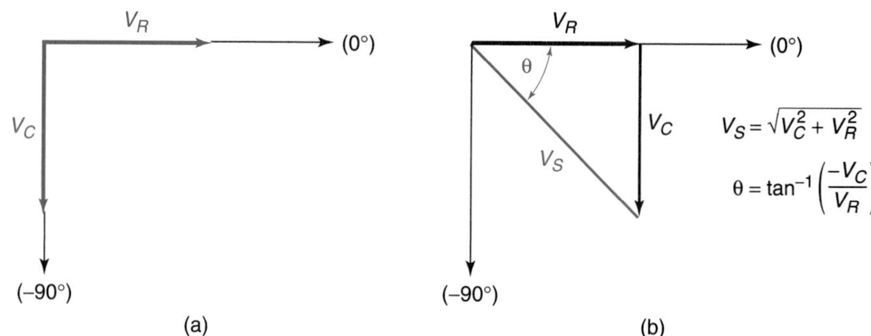

Figure 13.5

The *phase angle* (θ) identified in Figure 13.5b represents the phase difference between V_S and V_R. Since V_R is in phase with the circuit current, *θ represents the phase difference between the source voltage and the circuit current.* As shown in Figure 13.5b, θ is one of three angles in a right triangle. In Appendix F, this angle is shown to have a value of

$$\theta = \tan^{-1}\left(\frac{y}{x}\right)$$

where

$\tan^{-1}$ = the *inverse tangent* of the fraction
y = the magnitude of the y-axis value
x = the magnitude of the x-axis value

Relating this equation to Figure 13.5b, the phase angle of the source voltage can be found as

$$\theta = \tan^{-1}\left(\frac{-V_C}{V_R}\right) \quad (13.2)$$

Note that the *negative* value of V_C in equation (13.2) reflects the fact that its phasor falls on the *negative* y-axis (as shown in Figure 13.4). Example 13.1 demonstrates the calculation of source voltage for a series *RC* circuit.

Example 13.1

Note: Here's a quick-check for the result of any geometric addition: The result should be greater than either value and less than the algebraic sum of the two. For example, the result in Example 13.1 is 18 V, which is greater than either component voltage and less than the algebraic sum of the two (25 V).

The component voltages shown in Figure 13.6 were measured using an ac voltmeter. Calculate the source voltage for the circuit.

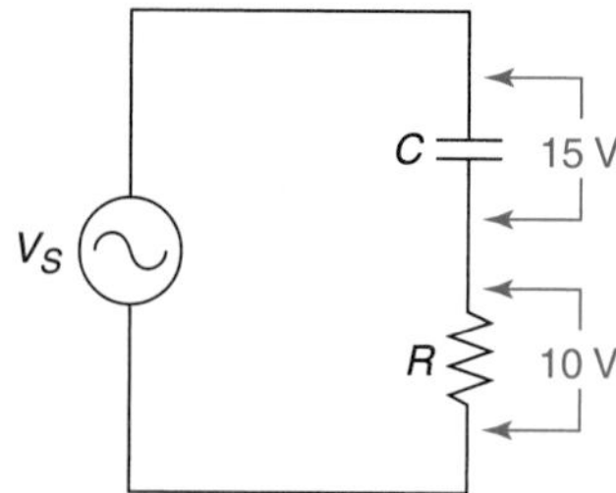

Figure 13.6

Solution: The source voltage has a magnitude of

$$V_S = \sqrt{V_C^2 + V_R^2} = \sqrt{(15\text{ V})^2 + (10\text{ V})^2} = 18\text{ V}$$

at a phase angle of

$$\theta = \tan^{-1}\left(\frac{-V_C}{V_R}\right) = \tan^{-1}\left(\frac{-15\text{ V}}{10\text{ V}}\right) = \tan^{-1}(-1.5) = -56.3°$$

Practice Problem 13.1

A circuit like the one shown in Figure 13.6 has measured values of $V_C = 20$ V and $V_R = 25$ V. Calculate the magnitude and phase angle of the source voltage.

The *negative phase angle* in Example 13.1 indicates that *the source voltage lags the circuit current* by 56.3°. For any *RC* circuit, the limits on the phase angle between the source voltage and circuit current are given as follows:

$$-90° < \theta < 0° \tag{13.3}$$

Remember: In a mathematical relationship involving two negative numbers, the less-negative number is considered to be the greater of the two. This is why θ is written as being greater than −90° in equation (13.3).

This range of values for θ makes sense when you consider that:

- The phase angle of a purely resistive circuit is 0°.
- The phase angle of the source voltage (relative to the circuit current) in a purely capacitive circuit is −90°.

A circuit containing both resistance and capacitance is neither purely resistive nor purely capacitive. Therefore, it must have a voltage phase angle that falls between these two extremes.

Series Impedance

◄ *OBJECTIVE 3*

In an *RC* circuit, the total impedance is made up of resistance (*R*) and capacitive reactance (X_C). As with series component voltages, these values must be added geometrically. This is the same approach we use to solve for Z_T in *RL* series circuits, with one important difference: the phase angle of X_C. The magnitude and phase angle of Z_T for a series *RC* circuit are found using

$$Z_T = \sqrt{X_C^2 + R^2} \tag{13.4}$$

and

$$\theta = \tan^{-1}\left(\frac{-X_C}{R}\right) \tag{13.5}$$

Here's why we use a *negative* value of X_C in equation (13.4). We know that we can solve for X_C using Ohm's law, as follows:

$$X_C = \frac{V_C\angle -90°}{I_C\angle 0°} = \frac{V_C}{I_C}\angle -90°$$

While this result does not include a magnitude, it does indicate that capacitive reactance has a phase angle of −90° (relative to the circuit current). Resistance, as always, has no phase angle (θ = 0°). As a result, the *R* and X_C phasors are plotted as shown in Figure 13.7. As you can see, X_C is written as a negative value in the equation to indicate its phase relationship with *R*.

Example 13.2 demonstrates the procedure for determining the total impedance in a series RC circuit.

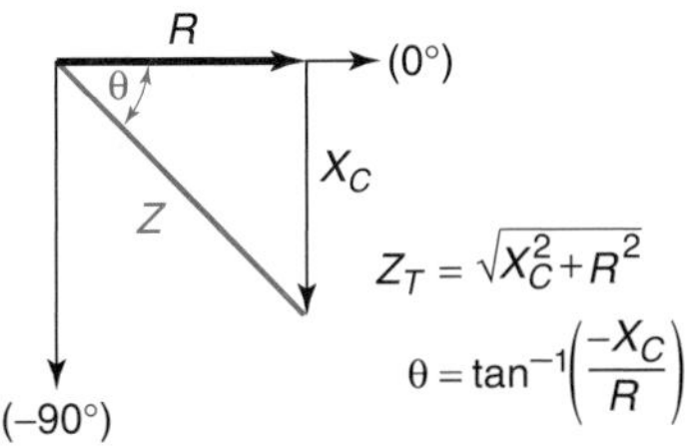

Figure 13.7

Example 13.2

Calculate the total impedance for the circuit shown in Figure 13.8.

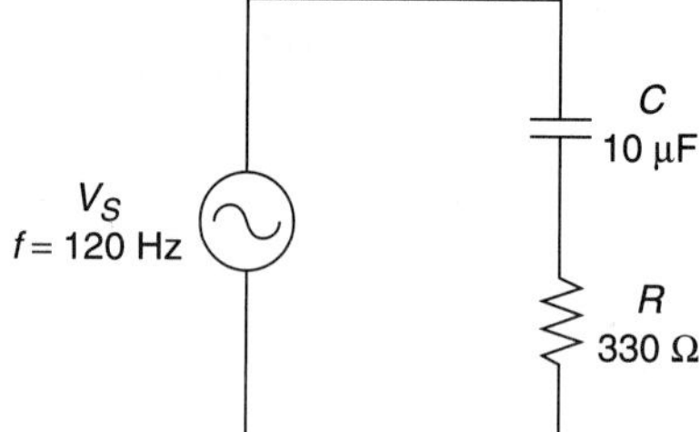

Figure 13.8

Solution: First, the value of X_C is found as

$$X_C = \frac{1}{2\pi fC} = \frac{1}{2\pi(120\text{ Hz})(10\text{ μF})} \cong 133\ \Omega$$

The circuit impedance can now be found as

$$Z_T = \sqrt{X_C^2 + R^2} = \sqrt{(133\ \Omega)^2 + (330\ \Omega)^2} \cong 356\ \Omega$$

at a phase angle of

$$\theta = \tan^{-1}\left(\frac{-X_C}{R}\right) = \tan^{-1}\left(\frac{-133\ \Omega}{330\ \Omega}\right) = \tan^{-1}(-0.403) = -22°$$

This result indicates that the circuit impedance (356 Ω) lags the circuit current by 22°.

Practice Problem 13.2

A circuit like the one shown in Figure 13.8 has the following values: $f = 300$ Hz, $C = 0.47$ μF, and $R = 1$ kΩ. Calculate the circuit impedance.

Resolving Series *RC* Circuit Phase Angles

Like numeric values, any group of phase angles associated with a given circuit must resolve themselves when the circuit analysis is complete. In other words, *the phase angles must combine in such a way that every circuit value conforms to the basic laws of circuit operation,* such as Ohm's law and Kirchhoff's laws. With this in mind, take a look at the circuit shown in Figure 13.9.

The circuit shown in the figure is basically the same one that we analyzed in Example 13.2. For the sake of discussion, a circuit current of 20 mA has been assumed, and Ohm's law has been used to calculate the resulting component voltages. According to equations (13.1) and (13.2),

$$V_S = \sqrt{V_C^2 + V_R^2} = \sqrt{(2.66\text{ V})^2 + (6.60\text{ V})^2} = 7.12\text{ V}$$

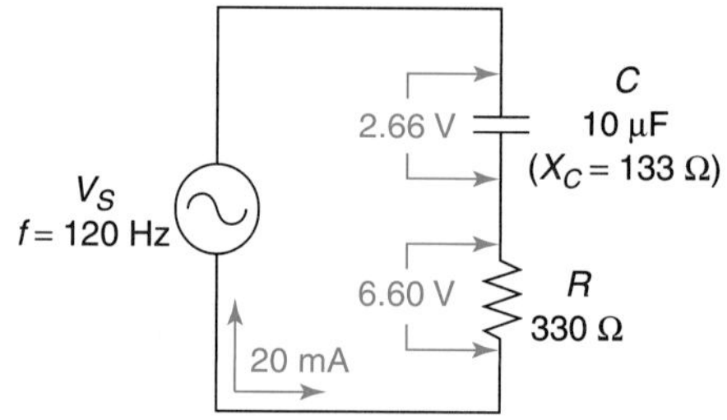

Figure 13.9

and

$$\theta = \tan^{-1}\left(\frac{-V_C}{V_R}\right) = \tan^{-1}\left(\frac{-2.66\ \text{V}}{6.6\ \text{V}}\right) = \tan^{-1}(-0.403) = -22°$$

Therefore, the source voltage has a value of 7.12 V$\angle -22°$. In Example 13.2, we calculated a total circuit impedance of 356 $\Omega\angle -22°$. When we use these two values, Ohm's law provides a circuit current of

$$I = \frac{V_S}{Z_T} = \frac{7.12\ \text{V}\angle -22°}{356\ \Omega\angle -22°} = 20\ \text{mA}\angle 0°$$

which agrees with our assumed value of circuit current. As you can see, the phase angles for the voltage, current, and impedance values all conform to basic circuit principles. This is always the case when a circuit is properly analyzed.

Calculating Series *RC* Circuit Current

When you use Ohm's law to calculate the total current in a series *RC* circuit, you may run into a situation where it appears that the circuit current has a *positive* phase angle. For example, consider the circuit shown in Figure 13.10. The circuit has a total impedance of

$$Z_T = \sqrt{X_C^2 + R^2} = \sqrt{(500\ \Omega)^2 + (150\ \Omega)^2} = 522\ \Omega$$

at a phase angle of

$$\theta = \tan^{-1}\left(\frac{-X_C}{R}\right) = \tan^{-1}\left(\frac{-500\ \Omega}{150\ \Omega}\right) = \tan^{-1}(-3.333) = -73.3°$$

With a circuit impedance of 522 $\Omega\angle -73.3°$, Ohm's law provides us with a circuit current of

$$I = \frac{V_S}{Z_T} = \frac{20\ \text{V}\angle 0°}{522\ \Omega\angle -73.3°} = 38.3\ \text{mA}\angle 73.3°$$

The positive phase angle in our result indicates that the circuit current *leads* the source voltage by 73.3°. This is the same thing as saying that the source voltage *lags* the circuit current by 73.3°. Therefore, we could express the phase relationship for this circuit in either of two ways:

$$\begin{aligned} V_S &= 20\ \text{V}\angle 0° \\ I &= 38.3\ \text{mA}\angle 73.3° \end{aligned} \qquad \text{or} \qquad \begin{aligned} V_S &= 20\ \text{V}\angle -73.3° \\ I &= 38.3\ \text{mA}\angle 0° \end{aligned}$$

The first set of values uses the source voltage as the phase angle reference, while the second uses the circuit current as the reference. Since current is constant throughout a

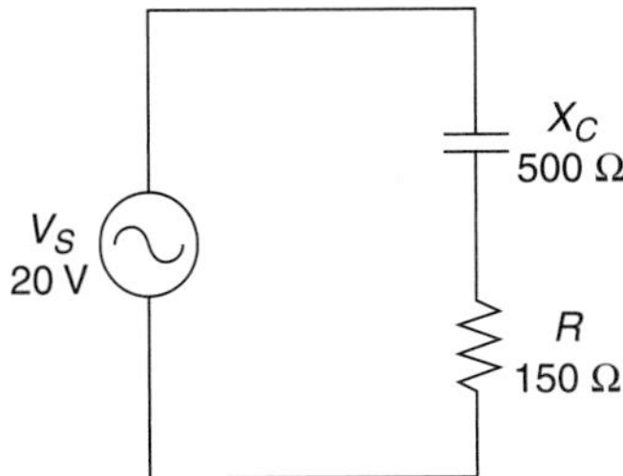

Figure 13.10

series circuit, using current as the 0° reference is the preferred method. Therefore, we adjust our results to be written as

$$V_S = 20\ \text{V}\angle{-73.3^\circ}$$
$$I = 38.3\ \text{mA}\angle 0^\circ$$

Series Circuit Analysis

OBJECTIVE 4 ➤ The basic analysis of a simple series *RC* circuit is performed as demonstrated in Example 13.3.

Example 13.3

Determine the voltage, current, and impedance values for the circuit shown in Figure 13.11.

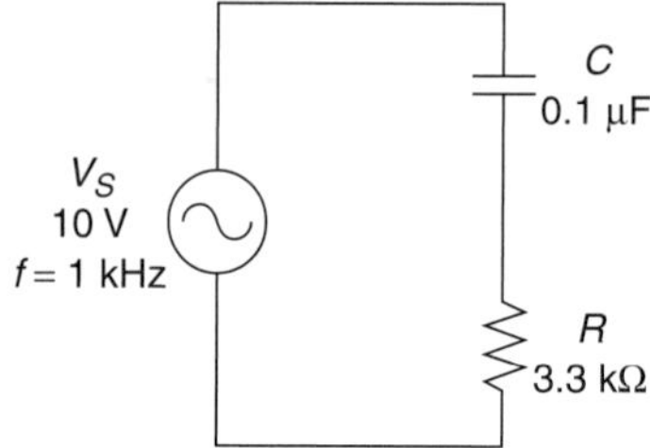

Figure 13.11

Solution: First, the value of X_C is found as

$$X_C = \frac{1}{2\pi fC} = \frac{1}{2\pi(1\ \text{kHz})(0.1\ \mu\text{F})} = 1.59\ \text{k}\Omega$$

The circuit impedance can now be found as

$$Z_T = \sqrt{X_C^2 + R^2} = \sqrt{(1.59\ \text{k}\Omega)^2 + (3.3\ \text{k}\Omega)^2} = 3.66\ \text{k}\Omega$$

at a phase angle of

$$\theta = \tan^{-1}\left(\frac{-X_C}{R}\right) = \tan^{-1}\left(\frac{-1.59\ \text{k}\Omega}{3.3\ \text{k}\Omega}\right) = \tan^{-1}(-0.4818) = -25.7^\circ$$

Using the values of Z_T and V_S, the circuit current is found as

$$I = \frac{V_S}{Z_T} = \frac{10\ \text{V}\angle 0^\circ}{3.66\ \text{k}\Omega\angle{-25.7^\circ}} = 2.73\ \text{mA}\angle 25.7^\circ$$

Since current is generally used as the 0° reference in a series circuit, the values of V_S and I are written as

$$V_S = 10\ \text{V}\angle{-25.7^\circ} \qquad \text{and} \qquad I = 2.73\ \text{mA}\angle 0^\circ$$

Now we can calculate V_C and V_R as follows:

$$V_C = I \cdot X_C = (2.73\ \text{mA})(1.59\ \text{k}\Omega\angle{-90^\circ}) = 4.34\ \text{V}\angle{-90^\circ}$$

and

$$V_R = I \cdot R = (2.73\ \text{mA})(3.3\ \text{k}\Omega) = 9.01\ \text{V}$$

Practice Problem 13.3

A circuit like the one shown in Figure 13.11 has the following values: $V_S = 12$ V, $f = 1.5$ kHz, $C = 0.22$ μF, and $R = 510$ Ω. Determine the voltage, current, and impedance values for the circuit.

There are always instances when we are interested only in calculating the component voltages in a series *RC* circuit. We can calculate the component voltages using the voltage-divider equation, just as we did with series *RL* circuits. For a series *RC* circuit, the voltage-divider equation is given as

$$V_n = V_S \frac{Z_n}{Z_T} \tag{13.6}$$

where

Z_n = the magnitude of R or X_C

Z_T = the geometric sum of R and X_C

Example 13.4 demonstrates the use of the voltage-divider equation in the analysis of a series *RC* circuit with two series capacitors. When a series circuit contains more than one capacitor and/or resistor, the analysis of the circuit requires a few more steps, but the technique is still the same.

Example 13.4

Calculate the values of Z_T, V_{C1}, V_{C2}, and V_R for the circuit shown in Figure 13.12.

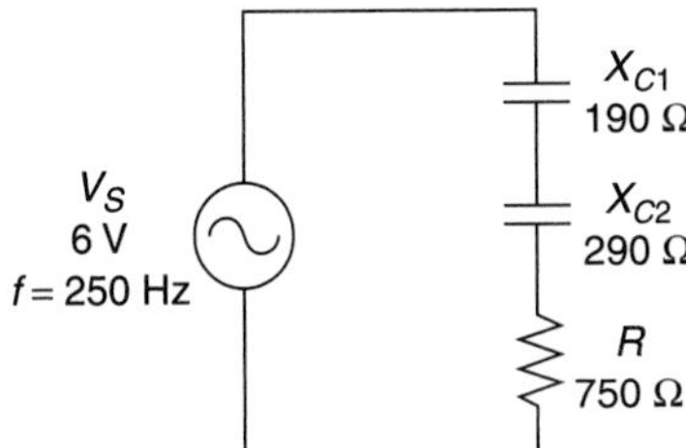

Figure 13.12

Solution: Since X_{C1} and X_{C2} are in phase, the total reactance equals the algebraic sum of their values, as follows:

$$X_{C(T)} = X_{C1} + X_{C2} = 190\ \Omega + 290\ \Omega = 480\ \Omega$$

Using the values of $X_{C(T)}$ and R, the magnitude of the circuit impedance is found to have a value of

$$Z_T = \sqrt{X_{C(T)}^2 + R^2} = \sqrt{(480\ \Omega)^2 + (750\ \Omega)^2} = 890\ \Omega$$

at a phase angle of

$$\theta = \tan^{-1}\left(\frac{-X_C}{R}\right) = \tan^{-1}\left(\frac{-4.80\ \Omega}{750\ \Omega}\right) = \tan^{-1}(-0.64) = -32.6°$$

We can now use the voltage-divider equation to calculate the magnitudes of the component voltages, as follows:

$$V_{C1} = V_S \frac{X_{C1}}{Z_T} = (6\ \text{V})\frac{190\ \Omega}{890\ \Omega} = 1.28\ \text{V}$$

$$V_{C2} = V_S \frac{X_{C2}}{Z_T} = (6\ \text{V})\frac{290\ \Omega}{890\ \Omega} = 1.96\ \text{V}$$

Note: Since the phase angles of V_S and Z_T are equal, they can be left out of the voltage-divider calculations without losing accuracy in the results.

and

$$V_R = V_S \frac{R}{Z_T} = (6\ \text{V})\frac{750\ \Omega}{890\ \Omega} = 5.06\ \text{V}$$

Practice Problem 13.4

A circuit like the one shown in Figure 13.12 has the following values: $V_S = 25$ V, $f = 2$ kHz, $C_1 = 47$ nF, $C_2 = 0.1$ μF, and $R = 750$ Ω. Calculate the values of Z_T, V_{C1}, V_{C2}, and V_R for the circuit.

Once again, we can verify our results by using the component voltages to calculate the value of the source voltage. Since V_{C1} and V_{C3} are in phase, their values can be added algebraically, as follows:

$$V_{C(T)} = V_{C1} + V_{C2} = 1.28 \text{ V} + 1.96 \text{ V} = 3.24 \text{ V}$$

Then, using the values of $V_{C(T)}$ and V_R, the magnitude of the source voltage can be found as

$$V_S = \sqrt{V_{C(T)}^2 + V_R^2} = \sqrt{(3.24 \text{ V})^2 + (5.06 \text{ V})^2} = 6.01 \text{ V}$$

Since our result is approximately equal to the given magnitude of V_S, we can assume that all our calculations were correct. When working with circuits containing multiple capacitors and/or resistors, just remember the following:

- Any two (or more) values that are in phase can be added algebraically.
- Any two (or more) values that are 90° out of phase must be added geometrically.

Series *RC* Circuit Frequency Response

OBJECTIVE 5 ➤ In Chapter 11, the term *frequency response* was used to describe any changes that occur in a circuit as a result of a change in operating frequency. For example, consider the circuit shown in Figure 13.13. The following describes the circuit's response to an *increase* in operating frequency:

- The increase in frequency causes X_C to decrease.
- The decrease in X_C causes Z_T to decrease and θ to become less negative.
- The decrease in Z_T causes I_T to increase.
- The decrease in X_C causes V_C to decrease.
- The decrease in V_C causes V_R to increase.

Example 13.5 verifies these cause-and-effect relationships.

Example 13.5

The voltage source in Figure 13.13 has the frequency limits shown. Calculate the circuit voltage, current, and impedance values at the frequency limits of the source.

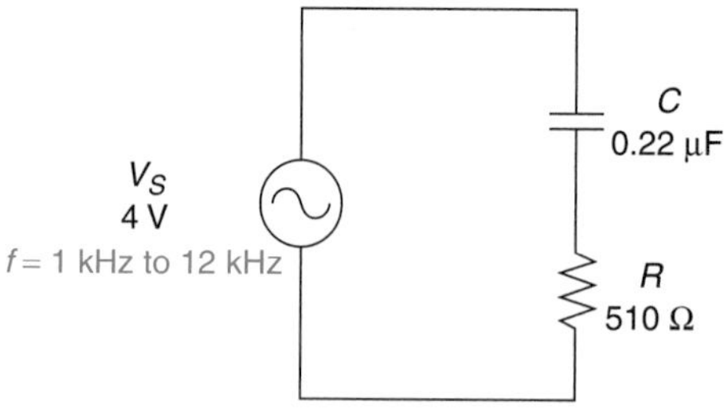

Figure 13.13

Solution: At $f = 1$ kHz, the reactance of the capacitor is found as

$$X_C = \frac{1}{2\pi fC} = \frac{1}{2\pi(1 \text{ kHz})(0.22 \text{ μF})} = 723 \text{ Ω}$$

The circuit impedance has a magnitude of

$$Z_T = \sqrt{X_C^2 + R^2} = \sqrt{(723 \text{ Ω})^2 + (510 \text{ Ω})^2} = 885 \text{ Ω}$$

at a phase angle of

$$\theta = \tan^{-1}\left(\frac{-X_C}{R}\right) = \tan^{-1}\left(\frac{-723\ \Omega}{510\ \Omega}\right) = \tan^{-1}(-1.418) = -54.8°$$

Now the magnitude of the circuit current is found as

$$I_T = \frac{V_S}{Z_T} = \frac{4\ \text{V}}{885\ \Omega} = 4.52\ \text{mA}$$

and the magnitudes of the component voltages are found as

$$V_C = I_T X_C = (4.52\ \text{mA})(723\ \Omega) = 3.27\ \text{V}$$

and

$$V_R = I_T R = (4.52\ \text{mA})(510\ \Omega) = 2.31\ \text{V}$$

When the operating frequency increases to 12 kHz, the reactance of the capacitor decreases to

$$X_C = \frac{1}{2\pi f C} = \frac{1}{2\pi(12\ \text{kHz})(0.22\ \mu\text{F})} = 60.3\ \Omega$$

This decrease in X_C gives us a circuit impedance of

$$Z_T = \sqrt{X_C^2 + R^2} = \sqrt{(60.3\ \Omega)^2 + (510\ \Omega)^2} \cong 514\ \Omega$$

at a phase angle of

$$\theta = \tan^{-1}\left(\frac{-X_C}{R}\right) = \tan^{-1}\left(\frac{-60.3\ \Omega}{510\ \Omega}\right) = \tan^{-1}(-0.1182) = -6.74°$$

The decreased circuit impedance increases the magnitude of the circuit current to

$$I_T = \frac{V_S}{Z_T} = \frac{4\ \text{V}}{514\ \Omega} = 7.78\ \text{mA}$$

Finally, the magnitudes of the component voltages at this operating frequency can be found as

$$V_C = I_T X_C = (7.78\ \text{mA})(60.3\ \Omega) = 469\ \text{mV}$$

and

$$V_R = I_T R = (7.78\ \text{mA})(510\ \Omega) = 3.97\ \text{V}$$

Practice Problem 13.5

A circuit like the one shown in Figure 13.13 has the following values: $V_S = 22$ V, $f = 1.5$ kHz to 6 kHz, $C = 10$ nF, and $R = 2$ kΩ. Calculate the circuit voltage, current, and impedance values at the frequency limits of the source.

When you compare the values obtained in Example 13.5 with our earlier statements, you'll see that the circuit responded to the change in frequency exactly as described. For future reference, the characteristics of series *RC* circuits are compared to those of series *RL* circuits in Figure 13.14.

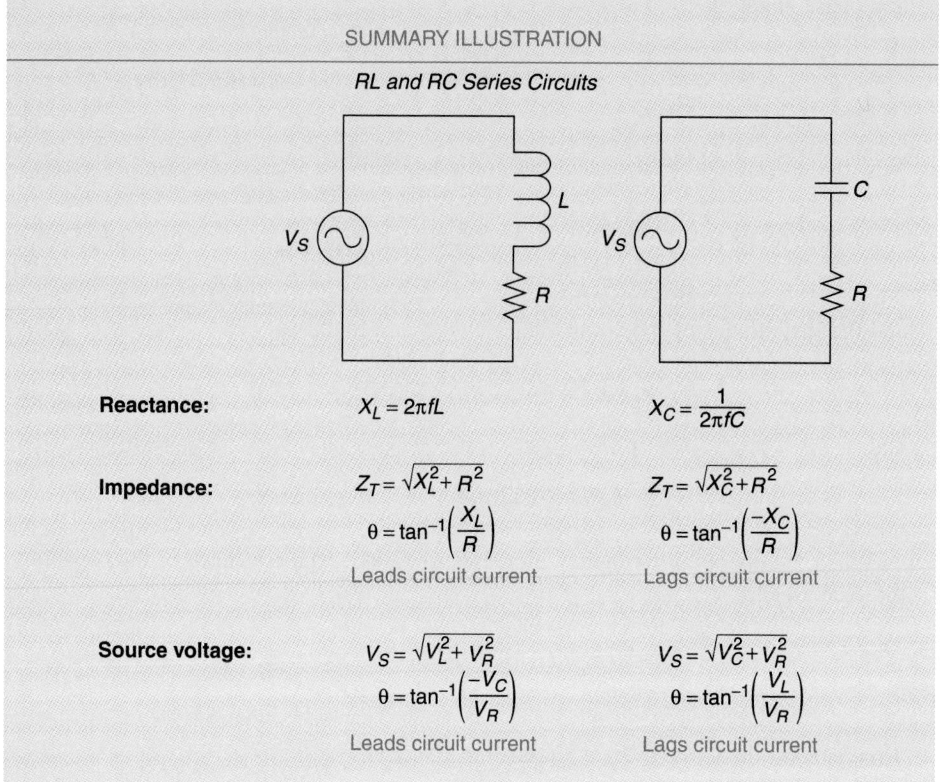

SUMMARY ILLUSTRATION
RL and RC Series Circuits
V_S
L
R
V_S
C
R
Reactance:
$X_L = 2\pi fL$
$X_C = \frac{1}{2\pi fC}$
Impedance:
$Z_T = \sqrt{X_L^2 + R^2}$
$\theta = \tan^{-1}\left(\frac{X_L}{R}\right)$
Leads circuit current
$Z_T = \sqrt{X_C^2 + R^2}$
$\theta = \tan^{-1}\left(\frac{-X_C}{R}\right)$
Lags circuit current
Source voltage:
$V_S = \sqrt{V_L^2 + V_R^2}$
$\theta = \tan^{-1}\left(\frac{-V_C}{V_R}\right)$
Leads circuit current
$V_S = \sqrt{V_C^2 + V_R^2}$
$\theta = \tan^{-1}\left(\frac{V_L}{V_R}\right)$
Lags circuit current

Figure 13.14

Section Review

1. What are the three fundamental characteristics of every series circuit?
2. In a series *RC* circuit, what is the phase relationship between V_C and V_R?
3. In a series *RC* circuit, what does the phase angle of the source voltage represent?
4. In a series *RC* circuit, what is the phase relationship between capacitive reactance and resistance?
5. In a series *RC* circuit, what does the phase angle of the circuit impedance represent?
6. When using Ohm's law to calculate the circuit current in a series *RC* circuit, what is indicated by a *positive* phase angle in the result? How is this phase angle normally resolved?
7. How is the voltage-divider equation modified for use in series *RC* circuits?
8. Describe the effect that an *increase* in operating frequency has on the voltage, current, and impedance values in a series *RC* circuit.

Critical Thinking

9. Predict the effect that a *decrease* in operating frequency has on the voltage, current, and impedance values in a series *RC* circuit.

13.2 Power Characteristics and Calculations

OBJECTIVE 6 ➤ Like *RL* circuits, *RC* circuits have significant values of apparent power, reactive power, and true power. Table 13.1 describes these values in terms of *RC* circuits.

The values of resistive and reactive power for a series *RC* circuit are calculated as shown in Example 13.6.

TABLE 13.1 Power Values in *RC* Circuits

Value	*Definition*
Resistive power (P_R)	The power dissipated by the resistance in an *RC* circuit. Also known as *true power.*
Reactive power (P_X)	The value found using $P = I^2 X_C$. The energy stored in the capacitor's electrostatic field, measured in volt-amperes-reactive (VAR). Also known as *imaginary power.*
Apparent power (P_{APP})	The combination of resistive (true) power and reactive (imaginary) power. Measured in volt-amperes (VA).
Power factor (PF)	The ratio of resistive power to apparent power.

Example 13.6

The circuit in Figure 13.15 has the values shown. Calculate the values of resistive and reactive power for the circuit.

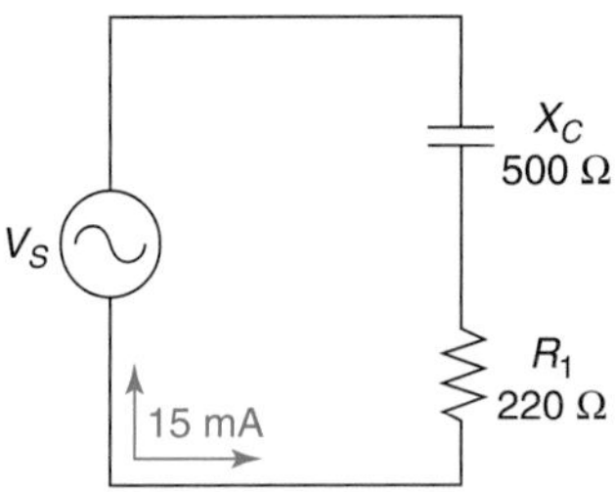

Figure 13.15

Solution: Using the values shown, the reactive power is found as

$$P_X = I^2 X_C = (15 \text{ mA})^2 (500 \ \Omega) = 112.5 \text{ mVAR}$$

and the resistive power is found as

$$P_R = I^2 R_1 = (15 \text{ mA})^2 (220 \ \Omega) = 49.5 \text{ mW}$$

Practice Problem 13.6

A circuit like the one shown in Figure 13.15 has the following values: $I = 22$ mA, $X_C = 200 \ \Omega$, and $R = 180 \ \Omega$. Calculate the values of P_X and P_R for the circuit.

You may be wondering why we didn't consider the dielectric resistance of the capacitor in the resistive power (P_R) calculation. The reason is that the dielectric resistance of any capacitor is usually much greater than its reactance. As a result, the dielectric resistance can be ignored in most situations.

Calculating Apparent Power (P_{APP})

Like voltage and impedance, apparent power in a series *RC* circuit equals the phasor sum of resistive power and reactive power. This relationship can be explained with the aid of Figure 13.16.

The graph in Figure 13.16a shows the phase relationship between X_C and *R*. As indicated, X_C has a $-90°$ phase angle relative to *R*. If we consider this phase angle in our P_X equation, we get

$$P_X = I^2 X_C = (I^2 \angle 0°)(X_C \angle -90°) = (I^2 X_C) \angle -90°$$

While the result does not commit to a magnitude, it does indicate that reactive power has a phase angle of $-90°$ (relative to the circuit current). As always, resistance and circuit current are in phase, so P_R has a phase angle of 0°. Since reactive power lags circuit

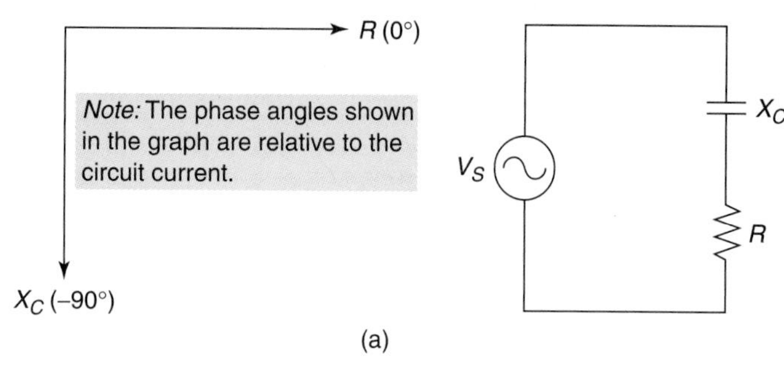

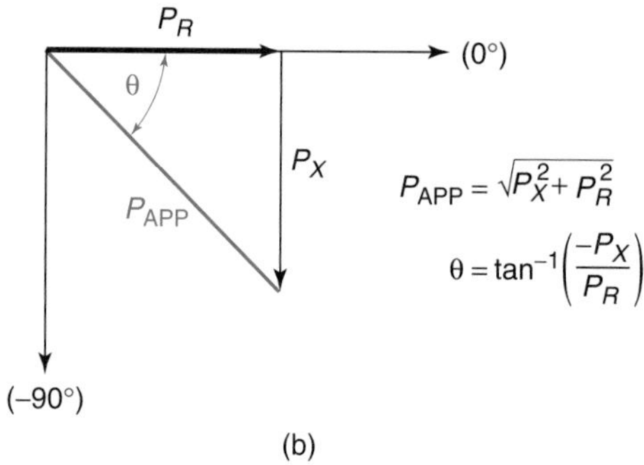

Figure 13.16

current by 90°, it also lags resistive power by the same angle. Therefore, apparent power equals the geometric sum of P_X and P_R (as shown in Figure 13.16b). By formula,

$$P_{APP} = \sqrt{P_X^2 + P_R^2} \tag{13.7}$$

where

P_{APP} = the apparent power, measured in volt-amperes

Earlier in this chapter, you were shown that

$$\theta = \tan^{-1}\left(\frac{-X_C}{R}\right)$$

Since P_X and P_R are in phase with X_C and R (respectively), the phase angle of P_{APP} can be found as

$$\theta = \tan^{-1}\left(\frac{-P_X}{P_R}\right) \tag{13.8}$$

Example 13.7 demonstrates the calculation of apparent power for a series *RC* circuit.

Example 13.7

Calculate the value of apparent power for the circuit in Example 13.6.

Solution: In Example 13.6, we calculated the following magnitudes: P_X = 112.5 mVAR, and P_R = 49.5 mW. Using these values, P_{APP} is found to have a magnitude of

$$P_{APP} = \sqrt{P_X^2 + P_R^2} = \sqrt{(112.5 \text{ mVAR})^2 + (49.5 \text{ mW})^2} = 122.9 \text{ mVA}$$

at a phase angle of

$$\theta = \tan^{-1}\left(\frac{-P_X}{P_R}\right) = \tan^{-1}\left(\frac{-112.5 \text{ mVAR}}{49.5 \text{ mW}}\right) = \tan^{-1}(-2.273) = -66.3°$$

Practice Problem 13.7

Calculate the value of P_{APP} for the circuit described in Practice Problem 13.6.

Power Factor

The **power factor (PF)** for a series *RC* circuit is *the ratio of resistive power to apparent power.* As with series *RL* circuits, it defines how much power is actually being dissipated in a circuit compared to the total circuit power. The power factor (PF) for a series *RC* circuit is found as

$$\text{PF} = \frac{P_R}{P_{APP}} \tag{13.9}$$

In Chapter 11, you were shown that

$$\cos\theta = \frac{P_R}{P_{APP}}$$

This relationship applies to *RC* circuits as well. If we transpose the equation, we get the following useful relationship:

$$P_R = P_{APP}\cos\theta \tag{13.10}$$

Example 13.8 demonstrates an application of this relationship.

Example 13.8

An *RC* circuit has measured value of $V_S = 12\ \text{V}_{ac}$, $I_S = 4.6$ mA, and $\theta = -30°$. Determine the amount of power that is dissipated by the circuit.

Solution: The value of apparent power is found as

$$P_{APP} = V_S I_S = (12\ \text{V}_{ac})(4.6\ \text{mA}) = 55.2\ \text{mVA}$$

and the power dissipated by the circuit is found as

$$P_R = P_{APP}\cos\theta = (55.2\ \text{mVA})(\cos -30°) = (55.2\ \text{VA})(0.866) = 47.8\ \text{mW}$$

Practice Problem 13.8

An *RC* circuit has measured value of $V_S = 8\ \text{V}_{ac}$, $I_S = 6.5$ mA, and $\theta = -28°$. Determine the amount of power that is dissipated by the circuit.

Note that equation (13.10) can also be used in the analysis of an *RL* circuit.

Section Review

1. Compare and contrast the following: *apparent power, resistive power,* and *reactive power.*
2. Why is the dielectric resistance of a capacitor normally ignored in the resistive power calculations for an *RC* circuit?
3. State and explain the phase relationship between *RC* circuit reactive power (P_X) and resistive power (P_R).
4. What is the power factor for an *RC* circuit? What does it tell you?
5. What (if anything) happens to the power factor of a series *RC* circuit when the circuit operating frequency increases? Explain your answer. Critical Thinking

13.3 Parallel *RC* Circuits

A *parallel RC circuit* contains *one or more resistors in parallel with one or more capacitors.* Several parallel *RC* circuits are shown in Figure 13.17. As you can see, none of the circuit branches contain more than one component. (If they did, we'd be dealing with a series-parallel *RC* circuit.)

Parallel Circuits: A Comparison

OBJECTIVE 7 ➤ As we did with series circuits, we're going to start by comparing parallel *RC* circuits to purely resistive and purely capacitive parallel circuits. Such a comparison is provided in Figure 13.18.

In Chapters 5 and 12, we established the fundamental characteristics of resistive and capacitive parallel circuits. As you know, these characteristics are as follows:

- All branch voltages equal the source voltage.
- The circuit current equals the sum of the branch currents.
- The total impedance (resistance or reactance) is lower than the lowest branch value.

The relationships listed for the resistive and capacitive circuits in Figure 13.18 should all seem familiar. The current and impedance relationships shown for the *RC* circuit are very similar to those for a parallel *RL* circuit. Specifically:

- The total current in a parallel *RC* circuit equals the geometric sum of the branch currents.
- The equation for total impedance involves the geometric sum of X_C and R (in the denominator of the equation).

Branch Currents

OBJECTIVE 8 ➤ As you know, current leads voltage by 90° in a capacitive circuit. At the same time, current and voltage are always in phase in a resistive circuit. When connected in parallel, a resistor and a capacitor produce the waveforms shown in Figure 13.19.

Since voltage is constant across all branches in a parallel circuit, the capacitor and resistor currents are referenced to a single voltage waveform. As shown in the figure,

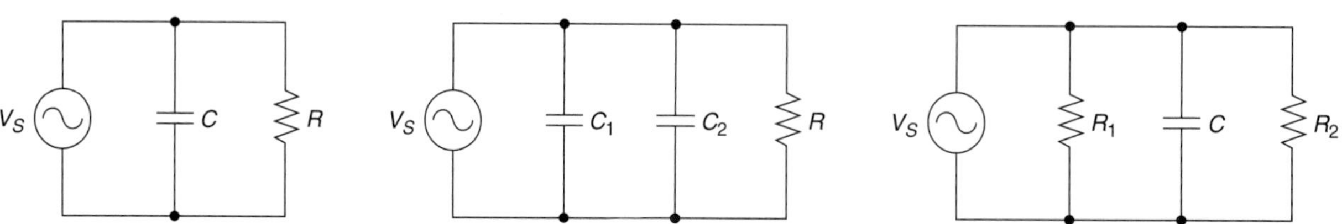

Figure 13.17 Parallel *RC* circuits.

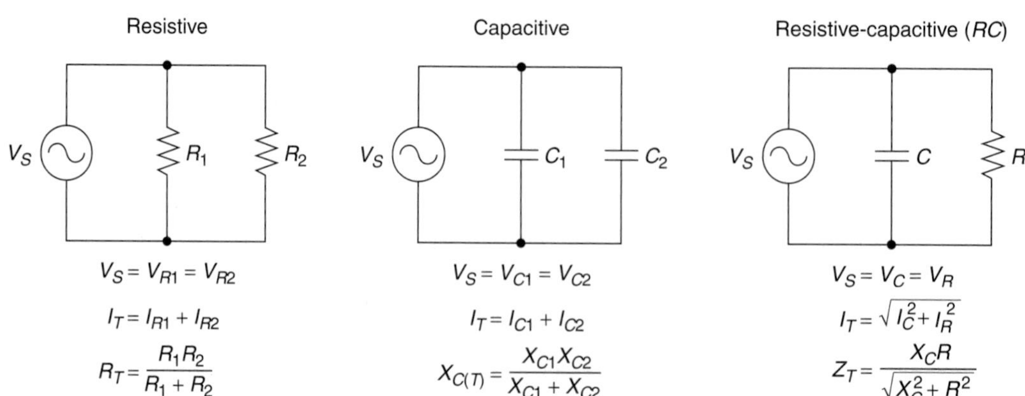

Figure 13.18 Comparisons of resistive, capacitive, and *RC* parallel circuits.

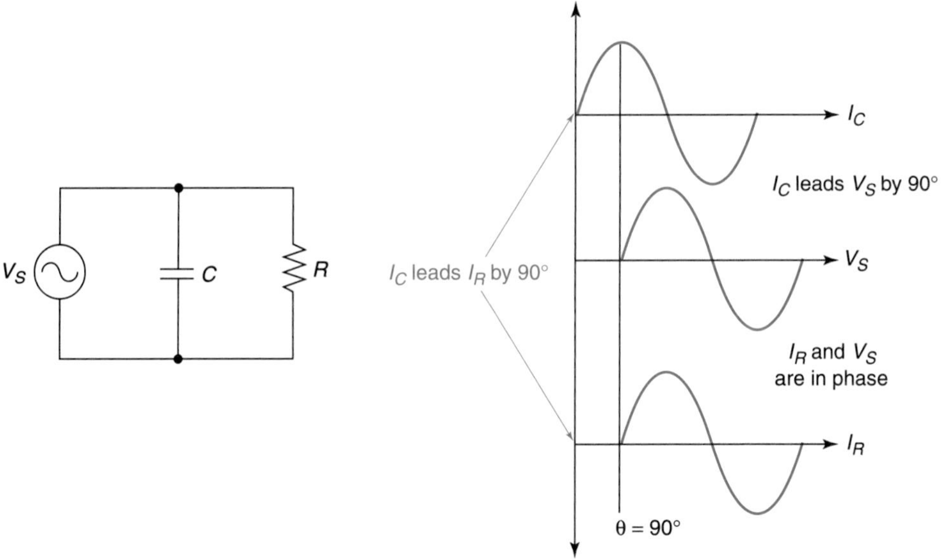

Figure 13.19 Parallel *RC* circuit waveforms.

capacitor current leads resistor current by 90°. This phase relationship holds true for every parallel *RC* circuit with an ac voltage source.

Since I_C leads I_R by 90°, the values of the component currents are represented using phasors (as shown in Figure 13.20a). By plotting the value of I_R on the *x*-axis (0°) and the value of I_C on the positive *y*-axis (90°), the phase relationship between the two is represented in the graph.

As in any parallel circuit, the sum of I_C and I_R must equal the total circuit current. The phase relationship between the two values requires that they be added geometrically. By formula,

$$I_T = \sqrt{I_C^2 + I_R^2} \tag{13.11}$$

Since voltage is equal across all branches in a parallel circuit, V_S is used as the reference for all phase angle measurements. Therefore, the value of θ for the circuit current indicates its phase relative to the source voltage. As shown in Figure 13.20b, this phase angle is found as

$$\theta = \tan^{-1}\left(\frac{I_C}{I_R}\right) \tag{13.12}$$

Example 13.9 demonstrates the calculations of total current for a parallel *RC* circuit.

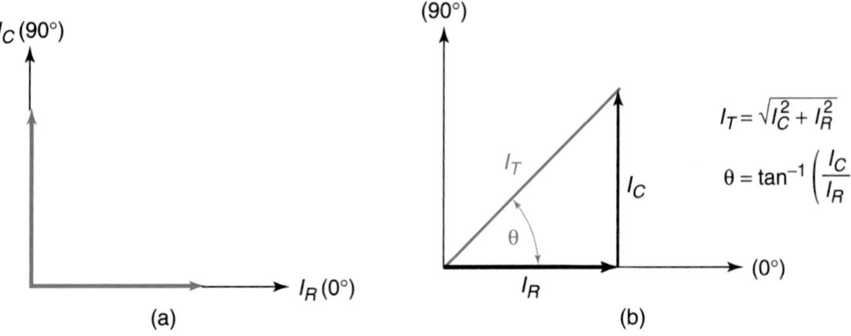

Figure 13.20 Phasor representations of resistor and capacitor current.

Example 13.9

The current values shown in Figure 13.21 were measured as shown. Calculate the total current for the circuit.

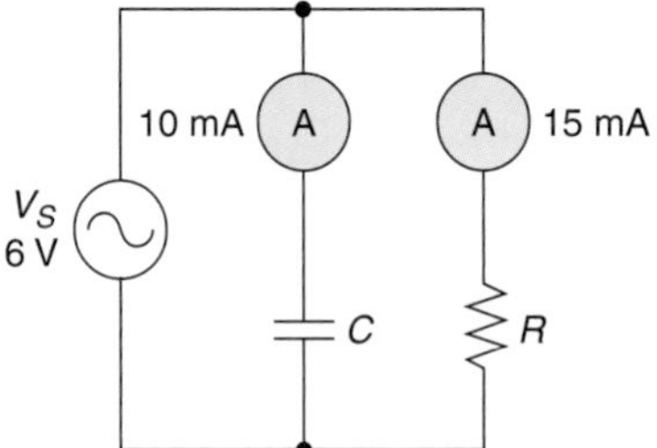

Figure 13.21

Solution: The circuit current has a magnitude of

$$I_T = \sqrt{I_C^2 + I_R^2} = \sqrt{(10 \text{ mA})^2 + (15 \text{ mA})^2} = 18 \text{ mA}$$

at a phase angle of

$$\theta = \tan^{-1}\left(\frac{I_C}{I_R}\right) = \tan^{-1}\left(\frac{10 \text{ mA}}{15 \text{ mA}}\right) = \tan^{-1}(0.667) = 33.7°$$

The positive value of θ indicates that the circuit current leads the source voltage by 33.7°.

Practice Problem 13.9

A circuit like the one shown in Figure 13.21 has the following values: $I_C = 25$ mA, and $I_R = 8$ mA. Calculate the total circuit current.

For any parallel *RC* circuit, the limits on the phase angle between circuit current and source voltage are given as follows:

$$0° < \theta < 90° \qquad (13.13)$$

This range of values for θ makes sense when you consider that I_T is the sum of I_R ($\theta = 0°$) and I_C ($\theta = 90°$). A parallel *RC* circuit is neither purely resistive nor purely capacitive. Therefore, it must have a phase angle that is greater than 0° but less than 90°.

Parallel-Circuit Impedance

The easiest way to calculate the total impedance in a parallel *RC* circuit begins with calculating the circuit current. Once the circuit current is known, Ohm's law states that the circuit impedance can be found as

$$Z_T = \frac{V_S}{I_T} \qquad (13.14)$$

As with any parallel circuit, the total impedance in a parallel *RC* circuit is always lower than any branch impedance value.

Phase Angle of Z_T

In our discussion of series *RC* circuits, it was shown that Z_T and V_S have equal phase angles (relative to the circuit current). In *parallel RC* circuits, Z_T and I_T have phase angles of equal magnitude but opposite signs: The current phase angle is *positive,* and the impedance phase angle is *negative.* This may seem confusing at first, but it makes sense if you apply Ohm's law to the circuit as follows:

TABLE 13.2 Parallel-Circuit Impedance Equations for Two-Component Resistive and Inductive Circuits

Method	Resistive Form	Capacitive Form
Reciprocal	$R_T = \dfrac{1}{\dfrac{1}{R_1} + \dfrac{1}{R_2}}$	$X_{C(T)} = \dfrac{1}{\dfrac{1}{X_{C1}} + \dfrac{1}{X_{C2}}}$
Product-over-sum	$R_T = \dfrac{R_1 R_2}{R_1 + R_2}$	$X_{C(T)} = \dfrac{X_{C1} X_{C2}}{X_{C1} + X_{C2}}$
Equal-value branches	$R_T = \dfrac{R}{n}$	$X_{CT} = \dfrac{X_C}{n}$
	(where n = the number of branches in the circuit)	

$$\begin{aligned} V_S\angle 0^\circ &= (I_T\angle\theta)(Z_T\angle -\theta) \\ &= I_T Z_T\angle(\theta - \theta) \\ &= I_T Z_T\angle 0^\circ \end{aligned}$$

Source voltage is the 0° reference for parallel circuits. The only way that the product of current and impedance can have a 0° phase angle is if the two angles are positive and negative equivalents. This holds true for any parallel *RC* (or *RL*) circuit.

Calculating Parallel-Circuit Impedance

In Chapters 5 and 12, you were shown how to calculate the total impedance (resistance or reactance) of purely resistive and purely capacitive parallel circuits. The commonly used equations for these parallel circuits are listed (as a reference) in Table 13.2.

Since X_C and R do not have the same phase angle, the equal-value branches approach cannot be used on a parallel *RC* circuit. At the same time, the phase angles of X_C and R make the reciprocal method long and tedious. As a result, the product-over-sum approach to calculating the circuit impedance is preferred for parallel *RC* circuits. The magnitude of Z_T is found as

$$Z_T = \frac{X_C R}{\sqrt{X_C^2 + R^2}} \tag{13.15}$$

Example 13.10 demonstrates the use of this equation.

Example 13.10

Calculate the magnitude of Z_T for the circuit shown in Figure 13.22.

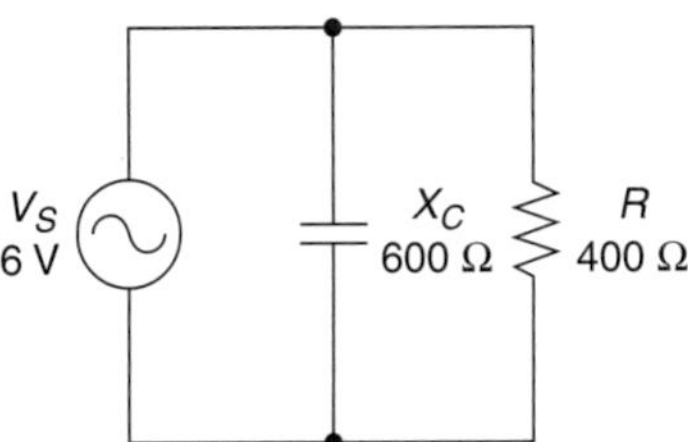

Figure 13.22

Solution: The magnitude of Z_T can be found as

$$Z_T = \frac{X_C R}{\sqrt{X_C^2 + R^2}} = \frac{(600\ \Omega)(400\ \Omega)}{\sqrt{(600\ \Omega)^2 + (400\ \Omega)^2}} \cong 333\ \Omega$$

Practice Problem 13.10

Calculate the value of Z_T for a circuit with values of $R = 150\ \Omega$ and $X_C = 300\ \Omega$.

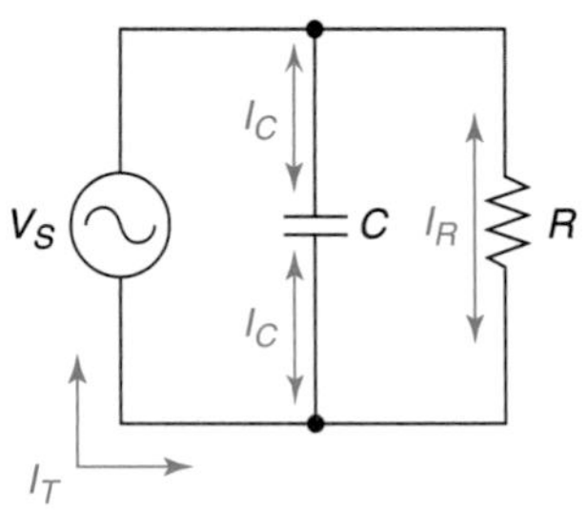

Figure 13.23

Calculating the Impedance Phase Angle

The phase angle of the circuit impedance for a parallel *RC* circuit can be found as

$$\theta = \tan^{-1}\left(\frac{R}{-X_C}\right) \tag{13.16}$$

As you can see, the fraction in equation (13.16) is the reciprocal of the one used to calculate the impedance phase angle in a series *RC* circuit. Example 13.11 applies this equation to the circuit shown in Figure 13.22.

Example 13.11

Calculate the phase angle of the circuit impedance for the circuit in Figure 13.22 (see Example 13.10).

Solution: The value of θ for the circuit impedance is found as

$$\theta = \tan^{-1}\left(\frac{R}{-X_C}\right) = \tan^{-1}\left(\frac{400\ \Omega}{-600\ \Omega}\right) = \tan^{-1}(-0.667) = -33.7°$$

Practice Problem 13.11

Calculate the phase angle of the circuit impedance in Practice Problem 13.10.

Parallel-Circuit Frequency Response

OBJECTIVE 9 ➤ The frequency response of a parallel *RC* circuit is quite different from that of a series *RC* circuit. For example, consider the circuit shown in Figure 13.23. The circuit's response to an *increase* in operating frequency is as follows:

- The increase in operating frequency causes X_C to decrease.
- The decrease in X_C causes I_C to increase.
- The increase in I_C causes the magnitude and phase angle of I_T to increase.
- The decrease in X_C causes the magnitude of Z_T to decrease. It also causes the phase angle of Z_T to become more negative.

Example 13.12 demonstrates the effects of an *increase* in operating frequency on parallel *RC* circuits.

Example 13.12

The voltage source in Figure 13.24 has the frequency limits shown. Calculate the circuit current and impedance values at the frequency limits of the source.

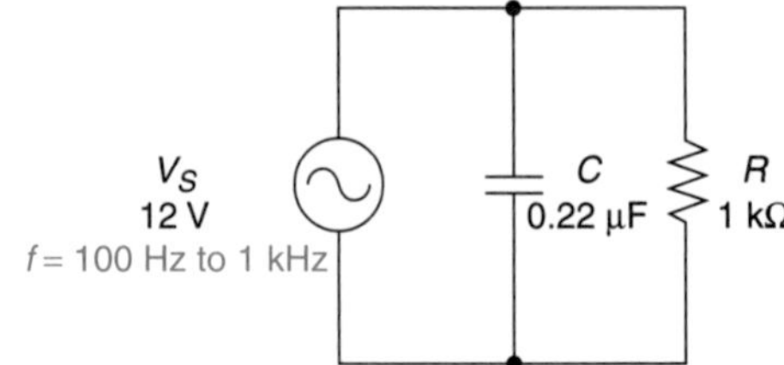

Figure 13.24

Solution: When the operating frequency is set to 100 Hz, the reactance of the capacitor is found as

$$X_C = \frac{1}{2\pi f C} = \frac{1}{2\pi(100\text{ Hz})(0.22\ \mu\text{F})} = 7.23\text{ k}\Omega$$

The magnitudes of the branch currents are found as

$$I_C = \frac{V_S}{X_C} = \frac{12\text{ V}}{7.23\text{ k}\Omega} = 1.66\text{ mA}$$

and

$$I_R = \frac{V_S}{R} = \frac{12\text{ V}}{1\text{ k}\Omega} = 12\text{ mA}$$

Combining these values, we get a circuit current magnitude of

$$I_T = \sqrt{I_C^2 + I_R^2} = \sqrt{(1.66\text{ mA})^2 + (12\text{ mA})^2} = 12.1\text{ mA}$$

at a phase angle of

$$\theta = \tan^{-1}\left(\frac{I_C}{I_R}\right) = \tan^{-1}\left(\frac{1.66\text{ mA}}{12\text{ mA}}\right) = \tan^{-1}(0.138) = 7.88°$$

Finally, the circuit impedance can be found as

$$Z_T = \frac{V_S}{I_T} = \frac{12\text{ V}\angle 0°}{12.1\text{ mA}\angle 7.88°} = 992\ \Omega\angle -7.88°$$

When the operating frequency increases to 1 kHz, the reactance of the capacitor decreases to

$$X_C = \frac{1}{2\pi f C} = \frac{1}{2\pi(1\text{ kHz})(0.22\ \mu\text{F})} = 723\ \Omega$$

The magnitudes of the branch currents are found as

$$I_C = \frac{V_S}{X_C} = \frac{12\text{ V}}{723\ \Omega} = 16.5\text{ mA}$$

and

$$I_R = \frac{V_S}{R} = \frac{12\text{ V}}{1\text{ k}\Omega} = 12\text{ mA}$$

Combining these values, we get a circuit current magnitude of

$$I_T = \sqrt{I_C^2 + I_R^2} = \sqrt{(16.5\text{ mA})^2 + (12\text{ mA})^2} = 20.4\text{ mA}$$

at a phase angle of

$$\theta = \tan^{-1}\left(\frac{I_C}{I_R}\right) = \tan^{-1}\left(\frac{16.5\text{ mA}}{12\text{ mA}}\right) = \tan^{-1}(1.375) = 54°$$

Finally, the circuit impedance is found as

$$Z_T = \frac{V_S}{I_T} = \frac{12\text{ V}\angle 0°}{20.4\text{ mA}\angle 54°} = 588\ \Omega\angle -54°$$

Practice Problem 13.12

A circuit like the one shown in Figure 13.24 has the following values: $V_S = 9$ V, $f = 120$ Hz to 6 kHz, $C = 1$ μF, and $R = 620\ \Omega$. Calculate the circuit current and impedance values at the frequency limits of the source.

When you compare the results in Example 13.12 with the statements made earlier, you'll see that the circuit responded to the increase in operating frequency exactly as described.

One Final Note

In this section, you have been introduced to parallel *RC* circuit relationships and analysis. Once again, we have a situation where all of the relationships discussed are similar to those given for *RL* circuits in Chapter 11. For future reference, the basic characteristics of parallel *RC* circuits are compared to those of parallel *RL* circuits in Figure 13.25.

Section Review

1. What are the three fundamental characteristics of any parallel circuit?
2. In a parallel *RC* circuit, what is the phase relationship between I_C and I_R?
3. In a parallel *RC* circuit, what does the current phase angle represent?
4. Why does the current in a parallel *RC* circuit have a *positive* phase angle?
5. What is the simplest approach to calculating the value of Z_T for a parallel *RC* circuit?

Critical Thinking

6. When $X_C = R$ in a parallel *RC* circuit, can the equal-value branches approach to calculating Z_T be used? Explain your answer.
7. Describe the response of a parallel *RC* circuit to a *decrease* in *operating frequency.*

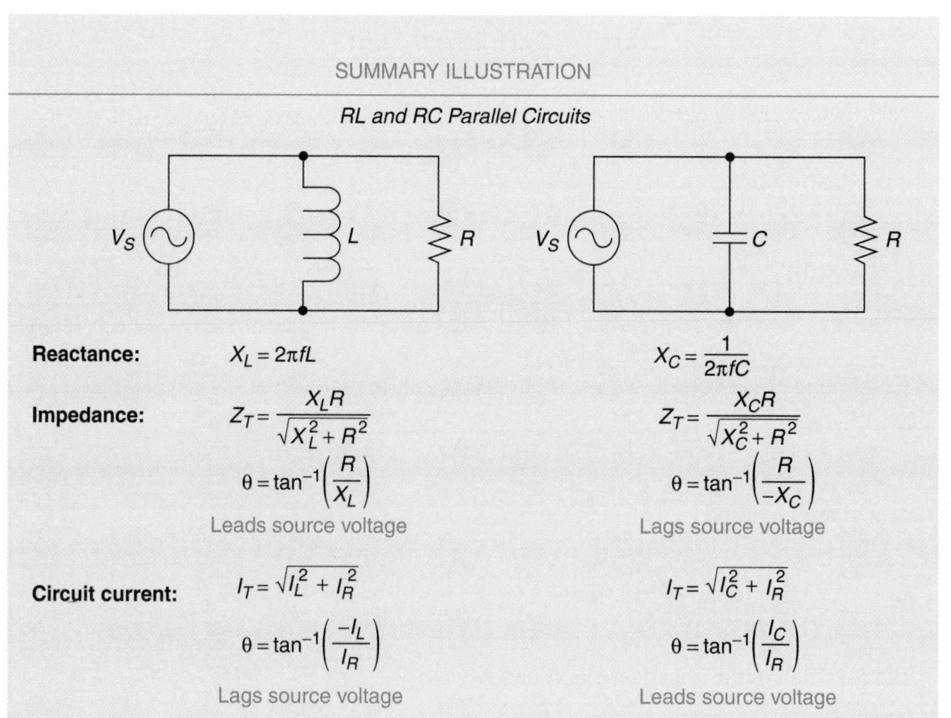

Figure 13.25

13.4 Series-Parallel *RC* Circuit Analysis

In Chapter 6, you learned that analyzing a series-parallel circuit is simply a matter of: ◄ *OBJECTIVE 10*

- Combining series components according to the rules of series circuits.
- Combining parallel components according to the rules of parallel circuits.

The analysis of a series-parallel *RC* circuit gets a bit complicated because of the phase angles involved. For example, consider the circuit shown in Figure 13.26a. Using the techniques outlined in the last section, the circuit was determined to have values of $X_C = 318.3\ \Omega$ and $Z_T = 264\ \Omega\angle -56°$. The circuit impedance is represented as a block (Z_P) in the equivalent circuit shown.

If we add a series resistor (R_S) to the circuit, we have the series-parallel circuit shown in Figure 13.26b. As the equivalent of this circuit shows, R_S is in series with Z_P. The phase difference between R_S and Z_P makes it difficult to add their polar notation values. However, *they can easily be added when converted to rectangular notation.* As discussed in Appendix F, a value in the form of $Z\angle\theta$ (polar notation) can be converted to the form of $X \pm jY$ (rectangular notation) using the relationships shown in Figure 13.27. When the value of Z_P is converted to rectangular form, it can be added to the value of R_S, as demonstrated in Example 13.13.

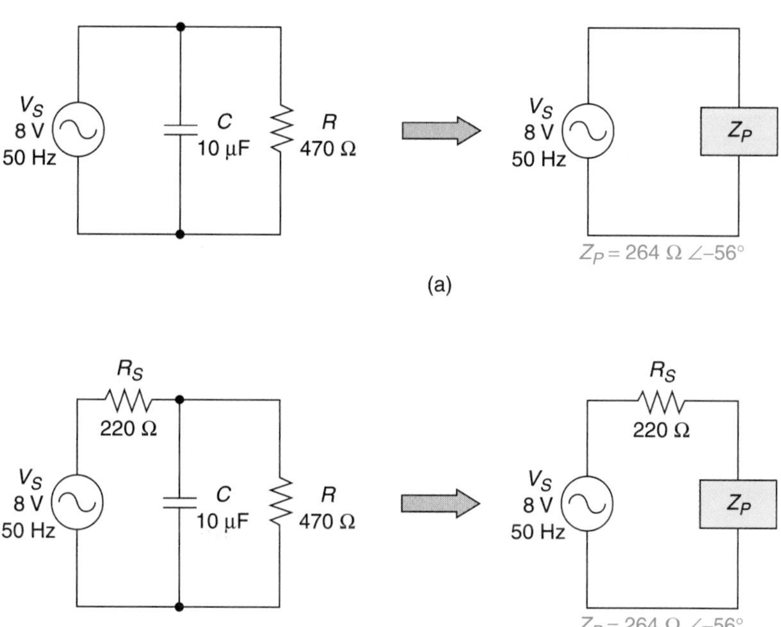

Figure 13.26

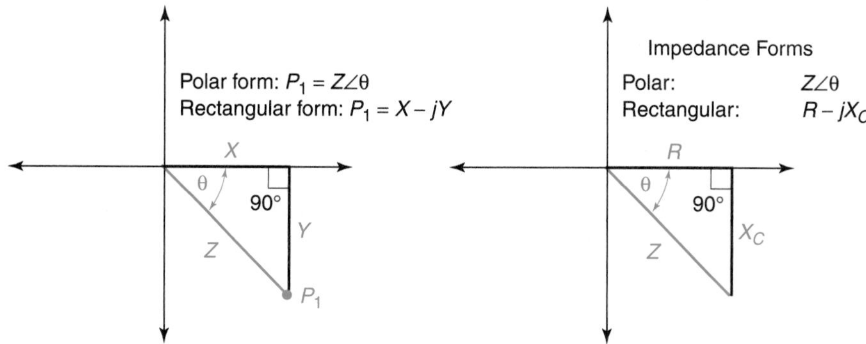

Figure 13.27 Polar-to-rectangular conversions.

Example 13.13

Determine the total impedance (in polar form) for the circuit shown in Figure 13.28a.

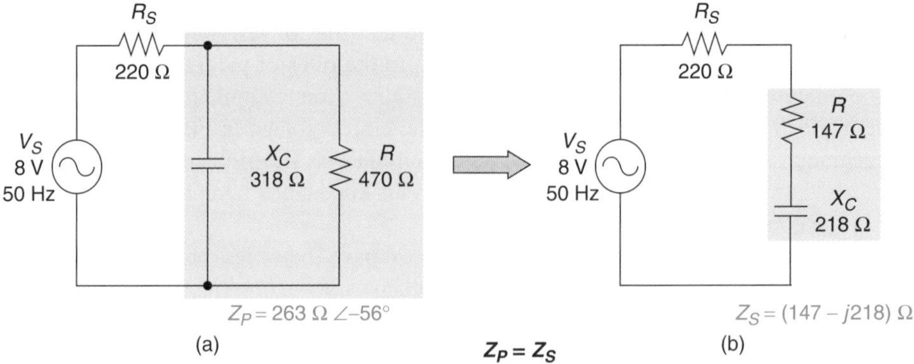

Figure 13.28

Solution: The parallel combination of R and C is shown to have a value of $Z_P = 264\ \Omega\angle-56°$. First, we convert this value to rectangular form as follows:

$$R = Z \cos \theta = (264\ \Omega)(\cos -56°) = (263\ \Omega)(0.559) = 147\ \Omega$$

and

$$X_C = Z \sin \theta = (264\ \Omega)(\sin -56°) = (263\ \Omega)(-0.829) = -218\ \Omega$$

These values indicate that Z_P equals the total impedance of a 147 Ω resistance in series with a 218 Ω capacitive reactance. In other words, it provides us with the series equivalent of Z_P shown in Figure 13.28b.

Using the equivalent circuit, the total resistance in the circuit is found as

$$R_T = R_S + R = 220\ \Omega + 147\ \Omega = 367\ \Omega$$

Now the magnitude of the circuit impedance is found as

$$Z_T = \frac{X_C R_T}{\sqrt{X_C^2 + R_T^2}} = \frac{(218\ \Omega)(367\ \Omega)}{\sqrt{(218\ \Omega)^2 + (367\ \Omega)^2}} \cong 187\ \Omega$$

and the phase angle of the circuit impedance is found as

$$\theta = \tan^{-1}\left(\frac{-X_C}{R_T}\right) = \tan^{-1}\left(\frac{-218\ \Omega}{367\ \Omega}\right) = \tan^{-1}(-0.594) = -30.7°$$

Thus, the total impedance of the original series-parallel circuit is $187\ \Omega\angle-30.7°$.

Practice Problem 13.13

A circuit like the one shown in Figure 13.28b has the following values: $R_S = 150\ \Omega$, $R = 348\ \Omega$, and $X_C = 80\ \Omega$. Calculate the total circuit impedance.

The rest of a typical circuit analysis involves solving for the component voltage and current values. Example 13.14 demonstrates the process from start to finish.

Example 13.14

Perform a typical analysis on the circuit shown in Figure 13.29.

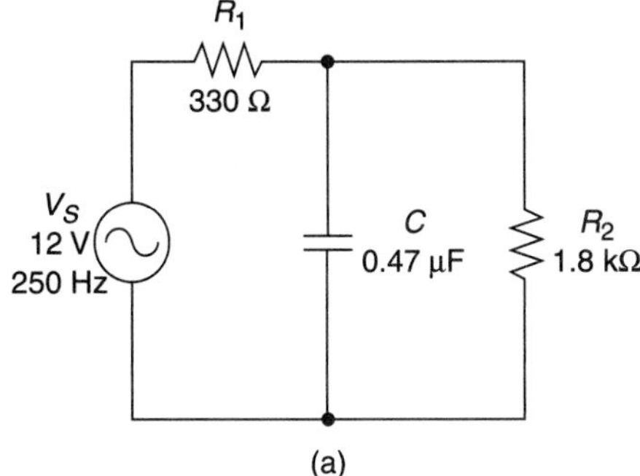

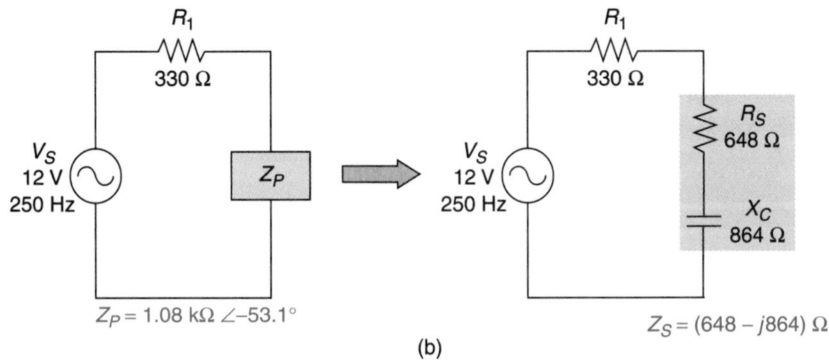

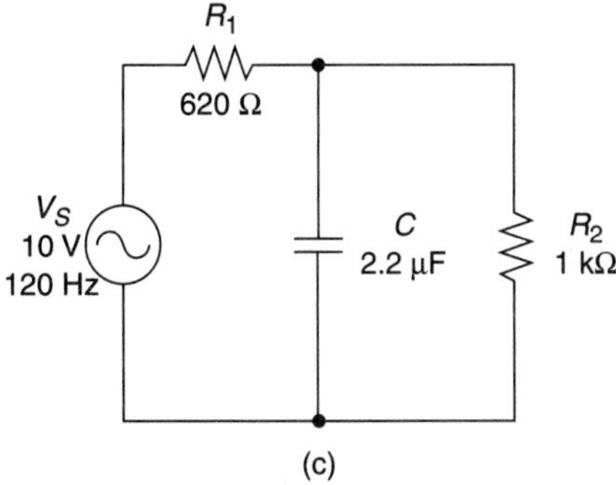

Figure 13.29

Solution: First, the reactance of the capacitor is found as

$$X_C = \frac{1}{2\pi fC} = \frac{1}{2\pi(250 \text{ Hz})(0.47 \text{ μF})} = 1.35 \text{ kΩ}$$

Now the parallel circuit impedance can be found as

$$Z_P = \frac{X_C R_2}{\sqrt{X_C^2 + R_2^2}} = \frac{(1.35 \text{ kΩ})(1.8 \text{ kΩ})}{\sqrt{(1.35 \text{ kΩ})^2 + (1.8 \text{ kΩ})^2}} = 1.08 \text{ kΩ}$$

at a phase angle of

$$\theta = \tan^{-1}\left(\frac{R_2}{-X_C}\right) = \tan^{-1}\left(\frac{1.8 \text{ kΩ}}{-1.35 \text{ kΩ}}\right) = \tan^{-1}(-1.333) = -53.1°$$

This impedance is represented as a block (Z_P) in Figure 13.29b. The series-equivalent of Z_P can now be found using

$$R = Z \cos \theta = (1.08 \text{ k}\Omega)(\cos -53.1°) = (1.08 \text{ k}\Omega)(0.6) = 648 \ \Omega$$

and

$$X_C = Z \sin \theta = (1.08 \text{ k}\Omega)(\sin -53.1°) = (1.08 \text{ k}\Omega)(-0.7997) = -864 \ \Omega$$

These values are included as Z_S in the series-equivalent circuit (see Figure 13.29b).

Once the series-equivalent circuit has been derived, the resistances in the circuit can be combined as follows:

$$R_T = R_S + R_1 = 648 \ \Omega + 330 \ \Omega = 978 \ \Omega$$

Now the circuit impedance can be found as

$$Z_T = \sqrt{X_C^2 + R_T^2} = \sqrt{(864 \ \Omega)^2 + (978 \ \Omega)^2} = 1.3 \text{ k}\Omega$$

at a phase angle of

$$\theta = \tan^{-1}\left(\frac{-X_C}{R_T}\right) = \tan^{-1}\left(\frac{-864 \ \Omega}{978 \ \Omega}\right) = \tan^{-1}(-0.8834) = -41.5°$$

Once the circuit impedance is determined, the circuit current can be found as

$$I_T = \frac{V_S}{Z_T} = \frac{12 \text{ V}}{1.3 \text{ k}\Omega\angle -41.5°} = 9.23 \text{ mA}\angle 41.5°$$

This result indicates that circuit current leads the source voltage by 41.5°. Since we are dealing with a series-equivalent circuit, the phase angles are adjusted to

$$V_S = 12 \text{ V}\angle -41.5° \qquad \text{and} \qquad I_T = 9.23 \text{ mA}\angle 0°$$

Now the voltage across R_1 can be found as

$$V_1 = I_T R_1 = (9.23 \text{ mA})(330 \ \Omega) = 3.05 \text{ V}$$

and the voltage across the parallel impedance (Z_P) can be found as

$$V_P = I_T Z_P = (9.23 \text{ mA}\angle 0°)(1.08 \text{ k}\Omega\angle -53.1°) \cong 9.97 \text{ V}\angle -53.1°$$

Now that we know the voltage across the parallel circuit, we can calculate the values of I_C and I_{R2} as follows:

$$I_C = \frac{V_P}{X_C} = \frac{9.97 \text{ V}\angle -53.1°}{1.35 \text{ k}\Omega\angle -90°} = 7.39 \text{ mA}\angle 36.9°$$

and

$$I_{R2} = \frac{V_P}{R_2} = \frac{9.97 \text{ V}\angle -53.1°}{1.8 \text{ k}\Omega\angle 0°} = 5.54 \text{ mA}\angle -53.1°$$

This completes the analysis of the circuit.

Practice Problem 13.14

Perform a typical analysis on the circuit shown in Figure 13.29c.

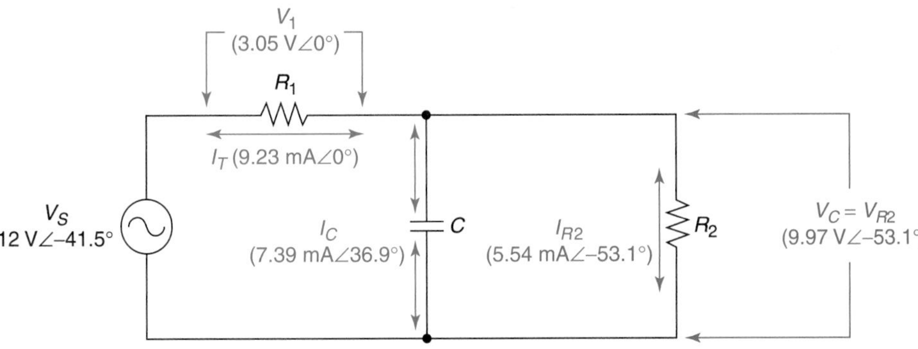

Figure 13.30 The voltages and currents calculated in Example 13.14.

Putting It All Together: Some Observations on Example 13.14

To make sense of the results in Example 13.14, we need to combine them and see how they fit together. For ease of discussion, Figure 13.30 includes the currents and voltages calculated in the example. (Note the source phase angle.) These values show that:

- $\boldsymbol{V_C = V_{R2}}$. This satisfies one of the basic rules of parallel components, namely, that the voltage across parallel components is always the same.
- $\boldsymbol{I_C}$ **leads** $\boldsymbol{I_R}$ **by 90°.** This can easily be proven by subtracting the phase angle of I_R from the phase angle of I_C. As you can see below, the phase difference is 90°.

$$\theta = 36.9° - (-53.1°) = 36.9° + 53.1° = 90°$$

- $\boldsymbol{I_{R2} + I_C = I_T}$. This satisfies Kirchhoff's current law. Since I_C and I_{R2} are 90° out of phase, they can be added geometrically, as follows:

$$I_T \cong \sqrt{I_C^2 + I_{R2}^2} = \sqrt{(7.39\text{ mA})^2 + (5.54\text{ mA})^2} = 9.236\text{ mA}$$

- $\boldsymbol{V_P + V_{R1} = V_S}$. This is Kirchhoff's voltage law for series circuits. By converting V_P and V_{R1} to rectangular form, we can add them together and then compare the results to the value of V_S. As you can see from the solution below, Kirchhoff's voltage law is satisfied (some accuracy is lost due to rounding):

$$\begin{aligned} V_{R1} &= (3.05 + j0.00)\text{ V} \\ V_P &= \underline{(5.99 - j7.97)\text{ V}} \\ & \ (9.04 - j7.97)\text{ V} = 12.05\text{ V}\angle{-41.4°} \cong V_S \end{aligned}$$

As you can see, the basic rules for circuit analysis still hold true. Regardless of the type of circuit, Kirchhoff's voltage and current laws, as well as the basic rules for series and parallel circuits, are valid. This is true for resistive, inductive, capacitive, *RC*, and *RL* circuits.

Section Review

1. How is the impedance of an *RC* circuit converted from polar form to rectangular form?
2. When a parallel *RC* impedance is converted from polar to rectangular form, what type of circuit conversion has taken place?
3. Refer to Figure 13.29a. What happens to the current through R_1 if the operating frequency *increases*? Explain your answer. Critical Thinking

---

Key Terms

The following terms were introduced in this chapter on the pages indicated:

Power factor (PF), 391

Resistive-capacitive (*RC*) circuit, 378

Practice Problems

1. The component voltages shown in Figure 13.31a were taken using an ac voltmeter. Determine the magnitude and phase angle of V_S for the circuit.
2. The component voltages shown in Figure 13.31b were taken using an ac voltmeter. Determine the magnitude and phase angle of V_S for the circuit.

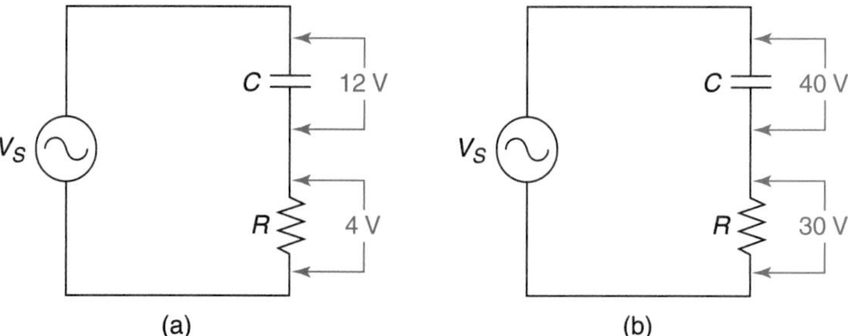

Figure 13.31

3. Calculate the magnitude and phase angle of the circuit impedance in Figure 13.32a.
4. Calculate the magnitude and phase angle of the circuit impedance in Figure 13.32b.

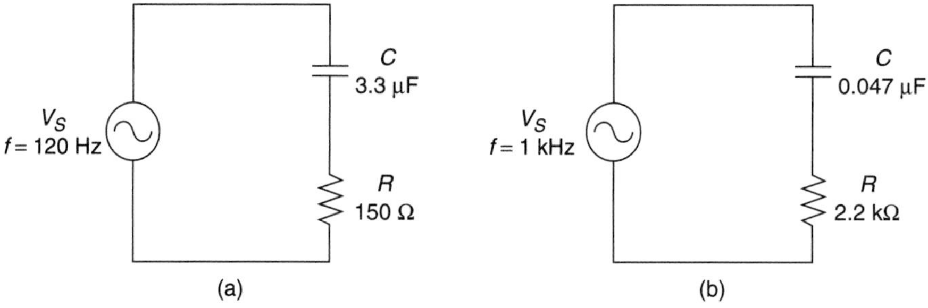

Figure 13.32

5. Calculate the values of Z_T, I_T, V_C, V_R, and θ for the circuit shown in Figure 13.33a.
6. Calculate the values of Z_T, I_T, V_C, V_R, and θ for the circuit shown in Figure 13.33b.

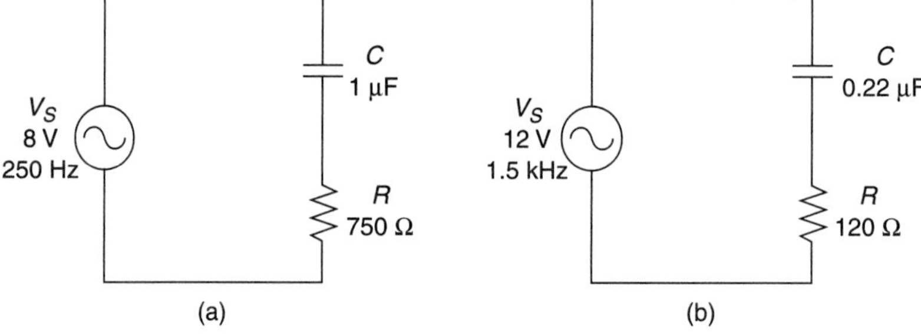

Figure 13.33

7. Calculate the values of Z_T, I_T, V_{C1}, V_{C2}, V_R, and θ for the circuit shown in Figure 13.34a.
8. Calculate the values of Z_T, I_T, V_{C1}, V_{C2}, V_{R1}, V_{R2} and θ for the circuit shown in Figure 13.34b.

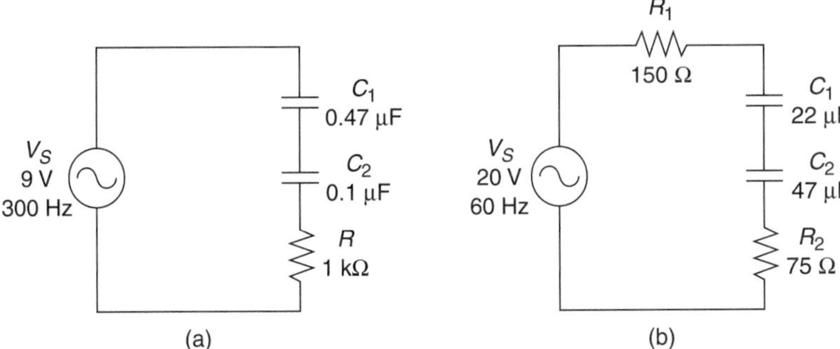

Figure 13.34

9. The voltage source in Figure 13.35a has the frequency limits shown. Calculate the range of Z_T, V_C, V_R, and θ values for the circuit.
10. The voltage source in Figure 13.35b has the frequency limits shown. Calculate the range of Z_T, V_C, V_R, and θ values for the circuit.

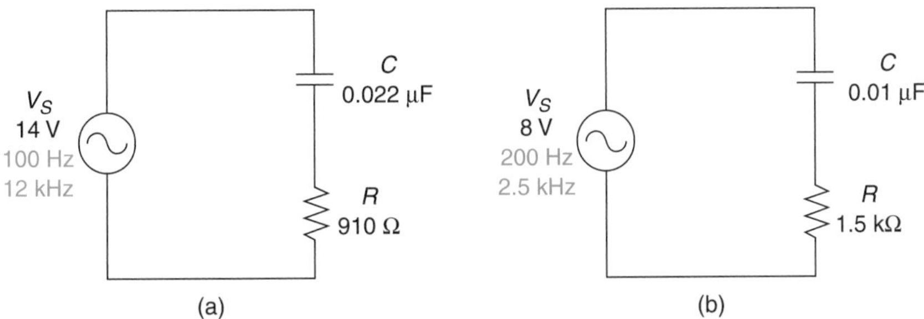

Figure 13.35

11. Refer to Figure 13.33a. Calculate the values of resistive and reactive power for the circuit shown.
12. Refer to Figure 13.33b. Calculate the values of resistive and reactive power for the circuit shown.
13. Calculate the magnitude and phase angle of apparent power for the circuit in problem 11.
14. Calculate the magnitude and phase angle of apparent power for the circuit in problem 12.
15. Refer to Figure 13.32a. The current for this circuit was measured at 100 μA. Determine the values of P_{APP}, $\cos\theta$, and P_R for the circuit.
16. Refer to Figure 13.32b. The current for this circuit was measured at 50 mA. Determine the values of P_{APP}, $\cos\theta$, and P_R for the circuit.
17. Determine the value of reactive power for the circuit in problem 15.
18. Determine the value of reactive power for the circuit in problem 16.
19. Calculate the impedance, voltage, and current values for the circuit shown in Figure 13.36a.

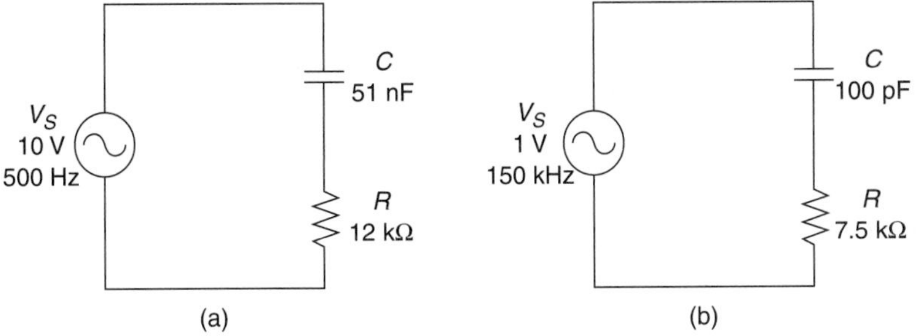

Figure 13.36

20. Calculate the impedance, voltage, and current values for the circuit shown in Figure 13.36b.

21. A parallel *RC* circuit has values of $I_C = 25$ mA and $I_R = 60$ mA. Calculate the magnitude and phase angle of the circuit current.

22. A parallel *RC* circuit has values of $I_C = 320$ mA and $I_R = 140$ mA. Calculate the magnitude and phase angle of the circuit current.

23. Calculate the total impedance for the circuit shown in Figure 13.37a.

24. Calculate the total impedance for the circuit shown in Figure 13.37b.

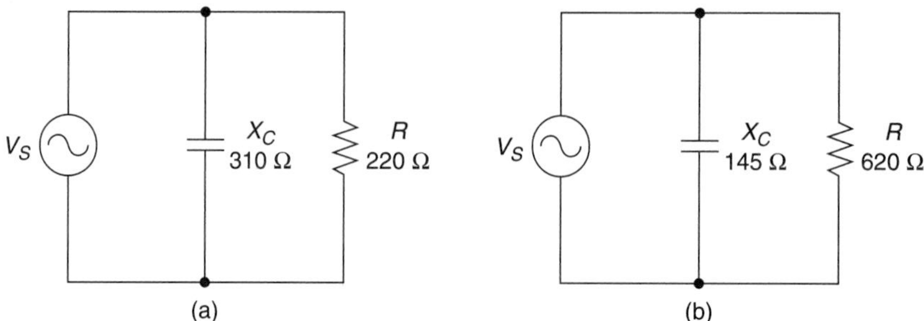

Figure 13.37

25. Calculate the magnitude and phase angle of the circuit impedance in Figure 13.38a.

26. Calculate the magnitude and phase angle of the circuit impedance in Figure 13.38b.

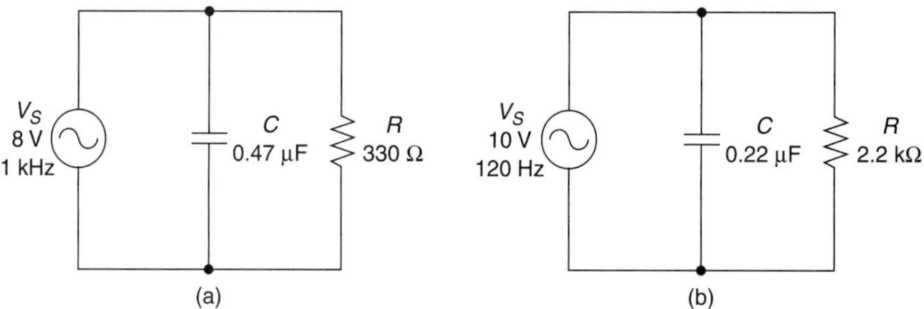

Figure 13.38

27. Calculate the component and circuit current values for the circuit shown in Figure 13.39a.

28. Calculate the component and circuit current values for the circuit shown in Figure 13.39b.

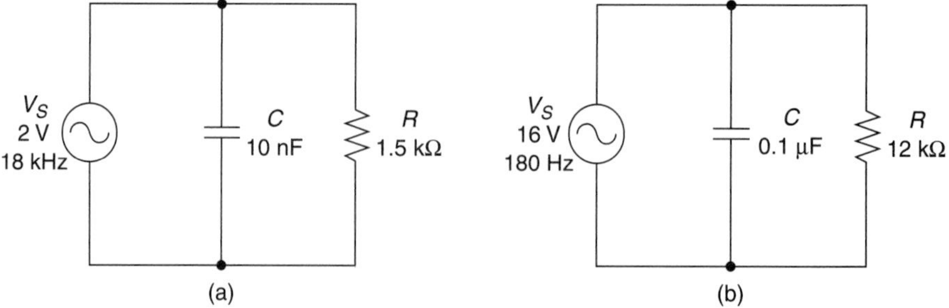

Figure 13.39

29. The voltage source in Figure 13.40a has the frequency limits shown. Calculate the circuit current and impedance values at the frequency limits of the source.

30. The voltage source in Figure 13.40b has the frequency limits shown. Calculate the circuit current and impedance values at the frequency limits of the source.

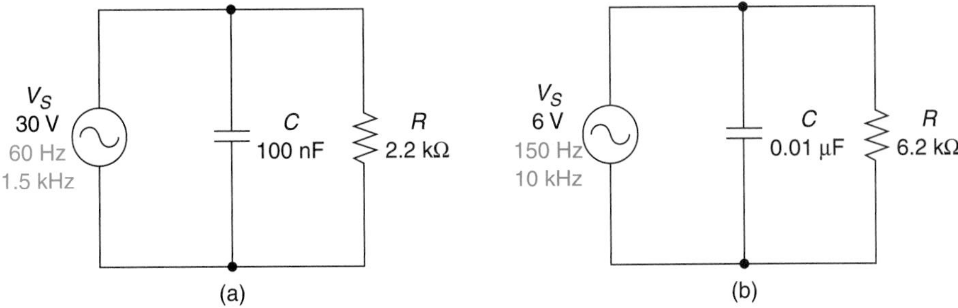

Figure 13.40

31. Refer to Problem 25. Express the circuit impedance in rectangular form.

32. Refer to Problem 26. Express the circuit impedance in rectangular form.

33. Calculate the component voltage and current values for the circuit shown in Figure 13.41a.

34. Calculate the component voltage and current values for the circuit shown in Figure 13.41b.

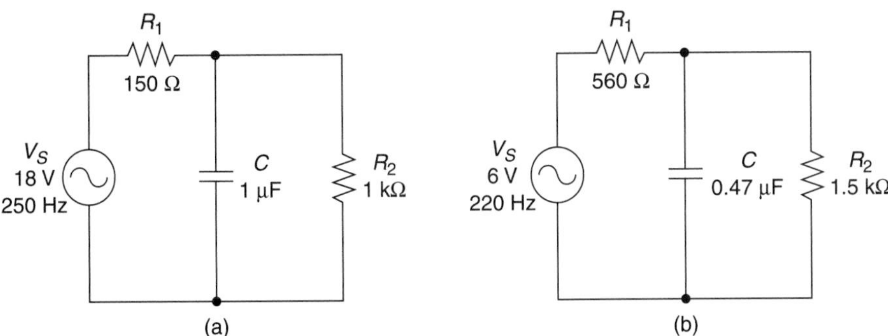

Figure 13.41

Looking Back

These problems relate to material presented in earlier chapters. The chapters are identified in brackets.

35. Convert each of the following values to engineering notation form: [Introduction]
    a. 77,350 Hz = ______
    b. 0.00047 F = ______
    c. 560,000 Ω = ______
    d. 0.00000051 F = ______

36. A two-resistor parallel circuit has the following values: $R_1 = 330\ \Omega$, $R_2 = 220\ \Omega$, and $V_S = 6$ V. Calculate the change in source power (P_S) that occurs if R_2 is replaced with a 100 Ω resistor. [Chapter 5]

37. What is the approximate resistance of a 100 ft length of 40 gauge copper wire? [Chapter 2]

38. A sine wave has the following values: $V_{pk} = 12$ V, and $f = 100$ Hz. What values of θ correspond to an instantaneous value of 6 V? [Chapter 9]

Pushing the Envelope

39. Determine the potentiometer setting that will provide an rms value of 20 mA in Figure 13.42. [*Hint:* Transpose equation (13.4) to provide an equation for *R*.]

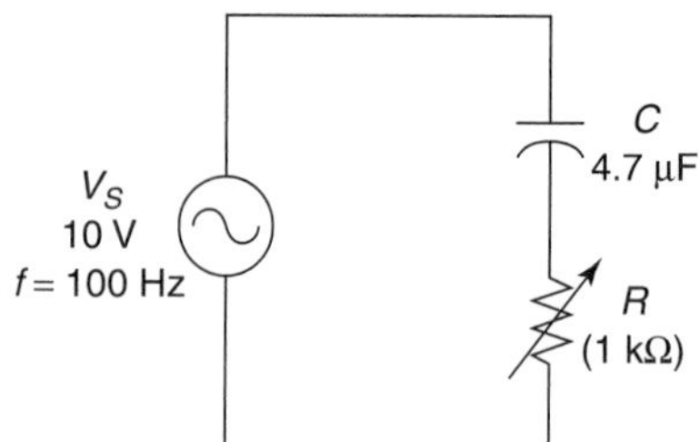

Figure 13.42

40. Determine the Q of the capacitor shown in Figure 13.43.

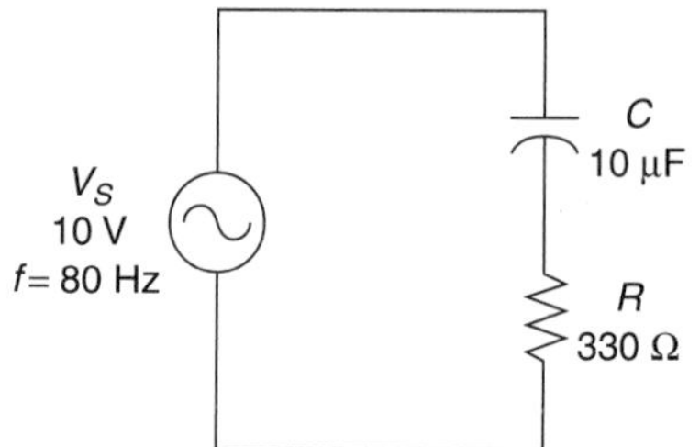

Figure 13.43 Capacitor leakage current = 10 μA @ V_C = 50 V

Answers to the Example Practice Problems

13.1 32 V$\angle -38.7°$

13.2 1.51 k$\Omega\angle -48.5°$

13.3 Z_T = 702 $\Omega\angle -43.4°$, I = 17.1 mA, V_S = 12 V$\angle -43.4°$, V_C = 8.24 V, V_R = 8.72 V

13.4 Z_T = 2.6 k$\Omega\angle -73.2°$, V_{C1} = 16.25 V, V_{C2} = 7.65 V, V_R = 7.21 V

13.5 When f = 1.5 kHz: Z_T = 10.8 k$\Omega\angle -79.3°$, I_T = 2.04 mA, V_C = 21.6 V, V_R = 4.08 V
When f = 6 kHz: Z_T = 3.32 k$\Omega\angle -53°$, I_T = 6.63 mA, V_C = 17.6 V, V_R = 13.3 V

13.6 96.8 mVAR, 87.1 mW

13.7 130 mVA$\angle -48°$

13.8 45.9 mW

13.9 26.2 mA$\angle 72.3°$

13.10 134 Ω

13.11 $-26.6°$

13.12 For f = 120 Hz: I_T = 16 mA$\angle 25°$, Z_T = 563 $\Omega\angle -25°$
For f = 6 kHz: I_T = 340 mA$\angle 87.6°$, Z_T = 26.4 $\Omega\angle -87.6°$

13.13 504 $\Omega\angle -9.13°$

13.14 X_C = 603 Ω, Z_P = 516 $\Omega\angle -58.9°$ = (267 − j442) Ω, R_T = 887 Ω, Z_T = 991 $\Omega\angle -26.5°$, I_T = 10.1 mA, V_S = 10 V$\angle -26.5°$, V_{R1} = 6.26 V, V_P = 5.22 V$\angle -58.9°$, I_C = 8.66 mA$\angle 31.1°$, I_{R2} = 5.22 mA$\angle -58.9°$

Multisim File References

Figure	File Reference
13.8	ETF13_8.msm
13.13	ETF13_13.msm
13.21	ETF13_21.msm
13.29	ETF13_29.msm

chapter 14

RLC Circuits

Objectives

After studying the material in this chapter, you should be able to:

1. Discuss the relationship between the component voltages and the operating characteristics of a series *LC* circuit.
2. Discuss the relationship between the reactance values and the operating characteristics of a series *LC* circuit.
3. Calculate the reactance and voltage values for a given series *LC* circuit.
4. Discuss the relationship between the branch currents and the operating characteristics of a parallel *LC* circuit.
5. Calculate the magnitude of the current through a parallel *LC* circuit.
6. Calculate the parallel equivalent reactance for any parallel *LC* circuit.
7. Describe *resonance,* and calculate the *resonant frequency* of a given *LC* circuit.
8. Describe the effects of *stray capacitance, stray inductance,* and *oscilloscope input capacitance* on resonant frequency.
9. Describe and analyze the operation of series and parallel resonant *LC* circuits.
10. Describe and analyze the operation of any series *RLC* circuit.
11. Describe and analyze the operation of any parallel *RLC* circuit.
12. Determine the impedance, voltage, and current values for a series-parallel *RLC* circuit.

Now that we have discussed *RL* and *RC* circuits, it is time to combine them into *RLC* (resistive-inductive-capacitive) circuits. We will start our discussion of *RLC* circuits with the operating characteristics of *LC* (inductive-capacitive) circuits.

14.1 Series *LC* Circuits

OBJECTIVE 1 ➤ When an inductor is connected in series with a capacitor, the circuit has the following current and voltage characteristics:

- The inductor and capacitor currents are equal ($I_L = I_C$).
- The inductor and capacitor voltages are 180° out of phase.

The first principle needs no clarification. The second can be explained with the help of Figure 14.1. Since $I_L = I_C$, the inductor and capacitor voltages are referenced to circuit current. As the voltage waveforms show:

- Inductor voltage *leads* circuit current by 90°.
- Capacitor voltage *lags* circuit current by 90°.

Therefore, there is a 180° difference in phase between V_L and V_C. Since they are 180° out of phase, the source voltage equals the difference between V_L and V_C, as follows:

$$V_S = V_L - V_C \tag{14.1}$$

Voltage Relationships and Phase Angles

There are three possible relationships between V_L and V_C in any series *LC* circuit:

- $V_L > V_C$
- $V_L < V_C$
- $V_L = V_C$

Depending on the relationship between V_L and V_C, the source sees the circuit as being inductive, capacitive, or a short circuit. This principle is illustrated in Figure 14.2. The positive *x*-axis is used to represent the circuit current. V_L is plotted on the positive *y*-axis, and V_C is plotted on the negative *y*-axis. As shown, the voltage phasors are 180° out of phase.

The sum of V_L and V_C determines the phase angle of V_S (relative to the circuit current). For example, Figure 14.3a shows the result phasor when $V_L > V_C$. In this case, V_L = 6 V, and V_C = 2 V. The result is a 4 V phasor that lies on the positive *y*-axis. This means that the source sees the circuit as being *inductive,* and V_S leads I_T by 90°.

When $V_L < V_C$, the result phasor falls on the negative *y*-axis, as shown in Figure 14.3b. In this case, V_L = 1 V, and V_C = 4 V. The result is a 3 V phasor that falls on the negative *y*-axis. This means the source sees the circuit as being *capacitive,* and V_S lags I_T by 90°.

When $V_L = V_C$, the *LC* circuit shorts out the voltage source. This "special case" condition is discussed in Section 14.3. For now, we will limit our discussion to cases where

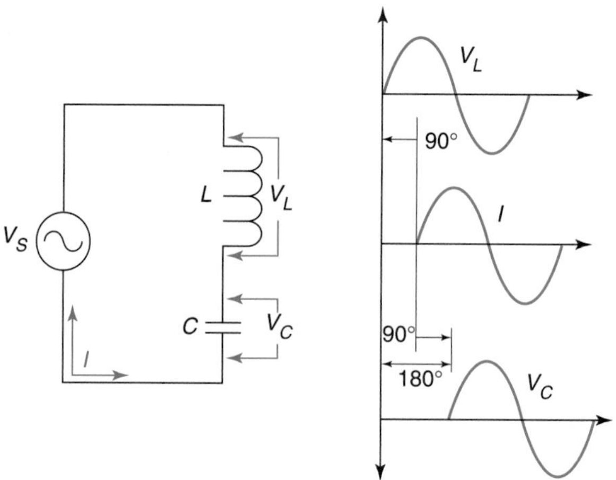

Figure 14.1 Series *LC* phase relationships.

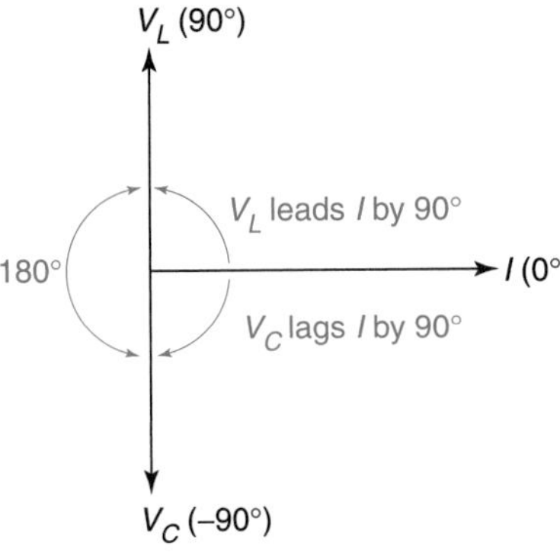

Figure 14.2 Phasor representations of V_L and V_C.

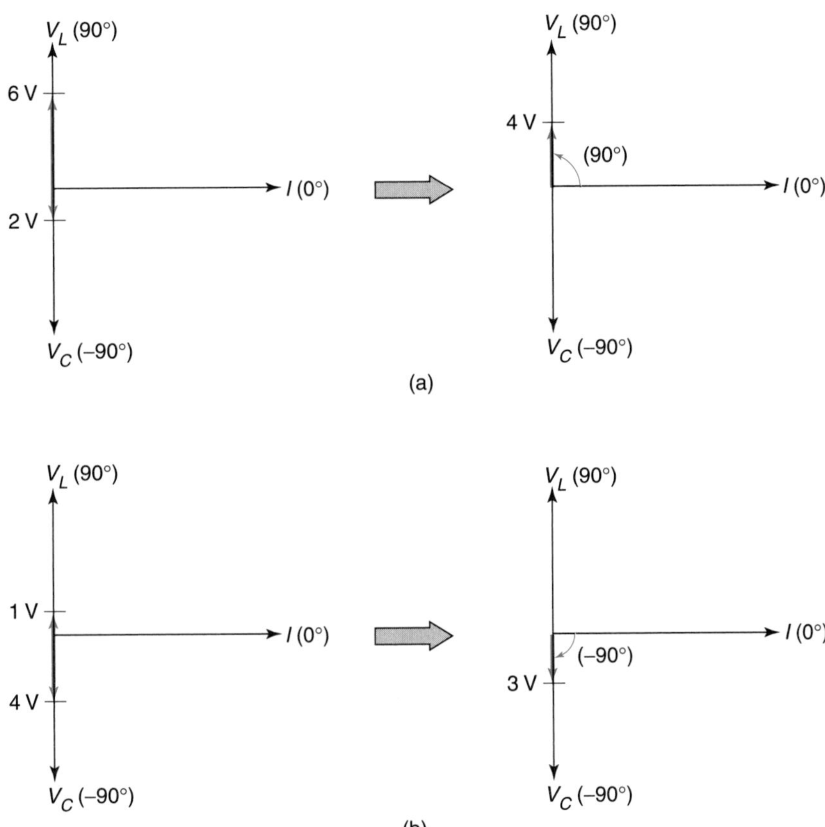

Figure 14.3 Examples of phasor combinations.

$V_L \neq V_C$. The effect of V_L and V_C on the voltage phase angle in a series LC circuit is demonstrated further in Example 14.1.

Example 14.1

Determine the value of V_S for the circuit shown in Figure 14.4.

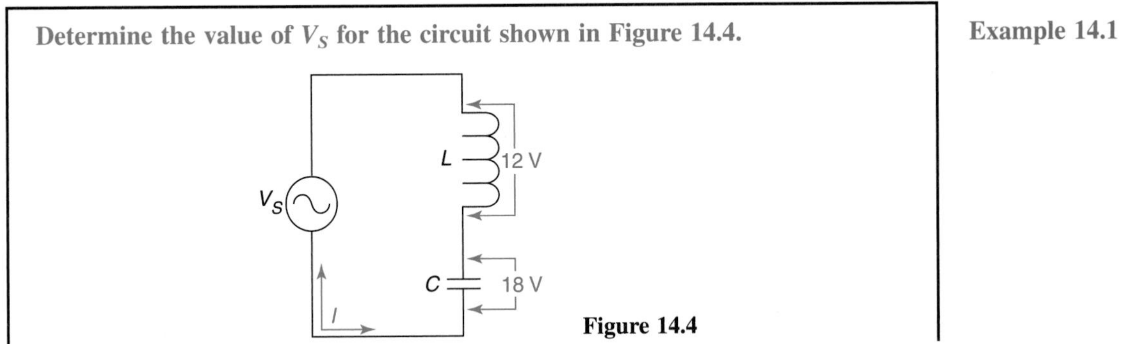

Figure 14.4

Solution: The source voltage is found as

$$V_S = V_L - V_C = 12\text{ V} - 18\text{ V} = -6\text{ V}$$

Since the result is *negative,* we know that V_S *lags* the circuit current by 90°. Therefore, $V_S = 6\text{ V}\angle -90°$.

Practice Problem 14.1

A circuit like the one shown in Figure 14.4 has values of $V_L = 1$ V and $V_C = 1.5$ V. Determine the value and phase angle of the source voltage.

The result in Example 14.1 indicates that the source sees the circuit as being capacitive in nature, because $V_C > V_L$. Example 14.2 demonstrates the phase relationship between V_S and circuit current when $V_L > V_C$.

Example 14.2

Determine the value of V_S for the circuit shown in Figure 14.5.

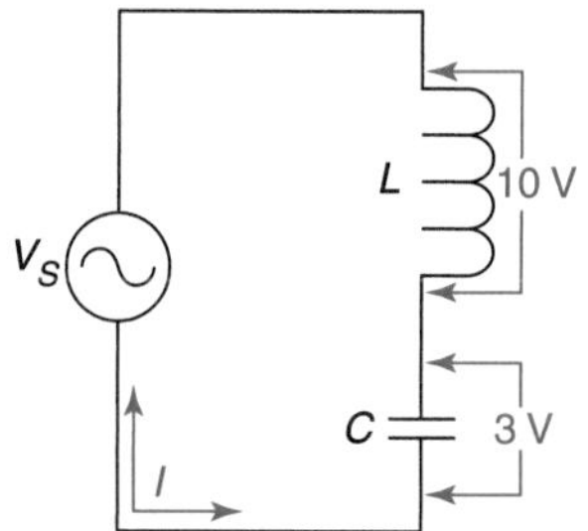

Figure 14.5

Solution: The value of the source voltage is found as

$$V_S = V_L - V_C = 10\text{ V} - 3\text{ V} = 7\text{ V}$$

Since the result is *positive,* we know that V_S *leads* the circuit current by 90°. Therefore, $V_S = 7\text{ V}\angle 90°$, and the source sees the circuit as being *inductive.*

Series Reactance (X_S)

OBJECTIVE 2 ➤ When an inductor and capacitor are connected in series, the *net series reactance* (X_S) is found as the difference between X_L and X_C. By formula,

$$X_S = X_L - X_C \quad (14.2)$$

The basis for this relationship is illustrated in Figure 14.6.

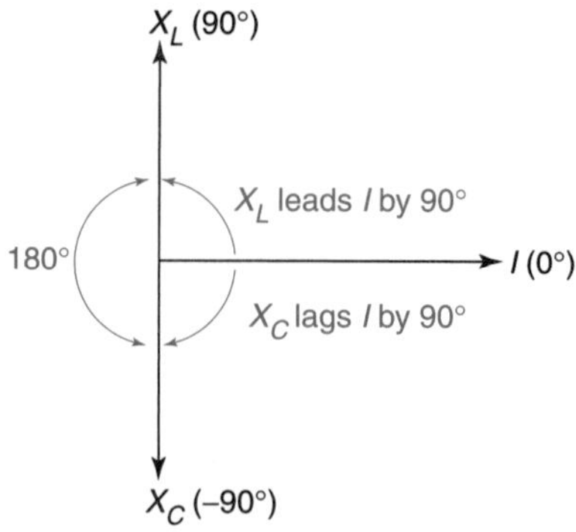

Figure 14.6

As you know, the phase angle of X_L in a series circuit is 90° (relative to circuit current). By the same token, the phase angle of X_C in a series circuit is −90° (relative to the circuit current). When X_L and X_C are plotted as shown in Figure 14.6, we can see that the phase angle between the reactance phasors is 180°.

As was the case with component voltages, the result of equation (14.2) indicates whether the circuit is seen by the source as being inductive, capacitive, or a short circuit. The case of $X_L = X_C$ is discussed in Section 14.3. For now, we're going to limit our discussion to cases where $X_L \neq X_C$. Example 14.3 demonstrates the phase relationship between X_S and circuit current when $X_L > X_C$.

Example 14.3

Determine the value of X_S for the circuit shown in Figure 14.7.

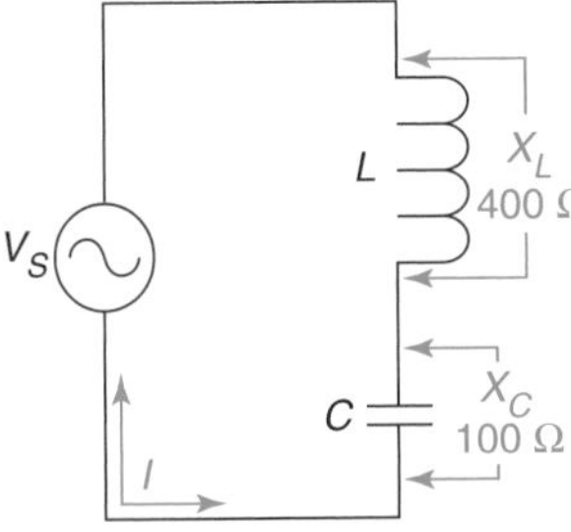

Figure 14.7

Solution: The net series reactance is found as

$$X_S = X_L - X_C = 400\ \Omega - 100\ \Omega = 300\ \Omega$$

Since the result is *positive,* we know that X_S leads the circuit current by 90°. Therefore, $X_S = 300\ \Omega\angle 90°$.

Practice Problem 14.3

A circuit like the one shown in Figure 14.7 has values of $X_L = 80\ \Omega$ and $X_C = 55\ \Omega$. Determine the value of X_S.

The result in Example 14.3 indicates that the source sees the circuit as being *inductive* in nature, because the value of X_L is greater than the value of X_C. Since the circuit is inductive in nature, the source voltage leads the circuit current by 90°. When $X_L < X_C$, the source sees the circuit as being *capacitive,* and V_S lags the circuit current by 90°.

Putting It All Together: Basic Series *LC* Circuit Analysis

◄ *OBJECTIVE 3*

In this section, we have established the basic voltage and reactance characteristics of series *LC* circuits. These characteristics are summarized in Table 14.1.

The relationships listed in Table 14.1 for $X_C > X_L$ are verified in Example 14.4.

TABLE 14.1 Series *LC* Circuit Conditions When $X_L \neq X_C$

Reactance Relationship	*Circuit Characteristics*
$X_L > X_C$	X_S leads circuit current by 90°. The source "sees" the circuit as inductive. V_S leads circuit current by 90°.
$X_C > X_L$	X_S lags circuit current by 90°. The source "sees" the circuit as capacitive. V_S lags circuit current by 90°.

Example 14.4

Determine the magnitudes and phase angles of V_L, V_C, and V_S for the circuit shown in Figure 14.8.

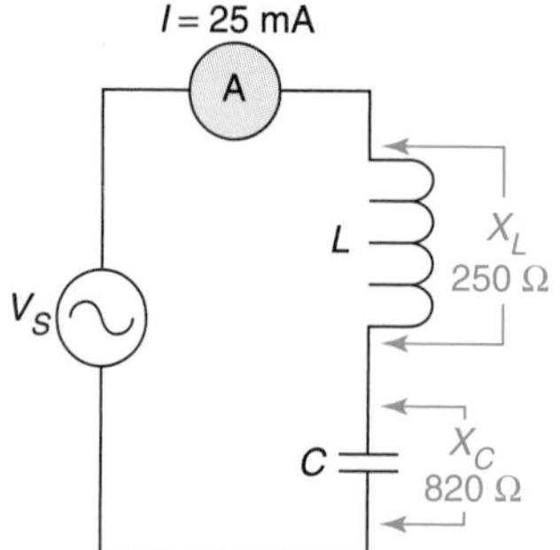

Figure 14.8

Solution: The component voltages can be found as

$$V_L = I \cdot X_L = (25 \text{ mA}\angle 0°)(250 \ \Omega\angle 90°) = 6.25 \text{ V}\angle 90°$$

and

$$V_C = I \cdot X_C = (25 \text{ mA}\angle 0°)(820 \ \Omega\angle -90°) = 20.5 \text{ V}\angle -90°$$

Using the calculated values of V_L and V_C,

$$V_S = V_L - V_C = 6.25 \text{ V} - 20.5 \text{ V} = -14.25 \text{ V}$$

As you can see, the result is *negative*. Therefore, $V_S = 14.25 \text{ V}\angle -90°$. The phase angle of V_S indicates that the source voltage lags the circuit current by 90°. Therefore, the circuit is *capacitive* in nature.

Practice Problem 14.4

A circuit like the one shown in Figure 14.8 has the following values: $X_L = 1.1 \text{ k}\Omega$, $X_C = 3.3 \text{ k}\Omega$, and $I = 4.5$ mA. Calculate V_L, V_C, and V_S for the circuit.

Section Review

1. List the current and voltage characteristics of a series *LC* circuit.
2. Describe the phase relationship between series values of V_L and V_C.
3. Describe the phase angle between source voltage and circuit current when:
   a. $V_L > V_C$ b. $V_C > V_L$
4. State and explain the phase relationship between series values of X_L and X_C.
5. Describe the phase angle between net series reactance and circuit current when:
   a. $X_L > X_C$ b. $X_L < X_C$
6. Describe the phase relationship between source voltage and circuit current in a series *LC* circuit when:
   a. $X_L > X_C$ b. $X_L < X_C$

14.2 Parallel *LC* Circuits

OBJECTIVE 4 ➤ When an inductor and a capacitor are connected in parallel, the circuit has the following current and voltage characteristics:

- The inductor and capacitor voltages are equal ($V_L = V_C$).
- The inductor and capacitor currents are 180° out of phase.

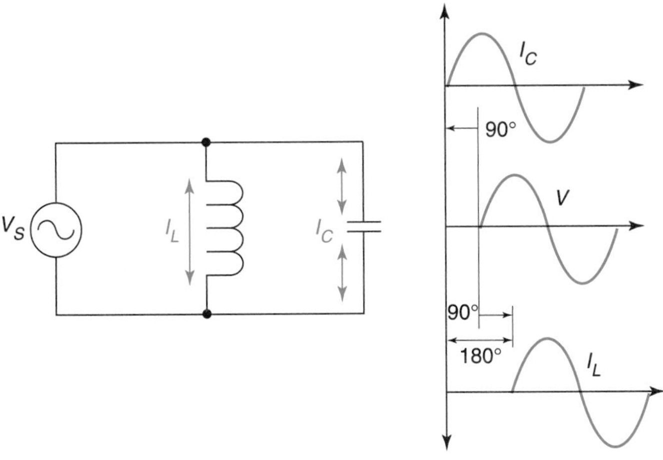

Figure 14.9 Inductor and capacitor phase relationships.

The first of these characteristics should be familiar by now. The second can be explained with the help of Figure 14.9. As the waveforms show:

- I_C leads V_S by 90°.
- I_L lags V_S by 90°.

Therefore, the current waveforms are 180° out of phase. Since they are 180° out of phase, the total current equals the difference between I_C and I_L. By formula, ◄ *OBJECTIVE 5*

$$I_T = I_C - I_L \qquad (14.3)$$

Current Relationships and Phase Angles

There are three possible relationships between I_C and I_L in any parallel *LC* circuit:

- $I_C > I_L$
- $I_C < I_L$
- $I_C = I_L$

Depending on the relationship between I_C and I_L, the source sees the circuit as being inductive, capacitive, or an open circuit. This principle is illustrated in Figure 14.10. The positive *x*-axis of the graph is used to represent the source voltage (V_S). Capacitor current leads V_S by 90°, so I_C is represented on the positive *y*-axis. Inductor current lags V_S by 90°, so I_L is represented on the negative *y*-axis. As shown, these current phasors are 180° out of phase.

When I_C and I_L are combined, the resulting phasor is inductive, capacitive, or zero, depending on the relationship between the two values. For example, Figure 14.11a shows the result when $I_C > I_L$. In this case,

$$I_T = I_C - I_L = 8\text{ mA} - 5\text{ mA} = 3\text{ mA}$$

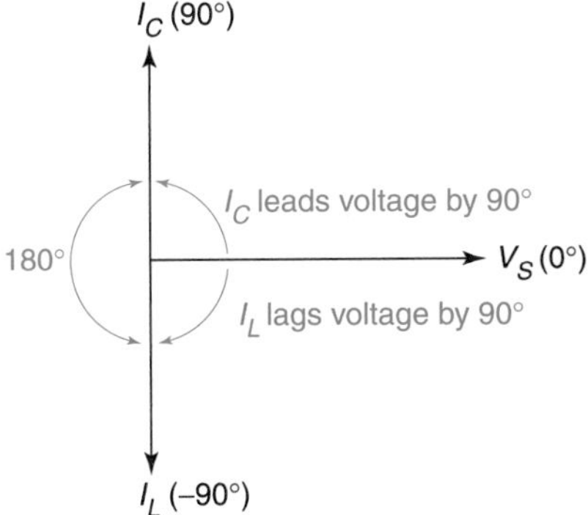

Figure 14.10 Phasor representations of I_C and I_L.

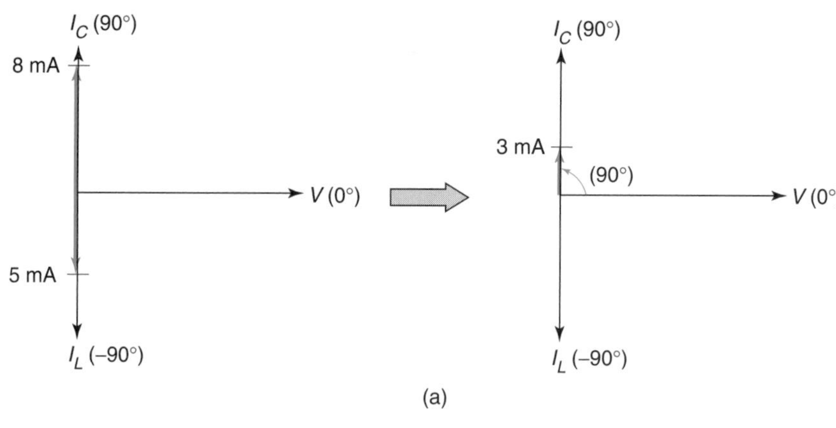

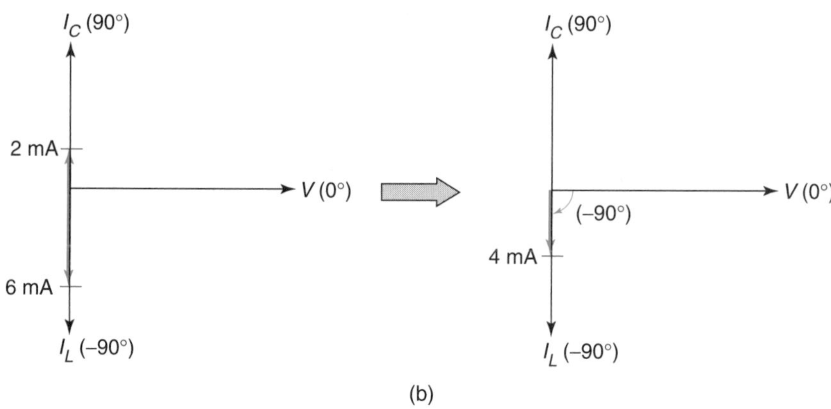

Figure 14.11 Examples of current phasor combinations.

The positive value of I_T indicates that the result is a 3 mA phasor that lies on the positive y-axis. Therefore, the circuit is *capacitive,* and I_T *leads* V_S by 90°.

When $I_C < I_L$, the resulting phasor falls on the negative y-axis, as shown in Figure 14.11b. In this case,

$$I_T = I_C - I_L = 2 \text{ mA} - 6 \text{ mA} = -2 \text{ mA}$$

The negative value of I_T indicates that the result is a 2 mA phasor that lies on the negative y-axis. Therefore, the circuit is *inductive,* and I_T *lags* V_S by 90°.

When $I_C = I_L$, the phasors have equal magnitudes and are in opposite directions. In this case, the sum of the phasors is 0 A, and the parallel LC circuit acts as an open. Note that $I_C = I_L$ is a special case that is discussed in Section 14.3.

Parallel Reactance (X_P)

OBJECTIVE 6 ➤ Because of the unique phase relationship between I_C and I_L, the total reactance of a parallel LC circuit can be greater than either (or both) of the branch reactance values. For example, the reactance values for the circuit shown in Figure 14.12 can be found as

$$X_L = \frac{V_S}{I_L} = \frac{6 \text{ V}}{40 \text{ mA}\angle -90°} = 150 \ \Omega\angle 90°$$

and

$$X_C = \frac{V_S}{I_C} = \frac{6 \text{ V}}{12 \text{ mA}\angle 90°} = 500 \ \Omega\angle -90°$$

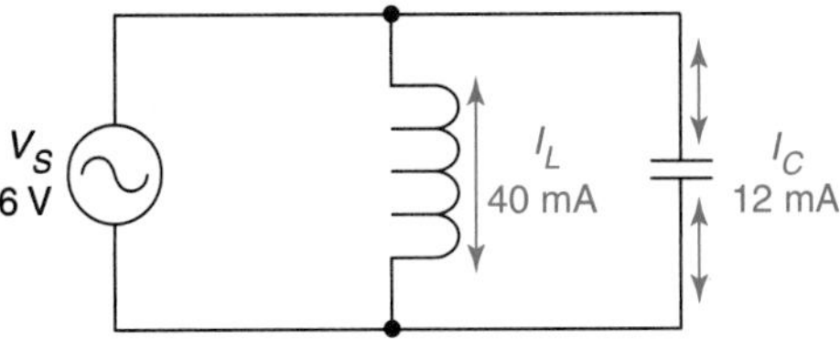

Figure 14.12

The total circuit current is found as

$$I_T = I_C - I_L = 12 \text{ mA} - 40 \text{ mA} = -28 \text{ mA} = 28 \text{ mA}\angle -90°$$

If we divide the source voltage by the total current, we get a total circuit reactance of

$$X_P = \frac{V_S}{I_T} = \frac{6 \text{ V}}{28 \text{ mA}\angle -90°} = 214.3 \ \Omega\angle 90°$$

As you can see, this value is greater than the value of $X_L = 150 \ \Omega$ calculated for the circuit.

When a capacitor and an inductor are connected in parallel, the total parallel reactance (X_P) can be found using

$$X_P = \frac{1}{\dfrac{1}{X_L} + \dfrac{1}{X_C}} \qquad (14.4)$$

or

$$X_P = \frac{X_L X_C}{X_L + X_C} \qquad (14.5)$$

Example 14.5 demonstrates the use of both equations.

Example 14.5

Calculate the value of X_P for the circuit in Figure 14.13.

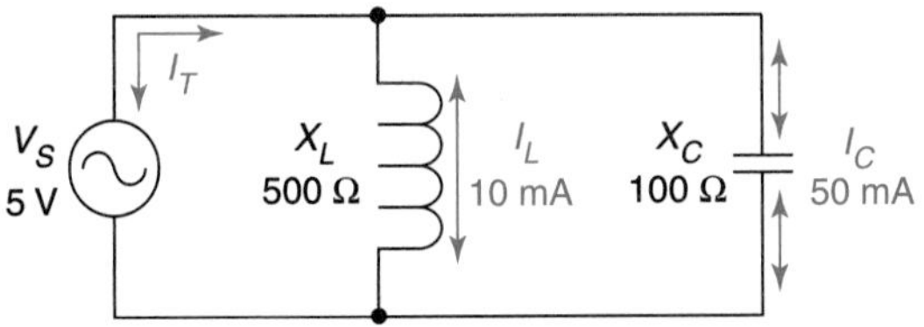

Figure 14.13

Solution: The value of X_P can be found as

$$X_P = \frac{1}{\dfrac{1}{X_L} + \dfrac{1}{X_C}} = \frac{1}{\dfrac{1}{500 \ \Omega} + \dfrac{1}{-100 \ \Omega}} = \frac{1}{\dfrac{1}{500 \ \Omega} - \dfrac{1}{100 \ \Omega}}$$
$$= -125 \ \Omega = 125 \ \Omega\angle -90°$$

or

$$X_P = \frac{X_L X_C}{X_L + X_C} = \frac{(500 \ \Omega)(-100 \ \Omega)}{500 \ \Omega + (-100 \ \Omega)} = -125 \ \Omega = 125 \ \Omega\angle -90°$$

Since the phase angle is $-90°$, the source sees the circuit as being *capacitive* in nature, and I_T leads V_S by 90°.

Practice Problem 14.5

A circuit like the one shown in Figure 14.13 has the following values: $V_S = 8$ V, $I_L = 20$ mA, and $I_C = 4$ mA. Calculate the value of X_P using Ohm's law, equation (14.4), and equation (14.5).

Putting It All Together: Basic Parallel *LC* Circuit Characteristics

The relationships among the currents, voltages, and reactances in a parallel *LC* circuit are summarized in Table 14.2. The relationships listed can be verified by reviewing the values and calculations in Example 14.5 and your results from Practice Problem 14.5.

TABLE 14.2 Parallel *LC* Circuit Conditions When $X_L \neq X_C$

Reactance Relationship	*Circuit Characteristics*
$X_L > X_C$	$I_C > I_L$. X_P has a negative phase angle. The circuit is *capacitive* in nature. Circuit current leads V_S by 90°.
$X_C > X_L$	$I_L > I_C$. X_P has a positive phase angle. The circuit is *inductive* in nature. Circuit current lags V_S by 90°.

An Important Point

Because of the unique phase relationships in a parallel *LC* circuit:

- The total circuit current is lower than the highest valued branch current.
- The value of X_P is greater than one (or both) of the branch reactance values.

These relationships can be verified using the following values from Example 14.5:

$$I_L = 10 \text{ mA} \qquad I_C = 50 \text{ mA} \qquad I_T = 40 \text{ mA}$$
$$X_L = 500\ \Omega \qquad X_C = 100\ \Omega \qquad X_P = 125\ \Omega$$

As you can see, $I_L < I_T < I_C$, and $X_L > X_P > X_C$. Note that total current is never the highest-valued current in the circuit. By the same token, X_P is never the lowest-valued reactance in the circuit.

Section Review

1. List the current and voltage characteristics of a parallel *LC* circuit.
2. Explain the phase relationship between I_L and I_C in a parallel *LC* circuit.
3. Describe the phase angle of circuit current (relative to source voltage) when:
   a. $I_L > I_C$ b. $I_C > I_L$
4. State and explain the phase relationship between parallel values of X_L and X_C.
5. Describe the phase angle of X_P (relative to source voltage) when:
   a. $X_L > X_C$ b. $X_C > X_L$
6. Describe the phase relationship between circuit current and source voltage in a parallel *LC* circuit when:
   a. $X_L > X_C$ b. $X_C > X_L$

Critical Thinking

7. What effect would a change in operating frequency have on the value of X_P in a parallel *LC* circuit? Explain your answer.

14.3 Resonance

◀ OBJECTIVE 7

In Chapters 11 and 13, you were shown that:

- Inductive reactance (X_L) varies directly with operating frequency.
- Capacitive reactance (X_C) varies inversely with operating frequency.

As you know, inductive and capacitive reactance can be found as

$$X_L = 2\pi f L \qquad \text{and} \qquad X_C = \frac{1}{2\pi f C}$$

For every LC circuit, there is an operating frequency where $X_L = X_C$, as shown in Figure 14.14. *When $X_L = X_C$, an LC circuit is said to be operating at* **resonance.** The *frequency at which this occurs* in a given LC circuit is referred to as its **resonant frequency (f_r).** The resonant frequency of a given LC circuit is found as

$$f_r = \frac{1}{2\pi\sqrt{LC}} \tag{14.6}$$

where

f_r = the frequency at which $X_L = X_C$. The use of equation (14.6) is demonstrated in Example 14.6.

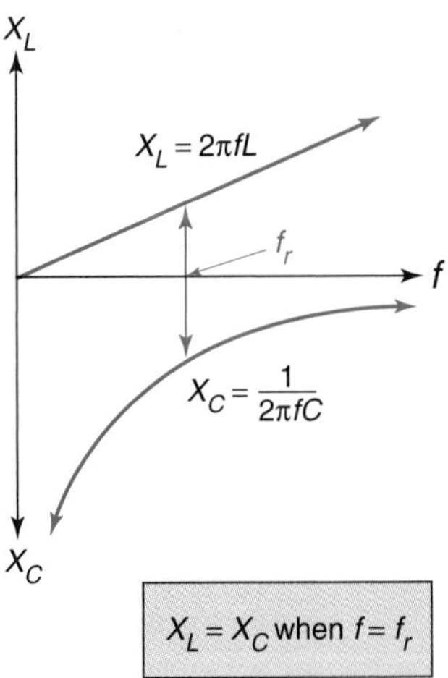

Figure 14.14

Example 14.6

Determine the resonant frequency for the series LC circuit shown in Figure 14.15.

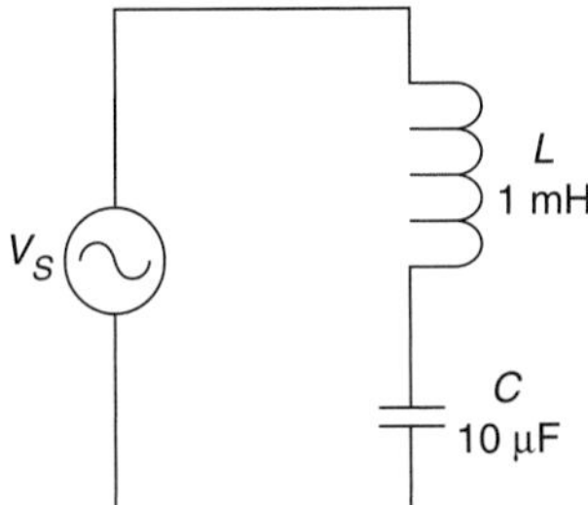

Figure 14.15

Solution: Using the values of L and C shown, the resonant frequency of the circuit can be found as

$$f_r = \frac{1}{2\pi\sqrt{LC}} = \frac{1}{2\pi\sqrt{(1\text{ mH})(10\text{ μF})}} = \frac{1}{2\pi\sqrt{10 \times 10^{-9}}}\text{ Hz} = 1.59\text{ kHz}$$

Practice Problem 14.6

A circuit like the one shown in Figure 14.15 has values of $L = 3.3$ mH and $C = 0.22$ μF. Calculate the resonant frequency of the circuit.

The values in Example 14.6 can be used to demonstrate the relationship among f_r, X_L, and X_C. At an operating frequency of $f_r = 1.59$ kHz, the circuit values of X_L and X_C can be found as

$$X_L = 2\pi f L = 2\pi(1.59\text{ kHz})(1\text{ mH}) \cong 10\ \Omega$$

and

$$X_C = \frac{1}{2\pi f C} = \frac{1}{2\pi(1.59\text{ kHz})(10\text{ μF})} = 10\ \Omega$$

As you can see, $X_L = X_C$ when the circuit is operated at its resonant frequency. Note that the principles and relationships described here apply to *parallel LC* circuits as well. That is, every parallel circuit:

- Has a resonant frequency, found as $f_r = \dfrac{1}{2\pi\sqrt{LC}}$.
- Has equal values of X_L and X_C when operated at its resonant frequency.

Factors Affecting the Value of f_r

OBJECTIVE 8 ➤ Equation (14.6) indicates that a change in either L or C causes a change in the value of f_r. Changing component values is not the only way that f_r can be affected. We will now take a brief look at some other factors that can affect the resonant frequency of an *LC* circuit.

Stray inductance Any time that alternating current passes through a conductor, a magnetic field is generated around that conductor. Most circuits contain lengths of wire, and these wires all have some amount of inductance, which is referred to as *stray inductance.* Stray inductance can also be generated by the leads on a component.

Sources of stray inductance can affect the resonant frequency of a low-inductance *LC* circuit. Stray inductance is rarely a factor in the operation of low-frequency circuits. However, as the operating frequency of a circuit increases, the values of stray inductance within the circuit can have a significant impact on the circuit operation.

Stray capacitance Whenever you have two conductors that are separated by an insulator, you have some amount of capacitance. Any two wires (conductors) separated by air (a dielectric) have some amount of capacitance between them, which is referred to as stray capacitance. Individual components can also be a source of stray capacitance. This stray capacitance can also affect the resonant frequency of a low-capacitance *LC* circuit.

Oscilloscope input capacitance Oscilloscopes have a measurable amount of *input capacitance.* Oscilloscope input capacitance can affect the resonant frequency of an *LC* circuit under test. This point is illustrated in Figure 14.16. When the oscilloscope is connected as shown, the total capacitance of the circuit *increases.* This causes a *decrease* in f_r, as given in equation (14.6). For this reason, a measured value of f_r is often lower than its calculated value.

A Practical Consideration: The input capacitance of an oscilloscope is provided on the specification page of its operator's manual. Oscilloscope input capacitance is typically in the low pF range.

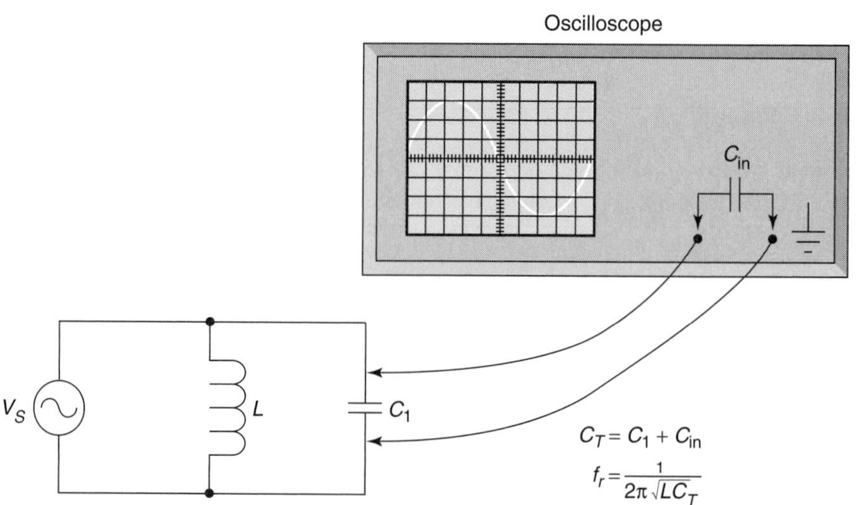

Figure 14.16 Oscilloscope input capacitance.

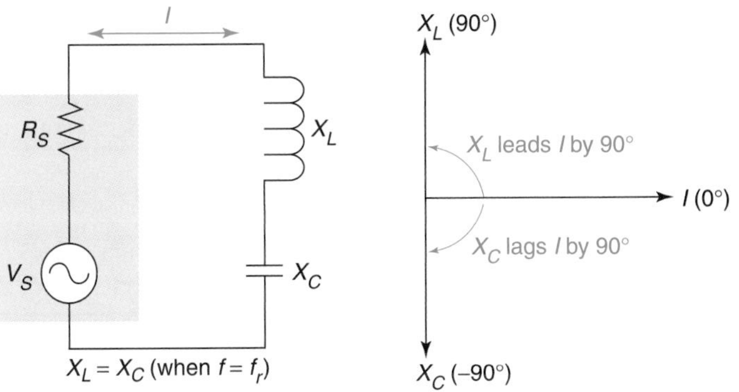

Figure 14.17
Series resonant circuit characteristics.

Series Resonant *LC* Circuits

◄ *OBJECTIVE 9*

When a series *LC* circuit is operated at resonance, it has the following characteristics:

- The total reactance of the series resonant circuit is 0 Ω.
- The voltage across the series *LC* circuit is 0 V.
- The circuit current and voltage are in phase; that is, the circuit is resistive in nature.

These characteristics can be explained using the circuit and graph shown in Figure 14.17.

Assuming that the circuit shown in Figure 14.17 is operating at resonance, the values of X_L and X_C are equal, and the total series reactance can be found as

$$X_T = X_L - X_C = 0\ \Omega$$

Since $X_T = 0\ \Omega$, the voltage across the series combination of L and C (V_{LC}) must be 0 V. This means that the source voltage in Figure 14.17 is dropped across the internal resistance of the source. Because V_S is dropped across R_S:

- The circuit is resistive in nature.
- The source voltage and circuit current are in phase.

The fact that $V_{LC} = 0$ V may lead you to believe that V_L and V_C each have a value of 0 V. However, this is not the case. V_L and V_C each have some measurable value, which may be much greater than the value of source voltage. However, the *sum* of the component voltages is 0 V (because of their phase relationship). This point is demonstrated in Example 14.7.

Example 14.7

Calculate the values of V_L, V_C, and V_{LC} for the circuit shown in Figure 14.18.

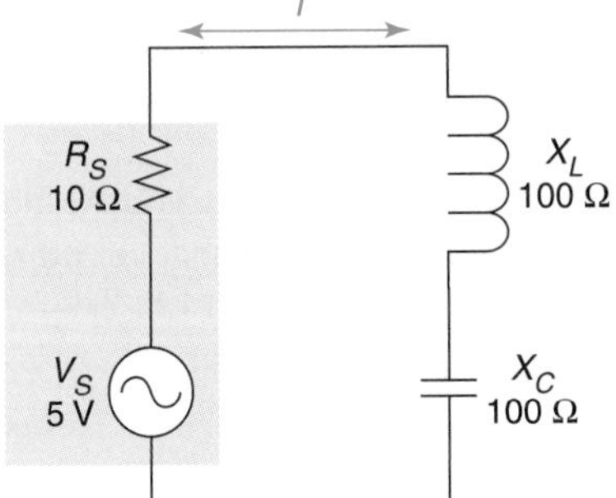

Figure 14.18

Solution: Since $X_L = X_C$, the total circuit reactance is 0 Ω, and R_S provides the only opposition to circuit current. Therefore,

$$I = \frac{V_S}{R_S} = \frac{5\text{ V}}{10\ \Omega} = 500\text{ mA}$$

Using $I = 500$ mA,

$$V_L = I \cdot X_L = (500\text{ mA})(100\ \Omega\angle 90^\circ) = 50\text{ V}\angle 90^\circ$$

and

$$V_C = I \cdot X_C = (500 \text{ mA})(100 \ \Omega\angle -90°) = 50 \text{ V}\angle -90°$$

Since these voltages are equal in magnitude and 180° out of phase, their sum is 0 V.

Practice Problem 14.7

A circuit like the one shown in Figure 14.18 has the following values: $V_S = 8$ V, $R_S = 40\ \Omega$, $X_L = 1$ kΩ, and $X_C = 1$ kΩ. Calculate V_L, V_C, and V_{LC} for the circuit.

As you can see, the circuit has values of V_L and V_C that are each ten times the value of the source voltage. However, their sum is 0 V because of the 180° phase relationship between them.

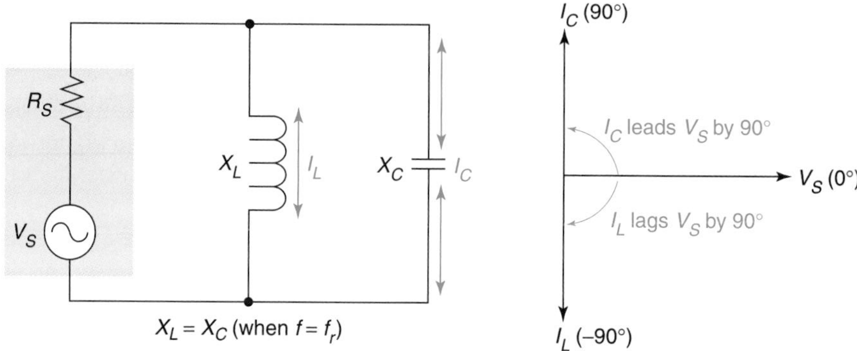

Figure 14.19 Parallel resonant circuit characteristics.

Parallel Resonant *LC* Circuits

When a parallel *LC* circuit is operated at resonance, it has the following characteristics:

- The sum of the currents through the parallel *LC* circuit is 0 A.
- The circuit has infinite reactance; that is, it acts as an open.

These characteristics can be explained using the circuit and graph shown in Figure 14.19.

Assuming that the circuit shown in Figure 14.19 is operating at resonance, $X_L = X_C$. Since the branch reactances and voltages are equal, so are the branch currents. As shown in the graph, I_L and I_C are 180° out of phase. Therefore, the total current in the circuit can be found as

$$I_T = I_C - I_L = 0 \text{ A}$$

Since the net current through the parallel resonant circuit is 0 A, the circuit is acting as an open. In effect, the total reactance of the parallel resonant circuit is infinite. These circuit characteristics are demonstrated further in Example 14.8.

Example 14.8

Calculate the values of I_C, I_L, I_T, and X_T for the parallel resonant circuit shown in Figure 14.20.

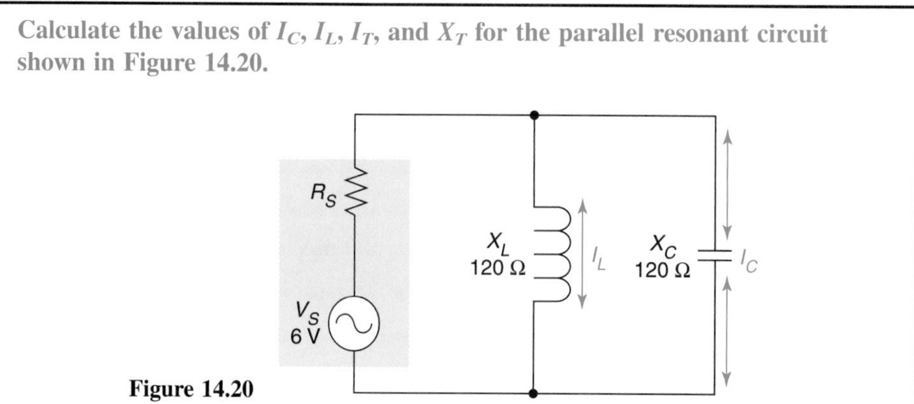

Figure 14.20

Solution: The value of capacitor current is found as

$$I_C = \frac{V_S}{X_C} = \frac{6\text{ V}}{120\ \Omega\angle -90^\circ} = 50\text{ mA}\angle 90^\circ$$

and value of the inductor current is found as

$$I_L = \frac{V_S}{X_L} = \frac{6\text{ V}}{120\ \Omega\angle 90^\circ} = 50\text{ mA}\angle -90^\circ$$

Since the branch currents are equal in magnitude and 180° out of phase, the net current through the circuit is 0 A. Using this value, the total reactance of the circuit approaches infinity, as follows:

$$X_T = \frac{V_S}{I_T} = \frac{6\text{ V}}{0\text{ A}} \rightarrow \infty\Omega$$

The arrow at the end of the calculation indicates that the value of X_T *approaches* infinity.

Practice Problem 14.8

A circuit like the one shown in Figure 14.20 has the following values: $V_S = 10$ V, $X_L = 40\ \Omega$, and $X_C = 40\ \Omega$. Calculate I_C, I_L, I_T, and X_T for the circuit.

Series Versus Parallel Resonance: A Comparison

In this section, you have been introduced to the characteristics of series and parallel resonant *LC* circuits. Figure 14.21 provides a summary comparison of the characteristics of these two types of circuits.

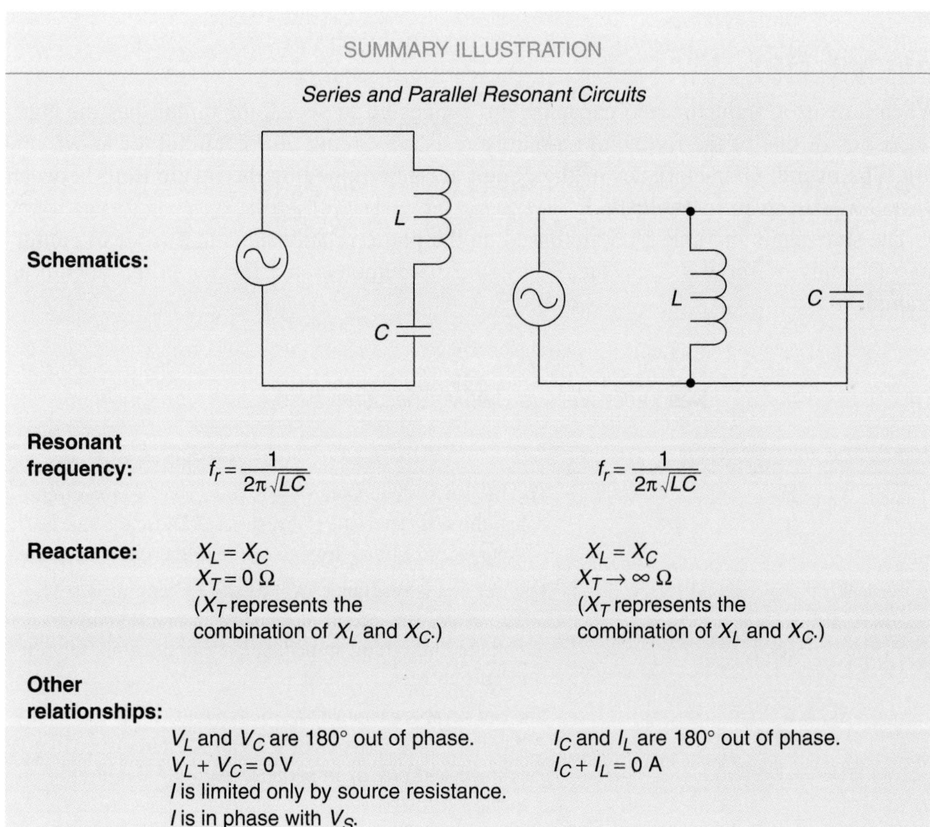

SUMMARY ILLUSTRATION
Series and Parallel Resonant Circuits
Schematics:
L
C
L
C
Resonant frequency:
$f_r = \frac{1}{2\pi\sqrt{LC}}$
$f_r = \frac{1}{2\pi\sqrt{LC}}$
Reactance:
$X_L = X_C$
$X_T = 0\ \Omega$
(X_T represents the combination of X_L and X_C.)
$X_L = X_C$
$X_T \rightarrow \infty\ \Omega$
(X_T represents the combination of X_L and X_C.)
Other relationships:
V_L and V_C are 180° out of phase.
$V_L + V_C = 0$ V
I is limited only by source resistance.
I is in phase with V_S.
I_C and I_L are 180° out of phase.
$I_C + I_L = 0$ A

Figure 14.21

One Final Note

The frequency response characteristics of series and parallel *LC* circuits are more complicated than represented in this section. In Chapter 15, we will discuss the frequency response of *LC* (and other) circuits. As you will see, values like inductor *Q*, capacitor *Q*, and circuit resistance all affect the frequency response of *LC* circuits.

Section Review

1. What is the *resonant frequency* of an *LC* circuit? How is it calculated?
2. How do changes in the values of *L* and *C* affect the resonant frequency of an *LC* circuit?
3. What is *stray inductance?*
4. What is *stray capacitance?*
5. How does the input capacitance of an oscilloscope affect the resonant frequency of an *LC* circuit under test?
6. List the basic characteristics of a *series resonant LC* circuit.
7. List the basic characteristics of a *parallel resonant LC* circuit.

14.4 Series and Parallel *RLC* Circuits

In this section, we will look at the operation and analysis of basic resistive-inductive-capacitive (*RLC*) circuits. Most of the principles covered here will probably seem familiar. For all practical purposes, we are merely combining circuit operating principles that were covered in earlier sections and chapters.

Series *RLC* Circuits

OBJECTIVE 10 ➤ When a resistor, inductor, and capacitor are connected in series, the circuit has the characteristics of one of the following: a purely resistive circuit, an *RC* circuit, or an *RL* circuit. The overall characteristics of the circuit are determined by the relationship between X_L and X_C, as given in Table 14.3.

The statements in Table 14.3 are based on the phase relationships that we learned during our study of series *LC* circuits. They will be demonstrated further in the upcoming examples.

TABLE 14.3 Reactance Versus Circuit Characteristics

Reactance Relationship	*Resulting Circuit Characteristics*
$X_L > X_C$	The net series reactance (X_S) is *inductive*, so the circuit has the characteristics of a series *RL* circuit: source voltage and circuit impedance lead the circuit current.
$X_L = X_C$	The net series reactance (X_S) of the *LC* circuit is 0 Ω. Therefore, the circuit is *resistive* in nature: source voltage and circuit impedance are both in phase with circuit current.
$X_C > X_L$	The net series reactance (X_S) is *capacitive*, so the circuit has the characteristics of a series *RC* circuit: source voltage and circuit impedance both lag the circuit current.

Total Series Impedance

We know that the net series reactance of an LC circuit is at a 90° angle to R when its value is inductive or capacitive. This means that the circuit impedance is found as the geometric sum of X_S and R. By formula,

$$Z_T = \sqrt{X_S^2 + R^2} \tag{14.7}$$

where

X_S = the net series reactance, found as $X_S = X_L - X_C$

The phase angle of the circuit impedance is found as

$$\theta = \tan^{-1}\left(\frac{X_S}{R}\right) \tag{14.8}$$

When $X_L > X_C$, the circuit impedance of a series RLC circuit has a positive phase angle, which indicates that the circuit has the characteristics of a series RL circuit. This relationship is demonstrated in Example 14.9.

Example 14.9

Determine the circuit impedance in Figure 14.22.

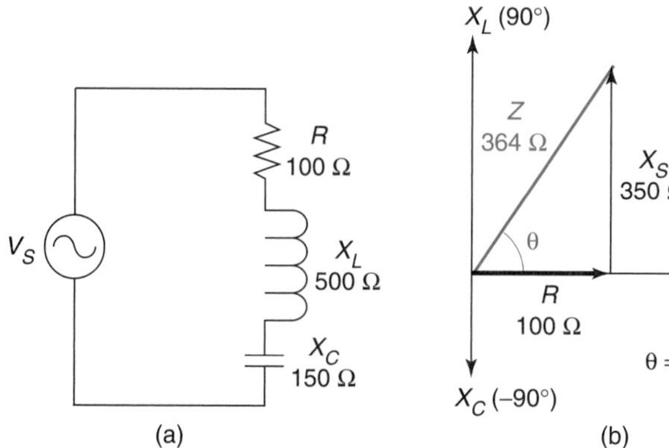

Figure 14.22

Solution: First, the net series reactance is found as

$$X_S = X_L - X_C = 500\ \Omega - 150\ \Omega = 350\ \Omega$$

The circuit impedance has a magnitude of

$$Z_T = \sqrt{X_S^2 + R^2} = \sqrt{(350\ \Omega)^2 + (100\ \Omega)^2} = 364\ \Omega$$

at a phase angle of

$$\theta = \tan^{-1}\left(\frac{X_S}{R}\right) = \tan^{-1}\left(\frac{350\ \Omega}{100\ \Omega}\right) = 74°$$

Since the phase angle of Z_T is *positive,* we know the circuit has the phase characteristics of a series RL circuit.

Practice Problem 14.9

A circuit like the one shown in Figure 14.22 has the following values: $R = 330\ \Omega$, $X_L = 840\ \Omega$, and $X_C = 600\ \Omega$. Calculate the circuit impedance.

When $X_C > X_L$, the circuit impedance of a series *RLC* circuit has a negative phase angle, which indicates that the circuit has the characteristics of a series *RC* circuit. This relationship is demonstrated in Example 14.10.

Example 14.10

Determine the magnitude and phase angle of the circuit impedance in Figure 14.23.

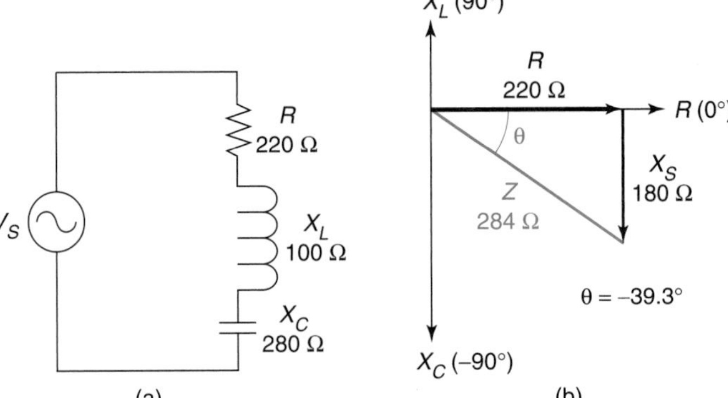

Figure 14.23

Solution: First, the net series reactance is found as

$$X_S = X_L - X_C = 100\ \Omega - 280\ \Omega = -180\ \Omega$$

The circuit impedance has a magnitude of

$$Z_T = \sqrt{X_S^2 + R^2} = \sqrt{(-180\ \Omega)^2 + (220\ \Omega)^2} = 284\ \Omega$$

at a phase angle of

$$\theta = \tan^{-1}\left(\frac{X_S}{R}\right) = \tan^{-1}\left(\frac{-180\ \Omega}{220\ \Omega}\right) = -39.3°$$

Since the phase angle of Z_T is *negative,* we know the circuit has the phase characteristics of a series *RC* circuit.

Practice Problem 14.10

A circuit like the one shown in Figure 14.23 has the following values: $R = 1\ \text{k}\Omega$, $X_L = 300\ \Omega$, and $X_C = 1.5\ \text{k}\Omega$. Calculate the circuit impedance.

When a series *RLC* is operating at resonance, the values of inductive and capacitive reactance are equal. When $X_L = X_C$, the net series reactance (X_S) is 0 Ω. With a value of $X_S = 0\ \Omega$, the circuit impedance is found as

$$Z_T = \sqrt{X_S^2 + R^2} = \sqrt{(0\ \Omega)^2 + R^2} = \sqrt{R^2} = R$$

As these values indicate, a series *RLC* circuit operating at resonance has no phase angle and is purely resistive in nature.

Series Circuit Frequency Response

You have seen how the values of reactance determine the operating characteristics of a series *RLC* circuit. Since reactance varies with operating frequency, the operating frequency of a series *RLC* circuit plays a role in its circuit characteristics. For example, consider the frequency graphs shown in Figure 14.24. The graph in Figure 14.24a illustrates the relationship between X_L and X_C when the operating frequency (f_o) is greater than the circuit's resonant frequency. As shown, $X_L > X_C$ whenever $f_o > f_r$. This indicates that a

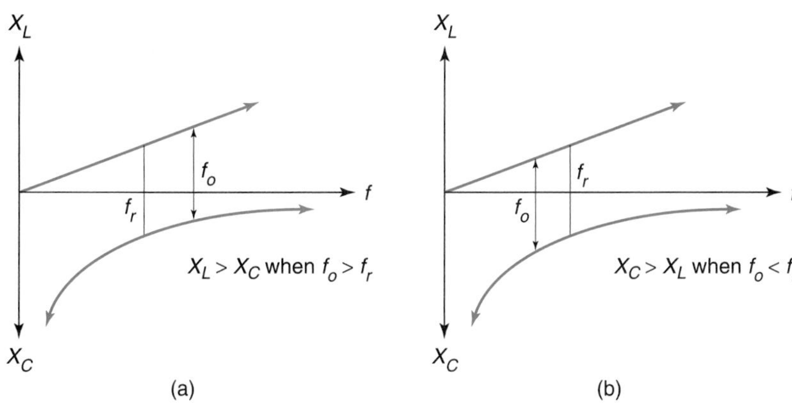

Figure 14.24 Reactance versus frequency.

series *RLC* circuit is *inductive* in nature when operated above its resonant frequency. By the same token, the graph in Figure 14.24b shows that $X_C > X_L$ whenever $f_o < f_r$. This indicates that a series *RLC* circuit is *capacitive* in nature when operated below its resonant frequency. Of course, we already know that a series *RLC* circuit is resistive in nature when $f_o = f_r$. The relationship between operating frequency and the characteristics of a series *RLC* circuit is summarized in Table 14.4.

We will discuss the effects of operating frequency on circuit operation in greater detail in Chapter 15. For now, we merely want to establish that f_o is one of the determining factors in the operating characteristics of a series *RLC* circuit.

Series Voltages

Until now, we have ignored the voltage relationships in series *RLC* circuits. As you know, current is considered to be the 0° reference in any series circuit. Therefore, the voltage phase angles in a series *RLC* circuit can be determined as follows:

$$\begin{aligned} V_L &= I \cdot X_L \\ &= (I\angle 0°)(X_L\angle 90°) \\ &= (I \cdot X_L)\angle 90° \end{aligned} \qquad \begin{aligned} V_R &= I \cdot R \\ &= (I\angle 0°)(R\angle 0°) \\ &= (I \cdot R)\angle 0° \end{aligned} \qquad \begin{aligned} V_C &= I \cdot X_C \\ &= (I\angle 0°)(X_C\angle -90°) \\ &= (I \cdot X_C)\angle -90° \end{aligned}$$

These calculations indicate that the component voltages in a series *RLC* circuit have the phase relationship illustrated in Figure 14.25.

With the phase relationships shown, the net reactive voltage (V_{LC}) equals the difference between V_L and V_C, as follows:

$$V_{LC} = V_L - V_C \tag{14.9}$$

TABLE 14.4 Effects of Operating Frequency on Series *RLC* Circuit Characteristics

Operating Frequency	*Reactance*	*Other Characteristics*
$f_o < f_r$	$X_C > X_L$	The circuit is *capacitive* in nature, meaning that it has the operating characteristics of a series *RC* circuit.
$f_o = f_r$	$X_L = X_C$	The reactances cancel out, meaning that the circuit has the operating characteristics of a purely *resistive* circuit.
$f_o > f_r$	$X_L > X_C$	The circuit is *inductive* in nature, meaning that it has the operating characteristics of a series *RL* circuit.

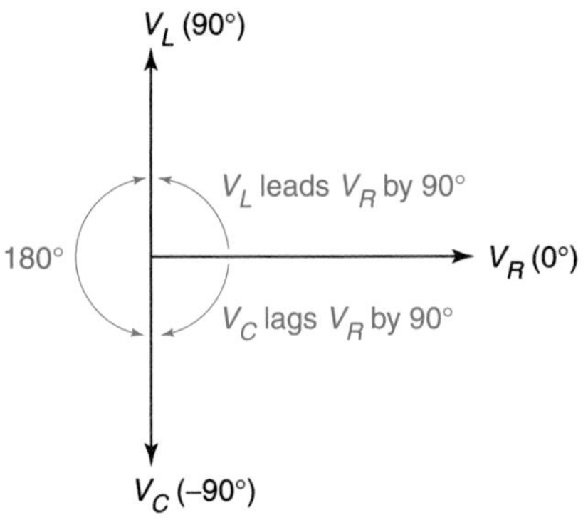

Figure 14.25 Voltage phase relationships in a series *RLC* circuit.

The magnitude of the source voltage equals the geometric sum of V_{LC} and V_R. By formula,

$$V_S = \sqrt{V_{LC}^2 + V_R^2} \tag{14.10}$$

where

V_S = the source voltage
V_{LC} = the net reactive voltage
V_R = the voltage across the resistor

Finally, the phase angle of the source voltage (relative to the circuit current) is found as

$$\theta = \tan^{-1}\left(\frac{V_{LC}}{V_R}\right) \tag{14.11}$$

As is the case in any series resistive-reactive circuit, the phase angle of the source voltage equals the phase angle of the circuit impedance.

The typical analysis of a series *RLC* circuit involves determining the values of reactance, impedance, current, and voltage. A typical analysis problem is provided in Example 14.11.

Example 14.11

Perform a typical analysis of the circuit shown in Figure 14.26.

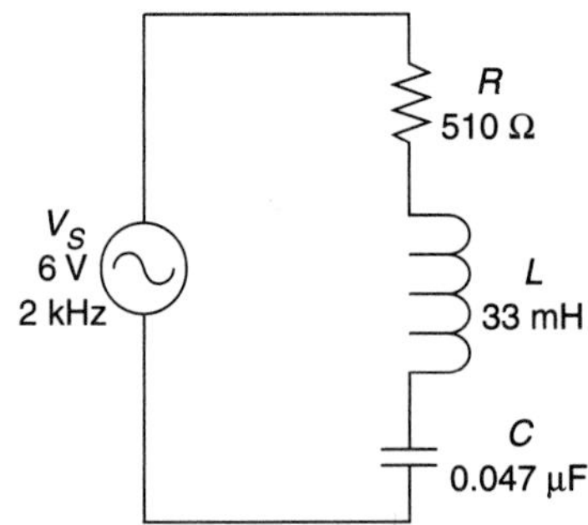

Figure 14.26

Solution: The value of inductive reactance is found as

$$X_L = 2\pi fL = 2\pi(2\text{ kHz})(33\text{ mH}) = 415\ \Omega$$

and the value of capacitive reactance is found as

$$X_C = \frac{1}{2\pi fC} = \frac{1}{2\pi(2\text{ kHz})(0.047\ \mu\text{F})} = 1.69\text{ k}\Omega$$

The net series reactance for the circuit is found as

$$X_S = X_L - X_C = 415\ \Omega - 1.69\ \text{k}\Omega \cong -1.28\ \text{k}\Omega$$

The circuit impedance has a magnitude of

$$Z_T = \sqrt{X_S^2 + R^2} = \sqrt{(-1.28\ \text{k}\Omega)^2 + (510\ \Omega)^2} = 1.38\ \text{k}\Omega$$

at a phase angle of

$$\theta = \tan^{-1}\left(\frac{X_S}{R}\right) = \tan^{-1}\left(\frac{-1.28\ \text{k}\Omega}{510\ \Omega}\right) = -68.3°$$

Since the phase angle of Z_T is *negative,* we know the circuit has the phase characteristics of a series *RC* circuit. Once we know the value of the circuit impedance, the magnitude of the circuit current can be found as

$$I = \frac{V_S}{Z_T} = \frac{6\ \text{V}}{1.38\ \text{k}\Omega} = 4.35\ \text{mA}$$

Remember, the circuit current in any series circuit is assumed to have a phase angle of 0°. Once the magnitude of the circuit current is known, the component voltages can be found as

$$V_R = I \cdot R = (4.35\ \text{mA}\angle 0°)(510\ \Omega\angle 0°) = 2.22\ \text{V}\angle 0°$$
$$V_L = I \cdot X_L = (4.35\ \text{mA}\angle 0°)(415\ \Omega\angle 90°) = 1.81\ \text{V}\angle 90°$$

and

$$V_C = I \cdot X_C = (4.35\ \text{mA}\angle 0°)(1.69\ \text{k}\Omega\angle -90°) = 7.35\ \text{V}\angle -90°$$

To verify our calculated component voltages, we can use them to calculate the value of the source voltage. If our result matches the known source voltage, then our calculations are correct. First, the net reactive voltage can be found as

$$V_{LC} = V_L - V_C = 1.81\ \text{V} - 7.35\ \text{V} = -5.54\ \text{V}$$

This result indicates that the net reactive voltage is capacitive in nature. Combining V_{LC} with V_R, we get a magnitude of

$$V_S = \sqrt{V_{LC}^2 + V_R^2} = \sqrt{(-5.54\ \text{V})^2 + (2.22\ \text{V})^2} = 5.97\ \text{V}$$

at a phase angle of

$$\theta = \tan^{-1}\left(\frac{V_{LC}}{V_R}\right) = \tan^{-1}\left(\frac{-5.54\ \text{V}}{2.22\ \text{V}}\right) = -68.2°$$

We know that the output from the voltage source in Figure 14.26 is 6 V. We also know that the source voltage has the same phase angle as the circuit impedance (−68.3°). Our calculated source voltage is approximately equal to 6 V∠−68.3°. This validates our results.

Practice Problem 14.11

A circuit like the one shown in Figure 14.26 has the following values: $V_S = 9$ V, $f = 10$ kHz, $R = 1.5$ kΩ, $L = 22$ mH, and $C = 33$ nF. Perform a typical analysis of the circuit.

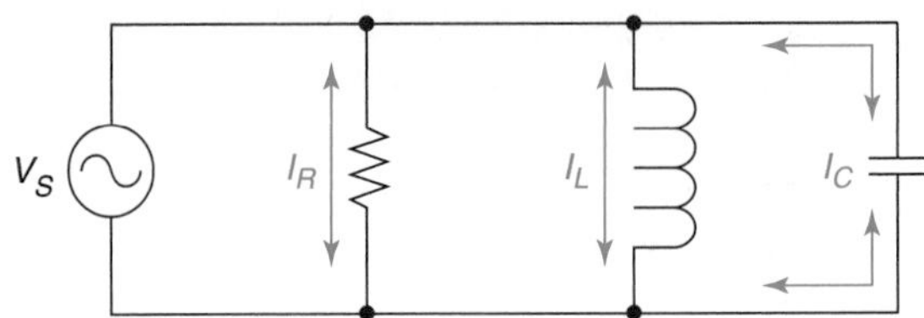

Figure 14.27 Parallel *RLC* circuit currents.

Parallel *RLC* Circuits

OBJECTIVE 11 ➤ When a resistor, capacitor, and inductor are connected in parallel, the circuit has the characteristics of a purely resistive circuit, a parallel *RL* circuit, or a parallel *RC* circuit. For example, look at the circuit shown in Figure 14.27. The overall characteristics of the circuit are determined by the relationship between I_L and I_C, as given in Table 14.5.

The statements in Table 14.5 are based on the phase relationships shown in Figure 14.28a. Since I_L and I_C are 180° out of phase, the net reactive current (I_{LC}) is equal to the difference between the two. This means that:

- I_{LC} parallels the negative *y*-axis when $I_L > I_C$ (see Figure 14.28b). In this case, I_{LC} is *inductive* in nature, and the circuit has the characteristics of a parallel *RL* circuit.
- I_{LC} has a value of 0 A when $I_L = I_C$ (see Figure 14.28c). In this case, the total circuit current equals the current through the resistor, meaning that the circuit is purely resistive.
- I_{LC} parallels the positive *y*-axis when $I_C > I_L$ (see Figure 14.28d). In this case, I_{LC} is *capacitive* in nature, and the circuit has the characteristics of a parallel *RC* circuit.

These relationships will be demonstrated further in upcoming examples.

Total Parallel Current

Figures 14.28(b) and (d) indicate that I_{LC} is at a 90° angle to I_R when its value is inductive or capacitive. In either case, the circuit current is found as the geometric sum of I_{LC} and I_R. By formula,

$$I_T = \sqrt{I_{LC}^2 + I_R^2} \tag{14.12}$$

where I_{LC} is the net reactive current, found as $I_{LC} = I_C - I_L$. The phase angle of the circuit current (relative to the source voltage) is found as

$$\theta = \tan^{-1}\left(\frac{I_{LC}}{I_R}\right) \tag{14.13}$$

TABLE 14.5 Reactive Current Versus Circuit Characteristics

Current Relationship	*Resulting Circuit Characteristics*
$I_L > I_C$	The net reactive current is *inductive,* so the circuit has the characteristics of a parallel *RL* circuit: source voltage leads the circuit current and lags the circuit impedance.
$I_L = I_C$	The resonant *LC* circuit has a net current of 0 A, so the circuit is *resistive* in nature: source voltage, current, and impedance are all in phase.
$I_C > I_L$	The net reactive current is *capacitive,* so the circuit has the characteristics of a parallel *RC* circuit: source voltage lags the circuit current and leads the circuit impedance.

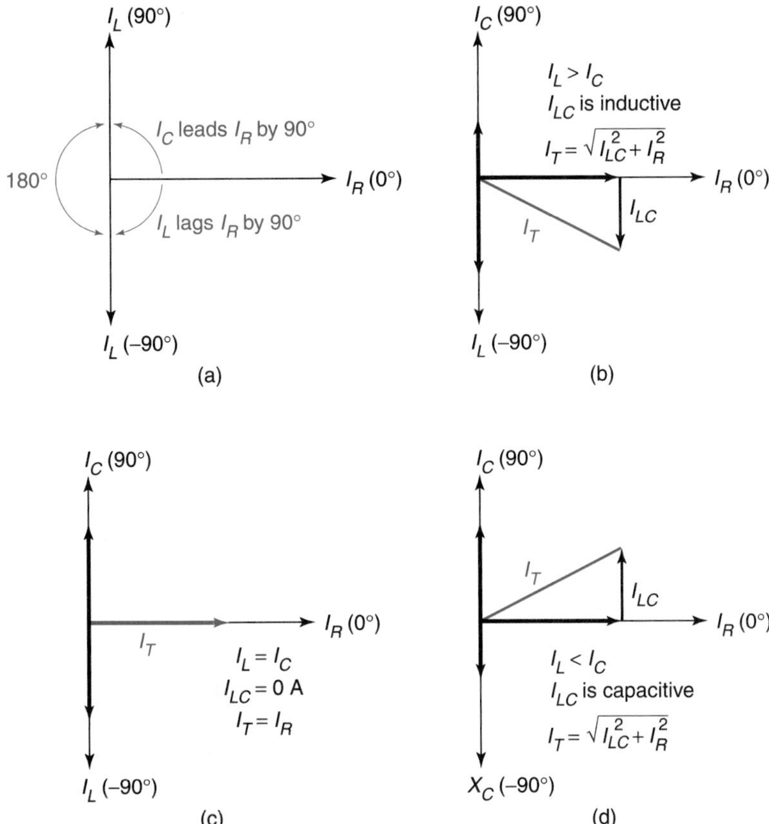

Figure 14.28 The relationship among I_L, I_C, and I_{LC}.

When $I_L > I_C$, the total current in a parallel *RLC* circuit has a negative phase angle, which indicates that the circuit has the characteristics of a parallel *RL* circuit. This point is demonstrated in Example 14.12.

Example 14.12

Determine the value of the circuit current in Figure 14.29.

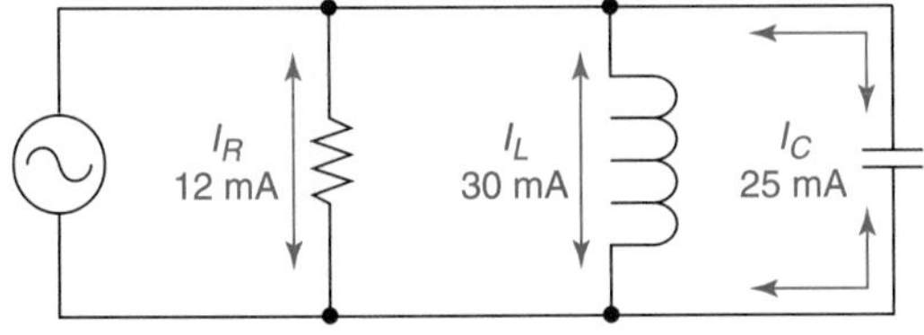

Figure 14.29

Solution: First, the net reactive current is found as

$$I_{LC} = I_C - I_L = 25\text{ mA} - 30\text{ mA} = -5\text{ mA}$$

The circuit current has a magnitude of

$$I_T = \sqrt{I_{LC}^2 + I_R^2} = \sqrt{(-5\text{ mA})^2 = (12\text{ mA})^2} = 13\text{ mA}$$

at a phase angle of

$$\theta = \tan^{-1}\left(\frac{I_{LC}}{I_R}\right) = \tan^{-1}\left(\frac{-5\text{ mA}}{12\text{ mA}}\right) = -22.6°$$

The negative phase angle of I_T indicates that the circuit is inductive in nature. Therefore, it has the phase characteristics of a parallel *RL* circuit.

Practice Problem 14.12

A circuit like the one shown in Figure 14.29 has the following values: $I_R = 30$ mA, $I_L = 50$ mA, and $I_C = 10$ mA. Calculate the value of the circuit current.

When $I_C > I_L$, the circuit current in a parallel *RLC* circuit has a positive phase angle, which indicates that the circuit has the characteristics of a parallel *RC* circuit. This point is demonstrated in Example 14.13.

Example 14.13

Determine the circuit current in Figure 14.30.

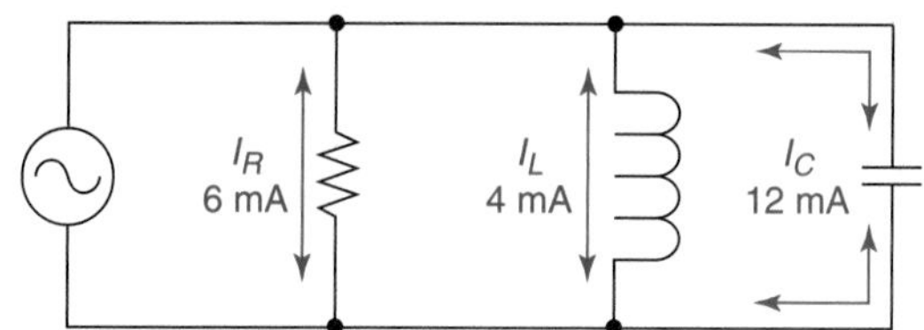

Figure 14.30

Solution: First, the net reactive current is found as

$$I_{LC} = I_C - I_L = 12 \text{ mA} - 4 \text{ mA} = 8 \text{ mA}$$

The circuit current has a magnitude of

$$I_T = \sqrt{I_{LC}^2 + I_R^2} = \sqrt{(8 \text{ mA})^2 + (6 \text{ mA})^2} = 10 \text{ mA}$$

at a phase angle of

$$\theta = \tan^{-1}\left(\frac{I_{LC}}{I_R}\right) = \tan^{-1}\left(\frac{8 \text{ mA}}{6 \text{ mA}}\right) = 53.1°$$

The positive phase angle of I_T indicates that the circuit is capacitive in nature. Therefore, it has the phase characteristics of a parallel *RC* circuit.

Practice Problem 14.13

A circuit like the one shown in Figure 14.30 has the following values: $I_R = 20$ mA, $I_L = 12$ mA, and $I_C = 40$ mA. Calculate the value of the circuit current.

When a parallel *RLC* circuit is operating at resonance, the values of I_L and I_C are equal. Therefore, the net reactive current is 0 A. With a value of $I_{LC} = 0$ A, the magnitude of the circuit current is found as

$$I_T = \sqrt{I_{LC}^2 + I_R^2} = \sqrt{(0 \text{ mA})^2 + I_R^2} = \sqrt{I_R^2} = I_R$$

As this value indicates, a parallel *RLC* circuit operating at resonance has no phase angle and is purely resistive in nature.

Parallel Circuit Frequency Response

You have seen how changes in operating frequency affect the operation of a series *RLC* circuit. The effects of operating frequency on parallel *RLC* circuit operation can be described using the graphs shown in Figure 14.31. The curves represent the relationship between component current and operating frequency (assuming that component voltage remains constant).

The graph in Figure 14.31a illustrates the relationship between I_L and I_C when the operating frequency (f_o) is greater than the circuit's resonant frequency. As shown, $I_C > I_L$ whenever $f_o > f_r$. Therefore, a parallel *RLC* circuit is *capacitive* in nature when operated

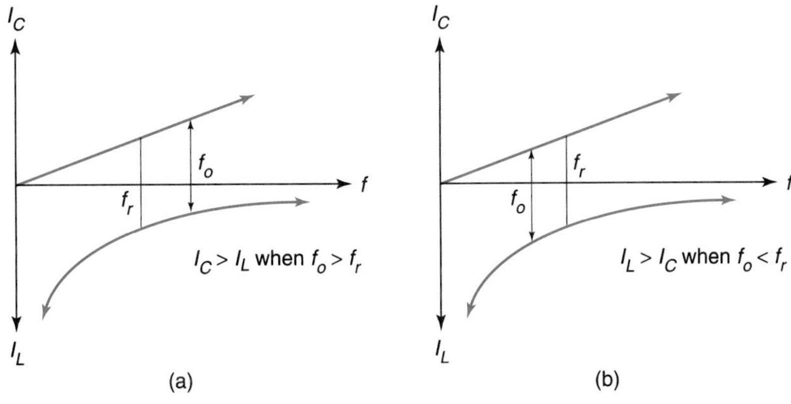

Figure 14.31 Reactive branch current versus frequency.

TABLE 14.6 Effects of Operating Frequency on Parallel *RLC* Circuit Characteristics

Operating Frequency	*Reactive Current*	*Other Characteristics*
$f_o > f_r$	$I_C > I_L$	The net reactive current is *capacitive* in nature. As a result, the circuit has the operating characteristics of a parallel *RC* circuit.
$f_o = f_r$	$I_C = I_L$	The currents cancel, so the net reactive current is 0 A. As a result, $I_T = I_R$, and the circuit has the operating characteristics of a purely *resistive* circuit.
$f_o < f_r$	$I_L > I_C$	The net reactive current is *inductive* in nature. As a result, the circuit has the operating characteristics of a parallel *RL* circuit.

above its resonant frequency. By the same token, the graph in Figure 14.31b indicates that $I_L > I_C$ whenever $f_o < f_r$. This means that a parallel *RLC* circuit is *inductive* in nature when operated below its resonant frequency. These relationships are summarized in Table 14.6.

Parallel Circuit Analysis

The analysis of a parallel *RLC* circuit normally begins with determining the branch currents and the total circuit current. Then, the values of V_S and I_T are used to calculate the circuit impedance. A typical analysis of a parallel *RLC* circuit is demonstrated in Example 14.14.

Example 14.14

Perform a typical analysis on the circuit shown in Figure 14.32.

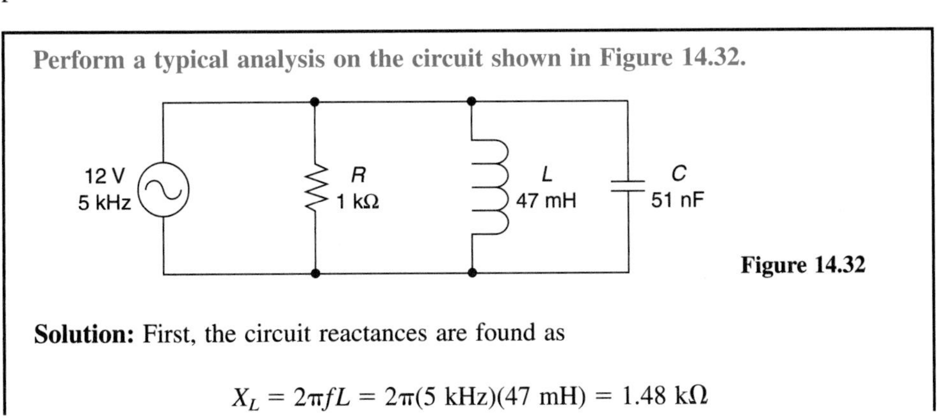

Figure 14.32

Solution: First, the circuit reactances are found as

$$X_L = 2\pi fL = 2\pi(5\text{ kHz})(47\text{ mH}) = 1.48\text{ k}\Omega$$

and

$$X_C = \frac{1}{2\pi fC} = \frac{1}{2\pi(5\text{ kHz})(51\text{ nF})} = 624\ \Omega$$

The reactive currents can now be found as

$$I_L = \frac{V_S}{X_L} = \frac{12\text{ V}\angle 0^\circ}{1.48\text{ k}\Omega\angle 90^\circ} = 8.11\text{ mA}\angle -90^\circ$$

and

$$I_C = \frac{V_S}{X_C} = \frac{12\text{ V}\angle 0^\circ}{624\ \Omega\angle -90^\circ} = 19.23\text{ mA}\angle 90^\circ$$

Since the currents are 180° out of phase, the net reactive current is found as

$$I_{LC} = I_C - I_L = 19.23\text{ mA} - 8.11\text{ mA} = 11.12\text{ mA}$$

The positive value of I_{LC} indicates that it is *capacitive* in nature. The value of I_R is found as

$$I_R = \frac{V_S}{R} = \frac{12\text{ V}}{1\text{ k}\Omega} = 12\text{ mA}$$

The circuit current has a magnitude of

$$I_T = \sqrt{I_{LC}^2 + I_R^2} = \sqrt{(11.12\text{ mA})^2 + (12\text{ mA})^2} = 16.36\text{ mA}$$

at a phase angle of

$$\theta = \tan^{-1}\left(\frac{I_{LC}}{I_R}\right) = \tan^{-1}\left(\frac{11.12\text{ mA}}{12\text{ mA}}\right) = 42.8^\circ$$

Finally, the circuit impedance can be found as

$$Z_T = \frac{V_S}{I_T} = \frac{12\text{ V}\angle 0^\circ}{16.36\text{ mA}\angle 42.8^\circ} = 733\ \Omega\angle -42.8^\circ$$

Practice Problem 14.14

A circuit like the one shown in Figure 14.32 has the following values: $V_S = 10$ V, $f = 8$ kHz, $R = 750\ \Omega$, $L = 10$ mH, and $C = 0.1\ \mu$F. Perform a typical analysis on the circuit.

Power Calculations

We have ignored power calculations throughout this section, because they haven't changed. Whether you are dealing with a series or a parallel *RLC* circuit, power is calculated just as it is for any resistive-reactive circuit. The values of apparent, reactive, and resistive power can be found using the relationships introduced earlier in the text.

Section Review

1. Describe the phase relationships for a series *RLC* circuit under the following conditions:
   a. $X_L > X_C$
   b. $X_L = X_C$
   c. $X_C > X_L$

2. What values are used to plot the impedance triangle for a series *RLC* circuit?
3. Describe the phase relationships for a series *RLC* circuit under the following conditions:
   a. $f_o > f_r$
   b. $f_o = f_r$
   c. $f_o < f_r$
4. Describe the phase relationships for a parallel *RLC* circuit under the following conditions:
   a. $I_C > I_L$
   b. $I_C = I_L$
   c. $I_L > I_C$
5. What values are used to plot the current triangle for a parallel *RLC* circuit?
6. Describe the phase relationships for a parallel *RLC* circuit under the following conditions:
   a. $f_o > f_r$
   b. $f_o = f_r$
   c. $f_o < f_r$
7. Refer to Example 14.14. If the circuit *operating frequency* is cut in half, what happens to each of the values calculated in the example? Critical Thinking
8. Refer to Example 14.14. What would happen to the value of *apparent power* for the circuit if the operating frequency were to double? Explain your answer.

14.5 Series-Parallel *RLC* Circuit Analysis

Series-parallel *RLC* circuits can be easy or difficult to analyze, depending on whether the values of *L* and *C* can be combined directly. For example, consider the circuit shown in Figure 14.33a. The values of *L* and *C* in that circuit can be combined and simplified easily. ◄ *OBJECTIVE 12*

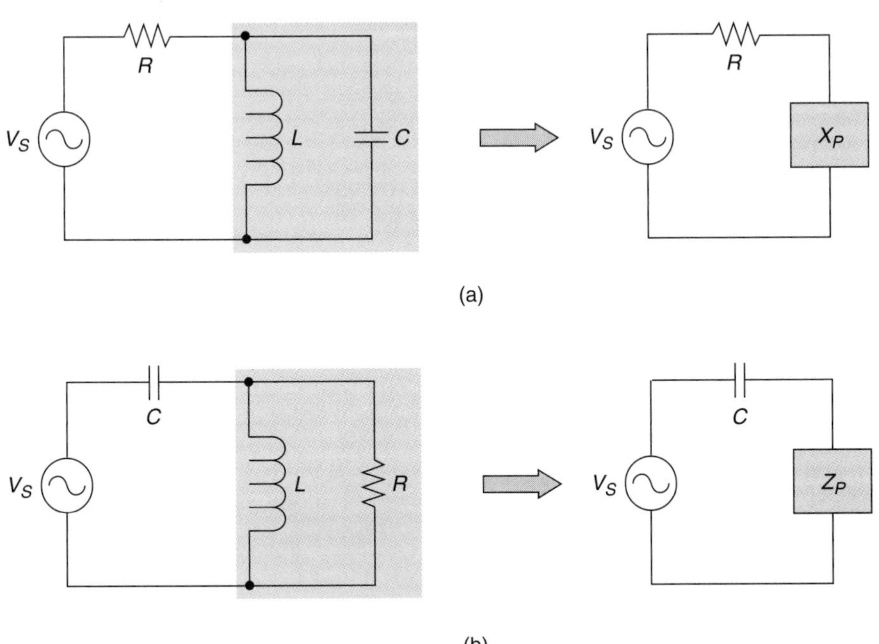

Figure 14.33

Series-parallel *RLC* circuit analysis becomes more involved when the reactive components cannot be combined directly. For example, consider the circuit shown in Figure 14.33b. Because of the component connections, the reactive components cannot be combined directly. In this case, the parallel *RL* circuit must be combined into a parallel equivalent (Z_P), then the circuit can be treated as a series circuit.

In this section, we are going to work through several circuit analysis problems. These will include circuits that are similar to those shown in Figures 14.33. While these circuits do not represent every possible series-parallel *RLC* combination, they should provide a basis for analyzing most circuit configurations.

Example 14.15

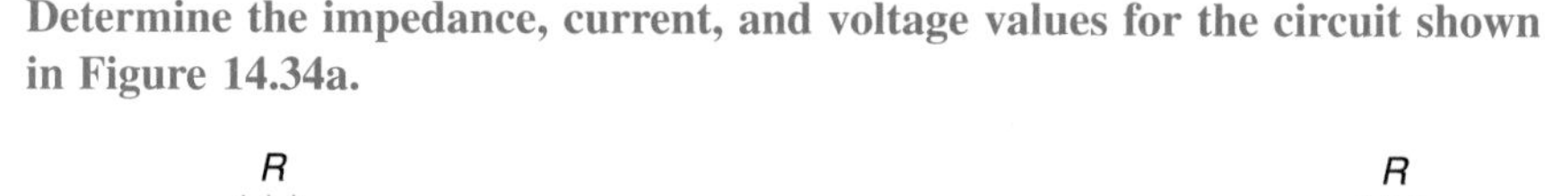

Determine the impedance, current, and voltage values for the circuit shown in Figure 14.34a.

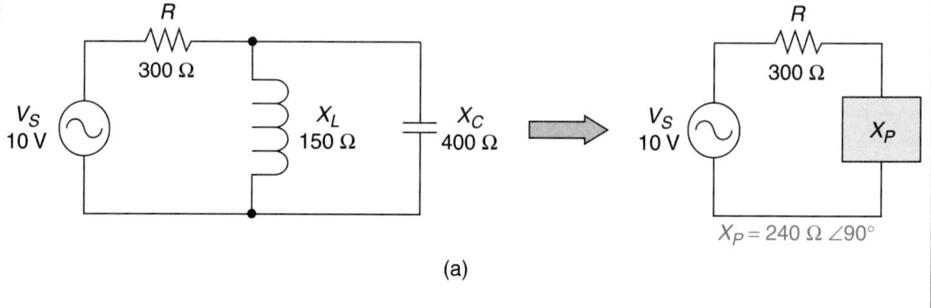

(a)

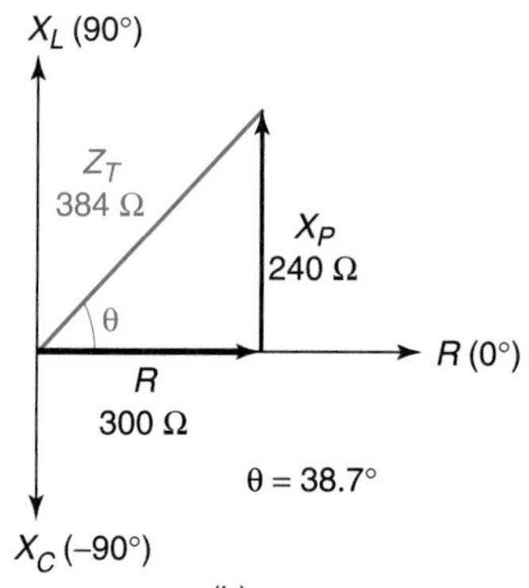

(b)

Figure 14.34

Solution: First, combine the reactances into a single parallel-equivalent value, as follows:

$$X_P = \frac{X_L X_C}{X_L + X_C} = \frac{(150\ \Omega)(-400\ \Omega)}{150\ \Omega + (-400\ \Omega)} = 240\ \Omega = 240\ \Omega\angle 90^\circ$$

Since the phase angle of X_P is positive, it is *inductive* in nature. The circuit impedance has a magnitude of

$$Z_T = \sqrt{X_P^2 + R^2} = \sqrt{(240\ \Omega)^2 + (300\ \Omega)^2} = 384\ \Omega$$

at a phase angle of

$$\theta = \tan^{-1}\left(\frac{X_P}{R}\right) = \tan^{-1}\left(\frac{240\ \Omega}{300\ \Omega}\right) = 38.7^\circ$$

Once the circuit impedance is known, the circuit current can be found as

$$I_T = \frac{V_S}{Z_T} = \frac{10\text{ V}}{384\ \Omega\angle 38.7^\circ} = 26\text{ mA}\angle -38.7^\circ$$

The phase angle of I_T indicates that the circuit current lags the source voltage by 38.7°. Therefore, the values of circuit current and source voltage can be written as

$$I_T = 26 \text{ mA}\angle 0° \qquad \text{and} \qquad V_S = 10 \text{ V}\angle 38.7°$$

Once the value of circuit current is known, the voltage across the parallel *LC* circuit can be found as

$$V_{LC} = I_T X_P = (26 \text{ mA}\angle 0°)(240 \ \Omega\angle 90°) = 6.24 \text{ V}\angle 90°$$

The phase angle of V_{LC} makes sense when you remember that X_P is inductive in nature. Using the calculated value of V_{LC}, the values of I_C and I_L can be found as

$$I_C = \frac{V_{LC}}{X_C} = \frac{6.24 \text{ V}\angle 90°}{400 \ \Omega\angle -90°} = 15.6 \text{ mA}\angle 180°$$

and

$$I_L = \frac{V_{LC}}{X_L} = \frac{6.24 \text{ V}\angle 90°}{150 \ \Omega\angle 90°} = 41.6 \text{ mA}\angle 0°$$

The phase angles for I_C and I_L indicate that these two currents are 180° out of phase, which is consistent with our discussions on parallel *LC* circuits.

Finally, the voltage across the resistor is found as

$$V_R = I_T R = (26 \text{ mA}\angle 0°)(300 \ \Omega) = 7.8 \text{ V}\angle 0°$$

Practice Problem 14.15

A circuit like the one shown in Figure 14.34a has the following values: $V_S = 18$ V, $R = 1.2$ kΩ, $X_L = 3$ kΩ, and $X_C = 1$ kΩ. Determine the impedance, current, and voltage values for the circuit.

Example 14.16

Determine the impedance, current, and voltage values for the circuit shown in Figure 14.35a.

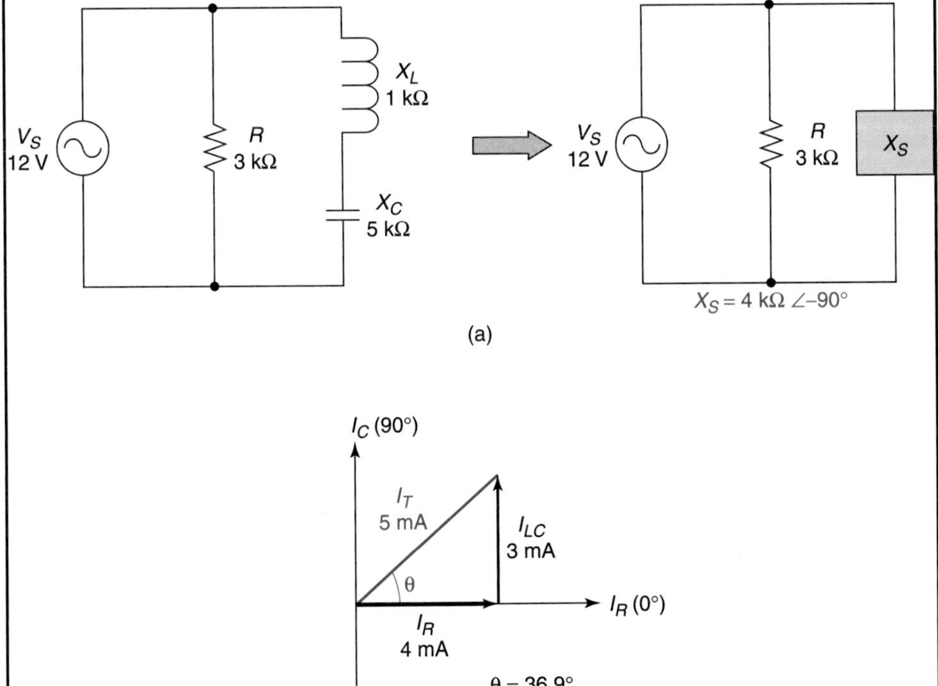

Figure 14.35

Solution: First, the net series reactance (X_S) is found as

$$X_S = X_L - X_C = 1\ \text{k}\Omega - 5\ \text{k}\Omega = -4\ \text{k}\Omega = 4\ \text{k}\Omega\angle -90^\circ$$

This value is included in the parallel-equivalent circuit shown in Figure 14.35a. The branch currents in the circuit can be found as follows:

$$I_R = \frac{V_S}{R} = \frac{12\ \text{V}\angle 0^\circ}{3\ \text{k}\Omega} = 4\ \text{mA}\angle 0^\circ$$

and

$$I_{LC} = \frac{V_S}{X_S} = \frac{12\ \text{V}\angle 0^\circ}{4\ \text{k}\Omega\angle -90^\circ} = 3\ \text{mA}\angle 90^\circ$$

The phase angle of I_{LC} indicates that it leads the source voltage by 90°. The circuit current has a magnitude of

$$I_T = \sqrt{I_{LC}^2 + I_R^2} = \sqrt{(3\ \text{mA})^2 + (4\ \text{mA})^2} = 5\ \text{mA}$$

at a phase angle of

$$\theta = \tan^{-1}\left(\frac{I_{LC}}{I_R}\right) = \tan^{-1}\left(\frac{3\ \text{mA}}{4\ \text{mA}}\right) = 36.9^\circ$$

With a net reactive current of 3 mA, the values of V_L and V_C can be found as

$$V_L = I_{LC}X_L = (3\ \text{mA}\angle 90^\circ)(1\ \text{k}\Omega\angle 90^\circ) = 3\ \text{V}\angle 180^\circ$$

and

$$V_C = I_{LC}X_C = (3\ \text{mA}\angle 90^\circ)(5\ \text{k}\Omega\angle -90^\circ) = 15\ \text{V}\angle 0^\circ$$

The phase angles for V_L and V_C indicate that these voltages are 180° out of phase, which is consistent with our discussion on series *LC* circuits.

Finally, the circuit impedance is found as

$$Z_T = \frac{V_S}{I_T} = \frac{12\ \text{V}\angle 0^\circ}{5\ \text{mA}\angle 36.9^\circ} = 2.4\ \text{k}\Omega\angle -36.9^\circ$$

Practice Problem 14.16

A circuit like the one shown in Figure 14.35a has the following values: $V_S = 6$ V, $R = 560\ \Omega$, $X_L = 2\ \text{k}\Omega$, and $X_C = 1.8\ \text{k}\Omega$. Determine the impedance, current, and voltage values for the circuit.

In the last two examples, we were able to directly combine the *LC* circuit into a single equivalent reactance and then use that reactance to analyze the circuit. When the values of inductance and capacitance cannot be combined directly, the analysis of the circuit gets more involved. This point is demonstrated in Example 14.17.

Example 14.17

Determine the impedance, current, and voltage values for the circuit shown in Figure 14.36a.

Solution: First, combine X_L and R into a single parallel-equivalent impedance (Z_P). This impedance has a magnitude of

$$Z_P = \frac{R \cdot X_L}{\sqrt{R^2 + X_L^2}} = \frac{(300\ \Omega)(200\ \Omega)}{\sqrt{(300\ \Omega)^2 + (200\ \Omega)^2}} = 166\ \Omega$$

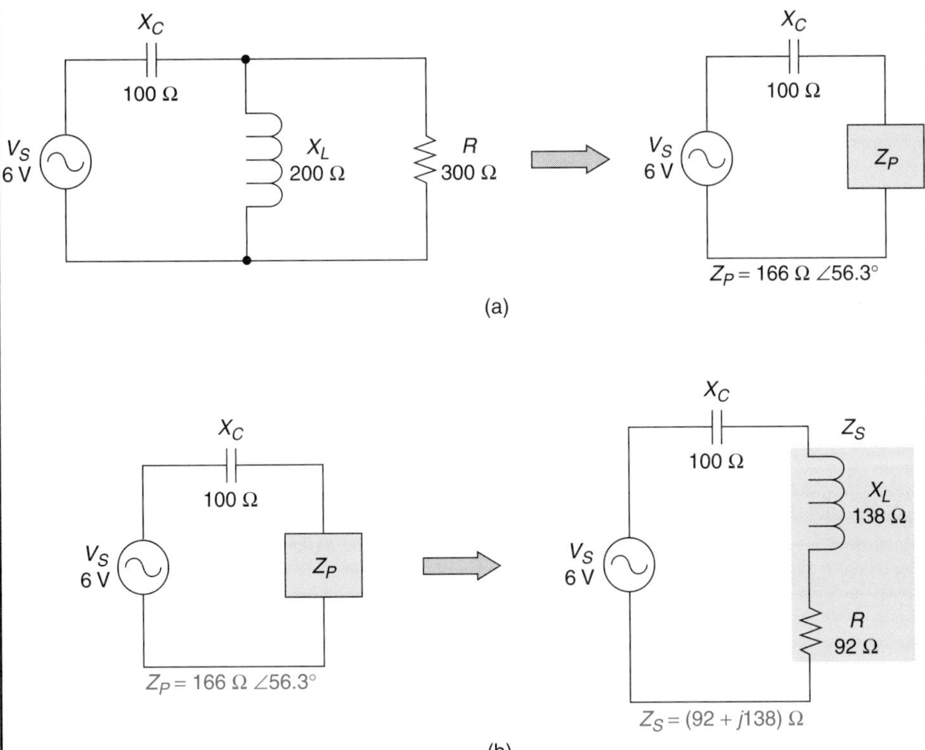

Figure 14.36

at a phase angle of

$$\theta = \tan^{-1}\left(\frac{R}{X_L}\right) = \tan^{-1}\left(\frac{300\ \Omega}{200\ \Omega}\right) = 56.3°$$

This value has been included in the series-equivalent circuit shown in Figure 14.36a.

Next, convert Z_P into a series-equivalent impedance. This is accomplished by converting Z_P to rectangular form, as follows:

$$R = Z_P \cos\theta = (166\ \Omega)(\cos 56.3°) = (166\ \Omega)(0.5548) = 92.1\ \Omega$$

and

$$X_S = Z_P \sin\theta = (166\ \Omega)(\sin 56.3°) = (166\ \Omega)(0.8320) = 138.1\ \Omega$$

Since X_S has a positive phase angle, it is represented as an inductive reactance in the series-equivalent circuit (see Figure 14.36b).

We now have a simple series *RLC* circuit, and the circuit analysis proceeds using the techniques discussed earlier in this chapter. Using these techniques, the following values are calculated for the series-equivalent circuit:

$$X_{LC} = 38\ \Omega \qquad Z_T = 99.5\ \Omega\angle 22.4° \qquad I_T = 60.3\ \text{mA}\angle 0°$$

$$V_S = 6\ \text{V}\angle 22.4° \qquad V_C = 6\ \text{V}\angle -90° \qquad V_{RL} = 10\ \text{V}\angle 56.3°$$

Note that the phase angle of the source voltage (V_S) is a result of setting the current phase angle to 0°. V_{RL} represents the voltage across the series *RL* circuit and, therefore, across the original parallel *RL* circuit. Using V_{RL}, the values of the branch currents in the original circuit can be found as

$$I_L = \frac{V_{RL}}{X_L} = \frac{10\ \text{V}\angle 56.3°}{200\ \Omega\angle 90°} = 50\ \text{mA}\angle -33.7°$$

and

$$I_R = \frac{V_{RL}}{R} = \frac{10\text{ V}\angle 56.3°}{300\ \Omega} = 33.3\text{ mA}\angle 56.3°$$

Since we had already calculated the current through and the voltage across the capacitor, the analysis of the circuit is complete.

Practice Problem 14.17

A circuit like the one shown in Figure 14.36a has the following values: $V_S = 8$ V, $X_C = 300\ \Omega$, $X_L = 800\ \Omega$, and $R = 150\ \Omega$. Determine the values of impedance, current, and voltage for the circuit.

Let's take a moment to look at the results from Example 14.17. To aid in our discussion, the calculated and given values are all shown in Figure 14.37.

Note that I_L lags I_R by 90°, as is always the case for a parallel *RL* circuit. If we have calculated I_L and I_R correctly, the sum of their values should equal the total circuit current. The sum of I_L and I_R can be found as follows:

$$\begin{aligned} I_L &= 50\text{ mA}\angle -33.7° = 41.6 - j27.7\text{ mA} \\ I_R &= 33.3\text{ mA}\angle 56.3° = \underline{18.5 + j27.7\text{ mA}} \\ &\qquad\qquad\qquad\quad 60.1 + j0 \quad \text{mA} \cong 60\text{ mA}\angle 0° \end{aligned}$$

By the same token, the sum of V_{RL} and V_C should equal the source voltage, as follows:

$$\begin{aligned} V_{RL} &= 9.96\text{ V}\angle 56.3° = 5.53 + j8.29\text{ V} \\ V_C &= 6\text{ V}\angle -90° \quad = \underline{0.00 - j6.00\text{ V}} \\ &\qquad\qquad\qquad\quad 5.53 + j2.29\text{ V} \cong 6\text{ V}\angle 22.4° \end{aligned}$$

Since all the values add up correctly, we can assume that all our calculations were correct.

One Final Note

It would be impossible to show examples for every type of series-parallel *RLC* circuit. However, you have seen in this section that analyzing such a circuit is a matter of:

1. Simplifying the circuit.
2. Solving for the overall circuit values.
3. Using the overall circuit values to solve for the individual component values.

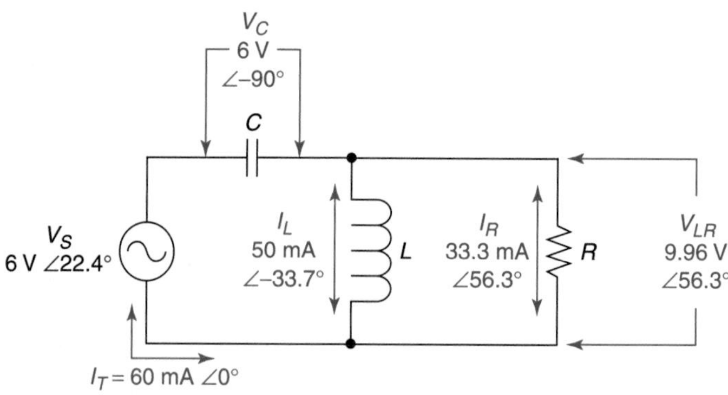

Figure 14.37

Section Review

1. Describe, in your own words, the approach you would take to analyze a series-parallel *RLC* circuit with values of *L* and *C* that can be combined directly.
2. Describe, in your own words, the approach you would take to analyze a series-parallel *RLC* circuit with values of *L* and *C* that cannot be combined directly.
3. Refer to Figure 14.36a (see Example 14.17). Determine the effect of an *increase* in operating frequency on each of the values calculated in the example. Critical Thinking

Key Terms

The following terms were introduced in this chapter on the pages indicated:

Resonance, 419
Resonant frequency (f_r), 419

Practice Problems

1. A series *LC* circuit has values of $V_L = 10$ V and $V_C = 3$ V. Determine the magnitude and phase angle of the voltage source.
2. A series *LC* circuit has values of $V_L = 6$ V and $V_C = 15$ V. Determine the magnitude and phase angle of the voltage source.
3. Determine the magnitude and phase angle of X_S for the circuit shown in Figure 14.38a.
4. Determine the magnitude and phase angle of X_S for the circuit shown in Figure 14.38b.

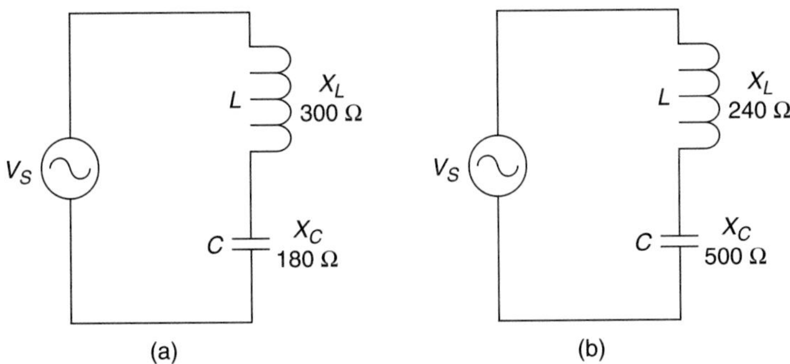

Figure 14.38

5. Determine the magnitude and phase angle of X_S for the circuit shown in Figure 14.39a.
6. Determine the magnitude and phase angle of X_S for the circuit shown in Figure 14.39b.

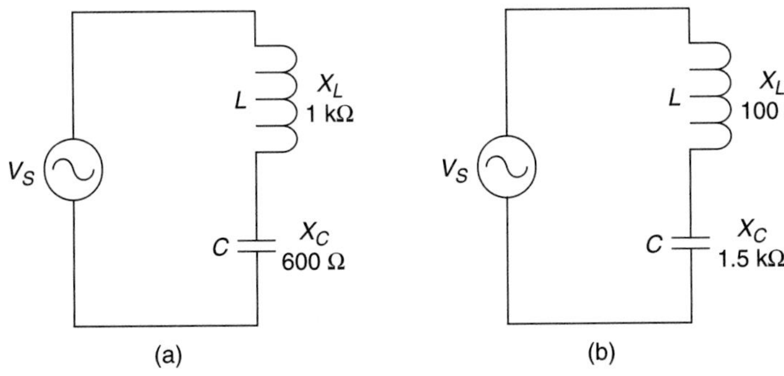

Figure 14.39

7. Determine the values of V_L, V_C, and V_S for the circuit shown in Figure 14.40a.
8. Determine the values of V_L, V_C, and V_S for the circuit shown in Figure 14.40b.

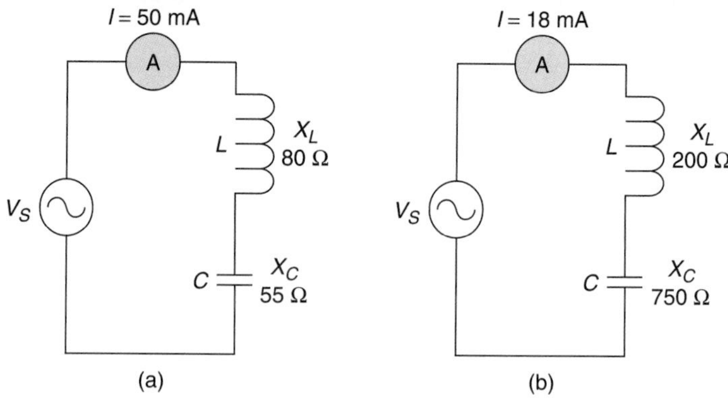

Figure 14.40

9. Determine the values of V_L, V_C, and V_S for the circuit shown in Figure 14.41a.

10. Determine the values of V_L, V_C, and V_S for the circuit shown in Figure 14.41b.

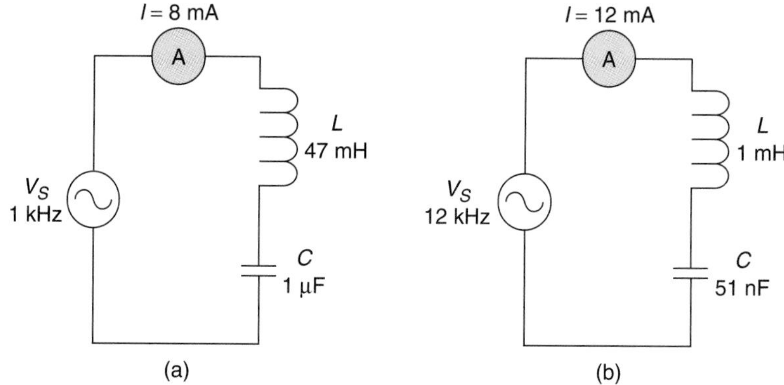

Figure 14.41

11. Determine the magnitude and phase angle of the circuit current shown in Figure 14.42a.

12. Determine the magnitude and phase angle of the circuit current shown in Figure 14.42b.

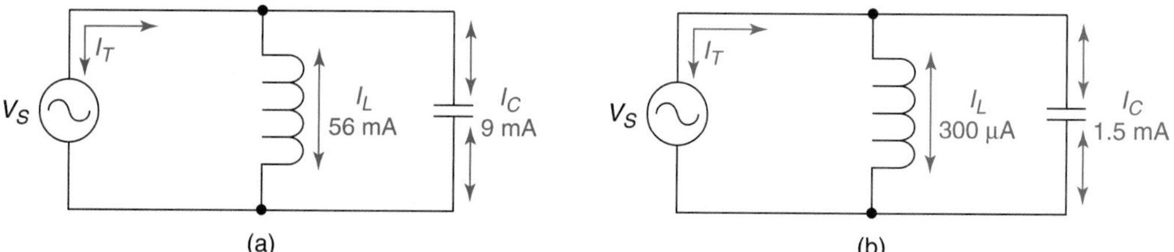

Figure 14.42

13. Determine the parallel equivalent reactance for the circuit shown in Figure 14.43a.

14. Determine the parallel equivalent reactance for the circuit shown in Figure 14.43b.

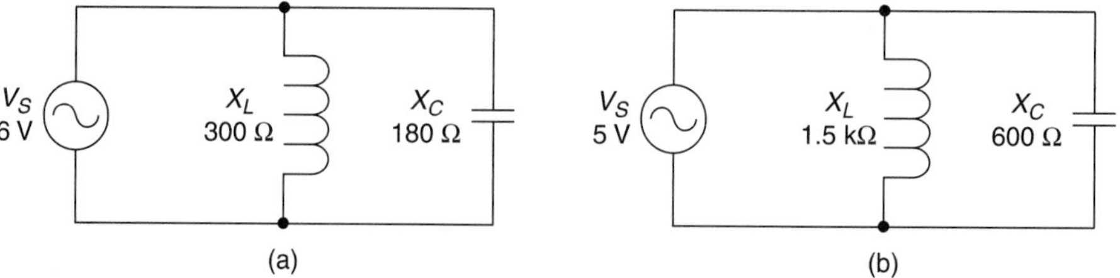

Figure 14.43

15. Determine the resonant frequency for the circuit shown in Figure 14.44a.

16. Determine the resonant frequency for the circuit shown in Figure 14.44b.

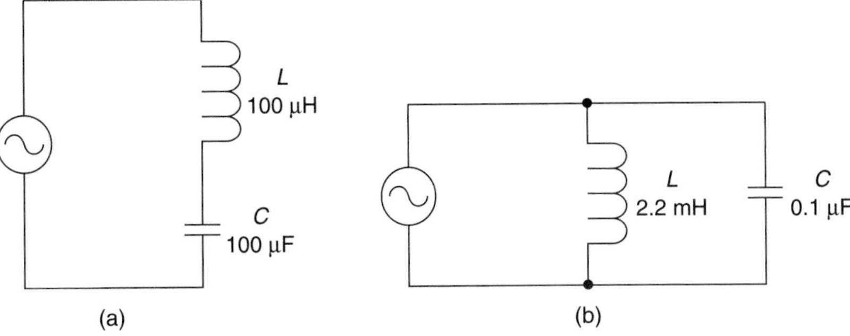

Figure 14.44

17. Calculate the values of X_L and X_C for the circuit shown in Figure 14.41a when it is operating at its resonant frequency.

18. Calculate the values of X_L and X_C for the circuit in Figure 14.41b when it is operating at its resonant frequency.

19. Calculate the values of V_L, V_C, and V_{LC} for the circuit shown in Figure 14.45a.

20. Calculate the values of V_L, V_C, and V_{LC} for the circuit shown in Figure 14.45b.

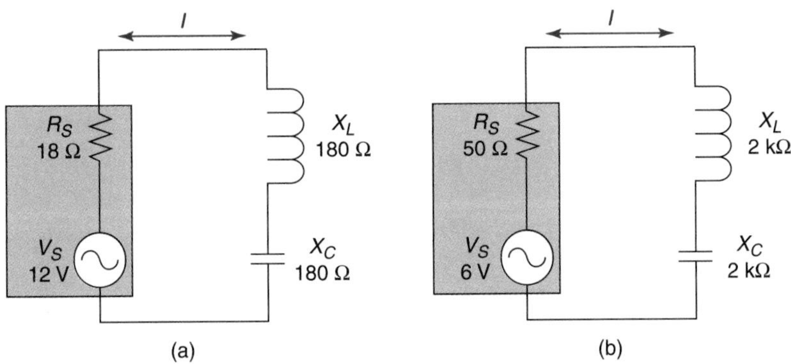

Figure 14.45

21. Determine the magnitude and phase angle of the circuit impedance in Figure 14.46a.

22. Determine the magnitude and phase angle of the circuit impedance in Figure 14.46b.

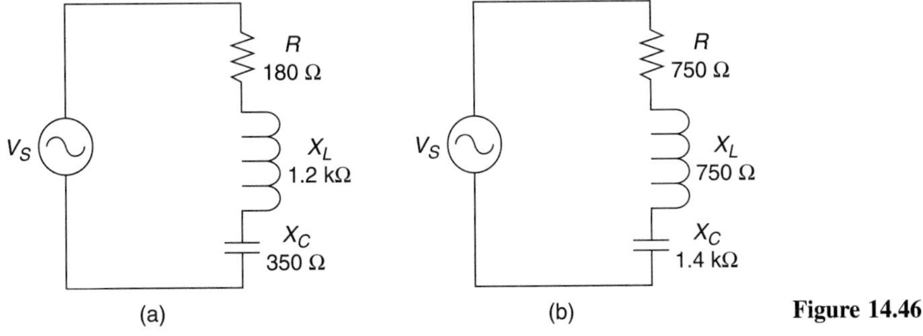

Figure 14.46

23. Determine the magnitude and phase angle of the source voltage in Figure 14.47a.

24. Determine the magnitude and phase angle of the source voltage in Figure 14.47b.

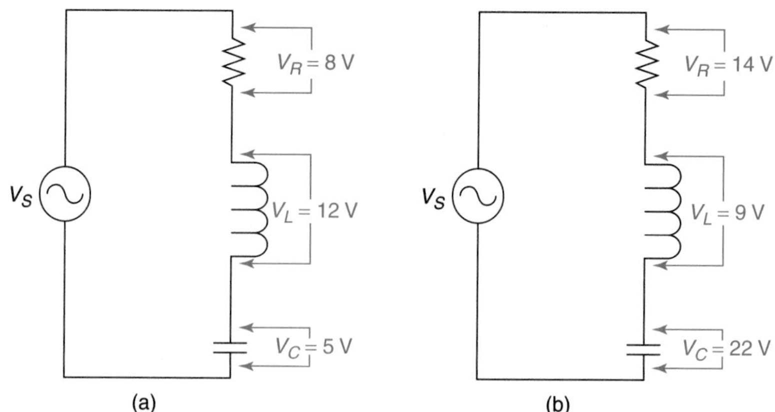

Figure 14.47

25. Calculate the impedance, current, and voltage values for the circuit shown in Figure 14.48a.

26. Calculate the impedance, current, and voltage values for the circuit shown in Figure 14.48b.

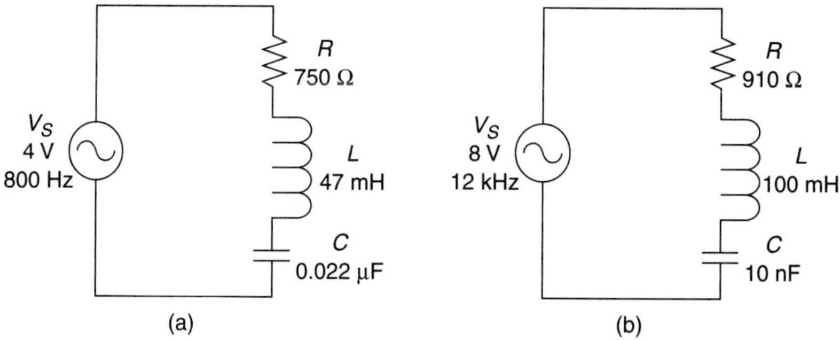

Figure 14.48

27. Determine the magnitude and phase angle of the circuit current shown in Figure 14.49a.

28. Determine the magnitude and phase angle of the circuit current shown in Figure 14.49b.

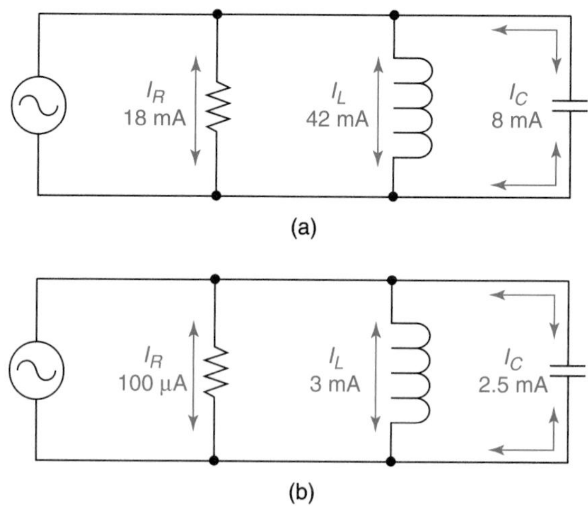

Figure 14.49

29. Calculate the impedance and current values for the circuit shown in Figure 14.50a.

30. Calculate the impedance and current values for the circuit shown in Figure 14.50b.

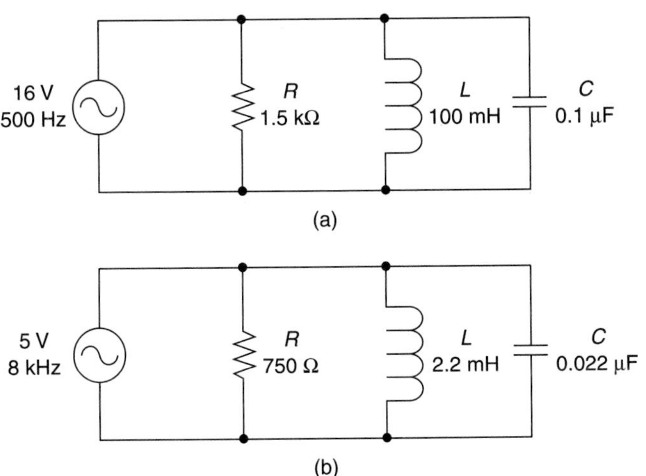

Figure 14.50

31. Calculate the impedance, current, and voltage values for the circuit shown in Figure 14.51a.
32. Calculate the impedance, current, and voltage values for the circuit shown in Figure 14.51b.

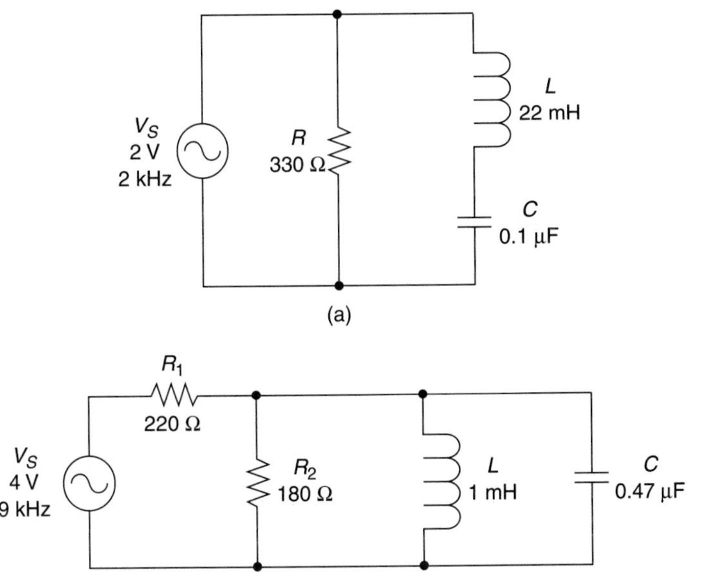

Figure 14.51

33. Calculate the impedance, current, and voltage values for the circuit shown in Figure 14.52a.
34. Calculate the impedance, current, and voltage values for the circuit shown in Figure 14.52b.

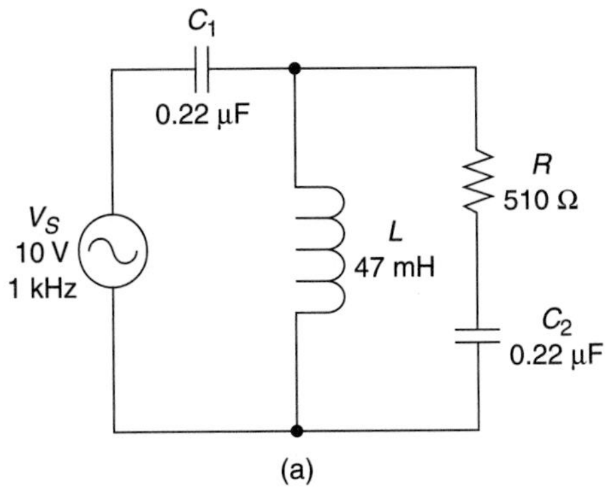

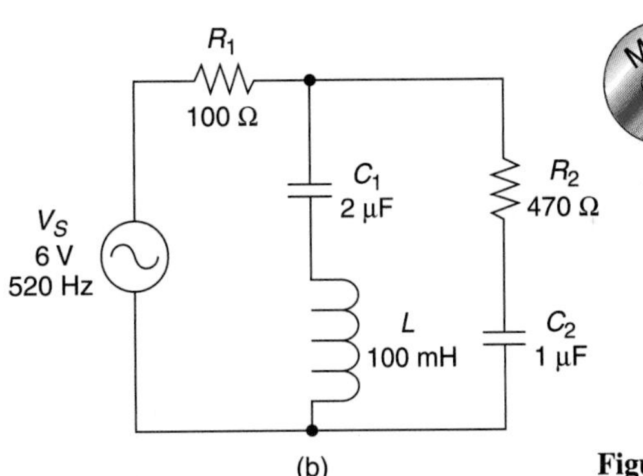

Figure 14.52

Looking Back

These problems relate to material presented in earlier chapters. The chapters are identified in brackets.

35. Convert each value to standard numeric form. [Introduction]
    a. 1.58 MHz = ______ c. 0.98 kΩ = ______
    b. 900 nF = ______ d. 650 μH = ______

36. A two-resistor series circuit has the following values: $R_1 = 330\ \Omega$, $R_2 = 220\ \Omega$, and $V_S = 6$ V. Calculate the change in source power (P_S) that occurs if R_1 is replaced with a 1 kΩ resistor. [Chapter 4]

37. A series *RC* circuit has the following values: $R = 1.5$ kΩ, $C = 1$ μF, and $f = 500$ Hz. Determine the change in θ that occurs if the operating frequency *doubles*. [Chapter 13]

38. A parallel *RL* circuit has the following values: $R = 100\ \Omega$, $L = 15$ mH, and $f = 6$ kHz. Calculate the change in apparent power (P_{APP}) that occurs if the value of *R* is changed to 330 Ω. [Chapter 11]

Answers to the Example Practice Problems

14.1 500 mV∠−90°

14.3 25 Ω∠90°

14.4 $V_L = 4.95$ V∠90°, $V_C = 14.85$ V∠−90°, $V_S = 9.9$ V∠−90°

14.5 500 Ω

14.6 5.91 kHz

14.7 $V_L = 200$ V∠90°, $V_C = 200$ V∠−90°, $V_{LC} = 0$ V

14.8 $I_C = 250$ mA∠90°, $I_L = 250$ mA∠−90°, $I_T = 0$ A, $X_T \rightarrow \infty\Omega$

14.9 408 Ω∠36°

14.10 1.56 kΩ∠−50.2°

14.11 $X_L = 1.38$ kΩ, $X_C = 482\ \Omega$, $X_S = 900\ \Omega$, $Z = 1.75$ kΩ∠31°, $I_T = 5.14$ mA, $V_R = 7.71$ V, $V_L = 7.09$ V, $V_C = 2.47$ V, $V_{LC} = 4.62$ V, $V_S = 9$ V∠31°

14.12 50 mA∠−53.1°

14.13 34.4 mA∠54.5°

14.14 $X_L = 503\ \Omega$, $I_L = 19.9$ mA∠−90°, $X_C = 199\ \Omega$, $I_C = 50.3$ mA ∠90°, $I_{LC} = 30.4$ mA∠90°, $I_R = 13.3$ mA, $I_T = 33.2$ mA∠63.4°, $Z_T = 301$ Ω∠−66.4°

14.15 $X_P = 1.5$ kΩ∠−90°, $Z_T = 1.92$ kΩ∠−51.3°, $I_T = 9.38$ mA, $V_S = 18$ V∠−51.3°, $V_{LC} = 14.1$ V∠−90°, $I_L = 4.7$ mA∠−180°, $I_C = 14.1$ mA∠0°, $I_{LC} = 9.4$ mA, $V_R = 11.3$ V

14.16 $I_{LC} = 30$ mA∠−90°, $I_R = 10.7$ mA∠0°, $I_T = 31.9$ mA∠−70.4°, $V_L = 60$ V∠0°, $V_C = 54$ V∠−180°, $V_{LC} = 6$ V, $Z_T = 188$ Ω∠70.4°

14.17 $Z_P = 147$ Ω∠10.6°, $X_S = 145\ \Omega$, $R_S = 27\ \Omega$, $X_{LC} = 273$ Ω∠−90°, $Z_T = 309$ Ω∠−62.1°, $I_T = 25.9$ mA, $V_S = 8$ V∠−62.1°, $V_C = 7.77$ V∠−90°, $V_{RL} = 3.8$ V∠10.6°, $I_L = 4.75$ mA∠−74.9°, $I_R = 25.3$ mA∠10.6°

Multisim File References

Figure	File Reference
14.26	ETF14_26.msm
14.32	ETF14_32.msm
14.49	ETF14_49.msm
14.50	ETF14_50.msm
14.51	ETF14_51.msm
14.52	ETF14_52.msm

chapter 15

Frequency Response and Passive Filters

Objectives

After studying the material in this chapter, you should be able to:

1. Describe *attenuation.*
2. Describe the relationship between circuit *amplitude* and *cutoff frequency.*
3. List the four primary filters, and identify their frequency response curves.
4. Identify (and solve for) each of the following on a frequency response curve: *bandwidth, center frequency, lower cutoff frequency,* and *upper cutoff frequency.*
5. Discuss the relationship among *filter Q, bandwidth,* and *center frequency.*
6. Describe and calculate the *average frequency* of a bandpass (or notch) filter.
7. Describe the relationship among *filter Q, center frequency,* and *average frequency.*
8. List and describe the commonly used logarithmic frequency scales.
9. Represent any power value in *bel (B)* or *decibel (dB)* form.
10. Convert *power gain* from standard numeric form to dB form, and vice versa.
11. Convert *voltage gain* from standard numeric form to dB form, and vice versa.
12. Convert *current gain* from standard numeric form to dB form, and vice versa.
13. Describe and analyze the operation of the *RC low-pass filter.*
14. Compare and contrast *bode plots* with *frequency response curves.*
15. Describe and analyze the operation of the *RL low-pass filter.*
16. Describe and analyze the operation of the *RC high-pass filter.*
17. Describe and analyze the operation of the *RL high-pass filter.*
18. Describe and analyze the operation of *series LC bandpass filters.*
19. Describe and analyze the operation of *shunt LC bandpass filters.*
20. Describe the operation of *LC notch filters.*

We touched on the topic of frequency response in Chapters 11, 13, and 14. Now, we will take a more in-depth look at frequency response and the operating characteristics of a group of circuits called *passive filters.* As you will learn, these circuits are designed for specific frequency response characteristics.

15.1 Frequency Response: Curves and Measurements

The subject of *frequency response* is complex, with many principles and concepts. In this section, we will begin by touching on several of the most general principles.

Attenuation

OBJECTIVE 1 ➤ Every circuit responds to a change in operating frequency to some extent. The changes may be too subtle to observe, or they may be dramatic. An example of the latter is illustrated in Figure 15.1.

The circuit represented in Figure 15.1a has input and output waveforms that are nearly identical. When the input frequency is increased by a factor of ten (as shown in Figure 15.1b), the amplitude of the circuit output drops significantly. *The signal loss caused by the frequency response of the circuit* is referred to as **attenuation.**

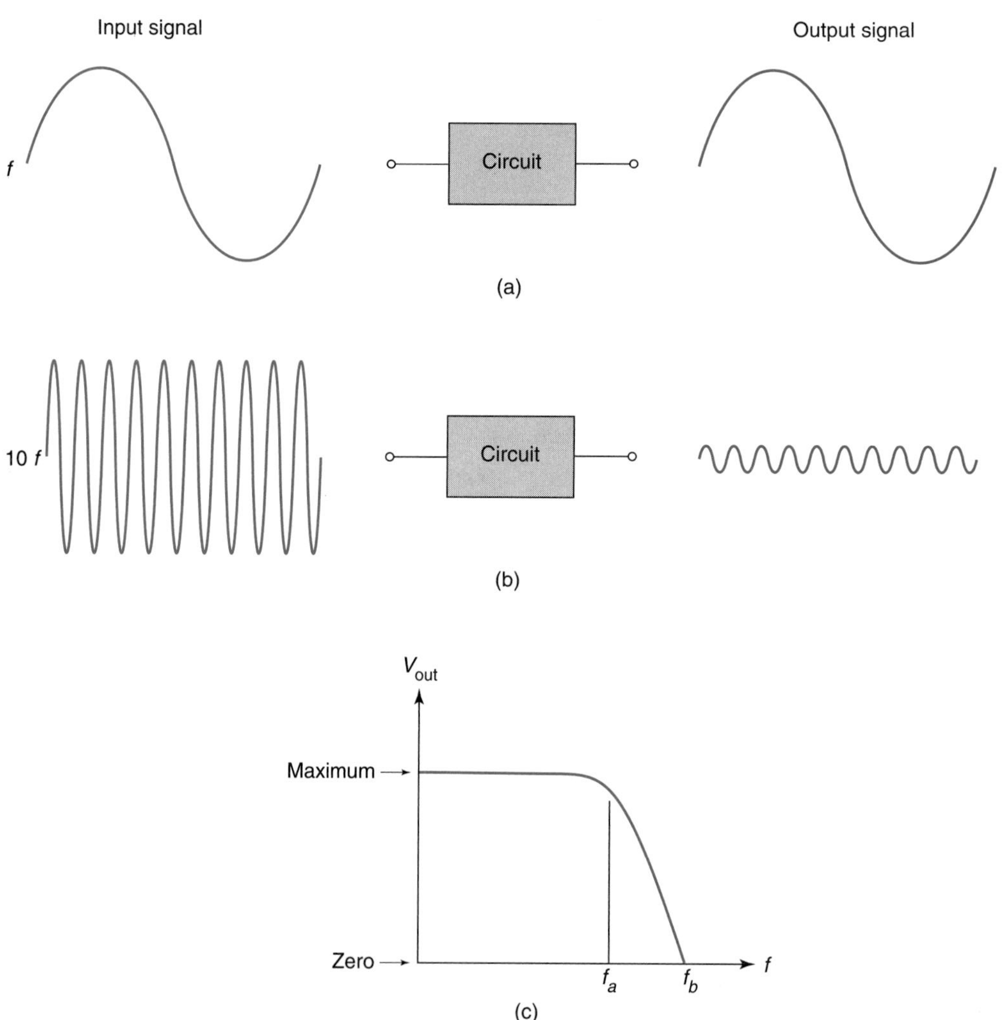

Figure 15.1 Attenuation

The frequency response described above can be graphed as shown in Figure 15.1c. As shown, the circuit output (V_{out}) stays relatively constant over the range of frequencies below f_a. As frequency increases above f_a, circuit attenuation causes the magnitude of V_{out} to drop. If the circuit operating frequency reaches f_b, the magnitude of V_{out} drops to zero.

Frequency Response Curves

The frequency response of a circuit is commonly described in terms of the effect that a change in frequency has on *the ratio of its output amplitude to its input amplitude.* For example, the curve shown in Figure 15.2a represents amplitude as a ratio of output voltage to input voltage. When $\frac{V_{out}}{V_{in}} = 1$, the ratio is at its maximum, and the input and output voltages are equal. When the circuit is operated at the frequency designated f_x, the output voltage is half the input voltage, and so on.

The ratio of a circuit's output amplitude to its input amplitude is referred to as **gain.** As such, the circuit represented by the curve shown in Figure 15.2a is described as having a maximum *voltage gain* of one (1).

A **frequency response curve** is *a graph that shows the effect that frequency has on circuit gain.* In most cases, these curves use a *power ratio* as the indicator of gain. For example, the curve shown in Figure 15.2b shows the relationship between frequency and **power gain** (*the ratio of output power to input power*).

Cutoff Frequency (f_C)

The frequency at which the power gain of a circuit drops to 50% of its maximum value is referred to as the **cutoff frequency (f_C).** This frequency is identified as shown in ◀ *OBJECTIVE 2*

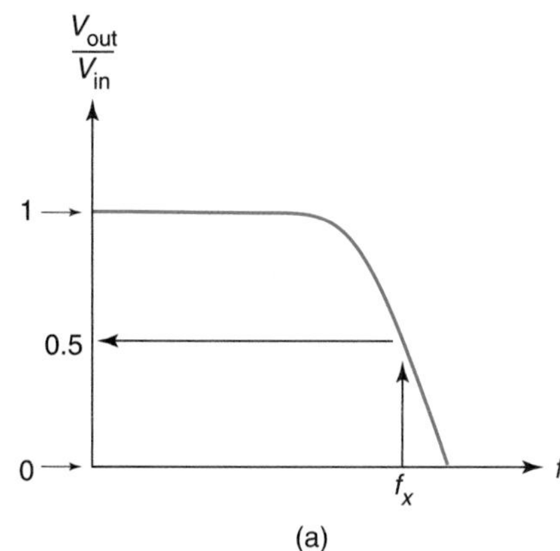

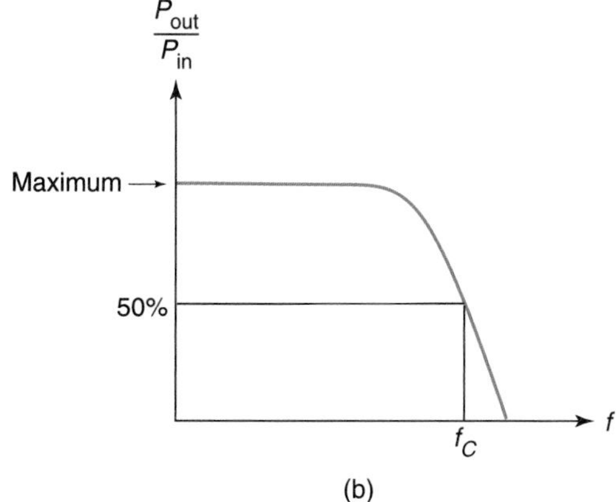

Figure 15.2 Amplitude measurements on frequency response curves.

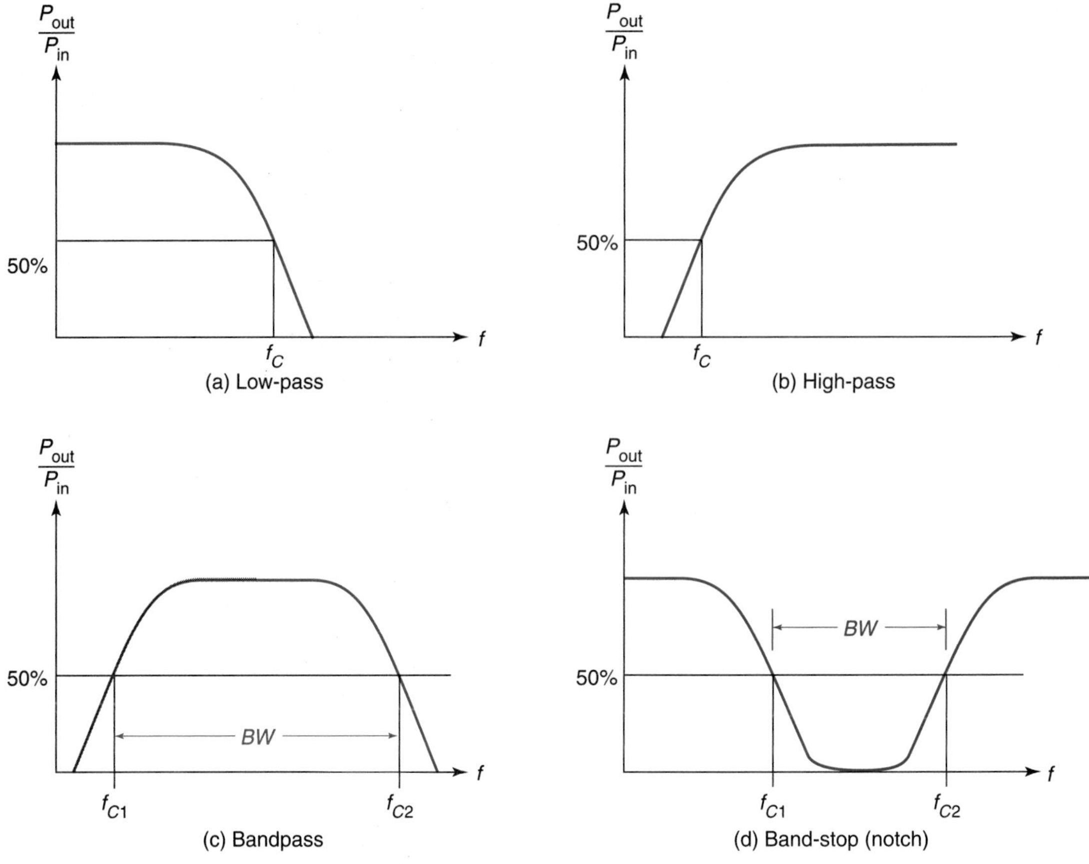

Figure 15.3 Filter frequency response curves.

Figure 15.2b. Note that the 50% standard for the power gain is an arbitrary value; that is, it is used only because it is agreed on by professionals in the field. The cutoff frequency of a component or circuit is determined by its resistive and reactive values (as will be demonstrated later in this chapter).

Filters

OBJECTIVE 3 ➤ Many circuits are designed to pass a specific range of frequencies while rejecting others. These circuits are generally referred to as *filters.* There are four basic types of filters:

- A **low-pass filter** is *designed to pass all frequencies below its cutoff frequency.*
- A **high-pass filter** is *designed to pass all frequencies above its cutoff frequency.*
- A **bandpass filter** is *designed to pass the band of frequencies between two cutoff frequencies* (f_{C1} *and* f_{C2}).
- A **band-stop filter,** or **notch filter,** is *designed to reject the band of frequencies between its cutoff frequencies.* That is, it passes the frequencies that are *below* f_{C1} and *above* f_{C2}.

The four characteristic filter response curves are shown in Figure 15.3.

Bandwidth and Center Frequency

OBJECTIVE 4 ➤ *The range, or band, of frequencies between the cutoff frequencies of a component or circuit* is referred to as its **bandwidth.** The concept of bandwidth is illustrated in Figures 15.3c and 15.3d. The bandwidth of a component or circuit equals the difference between its cutoff frequencies. By formula,

$$\text{BW} = f_{C2} - f_{C1} \tag{15.1}$$

where

BW = the bandwidth of the component or circuit, in Hz
f_{C2} = the upper cutoff frequency
f_{C1} = the lower cutoff frequency

The bandwidth of a component or circuit can be found as demonstrated in Example 15.1.

Example 15.1

A bandpass filter has the frequency response curve shown in Figure 15.4. Calculate the bandwidth of the circuit.

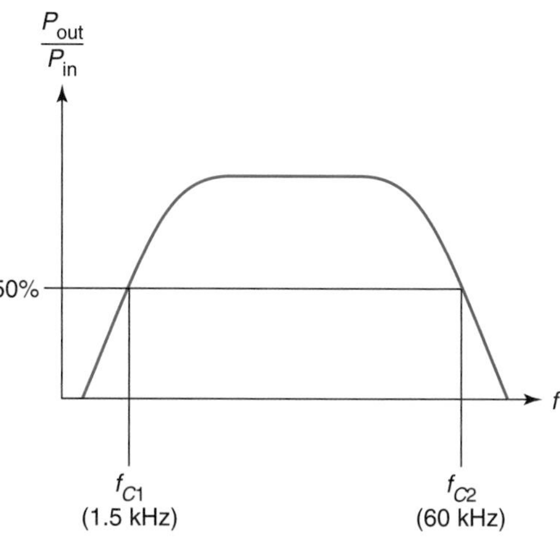

Figure 15.4

Solution: Using the cutoff frequencies shown, the bandwidth of the circuit is found as

$$BW = f_{C2} - f_{C1} = 60 \text{ kHz} - 1.5 \text{ kHz} = 58.5 \text{ kHz}$$

This result indicates that there is a 58.5 kHz spread between the cutoff frequencies of the circuit.

Practice Problem 15.1

A curve like the one shown in Figure 15.4 has the following values: f_{C2} = 88 kHz, and f_{C1} = 6.2 kHz. Calculate the bandwidth of the circuit represented by the curve.

Center Frequency (f_0)

To accurately describe the frequency response of a bandpass or notch filter, we need to know where the curve is located on the frequency spectrum. For example, consider the two bandpass curves shown in Figure 15.5. Each of these curves has a bandwidth of 50 kHz, yet clearly, they are not the same curve. To distinguish between them, we need to identify the *center frequency* of each curve.

Technically defined, **center frequency (f_0)** is *the frequency that equals the geometric average of the cutoff frequencies.* Therefore, center frequency is found as

$$f_0 = \sqrt{f_{C1} f_{C2}} \tag{15.2}$$

Example 15.2 shows how the value of f_0 is found for a component or circuit.

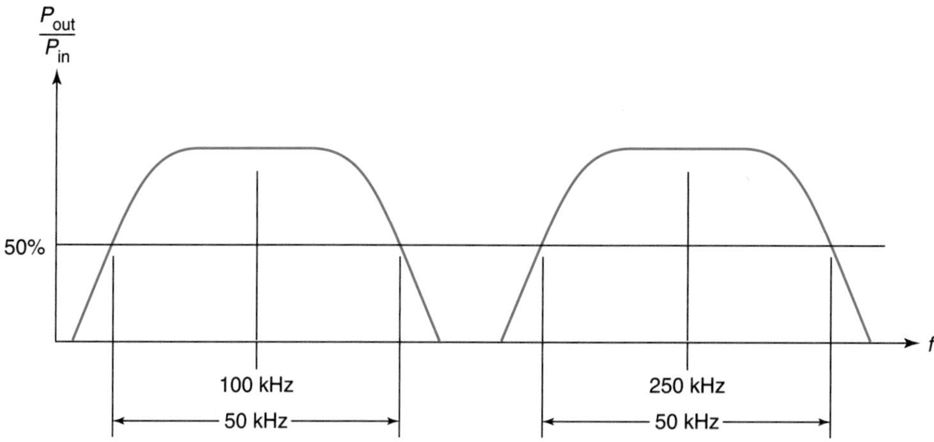

Figure 15.5 Bandpass curves with equal bandwidths.

Example 15.2

A bandpass filter has the following values: $f_{C1} = 5$ kHz, and $f_{C2} = 20$ kHz. Calculate the value of the circuit's center frequency.

Solution: The value of f_0 is found as

$$f_0 = \sqrt{f_{C1} f_{C2}} = \sqrt{(5 \text{ kHz})(20 \text{ kHz})} = 10 \text{ kHz}$$

Practice Problem 15.2

A band-stop filter has the following values: $f_{C1} = 300$ Hz and $f_{C2} = 120$ kHz. Calculate the value of the circuit's center frequency.

When we first hear the term *center frequency,* we tend to visualize a frequency that is halfway between the cutoff frequencies. However, this is not necessarily the case. Consider the results from Example 15.2. Obviously, 10 kHz is not halfway between 5 kHz and 20 kHz. However, it is at the *geometric center* of these two values, meaning that the ratio of f_0 to f_{C1} equals the ratio of f_{C2} to f_0. By formula,

$$\frac{f_0}{f_{C1}} = \frac{f_{C2}}{f_0} \tag{15.3}$$

If we apply this relationship to the values obtained in the example, we get

$$\frac{10 \text{ kHz}}{5 \text{ kHz}} = \frac{20 \text{ kHz}}{10 \text{ kHz}}$$

which is true. As you can see, the frequency ratios are equal. This is always the case for the geometric average of two values.

Frequency Calculations

When one cutoff frequency and the center frequency for a filter are known, the other cutoff frequency can be found using one of the following transposed forms of equation (15.3):

$$f_{C1} = \frac{f_0^2}{f_{C2}} \tag{15.4}$$

and

$$f_{C2} = \frac{f_0^2}{f_{C1}} \tag{15.5}$$

Example 15.3 demonstrates the first of these relationships.

A notch filter has the following values: $f_0 = 6$ kHz, and $f_{C2} = 120$ kHz. Calculate the value of the circuit's lower cutoff frequency.

Example 15.3

Solution: The filter's lower cutoff frequency is found as

$$f_{C1} = \frac{f_0^2}{f_{C2}} = \frac{(6\text{ kHz})^2}{120\text{ kHz}} = 300\text{ Hz}$$

Practice Problem 15.3

A bandpass filter has the following values: $f_0 = 500$ Hz, and $f_{C2} = 5$ kHz. Calculate the value of the filter's lower cutoff frequency.

If you compare the frequencies in Example 15.3 to the values given and calculated in Practice Problem 15.2, you'll see that the same curve was used for the two problems. The application of equation (15.5) is identical to the technique used in Example 15.3.

Filter Quality (Q)

◄ *OBJECTIVE 5*

You may recall that the Q of a coil or capacitor is a figure of merit that indicates how closely the characteristics of the component come to those of an *ideal* component. By the same token, the **quality (Q)** of a bandpass or notch filter *indicates how closely the characteristics of the circuit come to those of an ideal circuit.* The Q of a bandpass or notch filter equals the ratio of its center frequency to its bandwidth. By formula,

$$Q = \frac{f_0}{BW} \tag{15.6}$$

Example 15.4 demonstrates this relationship.

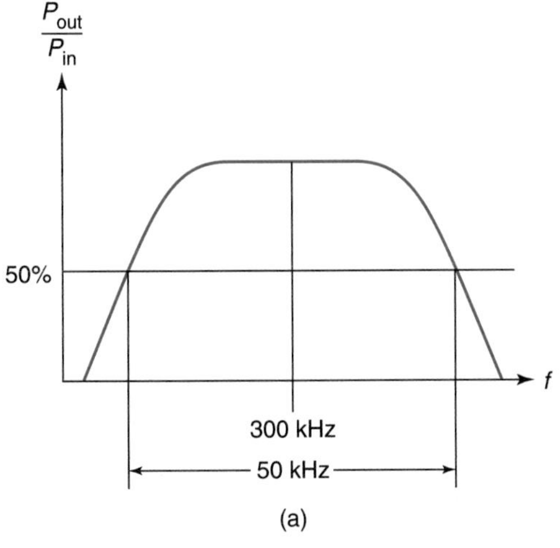

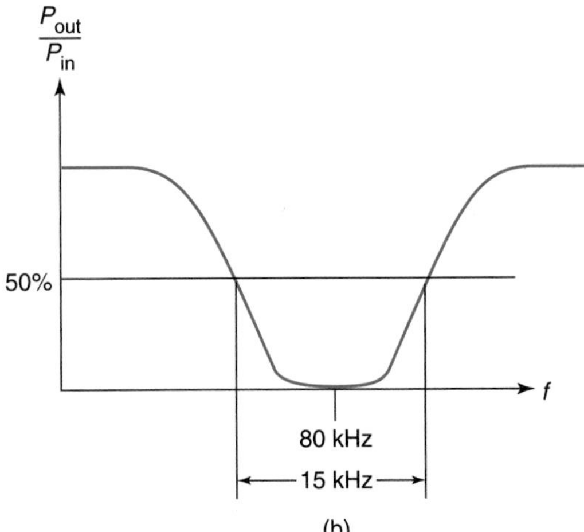

Figure 15.6

Determine the Q of the bandpass filter represented by the curve shown in Figure 15.6a.

Example 15.4

Solution: The Q of the filter is found as

$$Q = \frac{f_0}{BW} = \frac{300\ kHz}{50\ kHz} = 6$$

Note that Q has no unit of measure, because it is a ratio.

Practice Problem 15.4

Determine the Q of the notch filter represented in Figure 15.6b.

When working with a filter, the center frequency and Q of a filter are both determined using component values (as will be demonstrated later in this chapter). Then, these values are used to determine the bandwidth of the circuit, as follows:

$$BW = \frac{f_0}{Q} \tag{15.7}$$

Example 15.5 demonstrates a more common application of the relationship among Q, center frequency, and bandwidth.

Example 15.5

Through a series of calculations, a bandpass filter is determined to have the following values: $Q = 4.8$, and $f_0 = 96$ kHz. Calculate the circuit bandwidth.

Solution: Using these values of Q and f_0, the circuit bandwidth is found as

$$BW = \frac{f_0}{Q} = \frac{96 \text{ kHz}}{4.8} = 20 \text{ kHz}$$

Practice Problem 15.5

Using a series of calculations, a notch filter is found to have the following values: $Q = 3.3$, and $f_0 = 165$ kHz. Determine the circuit bandwidth.

Average Frequency (f_{ave})

OBJECTIVE 6 ➤ Every bandpass and notch filter has both a *center frequency* (f_0) and an *algebraic average frequency* (f_{ave}). This **average frequency** is *the frequency that lies halfway between the cutoff frequencies.* By formula,

$$f_{ave} = \frac{f_{C1} + f_{C2}}{2} \tag{15.8}$$

If we know the value of f_{ave} and the circuit bandwidth, we can solve for the two cutoff frequencies, as follows:

$$f_{C1} = f_{ave} - \frac{BW}{2} \tag{15.9}$$

and

$$f_{C2} = f_{ave} + \frac{BW}{2} \tag{15.10}$$

There is only one problem: Equation (15.8) assumes that we know the cutoff frequencies of the filter. When the cutoff frequencies of a filter are not known, we need to have another means of finding f_{ave}.

Determining the Value of f_{ave}

The values of f_0 and Q for a given filter can be used to determine the value of f_{ave}, as follows:

$$f_{ave} = f_0\sqrt{1 + \left(\frac{1}{2Q}\right)^2} \tag{15.11}$$

An application of this relationship is demonstrated in Example 15.6.

Example 15.6

A bandpass filter has the following values: $f_0 = 250$ kHz, and BW = 150 kHz. Determine the value of f_{ave} for the circuit.

Solution: First, the Q of the filter is found as

$$Q = \frac{f_0}{\text{BW}} = \frac{250 \text{ kHz}}{150 \text{ kHz}} = 1.67$$

The value of f_{ave} for the circuit can now be found as

$$f_{ave} = f_0 \sqrt{1 + \left(\frac{1}{2Q}\right)^2} = (250 \text{ kHz}) \sqrt{1 + \left(\frac{1}{[2][1.67]}\right)^2}$$
$$= (250 \text{ kHz})(1.044) = 261 \text{ kHz}$$

Practice Problem 15.6

A notch filter has the following values: $f_0 = 300$ kHz, and BW = 240 kHz. Determine the value of f_{ave} for the circuit.

Once the value of f_{ave} is known, the circuit cutoff frequencies can be determined as demonstrated in Example 15.7.

Example 15.7

Determine the cutoff frequencies for the circuit described in Example 15.6.

Solution: From Example 15.6, we know that $f_{ave} = 261$ kHz and BW = 150 kHz. Using these values, the cutoff frequencies of the circuit are found as

$$f_{C1} = f_{ave} - \frac{\text{BW}}{2} = 261 \text{ kHz} - \frac{150 \text{ kHz}}{2} = 186 \text{ kHz}$$

and

$$f_{C2} = f_{ave} + \frac{\text{BW}}{2} = 261 \text{ kHz} + \frac{150 \text{ kHz}}{2} = 336 \text{ kHz}$$

Practice Problem 15.7

Determine the cutoff frequencies for the notch filter described in Practice Problem 15.6.

Frequency Approximations

OBJECTIVE 7

When a bandpass or notch filter has a value of $Q \geq 2$, the center and average frequencies can be assumed to be approximately equal in value. By formula,

$$f_{ave} \cong f_0 \quad | \quad Q \geq 2 \tag{15.12}$$

The basis for this relationship can be seen by taking another look at equation (15.11). If we solve this equation for $Q = 2$, we get

$$f_{ave} = f_0 \sqrt{1 + \left(\frac{1}{2Q}\right)^2} = f_0 \sqrt{1 + \left(\frac{1}{4}\right)^2} = f_0(1.03)$$

This result indicates that there is only a 3% difference between f_{ave} and f_0 when $Q = 2$. Therefore, we can assume that the two frequencies are approximately equal.

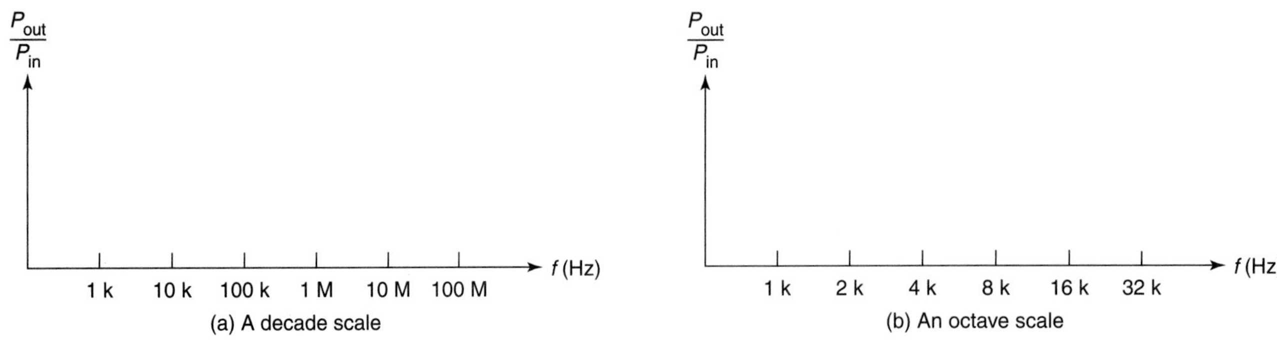

Figure 15.7 Logarithmic frequency scales.

Frequency Scales

OBJECTIVE 8 ➤

Up to this point, we have shown only the given and calculated frequencies on our response curves. In practice, however, frequency response curves are commonly plotted using frequency scales like those shown in Figure 15.7, which are referred to as *logarithmic scales.* A **logarithmic scale** is *one where the value of each increment is a whole-number multiple of the previous increment.* For example, the value of each increment shown in Figure 15.7a is ten times the value of the previous increment.

A frequency multiplier of ten is referred to as a **decade,** so Figure 15.7a is referred to as a **decade scale.** *A frequency multiplier of two* is referred to as an **octave,** so Figure 15.7b is referred to as an **octave scale.** Note that both decade and octave scales are used in practice. When a large frequency range is needed, a decade scale is used. When a more precise representation over a smaller frequency range is needed, an octave scale may be used.

Why logarithmic scales are preferred over algebraic scales.

In general, logarithmic scales are preferred for several reasons:

- An algebraic scale can be extremely long.
- When a logarithmic scale is used, the center frequency (which is the geometric average of the cutoff frequencies) falls in the physical center of the curve.

The first point can be seen when you compare Figure 15.8 to Figure 15.7b. Figure 15.8 is an algebraic scale with a constant (c) of 2 kHz between increments. To represent 32 kHz, this scale would have to extend well off the page.

The second point is illustrated in Figure 15.9a. As shown, the geometric center frequency (f_0) falls very close to the physical center of the curve. If an algebraic scale had been used, f_0 would have fallen very close to the left end of the scale, as shown in Figure 15.9b.

Critical Frequencies: Putting It All Together

Table 15.1 and Figure 15.10 summarize the critical frequencies and characteristics for the four main filter types. Remember that band-pass and notch filters have both a *geometric center frequency* (f_0) and an *algebraic average frequency* (f_{ave}). When a band-pass or notch filter has a value of $Q \geq 2$, these two frequencies are assumed to be approximately equal in value.

Figure 15.8

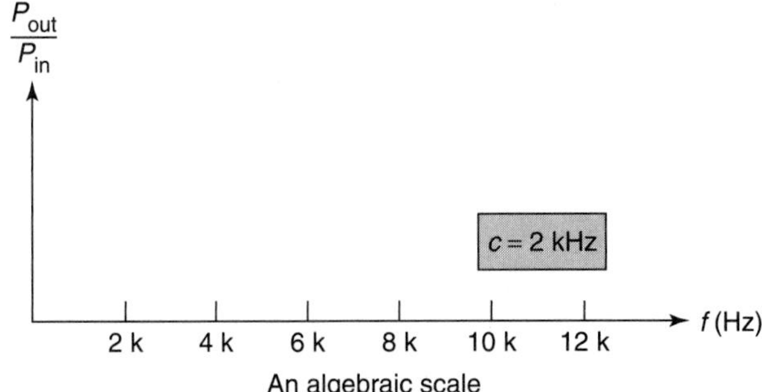

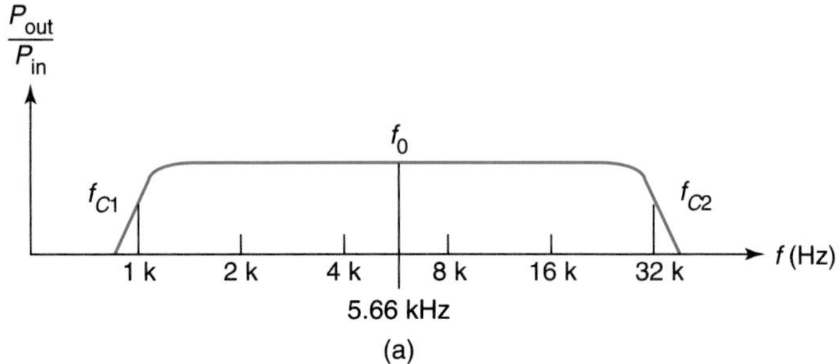

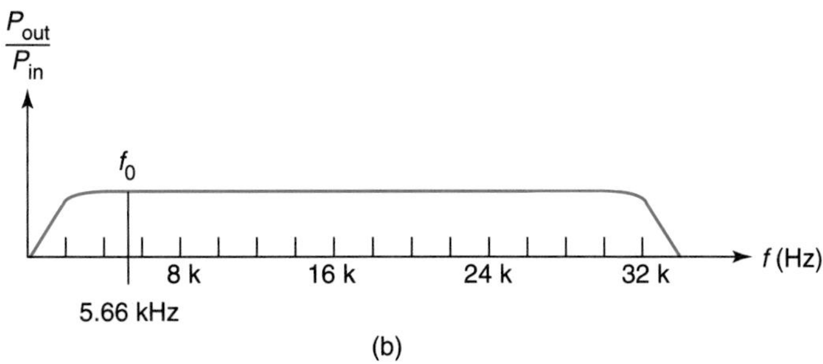

Figure 15.9

TABLE 15.1 Frequency Descriptions and Characteristics of Filters

Filter Type	*Characteristics*
High-pass	Passes all frequencies above its lower cutoff frequency (f_{C1}).
Low-pass	Passes all frequencies below its upper cutoff frequency (f_{C2}).
Bandpass	Passes all frequencies between its upper and lower cutoff frequencies. Normally described in terms of geometric center frequency (f_0) and bandwidth ($f_{C2} - f_{C1}$).
Notch (or band-stop)	Blocks (attenuates) all frequencies between its upper and lower cutoff frequencies. Normally described in terms of geometric center frequency (f_0) and bandwidth ($f_{C2} - f_{C1}$).

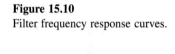

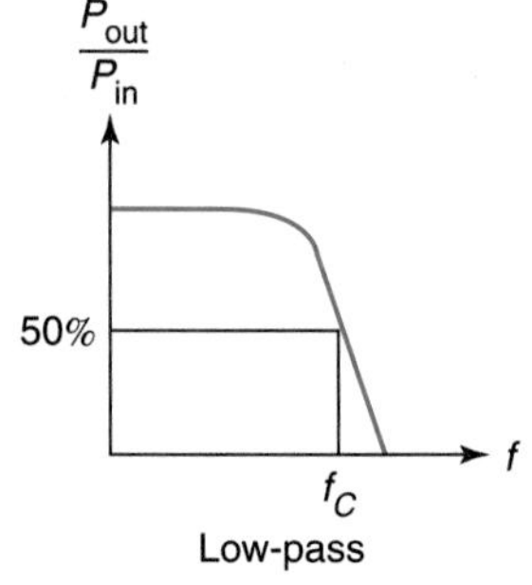

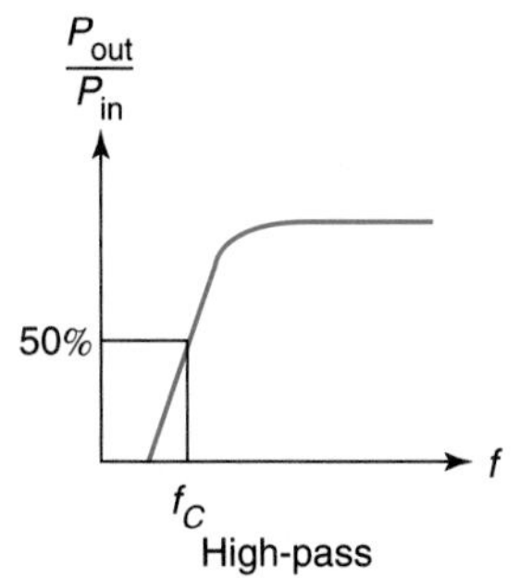

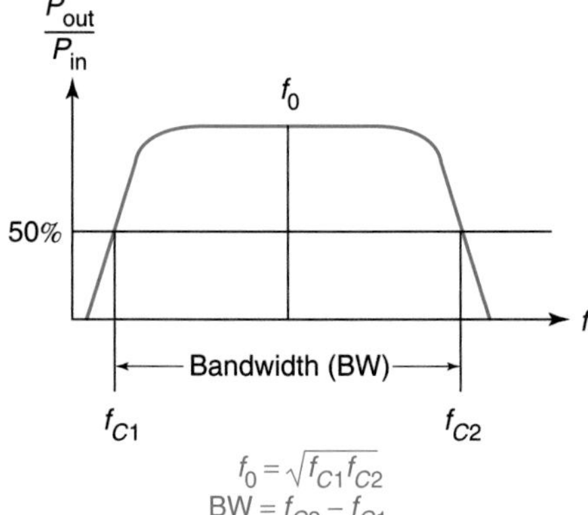

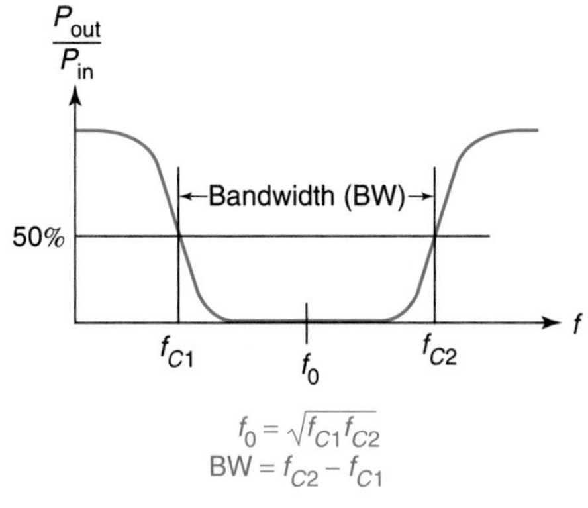

Figure 15.10
Filter frequency response curves.

Section Review

1. What is *attenuation*?
2. How is *amplitude* normally represented on a frequency response curve?
3. What is a *cutoff* frequency?
4. What is a *filter*?
5. List the four common types of filters, and describe the frequency response of each.
6. What is meant by the term *band*?
7. What is the *bandwidth* of a bandpass or notch filter?
8. What two values are required to accurately locate and describe the frequency response curve of a bandpass or notch filter?
9. What is the *center frequency* of a bandpass or notch filter?
10. Using *frequency ratios,* describe the relationship among f_{C1}, f_{C2}, and f_0.
11. What does the *quality* (Q) rating of a filter equal?
12. What is the *average frequency* (f_{ave}) of a bandpass or notch filter?
13. What determines the relationship between the values of f_{ave} and f_0 for a bandpass or notch filter?
14. Under what circumstance is it safe to assume that the values of f_{ave} and f_0 for a given filter are approximately equal?
15. What is a *logarithmic scale*?
16. What is a *decade scale*? An *octave scale*?
17. When would each of the scales mentioned in Question 16 be preferred over the other?
18. Why are *logarithmic* frequency scales preferred over *algebraic* frequency scales?

Critical Thinking

19. Why isn't the concept of *geometric center frequency* applied to low-pass filters?
20. What does the *algebraic average frequency* of every low-pass filter equal?

15.2 Amplitude Measurements: dB Power Gain

OBJECTIVE 9 ➤ Amplitude is normally expressed as a ratio of an output value (such as voltage or power) to a corresponding input value. In most cases, these ratios are represented using logarithmic values, called *decibels (dB)*. A **decibel (dB)** is *a logarithmic representation of a number.* Decibels allow us to easily represent very large and very small values. In this section, you will learn how to convert any numeric value to dB form, and vice versa. You will also learn how to perform basic mathematical operations involving dB values.

Bel Power Gain

At one time, power ratios were expressed using logarithmic values called *bels.* The **bel (B)** is a ratio of output power to input power, expressed as a common (base 10) logarithm. By formula,

$$A_{p(\mathrm{B})} = \log_{10} \frac{P_{out}}{P_{in}} \tag{15.13}$$

where

$A_{p(\mathrm{B})}$ = the *power gain* of a component or circuit, in bels

$\log_{10} \frac{P_{out}}{P_{in}}$ = the common log of the power ratio

The term *power gain* (A_p) is used in reference to amplifiers, which are capable of increasing the power level of their ac input signals. Therefore, amplifiers can have power ratios that are greater than one. The circuits we are going to be dealing with in this chapter have power ratios that are less than one ($P_{out} < P_{in}$) . However, we still use the term *gain* to describe their output to input power ratios. Note that the term **unity gain** is commonly used to refer to *a value of* $A_p = 1$. Using the symbol for power gain, equation (15.13) can be rewritten as

$$A_{p(B)} = \log_{10} A_p \qquad (15.14)$$

where $A_p = \dfrac{P_{out}}{P_{in}}$.

Decibels (dB)

Many years ago, industry decided that the bel was too large a unit of gain to be practical. As a result, the switch was made from bels to tenths of bels, or *decibels (dB)*. There are ten decibels in one bel, so

◄ *OBJECTIVE 10*

$$A_{p(dB)} = 10 \log \frac{P_{out}}{P_{in}} \qquad (15.15)$$

or

$$A_{p(dB)} = 10 \log A_p \qquad (15.16)$$

Example 15.8 demonstrates the use of this relationship.

Example 15.8

The output power of a certain bandpass filter is one-half the value of the input power. Calculate the dB power gain for the circuit.

Solution: The circuit power gain is found as

$$A_p = \frac{P_{out}}{P_{in}} = \frac{0.5\ P_{in}}{P_{in}} = 0.5$$

Now the dB power gain is found as

$$A_{p(dB)} = 10 \log A_p = 10 \log (0.5) = (10)(-0.3) = -3 \text{ dB}$$

Practice Problem 15.8

A bandpass filter is found to have values of $P_{in} = 1$ W and $P_{out} = 750$ mW. Calculate the dB power gain of the circuit.

As you can see, the dB power gain calculated in Example 15.8 was negative. Here are some important dB power gain relationships:

- A dB value is negative when $P_{out} < P_{in}$.
- A dB value is zero when $P_{out} = P_{in}$.
- A dB value is positive when $P_{out} > P_{in}$.

The first of these relationships was demonstrated in Example 15.8. The second relationship can be demonstrated as follows: When the values of P_{out} and P_{in} for a circuit are equal,

$$A_p = \frac{P_{out}}{P_{in}} = 1$$

and

$$A_{p(dB)} = 10 \log A_p = 10 \log (1) = (10)(0) = 0 \text{ dB}$$

The third relationship is demonstrated in Example 15.9.

Example 15.9

An amplifier is found to have a value of $A_p = 2$. Determine the dB power gain of the circuit.

Solution: The amplifier dB power gain is found as

$$A_{p(\text{dB})} = 10 \log A_p = 10 \log (2) = (10)(0.3) = 3 \text{ dB}$$

Practice Problem 15.9

An amplifier is found to have a value of $A_p = 100$. Determine the dB power gain of the circuit.

Reciprocal values of power gain have positive and negative dB values with equal magnitudes. This principle can be stated as follows:

$$\text{If } x \text{ dB} = y, \text{ then } -x \text{ dB} = \frac{1}{y}$$

For example, consider the values from Examples 15.8 and 15.9 listed here:

Example	*Power Gain*	*Power Gain (in dB)*
15.8	$0.5 \left(\frac{1}{2}\right)$	-3 dB
15.9	2.0	3 dB

Converting dB Power Gain to Standard Numeric Form

Any dB power gain value can be converted to standard numeric form using the following relationship:

$$A_p = \log^{-1}\left(\frac{A_{p(\text{dB})}}{10}\right) \tag{15.17}$$

where

$\log^{-1}$ represents the *inverse log* function

Earlier in this chapter, we defined A_p as the ratio of output power to input power. When the values of power gain and input power are known, the output power can be found as

$$P_{\text{out}} = A_p P_{\text{in}} \tag{15.18}$$

When we know the dB power gain and the input power of a circuit, we can solve for the output power as demonstrated in Example 15.10.

Example 15.10

A bandpass filter has maximum values of $A_{p(\text{dB})} = -1.8$ dB and $P_{\text{in}} = 2.5$ W. Determine the maximum output power for the filter.

Solution: The value of A_p for the circuit is found as

$$A_p = \log^{-1}\left(\frac{A_{p(\text{dB})}}{10}\right) = \log^{-1}\left(\frac{-1.8 \text{ dB}}{10}\right) = \log^{-1}(-0.18) = 0.661$$

Now the circuit output power can be found as

$$P_{\text{out}} = A_p P_{\text{in}} = (0.661)(2.5 \text{ W}) = 1.65 \text{ W}$$

Practice Problem 15.10

A bandpass filter has maximum values of $A_{p(\text{dB})} = -1.925$ dB and $P_{\text{in}} = 900$ mW. Determine the maximum output power for the filter.

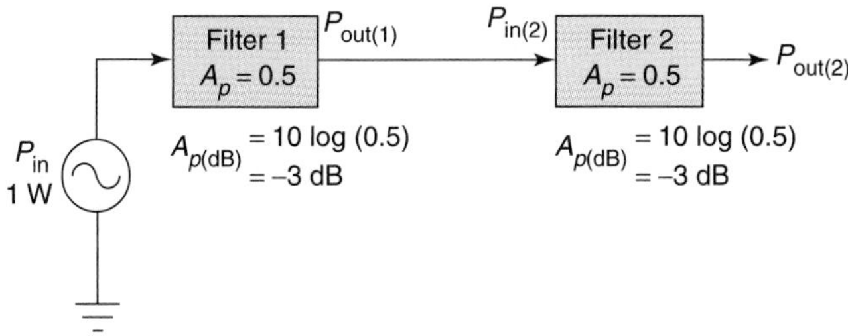

Figure 15.11 Cascaded filters.

Multistage Filter Gain

When filters are connected in series (as shown in Figure 15.11), they are said to be **cascaded.** *Each filter in the cascade* is referred to as a **stage.** *The combination of cascaded filters* is commonly referred to as an ***n*-stage filter,** where n is the number of stages. For example, the cascaded filter shown in Figure 15.11 would be referred to as a two-stage filter.

When the stage gains of a multistage filter are given in dB, the total output power can be found using the following procedure:

1. Determine the overall dB gain of the circuit.
2. Convert the dB gain to standard numeric form.
3. Multiply the input power by the overall gain to determine the total output power.

Note that the overall dB gain of the circuit equals the sum of the individual dB gains. By formula,

$$A_{p\text{T(dB)}} = A_{p1(\text{dB})} + A_{p2(\text{dB})} + \cdots + A_{pn(\text{dB})} \tag{15.19}$$

The output power from an n-stage filter can be calculated as demonstrated in Example 15.11.

Example 15.11

Determine the total output power for the circuit represented in Figure 15.12.

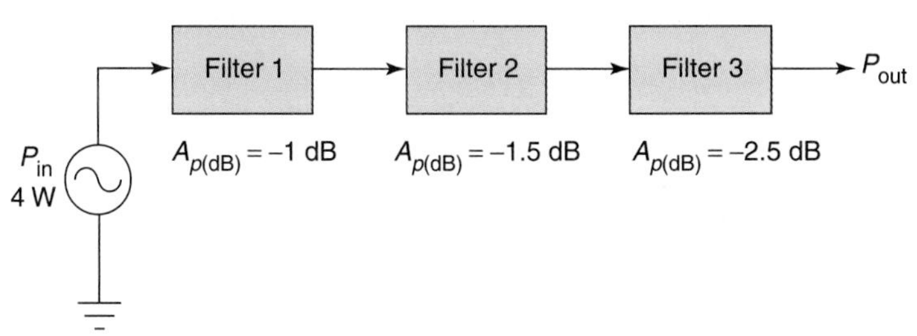

Figure 15.12

Solution: First, the sum of the stage gains (in dB) is found as

$$A_{p\text{T(dB)}} = (-1 \text{ dB}) + (-1.5 \text{ dB}) + (-2.5 \text{ dB}) = -5 \text{ dB}$$

Converting the result to standard numeric form, we get

$$A_p = \log^{-1}\left(\frac{A_{p\text{T(dB)}}}{10}\right) = \log^{-1}\left(\frac{-5 \text{ dB}}{10}\right) = \log^{-1}(-0.5) = 0.316$$

Finally, the output power from the circuit can be found as

$$P_{out} = A_p P_{in} = (0.316)(4\ \text{W}) = 1.26\ \text{W}$$

Practice Problem 15.11

Determine the value of P_{out} for the circuit shown in Figure 15.11.

The dBm Reference

The **dBm reference** represents *a power level* (rather than power gain) *that is referenced to* 1 *mW,* as shown in the following relationship:

$$P_{dBm} = 10 \log \frac{P_{out}}{1\ \text{mW}} \tag{15.20}$$

Values measured in dBm are actual power levels rather than power ratios, as demonstrated in Example 15.12.

Example 15.12

An amplifier has an output rated at 2 W. Determine the value of output power (in dBm) for the circuit.

Solution: The amplifier output power is converted to dBm as follows:

$$P_{dBm} = 10 \log \frac{P_{out}}{1\ \text{mW}} = 10 \log \frac{2\ \text{W}}{1\ \text{mW}} = 10 \log (2 \times 10^3) = 33\ \text{dBm}$$

As this calculation demonstrates, 2 W can be expressed as 33 dBm, meaning that it is 33 dB above 1 mW.

Practice Problem 15.12

An amplifier has an output that is rated at 12 W. Determine the value of output power (in dBm) for the circuit.

The dBm reference is most commonly used to identify a power limit. For example, a circuit may be rated for a maximum output power of 50 dBm. When this is the case, the maximum value of P_{out} can be determined using

$$P_{out} = (1\ \text{mW})\left(\log^{-1} \frac{P_{dBm}}{10}\right) \tag{15.21}$$

Using this relationship, we can easily convert a rating of 50 dBm, as follows:

$$P_{out} = (1\ \text{mW})\left(\log^{-1} \frac{P_{dBm}}{10}\right) = (1\ \text{mW})\left(\log^{-1} \frac{50}{10}\right) = (1\ \text{mW})(1 \times 10^5) = 100\ \text{W}$$

Section Review

1. What is a *bel*?
2. What is *power gain* (A_p)?
3. What is a *decibel* (dB)?
4. What characteristics do dB values share with bel values?
5. What is the procedure for calculating output power when input power and dB power gain are known?
6. What are the advantages of using dBs to represent gain?

7. How do you determine the overall dB gain of a multistage filter?
8. What is represented using the *dBm reference*?
9. What is the most common application for dBm values?

15.3 Amplitude Measurements: dB Voltage and Current Gain

In the last section, the decibel was defined in terms of *power gain*. In this section, you will see that *voltage gain* and *current gain* can also be expressed in decibels.

Voltage Gain

Voltage gain at the cutoff frequencies.

A bandpass response curve can be drawn as shown in Figure 15.13. The difference between this curve and those discussed earlier is that the cutoff frequencies are defined in terms of a *voltage ratio* (rather than a power ratio). *The ratio of circuit output voltage to input voltage* is generally referred to as **voltage gain (A_v).** As shown in the figure, the voltage gain of the curve equals 70.7% of its maximum value at the cutoff frequencies. By formula,

$$f = f_C \quad \text{when} \quad A_v = 0.707\, A_{v(\max)} \tag{15.22}$$

The 0.707 multiplier in equation (15.22) is based on the relationship between voltage and power. The maximum output power ($P_{\max}$) from a filter can be found as

$$P_{\max} = \frac{V_{\max}^2}{R_L}$$

where $V_{\max}$ is the maximum output voltage from the circuit. If the voltage gain of the circuit drops to 0.707 $A_{v(\max)}$, the output power drops to

$$P = \frac{(0.707\, V_{\max})^2}{R_L} = \frac{(0.707)^2\, V_{\max}^2}{R_L} = 0.5\frac{V_{\max}^2}{R_L} = 0.5\, P_{\max}$$

Assuming that the circuit input hasn't changed, the result above indicates that $A_p = 0.5\, A_{p(\max)}$. By definition, this occurs at the cutoff frequencies of the filter.

dB Voltage Gain

The dB voltage gain of a filter (or amplifier) is found using

$$A_{v(\text{dB})} = 20 \log \frac{v_{\text{out}}}{v_{\text{in}}} \tag{15.23}$$

◄ OBJECTIVE 11

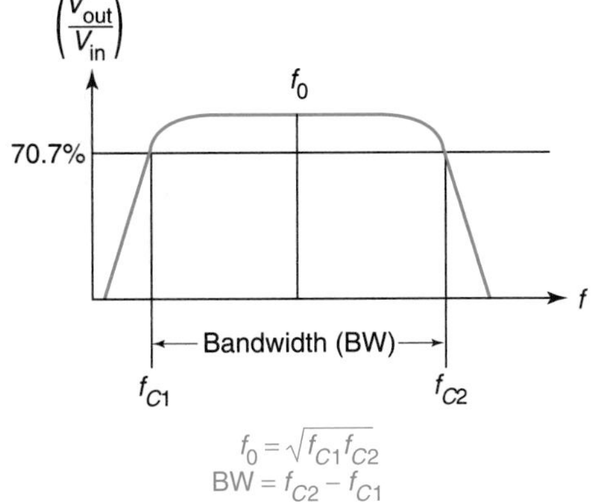

Figure 15.13
Voltage gain measurement on a frequency response curve.

or

$$A_{v(\text{dB})} = 20 \log A_v \tag{15.24}$$

where

A_v = the ratio of output voltage to input voltage

The calculation of dB voltage gain is demonstrated in Example 15.13.

Example 15.13

A filter has a value of $A_v = 0.707$. Calculate the value of $A_{v(\text{dB})}$ for the circuit.

Solution: Using equation (15.24), the dB voltage gain of the filter can be found as

$$A_{v(\text{dB})} = 20 \log A_v = 20 \log (0.707) = (20)(-0.15) = -3 \text{ dB}$$

Practice Problem 15.13

A filter has a value of $A_v = 0.5$. Calculate the value of $A_{v(\text{dB})}$ for the circuit.

The value of A_v used in Example 15.13 was selected to demonstrate further the relationship between voltage gain and power gain. Here is a comparison of the values from Examples 15.9 and 15.13:

Example	*Gain*	*Gain (in dB)*
15.9	$A_p = 0.5$	$A_{p(\text{dB})} = -3$ dB
15.13	$A_v = 0.707$	$A_{v(\text{dB})} = -3$ dB

As you can see, the two circuits have the same dB gain (−3 dB). This verifies the correlation between a power gain of 0.5 and a voltage gain of 0.707.

As you have seen, a conversion factor of 20 is used when calculating $A_{v(\text{dB})}$ in place of the 10 used when calculating $A_{p(\text{dB})}$. You may remember from your study of basic mathematics that

$$\log(x)^y = y \log x$$

Since power is proportional to the square of voltage, the multiplier for $A_{v(\text{dB})}$ is $10 \times 2 = 20$.

Converting dB Voltage Gain to Standard Numeric Form

Any dB voltage gain can be converted to standard numeric form using the following equation:

$$A_v = \log^{-1}\left(\frac{A_{v(\text{dB})}}{20}\right) \tag{15.25}$$

Example 15.14 demonstrates the process for converting dB voltage gain to standard numeric form.

Example 15.14

A filter is rated at $A_{v(\text{dB})} = -2.8$ dB. Calculate the value of A_v for the circuit.

Solution: Using equation (15.25), $A_{v(\text{dB})}$ is converted to standard numeric form as follows:

$$A_v = \log^{-1}\left(\frac{A_{v(\text{dB})}}{20}\right) = \log^{-1}\left(\frac{-2.8 \text{ dB}}{20}\right) = \log^{-1}(-0.14) = 0.724$$

This result tells us that $V_{out} = 0.724V_{in}$.

Practice Problem 15.14

A filter is rated at $A_{v(\text{dB})} = -3.5$ dB. Calculate the value of A_v for the circuit.

A *negative* dB voltage gain indicates that $A_v < 1$, as was demonstrated in Example 15.14. By the same token, a *positive* dB voltage gain indicates that $A_v > 1$. For example, an amplifier with a value of $A_{v(\text{dB})} = 2.8$ dB has a gain of

$$A_v = \log^{-1}\left(\frac{2.8 \text{ dB}}{20}\right) = \log^{-1}(0.14) = 1.38$$

This result illustrates another point: In Example 15.14, we found that an A_v of -2.8 dB equals 0.724. The result of 1.38 (above) is the reciprocal of 0.724. This demonstrates that the reciprocal relationships that we saw with dB power gain also apply to dB voltage gain.

Changes in dB Gain

Even though dB values of voltage and power gain are calculated differently, they change at the same rate. By formula,

$$\Delta A_{p(\text{dB})} = \Delta A_{v(\text{dB})} \tag{15.26}$$

where

$\Delta A_{p(\text{dB})}$ = the change in dB power gain
$\Delta A_{v(\text{dB})}$ = the change in dB voltage gain

For example, if the voltage gain of a filter drops by 3 dB, the power gain drops by the same amount.

Current Gain (A_i)

◄ OBJECTIVE 12

The ratio of circuit output current to input current is generally referred to as **current gain (A_i).** Current gain relationships are identical to those given earlier for voltage gain. For convenience, these relationships are rewritten here for current gain:

$$f = f_C \quad \text{when} \quad A_i = 0.707A_{i(\text{max})} \tag{15.27}$$

$$A_{i(\text{dB})} = 20 \log \frac{i_{\text{out}}}{i_{\text{in}}} \tag{15.28}$$

$$A_{i(\text{dB})} = 20 \log A_i \tag{15.29}$$

$$A_i = \log^{-1} \frac{A_{i(\text{dB})}}{20} \tag{15.30}$$

$$\Delta A_{p(\text{dB})} = \Delta A_{i(\text{dB})} \tag{15.31}$$

Since these relationships are identical to the voltage gain relationships given earlier, we do not need to elaborate on them any further at this point.

Section Review

1. What is *voltage gain* (A_v)?
2. What is the value of A_v (as a percentage) at the cutoff frequencies of a filter?
3. Compare the calculations of $A_{v(\text{dB})}$ and $A_{p(\text{dB})}$.
4. What is the relationship between $\Delta A_{v(\text{dB})}$ and $\Delta A_{p(\text{dB})}$?
5. What is *current gain* (A_i)?

15.4 RC and RL Low-Pass Filters

Simple *RL* and *RC* circuits can be used as low-pass and high-pass filters. In this section, we will look at *RL* and *RC* low-pass and high-pass circuits, their characteristics, and their analysis procedures. We will also introduce the *bode plot* (pronounced bō-dē plot).

RC Low-Pass Filters

OBJECTIVE 13 ➤ An *RC* low-pass filter is shown in Figure 15.14. The resistor is positioned directly in the signal path, and the capacitor is connected in parallel with the load. Since the capacitor and the load are in parallel, $V_L = V_C$. Note that *any component connected from a signal path to ground* is referred to as a **shunt component.**

The filtering action of this circuit is a result of the capacitor's response to an increase in frequency. As you know, a capacitor has near-infinite reactance when its operating frequency is 0 Hz. As shown in Figure 15.15a, the capacitor acts as an open circuit when $f_{in} = 0$ Hz. Therefore, it can be dropped from the equivalent circuit, and V_L equals the difference between V_S and V_R.

When f_{in} approaches infinity, X_C approaches 0 Ω, and the filter has the equivalent circuit shown in Figure 15.15b. As you can see, the capacitor is represented as a short circuit in parallel with the load, and $V_L = 0$ V.

Between the extremes represented in Figure 15.15 lies a range of frequencies over which V_L decreases from $(V_S - V_R)$ to 0 V. Our primary interest at this point is in establishing methods to determine:

- The maximum value of A_v for the circuit.
- The cutoff frequency (f_C) for the circuit.

Maximum Voltage Gain ($A_{v(max)}$)

The maximum gain of a low-pass filter occurs when the input frequency is 0 Hz. As shown in Figure 15.15a, the load forms a voltage divider with the filter resistor (R_F) when $f_{in} = 0$ Hz. Therefore, the maximum load voltage is found as

$$V_{L(\text{max})} = V_{\text{in}} \frac{R_L}{R_L + R_F} \tag{15.32}$$

As you know, voltage gain is the ratio of output voltage to input voltage. For the *RC* low-pass filter, the maximum voltage gain can be found as

$$A_{v(\text{max})} = \frac{R_L}{R_L + R_F} \tag{15.33}$$

Note that this equation can be derived by dividing both sides of equation (15.32) by V_{in}.

Equation (15.33) is important because it defines the limit on the voltage gain of an *RC* low-pass filter. Since $R_F + R_L > R_L$, the value of $A_{v(max)}$ must be less than one. A value

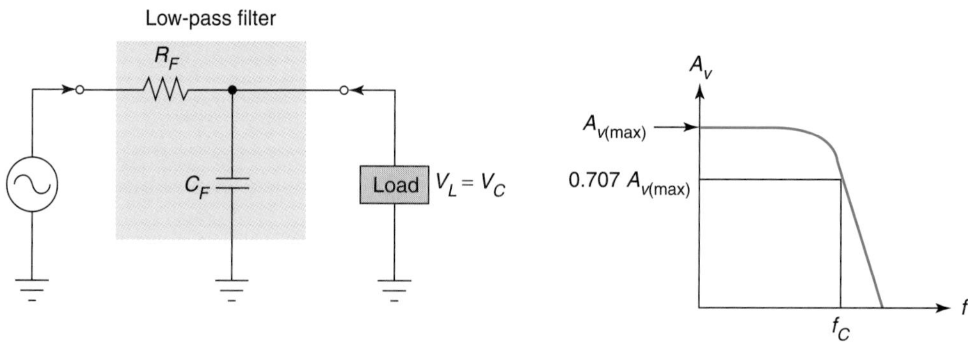

Figure 15.14 An *RC* low-pass filter and its frequency response curve.

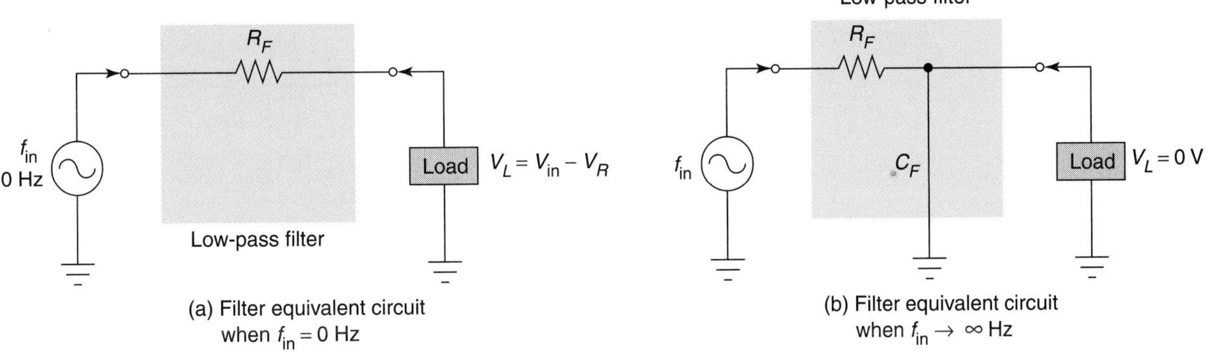

Figure 15.15 *RC* low-pass filter operating limits.

of $A_{v(\max)} \cong 1$ is achieved when $R_F << R_L$, so *RC* low-pass filters are normally designed using the lowest practical value of R_F.

Upper Cutoff Frequency (f_C)

The upper cutoff frequency for a given *RC* low-pass filter is determined by the circuit resistor and capacitor values. By formula,

$$f_C = \frac{1}{2\pi RC} \tag{15.34}$$

where

R = the total circuit resistance as seen by the capacitor

C = the value of the filter capacitor

As stated above, R is the total resistance *as seen by the capacitor.* This resistance can be identified as shown in Figure 15.16.

Even though only two resistors are shown in Figure 15.16a, the circuit has *three* resistance values: the filter resistor (R_F), the load resistance (R_L), and the source resistance (R_S). In Figure 15.16b, the source is replaced by a resistor representing R_S, and the capacitor is removed. The total resistance seen by the capacitor equals the Thevenin resistance (R_{TH}) across the open capacitor terminals. This resistance equals the parallel combination of ($R_S + R_F$) and R_L. By formula,

A Practical Consideration: The term *Thevenin resistance* is traditionally applied to load analysis. However, it can also be used to describe the resistance across any open pair of component terminals.

$$R_{TH} = (R_S + R_F) \parallel R_L \tag{15.35}$$

and

$$f_C = \frac{1}{2\pi R_{TH} C} \tag{15.36}$$

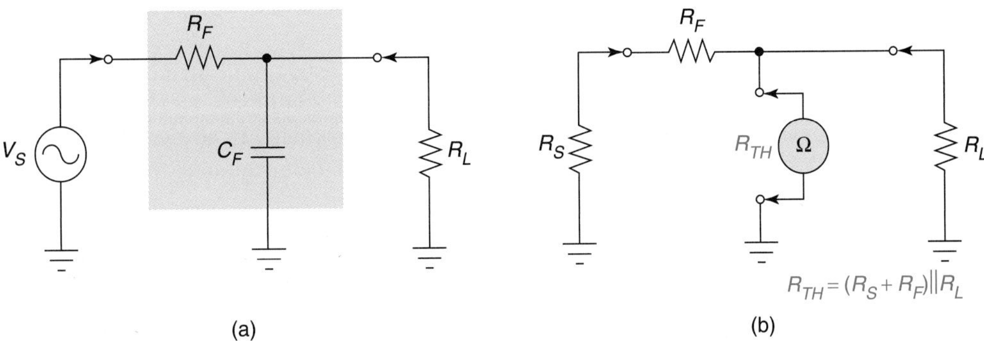

Figure 15.16 Filter resistance as seen by the capacitor.

Example 15.15 demonstrates the procedure for calculating the cutoff frequency of an *RC* low-pass filter.

Example 15.15

Determine the cutoff frequency for the circuit shown in Figure 15.17a.

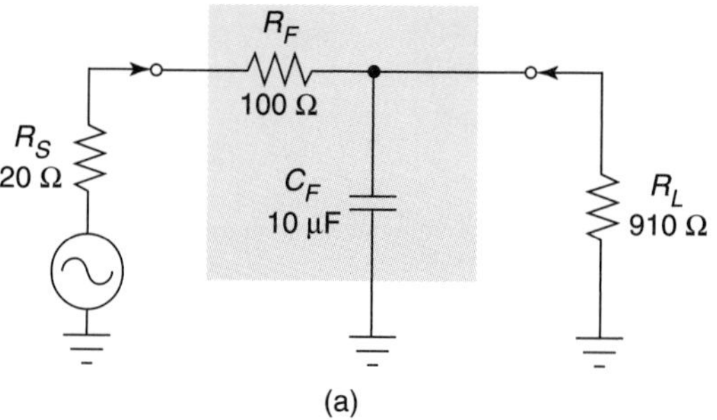

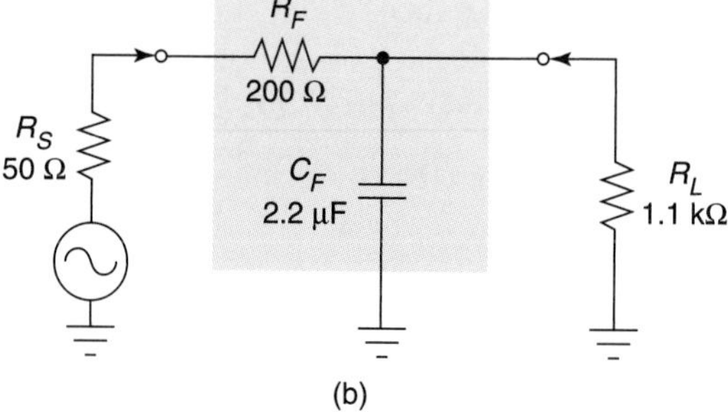

Figure 15.17

Solution: First, the value of R_{TH} is found as

$$R_{TH} = (R_S + R_F) \parallel R_L = (100\ \Omega + 20\ \Omega) \parallel 910\ \Omega = 106\ \Omega$$

Now the cutoff frequency of the circuit is found as

$$f_C = \frac{1}{2\pi R_{TH} C_F} = \frac{1}{2\pi(106\ \Omega)(10\ \mu\text{F})} = 150\ \text{Hz}$$

This result indicates that the voltage gain of the circuit drops to 70.7% of its maximum value when the operating frequency reaches 150 Hz.

Practice Problem 15.15

Determine the cutoff frequency for the circuit shown in Figure 15.17b.

Bode Plots

OBJECTIVE 14 ➤

A **bode plot** is *a normalized graph that represents frequency response as a change in gain (ΔA_v) versus operating frequency.* A frequency response curve and its equivalent bode plot are shown in Figure 15.18.

Bode plots are ideal representations of frequency response.

A bode plot is an ideal curve that assumes gain is constant until the cutoff frequency is reached. For example, ΔA_v is 0 dB from 0 Hz to 150 Hz in Figure 15.18b. At frequencies above 150 Hz, the bode plot shows the circuit gain changing at a constant rate. In contrast, the frequency response curve shows the voltage gain beginning to decrease well before the cutoff frequency is reached.

Filter Roll-Off Rate

As the input frequency to a low-pass filter increases above the value of f_C, *the circuit's gain drops at a relatively constant rate.* This drop in gain is referred to as **roll-off.** The

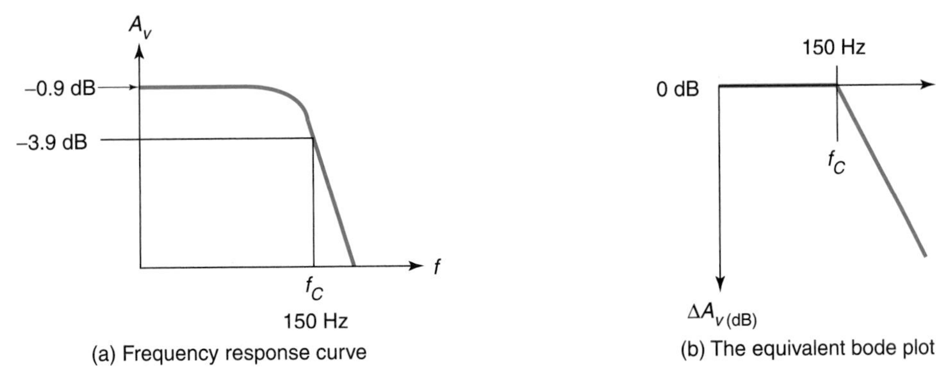

Figure 15.18 A frequency response curve and its equivalent bode plot.

roll-off rate of a filter is *the rate of change in gain, usually expressed in dB per octave or dB per decade.* For example, the bode plot in Figure 15.19a shows a decrease in gain of 6 dB whenever frequency doubles (6 dB per octave). The bode plot in Figure 15.19b shows a decrease in gain of 20 dB per decade.

There are two important points about the roll-off rates illustrated in Figure 15.19:

- The roll-off rates shown in the figure are equal. That is, a roll-off rate of 6 dB per octave equals a roll-off rate of 20 dB per decade.
- The roll-off rates shown apply to every single-stage *RC* low-pass filter and are independent of the component values. Even though the value of f_{C2} may vary from one *RC* low-pass filter to another, the roll-off rates do not.

Note: Roll-off rates are additive. Therefore, the overall roll-off rate for a multistage filter equals the sum of the stage roll-off rates. For example, the overall roll-off rate of a two-stage passive *RC* filter is 12 dB/octave (40 dB/decade).

RL Low-Pass Filters

An *RL* low-pass filter is shown in Figure 15.20. As you can see, the inductor is the series component and the resistor is the shunt component.

◀ *OBJECTIVE 15*

Circuit Operation

The filtering action of the circuit shown in Figure 15.20 is a result of the inductor's response to an increase in operating frequency. When f_{in} is 0 Hz, the inductive reactance is 0 Ω, but there is some amount of winding resistance (R_W). This winding resistance is shown in the equivalent circuit in Figure 15.21a. Note that the load voltage equals the difference between the source voltage and the voltage across R_W.

Theoretically, the operating frequency can become high enough for the inductor to act (effectively) as an open. In this case, the filter has the equivalent circuit shown in Figure 15.21b, and $V_L = 0$ V. Between these two extremes lies a range of frequencies over which V_L decreases from $(V_S - V_{RW})$ to 0 V.

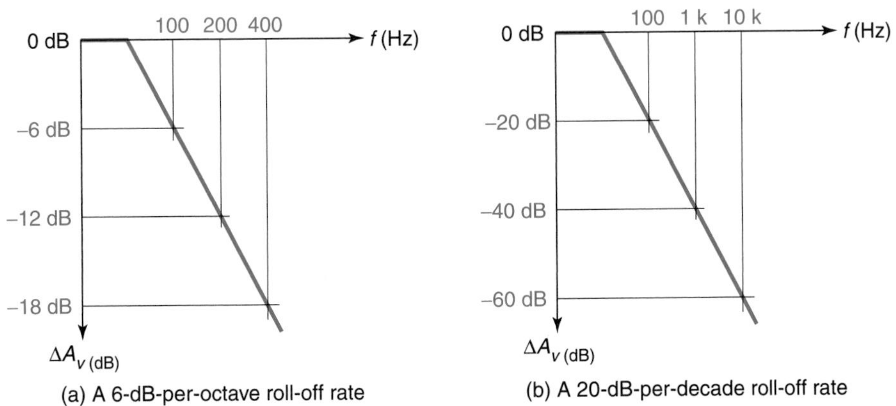

Figure 15.19 Filter roll-off rates.

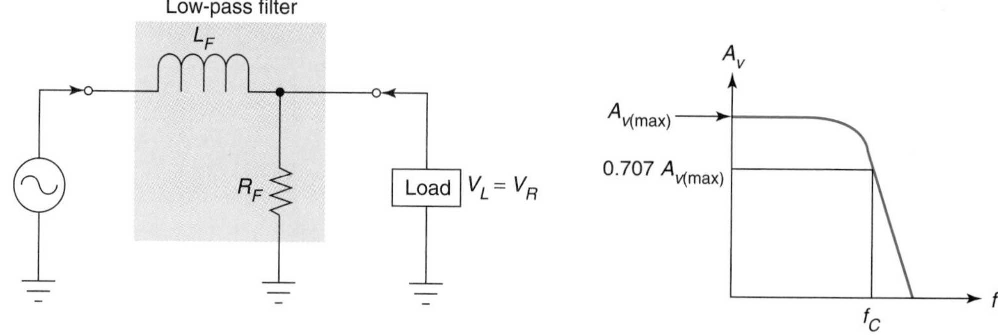

Figure 15.20 An *RL* low-pass filter and its frequency response curve.

Maximum Voltage Gain ($A_{v(max)}$)

The maximum gain of any low-pass filter occurs when the input frequency is 0 Hz. As shown in Figure 15.21a, the parallel combination of the load and R_F forms a voltage divider with R_W when $f_{in} = 0$ Hz. Therefore, the maximum load voltage is found as

$$V_{L(\max)} = V_{in}\frac{R_{EQ}}{R_{EQ} + R_W}\bigg| R_{EQ} = R_F \| R_L \tag{15.37}$$

As you know, voltage gain is the ratio of output voltage to input voltage. If we divide both sides of equation (15.37) by V_{in}, maximum voltage gain is shown to equal the resistance ratio. By formula,

$$A_{v(\max)} = \frac{R_{EQ}}{R_{EQ} + R_W}\bigg| R_{EQ} = R_F \| R_L \tag{15.38}$$

Equation (15.38) is important, because it defines the limit on the voltage gain of an *RL* low-pass filter. Since $R_{EQ} + R_W > R_{EQ}$, the value of $A_{v(\max)}$ must be less than one. A value of $A_{v(\max)} \cong 1$ is achieved when $R_W << R_L$.

Upper Cutoff Frequency (f_C)

The upper cutoff frequency for an *RL* low-pass filter is determined by the inductor, the parallel combination of R_F and R_L, and the source resistance (R_S). By formula,

$$f_C = \frac{R_{TH}}{2\pi L}\bigg| R_{TH} = R_S + (R_F \| R_L) \tag{15.39}$$

The use of this relationship is demonstrated in Example 15.16.

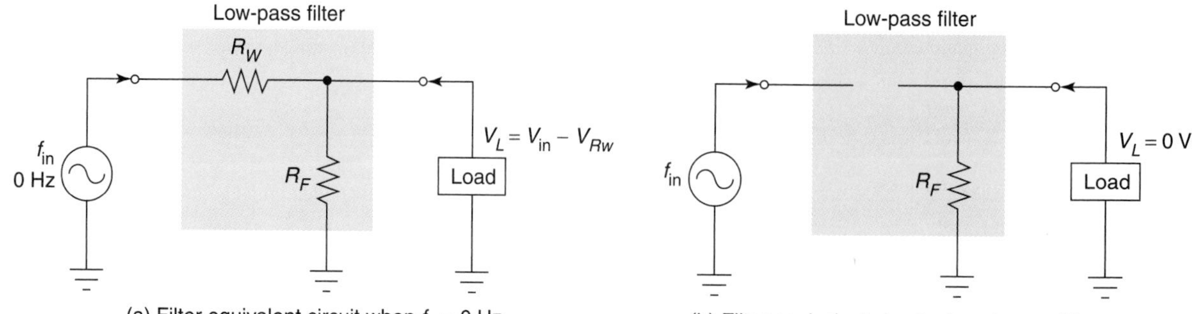

Figure 15.21 *RL* low-pass filter equivalent circuits.

Calculate the cutoff frequency for the circuit shown in Figure 15.22a.

Example 15.16

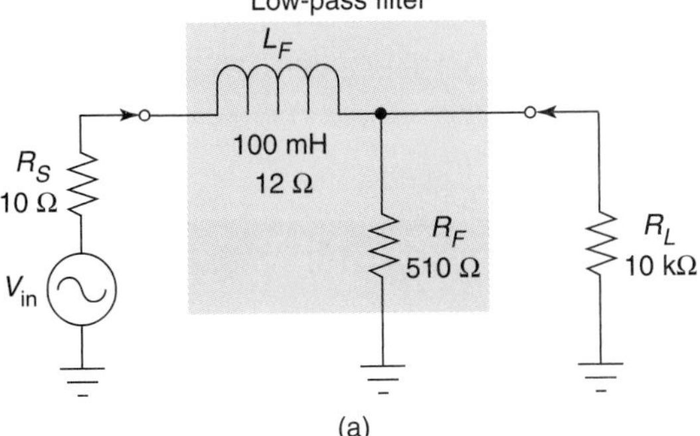

(a)

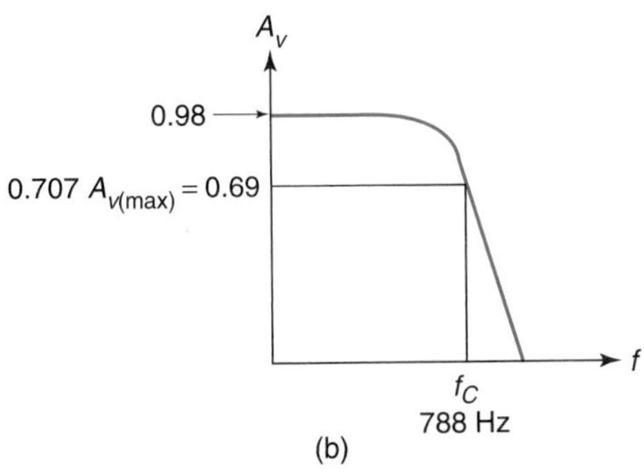

(b)

Figure 15.22

Solution: The value of R_{TH} is found as

$$R_{TH} = R_S + (R_F \| R_L) = 10\ \Omega + (510\ \Omega \| 10\ \text{k}\Omega) = 495\ \Omega$$

Now the filter cutoff frequency is found as

$$f_C = \frac{R_{TH}}{2\pi L} = \frac{495\ \Omega}{2\pi(100\ \text{mH})} = 788\ \text{Hz}$$

Practice Problem 15.16

A filter like the one shown in Figure 15.22a has the following values: $R_L = 680\ \Omega$, $R_F = 15\ \text{k}\Omega$, $R_S = 25\ \Omega$, $L = 470$ mH, and $R_W = 20\ \Omega$. Calculate the value of f_C for the circuit.

Roll-Off Rates

RL filter roll-off characteristics are identical to those of the *RC* low-pass filter. The gain remains relatively constant until the cutoff frequency is reached. Above f_C, it has a roll-off rate of 6 dB per octave (or 20 dB per decade). In fact, the bode plot for an *RL* low-pass filter is identical to that for an *RC* low-pass filter. As with the *RC* filter, the roll-off rate is independent of component values.

Section Review

1. What is a *shunt component*?
2. Why is the voltage gain of an *RC* circuit always less than one?
3. What is a *bode plot*?
4. What filter characteristic is normalized on a bode plot?
5. What is *roll-off*? What is the *roll-off rate* of a filter?

6. What is the relationship between the values of R and C in a low-pass filter and its roll-off rate?

7. What are the standard filter roll-off rates?

8. What characteristics do RL and RC low-pass filters share?

15.5 *RC* and *RL* High-Pass Filters

High-pass filters can be formed by reversing the positions of the resistive and reactive components in the RC and RL low-pass filters. The primary difference between a high-pass filter and its low-pass counterpart is component placement.

RC High-Pass Filters

OBJECTIVE 16 ➤ An RC high-pass filter is shown in Figure 15.23a. If you compare this circuit with the RC low-pass filter shown in Figure 15.14, you'll see that the filter capacitor and resistor positions are reversed between the two circuits.

The filtering action of the circuit in Figure 15.23 is a result of the capacitor's response to an increase in operating frequency. When f_{in} is 0 Hz, the reactance of the capacitor is infinite (for all practical purposes). This high reactance is represented as an open in the equivalent circuit shown in Figure 15.23b. Note that $V_L = 0$ V when $f_{in} = 0$ Hz.

As the filter input frequency increases, the reactance of the capacitor decreases. At some point, the reactance of the capacitor approaches 0 Ω, and it acts like a short circuit. As shown in Figure 15.23c, $V_L \cong V_{in}$ when $X_C \rightarrow 0\ \Omega$. Between the extremes shown in Figure 15.23 lies a range of frequencies over which V_L decreases from $V_L \cong V_{in}$ to $V_L = 0$ V.

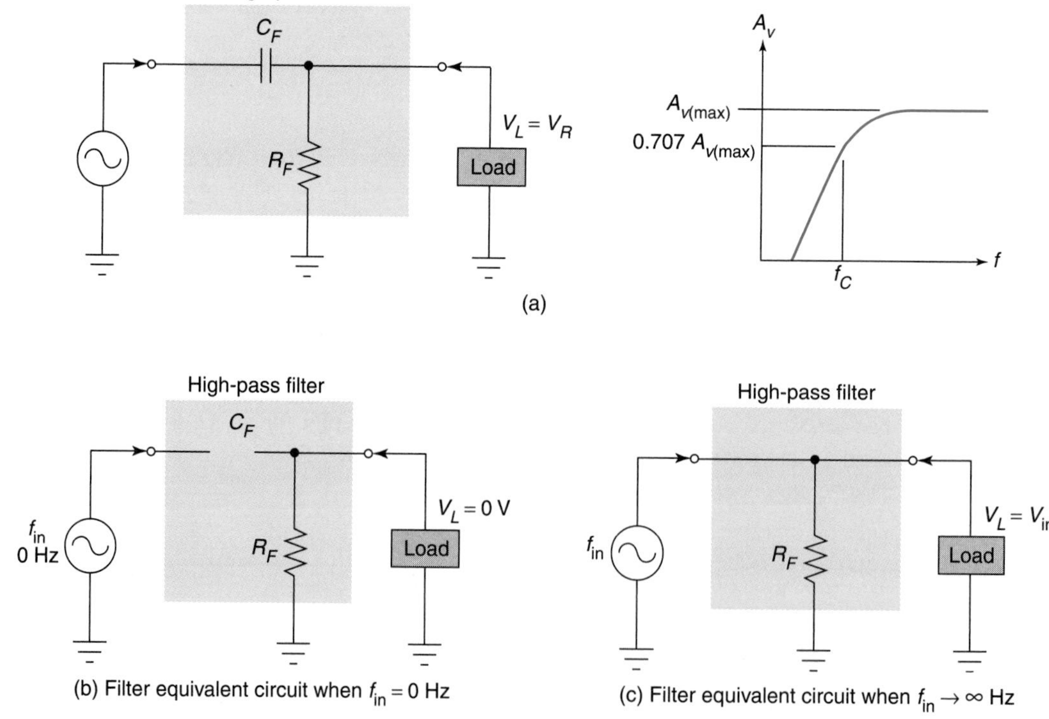

Figure 15.23 An RC high-pass filter and its frequency response curve.

Maximum Voltage Gain ($A_{v(max)}$)

The maximum gain of a high-pass filter occurs when $X_C \cong 0\ \Omega$. In this case, $V_L \cong V_{in}$ and

$$A_{v(\max)} \cong 1 \qquad (15.40)$$

Note that an approximation is used because the reactance of the capacitor never truly reaches 0 Ω.

Lower Cutoff Frequency (f_C)

The cutoff frequency for an *RC* high-pass filter is determined using the same relationship we established for the low-pass filter. By formula,

$$f_C = \frac{1}{2\pi R_{TH} C} \quad \Big| \quad R_{TH} = R_S + (R_F \| R_L)$$

We can use this equation, because the relationship among f_C, R, and C is not affected by the component positions. The cutoff frequency for an *RC* high-pass filter is calculated as demonstrated in Example 15.17.

Example 15.17

Calculate the cutoff frequency for the circuit shown in Figure 15.24a.

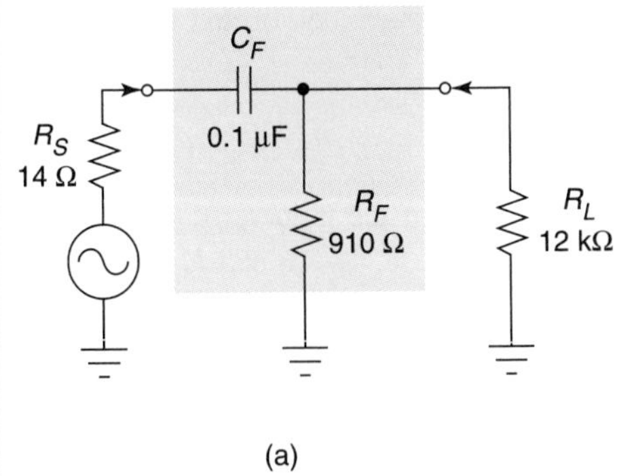

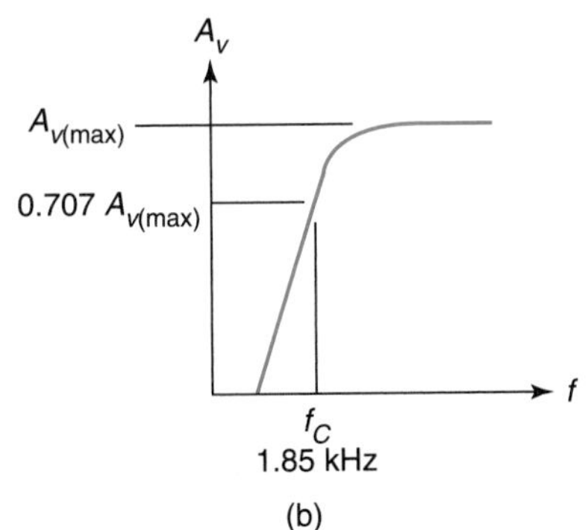

Figure 15.24

Solution: The circuit resistance equals the total circuit resistance as seen by the capacitor. The resistance seen by the capacitor equals the Thevenin resistance measured across the open capacitor terminals, found as

$$R_{TH} = R_S + (R_F \| R_L) = 14\ \Omega + (12\ \text{k}\Omega \| 910\ \Omega) = 860\ \Omega$$

The value of f_C is now found as

$$f_C = \frac{1}{2\pi R_{TH} C} = \frac{1}{2\pi(860\ \Omega)(0.1\ \mu\text{F})} = 1.85\ \text{kHz}$$

The circuit frequency response curve is shown in Figure 15.24b.

Practice Problem 15.17

A circuit like the one shown in Figure 15.24a has the following values: $C_F = 0.22\ \mu\text{F}$, $R_F = 15\ \text{k}\Omega$, $R_S = 12\ \Omega$, and $R_L = 750\ \Omega$. Calculate the circuit cutoff frequency.

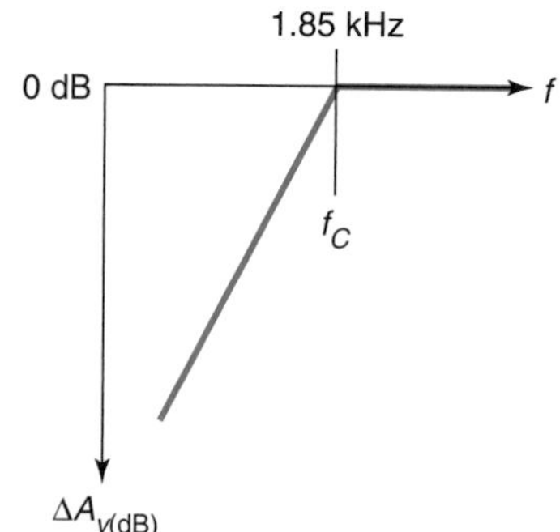

Figure 15.25 The bode plot for the circuit in Figure 15.24.

High-Pass Filter Bode Plots and Roll-Off Rates

The bode plot for a high-pass filter is simply a mirror image of the plot for a low-pass filter. The bode plot for the high-pass filter in Figure 15.24a is shown in Figure 15.25. As operating frequency decreases below the cutoff frequency, the dB voltage gain is shown to roll off at a constant rate. As is the case with *RC* low-pass filters, *RC* high-pass filters all have roll off rates of 6 dB per octave (20 dB per decade).

RL High-Pass Filters

OBJECTIVE 17 ➤ An *RL* high-pass filter is shown in Figure 15.26. If you compare this circuit with the *RL* low-pass filter shown in Figure 15.20, you'll see that the filter inductor and resistor positions are reversed between the two circuits.

Circuit Operation

As shown in Figure 15.27a, the extremely high reactance of the inductor causes it to act like an open circuit when $f_{\text{in}} >> f_C$. In this case, $V_L = V_{\text{in}} - V_{RF}$. When $f_{\text{in}} = 0$ Hz, the reactance of the inductor is approximately 0 Ω. In effect, the inductor acts like a short in parallel with the load. As a result, $V_L \cong 0$ V, as illustrated in Figure 15.27b. Between these two extremes lies a range of frequencies over which V_L decreases from $(V_{\text{in}} - V_{RF})$ to approximately 0 V.

Maximum Voltage Gain ($A_{v(\max)}$)

The maximum gain of an *RL* high-pass filter occurs when the operating frequency is high enough to cause the inductor to effectively act as an open ($X_L \to \infty\,\Omega$). As with the *RC* low-pass filter, maximum voltage gain is found as

$$A_{v(\max)} = \frac{R_L}{R_L + R_F} \tag{15.41}$$

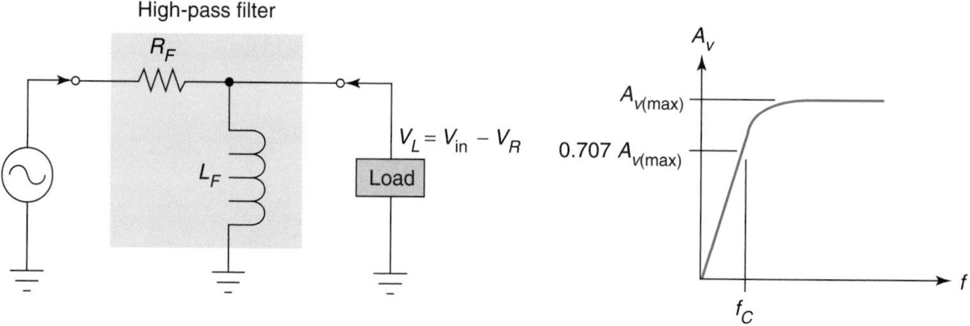

Figure 15.26 An *RL* high-pass filter and its frequency response curve.

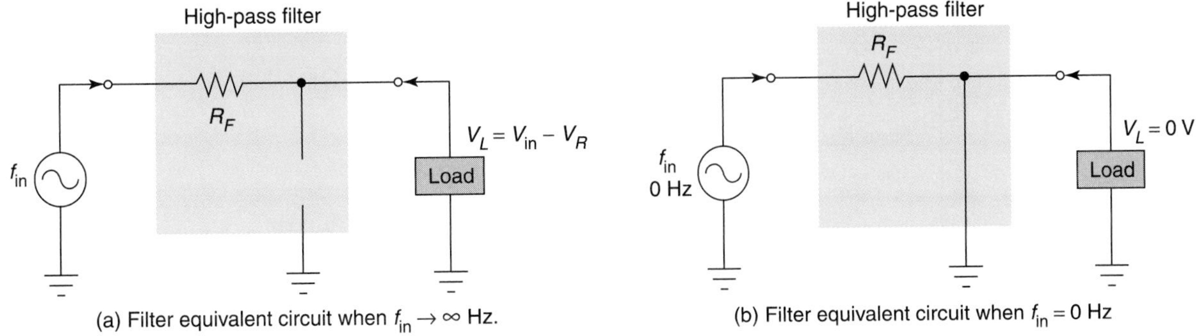

Figure 15.27 *RL* high-pass filter operating limits.

As equation (15.41) indicates, the voltage gain of the *RL* high-pass filter is always less than one. Also, as with the *RC* low-pass filter, maximum voltage gain is achieved by designing the filter so that $R_F << R_L$.

Lower Cutoff Frequency (f_C)

The lower cutoff frequency for an *RL* high-pass filter is determined by the inductor, the parallel combination of R_F and R_L, and the source resistance (R_S). By formula,

$$f_C = \frac{R_{TH}}{2\pi L} \quad | \quad R_{TH} = (R_F + R_S)\|R_L$$

This relationship was introduced earlier in the chapter. The value of R_{TH} is the resistance measured across the open inductor terminals. Example 15.18 demonstrates the cutoff frequency calculations for an *RL* high-pass filter.

Example 15.18

Calculate the cutoff frequency for the *RL* high-pass filter shown in Figure 15.28a.

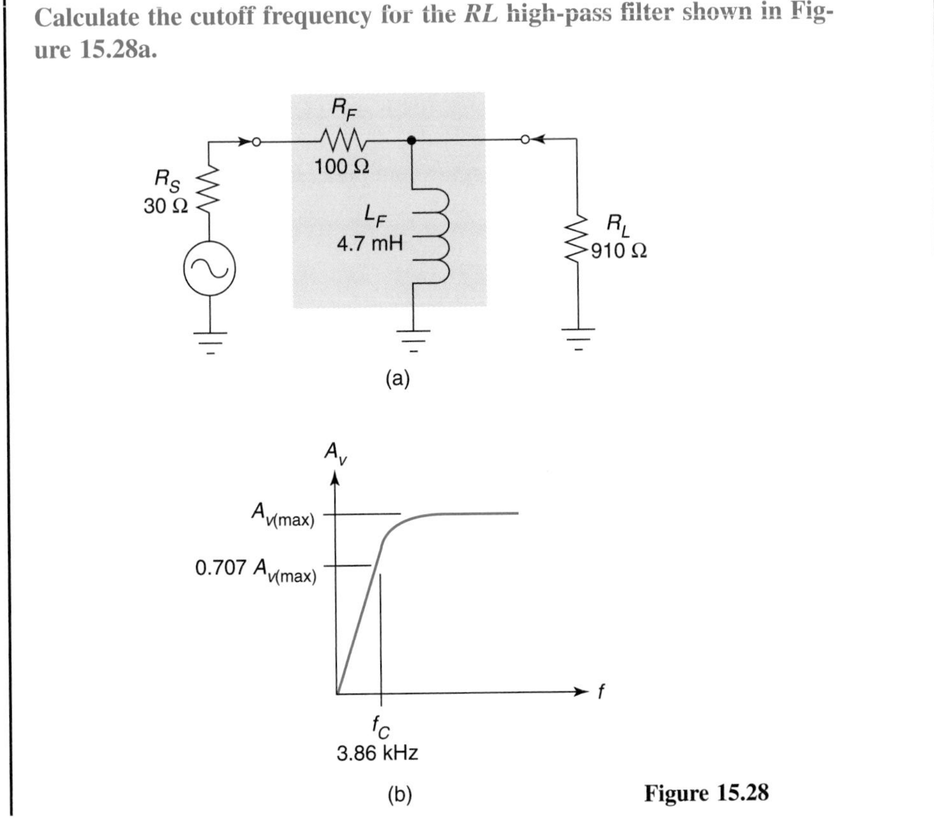

Figure 15.28

Solution: The value of R_{TH} is found as

$$R_{TH} = (R_S + R_F)\|R_L = (130\ \Omega)\|(910\ \Omega) = 114\ \Omega$$

Now the filter cutoff frequency is found as

$$f_C = \frac{R_{TH}}{2\pi L} = \frac{114\ \Omega}{2\pi(4.7\text{ mH})} = 3.86\text{ kHz}$$

The frequency response curve for this circuit is shown in Figure 15.28b.

Practice Problem 15.18

A circuit like the one shown in Figure 15.28a has the following values: $R_F = 20\ \Omega$, $R_S = 15\ \Omega$, $L_F = 1$ mH, and $R_L = 180\ \Omega$. Calculate the filter cutoff frequency.

As you can see, the analysis of an *RL* high-pass circuit is nearly identical to that of an *RC* low-pass filter.

Section Review

1. In terms of component placement, what is the difference between an *RC high-pass filter* and an *RC low-pass filter*?
2. Explain the filtering action of the *RC* high-pass filter shown in Figure 15.23a.
3. Compare *high-pass* filter roll-off to *low-pass* filter roll-off.
4. In terms of component placement, what is the difference between an *RL high-pass filter* and an *RL low-pass filter*?
5. Explain the filtering action of the *RL* high-pass filter shown in Figure 15.26a.

15.6 Bandpass and Notch Filters

The most common passive bandpass and notch filters are *LC* filters, like those shown in Figure 15.29. In each case, the filtering action is based on the *resonant characteristics* of the *LC* circuit. The basic characteristics of series and parallel resonant circuits are summarized in Figure 15.30.

Series *LC* Bandpass Filters

OBJECTIVE 18 ➤ The operation of a series *LC* bandpass filter is easiest to understand when the filter is represented as an equivalent circuit, like the one in Figure 15.31b. In this circuit, the net series reactance (X_S) of the filter is represented as a series component between the source and the load.

The operation of a series *LC* bandpass filter is based on the relationship between its input frequency and its resonant frequency, as follows:

- $f_{in} = f_r$. When the circuit is operating at resonance, the component reactances are equal, and $X_S = X_L - X_C = 0\ \Omega$. This means that, ignoring R_W, circuit current is determined by the source voltage and the load ($I_T = \frac{V_S}{R_L}$), the circuit is purely resistive, and the phase angle is 0°.
- $f_{in} < f_r$. When the input frequency drops below f_r, X_C increases, and X_L decreases. As a result, the net reactance is *capacitive,* and the impedance phase angle is negative. As f_{in} continues to decrease, X_S increases in magnitude, and the impedance phase angle becomes more negative. When f_{in} reaches 0 Hz, X_C and X_S approach infinity, and $I_T = 0$ A.

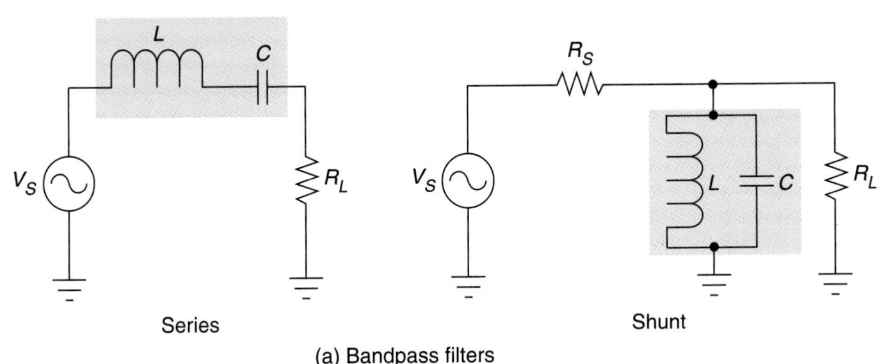

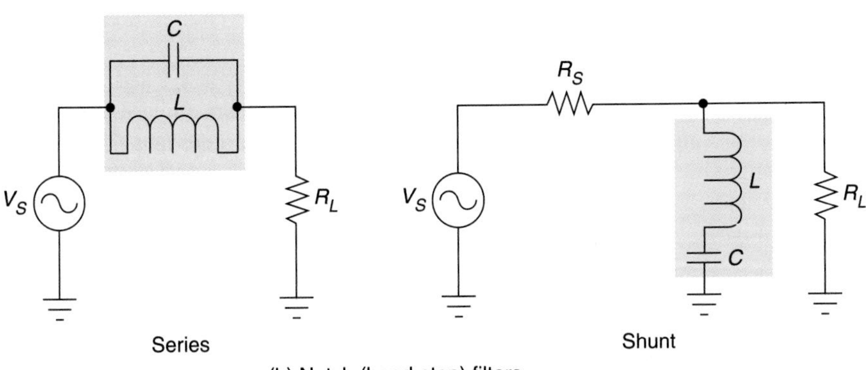

Figure 15.29
LC bandpass and notch filters.

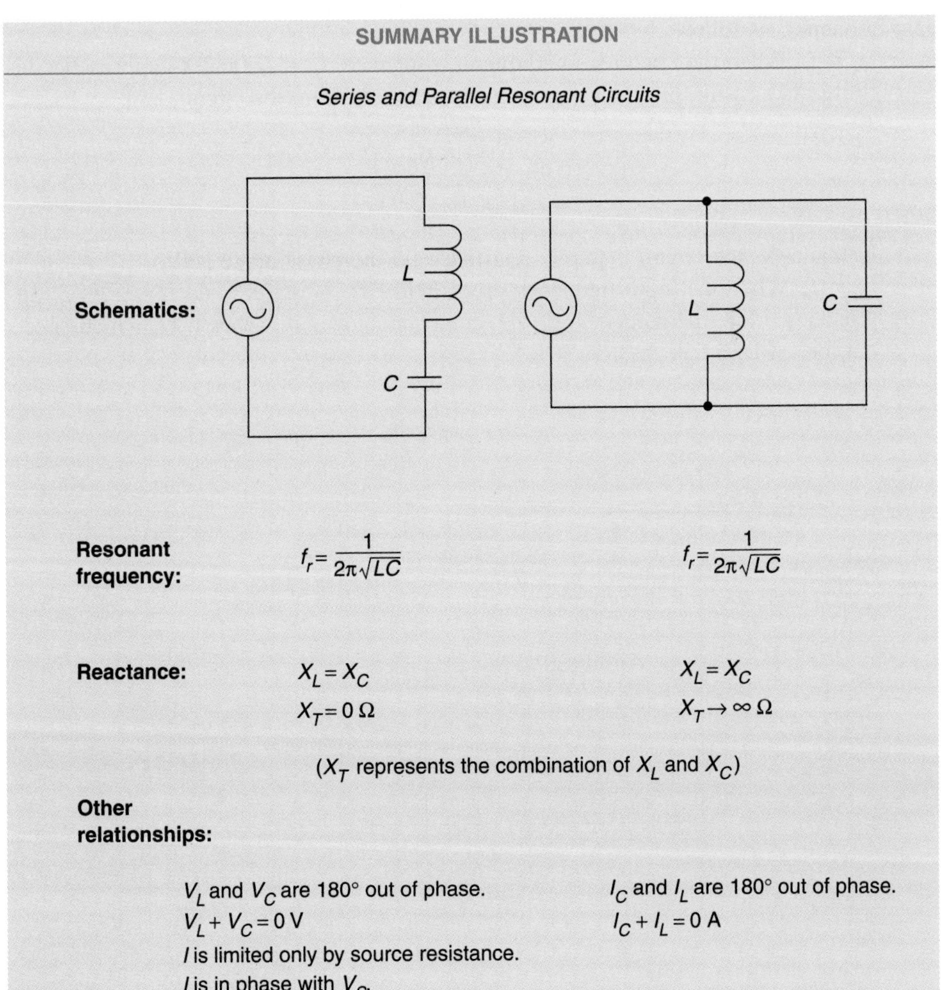

Figure 15.30

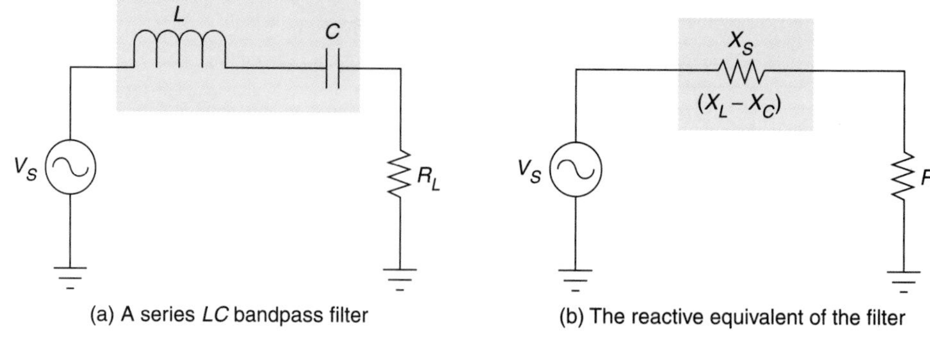

Figure 15.31 A series *LC* bandpass filter and its equivalent circuit.

- $f_{in} > f_r$. When the input frequency rises above f_r, X_C decreases, and X_L increases. As a result, the net reactance is *inductive,* and the impedance phase angle is positive. As f_{in} continues to increase, X_S increases in magnitude, and the impedance phase angle becomes more positive. As f_{in} approaches infinity, X_L and X_S approach infinity, and $I_T = 0$ A.

The curve in Figure 15.32 illustrates the frequency response of a series *LC* bandpass filter.

Shunt *LC* Bandpass Filters

OBJECTIVE 19 ➤ A shunt *LC* bandpass filter is shown in Figure 15.33 with its reactive equivalent circuit. Note that X_P represents the parallel combination of X_L and X_C. The operation of the shunt *LC* bandpass filter is based on the relationship between its input frequency and its resonant frequency, as follows:

- $f_{in} = f_r$. When the circuit is operating at resonance, $I_L = I_C$, and X_P approaches infinity. As a result, the filter is effectively removed from the circuit, and

$$V_L = V_{in} \frac{R_L}{R_L + R_S}$$

In this case, the circuit is purely resistive, and the phase angle is 0°.

- $f_{in} < f_r$. When the input frequency drops below f_r, X_L starts to decrease. As f_{in} approaches 0 Hz, the reactance of the shunt inductor approaches 0 Ω, which shorts out the load. In this case, $V_L \cong 0$ V.

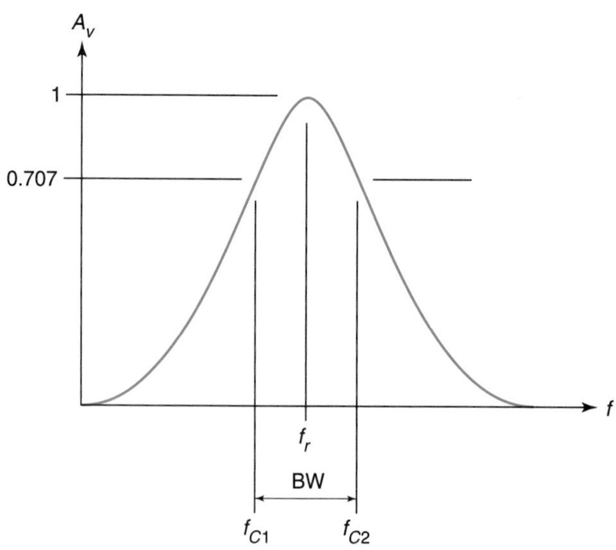

Figure 15.32 The frequency response curve of an *LC* bandpass filter.

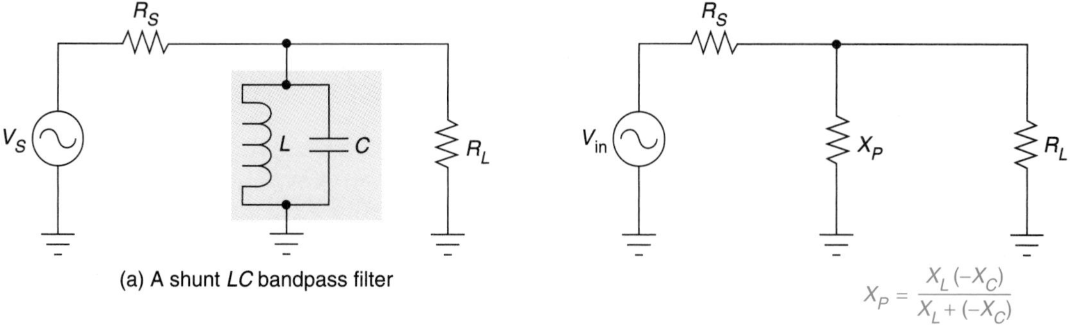

Figure 15.33 A shunt *LC* bandpass filter.

- $f_{in} > f_r$. When the input frequency rises above f_r, X_C starts to decrease. As f_{in} continues to increase, X_C approaches 0 Ω, which shorts out the load. Again, $V_L \cong 0$ V.

Note that the frequency response curve for the circuit shown in Figure 15.33 is identical to the one shown in Figure 15.32. Even though the two circuits operate differently, they produce the same overall frequency response.

Filter Quality (Q)

As you know, the Q of a component (or a circuit) is a figure of merit that indicates how close its characteristics come to those of the ideal component (or circuit). Inductors usually have much lower Q values than capacitors. Since the quality of a circuit can be no higher than its lowest-quality component, the Q of an *LC* filter is normally assumed to equal that of the inductor.

Series Filter Frequency Analysis

In Section 15.1, you were shown the relationship between bandwidth, center frequency (f_0), and filter quality (Q). This relationship was expressed as

$$\text{BW} = \frac{f_0}{Q}$$

The center frequency of an *LC* bandpass filter equals the resonant frequency of the circuit. Therefore, the bandwidth of a series *LC* bandpass filter can be found as

$$\text{BW} = \frac{f_r}{Q_L} \tag{15.42}$$

where

f_r = the resonant frequency of the *LC* filter
Q_L = the loaded Q of the filter

The **loaded *Q*** of a filter is *the quality of the circuit when a load is connected to its output terminals.* In earlier chapters, you were shown that the Q of an inductor equals the ratio of inductive reactance (X_L) to winding resistance (R_W). More precisely, it is the ratio of X_L to *the total series resistance.* For the circuit in Figure 15.34, the total series resistance equals the sum of the winding resistance, the load resistance, and the source resistance (not shown). Therefore, the loaded Q of the circuit is found as

$$Q_L = \frac{X_L}{R_S + R_W + R_L} \tag{15.43}$$

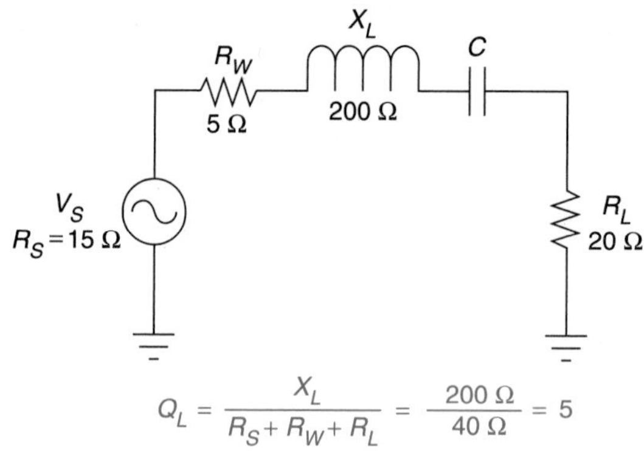

Figure 15.34 Loaded Q.

The calculation in Figure 15.34 shows that Q_L is significantly lower than the value of Q we would have obtained using only the inductor winding resistance.

Example 15.19 demonstrates the process used to solve for the loaded Q and bandwidth of a series LC bandpass filter.

Example 15.19

Determine the bandwidth of the series *LC* bandpass filter shown in Figure 15.35a.

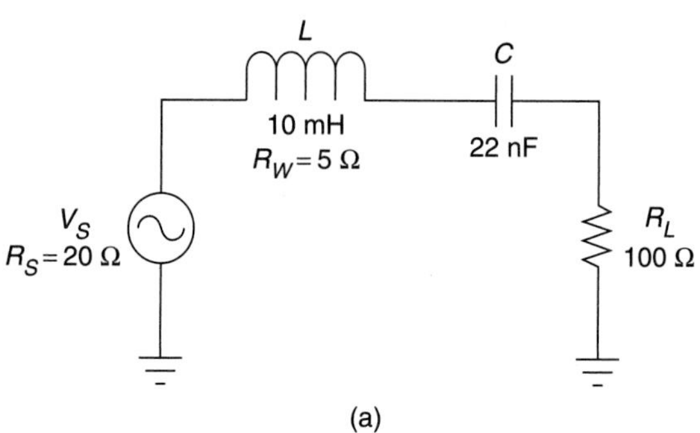

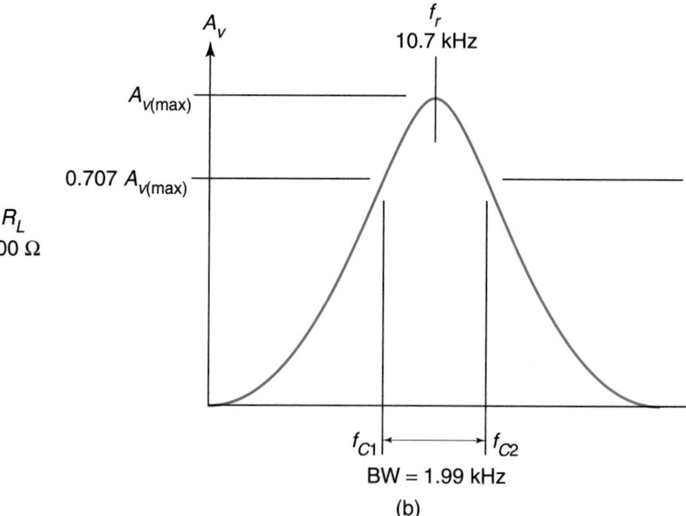

Figure 15.35

Solution: The resonant frequency of the circuit is found as

$$f_r = \frac{1}{2\pi\sqrt{LC}} = \frac{1}{2\pi\sqrt{(10\text{ mH})(22\text{ nF})}} = 10.7\text{ kHz}$$

At this operating frequency,

$$X_L = 2\pi f L = 2\pi(10.7\text{ kHz})(10\text{ mH}) = 672\ \Omega$$

and

$$Q_L = \frac{X_L}{R_S + R_W + R_L} = \frac{672\ \Omega}{125\ \Omega} = 5.38$$

Finally, the bandwidth of the circuit is found as

$$\text{BW} = \frac{f_r}{Q_L} = \frac{10.7\text{ kHz}}{5.38} = 1.99\text{ kHz}$$

The resulting frequency response curve for the circuit is shown in Figure 15.35b.

Practice Problem 15.19

A filter like the one shown in Figure 15.35a has the following values: $L = 22$ mH, $R_W = 8\ \Omega$, $R_S = 12\ \Omega$, $R_L = 330\ \Omega$, and $C = 4.7$ nF. Calculate the bandwidth of the filter.

Once we know the bandwidth and center frequency of a series bandpass filter, we have the information needed to calculate the circuit cutoff frequencies. Earlier, you were shown that the cutoff frequencies of a bandpass filter can be found using

$$f_{C1} = f_{\text{ave}} - \frac{\text{BW}}{2} \qquad \text{and} \qquad f_{C2} = f_{\text{ave}} + \frac{\text{BW}}{2}$$

where

f_{ave} = the frequency that lies half way between the cutoff frequencies

When $Q_L \geq 2$, the values of f_{ave} and f_r are approximately equal. In this case, the cutoff frequencies can be found using

$$f_{C1} = f_r - \frac{\text{BW}}{2} \tag{15.44}$$

and

$$f_{C2} = f_r + \frac{\text{BW}}{2} \tag{15.45}$$

For the circuit in Example 15.19,

$$f_{C1} = f_r - \frac{\text{BW}}{2} = 10.7\text{ kHz} - \frac{1.99\text{ kHz}}{2} \cong 9.71\text{ kHz}$$

and

$$f_{C2} = f_r + \frac{\text{BW}}{2} = 10.7\text{ kHz} + \frac{1.99\text{ kHz}}{2} \cong 11.70\text{ kHz}$$

When a filter has a value of $Q_L < 2$, then the value of f_{ave} must be calculated using

$$f_{\text{ave}} = f_r\sqrt{1 + \left(\frac{1}{2Q_L}\right)^2}$$

This relationship was established in Section 15.1.

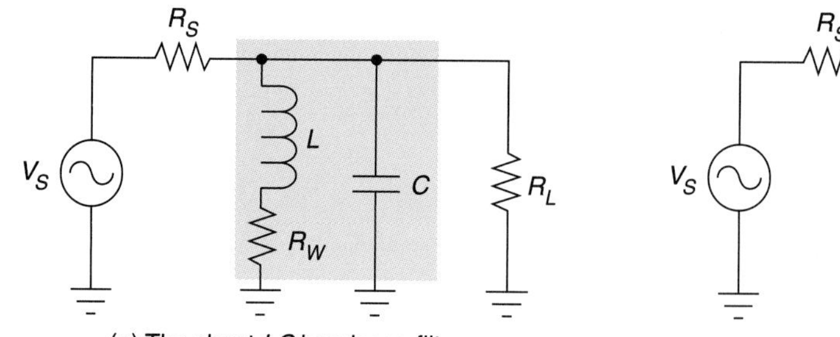

(a) The shunt *LC* bandpass filter

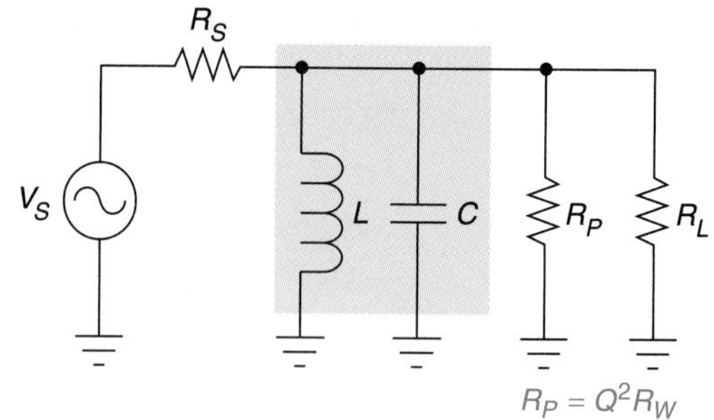

(b) Filter parallel resistance (R_P)

Figure 15.36 Loaded Q for a shunt LC filter.

Shunt Filter Frequency Analysis

The analysis of a shunt LC bandpass filter is nearly identical to that of a series filter. The primary difference lies with the calculation of Q_L. For example, consider the circuit shown in Figure 15.36a. To determine Q_L, we need to combine R_W with the other resistor values. This is done by representing the winding resistance of the coil as an *equivalent parallel resistance* (R_P), as shown in Figure 15.36b. This parallel resistance is found as

$$R_P = Q^2R_W \tag{15.46}$$

where

Q = the unloaded Q of the filter

From the viewpoint of the filter, the three resistors shown in Figure 15.36b are in parallel, and

$$Q_L = \frac{R_S \| R_P \| R_L}{X_L} \tag{15.47}$$

Example 15.20 demonstrates the process used to determine the loaded Q of a shunt bandpass filter.

Example 15.20

Determine the value of Q_L for the circuit shown in Figure 15.37.

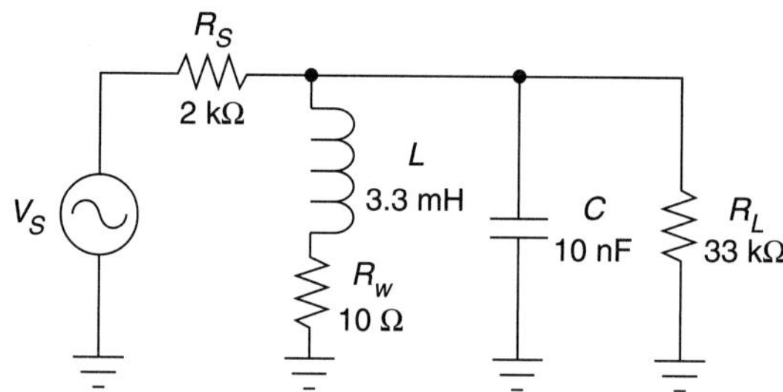

Figure 15.37

Solution: The resonant frequency of the circuit is found as

$$f_r = \frac{1}{2\pi\sqrt{LC}} = \frac{1}{2\pi\sqrt{(3.3\text{ mH})(10\text{ nF})}} = 27.7\text{ kHz}$$

At this operating frequency,

$$X_L = 2\pi fL = 2\pi(27.7\text{ kHz})(3.3\text{ mH}) = 574\ \Omega$$

and

$$Q = \frac{X_L}{R_W} = \frac{574\ \Omega}{10\ \Omega} = 57.4$$

The parallel equivalent (R_P) of the winding resistance is found as

$$R_P = Q^2 R_W = (57.4)^2(10\ \Omega) = 32.9\ \text{k}\Omega$$

and the loaded Q of the filter is found as

$$Q_L = \frac{R_S \| R_P \| R_L}{X_L} = \frac{2\ \text{k}\Omega \| 32.9\ \text{k}\Omega \| 33\ \text{k}\Omega}{574\ \Omega} = \frac{1.78\ \text{k}\Omega}{574\ \Omega} = 3.11$$

Practice Problem 15.20

A circuit like the one shown in Figure 15.37 has the following values: $R_S = 3.3\ \text{k}\Omega$, $L = 10$ mH, $R_W = 8\ \Omega$, $C = 22$ nF, and $R_L = 50\ \text{k}\Omega$. Calculate the value of Q_L for the filter.

Once the value of Q_L for a shunt bandpass filter has been determined, the rest of the frequency analysis proceeds just as it did for the series bandpass filter (see Example 15.19.)

Maximum Voltage Gain

Figure 15.36 shows a circuit similar to the one that we analyzed in Example 15.20, along with its resistive equivalent circuit. When the circuit is operated at resonance, the *LC* circuit is effectively an *open* circuit. If we view the *LC* circuit shown in Figure 15.23b as open, the remaining components in the circuit form a voltage divider. As such,

$$V_{L(\max)} = V_{\text{in}} \frac{R_{EQ}}{R_{EQ} + R_S} \tag{15.48}$$

where

$$R_{EQ} = R_P \| R_L$$

If we divide both sides of equation (15.48) by V_{in}, we get

$$\frac{V_{L(\max)}}{V_{\text{in}}} = \frac{R_{EQ}}{R_{EQ} + R_S}$$

or

$$A_{v(\max)} = \frac{R_{EQ}}{R_{EQ} + R_S} \tag{15.49}$$

Since $(R_{EQ} + R_S) > R_{EQ}$, the maximum voltage gain of the circuit is always less than one.

The Effects of Q on Filter Gain and Roll-Off

The effects of filter Q on bandpass frequency response are illustrated in Figure 15.38. As shown, a decrease in Q results in a decrease in $A_{v(\max)}$. As you learned earlier, $A_{v(\max)}$ is found as

$$A_{v(\max)} = \frac{R_{EQ}}{R_{EQ} + R_S} \Big| R_{EQ} = R_P \| R_L$$

Since $R_P = Q^2 R_W$, a decrease in Q results in a decrease in R_P. The decrease in R_P results in a decrease in R_{EQ}, and thus, the value of $A_{v(\max)}$. At the same time, a decrease in filter Q causes:

- The circuit bandwidth to increase.
- The roll-off rates of the filter to decrease.

The increase in bandwidth makes sense when you remember that bandwidth is inversely proportional to Q, as given by the equation shown in the figure.

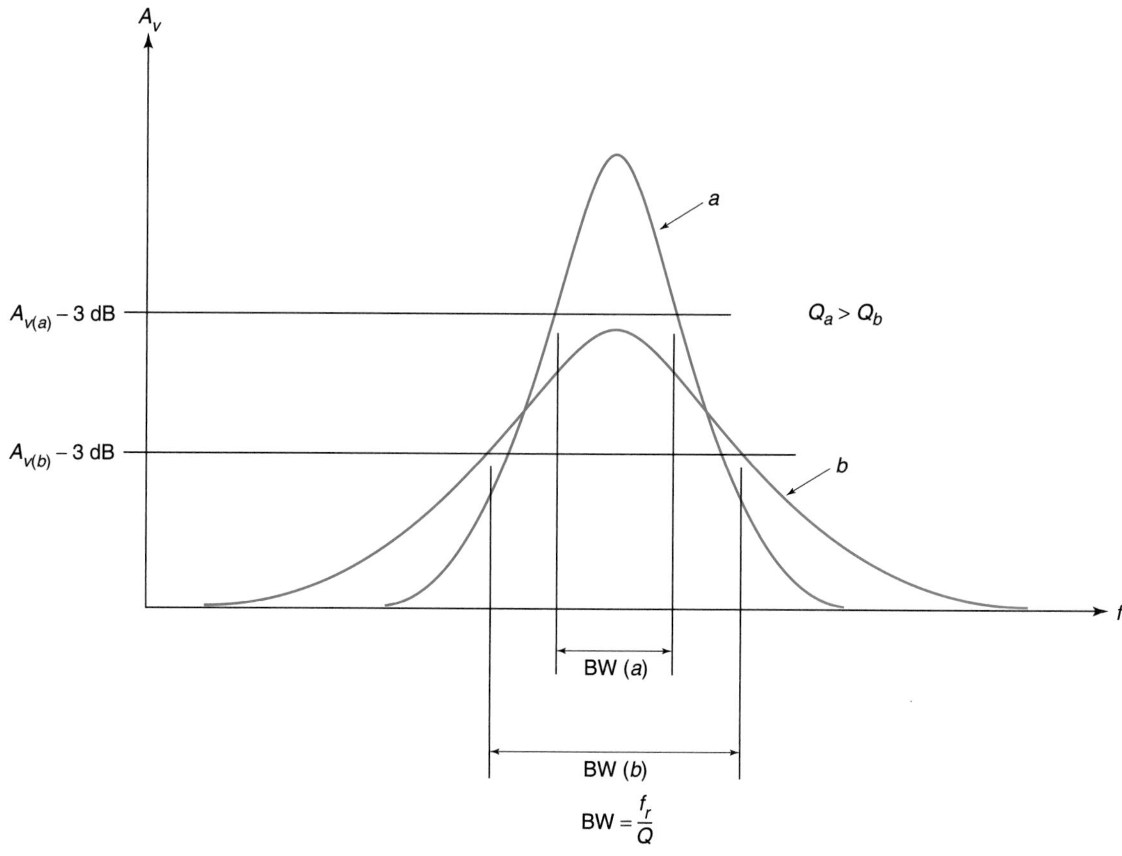

Figure 15.38 The effect of Q on circuit bandwidth.

The roll-off rate of a filter is a measure of its slope. As shown in Figure 15.38, the lower-Q curve (line b) intersects the higher-Q curve (line a). This indicates that the lower-Q curve has a more gradual slope, and thus, a lower roll-off rate.

Series *LC* Notch Filters

OBJECTIVE 20 ➤ A series *LC* notch filter is shown in Figure 15.39, along with its frequency response curve. The operation of the series *LC* notch filter is easy to understand when you consider the circuit response to each of the following conditions:

$$f_{\text{in}} = 0 \text{ Hz} \qquad f_{\text{in}} = f_r \qquad f_{\text{in}} \to \infty \text{ Hz}$$

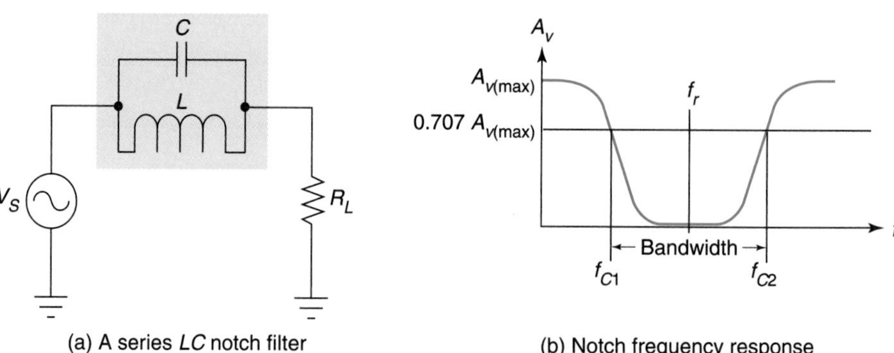

Figure 15.39 A series *LC* notch filter.

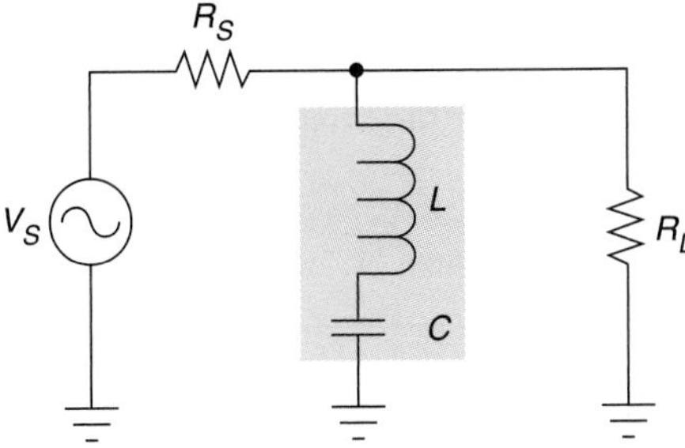

Figure 15.40 A shunt *LC* notch filter.

When $f_{in} = 0$ Hz, the inductor reactance is 0 Ω, V_S is coupled to the load via the inductor, and $V_L = V_{in}$. Note that $A_{v(max)}$ is always slightly less than one because of the voltage drop across the inductor winding resistance. As f_{in} increases, so does the impedance of the *LC* circuit. As the impedance of the *LC* circuit increases, circuit current and load voltage both decrease.

When $f_{in} = f_r$, the impedance of the *LC* filter is close to infinite. As a result, the load is effectively isolated from the source, and $V_L \cong 0$ V.

When $f_{in} \to \infty$ Hz (a theoretical value), the reactance of the filter capacitor is reduced to 0 Ω. In this case, the signal source is coupled directly to the load, and $V_L = V_{in}$. As f_{in} decreases from ∞ Hz toward f_r, the impedance of the *LC* circuit increases. As the impedance of the *LC* circuit increases, circuit current and load voltage both decrease.

Shunt *LC* Notch Filters

A shunt *LC* notch filter is shown in Figure 15.40. For all practical purposes, the frequency response curve of the circuit is identical to the one shown in Figure 19.39b.

The operation of the shunt *LC* notch filter is easy to understand when you consider the circuit response to each of the following conditions:

$$f_{in} = 0 \text{ Hz} \qquad f_{in} = f_r \qquad f_{in} \to \infty \text{ Hz}$$

When $f_{in} = 0$ Hz, the reactance of the capacitor is infinite, so the *LC* circuit acts as an open. In this case, the *LC* circuit is effectively removed from the circuit, and the maximum voltage gain can be found as

$$A_{v(max)} = \frac{R_L}{R_L + R_S}$$

As f_{in} starts to increase, the impedance of the *LC* circuit starts to decrease. This causes the voltage across the *LC* circuit (and therefore V_L) to decrease. When $f_{in} = f_r$, the *LC* circuit essentially shorts out the load. This means that $V_L \cong 0$ V when the circuit is operated at resonance.

When $f_{in} \to \infty$ Hz (a theoretical value), the reactance of the inductor is infinite. Therefore, the *LC* circuit acts as an open, just as it did when $f_{in} = 0$ Hz, and the load voltage is at its maximum value. As f_{in} decreases toward f_r, the impedance of the *LC* circuit starts to decrease. This causes the voltage across the *LC* circuit (and therefore V_L) to decrease.

Section Review

1. List the impedance, current, and voltage characteristics of a series resonant *LC* circuit.
2. List the impedance, current, and voltage characteristics of a parallel resonant *LC* circuit.
3. Describe the frequency response of the series *LC* bandpass filter shown in Figure 15.31.
4. Describe the frequency response of the shunt *LC* bandpass filter shown in Figure 15.33.

5. Which component in an LC filter has the greatest effect on the value of filter Q? Explain your answer.
6. Describe the effects of Q on LC filter gain and roll-off.
7. Describe the operation of the series LC notch filter shown in Figure 15.39.
8. Describe the operation of the shunt LC notch filter in Figure 15.40.

Key Terms

The following terms were introduced in this chapter on the pages indicated:

Attenuation, 448
Average frequency (f_{ave}), 454
Band-stop filter, 450
Bandpass filter, 450
Bandwidth, 450
Bel (B), 458
Bode plot, 468
Cascaded, 461
Center frequency (f_0), 451
Current gain (A_i), 465
Cutoff frequency (f_C), 449
dBm reference, 462
Decade, 456
Decade scale, 456
Decibel (dB), 458
Filter, 450
Frequency response curve, 449
Gain, 449
High-pass filter, 450
Loaded Q (Q_L), 479
Logarithmic scale, 456
Low-pass filter, 450
n-stage filter, 461
Notch filter, 450
Octave, 456
Octave scale, 456
Power gain (A_P), 449
Quality (Q), 453
Roll-off, 468
Roll-off rate, 469
Shunt component, 466
Stage, 461
Unity gain, 459
Voltage gain (A_v), 463

Practice Problems

1. Complete the table below.

	f_{C1}	f_{C2}	BW	f_0
a.	5 kHz	88 kHz	______	______
b.	640 Hz	1.8 kHz	______	______
c.	280 kHz	1.4 MHz	______	______

2. Complete the table below.

	f_{C1}	f_{C2}	BW	f_0
a.	2.2 kHz	63 kHz	______	______
b.	120 kHz	2 MHz	______	______
c.	400 Hz	800 Hz	______	______

3. Complete the table below.

	f_{C1}	f_{C2}	BW	f_0
a.	800 Hz	______	12 kHz	______
b.	5 kHz	______	______	10 kHz
c.	______	200 kHz	198 kHz	______

4. Complete the table below.

	f_{C1}	f_{C2}	BW	f_0
a.	400 Hz	______	39.6 kHz	______
b.	50 kHz	______	______	200 kHz
c.	______	400 kHz	50 kHz	______

5. A filter has values of BW = 18 kHz and f_0 = 120 kHz. Calculate the Q of the filter.
6. A filter has values of BW = 48 kHz and f_0 = 340 kHz. Calculate the Q of the filter.
7. A filter has values of f_0 = 52 kHz and Q = 6.8. Determine the circuit bandwidth.
8. A filter has values of f_0 = 650 Hz and Q = 2.5. Determine the circuit bandwidth.

9. Determine the value of f_{ave} for a filter with values of $f_0 = 88$ kHz and $Q = 1.2$.

10. Determine the value of f_{ave} for a filter with values of $f_0 = 340$ kHz and $Q = 1.8$.

11. Determine the value of f_{ave} for the filter described in Problem 7.

12. Determine the value of f_{ave} for the filter described in Problem 8.

13. A bandpass filter has a 330 kHz center frequency and an 80 kHz bandwidth. Determine the average frequency for the circuit.

14. A bandpass filter has a 460 kHz center frequency and an 240 kHz bandwidth. Determine the average frequency for the circuit.

15. Determine the cutoff frequencies for the circuit described in Problem 13.

16. Determine the cutoff frequencies for the circuit described in Problem 14.

17. Complete the table below.

P_{in}	P_{out}	A_P	$A_{p(dB)}$
a. 2 mW	80 mW	______	______
b. 48 mW	300 mW	______	______
c. 240 mW	60 mW	______	______

18. Complete the table below.

P_{in}	P_{out}	A_P	$A_{p(dB)}$
a. 60 mW	720 mW	______	______
b. 50 μW	30 mW	______	______
c. 40 mW	200 μW	______	______

19. The gain of a filter is −0.8 dB. Determine the output power when $P_{in} = 800$ mW.

20. The gain of a filter is −1.2 dB. Determine the output power when $P_{in} = 250$ mW.

21. The gain of a filter is −0.4 dB. Determine the output power when $P_{in} = 1.4$ W.

22. The gain of a filter is −1.95 dB. Determine the output power when $P_{in} = 1$ W.

23. Determine the output power for the circuit represented in Figure 15.41a.

24. Determine the output power for the circuit represented in Figure 15.41b.

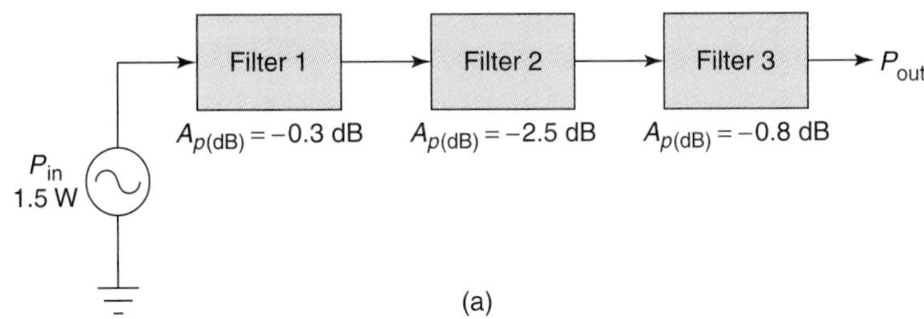

(a)

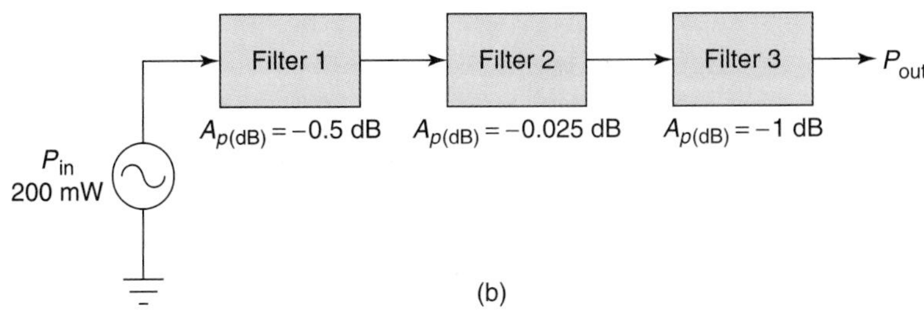

(b)

Figure 15.41

25. Express each of the following in dBm.
   a. 12 W
   b. 180 mW
   c. 2.5 W
   d. 300 μW

26. Express each of the following in dBm.

a. 640 mW
b. 7 W
c. 1 mW
d. 800 nW

27. Convert each of the following to a power value (in watts).

a. 3.5 dBm
b. 12 dBm
c. −1.2 dBm
d. −3.5 dBm

28. Convert each of the following to a power value (in watts).

a. 2 dBm
b. 7.2 dBm
c. 0.9 dBm
d. −2.2 dBm

29. Convert each of the following voltage gain (A_v) values to dB form.

a. 120
b. 40
c. 0.995
d. 0.978

30. Convert each of the following voltage gain (A_v) values to dB form.

a. 0.64
b. 0.75
c. 0.895
d. 0.159

31. Convert each of the following dB voltage gains to standard numeric form.

a. 1.2 dB
b. 4 dB
c. −3.6 dB
d. −2.5 dB

32. Convert each of the following dB voltage gains to standard numeric form.

a. 0.8 dB
b. −7.2 dB
c. −0.8 dB
d. −4.8 dB

33. Calculate the maximum voltage gain for the low-pass filter shown in Figure 15.42a.

34. Calculate the maximum voltage gain for the low-pass filter shown in Figure 15.42b.

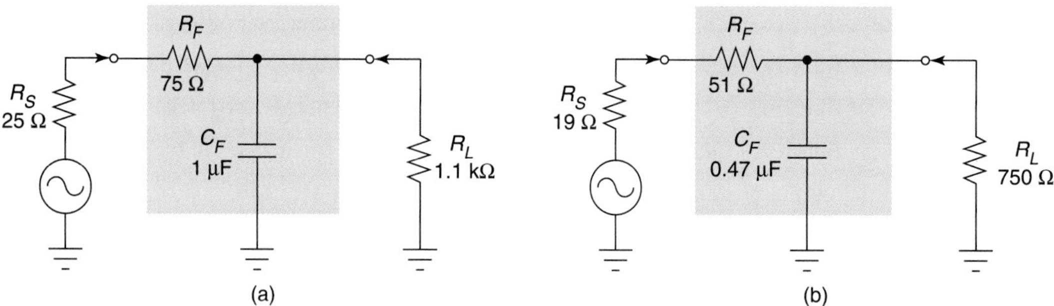

Figure 15.42

35. Calculate the maximum voltage gain for the low-pass filter shown in Figure 15.43a.

36. Calculate the maximum voltage gain for the low-pass filter shown in Figure 15.43b.

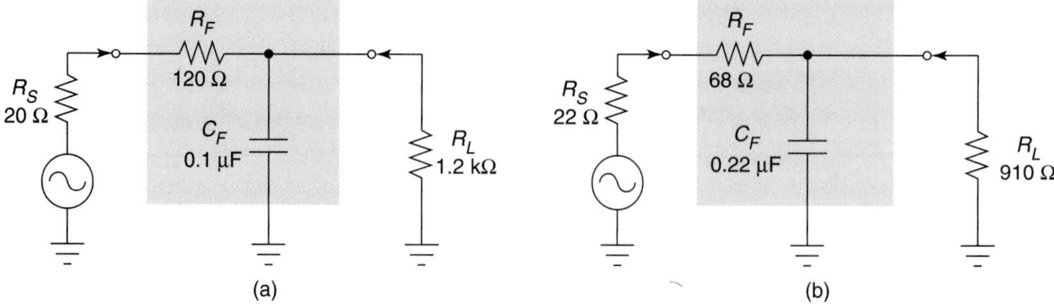

Figure 15.43

37. Calculate the cutoff frequency for the low-pass filter shown in Figure 15.42a.

38. Calculate the cutoff frequency for the low-pass filter shown in Figure 15.42b.
39. Calculate the cutoff frequency for the low-pass filter shown in Figure 15.43a.
40. Calculate the cutoff frequency for the low-pass filter shown in Figure 15.43b.
41. Calculate the cutoff frequency for the low-pass filter shown in Figure 15.44a, and plot its frequency response curve.
42. Calculate the cutoff frequency for the low-pass filter shown in Figure 15.44b, and plot its frequency response curve.

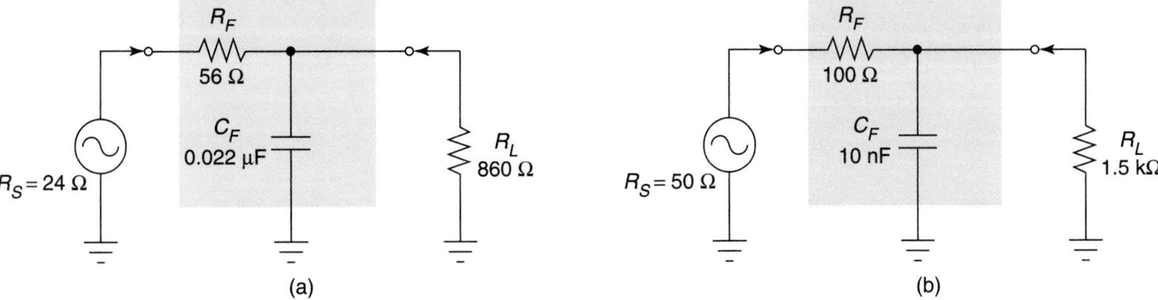

Figure 15.44

43. Calculate the maximum voltage gain for the low-pass filter shown in Figure 15.45a.
44. Calculate the maximum voltage gain for the low-pass filter shown in Figure 15.45b.

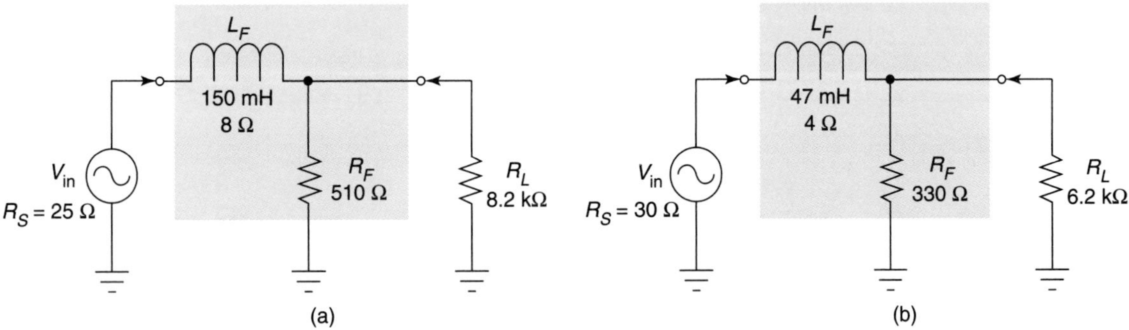

Figure 15.45

45. Calculate the maximum voltage gain for the low-pass filter shown in Figure 15.46a.
46. Calculate the maximum voltage gain for the low-pass filter shown in Figure 15.46b.

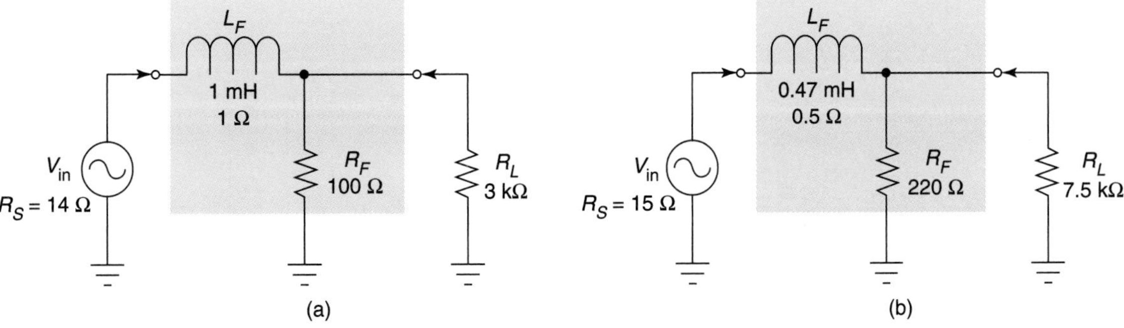

Figure 15.46

47. Calculate the cutoff frequency for the low-pass filter shown in Figure 15.45a.

48. Calculate the cutoff frequency for the low-pass filter shown in Figure 15.45b.

49. Calculate the cutoff frequency for the low-pass filter shown in Figure 15.46a.

50. Calculate the cutoff frequency for the low-pass filter shown in Figure 15.46b.

51. Calculate the cutoff frequency for the high-pass filter shown in Figure 15.47a.

52. Calculate the cutoff frequency for the high-pass filter shown in Figure 15.47b.

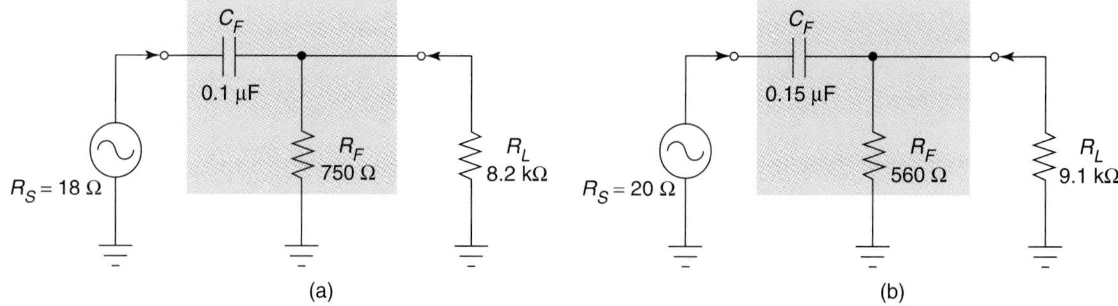

Figure 15.47

53. Calculate the maximum voltage gain and cutoff frequency for the high-pass filter shown in Figure 15.48a.

54. Calculate the maximum voltage gain and cutoff frequency for the high-pass filter shown in Figure 15.48b.

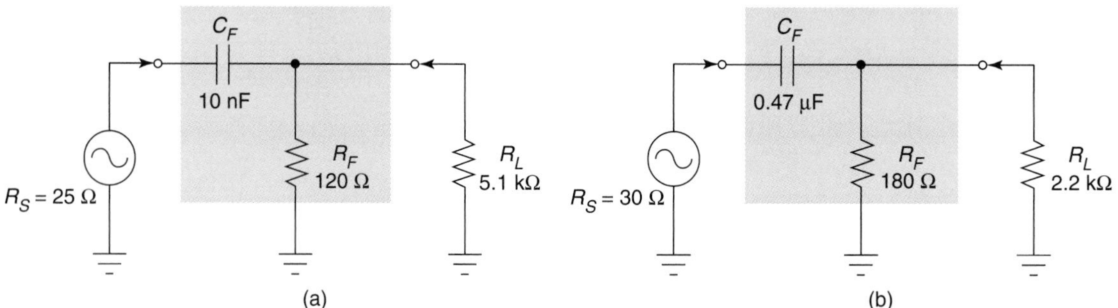

Figure 15.48

55. Calculate the cutoff frequency for the high-pass filter shown in Figure 15.49a.

56. Calculate the cutoff frequency for the high-pass filter shown in Figure 15.49b.

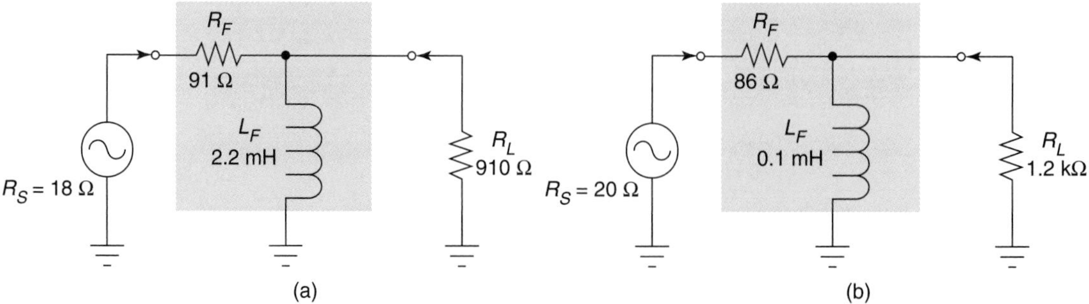

Figure 15.49

57. Determine the bandwidth of the bandpass filter shown in Figure 15.50a.

58. Determine the bandwidth of the bandpass filter shown in Figure 15.50b.

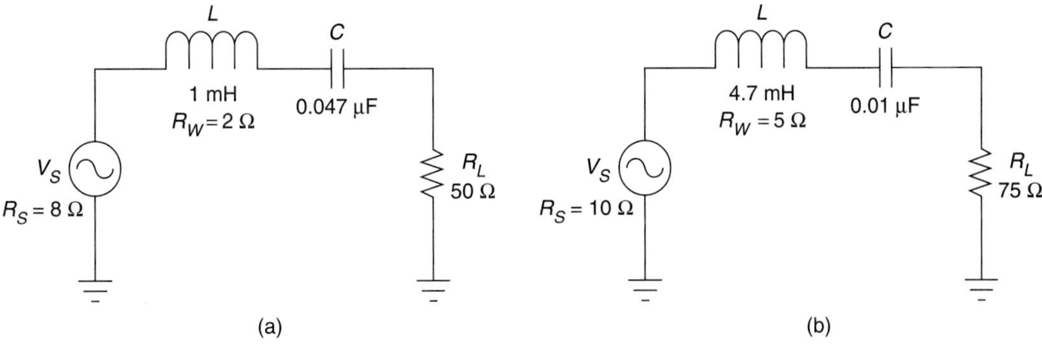

Figure 15.50

59. Determine the cutoff frequencies for the bandpass filter shown in Figure 15.50a.

60. Determine the cutoff frequencies for the bandpass filter shown in Figure 15.50b.

61. Determine the bandwidth and cutoff frequencies for the bandpass filter shown in Figure 15.51a.

62. Determine the bandwidth and cutoff frequencies for the bandpass filter shown in Figure 15.51b.

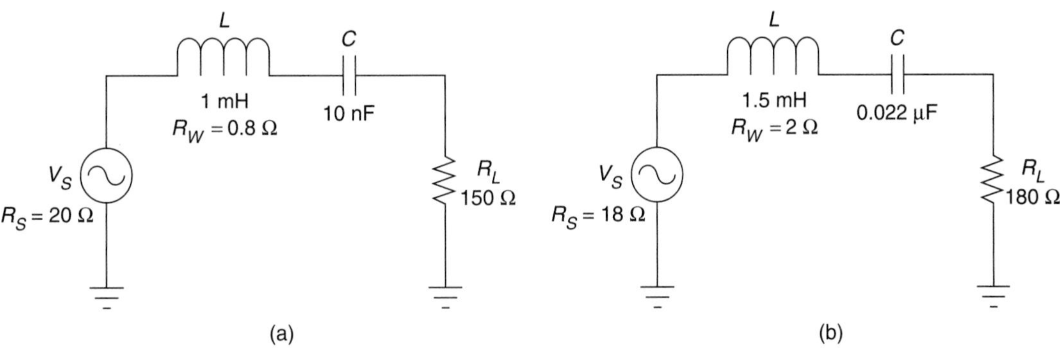

Figure 15.51

63. Determine the value of Q_L for the circuit shown in Figure 15.52a.

64. Determine the value of Q_L for the circuit shown in Figure 15.52b.

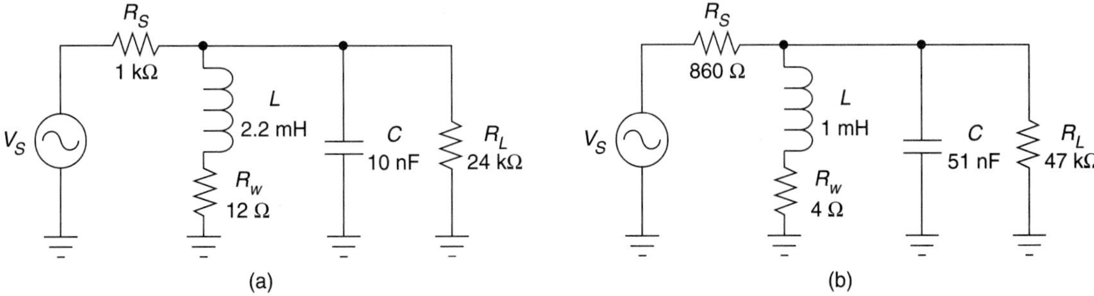

Figure 15.52

65. Calculate the bandwidth and cutoff frequencies for the bandpass filter shown in Figure 15.52a.

66. Calculate the bandwidth and cutoff frequencies for the bandpass filter shown in Figure 15.52b.

Looking Back

These problems relate to material presented in earlier chapters. The chapters are identified in brackets.

67. A sine wave has the following values: $T_C = 150\ \mu s$, and $V_{pk} = 12$ V. Determine the instantaneous voltage when $t = 50\ \mu s$. [Chapter 9]
68. Calculate the energy used (in kWh) to run a 750 W microwave oven for 20 minutes and a 1500 W space heater for 2 hours. [Chapter 3]
69. A 0.33 mH inductor is in series with a 200 Ω resistor. The components are connected to an 18 V_{ac}, 10 kHz source. Calculate the circuit phase angle. [Chapter 11]
70. A series *RC* circuit has a value of $Z_T = 372.02\ \Omega\angle -53.746°$. Determine the circuit values of *R* and X_C. [Chapter 13]
71. A magnet with a cross-sectional area of 1.8 cm^2 generates 900 μWb of flux at its poles. Calculate the flux density at the poles of the magnet. [Chapter 8]

Pushing the Envelope

72. An *RC* low-pass filter has the following *exact* values: $R_F = 100\ \Omega$, and $C_F = 10\ \mu F$. The filter is connected to a load that can vary over a range of 10 to 500 Ω. Calculate:
    a. The range of A_v values for the circuit.
    b. The range of f_C values for the circuit.

Answers to the Example Practice Problems

15.1	81.8 kHz	15.11	251 mW
15.2	6 kHz	15.12	40.8 dBm
15.3	50 Hz	15.13	−6 dB
15.4	5.33	15.14	0.668
15.5	50 kHz	15.15	355 Hz
15.6	323 kHz	15.16	229 Hz
15.7	203 kHz, 443 kHz	15.17	1.29 kHz
15.8	−1.25 dB	15.18	4.66 kHz
15.9	20 dB	15.19	2.53 kHz
15.10	578 mW	15.20	4.37

Multisim File References

Figure	File Reference
15.17	ETF15_17.msm
15.22	ETF15_22.msm
15.24	ETF15_24.msm
15.28	ETF15_28.msm
15.35	ETF15_35.msm
15.37	ETF15_37.msm

chapter 16

RL and *RC* Circuit Pulse Response

Objectives

After studying the material in this chapter, you should be able to:

1. Describe and explain the current and voltage waveforms found in an *RL* switching circuit.
2. Describe the *universal current curve* for an *RL* switching circuit.
3. Calculate the value of circuit current at any point on a *rise* or *decay curve* using the *curve equations.*
4. Calculate the *time constant* and *transition time* for a given series *RL* circuit.
5. Define and explain *RL time constant.*
6. Calculate the time required for the current in an *RL* switching circuit to reach a designated value.
7. Describe and explain the current and voltage waveforms found in an *RC* switching circuit.
8. Describe the *universal voltage current curve* for an *RC* switching circuit.
9. Calculate the value of capacitor voltage at any point on a rise or decay curve using the universal curve equations.
10. Calculate the time constant and transition time for a given series *RC* circuit.
11. Define and explain the *RC time constant.*
12. Calculate the time required for the current in an *RC* switching circuit to reach a designated value.

Any waveform that abruptly switches from one dc level to another (and back) is generally referred to as a *pulse.* A circuit with a pulse input is often referred to as a *switching circuit.* In this chapter, we discuss the switching (or pulse) response of *RL* and *RC* circuits.

16.1 Square Wave Response: *RL* Time Constants

In Chapter 9, you were introduced to rectangular waveforms. As a review, rectangular and square wave measurements are illustrated in Figure 16.1.

The waveforms shown in Figure 16.1 alternate between two dc levels that are identified as 0 V and $+V$. The time the waveform spends at the $+V$ level is the *pulse width* (PW), and the time it spends at the 0 V level is the *space width* (SW). The pulse width and space width combine to make up the *cycle time.* The square wave (see Figure 16.1b) is a special-case rectangular waveform that has equal PW and SW values.

Any circuit with a square (or rectangular) wave input is generally referred to as a **switching circuit,** since the input switches back and forth between dc levels. In this section, we are going to concentrate on the characteristics of series *RL* switching circuits.

Voltage and Current Transitions

In a rectangular waveform, *the change from either dc level to the other* is referred to as a **transition.** For example, the voltage and current transitions in a purely resistive circuit are shown in Figure 16.2.

The waveform transitions occur at the time increments shown at the bottom of the waveforms (t_1 through t_4). The transitions that occur at t_1 and t_3 are called *positive-going* transitions. The transitions at t_2 and t_4 are called *negative-going* transitions. Note that the V_R and current waveforms have the same shape as the V_S waveform. This is not the case when an inductor is added to the circuit.

RL Circuit Waveforms

OBJECTIVE 1 ➤ When a square wave is applied to a series *RL* circuit, the V_R and current waveforms are altered as shown in Figure 16.3. (We'll look at the V_L waveform in a moment.)

At t_1, the source voltage makes the transition from 0 V to $+V$. Since an inductor opposes any change in current, the circuit current cannot rise to its maximum value at the same rate as the source voltage (as it did in the purely resistive circuit). Rather, current rises to its maximum value over the time period shown (t_1 to t_2). Since V_R is always in phase with the circuit current, it also rises over the time period shown.

At t_2, the source voltage makes its negative transition, and once again, the inductor opposes the change in current. In this case, the current decay occurs over the time period shown (t_2 to t_3). As a result, V_R decays at the same rate over the same time period.

A Practical Consideration: When you study digital electronics, you'll learn how the pulse width and space width polarities can be the opposite of those shown here; that is, pulse width can be negative and space width can be positive.

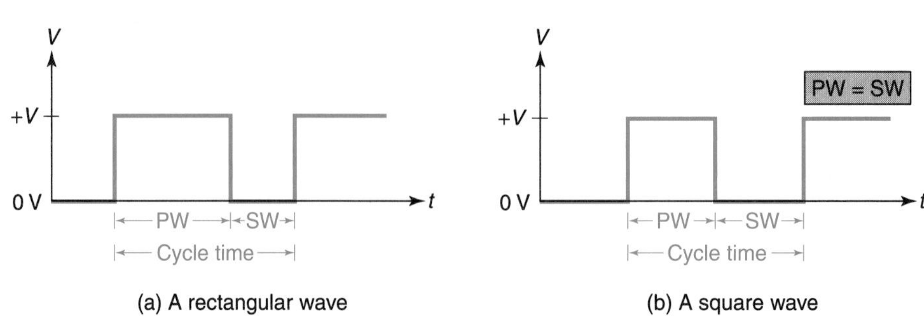

Figure 16.1 Rectangular and square wave measurements.

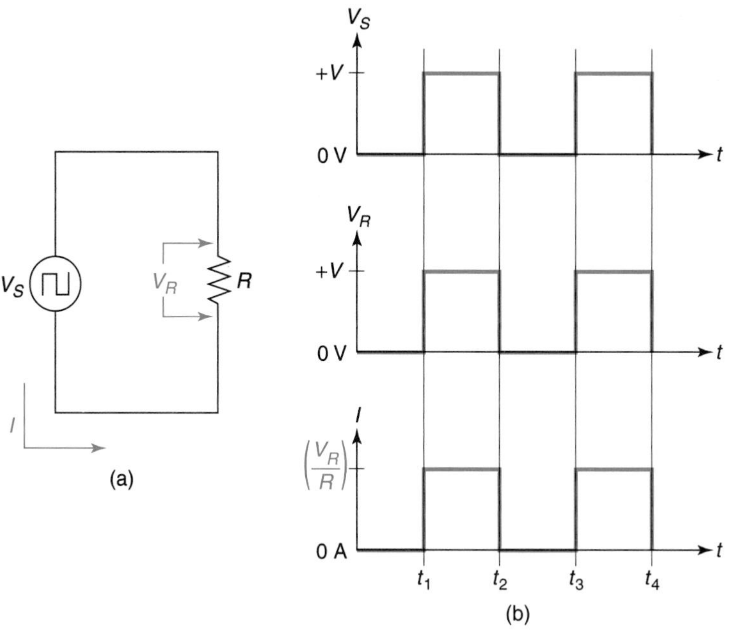

Figure 16.2 Voltage and current transitions in a purely resistive circuit.

Inductor Voltage (V_L)

The V_R and current waveforms shown in Figure 16.3 are easily explained in terms of the inductor's opposition to a change in current. This opposition slows the change in current that would normally result from a change in voltage, producing the waveforms shown. However, the waveform for the inductor voltage (V_L) requires a more thorough explanation. As Figure 16.4 shows, the V_L waveform has a shape that you haven't seen before. (For the sake of discussion, Figure 16.4 assumes a peak value of $V_S = +5$ V.)

The shape of the V_L waveform can be explained using Kirchhoff's voltage law. Since the source generates alternating dc voltages, the circuit operates according to dc

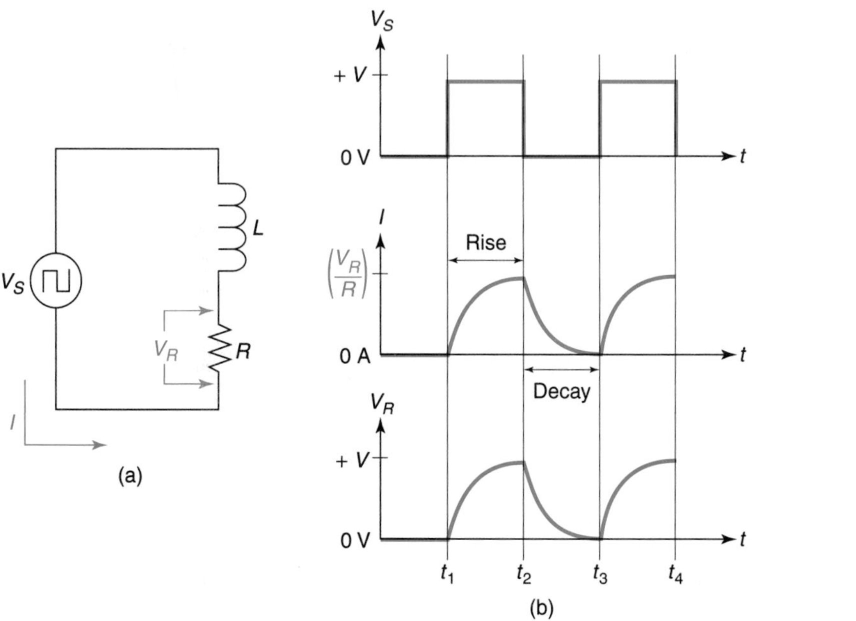

Figure 16.3 Transition delay in an *RL* circuit.

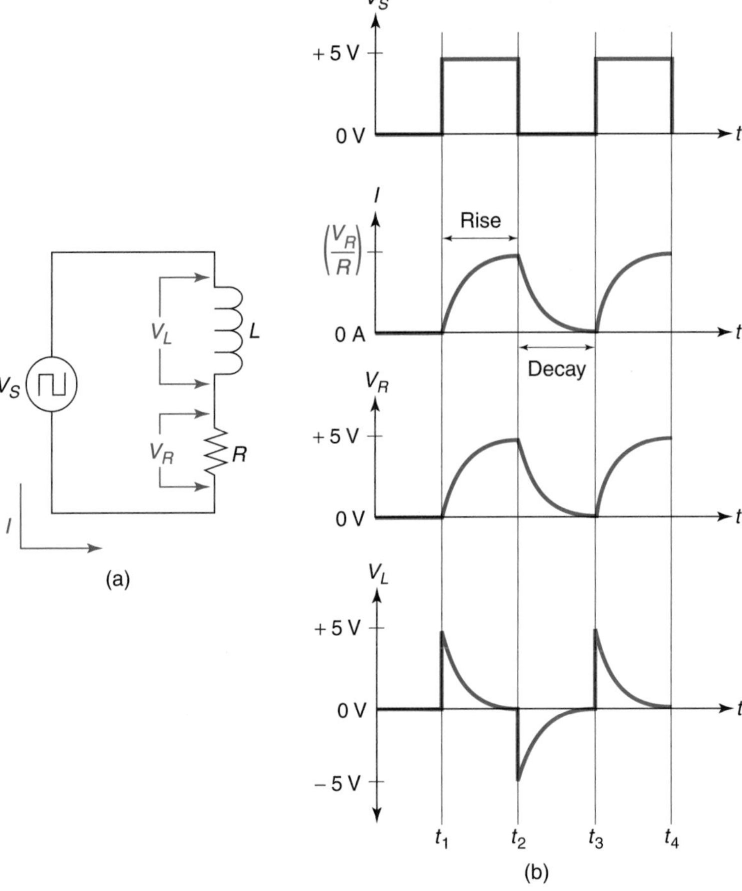

Figure 16.4 *RL* circuit waveforms.

principles. This means that the source voltage equals the *algebraic* sum of V_L and V_R. By formula,

$$V_S = V_L + V_R$$

or

$$V_L = V_S - V_R \tag{16.1}$$

With this relationship in mind, look at the voltage waveforms shown in Figure 16.4.

At t_1, the source makes the transition from 0 V to +5 V. At the instant of this transition $V_R = 0$ V, (since the rise in current is delayed by the inductor). Therefore, the value of V_L at t_1 is found as

$$V_L = V_S - V_R = +5\text{ V} - 0\text{ V} = +5\text{ V}$$

This value corresponds to the peak value of V_L shown at t_1 on the waveform.

Immediately before t_2, the source is still at +5 V. During the $t_1 \rightarrow t_2$ transition, V_R increases from 0 V to +5 V. At the same time, V_L decreases to

$$V_L = V_S - V_R = +5\text{ V} - 5\text{ V} = 0\text{ V}$$

At t_2, the source changes from +5 V to 0 V. But, at this instant, V_R still equals +5 V (since the current decay is delayed by the inductor). Therefore, the value of V_L at t_2 is found as

$$V_L = V_S - V_R = 0\text{ V} - 5\text{ V} = -5\text{ V}$$

This value corresponds to the abrupt transition from 0 V to −5 V that V_L makes at t_2. During the $t_2 \rightarrow t_3$ transition, V_R decays from +5 V to 0 V. At the same time, V_L increases to

$$V_L = V_S - V_R = 0\text{ V} - 0\text{ V} = 0\text{ V}$$

From then on, the cycle repeats itself.

More on *RL* Circuit Waveforms

Several more points need to be made regarding the waveforms shown in Figure 16.4. First, note that equation (16.1) holds true at any instant on the V_R and V_L waveforms. Since V_R and V_L change at the same rate, their sum always equals the supply voltage. For example, V_R reaches a value of 3 V at some point between t_1 and t_2. At that time, V_L has a value of

$$V_L = V_S - V_R = 5\text{ V} - 3\text{ V} = 2\text{ V}$$

This relationship between the circuit voltages holds true at every point on the voltage waveforms.

The second point we need to make deals with the rate at which current makes its transitions. As Figure 16.4 shows, the circuit current does not change at a *linear* (constant) rate. Rather, the *rate of change in current decreases* as the transition progresses. The more detailed current waveform shown in Figure 16.5 can be used to illustrate this concept. (For the sake of discussion, Figure 16.5 assumes that the current has a peak value of 100 mA.)

Each half of the waveform shown in Figure 16.5 is divided into *five equal time intervals.* For reasons you will be shown later, each time interval is identified using the Greek letter *tau* (τ). The estimated *change in current* (ΔI) during each of the first five time intervals is provided in Table 16.1. As you can see, the change in current *decreases* from one time interval to the next during the **rise** portion of the curve. The same principle holds true for the **decay** portion of the curve as well. The estimated values of ΔI for the decay portion of are listed in Table 16.2.

When you compare the values of ΔI in Tables 16.1 and 16.2, you can see that the two halves of the curve are *symmetrical.* That is, the rate of rise in any time interval matches the corresponding rate of decay. *This relationship holds true for every RL switching circuit.*

TABLE 16.1
Estimated Values of ΔI During the Rise Portion of the Curve Shown in Figure 16.5

Time Interval	*Estimated* ΔI
$t_0 \rightarrow t_1$	62 mA
$t_1 \rightarrow t_2$	22 mA
$t_2 \rightarrow t_3$	10 mA
$t_3 \rightarrow t_4$	2 mA
$t_4 \rightarrow t_5$	<2 mA

TABLE 16.2
Estimated Values of ΔI During the Decay Portion of the Curve Shown in Figure 16.5

Time Interval	*Estimated* ΔI
$t_5 \rightarrow t_6$	−62 mA
$t_6 \rightarrow t_7$	−22 mA
$t_7 \rightarrow t_8$	−10 mA
$t_8 \rightarrow t_9$	−2 mA
$t_9 \rightarrow t_{10}$	<−2 mA

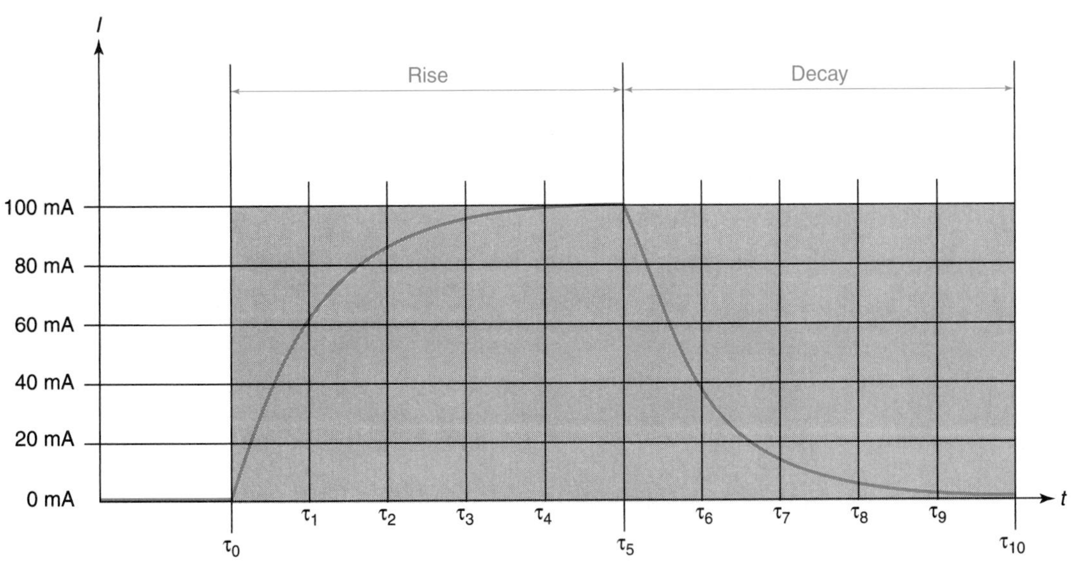

Figure 16.5 A close-up of the current waveform shown in Figure 16.4.

The Universal Curve

OBJECTIVE 2 ➤ The shape of the curve shown in Figure 16.5 is a function of the inductor's opposition to a change in circuit current. This means that the wave shape is *independent of the peak value of the circuit current.* For example, if we doubled the peak source voltage in Figure 16.4, the peak current generated through the circuit would also double. However, the shape of the curve would remain unchanged.

The current characteristics of an *RL* switching circuit can be represented using what is called a *universal curve.* A **universal curve** is *one that can be used to predict the operation of any specified type of circuit.* The **universal current curve** for an *RL* switching circuit is shown in Figure 16.6.

The *x*-axis of the universal curve is divided into the same time intervals that we labeled as τ earlier. The *y*-axis represents the ratio of circuit current (I_t) to the peak source current (I_{pk}). For example, at τ_1, the curve indicates that

$$\frac{I_t}{I_{pk}} \cong 0.62$$

At τ_2, the curve indicates that

$$\frac{I_t}{I_{pk}} \cong 0.85$$

and so on.

OBJECTIVE 3 ➤ The universal curve in Figure 16.6 is an **exponential curve,** meaning that *its rate of change is a function of a variable exponent.* For example, the rise portion of the universal curve shown in Figure 16.6 is actually a plot of the equation

$$\frac{I_t}{I_{pk}} = 1 - e^{-x} \qquad (16.2)$$

where

e = the base of the natural log system (approximately 2.71828)

x = the variable exponent

The only value on the right-hand side of the equation that is variable is x. Therefore, *the shape of the curve is a function of this variable exponent.* By definition, that makes the curve an *exponential curve.*

Note: Equations (16.2) and (16.3) can be used to solve the current ratio at any point in the waveform by using the following as the variable exponent:

$$x = t\frac{R}{L}$$

where

t = the time interval of interest from the start of the curve.

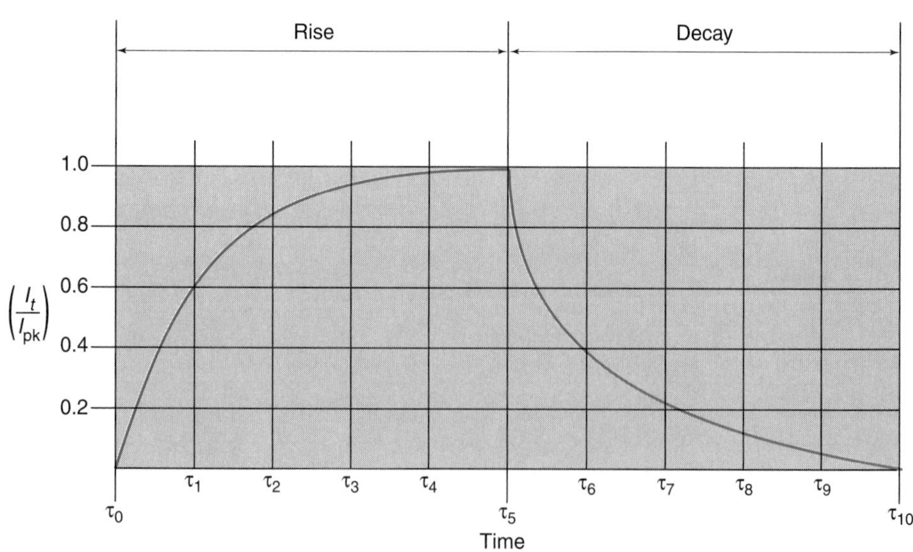

Figure 16.6 The universal current curve for an *RL* switching circuit.

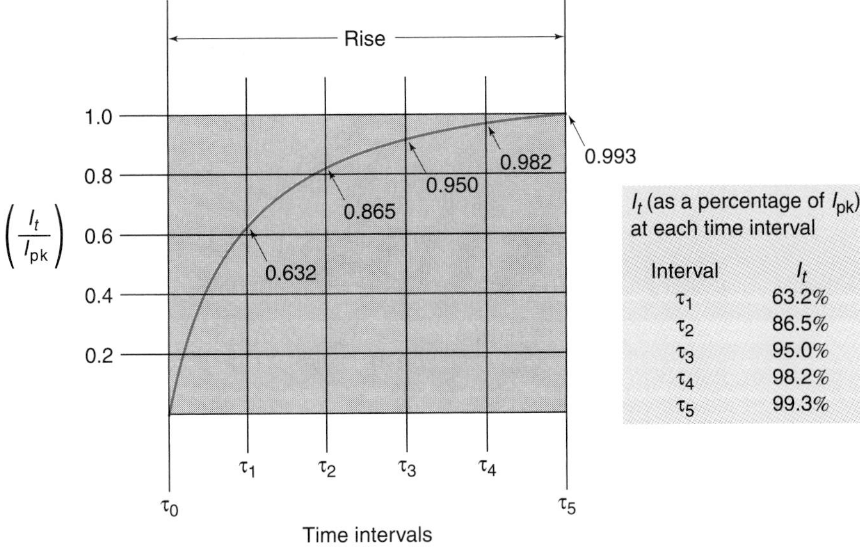

Figure 16.7 The calculated current ratio at each time interval.

Plotting the Universal Curve

If we solve equation (16.2) for *whole-number* values of $x = 1 \rightarrow 5$, we get the results listed in Table 16.3.

If we assume that the values of x listed in Table 16.3 correlate to the *time intervals* along the x-axis in Figure 16.7, then the curve forms a plot of the results. Using the values shown on the curve, we can express the value of I_t at each time interval as *a percentage of the peak circuit current* (I_{pk}). For example,

$$\frac{I_t}{I_{pk}} = 0.632$$

at τ_1. Therefore,

$$I_t = 0.632 I_{pk} \qquad \text{(at } \tau_1\text{)}$$

This equation indicates that I_t reaches 63.2% of its peak value at τ_1. Based on the percentages shown in the universal curve, we can say that *current increases from 0% to 99.3% of its peak value over the course of the five time intervals.*

TABLE 16.3
Results of Equation (16.2) for Increasing Values of x

Value of x	*Equation (16.2) Result*
1	0.632
2	0.865
3	0.950
4	0.982
5	0.993

The Decay Curve

The *decay curve* is simply a mirror image of the rise curve. Like the rise curve, it is an exponential curve. The exponential relationship used to plot the decay curve is

$$\frac{I_t}{I_{pk}} = e^{-x} \qquad (16.3)$$

If we solve this relationship for *whole-number* values of $x = 1 \rightarrow 5$, we get the results listed in Table 16.4.

The results listed in Table 16.4 are identified on the decay curve shown in Figure 16.8. Again, each time interval corresponds to a value of x from the table. As was the case with the rise curve, we can express the value of I_t at each time interval as *a percentage of the peak circuit current* (I_{pk}). As shown in the figure, *the current decreases from approximately 100% to 0.7% of its peak value over the course of the five time intervals.*

TABLE 16.4
Results of Equation (16.3) for Increasing Values of x

Value of x	*Equation (16.3) Result*
1	0.368
2	0.135
3	0.050
4	0.018
5	0.007

Figure 16.8
The decay curve.

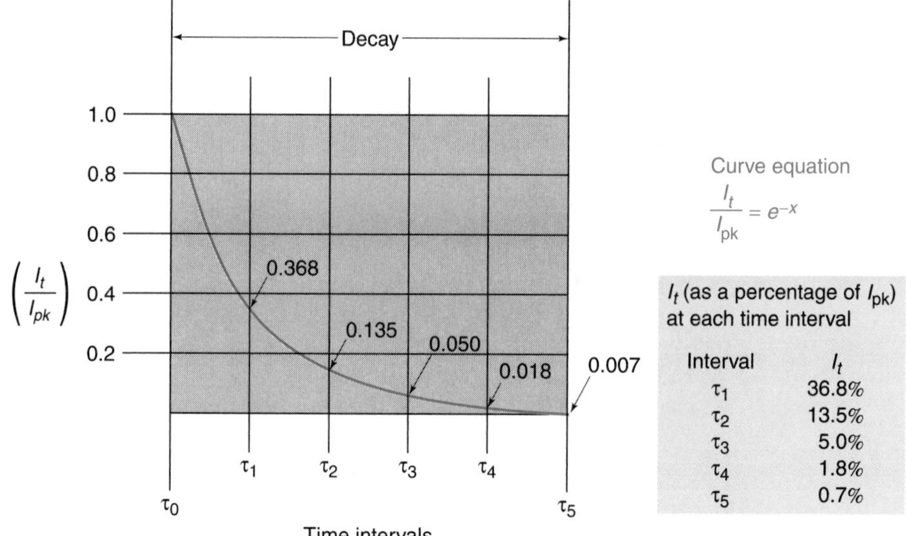

The *RL* Time Constant (τ)

➤*OBJECTIVE 4* Each time interval in the universal curve represents a real-time value that is referred to as a **time constant (τ)**. The term *constant* is used, because τ is independent of the circuit current magnitude and operating frequency.

For a given series *RL* switching circuit, the value of the *RL time constant* is found as

$$\tau = \frac{L}{R} \tag{16.4}$$

where

τ = the duration of each time interval, in seconds
L = the total series inductance
R = the total series resistance

Example 16.1 demonstrates the use of this equation.

Example 16.1

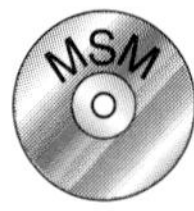

Determine the time constant for the circuit shown in Figure 16.9.

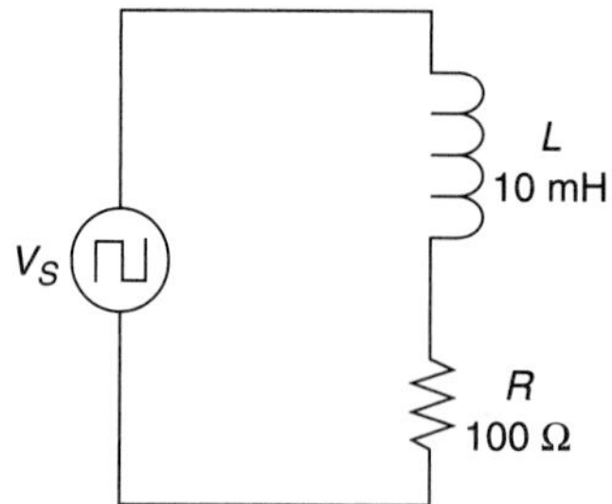

Figure 16.9

Solution: The time constant for the circuit is found as

$$\tau = \frac{L}{R} = \frac{10\text{ mH}}{100\ \Omega} = 100\ \mu\text{s}$$

Practice Problem 16.1

A circuit like the one shown in Figure 16.9 has values of $R = 3.3\text{ k}\Omega$ and $L = 33\text{ mH}$. Calculate the time constant for the circuit.

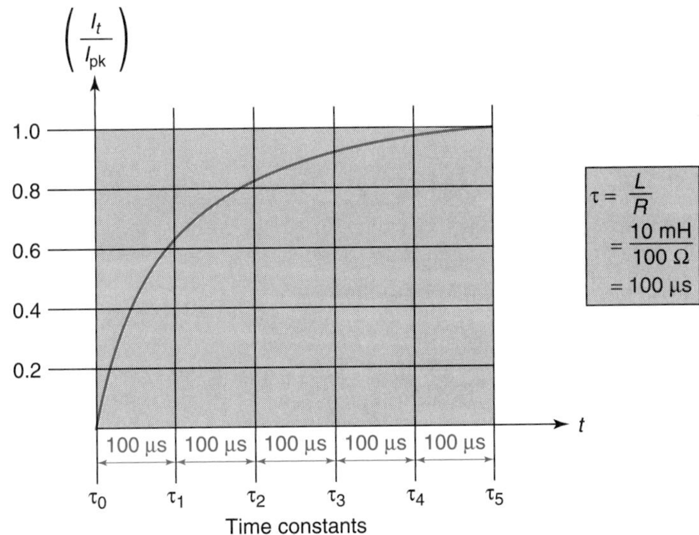

Figure 16.10 The rise curve for the circuit in Example 16.1.

Figure 16.10 shows the rise curve for the circuit in Example 16.1. As you can see, each time interval represents the value of the circuit time constant (100 μs). The total time required for the circuit current to complete its rise is approximately five time constants. By formula,

$$T = 5\tau \tag{16.5}$$

where

T = the time required for the circuit current to rise to its peak value, in seconds
τ = the time constant of the circuit, in seconds

Example 16.2 demonstrates the process for determining the total time required for the current in an RL circuit to rise from zero to its peak value.

Example 16.2

Determine the time required for I_t to reach its peak value in the circuit shown in Figure 16.11.

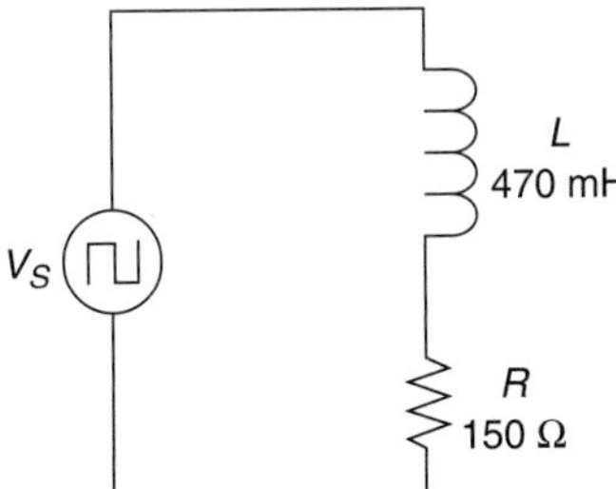

Figure 16.11

Solution: First, the circuit time constant is found as

$$\tau = \frac{L}{R} = \frac{470\text{ mH}}{150\ \Omega} = 3.13\text{ ms}$$

It takes five time constants for the rise portion of the curve to complete its transition. Therefore, the total time required for the transition is found as

$$T = 5\tau = (5)(3.13\text{ ms}) = 15.67\text{ ms}$$

Practice Problem 16.2

A circuit like the one shown in Figure 16.11 has values of $R = 33\ \Omega$ and $L =$ 100 mH. Determine the time required for the circuit current to reach its peak value.

Although Example 16.2 deals only with the rise curve of the circuit shown in Figure 16.11, its results apply equally to the decay curve. The value of τ for the decay curve is 3.13 ms, and it takes 15.67 ms for the decay curve to complete its transition.

Defining the Time Constant (τ)

OBJECTIVE 5 ➤ You have been shown how to calculate the value of the time constant for a given series *RL* circuit. Technically defined, an ***RL* time constant** is *the time required for the current in an RL switching circuit to increase (or decrease) by 63.2% of its maximum possible* ΔI. This definition can be explain with the aid of Figure 16.12.

The rise curve shown in the figure was derived using the rise equation. That is, the current ratio for each time interval was determined and then used to calculate the value of I_t.

At any time interval, the *maximum possible* ΔI equals *the difference between its value at the start of the time interval and its value at the end of the transition.* For example, the circuit current at τ_0 in Figure 16.12 is 0 mA. The current at the end of the transition is approximately 100 mA. Therefore, the maximum possible ΔI at τ_0 is found as

$$\Delta I_{max} = 100 \text{ mA} - 0 \text{ mA} = 100 \text{ mA}$$

During the first time constant, the circuit current increases by 63.2% of ΔI_{max}. Therefore, the current at τ_1 is found as

$$\begin{aligned} I_t &= I_{\tau_0} + 0.632(\Delta I_{max}) \\ &= 0 \text{ mA} + 0.632(100 \text{ mA}) \\ &= 0 \text{ mA} + 63.2 \text{ mA} \\ &= 63.2 \text{ mA} \end{aligned}$$

From τ_1 to τ_2, the current again changes by 63.2% of ΔI_{max}. However, the current at the start of the time interval has a value of $I_t = 63.2$ mA. Therefore, the value of ΔI_{max} at τ_1 is found as

$$\begin{aligned} \Delta I_{max} &= 100 \text{ mA} - 63.2 \text{ mA} \\ &= 36.8 \text{ mA} \end{aligned}$$

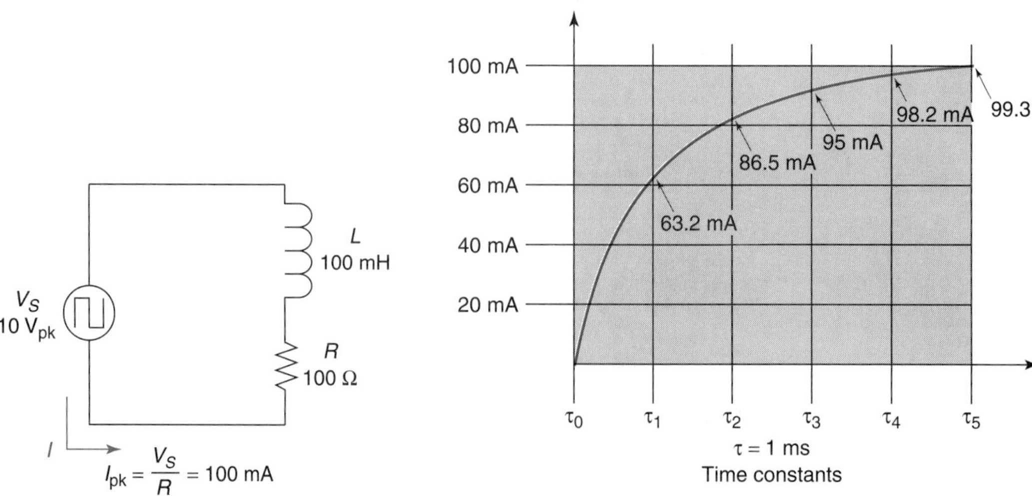

Figure 16.12 Current values at each time constant of an *RL* circuit.

Using this value of ΔI_{max}, the current at τ_2 is found as

$$\begin{aligned} I_t &= I_{\tau_1} + 0.632(\Delta I_{max}) \\ &= 63.2 \text{ mA} + 0.632(36.8 \text{ mA}) \\ &= 63.2 \text{ mA} + 23.3 \text{ mA} \\ &= 86.5 \text{ mA} \end{aligned}$$

As you can see, this value corresponds to the value shown on the curve when $t = \tau_2$. Continuing this series of calculations provides the circuit currents shown at each time interval in the curve (see Figure 16.12).

As the current calculations have demonstrated, *the current change from one time constant to the next equals 63.2% of the maximum possible current transition.* This is the basis for the definition of the time constant. The current relationship described here also applies to the decay curve of an *RL* switching circuit. However, in that case, the current during each time interval *decreases* by 63.2% of the maximum possible current transition.

Time Calculations

There are situations when it is desirable to know how long it will take for the current in an *RL* switching circuit to reach a designated value. For example, take a look at the **delay circuit** shown in Figure 16.13. This circuit uses the current rise of the *RL* circuit to provide a delay between the *positive-going* transitions of V_S and V_{out}.

The purpose served by a delay circuit like the one shown in Figure 16.13a is covered in a course on *digital* (switching) circuits. However, we can use the circuit to demonstrate the concept of a *signal delay.* Here's how the circuit works: The *switching circuit* is designed to switch between two output levels, 0 V and +5 V. The output from the circuit depends on the voltages at its signal inputs (A and B). The only time that the output from

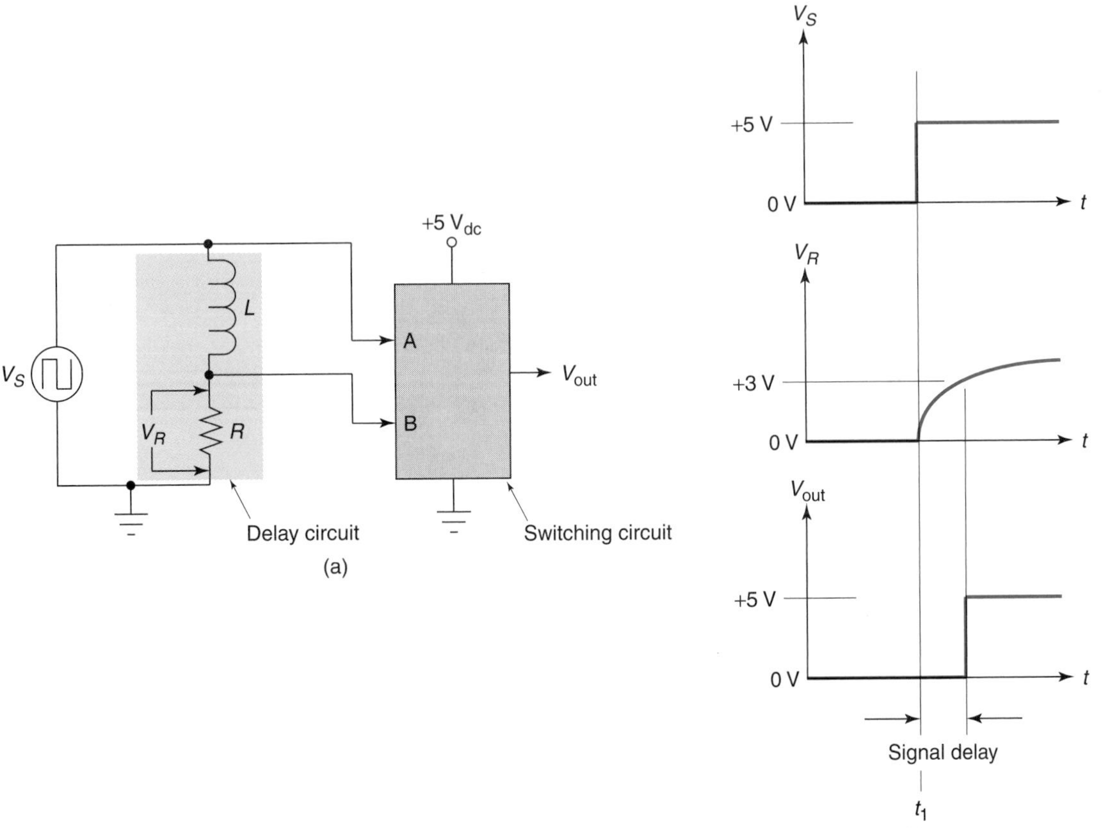

Figure 16.13 An *RL* delay circuit and waveforms.

the switching circuit is at +5 V is when *both of its input voltages* (V_A and V_B) are *greater than or equal to +3 V.*

At the instant V_S makes its *positive-going* transition from 0 V to +5 V, there is no voltage across the resistor. As the circuit current increases, V_R increases. When V_R reaches +3 V, the switching circuit has input values of $V_A = V_S = +5$ V and $V_B = V_R = +3$ V. These two input voltages cause V_{out} to make its positive-going transition. The time difference between the positive-going transitions of V_S and V_{out} is the *signal delay* in the waveforms.

OBJECTIVE 6 ➤ The analysis of a circuit like the one shown in Figure 16.13 requires that you know how long it takes for V_R to reach a desired level. The time required for V_R to reach a specified level can be found as

$$t = -\tau \ln\left(1 - \frac{V_R}{V_S}\right) \tag{16.6}$$

where

ln = the *natural log* function.

Equation (16.6) is derived in Appendix D. Its use is demonstrated in Example 16.3.

Example 16.3

The circuit shown in Figure 16.13a has values of $L = 330$ mH and $R = 1.5$ kΩ. Determine the duration of the signal delay.

Solution: The time constant of the *RL* circuit is found as

$$\tau = \frac{L}{R} = \frac{330 \text{ mH}}{1.5 \text{ k}\Omega} = 220 \ \mu\text{s}$$

V_{out} makes its transition when $V_R = +3$ V and $V_S = +5$ V. The time required for the V_R transition can be found as

$$t = -\tau \ln\left(1 - \frac{V_R}{V_S}\right) = -(220 \ \mu\text{s}) \ln\left(1 - \frac{3 \text{ V}}{5 \text{ V}}\right) = -(220 \ \mu\text{s}) \ln(0.4) \cong 202 \ \mu\text{s}$$

This result indicates that V_{out} makes its positive-going transition approximately 202 μs after V_S makes its positive-going transition.

Practice Problem 16.3

A circuit like the one shown in Figure 16.13 has the following values: $L = 470$ mH, $R = 4.7$ kΩ, and $V_S = +10 \ V_{pk}$. Determine the time required for V_R to reach a value of +8 V.

One Final Note

As your electronics education continues, you'll learn that the principles covered in this section apply to a wide variety of applications. For future reference, many of the principles and relationships introduced here are summarized in Figure 16.14.

Section Review

1. Define each of the following terms: *rectangular waveform, pulse width, space width, square wave, switching circuit.*
2. What is a *transition*?
3. Refer to Figure 16.3. What causes the shape of the I and V_R transitions?
4. Explain the shape of the V_L waveform in Figure 16.4.

5. What is the relationship between the *shape* of a current rise (or decay) curve and its peak current?
6. What is a *universal curve*? What purpose does it serve?
7. What is an *exponential curve*?
8. How is the value of the *time constant* in a series *RL* circuit found?
9. How many time constants are required to complete a rise (or decay) curve?
10. What is a *time constant*? Explain your answer.

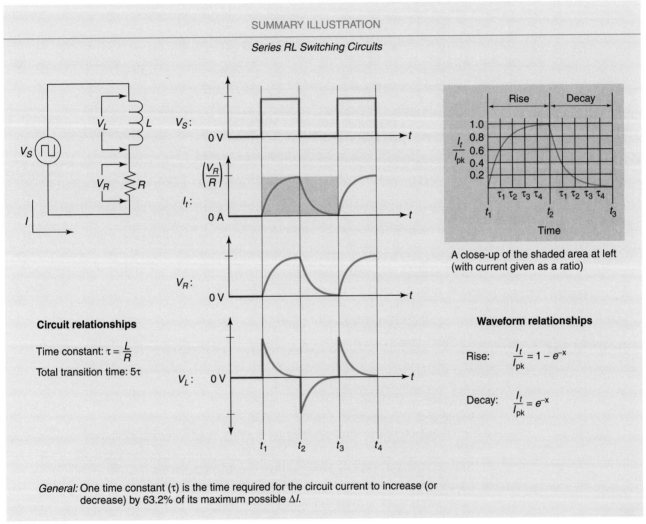

SUMMARY ILLUSTRATION

Series RL Switching Circuits

V_S V_L L V_R R I

V_S: 0 V t

I_t: $\left(\frac{V_R}{R}\right)$ 0 A t

V_R: 0 V t

V_L: 0 V t

t_1 t_2 t_3 t_4

Rise Decay

$\frac{I_t}{I_{pk}}$ 1.0 0.8 0.6 0.4 0.2

τ_1 τ_2 τ_3 τ_4 τ_1 τ_2 τ_3 τ_4

t_1 t_2 t_3

Time

A close-up of the shaded area at left (with current given as a ratio)

Circuit relationships

Time constant: $\tau = \frac{L}{R}$

Total transition time: 5τ

Waveform relationships

Rise: $\frac{I_t}{I_{pk}} = 1 - e^{-x}$

Decay: $\frac{I_t}{I_{pk}} = e^{-x}$

General: One time constant (τ) is the time required for the circuit current to increase (or decrease) by 63.2% of its maximum possible ΔI.

Figure 16.14

16.2 Square Wave Response: *RC* Time Constants

Earlier in this chapter, you were shown how an *RL* circuit responds to rectangular waveforms. In this section, we will discuss the response of an *RC* circuit to the same type of input.

RC Circuit Waveforms

◄ *OBJECTIVE 7*

When a square wave is applied to a purely resistive circuit, the voltage and current waveforms have identical shapes. When a square wave is applied to a series *RC* circuit, the

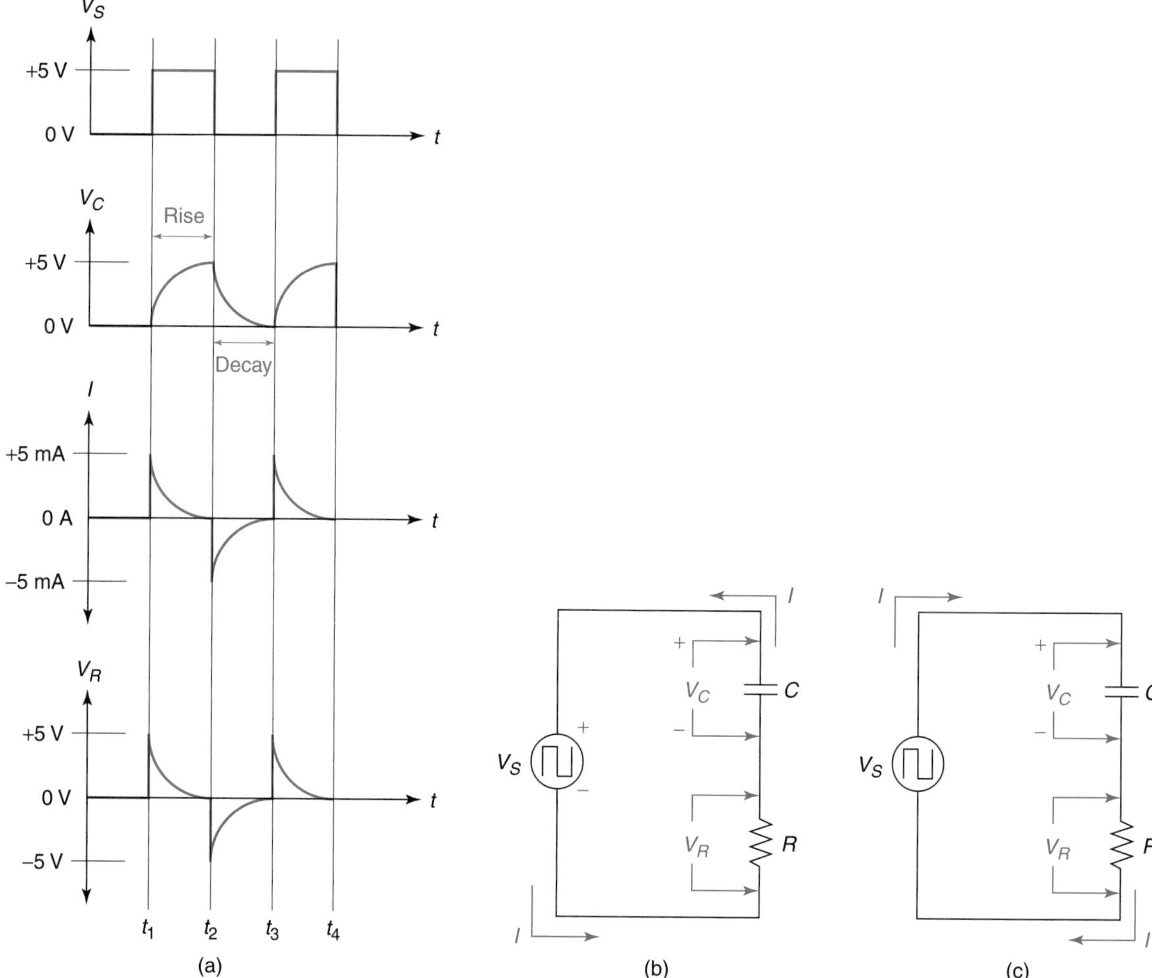

Figure 16.15

waveforms are altered as shown in Figure 16.15. These waveforms are a result of the charging of the capacitor.

The effects of capacitor charging on the *RC* circuit waveforms can be explained using the circuits shown in Figure 16.15. Since we are dealing with alternating dc voltages, the resistor voltage always equals the difference between the source voltage and the capacitor voltage. That is,

$$V_R = V_S - V_C$$

When the values of V_R and R are known, the circuit current can be found as

$$I_T = \frac{V_R}{R}$$

Combining these equations, we get

$$I_T = \frac{V_S - V_C}{R} \tag{16.7}$$

Using equation (16.7), let's take a look at the waveforms shown in Figure 16.15. At t_1, the source voltage makes a transition from 0 V to +5 V. At the instant of the transition, the capacitor has not yet begun to charge, and $V_C = 0$ V. Therefore, the circuit current at t_1 jumps to a value of

$$I_T = \frac{V_S - V_C}{R} = \frac{5\text{ V} - 0\text{ V}}{1\text{ k}\Omega} = 5\text{ mA}$$

This current begins to charge the capacitor (as shown in Figure 16.15b).

Since the charging current decreases as the capacitor charges, so does the rate at which the capacitor is charging. Eventually, the capacitor charges to the point where its voltage is approximately equal to the source voltage. At that time, there is approximately 0 V across the resistor, and the circuit current is reduced to near zero.

At t_2, the source voltage makes the transition back to 0 V. When this occurs, the capacitor begins to discharge through R and the source (as shown in Figure 16.15c). Because of the polarity of the capacitor voltage, the circuit current switches direction. The value of the circuit current at the moment the capacitor begins to discharge can be found as

$$I_T = \frac{V_S - V_C}{R} = \frac{0\text{ V} - 5\text{ V}}{1\text{ k}\Omega} = -5\text{ mA}$$

The minus sign in the result indicates that the current has reversed direction, as shown in the figure. The reversed circuit current results in a value of $V_R = -5$ V.

As the capacitor discharges, its plate-to-plate voltage decreases, as does the value of the discharge current. This is why the rate of change in the current curve decreases between t_2 and t_3. Eventually, the capacitor discharges completely. At that time, the circuit current is reduced to zero, as is the voltage across the capacitor.

One other point should be made regarding the waveforms shown in Figure 16.15. Even though we limited our discussion to those times immediately before and after the transitions, V_R and V_C change at the same rate. Therefore, their sum always equals the supply voltage (in keeping with Kirchhoff's voltage law).

The Universal Voltage Curve

The curve shown in Figure 16.16 should look very familiar. It is almost the same as the universal current curve for *RL* switching circuits. In this case, it is called the **universal voltage curve.** The universal current curve was the result of the inductor's opposition to ◄ *OBJECTIVE 8*

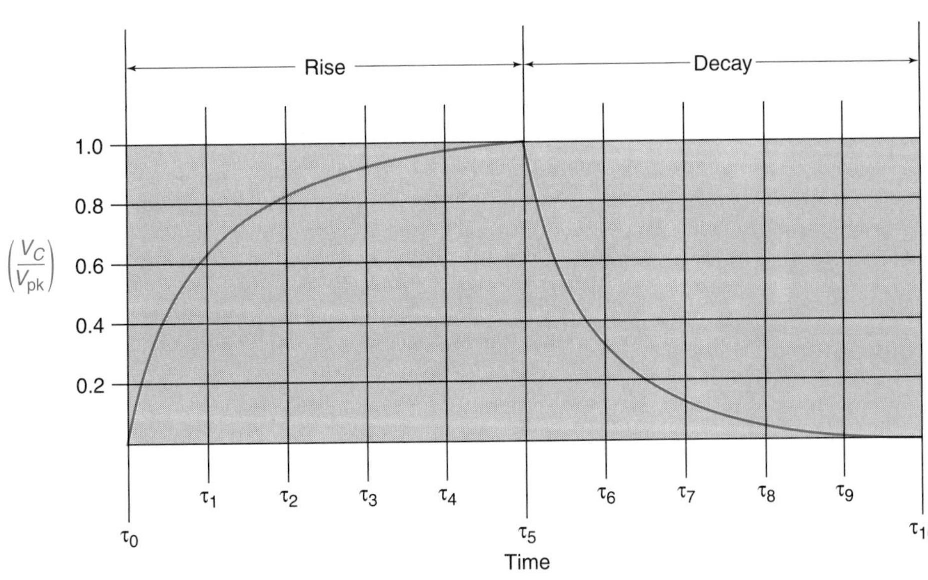

Figure 16.16 The universal voltage curve for a series *RC* switching circuit.

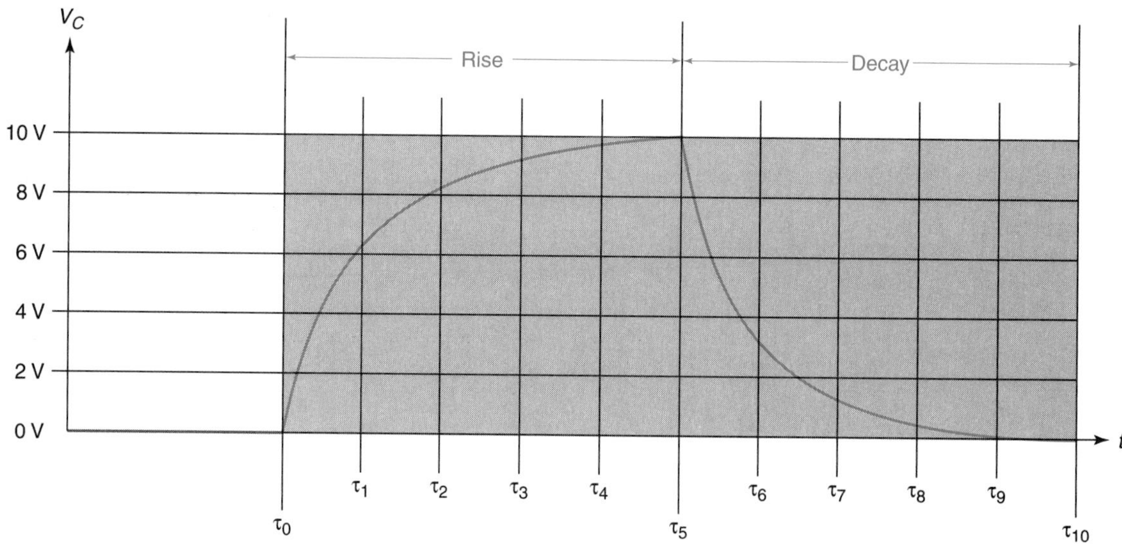

Figure 16.17

TABLE 16.5
Estimated Values of ΔV_C During the Rise Portion of the Curve Shown in Figure 16.17

Time Interval	*Estimated* ΔV_C
$t_0 \rightarrow t_1$	6.2 V
$t_1 \rightarrow t_2$	2.2 V
$t_2 \rightarrow t_3$	1.0 V
$t_3 \rightarrow t_4$	0.2 V
$t_4 \rightarrow t_5$	<0.2 V

a change in current. The universal voltage curve is caused by the capacitor's opposition to a change in voltage.

The x-axis of the universal voltage curve is divided into the same time intervals that we saw earlier. The y-axis, however, represents the ratio of capacitor voltage (V_C) to peak source voltage (V_{pk}). With this exception, the universal voltage curve and the universal current curve are identical.

As stated earlier, capacitor voltage does not change at a *linear* (constant) rate. Rather, the *rate of change* in V_C decreases as the transition progresses. This concept is illustrated further in Figure 16.17. For the sake of discussion, the V_C waveform in the figure is assumed to have a peak value of +10 V.

Each half of the waveform shown in Figure 16.17 is divided into *five equal time intervals.* Each time interval is identified using the Greek letter *tau*, τ. The estimated *change in voltage* (ΔV_C) during each of the first five time intervals is provided in Table 16.5. As you can see, the change in V_C *decreases* from one time interval to the next during the rising portion of the curve. The same principle also holds true for the *decay* portion of the curve. The estimated values of ΔV_C for the decay portion of the curve are listed in Table 16.6.

The values of ΔV_C in Tables 16.5 and 16.6 indicate that the two halves of the curve are *symmetrical.* That is, the rate of rise in any time interval matches the corresponding rate of decay. *This relationship holds true for every RC switching circuit.*

OBJECTIVE 9 ➤

The Curve Equations

We have already learned that a universal curve like the one shown in Figure 16.16 is an exponential curve, meaning that its rate of change is a function of a variable exponent. The relationship used to plot the rise portion of the curve is

$$\frac{V_C}{V_{pk}} = 1 - e^{-x} \tag{16.8}$$

where

e = the base of the natural log system (approximately 2.71828)
x = the variable exponent

The decay curve is simply a mirror image of the rise curve. The relationship used to plot the decay curve is

$$\frac{V_C}{V_{pk}} = e^{-x} \tag{16.9}$$

Note: Equations (16.8) and (16.9) can be used to solve the voltage ratio at any point in the waveform by using the following as the variable exponent:

$$x = \frac{t}{RC}$$

where

t = the time interval of interest from the start of the curve.

If you compare equations (16.8) and (16.9) to equations (16.2) and (16.3), you'll see why there are so many similarities between the universal voltage and current curves.

The *RC* Time Constant (τ)

◀ *OBJECTIVE 10*

Each time interval in the universal curve represents a real-time value that we have referred to as a *time constant* (τ). Remember, the term *constant* is used, because τ is independent of the circuit voltage (or current) magnitude and operating frequency. For a given series *RC* switching circuit, the value of the *RC* time constant is found as

$$\tau = RC \tag{16.10}$$

where

τ = the duration of each time interval, in seconds

C = the total series capacitance

R = the total series resistance

Example 16.4 demonstrates the use of this equation in determining the time constant for a series *RC* circuit with a square wave input.

TABLE 16.6
Estimated Values of ΔV_C During the Decay Portion of the Curve Shown in Figure 16.17

Time Interval	*Estimated* ΔV_C
$t_5 \rightarrow t_6$	−6.2 V
$t_6 \rightarrow t_7$	−2.2 V
$t_7 \rightarrow t_8$	−1.0 V
$t_8 \rightarrow t_9$	−0.2 V
$t_9 \rightarrow t_{10}$	<−0.2 V

Example 16.4

Determine the time constant for the circuit shown in Figure 16.18.

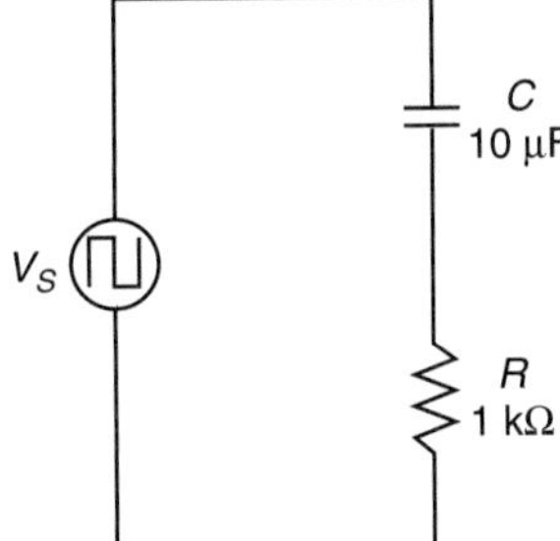

Figure 16.18

Solution: The time constant for the circuit is found as

$$\tau = RC = (1\ \text{k}\Omega)(10\ \mu\text{F}) = 10\ \text{ms}$$

Practice Problem 16.4

A circuit like the one shown in Figure 16.18 has values of $R = 220\ \Omega$ and $C = 0.1\ \mu$F. Calculate the time constant for the circuit.

Figure 16.19 shows the rise curve for the circuit in Example 16.4. As you can see, each time interval represents the value of the circuit time constant (10 ms). As shown in Figure 16.17, it takes five time constants (5τ) for capacitor voltage to reach its peak.

Defining the *RC* Time Constant

◀ *OBJECTIVE 11*

Earlier, you learned that the *RL time constant* is the time required for the current in an *RL* switching circuit to increase (or decrease) by 63.2% of its maximum possible ΔI. The ***RC* time constant** is defined as *the time required for the capacitor voltage in an RC switching circuit to increase (or decrease) by 63.2% of its maximum possible* ΔV. This definition can be explained with the aid of Figure 16.20.

At any time interval, the maximum possible ΔV equals the difference between its value at the start of the time interval and its value at the end of the transition. For example, assume that a capacitor voltage has an initial value of $V_C = 0$ V, and that the peak value of the applied signal is 100 V. The maximum possible ΔV at τ_0 is found as

$$\Delta V_{max} = 100 \text{ V} - 0 \text{ V} = 100 \text{ V}$$

During the first time constant, the capacitor voltage increases by 63.2% of ΔV_{max}. Therefore, the value of V_C at τ_1 is found as

$$V_C = V_{\tau 0} + 0.632(\Delta V_{max}) = 0 \text{ V} + 0.632(100 \text{ V}) = 0 \text{ V} + 63.2 \text{ V} = 63.2 \text{ V}$$

The relationship described here is demonstrated further in Example 16.5.

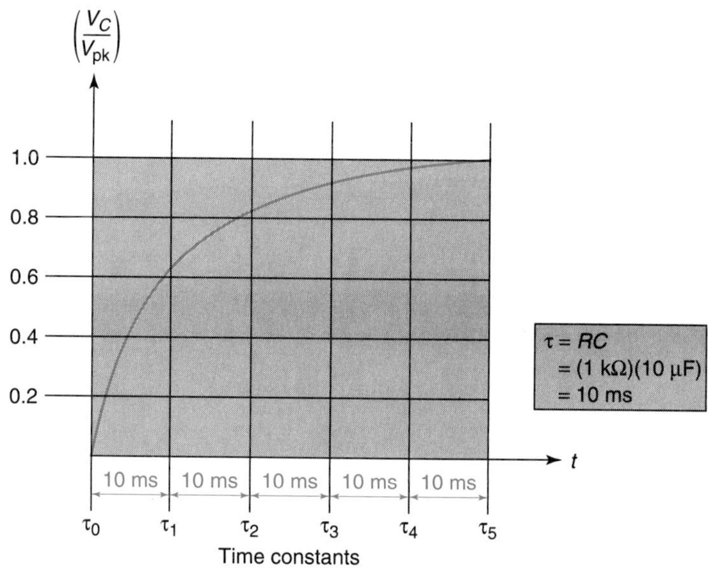

Figure 16.19

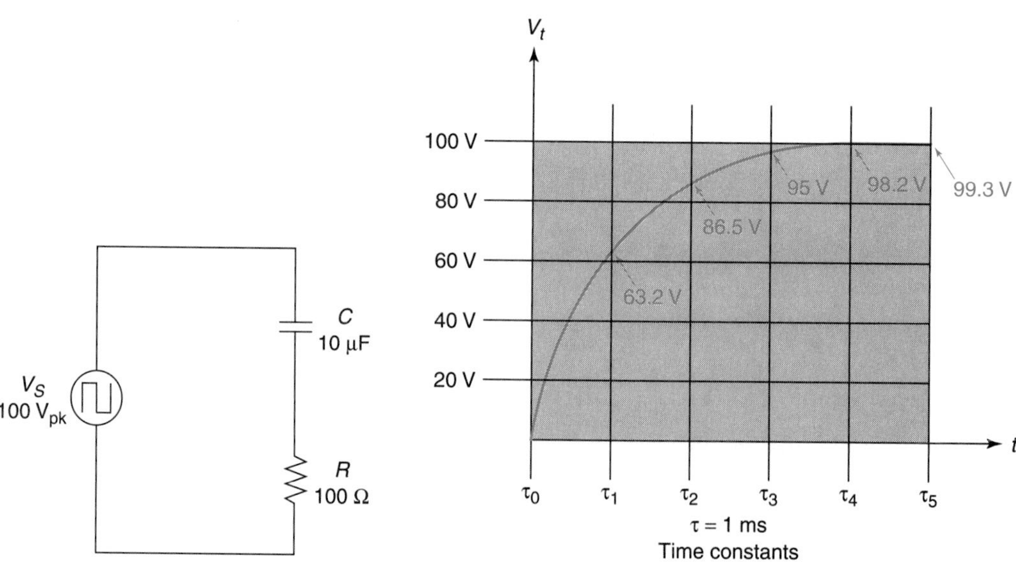

Figure 16.20

Example 16.5

The capacitor shown in Figure 16.20 has charged so that $V_C = 63.2$ V at the end of the first time constant. Calculate V_C after one additional time constant.

Solution: The value of ΔV_{max} is found as

$$\Delta V_{max} = 100 \text{ V} - 63.2 \text{ V} = 36.8 \text{ V}$$

Now we can solve for V_C at the second time constant, as follows:

$$V_C = V_{\tau 1} + 0.632(\Delta V_{max}) = 63.2 \text{ V} + 0.632(36.8 \text{ V}) = 63.2 \text{ V} + 23.3 \text{ V} = 86.5 \text{ V}$$

Comparing this result to the value shown at τ_2 in Figure 16.19, you can see that the values match.

Practice Problem 16.5

Assume that the capacitor shown in Figure 16.20 has charged to 86 V when the source makes the transition back to 0 V. Calculate V_C after one time constant. (*Hint:* The same rules apply for decay curves and rise curves.)

A Practical Application

The *RC* signal delay circuit shown in Figure 16.21 is similar to the *RL* signal delay circuit shown in Figure 16.13. The difference is that it uses the capacitor voltage rise of the *RC* circuit to provide a delay between the positive-going transitions of V_S and V_{out}.

The switching circuit is designed to switch between two output levels, 0 V and +5 V. The output from the circuit depends on the voltages at its signal inputs (A and B). The only time that the output from the switching circuit is at +5 V is when both of its input voltages (V_A and V_B) are greater than or equal to +3 V.

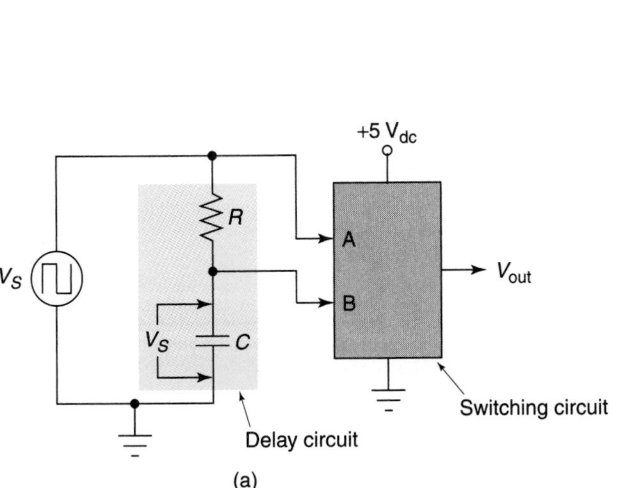

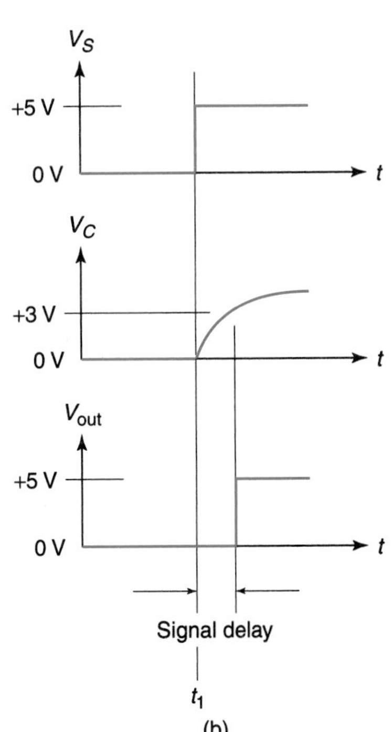

Figure 16.21 An *RC* delay circuit and waveforms.

At the instant V_S makes its positive-going transition from 0 V to +5 V, there is no voltage across the capacitor. As the capacitor charges, however, V_C increases. When V_C reaches +3 V, the switching circuit has input values of $V_A = V_S = +5$ V and $V_B = V_C = +3$ V. These two voltages meet the requirements for a +5 V output from the switching circuit, so V_{out} makes its positive-going transition. The time difference between the positive-going transitions of V_S and V_{out} is shown as the signal delay in the waveforms.

OBJECTIVE 12 ➤ The analysis of a circuit like the one shown in Figure 16.21 requires that you know how long it takes for V_C to reach its desired level. The time required for V_C to reach a specified level can be found as

$$t = -\tau \ln\left(1 - \frac{V_C}{V_S}\right) \qquad (16.11)$$

where

ln = the natural log function

The use of equation (16.11) is demonstrated in Example 16.6.

Example 16.6

Assume that the circuit shown in Figure 16.21 has values of $C = 10\ \mu F$ and $R = 330\ \Omega$. Determine the duration of the signal delay shown in the waveforms.

Solution: The time constant of the *RC* circuit is found as

$$\tau = RC = (330\ \Omega)(10\ \mu F) = 3.3 \text{ ms}$$

The time required for the V_C transition to +3 V is found as

$$t = -\tau \ln\left(1 - \frac{V_C}{V_S}\right) = -(3.3 \text{ ms}) \ln\left(1 - \frac{3\text{ V}}{5\text{ V}}\right) = -(3.3 \text{ ms}) \ln(0.4) = 3.02 \text{ ms}$$

This result indicates that V_{out} will make its positive-going transition 3.02 ms after V_S makes its positive-going transition.

Practice Problem 16.6

A circuit like the one shown in Figure 16.21 has the following values: $C = 2.2\ \mu F$, $R = 620\ \Omega$, and $V_S = +18\ V_{pk}$. Determine the time required for V_C to reach +10 V.

One Final Note

As your electronics education continues, you'll learn that the principles covered in this section are involved in a wide variety of applications. For future reference, many of the principles and relationships here are summarized in Figure 16.22.

Section Review

1. Refer to Figure 16.15. Explain the shape of the V_C and I transitions.
2. Explain the shape of the V_R waveform in Figure 16.15.
3. How is the value of the time constant in a series *RC* circuit found?
4. What is an *RC time constant*? Explain your answer.
5. Describe the operation of the *delay circuit* in Figure 16.21.

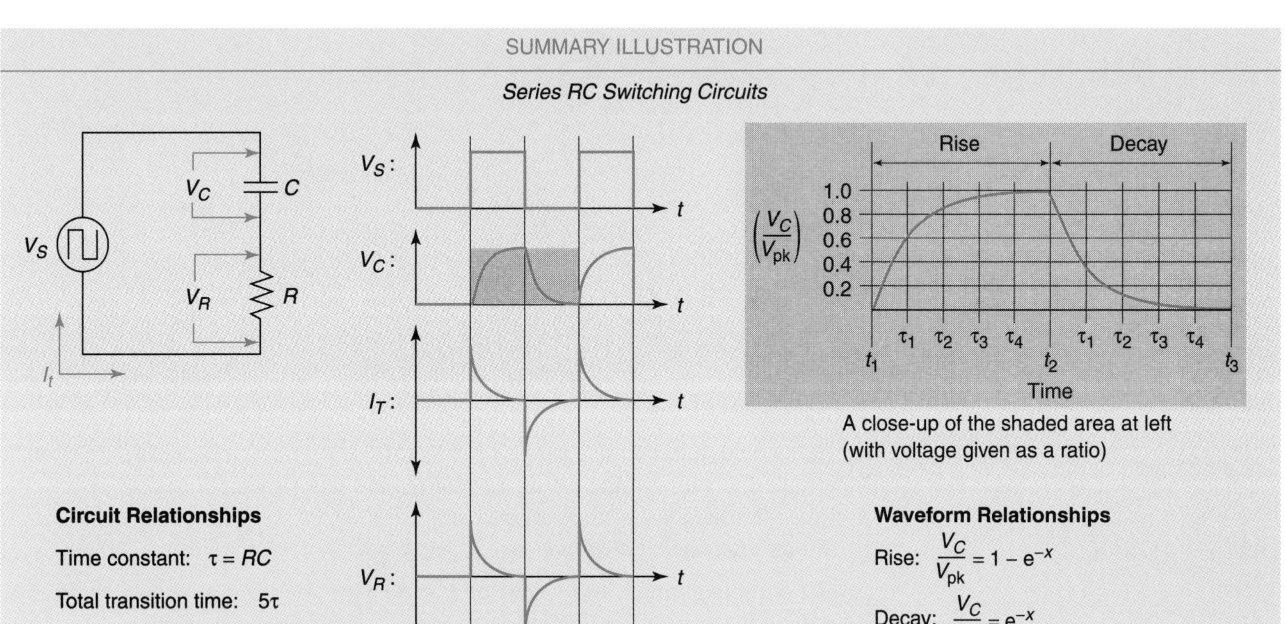

SUMMARY ILLUSTRATION
Series RC Switching Circuits
Circuit Relationships
Time constant: τ = RC
Total transition time: 5τ
Waveform Relationships
Rise: Vc/Vpk = 1 – e^-x
Decay: Vc/Vpk = e^-x
A close-up of the shaded area at left
(with voltage given as a ratio)
Rise
Decay
Time
General: One time constant (τ) is the time required for the capacitor voltage to increase (or decrease) by 63.2% of its maximum possible ΔV.

Figure 16.22

Key Terms

The following terms were introduced in this chapter on the pages indicated:

Decay, 497
Delay circuit, 503
Exponential curve, 498
Pulse, 494
RC time constant, 509
RL time constant, 502
Rise, 497
Switching circuit, 494
Time constant, 500
Transition, 494
Universal current curve, 498
Universal curve, 498
Universal voltage curve, 507

Practice Problems

1. Determine the time constant for the circuit shown in Figure 16.23a.
2. Determine the time constant for the circuit shown in Figure 16.23b.

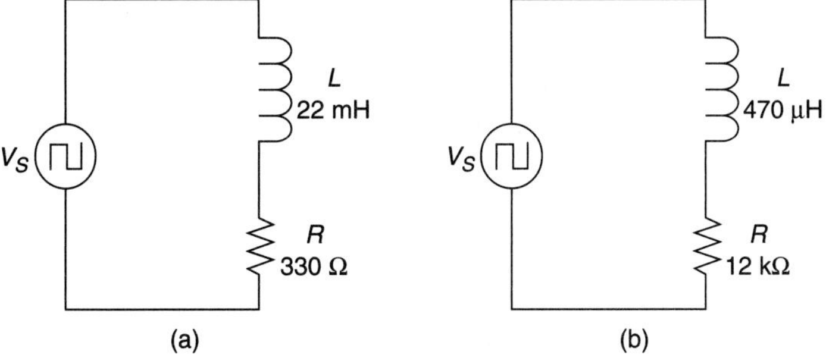

Figure 16.23

3. Determine the time required for the current in Figure 16.24a to reach its peak value.
4. Determine the time required for the current in Figure 16.24b to reach its peak value.

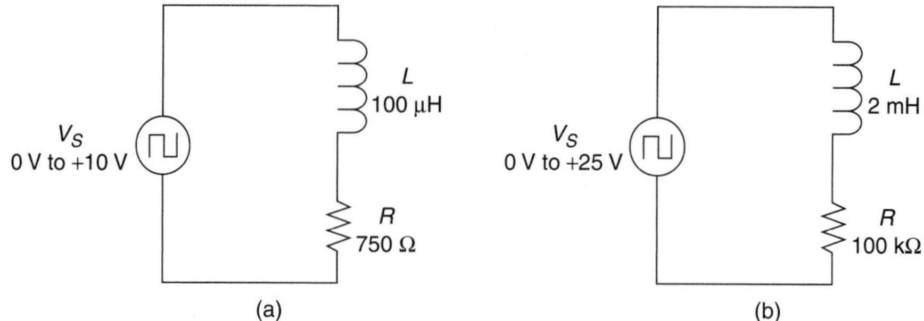

Figure 16.24

5. An RL switching circuit has values of $R = 150\ \Omega$, $L = 330$ mH, and $I_{pk} = 20$ mA. Determine the value of the circuit current (I_t) at the following times:
   a. 5 ms after the start of the current rise.
   b. 150 μs after the start of the current decay.
6. An RL switching circuit has values of $R = 33\ \Omega$, $L = 100\ \mu$H, and $I_{pk} = 75$ mA. Determine the value of the circuit current (I_t) at the following times:
   a. 1.4 μs after the start of the current rise.
   b. 2 μs after the start of the current decay.
7. Refer to Figure 16.24a. Determine the time required for V_R to reach +4 V on the rise curve.
8. Refer to Figure 16.24b. Determine the time required for V_R to reach +18 V on the rise curve.
9. Determine the time constant for the circuit shown in Figure 16.25a.
10. Determine the time constant for the circuit shown in Figure 16.25b.

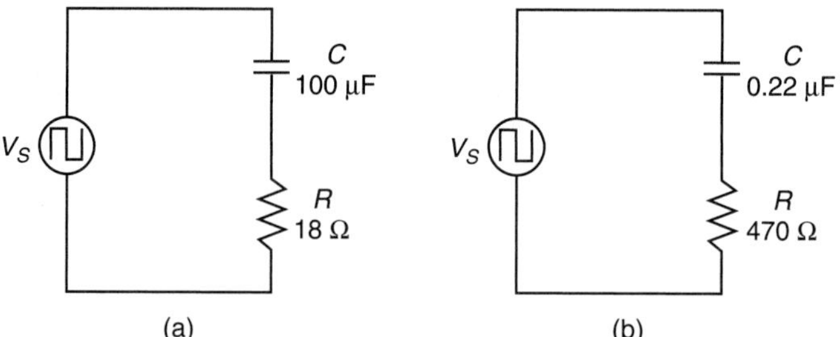

Figure 16.25

11. Determine the time required for V_C in Figure 16.26a to reach its peak value.
12. Determine the time required for V_C in Figure 16.26b to reach its peak value.

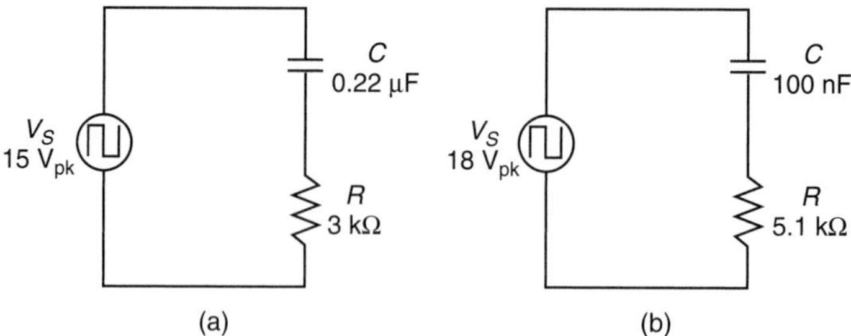

Figure 16.26

13. An RC switching circuit has values of $R = 750\ \Omega$, $C = 220\ \mu$F, and $V_{pk} = 18$ V. Determine the value of V_C at the following times:
    a. 100 ms after the start of the V_C rise cycle.
    b. 300 ms after the start of the V_C decay cycle.

14. An RC switching circuit has values of $R = 120\ \Omega$, $C = 47\ \mu F$, and $V_{pk} = 8$ V. Determine the value of V_C at the following times:
   a. 8 ms after the start of the V_C rise cycle.
   b. 4 ms after the start of the V_C decay cycle.

15. Refer to Figure 16.26a. Determine the time required for V_C to reach +6 V on its rise curve.

16. Refer to Figure 16.26b. Determine the time required for V_C to reach +15 V on its rise curve.

Looking Back

These problems relate to material presented in earlier chapters. The chapters are identified in brackets.

17. Convert the following values to scientific notation. [Introduction]
   a. 33.2 MHz
   b. 0.33 nF
   c. 430 kΩ
   d. 0.01 μH

18. A two-resistor parallel circuit has the following values: $R_1 = 330\ \Omega$, $R_2 = 220\ \Omega$, and $V_S = 6$ V. Calculate the change in source power (P_S) that occurs if R_1 is replaced with a 510 Ω resistor. [Chapter 5]

19. A series RC circuit has the following values: $R = 1.5\ k\Omega$, $C = 1\ \mu F$, and $f =$ 500 Hz. Determine the change in θ that occurs if the capacitor is replaced with a 3.3 μF component. [Chapter 13]

20. A series RL circuit has the following values: $R = 100\ \Omega$, $L = 15$ mH, and $f =$ 6 kHz. Calculate the change in apparent power (P_{APP}) that occurs if the operating frequency doubles. [Chapter 11]

Pushing the Envelope

21. Show that current in an RL circuit decreases by 63.2% of its maximum ΔI during each time interval on the decay curve using:
   a. The curve and curve equation in Figure 16.8.
   b. A series of calculations like those in Table 16.4.

Answers to the Example Practice Problems

16.1	10 μs	**16.4**	22 μs
16.2	15.2 ms	**16.5**	31.65 V
16.3	161 μs	**16.6**	1.1 ms

Multisim File References

Figure	File Reference
16.3	ETF16_3.msm
16.4	ETF16_4.msm
16.9	ETF16_9.msm
16.13	ETF16_13.msm
16.14	ETF16_14.msm
16.21	ETF16_21.msm

Part III

Discrete Electronic Devices and Circuits

chapter 17

Introduction to Solid State Components: Diodes

Objectives

After studying the material in this chapter, you should be able to:

1. Describe *covalent bonding.*
2. Compare and contrast *n-type* and *p-type* materials.
3. Describe what happens when a *pn-junction* is formed.
4. List the characteristics of a *forward-biased pn*-junction.
5. List the characteristics of a *reverse-biased pn*-junction.
6. Determine whether the diode in a circuit is *forward* or *reverse* biased.
7. List and contrast the ideal and practical *forward characteristics* of a diode.
8. Calculate the current and voltage values in a simple diode circuit.
9. List the main parameters of the *pn*-junction diode and explain how each limits the use of the component.
10. Describe the effects of diode *bulk resistance* and *reverse current* on circuit measurements.
11. Describe the effects of *temperature* on diode forward operation.
12. Obtain any parameter or characteristic value from a diode spec sheet.
13. Discuss the basic operating principles and characteristics of the *zener diode.*
14. List the main parameters of the zener diode and explain how each limits the use of the component.
15. Discuss the basic operating principles of the *light-emitting diode (LED).*
16. Given the device ratings, calculate the forward current through an LED circuit.

Electronic systems, such as radios, televisions, and computers, were originally constructed using vacuum tubes. Vacuum tube circuits can increase the strength of ac signals (amplify) and convert ac energy to dc energy (rectify). Unfortunately, tubes are large, fragile, and waste a tremendous amount of power.

During the 1940s, a team of scientists working for Bell Labs developed the *transistor,* the first solid-state device capable of amplifying an ac signal. The transistor is smaller, more rugged, and wastes much less power than a comparable vacuum tube. Solid-state components are made from elements that are classified as *semiconductors.* As you were told in Chapter 1, a semiconductor element is one that is neither a conductor nor an insulator, but rather, lies halfway between the two.

17.1 Semiconductors

Semiconductors are *atoms that contain four valence electrons,* and thus, neither good conductors nor good insulators. Three of the most common semiconductor materials are **silicon** (Si), **germanium** (Ge), and **carbon** (C). These atoms can be represented as shown in Figure 17.1. Of the semiconductors shown, silicon and germanium are used in the production of solid-state components. (Carbon is used mainly in the production of resistors and potentiometers.) Silicon is used more commonly than germanium in the production of solid-state components because it is more tolerant of heat. Although germanium is still used in some applications, they are relatively rare. For this reason, we will focus primarily on silicon throughout the upcoming chapters.

Charge and Conduction

When no outside force causes conduction, the number of electrons in a given atom equals the number of protons, and the net charge on the atom is zero. If an atom loses one valence electron, the net charge on the atom is positive. If an atom with an incomplete valence shell gains one electron, the net charge on the atom is negative.

As shown in Figure 17.2, there is an energy band above the **valence shell** (or *valence band*) called the **conduction band.** When an electron absorbs enough energy, it "jumps" from the valence band to the conduction band. This election is said to be in an *excited* state. An excited electron eventually gives up the energy that it absorbed and returns to its original energy level. The energy given up by the electron is in the form of *light* or *heat.*

Covalent Bonding

◄ *OBJECTIVE 1*

Covalent bonding is *the method by which some atoms complete their valence shells by "sharing" valence electrons with other atoms.* The covalent bonding of a group of silicon atoms is represented in Figure 17.3. As you can see, the center atom has eight valence electrons, four of its own plus one that belongs to each of the four surrounding atoms. These

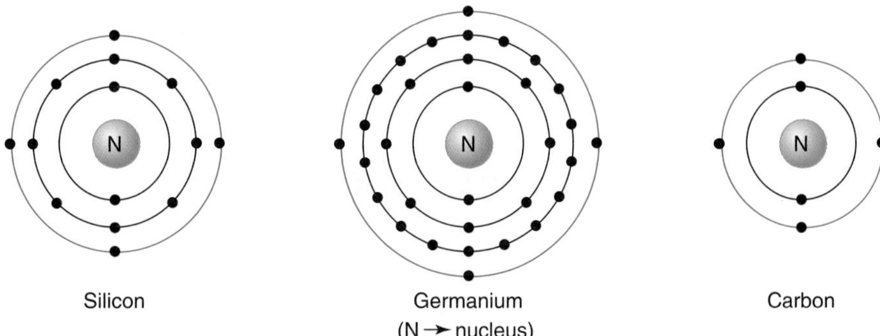

Figure 17.1 Semiconductor atoms.

Figure 17.2
Silicon energy gaps and levels.

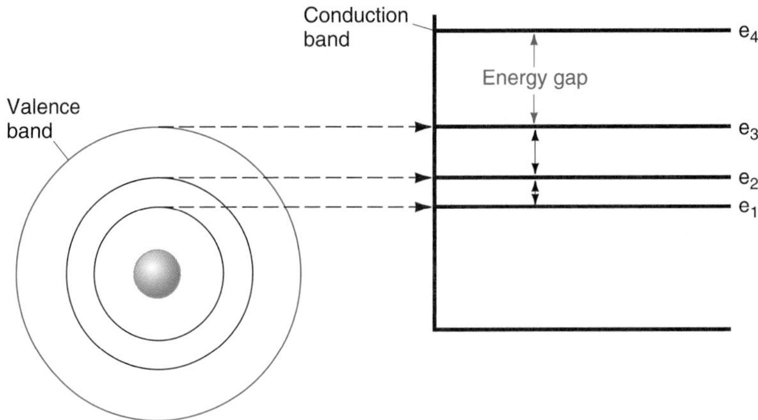

atoms, in turn, share electrons with four surrounding atoms, and so on. The results of this bonding are as follows:

- The atoms are held together, forming a solid substance.
- The atoms are all electrically stable, because their valence shells are complete. As a result, the bonded silicon atoms act as insulators.

When silicon atoms form covalent bonds, the resulting material is a silicon *crystal.* Carbon can crystallize in the same fashion as silicon, but it is too expensive to use in the production of solid-state components. (Carbon crystals are more commonly known as *diamonds.*)

Conduction

When a valence electron jumps to the conduction band, a gap is left in the covalent bond, as illustrated in Figure 17.4. This gap is referred to as a **hole.** For every conduction-band electron, there must exist a valence-band hole. This combination is referred to as an **electron-hole pair.** Within a few microseconds of becoming a free electron, that electron gives up its energy and falls into one of the holes in the covalent bond. This process is known as **recombination.**

The Relationship Between Conduction and Temperature

At room temperature, thermal energy (heat) causes the constant creation of electron-hole pairs. Thus, a semiconductor always has some number of free electrons, even when no voltage is applied to the element. As the temperature increases, so does the number of

Figure 17.3
Silicon covalent bonding.

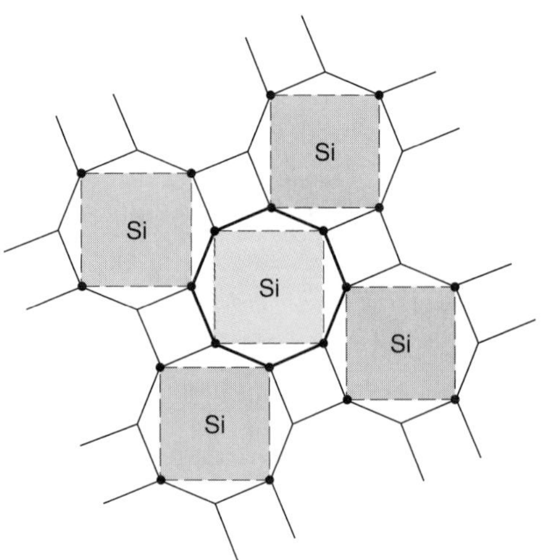

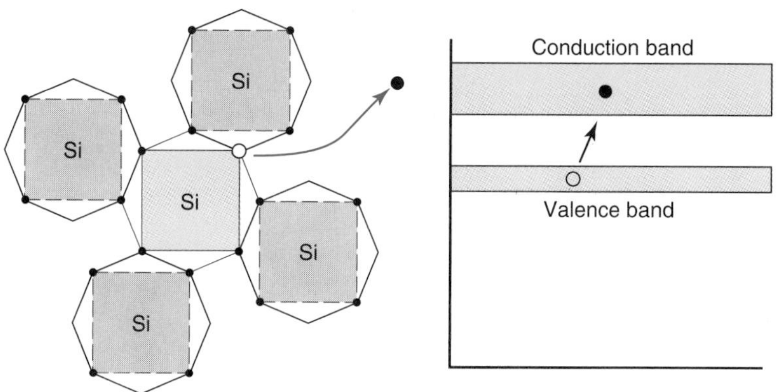

Figure 17.4 Generation of an electron-hole pair.

free electrons. The opposite is also true: as the temperature decreases, so does the number of free electrons. This is why the current through a semiconductor:

- Increases as the material warms up.
- Decreases as the material cools.

Section Review

1. How many valence electrons are there in a semiconductor?
2. What three semiconductor elements are most commonly used in electronics?
3. What forms of energy are given off by an electron that is falling into the valence band from the conduction band?
4. What is *covalent bonding*?
5. What are the effects of covalent bonding on semiconductor materials?
6. What is an *electron-hole pair*? What is *recombination*?
7. What is the relationship between *temperature* and *conductivity*?
8. Why aren't electron-hole pairs generated in a semiconductor when its temperature drops to absolute zero? Critical Thinking

---

17.2 Doping

As stated earlier, **intrinsic** (*pure*) silicon is a poor conductor. **Doping** is *the process of adding impurity atoms to intrinsic silicon to increase its conductivity.*

◄ OBJECTIVE 2

Two element types are commonly used for doping: *trivalent,* and *pentavalent.* A **trivalent** element is *one that has three valence electrons.* A **pentavalent** element is *one that has five valence electrons.* When trivalent atoms are added to intrinsic semiconductors, the resulting material is called a ***p*-type material.** When pentavalent impurity atoms are added, the resulting material is called an ***n*-type material.** The most commonly used doping elements are listed in Table 17.1. Note that the term *impurity* is used because these elements, when combined with intrinsic silicon, produce a material that is no longer pure.

TABLE 17.1 Commonly Used Doping Elements

Trivalent Impurity	*Pentavalent Impurity*
Aluminum (Al)	Phosphorus (P)
Gallium (Ga)	Arsenic (As)
Boron (B)	Antimony (Sb)
Indium (In)	Bismuth (Bi)

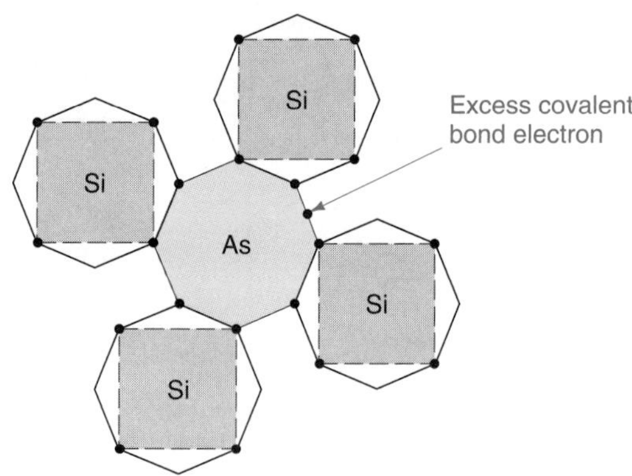

Figure 17.5 *N*-type material.

N-Type Materials

When pentavalent impurities are added to silicon, the resulting material contains too many electrons for the covalent bonds. Figure 17.5 shows four silicon atoms bonded with the arsenic atom, with each silicon atom sharing one arsenic electron. The fifth arsenic electron is not bound to any of the surrounding atoms, so it requires relatively little energy to enter the conduction band. When literally millions of arsenic atoms are added to pure silicon, the material then contains millions of electrons that are not part of the covalent bonds. All these electrons can be moved through the material with little difficulty.

Even though millions of electrons are not part of the covalent bonding, the *n*-type material is still electrically *neutral.* Each arsenic atom (and each silicon atom) still has the same number of protons as electrons; thus, the net charge on the material is zero.

Because *n*-type material has more conduction-band electrons than holes in the covalent bond, the electrons are called **majority carriers,** and the holes are called **minority carriers.** (The holes are caused by thermal energy excitation of electrons, as was discussed earlier.)

P-Type Materials

When intrinsic silicon is doped with a trivalent element, the resulting material is called a *p-type material.* As shown in Figure 17.6, doping with a trivalent element produces a valence-band hole. This hole results from the lack of a fourth valence electron in the doping element (aluminum, in this case). Since doping occurs on a larger scale, *valence band holes* are the majority carriers in a *p*-type material. Conduction band electrons (produced by thermal activity) are the minority carriers. Like an *n*-type material, however, a proton-electron balance exists within the *p*-type material, so it has a net charge of zero.

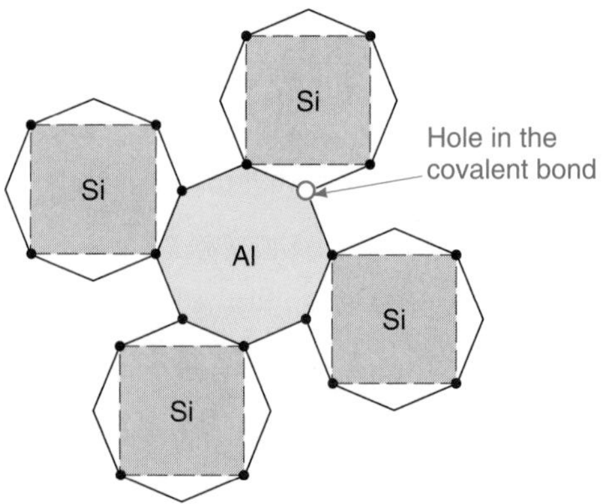

Figure 17.6 *P*-type material.

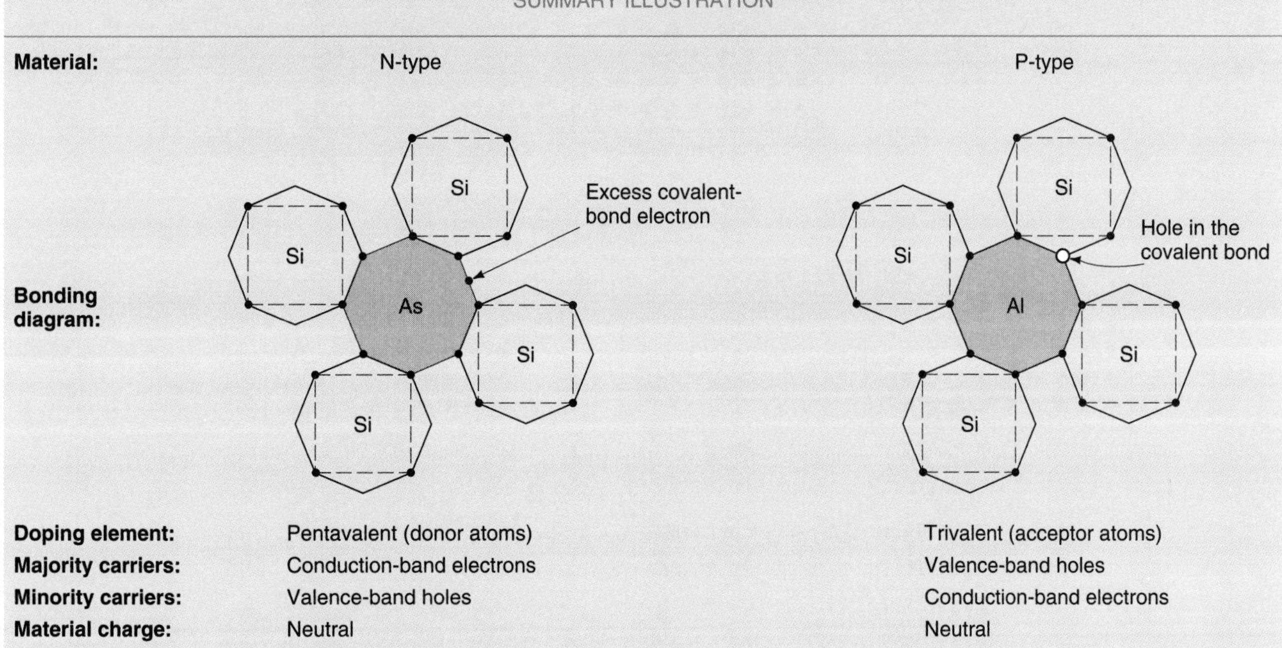

Figure 17.7

Figure 17.7 provides a comparison of n-type and p-type materials. Note that the terms *donor atom* and *acceptor atom* are used in the description of the pentavalent and trivalent doping elements. The meanings of these two terms will be made clear in the next section.

Section Review

1. What is *doping*?
2. What is an *impurity* element?
3. What are *trivalent* and *pentavalent* elements?
4. Despite their respective characteristics, *n-type* and *p-type* materials are still electrically neutral. Why?
5. In what ways are n-type and p-type materials similar? In what ways are they different?

17.3 The PN Junction

N-type and p-type materials become extremely useful when joined together to form what is called a *pn junction,* as illustrated in Figure 17.8a. Note that the excess electrons in the n-type material are represented using solid circles. The excess covalent bond holes in the p-type material are represented using open circles.

◄ OBJECTIVE 3

When the two materials are joined together, the free electrons from the n-type material begin to *diffuse* (wander) to the p-type material, as shown in Figure 17.8b. When a free electron crosses the *pn* junction, it fills one of the holes in the p-type material, completing the covalent bond, as illustrated in Figure 17.9. As the calculations in that figure demonstrate, the diffusion of an electron across the junction results in:

- One net *positive* charge in the n-type material.
- One net *negative* charge in the p-type material.

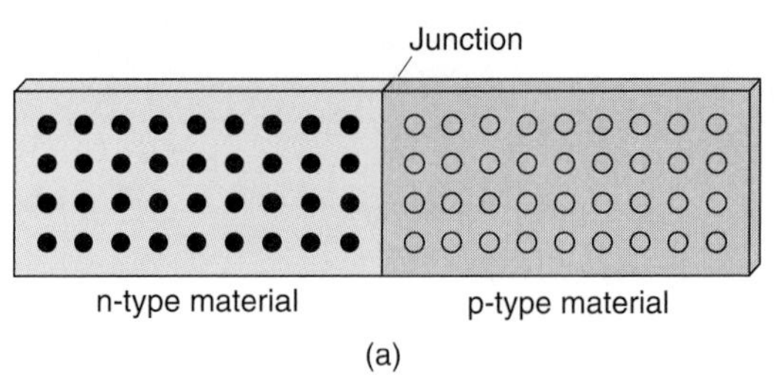

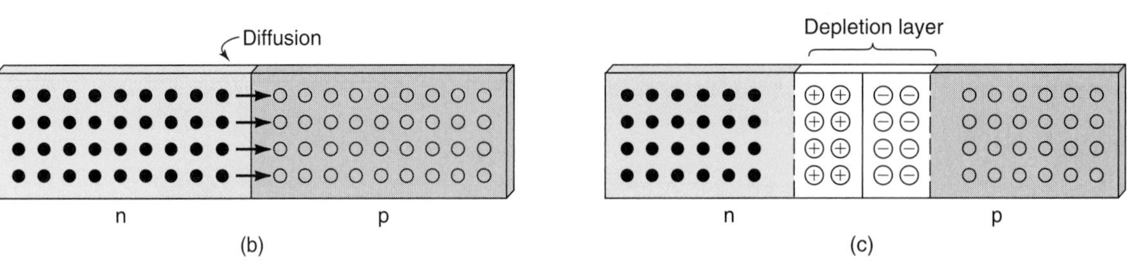

Figure 17.8 A *pn* junction.

Since the diffusion of electrons happens on a large scale, the junction ends up with a layer that is depleted of free-charge carriers. This **depletion layer** is represented in Figure 17.8c. Note that the overall charge of the layer is shown to be positive on the *n* side of the junction and negative on the *p* side.

With the buildup of (−) charges on the *p* side of the junction and of (+) charges on the *n* side, there is a natural difference of potential (voltage) across the junction. This difference of potential is referred to as the **barrier potential.** The barrier potential for silicon is approximately 0.7 V. The barrier potential for germanium is approximately 0.3 V.

Since the pentavalent atoms in the *n*-type material give up electrons, they are often referred to as **donor atoms.** The trivalent atoms, on the other hand, accept electrons and are referred to as **acceptor atoms.**

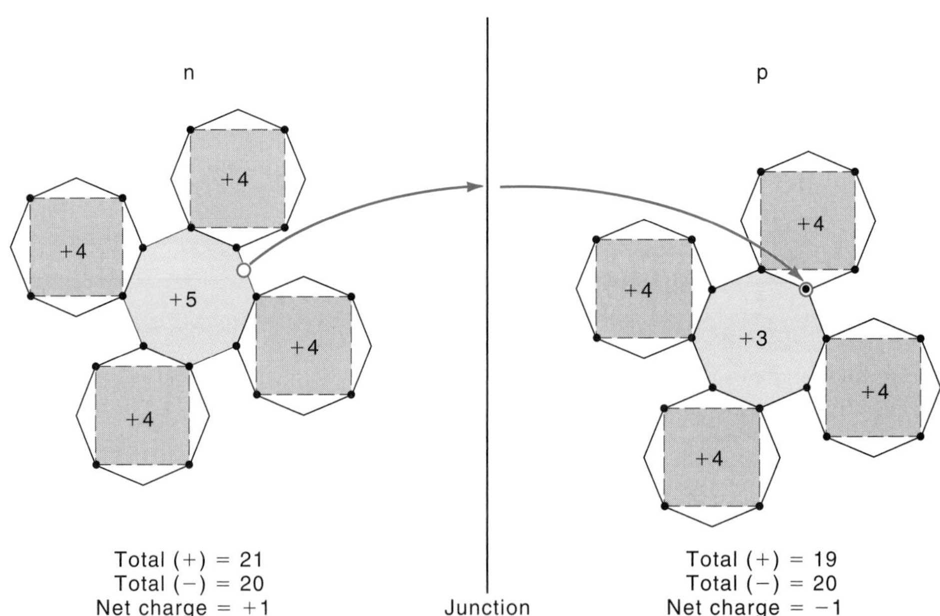

Figure 17.9 Charges resulting from electron diffusion.

Section Review

1. What is the overall charge on an *n-type* covalent bond that has just given up a conduction-band electron?
2. What is the overall charge on a *p-type* covalent bond that has just accepted an extra valence-band electron?
3. Describe the forming of the *depletion layer.*
4. What is *barrier potential*? What is its source?
5. What is the barrier potential for silicon? For germanium?

17.4 Bias

The value of a *pn* junction lies in our ability to vary the width of its depletion layer. By varying the width of the depletion layer, we can vary the resistance of the *pn* junction, and thus, the level of device current. The relationship between depletion layer width and junction current is summarized as follows:

Depletion Layer Width	*Junction Resistance*	*Junction Current*
Minimum	Minimum	Maximum
Maximum	Maximum	Minimum

Bias is *a potential applied to a pn junction to control the width of its depletion layer.* There are two types of bias: **forward bias** and **reverse bias**. A forward-biased *pn* junction has a minimum depletion layer width (i.e., low resistance). A reverse-biased *pn* junction has maximum depletion layer width (i.e., high resistance).

Forward Bias

◄ *OBJECTIVE 4*

A *pn* junction is forward biased when the applied voltage causes the *n*-type material to be *more negative* than the *p*-type material. The effects of forward bias are illustrated in Figure 17.10. When SW_1 is closed, a negative potential is applied to the *n*-type material, and a positive potential is applied to the *p*-type material. As a result:

- Conduction-band electrons in the *n*-type material are pushed toward the junction by the negative potential of the source.
- Conduction-band electrons in the *p*-type material (minority carriers) are drawn away from the junction by the positive potential of the source (leaving more valence-band holes near the junction).

Assuming that *V* is greater than the barrier potential of the junction, the electrons in the *n*-type material cross the depletion layer and recombine with the holes in the *p*-type

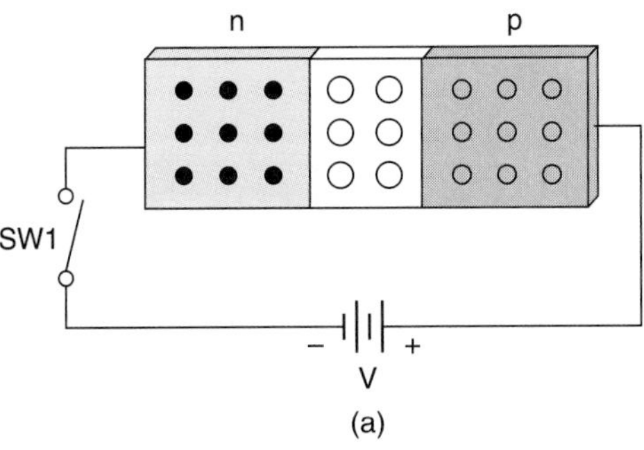

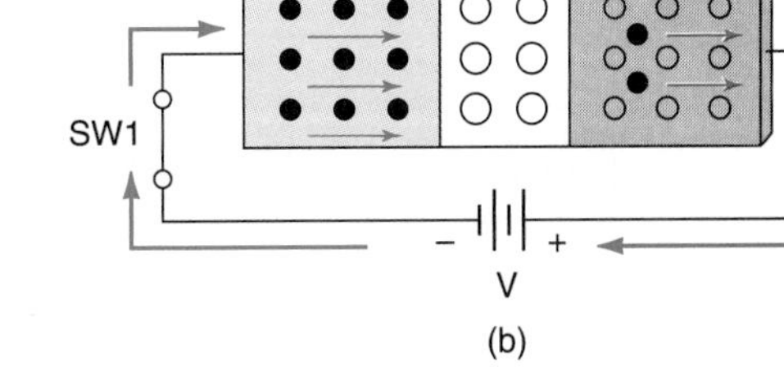

Figure 17.10 The effects of forward bias.

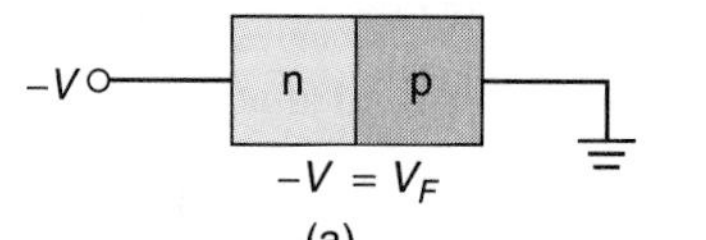

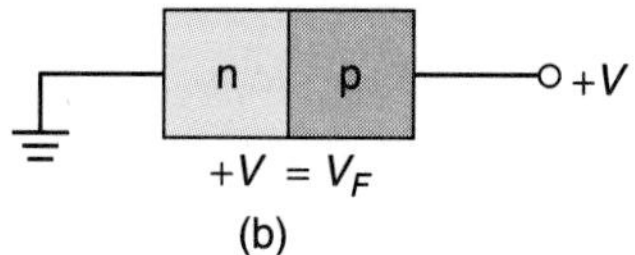

Figure 17.11 Forward biasing a *pn* junction.

material. This results in conduction through the junction, as illustrated in Figure 17.10b. Once a *pn* junction begins to conduct, it provides a slight opposition to current, referred to as **bulk resistance (R_B)**. Bulk resistance is typically around 25 Ω or less, so it is usually ignored in circuit calculations.

When a forward-biased *pn* junction begins to conduct, the forward voltage (V_F) across the junction is slightly greater than the barrier potential for the device. The value of V_F is approximated as

$$V_F \cong 0.7 \text{ V} \qquad \text{(for silicon)}$$

and

$$V_F \cong 0.3 \text{ V} \qquad \text{(for germanium)}$$

There are two ways a *pn* junction can be forward biased:

- By applying a potential to the *n*-type material that is *more negative* than the *p*-type material potential.
- By applying a potential to the *p*-type material that is *more positive* than the *n*-type material potential.

These two biasing methods are illustrated in Figure 17.11.

Reverse Bias

OBJECTIVE 5 ➤ A *pn* junction is reverse biased when the applied potential causes the *n*-type material to be *more positive* than the *p*-type material. When a *pn* junction is reverse biased, the depletion layer becomes wider, and the junction current is reduced to near zero. Reverse bias and its effects are illustrated in Figure 17.12. In Figure 17.12a, the junction is forward biased,

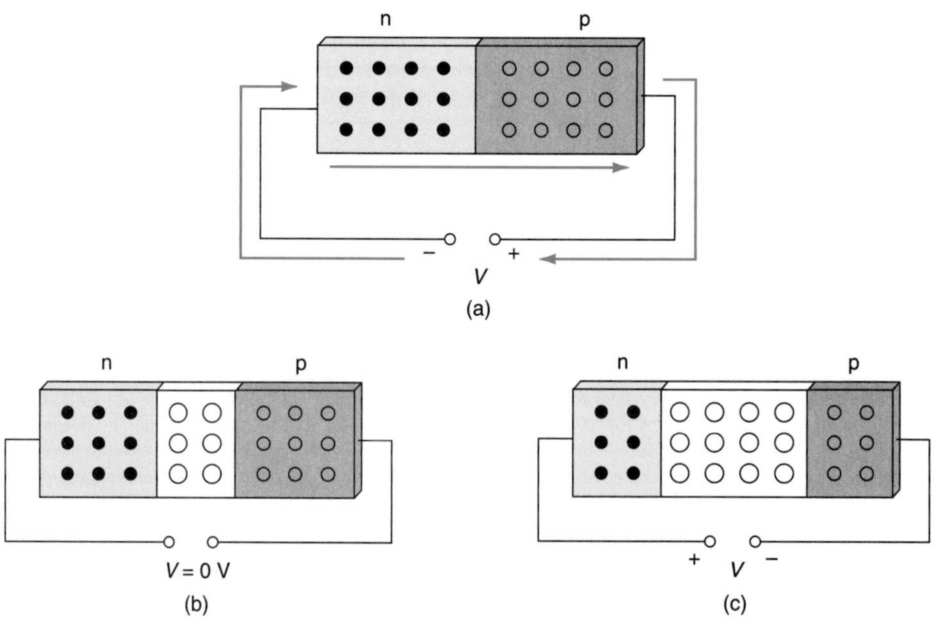

Figure 17.12 The effects of reverse bias.

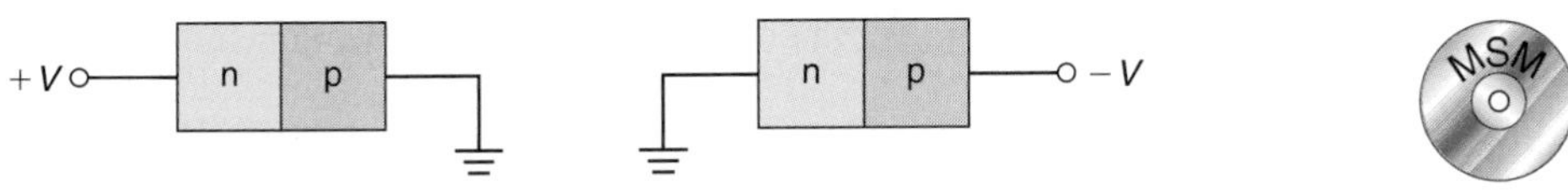

Figure 17.13 Some reverse-biased *pn* junctions.

allowing the current to pass with little opposition. If the biasing potential returns to zero, the depletion layer reforms, as illustrated in Figure 17.12b.

When we apply a voltage with the polarity shown in Figure 17.12c, the following occurs:

- The positive terminal of the source draws electrons from the *n*-type side of the *pn* junction, depleting it of free-charge carriers.
- The negative terminal of the source repels electrons toward the *p*-type material side of the junction. These electrons recombine with the valence-band holes near the junction, depleting it of free charge carriers.

As a result, the depletion layer widens on both sides of the junction. The overall effect of widening the depletion layer is to drastically increase junction resistance, and the junction current is reduced to near zero.

Just as there are two ways to forward bias a junction, there are two ways to reverse bias a junction:

- By applying a potential to the *n*-type material that makes it *more positive* than the *p*-type material.
- By applying a potential to the *p*-type material that makes it *more negative* than the *n*-type material.

The junctions shown in Figure 17.13 are both reverse biased as described above. The effects of forward and reverse biasing a *pn* junction are summarized in Figure 17.14.

A Final Note

Earlier, we said that both silicon and germanium were used to produce solid-state devices, but that silicon was, by far, the more common. There are several reasons for this:

- Silicon is more tolerant of heat.

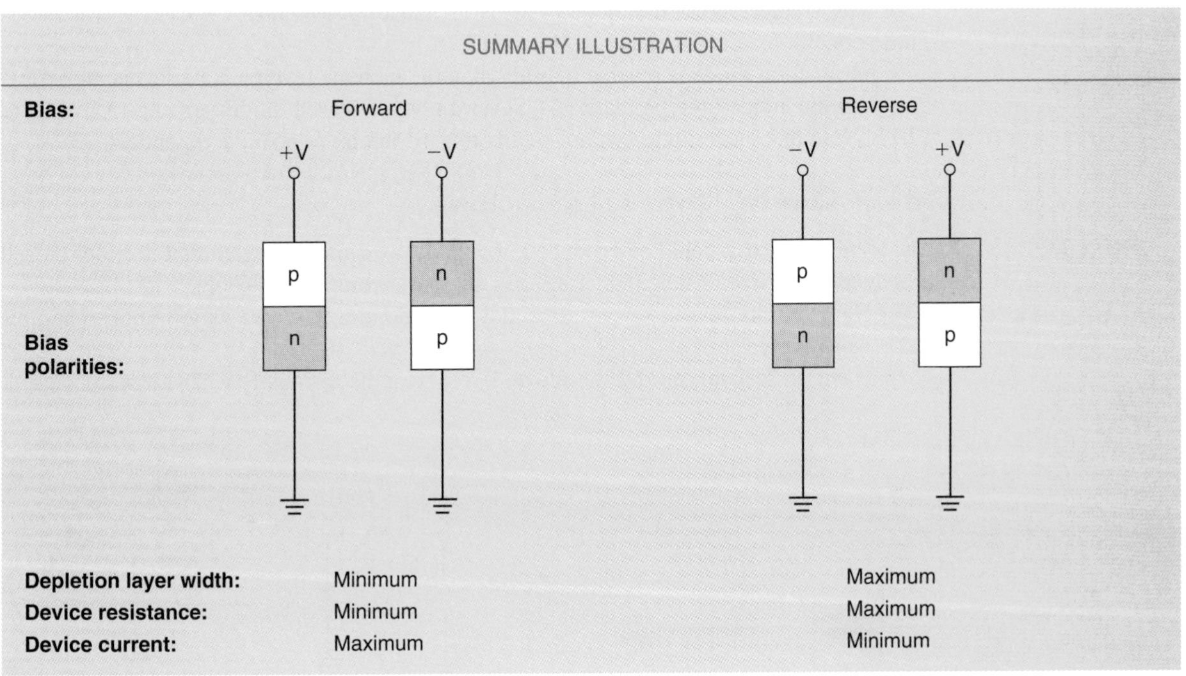

Figure 17.14

- Germanium oxide is water soluble, making germanium more difficult to process than silicon.
- Germanium allows more leakage current than silicon. (The impact of leakage current on circuit operation will become clear in later discussions.)

Section Review

1. What are the resistance and current characteristics of a *pn* junction when its depletion layer is at its maximum width? Minimum width?
2. What purpose is served by the use of *bias*?
3. What effect does *forward bias* have on the depletion layer width of a *pn* junction?
4. What are the approximate values of V_F for forward-biased silicon and germanium *pn* junctions?
5. What effect does *reverse bias* have on the depletion layer width of a *pn* junction?
6. Why is silicon more commonly used than germanium in the production of solid-state components?

17.5 PN Junction Diodes

A **diode** is *a two-electrode* (i.e., two-terminal) *component that acts as a one-way conductor.* The most basic type of diode is the ***pn*-junction diode,** which is nothing more than *a pn junction with a lead connected to each of the semiconductor materials.* The schematic symbol for the *pn*-junction diode is shown in Figure 17.15. The *n*-type material is called the **cathode,** and the *p*-type material is called the **anode.**

OBJECTIVE 6 ➤ As stated earlier, a *pn* junction conducts when the cathode (*n*-type material) is *more negative* than the anode (*p*-type material). Relating this characteristic to the schematic symbol of the diode, a diode is forward biased (conducts) when:

- The arrow points to the more negative of the diode potentials.
- The voltage across the diode exceeds its barrier potential (0.7 V).

These conditions are illustrated in Figure 17.16.

A *pn*-junction diode is reverse biased when the cathode (*n*-type material) is *more positive* than the anode (*p*-type material). Several reverse-biased diodes are shown in Figure 17.17. Note that each diode symbol points toward the more positive potential.

Ideal Diode Characteristics

OBJECTIVE 7 ➤ The ideal diode, if it could be produced, would act as a simple switch that is either closed (conducting) or open (nonconducting). As such, the component would:

- Have infinite resistance and drop the applied voltage across its terminals when reverse biased.
- Have no resistance, and therefore, 0 V across its terminals when forward biased.

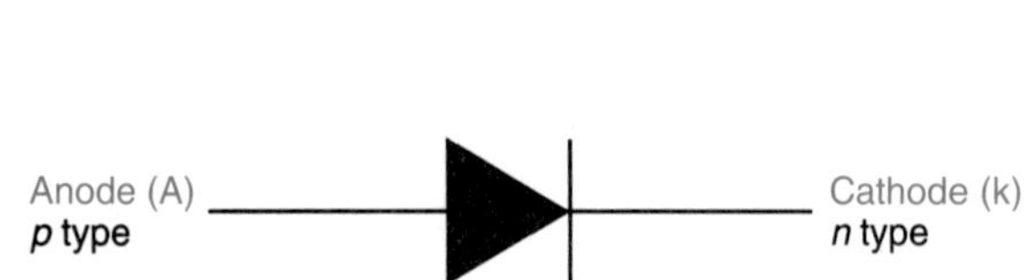

Figure 17.15 *PN*-junction diode schematic symbol.

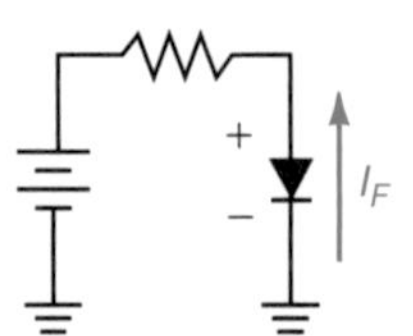

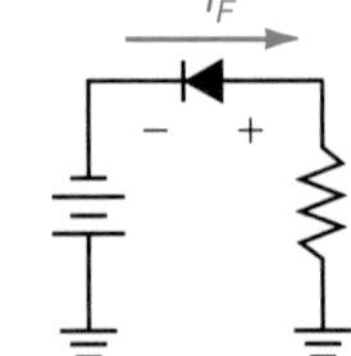

Figure 17.16 Forward-biased diodes.

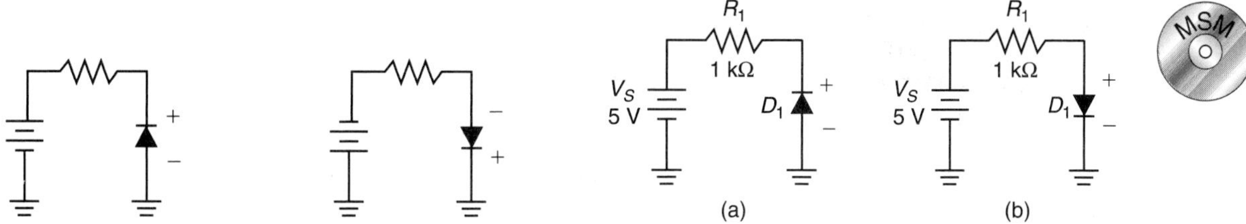

Figure 17.17 Reverse-biased diodes.

Figure 17.18

For example, the diode in Figure 17.18a is reverse biased. If we view the diode as an ideal component, it has infinite resistance. As a result:

- $I_T = 0$ A
- $V_{R1} = I_T R_1 = 0$ V
- $V_{D1} = V_S - V_{R1} = 5$ V

In Figure 17.18b, the diode is forward biased. If we again consider the diode to be ideal:

- $V_{D1} = 0$ V
- $V_{R1} = V_S - V_{D1} = 5$ V
- $I_T = \dfrac{V_{R1}}{R_1} = \dfrac{5\text{ V}}{1\text{ k}\Omega} = 5\text{ mA}$

It is important to remember that the characteristics described here are those of an *ideal* component. In practice, several additional factors must be considered.

The Practical Diode

As stated earlier, there is a slight voltage across a forward-biased *pn*-junction. This **forward voltage (V_F)** accounts for the difference between the characteristic curves shown in Figure 17.19. Note that the point where I_F suddenly increases is called the **knee voltage (V_K)**. As you can see, the ideal diode graph assumes that $V_K = 0$ V, while the practical diode graph assumes that $V_K = V_F \cong 0.7$ V. In practice, V_F may fall between 0.4 V and 1.1 V, depending on the type of diode and the value of diode current.

Using the curve shown in Figure 17.19b, we can make the following statements about the practical forward operating characteristics of a diode:

- Diode current remains at zero until the knee voltage is reached.
- Once the applied voltage reaches the value of V_K, the diode turns on and conducts.
- As long as the diode is conducting, $V_F \cong V_K$, regardless of the value of I_F.

Practical Circuit Analysis

◀ *OBJECTIVE 8*

The simplest practical analysis of a diode circuit can be demonstrated using the circuit shown in Figure 17.20. According to Kirchhoff's voltage law, the voltage across the

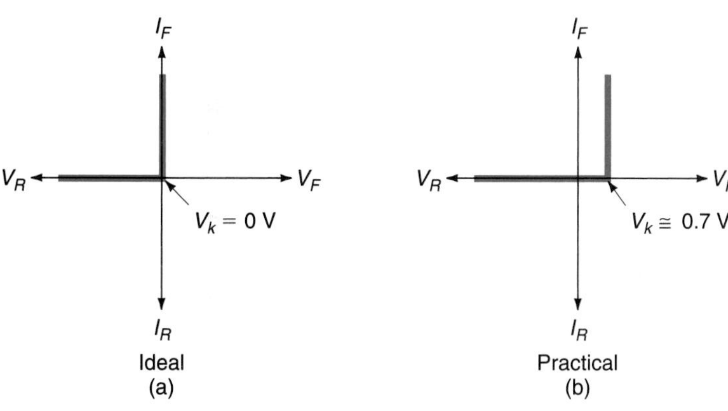

Figure 17.19 Diode characteristic curves.

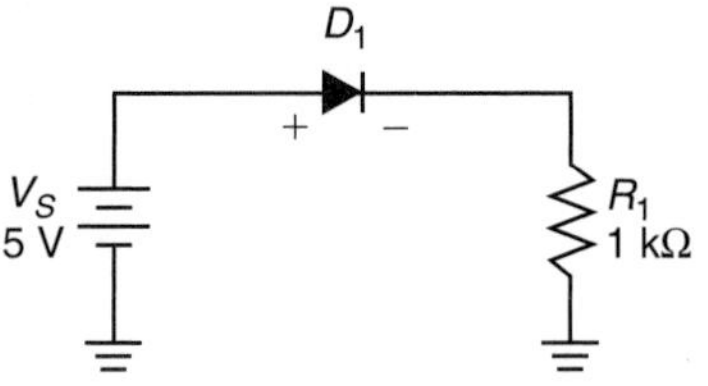

Figure 17.20

resistor must equal the difference between the source voltage and the voltage across the diode. By formula,

$$V_{R1} = V_S - V_F$$

If we assume that $V_F = 0.7$ V, we get

$$V_{R1} = V_S - 0.7 \text{ V} \tag{17.1}$$

According to Ohm's law,

$$I_T = \frac{V_{R1}}{R_1}$$

or

$$I_T = \frac{V_S - 0.7 \text{ V}}{R_1} \tag{17.2}$$

Using equation (17.2), the circuit current in Figure 17.20 is found as

$$I_T = \frac{V_S - 0.7 \text{ V}}{R_1} = \frac{5 \text{ V} - 0.7 \text{ V}}{1 \text{ k}\Omega} = 4.3 \text{ mA}$$

Example 17.1 further demonstrates the practical analysis of simple diode circuits.

Example 17.1

Determine the values of V_{R1} and I_T for the circuit shown in Figure 17.21.

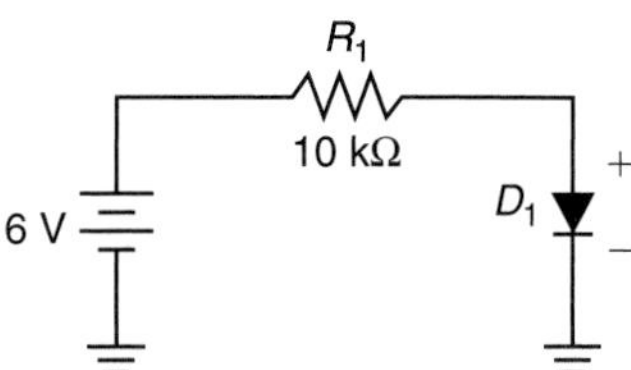

Figure 17.21

Solution: The voltage across the resistor is found as

$$V_{R1} = V_S - 0.7 \text{ V} = 6 \text{ V} - 0.7 \text{ V} = 5.3 \text{ V}$$

Using equation (17.2), the total circuit current is found as

$$I_T = \frac{V_S - 0.7 \text{ V}}{R_1} = \frac{6 \text{ V} - 0.7 \text{ V}}{10 \text{ k}\Omega} = 530 \ \mu A$$

Practice Problem 17.1

A circuit like the one shown in Figure 17.21 has a 5 V source and a 510 Ω resistor. Determine the values of V_{R1} and I_T for the circuit.

Section Review

1. What type of bias causes a *pn* junction diode to conduct?
2. What type of bias causes a *pn* junction diode to block current?
3. Draw the symbol for a *diode,* and label the terminals.
4. When analyzing a schematic diagram, how do you determine whether or not a diode is conducting?
5. When analyzing a schematic diagram, how do you determine the *direction* of diode current? Explain your answer.
6. What are the *ideal* forward characteristics of the *pn*-junction diode?
7. What are the *ideal* reverse characteristics of the *pn*-junction diode?
8. What diode characteristic is considered in the practical analysis of a diode circuit?

17.6 Diode Ratings

◄ *OBJECTIVE 9*

Assume that while troubleshooting a circuit, you find that a diode must be replaced. How can you determine whether or not a specific diode can be used in place of the faulty one? As you will see, several diode characteristics must be considered before substituting one diode for another.

Peak Reverse Voltage (V_{RRM})

Any insulator will break down and conduct if the applied voltage exceeds a specified value. *The maximum reverse voltage that won't force a diode to conduct* is called **peak reverse voltage (V_{RRM}).** When V_{RRM} is exceeded, the depletion layer may break down, allowing the diode to conduct in the reverse direction. Typical values of V_{RRM} range from a few volts to thousands of volts.

The effect that V_{RRM} has on the diode characteristic curve can be seen in Figure 17.22. Note that reverse current (I_R) is zero until the value of V_{RRM} (270 V, in this case) is exceeded. When V_R exceeds V_{RRM}, I_R increases rapidly as the depletion layer breaks down.

Peak reverse voltage is a very important **parameter** (*limit*). When considering whether or not to use a specific diode in a given application, the diode should have a V_{RRM} rating that is at least 20% greater than the maximum voltage it is expected to block.

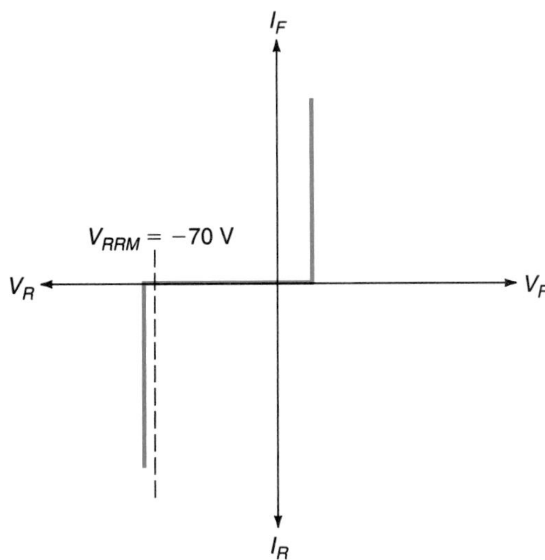

Figure 17.22

The value of V_{RRM} for a given diode can be obtained from the **specification (spec) sheet** (or data sheet) for the component. The spec sheet for a component lists all its important parameters and operating characteristics. Note that most *pn*-junction diode maximum reverse voltage ratings are multiples of 50 or 100 V. (Common values are 50 V, 100 V, 150 V, 200 V, and so on.)

Average Forward Current (I_0)

The **average forward current (I_0)** rating of a diode is its *maximum allowable value of dc forward current.* For example, the 1N4001 diode has an average forward current rating of 1 A. If the dc forward current exceeds 1 A, the diode may be destroyed from excessive heat. When you are considering whether or not to use a diode in a specific circuit, make sure that it has an I_0 rating that is at least 20% greater than the value of I_F for the circuit. The minimum acceptable I_0 rating for a diode is determined as demonstrated in Example 17.2.

Example 17.2

Determine the minimum I_0 rating that would be required for the diode shown in Figure 17.23.

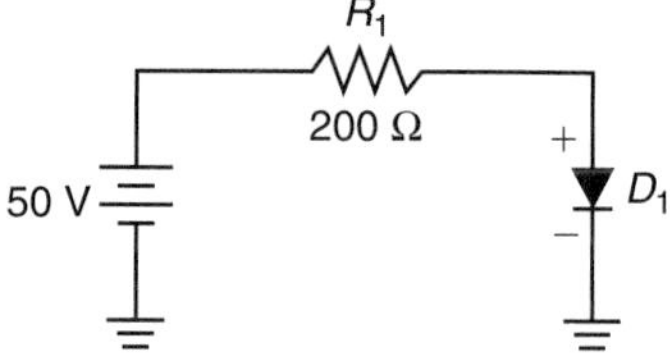

Figure 17.23

Solution: The forward diode current is found as

$$I_F = \frac{V_S - 0.7\text{ V}}{R_1} = \frac{50\text{ V} - 0.7\text{ V}}{200\ \Omega} = 246.5\text{ mA}$$

The diode must have an I_0 rating that is at least 20% greater than I_F. Therefore, the minimum acceptable I_0 rating for the component is found as

$$I_0 = (1.2)(246.5\text{ mA}) = 295.8\text{ mA} \qquad \text{(minimum)}$$

Practice Problem 17.2

A circuit like the one shown in Figure 17.23 has a 100 V source. If the resistor has a value of 51 Ω, what is the minimum allowable I_0 rating for the diode in the circuit?

A Practical Consideration: By using a diode with $I_0 >> I_F$ for a given application, you avoid pushing the diode to its current limit. As a result, the diode will last much longer.

Forward Power Dissipation ($P_{D(\max)}$)

The **forward power dissipation ($P_{D(\max)}$)** rating of a diode indicates the maximum possible power dissipation of the device when it is forward biased. Using the values of V_F and I_F, you can determine the required $P_{D(\max)}$ rating for a diode as demonstrated in Example 17.3.

Example 17.3

Calculate the minimum acceptable $P_{D(\max)}$ rating for any diode used in the circuit shown in Figure 17.24.

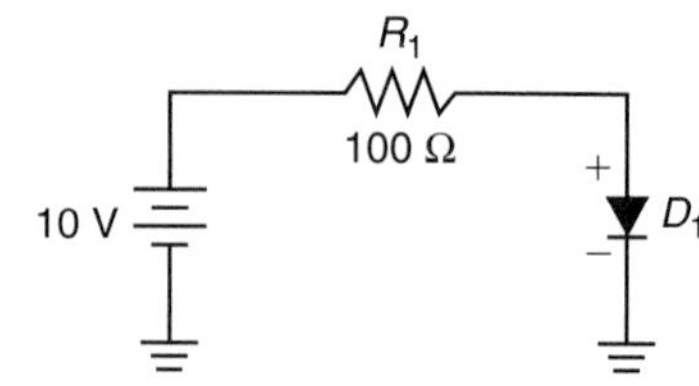

Figure 17.24

Solution: First, the total circuit current is found as

$$I_F = \frac{10\text{ V} - 0.7\text{ V}}{100\ \Omega} = 93\text{ mA}$$

Using $I_F = 93$ mA and $V_F = 0.7$ V, the power dissipation of the diode is found as

$$P_F = I_F V_F = (93\text{ mA})(0.7\text{ V}) = 65.1\text{ mW}$$

To provide a safety margin, the $P_{D(\max)}$ rating of the diode used should be at least 20% greater than the calculated value of P_F. Thus, the minimum value of $P_{D(\max)}$ is found as

$$P_{D(\max)} = 1.2P_F = (1.2)(65.1\text{ mW}) = 78.12\text{ mW} \qquad \text{(minimum)}$$

Practice Problem 17.3

A circuit like the one shown in Figure 17.24 has a 20 V source and a 68 Ω series resistor. What is the minimum required $P_{D(\max)}$ rating for any diode used in the circuit?

Diode Replacement

When replacing a faulty diode, three main parameters must be considered. Before substituting one diode for another, ask yourself:

- Is the V_{RRM} rating of the substitute component at least 20% greater than the maximum reverse voltage in the circuit?
- Is the I_0 rating of the substitute component at least 20% greater than the average (dc) value of I_F in the circuit?
- Is the $P_{D(\max)}$ rating of the substitute component at least 20% greater than the value of P_F required in the circuit?

If the answer to all of these questions is yes, then you can use the diode in the circuit.

Section Review

1. What is *peak reverse voltage*?
2. Why do you need to consider the V_{RRM} rating for a diode before attempting to use the diode in a specific circuit?
3. What is *average forward current*?
4. How do you determine whether or not the I_0 rating of a diode will be exceeded in a given circuit?
5. What is the *forward power dissipation* rating of a diode?
6. How do you determine whether or not a given diode has a $P_{D(\max)}$ rating that is high enough to allow the device to be used in a specific circuit?
7. A diode spec sheet contains a $P_{D(\max)}$ rating but no I_0 rating. How would you determine the value of I_0 for that component? Critical Thinking

17.7 Other Diode Characteristics

◄ *OBJECTIVE 10*

There are two diode characteristics that we have ignored to this point: *bulk resistance* (R_B), and *reverse current* (I_R). When these factors are taken into account, we get the diode characteristic curve shown in Figure 17.25. This illustration will be referred to throughout our discussion on R_B and I_R.

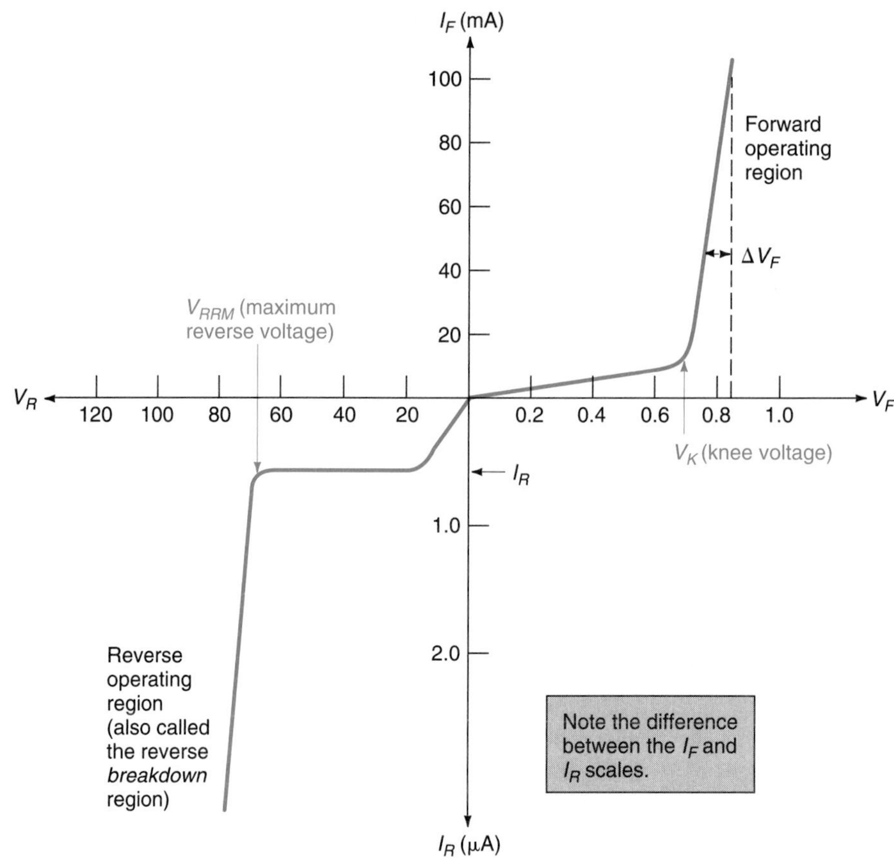

Figure 17.25 The true diode curve.

Bulk Resistance (R_B)

Earlier, you were told that *bulk resistance* is the natural resistance of the diode *p*-type and *n*-type materials. The effect of bulk resistance can be seen in the forward operation region of the curve. As Figure 17.25 illustrates, a change in forward voltage (ΔV_F) results from any change in the current passing through the bulk resistance of the diode.

Take a look at the diode equivalent circuit shown in Figure 17.26. The diode barrier potential is represented by the 0.7 V source, and its bulk resistance is represented by the series resistor, R_B. The current passing through the bulk resistance develops a voltage equal to $I_F R_B$. Taking this voltage into account, the value of V_F can be approximated more closely as

$$V_F = 0.7\text{ V} + I_F R_B \tag{17.3}$$

Note: The polarity shown for V_B in the figure was established earlier in the chapter. (See Figure 17.8.) The polarity signs for V_F indicate the polarity required to overcome V_B and generate a forward current (I_F) through the diode.

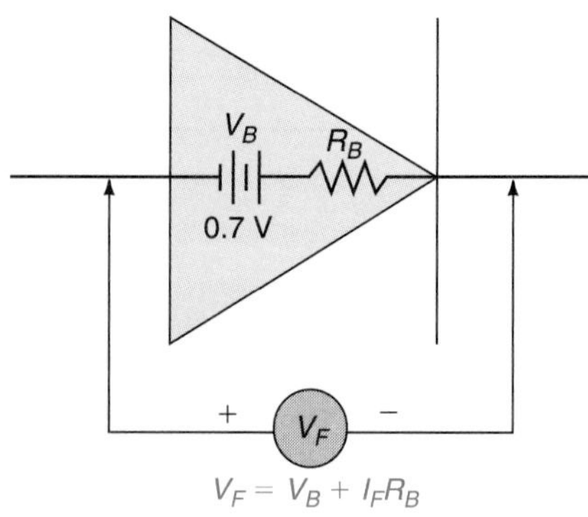

Figure 17.26 Diode equivalent circuit.

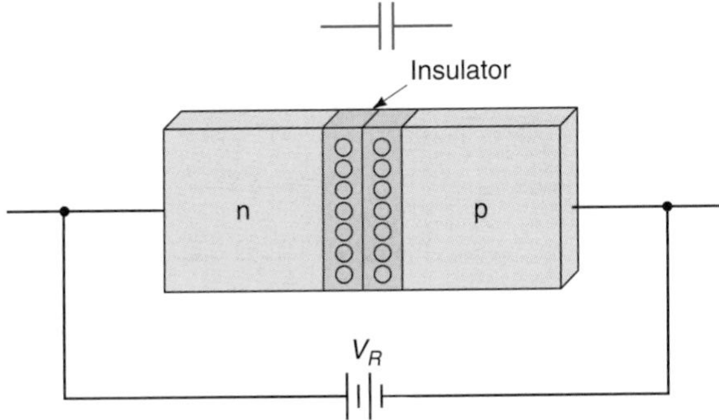

Figure 17.27 Diode capacitance.

As this relationship indicates, *V_F increases when I_F increases.* By the same token, *V_F decreases when I_F decreases.*

Reverse Current (I_R)

Ideally, there is no current through a reverse-biased diode. In reality, a very small amount of current passes through the diode when it is reverse biased. This current is referred to as **reverse current (I_R)** and is represented by the reverse operating region of the diode curve (see Figure 17.25). Reverse current is made of two independent components: *reverse saturation current* and *surface-leakage current.*

Reverse saturation current (I_S) is *a reverse current that is produced by thermal activity in the diode materials.* This current:

- Is temperature dependent; that is, it increases as temperature increases.
- Accounts for the major portion of diode reverse current.

Surface-leakage current (I_{SL}) is *a reverse current that is generated along the surface of the diode by the reverse voltage.* This current:

- Is not affected by changes in temperature.
- Accounts for only a small portion of the diode reverse current.

The total value of I_R for most diodes is typically in the microampere range or less. However, because I_S makes up the greatest portion of I_R, the value of I_R can increase significantly when there is a significant rise in operating temperature.

Diode Capacitance

An insulator placed between two closely spaced conductors forms a *capacitor.* When a diode is reverse biased, it forms a depletion layer (insulator) between two semiconductor materials (which are conductors by comparison). Therefore, a reverse-biased diode has some measurable amount of junction capacitance (as shown in Figure 17.27). Under normal circumstances, junction capacitance can be ignored because of its extremely low value (which is typically in the low picofarad range). However, when the high-frequency operation of a diode is analyzed, junction capacitance can become extremely important.

Temperature Effects on Diode Operation

Temperature has a significant effect on most of the diode characteristics discussed in this section. The effects of increased temperature on forward diode operation are illustrated in Figure 17.28. As you can see, there are two forward operation curves, one for 25°C and one for 100°C. Based on the graph shown in Figure 17.28, we can draw two conclusions: ◄ *OBJECTIVE 11*

- As temperature increases, I_F increases at a specified value of V_F. This is illustrated by the points labeled I_1 and I_2.
- As temperature increases, V_F decreases at a specified value of I_F. This is illustrated by the points labeled V_1 and V_2.

In practice, an increase in temperature usually results in both an *increase* in forward current (I_F) and a *decrease* in forward voltage (V_F).

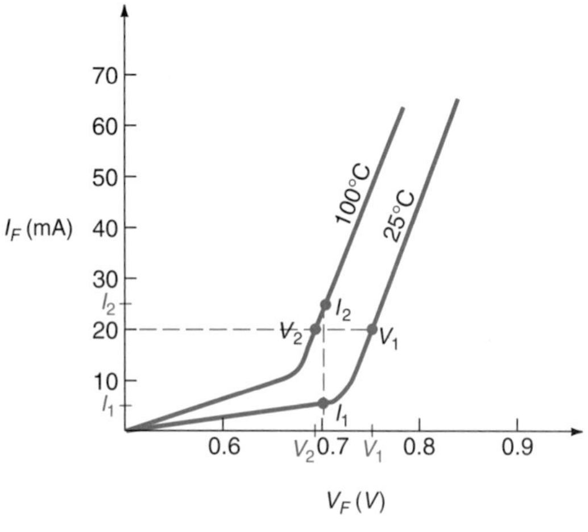

Figure 17.28 Temperature effects on diode forward operation.

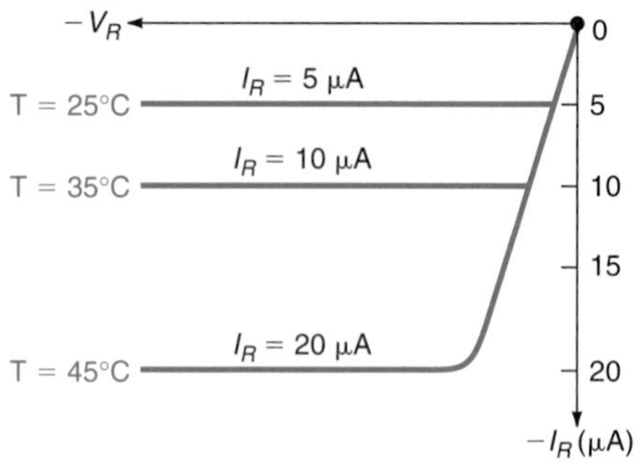

Figure 17.29 Temperature effects on diode reverse current.

Temperature affects diode reverse current in the same way that it affects forward current. The effect of temperature on I_R is illustrated in Figure 17.29.

Section Review

1. What is *bulk resistance*?
2. What is the relationship between I_F and V_F?
3. What is *reverse current*?
4. Which component of reverse current is affected by *temperature*?
5. Which component of reverse current is affected by the amount of *reverse bias*?
6. Which component of reverse current makes up a *majority* of that current?
7. Describe the reverse-biased diode as a capacitor.
8. What effect does an increase in temperature have on I_F and V_F?
9. What effect does an increase in temperature have on I_R?

Critical Thinking

10. The diode shown in Figure 17.30 allows 100 μA of reverse current. What voltage would be measured across the reverse-biased diode?

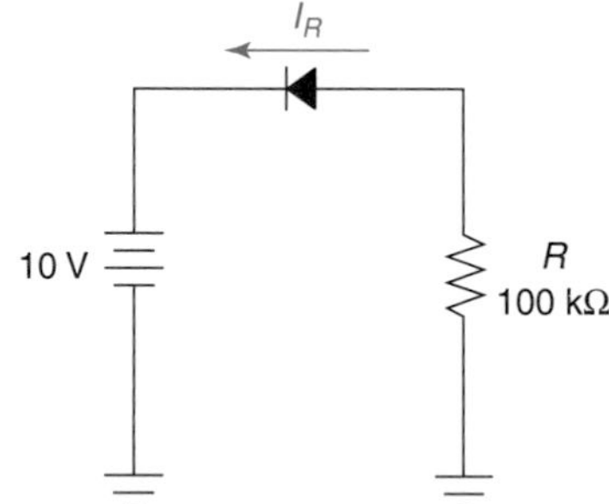

Figure 17.30

17.8 Diode Specifications

The specification sheet for any component lists its parameters and operating characteristics. Because parameters are limits, they are almost always designated as *maximum values.* Operating characteristics, on the other hand, may be designated as minimum, maximum, or typical values, depending on the rating.

Spec Sheet Organization

Figure 17.31 contains the spec sheet for the 1N4001–1N4007 (or 1N400X) series diodes. The *maximum ratings* table contains the diode parameters that must not be exceeded under any circumstances. If any maximum rating is exceeded, you will more than likely have to replace the diode. The *electrical characteristics* table contains the guaranteed operating characteristics of the device. These characteristics are guaranteed so long as the maximum ratings limits are observed. If any diode parameter is exceeded, however, the electrical characteristics values of the device cannot be guaranteed.

◄ *OBJECTIVE 12*

Diode Maximum Ratings

Many of the parameters listed in the maximum ratings table have already been covered in this chapter. Table 17.2 summarizes the parameters listed in Figure 17.31 and their meanings.

TABLE 17.2 Diode Parameters

Rating	*Description*
Peak repetitive reverse voltage (V_{RRM})	This is the maximum allowable V_R for the diode. This holds true for both dc and peak ac voltages.
RMS reverse voltage ($V_{R(\mathrm{rms})}$)	This rating is found by converting the peak repetitive voltage rating to an effective (rms) value.
Average half-wave rectified forward current (I_0)	This rating tells you the maximum average forward current the diode can handle. Note that the limit on forward current decreases when the temperature increases.
Peak surge current (I_{FSM})	This is the maximum *surge* (nonrepetitive) value of I_F that the diode can handle. With the average diode, this rating can be exceeded once. After that, you will have to replace the diode.
Operating and storage temperature range (T_J, T_{stg})	This one is self-explanatory. The diode can be used and stored at any temperature between −65°C and 175°C.

Note: The meaning of the term *half-wave rectified* is explained in Chapter 18.

Diode Identification

The terminals on most diodes can be identified as shown in Figure 17.32. Whenever possible, use the spec sheet diagram to identify the anode and cathode leads of a given diode.

Electrical Characteristics

As stated earlier, the values provided under *Electrical Characteristics* indicate the guaranteed operating characteristics of the diode. Table 17.3 provides a brief explanation of each electrical characteristic listed in Figure 17.31.

TABLE 17.3 Diode Electrical Characteristics

Rating	*Discussion*
Maximum forward voltage drop (V_F)	This is the maximum value that V_F will ever reach, determined at a temperature of 25°C.
Maximum full-cycle average forward voltage drop ($V_{F(AV)}$)	This is the maximum average forward voltage (V_F). This parameter is temperature dependent and was measured at 75°C.
Maximum reverse current (I_R)	These ratings are given for temperatures of 25°C and 100°C. The spec sheet lists this parameter *at rated dc voltages,* meaning that the rating is valid for all dc values of V_R at or below the component's V_{RRM} rating.
Maximum full-cycle average reverse current ($I_{R(AV)}$)	This is the maximum average value of I_R. Note that this rating is measured at a temperature of 75°C. At all temperatures below 75°C, the average value of I_R will be less than 30 μA.

1N4001 thru 1N4007

Designers▲Data Sheet

LEAD MOUNTED
SILICON RECTIFIERS

50-1000 VOLTS
DIFFUSED JUNCTION

"SURMETIC"▲ RECTIFIERS

. . . subminiature size, axial lead mounted rectifiers for general-purpose low-power applications.

Designers Data for "Worst Case" Conditions

The Designers▲ Data Sheets permit the design of most circuits entirely from the information presented. Limit curves – representing boundaries on device characteristics – are given to facilitate "worst case" design.

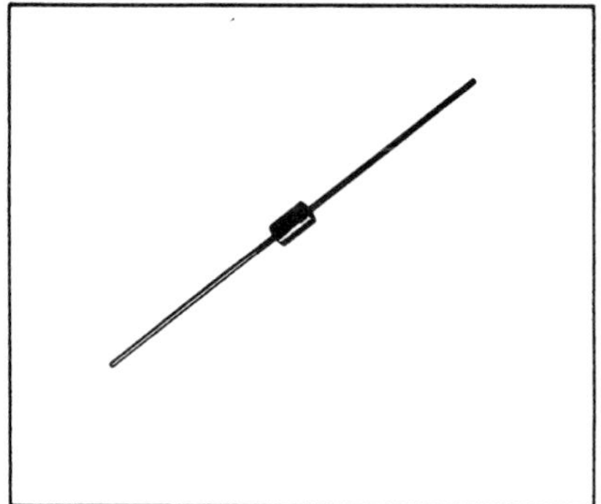

*MAXIMUM RATINGS

Rating	Symbol	1N4001	1N4002	1N4003	1N4004	1N4005	1N4006	1N4007	Unit
Peak Repetitive Reverse Voltage Working Peak Reverse Voltage DC Blocking Voltage	V_{RRM} V_{RWM} V_R	50	100	200	400	600	800	1000	Volts
Non-Repetitive Peak Reverse Voltage (halfwave, single phase, 60 Hz)	V_{RSM}	60	120	240	480	720	1000	1200	Volts
RMS Reverse Voltage	$V_{R(RMS)}$	35	70	140	280	420	560	700	Volts
Average Rectified Forward Current (single phase, resistive load, 60 Hz, see Figure 8, $T_A = 75^oC$)	I_O	1.0							Amp
Non-Repetitive Peak Surge Current (surge applied at rated load conditions, see Figure 2)	I_{FSM}	30 (for 1 cycle)							Amp
Operating and Storage Junction Temperature Range	T_J, T_{stg}	-65 to +175							^oC

*ELECTRICAL CHARACTERISTICS

Characteristic and Conditions	Symbol	Typ	Max	Unit
Maximum Instantaneous Forward Voltage Drop ($i_F = 1.0$ Amp, $T_J = 25^oC$) Figure 1	v_F	0.93	1.1	Volts
Maximum Full-Cycle Average Forward Voltage Drop ($I_O = 1.0$ Amp, $T_L = 75^oC$, 1 inch leads)	$V_{F(AV)}$	–	0.8	Volts
Maximum Reverse Current (rated dc voltage)	I_R			μA
$T_J = 25^oC$		0.05	10	
$T_J = 100^oC$		1.0	50	
Maximum Full-Cycle Average Reverse Current ($I_O = 1.0$ Amp, $T_L = 75^oC$, 1 inch leads	$I_{R(AV)}$	–	30	μA

*Indicates JEDEC Registered Data.

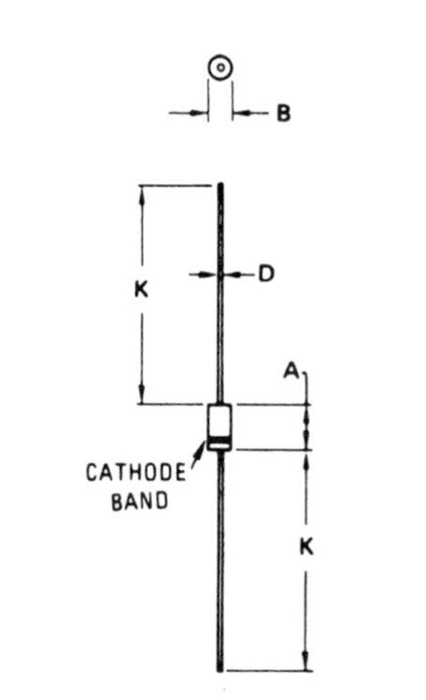

DIM	MILLIMETERS MIN	MILLIMETERS MAX	INCHES MIN	INCHES MAX
A	5.97	6.60	0.235	0.260
B	2.79	3.05	0.110	0.120
D	0.76	0.86	0.030	0.034
K	27.94	–	1.100	–

CASE 59-04
Does Not Conform to DO-41 Outline.

MECHANICAL CHARACTERISTICS

CASE: Void free, Transfer Molded
MAXIMUM LEAD TEMPERATURE FOR SOLDERING PURPOSES: 350°C, 3/8" from case for 10 seconds at 5 lbs. tension
FINISH: All external surfaces are corrosion-resistant, leads are readily solderable
POLARITY: Cathode indicated by color band
WEIGHT: 0.40 Grams (approximately)

▲Trademark of Motorola Inc.

DS 6015 R3

Figure 17.31 The Motorola 1N400X series specifications. (Copyright of Semiconductor Components Industries, LLC. Used by permission.)

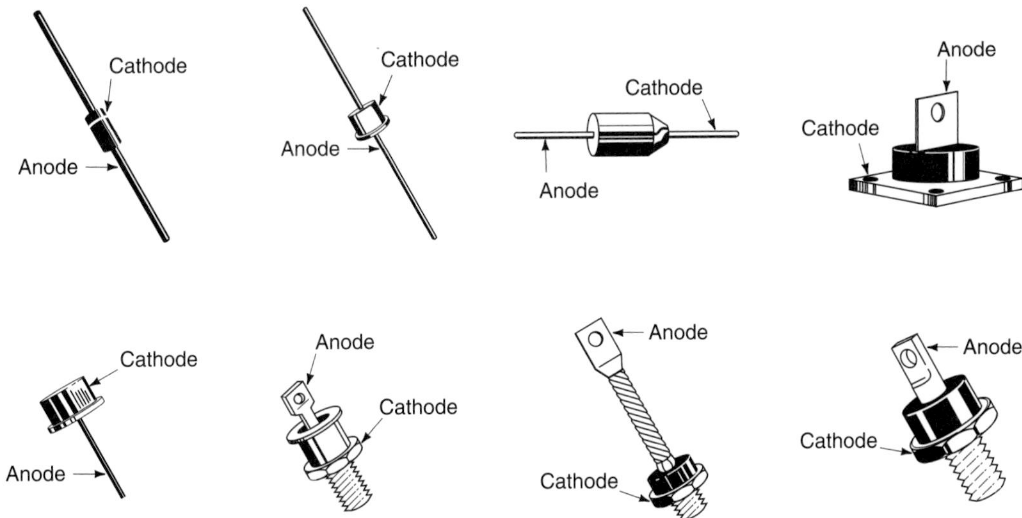

Figure 17.32 Common types of diodes. (Copyright of Semiconductor Components Industries, LCC. Used by permission.)

Many of the characteristics listed include both *typical* and *maximum* values. When analyzing the operation of the component in a circuit, the "typical" value is used. When determining circuit tolerances for circuit development purposes, the "maximum" value is used. When only one value is given, it is used for all applications.

Section Review

1. What is a *maximum rating*? Give an example.
2. What is an *electrical characteristic*? Give an example.
3. What is the difference between a *maximum rating* and an *electrical characteristic*?
4. Why does the limit on forward current *decrease* when temperature *increases*? Critical Thinking

17.9 Zener Diodes

◀ *OBJECTIVE 13*

A *pn*-junction diode operated in the *reverse breakdown* region of its operating curve is usually destroyed, but this is not the case for the *zener diode.* The **zener diode** is *a type of diode that is designed to work in the reverse breakdown region of its characteristic curve.* Figure 17.33 shows the reverse breakdown region of the zener diode characteristic curve. As illustrated, two things happen when the **reverse breakdown voltage (V_{BR})** is reached:

- The diode current (I_R) increases drastically.
- The diode reverse voltage (V_R) remains relatively constant.

In other words, *the voltage across a zener diode operated in this region is relatively constant over a wide range of device current values.* This makes the zener diode useful as a **voltage regulator**; *a circuit designed to maintain a constant voltage despite minor variations in circuit current or input voltage.*

Since the zener diode is normally operated in its reverse breakdown region, zener current is normally *in the direction of the arrow,* as shown in Figure 17.34. When the diode is conducting, the cathode is more positive than the anode. Thus, the arrow points in the direction of current.

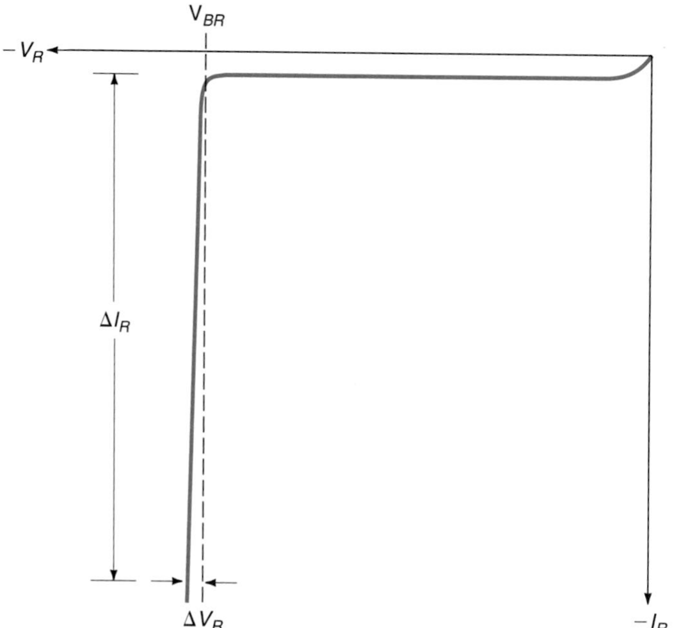

Figure 17.33 Reverse breakdown characteristics.

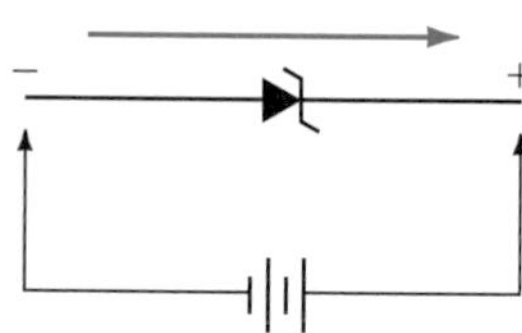

Figure 17.34 Zener current.

When reverse biased, the voltage across a zener diode is nearly constant and equal to its **zener voltage (V_Z)** rating. Zener diodes with V_Z ratings in the range of 1.8 V to several hundreds of volts and power dissipation ratings from 500 mW to 50 W are commonly available.

Zener Operating Characteristics

A zener diode maintains a near-constant reverse voltage for a range of reverse current values, as illustrated in Figure 17.35. The following zener current values are identified in the figure:

- **Zener knee current (I_{ZK})** is *the minimum current required to maintain voltage regulation (constant voltage).*
- **Maximum zener current (I_{ZM})** is *the maximum amount of current the diode can handle without being damaged or destroyed.*
- **Zener test current (I_{ZT})** is *the current level at which the V_Z rating of the diode is measured.*

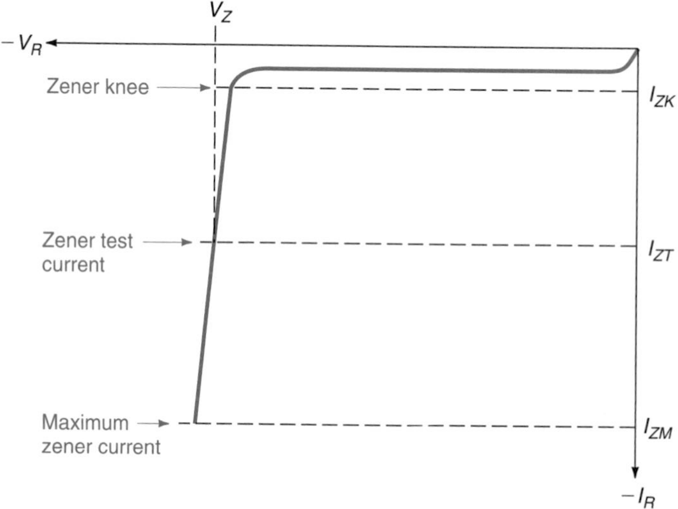

Figure 17.35 Zener reverse current values.

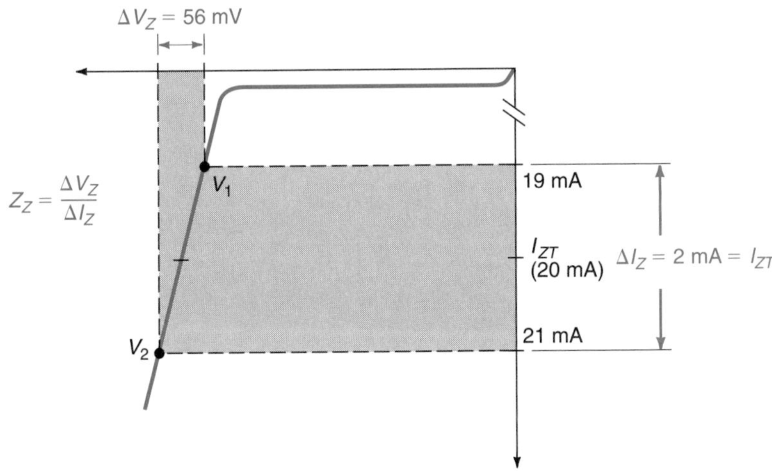

Figure 17.36 Determining zener impedance.

As long as the zener current is kept between the values of I_{ZK} and I_{ZM}, the voltage across the component (V_Z) is approximately equal to its **nominal** (*rated*) value.

Zener impedance (Z_Z) is *the zener diode's opposition to a change in current.* The value of Z_Z is measured as shown in Figure 17.36. The change in zener voltage (ΔV_Z) is measured for a specific change in zener current (ΔI_Z). Then, using these two values, the value of zener impedance is found using

$$Z_Z = \frac{\Delta V_Z}{\Delta I_Z}$$

Note that the value of Z_Z (at a specified value of ΔI_Z) is normally listed on the spec sheet for a zener diode.

Static reverse current (I_R) is *the reverse current through the diode when $V_R < V_Z$.* In other words, I_R is the reverse leakage current through the diode when it is off.

Section Review

1. What is the primary difference between zener diodes and *pn*-junction diodes?
2. What characteristic makes the zener diode useful as a *voltage regulator*?
3. How do you determine the direction of zener current in a schematic diagram?
4. What is the significance of the zener voltage rating?
5. Name and define the following: V_Z, Z_Z, I_{ZK}, I_{ZM}, I_{ZT}, and I_R.
6. Which zener rating limits the *total current* in a zener diode circuit?

17.10 Zener Diode Ratings

The ratings discussed in this section are listed on the spec sheet shown in Figure 17.37. We will refer to this figure throughout our study.

Maximum Ratings

OBJECTIVE 14

The two most important maximum ratings for the zener diode are *dc power dissipation* and the *power derating factor.* The dc power dissipation (P_D) rating is important because many zener diode spec sheets do not list the values of maximum zener current (I_{ZM}). When

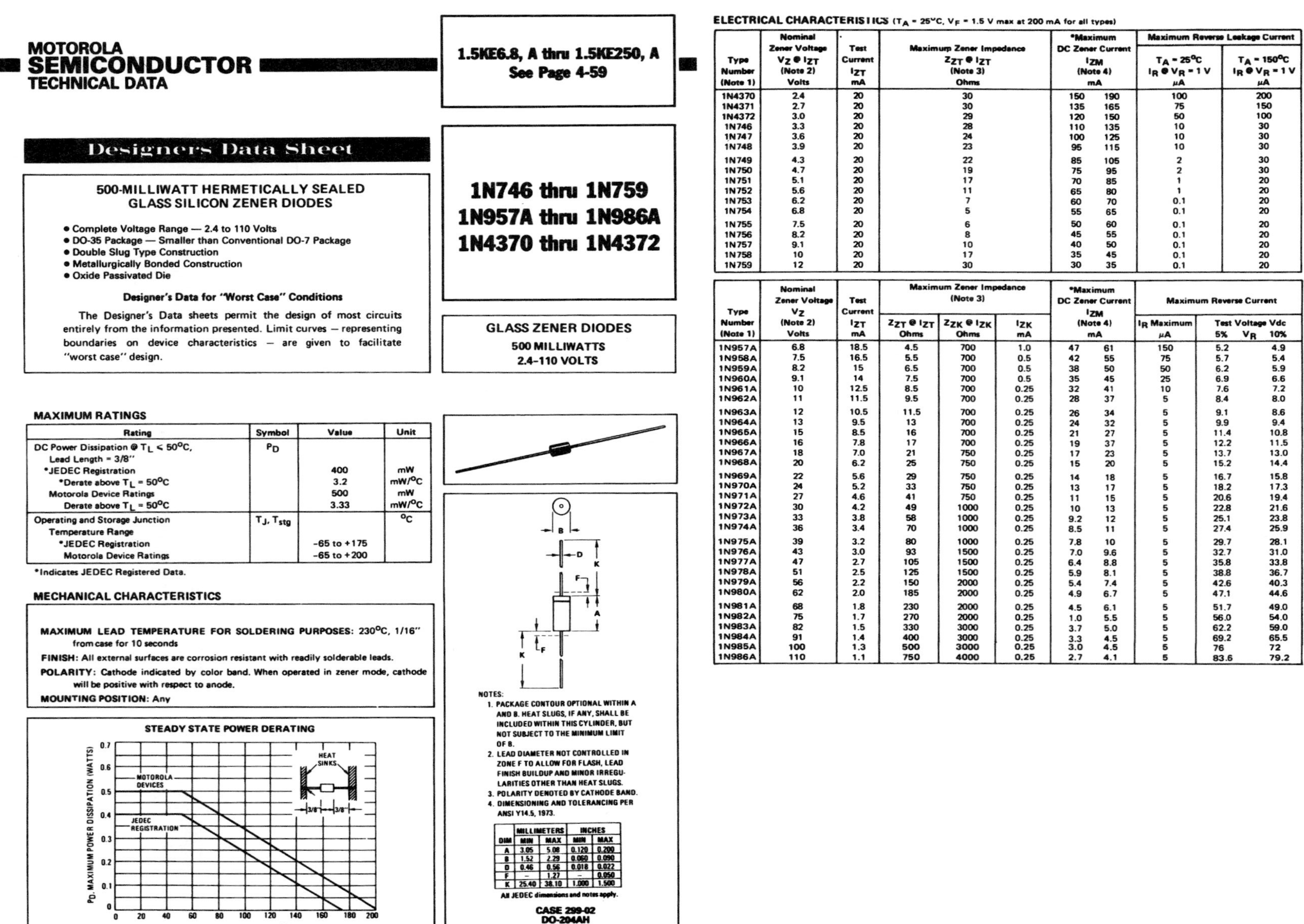

MOTOROLA
SEMICONDUCTOR
TECHNICAL DATA

1.5KE6.8, A thru 1.5KE250, A
See Page 4-59

Designers Data Sheet

1N746 thru 1N759
1N957A thru 1N986A
1N4370 thru 1N4372

500-MILLIWATT HERMETICALLY SEALED GLASS SILICON ZENER DIODES

- Complete Voltage Range — 2.4 to 110 Volts
- DO-35 Package — Smaller than Conventional DO-7 Package
- Double Slug Type Construction
- Metallurgically Bonded Construction
- Oxide Passivated Die

Designer's Data for "Worst Case" Conditions

The Designer's Data sheets permit the design of most circuits entirely from the information presented. Limit curves – representing boundaries on device characteristics – are given to facilitate "worst case" design.

GLASS ZENER DIODES
500 MILLIWATTS
2.4-110 VOLTS

MAXIMUM RATINGS

Rating	Symbol	Value	Unit
DC Power Dissipation @ $T_L \leq 50^oC$, Lead Length = 3/8"	P_D		
*JEDEC Registration		400	mW
*Derate above $T_L = 50^oC$		3.2	mW/^oC
Motorola Device Ratings		500	mW
Derate above $T_L = 50^oC$		3.33	mW/^oC
Operating and Storage Junction Temperature Range	T_J, T_{stg}		^oC
*JEDEC Registration		-65 to +175	
Motorola Device Ratings		-65 to +200	

*Indicates JEDEC Registered Data.

MECHANICAL CHARACTERISTICS

MAXIMUM LEAD TEMPERATURE FOR SOLDERING PURPOSES: 230°C, 1/16" from case for 10 seconds

FINISH: All external surfaces are corrosion resistant with readily solderable leads.

POLARITY: Cathode indicated by color band. When operated in zener mode, cathode will be positive with respect to anode.

MOUNTING POSITION: Any

STEADY STATE POWER DERATING

NOTES:
1. PACKAGE CONTOUR OPTIONAL WITHIN A AND B. HEAT SLUGS, IF ANY, SHALL BE INCLUDED WITHIN THIS CYLINDER, BUT NOT SUBJECT TO THE MINIMUM LIMIT OF B.
2. LEAD DIAMETER NOT CONTROLLED IN ZONE F TO ALLOW FOR FLASH, LEAD FINISH BUILDUP AND MINOR IRREGULARITIES OTHER THAN HEAT SLUGS.
3. POLARITY DENOTED BY CATHODE BAND.
4. DIMENSIONING AND TOLERANCING PER ANSI Y14.5, 1973.

DIM	MILLIMETERS MIN	MILLIMETERS MAX	INCHES MIN	INCHES MAX
A	3.05	5.08	0.120	0.200
B	1.52	2.29	0.060	0.090
D	0.46	0.56	0.018	0.022
F	–	1.27	–	0.050
K	25.40	38.10	1.000	1.500

All JEDEC dimensions and notes apply.

CASE 299-02
DO-204AH
GLASS

ELECTRICAL CHARACTERISTICS ($T_A = 25^oC$, $V_F = 1.5$ V max at 200 mA for all types)

Type Number (Note 1)	Nominal Zener Voltage V_Z @ I_{ZT} (Note 2) Volts	Test Current I_{ZT} mA	Maximum Zener Impedance Z_{ZT} @ I_{ZT} (Note 3) Ohms	*Maximum DC Zener Current I_{ZM} (Note 4) mA		Maximum Reverse Leakage Current $T_A = 25^oC$ I_R @ $V_R = 1$ V µA	$T_A = 150^oC$ I_R @ $V_R = 1$ V µA
1N4370	2.4	20	30	150	190	100	200
1N4371	2.7	20	30	135	165	75	150
1N4372	3.0	20	29	120	150	50	100
1N746	3.3	20	28	110	135	10	30
1N747	3.6	20	24	100	125	10	30
1N748	3.9	20	23	95	115	10	30
1N749	4.3	20	22	85	105	2	30
1N750	4.7	20	19	75	95	2	30
1N751	5.1	20	17	70	85	1	20
1N752	5.6	20	11	65	80	1	20
1N753	6.2	20	7	60	70	0.1	20
1N754	6.8	20	5	55	65	0.1	20
1N755	7.5	20	6	50	60	0.1	20
1N756	8.2	20	8	45	55	0.1	20
1N757	9.1	20	10	40	50	0.1	20
1N758	10	20	17	35	45	0.1	20
1N759	12	20	30	30	35	0.1	20

Type Number (Note 1)	Nominal Zener Voltage V_Z (Note 2) Volts	Test Current I_{ZT} mA	Maximum Zener Impedance (Note 3) Z_{ZT} @ I_{ZT} Ohms	Z_{ZK} @ I_{ZK} Ohms	I_{ZK} mA	*Maximum DC Zener Current I_{ZM} (Note 4) mA		Maximum Reverse Current I_R Maximum µA	Test Voltage Vdc V_R 5%	10%
1N957A	6.8	18.5	4.5	700	1.0	47	61	150	5.2	4.9
1N958A	7.5	16.5	5.5	700	0.5	42	55	75	5.7	5.4
1N959A	8.2	15	6.5	700	0.5	38	50	50	6.2	5.9
1N960A	9.1	14	7.5	700	0.5	35	45	25	6.9	6.6
1N961A	10	12.5	8.5	700	0.25	32	41	10	7.6	7.2
1N962A	11	11.5	9.5	700	0.25	28	37	5	8.4	8.0
1N963A	12	10.5	11.5	700	0.25	26	34	5	9.1	8.6
1N964A	13	9.5	13	700	0.25	24	32	5	9.9	9.4
1N965A	15	8.5	16	700	0.25	21	27	5	11.4	10.8
1N966A	16	7.8	17	700	0.25	19	37	5	12.2	11.5
1N967A	18	7.0	21	750	0.25	17	23	5	13.7	13.0
1N968A	20	6.2	25	750	0.25	15	20	5	15.2	14.4
1N969A	22	5.6	29	750	0.25	14	18	5	16.7	15.8
1N970A	24	5.2	33	750	0.25	13	17	5	18.2	17.3
1N971A	27	4.6	41	750	0.25	11	15	5	20.6	19.4
1N972A	30	4.2	49	1000	0.25	10	13	5	22.8	21.6
1N973A	33	3.8	58	1000	0.25	9.2	12	5	25.1	23.8
1N974A	36	3.4	70	1000	0.25	8.5	11	5	27.4	25.9
1N975A	39	3.2	80	1000	0.25	7.8	10	5	29.7	28.1
1N976A	43	3.0	93	1500	0.25	7.0	9.6	5	32.7	31.0
1N977A	47	2.7	105	1500	0.25	6.4	8.8	5	35.8	33.8
1N978A	51	2.5	125	1500	0.25	5.9	8.1	5	38.8	36.7
1N979A	56	2.2	150	2000	0.25	5.4	7.4	5	42.6	40.3
1N980A	62	2.0	185	2000	0.25	4.9	6.7	5	47.1	44.6
1N981A	68	1.8	230	2000	0.25	4.5	6.1	5	51.7	49.0
1N982A	75	1.7	270	2000	0.25	1.0	5.5	5	56.0	54.0
1N983A	82	1.5	330	3000	0.25	3.7	5.0	5	62.2	59.0
1N984A	91	1.4	400	3000	0.25	3.3	4.5	5	69.2	65.5
1N985A	100	1.3	500	3000	0.25	3.0	4.5	5	76	72
1N986A	110	1.1	750	4000	0.25	2.7	4.1	5	83.6	79.2

Figure 17.37 (Copyright of Semiconductor Components Industries, LLC. Used by permission.)

this is the case, you must calculate the value of I_{ZM} using the zener voltage rating (listed under *Electrical Characteristics*) and the dc power dissipation rating. Using these ratings, the value of I_{ZM} can be found using

$$I_{ZM} = \frac{P_D}{V_Z} \tag{17.4}$$

The use of this equation is demonstrated in Example 17.4.

Example 17.4

A 1N754 zener diode has a dc power dissipation rating of 500 mW and a zener voltage rating of 6.8 V. What is the value of I_{ZM} for the device?

Solution: Using equation (17.4), the value of I_{ZM} is found as

$$I_{ZM} = \frac{P_D}{V_Z} = \frac{500 \text{ mW}}{6.8 \text{ V}} = 73.5 \text{ mA}$$

If the total current through this diode exceeds 73.5 mA, you'll end up replacing the diode.

Practice Problem 17.4

A zener diode has a dc power dissipation rating of 1 W and a zener voltage rating of 27 V. What is the value of I_{ZM} for the device?

Power Derating Factor

The **power derating factor** (listed under *Maximum Ratings*) indicates *the decrease in the dc power dissipation capability of a device (per °C) at temperatures above a specified value*. Figure 17.37 shows a derating factor of 3.33 mW/°C for temperatures above 50°C. This means that you must decrease the dc power dissipation rating by 3.33 mW for every 1°C above 50°C. Example 17.5 demonstrates the use of this derating factor.

Example 17.5

A 1N746 is operated at a temperature of 75°C. What is the maximum dc power dissipation for the device at this temperature?

Solution: First, the total derating value is found as

$$\text{Derating value} = (3.33 \text{ mW/°C})(75°\text{C} - 50°\text{C}) = 83.25 \text{ mW}$$

We then subtract this derating value from the dc power dissipation rating to get the adjusted dc power dissipation limit, as follows:

$$P_D = 500 \text{ mW} - 83.25 \text{ mW} = 416.75 \text{ mW}$$

Thus, at 75°C, a 1N746 has a power dissipation limit of 416.75 mW.

Practice Problem 17.5

Determine the power dissipation rating of the 1N746 at 125°C. The dc power dissipation rating of the device is 500 mW and its derating factor is 3.33 mW/°C when operated above 50°C.

Electrical Characteristics

The electrical characteristics portion of the zener spec sheet shows the zener voltage, current, and impedance ratings for the entire group of diodes. If you look at the component numbers, you'll see that many end with the letter A. The letter indicates the *tolerance* of the ratings. In this case, the letter A indicates a ±5% tolerance in the ratings. When no letter follows the part number, the tolerance of the component is ±10%.

Note: The *Joint Electronic Device Engineering Council* (JEDEC) registered values are for military applications. By rating devices below their actual capabilities, JEDEC ratings provide a larger safety margin (which is required by military specification).

The maximum dc zener current (I_{ZM}) column needs a bit of explaining. First, you will notice that there are two sets of numbers. Those on the left correspond to the JEDEC registered value of 400 mW for dc power dissipation. Those on the right correspond to the manufacturer's device rating of 500 mW. The rated values of I_{ZM} are generally reduced by approximately 10% to keep you from driving the components to their absolute limits.

Component Substitution

When we were substituting one *pn*-junction diode for another, we were concerned only with whether or not the current and peak reverse voltage ratings were high enough to survive in the circuit. With zener diodes, the power rating of the substitute diode may be higher than needed, but the V_Z rating of the substitute must equal that of the component it is replacing. We cannot use a diode with a higher (or lower) V_Z rating than that of the original component.

Section Review

1. When a spec sheet does not list the value of I_{ZM}, how do you determine its value?
2. What is a *power derating factor*? How is it used?
3. Explain how you would determine whether or not one zener diode can be used in place of another.

17.11 Light-Emitting Diodes

OBJECTIVE 15 ➤ **Light-emitting diodes (LEDs)** are *diodes that emit light when biased properly.* The schematic symbol for the LED is shown in Figure 17.38. Although LEDs are available in various colors, such as infrared (which is not visible), red, green, yellow, orange, and blue, the schematic symbol is the same for all colors. The internal construction of a typical LED is illustrated in Figure 17.39.

Since LEDs have clear (or semiclear) cases, there is normally no label on the case to identify the leads. The leads are normally identified in one of the ways shown in Figure 17.40: the leads are of different lengths, one lead is flattened, or one side of the case is flattened.

LED Characteristics

LEDs have characteristic curves that are very similar to those for *pn*-junction diodes. However, they tend to have higher forward voltage (V_F) and lower reverse breakdown voltage (V_{BR}) ratings. The typical ranges for these values are as follows:

- Forward voltage: +1.2 to +4.3 V (typical).
- Reverse breakdown voltage: −3 to −10 V (typical).

Note: Reverse breakdown voltage (V_{BR}) is similar to *maximum reverse voltage* (V_{RRM}). This LED rating indicates the value of reverse voltage that can cause the reverse biased component to break down and conduct.

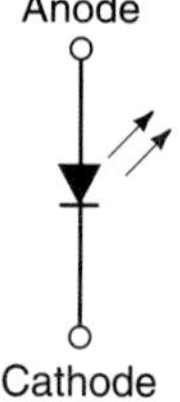

Figure 17.38 The LED schematic symbol.

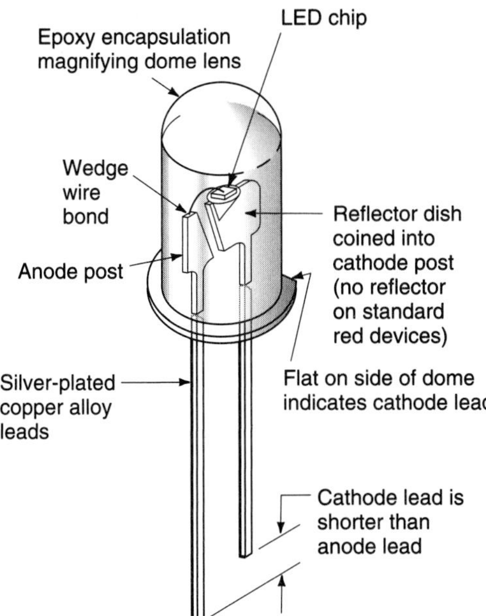

Figure 17.39 Construction features of the T-1 ¾ plastic LED lamp. (Courtesy of Agilent Technologies, Inc.)

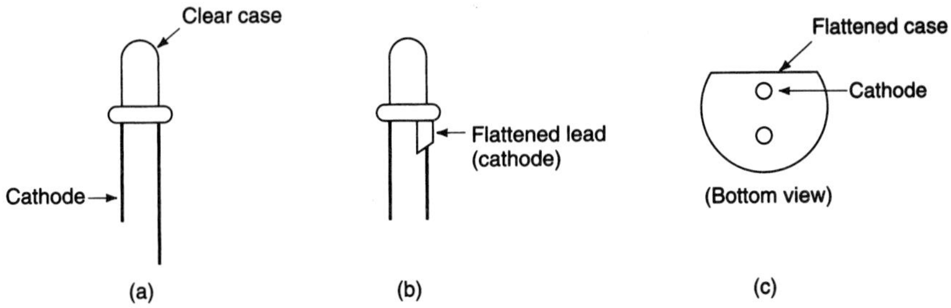

Figure 17.40

As you can see, LEDs typically have higher values of V_F than *pn*-junction diodes. Since the type of material used determines the color of light that is emitted, the forward voltage is not the same for all LEDs.

Current-Limiting Resistors

LEDs require the use of series *current-limiting resistors* because of their low maximum forward current ratings (typically 2 mA to 50 mA). A current-limiting resistor like the one shown in Figure 17.41 ensures that the maximum current rating of the LED will not be exceeded. The minimum value of R_S is determined using the following equation: ◄ *OBJECTIVE 16*

$$R_S = \frac{V_{\text{out(pk)}} - V_F}{I_F} \tag{17.5}$$

where

$V_{\text{out(pk)}}$ = the peak output voltage of the driving circuit
V_F = the *minimum* rated forward voltage for the LED
I_F = the *maximum* forward current rating for the LED

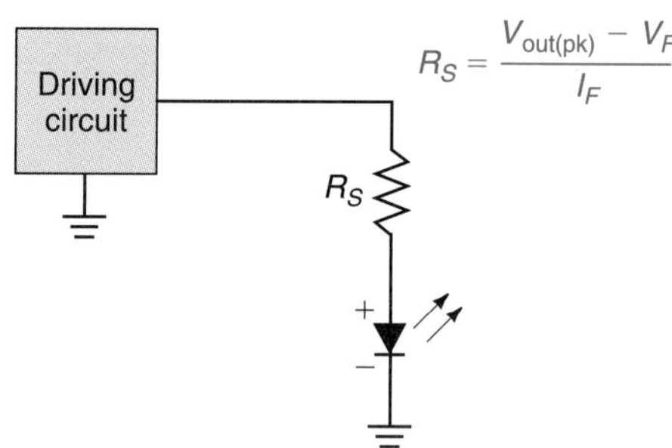

Figure 17.41 An LED with a current-limiting resistor.

To provide a safety margin, the value of I_F used in equation (17.5) should be limited to 80% of the rated value for the LED, as demonstrated in Example 17.6.

Example 17.6

The driving circuit shown in Figure 17.41 has a peak output of 8 V. The LED ratings are V_F = 1.8 to 2.0 V and I_F = 16 mA (maximum). What value of current-limiting resistor is needed in the circuit?

Solution: Using the 80% guideline, the LED forward current is found as

$$I_F = (0.8)(16 \text{ mA}) = 12.8 \text{ mA} \qquad \text{(maximum)}$$

Now the value of the current limiting resistor is found as

$$R_S = \frac{V_{out(pk)} - V_F}{I_F} = \frac{8 \text{ V} - 1.8 \text{ V}}{12.8 \text{ mA}} = 484.4 \ \Omega$$

The lowest standard-value resistor that has a value *greater than* 484.4 Ω is the 510 Ω resistor. This is the component that should be used.

Practice Problem 17.6

An LED has the following ratings: V_F = 1.4 to 1.8 V, and I_F = 12 mA (maximum). It is driven by a source with a peak value of 14 V. What is the lowest standard resistor value that can be used as R_S? (*Note:* A listing of the standard resistor values is provided Table 2.3, on page 48.)

Multicolor LEDs

Some LEDs will emit one color of light when biased in one direction, a second color when the polarity is reversed, and a third when the polarity is rapidly alternated. **Multicolor LEDs** *contain two diodes, one for each color, that are connected in reverse parallel.* When a voltage of either polarity is applied to the LED, one of the diodes is forward biased and emits light.

Multicolor LEDs are typically red when biased in one direction and green when biased in the other. If a multicolor LED is switched fast enough between the two polarities, the LED appears to produce a third color (in this case, yellow).

Section Review

1. How are the leads on an LED usually identified?
2. How do the electrical characteristics of LEDs differ from those of *pn*-junction diodes?
3. Why do LEDs need series current-limiting resistors?

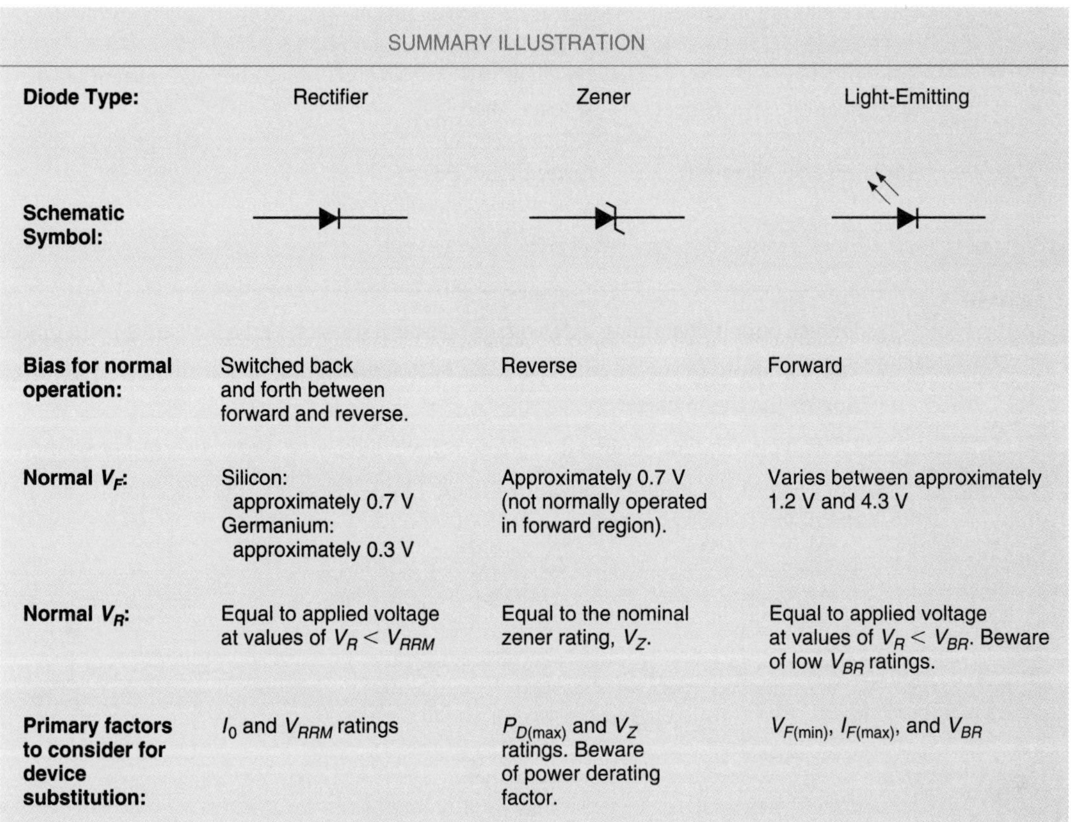

SUMMARY ILLUSTRATION

Diode Type:	Rectifier	Zener	Light-Emitting
Schematic Symbol:			
Bias for normal operation:	Switched back and forth between forward and reverse.	Reverse	Forward
Normal V_F:	Silicon: approximately 0.7 V Germanium: approximately 0.3 V	Approximately 0.7 V (not normally operated in forward region).	Varies between approximately 1.2 V and 4.3 V
Normal V_R:	Equal to applied voltage at values of $V_R < V_{RRM}$	Equal to the nominal zener rating, V_Z.	Equal to applied voltage at values of $V_R < V_{BR}$. Beware of low V_{BR} ratings.
Primary factors to consider for device substitution:	I_0 and V_{RRM} ratings	$P_{D(\max)}$ and V_Z ratings. Beware of power derating factor.	$V_{F(\min)}$, $I_{F(\max)}$, and V_{BR}

Figure 17.42

17.12 Diodes: A Comparison

The summary illustration in Figure 17.42 will help you to remember the key points about each diode covered in this chapter. If you have difficulty remembering any of the points listed, review the appropriate section.

Key Terms

The following terms were introduced in this chapter on the pages indicated:

Acceptor atom, 522
Anode, 526
Average forward current (I_0), 530
Barrier potential, 522
Bias, 523
Bulk resistance (R_B), 524
Carbon, 517
Cathode, 526
Conduction band, 517
Covalent bonding, 517
Depletion layer, 522
Diode, 526
Donor atom, 522
Doping, 519
Electron-hole pair, 518
Forward bias, 523
Forward power dissipation ($P_{D(\max)}$), 530
Forward voltage (V_F), 527
Germanium, 517
Hole, 518
Intrinsic, 519
Knee voltage (V_K), 527
Light-emitting diode (LED), 542
Majority carrier, 520
Maximum zener current (I_{ZM}), 538
Minority carrier, 520
Multicolor LED, 544
n-Type material, 519
Nominal, 539
Parameter, 529
Peak reverse voltage (V_{RRM}), 529
Pentavalent, 519
p-Type material, 519
pn-Junction diode, 526
Power derating factor, 541
Recombination, 518
Reverse bias, 523
Reverse breakdown voltage, 537
Reverse current (I_R), 533

Reverse saturation current (I_S), 533
Silicon, 517
Specification (spec) sheet, 530
Static reverse current (I_R), 539
Surface-leakage current (I_{SL}), 533
Trivalent, 519
Valence shell, 517
Voltage regulator, 537
Zener diode, 537
Zener impedance (Z_Z), 539
Zener knee current (I_{ZK}), 538
Zener test current (I_{ZT}), 538
Zener voltage (V_Z), 538

Practice Problems

1. Draw a circuit containing a dc voltage source, a resistor, and a forward-biased diode.
2. Add an arrow to the circuit you drew in Practice Problem 1 to indicate the direction of the diode current.
3. Draw a circuit containing a dc voltage source, a resistor, and a reverse-biased diode.
4. For each of the circuits shown in Figure 17.43, determine the direction (if any) of the diode forward current.

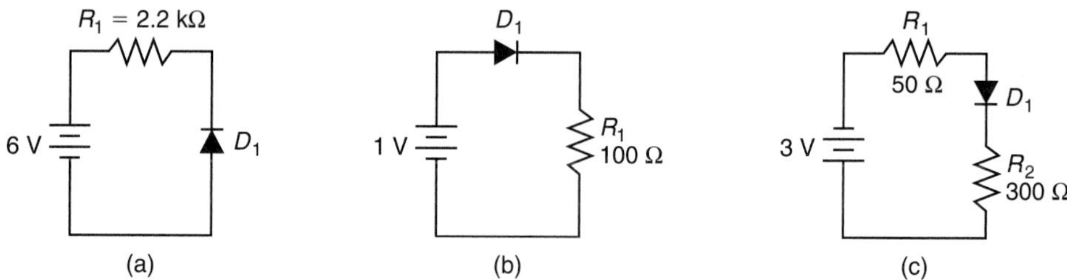

Figure 17.43

5. For each of the circuits shown in Figure 17.44, determine the direction (if any) of the diode forward current.

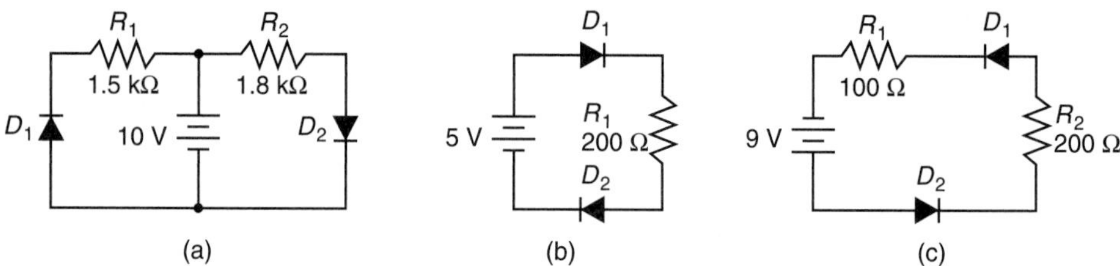

Figure 17.44

6. Determine the *ideal* voltage drop across each of the diodes shown in Figure 17.43.
7. Determine the *ideal* voltage drop across each of the components shown in Figure 17.44a.
8. Determine the values of V_{D1}, V_{R1}, and I_T for the circuit shown in Figure 17.43a.
9. Determine the values of V_{D1}, V_{R1}, and I_T for the circuit shown in Figure 17.43b.
10. Determine the values of V_{D1}, V_{R1}, V_{R2}, and I_T for the circuit shown in Figure 17.43c.
11. Determine the values of V_{D1}, V_{R1}, I_1, V_{D2}, V_{R2}, and I_2 for the circuit shown in Figure 17.44a.
12. Determine the values of V_{D1}, V_{D2}, V_{R1}, and I_T for the circuit shown in Figure 17.44b.
13. Determine the values of V_{D1}, V_{D2}, V_{R1}, V_{R2}, and I_T for the circuit shown in Figure 17.44c.
14. What is the minimum required peak reverse voltage rating for the diode shown in Figure 17.45a?

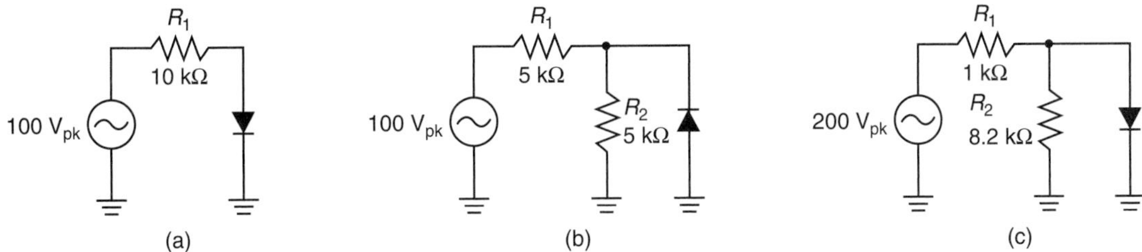

Figure 17.45

15. What is the minimum required peak reverse voltage rating for the diode shown in Figure 17.45b? (*Hint:* The resistors form a voltage divider.)
16. What is the minimum acceptable average forward current rating for the diode shown in Figure 17.46?

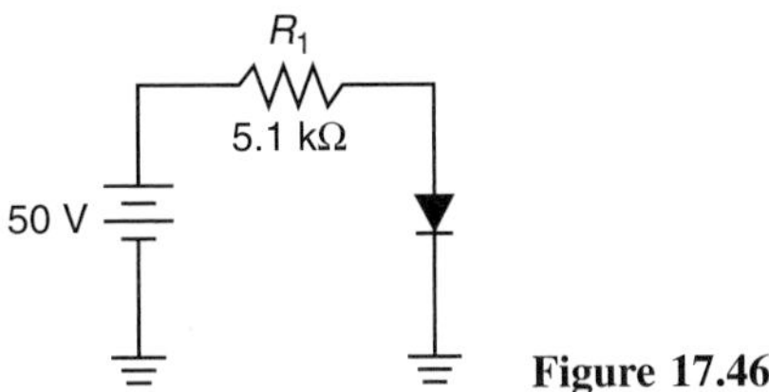

Figure 17.46

17. What is the minimum acceptable average forward power dissipation rating for the diode shown in Figure 17.46?
18. A diode has a $P_{D(\max)}$ rating of 1.2 W. What is the maximum allowable value of forward current for the device?
19. A diode has a $P_{D(\max)}$ rating of 750 mW. What is the limit on the value of average forward current for the device?
20. A silicon diode has a forward current of 10 mA and a bulk resistance (R_B) of 5 Ω. What is the V_F for the device?
21. A silicon diode has a forward current of 8.2 mA and a bulk resistance (R_B) of 12 Ω. What is the actual value V_F for the device?
22. A silicon diode has a bulk resistance of 20 Ω. At what value of I_F will the value of V_F equal 0.8 V?
23. Refer to Figure 17.45a. The diode in the circuit shown has a maximum rated value of $I_R = 10\ \mu A$ at 25°C. Assuming that I_R reaches its maximum value at each negative peak of the input cycle, what voltage will be measured across R_1 when I_R peaks? (Assume that $T = 25°C$.)
24. Refer to the diode spec sheet shown in Figure 17.47. In terms of maximum reverse voltage ratings, which of the diodes listed could be used in the circuit shown in Figure 17.45a?
25. Refer to the spec sheet shown in Figure 17.47. What is the maximum value of I_R for the 1N5398 at $T = 150°C$?
26. Refer to the spec sheet shown in Figure 17.47. What is the surge current rating for the 1N5391?
27. The chart shown in Figure 17.48 is called a *selector guide.* It is designed to help you find the appropriate diode for a given application by cross-referencing the I_O and V_{RRM} requirements for the component. Which of the diodes listed has the minimum acceptable ratings for use in a circuit with an average forward current of 24.5 A and a 225 V_{pk} source?

MOTOROLA
SEMICONDUCTOR
TECHNICAL DATA

1N5391 thru 1N5399

Designers Data Sheet

"SURMETIC" RECTIFIERS

. . . subminiature size, axial lead-mounted rectifiers for general-purpose, low-power applications.

Designers Data for "Worst Case" Conditions

The Designers Data Sheets permit the design of most circuits entirely from the information presented. Limits curves—representing boundaries on device characteristics—are given to facilitate "worst-case" design.

LEAD-MOUNTED SILICON RECTIFIERS

50-1000 VOLTS DIFFUSED JUNCTION

*MAXIMUM RATINGS

Rating	Symbol	1N5391	1N5392	1N5393	1N5395	1N5397	1N5398	1N5399	Unit
Peak Repetitive Reverse Voltage Working Peak Reverse Voltage DC Blocking Voltage	V_{RRM} V_{RWM} V_R	50	100	200	400	600	800	1000	Volts
Nonrepetitive Peak Reverse Voltage (Halfwave, Single Phase, 60 Hz)	V_{RSM}	100	200	300	525	800	1000	1200	Volts
RMS Reverse Voltage	$V_{R(RMS)}$	35	70	140	280	420	560	700	Volts
Average Rectified Forward Current (Single Phase, Resistive Load, 60 Hz, $T_L = 70°C$, 1/2" From Body)	I_O	← 1.5 →							Amp
Nonrepetitive Peak Surge Current (Surge Applied at Rated Load Conditions, See Figure 2)	I_{FSM}	← 50 (for 1 cycle) →							Amp
Storage Temperature Range	T_{stg}	← -65 to +175 →							°C
Operating Temperature Range	T_L	← -65 to +170 →							°C
DC Blocking Voltage Temperature	T_L	← 150 →							°C

*ELECTRICAL CHARACTERISTICS

Characteristic and Conditions	Symbol	Typ	Max	Unit
Maximum Instantaneous Forward Voltage Drop ($i_F = 4.7$ Amp Peak, $T_L = 170°C$, 1/2 Inch Leads)	v_F	–	1.4	Volts
Maximum Reverse Current (Rated dc Voltage) ($T_L = 150°C$)	I_R	250	300	μA
Maximum Full-Cycle Average Reverse Current (1) ($I_O = 1.5$ Amp, $T_L = 70°C$, 1/2 Inch Leads)	$I_{R(AV)}$	–	300	μA

*Indicates JEDEC Registered Data.

NOTE 1: Measured in a single-phase, halfwave circuit such as shown in Figure 6.25 of EIA RS-282, November 1963. Operated at rated load conditions $I_O = 1.5$ A, $V_r = V_{RWM}$, $T_L = 70°C$.

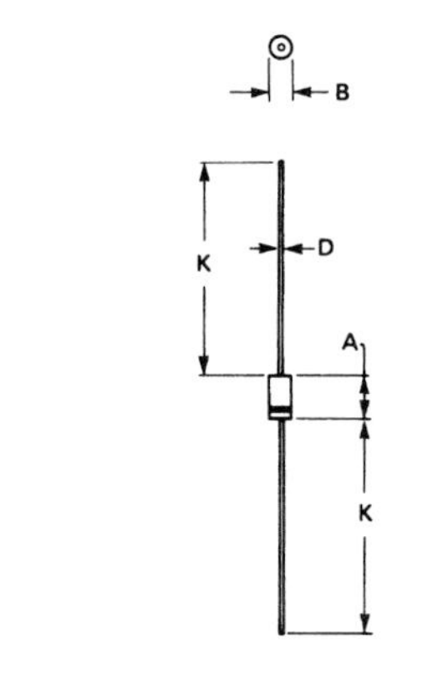

NOTES:
1. ALL RULES AND NOTES ASSOCIATED WITH JEDEC DO-41 OUTLINE SHALL APPLY.
2. POLARITY DENOTED BY CATHODE BAND.
3. LEAD DIAMETER NOT CONTROLLED WITHIN "F" DIMENSION.

DIM	MILLIMETERS MIN	MILLIMETERS MAX	INCHES MIN	INCHES MAX
A	5.97	6.60	0.235	0.260
B	2.79	3.05	0.110	0.120
D	0.76	0.86	0.030	0.034
K	27.94	—	1.100	—

CASE 59-04 PLASTIC

MECHANICAL CHARACTERISTICS

CASE: Transfer molded plastic

MAXIMUM LEAD TEMPERATURE FOR SOLDERING PURPOSES: 240°C, 1/8" from case for 10 seconds at 5 lbs. tension

FINISH: All external surfaces are corrosion-resistant, leads are readily solderable

POLARITY: Cathode indicated by color band

WEIGHT: 0.40 grams (approximately)

Figure 17.47 (Copyright of Semiconductor Components Industries, LLC. Used by permission.)

V_{RRM} (Volts)	I_O, AVERAGE RECTIFIED FORWARD CURRENT (Amperes) 1.0	1.5	3.0			6.0
	59-03 (DO-41) Plastic 	59-04 Plastic 	60-01 Metal 	267-03 Plastic 	267-02 Plastic 	194-04 Plastic 
50	†1N4001	**1N5391	1N4719	**MR500	1N5400	MR750
100	†1N4002	**1N5392	1N4720	**MR501	1N5401	MR751
200	†1N4003	1N5393 *MR5059	1N4721	**MR502	1N5402	MR752
400	†1N4004	1N5395 *MR5060	1N4722	**MR504	1N5404	MR754
600	†1N4005	1N5397 *MR5061	1N4723	**MR506	1N5406	MR756
800	†1N4006	1N5398	1N4724	MR508		MR758
1000	†1N4007	1N5399	1N4725	MR510		MR760
I_{FSM} (Amps)	30	50	300	100	200	400
T_A @ Rated I_O (°C)	75	$T_L = 70$	75	95	$T_L = 105$	60
T_C @ Rated I_O (°C)						
T_J (Max) (°C)	175	175	175	175	175	175

† Package Size: 0.120″ Max Diameter by 0.260″ Max Length.
* 1N5059 series equivalent Avalanche Rectifiers.
** Avalanche versions available, consult factory.

V_{RRM} (Volts)	I_O, AVERAGE RECTIFIED FORWARD CURRENT (Amperes) 12	20	24	25	30		40	50
	245A-02 (DO-203AA) Metal 		339-02 Plastic Note 1 	193-04 Plastic Note 2 	43-02 (DO-21) Metal 		42A-01 (DO-203AB) Metal 	43-04 Metal 
50	MR1120 1N1199,A,B	MR2000	MR2400	MR2500	1N3491	1N3659	1N1183A	MR5005
100	MR1121 1N1200,A,B	MR2001	MR2401	MR2501	1N3492	1N3660	1N1184A	MR5010
200	MR1122 1N1202,A,B	MR2002	MR2402	MR2502	1N3493	1N3661	1N1186A	MR5020
400	MR1124 1N1204,A,B	MR2004	MR2404	MR2504	1N3495	1N3663	1N1188A	MR5040
600	MR1126 1N1206,A,B	MR2006	MR2406	MR2506		Note 3	1N1190A	Note 3
800	MR1128	MR2008		MR2508		Note 3	Note 3	Note 3
1000	MR1130	MR2010		MR2510		Note 3	Note 3	Note 3
I_{FSM} (Amps)	300	400	400	400	300	400	800	600
T_A @ Rated I_O (°C)								
T_C @ Rated I_O (°C)	150	150	125	150	130	100	150	150
T_J (Max) (°C)	190	175	175	175	175	175	190	195

Note 1. Meets mounting configuration of TO-220 outline.
Note 2. Request Data Sheet for Mounting Information.
Note 3. Available on special order.

Figure 17.48 (Copyright of Semiconductor Components Industries, LLC. Used by permission.)

28. Refer to Figure 17.48. A circuit has an average forward current of 3.6 A and a 170 V_{pk} source. Which diode has the minimum acceptable ratings for use in this circuit?

29. Refer to Figure 17.48. A circuit has an average forward power dissipation (for the diode) of 2.8 W and a 470 V_{pk} source. Which diode has the minimum acceptable ratings for use in this circuit?

30. A zener diode spec sheet lists values of $I_{ZT} = 20$ mA. If a ΔI_Z of 1 mA results in a ΔV_Z of 25 mV, what is the value of zener impedance for the device?

31. Refer to Figure 17.49. In each circuit shown, determine whether or not the biasing voltage has the correct polarity for normal zener operation.

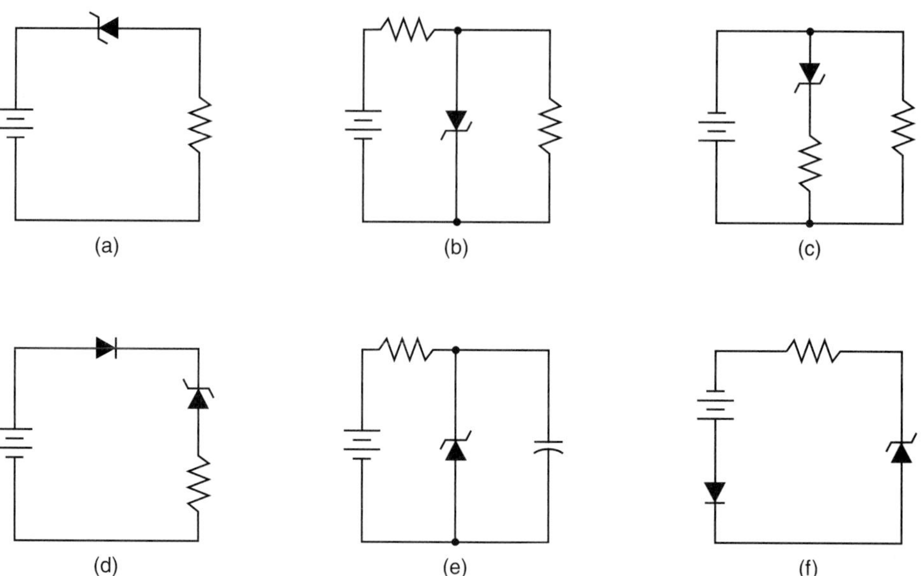

Figure 17.49

32. For each of the properly biased zener diodes shown in Figure 17.49, draw an arrow indicating the direction of the zener current.

33. A 6.8 V zener diode has a $P_{D(\text{max})}$ rating of 1 W. What is the value of I_{ZM} for the device?

34. A 24 V zener diode has a $P_{D(\text{max})}$ rating of 10 W. What is the value of I_{ZM} for the device?

35. A zener diode with a $P_{D(\text{max})}$ rating of 5 W has a derating factor of 8 mW/°C above 50°C. What is the maximum allowable value of P_D for the device at 120°C?

36. The MLL4678 zener diode has a $P_{D(\text{max})}$ rating of 250 mW and a derating factor of 1.67 mW/°C above 50°C. What is the maximum allowable value of P_D for the device at 150°C?

37. Refer to Figure 17.37. Which diode is rated at $V_Z = 27$ V?

38. Refer to Figure 17.37. What is the practical limit on the value of I_Z for the 1N957A?

39. Refer to Figure 17.37. Which of the diodes listed could be used in place of a 12 V zener diode that has a maximum operating current of 15 mA?

40. An LED is rated at $V_F = 1.5$ to 1.8 V and $I_F = 18$ mA (maximum). If the LED is driven by a 20 V_{pk} source, what standard value of current-limiting resistor is needed to protect the LED?

41. An LED is rated at $V_F = 1.6$ to 2.0 V and $I_F = 20$ mA (maximum). Determine the minimum standard resistor value that could be used as a current-limiting resistor if the LED is driven by a 32 V_{pk} source.

Looking Back

These problems relate to material presented in earlier chapters. The chapters are identified in brackets.

42. A sine wave has a peak value of 80 V. Calculate the instantaneous waveform voltage at $\theta = 60°$. [Chapter 9]

43. A sine wave has a peak value of 120 V and a cycle time of 20 ms. Calculate the instantaneous waveform voltage at $t = 5$ ms. [Chapter 9]

44. Derive the Thevenin equivalent of the circuit shown in Figure 17.50. [Chapter 7]

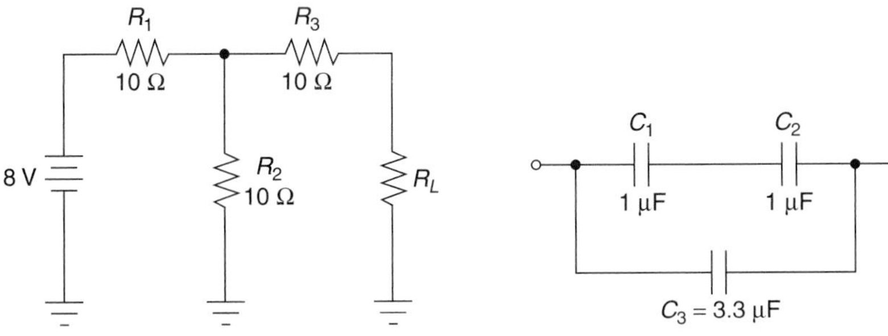

Figure 17.50 **Figure 17.51**

45. Calculate the total capacitance for the circuit shown in Figure 17.51. [Chapter 12]

46. A 100 V_{pk} source is connected to a 200 Ω load. Calculate the value of load power for the circuit. [Chapter 9]

47. A series LC circuit has values of $L = 3.3$ mH, $R_w = 2$ Ω, and $C = 10$ nF. Calculate the Q of the circuit when operated at resonance. [Chapter 14]

48. Calculate the value of resistance that will dissipate 100 mW when connected to a 20 V_{dc} source. [Chapter 3]

Pushing the Envelope

49. Calculate the total current for the circuit shown in Figure 17.52.

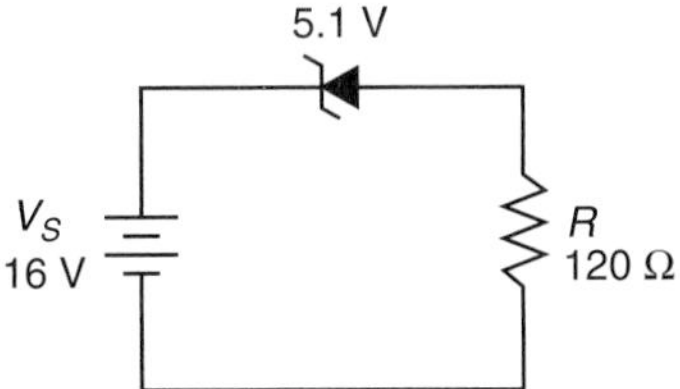

Figure 17.52

50. Calculate the power being dissipated by the zener diode shown in Figure 17.53.

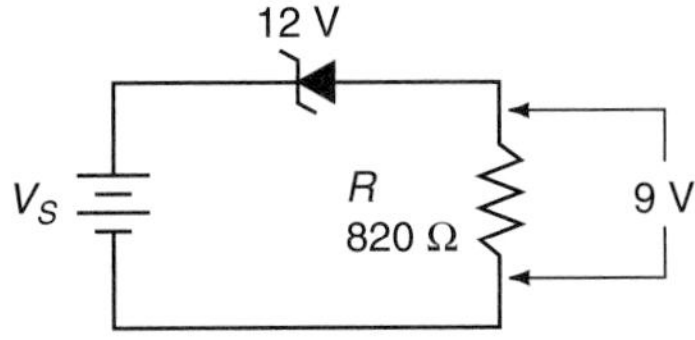

Figure 17.53

51. Explain how you could measure the value of R in Figure 17.54 without disconnecting the diode.

52. The spec sheet for the MLL755 zener diode is shown in Figure 17.55. What is the I_{ZM} for the device at 150°C?

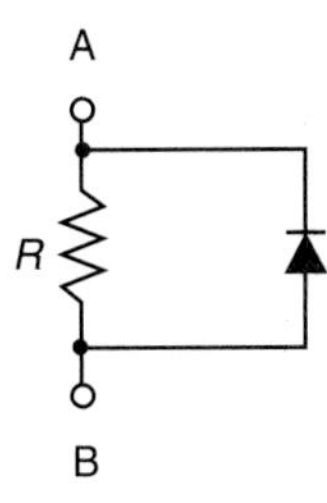

Figure 17.54

MOTOROLA
SEMICONDUCTOR
TECHNICAL DATA

MLL746 thru MLL759

MLL957A thru MLL986A

MLL4370 thru MLL4372

500 MILLIWATT HERMETICALLY SEALED GLASS SILICON ZENER DIODES

- Complete Voltage Range — 2.4 to 110 Volts
- Leadless Package for Surface Mount Technology
- Double Slug Type Construction
- Metallurgically Bonded Construction
- Nitride Passivated Die
- Available in 8 mm Tape and Reel
  T1 Cathode Facing Sprocket Holes
  T2 Anode Facing Sprocket Holes

LEADLESS GLASS ZENER DIODES

500 MILLIWATTS 2.4–110 VOLTS

MAXIMUM RATINGS

Rating	Symbol	Value	Unit
DC Power Dissipation @ $T_A \leq 50°C$ Derate above $T_A = 50°C$	P_D	500 3.3	mW mW/°C
Operating and Storage Junction Temperature Range	T_J, T_{stg}	−65 to +200	°C

MECHANICAL CHARACTERISTICS

CASE: Double slug type, hermetically sealed glass

MAXIMUM LEAD TEMPERATURE FOR SOLDERING PURPOSES: 230°C, for 10 seconds

FINISH: All external surfaces are corrosion resistant and readily solderable

POLARITY: Cathode indicated by color band. When operated in zener mode, cathode will be positive with respect to anode

MOUNTING POSITION: Any

STEADY STATE POWER DERATING

P_D, MAXIMUM POWER DISSIPATION (WATTS): 0, 0.2, 0.4, 0.6, 0.8, 1.0, 1.2, 1.4

T_A, LEAD TEMPERATURE (°C) AMBIENT: 0, 20, 40, 60, 80, 100, 120, 140, 160, 180, 200

P_D vs T_C

P_D vs T_A

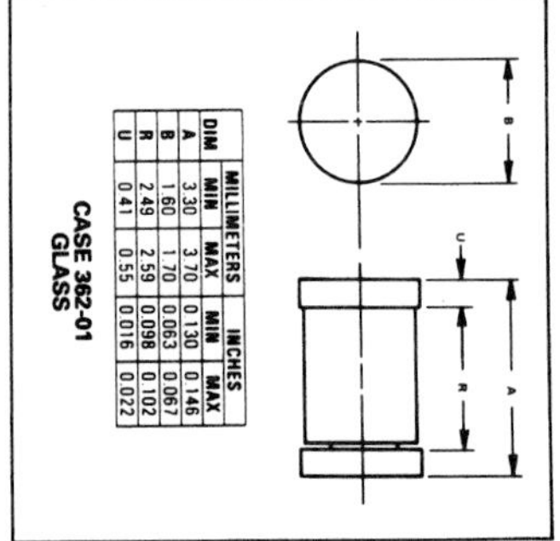

DIM	MILLIMETERS MIN	MILLIMETERS MAX	INCHES MIN	INCHES MAX
A	3.30	3.70	0.130	0.146
B	1.60	1.70	0.063	0.067
R	2.49	2.59	0.098	0.102
U	0.41	0.55	0.016	0.022

CASE 362-01 GLASS

ELECTRICAL CHARACTERISTICS (T_A = 25°C, V_F = 1.5 V Max @ 200 mA for all types)

Type Number (Note 1)	Nominal Zener Voltage V_Z @ I_{ZT} (Notes 1,2,3) Volts	Test Current I_{ZT} (Note 2) mA	Maximum Zener Impedance Z_{ZT} @ I_{ZT} (Note 4) Ohms	Maximum DC Zener Current I_{ZM} mA		Maximum Reverse Leakage Current T_A = 25°C I_R @ V_R = 1 V μA	T_A = 150°C I_R @ V_R = 1 V μA
MLL4370	2.4	20	30	150	190	100	200
MLL4371	2.7	20	30	135	165	75	150
MLL4372	3.0	20	29	120	150	50	100
MLL746	3.3	20	28	110	135	10	30
MLL747	3.6	20	24	100	125	10	30
MLL748	3.9	20	23	95	115	10	30
MLL749	4.3	20	22	85	105	2	30
MLL750	4.7	20	19	75	95	2	30
MLL751	5.1	20	17	70	85	1	20
MLL752	5.6	20	11	65	80	1	20
MLL753	6.2	20	7	60	70	0.1	20
MLL754	6.8	20	5	55	65	0.1	20
MLL755	7.5	20	6	50	60	0.1	20
MLL756	8.2	20	8	45	55	0.1	20
MLL757	9.1	20	10	40	50	0.1	20
MLL758	10	20	17	35	45	0.1	20
MLL759	12	20	30	30	35	0.1	20

Figure 17.55 (Copyright of Semiconductor Components Industries, LLC. Used by permission.)

53. The MLL756 zener diode cannot be used in the circuit shown in Figure 17.56. Why not? (*Note:* The temperature range shown is the normal operating temperature for the circuit.)

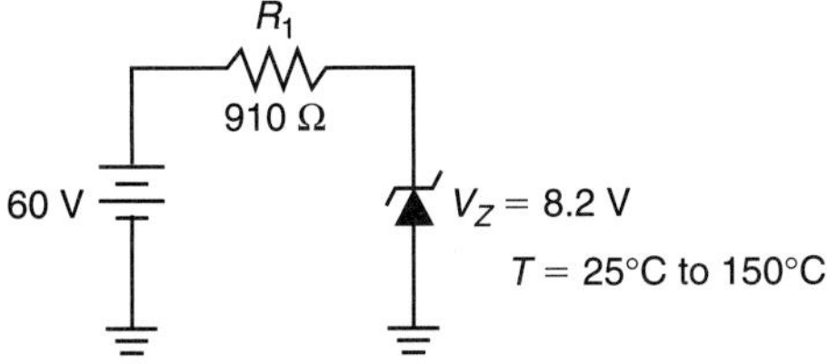

Figure 17.56

Answers to the Example Practice Problems

17.1 $V_{R_1} = 4.3$ V, $I_T = 8.43$ mA

17.2 The value of I_T for the circuit is 1.95 A. The rating for the diode would have to be 20% greater than this value, or 2.34 A.

17.3 $P_D = 198.7 \text{ mW} \times 1.2 = 238.4$ mW

17.4 $I_{ZM} = 37$ mA

17.5 P_D (at 125°C) = 252.5 mW

17.6 $R_S = 1.5$ kΩ minimum (standard value). The calculated value is 1313 Ω.

Multisim File References

Figure	File Reference
17.13	ETF17_13.msm
17.18	ETF17_18.msm
17.41	ETF17_41.msm

chapter 18

Basic Diode Circuits

Objectives

After studying the material in this chapter, you should be able to:

1. Describe the operation of the *half-wave rectifier.*
2. Calculate the peak and average load voltage and current for a positive half-wave rectifier.
3. Describe the operation of the *full-wave rectifier.*
4. Calculate the peak and average load voltage and current for any full-wave rectifier.
5. Describe the operation of the *bridge rectifier.*
6. Calculate the peak and average load voltage and current for a bridge rectifier.
7. Describe the effects of *filtering* on the output from a rectifier.
8. Describe the operation of *series clippers.*
9. Describe and analyze the operation of *shunt clippers.*
10. Describe and analyze the operation of *biased shunt clippers.*
11. Describe the effects of *negative clampers* and *positive clampers* on an input waveform.
12. Describe the circuit operation of a clamper.
13. Describe and analyze the operation of the *half-wave voltage doubler.*
14. Describe and analyze the operation of the *full-wave voltage doubler.*
15. Describe and analyze the operation of the *full-wave voltage tripler.*
16. Describe and analyze the operation of the *full-wave voltage quadrupler.*

It would take several volumes to discuss every diode application in modern electronics. In this chapter, we will concentrate on some of the most common diode circuits, including *rectifiers, clippers, clampers,* and *voltage multipliers.*

18.1 Half-Wave Rectifiers

Rectifiers are *diode circuits that convert ac into pulsating dc.* These circuits are used primarily in dc power supplies. A **dc power supply** is *a group of circuits used to convert the ac energy provided by the wall outlet into dc energy.* The power cord of any electronic system supplies the line power to the dc power supply, which then provides all the internal dc voltages needed for proper circuit operation.

A linear dc power supply can be broken down into the four circuit groups shown in Figure 18.1. The overall circuit operation can be described as follows:

- The 120 V_{ac} input is either stepped up or down by a transformer.
- The ac output from the transformer is converted into pulsating dc by a *diode rectifier.*
- The rectifier output is then applied to a *filter,* which reduces the variations in dc voltage.
- The output from the filter is applied to a *voltage regulator,* which maintains a constant output voltage

There are three types of rectifiers: *half-wave, full-wave,* and *bridge.* In this chapter, we will limit our discussion to the operation and analysis of rectifiers and basic filters. Voltage regulator operation is covered in detail in Chapter 25.

Positive and Negative Half-Wave Rectifiers

The **half-wave rectifier** is simply *a diode that is placed in series between a transformer and its load to eliminate either the negative or positive alternation of the input.* As shown in Figure 18.2: ◄ OBJECTIVE 1

- The output from a *positive half-wave rectifier* is a series of positive pulses.
- The output from a *negative half-wave rectifier* is a series of negative pulses.

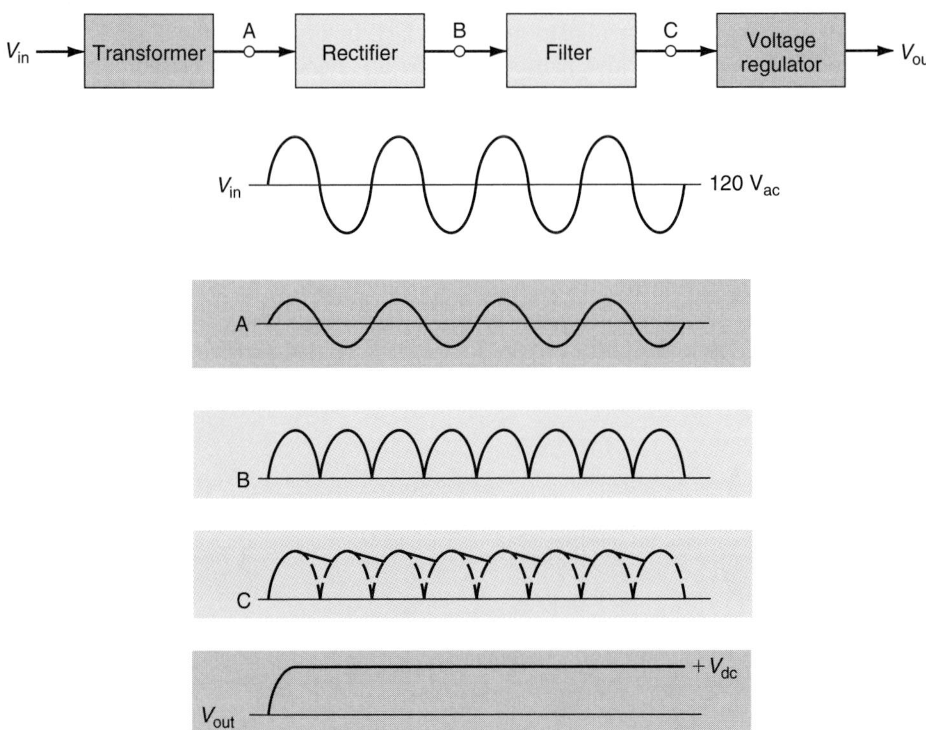

Figure 18.1 Basic power supply block diagram and waveforms.

Note: The half-wave rectifier requires few components and is inexpensive to produce. However, the circuit is not very efficient, so it is normally used for noncritical, low-current applications.

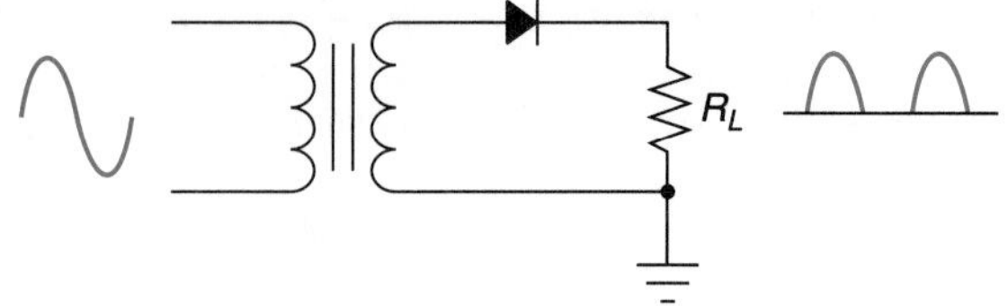

Positive half-wave rectifier

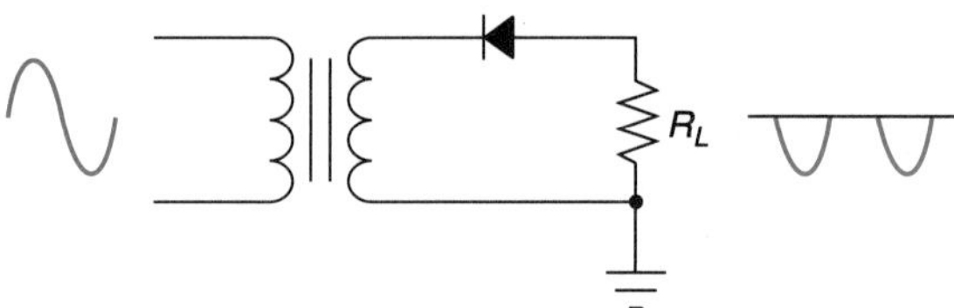

Negative half-wave rectifier

Figure 18.2 Half-wave rectifiers.

Basic Circuit Operation

A Practical Consideration: When we talk about *positive* and *negative* half-cycles of the input, we mean the polarity of the transformer secondary voltage measured from the top of the secondary to ground.

Figure 18.3 details the operation of a *positive half-wave rectifier* for one complete input cycle. During the positive half-cycle, D_1 is forward biased and conducts. This develops a voltage across R_L that is approximately equal to source voltage. During the negative half-cycle, D_1 is reverse biased and $I_T \cong 0$ A. Therefore, $V_L = 0$ V, and the voltage across the diode (V_{D1}) is approximately equal to V_S.

These relationships are illustrated further by the input/output waveforms shown in Figure 18.4. During T_1, the diode is forward biased, and $V_L \cong V_S$. During T_2, the diode is reverse biased, and $V_L \cong 0$ V.

Negative Half-Wave Rectifiers

Figure 18.5 shows a half-wave rectifier with the diode reversed. Unlike the positive half-wave rectifier, the anode in the negative half-wave rectifier is connected to the load, and the cathode is connected to the transformer secondary. In this circuit, the diode is forward biased (conducts) on the negative half-cycle of the input and reverse biased on the positive half-cycle. As a result, the positive half-cycle of the input is eliminated. Other than

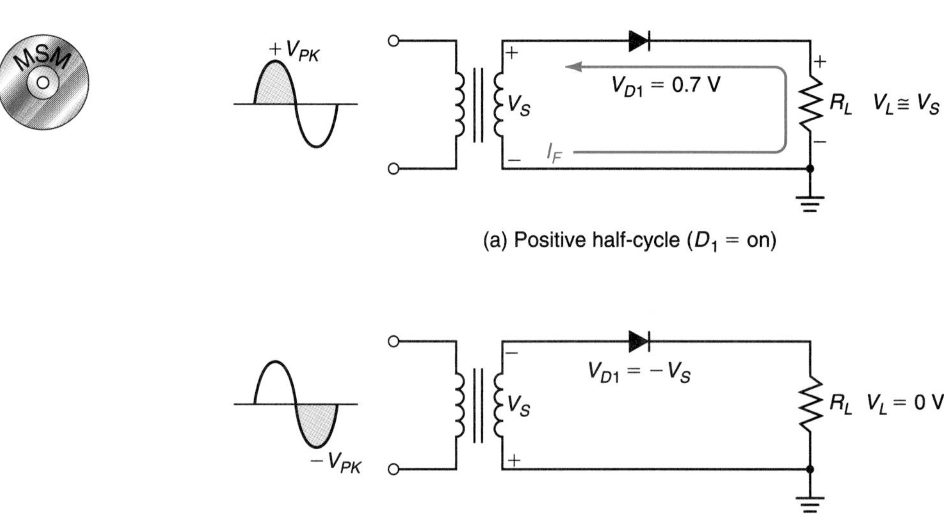

(b) Negative half-cycle (D_1 = off)

Figure 18.3 Ideal half-wave rectifier operation.

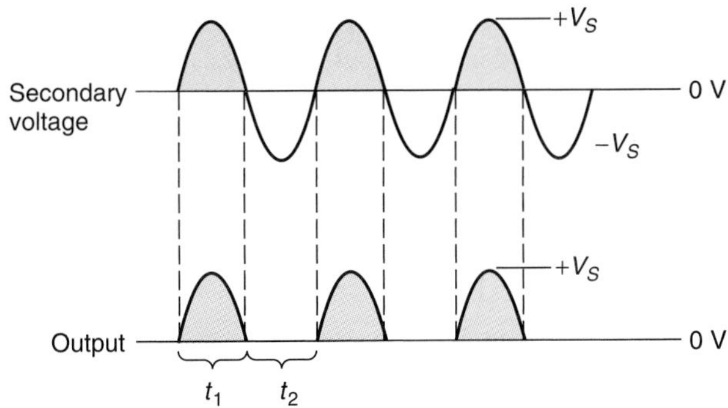

Figure 18.4 Input and output waveforms.

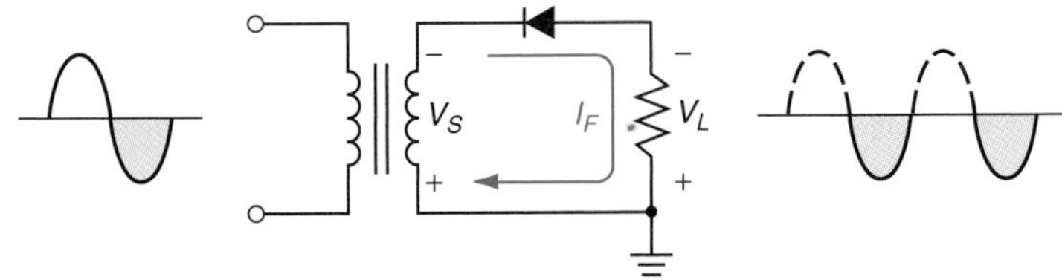

Figure 18.5 Negative half-wave rectifier operation.

that, however, the circuit works exactly like a positive half-wave rectifier. Positive and negative half-wave rectifiers are compared in Figure 18.6.

Calculating Load Voltage and Current Values

◄ *OBJECTIVE 2*

To this point, we have ignored the diode forward voltage (V_F) in the half-wave rectifier. When we take the value of V_F into account, the peak load voltage is found as

$$V_{L(\mathrm{pk})} = V_{S(\mathrm{pk})} - V_F \tag{18.1}$$

where

$V_{S(\mathrm{pk})}$ = the peak transformer secondary voltage

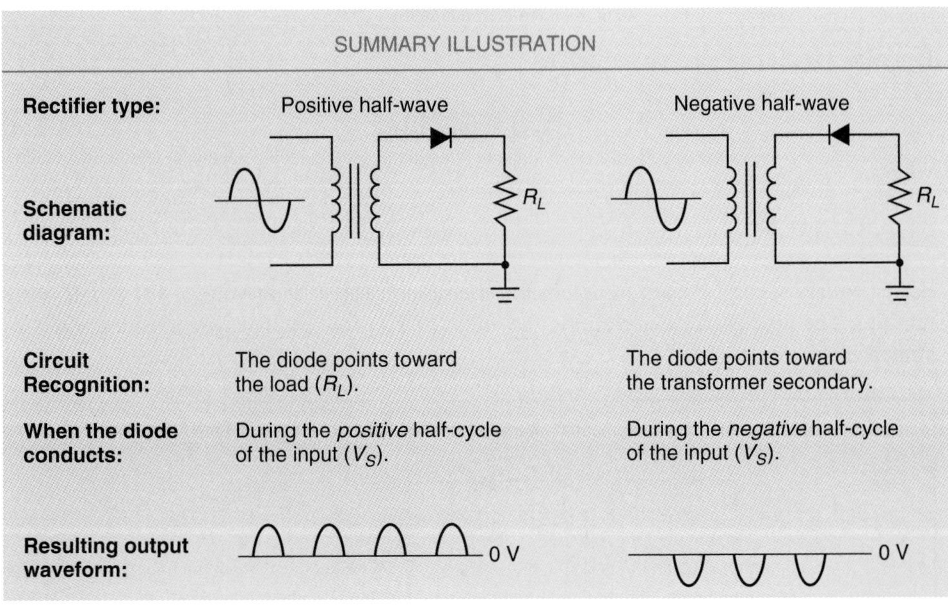

SUMMARY ILLUSTRATION
Rectifier type:
Positive half-wave
Negative half-wave
Schematic diagram:
R_L
R_L
Circuit Recognition:
The diode points toward the load (R_L).
The diode points toward the transformer secondary.
When the diode conducts:
During the *positive* half-cycle of the input (V_S).
During the *negative* half-cycle of the input (V_S).
Resulting output waveform:
0 V
0 V

Figure 18.6

More often than not, transformers are rated for a specific rms output voltage. For example, a 25 V transformer provides an output of 25 V_{rms} when supplied from a 120 V wall outlet. When a transformer is rated for a specific output voltage, simply divide the rated output voltage by 0.707 to obtain the value of $V_{S(pk)}$, as shown in Example 18.1.

Example 18.1

Determine the peak load voltage for the circuit shown in Figure 18.7.

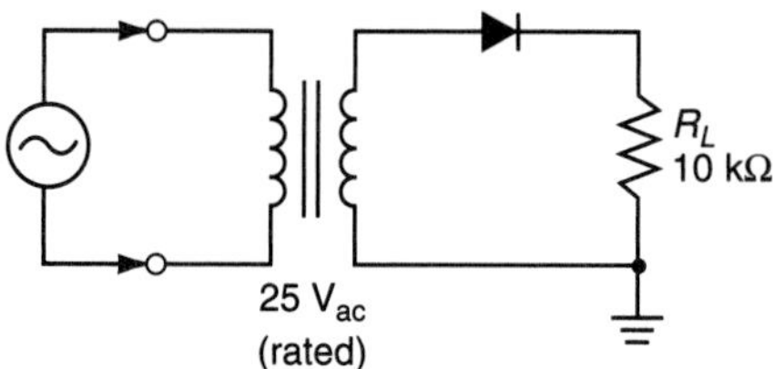

Figure 18.7

Solution: The transformer output is rated at 25 V. Because this is an rms value, the peak secondary voltage is found as

$$V_{S(pk)} = \frac{V_{S(rms)}}{0.707} = \frac{25\ V_{ac}}{0.707} = 35.4\ V$$

Now the value of $V_{L(pk)}$ is found as

$$V_{L(pk)} = V_{S(pk)} - 0.7\ V = 35.4\ V - 0.7\ V = 34.7\ V$$

Practice Problem 18.1

A 24 V_{ac} transformer is being used in a positive half-wave rectifier. What is the value of $V_{L(pk)}$ for the circuit?

Average Load Voltage and Current

The **average load voltage (V_{ave})** produced by a rectifier is *the dc average of the circuit output pulses.* As such, V_{ave} indicates the reading that you would get if you connected a dc voltmeter across the rectifier's load. For a half-wave rectifier, V_{ave} is found as

$$V_{ave} = \frac{V_{pk}}{\pi} \qquad (18.2)$$

Equation (18.2) can also be written as

$$V_{ave} = 0.318V_{pk} \qquad (18.3)$$

where

$0.318 \cong 1/\pi$.

Either equation can be used to determine the dc equivalent load voltage for a half-wave rectifier. The process for determining the value of V_{ave} for a half-wave rectifier is demonstrated in Example 18.2.

Example 18.2

Determine the value of V_{ave} for the circuit shown in Figure 18.8.

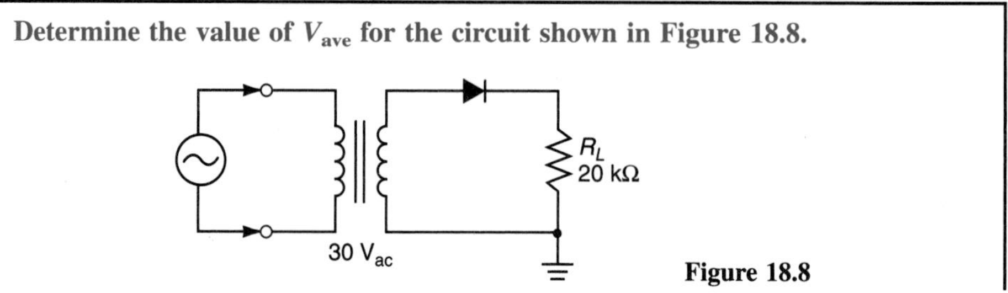

Figure 18.8

Solution: The peak secondary voltage is found as

$$V_{S(\text{pk})} = \frac{V_S}{0.707} = \frac{30\ \text{V}_{\text{ac}}}{0.707} = 42.4\ \text{V}$$

The peak load voltage is found as

$$V_{L(\text{pk})} = V_{S(\text{pk})} - 0.7\ \text{V} = 42.4\ \text{V} - 0.7\ \text{V} = 41.7\ \text{V}$$

and the average load voltage is found as

$$V_{\text{ave}} = \frac{V_{L(\text{pk})}}{\pi} = \frac{41.7\ \text{V}}{\pi} = 13.3\ \text{V}$$

Practice Problem 18.2

A half-wave rectifier like the one shown in Figure 18.8 has an output rated at 18 V_{ac}. What is the dc load voltage for the circuit?

When the value of V_{ave} is known, we can use Ohm's law to solve for the average load current (I_{ave}). For the circuit in Example 18.2,

$$I_{\text{ave}} = \frac{V_{\text{ave}}}{R_L} = \frac{13.3\ \text{V}}{20\ \text{k}\Omega} = 665\ \mu\text{A}$$

Knowing the value of I_{ave} is important when substituting one diode for another. You may recall that the maximum dc forward current that can be drawn through a diode is equal to the *average forward current* (I_0) rating of the device. When substituting one diode for another, you must make sure that the value of I_{ave} for the diode in the circuit is less than the I_0 rating of the substitute component.

Negative Half-Wave Rectifiers

The mathematical analysis of a negative half-wave rectifier is nearly identical to that for a positive half-wave rectifier. The only difference is that all the voltage polarities are reversed. Here is a simple approach for analyzing a negative half-wave rectifier:

- Analyze the circuit as if it were a positive half-wave rectifier.
- After completing your calculations, change all your calculated voltages from positive to negative.

Peak Inverse Voltage (PIV)

The maximum amount of reverse bias that will be applied to a diode in a given circuit is called the **peak inverse voltage (PIV).** For the half-wave rectifier, the PIV is found as

$$\text{PIV} = V_{S(\text{pk})} \qquad \text{(half-wave rectifier)} \qquad (18.4)$$

The basis for this equation can be seen by referring back to Figure 18.3. When the diode is reverse biased (see Figure 18.3b), no voltage is developed across the load. Therefore, all of V_S is dropped across the diode.

The PIV of a given circuit is important, because it determines the minimum allowable value of V_{RRM} for any diode used in the circuit.

Section Review

1. Describe the difference between the output waveforms of a *positive* and a *negative* half-wave rectifier.
2. How can you determine the output polarity of a half-wave rectifier from its schematic diagram?

3. Briefly explain the operation of a *positive* half-wave rectifier.

4. What type of meter is used to measure the *average load voltage* (V_{ave}) produced by a half-wave rectifier?

5. What diode parameter relates to the average load current (I_{ave}) in a half-wave rectifier? What is the relationship between the two?

6. What diode parameter relates to the peak inverse voltage (PIV) in a half-wave rectifier? What is the relationship between the two?

7. What component voltage does the PIV for a half-wave rectifier equal?

Critical Thinking

8. Based on the waveforms shown in Figure 18.4, why do you think the half-wave rectifier is viewed as being *inefficient?*

9. According to equation (9.5), the average value of one sine wave alternation is found as $V_{ave} = \frac{2V_{pk}}{\pi}$. Using this relationship and the output waveform shown in Figure 18.4, explain the relationship given in equation (18.2).

18.2 Full-Wave Rectifiers

OBJECTIVE 3 ➤ The **full-wave rectifier** consists of *two diodes that are connected to a center-tapped transformer,* as shown in Figure 18.9a. The result of this circuit construction is illustrated in Figure 18.9b. Note that the full-wave rectifier produces two half-cycles out for every one produced by the half-wave rectifier. As you will see, the center-tapped transformer plays a major role in the operation of a full-wave rectifier.

Basic Circuit Operation

Figure 18.10 illustrates the operation of the full-wave rectifier during one complete input cycle. During the positive half-cycle, D_1 is forward biased, and D_2 is reverse biased. Note the direction of current through the load (R_L). If we assume the diode to be *ideal,* the peak load voltage can be found as

$$V_{L(pk)} = \frac{V_{S(pk)}}{2} \tag{18.5}$$

Note that V_L is one-half the value of V_S, because the transformer is center tapped. (The voltage from one end of the transformer to the center tap is always one-half of the total secondary voltage.)

When the polarity of the input reverses, D_2 is forward biased, and D_1 is reverse biased. Note that *the direction of the load current has not changed,* even though the polarity of

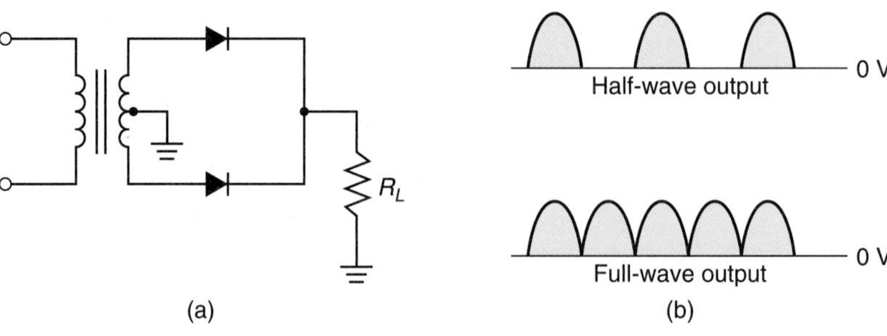

Figure 18.9 The full-wave rectifier.

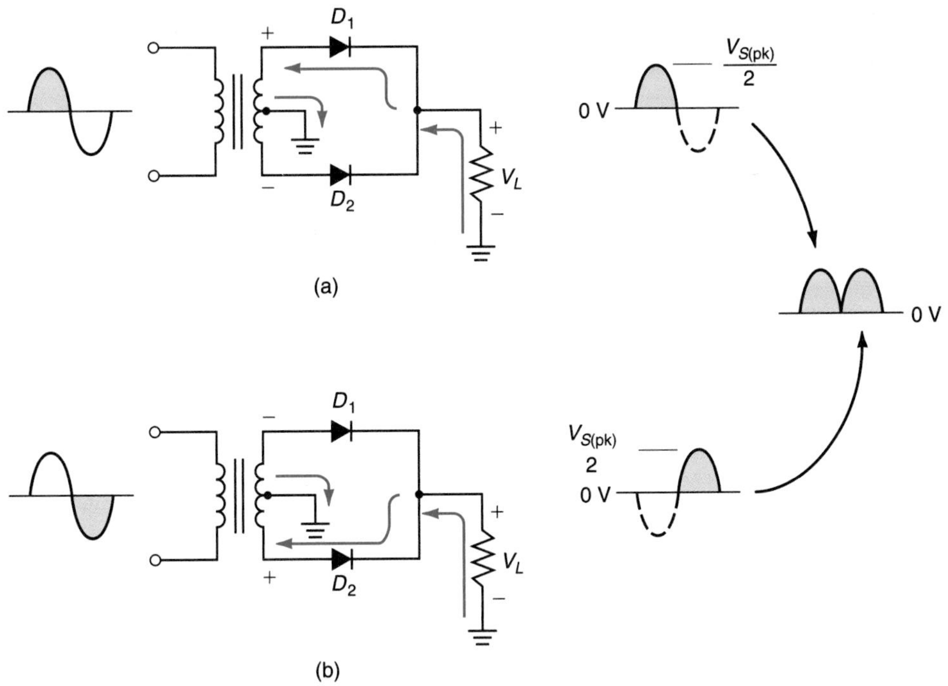

Figure 18.10
Full-wave operation.

the transformer secondary has. The combination of these two half-cycles gives the full-wave output waveform shown in the figure.

Calculating Load Voltage and Current Values

OBJECTIVE 4

When the diode forward voltage is taken into account, the peak load voltage for a full-wave rectifier is found as

$$V_{L(\text{pk})} = \frac{V_{S(\text{pk})}}{2} - 0.7\text{ V} \tag{18.6}$$

The full-wave rectifier produces twice as many output pulses (per input cycle) as the half-wave rectifier does. For this reason, the average load voltage for the full-wave rectifier is found as

$$V_{\text{ave}} = \frac{2V_{L(\text{pk})}}{\pi} \tag{18.7}$$

or

$$V_{\text{ave}} = 0.637V_{L(\text{pk})} \tag{18.8}$$

where

$0.637 \cong 2/\pi$.

The value of V_{ave} for a full-wave rectifier is calculated as shown in Example 18.3.

Example 18.3

Determine the dc load voltage for the circuit shown in Figure 18.11.

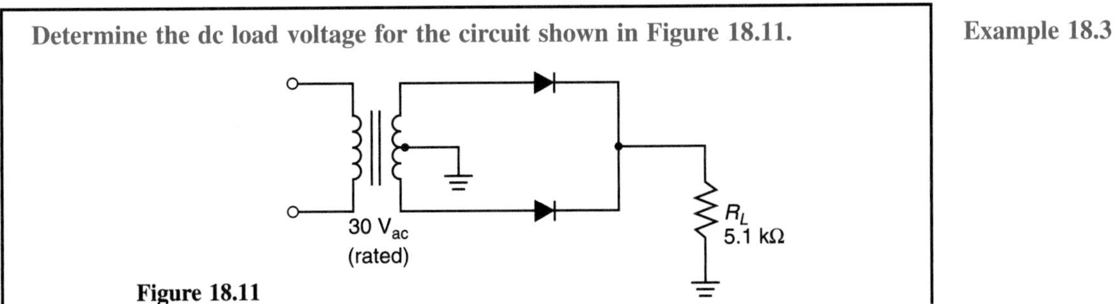

Figure 18.11

Solution: The value of $V_{S(pk)}$ is found as

$$V_{S(pk)} = \frac{30 \text{ V}_{ac}}{0.707} = 42.4 \text{ V}$$

The peak load voltage is now found as

$$V_{L(pk)} = \frac{V_{S(pk)}}{2} - 0.7 \text{ V} = 21.2 \text{ V} - 0.7 \text{ V} = 20.5 \text{ V}$$

Finally, the dc load voltage is found as

$$V_{ave} = \frac{2V_{L(pk)}}{\pi} = \frac{41 \text{ V}}{\pi} = 13.1 \text{ V}$$

Practice Problem 18.3

A full-wave rectifier is fed by a 24 V_{ac} center-tapped transformer. What is the dc load voltage for the circuit?

Once the peak and average load voltage values are known, the values of $I_{L(pk)}$ and I_{ave} can be determined using Ohm's law.

Negative Full-Wave Rectifiers

If we reverse the directions of the diodes in a positive full-wave rectifier, we have a negative full-wave rectifier. A negative full-wave rectifier and its output waveform are shown in Figure 18.12.

The mathematical analysis of a negative full-wave rectifier uses the same approach that we used for the positive half-wave rectifier, with one exception. The circuit values are calculated using the same relationships, but the calculated voltages are changed from positive to negative.

Peak Inverse Voltage

A Practical Consideration: It may seem confusing to have three different equations for PIV, however:

1. Equations (18.9) and (18.10) are effectively saying the same thing.
2. The 0.7 V drop included in equation (18.11) becomes important when dealing with low-voltage rectifiers (e.g., those typically found in personal computers).

When one of the diodes in a full-wave rectifier is reverse biased, the voltage across that diode is approximately equal to $V_{S(pk)}$. By formula,

$$\text{PIV} \cong V_{S(pk)} \tag{18.9}$$

For example, in Figure 18.13, the secondary develops peak values of +12 V and −12 V (measured from the center tap). With the polarities shown, D_1 is conducting, and D_2 is reverse biased. Ignoring the voltage across D_1, the cathode of D_2 is at +12 V, and its anode is at −12V. Thus, the total voltage across the diode is 24 V, the value of $V_{S(pk)}$.

The peak load voltage for the full-wave rectifier is approximately one-half the peak secondary voltage. Therefore, it is also accurate to say that

$$\text{PIV} \cong 2V_{L(pk)} \tag{18.10}$$

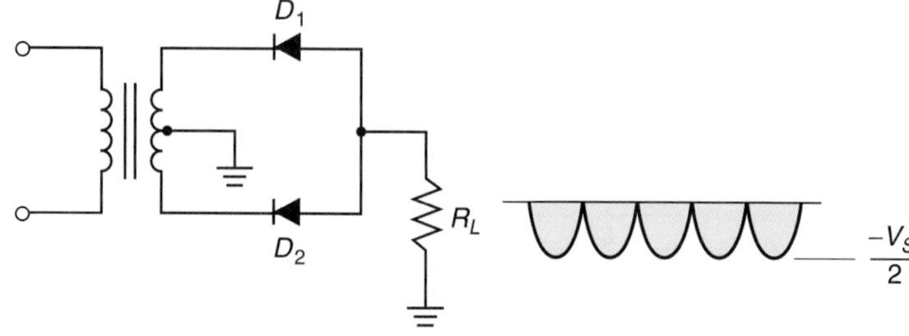

Figure 18.12
Negative full-wave rectifier.

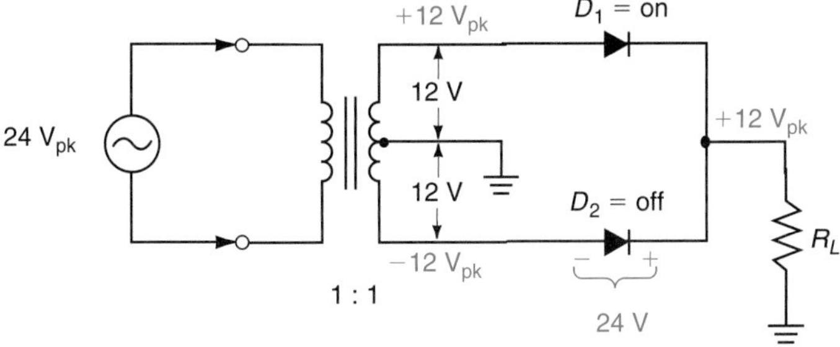

Figure 18.13 Full-wave rectifier PIV.

Up to this point, we have ignored the 0.7 V drop across the conducting diode. Therefore, a more accurate value of PIV can be obtained using

$$\text{PIV} = V_{S(\text{pk})} - 0.7\text{ V} \qquad (18.11)$$

Section Review

1. Briefly explain the operation of a full-wave rectifier.
2. How do you determine the PIV across each diode in a full-wave rectifier?
3. Briefly discuss the similarities and differences between the *half-wave rectifier* and the *full-wave rectifier*.
4. Why does a full-wave rectifier have twice the *efficiency* of a half-wave rectifier with the same peak load voltage? Critical Thinking

18.3 Full-Wave Bridge Rectifiers

The *bridge rectifier* is the most commonly used full-wave rectifier circuit for several reasons:

- It does not require the use of a center-tapped transformer.
- With the same source voltage, it produces nearly twice the peak output voltage of a full-wave center-tapped rectifier.

The bridge rectifier consists of four diodes connected as shown in Figure 18.14.

Circuit Operation

OBJECTIVE 5

As you know, the full-wave rectifier produces its output by alternating circuit conduction between two diodes. The **bridge rectifier** is *a full-wave rectifier that alternates conduction between two diode pairs,* as shown in Figure 18.15. When D_1 and D_3 are on, D_2 and D_4 are off, and vice versa. During the positive half-cycle, D_1 and D_3 are forward biased and conduct. Note the direction of the load current and the polarity of the load voltage.

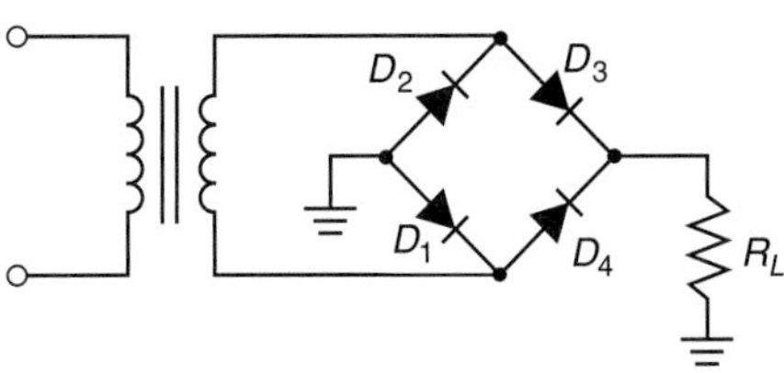

Figure 18.14 Bridge rectifier.

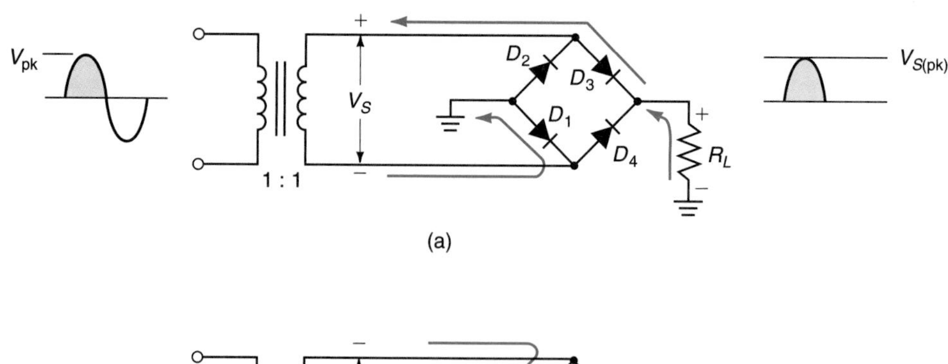

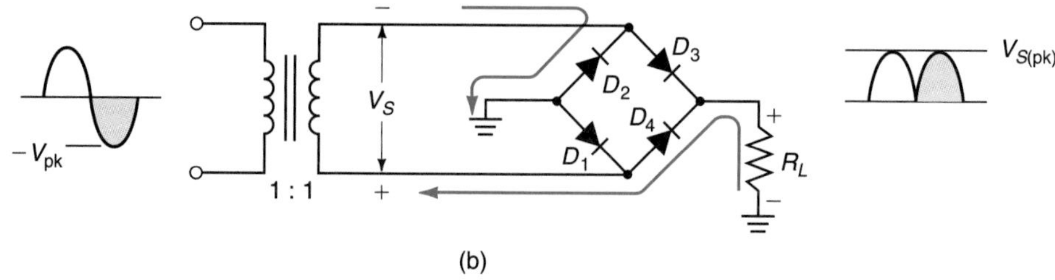

Figure 18.15 Bridge rectifier operation.

During the negative half-cycle, D_2 and D_4 conduct. Note that the direction of the load current has not changed, nor has the resulting polarity of the load voltage.

Calculating Load Voltage and Current Values

OBJECTIVE 6 ➤ Assuming the diodes to be *ideal* components, the peak output voltage from a bridge rectifier is approximately equal to V_S. For example, Figure 18.15a shows D_1 and D_3 conducting. If we assume these components are ideal conductors:

- The upper side of the load is connected directly to the top of the transformer secondary (via D_1).
- The lower side of the load is connected directly to the bottom of the transformer secondary (via D_2 and the ground connection).

Therefore, $V_{L(pk)} \cong V_{S(pk)}$. Taking into account the values of V_F across the two conducting diodes, a more accurate value of $V_{L(pk)}$ for the circuit is found as

$$V_{L(pk)} = V_{S(pk)} - 1.4 \text{ V} \tag{18.12}$$

The 1.4 V value represents the sum of the diode forward voltages. When the peak load voltage for a bridge rectifier is known, the average load values are found using the same relationships that we used for the full-wave rectifier. This point is demonstrated in Example 18.4.

Example 18.4

Determine the dc load voltage and current values for the circuit shown in Figure 18.16.

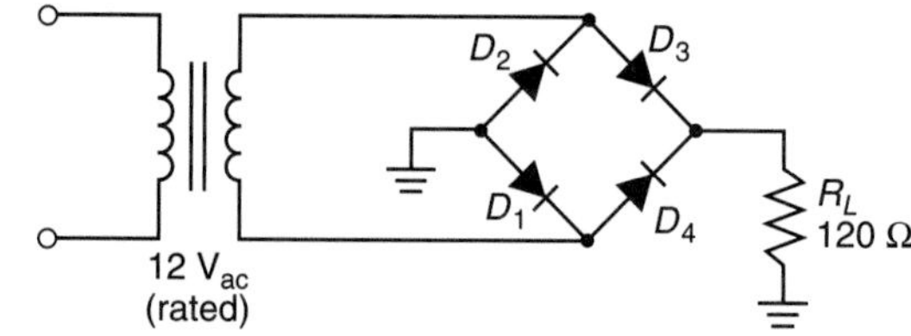

Figure 18.16

Solution: The peak secondary voltage is found as

$$V_{S(pk)} = \frac{12 \text{ V}_{ac}}{0.707} = 16.97 \text{ V}$$

The peak load voltage is now found as

$$V_{L(\text{pk})} = V_{S(\text{pk})} - 1.4 \text{ V} = 16.97 \text{ V} - 1.4 \text{ V} = 15.57 \text{ V}$$

The dc load voltage is found as

$$V_{\text{ave}} = \frac{2V_{L(\text{pk})}}{\pi} = \frac{(2)(15.57 \text{ V})}{\pi} = 9.91 \text{ V}$$

Finally, the dc load current is found as

$$I_{\text{ave}} = \frac{V_{\text{ave}}}{R_L} = \frac{9.91 \text{ V}}{120\ \Omega} = 82.6 \text{ mA}$$

Practice Problem 18.4

A bridge rectifier with a 1.2 kΩ load is fed by an 18 V_{ac} transformer. Determine the dc load voltage and current for the circuit.

Bridge Versus Full-Wave Rectifiers

The bridge rectifier provides twice the average output voltage and current values as a comparable full-wave rectifier. For example, the circuit in Example 18.4 has a 12 V_{ac} transformer and a 120 Ω load. If we were to calculate the peak and average values for a *full-wave* rectifier with the same transformer and load ratings, we would obtain the values listed in the Table 18.1.

As shown in Table 18.1, the output from the bridge rectifier is approximately twice the output from the comparable full-wave rectifier.

Peak Inverse Voltage (PIV)

The PIV of each diode in the bridge rectifier is approximately equal to V_S. Figure 18.17 helps to illustrate this point. If we assume that the conducting diodes are *ideal* components, they can be represented by wires (as shown in the figure).

The positive side of the secondary shown in Figure 18.17 is labeled A, and the negative side is labeled B. Connecting the common A points and the common B points gives us the circuit shown in Figure 18.17b. This equivalent circuit shows that the transformer secondary is connected across the parallel reverse-biased diodes. Thus, the PIV across

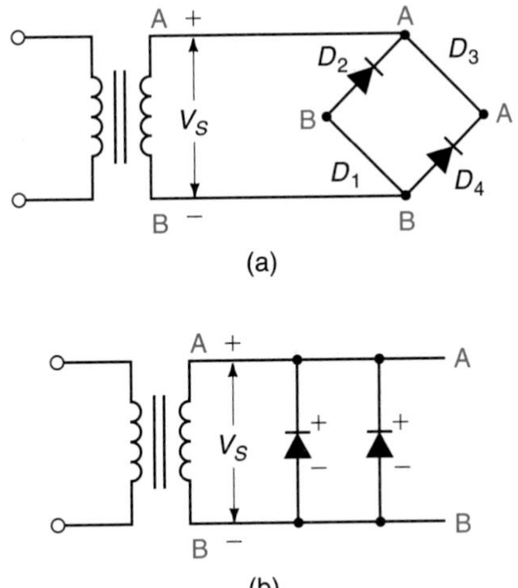

Figure 18.17 Bridge rectifier PIV.

Note: Since each "off" diode is in series with a conducting diode, the actual PIV equals ($V_{S(\text{pk})}$ − 0.7 V).

TABLE 18.1 The Peak and Average Values for the Circuit shown in Figure 18.16 and a Comparable Full-Wave Rectifier

Value	*Bridge Rectifier*	*Full-Wave Rectifier*
Peak load voltage ($V_{L(\text{pk})}$)	15.57 V	7.79 V
dc Load voltage (V_{ave})	9.91 V	4.96 V
dc Load current (I_{ave})	82.6 mA	41.3 mA

each diode is approximately equal to V_S. This same principle holds true for D_1 and D_3 when they are reverse biased.

Putting It All Together

The three commonly used rectifiers are the *half-wave, full-wave,* and *bridge* rectifiers. The output characteristics of these circuits are summarized in Figure 18.18.

The *half-wave rectifier* is the simplest of the three circuits. Even though it is commonly connected to the output of a transformer, this circuit can be connected directly to the ac line input. In other words, it does not require the use of a transformer for its operation.

The *full-wave rectifier* uses two diodes in conjunction with a center-tapped transformer. Since the center-tapped transformer is a vital part of the circuit operation, this rectifier cannot be connected directly to the ac line input.

The *bridge rectifier* is the most commonly used of the three circuits for several reasons:

- The peak and average output values are twice those of a comparable full-wave rectifier.
- The bridge rectifier (like the half-wave circuit) can be connected directly to an ac line input.

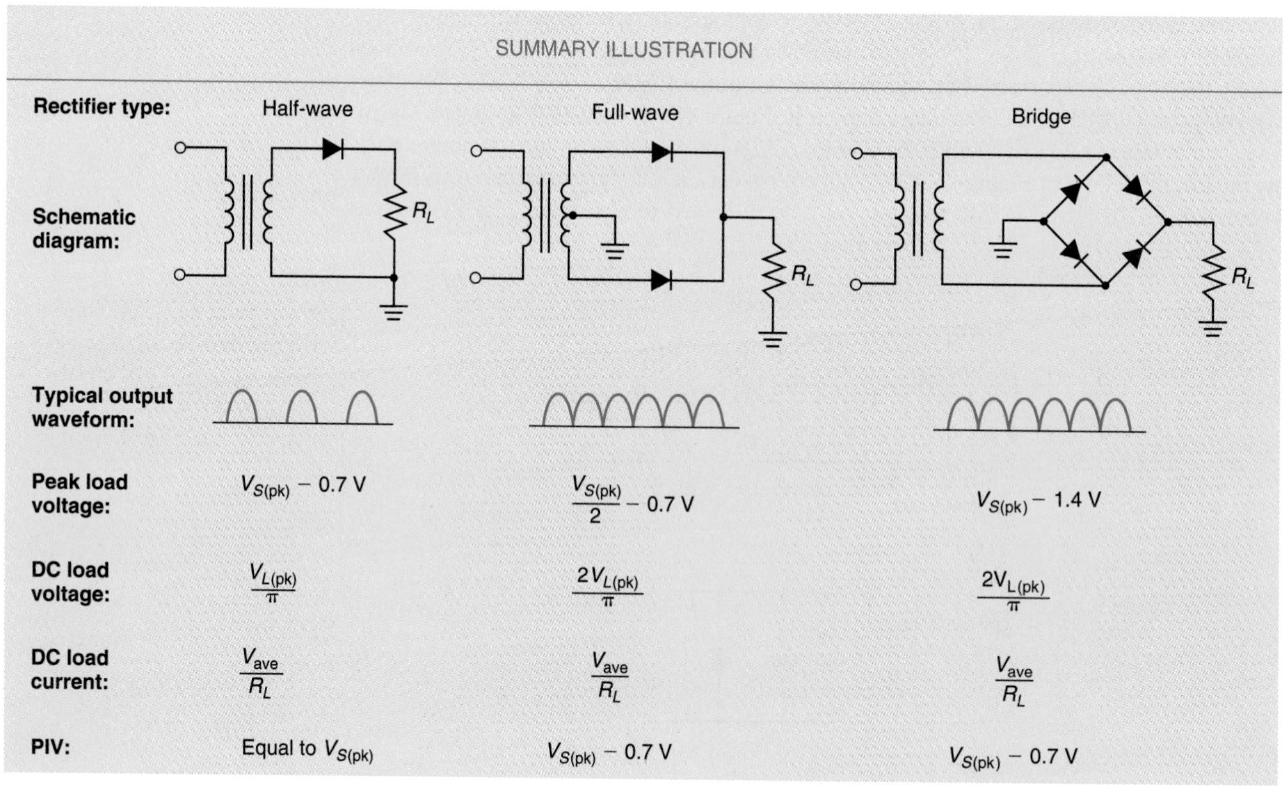

Figure 18.18

Section Review

1. Describe the operation of the *bridge rectifier.*
2. List the advantages that the bridge rectifier has over the full-wave rectifier.
3. Describe the method for determining the PIV for a diode in a bridge rectifier.

18.4 Filtered Rectifiers

The circuit that follows the rectifier in a power supply is the *filter* (see Figure 18.1). A **power supply filter** is *a circuit that reduces the variations in the rectifier output signal.*

◄ OBJECTIVE 7

The overall result of using a filter is illustrated in Figure 18.19. Note that voltage variations still remain after filtering; however, *the amplitude of those variations has been greatly reduced. The variation in the output voltage of a filter* is called **ripple voltage (V_R).** Power supplies are designed to produce as little ripple voltage as possible.

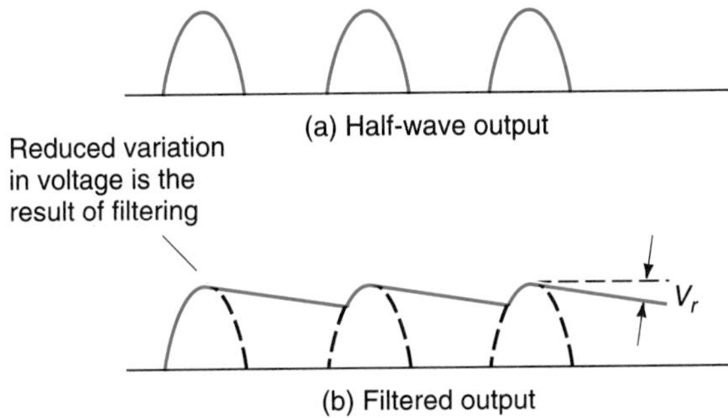

Figure 18.19 The effects of filtering on the output of a half-wave rectifier.

Basic Capacitive Filter

The most common type of filter is the **capacitive filter,** which is simply *a capacitor that is connected in parallel with the load.* A basic capacitive filter is shown in 18.20. The filtering action is based on the charge/discharge action of the capacitor. During the positive half-cycle, D_1 conducts, and the capacitor charges (as shown in Figure 18.20a). As the input starts to go negative, D_1 turns off, and the capacitor slowly discharges through the load resistance (as shown in Figure 18.20b). As the output from the rectifier drops below the charged voltage of the capacitor, the capacitor acts as the voltage source for the load.

The charge and discharge times of the filter capacitor are unequal, because there are two distinct *RC* time constants in the circuit. In Chapter 16, you were shown that the time constant of an *RC* circuit is found as

$$\tau = RC$$

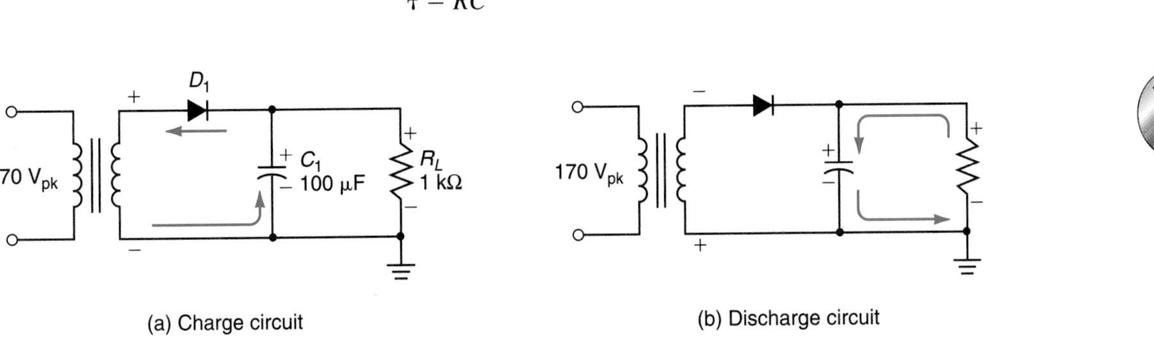

Figure 18.20 The basic capacitive filter.

where R and C are the total circuit resistance and capacitance values. You also learned that it takes five time constants (5τ) for the capacitor to fully charge or discharge.

In Figure 18.20a, the capacitor charges through the diode. For the sake of discussion, we'll assume that $R_{D1} = 5\ \Omega$. The time constant and capacitor charge times for the circuit are found as

$$\tau = R_{D1}C = 5\ \Omega \times 100\ \mu\text{F} = 500\ \mu\text{s}$$

and

$$T = 5\tau = (5)(500\ \mu\text{s}) = 2.5\ \text{ms}$$

The discharge path for the capacitor is through the load resistor (see Figure 18.20b). For this circuit, the time constant and capacitor charge times are found as

$$\tau = R_L C = (1\ \text{k}\Omega)(100\ \mu\text{F}) = 100\ \text{ms}$$

and

$$T = 5\tau = (5)(100\ \text{ms}) = 500\ \text{ms}$$

As you can see, the capacitor takes 200 times as long to discharge as it takes to charge. As a result, the capacitor charges almost instantly, yet barely starts to discharge before another charging voltage is provided by the rectifier.

A Practical Consideration: Ideally, there would be no ripple at the output of a filter. However, this can occur only under open-load conditions in any practical filter circuit. If the load is open, the capacitor does not have a discharge path, and the time constant of the RC circuit is (theoretically) infinite.

The amount of ripple voltage at the filter output is inversely proportional to the values of the filter capacitor and load resistance, as illustrated in Figure 18.21. As you can see:

- The lower the load resistance, the greater the capacitor discharge between half-cycles of the input.
- The lower the filter capacitance, the greater the capacitor discharge between half-cycles of the input.

Based on the waveforms shown in Figure 18.21b, you'd think that the best approach to filter design is to use an extremely high-value capacitor. However, the value of C affects not only the capacitor's discharge time but its charge time as well. This can result in a problem involving *surge current.*

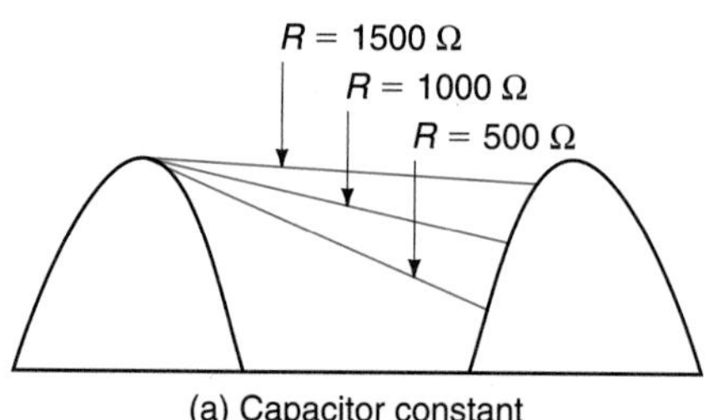

(a) Capacitor constant

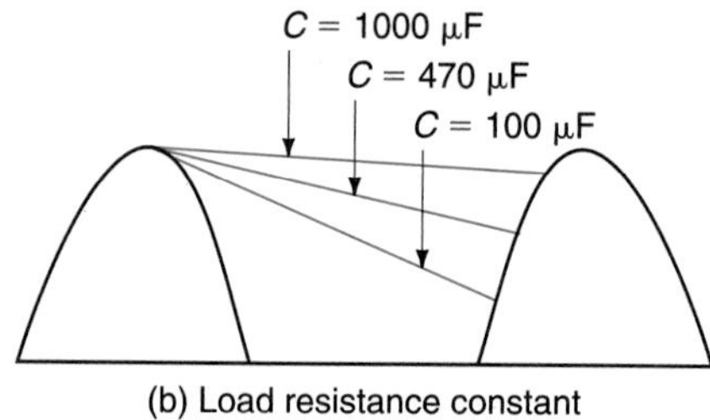

(b) Load resistance constant

Figure 18.21 The effects of load resistance and filter capacitance on ripple voltage.

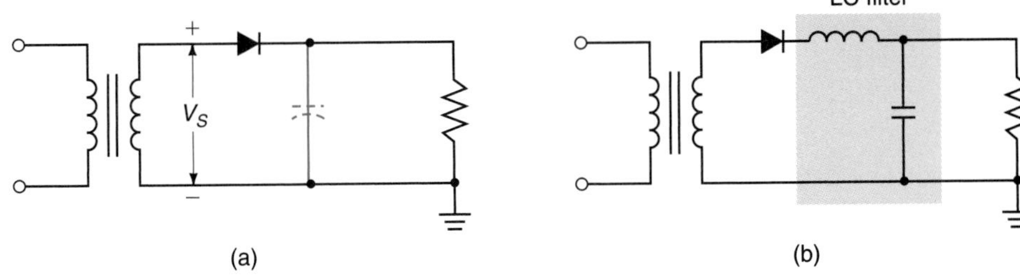

Figure 18.22

Surge Current

When you first turn on a power supply, the filter capacitor has no accumulated charge to oppose V_S. For the first instant, the discharged capacitor acts as a short circuit (as shown in Figure 18.22a). As a result, the current through the rectifier is limited only by:

- The winding resistance of the transformer secondary.
- The bulk resistance of the diode.

Since these resistances are usually very low, there is *a high initial current,* called **surge current.**

When the amount of surge current produced by a circuit is more than the rectifier diodes can handle, the problem can be resolved by using the *LC* filter shown in Figure 18.22b. The inductor opposes any rapid increase in current, and thus, limits the surge current. Surge current can also be limited by using a lower-value filter capacitor. However, there is one simple drawback to this approach: A lower-value capacitor charges in a shorter period of time, but the ripple voltage from the circuit increases (as shown in Figure 18.21b).

Filter Output Voltages

As shown in Figure 18.23, the output from a filter has a *peak value* (V_{pk}), an *average value* (V_{dc}), and a *ripple value* (V_r). The dc output voltage (V_{dc}) is shown to equal the peak voltage minus one-half the peak-to-peak value of the ripple voltage. By formula,

$$V_{dc} = V_{pk} - \frac{V_r}{2} \tag{18.13}$$

where

V_{pk} = the peak rectifier output voltage
V_r = the peak-to-peak value of ripple voltage

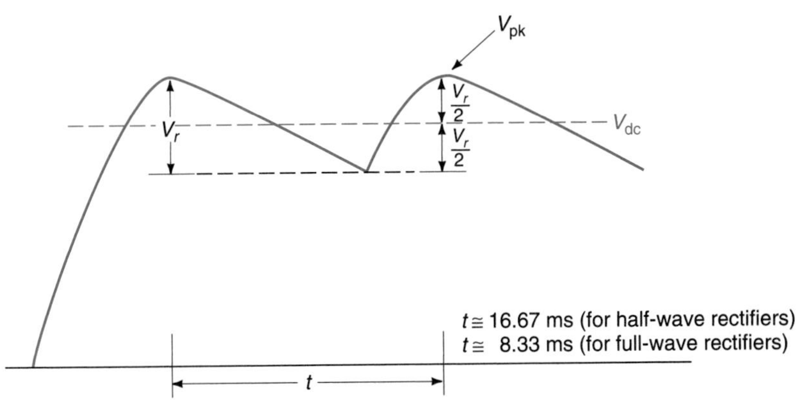

Figure 18.23

The ripple voltage at the filter output can be found as

$$V_r = \frac{I_L t}{C} \tag{18.14}$$

where

I_L = the dc load current
t = the time between charging peaks
C = the capacitance, in farads

The value of t used in equation (18.13) depends on the type of rectifier used (as indicated in Figure 18.23). Assuming a line frequency of 60 Hz, the time between the charging peaks for a half-wave rectifier can be found as

$$t = \frac{1}{f} = \frac{1}{60\text{ Hz}} = 16.67\text{ ms} \qquad \text{(half-wave rectifier)}$$

The full-wave rectifier produces two output peaks for every one produced by a half-wave rectifier. Therefore, the time between the charging peaks for a full-wave rectifier is found as

$$t = \frac{1}{f} = \frac{1}{120\text{ Hz}} = 8.33\text{ ms} \qquad \text{(full-wave rectifier)}$$

With one-half the time between charging peaks, the output from a filtered full-wave rectifier contains one-half the ripple of a comparable filtered half-wave rectifier. This point is demonstrated in Example 18.5.

Example 18.5

Determine the ripple output from each of the circuits shown in Figure 18.24. Assume that each circuit has a load current of 20 mA.

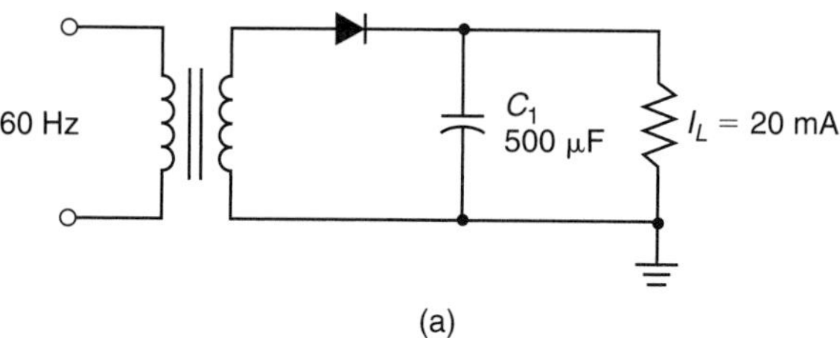

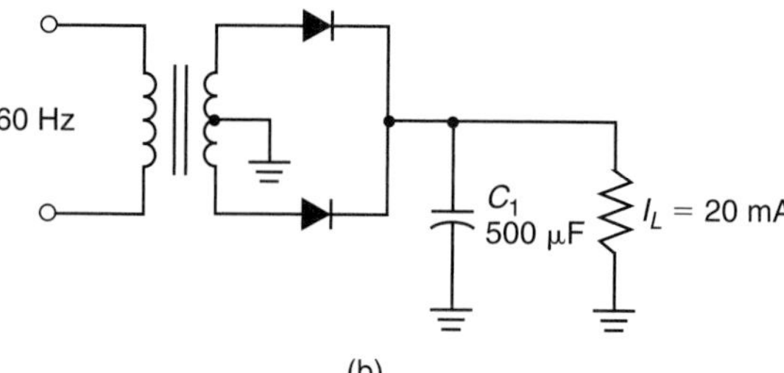

Figure 18.24

Solution: The half-wave rectifier shown in Figure 18.24a has a time of 16.67 ms between charging pulses. Therefore,

$$V_r = \frac{I_L t}{C} = \frac{(20\text{ mA})(16.67\text{ ms})}{500\ \mu\text{F}} \cong 667\text{ mV}_{PP}$$

For the full-wave rectifier shown in Figure 18.24b, $t = 8.33$ ms. Therefore,

$$V_r = \frac{I_L t}{C} = \frac{(20 \text{ mA})(8.33 \text{ ms})}{500\ \mu\text{F}} \cong 333 \text{ mV}_{PP}$$

As Example 18.5 demonstrates, the full-wave rectifier produces half as much ripple as the half-wave rectifier. This is a result of the shortened time period between capacitor charging pulses. Figure 18.25 compares the outputs from filtered half-wave and full-wave rectifiers.

Filter Effects on Rectifier Analysis

According to equation (18.13), you can find the value of V_{dc} for a filtered rectifier by subtracting $\frac{V_r}{2}$ from the peak rectifier output voltage. There is only one problem: According to equation (18.14), we need to know the value of the dc load current (I_L) to determine the value of V_r, and to determine the value of I_L, we need to know the value of V_{dc}.

What is the solution? If you look closely at Figure 18.23, you'll see that the final value of V_{dc} is very close to the value of V_{pk}. Thus, you can start your circuit calculations by assuming that

$$V_{dc} \cong V_{pk}$$

Using this assumed value of V_{dc}, you can calculate the approximate value of I_L. From there, you can find the approximate value of V_r and use it to calculate a more accurate value of V_{dc}.

Filter Effects on Diode PIV

For the full-wave and bridge rectifiers, the filter has no significant effect on the PIV across each diode. However, when filtered, the *half-wave rectifier* produces a diode PIV that is twice the secondary voltage. By formula,

$$\text{PIV} = 2V_{S(pk)} \qquad \text{(half-wave, filtered)} \qquad (18.15)$$

The basis for this relationship is illustrated in Figure 18.26. Assuming that the diode is an *ideal* component, C_1 charges to the value of $V_{S(pk)}$, as shown in Figure 18.26a when the diode conducts. When the polarity of V_S reverses as shown in Figure 18.26b, D_1 is reverse biased. For an instant, the reverse voltage across the diode is equal to $(V_{S(pk)} + V_C)$. Since these two voltages are equal, the peak inverse voltage across the diode is equal to $2V_{S(pk)}$.

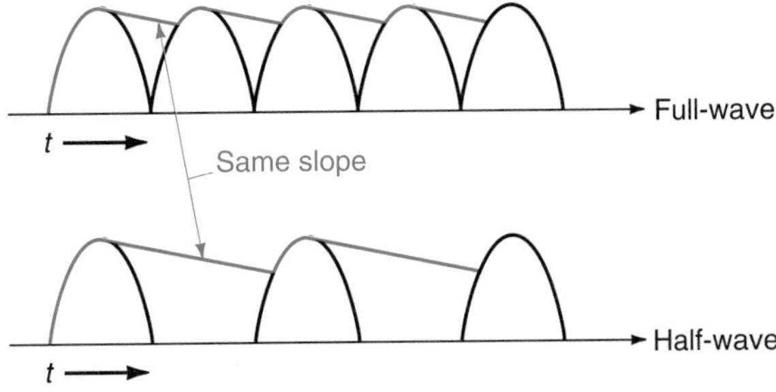

Figure 18.25 Full-wave versus half-wave ripple.

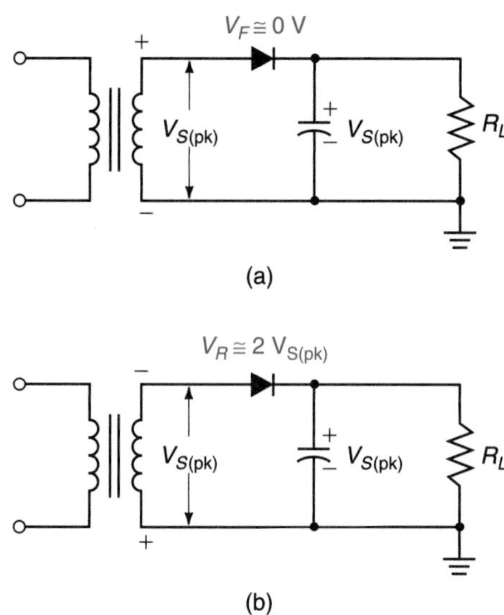

Figure 18.26 The effects of filtering on diode PIV in a half-wave rectifier.

Other Filter Types

There are several other types of filters, as shown in Figure 18.27. Each filter is designed to have high series impedance and low shunt impedance at the ripple frequency. These impedances combine to form a voltage divider at the ripple frequency, which greatly reduces the ripple amplitude.

Many filters use capacitors to oppose any changes in voltage and inductors to oppose any surge currents. For example, refer to the *LC* π filter shown in Figure 18.27c. The inductor opposes any rapid change in current while the capacitors short any voltage changes to ground.

One Final Note

Figure 18.1 showed the circuit groups that make up a basic dc power supply. In this chapter, you have been introduced to the basic principles of rectifier and filter operation. Most practical *voltage regulators* include components that we have yet to discuss. For that reason, voltage regulator circuits are covered in Chapter 25. In the next section, we will move on to discuss the operation of *clippers*.

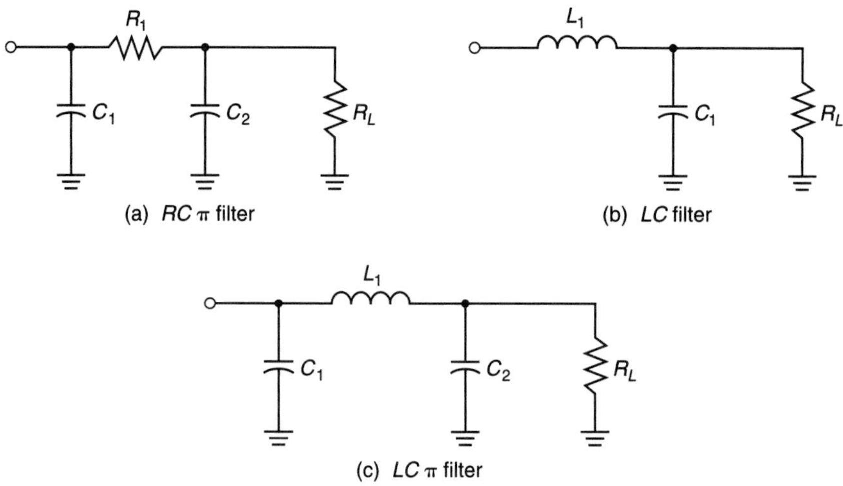

Figure 18.27 Other filter circuits.

Section Review

1. What are *filters* used for?
2. What is *ripple voltage*?
3. Describe the operation of the basic capacitive filter in terms of charge/discharge time constants.
4. What limits the value of a filter capacitor?
5. What causes *surge current* in a power supply?
6. Describe the relationship between V_r and V_{dc} for a *filtered rectifier.*
7. Why are *full-wave rectifiers* preferred over *half-wave rectifiers*?
8. Explain the effect that filtering has on the PIV across the diode in a half-wave rectifier.

18.5 Clippers

A **clipper** is *a diode circuit that eliminates a portion of its input signal.* For example, the *half-wave rectifier* is a clipper that eliminates one of the alternations of its ac input signal. Note that a clipper is often referred to as a **limiter,** because *it limits the voltage applied to its load.*

There are two types of clippers: *series clippers,* and *shunt clippers.* Each of these can be *positive* or *negative.* As shown in Figure 18.28:

- A **series clipper** is *a diode that is connected in series with its load.*
- A **shunt clipper** is *a diode that is connected in parallel with its load.*

Series Clippers

◄ OBJECTIVE 8

Series clipper operation is illustrated in Figure 18.29. Every series clipper has the same basic operating characteristics as a half-wave rectifier, so there is no need to go into any great detail on series clipper operation. Just note the following about the positive and negative series clippers shown:

- The *negative* series clipper eliminates the *negative alternation* of the input signal.
- The *positive* series clipper eliminates the *positive alternation* of the input signal.

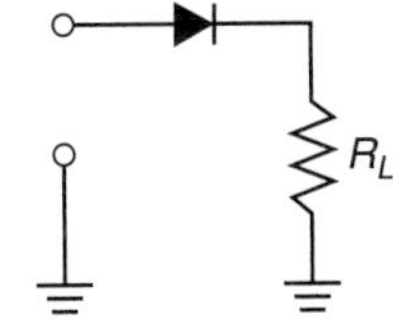

Negative series clipper

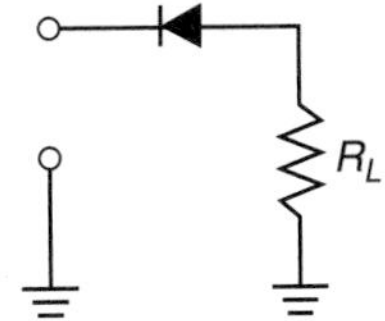

Positive series clipper

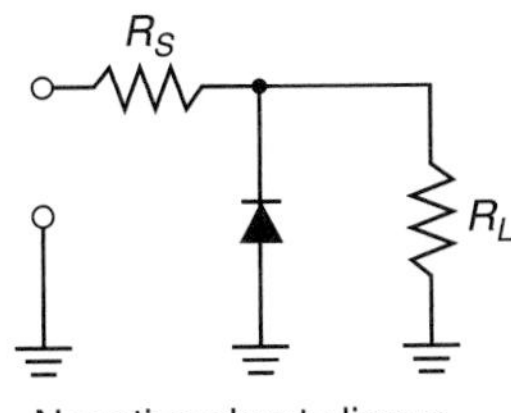

Negative shunt clipper

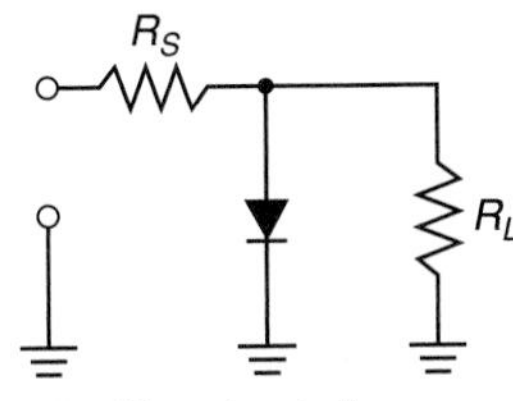

Positive shunt clipper

Figure 18.28

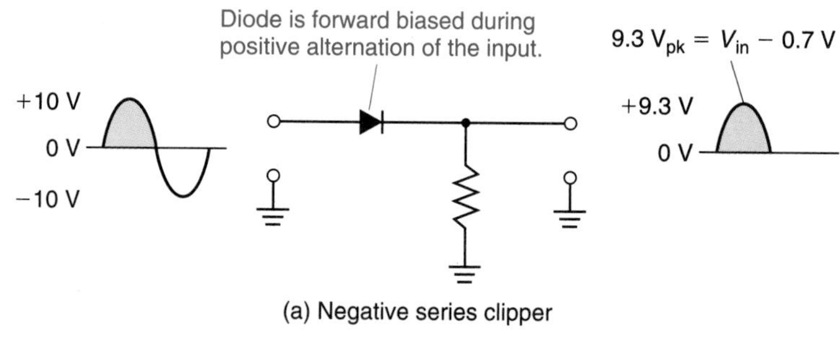

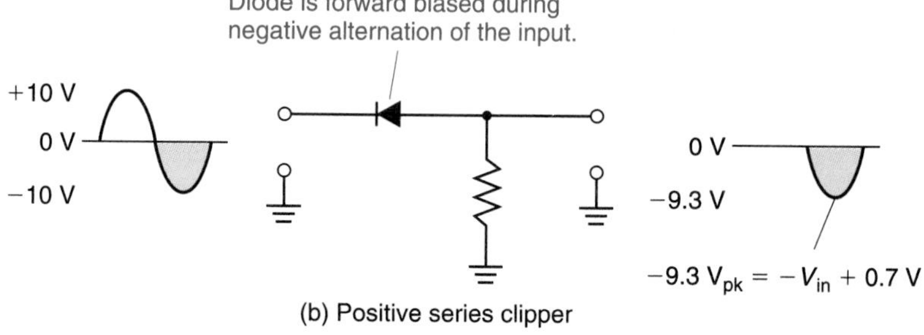

Figure 18.29 Series clipper operation.

The actual peak voltage delivered to the load (at the output of the clipper) is equal to the peak input voltage less the forward voltage drop of the diode. By formula,

$$V_L = V_{in(pk)} - 0.7\text{ V} \quad \text{(negative clipper)} \tag{18.16}$$

$$V_L = -V_{in(pk)} + 0.7\text{ V} \quad \text{(positive clipper)} \tag{18.17}$$

Shunt Clippers

OBJECTIVE 9 ➤ The operation of the shunt clipper is the exact opposite of the series clipper. The shunt clipper has an output when the diode is reverse biased, and shorts the output signal to ground when the diode is forward biased. The operation of a shunt clipper is illustrated in Figure 18.30.

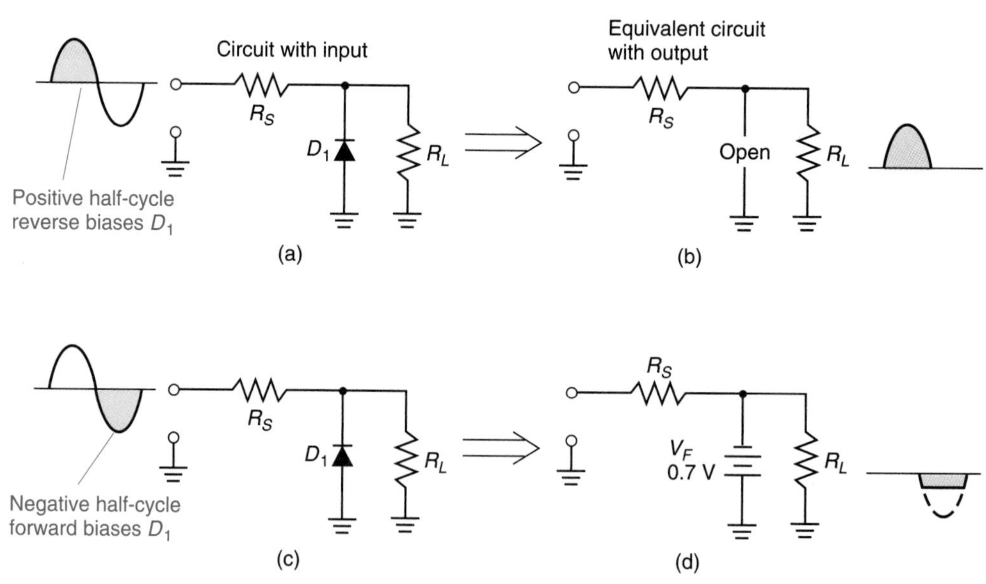

Figure 18.30 Shunt clipper operation.

The circuit shown in Figure 18.30 is a **negative shunt clipper.** When the diode is reverse biased, it acts like an open circuit (as shown in Figure 18.30b). Since R_S and R_L form a voltage divider, the load voltage can be found as

$$V_L = V_{in} \frac{R_L}{R_L + R_S} \tag{18.18}$$

When the diode is forward biased, it has a value of $V_F = -0.7$ V. Since the diode is in parallel with the load, V_L must also equal -0.7 V. These points are illustrated in Example 18.6.

Example 18.6

The negative shunt clipper shown in Figure 18.31 has peak input values of ±12 V. Solve for the value of V_L at the positive and negative input peaks.

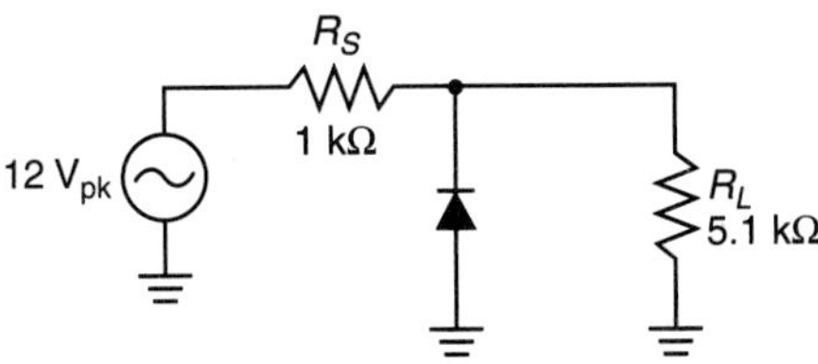

Figure 18.31

Solution: When $V_{in} = +12$ V, the diode is reverse biased and does not conduct. Therefore, the peak load voltage is found as

$$V_L = V_{in} \frac{R_L}{R_L + R_S} = (12\text{ V}) \frac{5.1\text{ k}\Omega}{6.1\text{ k}\Omega} = 10\text{ V}_{pk}$$

When $V_{in} = -12$ V, the diode is forward biased, and $V_{D1} = -0.7$ V. Since the load is in parallel with the diode,

$$V_L = V_{D1} = -0.7\text{ V}$$

Practice Problem 18.6

A negative shunt clipper has values of $R_L = 510\ \Omega$, $R_S = 100\ \Omega$, and $V_{in} = \pm 15\ \text{V}_{pk}$. Determine the value of V_L at the peak of each alternation of the input.

So far, we have focused on the negative shunt clipper. The **positive shunt clipper** works according to the same principles and is analyzed in the same fashion. This is illustrated in Example 18.7.

Example 18.7

The positive shunt clipper in Figure 18.32 has the input waveform shown. Determine the value of V_L for each of the input alternations.

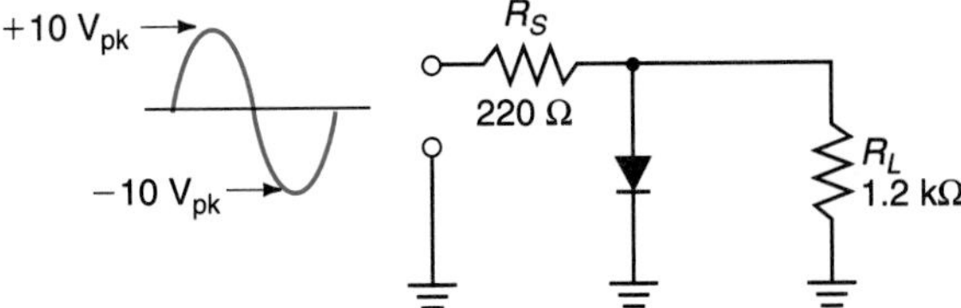

Figure 18.32

Solution: When the input is positive, the diode is forward biased. Therefore,

$$V_L = V_{D1} = 0.7\text{ V}$$

When the input is negative, the diode is reverse biased and effectively removed from the circuit. Therefore,

$$V_L = V_{\text{in}} \frac{R_L}{R_L + R_S} = (-10\ \text{V}_{\text{pk}}) \frac{1.2\ \text{k}\Omega}{1.42\ \text{k}\Omega} = -8.45\ \text{V}_{\text{pk}}$$

Using the values calculated, the output waveform for the circuit can be drawn as shown in Figure 18.33.

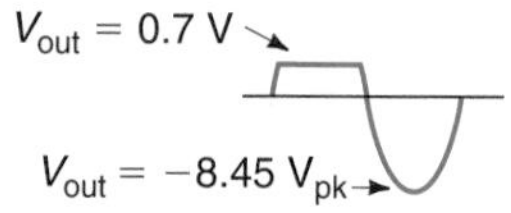

$V_{\text{out}} = 0.7$ V

$V_{\text{out}} = -8.45\ \text{V}_{\text{pk}}$

Figure 18.33

Practice Problem 18.7

A positive shunt clipper with values of $R_S = 100\ \Omega$ and $R_L = 1.1\ \text{k}\Omega$ has a $\pm 12\ \text{V}_{\text{pk}}$ input signal. Determine the value of V_L for each alternation of the input.

The Purpose Served by R_S

R_S is included in the shunt clipper as a *current-limiting resistor.* Its purpose becomes clear when you consider what would happen if the input signal forward biased the diode and R_S was not in the circuit.

Without R_S in the circuit, the forward-biased diode shorts the input signal to ground during the positive alternation of the source. The resulting excessive current may destroy the diode, damage the signal source, or both. In any practical application, the value of R_S is much lower than that of R_L. When $R_L >> R_S$, the load voltage is approximately equal to the value of V_{in} when the diode is reverse biased.

Biased Clippers

OBJECTIVE 10 ➤ A **biased clipper** is *a shunt clipper that uses a dc voltage source to bias its diode.* This allows the circuit to clip input voltages at values other than the normal diode V_F of 0.7 V. Figure 18.34 shows examples of positive- and negative-biased clippers. The *bias voltage* (V_B) is in series with the shunt diode, causing it to have a reference other than ground. The point at which the diode clips the waveform is equal to the sum of V_F and V_B.

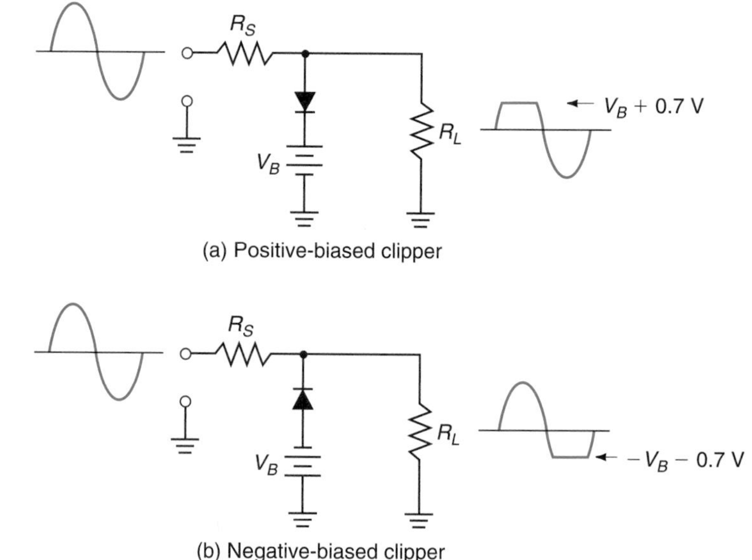

(a) Positive-biased clipper

(b) Negative-biased clipper

Figure 18.34

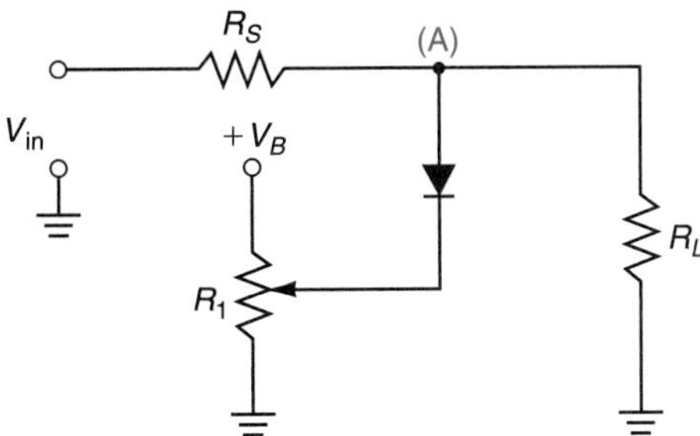

Figure 18.35

In practice, a potentiometer is normally used to provide an adjustable value of V_B, as shown in Figure 18.35. In this circuit, the biasing voltage ($+V_B$) is connected to the diode via the potentiometer (R_1). R_1 is adjusted to provide the desired clipping limit at point A. Note that the clipping voltage can never be greater than ($V_B + 0.7$ V) or equal to 0 V, because some voltage is always developed across the diode when it is conducting. By reversing the direction of the diode and the polarity of V_B, the circuit shown in Figure 18.35 can be modified to work as a negative-biased clipper. Figure 18.36 summarizes the various clipper configurations.

Transient Protection: A Clipper Application

Clippers are used in a wide variety of electronic systems. They are generally used to perform one of two functions:

- Altering the shape of a waveform.
- Circuit *transient* protection.

You have already seen one example of the first function in the half-wave rectifier; changing ac to pulsating dc.

A **transient** is *an abrupt current or voltage spike that has an extremely short duration.* Transients can do serious damage to a circuit. A clipper can help protect a circuit from transients, as illustrated in Figure 18.37.

The block labeled A in Figure 18.37a represents a digital circuit whose input voltage must not be allowed to go outside the range of 0 to +5 V. If the input goes more negative than −0.7 V, D_2 will be forward biased and short the negative transient to ground. If

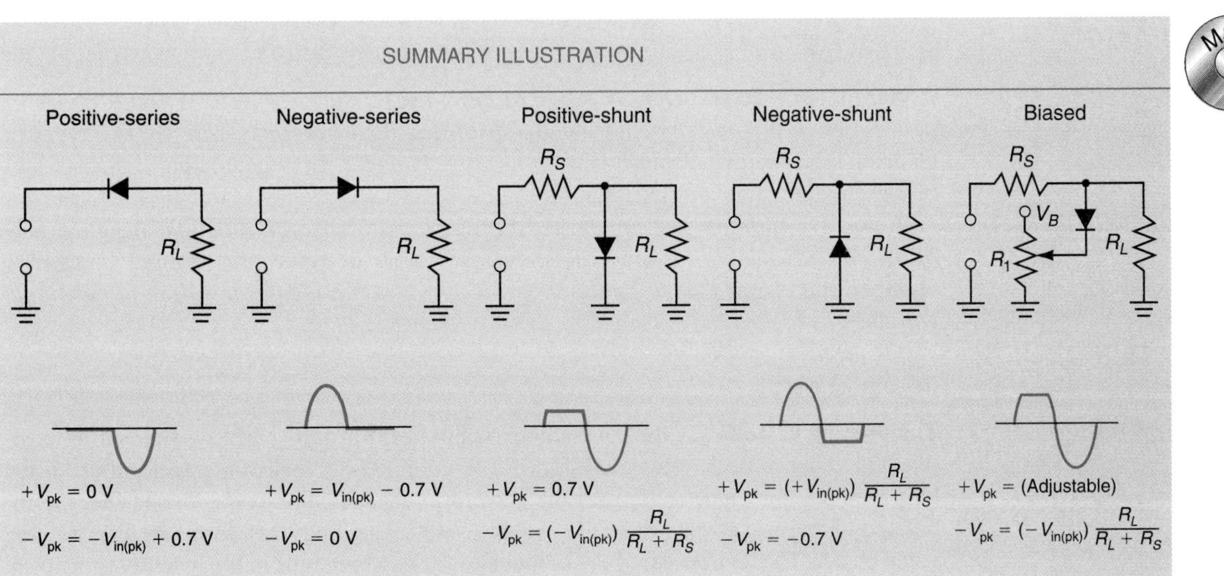

Figure 18.36

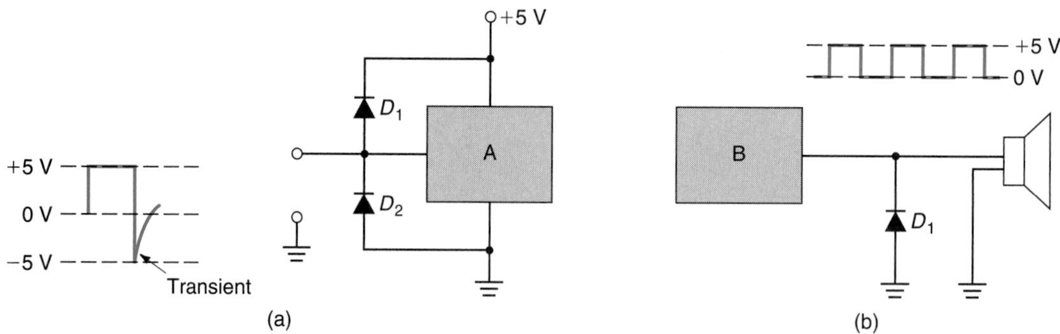

Figure 18.37

the input goes more positive than +5 V, D_1 will be forward biased and short the positive transient to the +5 V supply. In either case, the transient will bypass the input to circuit A.

The circuit shown in Figure 18.37b includes a clipper designed to protect the output of circuit B from transients produced when a square wave is used to drive a speaker (a common practice with computer sound effects). The speaker coil produces a counter emf that is greater in magnitude than the original voltage (and opposite in polarity). The clipping diode (D_1) conducts when the counter emf goes more negative than −0.7 V, shorting the transient to ground before it can harm the driving circuit.

Section Review

1. What purpose does a *clipper* serve?
2. What is another name for a clipper?
3. Discuss the differences between *series* clippers and *shunt* clippers.
4. What purpose is served by the *series resistor* (R_S) in the shunt clipper?
5. Describe the operation of a *biased clipper.*
6. What is a *transient*?

18.6 Clampers (DC Restorers)

A **clamper** (or **dc restorer**) is *a circuit designed to shift a waveform either above or below a given reference voltage without distorting the waveform.* There are two types of clampers: the *positive clamper,* and the *negative clamper.* The input/output characteristics of these two circuits are illustrated in Figure 18.38.

OBJECTIVE 11 ➤ A **positive clamper** is *a circuit that shifts its input waveform so that the negative peak of the output waveform is approximately equal to its* **dc reference voltage.** A **negative clamper** is *a circuit that shifts its input waveform so that the positive peak of the output waveform is approximately equal to its dc reference voltage.*

V_{ave} Shift

The average value (V_{ave}) for any symmetrical waveform falls halfway between its positive and negative peak values (as shown in Figure 18.38a). When a waveform is shifted by a clamper, the value of V_{ave} for the waveform changes. For example, consider the input and output waveforms shown in Figure 18.38b. The input waveform has a dc average value of 0 V. The dc average of the output is +10 V, which falls halfway between its peak values of +20 V and 0 V. The output waveform in Figure 18.38c has a dc average of −10 V, which falls halfway between its peak values of 0 V and −20 V.

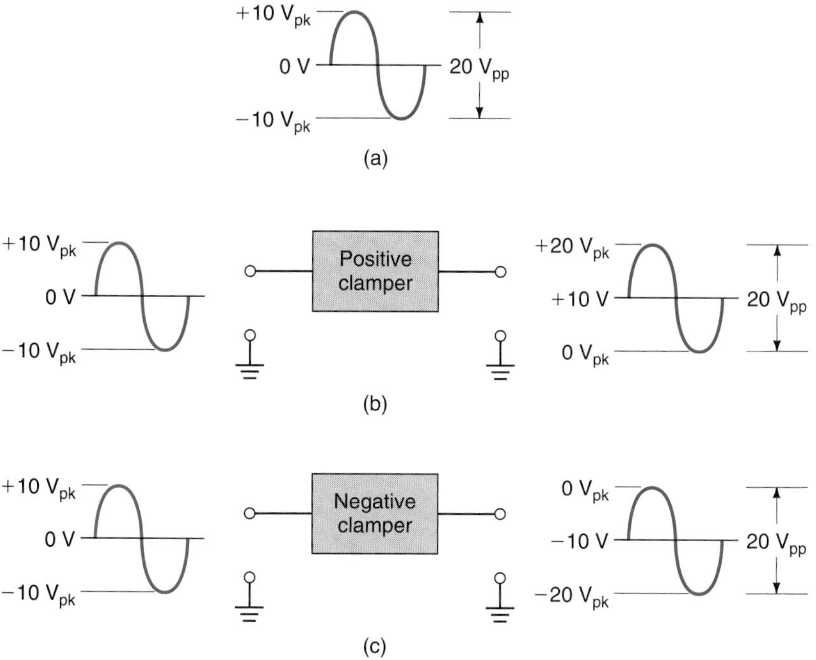

Figure 18.38

Clamper Operation

The schematic diagram of a positive clamper is shown in Figure 18.39. As you can see, this clamper is very similar to the shunt clipper, with the exception of the added capacitor (C_1). ◄ *OBJECTIVE 12*

Like the capacitive power supply filter, the clamper works on the basis of switching time constants. When the diode is forward biased, it provides a charging path for the capacitor, and the capacitor charge time is found as

$$T = 5\tau = 5(R_{D1}C_1) \tag{18.19}$$

where

R_{D1} = the bulk resistance of the diode

When the diode is reverse biased, the capacitor discharges through R_L. Thus, the capacitor discharge time is found as

$$T = 5\tau = 5(R_LC_1) \tag{18.20}$$

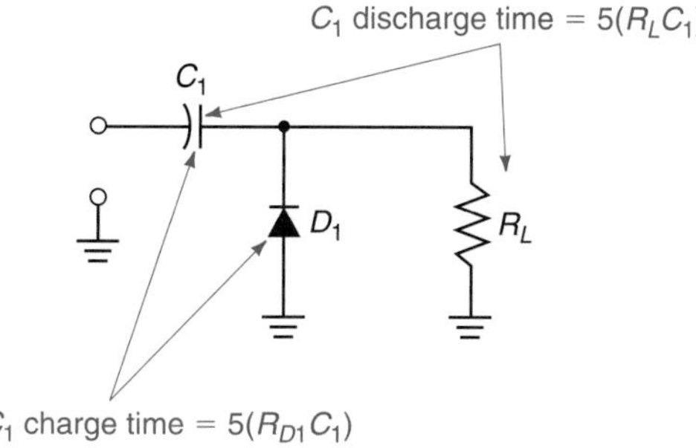

Figure 18.39 Clamper *charge* and *discharge* time constants.

Note the different resistance values that appear in equations (18.19) and (18.20). As Example 18.8 demonstrates, the difference between the capacitor charge and discharge times is significant.

Example 18.8

Determine the capacitor charge and discharge times for the circuit shown in Figure 18.39. Assume the circuit has values of $R_{D1} = 10\ \Omega$, $R_L = 10\ \text{k}\Omega$, and $C_1 = 1\ \mu\text{F}$.

Solution: The capacitor charges through the diode. The charge time is found as

$$T_C = 5(R_{D1}C_1) = (5)(10\ \Omega \times 1\ \mu\text{F}) = 50\ \mu\text{s}$$

The capacitor discharges through the resistor. Thus, the discharge time is found as

$$T_C = 5(R_LC_1) = (5)(10\ \text{k}\Omega \times 1\ \mu\text{F}) = 50\ \text{ms}$$

Practice Problem 18.8

A clamper has values of $R_{D1} = 8\ \Omega$, $R_L = 1.2\ \text{k}\Omega$, and $C_1 = 4.7\ \mu\text{F}$. Determine the charge and discharge times for the capacitor.

As shown in Example 18.8, it takes 1000 times as long for the capacitor to discharge as it does to charge. This is the basis of clamper operation.

The operation of a clamper is illustrated in Figure 18.40. For ease of discussion, we will look at the circuit's response to a square wave input. We will also assume that the diodes are ideal components.

Figure 18.40a shows the input waveform. When the input goes to its positive peak (+5 V), D_1 is forward biased. This provides a low-resistance current path for charging

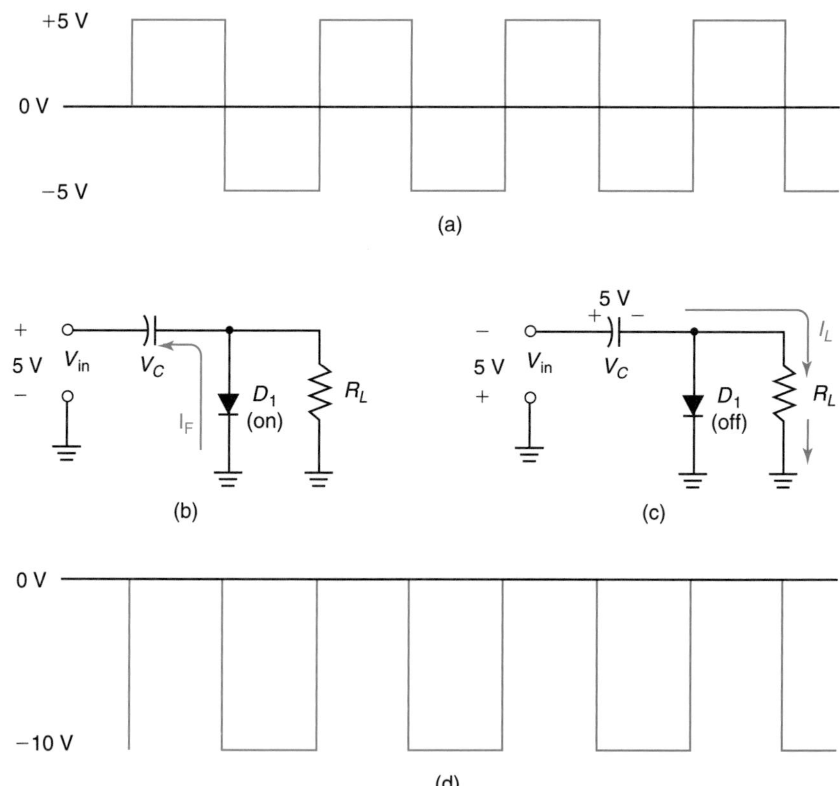

Figure 18.40

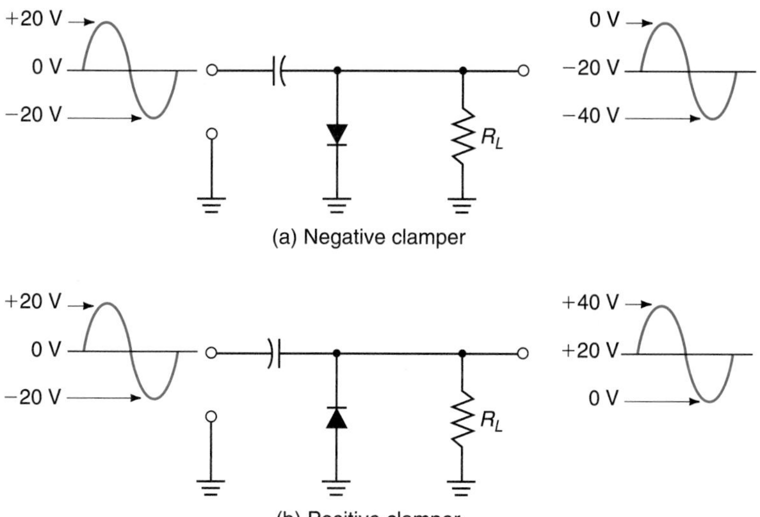

Figure 18.41

C_1, as shown in Figure 18.40b. Since the charge time is very short, C_1 rapidly charges to +5 V, leaving 0 V across the load.

When the input waveform begins to go negative, the output side of the capacitor (which is initially at 0 V) also begins to go negative. This turns D_1 off, and the capacitor is forced to discharge through the resistor, as shown in Figure 18.40c. Since the discharge time is very long, C_1 loses very little of its charge. With V_{in} and V_C having the values and polarities shown in Figure 18.40c, the load voltage is −10 V. Thus, with the input shown, the clamper provides an output square wave that varies between 0 V and −10 V, as shown in Figure 18.40d.

Negative Clampers Versus Positive Clampers

As with negative and positive clippers, the difference between negative and positive clampers is the direction of the diode. Figure 18.41 compares a positive and a negative clamper. It is important to remember that the capacitor (if polarized) must also be reversed to match the circuit input and output polarities.

Biased Clampers

A **biased clamper** is *a circuit that allows a waveform to be shifted so that it falls above or below a dc reference other than 0 V.* Several biased clampers are shown in Figure 18.42.

The circuits shown in Figures 18.42a and 18.42b simply use a dc supply and a potentiometer to establish a diode bias voltage (V_B). The circuit in Figure 18.42a shifts its input waveform *below* a reference voltage that is found as $V_{REF} = V_B + 0.7$ V. The circuit in Figure 18.42b shifts its input waveform *above* a reference voltage that is found as $V_{REF} = V_B - 0.7$ V.

The circuits shown in Figures 18.42c and 18.42d are referred to as *zener clampers.* A **zener clamper** is *a clamper that uses a zener diode to establish the pn-junction diode bias voltage.* The circuit shown in Figure 18.42c shifts its input waveform *above* a reference voltage of $V_{REF} = -V_Z + V_{D2}$. The circuit shown in Figure 18.42d shifts its input waveform *below* a reference voltage of $V_{REF} = -V_Z + V_{D2}$.

Section Review

1. In terms of input/output relationships, describe the difference between *positive clampers* and *negative clampers.*
2. What effect does a clamper have on the V_{ave} for a given input waveform?

3. Explain the operation of a clamper in terms of *switching time constants.*
4. What determines the *dc reference voltage* of a clamper?
5. How can you tell whether a given clamper is a positive or a negative clamper?

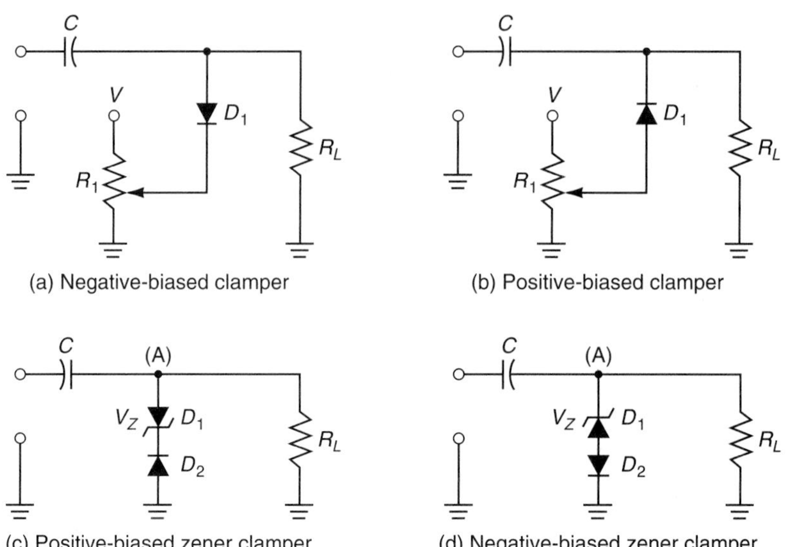

Figure 18.42

18.7 Voltage Multipliers

A **voltage multiplier** is *a circuit that provides a dc output that is a whole-number multiple of its peak input voltage.* For example, a **voltage doubler** is *a circuit that provides a dc output voltage that is twice its peak input voltage,* and so on. In this section, we will take a look at several types of voltage multipliers.

Half-Wave Voltage Doublers

OBJECTIVE 13 ➤ The **half-wave voltage doubler** is made up of two diodes and two capacitors. A half-wave voltage doubler is shown (along with its source and load) in Figure 18.43.

The operation of the half-wave voltage doubler is easier to understand if we assume that the diodes are ideal components. During the negative alternation of the input (shown in Figure 18.44a), D_1 is forward biased, and D_2 is reverse biased. If we represent D_1 as

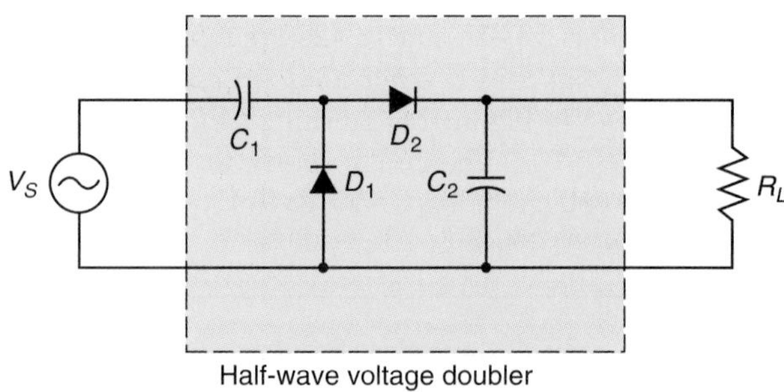

Figure 18.43 The half-wave voltage doubler.

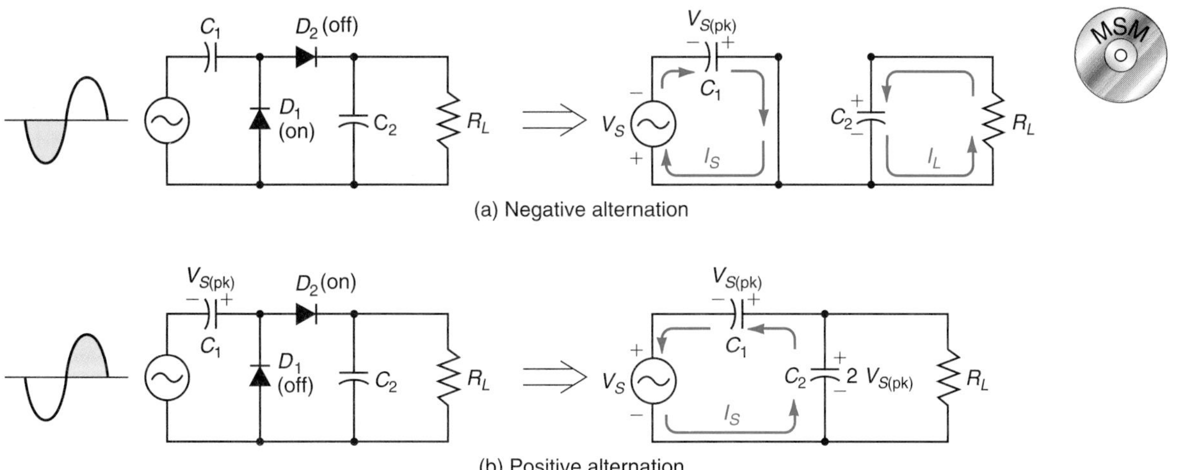

Figure 18.44

a short and D_2 as an open, we get the equivalent circuit shown. As you can see, two things happen during this time period:

- C_1 charges until its plate-to-plate voltage is equal to V_S.
- C_2 discharges through the load.

When the input polarity reverses, we have the circuit conditions that are shown in Figure 18.44b. During this alternation of the input cycle, D_1 is *off* and D_2 is *on*. The equivalent circuit in Figure 18.44b shows that C_1 (which is charged to the peak value of V_S) and the source voltage (V_S) act as series-aiding voltage sources. Thus, C_2 is charged to $2V_{S(pk)}$ by C_1 and the source. When V_S returns to its original polarity, D_2 is again turned off, and the only discharge path for C_2 is through the load. Normally, the time constant of this circuit is such that C_2 barely starts to discharge before the input reverses polarity again, recharging C_2.

The output waveform of the half-wave voltage doubler closely resembles that of a filtered half-wave rectifier. Typical input and output waveforms are shown in Figure 18.45. The output ripple from this circuit is calculated in the same manner as that from a filtered half-wave rectifier. High capacitor values and low current demands reduce the amount of ripple. Reversing the directions of the diodes in Figure 18.43 (along with the polarity of the two capacitors) produces a negative voltage doubler.

Full-Wave Voltage Doublers

◄ *OBJECTIVE 14*

The **full-wave voltage doubler** closely resembles the half-wave voltage doubler. It contains two diodes and two capacitors (C_1 and C_2), as shown in Figure 18.46a. Note the addition of the *filter capacitor* (C_3). This component reduces the output ripple from the voltage doubler.

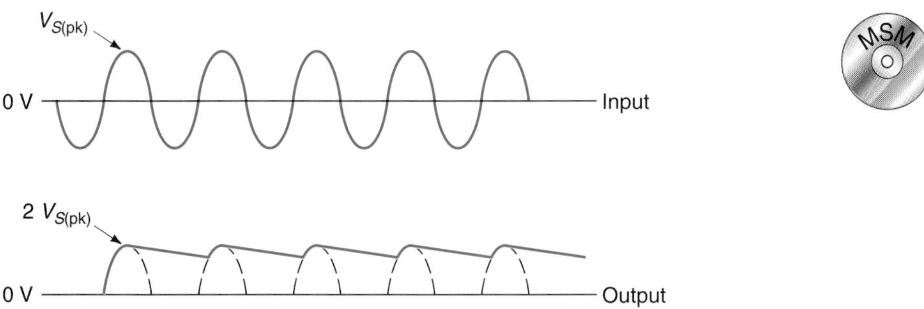

Figure 18.45

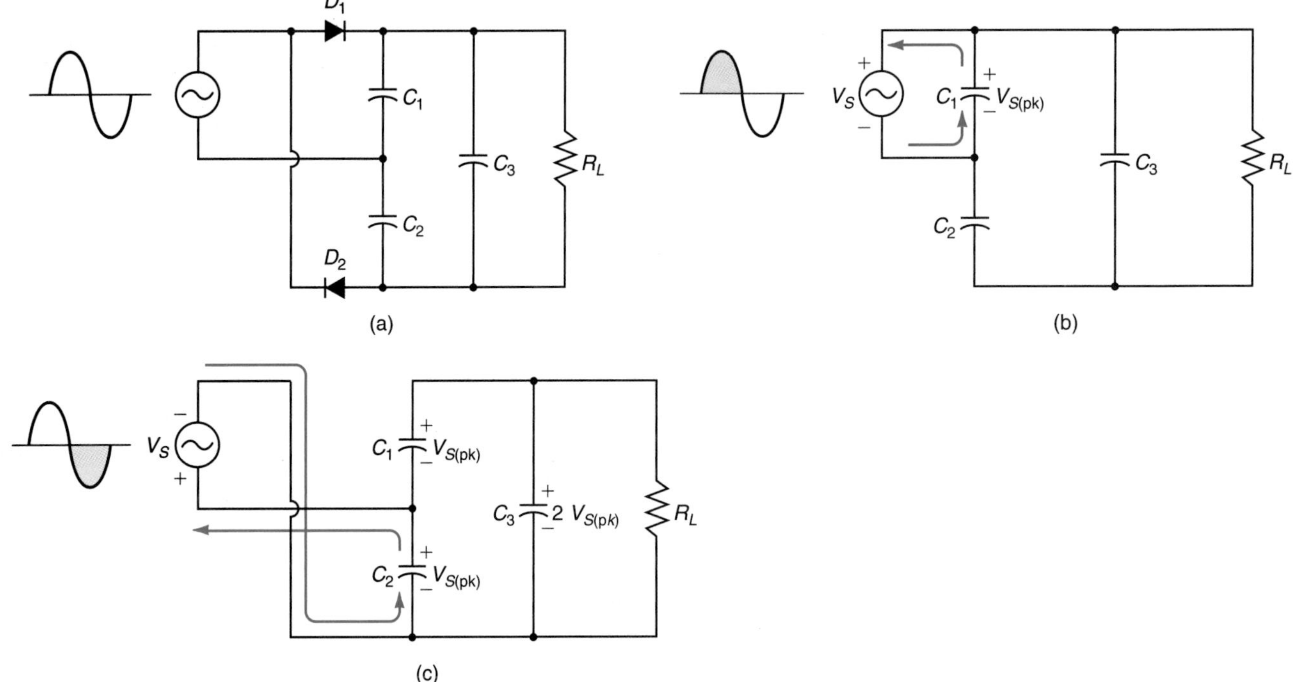

Figure 18.46 Full-wave voltage doubler operation.

During the positive half-cycle of the input, D_1 is forward biased, and D_2 is reverse biased. This gives us the equivalent circuit shown in Figure 18.46b. The equivalent circuit shows that C_1 charges to the value of $V_{S(pk)}$. When the input polarity reverses, D_1 is reverse biased, and D_2 is forward biased. This gives us the equivalent circuit shown in Figure 18.46c. As that circuit shows, C_2 now charges to the value of $V_{S(pk)}$. *Since* C_1 and C_2 are in series, the total voltage across the two components is $2V_{S(pk)}$.

With the added filter capacitor (C_3), there is very little ripple voltage. This is one of the advantages of using a full-wave voltage doubler in place of a half-wave doubler.

Voltage Triplers and Quadruplers

OBJECTIVE 15 ➤ **Voltage triplers** and **voltage quadruplers** are both variations on the basic half-wave voltage doubler. The schematic diagram for the voltage tripler is shown in Figure 18.47. As you can see, the circuit is very similar to the half-wave voltage doubler shown in Figure 18.43. In fact, the combination of D_1, D_2, C_1, and C_2 forms a half-wave doubler. The added components (D_3, C_3, and C_4) form the rest of the voltage tripler.

When V_S is negative (as shown in Figure 18.47a), D_1 and D_3 both conduct, allowing C_1 and C_3 to charge up to $V_{S(pk)}$. When V_S is positive (as shown in Figure 18.47b), D_2 is turned on, allowing C_2 to charge up to $2V_{S(pk)}$. The voltages across C_2 and C_3 now add up to $3V_{S(pk)}$. C_4 is a filter capacitor that is added to reduce the ripple voltage. Note that the value of C_4 is chosen so that it retains most of its plate-to-plate charge between charging cycles.

OBJECTIVE 16 ➤ The voltage quadrupler shown in Figure 18.48 is simply made up of two half-wave voltage doublers. The series combination of C_2 and C_4 charges the filter capacitor (C_5) until its plate-to-plate voltage is equal to $4V_{S(pk)}$. Again, the value of C_5 is chosen so that it retains most of its plate-to-plate charge between charging cycles.

Section Review

1. Describe the operation of the *half-wave voltage doubler.*
2. Describe the operation of the *full-wave voltage doubler.*
3. Why are full-wave voltage doublers preferred over half-wave voltage doublers?

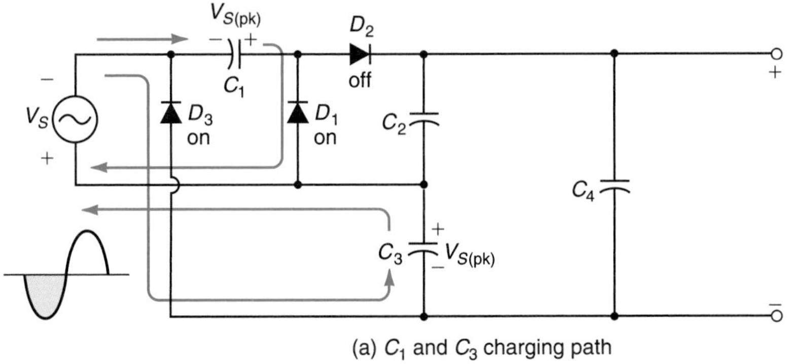

(a) C_1 and C_3 charging path

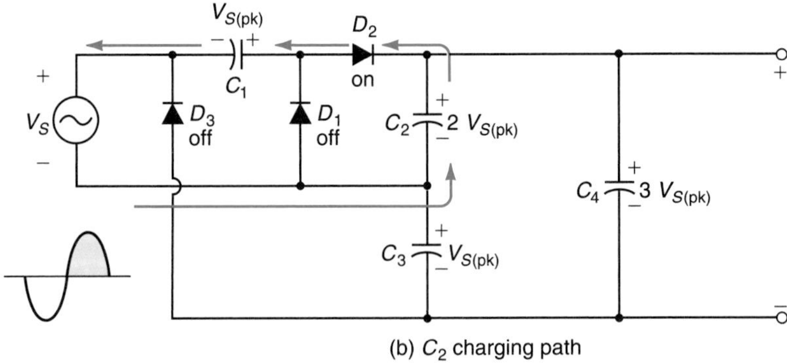

(b) C_2 charging path

Figure 18.47 Voltage tripler operation.

A Practical Consideration: When working with voltage multipliers, you should be extremely careful because of the high voltages that may be present within the circuit. All capacitors should be discharged with a shorting tool before any component is removed from the circuit.

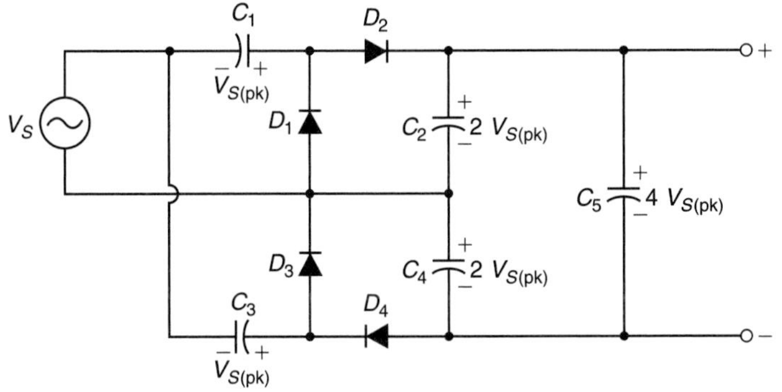

Figure 18.48 Voltage quadrupler.

MSM

4. What is the similarity between the *voltage tripler* and the *half-wave voltage doubler?*
5. What is the similarity between the *voltage quadrupler* and the *half-wave voltage doubler?*

Key Terms

The following terms were introduced in this chapter on the pages indicated:

Average load voltage (V_{ave}), 558
Biased clamper, 581
Biased clipper, 576
Bridge rectifier, 563
Capacitive filter, 567
Clamper, 578
Clipper, 573
DC power supply, 555
DC reference voltage, 578
DC restorer, 578

Full-wave rectifier, 560
Full-wave voltage doubler, 583
Half-wave rectifier, 555
Half-wave voltage doubler, 582
Limiter, 573
Negative clamper, 578
Negative shunt clipper, 575
Peak inverse voltage (PIV), 559
Positive clamper, 578
Positive shunt clipper, 575
Power supply filter, 567
Rectifier, 555
Ripple voltage (V_R), 567
Series clipper, 573
Shunt clipper, 573
Surge current, 569
Transient, 577
Voltage doubler, 582
Voltage multiplier, 582
Voltage quadrupler, 584
Voltage tripler, 584
Zener clamper, 581

Practice Problems

1. Determine the peak load voltage for the circuit shown in Figure 18.49.

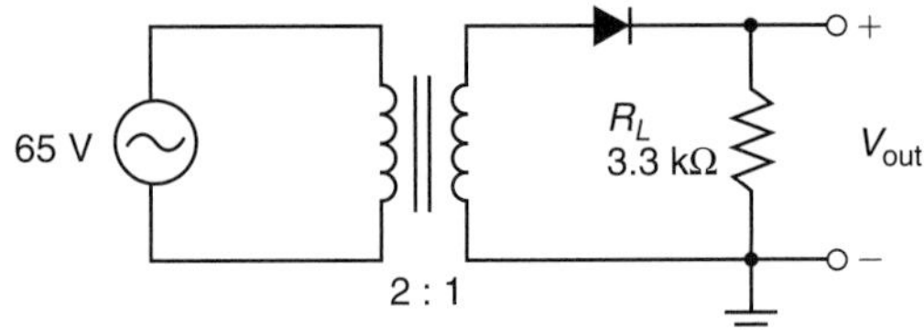

Figure 18.49

2. Determine the peak load voltage for the circuit shown in Figure 18.50.

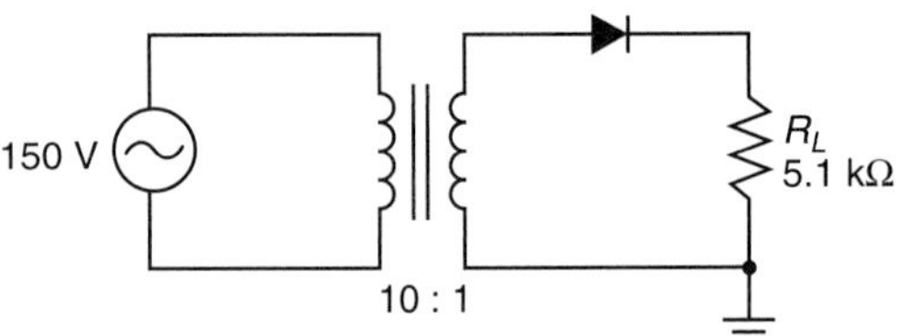

Figure 18.50

3. Determine the peak load voltage for the circuit shown in Figure 18.51.
4. Determine the peak load current for the circuit shown in Figure 18.51.

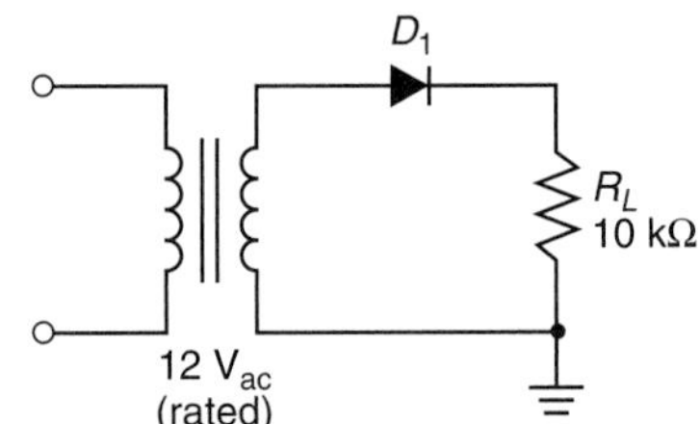

Figure 18.51

5. Determine the average (dc) load voltage for the circuit shown in 18.49.
6. Determine the average (dc) load voltage for the circuit shown in Figure 18.50.
7. Determine the average (dc) load voltage for the circuit shown in Figure 18.51.
8. Assume that the diode in Figure 18.49 is reversed. Determine the new values of $V_{L(pk)}$, V_{ave}, and I_{ave} for the circuit.
9. Assume that the diode in Figure 18.50 is reversed. Determine the new values of $V_{L(pk)}$, V_{ave}, and I_{ave} the circuit.
10. Assume that the diode in Figure 18.51 is reversed. Determine the new values of $V_{L(pk)}$, V_{ave}, and I_{ave} for the circuit.
11. Determine the PIV for the diode in Figure 18.50.

12. A negative half-wave rectifier with a 12 kΩ load is driven by a 20 V_{ac} transformer. Draw the schematic for the circuit and determine the following values: PIV, $V_{L(pk)}$, V_{ave}, and I_{ave}.

13. Determine the values of $V_{L(pk)}$, V_{ave}, and I_{ave} for the circuit shown in Figure 18.52.

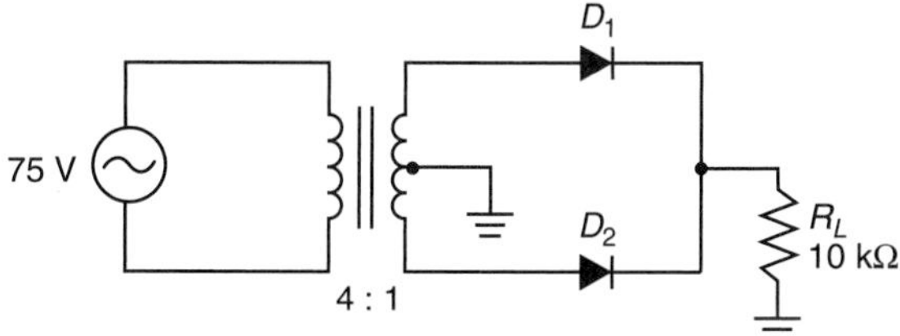

Figure 18.52

14. Determine the values of $V_{L(pk)}$, V_{ave}, and I_{ave} for the circuit shown in Figure 18.53.

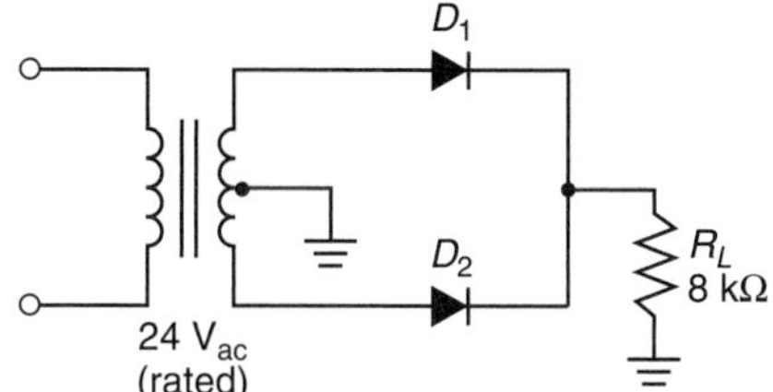

Figure 18.53

15. Determine the values of $V_{L(pk)}$, V_{ave}, and I_{ave} for the circuit shown in Figure 18.54.

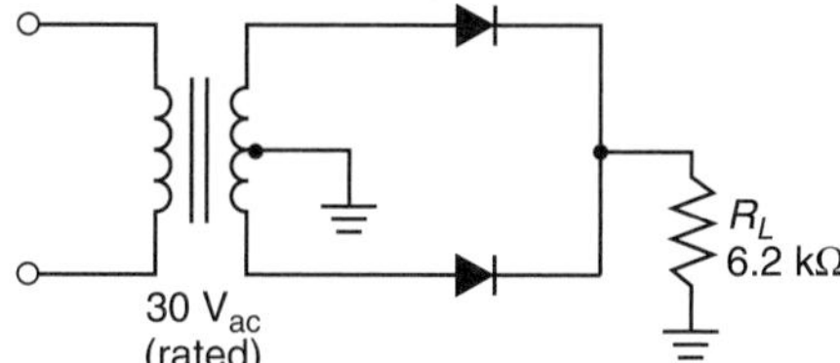

Figure 18.54

16. Determine the PIV of the circuit shown in Figure 18.52.
17. Determine the PIV of the circuit shown in Figure 18.53.
18. Determine the PIV of the circuit shown in Figure 18.54.
19. Assume that the diodes in Figure 18.52 are both reversed. Determine the new values of $V_{L(pk)}$, V_{ave}, and I_{ave} for the circuit.
20. A negative full-wave rectifier with a 910 Ω load is driven by a 16 V_{ac} transformer. Draw the schematic diagram of the circuit, and determine the following values: PIV, $V_{L(pk)}$, V_{ave}, and I_{ave}.
21. Determine the peak and average output voltage and current values for the circuit shown in Figure 18.55.

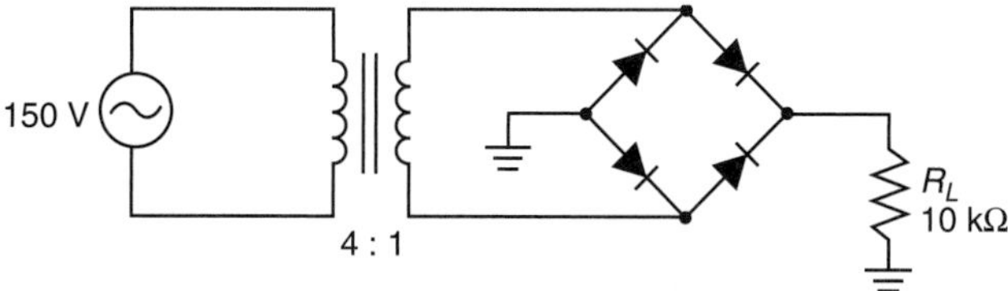

Figure 18.55

22. A bridge rectifier with a 1.2 kΩ load is driven by a 48 V_{ac} transformer. Draw the schematic diagram for the circuit, and calculate the dc load voltage and current values. Also, determine the PIV for each diode in the circuit.
23. What are the values of V_{dc} and V_r for the circuit shown in Figure 18.56a?
24. What are the values of V_{dc} and V_r for the circuit shown in Figure 18.56b?

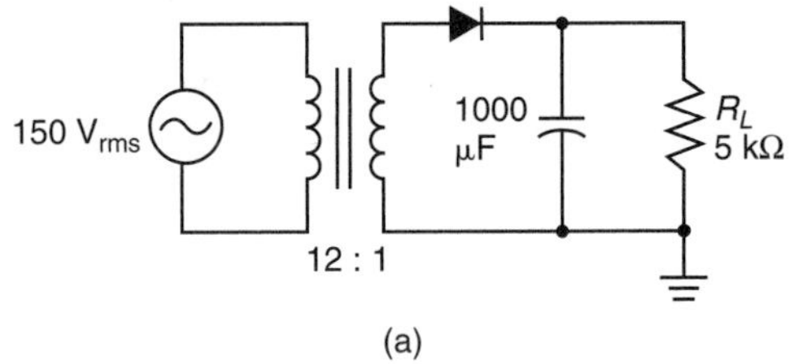

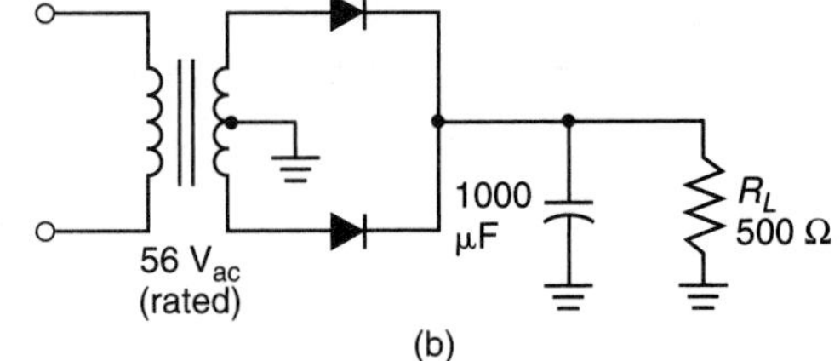

Figure 18.56

25. The circuit shown in Figure 18.57 has the following values: $V_S = 18\ \text{V}_{\text{ac}}$ (rated), $C = 470\ \mu\text{F}$, and $R_L = 820\ \Omega$. Determine V_r, V_{dc}, and I_L for this circuit.

26. The circuit shown in Figure 18.57 has the following values: $V_S = 24\ \text{V}_{\text{ac}}$ (rated), $C = 1200\ \mu\text{F}$, and $R_L = 200\ \Omega$. Determine V_r, V_{dc}, and I_L for the circuit.

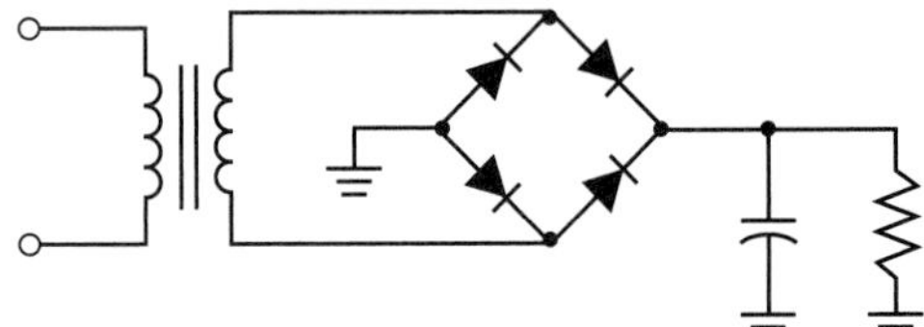

Figure 18.57

27. What is the PIV for the circuit described in Problem 26?

28. What is the PIV for the circuit shown in Figure 18.56a?

29. Determine the positive peak load voltage for the circuit shown in Figure 18.58a.

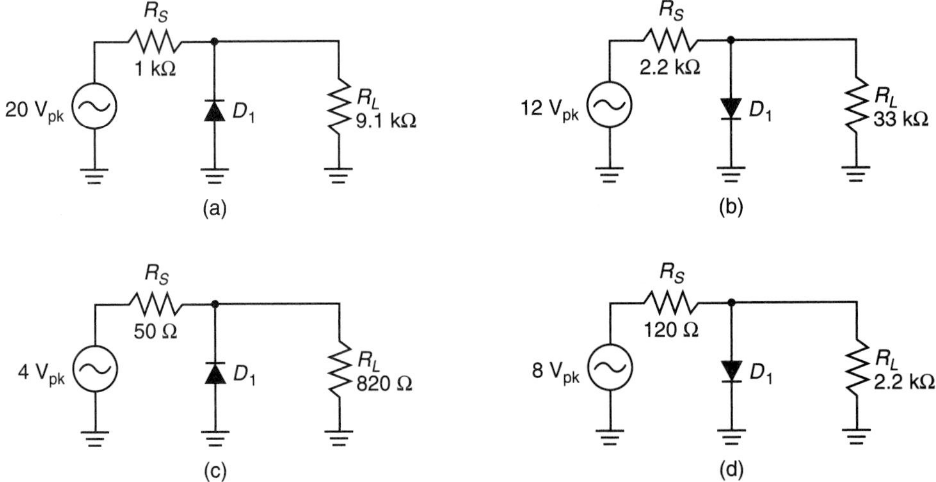

Figure 18.58

30. Draw the output waveform for the circuit shown in Figure 18.58a. Label the peak voltage values on the waveform.

31. Determine the negative peak load voltage for the circuit shown in Figure 18.58b.

32. Draw the output waveform for the circuit shown in Figure 18.58b. Label the peak voltage values on the waveform.

33. Determine the peak output voltage for the circuit shown in Figure 18.58c. Then, draw the waveform, and include the voltage values in the drawing.

34. Repeat Problem 5 for the circuit shown in Figure 18.58d.

35. The potentiometer in Figure 18.59a is set so that the anode of D_1 is at -2 V. Assuming that $V_S = 14\ V_{pp}$, determine the peak load voltages for the circuit, and draw the output waveform.

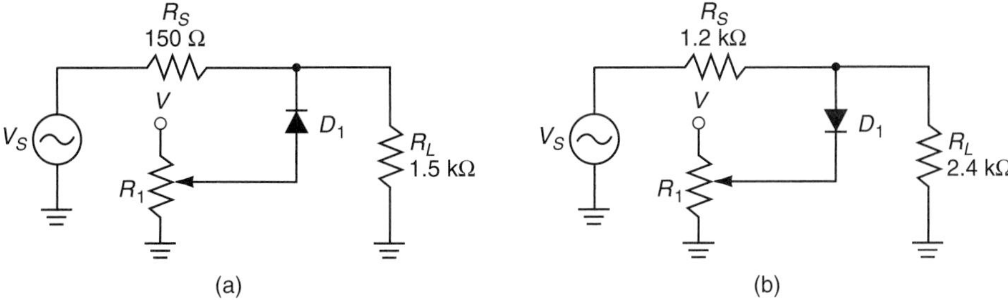

Figure 18.59

36. The potentiometer in Figure 18.59a is set so that the anode of D_1 is at 28 V. Assuming that $V_S = 24\ V_{pp}$, determine the peak load voltages for the circuit, and draw the output waveform.

37. The potentiometer in Figure 18.59b is set so that the cathode of D_1 is at $+4$ V. Assuming that $V_S = 22\ V_{pp}$, determine the peak load voltages for the circuit, and draw the output waveform.

38. The potentiometer in Figure 18.59b is set so that the cathode of D_1 is at $+2$ V. Assuming that $V_S = 4\ V_{pp}$, determine the peak load voltages for the circuit, and draw the output waveform. (Be careful on this one!)

39. A clamper with a 24 V_{pp} input shifts the waveform so that its peak voltages are 0 and -24 V. Determine the dc average of the output waveform.

40. A clamper with a 14 V_{pp} input shifts the waveform so that its peak voltages are 0 and $+14$ V. Determine the dc average of the output waveform.

41. The circuit shown in Figure 18.60a has values of $R_{D1} = 24\ \Omega$, $R_L = 2.2\ \text{k}\Omega$, and $C_1 = 4.7\ \mu\text{F}$. Determine the charge and discharge times for the capacitor.

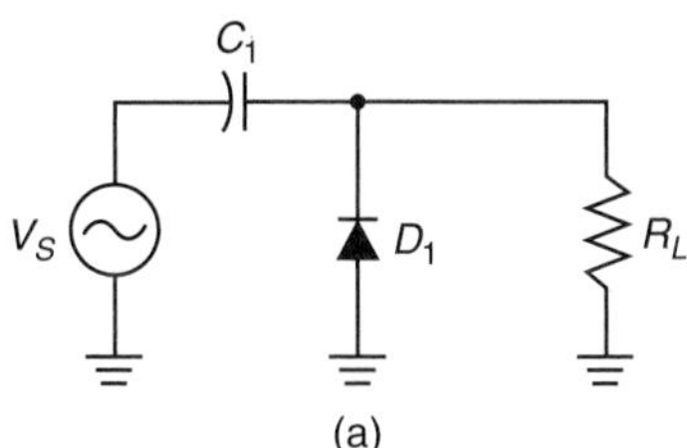

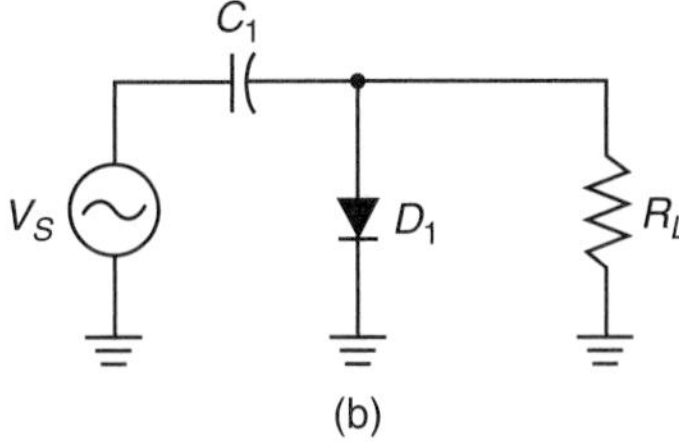

Figure 18.60

42. The circuit shown in Figure 18.60a has a 14 V_{pp} input signal. Draw the output waveform, and determine its peak voltage values.

43. The circuit shown in Figure 18.60a has values of $R_{D1} = 8\ \Omega$, $R_L = 1.2\ \text{k}\Omega$, and $C_1 = 33\ \mu\text{F}$. Determine the charge and discharge times for the capacitor.

44. The circuit shown in Figure 18.60b has values of $R_{D1} = 14\ \Omega$, $R_L = 1.5\ \text{k}\Omega$, and $C_1 = 1\ \mu\text{F}$. Determine the charge and discharge times for the capacitor.

45. The input to the circuit shown in Figure 18.60b is a 12 V_{pp} signal. Determine the peak load voltages for the circuit, and draw the output waveform.

46. The potentiometer in Figure 18.61a is set so that the anode of D_1 is at 13 V. The value of V_S is 9 V_{pp}. Draw the output waveform, and determine its peak voltage values.

47. The potentiometer in Figure 18.61b is set so that the cathode of D_1 is at $+6$ V. The value of V_S is 30 V_{pp}. Draw the output waveform for the circuit, and determine its peak voltage values.

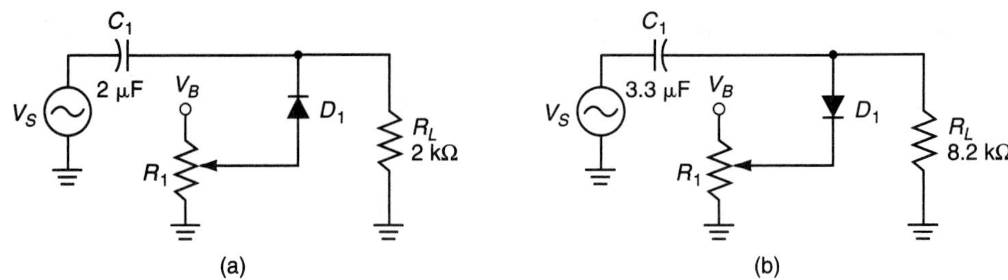

Figure 18.61

48. Draw the output waveform for the circuit shown in Figure 18.62a, and determine its peak voltage values.
49. Draw the output waveform for the circuit shown in Figure 18.62b, and determine its peak voltage values.

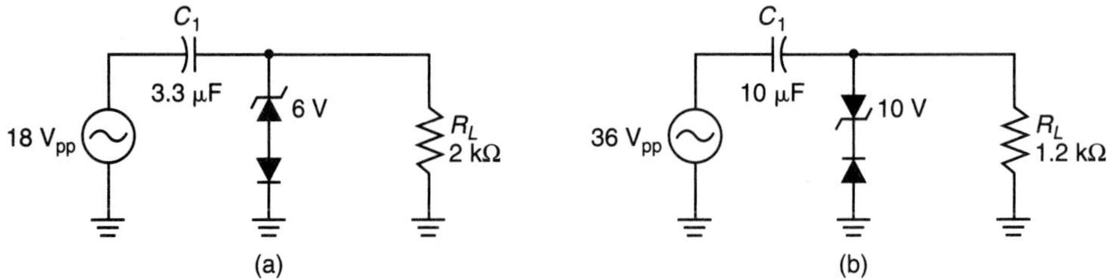

Figure 18.62

50. The circuit shown in Figure 18.63 has a 15 V_{pk} input signal. Determine V_{C1} and V_{C2} for the circuit. Assume that the diodes are ideal components.
51. The circuit shown in Figure 18.63 has a 48 V_{pk} input signal. Determine the values of V_{C1} and V_{C2} for the circuit. Assume that the diodes are ideal components.

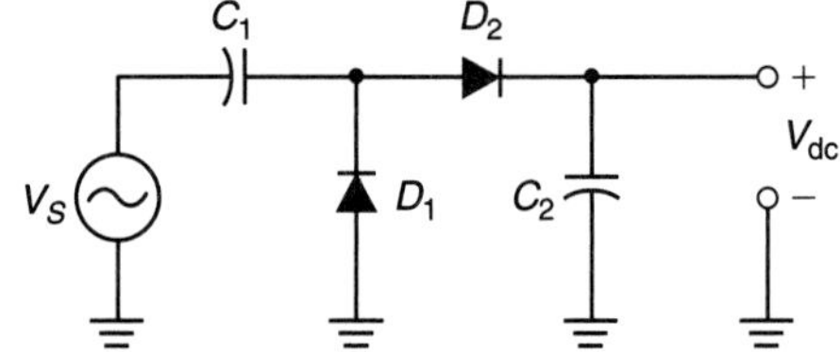

Figure 18.63

52. The circuit shown in Figure 18.64 has a 24 V_{pk} input signal. Determine V_{C1}, V_{C2}, and V_{C3} for the circuit. Assume that the diodes are ideal components.

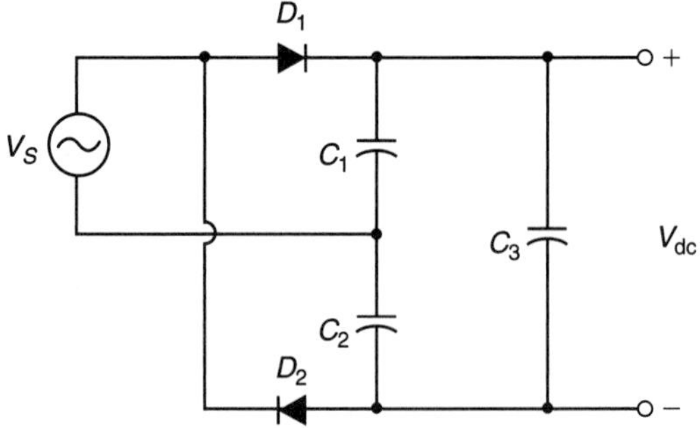

Figure 18.64

53. The circuit shown in Figure 18.64 has a 25 V_{ac} input signal. Determine V_{C1}, V_{C2}, and V_{C3} for the circuit. Assume that the diodes are ideal components. (Be careful on this one!)

54. Determine the dc output voltage for the circuit shown in Figure 18.65. Assume that the diodes are ideal components.

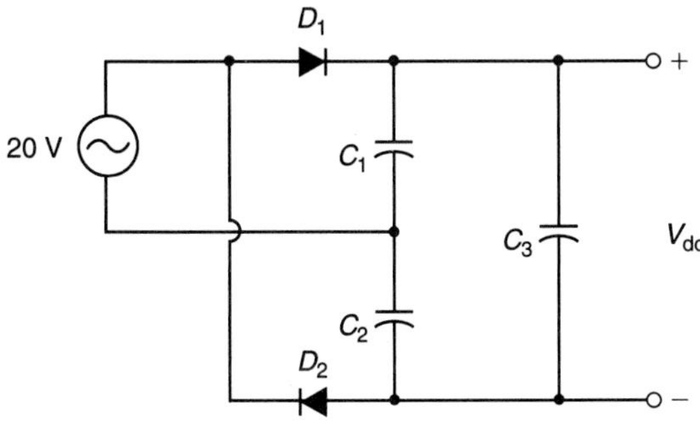

Figure 18.65

55. Determine the dc output voltage for the circuit shown in Figure 18.66. Assume that the diodes are ideal components.

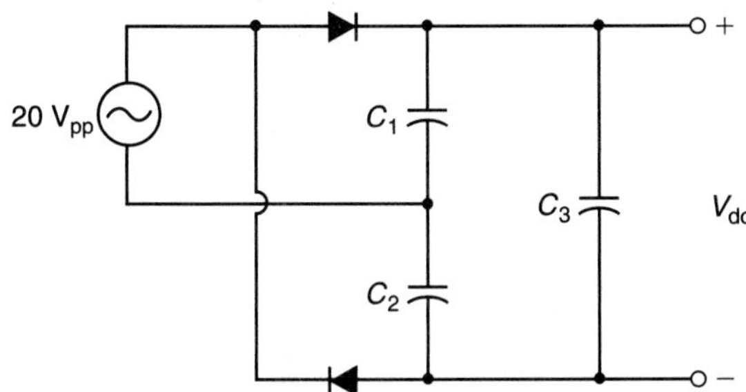

Figure 18.66

56. The circuit shown in Figure 18.67 has a 20 V_{pk} input signal. Determine the V_{C1}, V_{C2}, V_{C3}, and V_{C4} for the circuit. Assume that the diodes are ideal components.

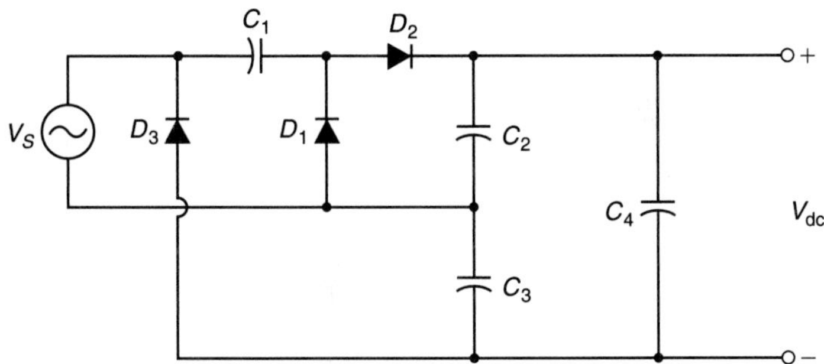

Figure 18.67

57. The circuit shown in Figure 18.67 has a 15 V_{ac} input signal. Determine V_{C1}, V_{C2}, V_{C3}, and V_{C4} for the circuit. Assume that the diodes are ideal components.

58. The circuit shown in Figure 18.68 has an 36 V_{ac} input signal. Determine V_{C1}, V_{C2}, V_{C3}, V_{C4}, and V_{C5} for the circuit. Assume that the diodes are ideal components.

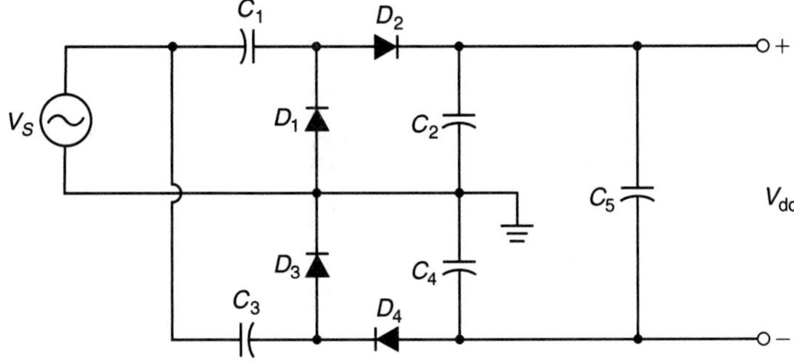

Figure 18.68

Looking Back

These problems relate to material presented in earlier chapters. The chapters are identified in brackets.

59. A 120 V_{pk} sine wave is applied to a 1.5 kΩ resistive load. Calculate the value of load power. [Chapter 9]

60. Derive the Norton equivalent of the circuit shown in Figure 18.69. [Chapter 7]

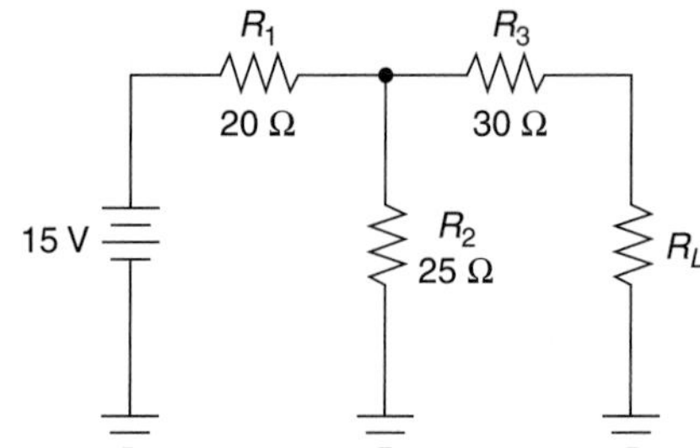

Figure 18.69

61. Calculate the total impedance for the circuit shown in Figure 18.70. [Chapter 11]

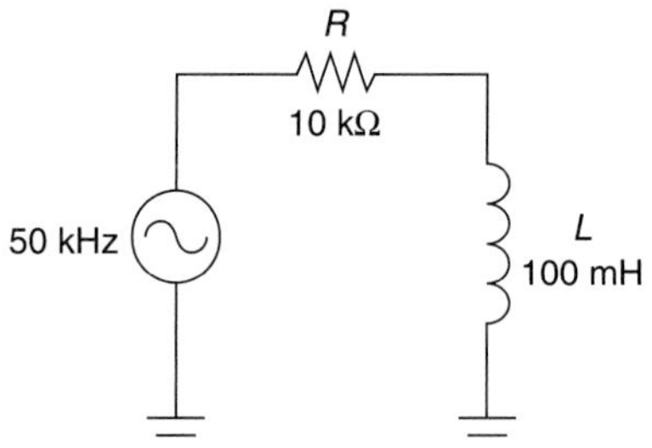

Figure 18.70

62. A parallel *LC* circuit has values of $L = 3.3$ mH, $R_W = 2\ \Omega$, and $C = 10$ nF. Calculate the *Q* of the circuit. [Chapter 14]

Pushing the Envelope

63. Explain the output waveform shown in Figure 18.71. (*Note:* There are no faulty components in the circuit.)

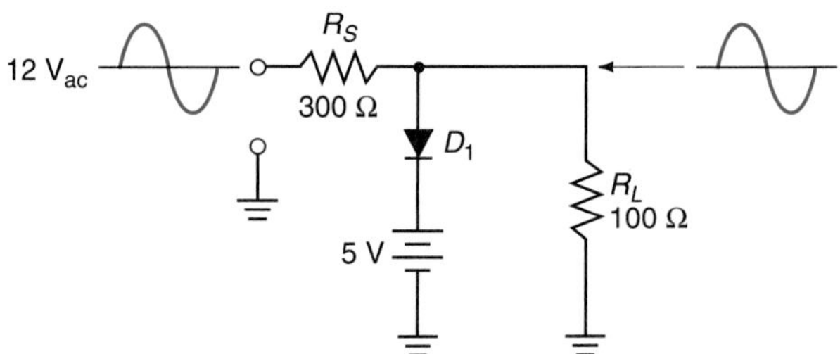

Figure 18.71

64. Figure 18.72 shows a half-wave voltage doubler with the diodes and capacitors reversed. Analyze the circuit, and show that it provides a negative dc output that is approximately twice the peak input voltage.

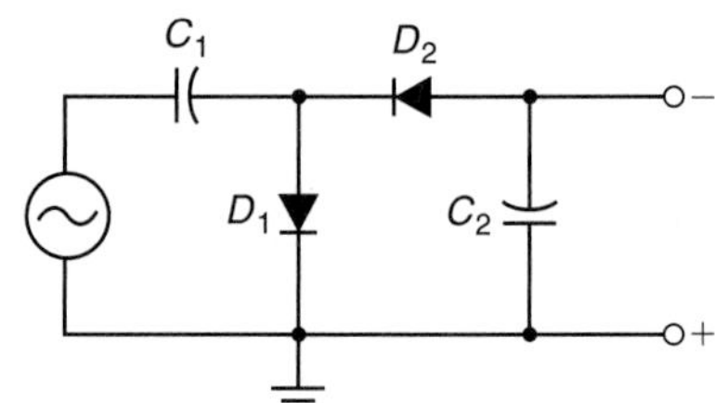

Figure 18.72

65. Schematics can be drawn in many different ways. For example, the circuit shown in Figure 18.73 is actually a voltage quadrupler. However, one of the components has accidentally been reversed. Which component is it? Explain your answer.

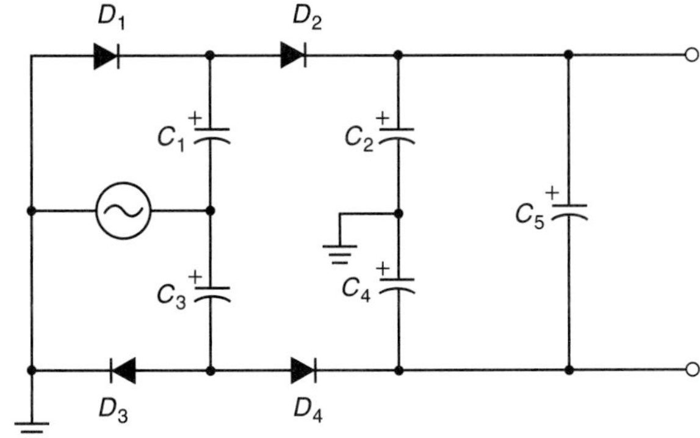

Figure 18.73

66. The circuit shown in Figure 18.74 is a zener clipper. Using your knowledge of diode operation, determine the shape of the circuit output waveform and its peak values.

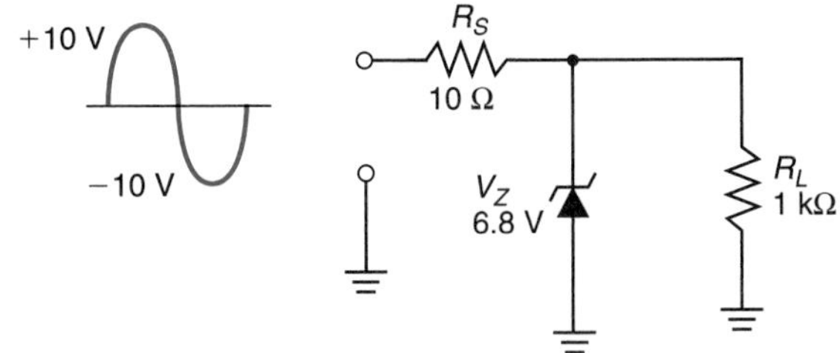

Figure 18.74

67. Repeat Problem 48 for the dual-zener clipper shown in Figure 18.75.

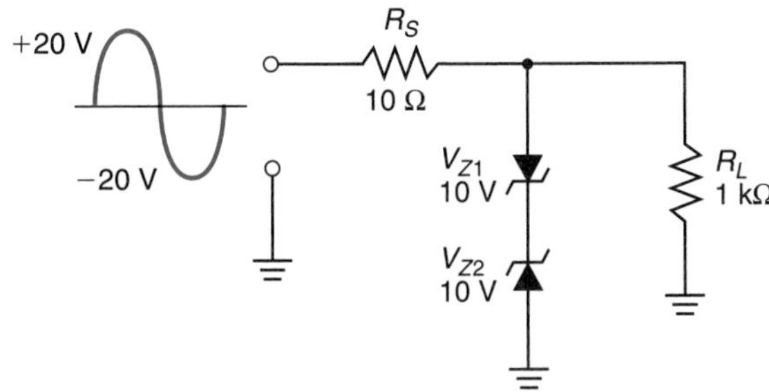

Figure 18.75

Answers to the Example Practice Problems

18.1 $V_L = 33.2\ \text{V}_{\text{pk}}$

18.2 $V_{\text{ave}} = 7.89$ V

18.3 $V_{\text{ave}} = 10.35\ \text{V}_{\text{dc}}$

18.4 $V_{\text{ave}} = 15.3\ \text{V}_{\text{dc}}$, $I_{\text{ave}} = 12.75$ mA

18.6 $12.54\ \text{V}_{\text{pk}}$, −0.7 V

18.7 0.7 V, $-11\ \text{V}_{\text{pk}}$

18.8 $T_C = 188$ ms, $T_D = 28.2$ ms

Multisim File References

Figure	File Reference	Figure	File Reference
18.3	ETF18_3.msm	18.43	ETF18_43.msm
18.10	ETF18_10.msm	18.44	ETF18_44.msm
18.16	ETF18_16.msm	18.45	ETF18_45.msm
18.20	ETF18_20.msm	18.48	ETF18_48.msm
18.32	ETF18_32.msm	18.50	ETF18_50.msm
18.36	ETF18_36.msm		

chapter 19

Bipolar Junction Transistor Operation and Biasing

Objectives

After studying the material in this chapter, you should be able to:

1. Name and identify each terminal of the *bipolar junction transistor* (*BJT*).
2. Identify the transistor terminal currents and voltages.
3. Describe the construction of a bipolar junction transistor.
4. Describe the characteristics of the *cutoff, saturation,* and *active* regions of operation.
5. State the relationship among the three transistor terminal currents.
6. Calculate any transistor terminal current given the other two currents, or *beta* (β) and one other terminal current.
7. Describe the *collector characteristic curve* of a BJT.
8. State the purpose of dc biasing circuits.
9. Describe the *dc load line* and *Q-point* of an amplifier.
10. Describe and analyze the operation of a *base-bias* circuit.
11. Describe the relationship between *beta, temperature,* and *Q-point shift.*
12. Describe and analyze the operation of a *voltage-divider bias* circuit.
13. Describe and analyze the operation of an *emitter-bias* circuit.
14. Describe and analyze the operation of a *collector-feedback bias* circuit.
15. Describe and analyze the operation of the *emitter-feedback bias* circuit.
16. Describe the differences between typical *pnp* biasing circuits and *npn* biasing circuits.
17. Describe the characteristics of *high-current, high-voltage,* and *high-power* transistors.
18. Describe the characteristics of *Darlington* transistors.
19. Describe the basic construction of *surface-mount components,* and list the advantages they have over other ICs.

The main building block of modern electronic systems is the *transistor.* The **transistor** is *a three-terminal device whose output current, voltage, and/or power are controlled by its input current.* In communications systems, it is used as the primary component in the **amplifier,** which is *a circuit used to increase the strength of an ac signal.* In digital computer systems, the transistor is used as a high-speed electronic switch capable of switching between two operating states (open and closed) at a rate of several billions of times per second.

There are two basic transistor types: the *bipolar junction transistor (BJT)*, and the *field-effect transistor (FET)*. The term *transistor* is generally used in reference to the BJT. The field-effect transistor is simply referred to as an FET. In this chapter, we will investigate BJT operation and biasing. FETs are discussed in Chapter 21.

19.1 Introduction to Bipolar Junction Transistors

The **bipolar junction transistor (BJT)** is a three-terminal component. The three terminals are called the **emitter,** the **collector,** and the **base.** The emitter and collector are made of the same type of semiconductor material (either *p*-type or *n*-type), while the base is made of the other type of material. The structure of the BJT is illustrated in Figure 19.1. ◄ *OBJECTIVE 1*

There are two types of BJTs. The ***npn* transistor** has *n*-type emitter and collector materials and a *p*-type base. The ***pnp* transistor** is constructed in the opposite manner: The transistor has *p*-type emitter and collector materials and an *n*-type base.

Figure 19.1 also shows the schematic symbols for the *npn* and *pnp* transistors. The arrow on the schematic symbol is significant for several reasons:

- *It identifies the component terminals.* The arrow is always drawn on the emitter terminal. The terminal farthest from the emitter is the collector, and the center terminal is the base.
- *The arrow always points toward the n-type material.* If the arrow points toward the base, the transistor is a *pnp* type. If it points toward the emitter, the transistor is an *npn* type.

Transistor Currents

The terminal currents of a transistor are illustrated in Figure 19.2. The emitter, collector, and base currents of the transistor are identified as I_E, I_C, and I_B, respectively. Under normal circumstances, I_E is slightly greater than I_C, and both currents are significantly greater than the base current (I_B). Note that the terminal currents for the *npn* and *pnp* transistors are in opposite directions. ◄ *OBJECTIVE 2*

The transistor is a *current-controlled device;* that is, the values of the collector and emitter currents are controlled primarily by the base current. Under normal circumstances,

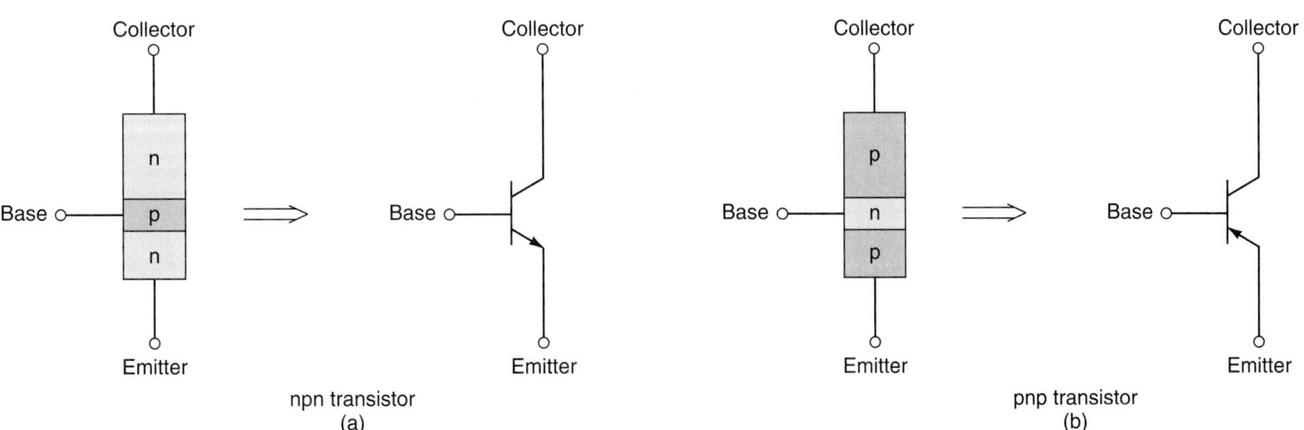

Figure 19.1 BJT construction and schematic symbols.

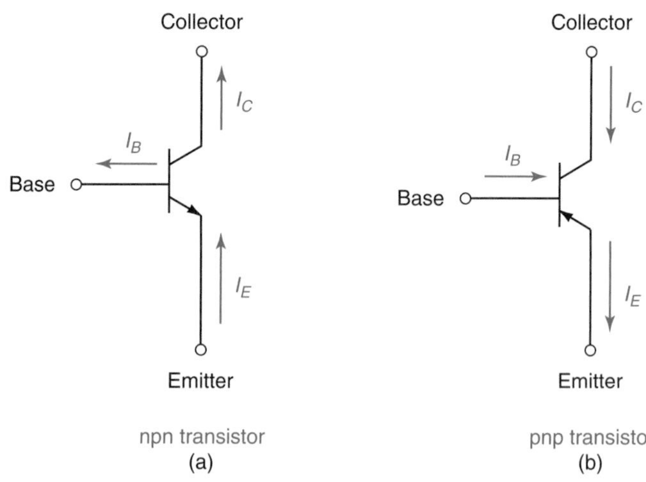

Figure 19.2 Transistor terminal currents.

the values of I_C and I_E vary directly with the value of I_B. An increase or decrease in I_B results in similar changes in I_C and I_E.

Normally, I_C is some multiple of I_B. *The factor by which current increases from the base to the collector* is referred to as the **forward current gain** and is represented by the Greek letter *beta* (β). When the values of I_B and β for a transistor are known, the value of I_C can be found as

$$I_C = \beta I_B \tag{19.1}$$

For example, if the transistor shown in Figure 19.2a has values of I_B = 50 mA and β = 120, I_C is found as

$$I_C = \beta I_B = (120)(50\ \mu\text{A}) = 6\ \text{mA}$$

Transistor Voltages

Several voltages are normally involved in any discussion of transistor operation. These voltages are listed in Table 19.1 and identified in Figure 19.3. Be sure that you can distinguish between these voltages, because it will make future discussions easier to follow.

TABLE 19.1 Transistor Voltages

Voltage Abbreviation	*Definition*
V_{CC}	*Collector supply voltage.* This is a power supply voltage applied directly or indirectly to the collector of the transistor.
V_{BB}	*Base supply voltage.* This is a dc voltage used to bias the base of the transistor. It may come directly from a dc supply or be applied indirectly to the base via a resistive circuit.
V_{EE}	*Emitter supply voltage.* This, again, is a dc biasing voltage.
V_C	The dc voltage *measured from the collector terminal to ground.*
V_B	The dc voltage *measured from the base terminal to ground.*
V_E	The dc voltage *measured from the emitter terminal to ground.*
V_{CE}	The dc voltage *measured across the collector and emitter terminals.*
V_{BE}	The dc voltage *measured across the base and emitter terminals.*
V_{CB}	The dc voltage *measured across the collector and base terminals.*

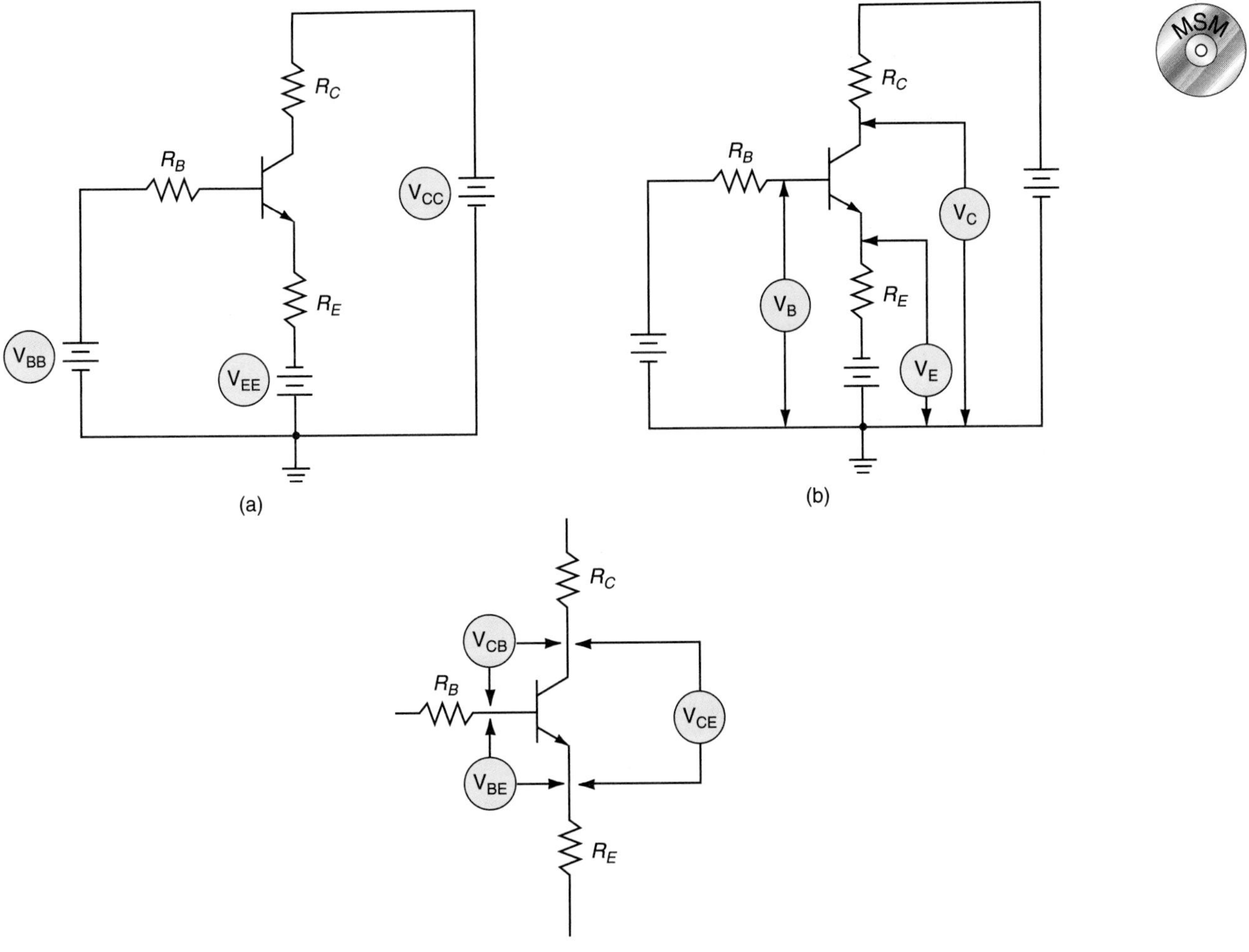

MSM

Figure 19.3 Transistor terminal voltages.

Section Review

1. What is the primary application for the transistor in communications electronics?
2. What is the primary application for the transistor in digital electronics?
3. What are the two basic types of transistors?
4. What are the terminals of a *bipolar junction transistor* (*BJT*) called?
5. What are the two types of BJTs? How do they differ from each other?
6. What does the arrow on the BJT schematic symbol indicate?
7. What is the "normal" relationship between I_C and I_B?
8. What is *forward current gain*?
9. Define the following: V_{CC}, V_{BB}, V_{EE}, V_C, V_E, V_B, V_{CE}, V_{BE}, and V_{CB}.

19.2 Transistor Construction and Operation

◄ *OBJECTIVE 3*

The three materials that form the transistor are joined so that they form two *pn* junctions, as shown in Figure 19.4. *The point at which the emitter and base are joined* is called the **base-emitter junction.** The **collector-base junction** is *the point at which the base and*

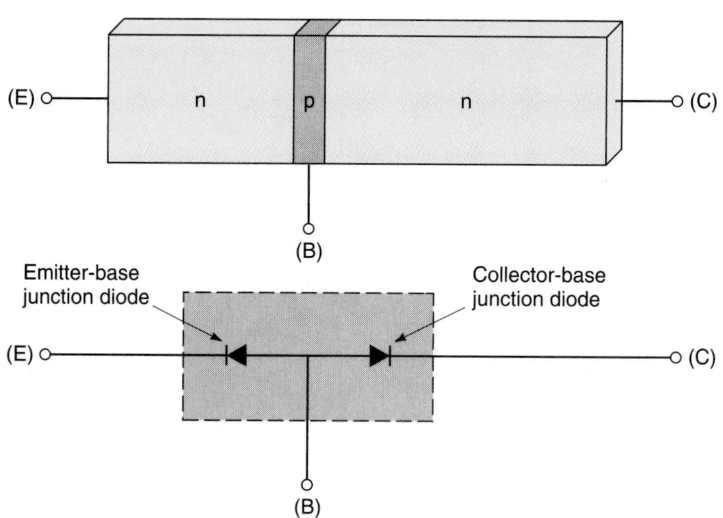

Figure 19.4 BJT construction.

collector are joined. The two junctions are normally operated using one (or more) of the following biasing combinations:

Base-Emitter Junction	*Collector-Base Junction*	*Operating Region*
Reverse biased	Reverse biased	Cutoff
Forward biased	Reverse biased	Active
Forward biased	Forward biased	Saturation

We will now look at all three of these operating conditions or regions. Though we will be concentrating on the *npn* transistor, all the principles covered apply equally to the *pnp* transistor.

Zero Bias

Figure 19.5 shows an *npn* transistor with no biasing potential applied. You may recall that an unbiased *pn* junction forms a depletion layer at room temperature. The two transistor depletion layers are shown in Figure 19.5a. Since both junctions are reverse biased, the values of I_E, I_C, and I_B equal zero in this operating condition.

Cutoff

OBJECTIVE 4 ➤ Two biasing voltages are connected to the transistor shown in Figure 19.5b. Note that both biasing sources in this figure have polarities that *reverse bias* their respective junctions.

With the polarities shown, the two depletion layers extend well into the emitter, base, and collector regions. Only *an extremely small amount of reverse current passes from the emitter to the collector,* and the transistor is said to be in **cutoff.**

Saturation

The opposite of cutoff is *saturation.* **Saturation** is *the operating condition where further increases in I_B do not cause an increase in I_C.* When a BJT is in saturation, I_C has reached its maximum possible value, as determined by V_{CC} and the total resistance in the collector-emitter circuit. This point is illustrated in Figure 19.6.

Assume for a moment that the transistor shown in Figure 19.6 has a value of $V_{CE} = 0$ V (an *ideal* value). If this is the case, the value of I_C depends completely on the values of

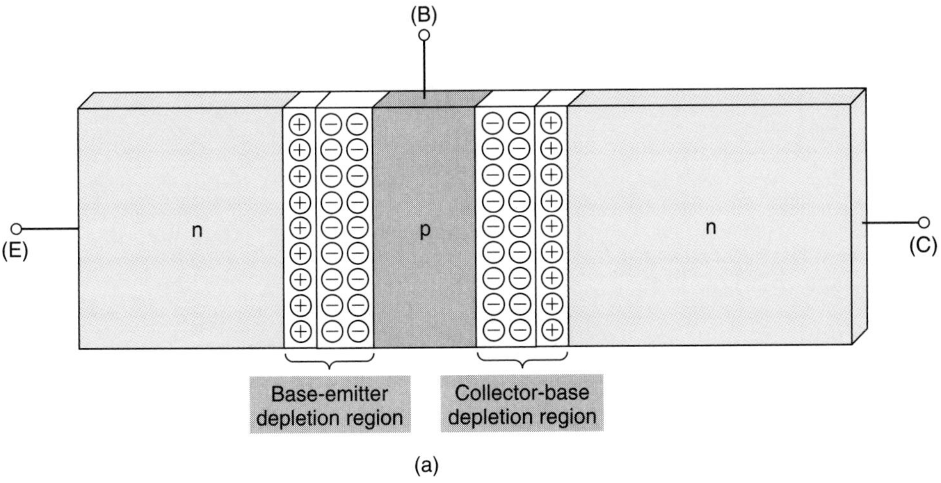

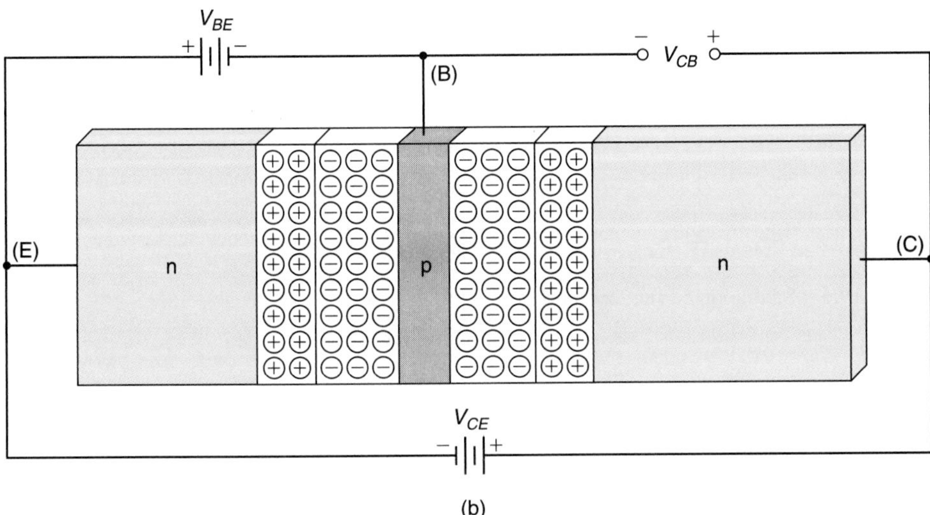

Figure 19.5 Zero biasing and cutoff.

V_{CC}, R_C (the collector resistor), and R_E (the emitter resistor). According to Ohm's law, this maximum value of I_C can be found as

$$I_{C(\max)} = \frac{V_{CC}}{R_C + R_E}$$

If I_B is increased to the point where I_C reaches this maximum value:

- Additional increases in I_B do not cause an increase in I_C, and the $I_C = \beta I_B$ relationship no longer holds true.
- Both of the transistor junctions become forward biased.

As shown in Figure 19.6, $V_{CE} \cong 0.3$ V (which is typical for a saturated transistor). With a value of $V_{BE} \cong 0.7$ V, the collector-base junction is biased to the difference between V_{CE} and V_{BE}, 0.4 V. This means that the collector-base junction is forward biased (even though it isn't fully on).

Active Operation

A transistor is said to be operating in its **active region** when *the base-emitter junction is forward biased and the collector-base junction is reverse biased.* In other words, the

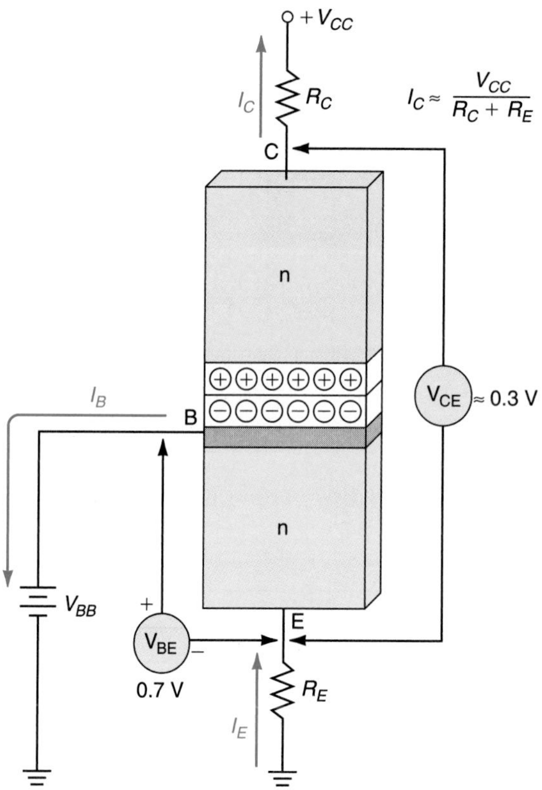

Figure 19.6 Saturation current conditions.

active region lies between cutoff and saturation. The biasing for active operation is illustrated in Figure 19.7.

When the base-emitter voltage (V_{BE}) is great enough to overcome the barrier potential of the junction (0.7 V), current is generated in the emitter and base regions. If the base and emitter regions acted as a normal diode, all of the emitter current would exit through the base. However, the base region is *very lightly doped,* so the resistance of the base material is much greater than that of the reverse-biased collector-base junction. Thus, *the vast majority of the emitter current continues through the reverse-biased collector-base junction to the collector circuit.*

As a reference, the characteristics of the three transistor operating regions are summarized in Figure 19.8.

Note: The idea of current through a reverse-biased *pn* junction should not seem that strange to you. The zener diode is designed for conduction through a reverse-biased junction.

Section Review

1. How are the two junctions of a transistor biased when the component is in:
   a. *Cutoff*?
   b. The *active region*?
   c. *Saturation*?
2. What is approximate the value of I_C when a transistor is in cutoff?
3. What controls the value of I_C when a transistor is saturated?
4. Describe the basic operation of a transistor biased for active-region operation.

Critical Thinking

5. How could you verify that the depletion layers shown in Figure 19.5a exist when there is no bias applied to a transistor?
6. When current passes through any resistance, *heat* is generated. Based on Figure 19.7a, which part of a BJT generates the greatest amount of heat?

Figure 19.7
Active operation.

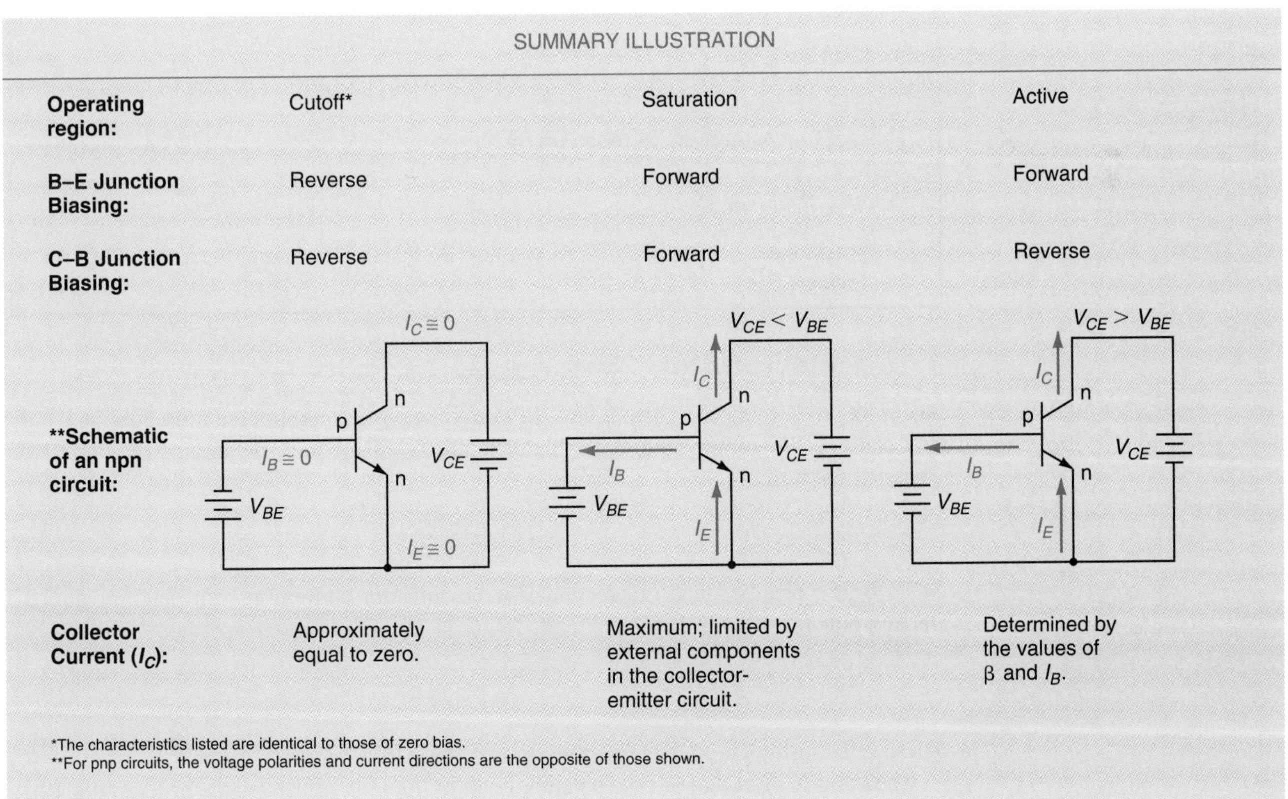

SUMMARY ILLUSTRATION

Operating region:	Cutoff*	Saturation	Active
B–E Junction Biasing:	Reverse	Forward	Forward
C–B Junction Biasing:	Reverse	Forward	Reverse
****Schematic of an npn circuit:**	$I_C \cong 0$, $I_B \cong 0$, $I_E \cong 0$, V_{BE}, V_{CE}	$V_{CE} < V_{BE}$, I_C, I_B, I_E, V_{BE}, V_{CE}	$V_{CE} > V_{BE}$, I_C, I_B, I_E, V_{BE}, V_{CE}
Collector Current (I_C):	Approximately equal to zero.	Maximum; limited by external components in the collector-emitter circuit.	Determined by the values of β and I_B.

*The characteristics listed are identical to those of zero bias.
**For pnp circuits, the voltage polarities and current directions are the opposite of those shown.

Figure 19.8

19.3 Transistor Currents

In this section, we will take a look at several current relationships and ratings and what they mean in common transistor applications.

Transistor Currents

In many transistor applications, the base current is varied to produce variations in I_C and I_E. A small change in I_B results in a larger change in the other terminal currents, as illustrated in Example 19.1.

Example 19.1

Determine the value of I_C for each value of I_B shown in Figure 19.9.

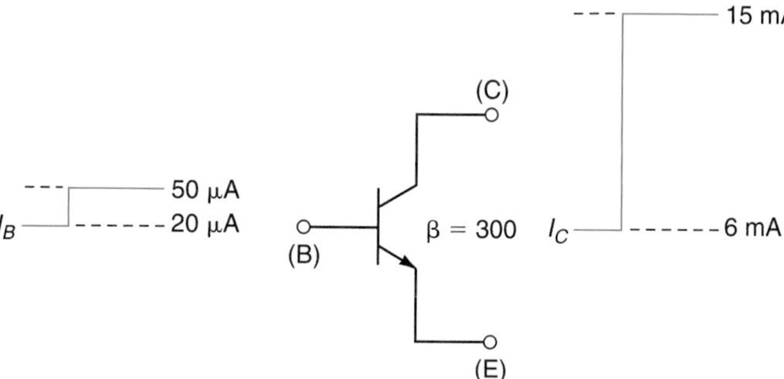

Figure 19.9

Solution: The base current has an initial value of 20 μA, and the transistor beta is 300. Using these values,

$$I_C = \beta I_B = (300)(20\ \mu\text{A}) = 6\ \text{mA}$$

When I_B increases to 50 μA, the collector current increases to

$$I_C = \beta I_B = (300)(50\ \mu\text{A}) = 15\ \text{mA}$$

Thus, a 30 μA increase in base current causes a 9 mA increase in collector current.

Practice Problem 19.1

A transistor has values of $I_B = 50\ \mu\text{A}$ and $\beta = 350$. Determine the value of I_C for the device.

The Relationship Among I_E, I_C, and I_B

OBJECTIVE 5 ➤ According to Kirchhoff's current law, the current leaving a component must equal the current entering the component. With this in mind, I_E must equal the sum of the other two currents. By formula,

$$I_E = I_C + I_B \tag{19.2}$$

Since I_C is typically much greater than I_B, the collector and emitter currents are approximately equal. By formula,

$$I_C \cong I_E \tag{19.3}$$

These relationships hold true for both *npn* and *pnp* transistors.

The validity of equation (19.3) can be seen by looking at the results in Example 19.1. When I_B was given as 50 μA, I_C was determined to be 15 mA. According to equation (19.2), I_E can be found as

$$I_E = I_C + I_B = 15 \text{ mA} + 50 \text{ μA} = 15.05 \text{ mA}$$

As you can see, the values of I_E and I_C are approximately equal for the transistor in Example 19.1.

DC Beta

The **dc beta** of a transistor is *the ratio of dc collector current to dc base current.* By formula,

$$\beta = \frac{I_C}{I_B} \tag{19.4}$$

This relationship is extremely important, because many transistor amplifier circuits have an input signal applied to the base and an output signal taken from the collector. Thus, *the dc beta rating represents the overall dc current gain of the transistor amplifier.*

We can use equations (19.1) and (19.2) to show how β affects the other terminal currents. As given in equation (19.1),

$$I_C = \beta I_B$$

If we combine this relationship with equation (19.2), we get

$$I_E = I_C + I_B = \beta I_B + I_B$$

or

$$I_E = (\beta + 1)I_B \tag{19.5}$$

As the next three examples illustrate, you can use beta and any one terminal current to determine the values of the other two terminal currents. ◄ *OBJECTIVE 6*

Example 19.2

Determine the values of I_C and I_E for the transistor shown in Figure 19.10.

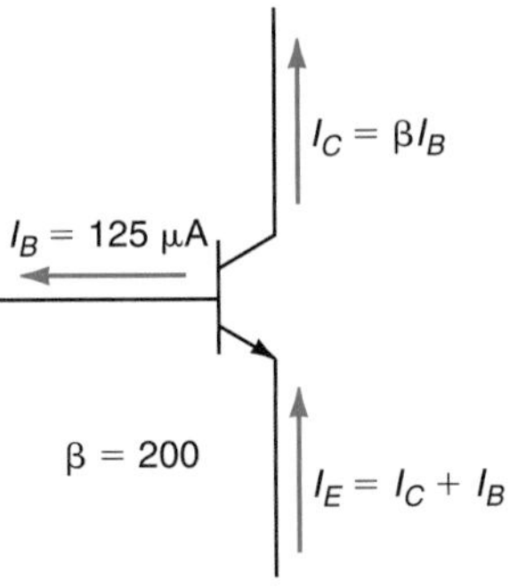

Figure 19.10

Solution: The value of I_C can be found as

$$I_C = \beta I_B = (200)(125 \text{ μA}) = 25 \text{ mA}$$

And the value of I_E is found as

$$I_E = I_C + I_B = 25 \text{ mA} + 125 \text{ μA} = 25.125 \text{ mA}$$

Practice Problem 19.2

The transistor shown in Figure 19.10 has values of $I_B = 50$ μA and $\beta = 400$. Determine the values of I_C and I_E for the device.

Example 19.3

Determine the values of I_C and I_B for the transistor shown in Figure 19.11.

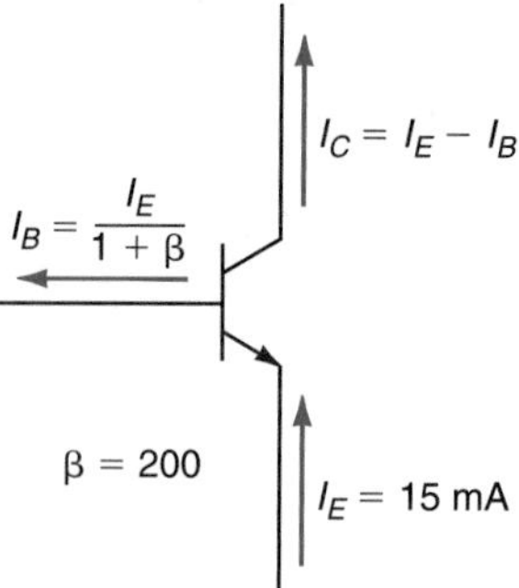

Figure 19.11

Solution: Equation (19.5) can be transposed to give us

$$I_B = \frac{I_E}{\beta + 1} = \frac{15 \text{ mA}}{201} = 74.6 \ \mu\text{A}$$

Now I_C can be found as

$$I_C = \beta I_B = (200)(74.6 \ \mu\text{A}) = 14.9 \text{ mA}$$

Practice Problem 19.3

The transistor shown in Figure 19.11 has values of $I_E = 12$ mA and $\beta = 140$. Determine the values of I_B and I_C for the device.

Example 19.4

Determine the values of I_B and I_E for the circuit shown in Figure 19.12.

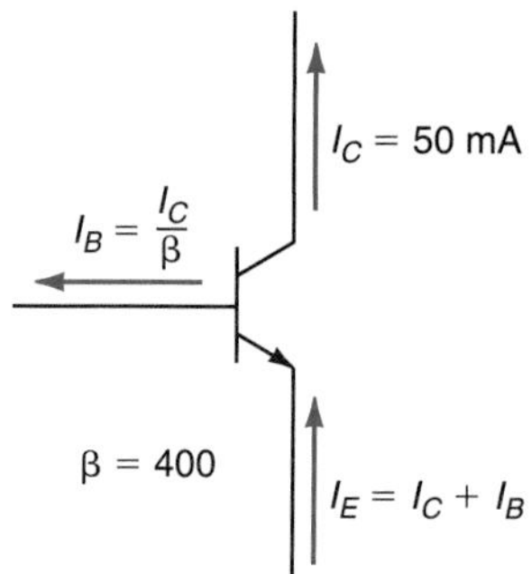

Figure 19.12

Solution: The base current can be found as

$$I_B = \frac{I_C}{\beta} = \frac{50 \text{ mA}}{400} = 125 \ \mu\text{A}$$

Now the emitter current can be found as

$$I_E = I_C + I_B = 50 \text{ mA} + 125 \ \mu\text{A} = 50.125 \text{ mA}$$

Practice Problem 19.4

The transistor shown in Figure 19.12 has values of $I_C = 80$ mA and $\beta = 170$. Determine the values of I_B and I_E for the device.

Because beta is a ratio of current values, it has no unit of measure. Typical beta ratings fall somewhere in the range of 50 to 300.

DC current gain The dc beta rating of a transistor is normally listed as **dc current gain (h_{FE})** on its specification sheet. Note that the label h_{FE} is normally used to represent dc beta, so we will be using this label in all our discussions from this point on. The origin of this label is discussed in Appendix E.

β_{dc} is normally identified as h_{FE} on transistor spec sheets.

DC Alpha

The **dc alpha** of a transistor is *the ratio of collector current to emitter current.* By formula,

$$\alpha = \frac{I_C}{I_E} \tag{19.6}$$

Since I_C is always less than I_E, the alpha rating of a transistor is always less than *unity* (1). Alpha can also be defined in terms of the transistor's beta rating, as follows:

$$\alpha = \frac{h_{FE}}{h_{FE} + 1} \tag{19.7}$$

Collector Characteristics Curves

◄ OBJECTIVE 7

The **collector characteristic curve** represents *the relationship between I_C and V_{CE} at a specified value of I_B*. Figure 19.13a shows a set of two collector curves. The collector curve is divided into four parts:

- The portion of the curve below the knee voltage (V_K) represents the *saturation* region of operation.
- The portion of the curve below the $I_B = 0$ A line represents the *cutoff* region of operation.
- The portion of the curve between V_K and the breakdown voltage (V_{BR}) represents the *active* region of operation.
- The portion of the curve to the right of V_{BR} represents the characteristics of the device when it is driven into breakdown.

Most amplifiers make use of the active region of operation (between V_K and V_{BR}). Note that the curves shown in Figure 19.13a are nearly flat in the area between V_K and V_{BR}. This indicates that I_C remains relatively constant over a wide range of V_{CE} values when the device is operated in its active region. This point can be demonstrated further using the circuit shown in Figure 19.14. According to Kirchhoff's voltage law, the value of V_{CE} for this circuit must equal the difference between V_{CC} and the voltage across the collector resistor (I_CR_C). By formula,

$$V_{CE} = V_{CC} - I_CR_C \tag{19.8}$$

According to equation (19.1), the circuit shown in Figure 19.14 has a value of

$$I_C = h_{FE}I_B = (100)(100\ \mu\text{A}) = 10\ \text{mA}$$

Using equation (19.8), we can now calculate a value of

$$V_{CE} = V_{CC} - I_CR_C = 10\ \text{V} - (10\ \text{mA})(470\ \Omega) = 10\ \text{V} - 4.7\ \text{V} = 5.3\ \text{V}$$

If we change the collector resistor to a value of 220 Ω, nothing has been done to affect the values of I_B and I_C. However, the value of V_{CE} changes to

$$V_{CE} = V_{CC} - I_CR_C = 10\ \text{V} - (10\ \text{mA})(220\ \Omega) = 10\ \text{V} - 2.2\ \text{V} = 7.8\ \text{V}$$

These calculations demonstrate that I_C is not controlled by the value of V_{CE} when the device is operated in its active region. The value of I_C varies primarily as a result of changes in I_B or h_{FE}.

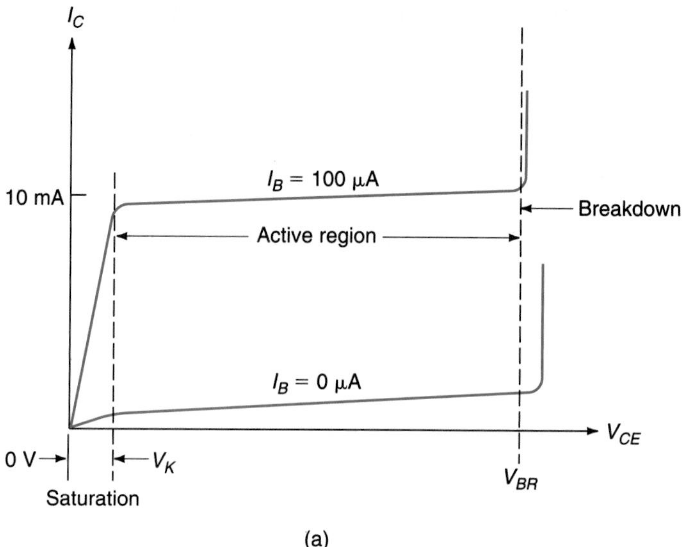

(a)

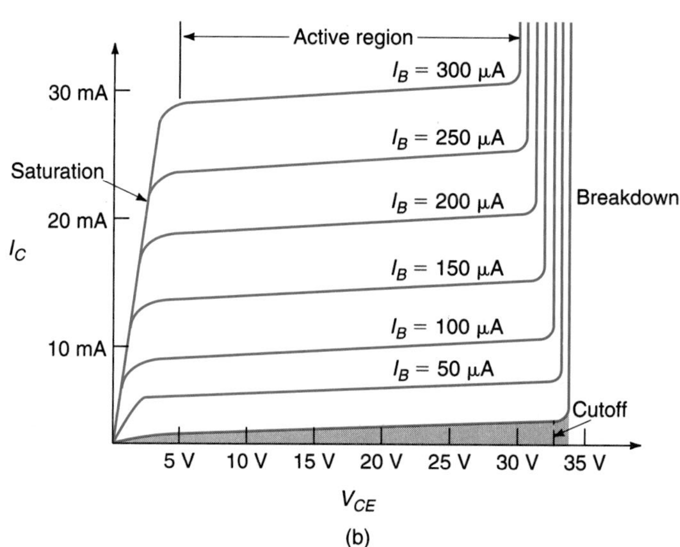

(b)

Figure 19.13 Collector characteristic curves.

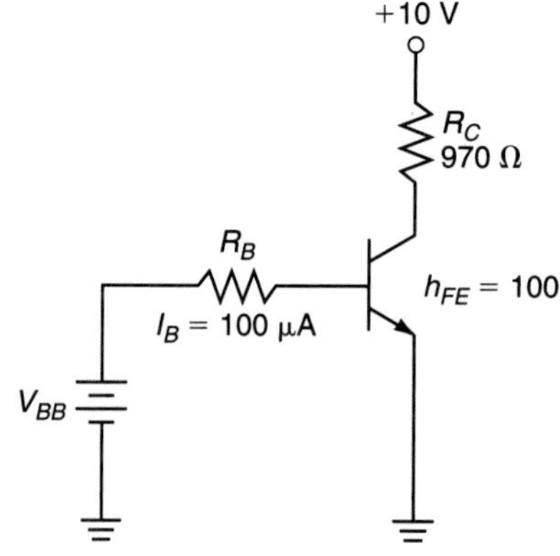

Figure 19.14

Transistor breakdown Transistor breakdown will occur if V_{CE} exceeds the breakdown voltage rating of the transistor. When breakdown occurs, I_C increases dramatically—until the transistor is destroyed by excessive heat. The far right-hand portion of each curve in Figure 19.13a represent the breakdown region. Note that collector curves are normally plotted as a *composite* of curves, each of which is plotted for a different value of I_B, as shown in Figure 19.13b.

Section Review

1. What is meant by the term *current-controlled device*?
2. What is meant by the term *current gain*?
3. What symbol is commonly used to represent dc current gain?
4. Under normal circumstances, what is the relationship between:
   a. Base current and collector current?
   b. Collector current and emitter current?
5. Why doesn't *beta* have any unit of measure?
6. What is *dc alpha*?
7. What is the limit on the value of alpha?
8. List and describe the four regions of the transistor collector curve.
9. What effect does V_{CE} have on I_C when a transistor is operated in its active region?

19.4 Introduction to Transistor Biasing

◄ OBJECTIVE 8

The ac operation of a transistor amplifier depends on its initial dc values of I_B, I_C, and V_{CE}. When I_B is varied around an initial value, I_C and V_{CE} vary around their initial values, as illustrated in Figure 19.15. The function of *dc biasing* is to set the initial values of I_B, I_C, and V_{CE}. Several bias methods are used to achieve different results, but each provides a means of setting the initial operating values of voltage and current.

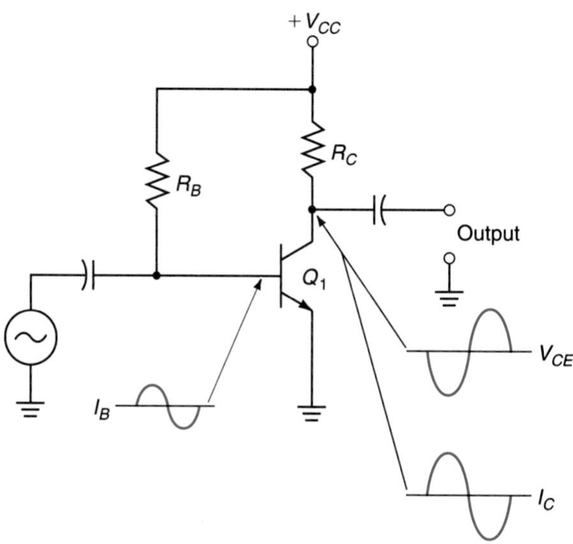

Figure 19.15 Typical amplifier operation.

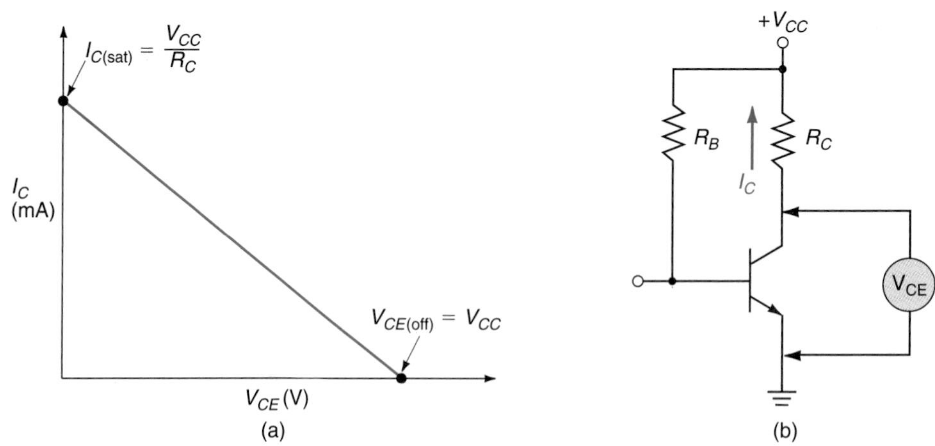

Figure 19.16 A generic dc load line.

The DC Load Line

OBJECTIVE 9 ➤ For every possible value of I_C, an amplifier has a corresponding value of V_{CE}. The **dc load line** for a given biasing circuit is *a graph that represents every possible combination of I_C and V_{CE} for the circuit.* A generic dc load line is shown in Figure 19.16a.

The ends of the load line are labeled $I_{C(sat)}$ and $V_{CE(off)}$. The $I_{C(sat)}$ point represents the *ideal* value of saturation current for the circuit. Ideally, a saturated transistor acts like a closed switch, and $V_{CE} = 0$ V. Therefore, the value of $I_{C(sat)}$ for the circuit shown in Figure 19.16b can be found as

$$I_{C(sat)} = \frac{V_{CC}}{R_C} \tag{19.9}$$

When the transistor is in cutoff, it acts like an open switch. Ideally, $I_C = 0$ A, and as a result, $I_C R_C = 0$ V. Since there is no voltage across the collector resistor, V_{CC} is dropped across the transistor, and

$$V_{CE(off)} = V_{CC} \tag{19.10}$$

Example 19.5 shows how these two equations are used to plot the dc load line for an amplifier.

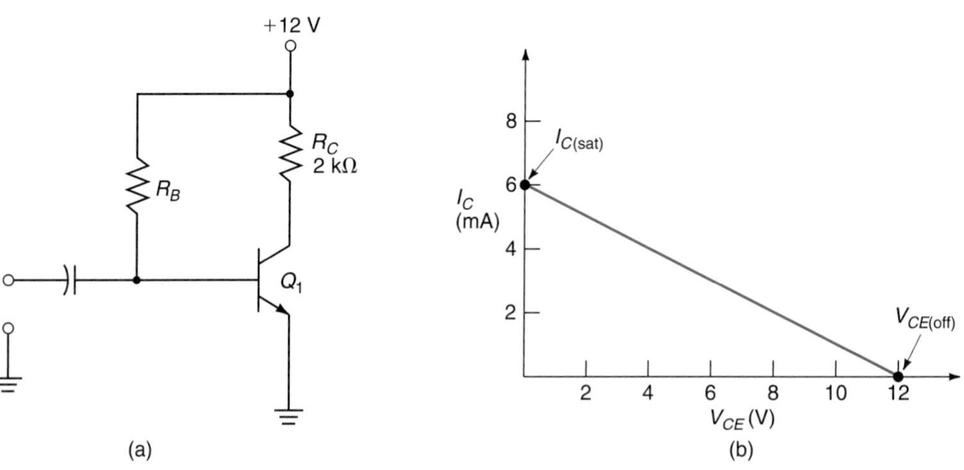

Figure 19.17

Example 19.5

Plot the dc load line for the circuit shown in Figure 19.17a.

Solution: $$V_{CE(\text{off})} = V_{CC} = 12\text{ V}$$

and

$$I_{C(\text{sat})} = \frac{V_{CC}}{R_C} = \frac{12\text{ V}}{2\text{ k}\Omega} = 6\text{ mA}$$

The plotted line is shown in Figure 19.17b.

Practice Problem 19.5

A circuit like the one shown in Figure 19.17a has values of $V_{CC} = +8$ V and $R_C = 1.1$ kΩ. Plot the dc load line for the circuit.

Note that R_B was not a part of the load line calculation in Example 19.5. This indicates that its value has no effect on load line end points.

The Q-Point

When a transistor has no input signal, it has constant values of I_C and V_{CE}. *The point on the dc load line that represents this combination of I_C and V_{CE}* is referred to as the ***Q*-point.** The letter *Q* comes from the word **quiescent,** meaning *at rest.*

For linear operation of an amplifier, it is desirable to have the *Q*-point centered on the load line. With a centered *Q*-point,

$$V_{CE} = \frac{V_{CC}}{2} \quad \text{(midpoint)} \quad (19.11)$$

and

$$I_C = \frac{I_{C(\text{sat})}}{2} \quad \text{(midpoint)} \quad (19.12)$$

as illustrated in Figure 19.18a. *An amplifier with a centered Q-point* is said to be **midpoint biased.**

Midpoint biasing allows optimum ac operation of the amplifier. This point is also illustrated in Figure 19.18b. When a sine wave is applied to the base of the transistor, I_C

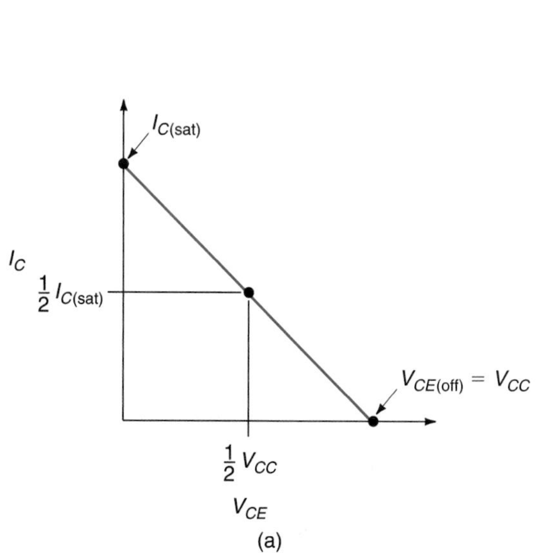

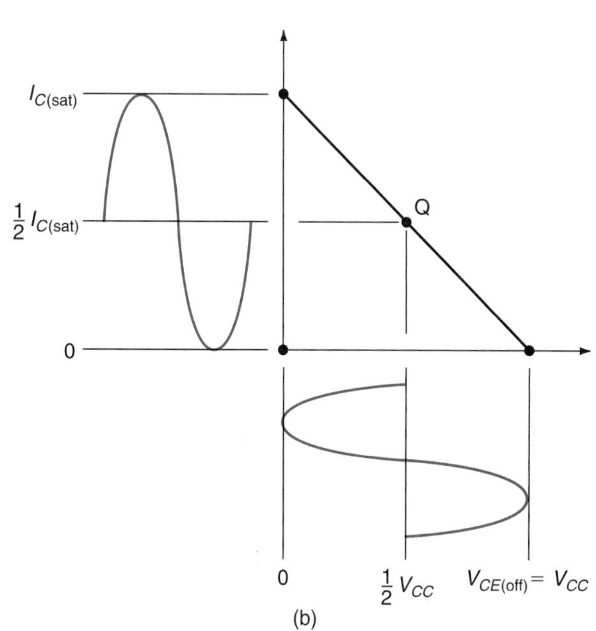

Figure 19.18 The Q-point for a midpoint-biased amplifier.

and V_{CE} both vary around their Q-point values. When the Q-point is centered, I_C and V_{CE} can both make the maximum possible transitions above and below their initial dc values.

Section Review

1. What purpose is served by a dc biasing circuit?
2. What is a *dc load line*?
3. What is represented by the point where the load line meets the I_C axis on the graph?
4. What is represented by the point where the load line meets the V_{CE} axis on the graph?
5. What does the *Q-point* of an amplifier represent?
6. What is meant by the term *quiescent*?
7. What does it mean when an amplifier is in its quiescent state?
8. What is *midpoint bias*?
9. Why is midpoint biasing desirable?

19.5 Base Bias

OBJECTIVE 10 ➤ *The simplest type of transistor biasing* is **base bias,** or **fixed bias.** A base-bias circuit is shown in Figure 19.19. The terminal currents are in the directions indicated. As the arrows in the figure indicate:

- Current enters the circuit through the transistor emitter (I_E).
- The current splits, with a portion leaving the transistor base (I_B) and the rest leaving the collector (I_C).
- The base current returns to the dc power supply via R_B.
- The collector current returns to the dc power supply via R_C.

Circuit Analysis

The analysis of a base-bias circuit begins with calculating the value of the transistor base current. According to Ohm's law, the base current in Figure 19.19 can be found as

$$I_B = \frac{V_{RB}}{R_B}$$

Note: The capacitors shown in Figure 19.19 are called *coupling capacitors.* The function of these components is discussed in Chapter 20.

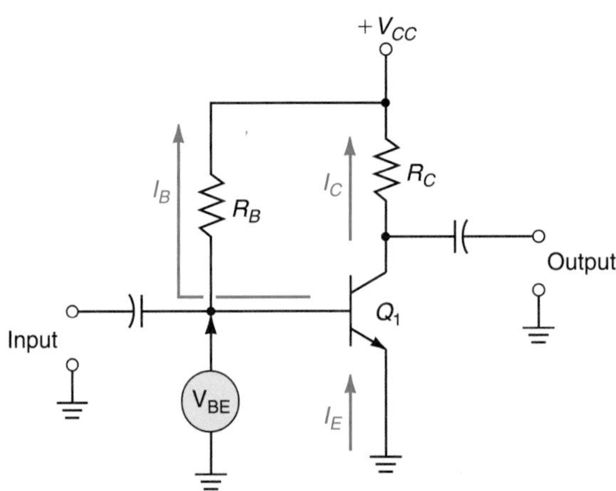

Figure 19.19 Base bias.

If we connect a meter from the transistor base to ground, as shown in Figure 19.19, we can see that the voltage from the transistor base to ground equals V_{BE} (0.7 V). Therefore, V_{RB} equals the difference between V_{CC} and V_{BE}. By formula,

$$V_{RB} = V_{CC} - V_{BE}$$

Substituting ($V_{CC} - V_{BE}$) into the base current equation gives us

$$I_B = \frac{V_{CC} - V_{BE}}{R_B} \tag{19.13}$$

Once the value of I_B is known, the values of I_C and V_{CE} can be found using these relationships established earlier in the chapter:

$$I_C = h_{FE}I_B$$

and

$$V_{CE} = V_{CC} - I_C R_C$$

The complete dc analysis of a base-bias circuit is demonstrated in Example 19.6.

Example 19.6

Determine the values of I_C and V_{CE} for the circuit shown in Figure 19.20.

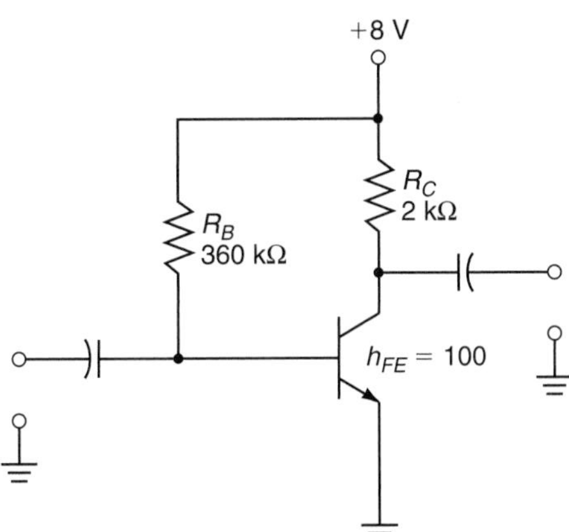

Figure 19.20

Solution: First, the circuit base current is found as

$$I_B = \frac{V_{CC} - V_{BE}}{R_B} = \frac{8\text{ V} - 0.7\text{ V}}{360\text{ k}\Omega} = 20.28\ \mu\text{A}$$

Next, the value of I_C is found as

$$I_C = h_{FE}I_B = (100)(20.28\ \mu\text{A}) = 2.03\text{ mA}$$

Finally, the value of V_{CE} is found as

$$V_{CE} = V_{CC} - I_C R_C = 8\text{ V} - (2.03\text{ mA})(2\text{ k}\Omega) = 3.94\text{ V}$$

Note that V_{CE} is very close to one-half the value of V_{CC}. This indicates that the amplifier is very nearly midpoint biased.

Practice Problem 19.6

A base-bias circuit like the one shown in Figure 19.20 has the following values: $V_{CC} = 12$ V, $R_C = 1$ kΩ, $h_{FE} = 150$, and $R_B = 500$ kΩ. Calculate the values of I_C and V_{CE}, and determine whether the circuit is midpoint biased.

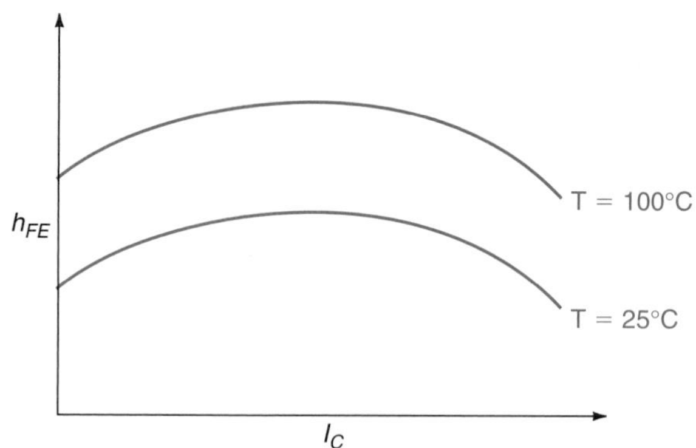

Figure 19.21 The relationship among h_{FE}, I_C, and temperature.

Q-Point Shift

OBJECTIVE 11 ➤ Even though they are easy to build and analyze, base-bias circuits are rarely used in applications that require stable midpoint-biased amplifiers. The reason is that they are extremely susceptible to a problem called *Q-point shift.* The term ***Q*-point shift** describes *a condition where a change in operating temperature causes a change in the Q-point values of I_C and V_{CE}.*

The curve shown in Figure 19.21 is called a **beta curve.** Note that the lower curve is measured at a value of $T = 25°C$, while the upper curve is measured at a value of $T = 100°C$. As the curves indicate, dc beta (h_{FE}) increases with temperature. When an increase in temperature causes h_{FE} to increase:

- I_C increases (because $I_C = h_{FE}I_B$).
- V_{CE} decreases (because $V_{CE} = V_{CC} - I_C R_C$).

Since these Q-point values are affected by the value of h_{FE}, the Q-point of the base-bias circuit can vary significantly with temperature.

Because they are susceptible to Q-point shift, base-bias circuits are used primarily in switching applications. When a transistor is used as a switch, it is constantly driven back and forth between saturation and cutoff, thus eliminating the need for midpoint bias. The characteristics and applications of the base-bias circuit are summarized in Figure 19.22.

A Few More Points

Throughout this section, we have referred to *the Q-point values of I_C and V_{CE}.* To simplify the upcoming discussions, we will use the labels I_{CQ} and V_{CEQ} to represent the Q-point values of I_C and V_{CE}.

Any biasing circuit that is susceptible to Q-point shift (like base bias) is referred to as a **beta-dependent circuit.** In the next two sections, we will discuss the operation of several **beta-independent circuits,** which are *circuits that provide values of I_{CQ} and V_{CEQ} that are relatively immune to changes in h_{FE}.*

Section Review

1. Describe the construction and current paths in a base-bias circuit.
2. What is the goal of the dc analysis of an amplifier?
3. List, in order, the steps taken to determine the Q-point values of I_C and V_{CE}.
4. Once you have calculated the values of I_C and V_{CE} for an amplifier, how can you determine whether or not the circuit is *midpoint biased*?
5. What is *Q-point shift*?

6. Demonstrate the following statement to be true: By increasing the value of R_C in Figure 19.22, it is possible to drive the transistor into saturation.

Critical Thinking

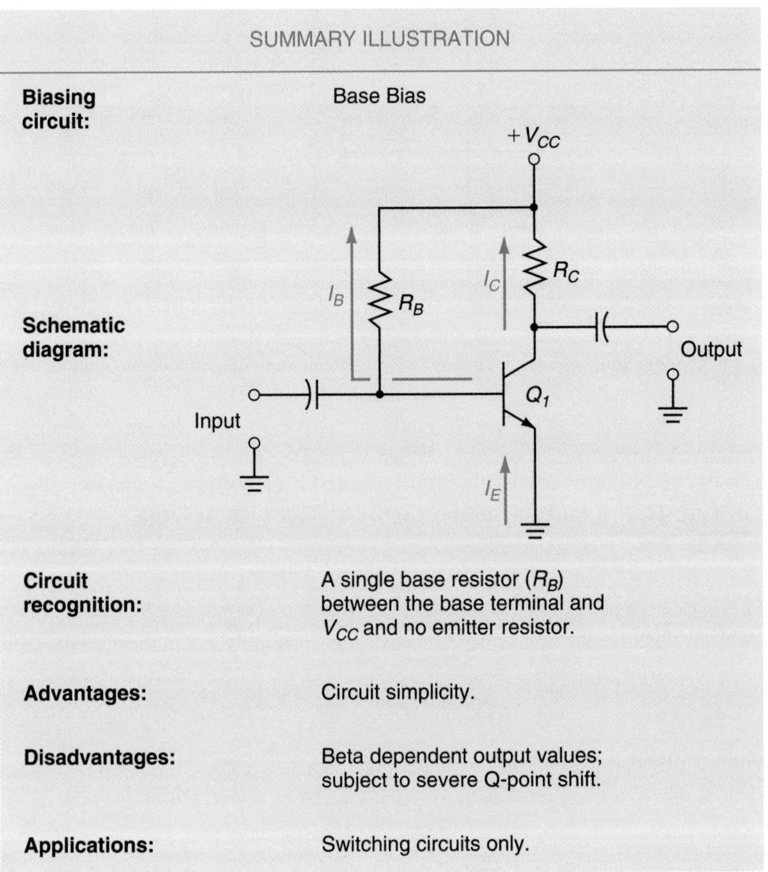

SUMMARY ILLUSTRATION
Biasing circuit:
Base Bias
Schematic diagram:
+Vcc
Rc
Ic
IB
RB
Output
Q1
Input
IE
Circuit recognition:
A single base resistor (RB) between the base terminal and Vcc and no emitter resistor.
Advantages:
Circuit simplicity.
Disadvantages:
Beta dependent output values; subject to severe Q-point shift.
Applications:
Switching circuits only.

Figure 19.22

19.6 Voltage-Divider Bias

The **voltage-divider bias** circuit is by far the most commonly used. This type of circuit, which is also referred to as **universal bias,** is shown in Figure 19.23. As you can see, the base of the transistor is connected to a voltage divider made up of R_1 and R_2. ◄ *OBJECTIVE 12*

The voltage divider in the base circuit sets the value of *base voltage* (V_B). Using the voltage-divider relationship, V_B is found as

$$V_B = V_{CC}\frac{R_2}{R_1 + R_2} \tag{19.14}$$

When the value of V_B is known, the value of *emitter voltage* (V_E) can be found using

$$V_E = V_B - 0.7\text{ V} \tag{19.15}$$

Note that the value of 0.7 V in equation (19.15) represents the base-emitter voltage (V_{BE}) of the transistor. Once the value of V_E is known, the emitter current (I_E) can be found as

$$I_E = \frac{V_E}{R_E} \tag{19.16}$$

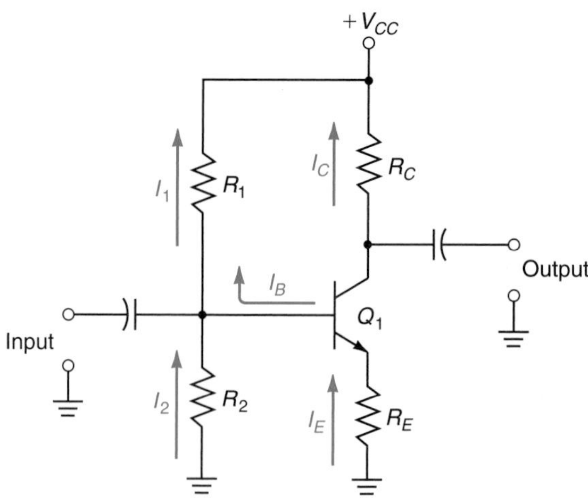

Figure 19.23 Voltage divider bias.

The sum of the voltages across the emitter resistor ($I_E R_E$), the transistor (V_{CE}), and the collector resistor ($I_C R_C$) must equal the supply voltage. By formula,

$$V_{CC} = I_E R_E + V_{CE} + I_C R_C$$

Since $I_C \cong I_E$ under normal operating circumstances, the above equation can be rewritten as

$$V_{CC} \cong I_C R_E + V_{CE} + I_C R_C$$

which, when transposed, gives us

$$V_{CE} \cong V_{CC} - I_C(R_C + R_E) \tag{19.17}$$

The values of I_{CQ} and V_{CEQ} for a voltage-divider bias circuit can be calculated as shown in Example 19.7.

Example 19.7

Determine the values of I_{CQ} and V_{CEQ} for the circuit shown in Figure 19.24.

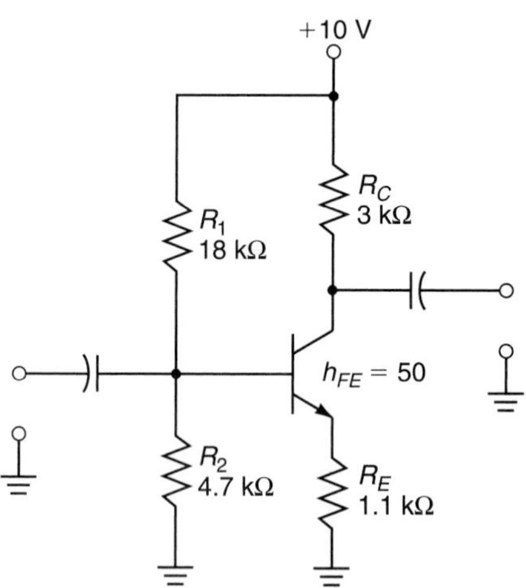

Figure 19.24

Solution: The voltage at the base of the transistor is found as

$$V_B = V_{CC}\frac{R_2}{R_1 + R_2} = (10\text{ V})\frac{4.7\text{ k}\Omega}{22.7\text{ k}\Omega} = 2.07\text{ V}$$

The voltage at the emitter of the transistor is found as

$$V_E = V_B - 0.7\text{ V} = 2.07\text{ V} - 0.7\text{ V} = 1.37\text{ V}$$

The value of I_{CQ} can now be approximated as

$$I_{CQ} \cong I_E = \frac{V_E}{R_E} = \frac{1.37\text{ V}}{1.1\text{ k}\Omega} = 1.25\text{ mA}$$

and the value of V_{CEQ} can be found as

$$V_{CEQ} \cong V_{CC} - I_{CQ}(R_C + R_E) = 10\text{ V} - (1.25\text{ mA})(4.1\text{ k}\Omega) = 4.88\text{ V}$$

Practice Problem 19.7

A circuit like the one shown in Figure 19.24 has the following values: $R_C = 620\ \Omega$, $R_E = 180\ \Omega$, $R_1 = 12\text{ k}\Omega$, $R_2 = 2.7\text{ k}\Omega$, and $V_{CC} = 10$ V. Determine the values of I_{CQ} and V_{CEQ} for the circuit.

Base Voltage: A Practical Consideration

According to equation (19.14), the value of V_B for a voltage-divider bias circuit can be found as

$$V_B = V_{CC}\frac{R_2}{R_1 + R_2}$$

This relationship is actually an approximation, because it does not take into account the input resistance of the transistor base terminal.

Base input resistance As shown in Figure 19.25a, I_2 and I_B combine at the junction of the voltage divider and the base of the transistor. If we trace the current paths from connection point to ground, we can see that R_2 is in parallel with the base-emitter circuit of the transistor. Just as current is increased by a factor of h_{FE} from base to emitter, R_E is increased by a factor of h_{FE} from emitter to base. As a result, there is a measurable amount of base input resistance. The value of this base input resistance can be found as

$$R_{\text{in(base)}} = h_{FE}R_E \qquad (19.18)$$

A Practical Consideration: Equation (19.5) indicates that current actually increases by a factor of $(h_{FE} + 1)$ from base to emitter. However, since h_{FE} is typically much greater than one, we normally assume that $I_E \cong h_{FE}I_B$.

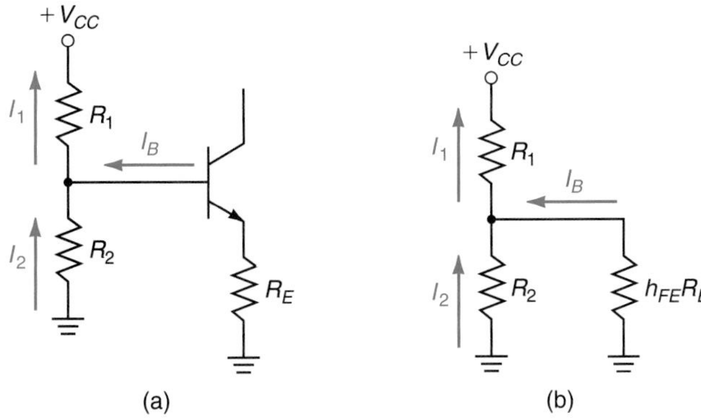

Figure 19.25

As shown in Figure 19.25b, this resistance is in parallel with R_2. Therefore, the equivalent resistance from the base of the transistor to ground (R_{EQ}) is the parallel combination of R_2 and $R_{in(base)}$. By formula,

$$R_{EQ} = R_2 \| h_{FE}R_E \tag{19.19}$$

Based on this relationship, a more accurate value of V_B can be found as

$$V_B = V_{CC}\frac{R_{EQ}}{R_1 + R_{EQ}} \Big| R_{EQ} = R_2 \| h_{FE}R_E \tag{19.20}$$

As shown in Figure 19.25b, the base input resistance ($h_{FE}R_E$) acts as a load on the voltage divider. If $h_{FE}R_E$ is greater than or equal to $10R_2$, we can ignore the loading effect of this resistance on the voltage divider. Otherwise, we must account for its effect on the value of V_B, as demonstrated in Example 19.8.

Example 19.8

MSM

Determine the values of I_{CQ} and V_{CEQ} for the amplifier shown in Figure 19.26.

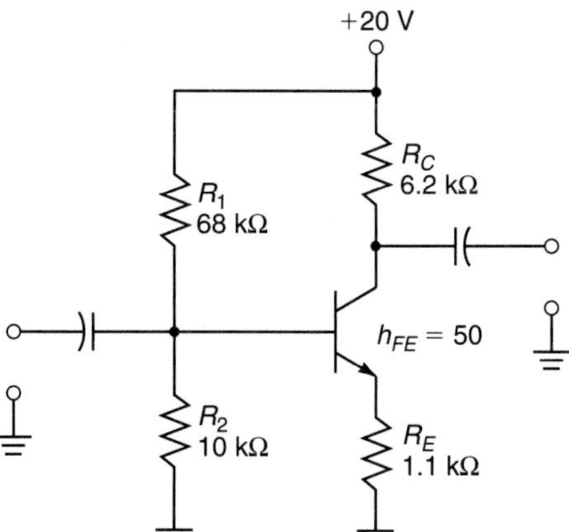

Figure 19.26

Solution: The base input resistance for the transistor is found as

$$R_{in(base)} = h_{FE}R_E = (50)(1.1\text{ k}\Omega) = 55\text{ k}\Omega$$

Because $R_{in(base)}$ is less than 10 times the value of R_2, we must determine V_B as follows:

$$R_{EQ} = R_2 \| h_{FE}R_E = 10\text{ k}\Omega \| 55\text{ k}\Omega = 8.46\text{ k}\Omega$$

and

$$V_B = V_{CC}\frac{R_{EQ}}{R_1 + R_{EQ}} = (20\text{ V})\frac{8.46\text{ k}\Omega}{76.46\text{ k}\Omega} = 2.21\text{ V}$$

The rest of the problem is solved as before. The emitter voltage is found as

$$V_E = V_B - V_{BE} = 2.21\text{ V} - 0.7\text{ V} = 1.51\text{ V}$$

and the collector current is found as

$$I_{CQ} \cong I_E = \frac{V_E}{R_E} = \frac{1.51\text{ V}}{1.1\text{ k}\Omega} = 1.37\text{ mA}$$

Finally, V_{CEQ} is found as

$$V_{CEQ} = V_{CC} - I_{CQ}(R_C + R_E) = 20\text{ V} - (1.37\text{ mA})(7.3\text{ k}\Omega) \cong 10\text{ V}$$

Practice Problem 19.8

Assume that the transistor shown in Figure 19.26 has a value of $h_{FE} = 80$. Recalculate the values of I_{CQ} and V_{CEQ}.

Saturation and Cutoff

When the transistor in a voltage-divider biased circuit is saturated, $V_{CE} \cong 0$ V, and the collector current equals the supply voltage divided by the total resistance between V_{CC} and ground. By formula,

$$I_{C(\text{sat})} = \frac{V_{CC}}{R_C + R_E} \tag{19.21}$$

When the transistor is in cutoff, all the supply voltage is dropped across the transistor, and

$$V_{CE(\text{off})} \cong V_{CC} \tag{19.22}$$

Applying these relationships to the circuit shown in Figure 19.26, we get

$$I_{C(\text{sat})} = \frac{V_{CC}}{R_C + R_E} = \frac{20\text{ V}}{7.3\text{ k}\Omega} = 2.74\text{ mA}$$

and

$$V_{CE(\text{off})} \cong V_{CC} = 20\text{ V}$$

Bias Stability

Earlier in this section, you were told that voltage-divider bias is the most commonly used biasing scheme. One reason is the fact that voltage-divider bias is a *beta-independent circuit;* that is, it provides values of I_{CQ} and V_{CEQ} that are relatively stable against changes in beta (h_{FE}). This can be seen in a relationship that was established early in this section:

$$I_{CQ} \cong I_E = \frac{V_E}{R_E}$$

As you can see, h_{FE} is not involved in the calculation of I_{CQ}. Therefore, the values of I_{CQ} and V_{CEQ} are not a function of h_{FE} and are relatively stable against any changes in transistor current gain.

At this point, we will move on to briefly discuss several other common biasing circuits. The characteristics of the voltage-divider bias circuit are summarized in Figure 19.27.

Section Review

1. List, in order, the steps required to determine the values of I_{CQ} and V_{CEQ} for a *voltage-divider bias* circuit.
2. What additional steps must be taken in the analysis of a voltage-divider bias circuit when $h_{FE}R_E < 10R_2$?
3. Why is voltage-divider bias relatively stable against changes in h_{FE}?

Critical Thinking

4. Would the $h_{FE}R_E < 10R_2$ guideline for ignoring the value of $h_{FE}R_E$ be considered valid in a biasing circuit using 1% tolerance components? Explain your answer.

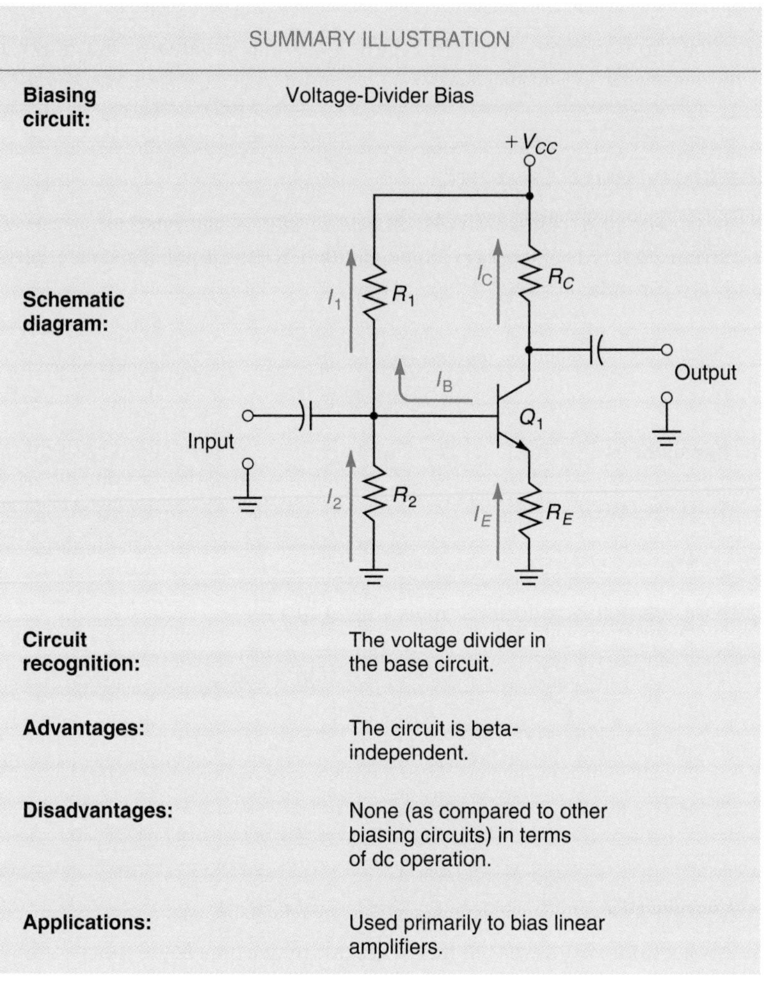
SUMMARY ILLUSTRATION
Biasing circuit:
Voltage-Divider Bias
+VCC
Schematic diagram:
I1
R1
IC
RC
IB
Output
Input
Q1
I2
R2
IE
RE
Circuit recognition:
The voltage divider in the base circuit.
Advantages:
The circuit is beta-independent.
Disadvantages:
None (as compared to other biasing circuits) in terms of dc operation.
Applications:
Used primarily to bias linear amplifiers.

Figure 19.27

19.7 Other Transistor Biasing Circuits

In this section, we will discuss several other beta-independent biasing circuits. While these circuits are not used as commonly as voltage-divider bias, they are used in a variety of applications. For this reason, you should be familiar with their operating principles and characteristics.

Emitter Bias

OBJECTIVE 13 ➤ An **emitter-bias** circuit *consists of several resistors and a dual-polarity power supply.* An emitter-bias circuit is shown in Figure 19.28. As the current arrows indicate:

- The circuit current originates at the emitter power supply ($-V_{EE}$), passes through the emitter resistor (R_E), and enters the transistor.
- A small portion of the emitter current leaves the transistor through the base terminal and passes through R_B to ground.
- Most of the emitter current passes through to the collector, then through the collector resistor (R_C) to the collector power supply (V_{CC}).

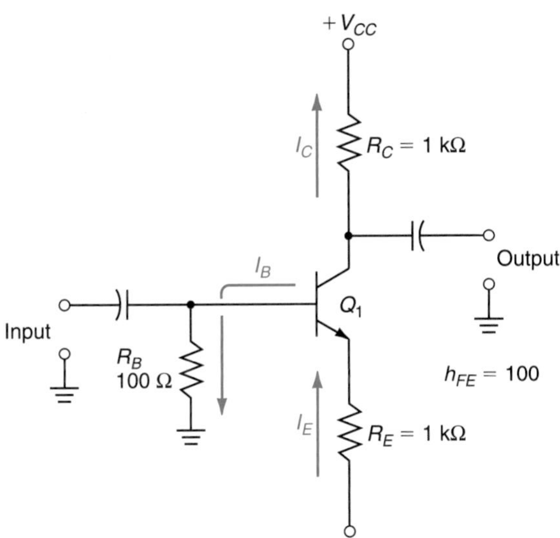

Figure 19.28 Emitter bias.

Circuit currents and voltages Emitter-bias circuits are always designed so that $V_B \cong 0$ V. For example, if we assume the circuit shown in Figure 19.28 has a value of $I_B = 100\ \mu$A, then

$$V_B = I_B R_B = (100\ \mu\text{A})(100\ \Omega) = 10\ \text{mV}$$

If we assume that the circuit shown in Figure 19.28 has a value of $V_B = 0$ V, then

$$V_E = 0\ \text{V} - V_{BE} = -V_{BE}$$

According to Kirchhoff's voltage law, the voltage across the emitter resistor (V_{RE}) must equal the difference between V_E and V_{EE}. Since $V_E = -V_{BE}$,

$$V_{RE} = V_{EE} - V_E = V_{EE} + V_{BE}$$

Using ($V_{EE} + V_{BE}$) and R_E, the value of the emitter current can be found as

$$I_E = \frac{|V_{EE} + V_{BE}|}{R_E}$$

Assuming that $I_{CQ} \cong I_E$ and $V_{BE} = 0.7$ V, this relationship can be rewritten as

$$I_{CQ} \cong \frac{|V_{EE} + 0.7\ \text{V}|}{R_E} \tag{19.23}$$

Note that the *absolute value* of ($V_{EE} + 0.7$ V) is used in equation (19.23) to obtain a *positive* value of I_{CQ}.

As stated earlier, the emitter-bias circuit has a value of $V_E \cong -V_{BE}$. If we assume that $V_{BE} = 0.7$ V, then $V_E = -0.7$ V. According to Kirchhoff's voltage law, the collector voltage (V_C) can be found as

$$V_C = V_{CC} - I_C R_C$$

Since $V_{CE} = V_C - V_E$, we can find the value of V_{CEQ} as

$$V_{CEQ} = V_{CC} - I_{CQ} R_C - V_E$$

or

$$V_{CEQ} = V_{CC} - I_{CQ}R_C + 0.7 \text{ V} \tag{19.24}$$

Example 19.9 demonstrates the Q-point analysis of an emitter-bias circuit.

Example 19.9

Determine the values of I_{CQ} and V_{CEQ} for the circuit shown in Figure 19.29.

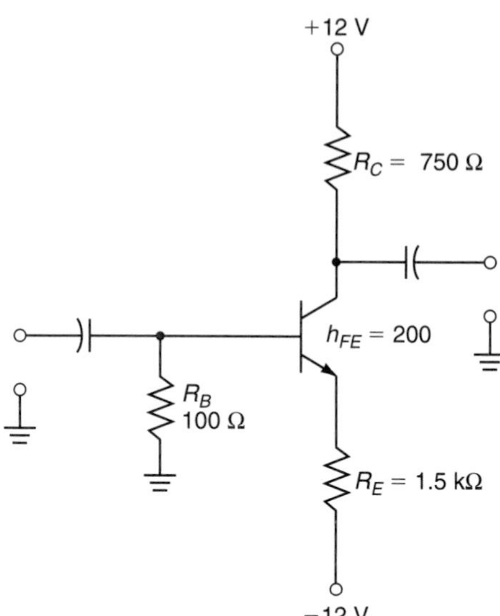

Figure 19.29

Solution: The value of I_{CQ} is found as

$$I_{CQ} \cong \frac{|V_{EE} + 0.7 \text{ V}|}{R_E} = \frac{|-12 \text{ V} + 0.7 \text{ V}|}{1.5 \text{ k}\Omega} = \frac{11.3 \text{ V}}{1.5 \text{ k}\Omega} = 7.53 \text{ mA}$$

and the value of V_{CEQ} is found as

$$V_{CEQ} = V_{CC} - I_{CQ}R_C + 0.7 \text{ V} = 12 \text{ V} - (7.53 \text{ mA})(750 \ \Omega) + 0.7 \text{ V} = 7.05 \text{ V}$$

Practice Problem 19.9

A circuit like the one shown in Figure 19.29 has the following values: $R_C = 1.5$ kΩ, $R_E = 3$ kΩ, $V_{CC} = +15$ V, and $V_{EE} = -15$ V. Determine the approximate values of I_{CQ} and V_{CEQ} for the circuit.

Saturation and cutoff The presence of two power supplies in the emitter-bias circuit affects the values of $I_{C(\text{sat})}$ and $V_{CE(\text{off})}$. When the two supply voltages are equal in magnitude (which is normally the case), the values of $I_{C(\text{sat})}$ and $V_{CE(\text{off})}$ can be found using

$$I_{C(\text{sat})} = \frac{2V_{CC}}{R_C + R_E} \tag{19.25}$$

and

$$V_{CE(\text{off})} = 2V_{CC} \tag{19.26}$$

Remember, these equations are valid only when the two supply voltages are equal in magnitude. When the supply voltages are *not* equal in magnitude, the difference between their values ($V_{CC} - V_{EE}$) is used in place of $2V_{CC}$ in equations (19.25) and (19.26).

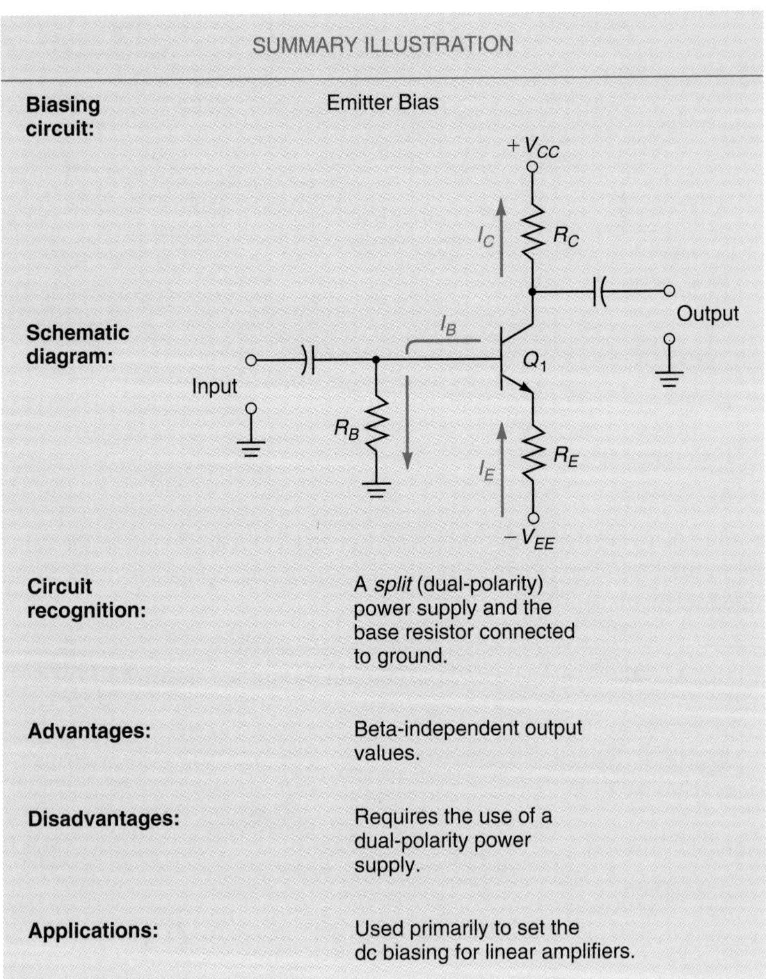
SUMMARY ILLUSTRATION
Biasing circuit:
Emitter Bias
+V_{CC}
I_C
R_C
Output
I_B
Schematic diagram:
Q_1
Input
R_B
I_E
R_E
−V_{EE}
Circuit recognition:
A *split* (dual-polarity) power supply and the base resistor connected to ground.
Advantages:
Beta-independent output values.
Disadvantages:
Requires the use of a dual-polarity power supply.
Applications:
Used primarily to set the dc biasing for linear amplifiers.

Figure 19.30

Emitter-bias characteristics The emitter-bias circuit is a beta-independent circuit, meaning that its output characteristics (I_{CQ} and V_{CEQ}) are not affected significantly by variations in beta (h_{FE}). This point is verified by equation (19.23), which shows that h_{FE} is not part of the I_{CQ} calculation. The characteristics of the emitter-bias circuit are summarized in Figure 19.30.

Collector-Feedback Bias

The term **feedback** is used to describe *a circuit that "feeds" a portion of the output voltage or current back to the input.* Among other benefits, the use of feedback helps to reduce the effects of beta variations on circuit Q-point values. ◄ *OBJECTIVE 14*

The **collector-feedback bias** circuit obtains its Q-point stability from the base resistor connection to the collector of the transistor (as shown in Figure 19.31a). The circuit currents are shown in Figure 19.31b. As you can see, the base current (I_B) path includes the collector resistor (R_C).

Circuit analysis The analysis of the collector-feedback bias circuit begins with calculating the value of I_B, as follows:

$$I_B = \frac{V_{CC} - V_{BE}}{R_B + h_{FE}R_C} \tag{19.27}$$

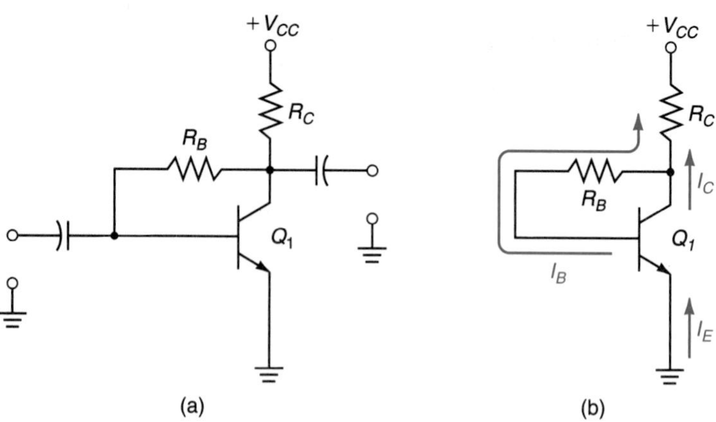

Figure 19.31 Collector feedback bias.

Once the value of I_B is known, the values of I_{CQ} and V_{CEQ} are found in the same fashion as they are in the base-bias circuit. The dc analysis of a collector-feedback bias circuit is demonstrated in Example 19.10.

Example 19.10

Determine the values of I_{CQ} and V_{CEQ} for the amplifier shown in Figure 19.32.

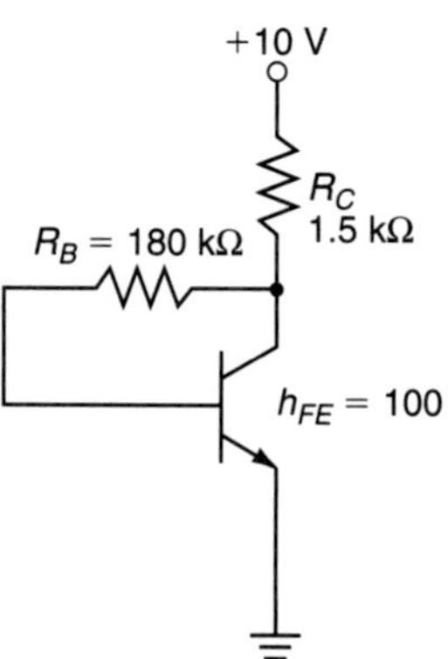

Figure 19.32

Solution: First, the value of I_B is found as

$$I_B = \frac{V_{CC} - V_{BE}}{R_B + h_{FE}R_C} = \frac{10\text{ V} - 0.7\text{ V}}{180\text{ k}\Omega + (100)(1.5\text{ k}\Omega)} = \frac{9.3\text{ V}}{330\text{ k}\Omega} = 28.2\ \mu\text{A}$$

Next, I_{CQ} is found as

$$I_{CQ} = h_{FE}I_B = (100)(28.2\ \mu\text{A}) = 2.82\text{ mA}$$

Finally, V_{CEQ} is found as

$$V_{CEQ} = V_{CC} - I_{CQ}R_C = 10\text{ V} - (2.82\text{ mA})(1.5\text{ k}\Omega) = 5.77\text{ V}$$

Practice Problem 19.10

A circuit like the one shown in Figure 19.32 has values of $R_C = 2\text{ k}\Omega$, $R_B = 240\text{ k}\Omega$, $h_{FE} = 120$, and $V_{CC} = +12$ V. Determine the values of I_{CQ} and V_{CEQ} for the circuit.

Q-Point stability The key to collector-feedback bias stability is the fact that *I_B and beta are inversely related* in this circuit. According to equation (19.27), I_B decreases when $h_{FE}R_C$ increases. With this in mind, consider the response of the amplifier shown in Figure 19.32 to an increase in temperature. If temperature increases, h_{FE} and I_C both

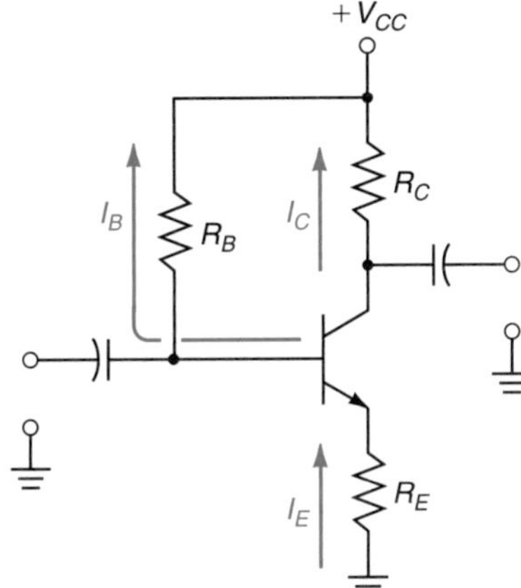

Figure 19.33 Emitter-feedback bias.

increase. The increase in h_{FE} causes I_B to *decrease,* partially offsetting the initial increase in I_C.

Emitter-Feedback Bias

◀ *OBJECTIVE 15*

Emitter-feedback bias works in the same basic fashion as the collector-feedback circuit. However, in this case, it is the emitter circuit that affects the value of I_B. The basic emitter-feedback bias circuit is shown in Figure 19.33. Note the directions of the circuit currents.

Circuit analysis The analysis of the emitter-feedback bias circuit begins with calculating the value of I_B, as follows:

$$I_B = \frac{V_{CC} - V_{BE}}{R_B + (h_{FE} + 1)\,R_E} \tag{19.28}$$

Once the value of I_B is known, the Q-point values of I_C and V_{CE} are found using

$$I_{CQ} = h_{FE}I_B$$

and

$$V_{CEQ} = V_{CC} - I_{CQ}(R_C + R_E)$$

Example 19.11 illustrates the process of determining the Q-point values for an emitter-feedback bias circuit.

Example 19.11

Determine the values of I_{CQ} and V_{CEQ} for the circuit shown in Figure 19.34.

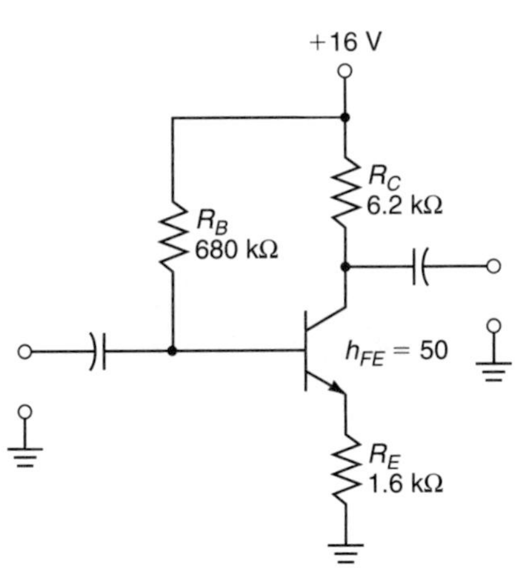

Figure 19.34

Solution: First, the value of I_B is found as

$$I_B = \frac{V_{CC} - V_{BE}}{R_B + (h_{FE} + 1)R_E} = \frac{16 \text{ V} - 0.7 \text{ V}}{680 \text{ k}\Omega + (51)(1.6 \text{ k}\Omega)} = 20.09 \ \mu\text{A}$$

The value of I_{CQ} is now found as

$$I_{CQ} = h_{FE}I_B = (50)(20.09 \ \mu\text{A}) = 1 \text{ mA}$$

Finally, the value of V_{CEQ} is found as

$$V_{CEQ} = V_{CC} - I_{CQ}(R_C + R_E) = 16 \text{ V} - (1 \text{ mA})(7.8 \text{ k}\Omega) = 8.2 \text{ V}$$

Practice Problem 19.11

A circuit like the one shown in Figure 19.34 has values of $V_{CC} = 16$ V, $R_B = 470$ kΩ, $R_C = 1.8$ kΩ, $R_E = 910$ Ω, and $h_{FE} = 100$. Determine the values of I_{CQ} and V_{CEQ} for the circuit.

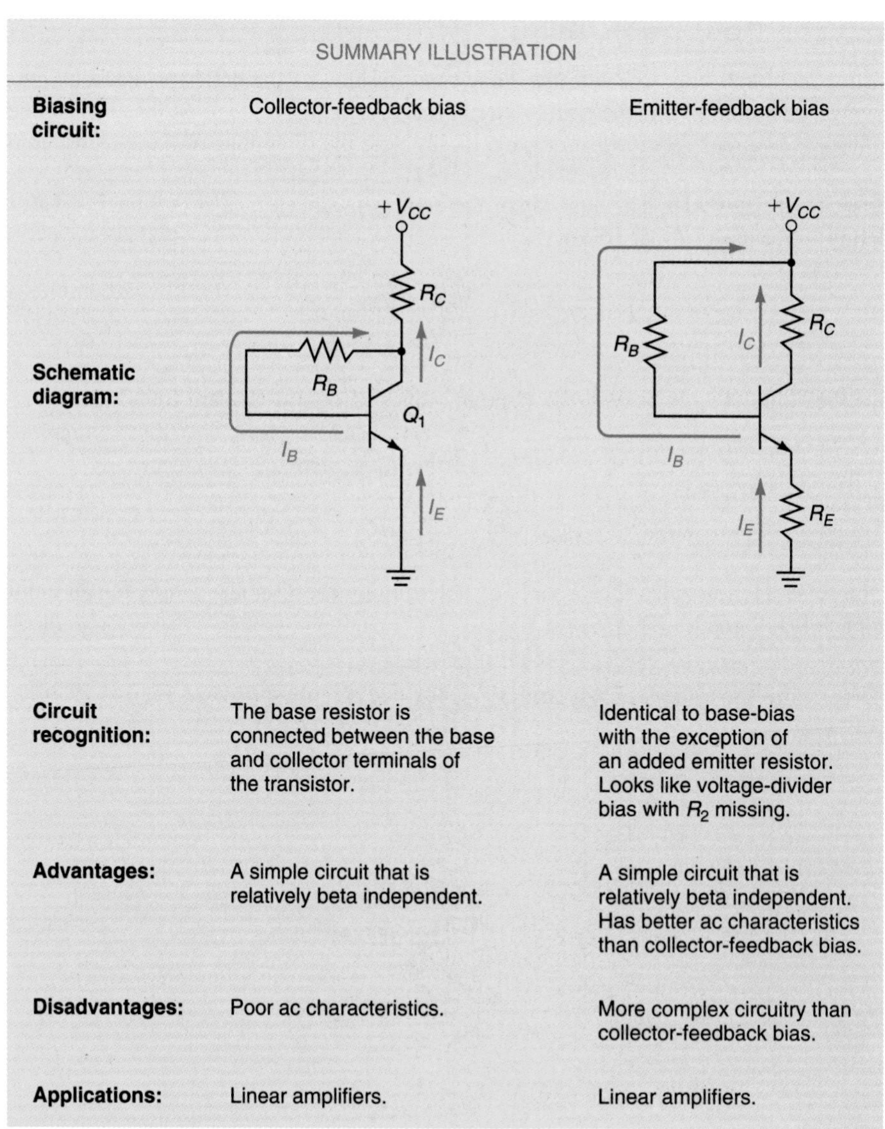

Figure 19.35

Q-Point stability Q-point stability in the emitter-feedback bias circuit is obtained just as it is in the collector-feedback bias circuit. If h_{FE} increases, the value of $(h_{FE} + 1)R_E$ also increases. This causes I_B to decrease, partially offsetting the effects of the increase in h_{FE} on I_{CQ}. This cause-and-effect relationship can be verified using equation (19.28).

One Final Note

While collector-feedback bias is the simpler of the feedback-bias circuits, emitter-feedback bias has ac characteristics that make it the more commonly used. The characteristics of the two feedback bias circuits are summarized in Figure 19.35.

Section Review

1. By design, what is the approximate value of V_B in an *emitter-bias* circuit?
2. Describe the construction and current characteristics of the emitter-bias circuit.
3. Why are the values of I_{CQ} and V_{CEQ} in an emitter-bias circuit relatively stable against changes in h_{FE}?
4. When do we use the value $2V_{CC}$ when determining the values of $I_{C(\text{sat})}$ and $V_{CE(\text{off})}$ for an emitter-bias circuit?
5. What is *feedback*? What is it used for?
6. Describe the construction and current characteristics of the *collector-feedback bias* circuit.
7. Explain why collector-feedback bias circuits are relatively stable against changes in *beta.*
8. Describe the construction and current characteristics of the *emitter-feedback bias* circuit.
9. Which transistor terminal current, if any, is *beta dependent* in an emitter-bias circuit? Critical Thinking

19.8 Related Topics

In this section, we will look at some topics that relate to transistors and biasing circuits, including transistor specification sheets, *npn* versus *pnp* transistors, and some common transistor packages.

Transistor Specification Sheets

For our discussion on transistor ratings, we will refer to the spec sheet for the Motorola 2N3903/2N3904 shown in Figure 19.36.

Maximum ratings The first three ratings shown in the maximum ratings chart are device breakdown voltage ratings. Every transistor has three breakdown voltage ratings. These ratings indicate the maximum *reverse* voltages that the transistor can withstand. For the 2N3904, these voltage ratings are

$$V_{CBO} = 60\text{ V} \qquad V_{CEO} = 40\text{ V} \qquad V_{EBO} = 6\text{ V}$$

These voltages are identified in Figure 19.37.

The value of V_{CBO} is important, because the collector-base junction is reverse biased for active-region operation, as shown in Figure 19.38. In this circuit, $V_{CE} = 40$ V, and $V_{BE} = 0.75$ V. The value of V_{CB} equals the difference between the other two voltages, or

A Practical Consideration: The "O" in the V_{CEO}, V_{EBO}, and V_{CBO} ratings indicate that the third terminal is open when the rating is measured. For example, V_{EBO} is measured with the collector terminal open (which ensures that the device is in cutoff when the rating is measured).

MAXIMUM RATINGS

Rating	Symbol	Value	Unit
Collector-Emitter Voltage	V_{CEO}	40	Vdc
Collector-Base Voltge	V_{CBO}	60	Vdc
Emitter-Base Voltage	V_{EBO}	6.0	Vdc
Collector Current — Continuous	I_C	200	mAdc
Total Device Dissipation @ T_A = 25°C Derate above 25°C	P_D	625 5.0	mW mW/°C
*Total Device Dissipation @ T_C = 25°C Derate above 25°C	P_D	1.5 12	Watts mW/°C
Operating and Storage Junction Temperature Range	T_J, T_{stg}	−55 to +150	°C

***THERMAL CHARACTERISTICS**

Characteristic	Symbol	Max	Unit
Thermal Resistance, Junction to Ambient	$R_{\theta JA}$	200	°C/W
Thermal Resistance, Junction to Case	$R_{\theta JC}$	83.3	°C/W

*Indicates Data in addition to JEDEC Requirements.

2N3903
2N3904★

CASE 29-04, STYLE 1
TO-92 (TO-226AA)

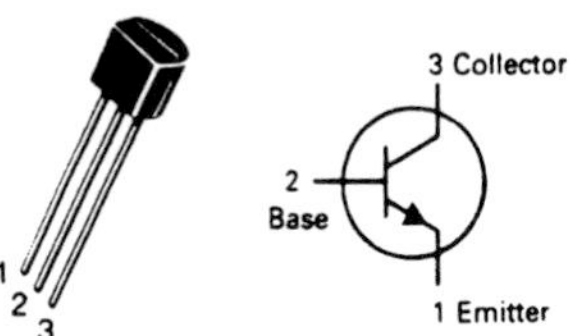

GENERAL PURPOSE TRANSISTORS

NPN SILICON

★This is a Motorola designated preferred device.

ELECTRICAL CHARACTERISTICS (T_A = 25°C unless otherwise noted.)

Characteristic		Symbol	Min	Max	Unit
OFF CHARACTERISTICS					
Collector-Emitter Breakdown Voltage(1) (I_C = 1.0 mAdc, I_B = 0)		$V_{(BR)CEO}$	40	—	Vdc
Collector-Base Breakdown Voltage (I_C = 10 μAdc, I_E = 0)		$V_{(BR)CBO}$	60	—	Vdc
Emitter-Base Breakdown Voltage (I_E = 10 μAdc, I_C = 0)		$V_{(BR)EBO}$	6.0	—	Vdc
Base Cutoff Current (V_{CE} = 30 Vdc, V_{BE} = 3.0 Vdc)		I_{BL}	—	50	nAdc
Collector Cutoff Current (V_{CE} = 30 Vdc, V_{BE} = 3.0 Vdc)		I_{CEX}	—	50	nAdc
ON CHARACTERISTICS					
DC Current Gain(1) (I_C = 0.1 mAdc, V_{CE} = 1.0 Vdc)	2N3903 2N3904	h_{FE}	20 40	— —	—
(I_C = 1.0 mAdc, V_{CE} = 1.0 Vdc)	2N3903 2N3904		35 70	— —	
(I_C = 10 mAdc, V_{CE} = 1.0 Vdc)	2N3903 2N3904		50 100	150 300	
(I_C = 50 mAdc, V_{CE} = 1.0 Vdc)	2N3903 2N3904		30 60	— —	
(I_C = 100 mAdc, V_{CE} = 1.0 Vdc)	2N3903 2N3904		15 30	— —	
Collector-Emitter Saturation Voltage(1) (I_C = 10 mAdc, I_B = 1.0 mAdc) (I_C = 50 mAdc, I_B = 5.0 mAdc)		$V_{CE(sat)}$	 — —	 0.2 0.3	Vdc
Base-Emitter Saturation Voltage(1) (I_C = 10 mAdc, I_B = 1.0 mAdc) (I_C = 50 mAdc, I_B = 5.0 mAdc)		$V_{BE(sat)}$	 0.65 —	 0.85 0.95	Vdc

Figure 19.36 The 2N3903-3904 specification sheet. (Copyright of Semiconductor Component Industries, LLC. Used by permission.)

39.25 V. If this voltage is greater than the maximum V_{CB} rating of the transistor, it may be damaged or destroyed.

The I_C rating is almost self-explanatory. It is the maximum allowable continuous value of I_C. For the 2N3904, the I_C rating is 200 mA.

The total device dissipation (P_D) rating is the same type of rating that is used for the zener and *pn*-junction diode. The 2N3904 has a P_D rating of 625 mW when the ambient temperature (T_A) is 25°C. If the case temperature (T_C) is held to 25°C (by a cooling fan

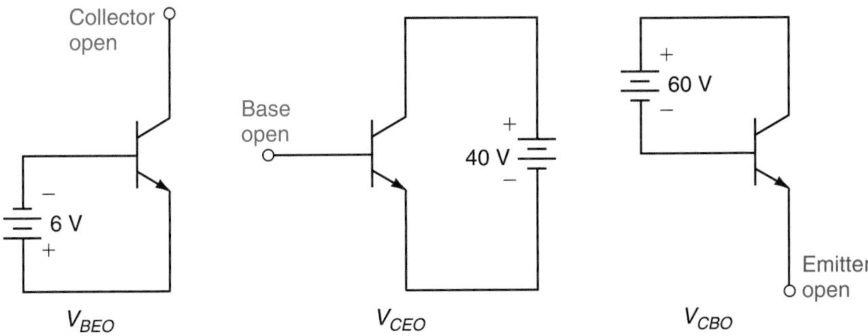

Figure 19.37 BJT breakdown voltage ratings.

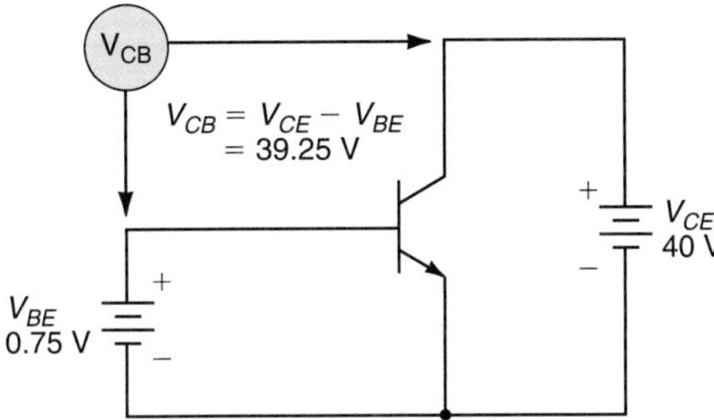

Figure 19.38 Collector-base junction biasing.

or a heat sink), the rating increases to 1.5 W. As always, both ratings must be derated as temperature increases.

Thermal characteristics Transistor thermal ratings are used primarily in circuit development applications, so we will not discuss them in depth. However, be aware that these ratings are used in determining the type of component cooling that is required for a given device in a specified application.

Off characteristics The *off characteristics* describe the operation of the transistor when it is in cutoff. The **collector cutoff current (I_{CEX})** rating indicates *the maximum value of I_C when the device is in cutoff.* The **base cutoff current (I_{BL})** rating indicates *the maximum base current present when the emitter-base junction is in cutoff.* In the case of the 2N3904, I_{BL} and I_{CEX} each have a maximum value of 50 nA.

On characteristics The *on characteristics* describe the dc operating characteristics for both the active and saturation regions of operation.

The **dc current gain (h_{FE})** rating of the transistor is the value of dc beta (β). As you can see, the values of h_{FE} are measured at different values of I_C, which indicates that dc current gain is affected by I_C. This relationship is consistent with the curves shown in Figure 19.21.

The **collector-emitter saturation voltage** rating indicates *the value of V_{CE} when the transistor is operated in saturation.* The **base-emitter saturation voltage** rating indicates *the value of V_{BE} when the transistor is operated in saturation.*

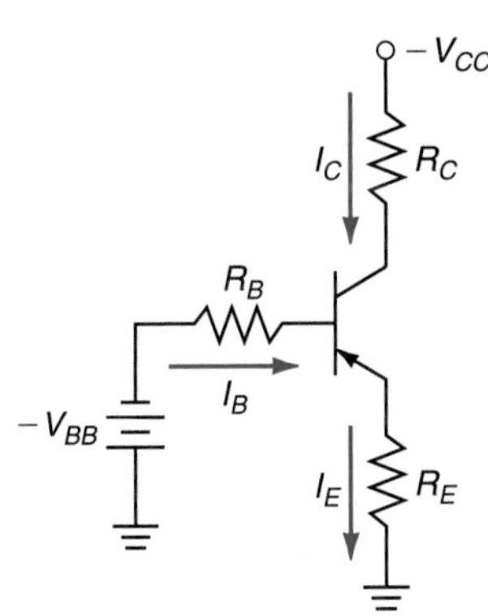

Figure 19.39
PNP voltage polarities and current directions.

PNP Versus NPN Transistors

◀ *OBJECTIVE 16*

So far, we have concentrated on the *npn* transistor. The *pnp* transistor has the same basic operating characteristics; the primary difference is the polarity of the biasing voltages and the direction of the terminal currents. These differences are illustrated in Figure 19.39. As you can see, V_{CC} and V_{BB} are both negative supply voltages, and both generate currents that enter the transistor through their respective terminals. The combination of I_C and I_B

exits the transistor through the emitter terminal. Note that all the transistor ratings discussed in this chapter apply equally to the *pnp* transistor, as do all of the current and voltage relationships.

Discrete Transistors

OBJECTIVE 17 ➤ The 2N3904 is a *small-signal* discrete transistor. This means that the component is designed for use in general, low-power applications. Other types of discrete transistors are designed for use in specific applications. Some of these discrete components are identified as follows:

- **High-voltage transistors** have unusually high reverse breakdown voltage ratings. They are used in circuits with high supply voltages, such as television CRT (cathode ray tube) control circuits.
- **High-current transistors** have very high maximum I_C ratings. Obviously, these devices are used in applications with high current demands, such as current regulator circuits.
- **High-power transistors** are designed for use in high-power circuits, such as regulated linear and switching power supplies. As you may have determined, these are devices with extremely high power dissipation ratings.

Darlington Transistors

OBJECTIVE 18 ➤ A **Darlington pair** is *a two-transistor configuration designed for high current gain.* The transistors in a Darlington pair are connected as shown in Figure 19.40a. For the sake of discussion, we will refer to Q_1 as the *input transistor* and Q_2 as the *output transistor.*

While a Darlington pair can be constructed using individual components, the configuration is most often produced as a single component. For example, the Darlington transistor in Figure 19.40b contains the transistors shown, with single base, collector, and emitter leads.

Figure 19.41 shows the fundamental voltage and current relationships in a Darlington pair. As illustrated in Figure 19.41a, there are two junctions between the base of Q_1 and the emitter of Q_2. Therefore,

$$V_E = V_B - 1.4 \text{ V} \tag{19.29}$$

where

1.4 V = the drop across the two emitter-base junctions

As you know, the emitter current for a given transistor can be approximated using $I_E \cong h_{FE}I_B$ (when $h_{FE} >> 1$). Applying this approximation to the Darlington pair shown in Figure 19.41b,

$$I_{E1} \cong h_{FE1}I_{B1}$$

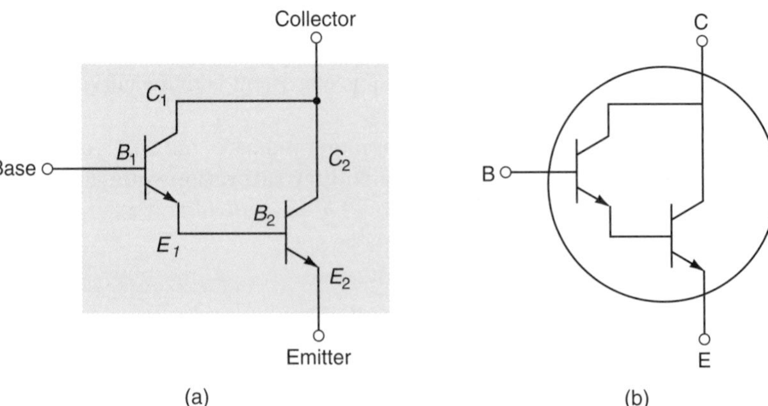

Figure 19.40

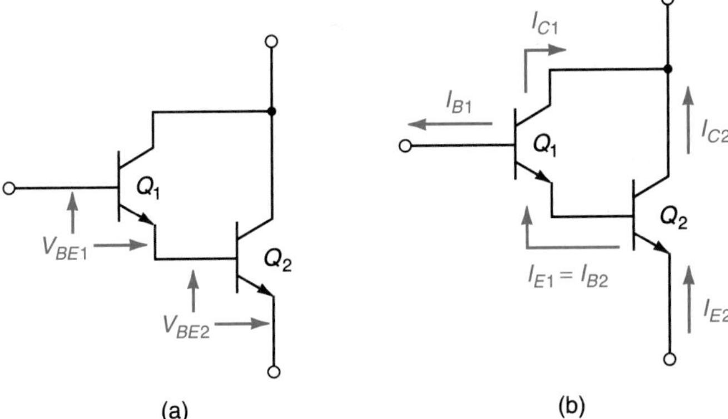

Figure 19.41

Since this current is drawn through the base of the output transistor, $I_{B2} = I_{E1}$, and the output emitter current can be found as

$$
\begin{aligned}
I_{E2} &= h_{FE2}I_{B2} \\
&= h_{FE2}(h_{FE1}I_{B1})
\end{aligned}
$$

or

$$I_{E2} = h_{FE1}h_{FE2}I_{B1} \tag{19.30}$$

This relationship demonstrates that the overall current gain of a Darlington pair is approximately equal to the product of the individual current gains. For example, if you were to construct a Darlington pair using two transistors rated at $h_{FE} = 100$, the overall dc current gain of the configuration would be (100)(100) = 10,000. Note that Darlington transistors typically have h_{FE} ratings in the tens of thousands.

Integrated Transistors

Integrated transistors come in packages that house more than one transistor in integrated form. Each transistor in an integrated circuit (IC) has the same types of maximum ratings and electrical characteristics as individual (or discrete) transistors. The primary difference is that integrated transistors tend to have lower maximum power and current ratings than their discrete-component counterparts.

Surface-Mount Components

The size advantage of integrated transistors has been taken one step further with the development of surface-mount technology. *Surface-mount components* (SMCs) are much smaller and lighter than other types of components. The term *surface mount* is derived from the fact that these ICs are mounted directly onto the surface of a PC board rather than soldered into holes or fitted into IC sockets. ◀ *OBJECTIVE 19*

Figure 19.42a shows a standard *dual in-line package* (DIP) and a surface-mount package. The DIP has, for a number of years, been one of the most commonly used types of IC package. If you compare the maximum A, B, and C dimensions of the two, you can see that the SMC covers roughly one-third the area of its DIP counterpart. This means that more of these components can be mounted in a given area of a PC board.

Figure 19.42a, shows that the pins of the SMC are flattened at the bottom, allowing it to be soldered directly onto PC board footprints. These footprints, which are represented in Figure 19.42b, are typically no larger than 1 mm^2. Because they are extremely sensitive to heat, great care must be taken when soldering SMCs onto a PC board.

CASE 751B-03 (SO-16) **PLASTIC**

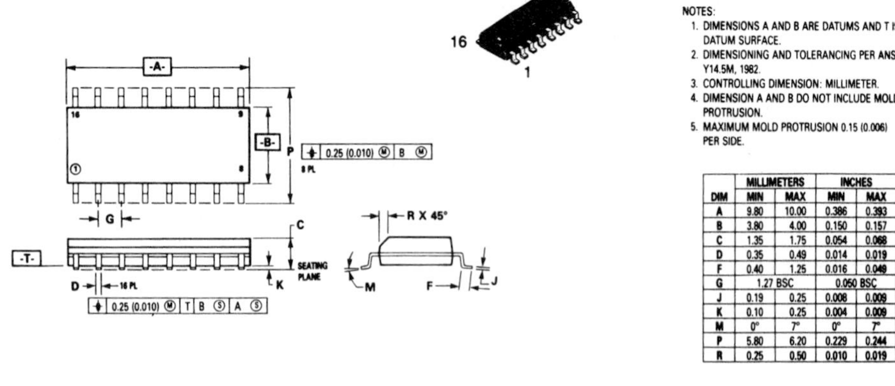

DIM	MILLIMETERS MIN	MILLIMETERS MAX	INCHES MIN	INCHES MAX
A	9.80	10.00	0.386	0.393
B	3.80	4.00	0.150	0.157
C	1.35	1.75	0.054	0.068
D	0.35	0.49	0.014	0.019
F	0.40	1.25	0.016	0.049
G	1.27 BSC		0.050 BSC	
J	0.19	0.25	0.008	0.009
K	0.10	0.25	0.004	0.009
M	0°	7°	0°	7°
P	5.80	6.20	0.229	0.244
R	0.25	0.50	0.010	0.019

CASE 646-06 (14-PIN DIP) **PLASTIC**

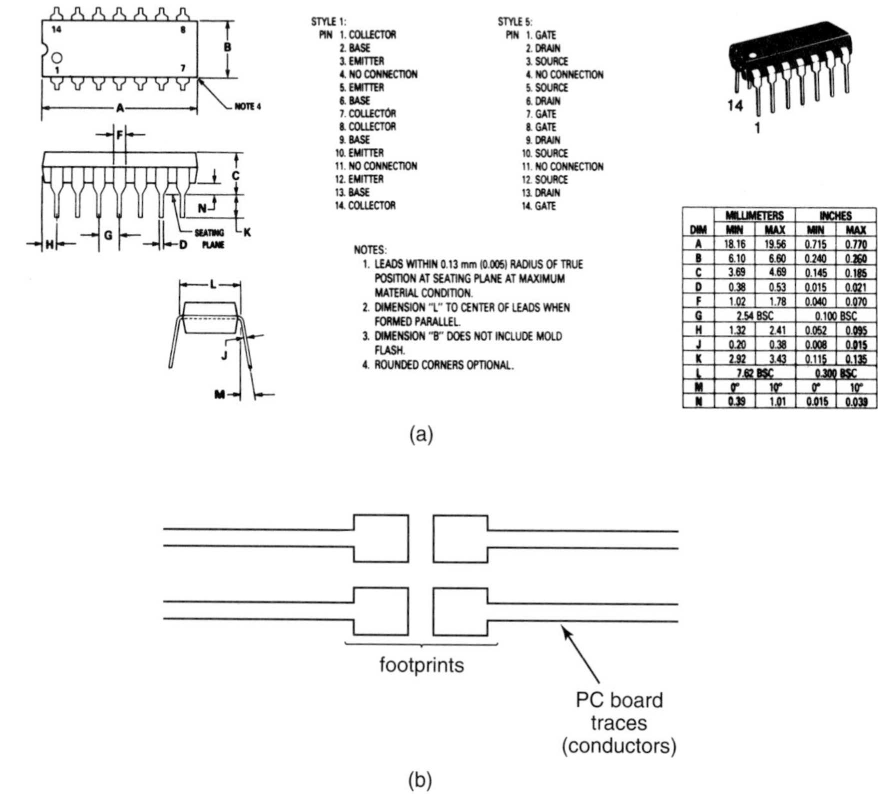

DIM	MILLIMETERS MIN	MILLIMETERS MAX	INCHES MIN	INCHES MAX
A	18.16	19.56	0.715	0.770
B	6.10	6.60	0.240	0.260
C	3.69	4.69	0.145	0.185
D	0.38	0.53	0.015	0.021
F	1.02	1.78	0.040	0.070
G	2.54 BSC		0.100 BSC	
H	1.32	2.41	0.052	0.095
J	0.20	0.38	0.008	0.015
K	2.92	3.43	0.115	0.135
L	7.62 BSC		0.300 BSC	
M	0°	10°	0°	10°
N	0.39	1.01	0.015	0.039

(b)

Figure 19.42 SMC dimensions. (Copyright of Semiconductor Components Industries, LLC. Used by permission.)

Section Review

1. What do the *off characteristics* of a transistor indicate?
2. What do the *on characteristics* of a transistor indicate?
3. What label is commonly used to represent *dc beta*?
4. What is *collector-emitter saturation voltage*?
5. What are the primary differences between *pnp* and *npn* transistor circuits?
6. Which transistor is used primarily in systems with *positive* dc supply voltages?
7. Which transistor is used primarily in systems with *negative* dc supply voltages?
8. What are *high-voltage transistors*? When are they typically used?

9. What are *high-current transistors*? When are they typically used?

10. What are *high-power transistors*? When are they typically used?

11. Describe the differences between SMCs and standard *DIP* ICs.

Key Terms

The following terms were introduced in this chapter on the pages indicated:

Active region, 599
Amplifier, 595
Base, 595
Base bias, 610
Base cutoff current (I_{BL}), 627
Base-emitter junction, 597
Base-emitter saturation voltage, 627
Beta (β), 596
Beta curve, 612
Beta-dependent circuit, 612
Beta-independent circuit, 612
Bipolar junction transistor (BJT), 595
Collector, 595
Collector-base junction, 597
Collector characteristic curve, 605
Collector cutoff current (I_{CEX}), 627
Collector-emitter saturation voltage, 627
Collector-feedback bias, 621
Cutoff, 598
Darlington pair, 628
DC Alpha, 605
DC Beta, 603
DC current gain (h_{FE}), 605
DC load line, 608
Emitter, 595
Emitter bias, 618
Emitter-feedback bias, 623
Feedback, 621
Fixed bias, 610
Forward current gain, 596
High-current transistor, 628
High-power transistor, 628
High-voltage transistor, 628
Integrated transistor, 629
Midpoint bias, 609
npn Transistor, 595
pnp Transistor, 595
Q-Point, 609
Q-point shift, 612
Quiescent, 609
Saturation, 598
Transistor, 595
Universal bias, 613
Voltage-divider bias, 613

Practice Problems

1. A BJT has values of $\beta = 320$ and $I_B = 12\ \mu A$. Determine I_C for the device.
2. A BJT has values of $\beta = 400$ and $I_B = 30\ \mu A$. Determine I_C for the device.
3. A BJT has values of $\beta = 254$ and $I_B = 1.01$ mA. Determine I_C for the device.
4. A BJT has values of $\beta = 144$ and $I_B = 82\ \mu A$. Determine I_C for the device.
5. A BJT has values of $I_B = 20\ \mu A$ and $I_C = 1.1$ mA. Determine I_E for the device.
6. A BJT has values of $I_B = 1.1$ mA and $I_C = 344$ mA. Determine I_E for the device.
7. A BJT has values of $I_B = 35\ \mu A$ and $\beta = 100$. Determine I_C and I_E for the device.
8. A BJT has values of $I_B = 150\ \mu A$ and $\beta = 400$. Determine I_C and I_E for the device.
9. A BJT has values of $I_B = 48\ \mu A$ and $\beta = 119$. Determine I_C and I_E for the device.
10. A BJT has values of $I_C = 12$ mA and $\beta = 440$. Determine I_B and I_E for the device.
11. A BJT has values of $I_C = 50$ mA and $\beta = 400$. Determine I_B and I_E for the device.
12. A BJT has values of $I_E = 65$ mA and $\beta = 380$. Determine I_B and I_C for the device.
13. A BJT has values of $I_E = 120$ mA and $\beta = 60$. Determine I_B and I_C for the device.
14. A BJT has a value of $\beta = 426$. Determine the value of α for the device.
15. A BJT has a value of $\beta = 350$. Determine the value of α for the device.
16. Refer to Figure 19.43. What is the approximate value of I_C when I_B is 40 mA?
17. Refer to Figure 19.43. What is the maximum value of V_{CE} when the device is saturated?

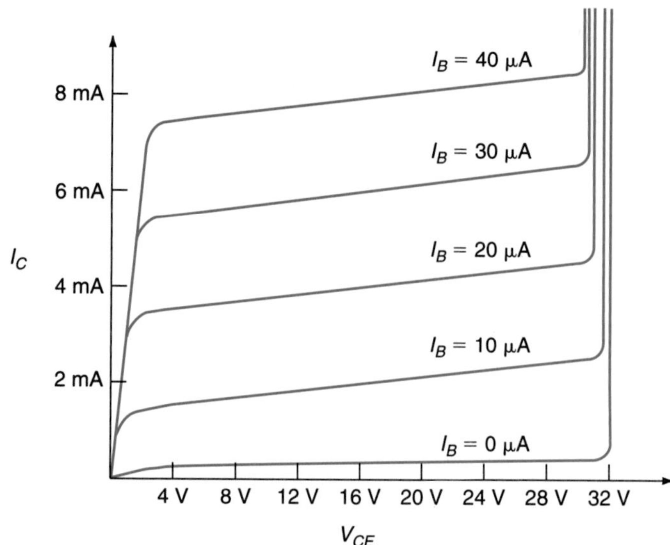

Figure 19.43

18. For each circuit shown in Figure 19.44, identify the type of transistor, and indicate the directions of the terminal currents.

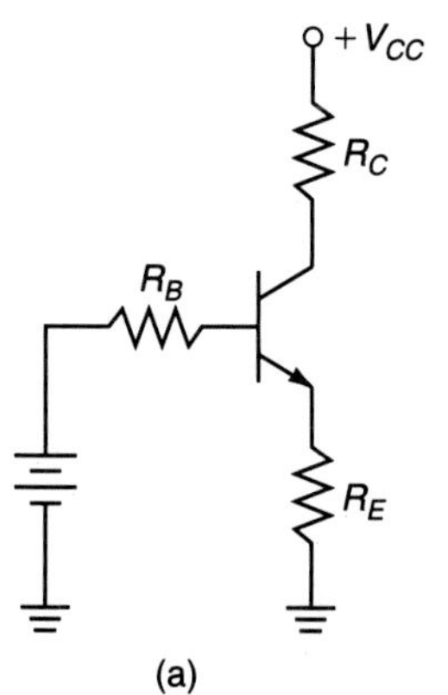

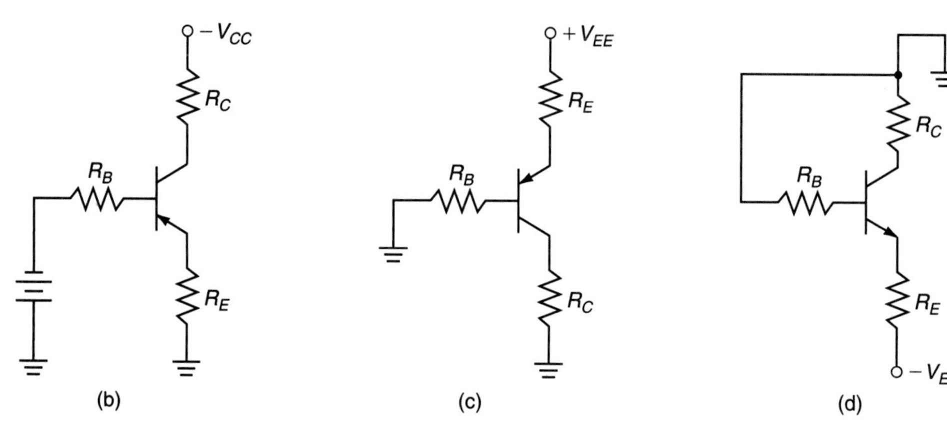

Figure 19.44

19. The circuit shown in Figure 19.45 has values of $V_{CC} = 18$ V and $R_C = 3.3$ kΩ. Plot the dc load line for the circuit.

20. The circuit shown in Figure 19.45 has values of $V_{CC} = 24$ V and $R_C = 9.1$ kΩ. Plot the dc load line for the circuit.

21. The circuit shown in Figure 19.45 has values of $V_{CC} = 14$ V and $R_C = 1$ kΩ. Plot its dc load line. Then, use equation (19.8) to verify the load line values of V_{CE} for $I_C = 2$ mA, $I_C = 8$ mA, and $I_C = 10$ mA.

22. The circuit shown in Figure 19.45 has values of $V_{CC} = 20$ V and $R_C = 2.4$ kΩ. Plot its dc load line. Then, use equation (19.8) to verify the load line values of V_{CE} for $I_C = 1$ mA, $I_C = 5$ mA, and $I_C = 7$ mA.

23. The circuit shown in 19.45 has values of $V_{CC} = 8$ V and $R_C = 1$ kΩ. Plot the dc load line for the circuit. Then, use your dc load line to determine the midpoint-bias values of I_C and V_{CE}.

24. The circuit shown in Figure 19.45 has values of $V_{CC} = 36$ V and $R_C = 36$ kΩ. Plot the dc load line for the circuit. Then, use your dc load line to determine the midpoint-bias values of I_C and V_{CE}.

25. Determine the Q-point values of I_C and V_{CE} for the circuit shown in Figure 19.46a.

26. Determine the Q-point values of I_C and V_{CE} for the circuit shown in Figure 19.46b.

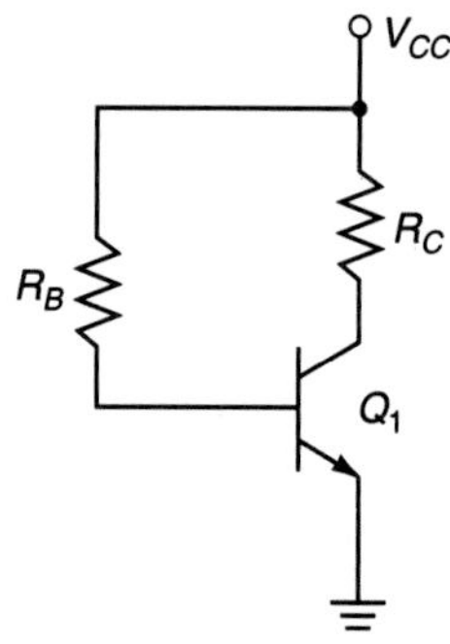

Figure 19.45

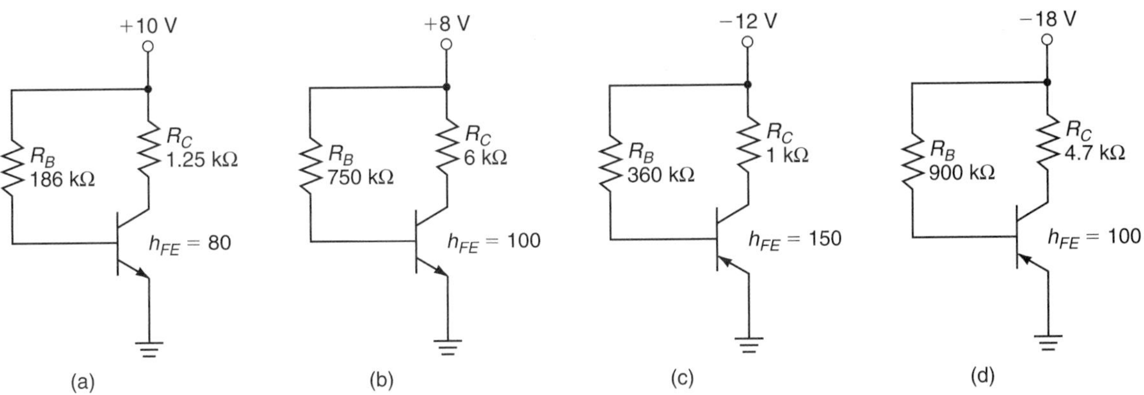

Figure 19.46

27. Determine the Q-point values of I_C and V_{CE} for the circuit shown in Figure 19.46c.
28. Determine the Q-point values of I_C and V_{CE} for the circuit shown in Figure 19.46d.
29. Determine whether or not the circuit shown in Figure 19.46a is midpoint biased.
30. Determine whether or not the circuit shown in Figure 19.46b is midpoint biased.
31. Determine whether or not the circuit shown in Figure 19.46c is midpoint biased.
32. Determine whether or not the circuit shown in Figure 19.46d is midpoint biased.
33. The value of h_{FE} shown in Figure 19.46a is measured at 25°C. At 100°C, the value of h_{FE} for the transistor is 119. Calculate I_C and V_{CE} for the circuit when at 100°C.
34. The value of h_{FE} shown in Figure 19.46b is measured at 25°C. At 100°C, the value of h_{FE} for the transistor is 119. Calculate I_C and V_{CE} for the circuit when it is operated at 100°C.
35. Determine the value of $R_{in(base)}$ for the circuit shown in Figure 19.47a.
36. Determine the value of $R_{in(base)}$ for the circuit shown in Figure 19.47b.
37. Determine the values of I_{CQ} and V_{CEQ} for the circuit shown in Figure 19.47a.
38. Determine the values of I_{CQ} and V_{CEQ} for the circuit shown in Figure 19.47b.
39. Determine the values of I_{CQ}, V_{CEQ}, and I_B for the circuit shown in Figure 19.47c.
40. Determine the values of I_{CQ}, V_{CEQ}, and I_B for the circuit shown in Figure 19.47d.
41. Determine the values of $I_{C(sat)}$ and $V_{CE(off)}$ for the circuit shown in Figure 19.47a. Then, using your answers from Problem 37, determine whether or not the circuit is midpoint biased.

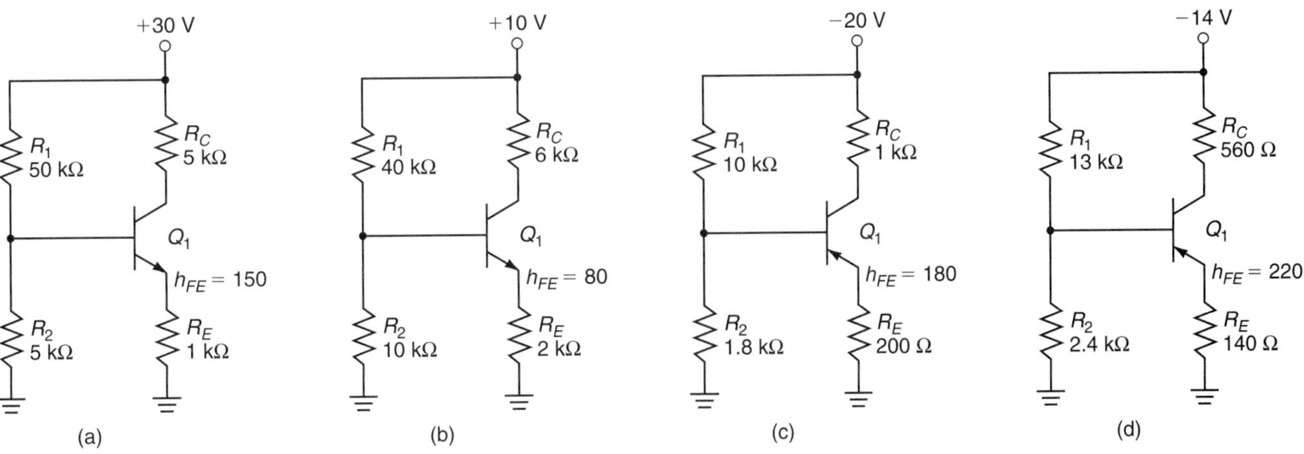

Figure 19.47

42. Determine the values of $I_{C(sat)}$ and $V_{CE(off)}$ for the circuit shown in Figure 19.47b. Then, using your answers from Problem 38, determine whether or not the circuit is midpoint biased.

43. Determine the values of $I_{C(sat)}$ and $V_{CE(off)}$ for the circuit shown in Figure 19.47c. Then, using your answers from Problem 39, determine whether or not the circuit is midpoint biased.

44. Determine the values of $I_{C(sat)}$ and $V_{CE(off)}$ for the circuit shown in Figure 19.47d. Then, using your answers from Problem 40, determine whether or not the circuit is midpoint biased.

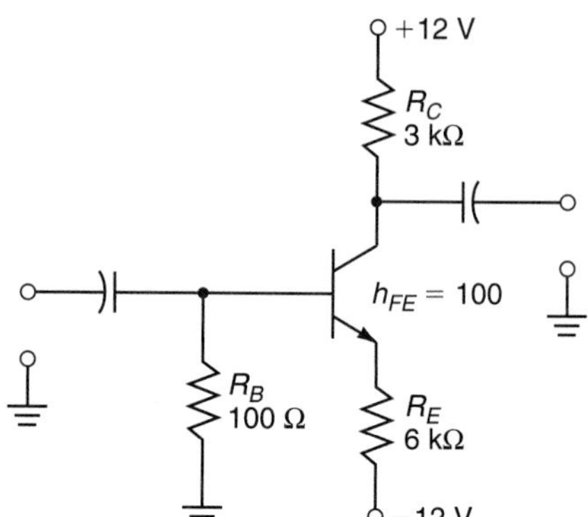

Figure 19.48

45. Determine the values of I_{CQ} and V_{CEQ} for the circuit shown in Figure 19.48.

46. For the circuit shown in Figure 19.48, double the values of R_C and R_E. Now recalculate the output values for the circuit. How do these values compare with those obtained in Problem 45?

47. Calculate the values of $I_{C(sat)}$ and $V_{CE(off)}$ for the circuit shown in Figure 19.48. Use the original component values shown.

48. For the circuit shown in Figure 19.48, double the values of R_C and R_E. Now recalculate values of $I_{C(sat)}$ and $V_{CE(off)}$. How do these values compare with those obtained in Problem 47?

49. Determine the values of I_{CQ}, V_{CEQ}, $I_{C(sat)}$, and $V_{CE(off)}$ for the circuit shown in Figure 19.49a.

50. Determine the values of I_{CQ}, V_{CEQ}, $I_{C(sat)}$, and $V_{CE(off)}$ for the circuit shown in Figure 19.49b.

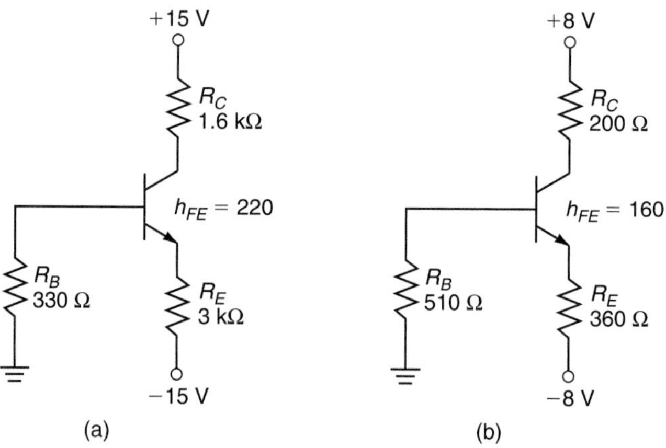

Figure 19.49

51. Determine the values of I_{CQ} and V_{CEQ} for the circuit shown in Figure 19.50a.

52. Determine the values of I_{CQ} and V_{CEQ} for the circuit shown in Figure 19.50b.

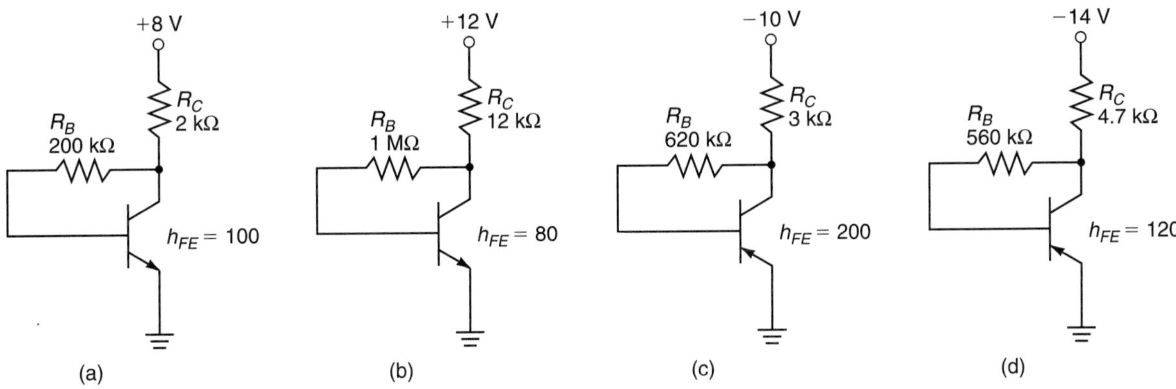

Figure 19.50

53. Determine the values of I_{CQ} and V_{CEQ} for the circuit shown in Figure 19.50c.

54. Determine the values of I_{CQ} and V_{CEQ} for the circuit shown in Figure 19.50d.

55. Determine the values of I_{CQ} and V_{CEQ} for the circuit shown in Figure 19.51a.

56. Determine the values of I_{CQ} and V_{CEQ} for the circuit shown in Figure 19.51b.

57. Determine the values of I_{CQ} and V_{CEQ} for the circuit shown in Figure 19.51c.

58. Determine the values of I_{CQ} and V_{CEQ} for the circuit shown in Figure 19.51d.

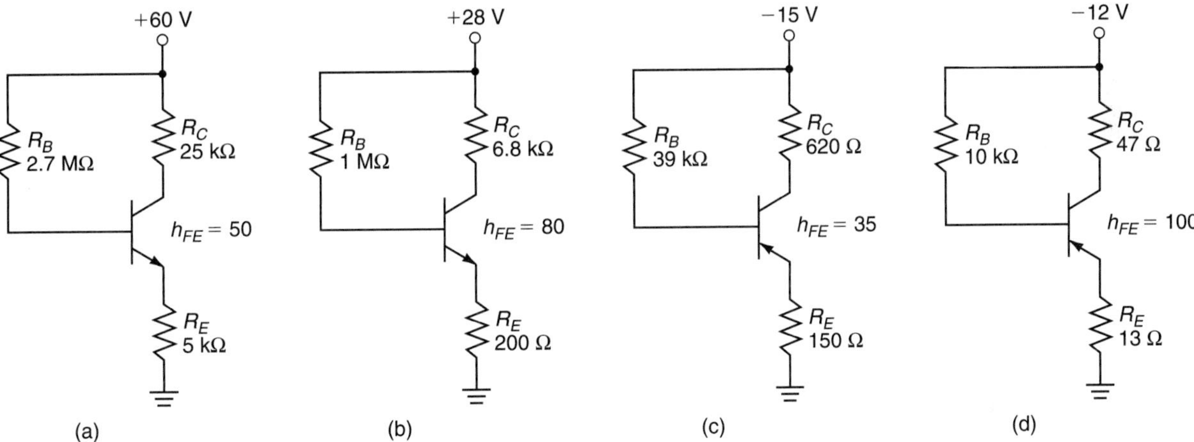

Figure 19.51

Looking Back

These problems relate to material presented in earlier chapters. The chapters are identified in brackets.

59. A series resistive circuit has a source resistance (R_S) that can be varied between 10 and 100 Ω and values of R_L = 910 Ω and V_S = 8 V. Determine its maximum value of load power. [Chapter 4]

60. Determine the total resistance for the parallel circuit shown in Figure 19.52. [Chapter 5]

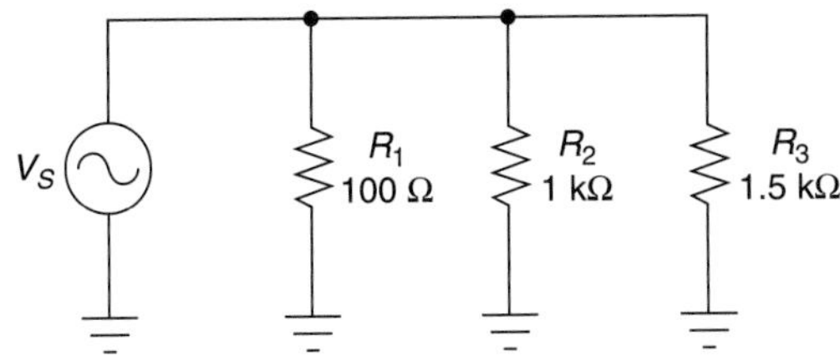

Figure 19.52

61. Determine the value of the bleeder current for the circuit shown in Figure 19.53. [Chapter 6]

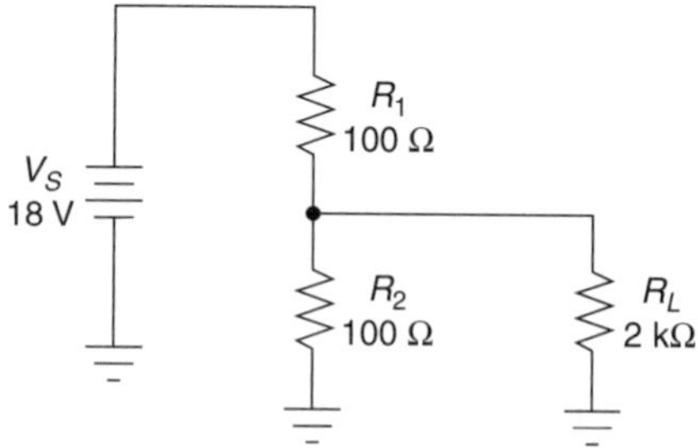

Figure 19.53

62. Determine the value of V_3 for the circuit shown in Figure 19.54. [Chapter 7]

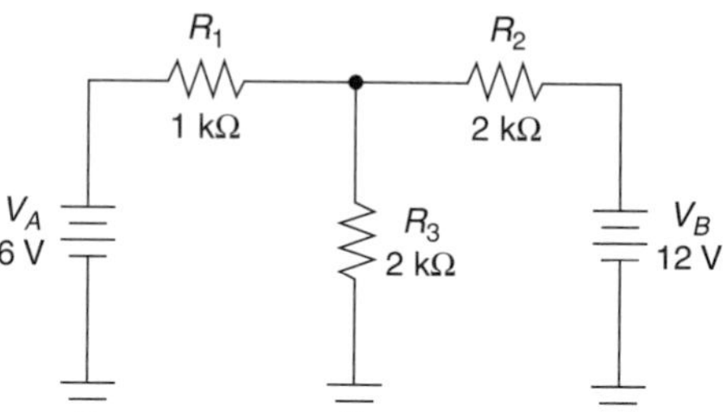

Figure 19.54

63. Calculate the reactance of a 47 μF capacitor when it is operated at 100 Hz. [Chapter 12]

64. A parallel *LC* circuit has values of $L = 33$ mH, $R_W = 2\ \Omega$, and $C = 100$ nF. Calculate the Q of the circuit when operated at resonance. [Chapter 14]

Pushing the Envelope

65. Starting with the defining equation for α, prove that $\alpha = \dfrac{\beta}{\beta + 1}$.

66. Starting with the defining equation for β, prove that $\beta = \dfrac{\alpha}{1 - \alpha}$.

67. Using equation (19.27), prove the following relationship to be true: A collector-feedback bias circuit is midpoint biased when $R_B = h_{FE}R_C$. (*Hint:* Idealize the emitter-base junction diode in the equation.)

68. The 2N3904 transistor cannot be used for the circuit shown in Figure 19.55. Why not? (*Hint:* Consider the dc load line for the amplifier and the maximum ratings shown in the 2N3904 spec sheet.)

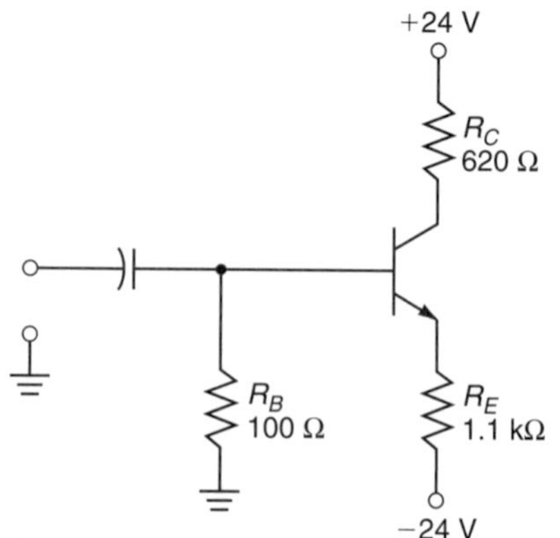

Figure 19.55

Answers to the Example Practice Problems

19.1 $I_C = 17.5$ mA

19.2 $I_C = 20$ mA, $I_E = 20.05$ mA

19.3 $I_B = 85.11\ \mu$A, $I_C = 11.91$ mA

19.4 $I_B = 470.6\ \mu$A, $I_E = 80.47$ mA

19.5 8 V, 7.27 mA

19.6 3.39 mA, 8.61 V. The circuit is *not* midpoint biased.

19.7 6.33 mA, 4.94 V

19.8 1.48 mA, 9.2 V

19.9 4.77 mA, 8.55 V

19.10 2.83 mA, 6.34 V

19.11 2.72 mA, 8.63 V

Multisim File References

Figure	File Reference
19.3	ETF19_3.msm
19.13b	ETF19_13b.msm
19.24	ETF19_24.msm
19.26	ETF19_26.msm
19.28	ETF19_28.msm
19.31	ETF19_31.msm
19.33	ETF19_33.msm
19.39	ETF19_39.msm

chapter 20

BJT Amplifiers

Objectives

After studying the material in this chapter, you should be able to:

1. List the three fundamental properties of amplifiers.
2. Discuss the concept of *gain*.
3. Draw and discuss the general model of a voltage amplifier.
4. Discuss the effects that amplifier input and output impedance have on the *effective* voltage gain of the circuit.
5. Describe the *ideal* voltage amplifier.
6. List, compare, and contrast the three BJT amplifier configurations.
7. Explain the input/output voltage phase relationship of a *common-emitter* amplifier.
8. Calculate the *ac emitter resistance* of a transistor.
9. Derive the *ac equivalent* for a common-emitter amplifier.
10. Describe the two primary roles that capacitors play in amplifiers.
11. Calculate the values of A_v, A_i, and Z_{in} for a common-emitter amplifier.
12. Calculate the gain and impedance values of a *common-collector* amplifier (*emitter follower*).
13. Calculate the gain and impedance values of a *common-base* amplifier.
14. Describe and analyze the operation of a *class B amplifier*.
15. Explain how *class AB* operation eliminates crossover distortion and thermal runaway.
16. Describe the effect that *swamping* has on common-emitter amplifier voltage gain.
17. List the advantages of using a *Darlington transistor* in an amplifier.
18. Determine the power dissipation requirements for the transistor(s) in any class A or class B amplifier.

When you cannot hear the output from your stereo, you turn up the volume. What you are actually doing is making a weak signal stronger; that is, you are increasing its power level. *The process of increasing the power of a signal* is referred to as **amplification.** *The circuits used to provide amplification* are referred to as **amplifiers.** Amplifiers are some of the most widely used circuits that you will encounter.

20.1 Amplifier Properties

All amplifiers have three fundamental properties: *gain, input impedance,* and *output impedance.* These properties can be combined to form a general amplifier model like the one shown in Figure 20.1. Note that the diamond shape in the model is used (in this case) to represent the gain of the circuit. We'll modify this symbol slightly after we discuss the concept of gain. ◄ *OBJECTIVE 1*

Amplifier Gain

All amplifiers exhibit a property called *gain.* The **gain** of an amplifier is *a value that indicates the magnitude relationship between the circuit's input and output signals.* If the gain of an amplifier is 100, then the output signal is 100 times greater than the input signal. There are three types of gain: *voltage gain, current gain,* and *power gain.* Gain is represented using the letter A, as shown in Table 20.1. Note that the subscript in each symbol identifies the type of gain. ◄ *OBJECTIVE 2*

Traditionally, gain is defined as *the ratio of an output value to its corresponding input value.* For example, voltage gain can be defined as the ratio of ac output voltage to ac input voltage. By formula,

$$A_v = \frac{v_{out}}{v_{in}} \tag{20.1}$$

Note: Amplifiers contain both *dc* and *ac* values. To distinguish between the two, lower-case letters are commonly used to represent the *ac* values.

where

v_{out} = the ac output voltage from the amplifier

v_{in} = the ac input voltage to the amplifier

The calculation of voltage gain, using input and output values, is demonstrated in Example 20.1.

TABLE 20.1 Gain Symbols

Type of Gain	*Symbol*
Voltage	A_v
Current	A_i
Power	A_p

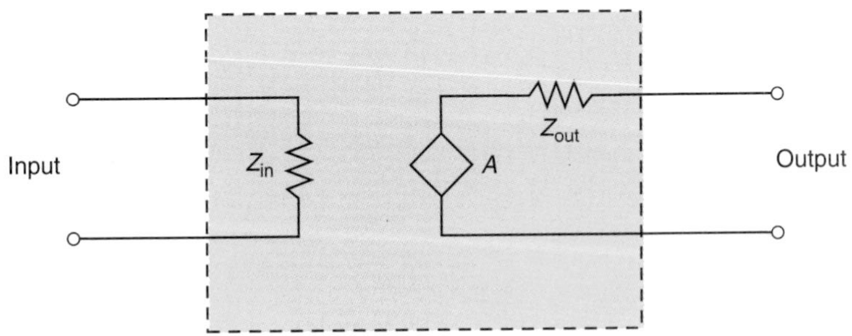

Figure 20.1 The general amplifier model.

Example 20.1

Figure 20.2 contains the generic symbol for an amplifier. Calculate the voltage gain (A_v) for the amplifier represented in the figure.

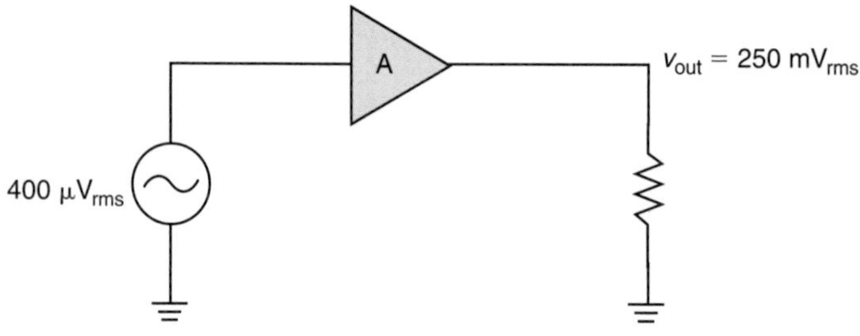

Figure 20.2

Solution: Using the given values of v_{in} and v_{out}, the amplifier voltage gain is found as

$$A_v = \frac{v_{out}}{v_{in}} = \frac{250 \text{ mV}}{400 \ \mu\text{V}} = 625$$

Practice Problem 20.1

An amplifier like the one represented in Figure 20.2 has the following values: v_{out} = 72 mV, and v_{in} = 300 μV. Calculate the voltage gain (A_v) of the circuit.

The result in Example 20.1 indicates that the ac output voltage is 625 times greater than the ac input voltage. Note that voltage gain has no units, because it is a ratio of an output value to an input value.

Under normal circumstances, the gain of an amplifier is a constant that is determined by circuit component values. Thus, a change in circuit input amplitude normally results in a change in output amplitude (not in a change in gain). When the value of v_{in} for an amplifier changes, the new value of v_{out} is found as

$$v_{out} = A_v v_{in} \tag{20.2}$$

This relationship is demonstrated in Example 20.2.

Example 20.2

The input to the circuit shown in Figure 20.2 changes to 240 μV. Calculate the resulting value of v_{out}.

Solution: In Example 20.1, A_v was found to be 625. When the input amplitude changes, the output amplitude changes to

$$v_{out} = A_v v_{in} = (625)(240 \ \mu\text{V}) = 150 \text{ mV}$$

Practice Problem 20.2

Refer to Practice Problem 20.1. Using the value of A_v found in that problem, calculate the value of v_{out} for v_{in} = 360 μV.

Like voltage gain, current and power gain can be defined as output-to-input ratios. These ratios are

$$A_i = \frac{i_{out}}{i_{in}} \tag{20.3}$$

and

$$A_p = \frac{P_{out}}{P_{in}} \tag{20.4}$$

Since they are ratios, current and power gain (like voltage gain) have no units.

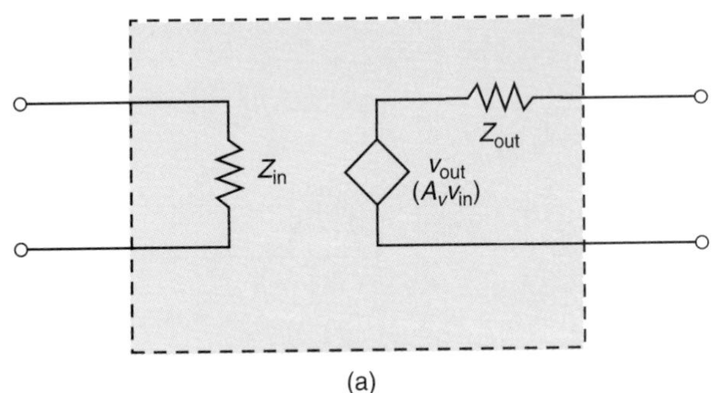

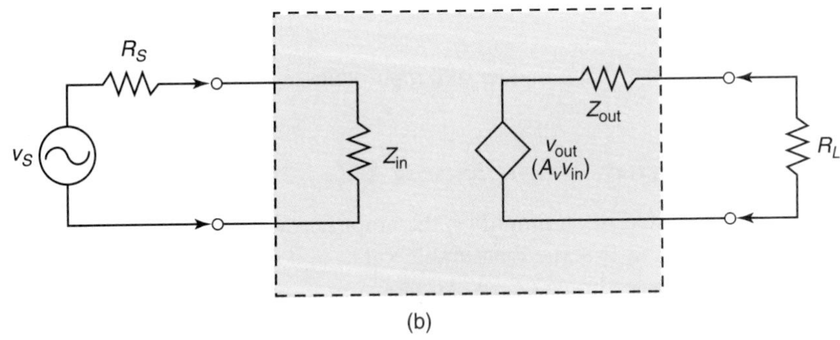

Figure 20.3

A Practical Consideration: The Z_{in} and Z_{out} labels in Figure 20.3 are used because they represent transistor ratings listed on most spec sheets. Though listed as *impedances,* they can be combined algebraically with R_S and R_L. *Remember:* impedance *can* be purely resistive.

The General Voltage Amplifier Model

◄ *OBJECTIVE 3*

Now that we have defined gain, we need to modify the amplifier model as shown in Figure 20.3a. As you can see, the diamond shape from Figure 20.1 is actually used to represent v_{out} as a voltage source (defined as $A_v v_{in}$).

When we add a signal source and a load to the amplifier model, we obtain the circuit shown in Figure 20.3b. The input circuit consists of V_S, R_S, and Z_{in} (the amplifier input impedance). The output circuit consists of v_{out}, Z_{out} (the amplifier output impedance), and R_L. As you can see, the input and output circuits are nearly identical in terms of the type of components they contain.

Amplifier Input Impedance (Z_{in})

◄ *OBJECTIVE 4*

When an amplifier is connected to a signal source, the source sees the amplifier as a load. For example, the amplifier in Figure 20.4 acts as a 1.5 kΩ load that is in series with the source resistance (R_S). If we assume that the **amplifier input impedance (Z_{in})** is purely resistive, the signal voltage at the amplifier input is found as

$$v_{in} = v_S \frac{Z_{in}}{R_S + Z_{in}} \qquad (20.5)$$

Figure 20.4

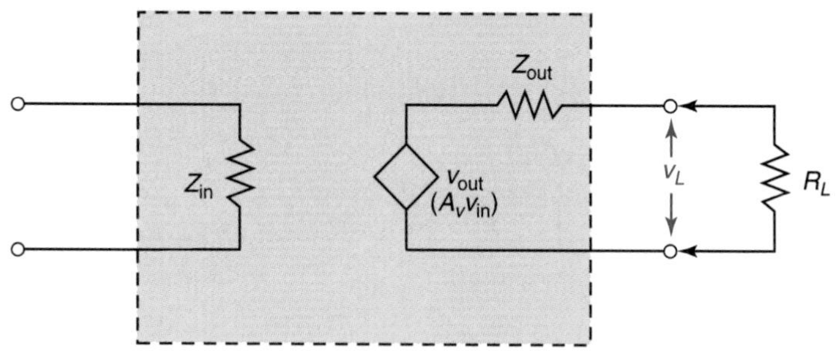

Figure 20.5

Since R_S and Z_{in} form a voltage divider, the input voltage to the amplifier must be lower than the rated value of the source.

Amplifier Output Impedance (Z_{out})

When a load is connected to an amplifier, the amplifier acts as the source for that load. As with any source, there is some measurable value of source impedance—in this case, the **amplifier output impedance (Z_{out})**. As shown in Figure 20.5, the output impedance of the amplifier forms a voltage divider with the load. Therefore, the load voltage is found using

$$v_L = v_{out}\frac{R_L}{R_L + Z_{out}} \tag{20.6}$$

The Combined Effects of the Input and Output Circuits

Together, the input and output circuits can cause a fairly significant reduction in the effective voltage gain of an amplifier. For example, consider the circuit shown in Figure 20.6. To see the combined effects of the input and output circuits, we need to calculate the value of v_L for the circuit.

The first step is to determine the value of v_{in}, as follows:

$$v_{in} = v_S\frac{Z_{in}}{R_S + Z_{in}} = (15\text{ mV})\frac{980\ \Omega}{1\text{ k}\Omega} = 14.7\text{ mV}$$

Now, the value of v_{out} is found as

$$v_{out} = A_v v_{in} = (340)(14.7\text{ mV}) = 5\text{ V}$$

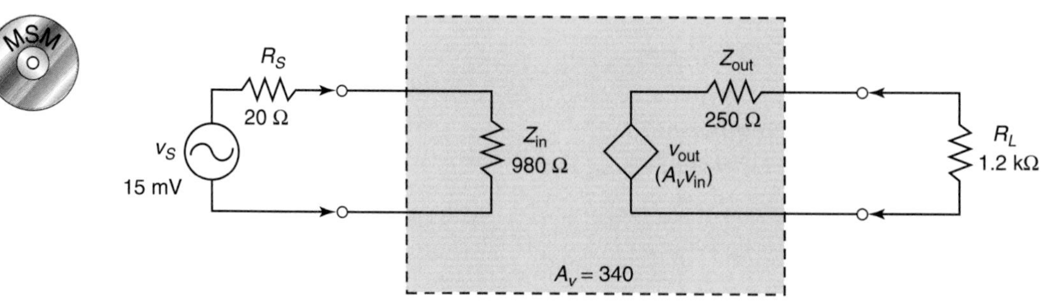

Figure 20.6

and the load voltage is found as

$$v_L = v_{\text{out}} \frac{R_L}{R_L + Z_{\text{out}}} = (5 \text{ V}) \frac{1.2 \text{ k}\Omega}{1.45 \text{ k}\Omega} = 4.14 \text{ V}$$

The **effective voltage gain** of the amplifier equals *the ratio of load voltage to source voltage.* By formula,

$$A_{v(\text{eff})} = \frac{v_L}{v_S} \tag{20.7}$$

Therefore, the effective voltage gain of the circuit can be found as

$$A_{v(\text{eff})} = \frac{v_L}{v_S} = \frac{4.14 \text{ V}}{15 \text{ mV}} = 276$$

As you can see, the input and output circuits have reduced the voltage gain of this amplifier from 340 to an effective value of 276.

How to reduce the effects of the input and output circuits on amplifier voltage gain.

The effects that the input and output circuits have on the voltage gain of an amplifier are significantly reduced by designing the circuit so that:

- $Z_{\text{in}} >> R_S$.
- $Z_{\text{out}} << R_L$.

For example, consider the circuit shown in Figure 20.7. This is essentially the same circuit as the one shown in Figure 20.6, but we have changed the values of Z_{in} and Z_{out}. For this circuit,

$$v_{\text{in}} = v_S \frac{Z_{\text{in}}}{Z_{\text{in}} + R_S} = (15 \text{ mV}) \frac{8 \text{ k}\Omega}{8.02 \text{ k}\Omega} = 14.96 \text{ mV}$$

$$v_{\text{out}} = A_v v_{\text{in}} = (340)(14.96 \text{ mV}) = 5.1 \text{ V}$$

$$v_L = v_{\text{out}} \frac{R_L}{R_L + Z_{\text{out}}} = (5.1 \text{ V}) \frac{1.2 \text{ k}\Omega}{1.22 \text{ k}\Omega} = 5 \text{ V}$$

and

$$A_{v(\text{eff})} = \frac{v_L}{v_S} = \frac{5 \text{ V}}{15 \text{ mV}} = 333.3$$

In most practical circuits, little can be done to change the resistance of the signal source or load. However, the values of Z_{in} and Z_{out} are determined by the design of a given voltage amplifier.

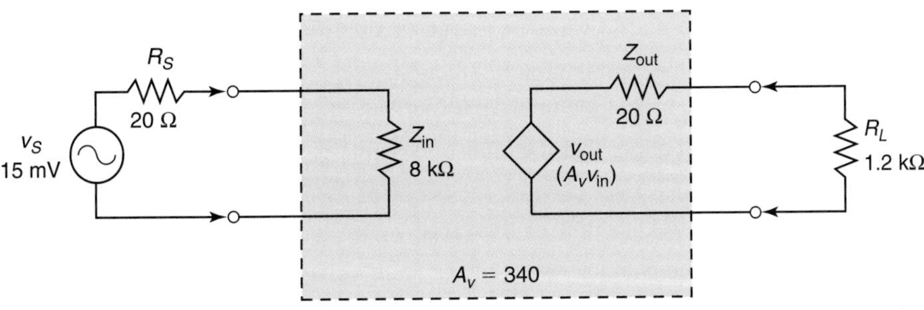

Figure 20.7

Figure 20.8

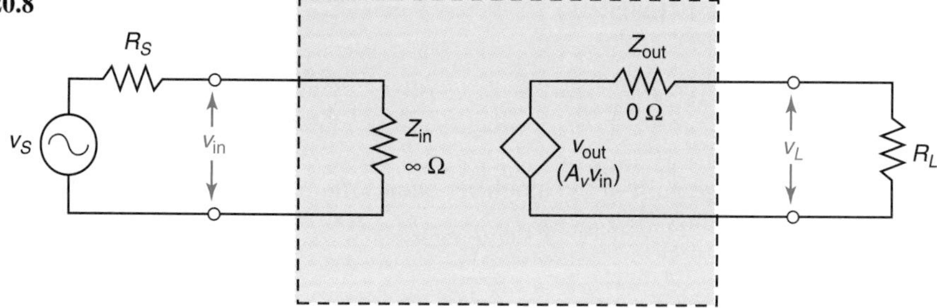

The Ideal Voltage Amplifier

OBJECTIVE 5 ➤ The **ideal voltage amplifier,** if it could be produced, would have the following characteristics (among others):

- Infinite gain (if needed).
- Infinite input impedance.
- Zero output impedance.

The first of these characteristics needs little explanation. An ideal amplifier would be capable of providing any value of gain. The impedance characteristics of an ideal voltage amplifier are illustrated in Figure 20.8. With infinite input impedance, there would be no current in the input circuit and, therefore, no voltage dropped across the source resistance (R_S). With no voltage dropped across R_S,

$$v_{in} = v_S \qquad \text{(for the ideal amplifier)}$$

With zero output impedance, there would be no voltage divider in the output circuit of the amplifier. Therefore,

$$v_L = v_{out} \qquad \text{(for the ideal amplifier)}$$

Since there would be no reduction of voltage by either the input or output circuits, the effective voltage gain of the circuit would equal the calculated value of A_v.

Values of $Z_{in} = \infty\Omega$ and $Z_{out} = 0\ \Omega$ have not yet been achieved in practical circuits. However, it is possible to "effectively" achieve them through proper circuit design (as stated earlier in this section). When a voltage amplifier is designed so that $Z_{in} >> R_S$ and $Z_{out} << R_L$, the circuit can come very close to having the impedence characteristics of an ideal voltage amplifier.

Section Review

1. What is *amplification*?
2. What is *gain*? What are the three types of gain?
3. Define *voltage gain* as a ratio.
4. What determines the voltage gain of an amplifier?
5. What effect does the combination of *source resistance* (R_S) and *amplifier input impedance* (Z_{in}) have on the value of v_{in} for an amplifier?
6. What effect does the combination of *amplifier output impedance* (Z_{out}) and *load resistance* (R_L) have on the value of v_L for an amplifier?
7. Explain the overall effect that the input and output circuits have on the voltage gain of an amplifier.
8. How is the effect in Question 7 minimized?
9. List and explain the characteristics of the *ideal* voltage amplifier.

Critical Thinking

10. How do you think the gain of an ideal voltage amplifier would be affected by a change in *operating frequency*?

20.2 BJT Amplifier Configurations

Now that we have established the gain, input impedance, and output impedance characteristics of the ideal voltage amplifier, we will take a brief look at several types of BJT amplifiers to see how they compare with the ideal amplifier. There are three BJT amplifier configurations, each having a unique combination of characteristics.

◀ *OBJECTIVE 6*

The Common-Emitter Amplifier

The **common-emitter (CE) amplifier** is the most widely used BJT amplifier. A typical CE amplifier is shown in Figure 20.9. As you can see, the input is applied to the base of the transistor, and the output is taken from the collector. The term *common emitter* is used for two reasons:

- The emitter terminal of the transistor is common to both the input and the output circuits.
- The emitter terminal of the transistor is normally set at 0 V_{ac}, and thus is at ac ground (or *ac common*). The ac ground is provided by the *bypass capacitor* (C_B), which is connected to the emitter terminal of the transistor.

To discuss the characteristics of the CE (or any other) amplifier, we need to establish some boundaries. For the sake of comparison, we'll classify gain and impedance values as being *low, midrange,* or *high*. These classifications are broken down as shown in Table 20.2. Note that the ranges listed are open to debate and should not be taken as standard values. They are provided merely as a basis for comparison.

Typical CE amplifier characteristics

Using the ranges given in Table 20.2, we can classify the CE amplifier as an amplifier that typically has:

- Midrange values of voltage and current gain.
- High power gain.
- Midrange input and output impedance values.

The CE amplifier is also the only BJT amplifier that produces a 180° voltage phase shift from its input to its output (as shown in Figure 20.9). The basis of this voltage phase shift is discussed later in this chapter.

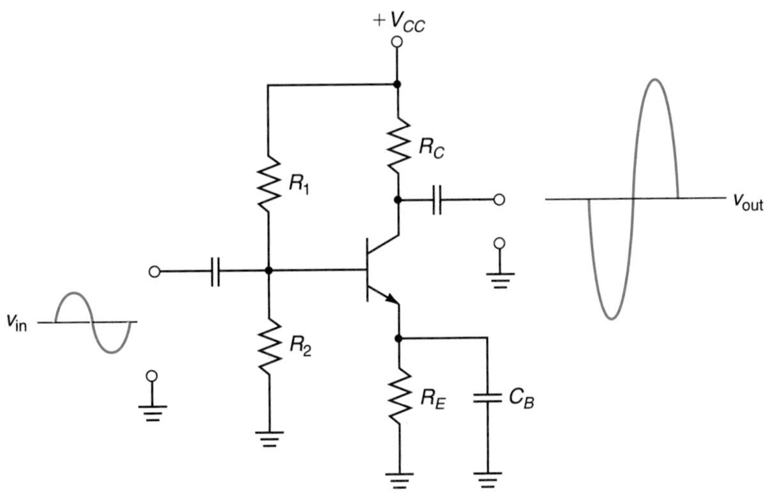

Figure 20.9 A common-emitter amplifier.

TABLE 20.2 Property Ranges

Property	*Low*	*Midrange*	*High*
Gain	<100	100–1000	>1000
Impedance	<1 kΩ	1–10 kΩ	>10 kΩ

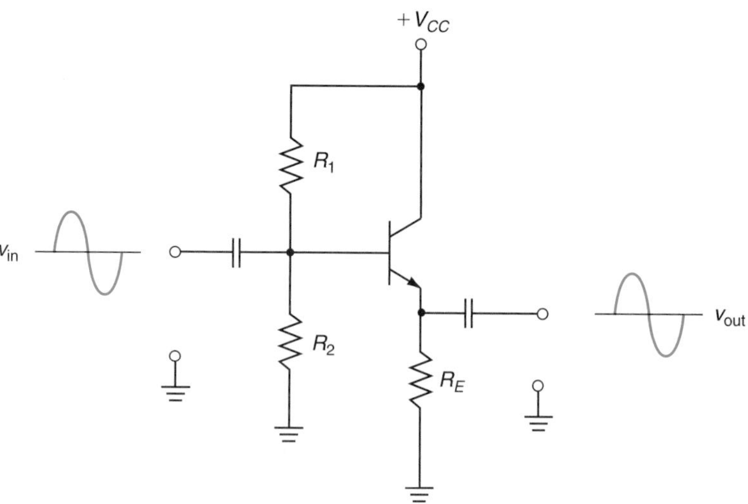

Figure 20.10 A common-collector amplifier.

The Common-Collector Amplifier (Emitter Follower)

A typical **common-collector (CC) amplifier** is shown in Figure 20.10. As you can see, the input is applied to the base of the transistor, and the output is taken from the emitter. In this case, the collector terminal of the transistor is part of both the input and output circuits and provides the ac ground (or common). The CC amplifier typically has the following characteristics:

Typical CC amplifier characteristics

- Midrange current gain.
- Extremely low voltage gain (slightly less than 1).
- High input impedance and low output impedance.

The most unique characteristic here is the extremely low voltage gain of this configuration. If we were to assume that the circuit shown in Figure 20.10 has a value of $A_v = 1$, the output waveform would be identical to the input waveform (as shown).

Since the ac signal at the emitter closely "follows" the ac signal at the base, the CC amplifier is commonly referred to as an *emitter follower.* Note that this circuit is used for its current gain and impedance characteristics.

The Common-Base Amplifier

The **common-base (CB) amplifier** is the least often used BJT amplifier configuration. A typical CB amplifier is shown in Figure 20.11. As you can see, the input is applied to the emitter of the transistor, and the output is taken from the collector.

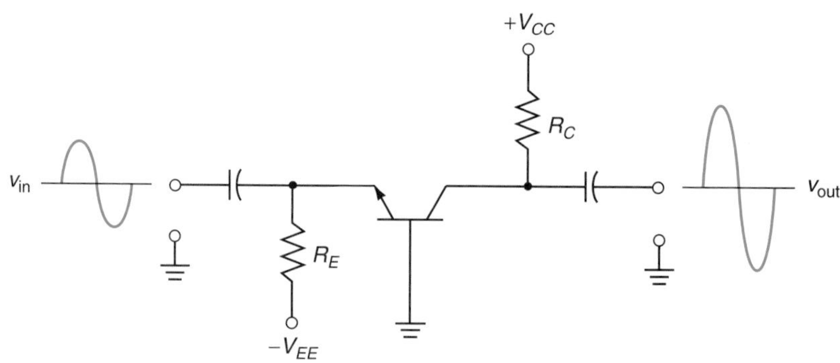

Figure 20.11 A common-base amplifier.

TABLE 20.3 A Comparison of CE, CC, and CB Circuit Characteristics

Amplifier Type	A_v	A_i	A_p	Z_{in}	Z_{out}
CE	Midrange	Midrange	High	Midrange	Midrange
CC	<1	Midrange	$<A_i$	High	Low
CB	Midrange	<1	$<A_v$	Low	High

The CB amplifier typically has the following characteristics:

Typical CB amplifier characteristics

- Midrange voltage gain.
- Extremely low current gain (slightly less than 1).
- Low input impedance and high output impedance.

If you compare the CB amplifier to the ideal voltage amplifier, you'll see one of the reasons why it is rarely used: Its impedance characteristics are exact opposites of those for the ideal voltage amplifier. In other words, the effective voltage gain of this circuit is nowhere near its ideal value of A_v.

Comparing the BJT Amplifier Configurations

For the sake of comparison, the gain and impedance characteristics of CE, CC, and CB amplifiers are listed in Table 20.3. Note the power gain entries in the table for the CC and CB amplifiers. These entries are based on the following relationship:

$$A_p = A_v A_i \tag{20.8}$$

Just as power equals the product of voltage and current, A_p equals the product of A_v and A_i. The CC amplifier has a value of $A_v < 1$, so its power gain must be lower than its current gain. By the same token, the CB amplifier has a value of $A_i < 1$, so its power gain must be lower than its voltage gain.

If you compare the values given in Table 20.3, you'll see that CC and CB amplifiers are nearly opposites in terms of their gain and impedance characteristics. At the same time, the CE amplifier is shown to be a "middle-of-the-road" circuit.

One Final Note

The material in this section has been presented to introduce you to the different types of BJT amplifiers. Throughout the remainder of this chapter, we will discuss the various configurations in greater detail. As you continue reading, keep in mind the relationships described in Table 20.3 and our discussion on amplifier gain and impedance characteristics.

Section Review

1. What is the basis for the term *common emitter*?
2. What are the gain and impedance characteristics of CE amplifiers?
3. In terms of input and output ac voltages, how is the CE amplifier unique among the BJT amplifiers?
4. What is the basis for the term *common collector*?
5. What are the gain and impedance characteristics of CC amplifiers?
6. What is another name commonly used for the CC amplifier? What is the basis of this name?
7. What is the basis for the term *common base*?
8. What are the gain and impedance characteristics of CB amplifiers?

20.3 Common-Emitter Amplifiers: Fundamental Concepts

In a common-emitter amplifier, the input signal is applied to the base, the output signal is taken from the collector, and the emitter is common to both the input and the output circuits. A CE amplifier is shown in Figure 20.12.

In this section, we're going to take a look at some general topics that are a part of CE amplifier operation and analysis. Then, in the next section, we will discuss CE amplifier gain and impedance characteristics and analysis.

Input/Output Phase Relationships

OBJECTIVE 7 ➤ As stated earlier, the CE amplifier is unique, because the input and output signal *voltages* are 180° out of phase. The source of this phase relationship is illustrated in Figure 20.13. As the input (base) current changes by ±5 μA, the output current changes by ±500 μA ($A_i = 100$). Since the output is taken from the transistor collector, the output voltage equals v_C, which is found as

$$v_C = V_{CC} - i_C R_C$$

As this equation illustrates, V_C decreases when collector current (I_C) increases. Thus, output current and output voltage are 180° out of phase. Since the input voltage (and current) are in phase with the output current, the input and output voltages must be 180° out of phase.

AC Emitter Resistance (r'_e)

OBJECTIVE 8 ➤ The emitter-base junction of a transistor has a dynamic resistance, called **ac emitter resistance (r'_e)**. As you will learn later, this resistance is used in some gain and impedance calculations. For a small-signal amplifier, the value of the ac emitter resistance can be approximated using

Note: The derivation of equation (20.9) is shown in Appendix D.

$$r'_e = \frac{25 \text{ mV}}{I_E} \qquad (20.9)$$

where

r'_e = the ac emitter resistance
I_E = the dc emitter current

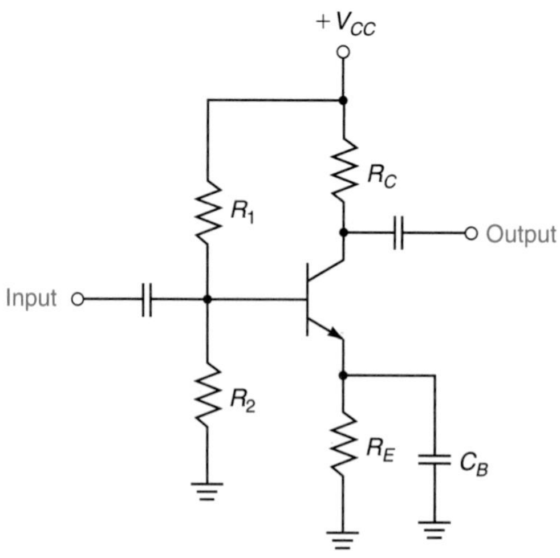

Figure 20.12

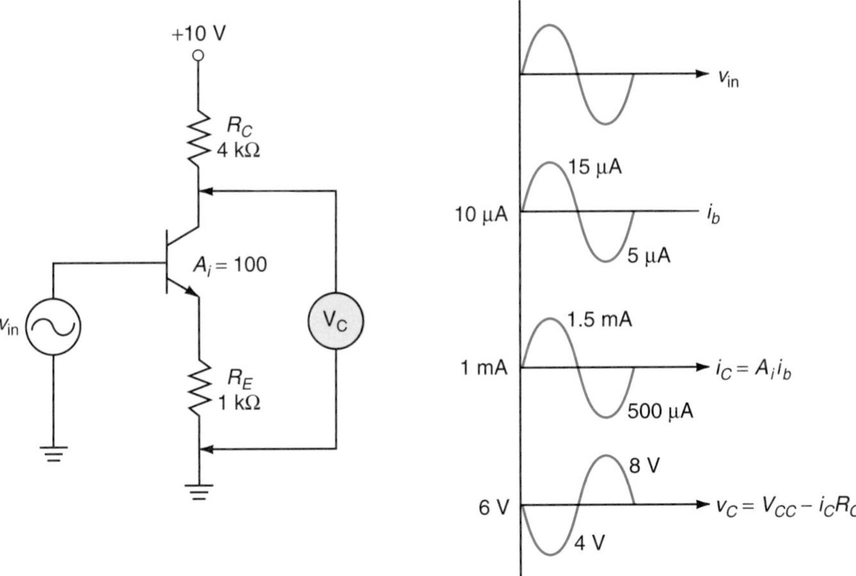

Figure 20.13

For example, if a CE amplifier has a value of $I_E = 2$ mA, the ac emitter resistance of the transistor can be found as

$$r'_e = \frac{25 \text{ mV}}{I_E} = \frac{25 \text{ mV}}{2 \text{ mA}} = 12.5\ \Omega$$

Small-Signal Current Gain (β_{ac})

The small-signal current gain (β_{ac}) of a transistor is different than its dc current gain (β_{dc}). This is because β_{dc} is measured with *fixed* values of I_B and I_C, while β_{ac} is measured using *changing* current values. The difference between these measurement techniques is illustrated in Figure 20.14. In Figure 20.14b, I_B is varied to cause a change in I_C around the Q-point. The change in I_C (ΔI_C) is then divided by the change in I_B (ΔI_B) to obtain the value of β_{ac}. By formula,

$$\beta_{ac} = \frac{\Delta I_C}{\Delta I_B} \tag{20.10}$$

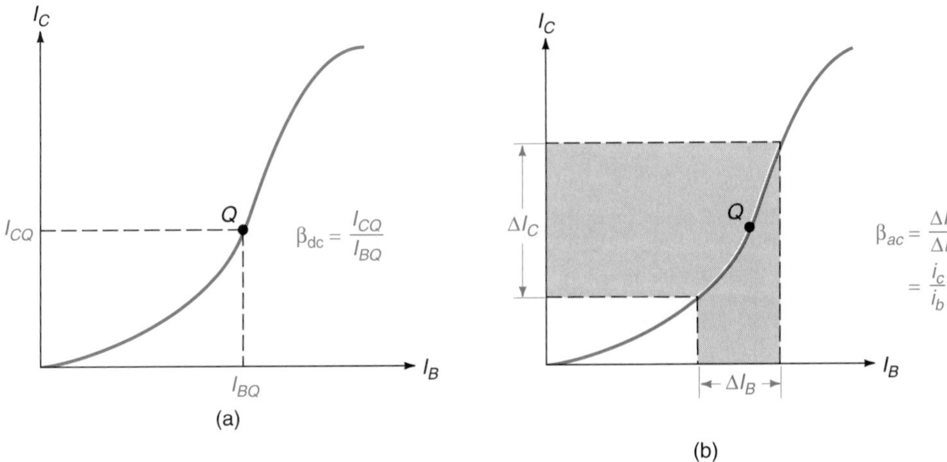

Figure 20.14 Determining the value of β_{ac}.

which is often written as

$$h_{fe} = \frac{i_c}{i_b} \qquad (20.11)$$

where

h_{fe} = the small-signal current gain of the transistor
i_c = the change in collector current
i_b = the change in base current

h_{FE} = dc beta
h_{fe} = ac beta

As you know, spec sheets label dc beta as h_{FE}. To distinguish the small-signal current gain, spec sheets use the label h_{fe} to represent β_{ac}.

AC Equivalent Circuits

OBJECTIVE 9 ➤

The small-signal operation of any amplifier is easier to understand when we represent the original circuit using an **ac equivalent circuit.** This circuit, which is used in many ac calculations, is easily derived using a two-step process:

1. Short circuit all capacitors.
2. Replace all dc sources with a ground symbol.

For reasons that we will discuss in a moment, most CE amplifiers contain one or more capacitors that are selected to have an *ideal* value of $X_C = 0\ \Omega$ at the circuit operating frequency. The first step in deriving the ac equivalent circuit is to represent these capacitors as short circuits. The second step is based on the fact that dc sources have extremely low internal resistance values. The ac equivalent of the amplifier shown in Figure 20.15 was derived using these steps.

You need to be absolutely clear on one point: *The ac equivalent circuit is based on the ac characteristics of the amplifier and is not a part of the dc analysis of the amplifier.*

Coupling and Bypass Capacitors

OBJECTIVE 10 ➤

Most CE amplifiers contain both *coupling* and *bypass capacitors,* as shown in Figure 20.16. A **coupling capacitor** is *a capacitor that is connected between amplifier stages to provide dc isolation between the stages while allowing an ac signal to pass without distortion.* The dc isolation characteristic of the coupling capacitor can be seen by comparing the dc voltages shown on the two sides of C_{C2} in Figure 20.16. The dc voltage on the left-hand side of C_{C2} equals the value of V_C from the first stage (5.6 V). The dc voltage on the right-hand side of C_{C2} equals the value dc biasing voltage for the second

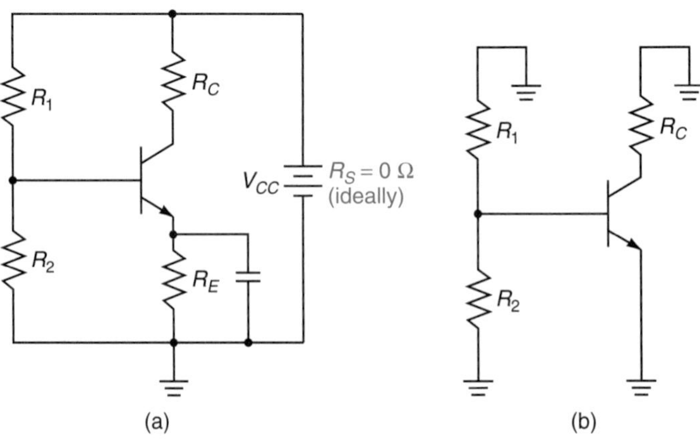

Figure 20.15 A CE amplifier and its ac equivalent circuit.

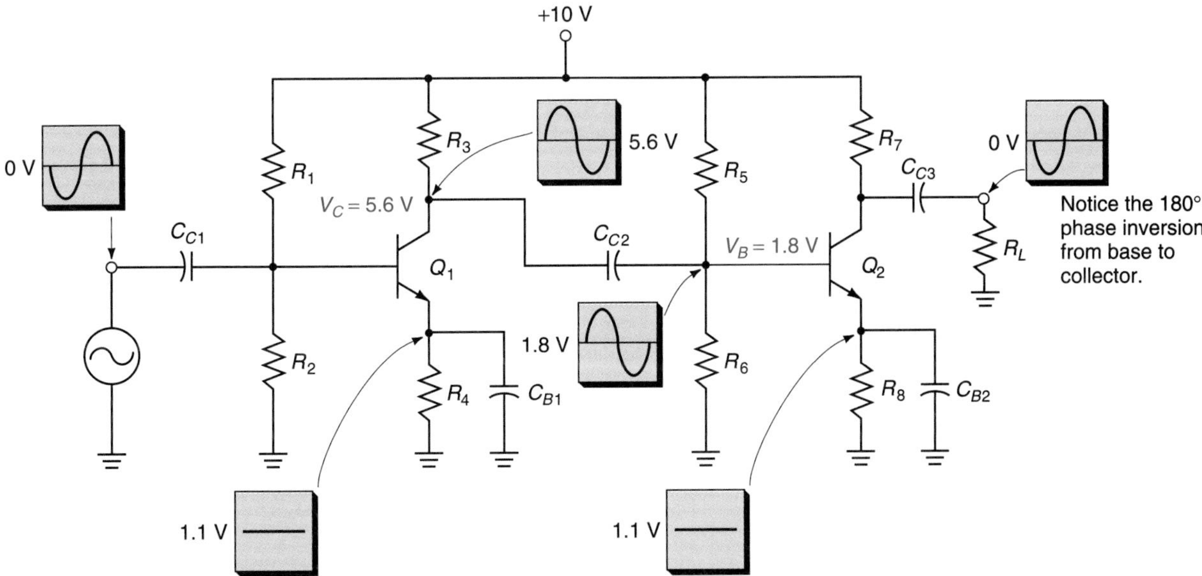

Figure 20.16 A two-stage CE amplifier.

stage (1.8 V). At the same time, the signal at the collector of Q_1 is coupled to the base of Q_2.

A **bypass capacitor** is *a capacitor that is used to establish an ac ground at a specific point in a circuit.* In Figure 20.16, there are two bypass capacitors (C_{B_1} and C_{B_2}). A bypass capacitor and its ac equivalent are shown in Figure 20.17. In the ac equivalent circuit, a wire that represents the shorting effect of the capacitor shorts the emitter resistor.

The effect of a bypass capacitor on amplifier operation can be seen in Figure 20.16. The input to each amplifier stage causes a change in the emitter and the collector currents. The change in collector current results in the collector waveforms shown. However, any change in emitter current (ΔI_E) does not produce a change in emitter voltage, because ΔI_E bypasses the emitter resistor via the capacitor. At the same time, the bypass capacitor acts as a *dc open,* so it does not affect the dc biasing of the transistor. As you will see, establishing this ac ground at the transistor emitter increases the circuit voltage gain.

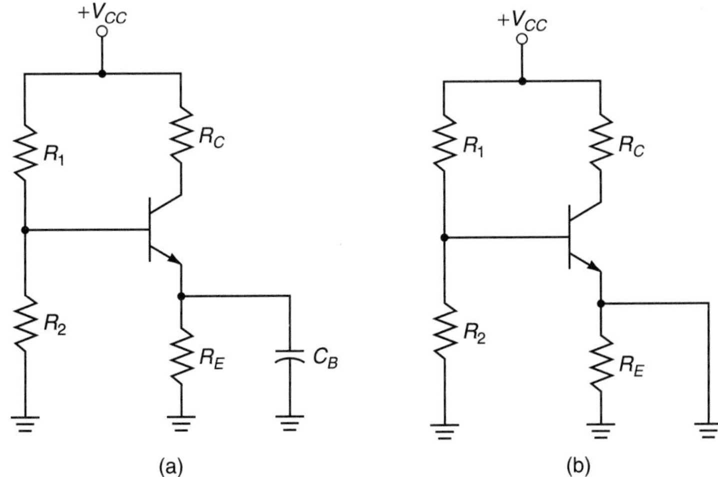

Figure 20.17 The effects of a bypass capacitor.

Section Review

1. Explain the input/output voltage phase relationship of the *CE amplifier.*
2. What is r'_e?
3. What is the difference between h_{FE} and h_{fe}?
4. Describe how h_{FE} and h_{fe} are measured.
5. What steps are taken to derive the *ac equivalent circuit* for a given amplifier?
6. What effects do the ac characteristics of an amplifier have on the dc characteristics of the circuit?
7. What two purposes are served by capacitors in multistage amplifiers?
8. Why is *dc isolation* between amplifier stages important?
9. What effect does a *coupling capacitor* have on the ac signal that is coupled from one amplifier stage to the next?
10. Why don't you see a signal voltage at the emitter terminal of an amplifier with a *bypass capacitor*?

Critical Thinking

11. List, in order, the calculations you would need to make to determine the value of r'_e for a voltage-divider biased CE amplifier.

20.4 Common-Emitter Amplifiers: Gain and Impedance Characteristics

OBJECTIVE 11 ➤ Earlier in this chapter, you were shown that amplifier *gain* and *impedance* are the primary values of interest in any circuit analysis. In this section, you will be shown how to calculate these values for a common-emitter amplifier.

Voltage Gain (A_v)

The **voltage gain (A_v)** of an amplifier is *the factor by which a signal voltage increases from the input of an amplifier to its output,* given as

$$A_v = \frac{v_{out}}{v_{in}}$$

The value of A_v for a common-emitter amplifier equals the ratio of total ac collector resistance to total ac emitter resistance. For the circuit shown in Figure 20.18a, this ratio is found as

$$A_v = \frac{r_C}{r'_e} \quad (20.12)$$

where

r_C = the total ac resistance of the collector circuit

r'_e = the ac emitter resistance of the transistor

The values in equation (20.12) can be identified using the ac equivalent circuit shown in Figure 20.18b. The bypass capacitor effectively shorts out the emitter resistor, so the only ac resistance in the emitter circuit is that of the transistor emitter, r'_e. The equivalent circuit also shows R_C and R_L as being in *ac parallel.* Therefore, the total ac resistance in the collector circuit is found as

$$r_C = R_C \| R_L \quad (20.13)$$

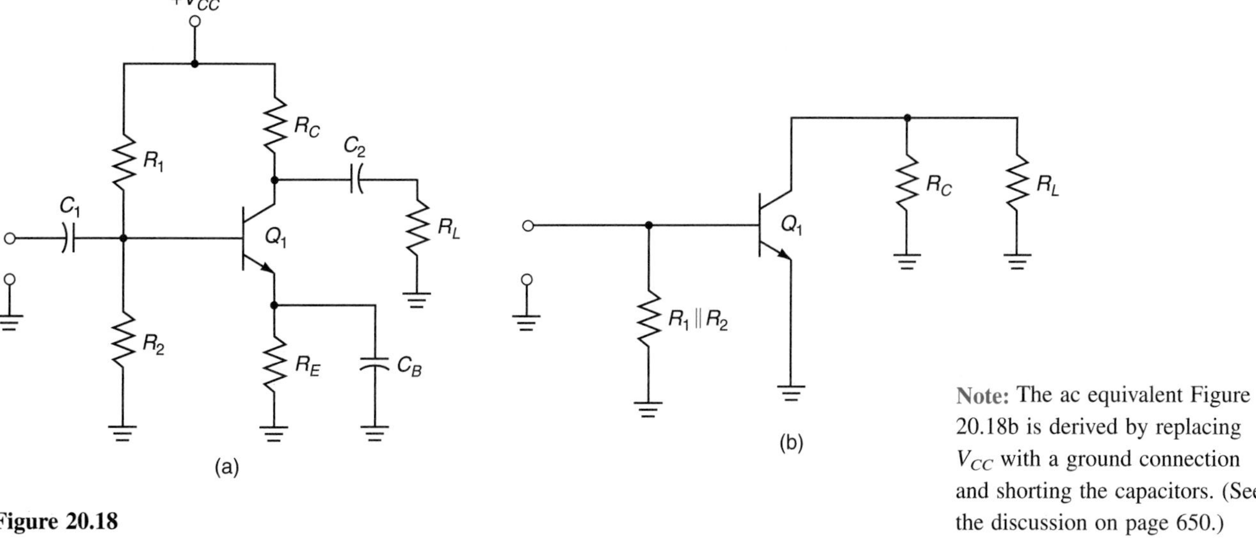

Figure 20.18

Note: The ac equivalent Figure 20.18b is derived by replacing V_{CC} with a ground connection and shorting the capacitors. (See the discussion on page 650.)

Example 20.3 demonstrates the series of calculations required to determine the voltage gain of a common-emitter amplifier.

Example 20.3

Determine the voltage gain (A_v) of the circuit shown in Figure 20.19.

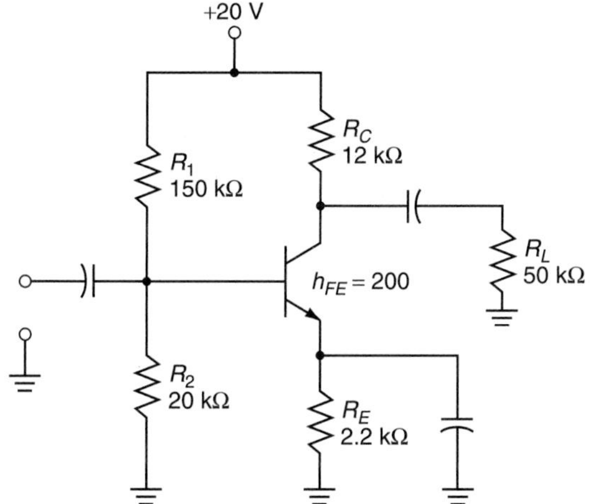

Figure 20.19

Solution: The value of A_v is found using the following steps:

Step 1: V_B is found as $$V_B = V_{CC}\frac{R_2}{R_1 + R_2} = (20\text{ V})\frac{20\text{ k}\Omega}{170\text{ k}\Omega} = 2.35\text{ V}$$

Step 2: V_E is found as $$V_E = V_B - 0.7\text{ V} = 2.35\text{ V} - 0.7\text{ V} = 1.65\text{ V}$$

Step 3: I_E is found as $$I_E = \frac{V_E}{R_E} = \frac{1.65\text{ V}}{2.2\text{ k}\Omega} = 750\ \mu\text{A}$$

Step 4: r'_e is found as $$r'_e = \frac{25\text{ mV}}{I_E} = \frac{25\text{ mV}}{750\ \mu\text{A}} = 33.3\ \Omega$$

Step 5: r_C is found as $$r_C = R_C \| R_L = 12\text{ k}\Omega \| 50\text{ k}\Omega = 9.7\text{ k}\Omega$$

Step 6: A_v is found as $$A_v = \frac{r_C}{r'_e} = \frac{9.7\text{ k}\Omega}{33.3\ \Omega} = 291$$

Therefore, the ratio of output signal voltage to input signal voltage for this circuit is 291.

> ***Practice Problem 20.3***
>
> An amplifier like the one shown in Figure 20.19 has the following values: $V_{CC} =$ 30 V, $R_1 = 51$ kΩ, $R_2 = 5.1$ kΩ, $R_C = 5.1$ kΩ, $R_L = 10$ kΩ, $R_E = 910$ Ω, and $h_{FE} = 250$. Determine the voltage gain of the amplifier.

Amplifier Input Impedance (Z_{in})

The input impedance of an amplifier is determined using its ac equivalent circuit. As Figure 20.20 shows, the input circuit of an amplifier contains the parallel combination of R_1, R_2, and the base of the transistor. Therefore, the amplifier input impedance is found as

$$Z_{in} = R_1 \| R_2 \| Z_{in(base)} \tag{20.14}$$

where

Z_{in} = the input impedance to the amplifier

$Z_{in(base)}$ = the input impedance to the transistor base

You may recall that the dc input resistance of a transistor base is found as

$$R_{in(base)} = h_{FE}R_E$$

The input impedance of the base can be determined in the same manner. The only differences are as follows:

- The ac emitter resistance (r'_e) is used in place of R_E.
- The value of ac beta (h_{fe}) is used in place of dc beta (h_{FE}).

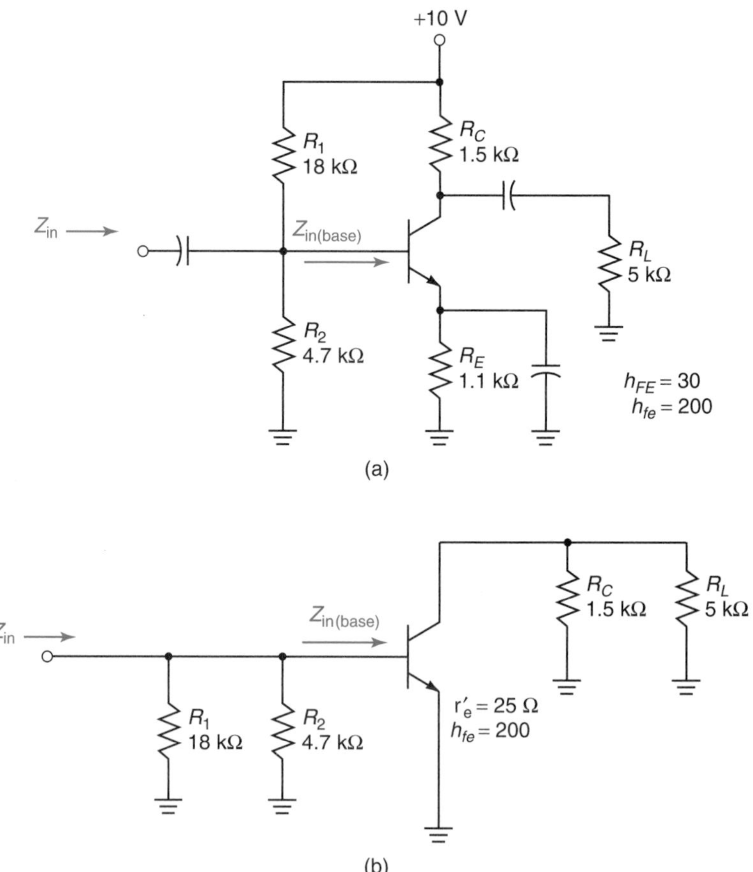

Figure 20.20 A common-emitter amplifier and its ac equivalent circuit.

Therefore, the value of $Z_{\text{in(base)}}$ can be found as

$$Z_{\text{in(base)}} = h_{fe}r'_e \qquad (20.15)$$

Since the input impedance of the transistor is in ac parallel with the base biasing resistors, the total input impedance to the amplifier can be found as demonstrated in Example 20.4.

Example 20.4

Determine the input impedance to the amplifier represented by the equivalent circuit shown in Figure 20.20b.

Solution: The base input impedance is found as

$$Z_{\text{in(base)}} = h_{fe}r'_e = (200)(25\ \Omega) = 5\ \text{k}\Omega$$

and the amplifier input impedance is found as

$$Z_{\text{in}} = R_1 \| R_2 \| Z_{\text{in(base)}} = 18\ \text{k}\Omega \| 4.7\ \text{k}\Omega \| 5\ \text{k}\Omega = 2.14\ \text{k}\Omega$$

Practice Problem 20.4

Determine Z_{in} for the amplifier described in Practice Problem 20.3. Assume that the transistor has a value of $h_{fe} = 200$.

Amplifier Current Gain (A_i)

As you were told earlier in the chapter, amplifier **current gain (A_i)** is found as

$$A_i = \frac{i_{\text{out}}}{i_{\text{in}}}$$

where

i_{in} = the ac source current

i_{out} = the ac load current

You were also told that the small-signal current gain of the *transistor* in a common-emitter amplifier is found as

$$h_{fe} = \frac{i_c}{i_b}$$

If you refer back to Figure 20.20b, you'll see that the input of the common-emitter amplifier contains a current divider made of R_1, R_2, and the base of the transistor. Since a portion of the ac source current passes through the biasing resistors, $i_b < i_{\text{in}}$. At the same time, the collector circuit of the amplifier forms another current divider. Since a portion of the ac collector current passes through the collector resistor, $i_{\text{out}} < i_c$. Thus, the overall current gain of the amplifier (A_i) is significantly lower than the current gain of the transistor (h_{fe}). The overall current gain of a CE amplifier can be calculated using

$$A_i = h_{fe}\left(\frac{Z_{\text{in}}r_C}{Z_{\text{in(base)}}R_L}\right) \qquad (20.16)$$

Note: The derivation of equation (20.16) is shown in Appendix D.

where

A_i = the current gain of the CE amplifier

h_{fe} = the current gain of the transistor

Note that the fraction in the equation represents the reduction factor introduced by the biasing and output components. The use of equation (20.16) is demonstrated in Example 20.5.

Example 20.5

Calculate the value of A_i for the circuit shown in Figure 20.20b.

Solution: For the circuit shown,

$$Z_{\text{in(base)}} = h_{fe}r'_e = (200)(25\ \Omega) = 5\ \text{k}\Omega$$

$$Z_{\text{in}} = R_1 \| R_2 \| Z_{\text{in(base)}} = 18\ \text{k}\Omega \| 4.7\ \text{k}\Omega \| 5\ \text{k}\Omega = 2.14\ \text{k}\Omega$$

and

$$r_C = R_C \| R_L = 1.5\ \text{k}\Omega \| 5\ \text{k}\Omega = 1.15\ \text{k}\Omega$$

Now the amplifier current gain can be found as

$$A_i = h_{fe}\left(\frac{Z_{\text{in}}r_C}{Z_{\text{in(base)}}R_L}\right) = (200)\left[\frac{(2.14\ \text{k}\Omega)(1.15\ \text{k}\Omega)}{(5\ \text{k}\Omega)(5\ \text{k}\Omega)}\right] = (200)(0.0984) = 19.68$$

As you can see, the value of A_i for this circuit is significantly lower than the value of h_{fe}.

Practice Problem 20.5

A common-emitter amplifier has the following values: $h_{fe} = 300$, $r_C = 2.15\ \text{k}\Omega$, $R_L = 6.2\ \text{k}\Omega$, $Z_{\text{in(base)}} = 8.2\ \text{k}\Omega$, and $Z_{\text{in}} = 3.8\ \text{k}\Omega$. Calculate the value of A_i for the circuit.

Power Gain (A_p)

As stated earlier in the chapter, **power gain (A_p)** is found as *the product of amplifier voltage gain and current gain.* By formula,

$$A_p = A_v A_i$$

This is consistent with calculating component power dissipation as the product of component voltage and current.

Section Review

1. How is the *voltage gain* (A_v) of a common-emitter amplifier defined in terms of circuit resistance values?
2. How is the *ac collector resistance* of a common-emitter amplifier determined?
3. List, in order, the steps usually required to calculate the value of A_v for a common-emitter amplifier.
4. What is the relationship between the values of A_i and h_{fe}?
5. What is the relationship between *power gain* (A_p) and the other amplifier gains?

20.5 The Emitter Follower (Common-Collector Amplifier)

The **emitter follower** is *a current amplifier with a voltage gain that is less than one.* The input signal is applied to the base of the transistor and the output signal is taken from the emitter. As shown in Figure 20.21, the circuit has no collector resistor and a

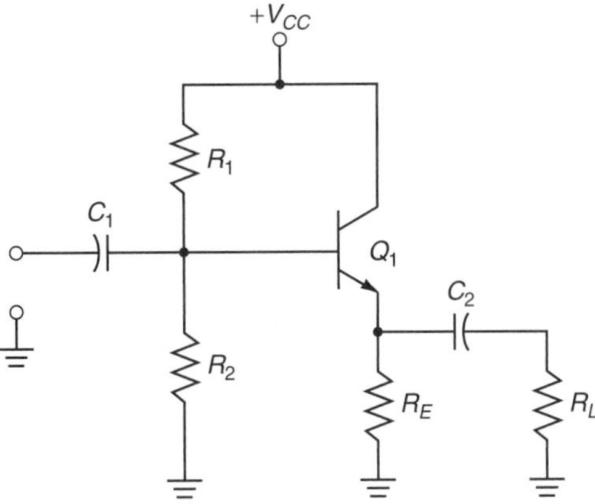

Figure 20.21 The emitter follower (common-collector amplifier).

load that is connected to the transistor emitter. These are two of the circuit recognition features for the emitter follower.

DC Characteristics

The lack of a collector resistor in the emitter follower affects the dc analysis of the circuit. For the circuit shown in Figure 20.21,

$$V_B = V_{CC}\frac{R_2}{R_1 + R_2}$$

$$V_E = V_B - 0.7\text{ V}$$

and

$$I_E = \frac{V_E}{R_E}$$

Don't forget: If $h_{FE}R_E < 10R_2$, the base input resistance must be considered when calculating the value of V_B. (See the discussion on pages 615–616.)

These relationships are identical to those used for analyzing any voltage-divider biased circuit. However, the value of V_{CEQ} for the circuit shown in Figure 20.21 is found using

$$V_{CEQ} = V_{CC} - V_E \tag{20.17}$$

Note that this relationship reflects the lack of a collector resistor in the circuit.

Saturation and Cutoff

The lack of a collector resistor in the emitter follower also affects the value of $I_{C(\text{sat})}$ for the circuit. When the transistor is saturated, $V_{CE} \cong 0$ V, and $V_E \cong V_{CC}$. Therefore, the value of $I_{C(\text{sat})}$ is found as

$$I_{C(\text{sat})} \cong \frac{V_{CC}}{R_E} \tag{20.18}$$

As with all the biasing circuits we have covered so far, the total applied voltage is dropped across the transistor when it is in cutoff. Therefore, $V_{CE(\text{off})}$ is found as

$$V_{CE(\text{off})} \cong V_{CC}$$

Small-Signal Characteristics and Analysis

OBJECTIVE 12

Figure 20.22 shows a typical emitter follower and its ac equivalent. Note that r_E in Figure 20.22b represents the total ac resistance in the emitter circuit. This resistance is found as

$$r_E = R_E \| R_L \tag{20.19}$$

Note: The *dc forward current gain of the transistor in an emitter follower* is designated as h_{FC}. The value of h_{FC} is found as

$$h_{FC} = h_{FE} + 1$$

However, when $h_{FE} >> 1$, we assume that $h_{FC} \cong h_{FE}$.

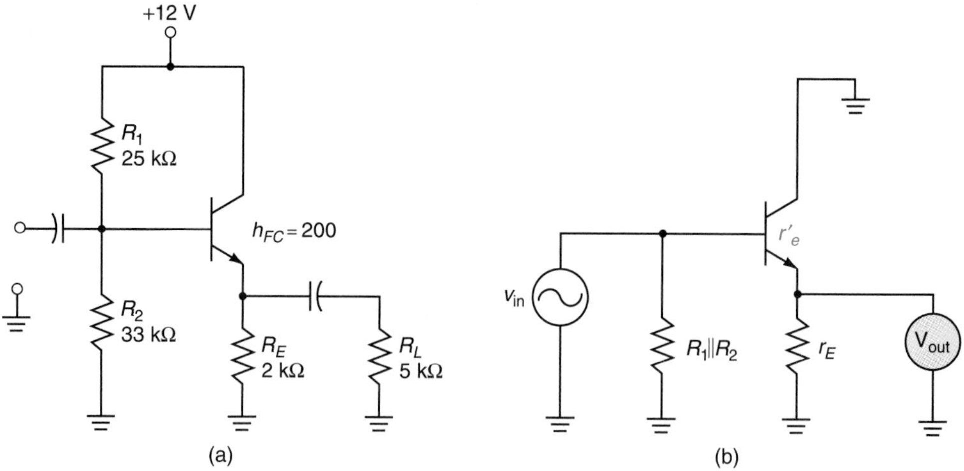

Figure 20.22 A typical emitter follower and its equivalent circuit.

The low voltage gain of the emitter follower results from the output circuit. As indicated in Figure 20.22b, this circuit contains a voltage divider made up of r'_e and r_E. Ignoring the slight voltage across the emitter-base junction of the transistor,

$$v_{out} = v_{in}\,\frac{r_E}{r'_e + r_E}$$

or

$$\frac{v_{out}}{v_{in}} = \frac{r_E}{r'_e + r_E}$$

Finally, since $A_v = \dfrac{v_{out}}{v_{in}}$, we can calculate the voltage gain of the circuit as

$$A_v = \frac{r_E}{r'_e + r_E} \qquad (20.20)$$

In most practical applications, $r_E >> r'_e$. When this is the case,

$$A_v \cong 1 \qquad \text{(when } r_E >> r'_e\text{)} \qquad (20.21)$$

The greater the difference between r_E and r'_e, the closer the amplifier voltage gain comes to one. In practice, the voltage gain of the emitter follower is usually between 0.8 and 0.999. The calculation of A_v is demonstrated in Example 20.6.

Example 20.6

Determine the value of A_v for the circuit shown in Figure 20.22a.

Solution: First, the ac resistance of the emitter circuit is found as

$$r_E = R_E \| R_L = 2\text{ k}\Omega \| 5\text{ k}\Omega = 1.43\text{ k}\Omega$$

Using established procedures, I_E is found to be approximately 3.1 mA, and r'_e is found as

$$r'_e = \frac{25\text{ mV}}{I_E} = \frac{25\text{ mV}}{3.1\text{ mA}} = 8.1\ \Omega$$

Now the voltage gain of the amplifier can be found as

$$A_v = \frac{r_E}{r'_e + r_E} = \frac{1.43\text{ k}\Omega}{8.1\ \Omega + 1.43\text{ k}\Omega} = 0.9944$$

Practice Problem 20.6

A circuit like the one shown in Figure 20.22a has the following values: $V_{CC} = +18$ V, $R_E = 910\ \Omega$, $R_1 = 16\text{ k}\Omega$, $R_2 = 22\text{ k}\Omega$, $R_L = 4\text{ k}\Omega$, and $h_{FC} = 200$. Determine the value of A_v for the circuit.

Current Gain

Earlier, we saw that the current gain of a CE amplifier is significantly lower than h_{fe} because of the current divisions that occur in both the input and the output circuits. The same principle holds true for the emitter follower. The current gain of an emitter follower is found as

$$A_i = h_{fc}\left(\frac{Z_{in}r_E}{Z_{in(base)}R_L}\right) \qquad (20.22)$$

where

h_{fc} = the *base-to-emitter current gain*

Note that $h_{fc} = h_{fe} + 1 \cong h_{fe}$.

Input Impedance (Z_{in})

As with the common-emitter amplifier, the input impedance of the emitter follower is found as the parallel equivalent of the base resistors and the transistor input impedance. By formula,

$$Z_{in} = R_1 \| R_2 \| Z_{in(base)}$$

In the case of the emitter follower, the base input impedance is equal to the product of the device current gain and the total ac emitter resistance. By formula,

$$Z_{in(base)} = h_{fc}(r'_e + r_E) \qquad (20.23)$$

where

h_{fc} = the base-to-emitter current gain of the transistor
r'_e = the ac resistance of the transistor emitter
$r_E = R_E \| R_L$

Example 20.7 illustrates the procedure used to determine the input impedance of an emitter follower.

Example 20.7

Determine the input impedance of the amplifier shown in Figure 20.22a. Assume the transistor has a value of $h_{fc} \cong h_{fe} = 220$.

Solution: From Example 20.6, we know that $r_E = 1.43\text{ k}\Omega$ and $r'_e = 8.1\ \Omega$. Therefore,

$$Z_{in(base)} = h_{fc}(r'_e + r_E) = (220)(8.1\ \Omega + 1.43\text{ k}\Omega) = 316.4\text{ k}\Omega$$

and

$$Z_{in} = R_1 \| R_2 \| Z_{in(base)} = 25\text{ k}\Omega \| 33\text{ k}\Omega \| 316.4\text{ k}\Omega = 13.6\text{ k}\Omega$$

Practice Problem 20.7

Assume that the amplifier described in Practice Problem 20.6 has a current gain of $h_{fc} = 240$ and a 2 kΩ load. Determine the input impedance of the circuit.

When $r_E >> r'_e$, as was the case in Example 20.6, equation (20.23) can be simplified to

$$Z_{in(base)} \cong h_{fc} r_E \qquad \text{(when } r_E >> r'_e\text{)} \tag{20.24}$$

Output Impedance (Z_{out})

When a load is connected to an amplifier, the output impedance of the circuit acts as the source impedance for that load. The output impedance of an emitter follower is found as

$$Z_{out} = R_E \| \left(r'_e + \frac{R'_{in}}{h_{fc}} \right) \tag{20.25}$$

where

$R'_{in} = R_1 \| R_2 \| R_S$

$R_S =$ the resistance of the input signal source

The use of this equation is demonstrated in Example 20.8.

Example 20.8

Determine the output impedance (Z_{out}) of the amplifier shown in Figure 20.23.

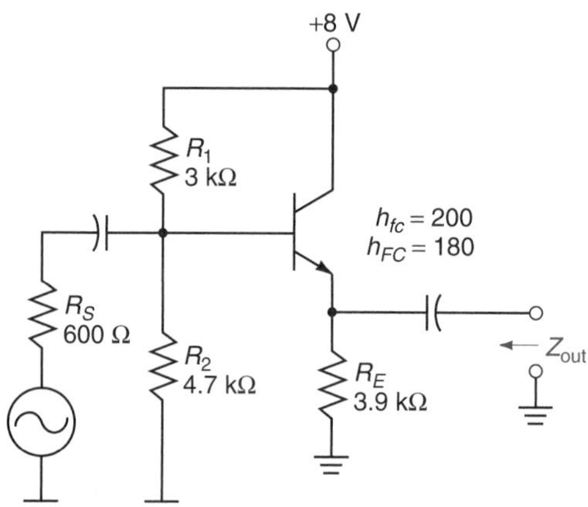

Figure 20.23

Solution: Using the established approach, the circuit is found to have a value of $I_E = 1.07$ mA, and

$$r'_e = \frac{25 \text{ mV}}{I_E} = \frac{25 \text{ mV}}{1.07 \text{ mA}} = 23.4\ \Omega$$

The value of R'_{in} is now found as

$$R'_{in} = R_1 \| R_2 \| R_S = 3\text{ k}\Omega \| 4.7\text{ k}\Omega \| 600\ \Omega = 452\ \Omega$$

and the output impedance of the amplifier is found as

$$Z_{out} = R_E \| \left(r'_e + \frac{R'_{in}}{h_{fc}} \right) = 3.9\text{ k}\Omega \| \left(23.4\ \Omega + \frac{452\ \Omega}{200} \right)$$

$$= 3.9\text{ k}\Omega \| 25.7\ \Omega = 25.5\ \Omega$$

Practice Problem 20.8

A circuit like the one shown in Figure 20.23 has the following values: $V_{CC} = +12$ V, $R_E = 2$ kΩ, $R_1 = 2.5$ kΩ, $R_2 = 3.3$ kΩ, $R_S = 500$ Ω, $h_{FC} = 180$, and $h_{fc} = 200$. Calculate the output impedance for the amplifier.

Power Gain (A_p)

As with the CE amplifier, the power gain (A_p) of an emitter follower is found as the product of current gain (A_i) and voltage gain (A_v). By formula,

$$A_p = A_v A_i$$

Because the value of A_v is always slightly less than one, the power gain of the emitter following is always *slightly less than the value of the circuit current gain* (A_i).

Section Review

1. What are the circuit recognition features of the *emitter follower*?
2. What is the limit on the value of *voltage gain* (A_v) for an emitter follower?
3. List, in order, the steps that you would take to determine the value of A_v for an emitter follower.
4. How is the value of *current gain* (A_i) for an emitter follower determined?
5. List, in order, the steps you would take to determine the value of *input impedance* (Z_{in}) for an emitter follower.
6. List, in order, the steps you would take to determine the value of *output impedance* (Z_{out}) for an emitter follower.
7. What is the limit on the value of *power gain* (A_p) for an emitter follower?
8. What is the relationship between the values of V_{CE} and V_E when an emitter follower is midpoint biased? Critical Thinking
9. Some emitter followers contain a *collector resistor* (R_C) in parallel with a *bypass capacitor* (to establish an ac ground at the transistor collector). Which of the dc analysis equations must be modified for this circuit?

20.6 The Common-Base Amplifier

The common-base amplifier provides voltage gain with no current gain. The input signal is applied to the transistor emitter, and the output is taken from the collector, as shown in Figure 20.24. Despite its unfamiliar appearance, the circuit in Figure 20.24 is an emitter-biased amplifier.

AC Analysis

◀ *OBJECTIVE 13*

The ac equivalent circuit for the common-base amplifier is shown in Figure 20.25. We will use this circuit for our discussion on the circuit's ac characteristics.

The input signal is applied across the emitter-base junction, so

$$v_{in} = i_e r'_e$$

The output signal is developed across the ac collector resistance, so

$$v_{out} = i_c r_C \quad \Big| \quad r_C = R_C \| R_L$$

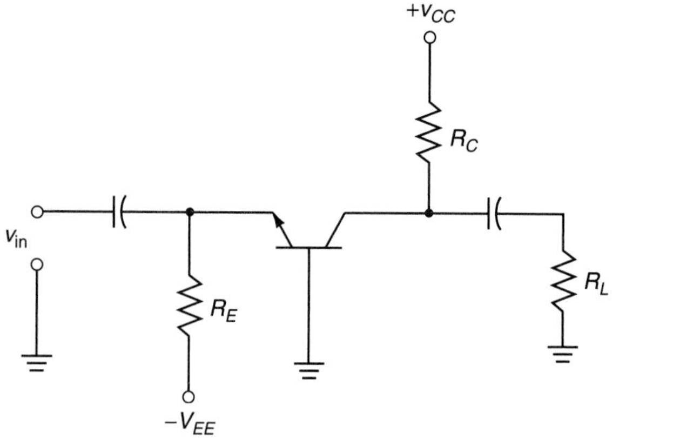

Figure 20.24 A common-base amplifier.

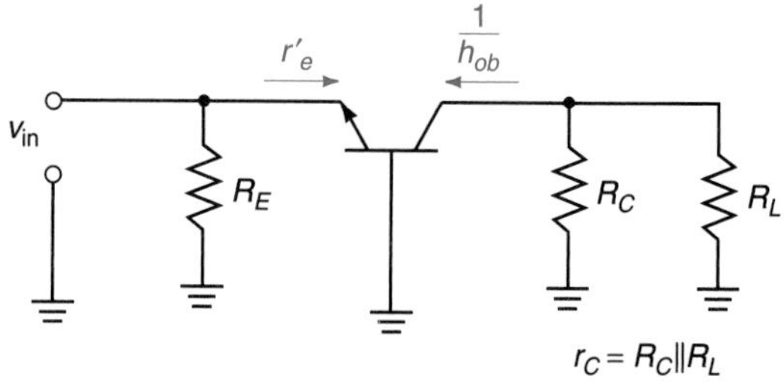

Figure 20.25 The common-base ac equivalent circuit.

Therefore, we can calculate the circuit voltage gain using

$$A_v = \frac{v_{\text{out}}}{v_{\text{in}}} = \frac{i_c r_C}{i_e r'_e}$$

Since $i_c \cong i_e$, we can calculate the circuit voltage gain using

$$A_v = \frac{r_C}{r'_e} \tag{20.26}$$

The emitter current is approximately equal to the collector current, so the current gain is generally assumed to have a value of

$$A_i \cong 1 \tag{20.27}$$

The input impedance to the amplifier is the parallel combination of r'_e and R_E. By formula,

$$Z_{\text{in}} = r'_e \| R_E \tag{20.28}$$

When $R_E >> r'_e$, which is normally the case, the value of Z_{in} can be approximated as

$$Z_{\text{in}} \cong r'_e \tag{20.29}$$

The output impedance of the amplifier is the parallel combination of R_C and the output impedance of the transistor collector $\left(\frac{1}{h_{ob}}\right)$. By formula,

Note: $\boldsymbol{h_{ob}}$ is *the output admittance rating of the transistor collector.* The reciprocal of this value $\left(\frac{1}{h_{ob}}\right)$ represents the output impedance of the transistor. The various *h*-parameters are discussed in detail in Appendix E.

$$Z_{\text{out}} = R_C \| \frac{1}{h_{ob}} \tag{20.30}$$

Since $\frac{1}{h_{ob}}$ is typically very high (in the hundreds of kilohms), the output impedance is approximated as

$$Z_{\text{out}} \cong R_C \tag{20.31}$$

Summary

The ac characteristics of a common-base amplifier can be summarized as follows:

- Relatively high voltage gain.
- Current gain that is slightly less than one.

- Low input impedance.
- High output impedance.
- Input and output signal voltages and currents that are in phase.

Section Review

1. What are the gain characteristics of a common-base amplifier?
2. What are the input/output characteristics of a common-base amplifier?

20.7 Class B Amplifiers

All the amplifiers that we have covered to this point have been *class A amplifiers.* A **class A amplifier** *contains a single active device* (such as a transistor) *that conducts during the entire* 360° *of the input signal cycle.* The primary drawback of class A operation is that it has a maximum theoretical efficiency of 25%. This means that the amplifier uses at least 75% of the power drawn from the dc power supply. This due partly to the fact that the components in the amplifier dissipate power even when there is no active input signal.

Note: The maximum theoretical efficiency ratings of class A and class B amplifiers are derived in Appendix D.

The **class B amplifier** was developed to improve on the low efficiency rating of the class A amplifier. Unlike the class A amplifier, the class B amplifier consumes very little power when it has no input signal. The maximum theoretical efficiency rating of a class B amplifier is approximately 78.5%.

Figure 20.26 shows the most commonly used class B amplifier. This circuit is referred to as a **complementary-symmetry amplifier,** or **push-pull emitter follower.** The circuit recognition feature is the use of **complementary transistors** (*a pair, one npn and one pnp, with matched characteristics*).

Class B Operation Overview

◄ OBJECTIVE 14

The term *push-pull* comes from the fact that the two transistors in a class B amplifier conduct on alternating half-cycles of the input signal. For example, consider the circuit shown

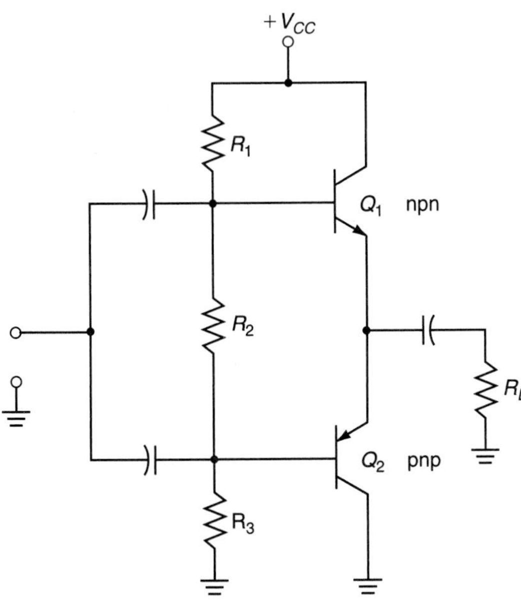

Figure 20.26 Class B complementary-symmetry amplifier.

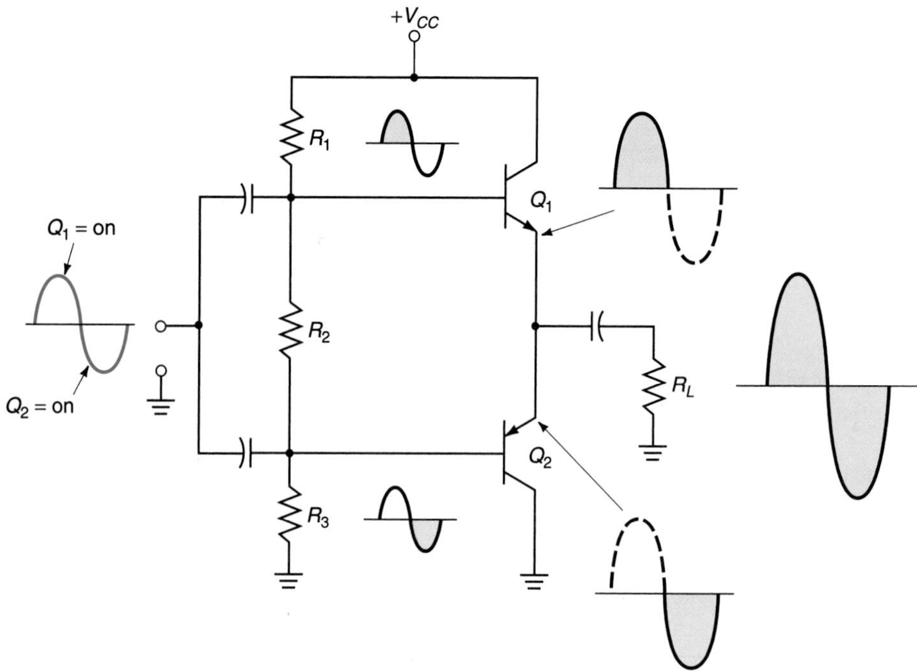

Figure 20.27 Class B operation.

in Figure 20.27. When the amplifier is in its *quiescent* state (no signal input), both transistors are biased in *cutoff.* When the input signal goes positive, Q_1 is biased above cutoff, and the transistor starts to conduct. During this time, however, Q_2 is still biased in cutoff. When the input goes into its negative half-cycle, Q_1 is driven back into cutoff, and conduction through Q_2 starts to increase. The fact that both transistors are never fully conducting at the same time is the key to the high efficiency rating of this amplifier.

Because both transistors are biased in cutoff in their quiescent state, class B amplifiers are subject to *crossover distortion,* which appears as flat lines between the output alternations (as shown in Figure 20.28). **Crossover distortion** is *distortion caused by both transistors being in cutoff for a short time between the positive and negative alternations*

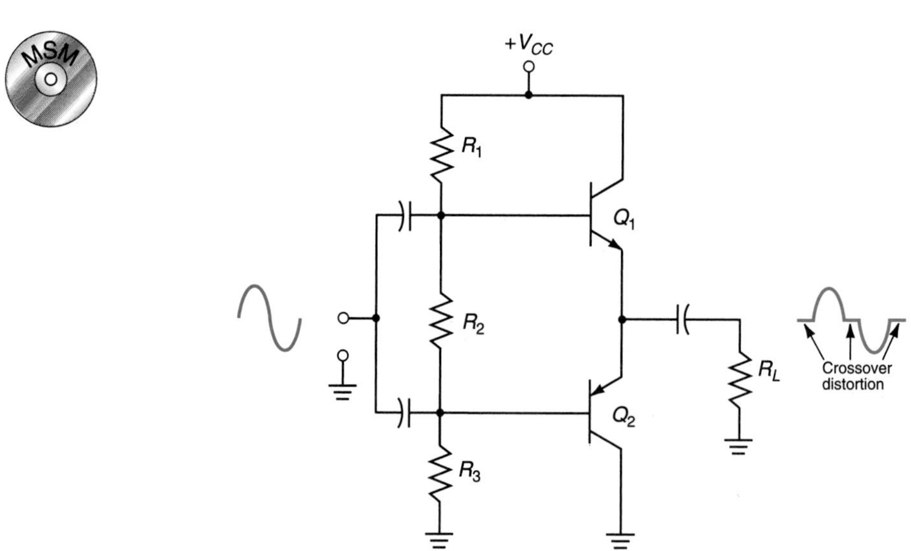

Figure 20.28 Crossover distortion.

of the amplifier input signal. Crossover distortion is prevented by the use of *diode bias* (which is discussed later in this section).

DC Operating Characteristics

It was stated earlier that both transistors in the class B amplifier are biased in cutoff. Assuming that the transistors are *matched,* they have equal values of V_{CE} when biased in cutoff. Therefore, each transistor drops one-half of the supply voltage, and the value of V_{CEQ} for either transistor can be found as

$$V_{CEQ} = \frac{V_{CC}}{2} \tag{20.32}$$

Since the transistors are biased in cutoff, the value of I_{CQ} is approximately equal to the *collector leakage current* (I_{CBO}) rating of the transistors. Because transistor I_{CBO} ratings are typically in the nanoamp range, we generally assume that

$$I_{CQ} \cong 0 \text{ A} \tag{20.33}$$

AC Operating Characteristics

The complementary-symmetry amplifier is basically an emitter follower, so many of the ac relationships are similar (or identical) to those introduced earlier in the chapter. For example, the current gain of an emitter follower is found as

$$A_i = h_{fc}\left(\frac{Z_{\text{in}} r_E}{Z_{\text{in(base)}} R_L}\right) \quad | \quad r_E = R_E \| R_L$$

Since the load acts as the emitter resistor for the class B amplifier, $r_E = R_L$, and the A_i equation simplifies to

$$A_i = h_{fc}\left(\frac{Z_{\text{in}}}{Z_{\text{in(base)}}}\right) \tag{20.34}$$

The voltage gain of the class B amplifier is slightly less than one (as it is with any emitter follower). A close approximation of A_v can be found using

$$A_v = \frac{R_L}{R_L + r'_e} \tag{20.35}$$

As with any amplifier, the power gain equals the product of A_v and A_i. By formula,

$$A_p = A_v A_i$$

Amplifier impedance If you look at the class B amplifier, you will notice that the load resistor is connected to the emitters of the two transistors. Because there is no emitter resistor and the load resistor is not bypassed, the transistor input impedance is found as

$$Z_{\text{in(base)}} = h_{fc}(r'_e + R_L) \tag{20.36}$$

The output impedance of the class B amplifier is solved in a similar way to that of the class A emitter follower. The only difference is that there is no emitter resistor. Therefore, the amplifier output impedance is found as

$$Z_{\text{out}} = r'_e + \frac{R'_{\text{in}}}{h_{fc}} \tag{20.37}$$

where

$$R'_{\text{in}} = R_1 \| R_3 \| R_S$$

Note that R_2 is not considered in the calculation of R'_{in}, because it is bypassed by the input coupling capacitors. (These capacitors short circuit the input signal around R_2.)

Power Calculations

Class B amplifier output power can be calculated using any of the standard power relationships. For example,

$$P_L = \frac{V_L^2}{R_L}$$

As you know $V_{rms} = 0.707V_{pk}$, and $V_{pk} = \frac{V_{PP}}{2}$. Based on these relationships, amplifier output power can be found as

$$P_L = \frac{(0.3535V_{PP})^2}{R_L} = \frac{0.125V_{PP}^2}{R_L}$$

or

$$P_L = \frac{V_{PP}^2}{8R_L} \tag{20.38}$$

Equation (20.38) is important, because it provides a means of determining the maximum possible load power for a given class B amplifier.

The maximum undistorted output that can be provided by an amplifier is referred to as its **compliance (PP).** The compliance of a class B amplifier is approximately twice the value of V_{CEQ}. By formula,

$$PP = 2V_{CEQ} \tag{20.39}$$

Since $V_{CEQ} \cong \frac{V_{CC}}{2}$, equation (20.39) can be rewritten as

$$PP \cong V_{CC} \tag{20.40}$$

The maximum possible load power for a class B amplifier is determined as shown in Example 20.9.

Example 20.9

Determine the maximum load power for the circuit shown in Figure 20.29.

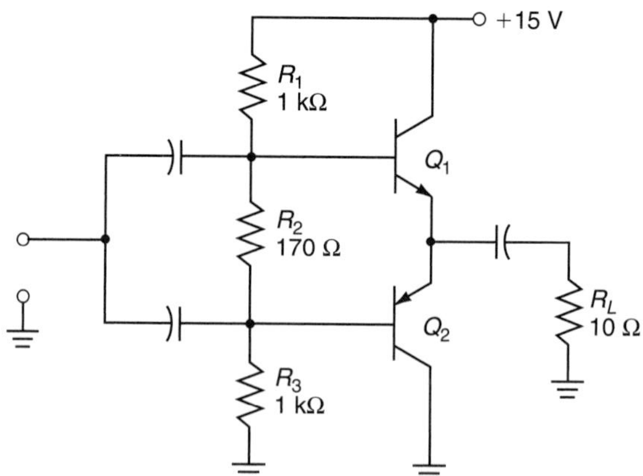

Figure 20.29

Solution: The compliance of the amplifier is found as

$$PP \cong V_{CC} = 15 \text{ V}$$

and the maximum possible load power is found as

$$P_{L(\max)} = \frac{PP^2}{8R_L} = \frac{(15\text{ V})^2}{(8)(10\ \Omega)} = 2.81\text{ W}$$

Practice Problem 20.9

A circuit like the one shown in Figure 20.29 has values of $V_{CC} = +12$ V and $R_L =$ 8 Ω. Determine the maximum value of load power for the circuit.

The class B amplifier relationships we have established up to this point are summarized in Figure 20.30.

Diode Bias

As shown in Figure 20.31, **diode bias** uses *a pair of diodes in place of the resistor between the transistor base terminals.* These diodes, called **compensating diodes,** are chosen to match the base-emitter junction characteristics of the two transistors. As you will see, the diodes prevent *crossover distortion* and *thermal runaway* when they are properly matched to the amplifier transistors.

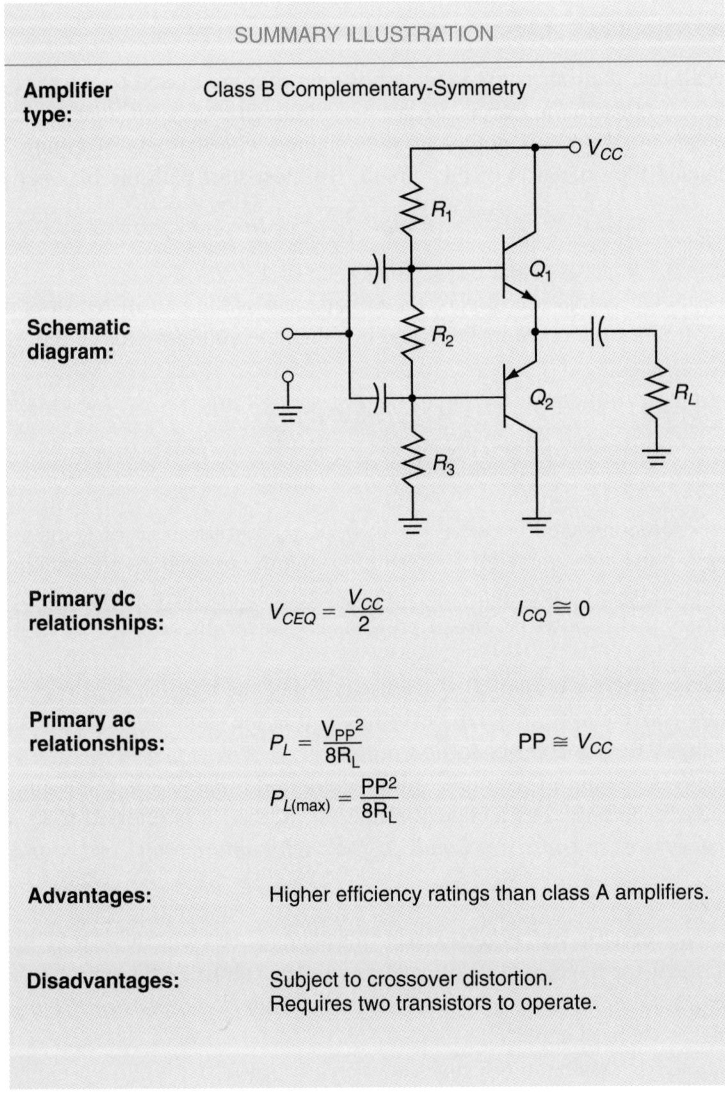

SUMMARY ILLUSTRATION
Amplifier type:
Class B Complementary-Symmetry
Schematic diagram:
V_{CC}
R_1
Q_1
R_2
Q_2
R_L
R_3
Primary dc relationships:
$V_{CEQ} = \frac{V_{CC}}{2}$
$I_{CQ} \cong 0$
Primary ac relationships:
$P_L = \frac{V_{PP}^2}{8R_L}$
$PP \cong V_{CC}$
$P_{L(\max)} = \frac{PP^2}{8R_L}$
Advantages:
Higher efficiency ratings than class A amplifiers.
Disadvantages:
Subject to crossover distortion.
Requires two transistors to operate.

Figure 20.30

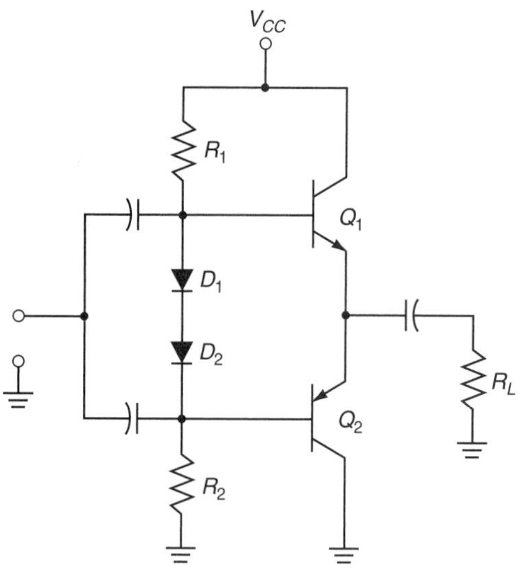

Figure 20.31 Diode bias.

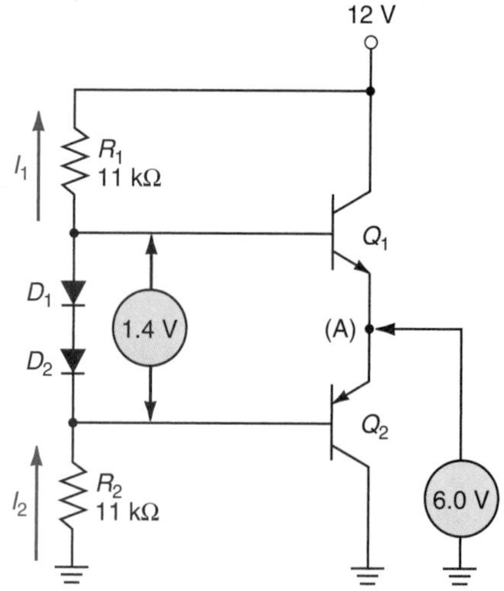

Figure 20.32

Diode Bias DC Characteristics

You may recall that the transistors in a class B amplifier are biased in cutoff ($I_{CQ} \cong 0$ A). When diode bias is used, the transistors are biased *just above cutoff;* that is, there is some measurable value of I_{CQ}. This point can be explained with the help of Figure 20.32.

To understand the operation of this circuit, we must start with the bias network. Note that:

- $R_1 = R_2$.
- There is 0.7 V across each diode, for a total of 1.4 V.

Since there is 1.4 V across the diodes, the voltage across the pair of resistors must equal (V_{CC} − 1.4 V). Since the resistors are equal in value, the voltage across either resistor can be found as

$$V_R = \frac{V_{CC} - 1.4\text{ V}}{2} \tag{20.41}$$

The diodes are connected between the base of Q_1 and the base of Q_2, so the sum of V_{BE1} and V_{BE2} must also equal 1.4 V. Assuming that the transistors are matched, there is 0.7 V across each base-emitter junction, and both transistors have some low value of I_{CQ}. As you will see, this is how the circuit prevents crossover distortion.

OBJECTIVE 15 ➤ **Eliminating crossover distortion** Since the transistors in Figure 20.32 are biased slightly above cutoff, they are never fully off at the same time. As a result, the circuit is not subject to crossover distortion problems.

Because I_{CQ} has some measurable value when diode bias is used, the amplifier cannot technically be classified as a class B amplifier. Rather, it is referred to as a **class AB amplifier,** because *each transistor conducts for slightly more than 180° of the input signal cycle.*

Eliminating thermal runaway In Chapter 19, you were told that *Q*-point stability is critical for the biasing circuit of any linear amplifier. **Thermal runaway** is *a phenomenon where an increase in temperature can result in a transistor driving itself into saturation.* Here's what happens:

- An increase in temperature causes an increase in transistor current gain (h_{FE}).
- The increase in current gain causes collector current to increase.
- The increase in I_C causes further increases in temperature and h_{FE}.

Left unchecked, these changes can have a snowball effect, taking the transistor into saturation.

For diode bias to eliminate the problem of thermal runaway, two conditions must be met:

- The diodes and the transistor base-emitter junctions must be very nearly perfectly matched.
- The diodes and the transistors must be in **thermal contact,** meaning that they must be *in physical contact with each other or a common surface so that they have equal operating temperatures.*

As long as these conditions are met, here's what happens when the temperature increases:

- The increase in temperature causes the 1.4 V drop across the diodes to decrease.
- The decrease in diode voltage causes the base-emitter junction voltages of the transistors to decrease.
- The decrease in V_{BE} for each transistor reduces I_B, which reduces the value of I_C.

This decrease in I_C prevents thermal runaway from occurring.

Note that the diodes can be placed in thermal contact with the power transistors by attaching them to the heat-sink tab of a power transistor or attaching them directly to the transistor heat sink (if one is present).

Section Review

1. What advantage does the class B amplifier have over the class A amplifier?
2. What are *complementary transistors*?
3. Describe the operating characteristics of the class B amplifier.
4. What are the typical values of I_{CQ} and V_{CEQ} for a class B amplifier?
5. What is *compliance*? What does the compliance of a class B amplifier equal?
6. Explain how *diode bias* develops the proper values of V_B for the amplifier transistors.
7. What is *class AB* operation?
8. How does the class AB amplifier eliminate *crossover distortion*?
9. How does the class AB amplifier eliminate *thermal runaway*?

20.8 Related Topics

This section contains several miscellaneous topics that relate to the material presented in this chapter.

Amplifier Swamping

◄ OBJECTIVE 16

Amplifier **swamping** *reduces variations in voltage gain by increasing the ac resistance of the emitter circuit.* A swamped amplifier is shown in Figure 20.33a.

The swamped, or *gain-stabilized,* amplifier has a higher ac emitter resistance, because only part of the dc emitter resistance is bypassed. The unbypassed emitter resistance (r_E) is part of the ac equivalent circuit, as shown in Figure 20.33b. The voltage gain for this amplifier is found as

$$A_v = \frac{r_C}{r'_e + r_E} \qquad (20.42)$$

The effect on gain stability The voltage gain of a CE amplifier can be affected by changes in temperature. You may recall that the dc current gain of a transistor

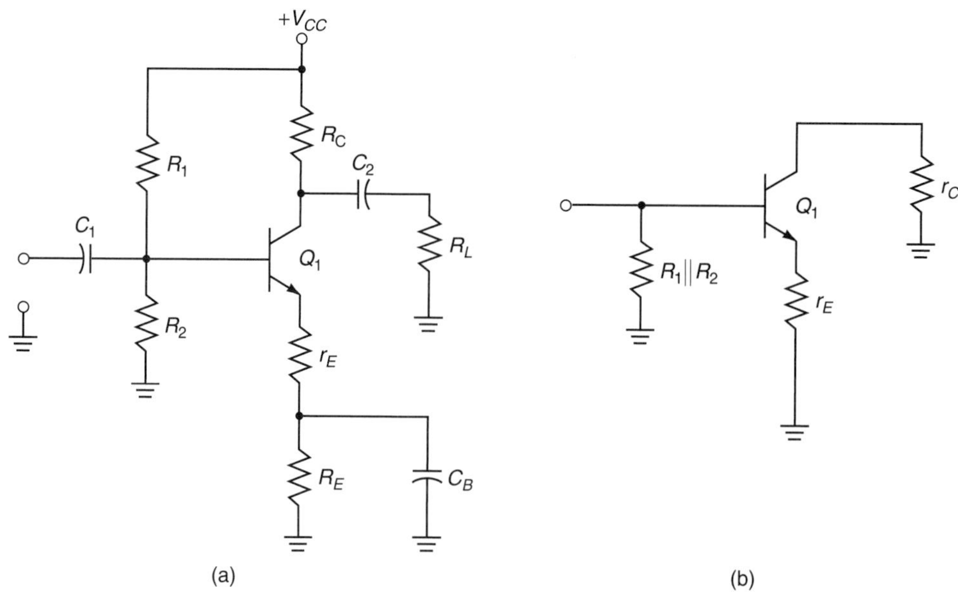

Figure 20.33 A gain-stabilized common-emitter amplifier and its ac equivalent.

(h_{FE}) varies with temperature. Any change in h_{FE} can affect the value of I_E and, therefore, the values of r'_e and A_v.

Swamping improves the stability of A_v by reducing the effect of changes in r'_e. When $r_E >> r'_e$, any change in r'_e caused by a change in temperature has little effect on the overall voltage gain of the amplifier. At the same time, the addition of r_E reduces the overall voltage gain of the amplifier. Therefore, the added gain stability is obtained at the cost of reduced voltage gain.

The effect on base input impedance The addition of a swamping resistor to a common-emitter amplifier increases the base input impedance of its transistor. You may recall that base input impedance equals the product of small signal current gain (h_{fe}) and the total ac emitter resistance. The ac emitter resistance of a swamped amplifier equals ($r'_e + r_E$). Therefore,

$$Z_{\text{in(base)}} = h_{fe}(r'_e + r_E) \tag{20.43}$$

Darlington Amplifiers

OBJECTIVE 17 ➤ In Chapter 18, you were introduced to the *Darlington pair,* a two-transistor device with extremely high current gain. A Darlington emitter follower is shown in Figure 20.34.

Just as a Darlington transistor has extremely high dc current gain (h_{FE}), it also has extremely high small-signal current gain (h_{fe}). For example, the Darlington transistor shown in Figure 20.34 has a rating of $h_{fe} = 20{,}000$ at $I_C = 10$ mA.

The high small-signal current gain of the Darlington transistor also results in high base input impedance. For example, let's assume that the amplifier shown in Figure 20.34 has a value of $I_E = 10$ mA. Using this value,

$$r'_e = \frac{25 \text{ mV}}{I_E} = \frac{25 \text{ mV}}{10 \text{ mA}} = 2.5\ \Omega$$

and

$$Z_{\text{in(base)}} = h_{fc}(r'_e + r_E) = (20{,}000)(2.5\ \Omega + 1.1\ \text{k}\Omega) \cong 22\ \text{M}\Omega$$

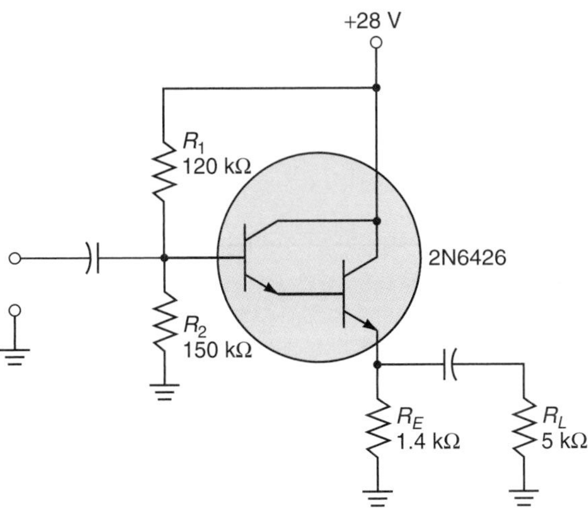

Figure 20.34 A Darlington emitter follower.

Transistor Power Requirements

◄ *OBJECTIVE 18*

When considering a transistor for a specific amplifier application, you must make sure that the power dissipation rating of the transistor is sufficient for that application. How do you determine the amount of power that a transistor will be required to dissipate in a specific circuit? For class A amplifiers, use the equation

$$P_D = V_{CEQ}I_{CQ} \tag{20.44}$$

For class B and class AB amplifiers, use the equation

$$P_D = \frac{V_{PP}^2}{40R_L} \tag{20.45}$$

The derivations of these relationships are provided in Appendix D. Once you calculate the value of P_D for a specific circuit, it is a good idea (as always) to use $1.2P_D$ as the minimum acceptable power dissipation rating for the transistor.

Section Review

1. How does *swamping* reduce the effect of variations in the value of r'_e?
2. What effect does swamping have on the values of A_v and $Z_{in(base)}$ of an amplifier?
3. What are the advantages of using a *Darlington transistor* in an emitter follower?

Key Terms

The following terms were introduced in this chapter on the pages indicated:

AC emitter resistance (r'_e), 648
AC equivalent circuit, 650
Amplification, 639
Amplifier, 639
Amplifier input impedance (Z_{in}), 641
Amplifier output impedance (Z_{out}), 642
Bypass capacitor, 651
Class A amplifier, 663
Class AB amplifier, 668
Class B amplifier, 663
Common-base (CB) amplifier, 646
Common-collector (CC) amplifier, 646
Common-emitter (CE) amplifier, 645
Compensating diodes, 667
Complementary-symmetry amplifier, 663
Complementary transistors, 663
Compliance (PP), 666
Coupling capacitor, 650
Crossover distortion, 664
Current gain (A_i), 655
Diode bias, 662

Effective voltage gain ($A_{v(\text{eff})}$), 643
Emitter follower, 656
Gain, 639
Ideal voltage amplifier, 644
Power gain (A_p), 656
Push-pull emitter follower, 663
Swamping, 669
Thermal contact, 669
Thermal runaway, 668
Voltage gain (A_v), 652

Practice Problems

1. Complete the following table.

v_{in}	v_{out}	A_v
1.2 mV	300 mV	_____
200 μV	18 mV	_____
24 mV	2.2 V	_____
800 μV	140 mV	_____

2. Complete the following table.

v_{in}	v_{out}	A_v
38 mV	600 mV	_____
6 mV	9.2 V	_____
500 μV	88 mV	_____
48 mV	48 mV	_____

3. For each combination of v_{in} and A_v, calculate the value of v_{out}.

v_{in}	v_{out}	A_v
240 μV	540	_____
1.4 mV	300	_____
24 mV	440	_____
800 μV	720	_____

4. Calculate the value of v_{in} for the circuit shown in Figure 20.35a.

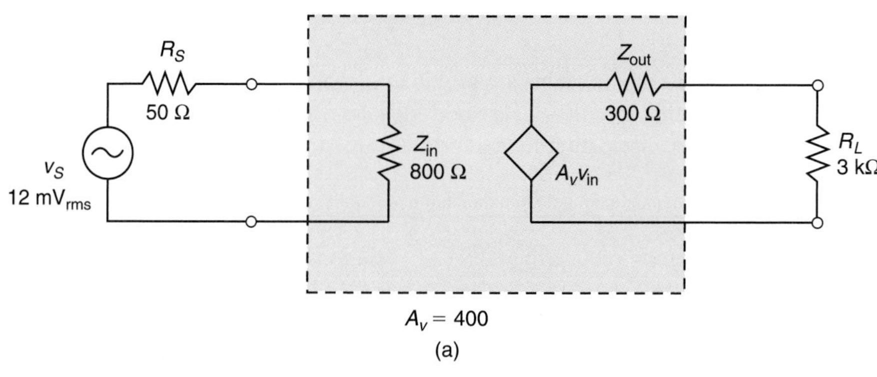

(a)

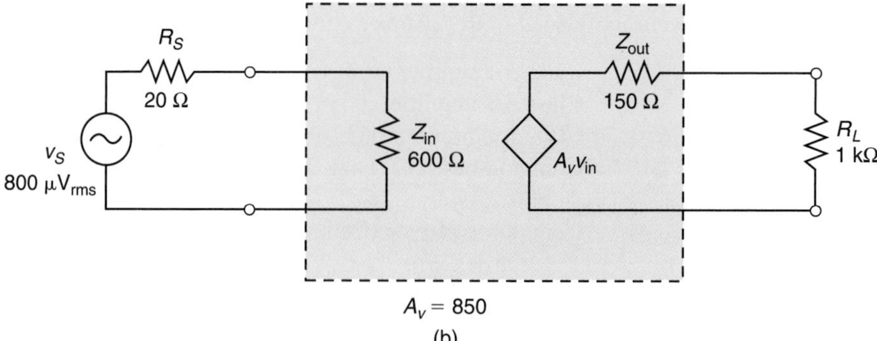

(b)

Figure 20.35

5. Calculate the value of v_{in} for the circuit shown in Figure 20.35b.
6. Calculate the value of v_L for the circuit shown in Figure 20.35a.
7. Calculate the value of v_L for the circuit shown in Figure 20.35b.
8. Calculate the effective voltage gain of the amplifier shown in Figure 20.35a.
9. Calculate the effective voltage gain of the amplifier shown in Figure 20.35b.
10. Calculate the effective voltage gain of the amplifier shown in Figure 20.36a.
11. Calculate the effective voltage gain of the amplifier shown in Figure 20.36b.

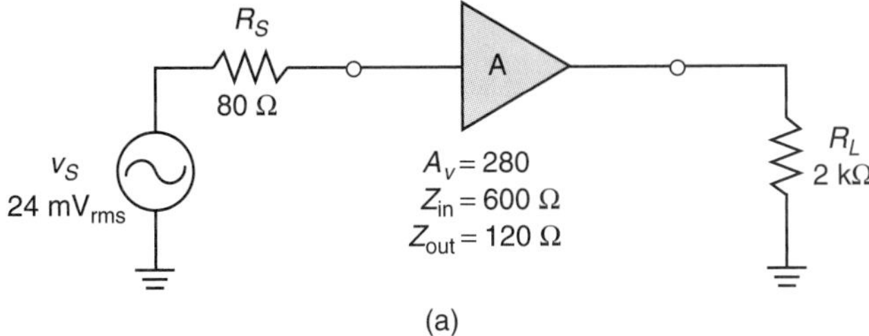

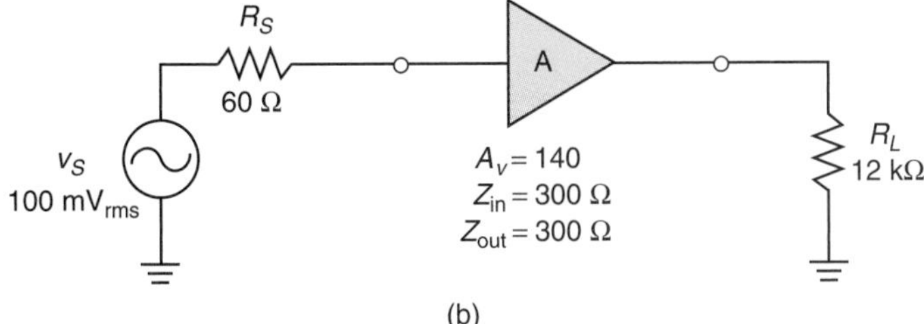

Figure 20.36

12. A CE amplifier has an emitter current of 12 mA. Determine the value of r'_e for the circuit.
13. A CE amplifier has an emitter current of 10 mA. Determine the value of r'_e for the circuit.
14. A CE amplifier has values of $V_E = 2.2$ V and $R_E = 910\ \Omega$. Determine the value of r'_e for the circuit.
15. A CE amplifier has values of $V_E = 12$ V and $R_E = 4.7\ \text{k}\Omega$. Determine the value of r'_e for the circuit.
16. Determine the value of r'_e for the circuit shown in Figure 20.37.
17. Determine the value of r'_e for the circuit shown in Figure 20.38.

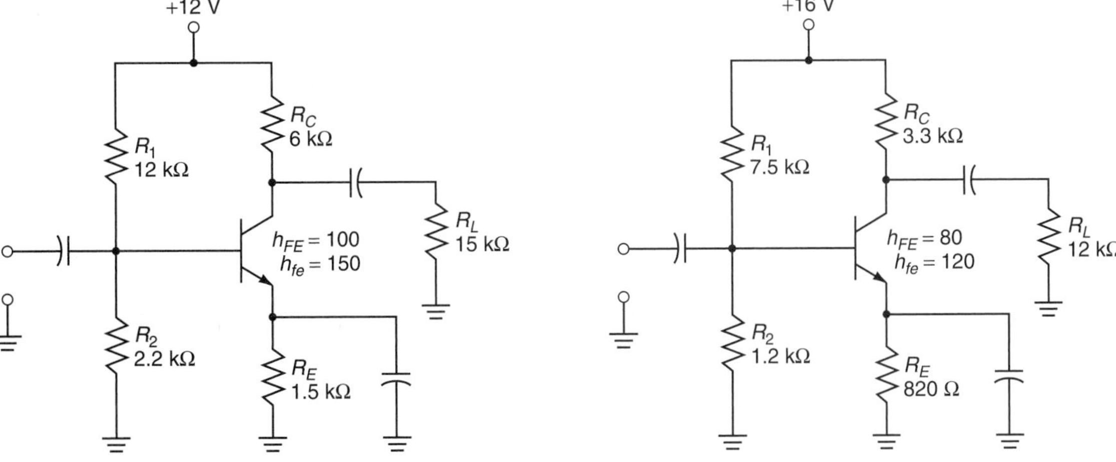

Figure 20.37

Figure 20.38

18. Derive the ac equivalent circuit for the amplifier shown in Figure 20.37. Include all component values.

19. Derive the ac equivalent circuit for the amplifier shown in Figure 20.38. Include all component values.

20. A CE amplifier has values of $r_C = 2.2\text{ k}\Omega$ and $r'_e = 22.8\ \Omega$. Determine the value of A_v for the circuit.

21. A CE amplifier has values of $r_C = 4.7\text{ k}\Omega$ and $r'_e = 32\ \Omega$. Determine the value of A_v for the circuit.

22. Determine the value of A_v for the amplifier shown in Figure 20.37.

23. Determine the value of A_v for the amplifier shown in Figure 20.38.

24. Determine the value of A_v for the amplifier shown in Figure 20.39.

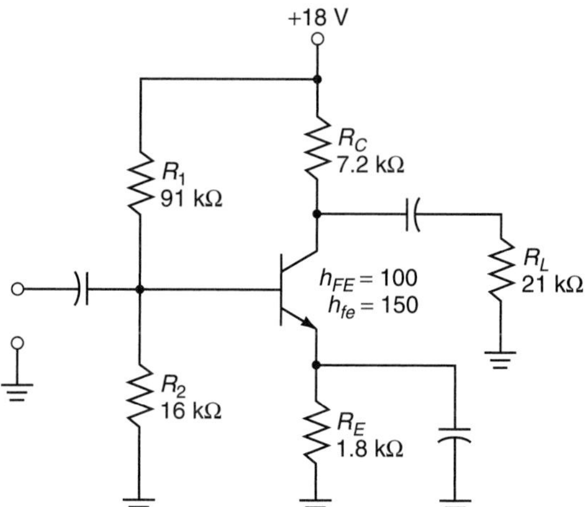

Figure 20.39

25. Determine the value of A_v for the amplifier shown in Figure 20.40.

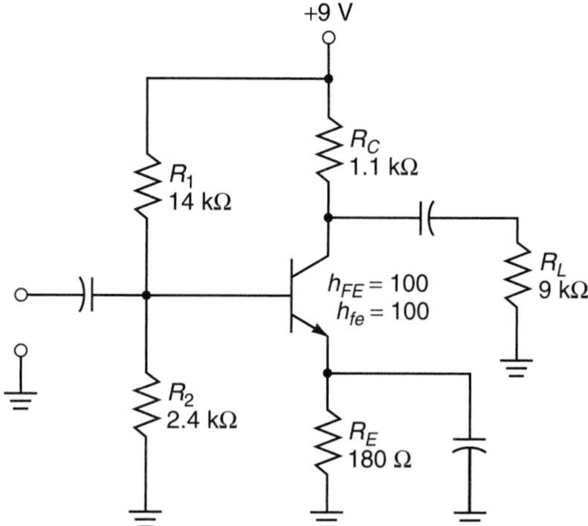

Figure 20.40

26. Determine the value of A_p for the circuit in Figure 20.37. Assume that the circuit has a value of $A_i = 11$.

27. Determine the value of A_p for the circuit shown in Figure 20.38. Assume that the circuit has a value of $A_i = 9.3$.

28. Determine the value of A_p for the circuit shown in Figure 20.39. Assume that the circuit has a value of $A_i = 7.5$.

29. Determine the values of I_{CQ} and V_{CEQ} for the circuit in Figure 20.41.

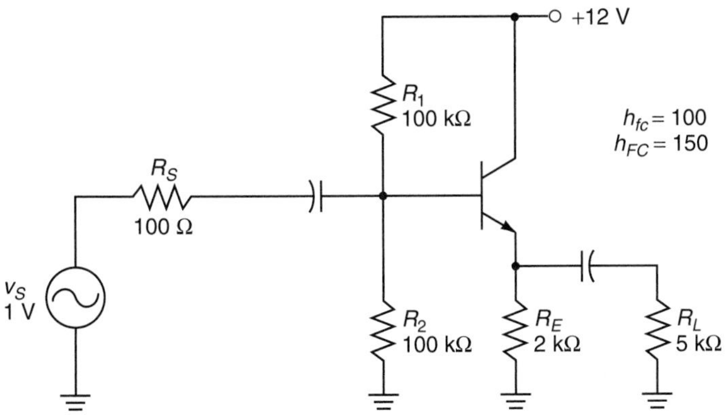

Figure 20.41

30. Determine the values of I_{CQ} and V_{CEQ} for the circuit shown in Figure 20.42.

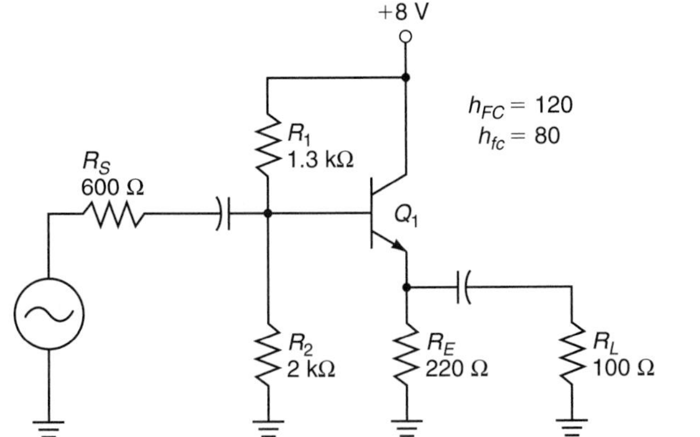

Figure 20.42

31. Determine the value of Z_{in} for the circuit shown in Figure 20.41.
32. Determine the value of Z_{in} for the circuit in Figure 20.42.
33. Determine the values of A_v, A_i, and A_p for the circuit shown in Figure 20.41.
34. Determine the values of A_v, A_i, and A_p for the circuit shown in Figure 20.42.
35. Determine the value of Z_{out} for the circuit shown in Figure 20.41.
36. Determine the value of Z_{out} for the circuit shown in Figure 20.42.
37. Calculate the values of A_v, A_i, A_p, Z_{in}, and Z_{out} for the amplifier shown in Figure 20.43.

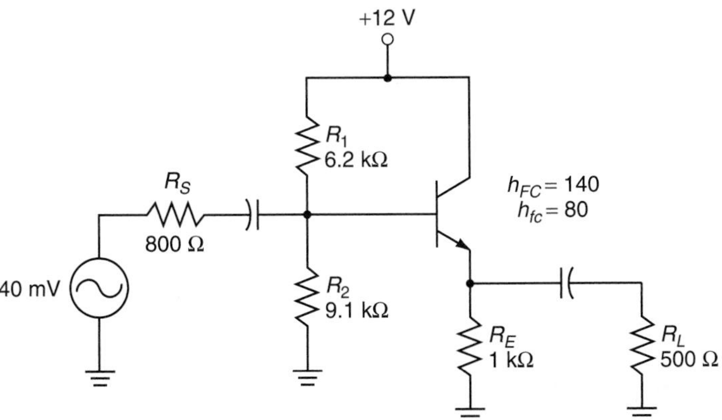

Figure 20.43

38. Calculate the values of A_v, A_i, A_p, Z_{in}, and Z_{out} for the amplifier shown in Figure 20.44.

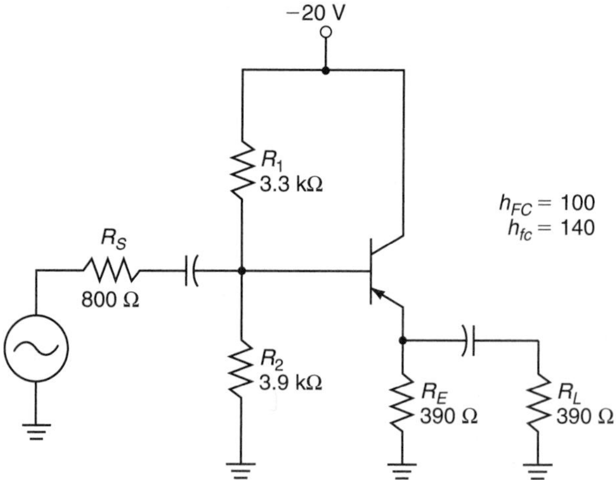

Figure 20.44

39. Determine the value of A_v for the amplifier shown in Figure 20.45.

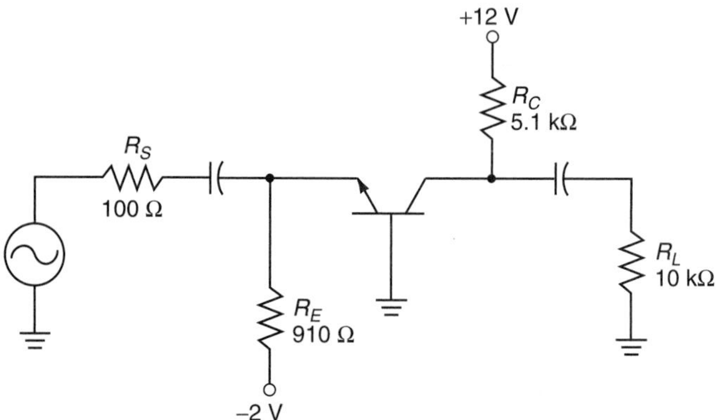

Figure 20.45

40. Determine the value of A_v for the amplifier shown in Figure 20.46.

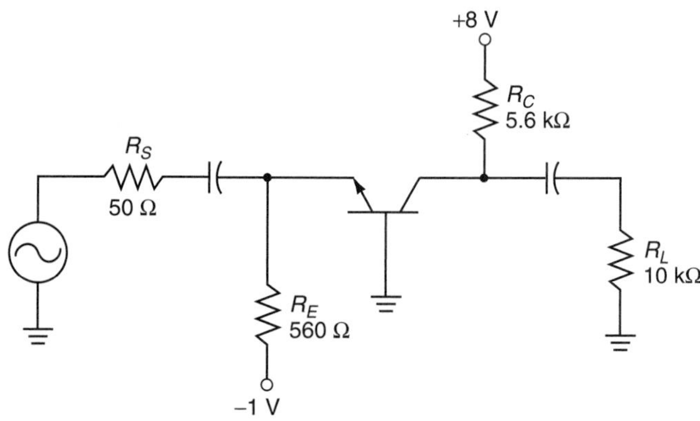

Figure 20.46

41. Determine the values of Z_{in} and Z_{out} for the amplifier shown in Figure 20.45.

42. Determine the values of Z_{in} and Z_{out} for the amplifier shown in Figure 20.46.

43. Determine the values of V_{CEQ}, V_{B1}, and V_{B2} for the amplifier shown in Figure 20.47.

44. Determine the values ofV_{CEQ}, V_{B1}, and V_{B2} for the amplifier shown in Figure 20.48.

45. Calculate the values of PP and $P_{L(max)}$ for the amplifier shown in Figure 20.47.

46. Calculate the values of PP and $P_{L(max)}$ for the amplifier shown in Figure 20.48.

47. Calculate the value of P_D for the transistor shown in Figure 20.37.

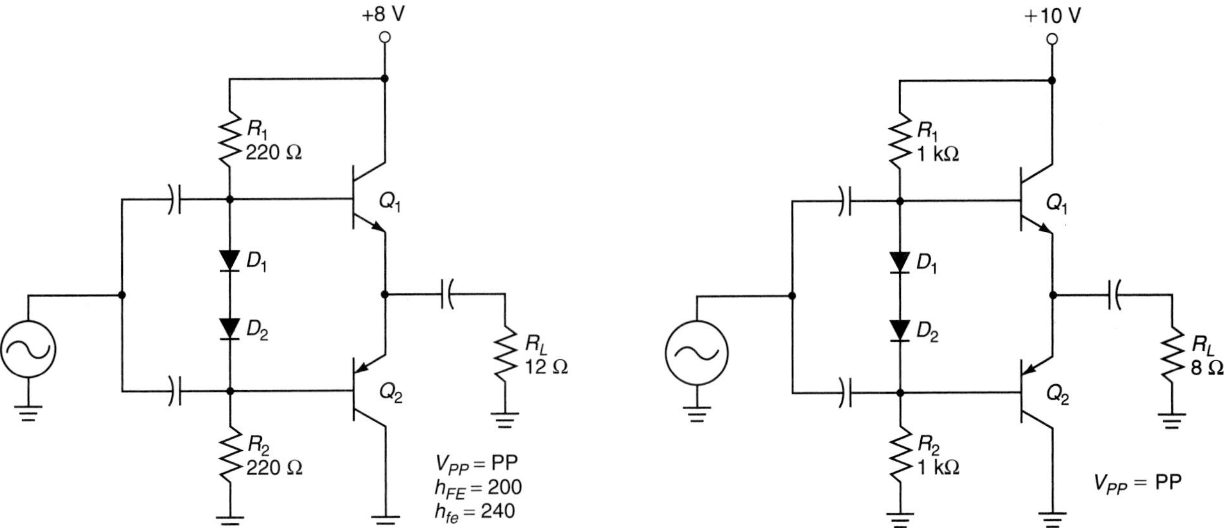

Figure 20.47

Figure 20.48

48. Calculate the value of P_D for the transistor shown in Figure 20.46.
49. Calculate the value of P_D for the transistors shown in Figure 20.47.
50. Calculate the value of P_D for the transistors shown in Figure 20.48.

Looking Back

These problems relate to material presented in earlier chapters. The chapters are identified in brackets.

51. Convert each value to engineering notation. [Introduction]
    a. 3.2×10^{-4} A = ______
    b. 0.89×10^{9} Ω = ______
52. Calculate the conductance of a 510 kΩ resistor. [Chapter 1]
53. Using Table 2.1, calculate the resistance of a 1200 ft length of 14 gauge wire. [Chapter 2]
54. A variable inductor has an adjusted value of 120 mH. Determine the value of the component if it is adjusted so that the effective length of the component doubles. [Chapter 10]
55. Determine the average value of the waveform shown in Figure 20.49. [Chapter 9]
56. Draw the output waveform for the circuit shown in Figure 20.50. Determine the peak output values, and include them in the figure. [Chapter 18]

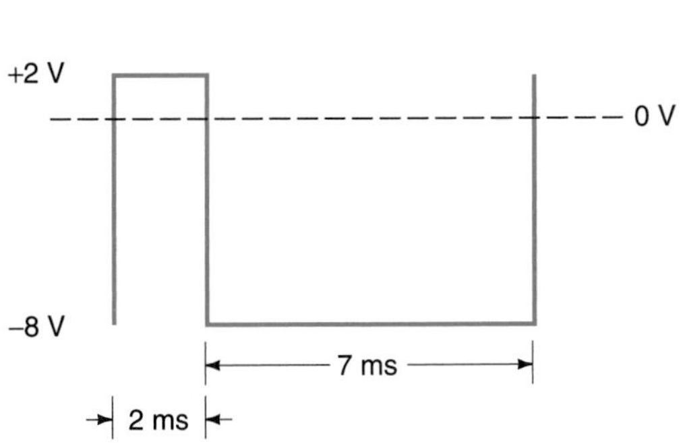

Figure 20.49

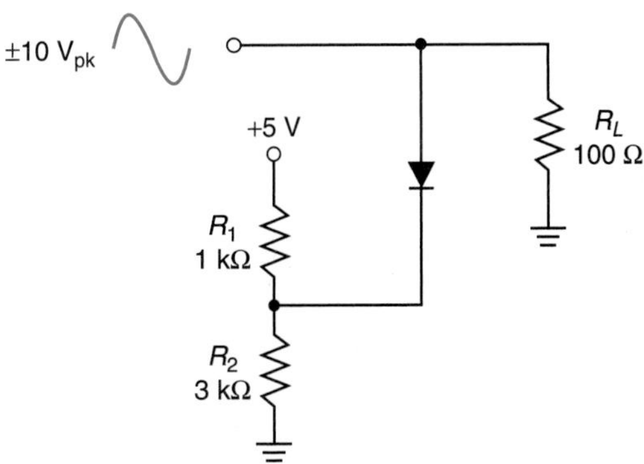

Figure 20.50

Answers to the Example Practice Problems

20.1	240	**20.6**	0.9967
20.2	86.4 mV	**20.7**	8.4 kΩ
20.3	301	**20.8**	9.95 Ω
20.4	1.5 kΩ	**20.9**	2.25 W
20.5	48.21		

Multisim File References

Figure	File Reference
20.6	ETF20_6.msm
20.12	ETF20_12.msm
20.28	ETF20_28.msm

chapter 21

Field-Effect Transistors and Circuits

Objectives

After studying the material in this chapter, you should be able to:

1. List the types of field-effect transistors (FETs).
2. Explain the relationship among JFET channel width, gate-source voltage (V_{GS}), and drain current (I_D).
3. State the relationship between drain-source voltage (V_{DS}) and drain current (I_D).
4. Describe the gate input impedance characteristics of the JFET.
5. Determine the range of *Q*-point values for a given JFET biasing circuit.
6. Describe and analyze the ac operation of the *common-source* amplifier.
7. Describe and analyze the ac operation of *common-drain* and *common-gate* amplifiers.
8. Identify the two types of MOSFETs and describe the construction of each.
9. Describe the devices and precautions used to protect MOSFETs.
10. Describe the *depletion-mode* operation of the D-MOSFET.
11. Describe the *enhancement-mode* operation of the D-MOSFET.
12. Describe and analyze zero-bias operation.
13. Define *threshold voltage* ($V_{GS(th)}$), and discuss its significance.
14. Calculate the value of I_D for an E-MOSFET, given the component ratings and circuit voltages.
15. List the FET biasing circuits that can and cannot be used with E-MOSFETs.
16. List the advantages that CMOS logic circuits have over comparable BJT logic circuits.
17. Describe the characteristics of VMOS and LDMOS devices.
18. Describe the purpose served by a *dual-gate* MOSFET.

You may recall that the bipolar junction transistor (BJT) is a current-controlled device, meaning that its output characteristics are varied by a change in input current. The **field-effect transistor (FET)** is *a voltage-controlled device.* The output characteristics of the FET are controlled by its input voltage, not its input current.

OBJECTIVE 1 ➤ There are two basic types of FETs: the **junction FET (JFET)**, and the **metal-oxide-semiconductor FET (MOSFET)**. As you will be shown, the operating principles of these two components vary, as do their applications and limitations.

21.1 Introduction to JFETs

Unlike the BJT, the JFET has only two materials: a single *n*-type material, and a single *p*-type material. JFET construction is illustrated in Figure 21.1. As you can see, the device has three terminals, labeled *source, drain,* and *gate.* The **source** can be viewed as *the counterpart of the BJT emitter,* the **drain** as *the counterpart of the collector,* and the **gate** as *the counterpart of the base.*

The material that connects the source to the drain is referred to as the **channel**. A given JFET is identified by its channel material. Thus the components in Figure 21.1 are referred to as the *p-channel JFET* and the *n-channel JFET.* Note that the blocks on either side of the channel represent the gate material. Though drawn as separate blocks, the gate material actually surrounds the channel in the same way that a belt surrounds your waist.

The JFET schematic symbols are shown in Figure 21.2. Note that the arrow points *in* on the *n*-channel JFET symbol and points *out* on the *p*-channel JFET symbol. As shown in Figure 21.3:

- *N*-channel JFETs normally require positive supply voltages.
- *P*-channel JFETs normally require negative supply voltages.

Note the directions of the circuit currents shown in Figure 21.3.

Throughout this chapter, we will concentrate on *n*-channel JFET operation. Except for the current directions and voltage polarities, the principles covered apply equally to the *p*-channel JFET.

Operation Overview

OBJECTIVE 2 ➤ JFET operation is based on *varying the channel width to control the drain current.* There are two ways that channel width can be varied. The first is to reverse bias the gate-source junction. In Figure 21.4, the voltage across the gate-source junction (V_{GS}) reverse biases the junction, causing a depletion layer to form. As you know, the depletion layer acts

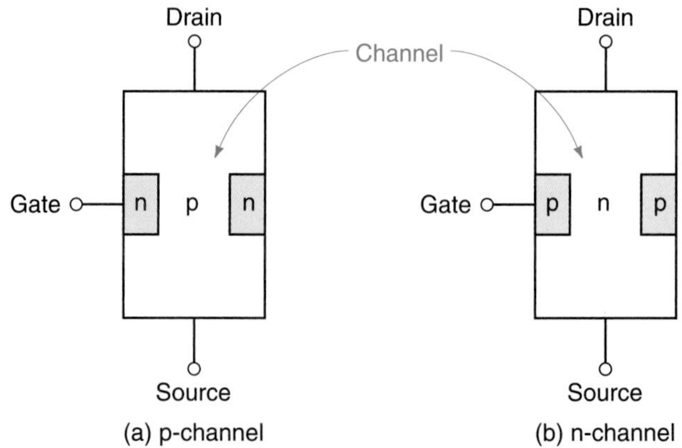

Figure 21.1 JFET construction.

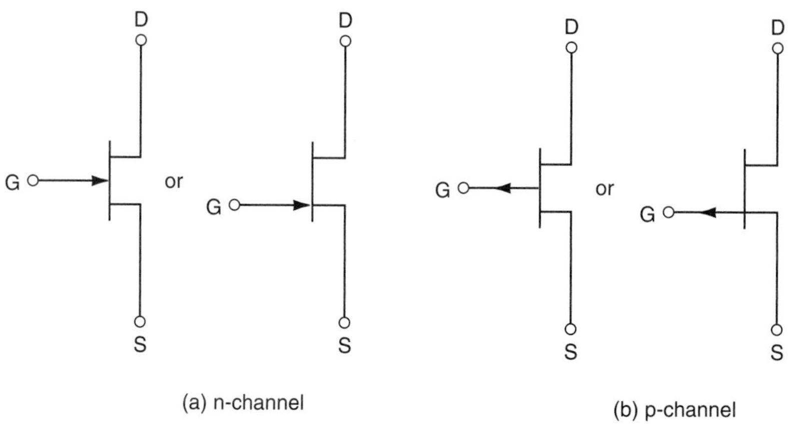

Figure 21.2 JFET schematic symbols.

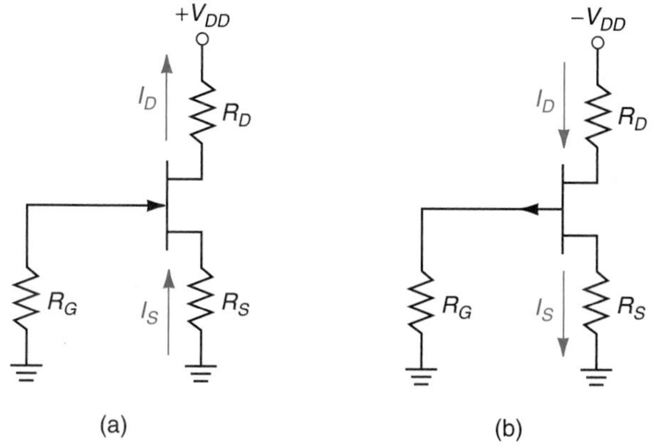

Figure 21.3 JFET supply voltages and circuit currents.

as an insulator. When the size of the depletion layer increases, it decreases the cross-sectional area of the channel, causing channel conduction to decrease. The greater the magnitude of V_{GS}, the greater the size of the depletion layer, and the lower the device current (as illustrated in Figure 21.4). In other words, as V_{GS} becomes more negative, drain current (I_D) decreases.

Figure 21.5 shows another means by which channel width can be varied. In Figure 21.5a, $V_{GS} = 0$ V, but a small depletion layer still surrounds the gate. This depletion layer is a result of the relationship between V_{GS} and V_{DS}. Since the *p*-type gate is more negative than the *n*-type drain, a small depletion layer forms around the gate. The greater the

◄ *OBJECTIVE 3*

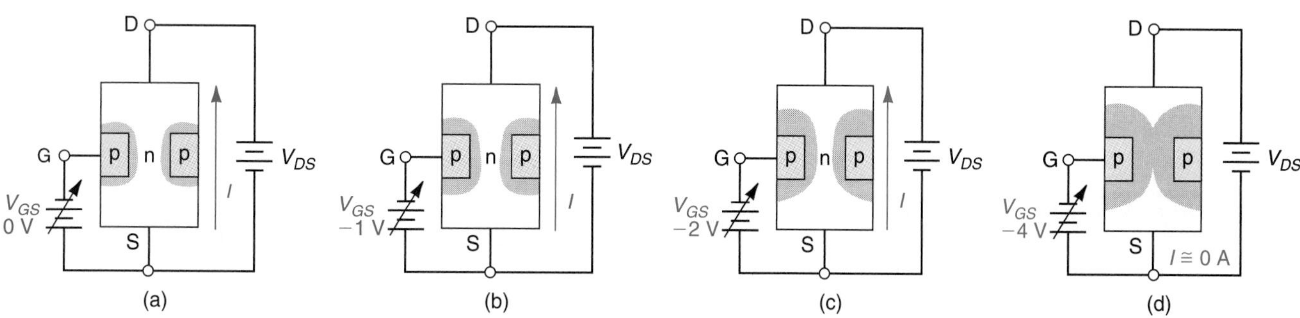

Figure 21.4 The relationship between V_{GS} and I_D.

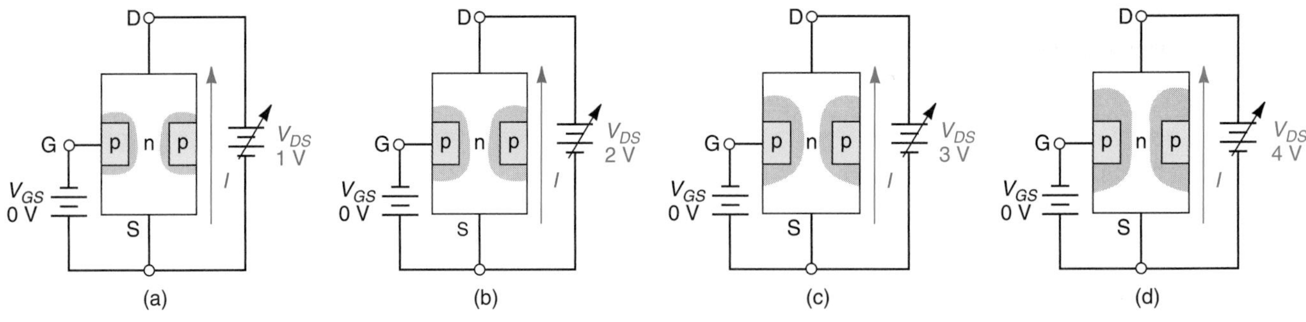

Figure 21.5 The effect of varying V_{DS} as V_{GS} remains constant.

magnitude of V_{DS}, the greater the size of the depletion layer. Thus, the second method for increasing the size of the depletion layer is to hold V_{GS} constant while increasing V_{DS} (as shown in Figure 21.5).

Pinch-Off Voltage (V_P)

It would seem that the increasing depletion layer width in Figure 21.5 would reduce the value of I_D. However, V_{DS} increases (initially) at a greater rate than the resistance of the channel, so I_D increases. Eventually, *a point is reached where further increases in V_{DS} are offset by proportional increases in channel resistance.* The value of V_{DS} at which this occurs is called the **pinch-off voltage (V_P)**. As shown in Figure 21.6a, the value of I_D levels off (becomes constant) as V_{DS} increases beyond the value of V_P.

The portion of the curve to the left of V_P (where $V_{DS} < V_P$) is referred to as the **ohmic region**. As V_{DS} increases through the ohmic region, I_D increases. When V_{DS} reaches the value of V_P, drain current levels off and remains constant for all values of V_{DS} between V_P and the breakdown voltage (V_{BR}). Note that *the region of operation between V_P and*

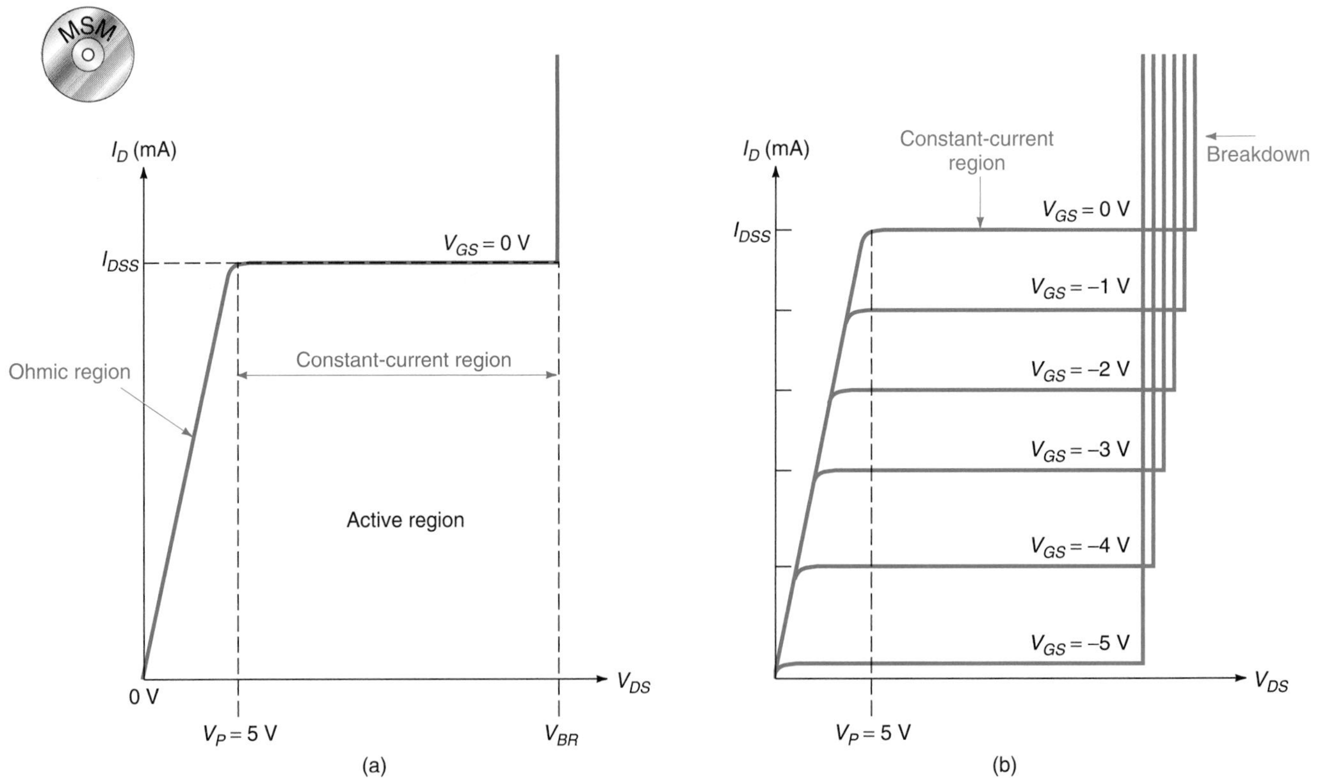

Figure 21.6 JFET drain curves.

V_{BR} is called the **constant-current region.** As long as V_{DS} remains within this range, I_D remains constant at a given value of V_{GS}.

Shorted-Gate Drain Current (I_{DSS})

The maximum possible value of drain current at a given value of V_{DS} is referred to as the **shorted-gate drain current (I_{DSS}).** The I_{DSS} rating for a given JFET is measured under the following conditions:

$$V_{GS} = 0 \text{ V} \qquad V_{DS} = V_P$$

The current through a given JFET cannot exceed the value of I_{DSS}. In this respect, I_{DSS} can be viewed as being the JFET equivalent of $I_{C(\text{sat})}$ for a BJT circuit.

Gate-Source Cutoff Voltage ($V_{GS(\text{off})}$)

The relationship among V_{GS}, V_{DS}, and I_{DSS} is illustrated in Figure 21.6b. When V_{GS} is more negative than 0 V:

- The JFET pinches off at some voltage that is less than its pinchoff (V_P) rating.
- The constant-current value of I_D drops below the value of I_{DSS}.

The value of V_{GS} that causes I_D to drop to approximately zero is called the **gate-source cutoff voltage ($V_{GS(\text{off})}$).** For conduction to occur through the device, V_{GS} must be less negative than $V_{GS(\text{off})}$. Note that the $V_{GS(\text{off})}$ rating for a JFET always has the same magnitude as its V_P rating. For example, if a JFET has a rating of $V_{GS(\text{off})} = -8$ V, then $V_P = 8$ V.

JFET Biasing

The gate-source junction of a JFET is never allowed to become forward biased, because the gate material is not constructed to handle any significant amount of current. If the junction is ever allowed to become forward biased, the gate current may destroy the component.

Since the gate is always reverse biased, a JFET has extremely high characteristic gate impedance, typically in the high megohm (MΩ) range. The high gate impedance of the JFET has led to its extensive use in microcomputer circuits. The low gate current requirement of the component makes it perfect for use in ICs, where thousands of transistors must be etched onto a single piece of silicon. The low gate input current helps the IC to remain relatively cool, allowing more components to be placed in a smaller physical area.

◀ *OBJECTIVE 4*

Component Control

It was stated earlier that the JFET is a voltage-controlled device, while the BJT is a current-controlled device. As you know, BJT collector current is controlled by its base current. In contrast, JFET drain current is controlled by its gate-source voltage. The value of I_D for a given JFET is found as

$$I_D = I_{DSS}\left(1 - \frac{V_{GS}}{V_{GS(\text{off})}}\right)^2 \qquad (21.1)$$

A Practical Consideration: Equation (21.1) works only when

$$|V_{GS}| < |V_{GS(\text{off})}|$$

If the above condition is not fulfilled, the equation will give you a false value of I_D. Just remember, if V_{GS} is more negative (or equal to) $V_{GS(\text{off})}$, I_D will be zero.

where

I_{DSS} = the shorted-gate drain current
V_{GS} = the gate-source voltage
$V_{GS(\text{off})}$ = the gate-source cutoff voltage

The use of equation (21.1) is demonstrated in Example 21.1.

Example 21.1

Determine the value of drain current for the circuit shown in Figure 21.7.

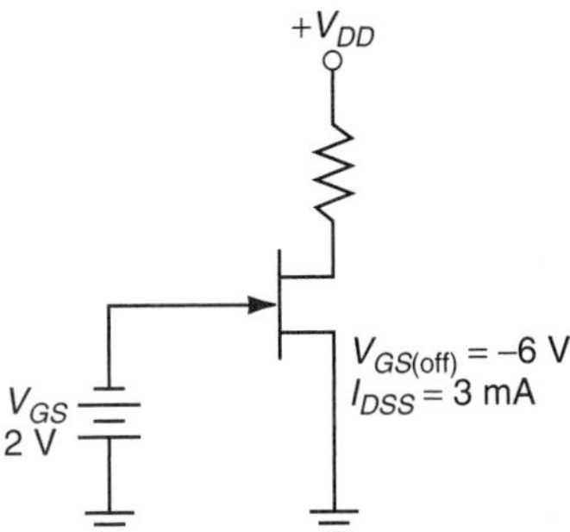

Figure 21.7

Solution: The drain current for the circuit is found as

$$I_D = I_{DSS}\left(1 - \frac{V_{GS}}{V_{GS(\text{off})}}\right)^2 = (3\text{ mA})\left(1 - \frac{-2\text{ V}}{-6\text{ V}}\right)^2 = (3\text{ mA})(0.444) = 1.33\text{ mA}$$

Practice Problem 21.1

A JFET has the following ratings: $I_{DSS} = 12$ mA, and $V_{GS(\text{off})} = -6$ V. Determine the value of drain current for the device when $V_{GS} = -3$ V.

If you examine equation (21.1) closely, you'll notice that V_{GS} is the only variable on the right-hand side. Since I_{DSS} and $V_{GS(\text{off})}$ are component ratings that do not change for a given JFET, the value of I_D is a function of V_{GS}. In other words, I_D changes as a result of changes in V_{GS}. Thus, the JFET is a voltage-controlled device.

Transconductance Curves

The **transconductance curve** for a JFET *represents every possible combination of V_{GS} and I_D for the device.* The transconductance curve for a specified JFET is plotted as demonstrated in Example 21.2.

Example 21.2

Plot the transconductance curve for a JFET having values of $V_{GS(\text{off})} = -6$ V and $I_{DSS} = 3$ mA.

Solution: The end points for the curve are defined by $V_{GS(\text{off})}$ and I_{DSS}. We know that $I_D = 0$ mA when $V_{GS} = V_{GS(\text{off})}$. Therefore, one end of the curve has a value of (−6 V, 0 mA). We also know that $I_D = I_{DSS}$ when $V_{GS} = 0$ V. Therefore, the other end of the curve has a value of (0 V, 3 mA). We now choose three other values of V_{GS} (−1 V, −3 V, and −5 V) and calculate the corresponding value of I_D for each:

At $V_{GS} = -1$ V:

$$I_D = I_{DSS}\left(1 - \frac{V_{GS}}{V_{GS(\text{off})}}\right)^2 = (3\text{ mA})\left(1 - \frac{-1\text{ V}}{-6\text{ V}}\right) = 2.08\text{ mA}$$

At $V_{GS} = -3$ V:

$$I_D = I_{DSS}\left(1 - \frac{V_{GS}}{V_{GS(\text{off})}}\right)^2 = (3\text{ mA})\left(1 - \frac{-3\text{ V}}{-6\text{ V}}\right) = 0.75\text{ mA}$$

At $V_{GS} = -5$ V:

$$I_D = I_{DSS}\left(1 - \frac{V_{GS}}{V_{GS(\text{off})}}\right)^2 = (3\text{ mA})\left(1 - \frac{-5\text{ V}}{-6\text{ V}}\right) = 0.083\text{ mA}$$

The following combinations of V_{GS} and I_D are used to plot the curve shown in Figure 21.8.

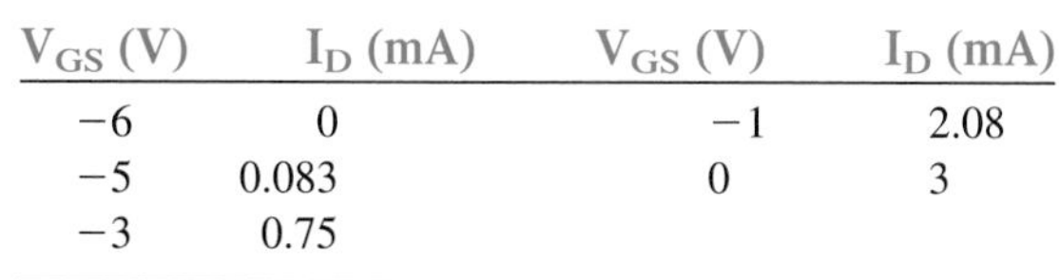

V_{GS} (V)	I_D (mA)	V_{GS} (V)	I_D (mA)
−6	0	−1	2.08
−5	0.083	0	3
−3	0.75		

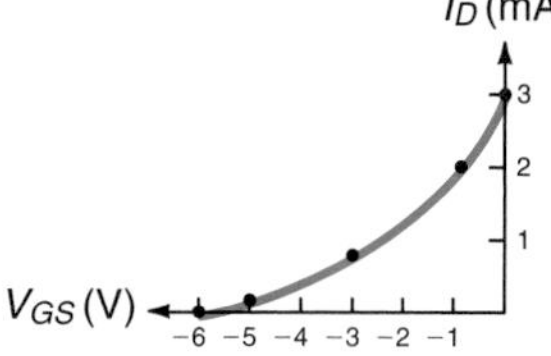

Figure 21.8

Practice Problem 21.2

A JFET has parameters of $V_{GS(\text{off})} = -20$ V and $I_{DSS} = 12$ mA. Plot the transconductance curve for the device using V_{GS} values of 0 V, −5 V, −10 V, −15 V, and −20 V.

The transconductance curve for a given JFET is used in both the dc and ac analyses of any amplifier using the device. It should be noted that most JFET spec sheets list more than one value of $V_{GS(\text{off})}$ and I_{DSS}. For example, the spec sheet for the 2N5457 lists the following:

$$V_{GS(\text{off})} = -0.5 \text{ V to } -6 \text{ V} \qquad I_{DSS} = 1 \text{ mA to } 5 \text{ mA}$$

When a range of values is given, you must use the two minimum values to plot one curve and the two maximum values to plot another curve on the same graph.

One Final Note

In this section, we have used *n*-channel JFETs in all the examples, because they are used far more commonly than *p*-channel JFETs. Just remember that all the *n*-channel JFET principles apply equally to *p*-channel JFETs. The only differences are the voltage polarities and the direction of the drain and the source currents. The *n*-channel and *p*-channel JFETs are contrasted in Figure 21.9.

Section Review

1. What is a field-effect transistor (FET)? What are the two types of FETs?
2. What are the three terminals of a *junction FET* (*JFET*)?
3. Describe the physical relationship between the JFET gate and channel.
4. What are the two types of JFETs?
5. What is the relationship between channel width and drain current?
6. What is the relationship between V_{GS} and channel width?
7. What is the relationship between V_{GS} and drain current?
8. What is *pinch-off voltage* (V_P)?
9. What effect does an increase in V_{DS} have on I_D when $V_{DS} < V_P$?
10. What effect does an increase in V_{DS} have on I_D when $V_{DS} \geq V_P$?
11. What is the operating region above V_P called?
12. What is V_{DSS}?
13. What is the relationship between I_D and I_{DSS}?
14. What is $V_{GS(\text{off})}$?

15. What is the primary restriction on the value of V_{GS}?

16. Why do JFETs typically have extremely high gate input impedance?

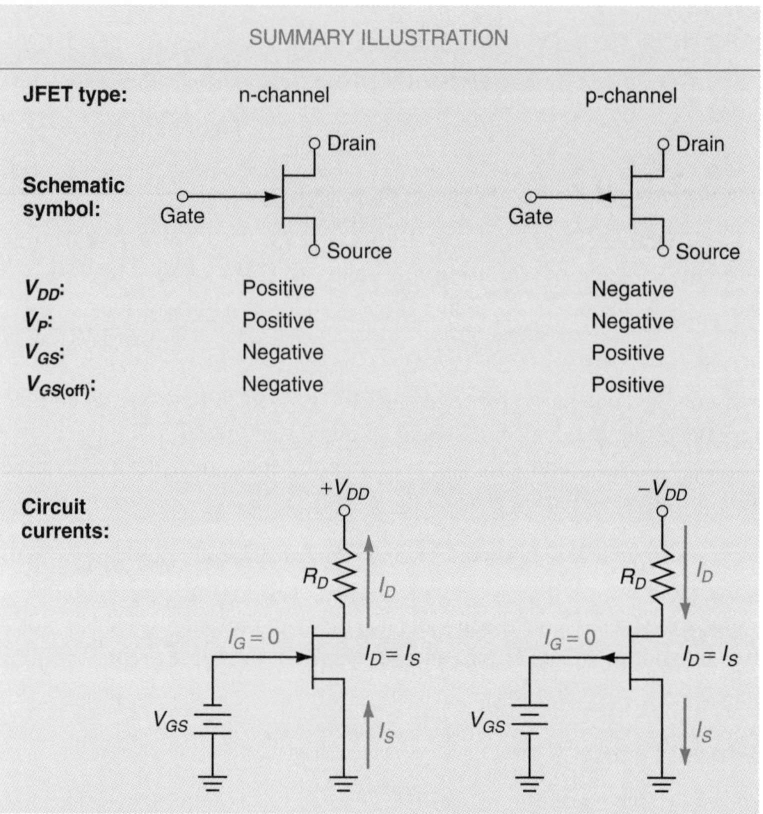

Figure 21.9

21.2 JFET Biasing Circuits

JFET biasing circuits are very similar to BJT biasing circuits. The main difference is the operation of the active components themselves.

OBJECTIVE 5 ➤

Gate Bias

Gate bias is *the JFET counterpart of base bias.* The gate-bias circuit is shown in Figure 21.10. The gate supply voltage ($-V_{GG}$) is used to reverse bias the gate-source junction. Since there is no gate current, there is no voltage across R_G, and

$$V_{GS} = V_{GG} \tag{21.2}$$

Using V_{GG} as V_{GS} in equation (21.1) allows us to calculate the value of I_D. Once the value of I_D is known, V_{DS} can be found as

$$V_{DS} = V_{DD} - I_D R_D \tag{21.3}$$

Example 21.3 demonstrates the complete Q-point analysis of a simple gate-bias circuit.

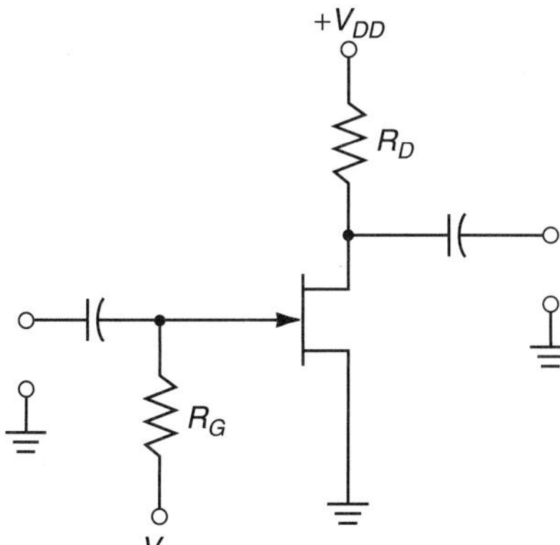

Figure 21.10 Gate bias.

Example 21.3

The JFET shown in Figure 21.11 has values of $V_{GS(\text{off})} = -8$ V and $I_{DSS} =$ 16 mA. Determine the Q-point values of I_D and V_{DS} for the circuit.

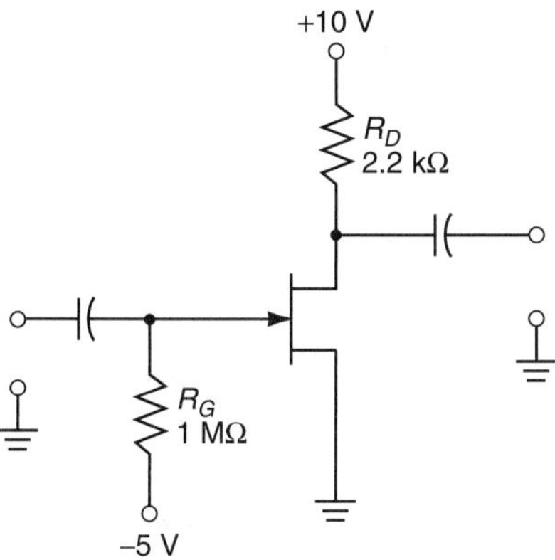

Figure 21.11

Solution: Since there is no current through the gate circuit, none of V_{GG} is dropped across the gate resistor, and

$$V_{GS} = V_{GG} = -5 \text{ V}$$

Using $V_{GS} = -5$ V and the device ratings,

$$I_D = I_{DSS}\left(1 - \frac{V_{GS}}{V_{GS(\text{off})}}\right)^2 = (16 \text{ mA})\left(1 - \frac{-5 \text{ V}}{-8 \text{ V}}\right)^2 = 2.25 \text{ mA}$$

and

$$V_{DS} = V_{DD} - I_D R_D = 10 \text{ V} - (2.25 \text{ mA})(2.2 \text{ k}\Omega) = 5.05 \text{ V}$$

Practice Problem 21.3

Assume that the JFET shown in Figure 21.11 has values of $V_{GS(\text{off})} = -10$ V and $I_{DSS} = 12$ mA. Determine the Q-point values of I_D and V_{DS} for the circuit.

Example 21.3 assumed only one value for I_{DSS} and $V_{GS(\text{off})}$. The fact that a given JFET can have a range of values for $V_{GS(\text{off})}$ and I_{DSS} leads us to a major problem with the gate-bias circuit: *Gate bias does not provide a stable Q-point from one JFET to another.* This problem is illustrated in Example 21.4.

Example 21.4

Determine the range of Q-point values for the circuit shown in Figure 21.12. Assume the JFET has ranges of $V_{GS(\text{off})} = -1$ to -7 V and $I_{DSS} =$ 2 to 9 mA.

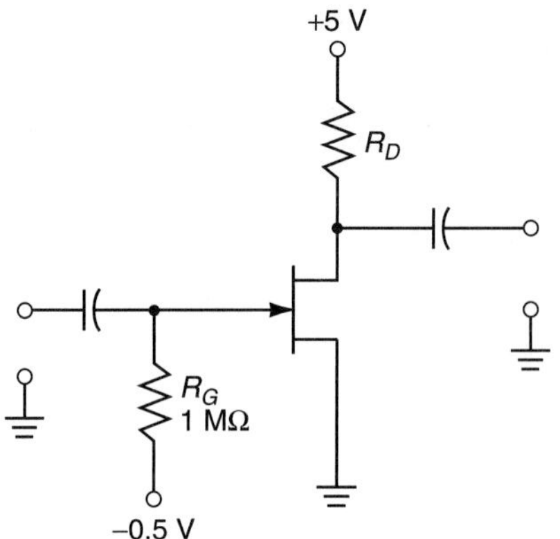

Figure 21.12

A Practical Consideration: When only a maximum curve can be plotted, the Q-point can fall anywhere between the points where the bias line intersects the curve and either axis of the graph.

Solution: Using the established procedure, the two transconductance curves for the JFET are plotted as shown in Figure 21.13. Locating the point representing $V_{GS} = -0.5$ V on the x-axis of the graph, a vertical line, called a **bias line,** is drawn from this point through the two transconductance curves as shown. The points where the line intersects the curves represent the maximum and minimum Q-points. The line between these two points represents all the possible Q-points for the circuit shown in Figure 21.12.

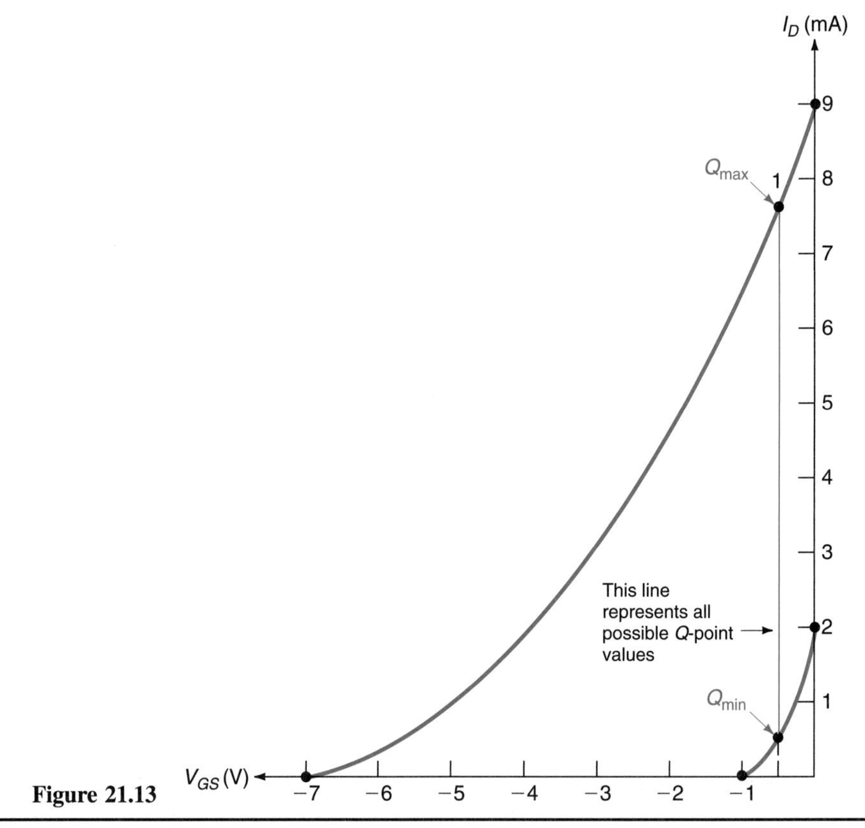

Figure 21.13

Figure 21.13 helps to illustrate the primary drawback of gate bias. Even with V_{GS} fixed at −0.5 V, the graph indicates that the Q-point value of I_D can fall anywhere between approximately 0.5 mA and 7.8 mA. Because of this Q-point instability, gate bias is used primarily in switching applications.

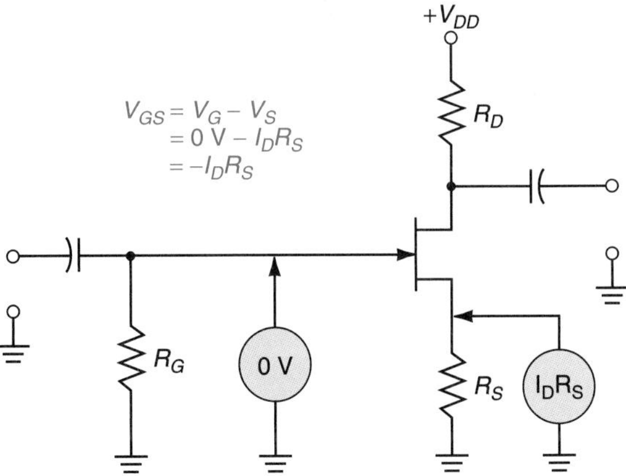

Figure 21.14 Self-bias.

Self-Bias

Self-bias is *a type of JFET that uses a source resistor to help reverse bias the JFET gate.* A self-bias circuit is shown in Figure 21.14. Note that the gate is returned to ground via R_G, and a resistor has been added to the source circuit (R_S). Figure 21.14 helps to explain the role that the source resistor plays in reverse biasing the gate.

In any JFET circuit, $I_D = I_S$ because there is no current through the gate. Therefore,

$$V_S = I_DR_S \tag{21.4}$$

Since there is no gate current, there is no voltage across R_G, and the self-bias circuit has a value of

$$V_G = 0\text{ V} \tag{21.5}$$

These relationships show that the source terminal is at some positive potential (due to the voltage across R_S) and that the gate is at ground. As a result, the necessary *negative* value of V_{GS} is established as follows:

$$V_{GS} = V_G - V_S = 0\text{ V} - I_DR_S$$

or

$$V_{GS} = -I_DR_S \tag{21.6}$$

If equation (21.6) seems confusing, just remember that there is no rule stating that the voltage at the gate terminal must be negative, only that the gate terminal must be more negative than the source terminal. Figure 21.15 helps to illustrate this point. In

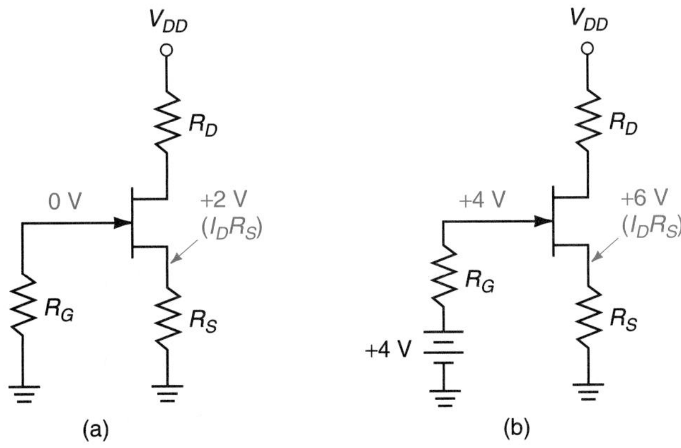

Figure 21.15 Two circuits with $V_{GS} = -2$ V.

each circuit shown, the gate is 2 V more negative than the source. Therefore, each has a value of $V_{GS} = -2$ V.

Equation (21.6) is the *bias line equation* for the self-bias circuit. To plot the dc bias line for a self-bias circuit:

1. Plot the minimum and maximum transconductance curves for the JFET used in the circuit.
2. Choose any value of V_{GS}, and determine the corresponding value of I_D using

$$I_D = \frac{-V_{GS}}{R_S} \tag{21.7}$$

3. Plot the point determined by equation (21.7), and draw a line from this point to the graph origin (the [0, 0] point).
4. The points where the bias line crosses the two transconductance curves define the limits of the Q-point operation of the circuit.

This procedure is demonstrated in Example 21.5.

Example 21.5

Determine the range of Q-point values for the circuit shown in Figure 21.16.

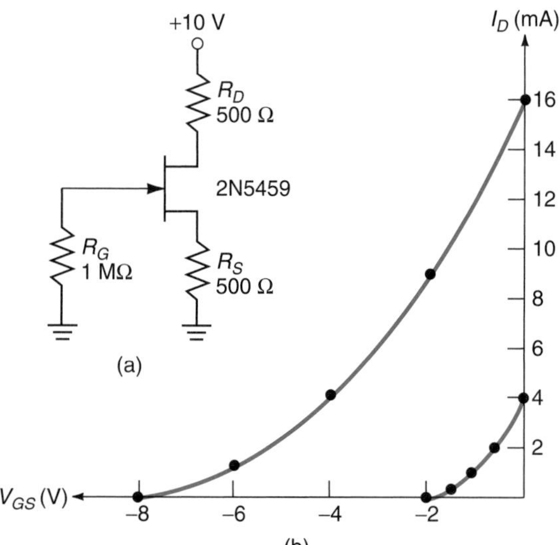

Figure 21.16

Solution: The 2N5459 spec sheet lists the following values for $V_{GS(\text{off})}$ and I_{DSS}:

$$V_{GS(\text{off})} = -2 \text{ V to } -8 \text{ V} \qquad I_{DSS} = 4 \text{ mA to } 16 \text{ mA}$$

Using these two sets of values, the transconductance curves are plotted as shown in Figure 21.16b. The value of $V_{GS} = -4$ V is chosen at random to calculate one possible value of I_D as follows:

$$I_D = \frac{-V_{GS}}{R_S} = \frac{4 \text{ V}}{500 \ \Omega} = 8 \text{ mA}$$

This point (−4, 8) is plotted on the graph, and a line is drawn from the point to the graph origin. The plot of the line is shown in Figure 21.17. The points where the bias line intersects the transconductance curves are labeled and used to obtain the maximum and minimum Q-point values listed in Figure 21.17.

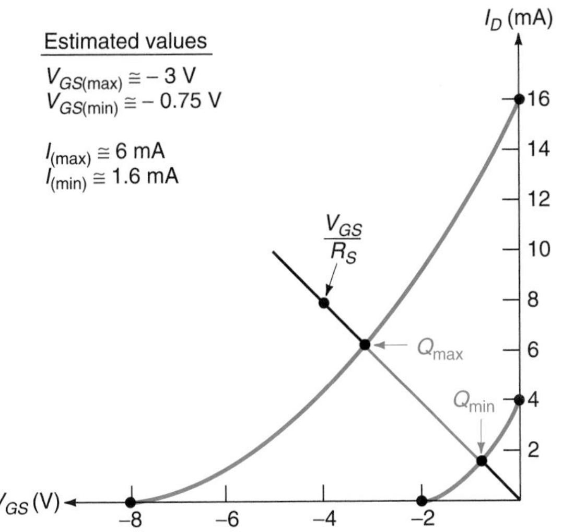

Figure 21.17

Practice Problem 21.5

A JFET with parameters of $V_{GS(off)} = -5$ to -10 V and $I_{DSS} = 5$ to 10 mA is used in a self-bias circuit with a value of $R_S = 2$ kΩ. Plot the transconductance curves and bias line for the circuit.

A Practical Consideration: Q-point instability is most evident when replacing one JFET with another. For example, if you were to replace the 2N5459 in Figure 21.16 with another, the Q-point value of I_D could shift to any point within the range shown in Figure 21.17.

As Example 21.5 demonstrates, self-bias produces a more stable Q-point value of I_D than gate bias. However, the variations in the Q-point value of I_D still leave a lot to be desired (in terms of circuit stability). As you will see, *voltage-divider bias* provides a far more stable Q-point than either self-bias or gate bias.

Voltage-Divider Bias

The voltage-divider biased JFET amplifier is very similar to its BJT counterpart. This biasing circuit is shown in Figure 21.18. The gate voltage is established by the voltage-divider as follows:

$$V_G = V_{DD} \frac{R_2}{R_1 + R_2} \tag{21.8}$$

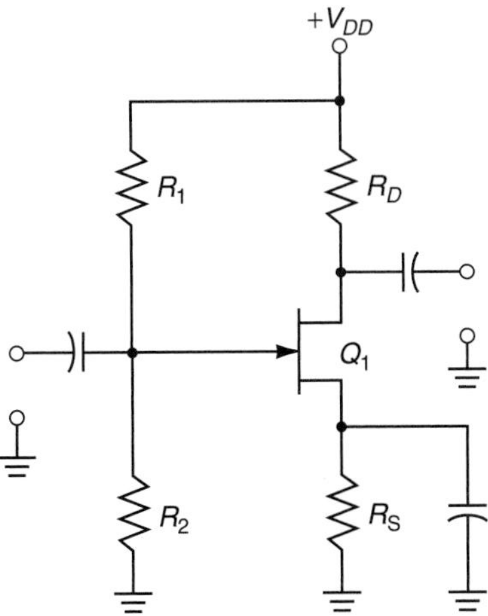

Figure 21.18 Voltage-divider bias.

To solve for V_{GS}, we must first plot the dc bias line for the circuit. The method used to plot the dc bias line for the voltage-divider bias circuit is a bit strange—easy, but strange. The exact process is as follows:

1. As usual, plot the transconductance curves for the specific JFET.
2. Calculate the value of V_G using equation (21.8).
3. *On the positive x-axis of the graph,* plot a point at the value of V_G.
4. Solve for I_D at $V_{GS} = 0$ V using

$$I_D = \frac{V_G}{R_S}$$

(*Note:* $V_S = V_G$ when $V_{GS} = 0$ V. Therefore, this equation gives us the value of I_D at $V_{GS} = 0$ V.)

5. Locate the point on the *y*-axis that corresponds to the value found in step 4.
6. Draw a line from the V_G point through the point plotted on the *y*-axis and both transconductance curves. The curve intersection points represent the *Q*-point limits of the circuit.

Example 21.6 demonstrates this procedure.

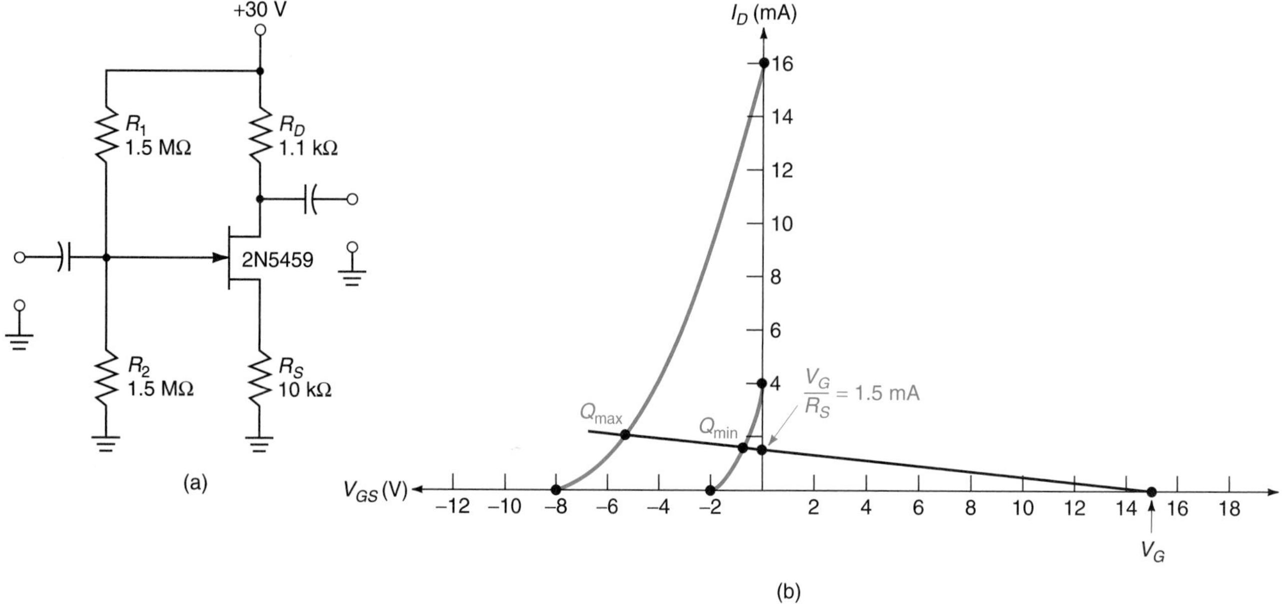

Figure 21.19

Example 21.6

Plot the dc bias line for the circuit shown in Figure 21.19a.

Solution: Figure 21.19a contains the 2N5459, the JFET whose transconductance curves were plotted in Example 21.5. These curves can be seen in Figure 21.19b. The value of V_G for the circuit is calculated as

$$V_G = V_{DD}\frac{R_2}{R_1 + R_2} = (30 \text{ V})\frac{1.5 \text{ M}\Omega}{3 \text{ M}\Omega} = 15 \text{ V}$$

This point is now plotted on the positive *x*-axis, as shown in Figure 21.19b. The value of V_G is also used to find the *y*-axis intersect point, as follows:

$$I_D = \frac{V_G}{R_S} = \frac{15 \text{ V}}{10 \text{ k}\Omega} = 1.5 \text{ mA}$$

A line is now drawn through the two points and the two curves to establish the Q-points shown.

Practice Problem 21.6

Assume that the circuit shown in Figure 21.19a has been changed so that V_{DD} = 36 V, R_1 = 10 MΩ, R_2 = 3.3 MΩ, R_D = 1.8 kΩ, and R_S = 3 kΩ. Plot the dc bias line for the amplifier.

The bias line shown in Figure 21.19 indicates that the circuit provides a highly stable Q-point value of I_D. A close-up view of the bias line (see Figure 21.20) shows that V_{GS} varies between approximately −5 and −0.5 V, but that I_D varies between 2 mA and approximately 1.5 mA. Thus, voltage-divider bias provides a much more stable Q-point than either gate bias or self-bias.

Just as we are interested in knowing the Q-point value of V_{CE} for a BJT biasing circuit, we are also interested in knowing the value of V_{DS} for any JFET biasing circuit. For the voltage-divider bias circuit, V_{DS} is found as

$$V_{DS} = V_{DD} - I_D(R_D + R_S) \tag{21.9}$$

Even though voltage-divider bias gives us a stable value of I_D, the same cannot be said for V_{DS}. For example, the range of V_{DS} for the circuit in Example 21.6 can be found as follows:

When I_D = 2 mA:

$$V_{DS} = V_{DD} - I_D(R_D + R_S) = 30\text{ V} - (2\text{ mA})(11.1\text{ k}\Omega) = 7.8\text{ V}$$

When I_D = 1.55 mA:

$$V_{DS} = V_{DD} - I_D(R_D + R_S) = 30\text{ V} - (1.6\text{ mA})(11.1\text{ k}\Omega) = 12.24\text{ V}$$

Summary

Gate bias, self-bias, and voltage-divider bias are the three most commonly used JFET biasing circuits. Each of these circuits has its own type of dc bias line, advantages, and

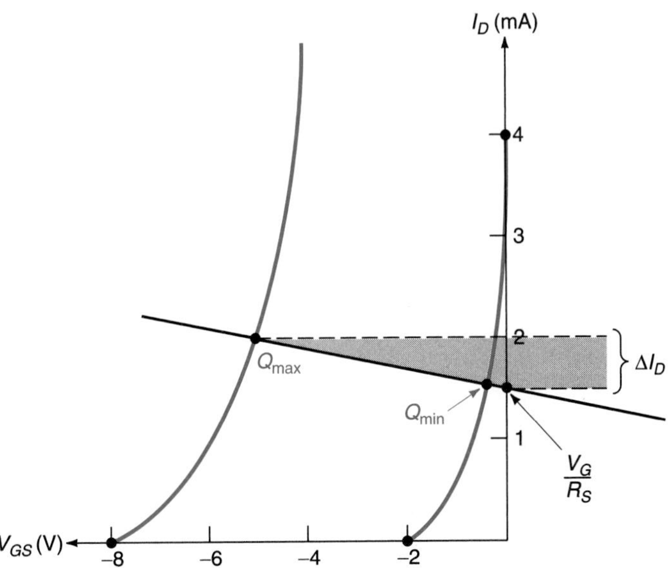

Figure 20.20 A close-up view of the bias line shown in Figure 21.19.

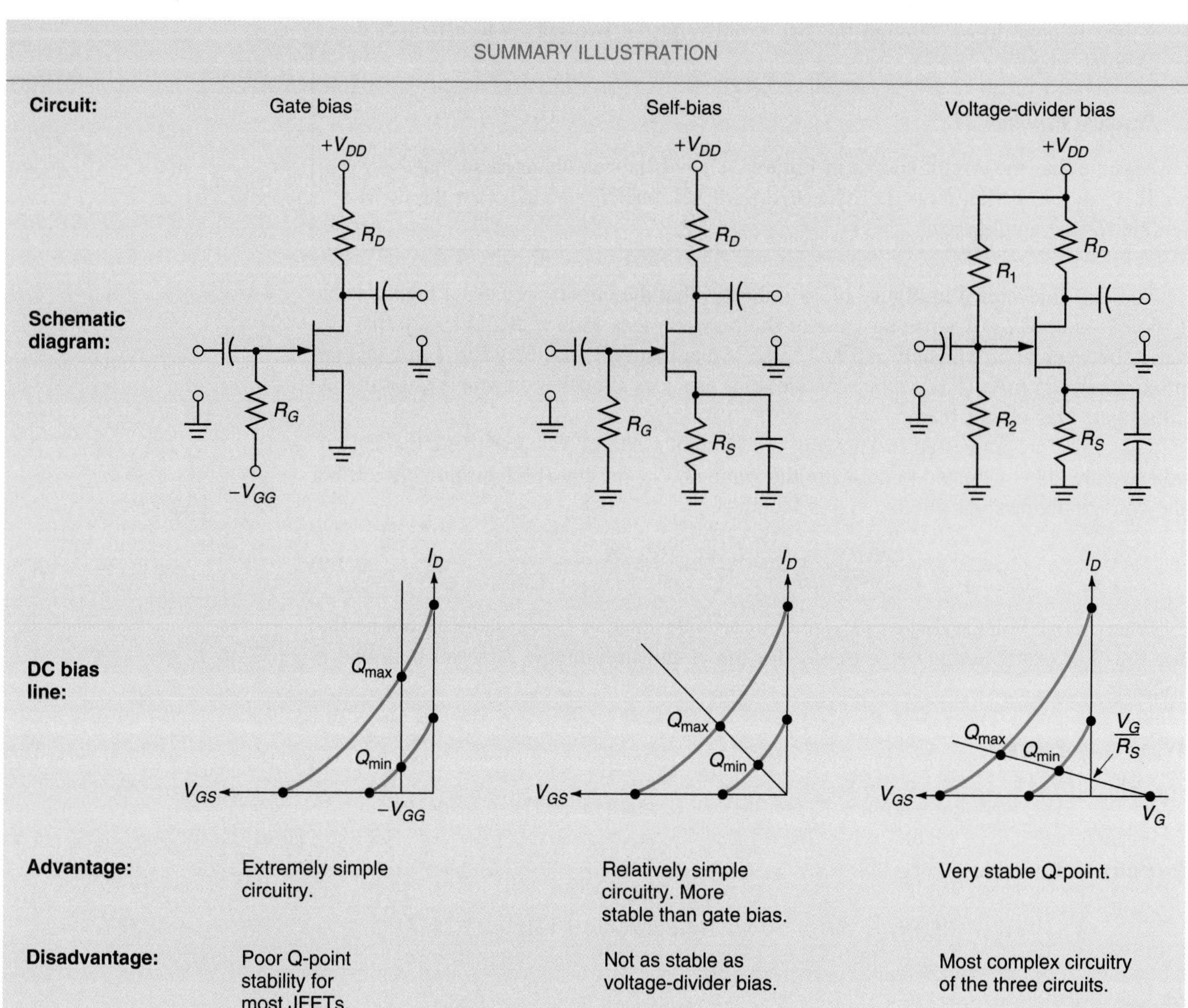

Figure 21.21

disadvantages. The characteristics of these three dc biasing circuits are summarized in Figure 21.21.

Section Review

1. What is the primary disadvantage of gate bias?
2. How does *self-bias* produce a negative value of V_{GS}?
3. What is the process used to plot the dc bias line for a self-bias circuit?
4. What is the process used to plot the dc bias line for the voltage-divider bias circuit?

Critical Thinking

5. Designing for $I_D \cong 0$ A is common for switching circuits. What change would you make to the circuit in Example 21.5 to ensure that $I_D \cong 0$ A for any 2N5458 used in the circuit?
6. Which BJT biasing circuit could be viewed as being the counterpart of a self-bias circuit? Explain your answer.

21.3 Common-Source Amplifiers

In many ways, the ac operation of a JFET amplifier is similar to that of a BJT amplifier. However, there are differences that warrant discussion, such as the effects of changes in JFET transconductance. In this section, we will discuss the operation of the **common-source (CS) amplifier,** which is *the JFET counterpart of the BJT common-emitter amplifier.*

Operation Overview

In general terms, a common-source amplifier has several of the characteristics of a common-emitter amplifier. For example, the common-source amplifier shown in Figure 21.22 provides voltage gain for its input signal and a 180° signal voltage phase shift from input to output. At the same time, there are significant differences in the operating principles of the common-source and common-emitter amplifiers. ◄ *OBJECTIVE 6*

The small-signal operation of the common-source amplifier can be explained with the help of Figure 21.23. The initial dc operating values are represented as straight-line values in the blocks. Here's what happens when the input signal rises to its positive peak value ($+2\ V_{pk}$):

1. V_G increases to $+10$ V.
2. V_{GS} decreases to -2 V ($V_{GS} = V_G - V_S = 10\text{ V} - 12\text{ V} = -2\text{ V}$).
3. The decrease in V_{GS} causes an increase in I_D (as shown in the transconductance curve).
4. The increase in I_D causes V_D to decrease ($V_D = V_{DD} - I_DR_D$).

Thus, when the input voltage increases, the output voltage decreases. When the input signal decreases to its negative peak value ($-2\ V_{pk}$):

1. V_G decreases to $+6$ V.
2. V_{GS} increases to -6 V ($V_{GS} = V_G - V_S = 6\text{ V} - 12\text{ V} = -6\text{ V}$).
3. The increase in V_{GS} causes a decrease in I_D.
4. The decrease in I_D causes V_D to increase.

As you can see, the small-signal operation of the amplifier is based on changing values of V_{GS} and I_D. When you start dealing with *changing values* of V_{GS} and I_D, you get into the area of transconductance.

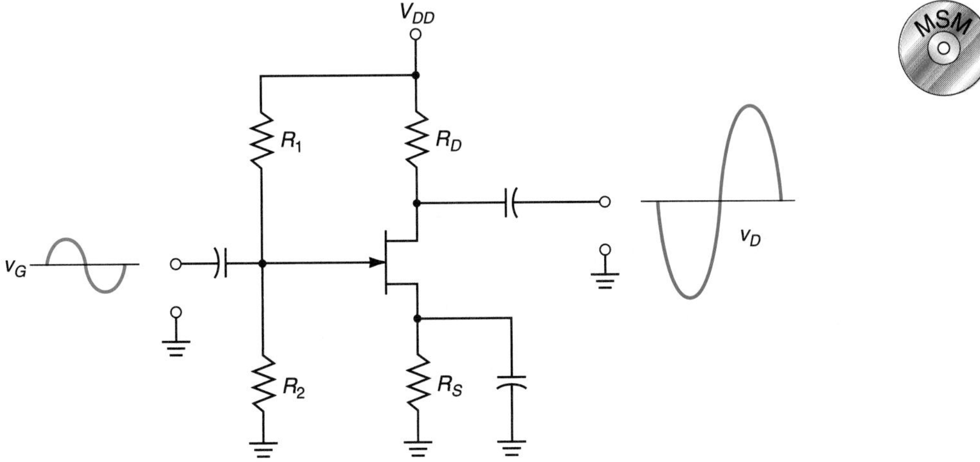

Figure 21.22

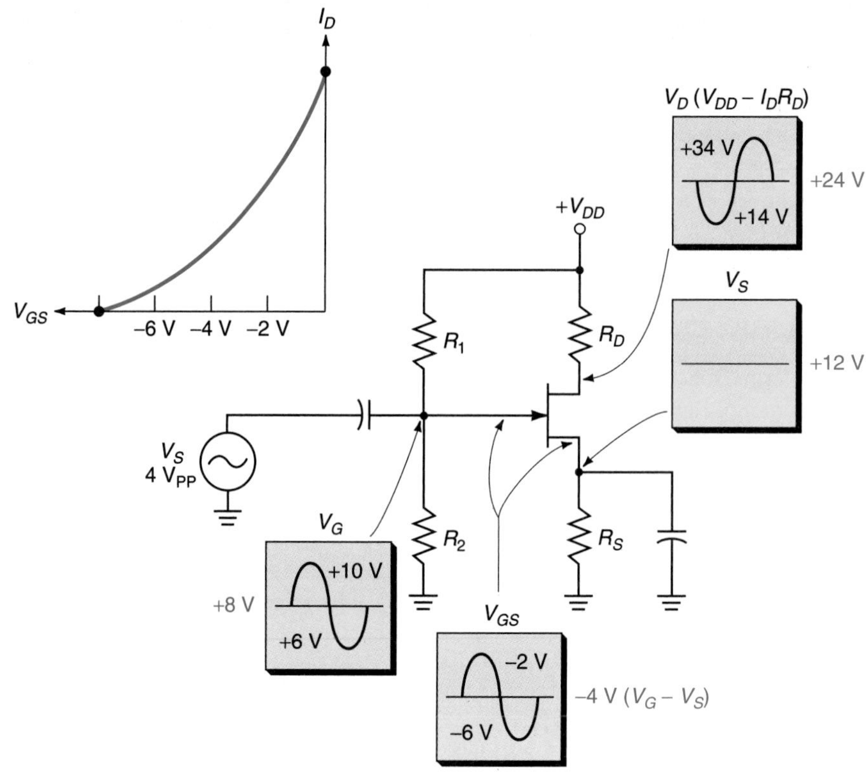

Figure 21.23 Common-source amplifier operation.

Transconductance

We have been plotting transconductance curves throughout this chapter as graphs of V_{GS} versus I_D. **Transconductance** is actually *a ratio of a change in drain current to a change in gate-source voltage.* By formula,

$$g_m = \frac{\Delta I_D}{\Delta V_{GS}} \tag{21.10}$$

where

g_m = the transconductance of the JFET at a given value of V_{GS}
ΔI_D = the change in drain current
ΔV_{GS} = the change in gate-source voltage

As a rating, transconductance is measured in microsiemens (μS) or micromhos (μmhos). You may recall that conductance is measured in siemens or mhos.

The transconductance of a JFET varies from one point on the transconductance curve to another. For example, refer to the 2N5459 transconductance curve shown in Figure 21.24. Two sets of points are selected on the curve, each marking a ΔV_{GS} of 1 V. For each ΔV_{GS}, the corresponding change in drain current (ΔI_D) is shown. Using equation (21.10), we can calculate the value of g_m that corresponds to each change in V_{GS}, as follows:

$$g_m = \frac{\Delta I_D}{\Delta V_{GS}} = \frac{1.25 \text{ mA}}{1 \text{ V}} = 1250\ \mu\text{S} \quad \text{and} \quad g_m = \frac{\Delta I_D}{\Delta V_{GS}} = \frac{2.75 \text{ mA}}{1 \text{ V}} = 2750\ \mu\text{S}$$

As you can see, the value of g_m is not constant across the transconductance curve. The value of g_m at a specified value of V_{GS} can be found using

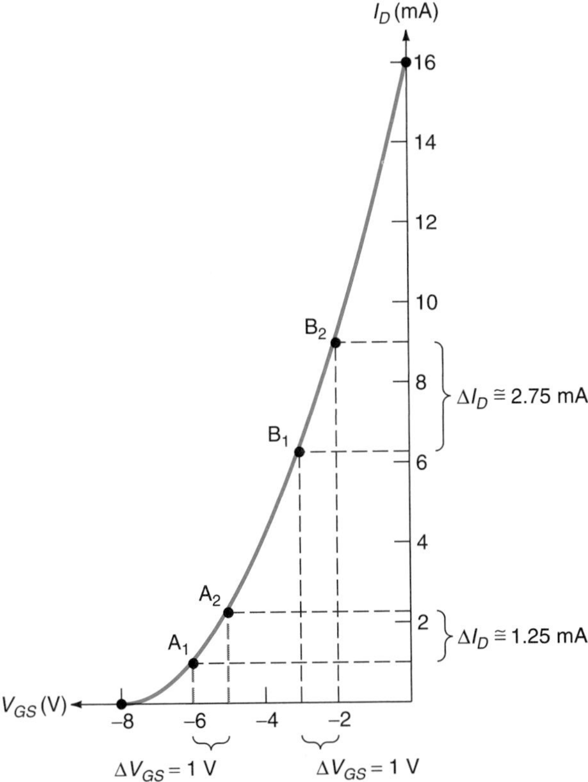

Figure 21.24 The 2N5459 maximum-ratings transconductance curve.

$$g_m = g_{m0}\left(1 - \frac{V_{GS}}{V_{GS(\text{off})}}\right) \tag{21.11}$$

where

g_m = the value of transconductance at the specific value of V_{GS}

g_{m0} = the maximum value of g_m, measured at $V_{GS} = 0$ V

Example 21.7 demonstrates the use of this relationship.

Example 21.7

The 2N5459 has the following maximum ratings: $g_{m0} = 6000\ \mu S$ and $V_{GS(\text{off})} = -8$ V. Determine the values of g_m for this device at $V_{GS} = -3$ V and $V_{GS} = -5$ V.

Solution: At $V_{GS} = -3$ V,

$$g_m = g_{m0}\left(1 - \frac{V_{GS}}{V_{GS(\text{off})}}\right) = (6000\ \mu\text{S})\left(1 - \frac{-3\text{ V}}{-8\text{ V}}\right) = 3750\ \mu\text{S}$$

At $V_{GS} = -5$ V,

$$g_m = g_{m0}\left(1 - \frac{V_{GS}}{V_{GS(\text{off})}}\right) = (6000\ \mu\text{S})\left(1 - \frac{-5\text{ V}}{-8\text{ V}}\right) = 2250\ \mu\text{S}$$

Practice Problem 21.7

The 2N5486 has maximum values of $g_{m0} = 8000\ \mu S$ and $V_{GS(\text{off})} = -6$ V. Using these values, determine the values of g_m at $V_{GS} = -2$ V and $V_{GS} = -4$ V.

Figure 21.25
A common-source amplifier and its ac equivalent circuit.

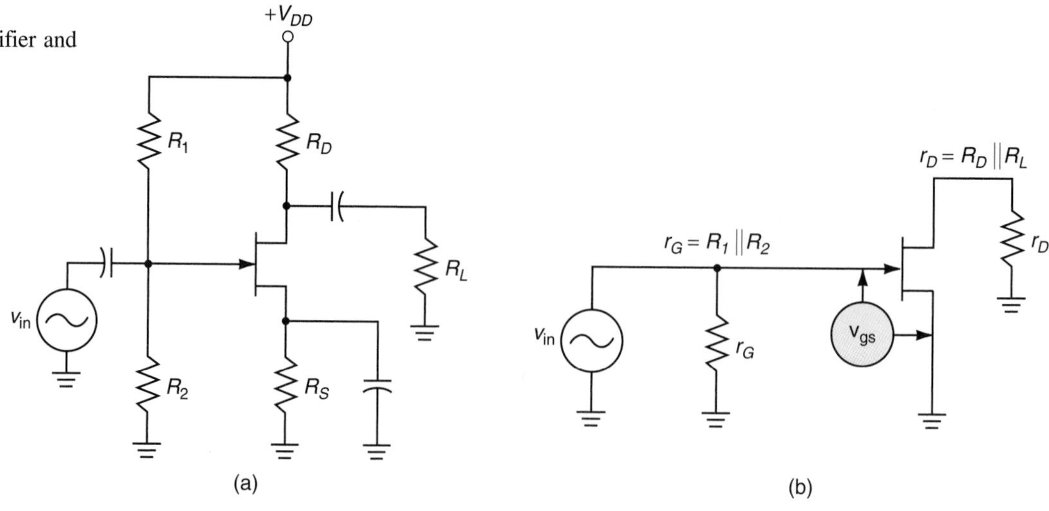

Amplifier Voltage Gain

The ac operation of a JFET amplifier is closely related to the value of g_m at the biasing value of V_{GS}. Figure 21.25 shows a basic common-source amplifier and its ac equivalent. Note that the biasing resistors have been replaced in Figure 21.25b with a parallel equivalent resistor, r_G. Also, R_D and R_L have been replaced by their parallel equivalent resistor, r_D.

The voltage gain of the common-source amplifier is found as

$$A_v = g_m r_D \tag{21.12}$$

As you know, the Q-point stability can be a problem with JFET amplifiers. Equation (21.12) points out another problem. The value of A_v depends on the value of g_m. In turn, the value of g_m depends on V_{GS}, which can vary from one JFET to another. This leads to the conclusion that the voltage gain of a common-source amplifier can also be relatively unstable, as illustrated in Example 21.8.

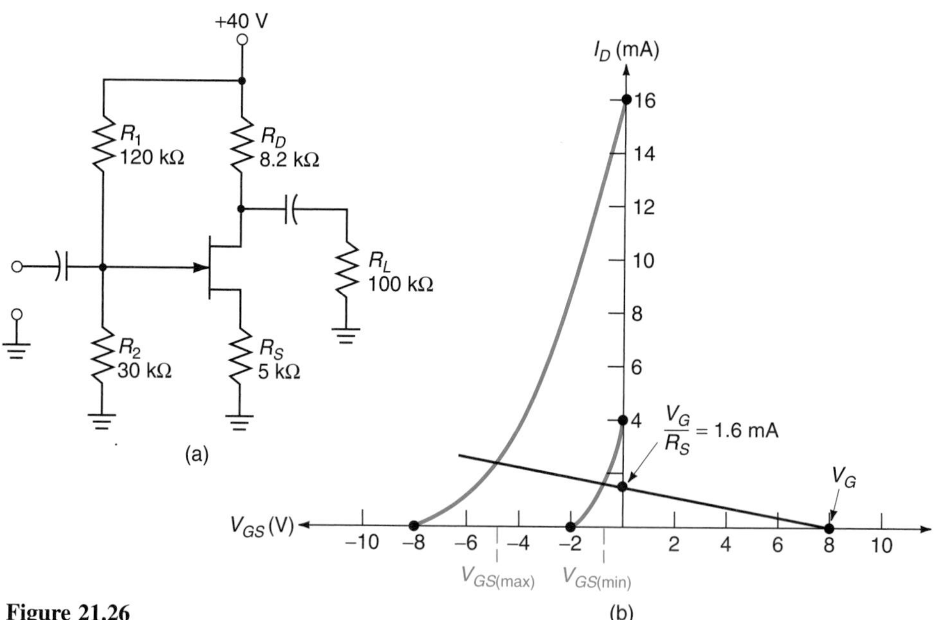

Figure 21.26

Example 21.8

Determine the maximum and minimum values of A_v for the amplifier shown in Figure 21.26. Assume that the 2N5459 has values of g_{m0} = 6000 µS at $V_{GS(\text{off})}$ = −8 V and of g_{m0} = 2000 µS at $V_{GS(\text{off})}$ = −2 V.

Solution: The transconductance curves for the 2N5459 shown in Figure 21.26b were plotted in Example 21.5. The bias line was plotted using the procedure demonstrated in Example 21.6. As shown, the maximum and minimum values of V_{GS} are approximately -5 V and -1 V, respectively. Using the *maximum* values of $V_{GS(\text{off})}$, V_{GS}, and g_{m0},

$$g_m = g_{m0}\left(1 - \frac{V_{GS}}{V_{GS(\text{off})}}\right) = (6000\ \mu\text{S})\left(1 - \frac{-5\text{ V}}{-8\text{ V}}\right) = 2250\ \mu\text{S}$$

Using the *minimum* values of $V_{GS(\text{off})}$, V_{GS}, and g_{m0},

$$g_m = g_{m0}\left(1 - \frac{V_{GS}}{V_{GS(\text{off})}}\right) = (2000\ \mu\text{S})\left(1 - \frac{-1\text{ V}}{-2\text{ V}}\right) = 1000\ \mu\text{S}$$

The value of r_D is found as

$$r_D = R_D \| R_L = 7.58\text{ k}\Omega$$

Using the maximum value of g_m, we get

$$A_v = g_m r_D = (2250\ \mu\text{S})(7.58\text{ k}\Omega) = 17.06 \qquad \text{(maximum)}$$

Using the minimum value of g_m, we get

$$A_v = g_m r_D = (1000\ \mu\text{S})(7.58\text{ k}\Omega) = 7.58 \qquad \text{(minimum)}$$

Practice Problem 21.8

A JFET has the following ratings: $g_{m0} = 4000\ \mu$S at $V_{GS(\text{off})} = -6$ V and $g_{m0} = 2000\ \mu$S at $V_{GS(\text{off})} = -2$ V. The component is used in a common-source amplifier with values of $R_D = 3.3$ kΩ, $R_L = 4.7$ kΩ, and $V_{GS} = -750$ mV to -3.5 V. Calculate the range of A_v for the circuit.

JFET Swamping

Just as a BJT amplifier can be *swamped* to reduce the effects of variations in r'_e, a JFET amplifier can be swamped to reduce the effects of variations in g_m. A swamped JFET is shown in Figure 21.27a. As you can see, the source resistance is only partially

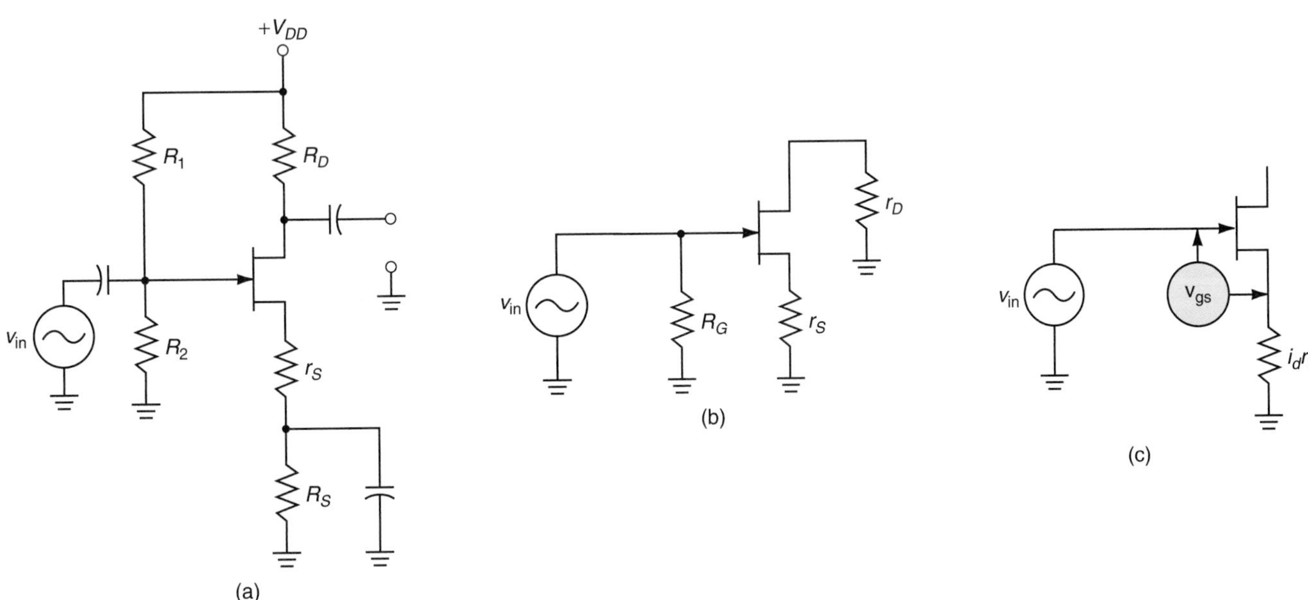

Figure 21.27 A swamped common-source amplifier.

bypassed (just as the emitter resistance is partially bypassed in the swamped BJT amplifier).

The voltage gain of the swamped JFET amplifier is found as

$$A_v = \frac{r_D}{r_S + \left(\frac{1}{g_m}\right)} \tag{21.13}$$

As long as $r_S >> \frac{1}{g_m}$ (which is generally the case), the resistor swamps out the effects of variations in g_m. This point is illustrated in Example 21.9.

Example 21.9

Determine the maximum and minimum gain values for the circuit shown in Figure 21.28.

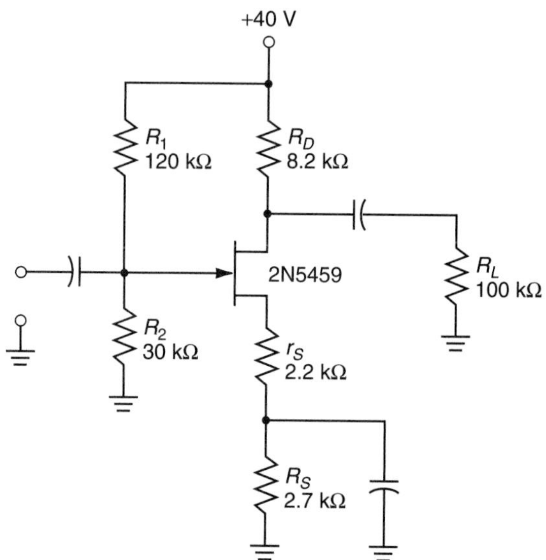

Figure 21.28

Solution: The circuit shown in Figure 21.28 is nearly identical to the one shown in Figure 21.26. The difference of 100 Ω in the source circuit is not enough to change significantly the values obtained in Example 21.8, so we will use those values here.

Using the values from Figure 21.28 and Example 21.8, the maximum value of A_v is found as

$$A_v = \frac{r_D}{r_S + \left(\frac{1}{g_m}\right)} = \frac{7.58\text{ k}\Omega}{2.2\text{ k}\Omega + 444\ \Omega} = 2.87$$

and the minimum value of A_v is found as

$$A_v = \frac{r_D}{r_S + \left(\frac{1}{g_m}\right)} = \frac{7.58\text{ k}\Omega}{2.2\text{ k}\Omega + 1\text{ k}\Omega} = 2.37$$

Practice Problem 21.9

A swamped JFET amplifier has values of $r_D = 6.6$ kΩ, $r_S = 1.5$ kΩ, and $g_m =$ 2000 to 4000 μS. Determine the range of A_v values for the circuit.

As you can see, swamping improves the stability of A_v. However, there is one drawback. As with swamped BJT amplifiers, swamping results in reduced voltage gain. This can be seen by comparing the values of A_v for Examples 21.8 and 21.9.

Amplifier Input Impedance

Due to the extremely high input impedance of the JFET, the overall input impedance to a common source amplifier is higher than that of a similar common emitter amplifier. For the gate-bias and self-bias circuits, the amplifier input impedance is found as

$$Z_{in} \cong R_G \qquad (21.14)$$

For the voltage-divider biased amplifier, Z_{in} is found as

$$Z_{in} \cong R_1 \| R_2 \qquad (21.15)$$

In a voltage-divider biased BJT amplifier, the bias resistors are limited to relatively low values to minimize the loading effects of the transistor base. This is not the case with JFET amplifiers. Since there is no gate current, there is no loading on the voltage divider in the gate circuit, allowing high-value biasing resistors to be used. As a result, a common-source amplifier typically has much greater input impedance than a comparable common-emitter amplifier.

Section Review

1. Explain the common-source amplifier circuit response to an input signal.
2. What is *transconductance*?
3. How does the value of g_m relate to the position of a circuit Q-point on the transconductance curve?
4. What is the relationship between the values of A_v and V_{GS} for a common-source amplifier?
5. What purpose is served by *swamping* a JFET amplifier?
6. How does JFET amplifier input impedance compare to that of a comparable BJT amplifier? Explain.

21.4 Common-Drain and Common-Gate Amplifiers

OBJECTIVE 7

The *common-drain* and *common-gate configurations* are the JFET counterparts of the common-collector and common-base BJT amplifiers. In this section, we will take a brief look at each of these amplifiers.

The Common-Drain Amplifier (Source Follower)

The **common-drain amplifier** is *an amplifier that accepts an input signal at its gate and provides an output signal at its source.* The circuit input and output signals are in phase and nearly equal in amplitude. For this reason, the common-drain amplifier is referred to as a **source follower.** A source follower is shown in Figure 21.29. The characteristics of the circuit are summarized as follows:

Input impedance:	High
Output impedance:	Low
Gain:	$A_v < 1$

The voltage gain (A_v) of the source follower is found using

$$A_v = \frac{r_S}{r_S + \left(\frac{1}{g_m}\right)} \qquad (21.16)$$

where

$r_S = R_S \| R_L$

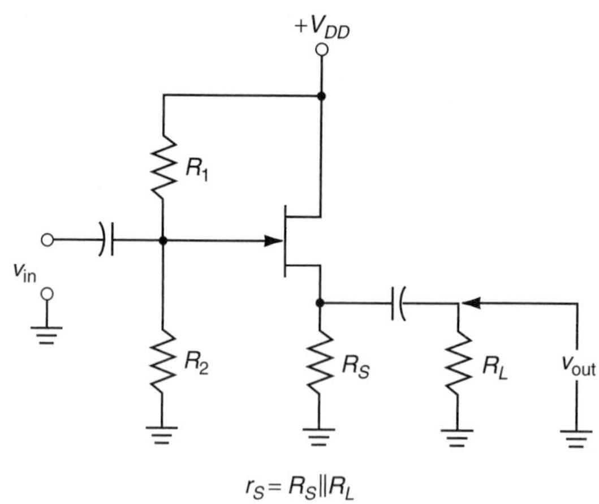

Figure 21.29 The source follower.

The basis for equation (21.16) is easy to see if you compare the source follower to the swamped common-source amplifier shown in Figure 21.28. Both circuits have some value of unbypassed source resistance. Thus, the voltage gains of the two circuits are calculated in the same basic fashion. The primary difference is that the output voltage from the source follower appears across r_S, so this value is used in place of r_D in the swamped amplifier equation. Since r_S appears in both the numerator and the denominator of equation (21.16), source follower voltage gain is always less than one.

As with the common-source amplifier, the total input impedance to the source follower equals the parallel equivalent resistance of the input circuit. When voltage-divider bias is used, Z_{in} is found as

$$Z_{in} = R_1 \| R_2$$

When gate bias or self-bias is used,

$$Z_{in} \cong R_G$$

Don't Forget: Output impedance is measured with the load disconnected. Thus, r_S is replaced by R_S in Figure 21.30 and equation (21.17).

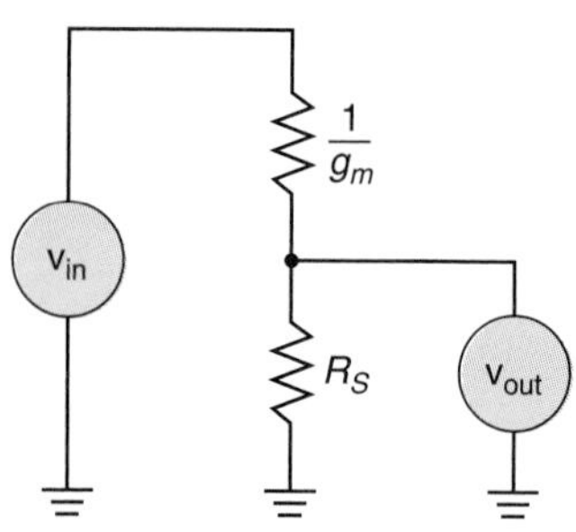

Figure 21.30

Again, the high gate input impedance eliminates it from any Z_{in} calculations.

The output impedance of the amplifier is determined with a little help from equation (21.16). If you look closely at this equation, you will notice that the source circuit is essentially a voltage divider. As a result, the circuit can be represented as shown in Figure 21.30. This is a resistive equivalent of the source-follower output circuit. To the load, the circuit appears as two resistors that are connected in parallel. Thus, the output impedance of the amplifier is given as

$$Z_{out} = R_S \| \frac{1}{g_m} \qquad (21.17)$$

So, how does equation (21.17) show that the value of Z_{out} is always low? The normal range of g_m is 1000 μS or greater. For $g_m = 1000$ μS,

$$\frac{1}{g_m} = \frac{1}{0.001} = 1\text{ k}\Omega$$

Since we used the *lowest* typical value of g_m, $\frac{1}{g_m}$ is typically lower than 1 kΩ. This means that Z_{out} is also typically less than 1 kΩ. Example 21.10 demonstrates the analysis of a source-follower circuit.

Example 21.10

Determine the maximum and minimum values of A_v and Z_{out} for the circuit shown in Figure 21.31. Also, determine the value of Z_{in} for the circuit.

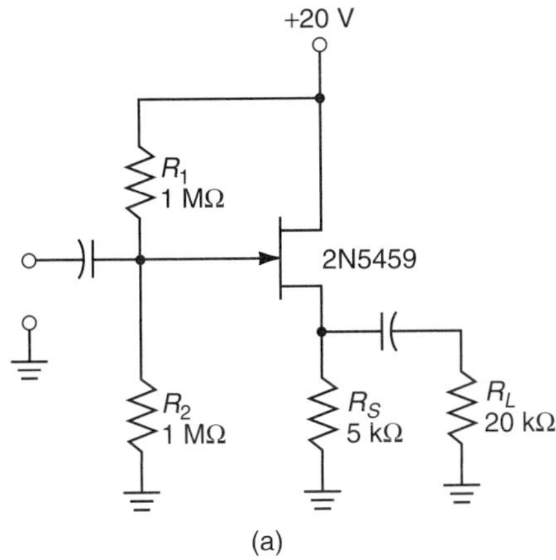

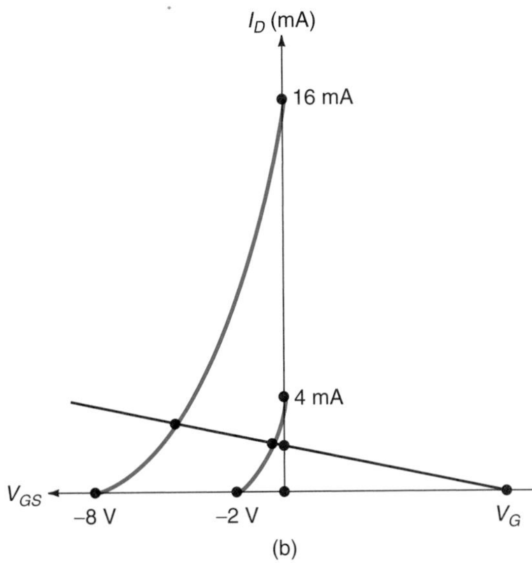

Figure 21.31

Solution: Using the dc bias line shown in Figure 21.31b, we determine the minimum and maximum values of V_{GS} to be approximately -0.5 V and -5 V, respectively. The 2N5459 has values stated on the spec sheet of $g_{m0} = 6000\ \mu\text{S}$ at $V_{GS(\text{off})} = -8$ V and $g_{m0} = 2000\ \mu\text{S}$ at $V_{GS(\text{off})} = -2$ V. These values are used with our minimum and maximum values of V_{GS} to determine the minimum and maximum values of g_m, as follows:

$$g_m = g_{m0}\left(1 - \frac{V_{GS}}{V_{GS(\text{off})}}\right) = (6000\ \mu\text{S})\left(1 - \frac{-5\text{ V}}{-8\text{ V}}\right) = 2250\ \mu\text{S} \qquad \text{(maximum)}$$

and

$$g_m = g_{m0}\left(1 - \frac{V_{GS}}{V_{GS(\text{off})}}\right) = (2000\ \mu\text{S})\left(1 - \frac{-0.5\text{ V}}{-2\text{ V}}\right) = 1500\ \mu\text{S} \qquad \text{(minimum)}$$

The value of r_S is now found as

$$r_S = R_S \| R_L = 5\text{ k}\Omega \| 20\text{ k}\Omega = 4\text{ k}\Omega$$

Using the maximum value of g_m, the maximum value of A_v is found as

$$A_v = \frac{r_S}{r_S + \left(\frac{1}{g_m}\right)} = \frac{4\text{ k}\Omega}{4\text{ k}\Omega + 444\ \Omega} = 0.9001 \qquad \text{(maximum)}$$

Using the minimum value of g_m, the minimum value of A_v is found as

$$A_v = \frac{r_S}{r_S + \left(\frac{1}{g_m}\right)} = \frac{4\text{ k}\Omega}{4\text{ k}\Omega + 667\ \Omega} = 0.8571 \qquad \text{(minimum)}$$

Using the maximum and minimum values of g_m, the corresponding values of Z_{out} are found as

$$Z_{out} = R_S \| \frac{1}{g_m} = 5\text{ k}\Omega \| 667\ \Omega = 588\ \Omega \qquad \text{(maximum)}$$

and

$$Z_{out} = R_S \| \frac{1}{g_m} = 5\text{ k}\Omega \| 444\ \Omega = 408\ \Omega \qquad \text{(minimum)}$$

Finally, Z_{in} is found as

$$Z_{in} = R_1 \| R_2 = 500\text{ k}\Omega$$

Practice Problem 21.10

A JFET has the following ratings: $g_{m0} = 4000\ \mu\text{S}$ at $V_{GS(\text{off})} = -6$ V and $g_{m0} = 2000\ \mu\text{S}$ at $V_{GS(\text{off})} = -1$ V. The component is used in a source follower with values of $R_S = 3.3\text{ k}\Omega$, $R_L = 18\text{ k}\Omega$, and $V_{GS} = -500$ mV to -3.0 V. Calculate the range of A_v and Z_{out} for the circuit.

Example 21.10 serves to demonstrate several things. First, source followers have high input impedance and low output impedance. Even when Z_{out} was at its maximum value, the ratio of Z_{in} to Z_{out} was close to 850:1. This example also demonstrates that there is a relationship among g_m, A_v, and Z_{out}. When g_m is at its maximum value, A_v is at its maximum value and Z_{out} is at its minimum value. The reverse also holds true. Based on these observations, we can state the following relationships for the source follower:

- A_v varies directly with g_m.
- Z_{out} varies inversely with g_m.

As always, the value of Z_{in} is not related to the value of g_m.

The Common-Gate Amplifier

The **common-gate amplifier** *is an amplifier that accepts an input signal at its source terminal and provides an output signal at its drain terminal.* The common-gate amplifier is shown in Figure 21.32. The characteristics of the common-gate circuit are summarized as follows:

Input impedance:	Low
Output impedance:	High (compared to Z_{in})
Gain:	$A_v > 1$

Like the common-source amplifier, the voltage gain of the common-gate amplifier equals the product of its transconductance and ac drain resistance. By formula,

$$A_v = g_m r_D$$

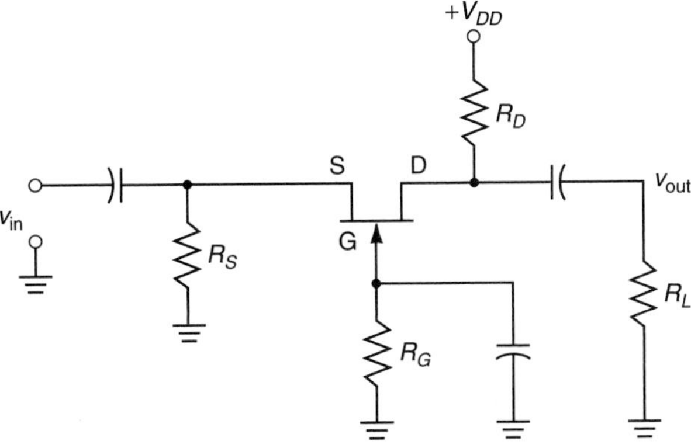

Figure 21.32 A common-gate amplifier.

Since the input signal is applied to its source, the input impedance of the common-gate amplifier is found as

$$Z_{in} = R_S \| \frac{1}{g_m} \tag{21.18}$$

This relationship makes sense, because the input impedance of the common-gate amplifier is measured at the same terminal (the source) as the output impedance of the source follower.

Figure 21.33 shows that the output impedance is made up of the parallel combination of R_D and the resistance of the JFET drain. By formula,

$$Z_{out} = R_D \| r_d \tag{21.19}$$

where

r_d = the resistance of the JFET terminal

The resistance of the JFET drain can be calculated using its *output admittance* (y_{os}) rating, as follows:

$$r_d = \frac{1}{y_{os}} \tag{21.20}$$

The value of r_d is typically much greater than R_D, so the value of Z_{out} can normally be approximated as

$$Z_{out} \cong R_D \tag{21.21}$$

Summary

You have now been shown the gain and impedance characteristics of the common-source, source-follower, and common-gate amplifiers. These three amplifiers and their characteristics are summarized in Figure 21.34.

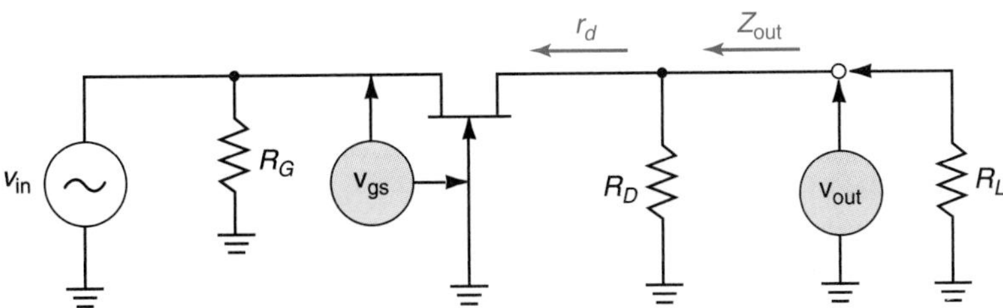

Figure 21.33 The ac equivalent circuit for a common-gate amplifier.

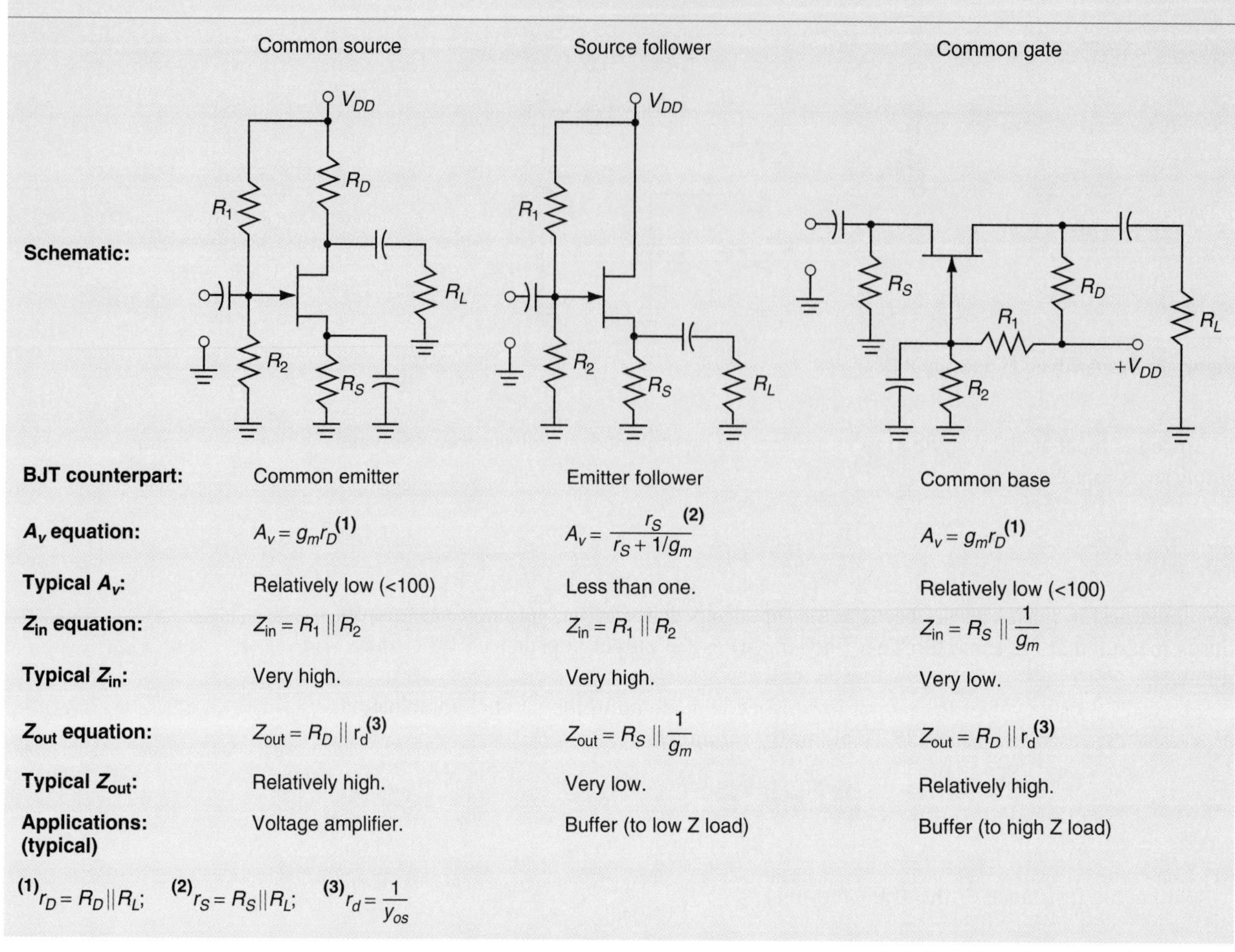

SUMMARY ILLUSTRATION
Common source
Source follower
Common gate
Schematic:
V_{DD}
R_D
R_1
R_L
R_2
R_S
$+V_{DD}$
BJT counterpart: Common emitter | Emitter follower | Common base
A_v equation: $A_v = g_m r_D^{(1)}$ | $A_v = \frac{r_S}{r_S + 1/g_m}^{(2)}$ | $A_v = g_m r_D^{(1)}$
Typical A_v: Relatively low (<100) | Less than one. | Relatively low (<100)
Z_{in} equation: $Z_{in} = R_1 \| R_2$ | $Z_{in} = R_1 \| R_2$ | $Z_{in} = R_S \| \frac{1}{g_m}$
Typical Z_{in}: Very high. | Very high. | Very low.
Z_{out} equation: $Z_{out} = R_D \| r_d^{(3)}$ | $Z_{out} = R_S \| \frac{1}{g_m}$ | $Z_{out} = R_D \| r_d^{(3)}$
Typical Z_{out}: Relatively high. | Very low. | Relatively high.
Applications: (typical) Voltage amplifier. | Buffer (to low Z load) | Buffer (to high Z load)
$^{(1)}r_D = R_D \| R_L$; $^{(2)}r_S = R_S \| R_L$; $^{(3)}r_d = \frac{1}{y_{os}}$

Figure 21.34

Section Review

1. List the gain and impedance characteristics of the source follower.
2. List, in order, the steps taken to determine the range of A_v values for a source follower.
3. List the gain and impedance characteristics of the common-gate amplifier.

21.5 Introduction to MOSFETs: D-MOSFET Operation and Biasing

The primary drawback to JFETs is that the gate must be reverse biased. Recall that *the reverse bias applied to the gate is varied to deplete the channel of free carriers,* thus reducing the size of the channel. This type of operation is referred to as **depletion-mode operation.**

The metal-oxide-semiconductor FET, or *MOSFET,* is a device that can be operated in the **enhancement mode.** This means that *the input signal can be used to increase the effective size of the channel.* This also means that the device is not restricted to operating with its gate reverse biased.

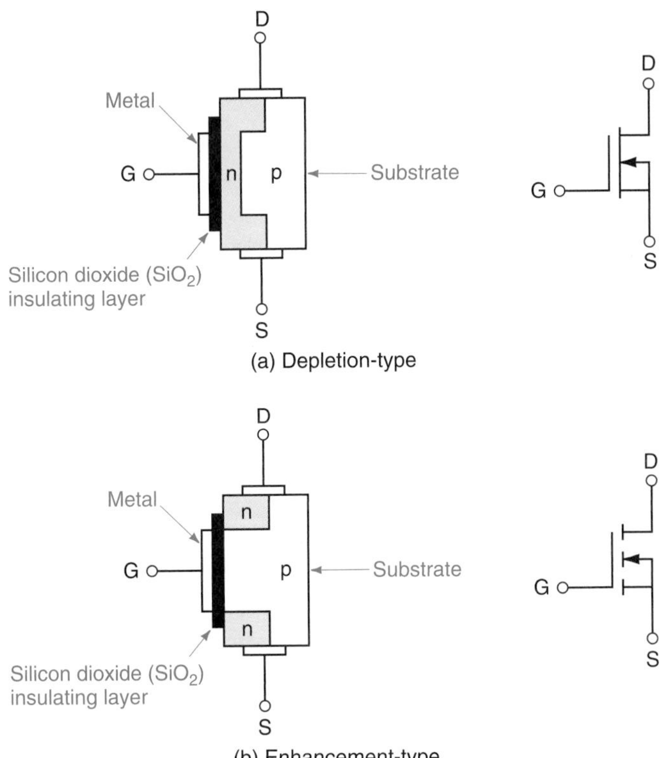

Figure 21.35 *n*-Channel MOSFETs.

MOSFET Construction and Handling

There are two basic types of MOSFETs. **Depletion-type MOSFETs (D-MOSFETs)** are *MOSFETs that can be operated in both the depletion and enhancement modes.* **Enhancement-type MOSFETs (E-MOSFETs)** are *MOSFETs that are restricted to enhancement-mode operation.* The primary difference between these types is their physical construction, as illustrated in Figure 21.35. As you can see, the D-MOSFET has a physical channel (shaded area) between the source and drain terminals. The E-MOSFET, however, depends on the gate voltage to form a channel between the source and drain terminals. This point will be discussed in detail later in this chapter. ◀ *OBJECTIVE 8*

Each of the MOSFETs shown in Figure 21.35 has a silicon dioxide (SiO_2) insulating layer between its gate and the rest of the component. Since the gate terminal is a metal conductor, going from gate to substrate you have metal-oxide-semiconductor, which is where the component name comes from.

The foundation of the MOSFET is called the **substrate.** Note that an *n*-channel MOSFET has a *p*-material substrate, and that a *p*-channel MOSFET has an *n*-material substrate. As with the JFET, an arrow pointing inward represents an *n*-channel device, and an arrow pointing outward represents a *p*-channel device. Figure 21.36 shows the *p*-channel D- and E-MOSFET symbols.

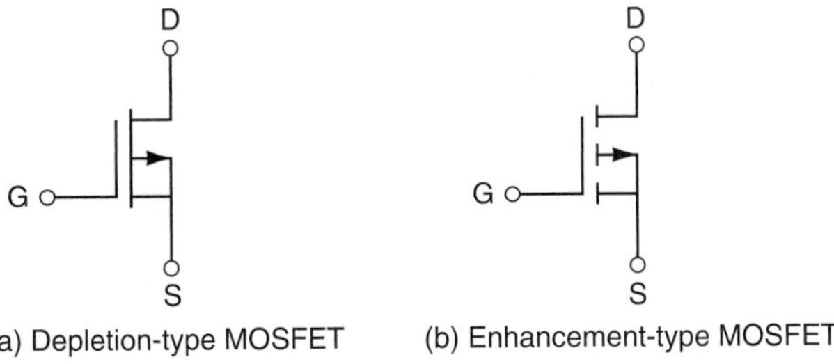

Figure 21.36 *p*-Channel MOSFET symbols.

OBJECTIVE 9 ➤

The layer of SiO_2 that insulates the gate from the channel is extremely thin and can be destroyed easily by static electricity. In fact, the static electricity generated by the human body can be sufficient to ruin a MOSFET, so some precautions must be taken to protect the component.

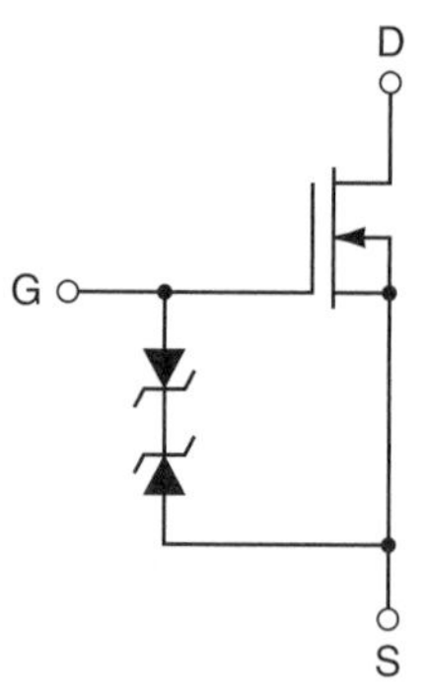

Figure 21.37
MOSFET static protection.

Many MOSFETs are manufactured with protective diodes etched between the gate and the source, as shown in Figure 21.37. These diodes are configured so that they will conduct in either direction when a predetermined voltage is reached, thus protecting the device.

When MOSFETs without protected inputs are used, some basic handling precautions must be taken to prevent the components from being damaged:

- Store the devices with the leads shorted together or in conductive foam. Never store MOSFETs in Styrofoam. (Styrofoam is the best static electricity generator ever devised.)
- Do not handle MOSFETs unless you need to. When you do handle any MOSFET, hold the component by the case, not by the leads.
- Do not install or remove any MOSFET while power is applied to a circuit. Also, make sure that any signal source is removed from a MOSFET circuit before turning the supply voltage off or on.

D-MOSFETs

OBJECTIVE 10 ➤

When operating in the depletion mode, the characteristics of the D-MOSFET are very similar to those of the JFET. The operation of the D-MOSFET is illustrated in Figure 21.38.

Figure 21.38a shows the D-MOSFET operating conditions when $V_{GS} = 0$ V (depletion mode). As is the case with JFETs, $I_D = I_{DSS}$ when $V_{GS} = 0$ V. When V_{GS} is negative (see Figure 21.38b), the gate-source junction is reverse biased, reducing the width of the channel and increasing its resistance. Thus, $I_D < I_{DSS}$ when V_{GS} is negative. This is exactly the same as the normal JFET operating state.

OBJECTIVE 11 ➤

Figure 21.38c shows the enhancement mode of operation. When V_{GS} is positive, the channel is widened, reducing its resistance. The positive gate voltage attracts minority carriers (electrons) in the *p*-type material to the *pn* junction, effectively increasing the width of the *n*-type material, and thus, the channel. When the channel width is increased, I_D can be greater than I_{DSS}, which is not possible with the JFET. The maximum allowable value of I_D is listed on the spec sheet of a given MOSFET.

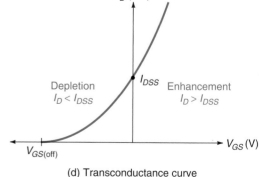

Figure 21.38
D-MOSFET operation.

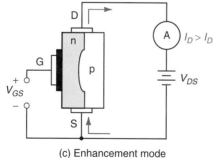

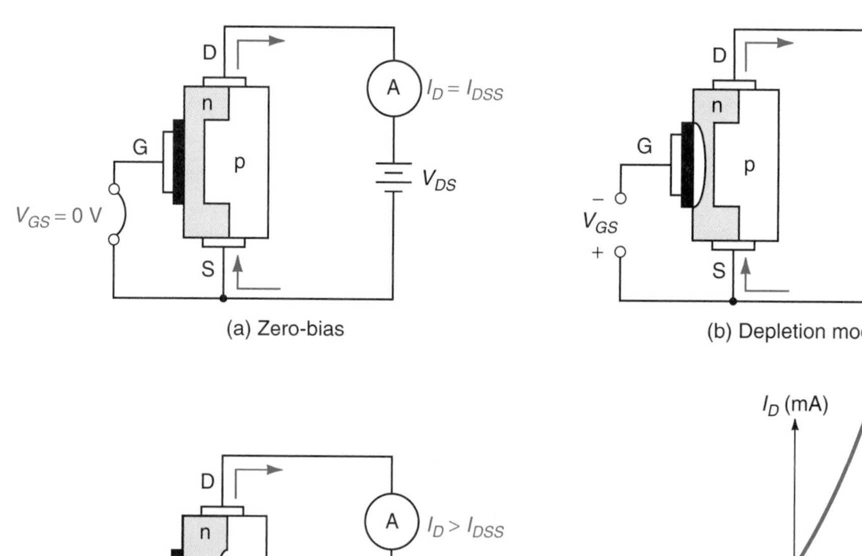

The combination of these three operating states is represented by the D-MOSFET transconductance curve shown in Figure 21.38d. The D-MOSFET uses the same transconductance equation as the JFET. As given earlier in the chapter,

$$I_D = I_{DSS}\left(1 - \frac{V_{GS}}{V_{GS(\text{off})}}\right)^2$$

A D-MOSFET transconductance curve is plotted in Example 21.11.

Example 21.11

A D-MOSFET has parameters of $V_{GS(\text{off})} = -6$ V and $I_{DSS} = 1$ mA. Plot the transconductance curve for the device.

Solution: From the component ratings given, we can determine two of the points on the curve, as follows:

$$V_{GS} = -6\text{ V}, I_D = 0\text{ mA} \qquad V_{GS} = 0\text{ V}, I_D = 1\text{ mA}$$

Now we will use equation (21.1) to determine the coordinates of several more points. When $V_{GS} = -3$ V,

$$I_D = I_{DSS}\left(1 - \frac{V_{GS}}{V_{GS(\text{off})}}\right)^2 = (1\text{ mA})\left(1 - \frac{-3\text{ V}}{-6\text{ V}}\right)^2 = 0.25\text{ mA}$$

When $V_{GS} = -1$ V,

$$I_D = I_{DSS}\left(1 - \frac{V_{GS}}{V_{GS(\text{off})}}\right)^2 = (1\text{ mA})\left(1 - \frac{-1\text{ V}}{-6\text{ V}}\right)^2 = 0.694\text{ mA}$$

When $V_{GS} = +1$ V,

$$I_D = I_{DSS}\left(1 - \frac{V_{GS}}{V_{GS(\text{off})}}\right)^2 = (1\text{ mA})\left(1 - \frac{+1\text{ V}}{-6\text{ V}}\right)^2 = 1.36\text{ mA}$$

When $V_{GS} = +3$ V,

$$I_D = I_{DSS}\left(1 - \frac{V_{GS}}{V_{GS(\text{off})}}\right)^2 = (1\text{ mA})\left(1 - \frac{+3\text{ V}}{-6\text{ V}}\right)^2 = 2.25\ mA$$

Using these values, we can plot the transconductance curve shown in Figure 21.39.

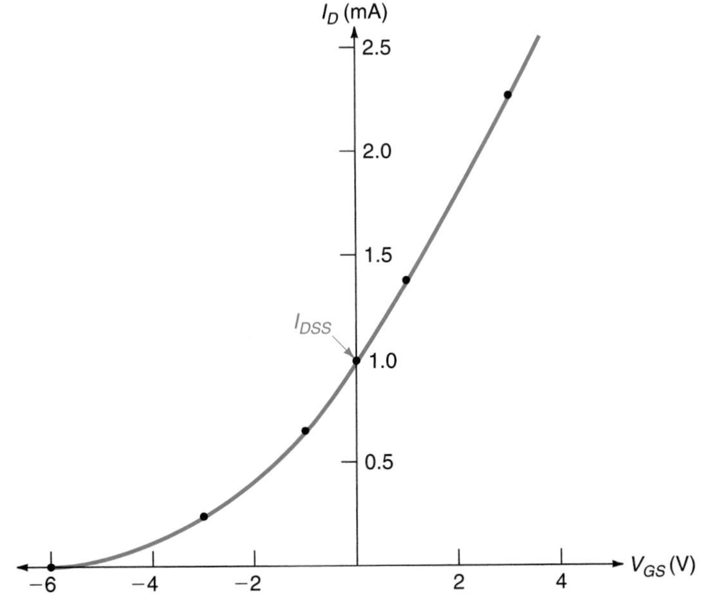

Figure 21.39

Practice Problem 21.11

A D-MOSFET has values of $V_{GS(\text{off})} = -4$ to -6 V and $I_{DSS} = 8$ to 12 mA. Plot the minimum and maximum transconductance curves for the device. Use the following values for V_{GS}: -4 V, -2 V, 0 V, 2 V, and 4 V.

Transconductance

The value of g_m for a D-MOSFET is found in the same way that it is for a JFET. As given earlier in the chapter,

$$g_m = g_{m0}\left(1 - \frac{V_{GS}}{V_{GS(\text{off})}}\right)$$

Just as I_D can be greater than I_{DSS} in the enhancement mode, g_m can be greater than g_{m0}.

D-MOSFET Biasing Circuits

D-MOSFET biasing circuits are the same as those used for JFETs, and have the same overall characteristics as JFETs when used in the common-source, common-drain, and common-gate configurations. All the dc and ac relationships covered for JFET circuits hold true for D-MOSFET circuits.

OBJECTIVE 12 ➤ The primary difference between the D-MOSFET and the JFET is the fact that the D-MOSFET is not limited to negative values of V_{GS}. In fact, one common biasing circuit, called **zero bias,** sets V_{GS} to 0 V. As shown in Figure 21.40, a zero-bias circuit is nearly identical to the JFET self-bias circuit, except that there is no source resistor. With both V_S and V_G at 0 V:

$$V_{GS} = 0 \text{ V} \qquad \text{and} \qquad I_D = I_{DSS}$$

When the circuit is designed, the value of the drain resistor (R_D) is selected so that it drops approximately one-half the supply voltage, leaving $V_{DS} \cong \frac{1}{2}V_{DD}$.

D-MOSFET Input Impedance

As a result of the insulating (SiO_2) layer, the gate impedance of a D-MOSFET is extremely high. For example, one MOSFET has a maximum gate current of 10 pA when $V_{GS} = 35$ V. This translates to a gate input impedance of 3.5×10^{12} Ω.

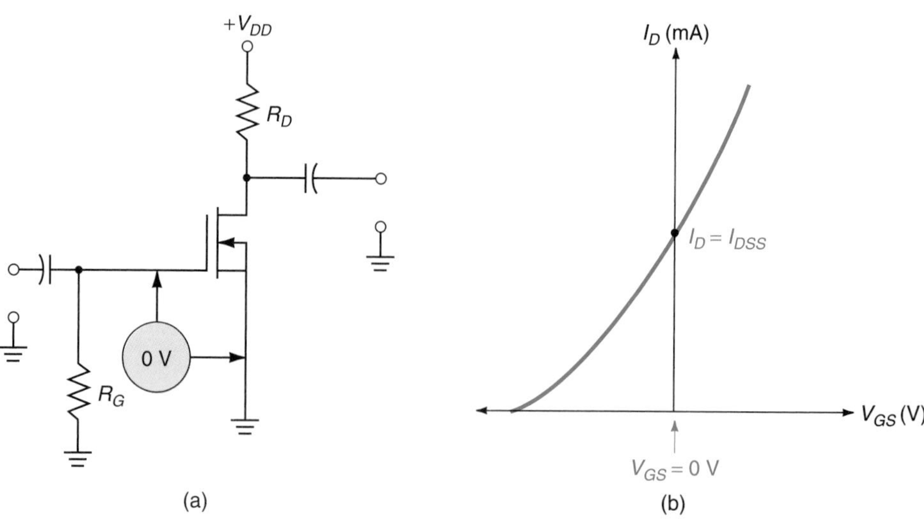

Figure 21.40 Zero bias.

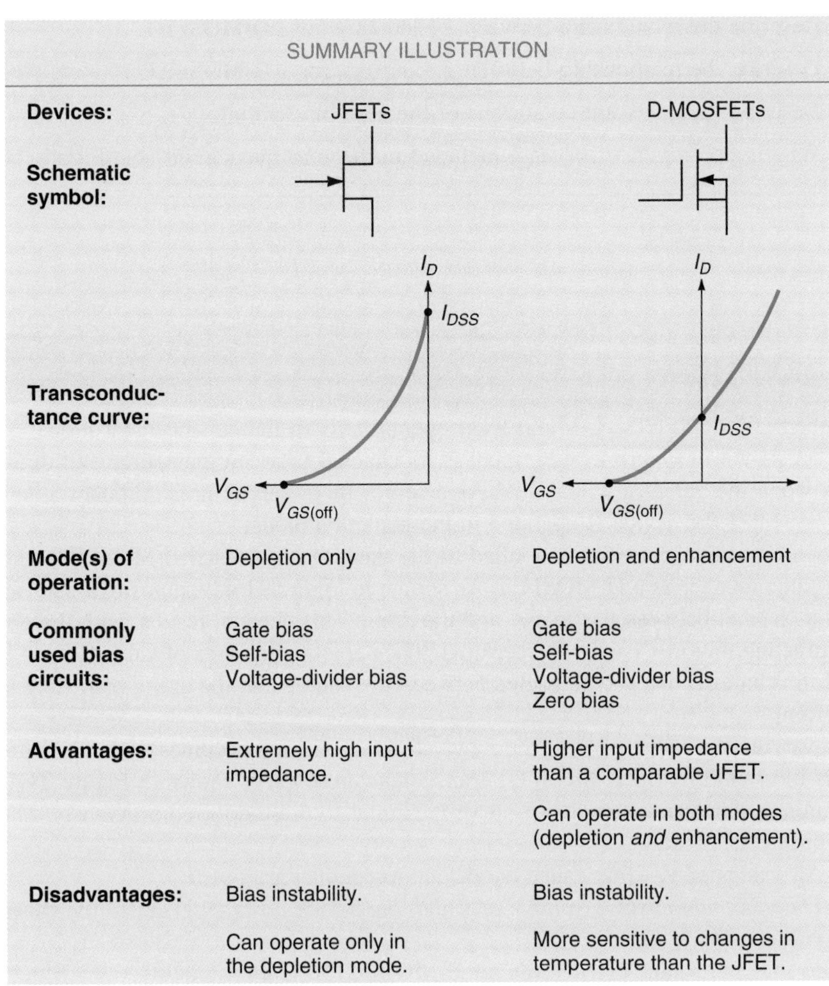

Figure 21.41

D-MOSFETs Versus JFETs

Figure 21.41 compares many of the characteristics of JFETs and D-MOSFETs.

Section Review

1. What is a *MOSFET*?
2. What is *depletion-mode* operation?
3. What is *enhancement-mode* operation?
4. In terms of operating modes, what is the difference between a D-MOSFET and an E-MOSFET?
5. In terms of physical construction, what is the difference between a D-MOSFET and an E-MOSFET?
6. Discuss the means by which MOSFET inputs may be internally protected from static electricity.
7. What precautions should be observed when handling MOSFETs that are not internally protected from static electricity?
8. Describe the depletion-mode relationship between V_{GS} and I_D for a D-MOSFET.
9. Describe the enhancement-mode relationship between V_{GS} and I_D for a D-MOSFET.

10. Describe the relationship between I_D and I_{DSS} for a D-MOSFET.

11. Describe the relationship between g_m and g_{m0} for a D-MOSFET.

12. Describe the quiescent conditions of the zero-bias circuit.

Critical Thinking

13. Why is the *gate resistor* (R_G) included in the zero bias circuit (see Figure 21.40)?

21.6 E-MOSFETs

Note: In this section, we are focusing on the n-channel E-MOSFET. Its p-channel counterpart operates according to the same principles. The only operating differences (as usual) are the voltage polarities and current directions.

Enhancement MOSFETs (E-MOSFETs) operate *only* in the enhancement mode ($V_{GS} >$ 0 V). As shown in Figure 21.42a, there is no channel between the source and drain materials when the value of $V_{GS} = 0$ V. As a result, there can be no significant amount of drain current if the device is operated in the depletion mode.

When a positive potential is applied to the gate, the n-channel E-MOSFET responds in the same manner as a D-MOSFET. As shown in Figure 21.42b, a *positive* gate voltage forms a channel between the source and the drain. This channel results from the positive V_{GS} attracting minority carriers (electrons), thus depleting the area of valence-band holes. This forms an effective n-type bridge between the source and the drain, providing a path for current.

OBJECTIVE 13 ➤

By varying the value of V_{GS}, the newly formed channel becomes wider or narrower. This is illustrated by the E-MOSFET transconductance curve shown in Figure 21.43. As you can see, this transconductance curve is similar to those covered previously. The primary differences are:

- All values of V_{GS} that cause the device to conduct are *positive.*
- *The minimum positive voltage at which the device turns on* is called the **threshold voltage, ($V_{GS(th)}$).**

OBJECTIVE 14 ➤

Note that the value of I_{DSS} for an E-MOSFET is approximately 0 A. Since I_{DSS} for the E-MOSFET is near zero, the standard transconductance formula will not work. To determine the value of I_D at a given value of V_{GS} you must use the following relationship:

$$I_D = k[V_{GS} - V_{GS(th)}]^2 \tag{21.22}$$

where k is a constant for the MOSFET, found as

$$k = \frac{I_{D(\text{on})}}{[V_{GS(\text{on})} - V_{GS(th)}]^2} \tag{21.19}$$

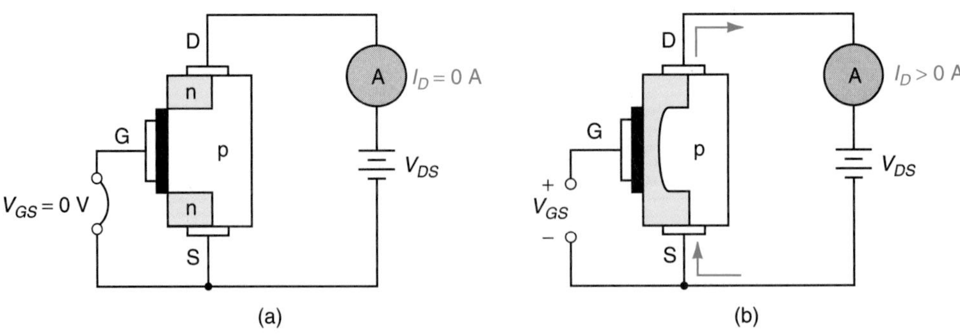

Figure 21.42 E-MOSFET operation.

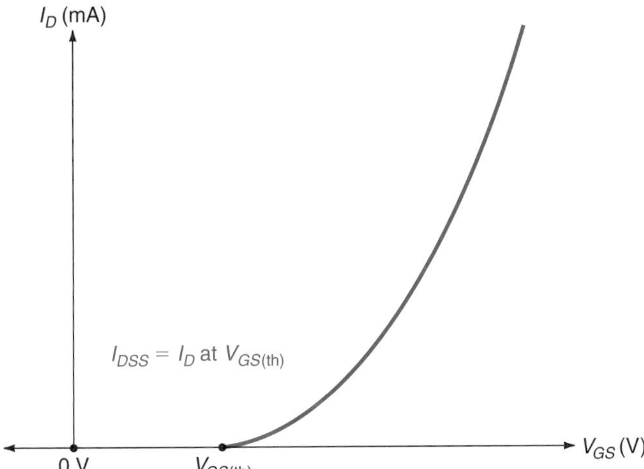

Figure 21.43 E-MOSFET transconductance curve.

The values used in equation (21.23) to determine the value of k are obtained from the spec sheet of the E-MOSFET being used. For example, the spec sheet for the 3N169 E-MOSFET lists the following parameters:

$$I_{D(\text{on})} = 10 \text{ mA} \qquad \text{(minimum)}$$
$$V_{GS(th)} = 0.5 \text{ V} \qquad \text{(minimum)}$$

According to the spec sheet, the $I_{D(\text{on})}$ rating is measured at $V_{GS} = 10$ V. Using this value with the two ratings listed above, the value of k can be found as

$$k = \frac{I_{D(\text{on})}}{[V_{GS(\text{on})} - V_{GS(th)}]^2} = \frac{10 \text{ mA}}{[10 \text{ V} - 0.5 \text{ V}]^2} = 1.11 \times 10^{-4} \text{ mA/V}^2$$

The entire process for finding I_D is demonstrated in Example 21.12.

Example 21.12

Determine the value of I_D for the circuit shown in Figure 21.44.

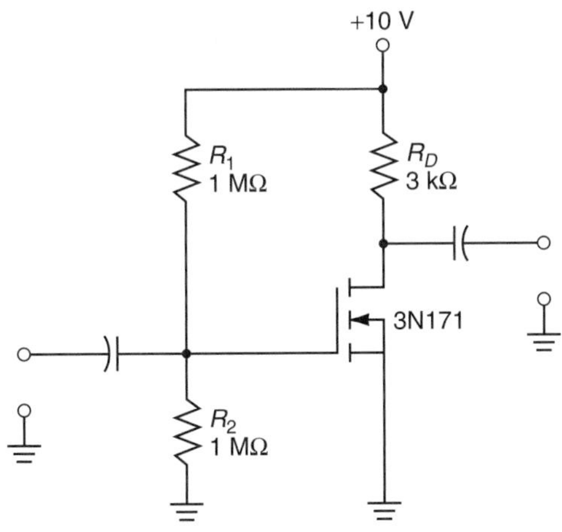

Figure 21.44

Solution: The spec sheet for the 3N171 provides the following ratings: $I_{D(\text{on})} = 10$ mA at $V_{GS} = 10$ V and $V_{GS(th)} = 1.5$ V. Using these ratings, the value of k is found as

$$k = \frac{I_{D(\text{on})}}{[V_{GS(\text{on})} - V_{GS(th)}]^2} = \frac{10 \text{ mA}}{[10 \text{ V} - 1.5 \text{ V}]^2} = 1.38 \times 10^{-4} \text{ mA/V}^2$$

The circuit does not contain a source resistor, so $V_{GS} = V_G$. The value of V_G is found as

$$V_G = V_{DD}\frac{R_2}{R_1 + R_2} = (10\text{ V})\frac{1\text{ M}\Omega}{2\text{ M}\Omega} = 5\text{ V}$$

Using the value of $V_{GS} = 5$ V in equation (21.22), the value of I_D is found as

$$I_D = k[V_{GS} - V_{GS(th)}]^2 = (1.38 \times 10^{-4}\text{ mA/V}^2)(5\text{ V} - 1.5\text{ V})^2 = 1.69\text{ mA}$$

Practice Problem 21.12

An E-MOSFET has values of $I_{D(\text{on})} = 14$ mA at $V_{GS} = 12$ V and $V_{GS(th)} = 2$ V. Determine I_D at $V_{GS} = 16$ V.

E-MOSFET Biasing Circuits

OBJECTIVE 15 ➤ Several of the biasing circuits used for JFETs and D-MOSFETs cannot be used to bias E-MOSFETs. For example, zero-bias and self-bias circuits produce values of $V_{GS} \leq 0$ V. Therefore, neither of these circuits can be used to bias an E-MOSFET. However, voltage-divider bias can be used, as can gate-bias (with a *positive* V_{GG}).

There is one bias circuit that is unique to E-MOSFETs. **Drain-feedback bias** is *the MOSFET counterpart of BJT collector-feedback bias.* A drain-feedback bias circuit is shown (along with its primary relationships) in Figure 21.45. Note that the circuit provides a value of $V_{GS} = V_{DS}$. The $I_{D(\text{on})}$ rating of a given E-MOSFET is measured with $V_{GS} = V_{DS}$, so drain-feedback bias always provides a Q-point value of $I_D = I_{D(\text{on})}$.

Section Review

1. How does a positive value of V_{GS} develop a channel through an *E-MOSFET*?
2. What is *threshold voltage* ($V_{GS(th)}$)?
3. When $V_{GS} = V_{GS(th)}$, what is the value of I_D for an E-MOSFET?
4. List, in order, the steps required to determine the value of I_D for an E-MOSFET at a given value of V_{GS}.
5. Which D-MOSFET biasing circuits can be used with E-MOSFETs?
6. Which D-MOSFET biasing circuits *cannot* be used with E-MOSFETs?

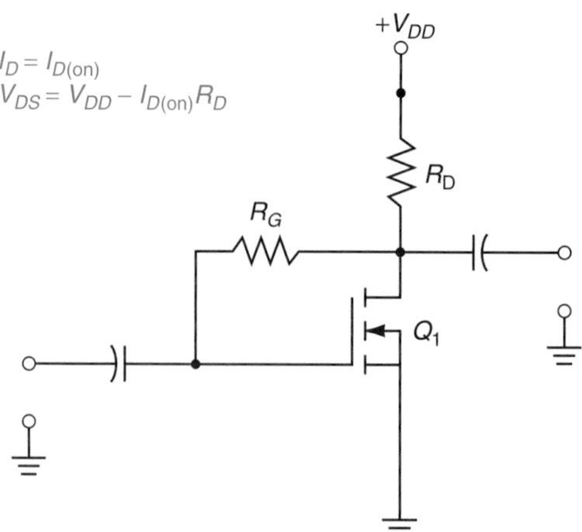

Figure 21.45 Drain-feedback bias.

21.7 Complementary MOSFETs: A MOSFET Application

The main contribution to electronics made by MOSFETs can be found in the area of digital electronics. Digital circuits are designed to respond to *rectangular waveforms.* As you know, these waveforms switch back and forth between two dc levels. In digital applications, these dc levels are referred to as **logic levels.**

A group of circuits that have similar operating characteristics is referred to as a **logic family.** All the circuits in a given logic family respond to the same voltage levels, have similar speed and power dissipation characteristics, and can be connected directly together. One such logic family is **complementary MOS (CMOS).** This logic family is made up entirely of MOSFETs. The basic CMOS *inverter* is shown in Figure 21.46a. An **inverter** is *a digital circuit that converts each logic level to the other.* For the circuit shown in Figure 21.46a, a 0 V input results in a +5 V output, and vice versa.

The operation of a CMOS inverter is easiest to understand if we look at the transistors as individual switches, as shown in Figures 21.46b and 21.46c. Note that Q_1 is a *p*-channel MOSFET, and that Q_2 is an *n*-channel MOSFET. When the input to the inverter is *low* (0 V), $V_{GS(Q1)} = -5$ V. As a result, Q_1 is on (conducting). In contrast, $V_{GS(Q2)} = 0$ V, and Q_2 is off (not conducting). With Q_1 on and Q_2 off, V_{out} is ideally at the source potential of Q_1, or +5 V (high). This is the inverse of the low input state.

When the input to the inverter is *high* (+5 V), $V_{GS(Q1)} = 0$ V, and Q_1 is off. $V_{GS(Q2)}$ is now +5 V, because its source is tied to ground. Q_2 is an *n*-channel device, so the positive V_{GS} causes it to conduct. With Q_1 off and Q_2 on, V_{out} is ideally at the source potential of Q_2, or 0 V (low). This is the inverse of the high input state.

There are several reasons that CMOS circuits work so well for digital applications: ◄ *OBJECTIVE 16*

- CMOS circuits are less complex than BJT logic circuits. As a result, more CMOS circuits can be etched onto an IC.
- CMOS circuits draw much less supply current than BJT logic circuits and, thus, run cooler.
- CMOS circuits need almost no input current. This means that a near-infinite number of CMOS loads can be driven in parallel by a single CMOS source. (BJT logic circuits are usually restricted to no more than 10 loads per BJT source.)

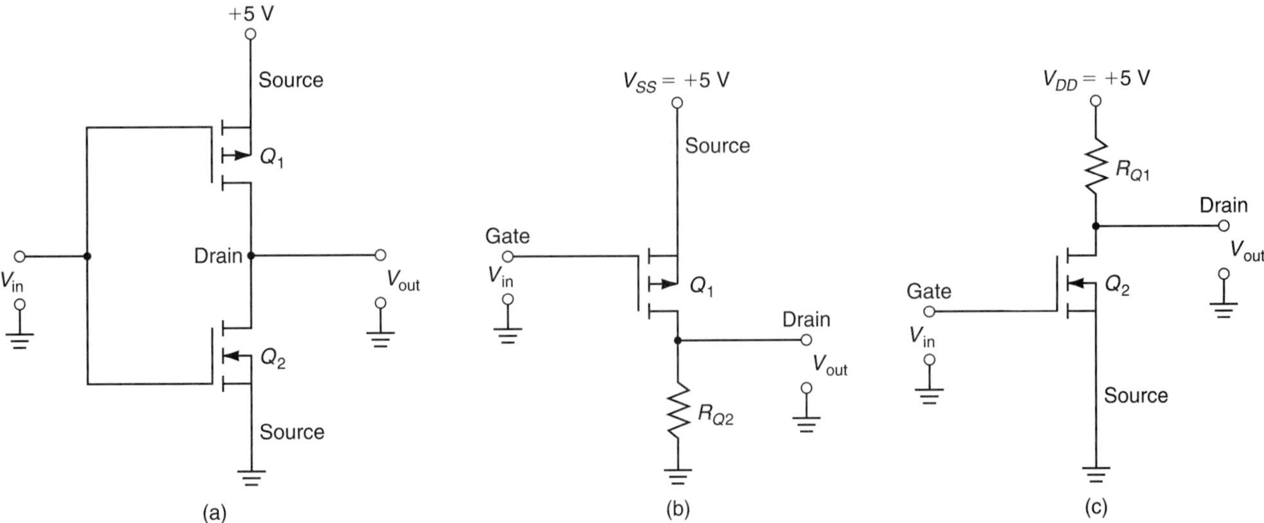

Figure 21.46 Complementary MOS inverter.

Section Review

1. What are *logic levels*?
2. What is a *logic family*?

3. What is an *inverter*?
4. Describe the physical construction of the CMOS inverter.
5. Describe the operation of the CMOS inverter.

21.8 Related Topics

In this section, we will briefly discuss some topics that relate to the material presented in this chapter.

Power MOSFETs

Advances in engineering have produced a variety of MOSFETs that are designed specifically for high-power applications. In this section, we will take a brief look at some of these devices.

OBJECTIVE 17 ➤ **V-MOSFET** Because of its physical construction, a **V-MOSFET (VMOS)** can handle much higher drain currents than a standard MOSFET. As shown in Figure 21.47a, the *n*-type materials are labeled n^+ (lightly doped) and n^- (heavily doped). Note that this is an enhancement-type MOSFET, as there is no physical channel connecting the source and drain terminals.

When a positive gate voltage is applied to the device, an *n*-type channel forms around the V-shaped gate, as shown in Figure 21.47b. The shape of this gate results in a wider channel and, therefore, higher drain currents at a given value of V_{GS}.

A VMOS device is not susceptible to thermal runaway because it has a *positive* temperature coefficient. This means that the channel resistance increases when temperature increases. Because of its higher drain current ratings and positive temperature coefficients, VMOS devices can be used in many high-power applications where the standard MOSFET cannot be used.

Lateral double-diffused MOSFET Another type of power MOSFET is the **lateral double-diffused MOSFET (LDMOS).** This type of E-MOSFET uses a very small channel region and a heavily doped *n*-type substrate (n^-) to obtain high drain current and low channel resistance ($r_{d(\text{on})}$). LDMOS devices have typical $r_{d(\text{on})}$ values in the range of 2 Ω or less. With this low channel resistance, it can handle currents as high as 20 A. The basic construction of this E-MOSFET is shown in Figure 21.48.

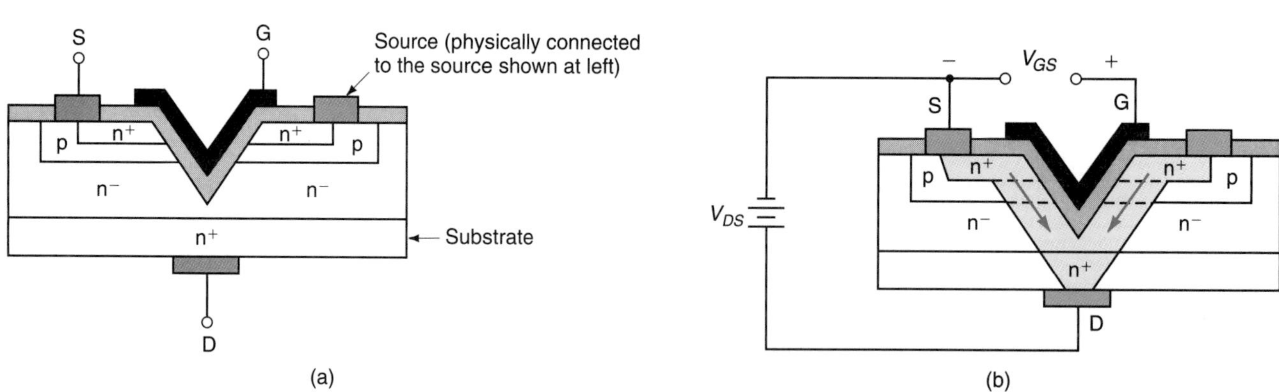

Figure 21.47 V-MOSFET (VMOS) construction.

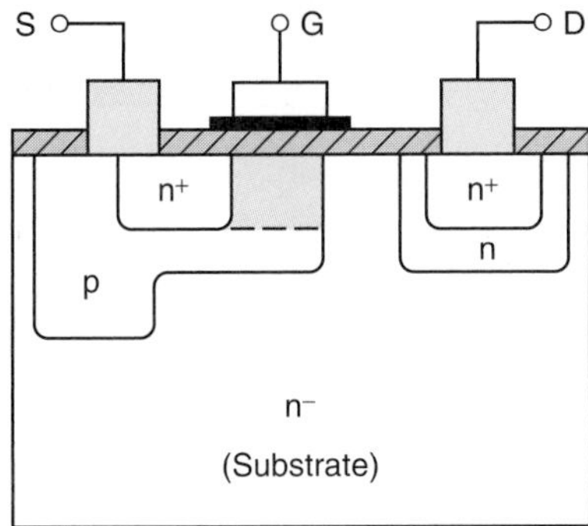

Figure 21.48 Lateral double-diffused MOSFET (LDMOS).

Cascode Amplifiers

Both JFETs and MOSFETs have relatively high input capacitance, which can adversely affect their high-frequency operation. The *cascode amplifier* was developed to overcome this shortcoming. A **cascode amplifier** is *a circuit that consists of a common-source amplifier in series with a common-gate amplifier,* as shown in Figure 21.49. Q_1 is a common-gate amplifier that is voltage-divider biased. Q_2 is a CS circuit and is self-biased. The input to the circuit is applied to the gate of Q_2, and the output is taken from the drain of Q_1. Therefore, input to output, the two MOSFETs are in series.

Connecting the two components in series has the same effect on the amplifier's input capacitance as connecting any two capacitances in series: *The total series capacitance is lower than either individual capacitance.* This makes the cascode amplifier better suited for high-frequency applications.

A simple cascode amplifier can be constructed using a **dual-gate MOSFET,** which, as its name implies, has two gate terminals. The schematic symbol of the dual-gate MOSFET is shown in Figure 21.50. This circuit is the equivalent of that shown in Figure 21.49. Gate 1 of the MOSFET is connected to act as a common-gate amplifier. Gate 2 is connected to act as the common-source amplifier. This amplifier has the same capacitance characteristics and high-frequency capabilities as the amplifier shown in Figure 21.49.

◀ *OBJECTIVE 18*

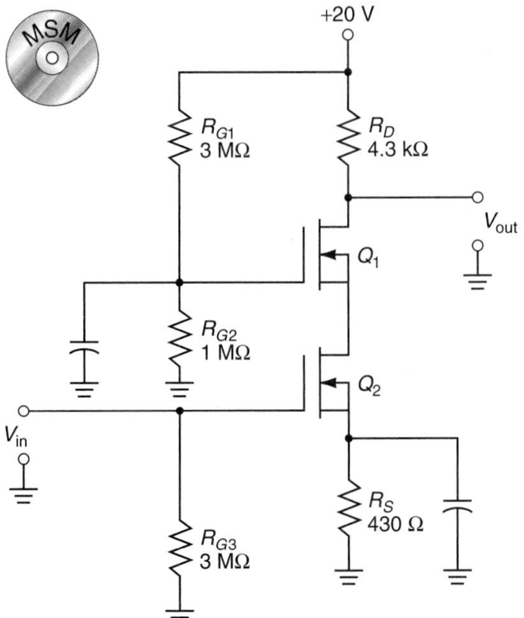

Figure 21.49 Cascode amplifier.

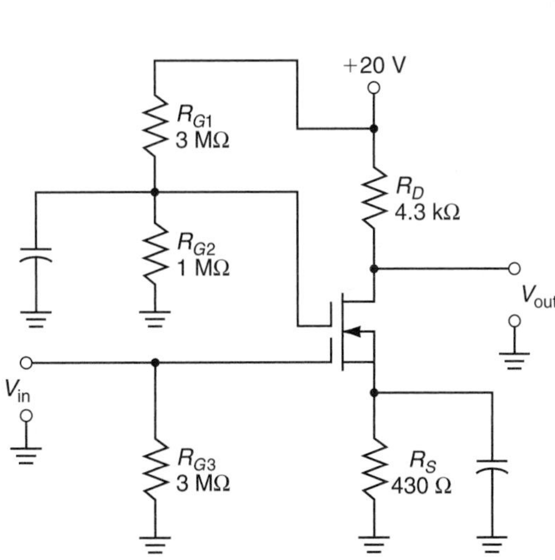

Figure 21.50 Dual-gate MOSFET cascode amplifier.

Figure 21.51
MOSFET power driver. (Copyright of Semiconductor Components Industries, LLC. Used by permission.)

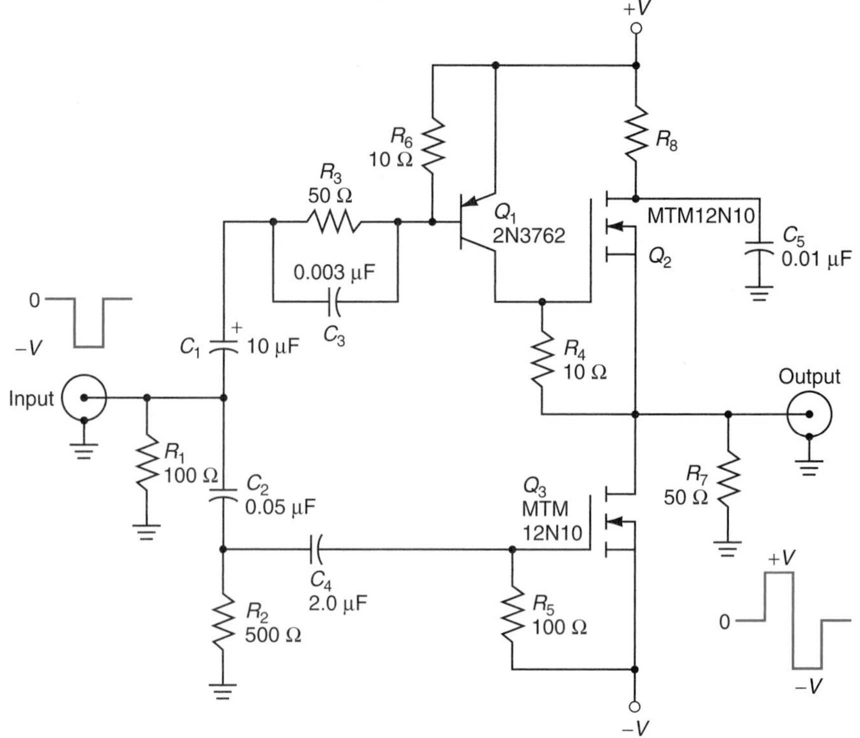

Power MOSFET Drivers

Many digital communications systems require power amplifiers that can produce high-speed, high-current digital outputs. These signals can be produced using a power MOSFET driver circuit like the one shown in Figure 21.51.

The high-current capability of the driver is provided by the low channel resistance [$r_{d(on)}$] of the power MOSFETs (typically 2 Ω or less). Thus, depending on the values of +V, −V, and R_8, the MOSFET pair can output currents as high as 20 to 35 A. The high-speed capability of the circuit is a result of the extremely low circuit resistance values. The low-value resistors allow the capacitors in the circuit to charge and discharge very rapidly, improving the frequency response of the driver.

Section Review

1. Describe the physical construction of a VMOS device.
2. Describe the physical construction of an LDMOS device.
3. What is a *cascode* amplifier?
4. How does the cascode amplifier reduce input capacitance?
5. Describe the dual-gate MOSFET cascode amplifier.
6. What purpose does the power MOSFET driver serve in digital communications?

Key Terms

The following terms were introduced in this chapter on the pages indicated:

Bias line, 688
Cascode amplifier, 717
Channel, 680
Common-drain amplifier, 701
Common-gate amplifier, 704
Common-source (CS) amplifier, 695
Complementary MOS (CMOS), 715

Constant-current region, 683
Depletion-mode operation, 706
Depletion-type MOSFET (D-MOSFET), 707
Drain, 680
Drain-feedback bias, 714
Dual-gate MOSFET, 717
Enhancement-mode operation, 706
Enhancement-type MOSFET (E-MOSFET), 707
Field-effect transistor (FET), 680
Gate, 680
Gate bias, 686
Gate-source cutoff voltage ($V_{GS(\text{off})}$), 682
Inverter, 715
Junction field-effect transistor (JFET), 680
Lateral double-diffused MOS (LDMOS), 716
Logic family, 715
Logic levels, 715
Metal-oxide-semiconductor FET (MOSFET), 680
Ohmic region, 682
Pinch-off voltage (V_P), 682
Self-bias, 689
Shorted-gate drain current (I_{DSS}), 683
Source, 680
Source follower, 701
Substrate, 707
Threshold voltage ($V_{GS(th)}$), 712
Transconductance, 696
Transconductance curve, 684
V-MOSFET (VMOS), 716
Zero bias, 710

Practice Problems

1. A JFET amplifier has values of $I_{DSS} = 8$ mA, $V_{GS(\text{off})} = -12$ V, and $V_{GS} = -6$ V. Determine the value of I_D for the circuit.
2. A JFET amplifier has values of $I_{DSS} = 16$ mA, $V_{GS(\text{off})} = -5$ V, and $V_{GS} = -4$ V. Determine the value of I_D for the amplifier.
3. A JFET amplifier has values of $I_{DSS} = 14$ mA, $V_{GS(\text{off})} = -6$ V, and $V_{GS} = 0$ V. Determine the value of I_D for the amplifier.
4. A JFET amplifier has values of $I_{DSS} = 10$ mA, $V_{GS(\text{off})} = -8$ V, and $V_{GS} = -5$ V. Determine the value of I_D for the amplifier.
5. A JFET amplifier has values of $I_{DSS} = 12$ mA, $V_{GS(\text{off})} = -10$ V, and $V_{GS} = -4$ V. Determine the value of I_D for the amplifier.
6. The JFET amplifier described in Problem 5 has a value of $V_{GS} = -14$ V. Determine the value of I_D for the circuit.
7. The 2N5484 has values of $V_{GS(\text{off})} = -0.3$ to -3 V and $I_{DSS} = 1.0$ to 5.0 mA. Plot the minimum and maximum transconductance curves for the device.
8. The 2N5485 has values of $V_{GS(\text{off})} = -0.5$ to -4.0 V and $I_{DSS} = 4$ to 10 mA. Plot the minimum and maximum transconductance curves for the device.
9. The 2N5457 has values of $V_{GS(\text{off})} = -0.5$ to -6.0 V and $I_{DSS} = 1$ to 5 mA. Plot the minimum and maximum transconductance curves for the device.
10. The 2N5458 has values of $V_{GS(\text{off})} = -1$ to -7 V and $I_{DSS} = 2$ to 9 mA. Plot the minimum and maximum transconductance curves for the device.
11. The 2N3437 has parameters of $V_{GS(\text{off})} = -5$ V (maximum) and $I_{DSS} = 0.8$ to 4.0 mA. Plot the maximum transconductance curve for the device.
12. The 2N3438 has parameters of $V_{GS(\text{off})} = -2.5$ V (maximum) and $I_{DSS} = 0.2$ to 1.0 mA. Plot the maximum transconductance curve for the device.
13. Determine the values of I_D and V_{DS} for the amplifier shown in Figure 21.52a.
14. Determine the values of I_D and V_{DS} for the amplifier shown in Figure 21.52b.
15. The 2N5486, which has values of $V_{GS(\text{off})} = -2$ to -6 V and $I_{DSS} = 8$ to 20 mA, is used in the circuit shown in Figure 21.52a. Determine the range of I_D values for the circuit.
16. A JFET with values of $V_{GS(\text{off})} = -5$ to -10 V and $I_{DSS} = 4$ to 8 mA is used in the circuit shown in Figure 21.52b. Determine the range of I_D values for the circuit.

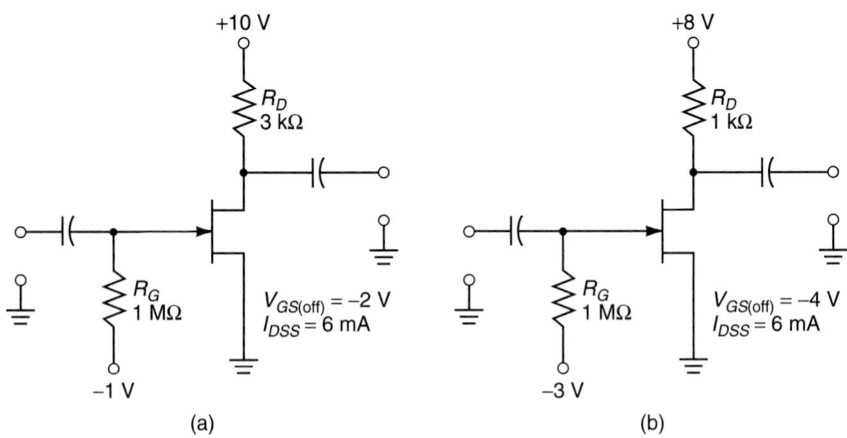

Figure 21.52

17. The 2N3437 described in Problem 11 is used in the circuit shown in Figure 21.53. Determine the range of I_D values for the circuit.

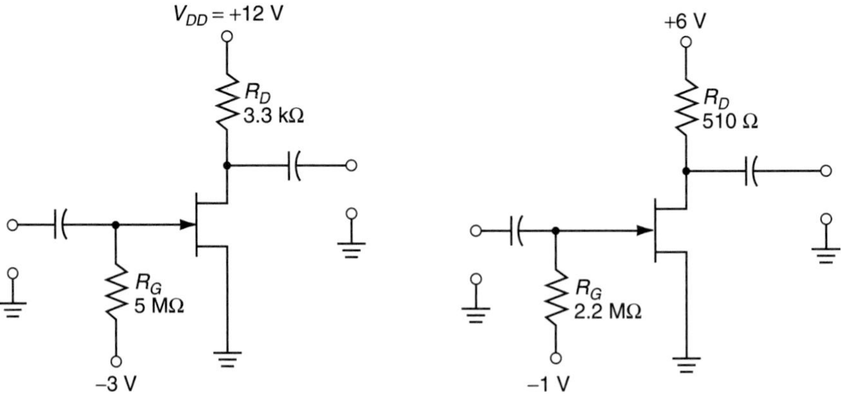

Figure 21.53 **Figure 21.54**

18. The 2N3438 described in Problem 12 is used in the circuit shown in Figure 21.54. Determine the range of I_D values for the circuit.
19. Determine the ranges of V_{GS}, I_D, and V_{DS} for the circuit shown in Figure 21.55. The 2N5484 is described in Problem 7.
20. Determine the ranges of V_{GS}, I_D, and V_{DS} for the circuit shown in Figure 21.56. The 2N5485 is described in Problem 8.

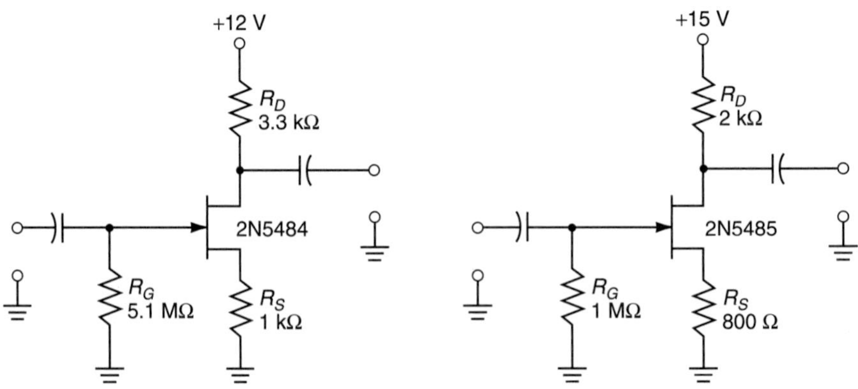

Figure 21.55 **Figure 21.56**

21. Determine the ranges of V_{GS}, I_D, and V_{DS} for the circuit shown in Figure 21.57. The 2N5458 is described in Problem 10.

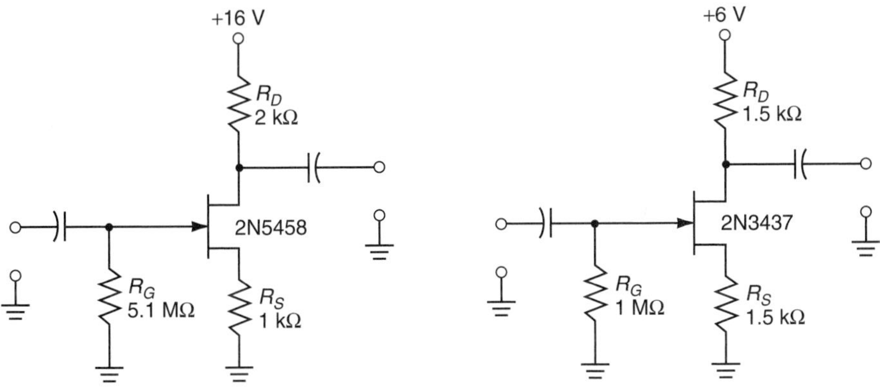

Figure 21.57 **Figure 21.58**

22. Determine the ranges of V_{GS}, I_D, and V_{DS} for the circuit shown in Figure 21.58. The 2N3437 is described in Problem 11.

23. Determine the ranges of I_D and V_{DS} for the circuit shown in Figure 21.59.

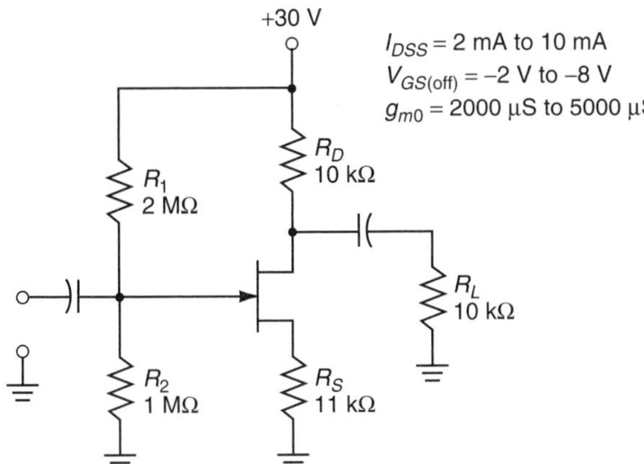

Figure 21.59

24. Determine the ranges of I_D and V_{DS} for the circuit shown in Figure 21.60.

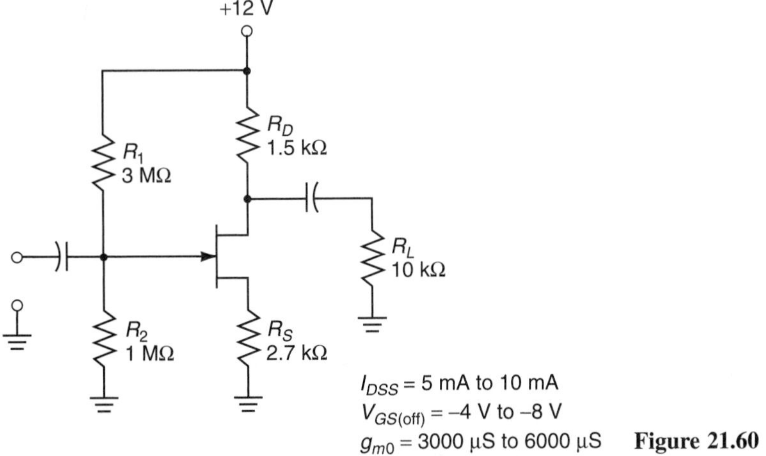

Figure 21.60

25. Determine the ranges of I_D and V_{DS} for the circuit shown in Figure 21.61.

26. Determine the ranges of I_D and V_{DS} for the circuit shown in Figure 21.62.

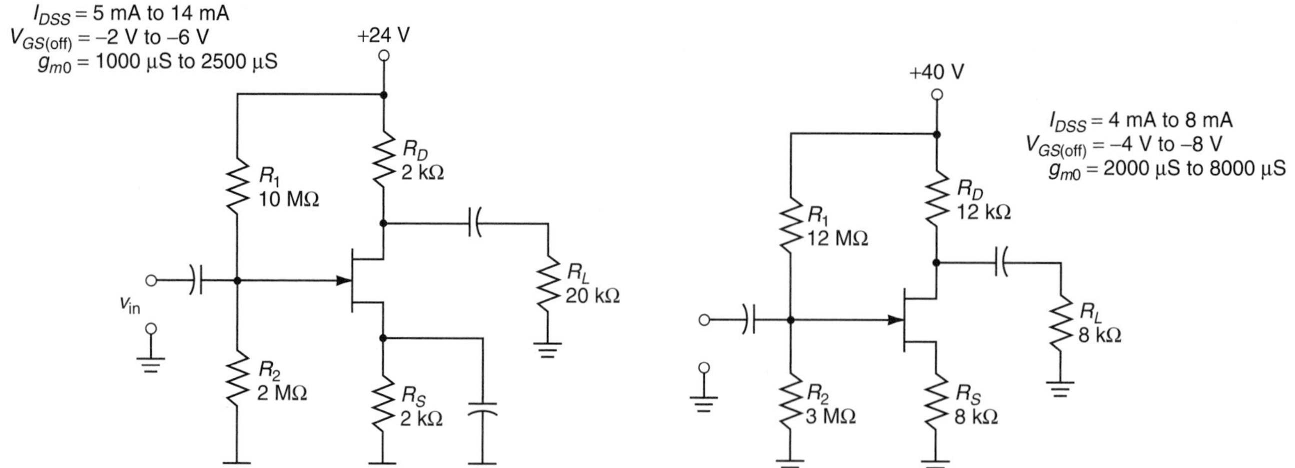

Figure 21.61

Figure 21.62

27. The 2N5484 has values of g_{m0} = 3000 to 6000 μS and $V_{GS(\text{off})} = -0.3$ to -3.0 V. Determine the range of g_m when $V_{GS} = -0.2$ V.
28. The 2N5485 has values of g_{m0} = 3500 to 7000 μS and $V_{GS(\text{off})} = -0.5$ to -4.0 V. Determine the range of g_m when $V_{GS} = -0.4$ V.
29. The 2N3437 has maximum values of $V_{GS(\text{off})} = -5$ V and g_{m0} = 6000 μS. Determine the maximum values of g_m at $V_{GS} = -1$ V, -2.5 V, and -4 V.
30. The 2N3438 has maximum values of $V_{GS(\text{off})} = -2.5$ V and g_{m0} = 4500 μS. Determine the maximum values of g_m at $V_{GS} = -1$ V, -1.5 V, and -2 V.
31. Determine the range of A_v values for the circuit shown in Figure 21.59.
32. Determine the range of A_v values for the circuit shown in Figure 21.60.
33. Determine the range of A_v values for the circuit shown in Figure 21.61.
34. Determine the range of A_v values for the circuit shown in Figure 21.62.
35. Determine the value of Z_{in} for the amplifier shown in Figure 21.59.
36. Determine the value of Z_{in} for the amplifier shown in Figure 21.60.
37. Determine the values of A_v, Z_{in}, and Z_{out} for the amplifier shown in Figure 21.63.
38. Determine the values of A_v, Z_{in}, and Z_{out} for the amplifier shown in Figure 21.64.

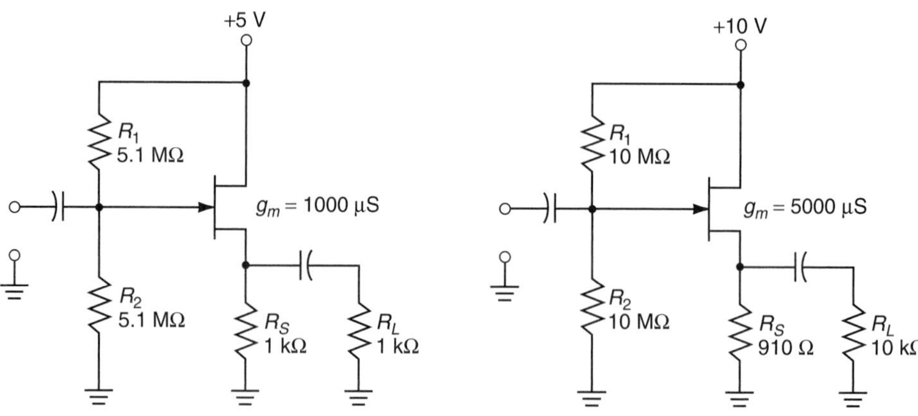

Figure 21.63

Figure 21.64

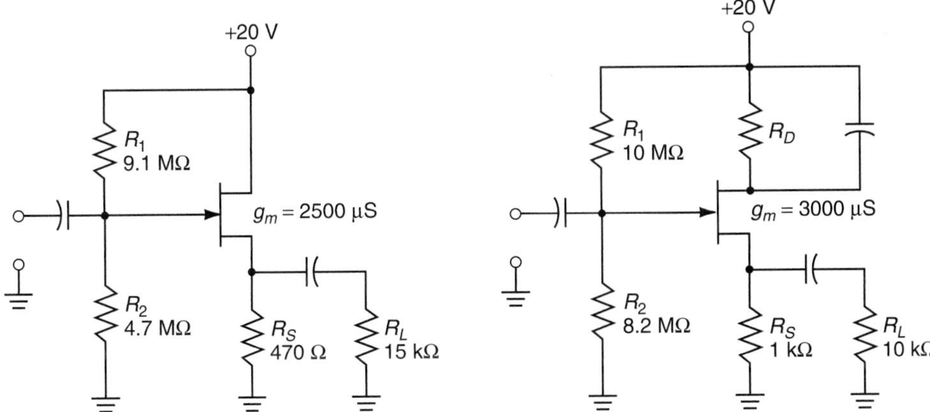

Figure 21.65

Figure 21.66

39. Determine the values of A_v, Z_{in}, and Z_{out} for the amplifier shown in Figure 21.65.

40. Determine the values of A_v, Z_{in}, and Z_{out} for the amplifier shown in Figure 21.66.

41. Determine the values of A_v, Z_{in}, and Z_{out} for the circuit shown in Figure 21.67.

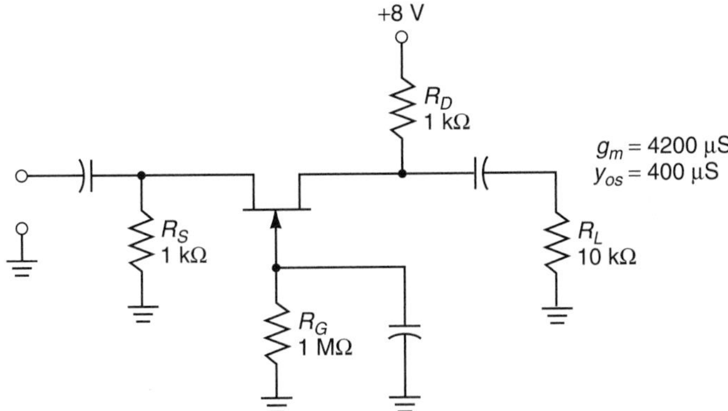

Figure 21.67

42. Determine the values of A_v, Z_{in}, and Z_{out} for the circuit shown in Figure 21.68.

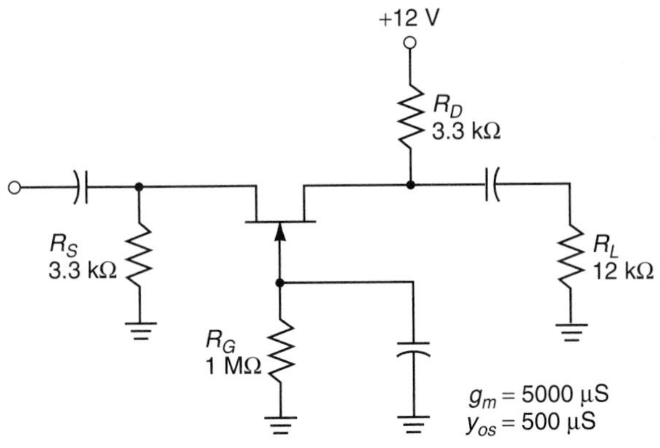

Figure 21.68

43. A D-MOSFET has values of $V_{GS(\text{off})} = -4$ V and $I_{DSS} = 8$ mA. Plot the transconductance curve for the device.

44. A D-MOSFET has values of $V_{GS(\text{off})} = -8$ V and $I_{DSS} = 12$ mA. Plot the transconductance curve for the device.

45. A D-MOSFET has values of $V_{GS(\text{off})} = -4$ to -6 V and $I_{DSS} = 6$ to 9 mA. Plot the minimum and maximum transconductance curves for the device.

46. A D-MOSFET has values of $V_{GS(\text{off})} = -4$ to -10 V and $I_{DSS} = 5$ to 10 mA. Plot the minimum and maximum transconductance curves for the device.

47. The D-MOSFET described in Problem 43 has a value of $g_{m0} = 2000$ μS. Determine the values of g_m at $V_{GS} = -2$ V, 0 V, and +2 V.

48. The D-MOSFET described in Problem 44 has a value of $g_{m0} = 3000$ μS. Determine the values of g_m at $V_{GS} = -5$ V, 0 V, and +5 V.

49. An E-MOSFET has ratings of $V_{GS(th)} = 4$ V and $I_{D(\text{on})} = 12$ mA at $V_{GS} = 8$ V. Determine the value of I_D for the device when $V_{GS} = 6$ V.

50. An E-MOSFET has ratings of $V_{GS(th)} = 2$ V and $I_{D(\text{on})} = 10$ mA at $V_{GS} = 8$ V. Determine the value of I_D for the device when $V_{GS} = +12$ V.

51. An E-MOSFET has ratings of $V_{GS(th)} = 1$ V and $I_{D(\text{on})} = 8$ mA at $V_{GS} = 4$ V. Determine the values of I_D for the device when $V_{GS} = 1$ V, 4 V, and 5 V.

52. An E-MOSFET has values of $V_{GS(th)} = 3$ V and $I_{D(\text{on})} = 2$ mA at $V_{GS} = 5$ V. Determine the values of I_D for the device when $V_{GS} = 3$ V, 5 V, and 8 V.

Looking Back

These problems relate to material presented in earlier chapters. The chapters are identified in brackets.

53. A base-bias circuit has values of $V_{CC} = 12$ V and $R_C = 510\ \Omega$. Calculate the values of $V_{CE(\text{off})}$ and $I_{C(\text{sat})}$ for the circuit. [Chapter 20]

54. A series LC circuit has values of $L = 1$ mH and $C = 0.1$ μF. Calculate the resonant frequency of the circuit. [Chapter 16]

55. Calculate the half-cycle average value of a 100 V_{pk} sine wave. [Chapter 9]

56. Draw the frequency response curve for the passive filter shown in Figure 21.69, and include its cutoff frequency. [Chapter 15]

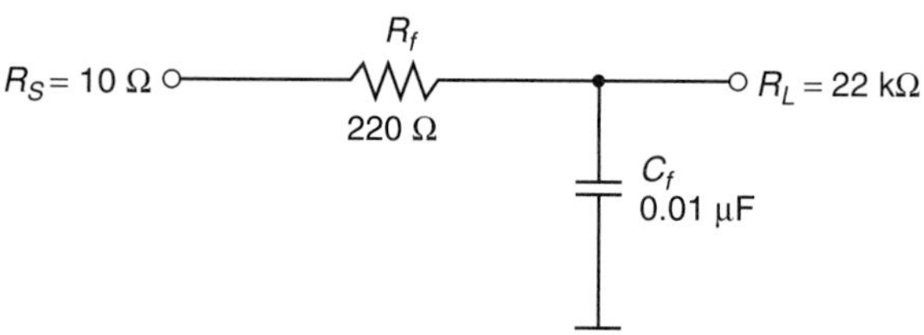

Figure 21.69

57. Calculate the average load voltage for the circuit shown in Figure 21.70. [Chapter 19]

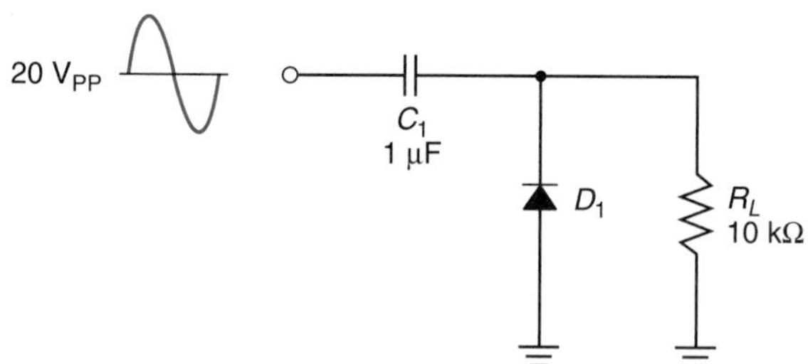

Figure 21.70

58. A rectangular waveform has the following values: PW = 10 μs, and $f = 20$ kHz. Calculate the duty cycle of the waveform. [Chapter 9]

Pushing the Envelope

59. The biasing circuit shown in Figure 21.71 is a current-source bias circuit. Determine how this circuit obtains a stable value of I_D.

60. The 2N5486 cannot be substituted for the JFET shown in Figure 21.72. Why not?

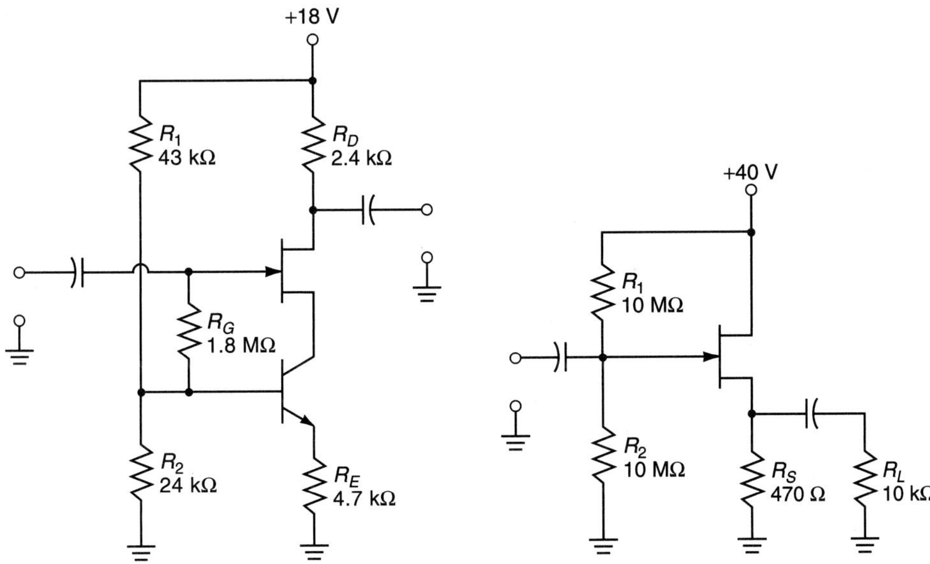

Figure 21.71

Figure 21.72

Answers to the Example Practice Problems

21.1 3 mA

21.2 See Figure 21.73.

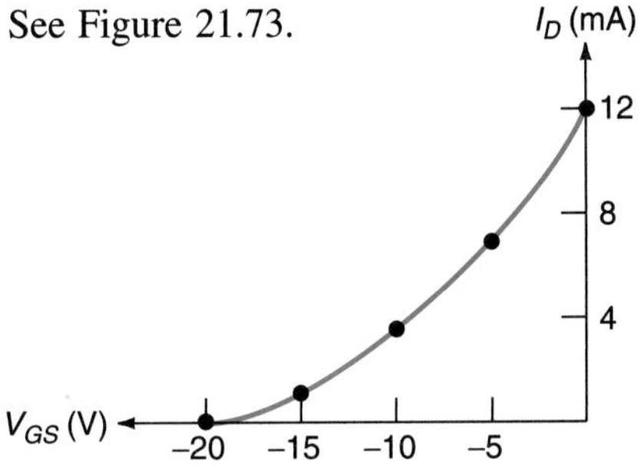

Figure 21.73

21.3 $I_D = 3$ mA, $V_{DS} = 3.4$ V

21.5 See Figure 21.74.

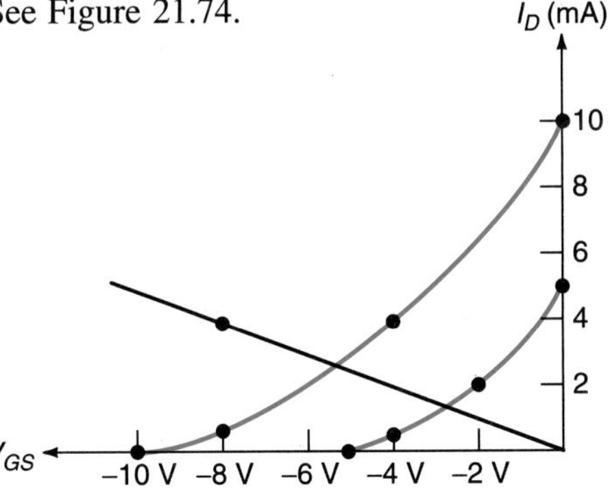

Figure 21.74

21.6 See Figure 21.75.

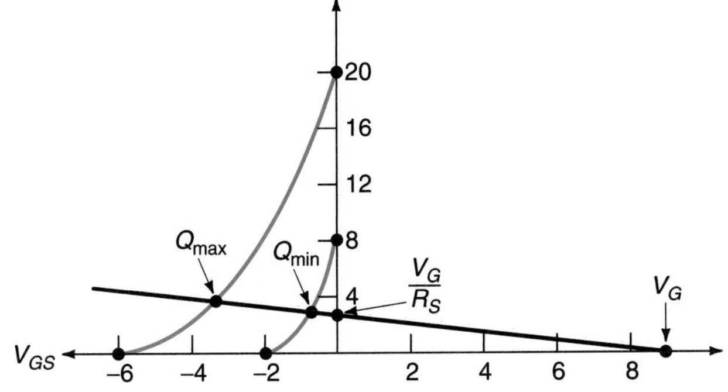

Figure 21.75

21.7 $g_m = 5333\ \mu S$ at $V_{GS} = -2$ V; $g_m = 2667\ \mu S$ at $V_{GS} = -4$ V

21.8 $A_v = 2.43$ to 3.23

21.9 $A_v = 3.3$ to 3.77

21.10 $A_v = 0.848$ to 0.736, $Z_{out} = 434\ \Omega$ to $767\ \Omega$

21.11 See Figure 21.76.

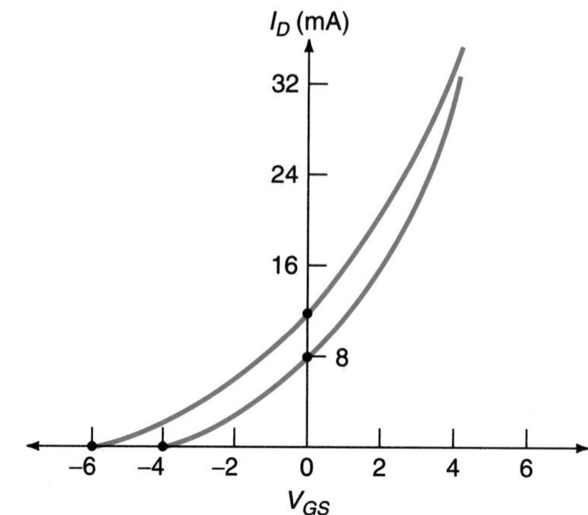

Figure 21.76

21.12 27.44 mA

Multisim File References

Figure	File Reference
21.6	ETF21_6.msm
21.7	ETF21_7.msm
21.10	ETF21_10.msm
21.22	ETF21_22.msm
21.43	ETF21_43.msm
21.49	ETF21_49.msm

chapter 22

Operational Amplifiers

Objectives

After studying the material in this chapter, you should be able to:

1. Describe the *operational amplifier* (op-amp).
2. Identify the input, output, and supply pins of a given op-amp.
3. State the purpose served by a *differential amplifier.*
4. Describe *open-loop voltage gain* and state its typical range of values.
5. Describe the effects of the *input polarities, supply voltages,* and *load resistance* on the output from an op-amp.
6. Describe the operation of a discrete differential amplifier.
7. Describe the various voltage and current ratings for an op-amp.
8. Discuss the relationship between *slew-rate* and op-amp operating frequency.
9. Describe an analyze the operation of an *inverting* amplifier.
10. Describe and analyze the operation of a *noninverting* amplifier.
11. Describe and analyze the frequency response of inverting and noninverting amplifiers.
12. Describe and analyze the operation of a *comparator.*
13. Describe and analyze the operation of an *integrator.*
14. Describe and analyze the operation of a *differentiator.*
15. Describe and analyze the operation of a *summing amplifier.*

Individual components, such as the 2N3904 BJT and the 2N5459 FET, are classified as **discrete components.** The term *discrete* indicates that *each physical package contains only one component.* For example, when you purchase a 2N3904, you are buying a single component that is housed in its own casing.

Over the years, advances in manufacturing technology have made it possible to produce entire circuits on a single piece of semiconductor material. This type of circuit, which is housed in a single casing, is referred to as an **integrated circuit (IC).** ICs range in complexity from simple circuits containing a few active and/or passive components to complex circuits containing hundreds of thousands of components. The more complex the internal circuitry of an IC, the more complex its function.

The major impacts of ICs all relate to their internal operation and relatively low manufacturing cost. Circuit operations that once took hundreds of discrete components to perform can now be accomplished with a single IC. At the same time, the cost of an IC is generally lower than that of a comparable discrete-component circuit. This has made electronic systems less expensive to produce.

22.1 Operational Amplifiers: An Overview

OBJECTIVE 1 ➤ The **operational amplifier (op-amp)** is *a high-gain dc amplifier that has high input impedance and low output impedance.* The internal circuitry, schematic symbol, and pin diagram for the 741 general-purpose operational amplifier are shown in Figure 22.1.

OBJECTIVE 2 ➤ The inputs to the op-amp are labeled *inverting* (−) and *noninverting* (+). Normally, an input signal is applied to one input, and the other input is wired to control the operating characteristics of the component. The particular application determines which input pin is used as the active input.

The op-amp has two dc power supply inputs, which are labeled $+V$ and $-V$. They are normally connected in one of two ways, as illustrated in Figure 22.2. Again, the specific wiring of these two pins depends on the particular application.

IC Identification

Many op-amps are identified using a seven-character ID code, as shown in Figure 22.3. The *prefix* identifies the manufacturer. The *designator* includes a number that indicates the type of op-amp and a letter that indicates the operating temperature range. The *suffix* identifies the type of package. Some commonly used temperature codes are listed in Table 22.1.

Op-Amp Packages

Op-amps are available in dual-in-line (DIP) packages, metal cans, and surface-mount packages (SMPs). The DIP and metal can packages are illustrated in Figure 22.4. Metal cans (type TO-5) are available with 8, 10, or 12 leads. Op-amp DIPs commonly have 8 or 14 pins. Of the three types, DIPs are still the most commonly used.

TABLE 22.1 Temperature Codes

Code	*Application*	*Temperature Range (°C)*
C	Commercial	0 to 70
I	Industrial	−25 to 85
M	Military	−55 to 125

Section Review

1. What is a *discrete* component?
2. What is an *integrated circuit*?
3. What is an *op-amp*?

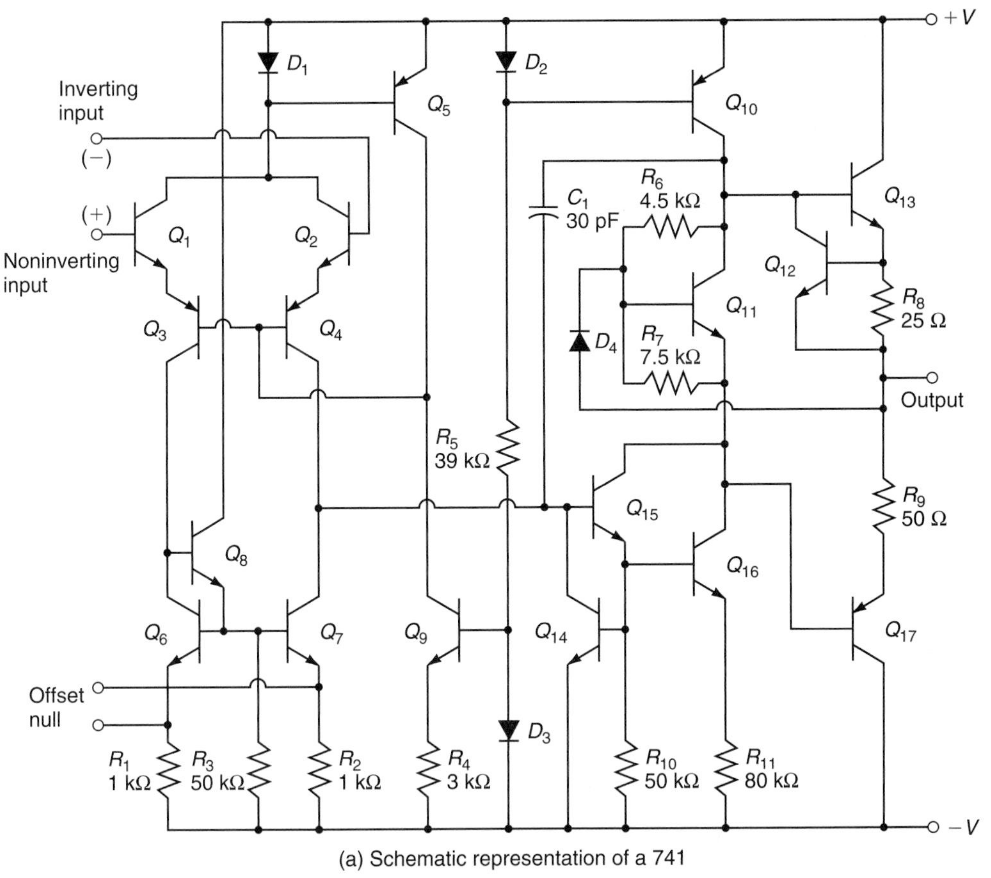

(a) Schematic representation of a 741

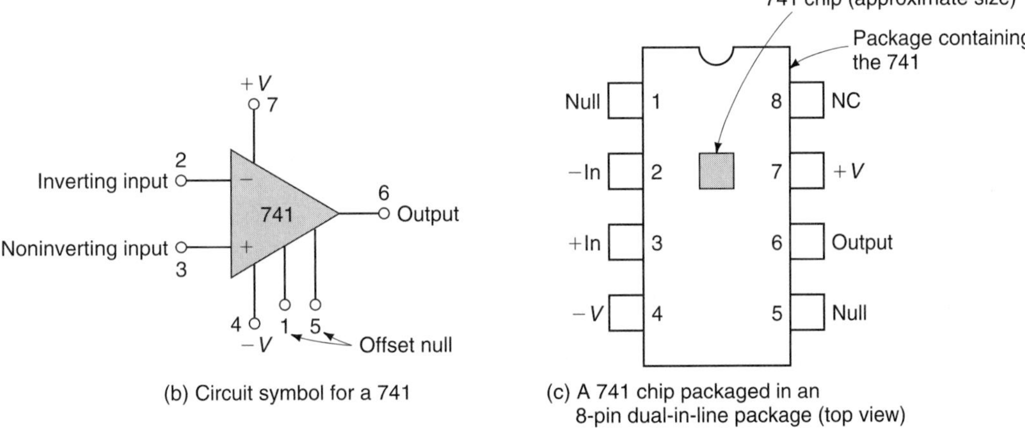

(b) Circuit symbol for a 741

(c) A 741 chip packaged in an 8-pin dual-in-line package (top view)

Figure 22.1 A 741 operational amplifier (op-amp).

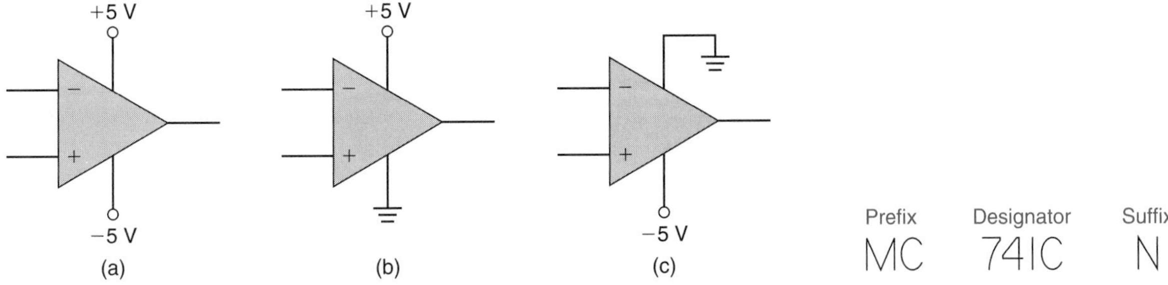

Figure 22.2 Op-amp supply voltages.

Figure 22.3 Op-amp ID code.

4. Draw the schematic symbol for an op-amp, and identify the signal and supply voltage inputs.
5. What are the three parts of the *op-amp identification code*? What does each part of the code tell you?
6. What are the three types of *op-amp packages*?

Note: NC stands for *not connected.* The pins labeled NC are not used.

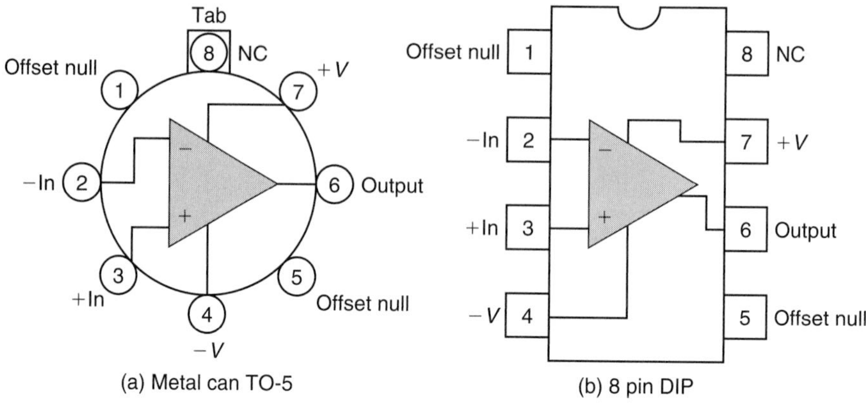

Figure 22.4 Op-amp packages.

22.2 Operation Overview

OBJECTIVE 3 ➤ The input stage of the op-amp is a **differential amplifier,** which is *a circuit that amplifies the difference of potential* (v_d) *across the input terminals.* As shown in Figure 22.5,

$$v_d = v_2 - v_1 \tag{22.1}$$

where

v_d = the voltage that will be amplified
v_1 = the voltage applied to the inverting input
v_2 = the voltage applied to the noninverting input

The output from the amplifier (for a given differential input) depends on four main factors: the *circuit gain, input polarity, supply voltages,* and *load resistance.* We will now look at each of these factors.

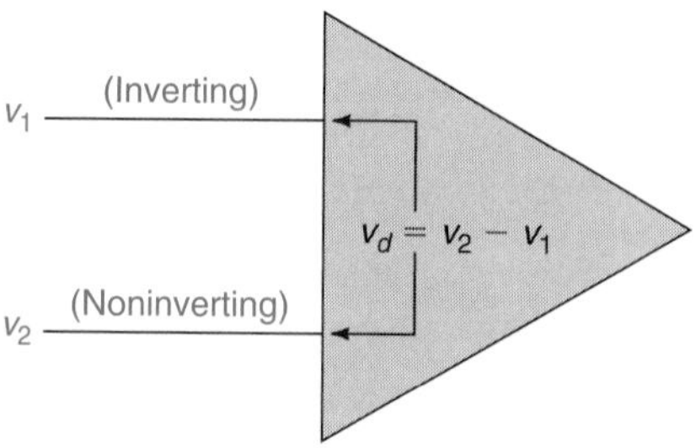

Figure 22.5

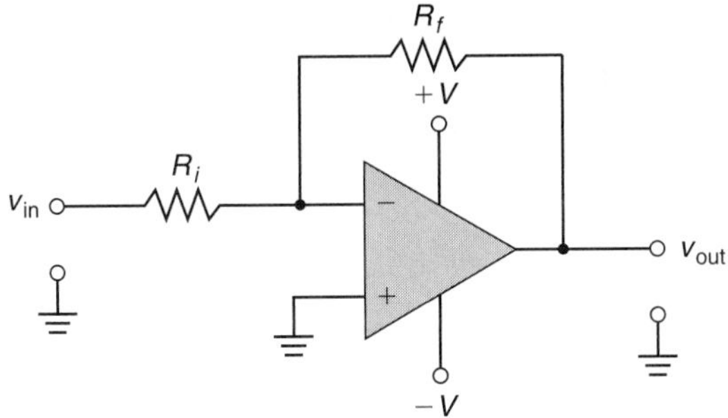

Figure 22.6 Op-amp feedback path.

Op-Amp Gain

The maximum possible gain provided by a given op-amp is referred to as its **open-loop voltage gain (A_{OL}).** The value of A_{OL} is generally greater than 10,000. For example, the Fairchild KA741 op-amp has an open-loop voltage gain of 200,000 (typical). ◄ *OBJECTIVE 4*

The term *open loop* indicates that there is no **feedback path** from the output to the op-amp input. Op-amp circuits usually contain one or more feedback paths. In Figure 22.6, the circuit containing R_f is the feedback path. Feedback reduces the effective gain of the op-amp. The effects of feedback on op-amp operation are discussed later in this chapter.

Input/Output Polarity

◄ *OBJECTIVE 5*

The polarity relationship between v_1 and v_2 determines the polarity of the op-amp output voltage. When v_1 is more negative than v_2, the op-amp output voltage swings toward $+V$. When v_1 is more positive than v_2, the output voltage swings toward $-V$. These relationships can be summarized as follows: *When the input voltage polarities, with respect to each other, match the polarity signs in the schematic symbol, the output voltage is positive; when they do not, the output is negative.* It is important to note that the output polarity is determined by the polarities of v_1 and v_2 *with respect to each other,* not by their polarities with respect to ground. Figure 22.7 illustrates this concept with several combinations of inputs.

Take a look at Figure 22.7e. When v_1 goes more positive than ground, the relative input polarities do not match the polarity signs, and the output goes negative. When v_1 goes more negative than ground, the relative input polarities match the polarity signs, and the output goes positive. Note the 180° phase shift between the op-amp input and output signals. This is where the term **inverting input** comes from. If you apply the same reasoning to the circuit shown in Figure 22.7f, you will see why the (+) input is referred to as the **noninverting input.**

There is a mathematical way to determine the polarity of the op-amp output voltage for a given set of input voltages. You may remember that equation (22.1) defines the differential input as

$$v_d = v_2 - v_1$$

When the result of this equation is positive, the op-amp output voltage is positive. When the result is negative, the output voltage is negative. These relationships are illustrated in Example 22.1.

Example 22.1

Determine the output voltage polarity for circuits (a) and (b) in Figure 22.7 using equation (22.1).

Solution: For circuit (a),

$$v_d = v_2 - v_1 = 4\text{ V} - 2\text{ V} = 2\text{ V}$$

Since the differential input (v_d) is positive, the op-amp output voltage is positive. For circuit (b),

$$v_d = v_2 - v_1 = 0\text{ V} - 2\text{ V} = -2\text{ V}$$

Since v_d is negative, the op-amp output voltage is negative.

Practice Problem 22.1

Using equation (22.1), determine the output voltage polarities for circuits (c) and (d) in Figure 22.7.

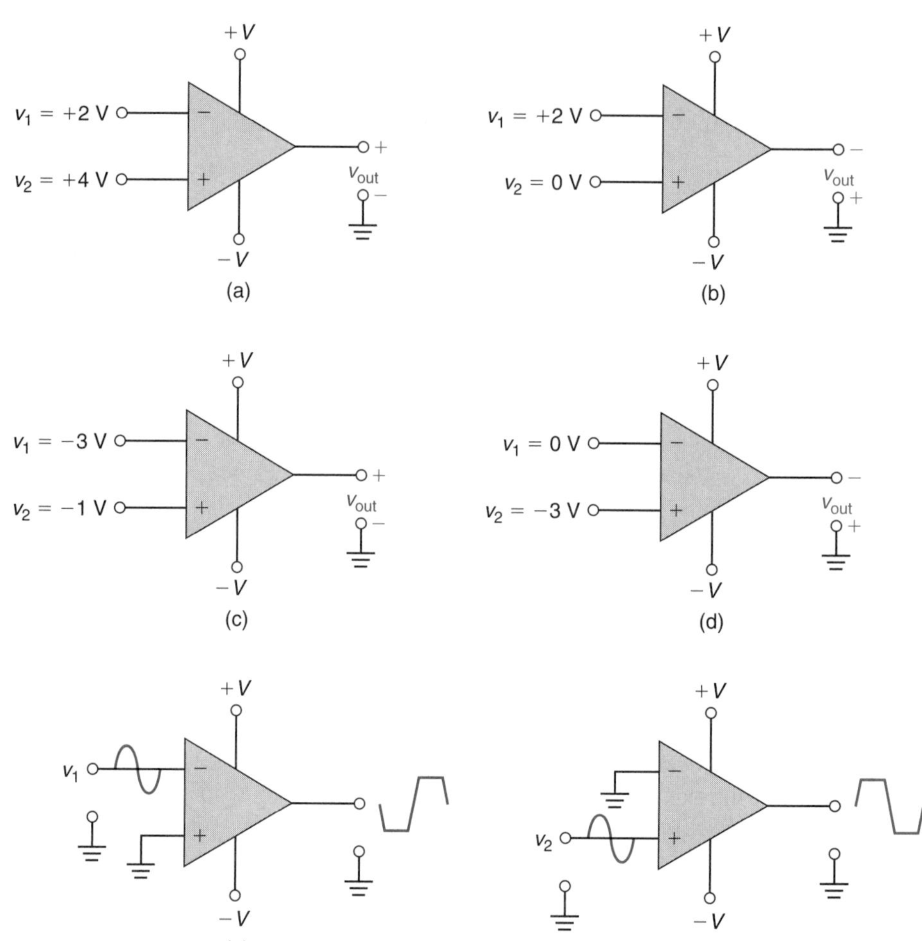

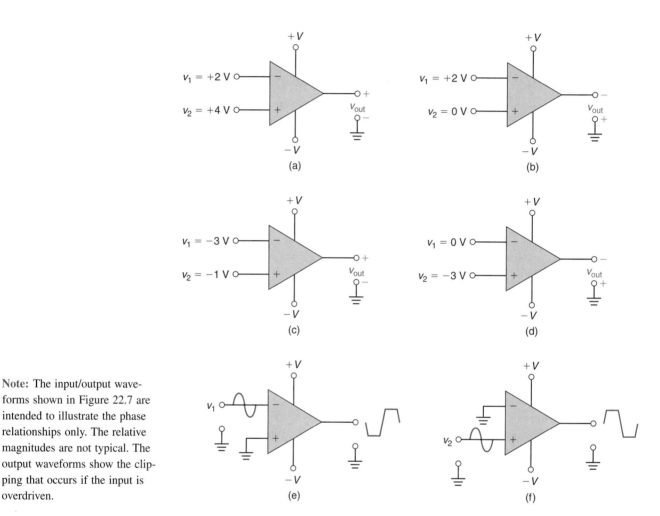

Note: The input/output waveforms shown in Figure 22.7 are intended to illustrate the phase relationships only. The relative magnitudes are not typical. The output waveforms show the clipping that occurs if the input is overdriven.

Figure 22.7 Op-amp input/output polarity relationships.

You may be wondering why we went to all the trouble of developing an observation method of analysis when the mathematical method seems so much simpler. The answer is simple. When you are analyzing a schematic, it is easier to use the mathematical method. However, when you are taking *actual* circuit measurements, the observation method is easier. You only need to determine the polarity relationship between the inputs, and then you can predict and observe the output voltage polarity.

Supply Voltages

The supply voltages ($+V$ and $-V$) determine the limits of the output voltage swing. For example, consider the circuits shown in Figure 22.8. Circuit (a) has supply voltages of

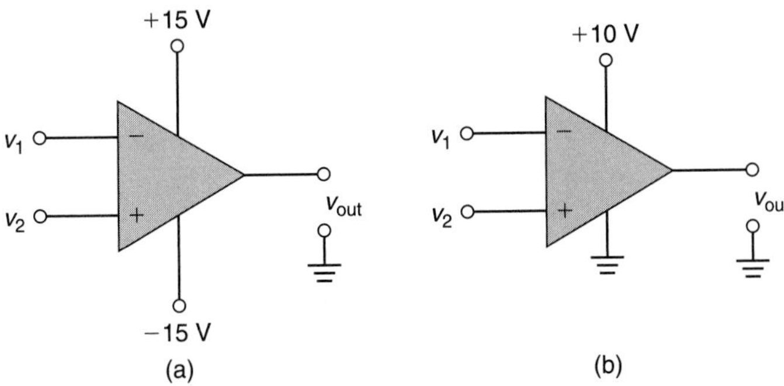

Figure 22.8 Op-amp supply voltages.

±15 V. Therefore, the output cannot go more positive than +15 V or more negative than −15 V. The output from circuit (b) would have limits of +10 V and ground.

In practice, the peak output voltage cannot reach either $+V$ or $-V$, because some voltage is dropped across the op-amp output circuit (see Figure 22.1a). The actual limits on v_{out} depend on the specific op-amp and the load resistance. For example, the spec sheet for the KA741 op-amp lists the following parameters:

Parameter	*Condition*	*Typical Value*[a]
Output voltage swing	$R_L \geq 10\ \text{k}\Omega$	±14 V
	$R_L \geq 2\ \text{k}\Omega$	±13 V

[a]$V_S = \pm 15$ V, $T_A = 25°$C.

If the load resistance is less than 2 kΩ, a graph like the one in Figure 22.9 can be used to determine the limits on the output voltage. If the source voltage is set to values other than ±15 V, a graph like the one in Figure 22.10 can be used to determine the limits on the output voltage. Generally, the following guidelines apply to any 741 op-amp:

Value of Load Resistance (kΩ)	*Max. (+) Output*	*Max. (−) Output*
>10	$(+V) - 1$ V	$(-V) + 1$ V
2 to 10	$(+V) - 2$ V	$(-V) + 2$ V

The following examples will help tie these concepts together.

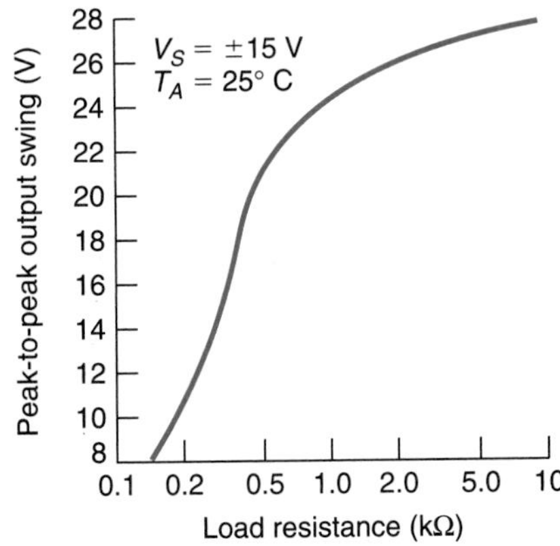

Figure 22.9 Output voltage as a function of load resistance.

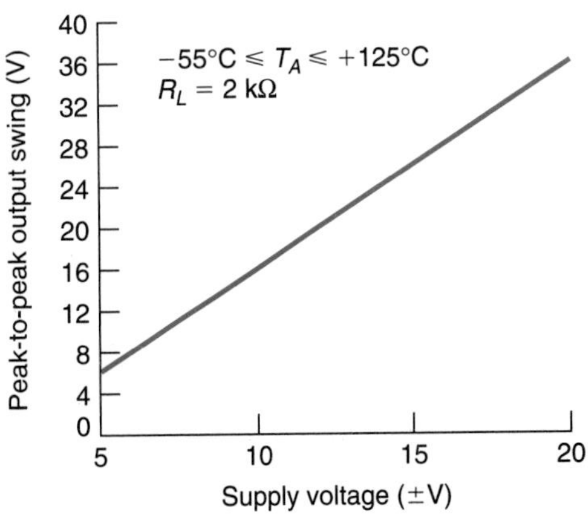

Figure 22.10 Output voltage as a function of supply voltage.

Example 22.2

Determine the maximum allowable value of v_{in} for the circuit shown in Figure 22.11. Assume that the voltage gain of the amplifier is 200.

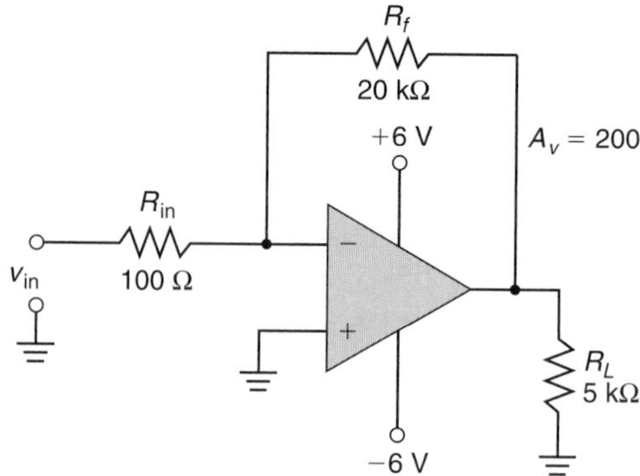

Figure 22.11

Solution: First, determine the maximum allowable peak output voltage values. Since the load resistance is between 2 and 10 kΩ, the peak output values are found as

$$V_{pk(+)} = (+V) - 2\text{ V} = +4 \qquad \text{(maximum)}$$

and

$$V_{pk(-)} = (-V) + 2\text{ V} = -4\text{ V} \qquad \text{(maximum)}$$

These results indicate that the maximum peak-to-peak output from the op-amp is 8 V. Using this value,

$$v_{in} = \frac{v_{out}}{A_v} = \frac{8\text{ V}_{PP}}{200} = 40\text{ mV}_{PP}$$

Practice Problem 22.2

An op-amp circuit has the following values: $+V = 12$ V, $-V = -12$ V, $A_v = 140$, and $R_L = 20$ kΩ. Determine the maximum allowable peak-to-peak input voltage for the amplifier.

Example 22.3

Determine the maximum peak output voltages for the circuit shown in Figure 22.12. Also, determine the maximum allowable value of v_{in} for the circuit. Assume that the gain of the amplifier is 121.

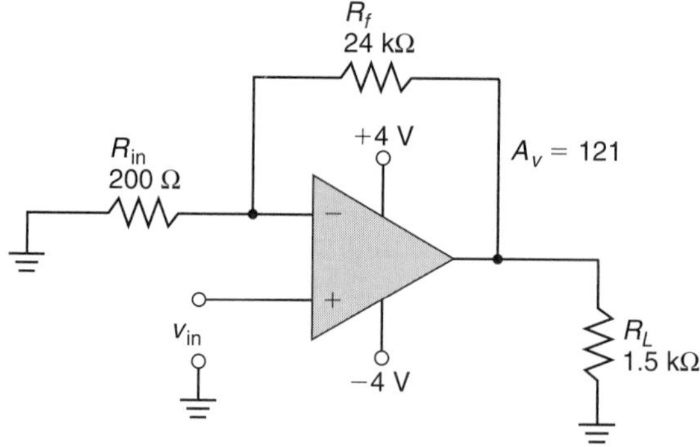

Figure 22.12

Solution: Since the load resistance is less than 2 kΩ, we must refer to the graph in Figure 22.9. The supply voltages are well below the maximum output value indicated in the graph for a 1.5 kΩ load (26 V_{PP} or ±13 V), so we can safely assume that the load will not restrict the output voltage swing. Thus,

$$V_{pk(+)} = (+V) - 2\text{ V} = +2\text{ V} \qquad \text{(maximum)}$$

and

$$V_{pk(-)} = (-V) + 2\text{ V} = -2\text{ V} \qquad \text{(maximum)}$$

Thus, the maximum output is 4 V_{PP}, and the maximum input is found as

$$v_{in} = \frac{v_{out}}{A_v} = \frac{4\text{ V}_{PP}}{121} = 33.06\text{ mV}_{PP}$$

Practice Problem 22.3

Determine the maximum allowable peak-to-peak input voltage for the amplifier described in Practice Problem 22.2 when R_L =1 kΩ. Use the curve in Figure 22.9 to determine the output voltage limits.

Section Review

1. What is a *differential amplifier*?
2. What determines the polarity of an op-amp output?
3. What is *open-loop voltage gain* (A_{OL})?
4. What is the typical range of A_{OL}?
5. Describe the *observation method* for determining the output voltage polarity for an op-amp.
6. Describe the *mathematical method* for determining the output voltage polarity for an op-amp.
7. What effect do the supply voltages of an op-amp have on its maximum peak-to-peak output voltage?
8. What effect does the load resistance of an op-amp have on its maximum peak-to-peak output voltage?
9. Using the component labels in Figure 22.1, explain why the op-amp output voltage cannot reach the values of $+V$ or $-V$. Critical Thinking

22.3 Differential Amplifiers and Op-Amp Specifications

Diodes, BJTs, and FETs all have parameters and electrical characteristics that affect their operation. Op-amps are no different. Earlier in this chapter, you were told that the input circuit to an op-amp is a differential amplifier. As you will see, this circuit determines many op-amp electrical characteristics.

The Basic Differential Amplifier

A *differential amplifier* is a circuit that accepts two inputs and produces an output that is proportional to the *difference* between those inputs. A basic differential amplifier is shown in Figure 22.13. Note that the input to Q_2 is the *noninverting* input (NI), and the input to Q_1 is the *inverting* input (I). ◄ *OBJECTIVE 6*

MSM

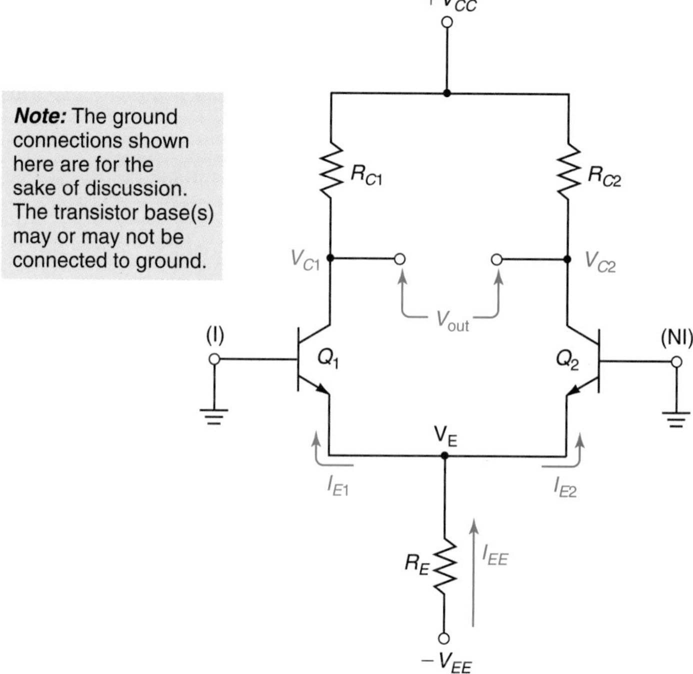
Note: The ground connections shown here are for the sake of discussion. The transistor base(s) may or may not be connected to ground.
+VCC
RC1
RC2
VC1
VC2
Vout
(I)
(NI)
Q1
Q2
VE
IE1
IE2
RE
IEE
−VEE

Figure 22.13 The basic differential amplifier.

Ideally, the Q_1 and Q_2 circuits have identical characteristics, so the quiescent values of I_E for the transistors are equal. When the two inputs are grounded as shown in Figure 22.13:

$$I_{E1} = I_{E2} \tag{22.2}$$

Since both emitter currents pass through R_E,

$$I_{E1} + I_{E2} = I_{EE} \tag{22.3}$$

where

$$I_{EE} = \frac{V_E - V_{EE}}{R_E} \tag{22.4}$$

Assuming negligible base current, $I_C \cong I_E$, and

$$I_{C1} + I_{C2} = I_{EE}$$

When $I_{C1} = I_{C2}$, the collector voltage for each transistor can be found using

$$V_C = V_{CC} - I_C R_C$$

and

$$V_{out} = V_{C1} - V_{C2} = 0 \text{ V} \qquad \text{(when } I_{C1} = I_{C2}\text{)}$$

When we apply a signal to the inverting input (Figure 22.14a), the positive alternation causes the current through Q_1 to increase. Since the value of I_{EE} is relatively fixed, the increase in I_{E1} results in a *decrease* in I_{E2}. The increase in I_{E1} and the decrease in I_{E2} cause V_{C1} to decrease and V_{C2} to increase. Assuming that the output signal (V_{out}) is taken from output 1 (with output 2 being the reference), we obtain the negative output

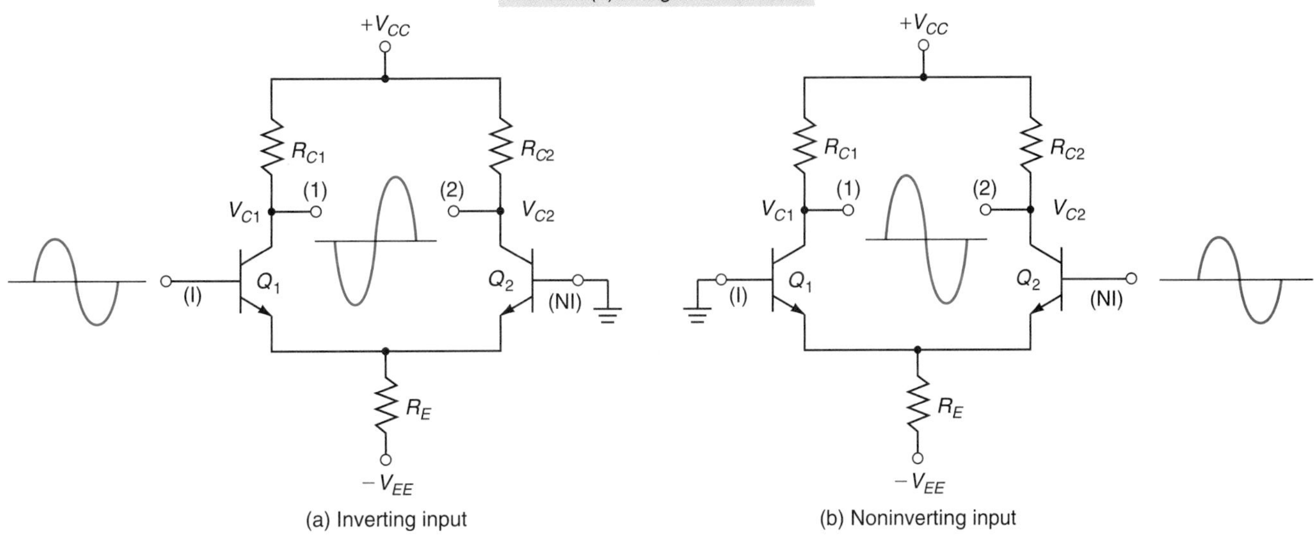

Figure 22.14

alternation shown in the figure. When the input goes negative, I_{E1} decreases and I_{E2} increases. As a result, V_{C1} increases, V_{C2} decreases, and v_{out} increases. Thus, the output voltage is 180° out of phase with the input.

In Figure 22.14b, a positive input causes I_{E2} to increase and I_{E1} to decrease. As you have been shown, these changes cause output 1 to go positive. A negative-going input causes I_{E2} to decrease and I_{E1} to increase. In this case, output 1 goes negative. Thus, the input and output voltages (in this case) are in phase.

Modes of Operation

There are three modes of operation for a differential amplifier:

- *Single-ended mode:* In single-ended mode, a signal is applied to only one input. The inactive input is normally connected directly (or via a resistor) to ground. Depending on which input is active, it is classified as either an inverting amplifier (Figure 22.14a) or a noninverting amplifier (Figure 22.14b).
- *Differential mode:* For differential operation, two signals are applied to the amplifier inputs. The output magnitude and polarity reflect the relationship between the two inputs.
- *Common mode:* This mode of operation occurs when identical signals are applied to both inputs. Ideally, with a differential input of $v_2 - v_1 = 0$ V, the circuit output also equals 0 V. The benefit of this mode is that any noise or undesired signal appearing at the two inputs simultaneously does not generate an output. In other words, common mode signals (ideally) cancel each other out.

Output Offset Voltage

Even though the transistors in the differential amplifier are closely matched, there are always some differences between the two. Any imbalance in V_{BE} between the transistors can result in an **output offset voltage,** which is *a positive or negative output voltage which appears at the output of an op-amp when no signal is applied to the inputs.* ◄ *OBJECTIVE 7*

One method of eliminating output offset voltage is to connect the op-amp's offset null pins as shown in Figure 22.15a. When the **offset null** is used, power is applied to the circuit, and the potentiometer is adjusted to eliminate the output offset.

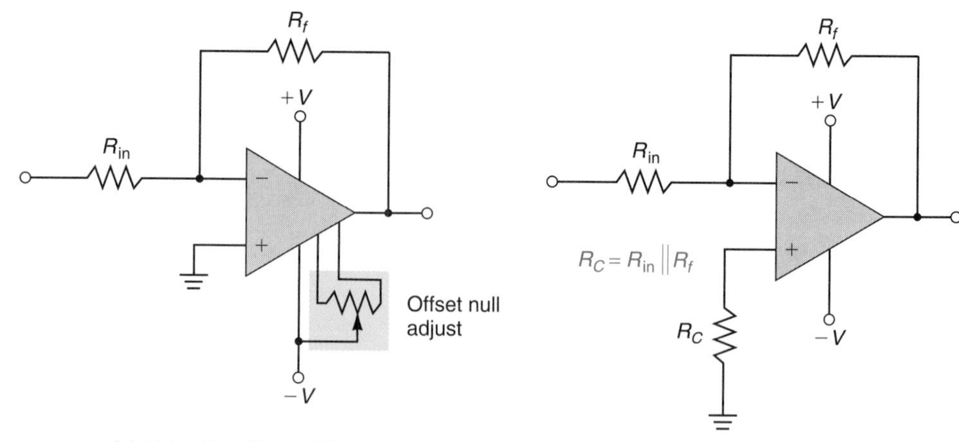

Figure 22.15 Reducing output offset voltage.

Input Offset Current

Input offset current is *a slight difference between the two input currents (i_1 and i_2) caused by a mismatch between the transistors.* This mismatch can also result in an output offset voltage. Figure 22.15b shows how a compensating resistor (R_C) between the noninverting input and ground can be used to compensate for any current mismatch. The current through R_C generates a proportional input voltage that offsets the imbalance caused by any difference in the input currents.

Input Bias Current

The inputs to an op-amp require some amount of dc biasing current for the BJTs in the differential amplifier. Therefore, an op-amp will not produce the expected output if either input is open. An **input bias current** path must *always* be provided for both op-amp inputs.

Common-Mode Rejection Ratio (CMRR)

Common-mode signals are *identical signals that are applied simultaneously to the two inputs of an op-amp.* Since the op-amp is designed to respond to the *difference* between two input signals, the ideal op-amp doesn't produce an output if there is no difference between the two inputs.

A measure of the ability of an op-amp to "ignore" common-mode signals is its **common-mode rejection ratio (CMRR).** Technically, CMRR is *the ratio of differential voltage gain to common-mode voltage gain.* For example, let's assume that the op-amp has a differential voltage gain of 1500 and a common-mode voltage gain of 0.01. The CMMR for the op-amp is found as

$$\text{CMRR} = \frac{A_v\ (\text{differential})}{A_v\ (\text{common mode})} = \frac{1500}{0.01} = 150{,}000\ (103.5\ \text{dB})$$

Note that CMMR is usually expressed in decibels.

Common-mode signals are usually undesired signals that are caused by external interference. The common-mode rejection ratio indicates the op-amp's ability to reject such unwanted signals.

Power Supply Rejection Ratio (PSRR)

Note: The ratings portion of the KA741 spec sheet can be found in Appendix C.

The **power supply rejection ratio (PSRR)** is *an op-amp rating that indicates the change in output voltage that occurs as a result of a change in supply voltage.* For example, the KA741 has a PSRR of 96 dB (typical). This rating indicates that any change in the power supply voltage is 96 dB greater than the resulting change in output voltage. For example, if the supply voltage changes by 1 V, the output voltage changes by approximately 16 μV.

Output Short-Circuit Current

Op-amps are protected internally from a shorted load. The **output short-circuit current** rating is *the maximum value of output current as measured with the load shorted.* For the KA741, this rating is 25 mA.

The output short-circuit current rating helps to explain why the op-amp output voltage drops when the load resistance decreases. For example, if a 50 Ω load is connected to the output of a KA741, the maximum possible output voltage is found as

$$V_L = I_{out}R_L = (25 \text{ mA})(50\ \Omega) = 1.25 \text{ V}$$

Slew Rate

◀ *OBJECTIVE 8*

The **slew rate** of an op-amp is *a measure of how fast the output voltage can change in response to a change at either signal input.* The slew rate of the KA741 is 0.5 V/μs (typical). This means that the output can change by a maximum of 0.5 V every microsecond. Since frequency is related to time, the slew rate can be used to determine the maximum operating frequency of the op-amp, as follows:

$$f_{max} = \frac{\text{slew rate}}{2\pi V_{pk}} \tag{22.5}$$

where

V_{pk} = the peak output voltage from the op-amp.

Example 22.4 demonstrates the use of this relationship.

Example 22.4

Determine the maximum operating frequency for the circuit shown in Figure 22.16. (Assume that 2 kΩ < R_L < 10 kΩ.)

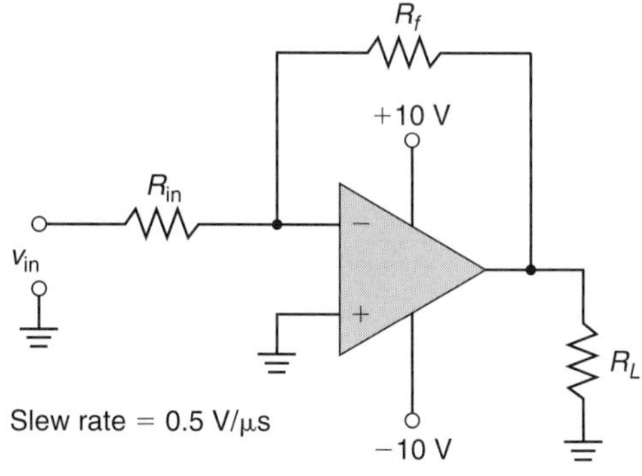

Figure 22.16

Solution: The maximum peak output voltage for this circuit is approximately 8 V. Using this value, the maximum operating frequency for the amplifier is found as

$$f_{max} = \frac{\text{slew rate}}{2\pi V_{pk}} = \frac{0.5 \text{ V/μs}}{2\pi(8 \text{ V})} = \frac{500 \text{ kHz}}{50.27} = 9.95 \text{ kHz}$$

Practice Problem 22.4

An op-amp with a slew rate of 0.4 V/μs has a 10 V_{pk} output. Determine the maximum operating frequency for the component.

Note: The slew rate in Example 22.4 converts to a frequency as follows:

$$\frac{0.5}{1 \text{ μs}} = \frac{0.5}{1 \times 10^{-6} \text{ s}} = 0.5 \times 10^6 \text{ Hz} = 500 \text{ kHz}$$

Note: The input and output waveforms may or may not be in phase as shown in Figure 22.17, depending on whether the component is being used as an *inverting* or *noninverting* circuit.

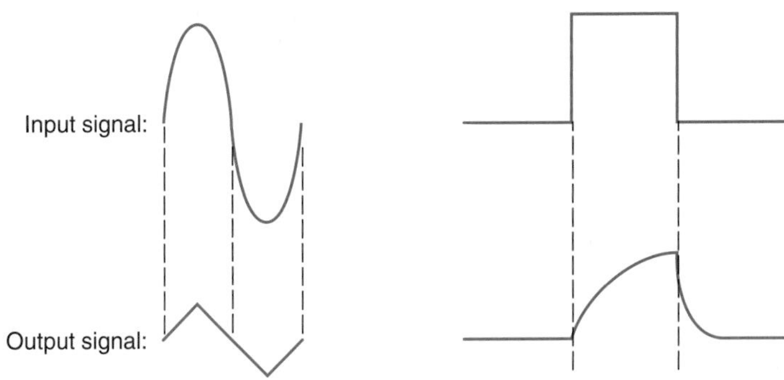

Figure 22.17 Slew-rate distortion.

When the maximum required output from the op-amp decreases, the maximum possible operating frequency increases. For example, if the peak output from the circuit shown in Figure 22.16 drops from 8 V_{pk} to 100 mV_{pk}, then

$$f_{max} = \frac{\text{slew rate}}{2\pi V_{pk}} = \frac{0.5 \text{ V}/\mu\text{s}}{2\pi(100 \text{ mV})} = 796 \text{ kHz}$$

This shows that the op-amp can be operated at a much higher frequency when it is used as a small-signal amplifier than when it is used as a large-signal amplifier.

The effects of operating an op-amp beyond its frequency limit (f_{max}) are illustrated in Figure 22.17. This figure shows two input waveforms and the distorted output waveforms. In both cases, the input waveform is changing at a rate that is greater than that at which the op-amp output can change.

Input/Output Resistance

As stated earlier in the chapter, op-amps typically have high input resistance and low output resistance. For the KA741, the input resistance is typically 2 MΩ. The output resistance of the component is 75 Ω up to frequencies of approximately 100 kHz. Output resistance then increases from 75 Ω to approximately 300 Ω (at a frequency of 1 MHz).

Other Op-Amp Characteristics

At this point, we will take a brief look at four additional electrical characteristics. These characteristics are found on most common op-amp spec sheets and are defined here for future reference:

- The **input voltage range** indicates *the maximum differential input to the op-amp.* If the voltage across the op-amp inputs exceeds this value, the component can be damaged. As shown on its spec sheet, the input voltage range of the KA741 can be as low as ±12 V.
- The **large-signal voltage gain** is simply *another name for open-loop voltage gain* (A_{OL}). The KA741 has a large-signal voltage gain rating of 200,000 (typical).
- The **supply current rating** indicates *the quiescent current that is drawn from the power supply.* For example, the KA741 draws a maximum of 2.8 mA from the dc power supply when it has no active input signal.
- The **power consumption rating** indicates *the amount of power that the op-amp will dissipate in its quiescent state.* For example, the power dissipation rating of the KA741 is 85 mW (maximum).

Section Review

1. Describe the differential amplifier's response to an input signal at its inverting input.
2. Describe the differential amplifier's response to an input signal at its noninverting input.

3. What is *output offset voltage*?
4. What effect does the *offset null* control have on the input offset voltage and output offset voltage? Explain your answer.
5. What is *input offset current*?
6. What is *input bias current*? What restriction does input bias current place on the wiring of an op-amp?
7. What are *common-mode* signals?
8. What is the *common-mode rejection ratio* (*CMRR*)? Why is a high CMRR considered desirable?
9. What is the *power supply rejection ratio* (*PSRR*)?
10. What is the *output short-circuit current* rating?
11. Discuss the relationship between *load resistance* and *maximum output voltage swing* in terms of the output short-circuit current rating.
12. What is *slew rate*? What circuit parameter is limited by the op-amp slew rate?
13. What is the result of trying to operate an op-amp beyond its slew-rate limit?

22.4 Inverting Amplifiers

◄ *OBJECTIVE 9*

The **inverting amplifier** is *the op-amp counterpart of the common-emitter and common-source circuits.* Like these discrete component circuits, the inverting amplifier produces a 180° voltage phase shift from input to output. At the same time, the inverting amplifier contains far fewer components than its discrete component counterparts. As shown in Figure 22.18a, the inverting amplifier contains an op-amp and two resistors.

The key to understanding inverting amplifier operation lies in the differential input circuit. If we assume that the amplifier in Figure 22.18a has a maximum output of ±5 V and an open-loop voltage gain of 200,000, the maximum possible differential input voltage is

$$v_d = \frac{v_{out}}{A_v} = \frac{5\text{ V}}{200{,}000} = 25\ \mu\text{V}$$

If the noninverting input is connected to ground (as shown in Figure 22.18b), the inverting input is always within ±25 μV of ground. As a result, the potential at the inverting input of the op-amp is referred to as **virtual ground.**

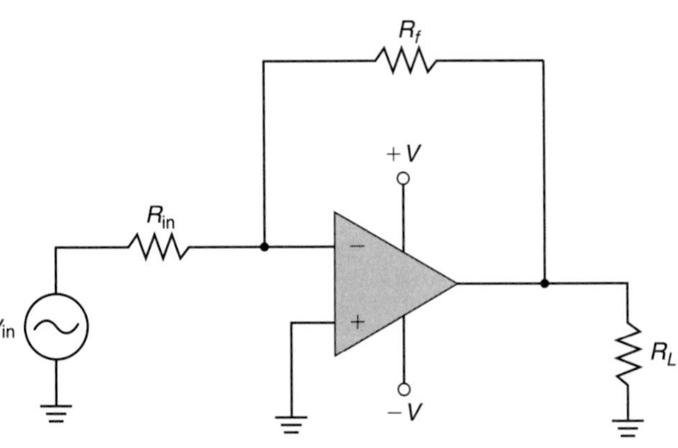

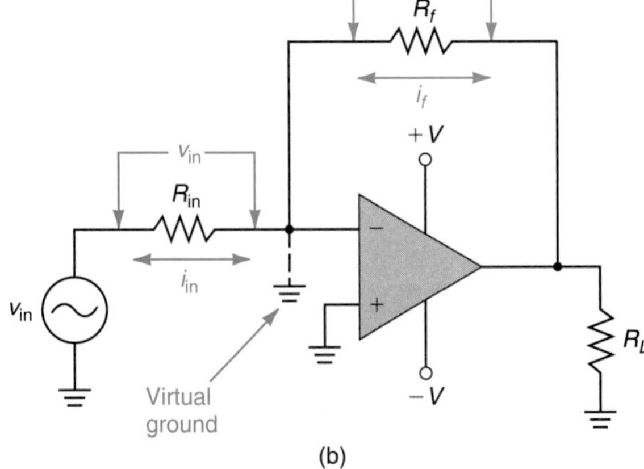

Figure 22.18 Inverting amplifier operation.

With the inverting input at virtual ground, the voltage across the *input resistor* (R_{in}) equals v_{in}. As shown in Figure 22.12b,

$$v_{in} = i_{in}R_{in} \tag{22.6}$$

At the same time, the voltage across the *feedback resistor* (R_f) equals v_{out}. By formula,

$$v_{out} = i_f R_f \tag{22.7}$$

Voltage gain equals the ratio of output voltage to input voltage, so the voltage gain of the inverting amplifier can be expressed as

$$A_v = \frac{i_f R_f}{i_{in} R_{in}} \tag{22.8}$$

As you know, the input resistance of an op-amp is extremely high (typically in the MΩ range). Since the values of R_{in} and R_f are typically much lower in value than the op-amp input resistance, $i_f \cong i_{in}$, and equation (22.8) can be rewritten as

$$A_v = \frac{R_f}{R_{in}} \tag{22.9}$$

Note: The value of A_v for an inverting amplifier is often written as a negative value to indicate the 180° voltage phase shift.

Thus, for the inverting amplifier, A_v is simply the ratio of R_f to R_{in}. The open-loop voltage gain (A_{OL}) is the gain of the op-amp when there is no feedback path. When a feedback path is present, the resulting voltage gain is referred to as the **closed-loop voltage gain (A_{CL}).**

Amplifier Input Impedance

While the op-amp has extremely high input impedance, the inverting amplifier does not. The reason for this can be seen by referring back to Figure 22.18b. As this figure shows, the voltage source "sees" an input resistance (R_{in}) that is going to (virtual) ground. Thus, the input impedance for the inverting amplifier is found as

$$Z_{in} \cong R_{in} \tag{22.10}$$

Amplifier Output Impedance

As shown in Figure 22.19, the output impedance of the inverting amplifier is the parallel combination of R_f and the output impedance of the op-amp itself. Since R_f is normally much greater than the value of Z_{out} for the op-amp, the output impedance of the circuit is usually assumed to be slightly lower than the Z_{out} rating of the op-amp.

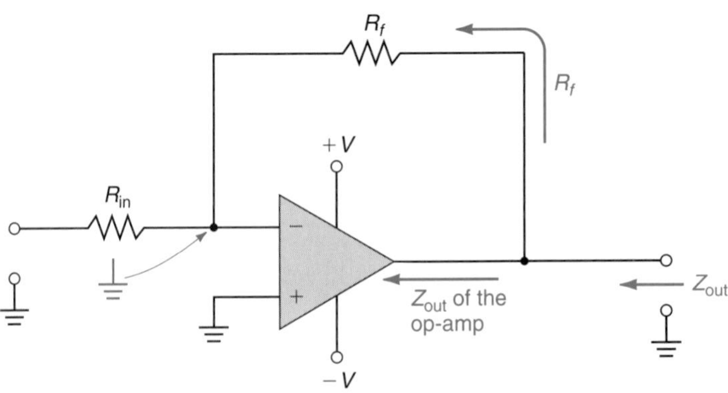

Figure 22.19 Inverting amplifier output impedance.

Amplifier CMRR

The CMRR of the inverting amplifier is typically much lower than that of the op-amp. The CMRR of the inverting amplifier is found as the ratio of *closed-loop gain* to *common-mode gain.* By formula,

$$\text{CMRR} = \frac{A_{CL}}{A_{CM}} \tag{22.11}$$

where

A_{CL} = the closed-loop voltage gain of the inverting amplifier
A_{CM} = the common-mode gain of the op-amp

Even though the CMRR of an inverting amplifier is always lower than that of the op-amp, it is still extremely high in most cases.

Inverting Amplifier Analysis

The complete analysis of an inverting amplifier is demonstrated in Example 22.5.

Example 22.5

Perform the complete analysis of the circuit shown in Figure 22.20.

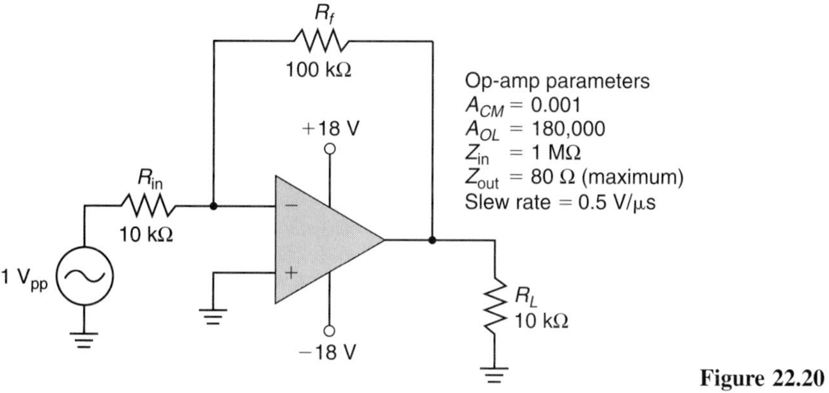

Figure 22.20

Solution: First, the closed-loop voltage gain (A_{CL}) of the circuit is found as

$$A_{CL} = \frac{R_f}{R_{\text{in}}} = \frac{100\ \text{k}\Omega}{10\ \text{k}\Omega} = 10$$

The circuit input impedance is found as

$$Z_{\text{in}} \cong R_{\text{in}} = 10\ \text{k}\Omega$$

The circuit output impedance is slightly lower than the output impedance of the op-amp, 80 Ω (maximum). The CMRR of the circuit is found as

$$\text{CMRR} = \frac{A_{CL}}{A_{CM}} = \frac{10}{0.001} = 10{,}000$$

To calculate $f_{\max}$, we need to determine the circuit's peak output voltage:

$$v_{\text{out}} = A_{CL}v_{\text{in}} = (10)(1\ \text{V}_{\text{PP}}) = 10\ \text{V}_{\text{PP}} = 5\ \text{V}_{\text{pk}}$$

Using this peak output voltage, the maximum operating frequency is found as

$$f_{\max} = \frac{\text{slew rate}}{2\pi V_{\text{pk}}} = \frac{0.5\ \text{V}/\mu\text{s}}{31.42\ \text{V}} = 15.9\ \text{kHz}$$

Practice Problem 22.5

An op-amp used in an inverting amplifier has the following ratings: $A_{CM} = 0.02$, $A_{OL} = 150{,}000$, $Z_{in} = 1.5\ \text{M}\Omega$, $Z_{out} = 50\ \Omega$ (maximum), and slew rate $= 0.75\ \text{V}/\mu\text{s}$. The supply voltages $= \pm 12$ V, $v_{in} = 50\ \text{mV}_{PP}$, $R_f = 250\ \text{k}\Omega$, and $R_{in} = 1\ \text{k}\Omega$. Perform a complete analysis of the circuit.

Section Review

1. Explain why the voltage gain of an inverting amplifier (A_{CL}) is equal to the ratio of R_f to R_{in}.
2. Why is the input impedance of an inverting amplifier generally less than that of the op-amp?
3. Why isn't R_f normally considered in the *output impedance* calculation for an inverting amplifier?
4. Why is the CMRR of an inverting amplifier lower than that of its op-amp?
5. List, in order, the steps involved in performing the complete analysis of an inverting amplifier.

Critical Thinking

6. Keeping equation (22.10) in mind, how would you go about increasing the A_{CL} of an inverting amplifier?
7. Refer to Example 22.5. What effect would doubling the value of R_{in} have on each of the values calculated in the example?

22.5 Noninverting Amplifiers

OBJECTIVE 10 ➤ The **noninverting amplifier** shares many characteristics of the inverting amplifier, with two exceptions:

- The noninverting amplifier has much higher circuit input impedance.
- The noninverting amplifier does not produce a 180° voltage phase shift from input to output.

The basic noninverting amplifier is shown in Figure 22.21. Note that the input is applied to the noninverting op-amp input, and that the input resistor (R_{in}) is returned to ground.

As shown in Figure 22.21, the input signal (v_{in}) is applied to the noninverting input. Since there is only a slight difference of potential between the signal inputs, v_{in} is effectively coupled from the noninverting input to the inverting input (as shown in the figure). As a result, v_{in} appears across R_{in}, and

$$v_{in} = i_{in}R_{in}$$

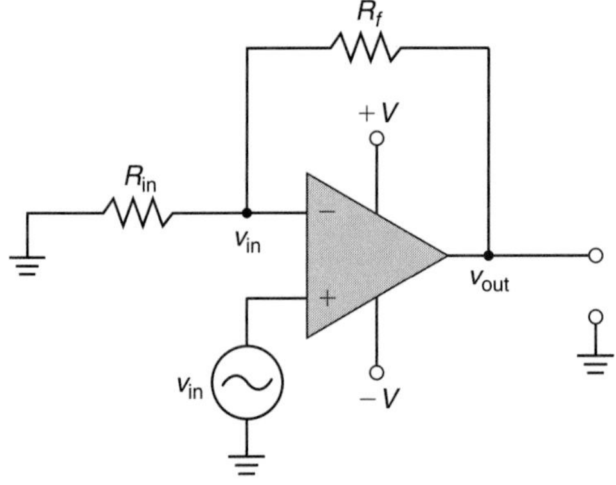

Figure 22.21 The noninverting amplifier.

As shown in Figure 22.21, the voltage across R_f equals $(v_{out} - v_{in})$. Therefore,

$$i_f = \frac{v_{out} - v_{in}}{R_f}$$

or

$$v_{out} = i_f R_f + v_{in}$$

Voltage gain equals the ratio of output voltage to input voltage, so the closed-loop voltage gain of the circuit can be expressed as

$$A_{CL} = \frac{i_f R_f + v_{in}}{i_{in} R_{in}}$$

Since $i_f \cong i_{in}$, the above relationship simplifes to

$$A_{CL} = \frac{R_f}{R_{in}} + 1 \qquad (22.12)$$

Amplifier Input and Output Impedance

Since the input signal is applied directly to the op-amp, the noninverting amplifier has extremely high input impedance. As with the inverting amplifier, the circuit's output impedance is slightly less than that of the op-amp.

Noninverting Amplifier Analysis

The complete analysis of the noninverting amplifier is demonstrated in Example 22.6.

Example 22.6

Perform the complete analysis of the noninverting amplifier shown in Figure 22.22.

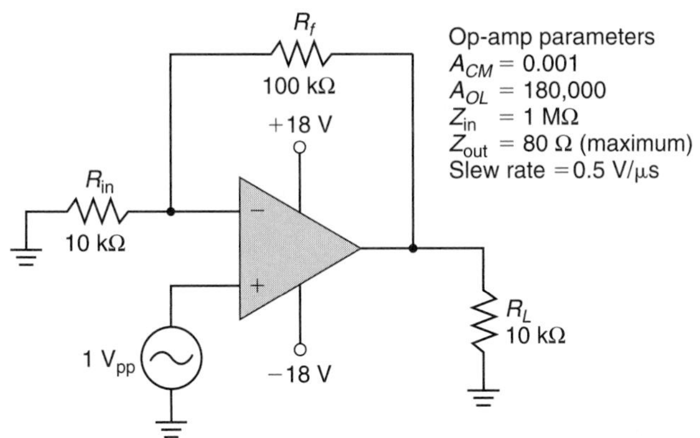

Figure 22.22

Solution: The closed-loop voltage gain (A_{CL}) of the circuit is found as

$$A_{CL} = \frac{R_f}{R_{in}} + 1 = \frac{100\text{ k}\Omega}{10\text{ k}\Omega} + 1 = 11$$

The circuit values of Z_{in} and Z_{out} are determined by the op-amp ratings: $Z_{in} \geq 1\text{ M}\Omega$, and $Z_{out} < 80\ \Omega$. The CMRR for the circuit is found as

$$\textbf{CMRR} = \frac{A_{CL}}{A_{CM}} = \frac{11}{0.001} = 11{,}000$$

To determine the value of f_{max}, we must first find the peak output voltage as

$$v_{out} = A_{CL}v_{in} = (11)(1\ V_{PP}) = 11\ V_{PP} = 5.5\ V_{pk}$$

The value of f_{max} is now found as

$$f_{max} = \frac{\text{slew rate}}{2\pi V_{pk}} = \frac{0.5\ V/\mu s}{34.56\ V} = 14.47\ kHz$$

Practice Problem 22.6

The circuit described in Practice Problem 22.5 is wired as a noninverting amplifier. Perform a complete analysis on the new circuit.

Figure 22.23 compares the characteristics of noninverting and inverting amplifiers.

The Voltage Follower

If we remove R_{in} and R_f from the noninverting amplifier and short the output to the inverting input, we have a **voltage follower.** This circuit, which is *the op-amp counterpart of the emitter and source follower,* is shown in Figure 22.24. You may recall that the characteristics of the emitter and source followers are:

- High Z_{in} and low Z_{out}.
- Input and output signals that are in phase.
- A_v that is less than 1.

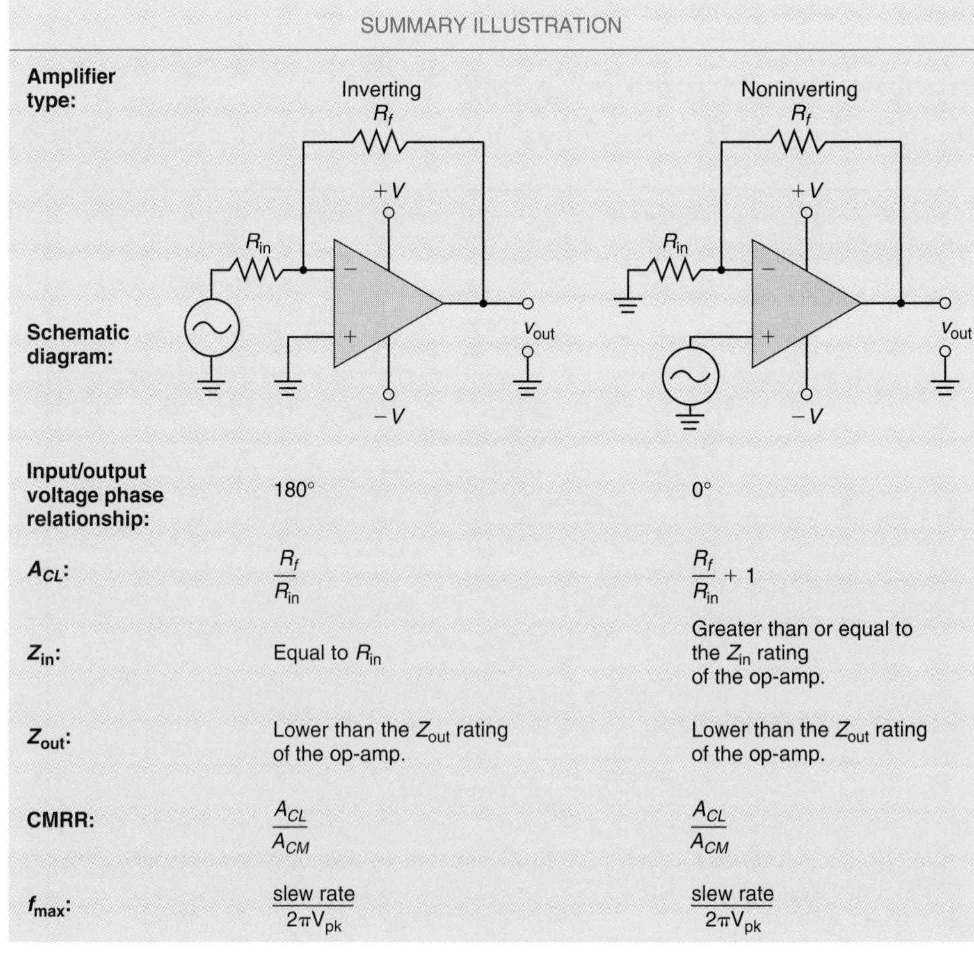

SUMMARY ILLUSTRATION
Amplifier type:
Inverting
Noninverting
R_f
R_{in}
$+V$
$-V$
V_{out}
Schematic diagram:
Input/output voltage phase relationship:
180°
0°
A_{CL}:
$\frac{R_f}{R_{in}}$
$\frac{R_f}{R_{in}} + 1$
Z_{in}:
Equal to R_{in}
Greater than or equal to the Z_{in} rating of the op-amp.
Z_{out}:
Lower than the Z_{out} rating of the op-amp.
Lower than the Z_{out} rating of the op-amp.
CMRR:
$\frac{A_{CL}}{A_{CM}}$
$\frac{A_{CL}}{A_{CM}}$
f_{max}:
$\frac{\text{slew rate}}{2\pi V_{pk}}$
$\frac{\text{slew rate}}{2\pi V_{pk}}$

Figure 22.23

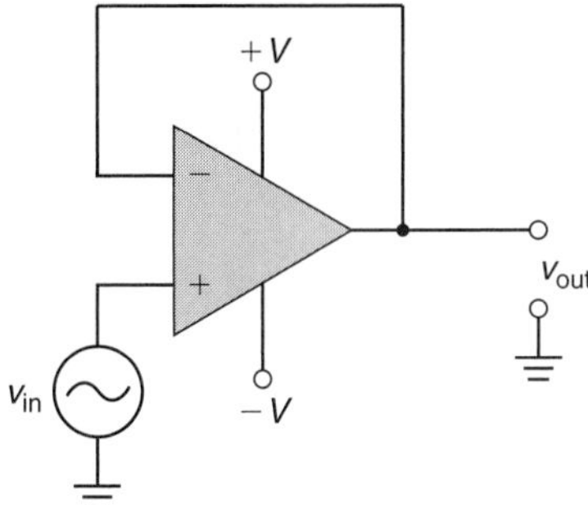

Figure 22.24 The voltage follower.

The first two characteristics are accomplished by using an op-amp in a noninverting circuit configuration. Since the voltage follower has no feedback resistor, its voltage gain is found as

$$A_{CL} = \frac{R_f}{R_{in}} + 1 = \frac{0\ \Omega}{R_{in}} + 1 = 1$$

The values of Z_{in}, Z_{out}, and f_{max} for the voltage follower are calculated using the same equations that we used for the basic noninverting amplifier. Since $A_{CL} = 1$, the CMMR for this circuit is found using

$$\text{CMMR} = \frac{1}{A_{CM}} \tag{22.13}$$

Section Review

1. How do you calculate the value of A_{CL} for a *noninverting amplifier*?
2. How do you determine the values of Z_{in} and Z_{out} for a noninverting amplifier?
3. How do you determine the CMRR for a noninverting amplifier?
4. Compare and contrast the inverting amplifier with the noninverting amplifier.
5. Describe the gain and impedance characteristics of the *voltage follower.*
6. Assume you have two circuits: a voltage follower, and a noninverting amplifier with a value of $A_{CL} = 10$. Assuming that the circuits have identical input signals, which will have the higher operating frequency? Explain your answer. Critical Thinking

22.6 Op-Amp Frequency Response

You have already been introduced to some of the frequency considerations involving op-amps. In this section, we will take a closer look at the effect of input frequency on the operation of an op-amp.

Frequency Versus Gain

◀ *OBJECTIVE 11*

The voltage gain of an op-amp remains stable between 0 Hz and some upper cutoff frequency (f_{C2}). This relationship is represented by the bode plot shown in Figure 22.25.

The op-amp is classified as a **dc amplifier,** because it has no lower cutoff frequency (f_{C1}). This means that the gain of the component remains relatively constant down to 0 Hz. As the input frequency increases, a point is reached at which the voltage gain starts to decrease. At the upper cutoff frequency (f_{C2}), the voltage gain is 3 dB lower than its

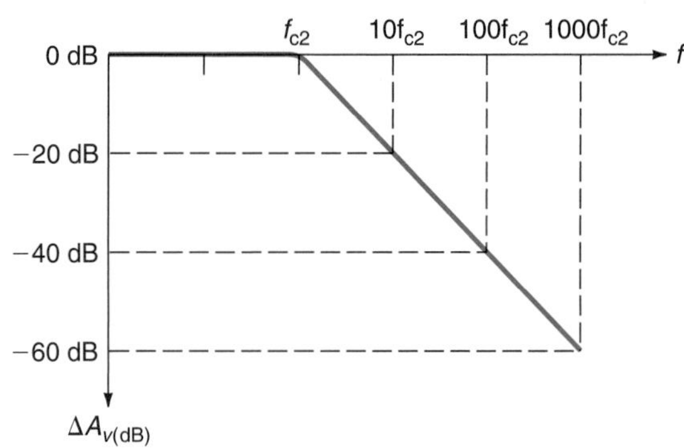

Figure 22.25 Op-amp frequency response.

midband (maximum) value. Note that the gain of the op-amp continues to decrease at the standard rate of 20 dB per decade above f_{C2}. Thus, increasing the operating frequency decreases the component gain.

There is another way to look at this frequency versus gain relationship: *Decreasing the voltage gain of an op-amp increases its maximum operating frequency.* This relationship can be demonstrated using the op-amp frequency response curve shown in Figure 22.26.

The maximum voltage gain on the curve is the open-loop voltage gain of the component, which is approximately 106 dB (200,000). If you want to operate the component so that $A_{v(\text{dB})} = A_{OL(\text{max})}$, you are limited to a maximum operating frequency of approximately 5 Hz. Above this frequency, A_{OL} begins to decrease.

Now, what if we use a feedback path that reduces the closed-loop gain of the amplifier to 60 dB? Figure 22.26 shows that $f_{C2} = 1$ kHz at $A_{CL} = 60$ dB. As you can see, decreasing the gain of the amplifier increases the value of f_{C2} of the device. In fact, each time we decrease A_{CL} by 20 dB, we increase the cutoff frequency by one decade. The maximum operating frequency of an op-amp is measured at $A_{CL} = 0$ dB (unity). This

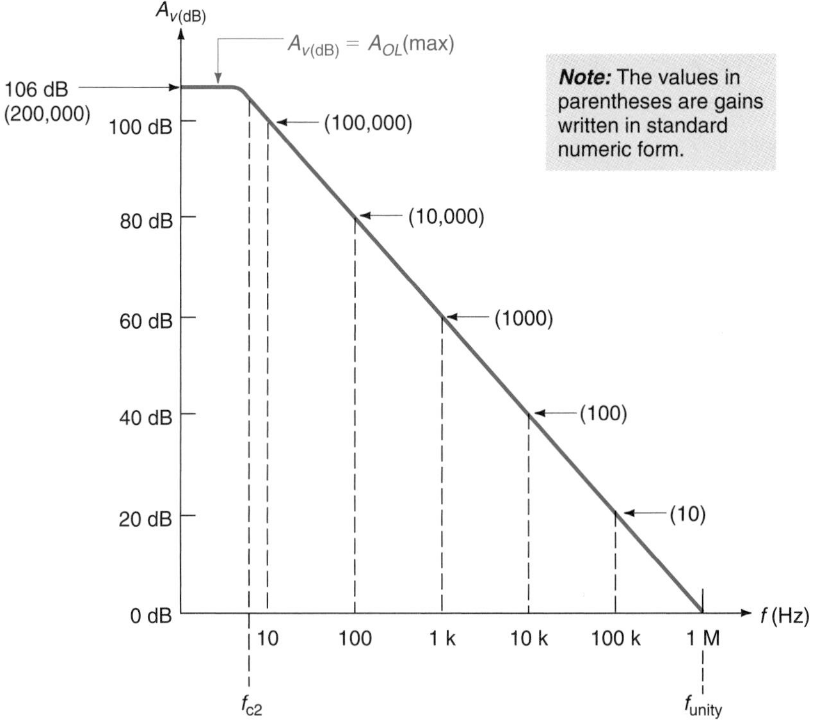

Figure 22.26

frequency is called the **unity-gain frequency (f_{unity}).** For the op-amp represented by the curve in Figure 22.26, f_{unity} is approximately equal to 1 MHz.

Since bandwidth increases as voltage gain decreases, we can make the following statements:

- The higher the gain of an op-amp, the narrower its bandwidth.
- The lower the gain of an op-amp, the wider its bandwidth.

Gain-Bandwidth Product

The **gain-bandwidth product** of an amplifier *is a constant that always equals the unity-gain frequency of its op-amp.* By formula,

$$A_{CL} f_{C2} = f_{unity} \quad (22.14)$$

The other two forms of equation (22.14) are

$$A_{CL} = \frac{f_{unity}}{f_{C2}} \quad (22.15)$$

and

$$f_{C2} = \frac{f_{unity}}{A_{CL}} \quad (22.16)$$

Examples 22.7 and 22.8 demonstrate the value of these equations in the frequency analysis of an op-amp circuit.

Example 22.7

The LM318 op-amp has a unity-gain frequency of 15 MHz. Determine the bandwidth of the LM318 when $A_{CL} = 500$ and the maximum value of A_{CL} at $f_{C2} = 200$ kHz.

Solution: When $A_{CL} = 500$, the value of f_{C2} is found as

$$f_{C2} = \frac{f_{unity}}{A_{CL}} = \frac{15 \text{ MHz}}{500} = 30 \text{ kHz}$$

Since the op-amp is capable of operating as a dc amplifier,

$$\text{BW} = f_{C2} = 30 \text{ kHz}$$

At $f_{C2} = 200$ kHz, the maximum value of A_{CL} is found as

$$A_{CL} = \frac{f_{unity}}{f_{C2}} = \frac{15 \text{ MHz}}{200 \text{ kHz}} = 75$$

Practice Problem 22.7

An op-amp has a gain-bandwidth product of 25 MHz. What is the BW of the device when $A_{CL} = 200$?

Example 22.8

We need to replace the op-amp in a circuit that has values of $A_{CL} = 500$ and BW = 80 kHz. All other factors being equal, can the op-amp represented by the curve in Figure 22.26 be used in this circuit?

Solution: The op-amp represented by Figure 22.26 has a value of $f_{unity} = 1$ MHz. Therefore, any product of $A_{CL} f_{C2}$ must be less than or equal to this value for the op-amp to be used. For our application,

$$A_{CL} f_{C2} = (500)(80 \text{ kHz}) = 40 \text{ MHz}$$

Since this value is greater than the f_{unity} rating of the substitute component, it cannot be used.

Practice Problem 22.8

We need to construct an amplifier with values of $A_{CL} = 52$ dB and BW = 10 kHz. Determine if an op-amp with a gain-bandwidth product of 5 MHz can be used in this application.

Some op-amp spec sheets list an f_{unity} rating; others will simply list a bandwidth rating. For example, the spec sheet for the KA741E op-amp lists a bandwidth rating of 1.5 MHz. This is the unity-gain frequency. When the operating curves for an op-amp are available, the gain versus frequency curve can be used to determine the value of f_{unity}. By selecting any frequency and multiplying it by the corresponding limit on voltage gain, you can obtain the value of f_{unity}.

Section Review

1. What is the relationship between op-amp operating frequency and voltage gain?
2. What is the *unity gain frequency* of an op-amp?
3. What is the relationship among f_{unity}, f_{C2}, and the voltage gain of an op-amp?
4. How can the gain versus frequency curve for an op-amp be used to determine f_{unity} for the device?

22.7 Op-Amp Circuits

In this section, you will be introduced to several common op-amp circuits to give you an idea of the versatility of these components.

Comparators

OBJECTIVE 12 ➤ A **comparator** is *an op-amp circuit that compares two voltages and provides an output indicating the relationship between them.* Generally, comparators are used to compare either:

- Two changing voltages to each other (as in comparing two sine waves).
- A changing voltage to a fixed reference voltage.

Figure 22.27 shows a comparator connected to a sinusoidal voltage source and a +10 V reference voltage (V_{ref}). When the voltage at the noninverting input exceeds the reference voltage, the output goes high and remains high until the input drops below +10 V. At that time, the output of the comparator goes low (as illustrated in the figure).

The output from a comparator is normally a dc voltage that indicates the polarity (or magnitude) relationship between the two input voltages. *A comparator used to compare a changing voltage to a fixed dc level* (as shown in Figure 22.27) is referred to as a **level detector.**

The most noticeable circuit recognition feature of the comparator is the lack of any feedback path. Therefore, the voltage gain of the circuit equals the open loop gain (A_{OL}) of the op-amp. With such a high gain, virtually any difference voltage at the inputs causes the output to go to one of the voltage extremes and then stay there until the difference voltage is removed. The polarity of the differential input voltage determines if the comparator output goes positive or negative, as demonstrated in Example 22.9.

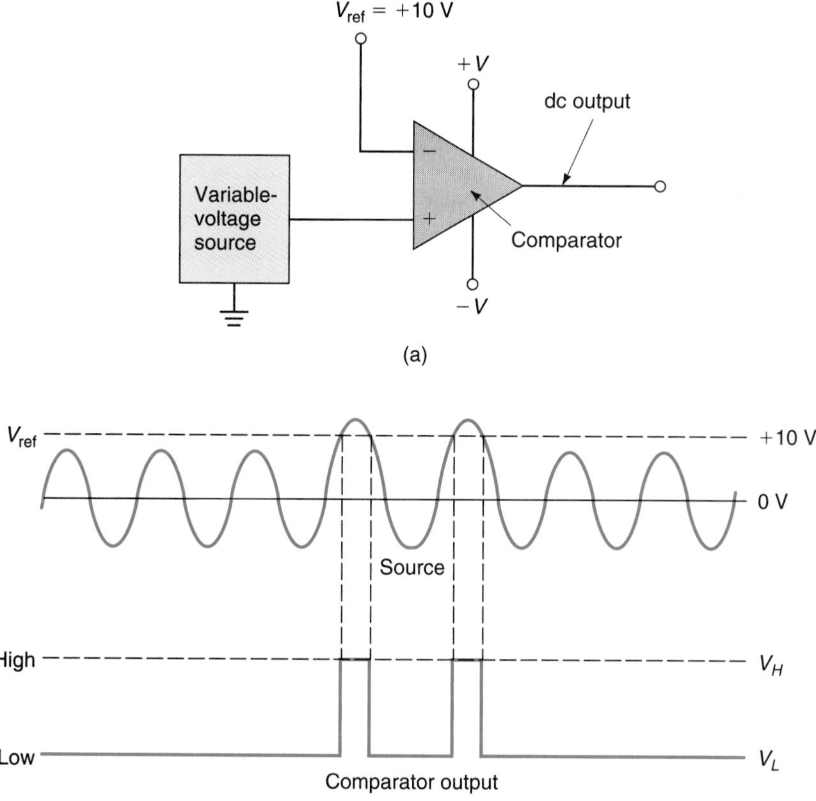

Figure 22.27 Basic comparator operation.

Example 22.9

Determine the output from the circuit shown in Figure 22.28 when the inverting input is at +5.01 V and +4.99 V.

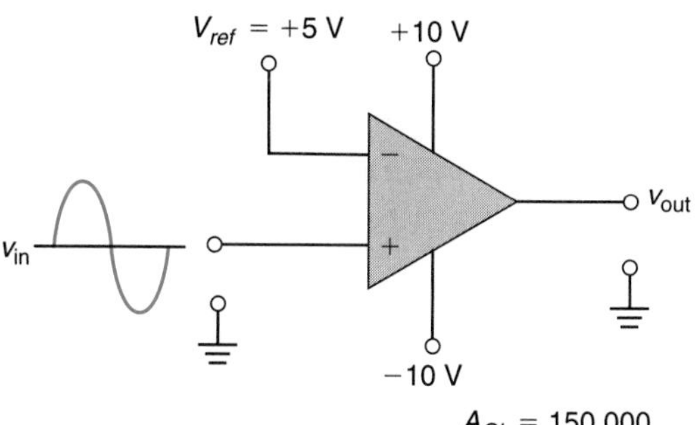

Figure 22.28

Solution: When the input voltage is at +5.01 V,

$$v_d = v_{in} - V_{ref} = 5.01\text{ V} - 5\text{ V} = +0.01\text{ V}$$

and

$$v_{out} = A_{OL}v_d = (150{,}000)(0.01\text{ V}) = 150\text{ V}$$

Since this is clearly beyond the output limits of the circuit, the output voltage is found as

$$v_{out} \cong (+V) - 1\text{ V} = +9\text{ V}.$$

When the input voltage is at +4.99 V,

$$v_d = v_{\text{in}} - V_{ref} = +4.99\text{ V} - 5\text{ V} = -0.01\text{ V}$$

and

$$v_{\text{out}} = A_{OL}v_d = (150{,}000)(-0.01\text{ V}) = -150\text{ V}$$

Again, this output is clearly impossible for the circuit shown. Thus, the output is found as

$$v_{\text{out}} \cong (-V) + 1\text{ V} = -9\text{ V}$$

Practice Problem 22.9

A comparator like the one shown in Figure 22.28 has its inverting input connected to a +2 V reference. The op-amp is connected to ±12 V supplies and has a value of A_{OL} = 70,000. Determine the output voltage for input voltages of +2.001 V and +1.999 V.

In many cases, a voltage divider is used to set the comparator reference voltage. For example, the reference voltage for the circuit shown in Figure 22.29 is found as

$$V_{ref} = +V\frac{R_2}{R_1 + R_2} = (+5\text{ V})\frac{30\text{ k}\Omega}{150\text{ k}\Omega} = +1\text{ V}$$

The circuit shown in Figure 22.29 contains a bypass capacitor in the voltage-divider circuit. This bypass capacitor is included to prevent the variations in v_{in} from being coupled to the voltage-divider circuit through the op-amp. The result is that the output of the circuit is much more reliable.

Circuit variations Up to this point, we have assumed that the comparator is used to provide a *positive* output when the input voltage is more *positive* than some *positive* reference voltage. However, this is not always the case. For example, take a look at the circuits shown in Figure 22.30. The input/output relationships for the circuits shown are as follows:

- *Figure 22.30a:* The circuit has a *positive* output when the input voltage is more *negative* than the *positive* reference voltage.
- *Figure 22.30b:* The circuit has a *positive* output when the input voltage is more *positive* than the *negative* reference voltage.
- *Figure 22.30c:* The circuit has a *positive* output when the input voltage is *positive.*
- *Figure 22.30d:* The circuit has a *positive* output when the input voltage is *negative.*

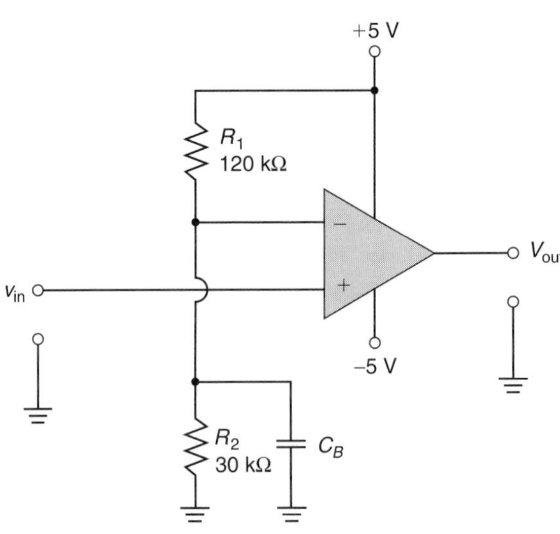

Figure 22.29 Comparator with a reference-setting circuit.

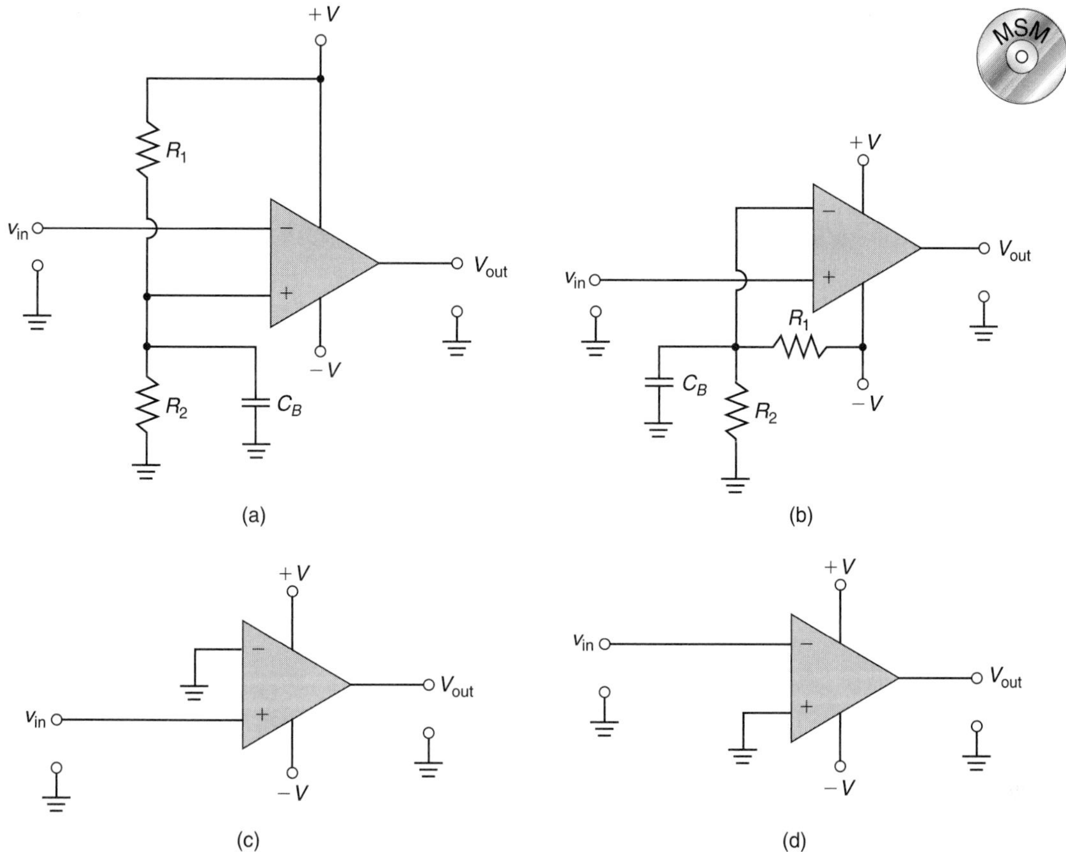

Figure 22.30 Comparator circuit variations.

Integrators

An **integrator** is *a circuit that converts a square wave into a triangular wave,* as shown in Figure 22.31. As you can see, the integrator contains a single op-amp and an *RC* circuit. ◄ *OBJECTIVE 13*

The integrator provides an output that is *proportional* to the area of its input waveform. The concept of waveform area is illustrated in Figure 22.32. The area of the waveform shown in Figure 22.32a is found as

$$A = Vt$$

where

A = the area of the waveform
V = the peak voltage of the waveform
t = the pulse width of the waveform

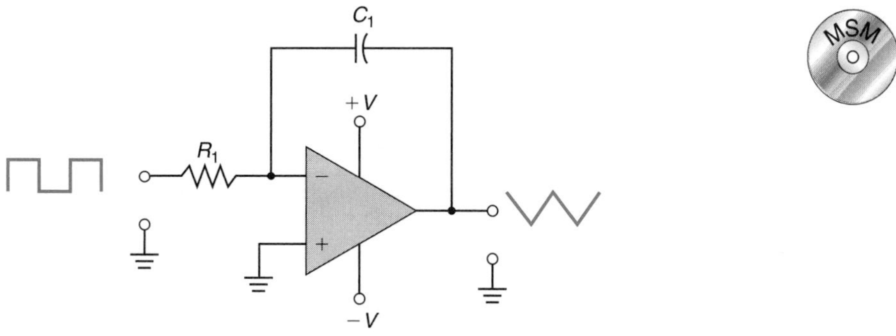

Figure 22.31 Op-amp integrator.

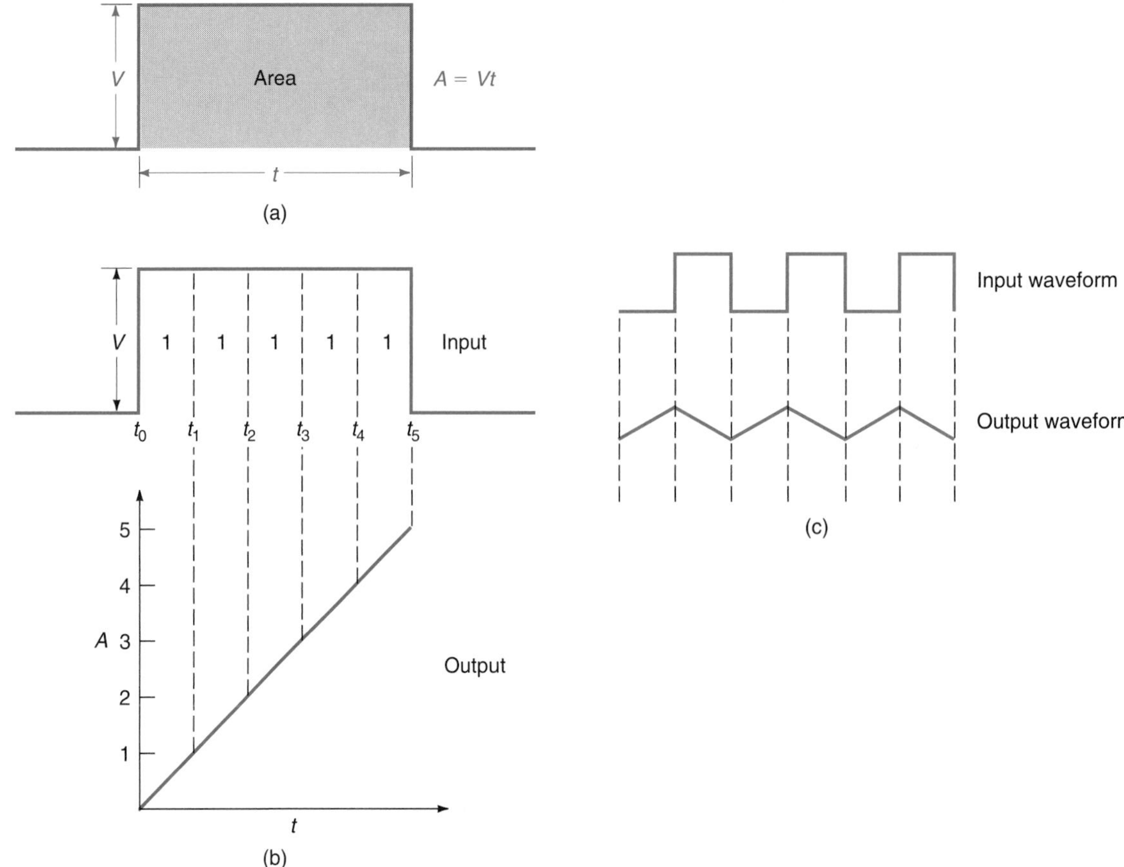

Figure 22.32

Figure 22.32b shows how the output of the integrator varies with the area of the input waveform. The input waveform has been divided into five equal sections. From t_0 to t_1, we have one unit of area. In response to this input condition, the output waveform of the integrator goes to 1. When t_2 is reached, the area of the waveform is now two units, and the integrator output goes to 2, and so on.

As the input cycles continue, the integrator produces the corresponding output waveform shown in Figure 22.32c. Note that the waveform is shifted 180°, because the input signal is applied to the inverting input of the op-amp (see Figure 22.31). As you can see, the integrator is acting as a square wave–to–triangular waveform converter.

Integrator operation The simplest integrator is an *RC* circuit like the one shown in Figure 22.33. With the square-wave input shown, the *RC* integrator would ideally have a linear triangular output. The only problem is that the capacitor does not charge/discharge at a linear rate but, rather, at an exponential rate. This produces the "actual" waveform shown in the figure.

By adding an op-amp to an *RC* integrator, we can obtain a linear output that is much closer to the ideal. The key to is to provide a constant-current charge path for the capacitor. This means that the rate at which the capacitor charges becomes constant, and a linear output is produced. Keeping this in mind, let's take a look at the circuit shown in Figure 22.34. The constant-current characteristic of this circuit is based on two well-known points:

- The inverting input to the op-amp is held at virtual ground.
- Virtually all of I_1 passes through the feedback path because of the op-amp's very high input impedance.

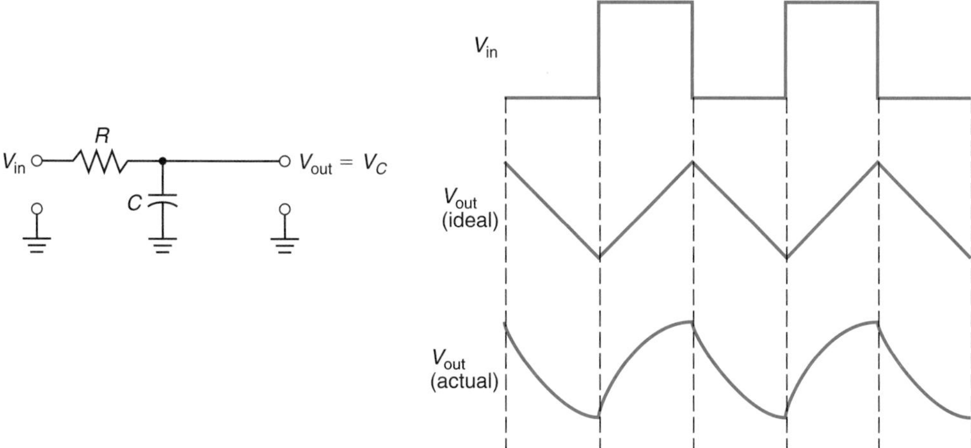

Figure 22.33 RC integrator.

Since the inverting input is held at virtual ground, the value of input current (I_1) is found as

$$I_1 = \frac{V_{in}}{R_1}$$

Assuming that V_{in} is constant over a period of time and R_1 is fixed, I_1 remains constant over the same period. As a result, the capacitor charges (and discharges) at a linear rate. This produces the ramp (ideal) output shown in Figure 22.35. Note that a *positive* input produces a *negative* ramp, because the integrator input is applied to the inverting op-amp input. By the same token, the transition to $-V$ produces the *positive* ramp shown in the figure.

Differentiators

If we reverse the integrator capacitor and resistor (as shown in Figure 22.36), we have a *differentiator.* The **differentiator** is *a circuit that provides an output that is proportional to the rate of change of its input signal.* The input/output relationship of the differentiator is illustrated in Figure 22.37.

◄ *OBJECTIVE 14*

The input signal for the differentiator is a triangular waveform. Between t_0 and t_2, the rate of change of the input is constant. The change is in a *positive* direction, so the inverting action of the op-amp produces a constant *negative* output voltage. Between t_2 and t_4, the rate of change is a *negative* constant, so the output switches to a constant *positive*

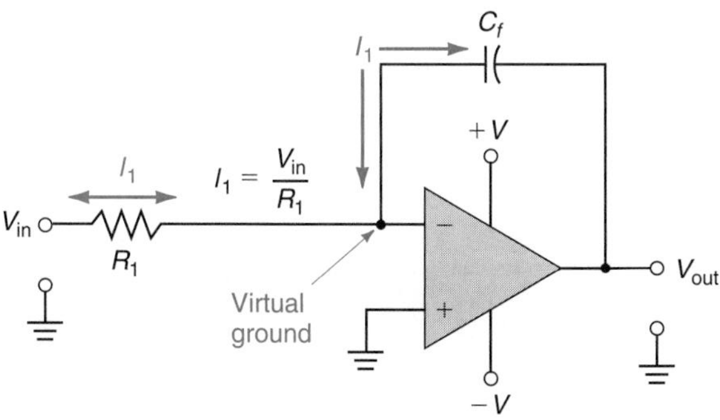

Figure 22.34 Op-amp integrator constant-current characteristics.

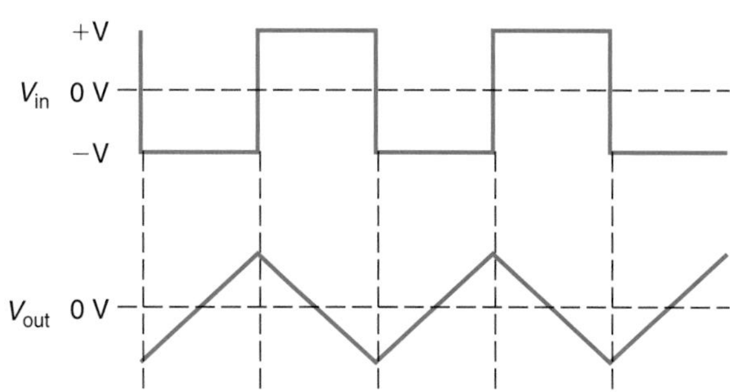

Figure 22.35 Op-amp integrator phase relationship.

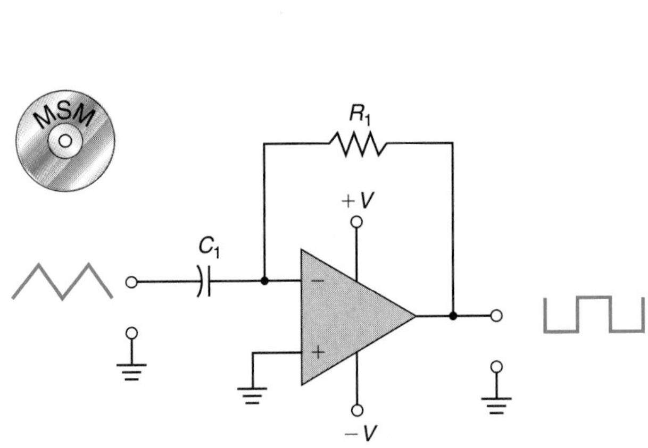

Figure 22.36 The differentiator.

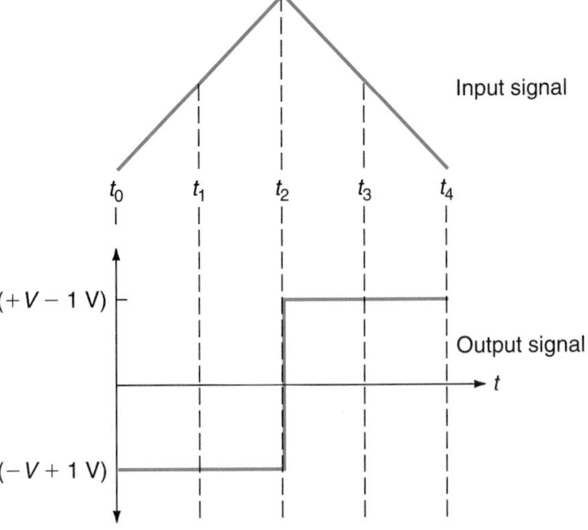

Figure 22.37 The input/output relationship of the differentiator.

voltage. Thus, with a *triangular* input waveform, the differentiator produces a *rectangular* output waveform.

Summing Amplifiers

OBJECTIVE 15 ➤ The **summing amplifier** is *an op-amp circuit that provides an output proportional to the sum of its inputs.* A basic summing amplifier is shown in Figure 22.38. To understand this circuit, we start by considering each input as an individual circuit. If V_1 were the only input, the output would be found as

$$V_{out} = -V_1 \frac{R_f}{R_1}$$

Similarly, if V_2 or V_3 were the only inputs, the respective outputs would be found as

$$V_{out} = -V_2 \frac{R_f}{R_2} \qquad \text{or} \qquad V_{out} = -V_3 \frac{R_f}{R_3}$$

When we have all three inputs, the output is found using the sum of these three equations:

$$V_{out} = \frac{-V_1 R_f}{R_1} + \frac{-V_2 R_f}{R_2} + \frac{-V_3 R_f}{R_3}$$

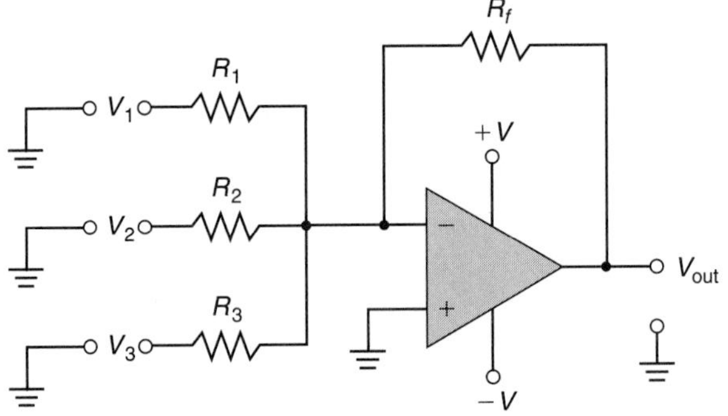

Figure 22.38 Summing amplifier.

Or, after factoring out the value of $-R_f$,

$$V_{out} = -R_f\left(\frac{V_1}{R_1} + \frac{V_2}{R_2} + \frac{V_3}{R_3}\right) \tag{22.17}$$

Example 22.10 demonstrates the procedure for determining the output from a summing amplifier.

Example 22.10

Determine the output voltage from the summing amplifier shown in Figure 22.39.

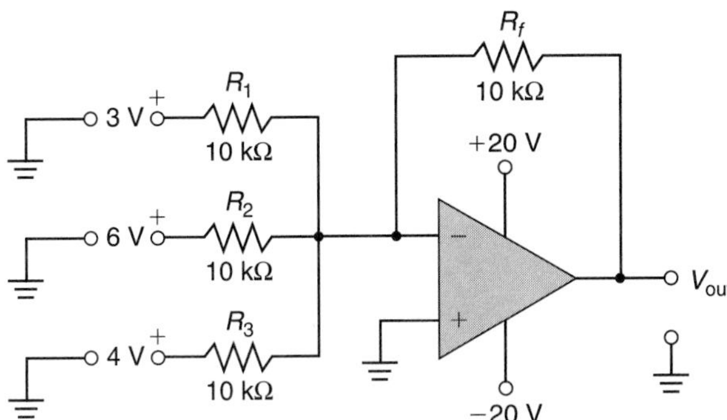

Figure 22.39

Solution: Using the values shown in the figure, the output voltage is found as

$$\begin{aligned} V_{out} &= -R_f\left(\frac{V_1}{R_1} + \frac{V_2}{R_2} + \frac{V_3}{R_3}\right) \\ &= (-10\text{ k}\Omega)\left(\frac{+3\text{ V}}{10\text{ k}\Omega} + \frac{+6\text{ V}}{10\text{ k}\Omega} + \frac{+4\text{ V}}{10\text{ k}\Omega}\right) = -13\text{ V} \end{aligned}$$

Practice Problem 22.10

A summing amplifier like the one shown in Figure 22.39 has the following values: $R_f = 2\text{ k}\Omega$, $R_1 = R_2 = R_3 = 1\text{ k}\Omega$, $V_1 = 1\text{ V}$, $V_2 = 500\text{ mV}$, and $V_3 = 1.5\text{ V}$. The supply voltages for the circuit are ± 18 V. Determine the value of V_{out} for the circuit.

If you compare the result in Example 22.10 with the input voltages, you will see that the output magnitude equals the sum of the input voltages. It is not always possible to have an output that is equal to the sum of the input voltages, because the sum of the inputs may exceed the maximum output voltage of the circuit. To solve this problem, the value of R_f is usually chosen so that the output voltage is *proportional* to the sum of the inputs. This point is illustrated in Example 22.11.

Example 22.11

Determine the output voltage from the circuit shown in Figure 22.40.

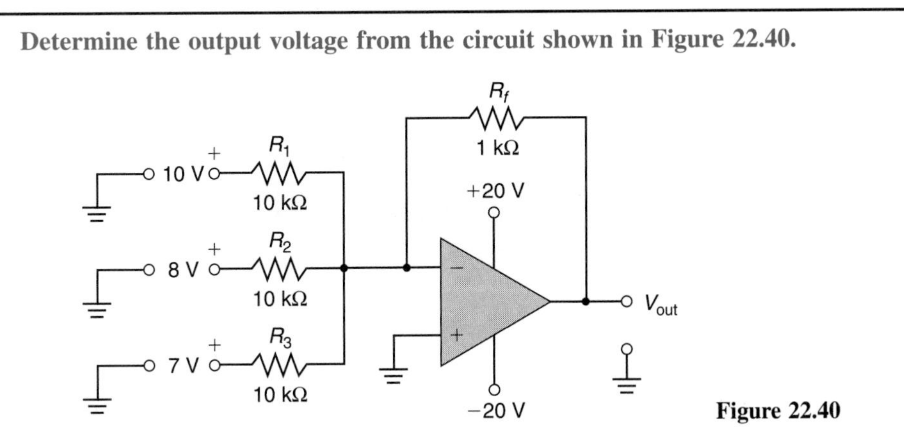

Figure 22.40

Solution: Using the values shown in the figure, the output voltage is found as

$$V_{out} = -R_f\left(\frac{V_1}{R_1} + \frac{V_2}{R_2} + \frac{V_3}{R_3}\right)$$

$$= (-1\text{ k}\Omega)\left(\frac{+10\text{ V}}{10\text{ k}\Omega} + \frac{+8\text{ V}}{10\text{ k}\Omega} + \frac{+7\text{ V}}{10\text{ k}\Omega}\right) = -2.5\text{ V}$$

Practice Problem 22.11

A summing amplifier like the one shown in Figure 22.40 has the following values: $R_f = 1\text{ k}\Omega$, $R_1 = R_2 = R_3 = 3\text{ k}\Omega$, $V_1 = 2$ V, $V_2 = 1$ V, and $V_3 = 6$ V. The supply voltages are ± 12 V. Determine the value of V_{out} for the circuit.

In Example 22.11, we solved the problem of overdriving the amplifier with +10 V, +8 V, and +7 V inputs by *decreasing* the value of R_f. Now the output magnitude is not equal to the sum of the inputs, but rather, is *proportional* to the sum of the inputs. In this case, it equals *one-tenth* the input sum. Note that the input resistors of a summing amplifier do not always have equal values. In other words, the inputs can be *weighted* so that certain inputs have more of an effect on the output.

Summing Amplifier Modifications

Figure 22.41 shows a summing amplifier designed to operate as an *averaging amplifier.* An **averaging amplifier** is *a circuit that produces an output equal to the average of its inputs.*

The input/output relationship for the circuit shown in Figure 22.41 can be found using

$$-V_{out} = \frac{R_f}{R_1}V_1 + \frac{R_f}{R_2}V_2 + \frac{R_f}{R_3}V_3$$

If you look at the figure, the feedback resistor is one-third the value of each input resistor. Therefore,

$$-V_{out} = \frac{1}{3}V_1 + \frac{1}{3}V_2 + \frac{1}{3}V_3$$

or

$$-V_{out} = \frac{V_1 + V_2 + V_3}{3}$$

This, by definition, is the average of the three input voltages. A summing amplifier is wired as an averaging amplifier when:

- All input resistors (R_1, R_2, etc.) are equal in value.
- The ratio of any input resistor to R_f equals the number of inputs.

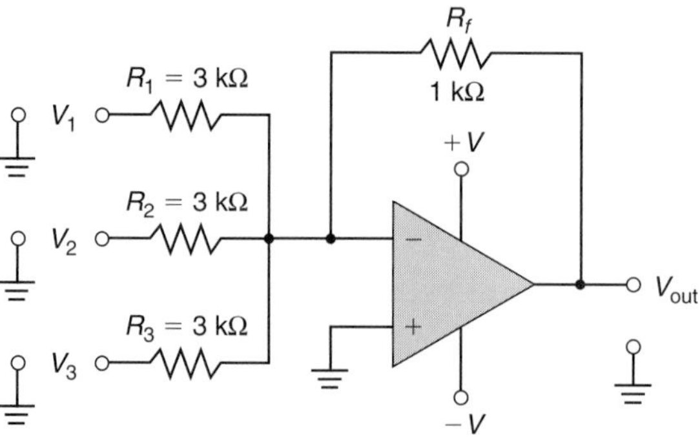

Figure 22.41 Averaging amplifier.

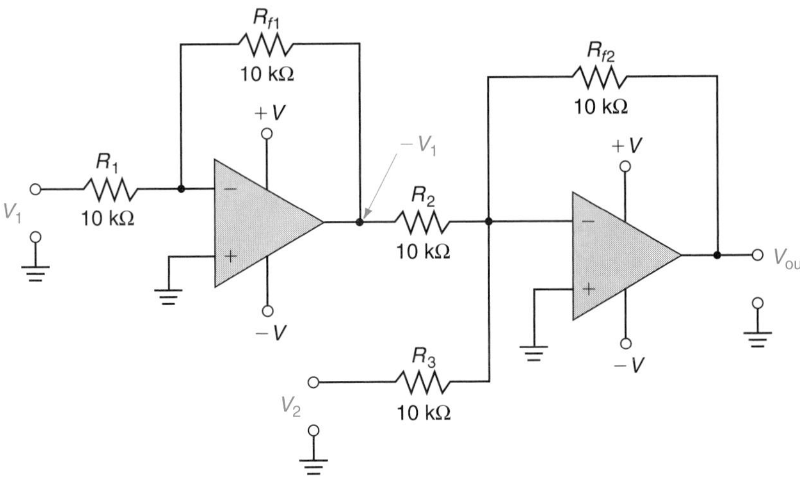

Figure 22.42 A subtractor.

For example, in Figure 22.41, all the input resistors are equal (3 kΩ). Also, the ratio of any input resistor to the feedback resistor (3 kΩ/1 kΩ) equals the number of inputs to the circuit (three).

Another variation on the summing amplifier is shown in Figure 22.42. This circuit, called a **subtractor,** provides an output that equals the difference between V_1 and V_2.

V_1 is applied to a unity-gain inverting amplifier. The output from this circuit equals $-V_1$. This output is then applied to a summing amplifier (also having unity gain) along with V_2. The output from the second op-amp is found as

$$-V_{out} = V_2 + (-V_1) = V_2 - V_1$$

As we have shown, the summing amplifier can be configured for a variety of applications. Summing amplifier characteristics are summarized in Figure 22.43.

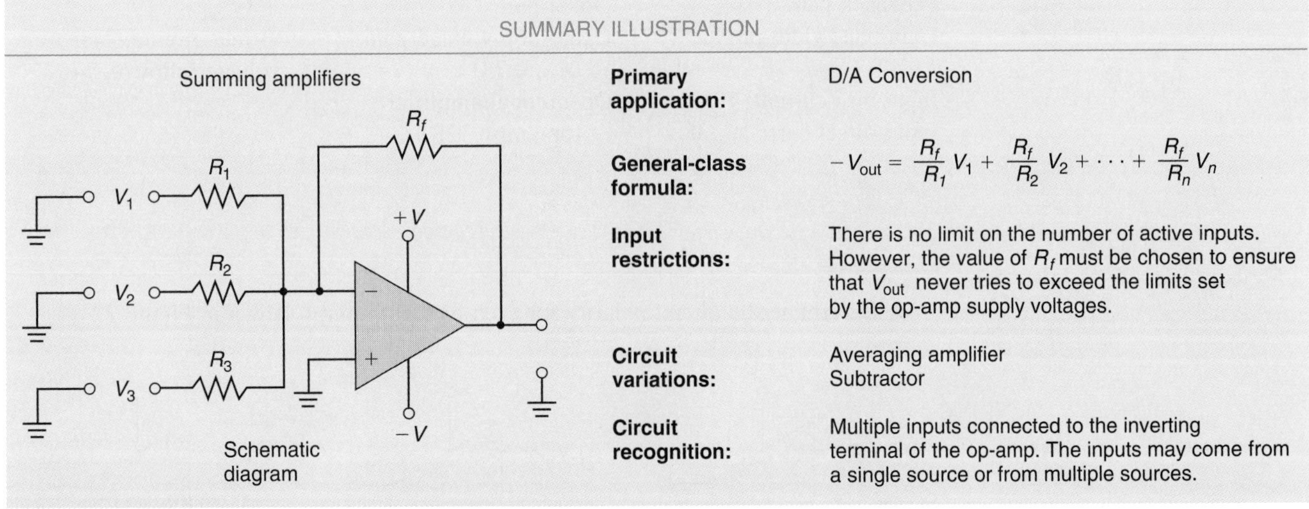

Figure 22.43

Section Review

1. What is a *comparator?*
2. What is the circuit recognition feature of a comparator?

3. What does the voltage gain of a comparator equal?
4. What is an *integrator*?
5. Describe the circuit operation of the op-amp integrator.
6. What is a *differentiator*?
7. Describe the circuit operation of the differentiator.
8. What is a *summing amplifier*?
9. Why is the output from a summing amplifier often *proportional* to (rather than equal to) the sum of its input voltages?
10. What is an *averaging* amplifier?
11. What are the resistor requirements of an averaging amplifier?
12. What is a *subtractor*?

Key Terms

The following terms were introduced in this chapter on the pages indicated:

Averaging amplifier, 758
Closed-loop voltage gain (A_{CL}), 742
Common-mode rejection ratio (CMRR), 738
Common-mode signals, 738
Comparator, 750
DC amplifier, 747
Differential amplifier, 730
Differentiator, 755
Discrete component, 728
Feedback path, 731
Gain-bandwidth product, 749
Input bias current, 738
Input offset current, 738
Input voltage range, 740
Integrated circuit (IC), 728
Integrator, 753
Inverting amplifier, 741
Inverting input, 731
Large-signal voltage gain, 740
Level detector, 750
Noninverting amplifier, 744
Noninverting input, 731
Offset null, 737
Open-loop voltage gain (A_{OL}), 731
Operational amplifier (op-amp), 728
Output offset voltage, 737
Output short-circuit current, 739
Power consumption rating, 740
Power supply rejection ratio (PSRR), 738
Slew rate, 739
Subtractor, 759
Summing amplifier, 756
Supply current rating, 740
Unity-gain frequency (f_{unity}), 749
Virtual ground, 741
Voltage follower, 746

Practice Problems

1. Determine the output polarity for each of the op-amps shown in Figure 22.44.

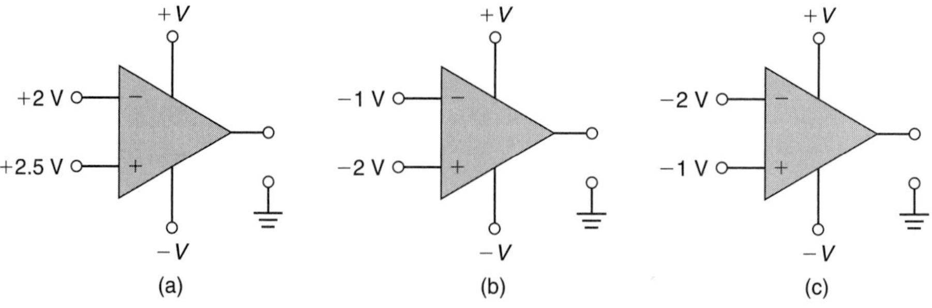

Figure 22.44

2. Determine the output polarity for each of the op-amps shown in Figure 22.45.

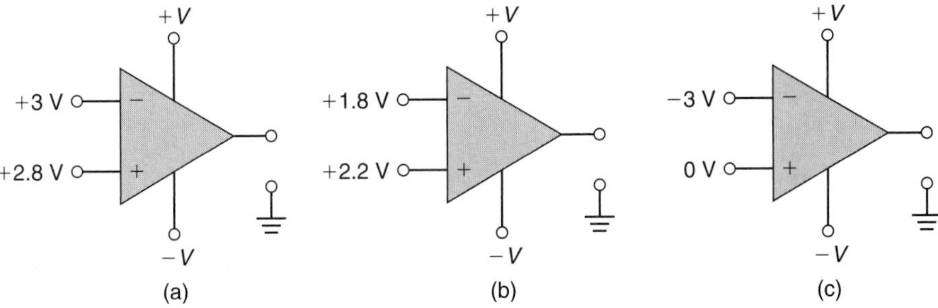

Figure 22.45

3. Determine the maximum peak-to-peak output voltage for the amplifier shown in Figure 22.46.

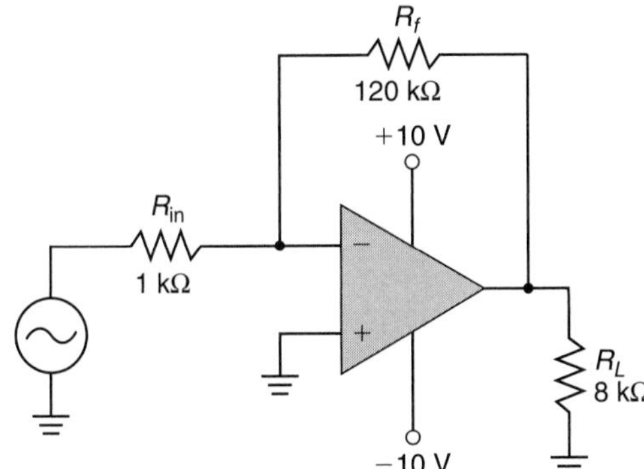

Figure 22.46

4. Determine the maximum peak-to-peak output voltage for the amplifier shown in Figure 22.47.

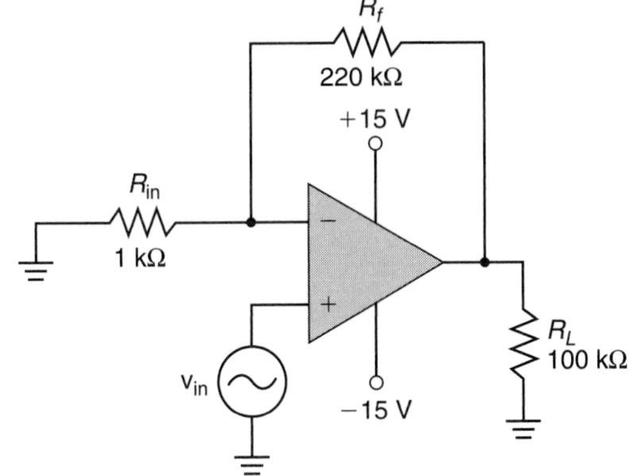

Figure 22.47

5. The amplifier in Problem 3 has a voltage gain of 120. Determine the maximum allowable peak-to-peak input voltage for the circuit.

6. The amplifier in Problem 4 has a voltage gain of 220. Determine the maximum allowable peak-to-peak input voltage for the circuit.

7. Determine the maximum allowable peak-to-peak input voltage for the amplifier shown in Figure 22.48.

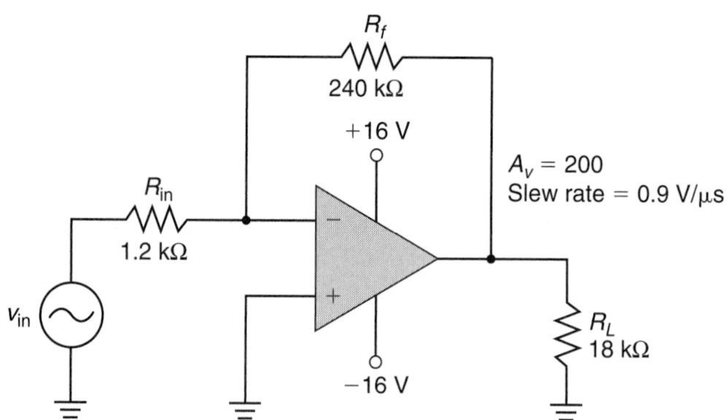

Figure 22.48

8. Determine the maximum allowable peak-to-peak input voltage for the amplifier shown in Figure 22.49.

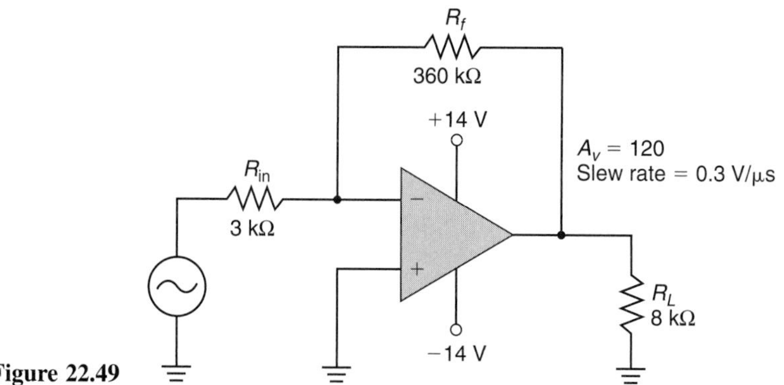

Figure 22.49

9. Determine the maximum operating frequency for the amplifier shown in Figure 22.48. Assume that the circuit has a 10 mV_{PP} input signal.
10. Determine the maximum operating frequency for the amplifier shown in Figure 22.49. Assume that the circuit has a 40 mV_{PP} input signal.
11. Determine the maximum operating frequency for the amplifier shown in Figure 22.50.
12. Determine the maximum operating frequency for the amplifier shown in Figure 22.51.

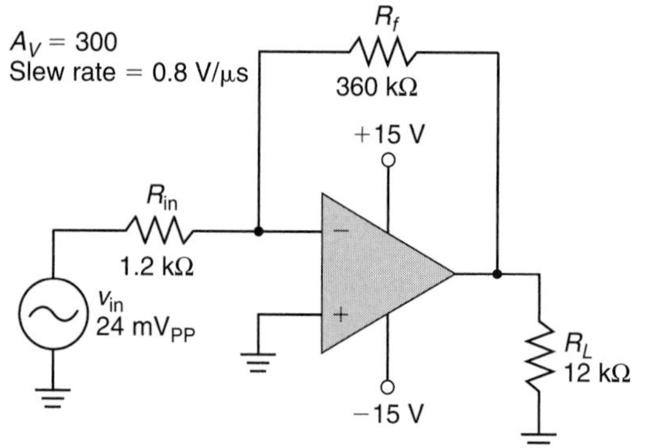

Figure 22.50

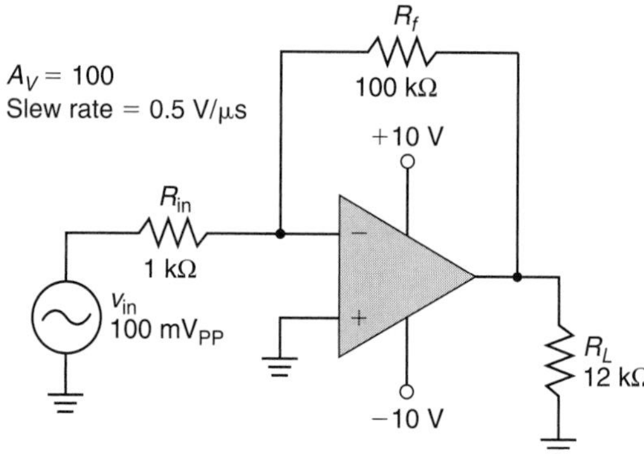

Figure 22.51

13. Perform the complete analysis of the amplifier shown in Figure 22.52.

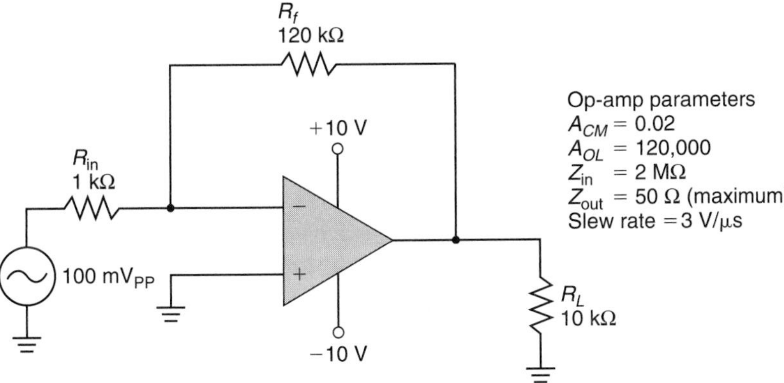

Figure 22.52

14. Perform the complete analysis of the amplifier shown in Figure 22.53.

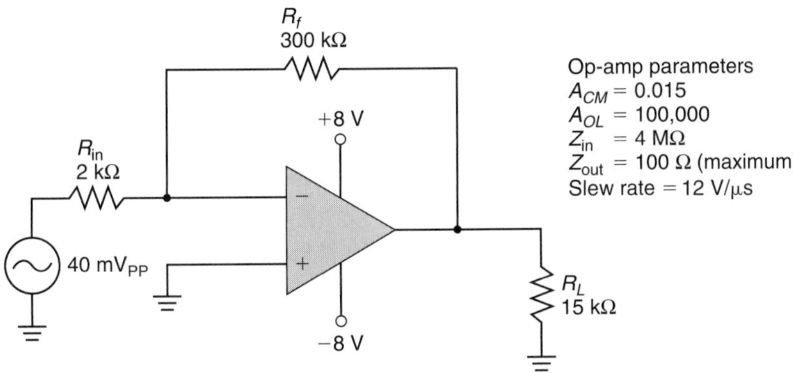

Figure 22.53

15. Perform the complete analysis of the amplifier shown in Figure 22.54.

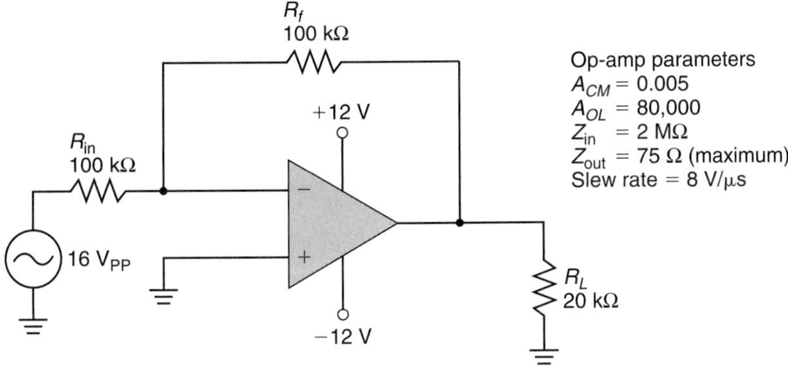

Figure 22.54

16. Perform the complete analysis of the amplifier shown in Figure 22.55.

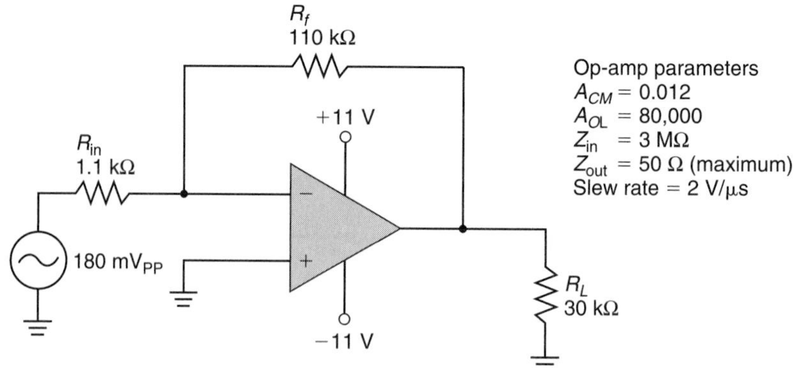

Figure 22.55

17. Perform the complete analysis of the amplifier shown in Figure 22.56.

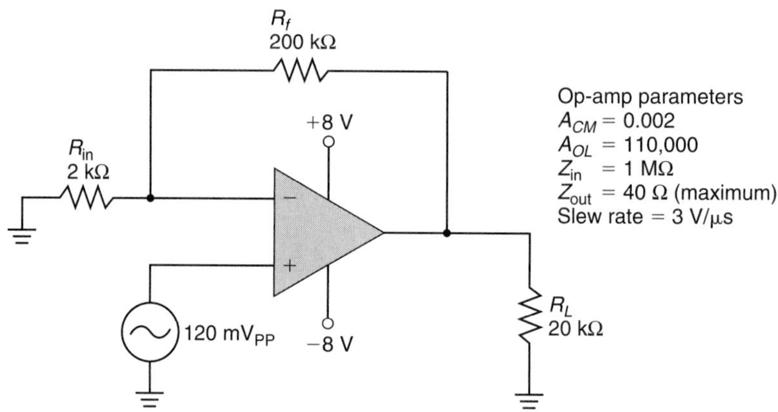

Figure 22.56

18. Perform the complete analysis of the amplifier shown in Figure 22.57.

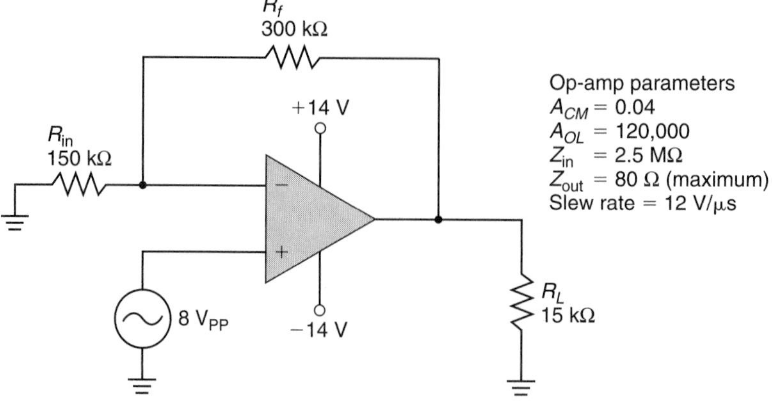

Figure 22.57

19. Perform the complete analysis of the amplifier shown in Figure 22.58.

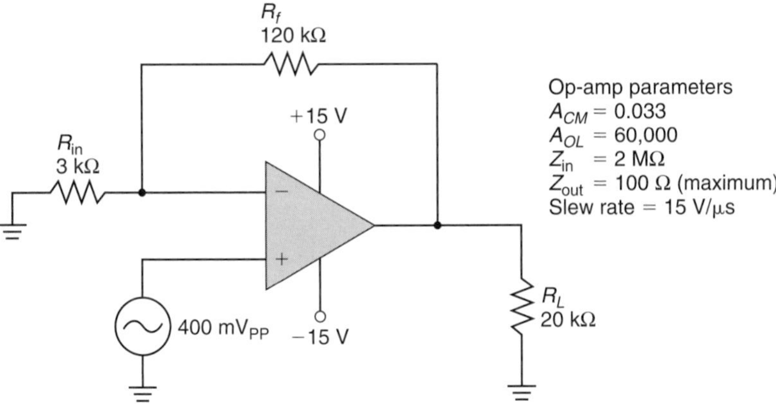

Figure 22.58

20. Perform the complete analysis of the amplifier shown in Figure 22.59.

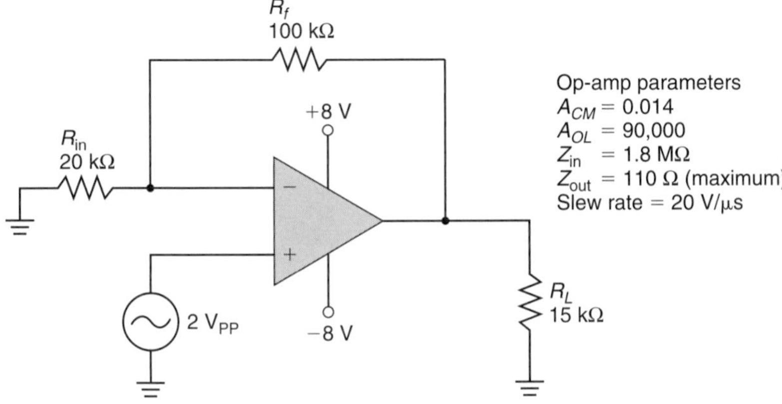

Figure 22.59

21. Perform the complete analysis of the voltage follower shown in Figure 22.60.

22. Perform the complete analysis of the voltage follower shown in Figure 22.61.

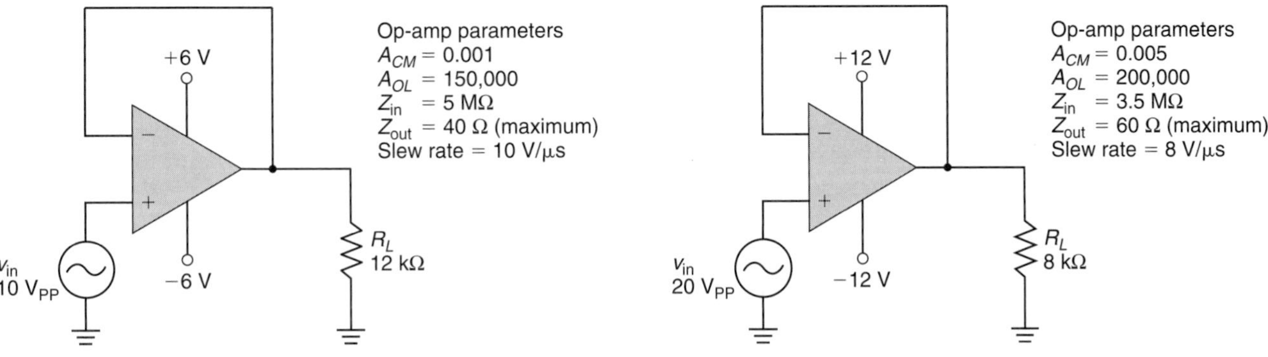

Figure 22.60 **Figure 22.61**

23. An op-amp has a unity-gain frequency of 12 MHz. Determine the bandwidth of the device when $A_{CL} = 400$.

24. An op-amp has a unity-gain frequency of 14 MHz. Determine the bandwidth of the device when $A_{CL} = 320$.

25. An op-amp has a unity-gain frequency of 25 MHz. Determine the bandwidth of the device when $A_{CL} = 42$ dB.

26. An op-amp has a unity-gain frequency of 1 MHz. Determine the bandwidth of the device when $A_{CL} = 20$ dB.

27. We need to construct an amplifier with values of $A_{CL} = 200$ and $f_{C2} = 120$ kHz. Can we use an op-amp with $f_{unity} = 28$ MHz for this application?

28. We need to construct an amplifier with values of $A_{CL} = 24$ dB and $f_{C2} = 40$ kHz. Can we use an op-amp with $f_{unity} = 1$ MHz for this application?

29. An op-amp circuit with $A_{CL} = 120$ has a measured value of $f_{C2} = 100$ kHz. What is the unity-gain frequency of the op-amp?

30. An op-amp circuit with $A_{CL} = 300$ has a measured value of $f_{C2} = 88$ kHz. What is the unity-gain frequency of the op-amp?

31. The circuit shown in Figure 22.62 has a measured f_{C2} of 250 kHz. What is the value of f_{unity} for the op-amp?

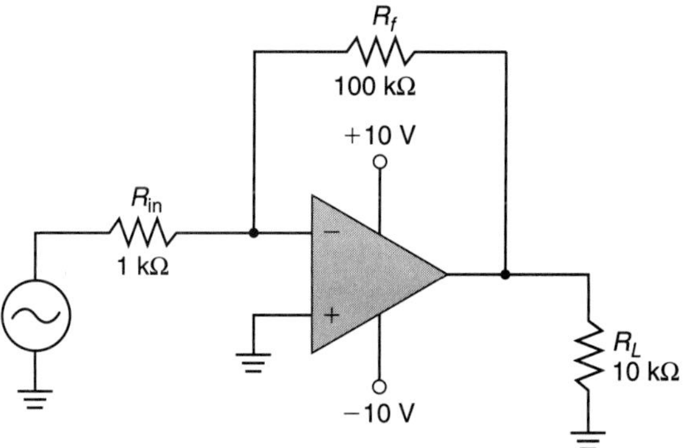

Figure 22.62

32. The circuit shown in Figure 22.63 has a measured f_{C2} of 100 kHz. What is the value of f_{unity} for the op-amp?

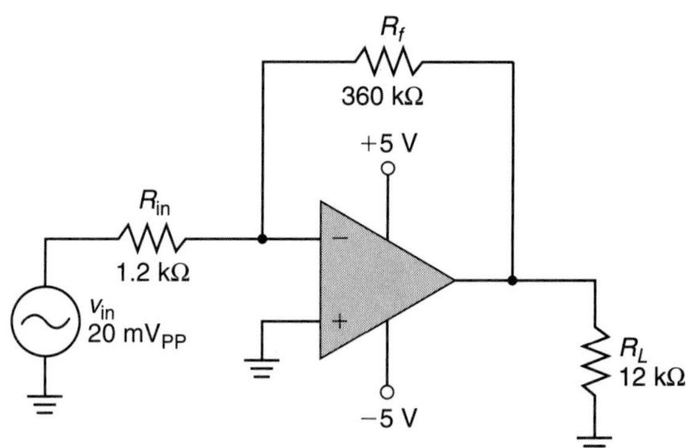

Figure 22.63

33. Determine the value of V_{ref} for the comparator shown in Figure 22.64.

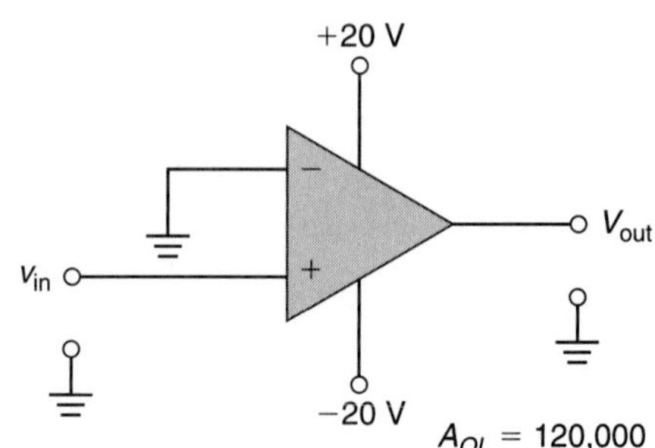

Figure 22.64

34. Determine the value of V_{ref} for the comparator shown in Figure 22.65.

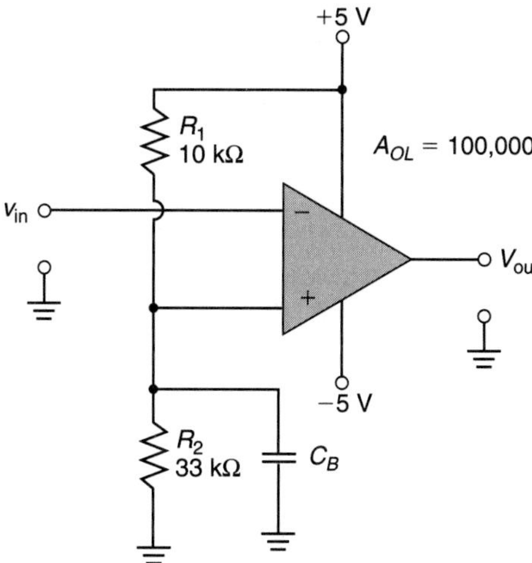

Figure 22.65

35. Determine the value of V_{ref} for the comparator shown in Figure 22.66.

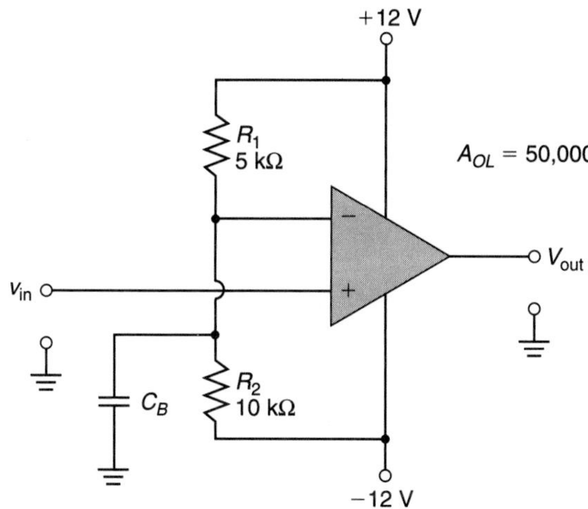

Figure 22.66

36. Determine the value of V_{ref} for the comparator shown in Figure 22.67.

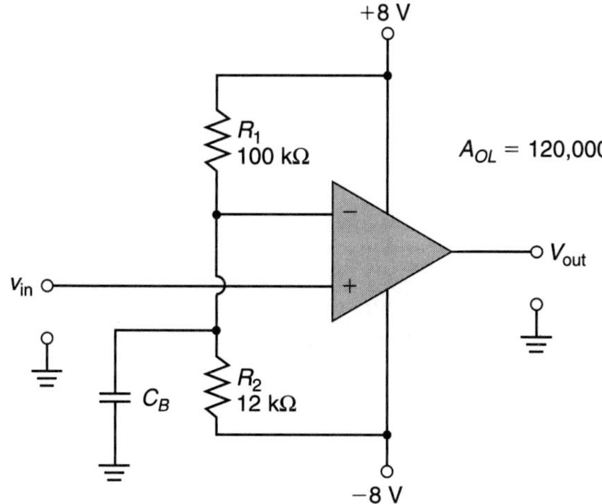

Figure 22.67

37. Determine the output for the summing amplifier shown in Figure 22.68a.

38. Determine the output for the summing amplifier shown in Figure 22.68b.

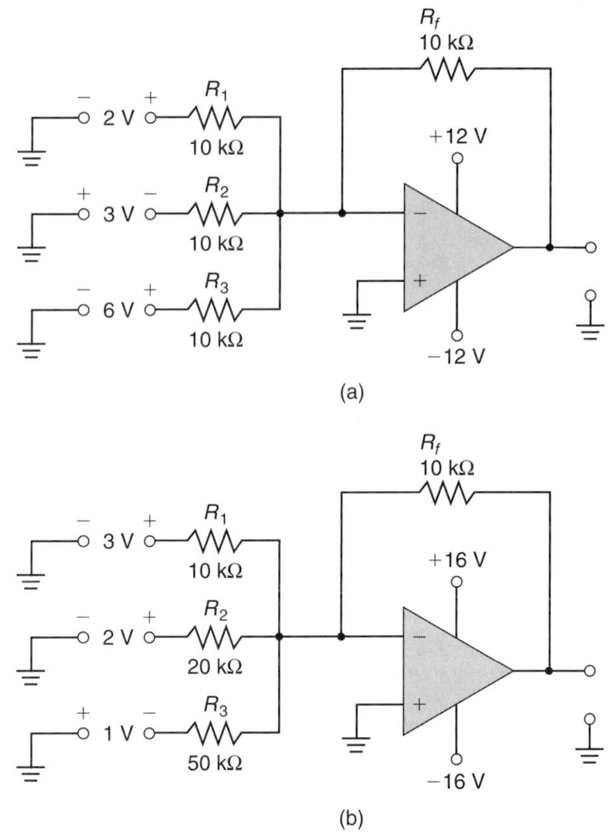

Figure 22.68

39. Determine the output for the summing amplifier in Figure 22.69.

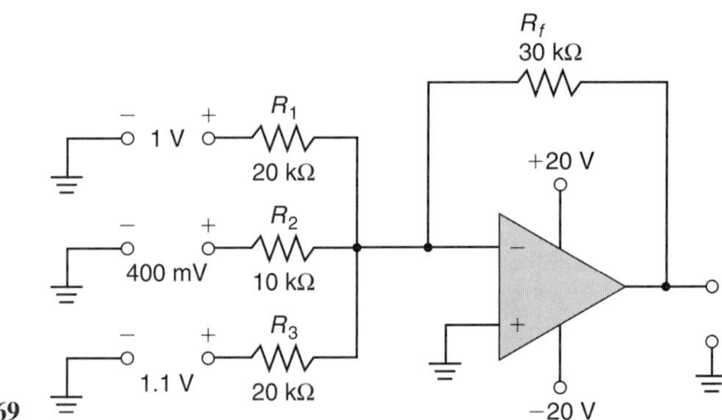

Figure 22.69

40. The feedback resistor in Figure 22.69 is changed to 24 kΩ. Determine its new output voltage.

41. A five-input summing amplifier has values of $R_1 = R_2 = R_3 = R_4 = R_5 = 15$ kΩ. What value of feedback resistor is required to produce an averaging amplifier?

42. A four-input summing amplifier has values of $R_1 = R_2 = R_3 = R_4 = 20$ kΩ. What value of feedback resistor is required to produce an averaging amplifier?

43. Refer back to Figure 22.42 (page 759). Assume that $V_1 = 4$ V and $V_2 = 6$ V. Determine the output voltage from the circuit.

44. Refer back to Figure 22.42 (page 759). Assume that R_{f2} in the circuit has been changed to 30 kΩ. Also, assume that the input voltages are $V_1 = 1$ V and $V_2 = 3$ V. Determine the output voltage from the circuit.

Looking Back

These problems relate to material presented in earlier chapters. The chapters are identified in brackets.

45. A 60 V_{PP} sine wave is applied to a 250 Ω resistive load. Calculate the value of load power. [Chapter 9]

46. Using Table 2.1, calculate the resistance of a 1500 ft length of 18 gauge wire. [Chapter 2]

47. A series circuit has values of $R = 200$ Ω and $L = 10$ mH. Calculate the phase angle for the circuit at 1.2 kHz. [Chapter 11]

48. A parallel LC circuit has values of $L = 300$ μH, $R_W = 3$ Ω, and $C = 510$ nF. Calculate the Q of the circuit. [Chapter 14]

49. A JFET has maximum values of $V_{GS(\text{off})}$ and -6 V and $I_{DSS} = 12$ mA. Calculate the maximum I_D when $V_{GS} = -5$ V. [Chapter 23]

50. An emitter follower has the following values: $r'_e = 10$ Ω, $r_E = 180$ Ω, and $A_i = 24$. Calculate the power gain of the circuit. [Chapter 22]

Pushing the Envelope

51. The op-amp in Figure 22.70 has the input and output voltages shown. Determine the CMRR of the device.

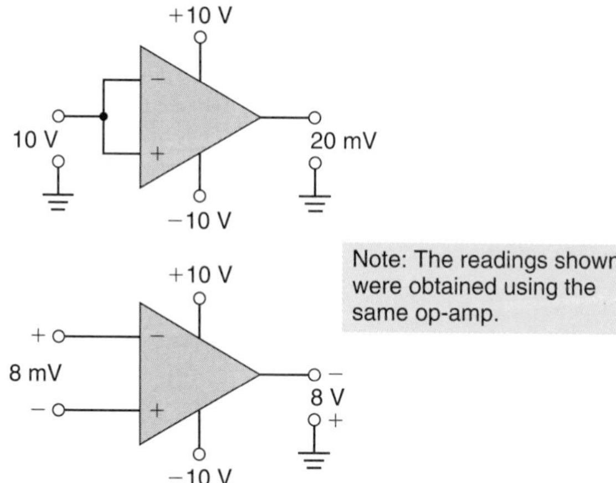

Figure 22.70

52. Figure 22.71 is the parts-placement diagram for an op-amp circuit. Analyze the circuit, and draw its schematic diagram. Then, perform a complete analysis of the circuit to determine its values of A_{CL}, Z_{in}, Z_{out}, CMRR, and f_{max}.

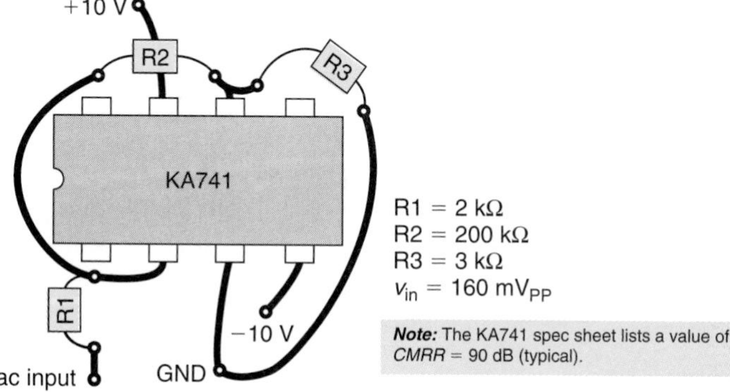

Figure 22.71

53. Calculate the voltage gain of the audio amplifier shown in Figure 22.72.

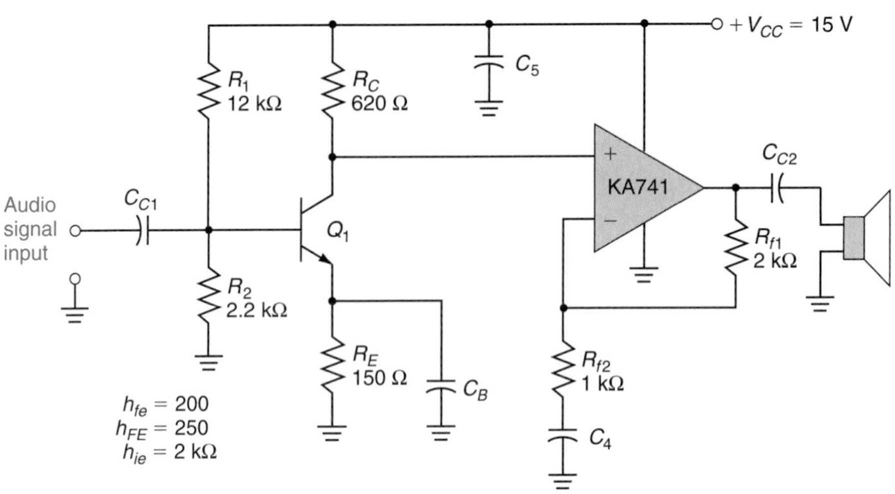

Figure 22.72

54. Design a circuit to solve the following equation:

$$-V_{\text{out}} = \frac{V_1 - V_2}{2} + \frac{V_3}{3}$$

The circuit is to contain only two op-amps.

Answers to the Example Practice Problems

22.1 Circuit (c): $v_d = +2$ V, positive; circuit (d): $v_d = -3$ V, negative

22.2 157 mV_{PP}

22.3 142.9 mV_{PP}

22.4 6.37 kHz

22.5 $A_{CL} = 250$, $Z_{\text{in}} = 1$ kΩ, $Z_{\text{out}} < 50$ Ω, CMRR = 12,500, $f_{\max} = 19.1$ kHz

22.6 $A_{CL} = 251$, $Z_{\text{in}} \geq 1.5$ MΩ, $Z_{\text{out}} < 50$ Ω, CMRR = 12,550, $f_{\max} = 19.02$ kHz

22.7 125 kHz

22.8 $A_{CL} f_{C2} = 3.98$ MHz; yes, it can be used.

22.9 $V_{\text{out}} = \pm 11$ V

22.10 −6 V

22.11 −3 V

Multisim File References

Figure	File Reference
22.13	ETF22_13.msm
22.19	ETF22_19.msm
22.24	ETF22_24.msm
22.29	ETF22_29.msm
22.30	ETF22_30.msm
22.31	ETF22_31.msm
22.36	ETF22_36.msm

chapter 23

Active Filters and Oscillators

Objectives

After studying the material in this chapter, you should be able to:

1. Compare and contrast the frequency-response characteristics of *ideal* and *practical* tuned amplifiers.
2. Discuss the *quality* (*Q*) factor of a tuned amplifier, the factors that affect its value, and its relationship to amplifier bandwidth.
3. Describe the frequency-response curves of the *low-pass, high-pass, bandpass,* and *band-stop* (*notch*) *filters.*
4. Compare and contrast the frequency-response curves of *Butterworth, Chebyshev,* and *Bessel filters.*
5. Perform the gain and frequency analysis of low-pass and high-pass active filters.
6. Perform the gain and frequency analysis of the two-stage bandpass, multiple-feedback bandpass, and notch filters.
7. State the function of the *oscillator.*
8. Describe *positive feedback,* how it is produced, and how it maintains oscillations after an oscillator is triggered.
9. Discuss the *Barkhausen criterion* and its effect on oscillator operation.
10. List the three requirements for proper oscillator operation.
11. Describe the operating characteristics of the *phase-shift* and *Wien-bridge* RC oscillators.
12. Describe the operation and perform the complete frequency analysis of a *Colpitts oscillator.*
13. Describe the operation of the *Hartley, Clapp,* and *Armstrong LC oscillators.*
14. Describe the operating principles of crystals and crystal-controlled oscillators.

Communications (and other) systems make extensive use of *tuned circuits.* A **tuned circuit** is *one that is designed to operate within a specific band of frequencies.* For example, a tuned amplifier may be designed to amplify only those frequencies that fall within ±20 kHz of 1000 kHz; that is, between 980 kHz and 1020 kHz. As long as the input signal is within this frequency range, it is amplified (or passed without being attenuated). If the input signal goes outside of this range, amplification is greatly reduced (or the signal is attenuated).

There are many types of tuned circuits. In this chapter, we will discuss *active filters* and *oscillators.* An **active filter** is *a tuned op-amp circuit that is designed to pass (or reject) a specific band of frequencies.* An **oscillator,** on the other hand, is *a circuit designed to generate a signal at a specific frequency.* In other words, active filters are designed to *respond to* specific frequencies, while oscillators are used to *generate* specific frequencies.

23.1 Tuned Amplifier Characteristics

OBJECTIVE 1 ➤ The *ideal* tuned amplifier would have zero ($-\infty$ dB) gain when operated outside of its *pass band.* For example, the amplifier represented by the ideal frequency response curve in Figure 23.1 would attenuate all frequencies below a *lower cutoff frequency* (f_{C1}) and above an *upper cutoff frequency* (f_{C2}). At all frequencies within its **pass band** (the range of frequencies between f_{C1} and f_{C2}), the circuit would provide a constant value of **mid-band gain ($A_{v(\text{mid})}$).** Thus, the circuit would pass all the frequencies within its pass band and stop all the others.

As you have seen, ideal characteristics are seldom achieved. A more practical frequency response curve is shown in Figure 23.1. Note that the gain of a practical filter decreases gradually at frequencies both above and below its geometric center frequency (f_0).

Roll-Off Rate Versus Bandwidth

Note: The material in this section is intended primarily as a review. A more detailed discussion of tuned circuit characteristics can be found in Section 15.1 (pp. 448–457).

Tuned amplifiers are designed for a specific bandwidth. For example, take a look at Figure 23.2, where we see a pass band (shaded) with two stop bands, A and B. A tuned amplifier, centered on 1000 kHz with a bandwidth of 40 kHz, passes all the frequencies within the pass band, while rejecting all those in the stop bands. If the bandwidth was increased to 100 kHz, the amplifier would pass a portion of the frequencies in both stop bands, which is not acceptable. This demonstrates the importance of having a specific

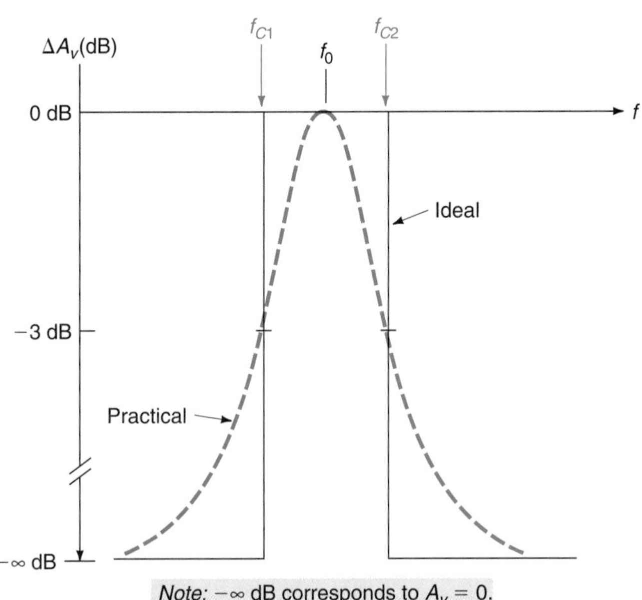

Figure 23.1 Ideal versus practical pass band characteristics.

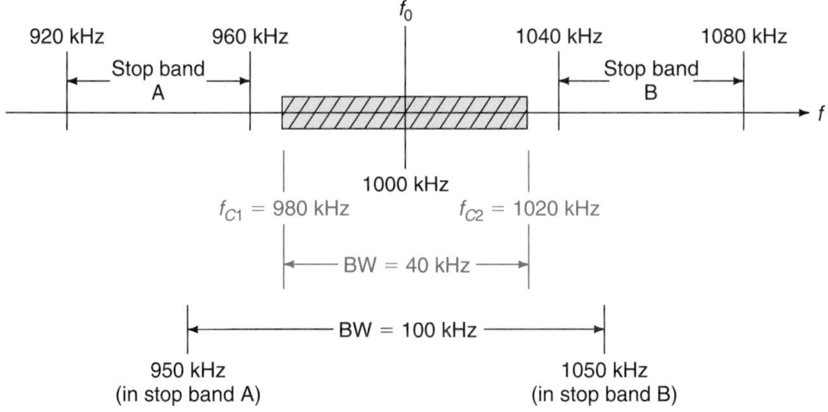

Figure 23.2

An Example: The frequency ranges shown in Figure 23.2 could represent the frequency ranges of three AM radio stations. If the tuned amplifier has a bandwidth of 40 kHz, it will pick up only the radio station to which it is tuned (the center station). If it has a bandwidth of 1000 kHz, the tuned amplifier will pick up the center station and portions of the other two station signals as well.

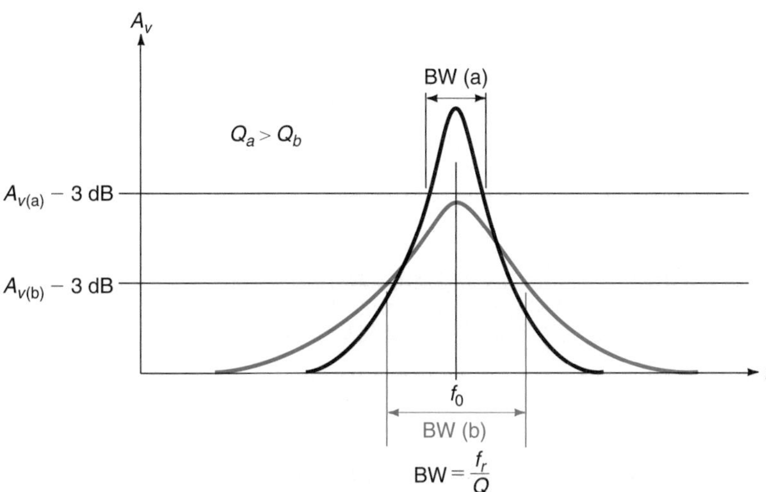

Figure 23.3 Bandwidth versus roll-off rate.

(and predictable) bandwidth. Keeping this in mind, let's take a look at the relationship between *roll-off rate* and *bandwidth.*

The lower the roll-off rate of an amplifier, the greater its bandwidth. Figure 23.3 shows the frequency-response curves of two tuned amplifiers with the same center frequency (f_0). As you can see, the circuit with the lower roll-off rate has a much greater bandwidth. The roll-off rate and bandwidth of a tuned amplifier are controlled by the quality (Q) of the circuit. ◄ *OBJECTIVE 2*

As with passive filters, the Q of a tuned amplifier is a figure of merit that equals the ratio of geometric center frequency (f_0) to bandwidth (BW). By formula,

$$Q = \frac{f_0}{\text{BW}} \tag{23.1}$$

Example 23.1 illustrates the relationship given in equation (23.1).

Example 23.1

What is the required Q of a tuned amplifier used for the application in Figure 23.2?

Solution: The pass band (shaded) in Figure 23.2 has a geometric center frequency of 1000 kHz and a bandwidth of 40 kHz. Using these two values,

$$Q = \frac{f_0}{\text{BW}} = \frac{1000 \text{ kHz}}{40 \text{ kHz}} = 25$$

Practice Problem 23.1

An amplifier with an f_0 of 1200 kHz has a BW of 20 kHz. Determine the Q of the amplifier.

The Q of a tuned amplifier actually depends on component values within the circuit (as does the value of f_0). Once these values have been calculated, the bandwidth of a tuned amplifier is found using

$$\text{BW} = \frac{f_0}{Q} \tag{23.2}$$

As we discuss various tuned amplifiers, you will be shown how to calculate the values of f_0 and Q for each. Once these values are known, determining the bandwidth of a tuned amplifier is simple.

Geometric Center Frequency

As you may recall, f_0 is the geometric average of f_{C1} and f_{C2}. By formula,

$$f_0 = \sqrt{f_{C1} f_{C2}} \tag{23.3}$$

When amplifier Q is greater than (or equal to) two, f_0 approaches the *algebraic average* of the cutoff frequencies (f_{ave}), which is found as

$$f_{\text{ave}} = \frac{f_{C1} + f_{C2}}{2} \tag{23.4}$$

These relationships were introduced in Chapter 15.

Section Review

1. What are the characteristics of the *ideal* tuned amplifier?
2. How does the frequency response of the *practical* tuned amplifier vary from that of the ideal tuned amplifier?
3. What is the relationship between the *roll-off rate* and the *bandwidth* of a tuned amplifier?
4. What is the *quality* (Q) rating of a tuned amplifier?
5. What is the relationship between the values of Q and BW for a tuned amplifier?
6. What determines the value of Q for a tuned amplifier?
7. What is the relationship between the values of Q and f_0 for a tuned amplifier?

23.2 Active Filters: An Overview

OBJECTIVE 3 ➤ In Chapter 15, you were introduced to *passive filters* that contain *RC*, *RL*, and *LC* circuits. An *active filter* is a tuned op-amp circuit. The frequency-response curves for the four basic types of active filters are shown in Figure 23.4. Remember that the concepts of Q, center frequency (f_0), and bandwidth (BW) are related primarily to the bandpass and notch filters. When we are dealing with low-pass and high-pass filters, we are concerned only with the value of the circuit cutoff frequency.

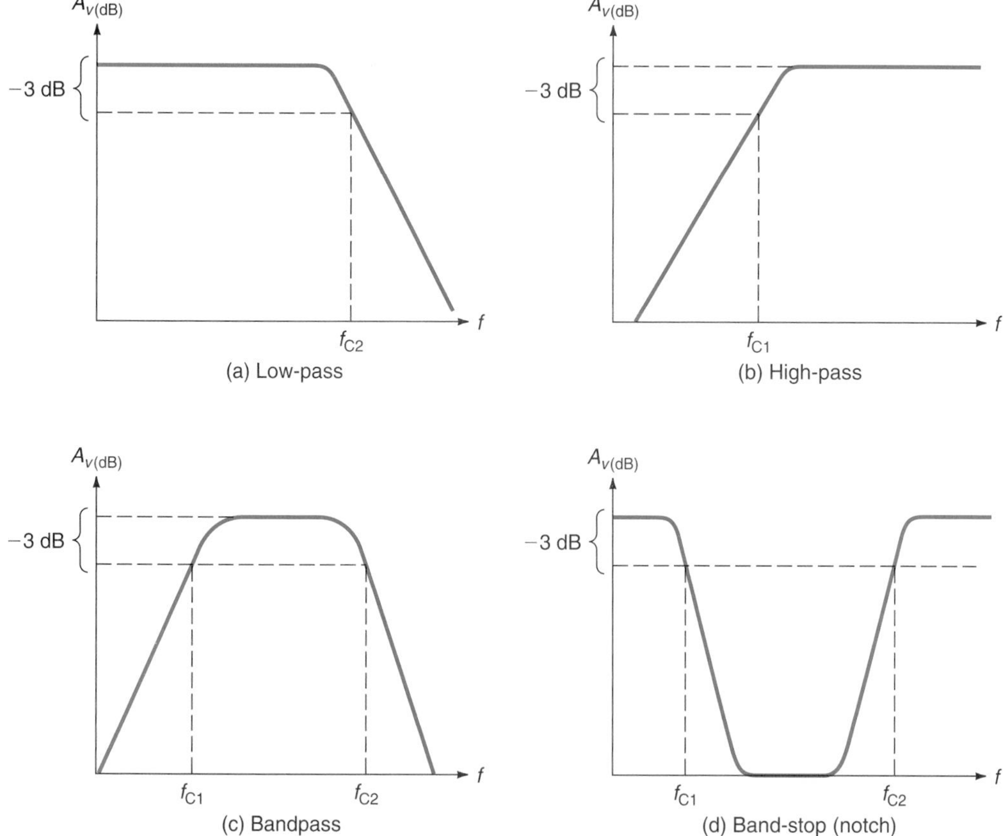

Figure 23.4 Active filter frequency-response curves.

General Terminology

Before we discuss any specific circuits, we need to introduce some common terms that are used to describe filters. A **pole** is simply *an RC circuit.* Thus, a *one-pole filter* contains one *RC* circuit, a *two-pole filter* contains two *RC* circuits, and so on. The term **order** is used to identify *the number of poles.* For example, a *first-order filter* contains one pole, a *second-order filter* contains two poles, and so on.

The more poles in an active filter, the higher the gain roll-off rate outside the pass band. One type of active filter (called a Butterworth filter) has a roll-off rate of 20 dB/decade per pole. The following table illustrates the relationship among order, poles, and gain roll-off for Butterworth filters:

Filter Type	*Number of Poles*[a]	*Total Gain Roll-Off*
First order	1	20 dB/decade
Second order	2	40 dB/decade
Third order	3	60 dB/decade

[a]*RC* circuits.

Butterworth, Chebyshev, and Bessel Filters

OBJECTIVE 4

The **Butterworth filter** has relatively flat response across its pass band, as shown in Figure 23.5. When we speak of *flat response,* we mean that the value of $A_{v(\text{dB})}$ remains relatively constant across the pass band. Butterworth filters are sometimes referred to as *maximally flat* or *flat-flat* filters.

The **Chebyshev filter** has a higher initial roll-off rate (per pole) than a Butterworth filter at frequencies just outside the pass band. As the operating frequency moves further

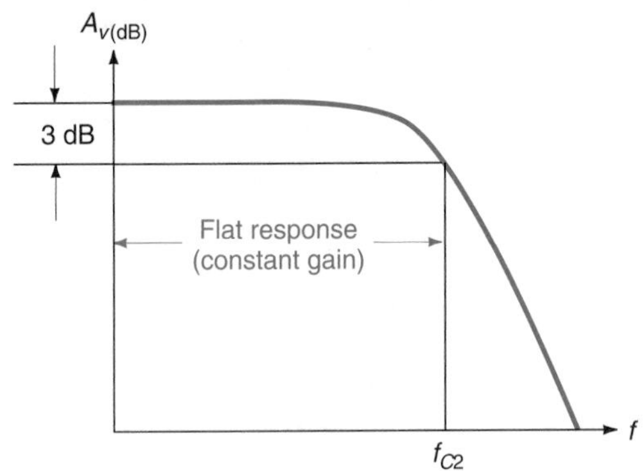

Figure 23.5 Butterworth low-pass response curve.

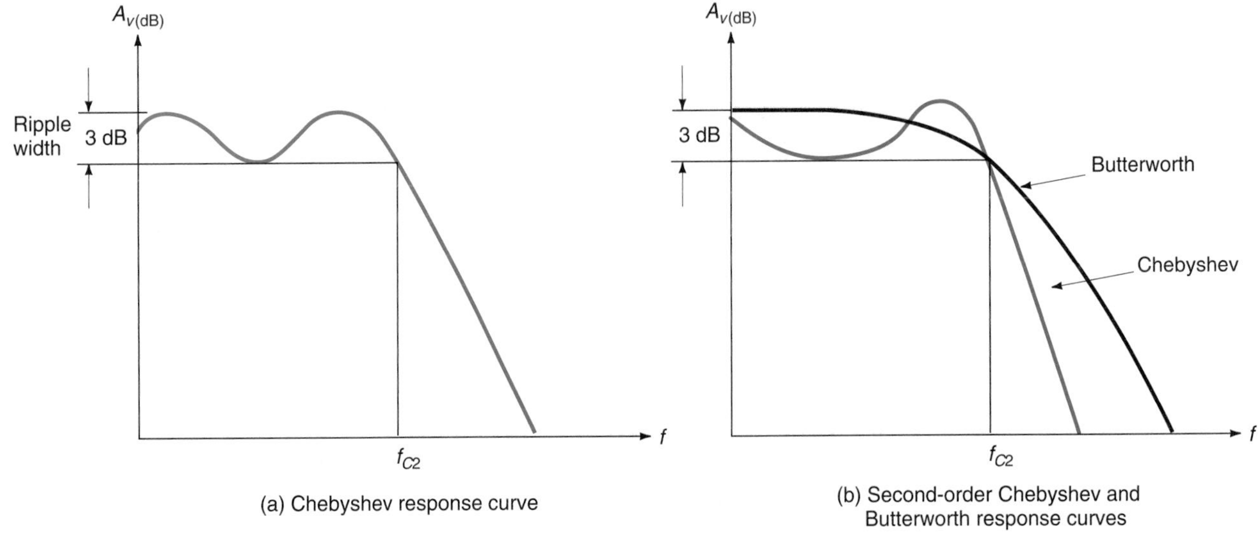

Figure 23.6 Chebyshev frequency response.

outside the pass band, the two filters have equal roll-off rates. Further, the gain of a Chebyshev filter is not constant across its pass band. Figure 23.6a shows the response curve of a first-order Chebyshev filter. The **ripple width** is the difference between the minimum and maximum values of midband gain. Figure 23.6b compares the response curves of second-order Chebyshev and Butterworth filters.

Since the Butterworth filter provides consistent gain across its pass band, it is the more commonly used of the two. Even so, the Butterworth filter has its own drawback: *The time delay produced by the Butterworth filter is not constant across its pass band.* This means that different frequencies experience different phase shifts (input to output). This can produce severe signal distortion.

The **Bessel filter** is *a circuit that is designed to provide a constant phase shift across its pass band.* The constant phase shift of the Bessel filter results in greater **fidelity** (i.e., *the ability to reproduce a waveform accurately*) than either the Butterworth or Chebyshev filter. However, this circuit has the disadvantage of a lower initial roll-off rate, as shown in Figure 23.7.

Of the three filter types, the Butterworth is the most commonly used (as mentioned earlier). For this reason, we will concentrate on this type of circuit in the upcoming section.

Section Review

1. What is an *active filter?*
2. Describe the frequency-response characteristics of *low-pass, high-pass, bandpass,* and *notch filters.*

3. What is a *pole*?
4. What is the relationship between the *order* of a filter and its *poles*?
5. Why is the number of poles in an active filter important?
6. What are the frequency response characteristics of a *Butterworth filter*?
7. Compare the frequency response of a Chebyshev filter to that of a Butterworth filter.
8. What is the inherent drawback of a Butterworth filter? What can result from this drawback?
9. What is a *Bessel filter*? What are its primary strengths and weaknesses?
10. Write a brief paragraph comparing the frequency-response characteristics of Butterworth, Chebyshev, and Bessel filters.

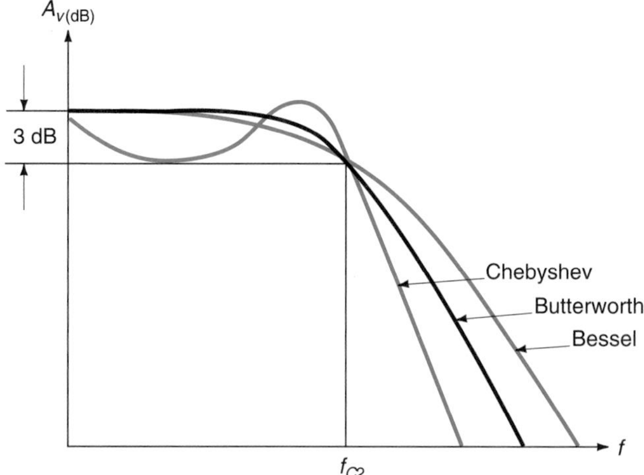

Figure 23.7 Chebyshev, Butterworth, and Bessel response curves.

23.3 Low-Pass and High-Pass Filters

The Single-Pole Low-Pass Filter

The single-pole low-pass Butterworth filter can be designed as either a high-gain circuit (Figure 23.8a) or as a voltage follower (Figure 23.8b). As you can see, the *RC* circuit at ◀ *OBJECTIVE 5*

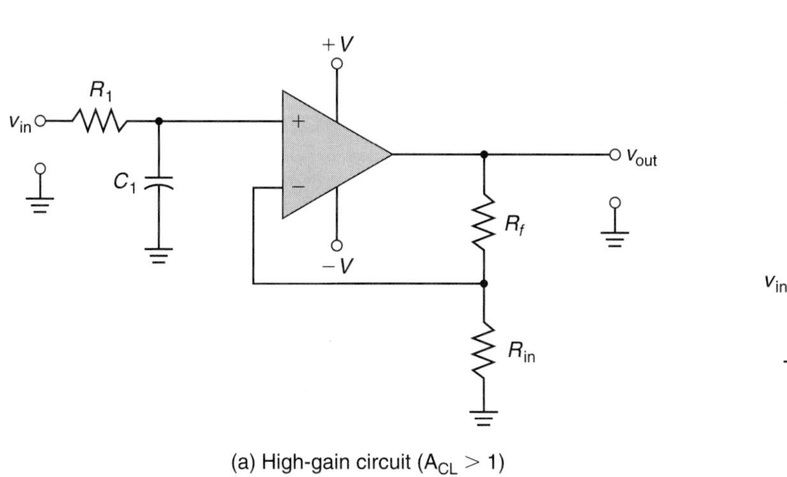

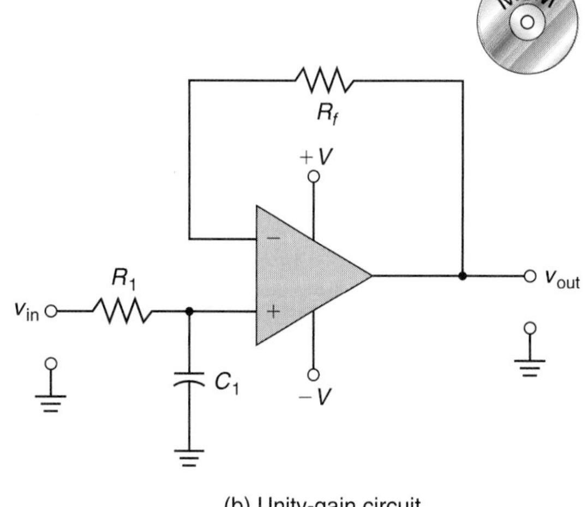

Figure 23.8 *RC* low-pass filters.

the input is simply a low-pass filter like those discussed in Chapter 15. This RC filter determines the upper cutoff frequency of the circuit, which is found as

$$f_{C2} = \frac{1}{2\pi RC} \tag{23.5}$$

Example 23.2

Note: Many of the active filter circuits contain op-amps that are drawn with the noninverting input at the *top* of the symbol (like the one in Figure 23.9). They are drawn like this to simplify the illustrations.

Determine the bandwidth and closed-loop gain of the single-pole low-pass filter shown in Figure 23.9.

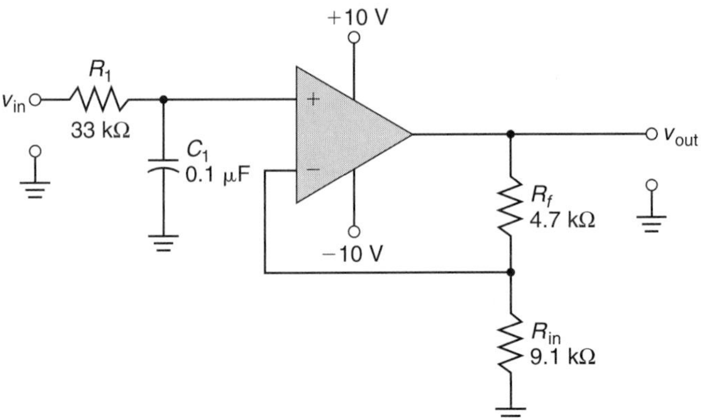

Figure 23.9

Solution: The upper cutoff frequency is found as

$$f_{C2} = \frac{1}{2\pi R_1 C_1} = \frac{1}{2\pi(33\text{ k}\Omega)(0.1\ \mu\text{F})} = 48.23\text{ Hz}$$

Since the op-amp is a dc amplifier (0 Hz), the bandwidth of the filter is equal to the value of f_{C2}. The circuit is a noninverting amplifier, so its value of A_{CL} is found using

$$A_{CL} = \frac{R_f}{R_{in}} + 1 = \frac{4.7\text{ k}\Omega}{9.1\text{ k}\Omega} + 1 = 1.52$$

Practice Problem 23.2

A single-pole low-pass filter like the one shown in Figure 23.9 has values of R_1 = 47 kΩ, C_1 = 0.033 μF, R_f = 10 kΩ, and R_{in} = 10 kΩ. Determine the values of f_{C2} and A_{CL} for the circuit.

Although the value of A_{CL} in Example 23.3 is relatively low, it can be made much higher by changing the ratio of R_f to R_{in}. When unity (0 dB) gain is desired, the circuit shown in Figure 23.8b is used with $R_f = R_1$. Above f_{C2}, circuit gain rolls off at a rate of 20 dB/decade (as with any single-pole low-pass filter).

The Two-Pole Low-Pass Filter

A two-pole low-pass filter has a roll-off rate of 40 dB/decade. Two commonly used two-pole low-pass filter configurations are shown in Figure 23.10. As the operating frequency increases beyond f_{C2}, each RC circuit reduces A_{CL} by 20 dB/decade, giving a total of 40 dB/decade. The cutoff frequency for the circuit is found as

$$f_{C2} = \frac{1}{2\pi\sqrt{R_1 R_2 C_1 C_2}} \tag{23.6}$$

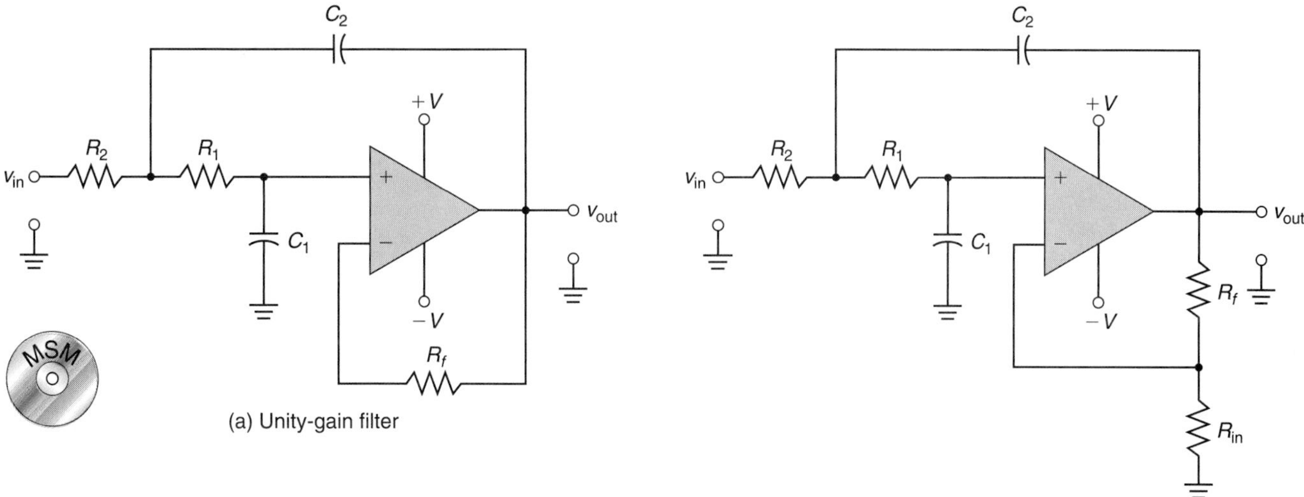

Figure 23.10 Common two-pole low pass filters.

There is a restriction on the closed-loop gain of a two-pole low-pass filter. For the filter to have a Butterworth response curve, A_{CL} can be no greater than 4 dB (1.586). The derivation of this value involves calculus and is not covered here. In this case, we will simply accept the value as valid. The variable-gain filter can be designed for any value of A_{CL} between 0 dB and 4 dB.

Note: The two-pole circuits in Figure 23.10 are also referred to as **Sallen-Key** filters (after their developers) and *VCVS* (voltage-controlled voltage-source) filters.

High-Pass Filters

Figure 23.11 shows several typical high-pass filters. The high-pass filter differs from the low-pass filter in that the resistor and capacitor positions are reversed. The value of f_{C1} for each circuit is found using the same equation that we used to find f_{C2} for the low-pass filter. We will not spend any time on these circuits, because their operating principles should be familiar to you by now.

Filter Gain Requirements

The gain requirements for a variety of Butterworth active filters are summarized in Table 23.1. The derivations of these gain requirements are beyond the scope of this text. Just be aware that almost every type of active filter has gain requirements that

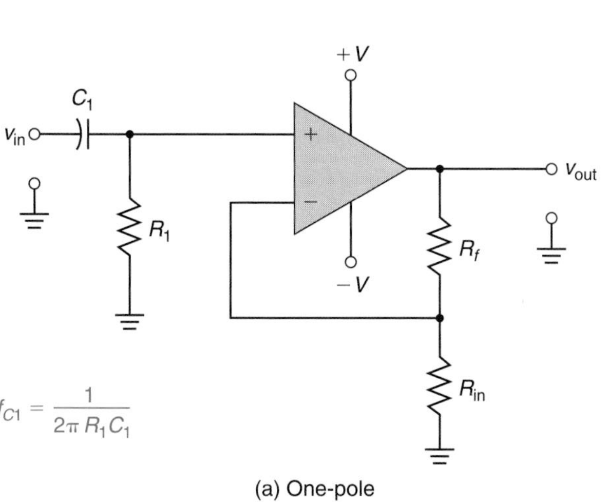

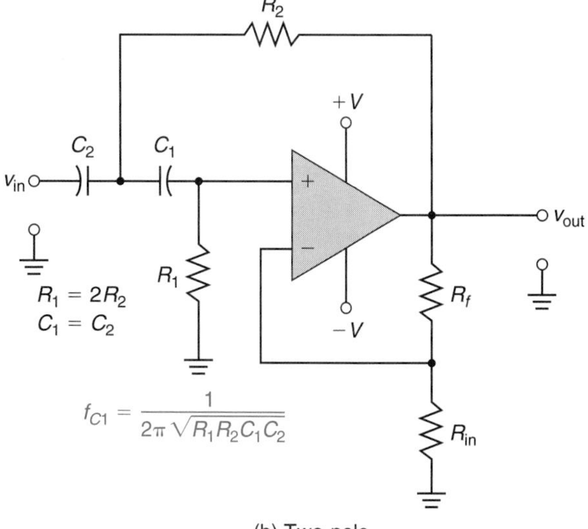

Figure 23.11 Typical high-pass filters.

Note that the maximum gain for low-pass and high-pass active filters increases by 2 dB for each pole added to the circuit. Also note that the approximated roll-off rate is equal to 20 dB/decade per pole.

TABLE 23.1 Butterworth Filter Gain Requirements

Number of Poles	*Maximum Overall Voltage Gain*	*Approximate Roll-Off Rate (dB/Decade)*
2	1.586 (4 dB)[a]	40
3	2 (6 dB)	60
4	2.58 (8 dB)	80
5	3.29 (10 dB)	100

[a]Decibel values are rounded to the nearest whole number.

must be fulfilled if the circuit is to have a Butterworth response curve. Any of the multiple filters described in Table 23.1 can be constructed using the appropriate number of two-pole and one-pole cascaded stages.

The component requirements for one-pole and two-pole Butterworth low-pass and high-pass active filters are summarized in Figures 23.12 and 23.13.

Section Review

1. In terms of their frequency response characteristics, what is the difference between *low-pass* and *high-pass* active filters?
2. In terms of component placement, what is the difference between low-pass and high-pass active filters?
3. Describe the dB gain and roll-off characteristics of low-pass and high-pass filters.

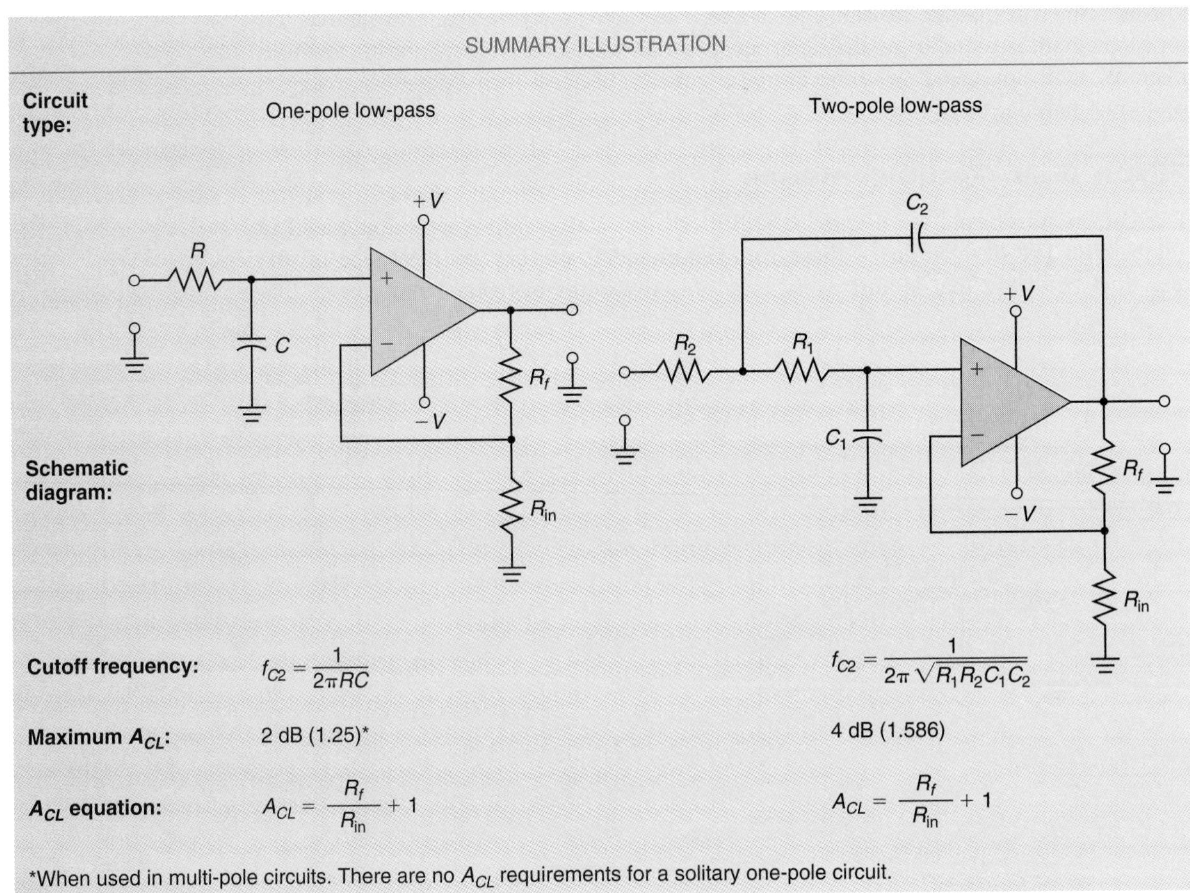

SUMMARY ILLUSTRATION
Circuit type:
One-pole low-pass
Two-pole low-pass
+V
R
C
−V
R_f
R_{in}
C_2
R_2
R_1
C_1
Schematic diagram:
Cutoff frequency:
$f_{C2} = \frac{1}{2\pi RC}$
$f_{C2} = \frac{1}{2\pi\sqrt{R_1R_2C_1C_2}}$
Maximum A_{CL}:
2 dB (1.25)*
4 dB (1.586)
A_{CL} equation:
$A_{CL} = \frac{R_f}{R_{in}} + 1$
$A_{CL} = \frac{R_f}{R_{in}} + 1$
*When used in multi-pole circuits. There are no A_{CL} requirements for a solitary one-pole circuit.

Figure 23.12

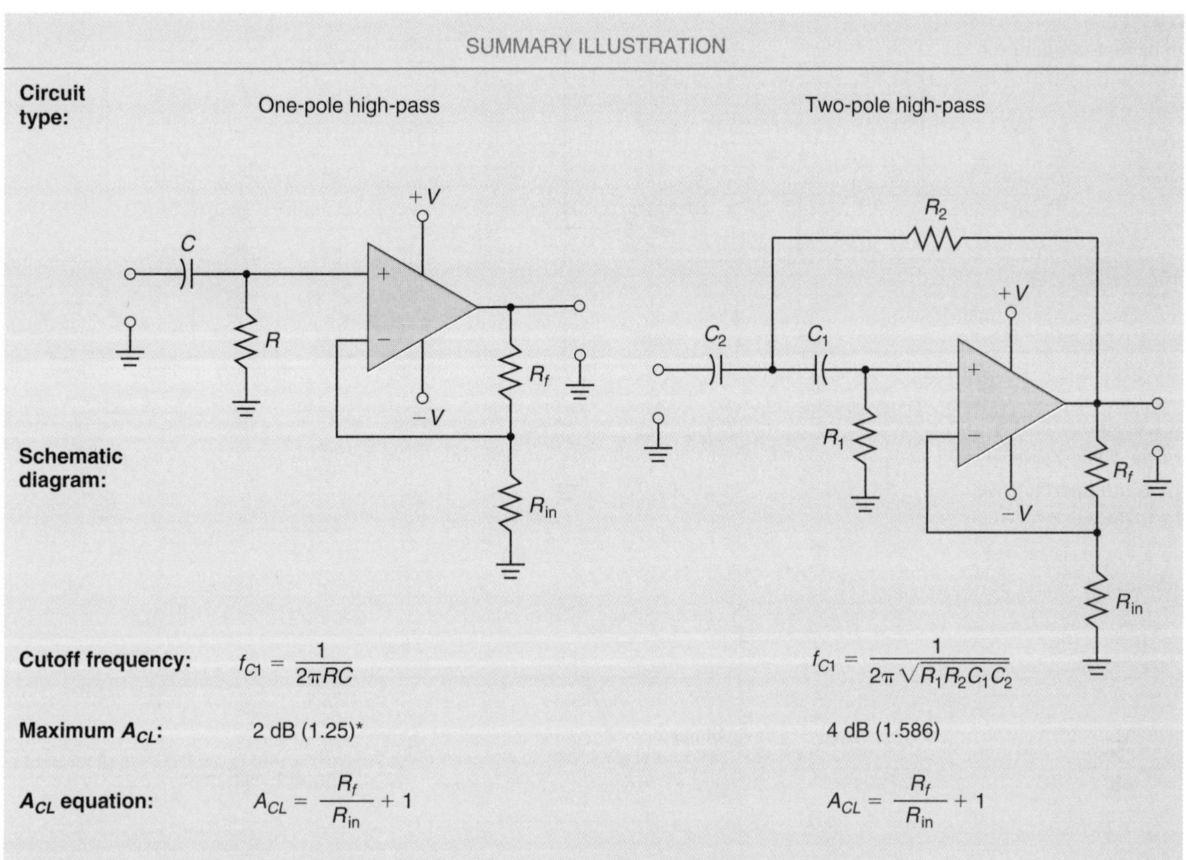

Figure 23.13

23.4 Bandpass and Notch Filters

The Two-Stage Bandpass Filter

A bandpass filter can be constructed by cascading a high-pass filter and a low-pass filter, as shown in Figure 23.14a. The first stage of the amplifier passes all frequencies that are below its value of f_{C2}. All frequencies passed by the first stage are then coupled to the second stage, which passes all frequencies above its value of f_{C1}. The result of this circuit action is illustrated in Figure 23.14b. ◄ *OBJECTIVE 6*

The frequency analysis of this circuit is relatively simple. Once you solve for the cutoff frequencies of each circuit, you can solve for other circuit characteristics, just as we did for passive filters in Chapter 15:

$$\text{BW} = f_{C2} - f_{C1}$$

$$f_0 = \sqrt{f_{C1}f_{C2}}$$

$$Q = \frac{f_0}{\text{BW}}$$

Example 23.3 demonstrates the complete frequency analysis of a two-stage bandpass filter.

Figure 23.14
A two-stage bandpass filter.

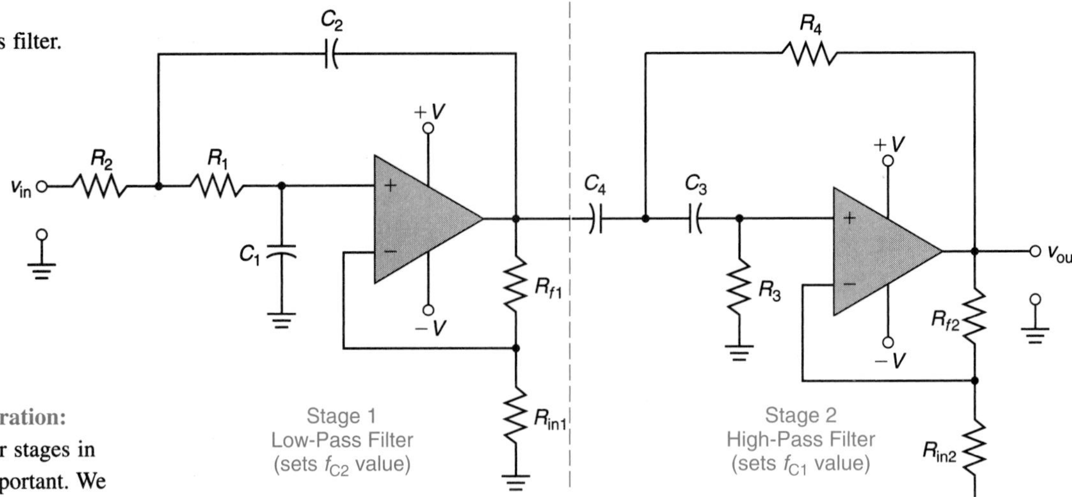

A Practical Consideration: The order of the filter stages in Figure 23.14 is unimportant. We could have a high-pass first stage connected to a low-pass second stage, and the results would be the same.

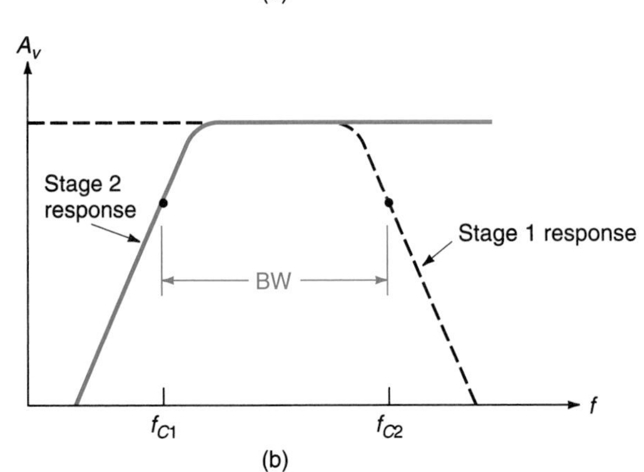

Figure 23.15

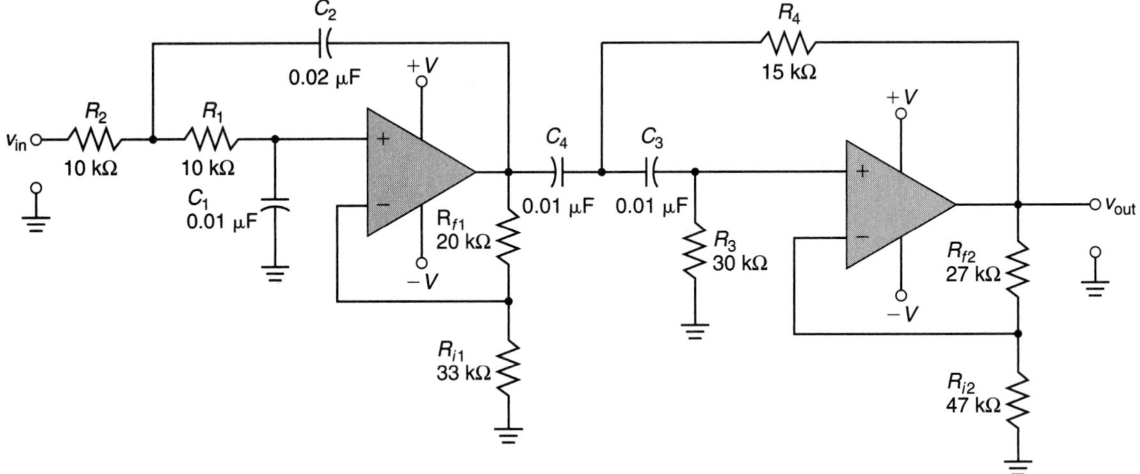

Example 23.3

Perform a complete frequency analysis of the amplifier shown in Figure 23.15.

Solution: The value of f_{C2} is determined by the first stage of the circuit (a low-pass filter), as follows:

$$f_{C2} = \frac{1}{2\pi\sqrt{R_1R_2C_1C_2}} = \frac{1}{2\pi\sqrt{(10\text{ k}\Omega)(10\text{ k}\Omega)(0.01\ \mu\text{F})(0.02\ \mu\text{F})}} = 1.13\text{ kHz}$$

The value of f_{C1} is determined by the second stage of the circuit (a high-pass filter), as follows:

$$f_{C1} = \frac{1}{2\pi\sqrt{R_3R_4C_3C_4}} = \frac{1}{2\pi\sqrt{(30\text{ k}\Omega)(15\text{ k}\Omega)(0.01\ \mu\text{F})(0.01\ \mu\text{F})}} = 750\text{ Hz}$$

Now we can solve for the amplifier bandwidth as

$$\text{BW} = f_{C2} - f_{C1} = 1.13\text{ kHz} - 750\text{ Hz} = 380\text{ Hz}$$

The geometric center frequency is found as

$$f_0 = \sqrt{f_{C1}f_{C2}} = \sqrt{(750\text{ Hz})(1.13\text{ kHz})} = 921\text{ Hz}$$

Finally, the filter Q is found as

$$Q = \frac{f_0}{\text{BW}} = \frac{921\text{ Hz}}{380\text{ Hz}} = 2.42$$

Practice Problem 23.3

A filter like the one shown in Figure 23.15 has values of $R_1 = R_2 = 12\text{ k}\Omega$, $C_1 = C_3 = C_4 = 0.01\ \mu\text{F}$, $C_2 = 0.02\ \mu\text{F}$, $R_3 = 39\text{ k}\Omega$, and $R_4 = 20\text{ k}\Omega$. Perform the frequency analysis of the filter.

The two-stage bandpass filter has the disadvantage of requiring two op-amps and a relatively large number of resistors and capacitors. The **multiple-feedback bandpass filter** requires fewer components (but is a bit more difficult to understand and analyze) than the two-stage filter.

Multiple-Feedback Bandpass Filters

The multiple-feedback bandpass filter, which is shown in Figure 23.16, derives its name from the fact that it has two feedback networks; one capacitive, and one resistive. The series capacitor (C_1) affects the low-frequency response of the filter. The shunt capacitor (C_2) affects its high-frequency response. This point is illustrated in Figure 23.17.

Let's begin our discussion of this circuit by assuming that $C_2 < C_1$ and that the capacitors act like open circuits until a *short-circuit frequency* is reached. (Granted, capacitor operation is more complicated than this, but the simplified view will help you to grasp the basic principles of how this circuit works.)

When the input frequency is below the short-circuit frequency for C_1, both capacitors act as open circuits. Since $C_1 > C_2$, C_2 is open as long as C_1 is open. This condition is

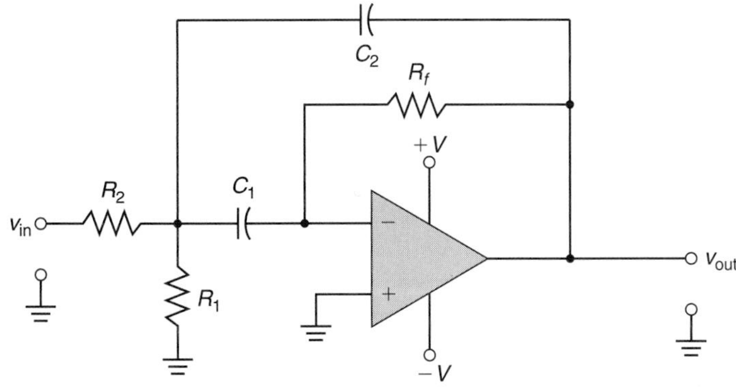

Figure 23.16 A multiple-feedback bandpass filter.

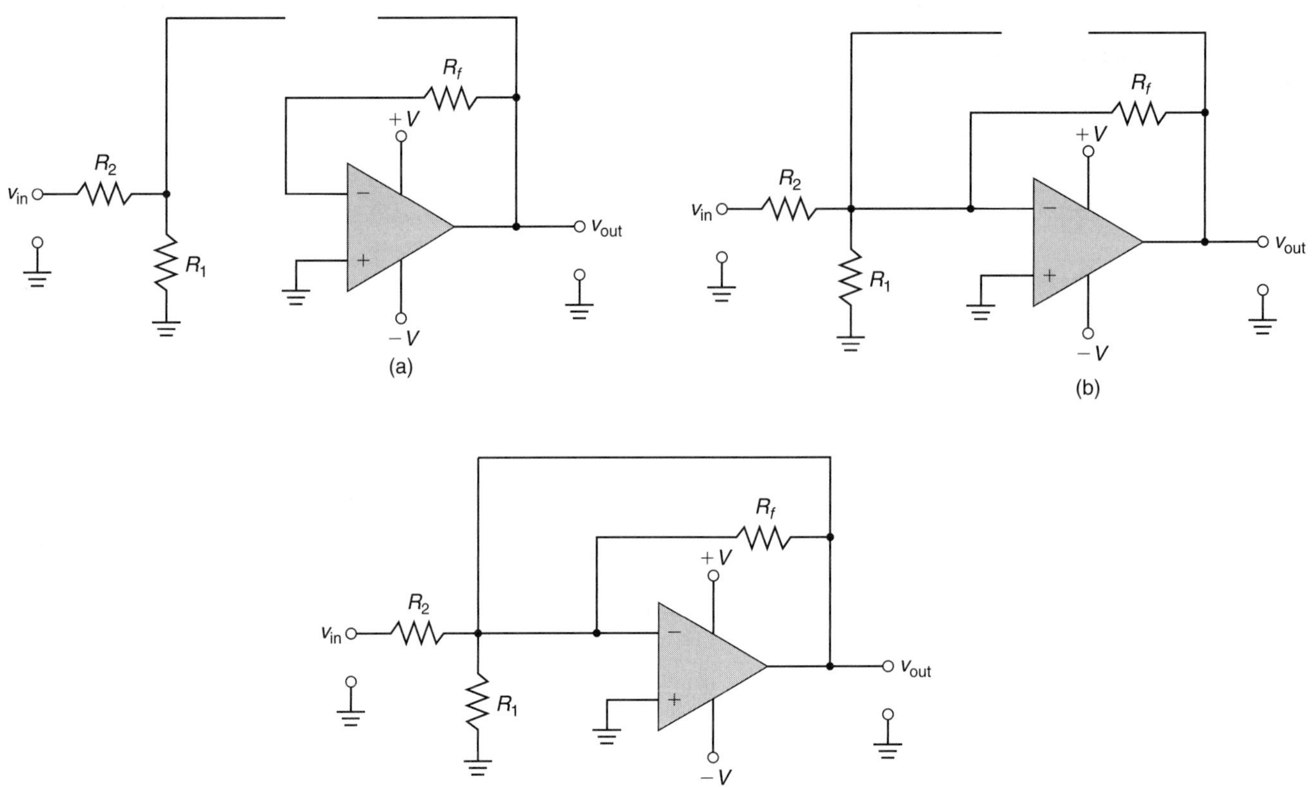

Figure 23.17 Bandpass filter operation.

illustrated in Figure 23.17a. In this case, there is no output from the circuit, because the input signal is blocked.

Now assume that the lower cutoff frequency for the filter (f_{C1}) equals the short-circuit frequency for C_1. When f_{C1} is reached, C_1 becomes a short circuit, while C_2 remains open. This gives us the equivalent circuit shown in Figure 23.17b. At this point, v_{in} is applied to the inverting input of the op-amp, and the filter acts as an inverting amplifier. Thus, at frequencies above f_{C1}, the filter has an output.

Now assume that f_{C2} equals the short-circuit frequency of C_2. When f_{C2} is reached, we have the equivalent circuit shown in Figure 23.17c. Since R_f is effectively shorted by C_2, the closed-loop voltage gain of the circuit is zero at frequencies above f_{C2}. If we put these three equivalent circuits together, we have a circuit with zero voltage gain below f_{C1}, relatively high voltage gain between f_{C1} and f_{C2}, and zero voltage gain above f_{C2}. This, by definition, is a bandpass filter.

Circuit Frequency Analysis

As you know, the cutoff frequency for a two-stage low- or high-pass active filter is found using

$$f_{C2} = \frac{1}{2\pi\sqrt{R_1 R_2 C_1 C_2}}$$

This equation is modified to provide us with the following equation for the geometric center frequency of a multiple-feedback bandpass filter:

$$f_0 = \frac{1}{2\pi\sqrt{(R_1 \| R_2) R_f C_1 C_2}} \tag{23.7}$$

Example 23.4 demonstrates the use of equation (23.7) in determining the geometric center frequency of a multiple-feedback filter.

Example 23.4

Determine the value of f_0 for the filter shown in Figure 23.18.

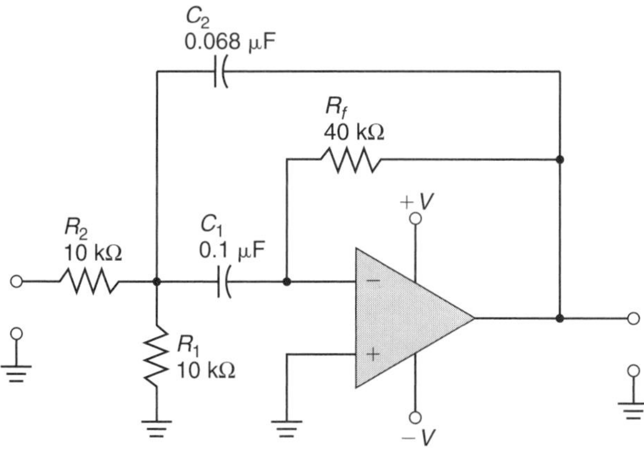

Figure 23.18

Solution: The center frequency of the circuit is found as

$$f_0 = \frac{1}{2\pi\sqrt{(R_1 \| R_2)R_f C_1 C_2}}$$

$$= \frac{1}{2\pi\sqrt{(10\text{ k}\Omega \| 10\text{ k}\Omega)(40\text{ k}\Omega)(0.1\ \mu\text{F})(0.068\ \mu\text{F})}} = 136.5\text{ Hz}$$

Practice Problem 23.4

A multiple-feedback filter has the following values: $R_1 = R_2 = 30\text{ k}\Omega$, $R_f = 68\text{ k}\Omega$, $C_1 = 0.22\ \mu\text{F}$, and $C_2 = 0.1\ \mu\text{F}$. Determine the value of f_0 of the circuit.

Once the center frequency of a multiple-feedback filter is known, we can calculate the value of Q for the circuit using the following equations:

$$C = \sqrt{C_1 C_2} \tag{23.8}$$

$$Q = \pi f_0 R_f C \tag{23.9}$$

Example 23.5 demonstrates the use of these two equations.

Example 23.5

Determine the values of Q and BW for the filter shown in Figure 23.18.

Solution: First, the value of C is found as

$$C = \sqrt{C_1 C_2} = \sqrt{(0.1\ \mu\text{F})(0.068\ \mu\text{F})} = 0.082\ \mu\text{F}$$

In Example 23.4, f_0 was found to be 136.5 Hz. Therefore,

$$Q = \pi f_0 R_f C = \pi(136.5\text{ Hz})(40\text{ k}\Omega)(0.082\ \mu\text{F}) = 1.41$$

and

$$\text{BW} = \frac{f_0}{Q} = \frac{136.5\text{ Hz}}{1.41} = 96.8\text{ Hz}$$

Practice Problem 23.5

Determine the values of Q and BW for the filter described in Practice Problem 23.4.

Once the values of center frequency, Q, and BW are known, f_{C1} and f_{C2} can be determined. When $Q \geq 2$, we can use the following approximations:

$$f_{C1} \cong f_0 - \frac{\text{BW}}{2} \quad \text{and} \quad f_{C2} \cong f_0 + \frac{\text{BW}}{2}$$

The majority of active filters have a Q value that is greater than two. If, however, Q is less than two, we must use two equations that are a bit more complex:

$$f_{C1} = f_0 \sqrt{1 + \left(\frac{1}{2Q}\right)^2} - \frac{\text{BW}}{2} \quad \text{and} \quad f_{C2} = f_0 \sqrt{1 + \left(\frac{1}{2Q}\right)^2} + \frac{\text{BW}}{2}$$

All of these relationships were introduced in Chapter 15.

Filter Gain

In most cases, the analysis of a multiple-feedback filter begins and ends with calculating the frequency response of the circuit. However, it is sometimes valuable to know the closed-loop voltage gain of the circuit. The following equation is used to find the closed-loop voltage gain of a multiple-feedback filter:

$$A_{CL} = \frac{R_f}{2R_{\text{in}}} \qquad (23.10)$$

where

R_{in} = the circuit series input resistor

Notch Filters

As you know, a notch filter is designed to block all frequencies that fall within its bandwidth. A multistage notch filter block diagram and frequency-response curve are shown in Figure 23.19. As you can see, the circuit is made up of a high-pass filter, a low-pass filter, and a summing amplifier. Note that in this case, the low-pass filter determines f_{C1}, and the high-pass filter determines f_{C2}.

When $f_{\text{in}} < f_{C1}$, the low-pass filter passes the input signal to the summing amplifier. Since the input frequency is below f_{C2}, the output from the summing amplifier equals the output from the low-pass filter. When $f_{\text{in}} > f_{C2}$, the high-pass filter passes the input signal to the summing amplifier. Since the input frequency is above f_{C1}, the output from the summing amplifier equals the output from the high-pass filter. Therefore, the notch filter passes all frequencies below f_{C1} and above f_{C2}.

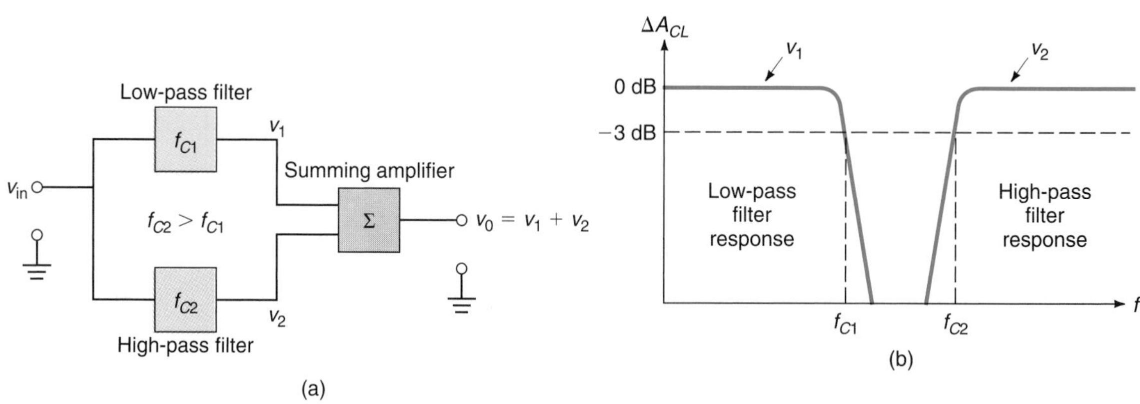

Figure 23.19 Multistage notch filter block diagram and its frequency-response curve.

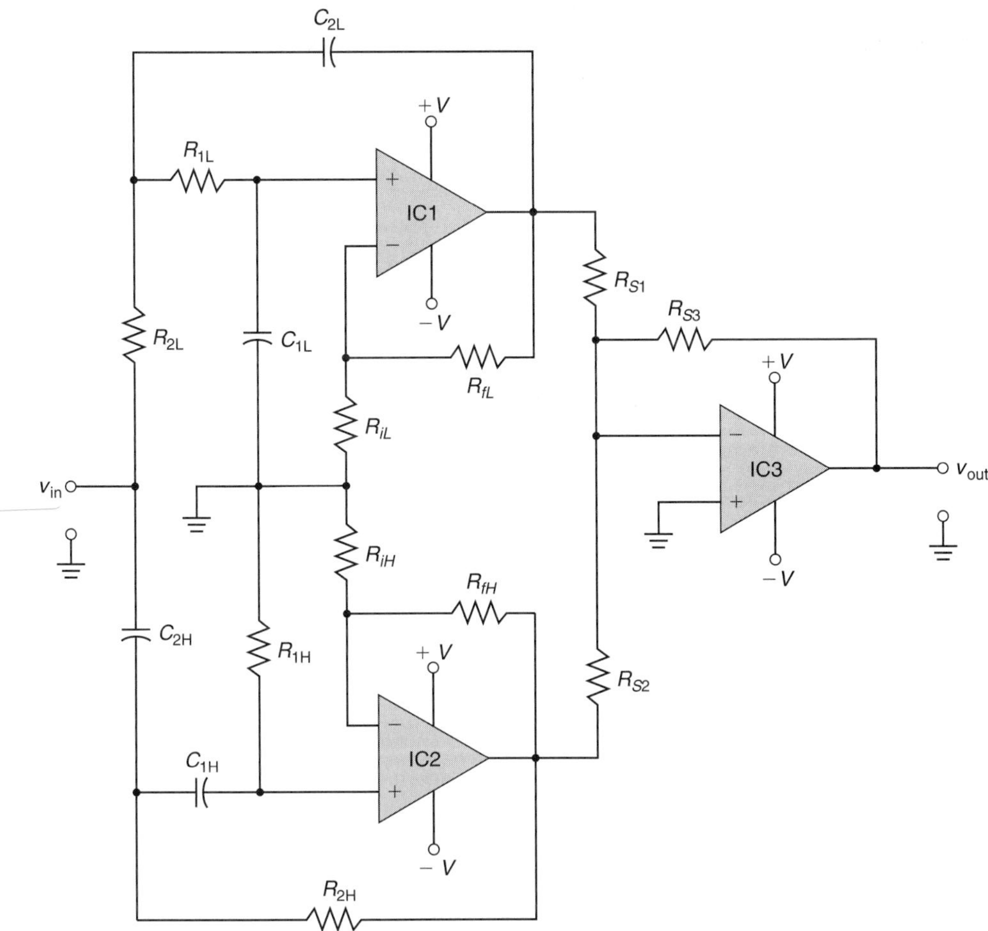

Figure 23.20

When the input frequency is between f_{C1} and f_{C2}, neither filter produces an output (ideally), and the output from the summing amplifier equals zero. In practice, of course, the exact output from the notch filter depends on how close the input frequency is to either f_{C1} or f_{C2}.

The circuit described above is constructed as shown in Figure 23.20. While the circuit may appear confusing at first, closer inspection shows that it is made of circuits that we have already discussed. The low-pass filter consists of the op-amp labeled *IC*1 and the components with the *L* subscript. The high-pass filter consists of the op-amp labeled *IC*2 and the components with the *H* subscript. Finally, the summing amplifier is made of the op-amp labeled *IC*3 and the components with the *S* subscript.

The Multiple-Feedback Notch Filter

The multiple-feedback notch filter shown in Figure 23.21 is very similar to its bandpass counterpart. However, the connection of v_{in} to the noninverting input of the op-amp (via the R_2–R_3 voltage divider) alters the overall circuit response, as follows:

- When $f_{in} < f_{C1}$, C_1 prevents the input voltage from reaching the inverting input. However, the signal does reach the noninverting input, and the circuit produces a noninverted output.
- When $f_{C2} > f_{in} > f_{C1}$, the input signal is applied to both inputs and is seen by the op-amp as a common-mode signal. Since op-amps reject common-mode input signals, there is little or no output when f_{in} falls within the stop-band of the filter.
- When $f_{in} > f_{C2}$, C_2 partially shorts out the signal at the inverting input, reducing its amplitude. However, v_{in} is still applied to the noninverting op-amp input, so the

A Practical Consideration: C_1 always produces a phase shift. Even though the phase shift may be slight, it still prevents v_{in} from appearing simultaneously at the two signal inputs of the op-amp. Therefore, the circuit still has a low-amplitude output signal when operated within its stop-band.

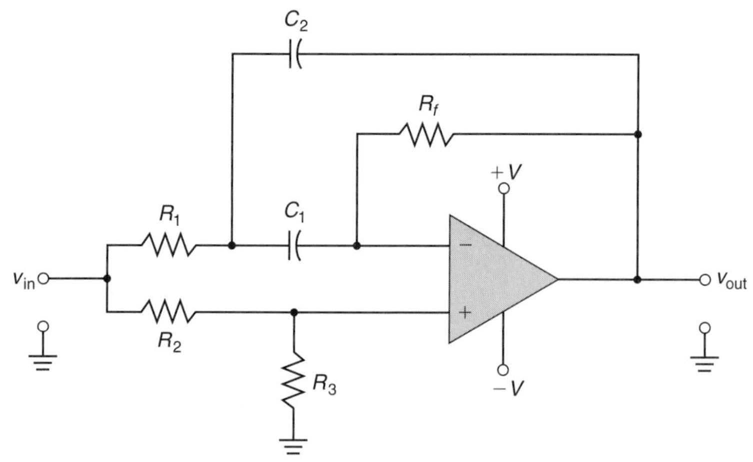

Figure 23.21 A multiple-feedback notch filter.

op-amp amplifies the difference between the two input signals. Therefore, the circuit has a measurable output.

The frequency analysis of the notch filter is very similar to that of the bandpass filter. The only equation modification is as follows:

$$f_0 = \frac{1}{2\pi\sqrt{R_1 R_f C_1 C_2}} \tag{23.11}$$

The gain of this circuit is dictated by C_2. Above f_{C2}, C_2 effectively shorts out R_f, so $A_{CL} \cong 1$. To have equal gain above and below the stop band, R_f and R_1 must be chosen so that the closed-loop gain is as close to unity as possible when $f_{in} < f_{C1}$.

Section Review

1. Describe how cascaded low-pass and high-pass filters can form a bandpass filter.
2. In terms of circuit construction and analysis, contrast the two-stage bandpass filter with the multiple-feedback bandpass filter.
3. Which capacitor in the multiple-feedback bandpass filter controls the high-frequency response of the circuit? Which one controls its low-frequency response?
4. Briefly describe the operation of the multiple-feedback bandpass filter.
5. List, in order, the steps taken to analyze the frequency response of a multiple-feedback bandpass filter.
6. How does the analysis of a $Q \geq 2$ multiple-feedback filter differ from that of a $Q < 2$ filter?
7. Describe the operation of the multistage notch filter shown in Figure 23.20.
8. Briefly describe the operation of the multiple-feedback notch filter.

Critical Thinking

9. Refer to Figure 23.14. If a fault causes f_{C2} to drop below the value of f_{C1}, what effect does it have on the circuit frequency response?

23.5 Introduction to Oscillators

OBJECTIVE 7 ➤ An *oscillator* is a circuit that produces an output waveform without an external signal source. As such, an oscillator functions as a *signal generator.* To understand the operation of oscillators, we have to briefly discuss the concept of *positive feedback.*

Positive Feedback

The term **feedback** describes *any connection that provides a signal path from the output of a circuit back to its input.* We have dealt with a variety of feedback circuits in our discussions on op-amp circuits.

Feedback is generally classified as being either *negative* or *positive.* **Negative feedback** provides *a feedback signal that is 180° out of phase with the circuit input signal.* One method of obtaining negative feedback is illustrated in Figure 23.22a. In the circuit shown, the amplifier (A) is providing a 180° voltage phase shift, but the feedback network (α) is not. The result is that the total voltage phase shift around the loop is 180°, and the feedback signal is out of phase with the input signal.

A practical example of a negative feedback path is shown in Figure 23.22b. As you know, the output from the inverting amplifier is 180° out of phase with the input signal. Since the feedback path (R_f) does not introduce a phase shift, the feedback signal is 180° out of phase with the input. As previous examples have demonstrated, the addition of a negative feedback path to an op-amp has the effect of:

- Decreasing amplifier gain (from A_{OL} to A_{CL}).
- Increasing the amplifier bandwidth.

◄ OBJECTIVE 8

Positive feedback provides *a feedback signal that is in phase with the circuit input.* One method of obtaining positive feedback is illustrated in Figure 23.23. In this case, the amplifier (A) and the feedback network (α) both introduce a 180° voltage phase shift into the loop. This results in a total phase shift of 360° (or 0°), and the feedback signal is in

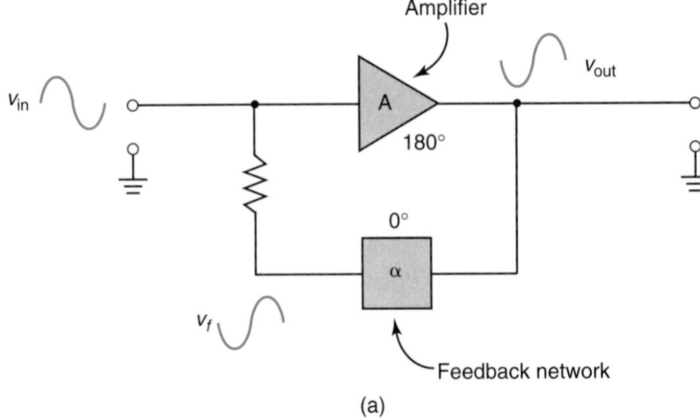

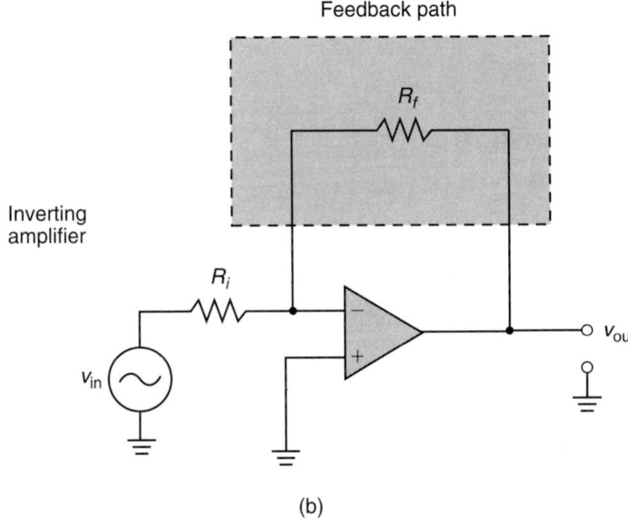

Figure 23.22 Negative feedback.

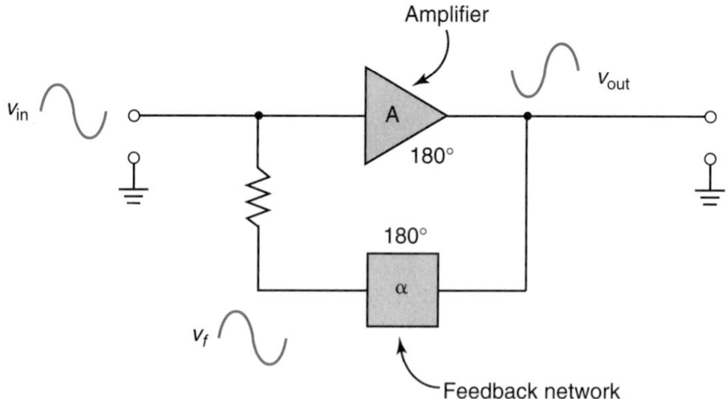

Figure 23.23 Positive feedback.

phase with the input. The same result can be achieved with the circuit configuration shown in Figure 23.24. Since neither component in the loop introduces a phase shift, the output and input signals are in phase.

Oscillators: The Basic Idea

The circuit shown in Figure 23.23, though useful in explaining positive feedback, has an input signal. This is inconsistent with our definition of an oscillator. However, by modifying the circuit as shown in Figure 23.25, we develop one that is very useful for describing the basic operating principles of an oscillator.

With the switch shown in Figure 23.25a closed, the circuit waveforms are as illustrated in the figure. Now assume that the switch is opened while the circuit is in operation. When this occurs, the only input to the amplifier is the feedback voltage (v_f). The amplifier responds to this signal in the same way that it did to the original input signal. In other words, v_f is amplified and sent to the output. Since the feedback network takes a portion of that output and sends it back to the input, the amplifier receives another input cycle, and another output cycle is produced. This process continues as long as power is applied to the circuit, and the circuit produces a sinusoidal output with no external signal source.

The feedback network in the oscillator *generates* an input to the amplifier that, in turn, *generates* an input to the feedback network. Since positive feedback produces this circuit action, it is often referred to as **regenerative feedback.** Regenerative feedback is the basis of operation for all oscillators. An oscillator needs only a trigger signal to start the oscillating circuit action. In fact, most oscillators provide their own trigger signals (as you will see later in this section).

The Barkhausen Criterion

OBJECTIVE 9 ➤ The amplifier in an oscillator provides voltage gain (A_v). The feedback network, on the other hand, *attenuates* its input signal. As you know, *attenuation* is a reduction in the

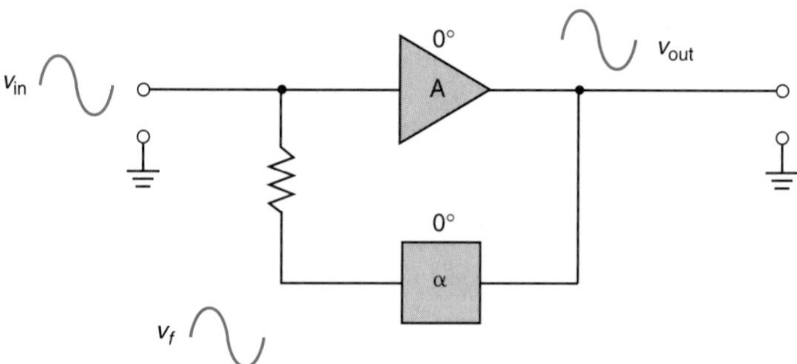

Figure 23.24

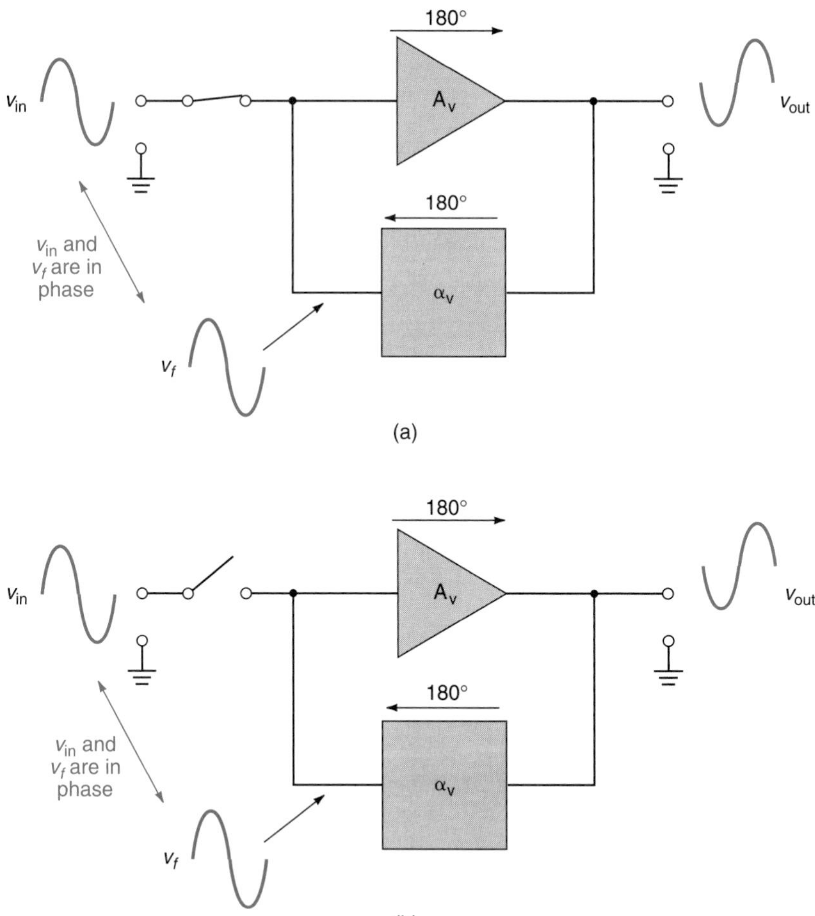

Figure 23.25

amplitude of a signal. Note that the Greek letter *alpha* (α) is commonly used to represent attenuation. (Don't confuse this with the alpha rating of a transistor.)

The **attenuation factor (α_v)** of the feedback network shown in Figure 23.25 equals *the ratio of the feedback voltage to the circuit output voltage.* By formula,

$$\alpha_v = \frac{v_f}{v_{out}} \tag{23.12}$$

Since the feedback network attenuates its input signal, the feedback voltage (v_f) is always less than the circuit output voltage (v_{out}), and

$$\alpha_v < 1 \tag{23.13}$$

For an oscillator to operate properly, the following relationship must be fulfilled:

$$\alpha_v A_v = 1 \tag{23.14}$$

This relationship is known as the **Barkhausen criterion.** The results of not fulfilling this criterion are as follows:

- If $\alpha_v A_v < 1$, the oscillations will fade out within a few cycles.
- If $\alpha_v A_v > 1$, the oscillator will drive itself into saturation and cutoff clipping.

These points are easiest to understand if we apply them to several circuits using different combinations of A_v and α_v. Since the input to an oscillator is the feedback voltage (v_f), the output voltage is found as

$$v_{out} = A_v v_f$$

The value of v_f depends on the values of α_v and v_{out}, as follows:

$$v_f = \alpha_v v_{out}$$

Now we will see what happens as each of the circuits shown in Figure 23.26 progresses through several cycles of operation. We will assume that the initial input to each amplifier is a 0.1 V_{pk} signal.

Note: The values of v_{out} and v_f for each cycle are found using the appropriate equations from page 791. The value of v_f at the end of each cycle is the value of v_{in} for the next cycle.

Starting with this input value, the circuit in Figure 23.26a progresses as follows:

Cycle	v_{in}	v_{out}	v_f
1	0.1 V_{pk}	10 V_{pk}	0.05 V_{pk}
2	0.05 V_{pk}	5 V_{pk}	0.025 V_{pk}
3	0.025 V_{pk}	2.5 V_{pk}	0.0125 V_{pk}

As you can see, when $\alpha_v A_v < 1$, the oscillations lose amplitude on each progressive cycle and will eventually fade out. This *loss of signal amplitude* is called **damping.**

When $\alpha_v A_v > 1$, the output amplitude of the circuit shown in Figure 23.26b increases with each cycle. This problem is illustrated as follows:

Cycle	v_{in}	v_{out}	v_f
1	0.1 V_{pk}	10 V_{pk}	1 V_{pk}
2	1 V_{pk}	100 V_{pk}	10 V_{pk}
3	10 V_{pk}	1000 V_{pk}	100 V_{pk}

For this circuit, it took only two cycles to start hitting some impossible values for v_{out}. The point, however, is that v_{out} is increasing from each cycle to the next. Eventually, the output starts to experience saturation and cutoff clipping. Thus, $\alpha_v A_v$ cannot be greater than one.

The only condition that provides a constant sinusoidal output from an oscillator is when $\alpha_v A_v = 1$, as it does in the circuit shown in Figure 23.26c. This results in the circuit operation as follows:

Cycle	v_{in}	v_{out}	v_f
1	0.1 V_{pk}	10 V_{pk}	0.1 V_{pk}
2	0.1 V_{pk}	10 V_{pk}	0.1 V_{pk}
3	0.1 V_{pk}	10 V_{pk}	0.1 V_{pk}

Obviously, this series of cycles will continue over and over. Thus, having a product of $\alpha_v A_v = 1$ causes the oscillator to have a consistent sinusoidal output, which is the only acceptable outcome.

OBJECTIVE 10 ➤

Now we have established the three requirements for an oscillator:

- Regenerative feedback.
- An initial input trigger to start the oscillations.
- $\alpha_v A_v = 1$ (fulfilling the Barkhausen criterion).

As long as all these conditions are fulfilled, we have an oscillator.

Section Review

1. What is an *oscillator*?
2. What is *positive feedback*?
3. In terms of phase shifts, how is positive feedback usually produced?
4. How does positive feedback maintain the oscillations started by a *trigger*?
5. What is the *Barkhausen criterion*?

6. What happens when $\alpha_v A_v > 1$?
7. What happens when $\alpha_v A_v < 1$?
8. What is *damping*?
9. List the three requirements for proper oscillator operation.

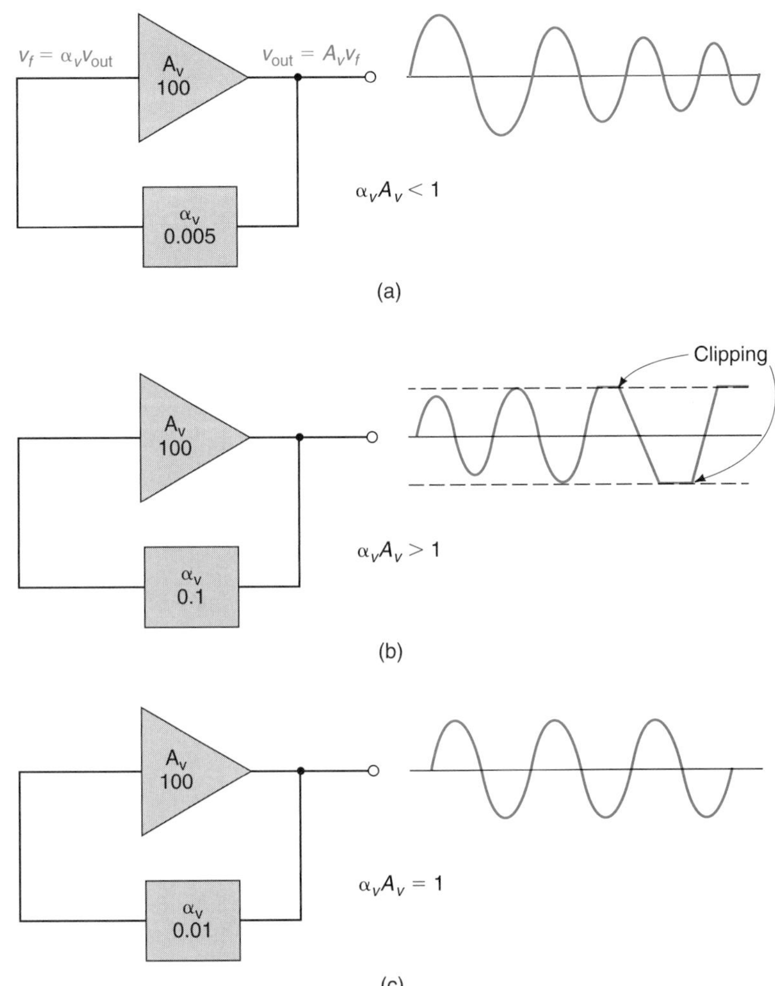

Figure 23.26 Effects of $\alpha_v A_v$ on oscillator operation.

23.6 Phase-Shift Oscillators

Probably the easiest oscillator to understand is the **phase-shift oscillator.** This circuit contains three *RC* circuits in its feedback network, as shown in Figure 23.27. You may recall that an *RC* circuit introduces a phase shift (θ), which is found as ◄ *OBJECTIVE 11*

$$\theta = \tan^{-1} \frac{-X_C}{R}$$

At this point, we are not interested in the exact value of θ. However, the fact that θ changes with X_C (and therefore, with frequency) means that a given *RC* circuit can be designed to produce a specific phase shift at a specific frequency.

In a phase-shift oscillator, the three *RC* circuits produce a combined phase shift of 180° at some specific operating frequency (f_0). This allows the circuit to oscillate at that

A Practical Consideration: A phase shift oscillator can also be constructed using shunt resistors and series capacitors; that is, by reversing the capacitor and resistor positions shown in Figure 23.27.

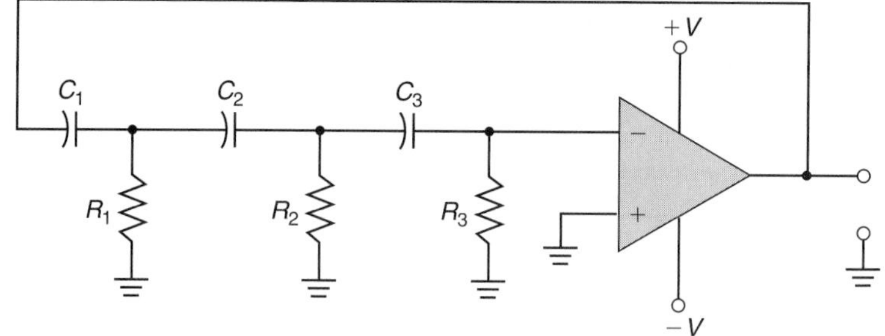

Figure 23.27 A phase-shift oscillator.

frequency, provided that the Barkhausen criterion has been met. For example, let's assume that the *RC* circuit shown in Figure 23.27 produces a 180° voltage phase shift at $f_0 = 10$ kHz and that $\alpha_v A_v = 1$ at the same frequency. As a result of these conditions, the circuit oscillates, generating a 10 kHz sinusoidal output.

Practical Considerations

Phase-shift oscillators are rarely used, because they are extremely unstable. Oscillator **stability** is *a measure of its ability to maintain an output that is constant in frequency and amplitude.* Phase-shift oscillators are extremely difficult to stabilize in terms of frequency. Though rarely used, they are introduced here because they are relatively easy to understand and serve as a valuable learning tool.

Another practical consideration involves the Barkhausen criterion. The relationship $\alpha_v A_v = 1$ holds true only for ideal circuits. In any practical circuit, the product of $\alpha_v A_v$ must be *slightly greater than* one, because there are always some losses in the amplifier and the feedback circuit. How much greater than one? Just enough to sustain oscillations.

Finally, we must address the question of how the oscillations start. Consider what happens when power is first applied to the circuit shown in Figure 23.27. Since this is an inverting amplifier, there is a 180° phase difference between the op-amp's input and output. This means that when power is applied, one of two things must happen:

- The input remains stable, and the output changes to the opposite-polarity extreme.
- The output remains stable, and the input voltage changes to the opposite-polarity voltage.

Which of these possibilities actually happens is of no consequence. Either way, a transition occurs between the output and the input. This transition is enough to trigger the oscillating process.

Section Review

1. Describe the construction of the phase-shift oscillator.
2. How is regenerative feedback produced by the phase-shift oscillator?
3. What is oscillator *stability*?
4. Why are phase-shift oscillators rarely used?
5. Why must $\alpha_v A_v$ be slightly greater than one in a practical oscillator?
6. How is a *trigger* produced in a practical oscillator?

23.7 The Wien-Bridge Oscillator

The **Wien-bridge oscillator** shown in Figure 23.28 is one of the more commonly used low-frequency *RC* oscillators. This circuit achieves regenerative feedback by producing no phase shift (0°) at its operating frequency. The Wien-bridge oscillator has two feedback paths: a *positive* feedback path (to the noninverting input), and a *negative* feedback path

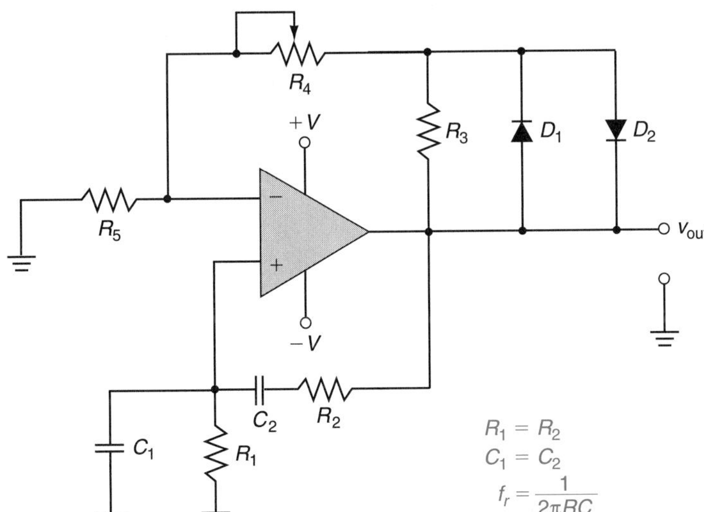

Figure 23.28 The Wien-bridge oscillator.

(to the inverting input). The positive feedback path is used to produce oscillations, while the negative feedback path is used to control the A_{CL} of the circuit.

The Positive Feedback Path

The positive feedback network contains two RC circuits. R_1C_1 forms a low-pass filter, while R_2C_2 forms a high-pass filter. Connected in series, they form a bandpass filter. When operating at midband, the filter introduces no phase shift (0°). This circuit is designed so that $R_1 = R_2$ and $C_1 = C_2$. Thus, the two filters have the same cutoff frequency. As a result, the circuit oscillates at this frequency. This concept is illustrated in Figure 23.29.

One final note on the positive feedback circuit: You will often see *trimmer potentiometers* added in series with R_1 and R_2. These trimmers allow the feedback circuit (and thus, the operating frequency) to be "fine-tuned."

The Negative Feedback Circuit

As you know, the closed-loop voltage gain of a noninverting amplifier is found as

$$A_{CL} = \frac{R_f}{R_{\text{in}}} + 1$$

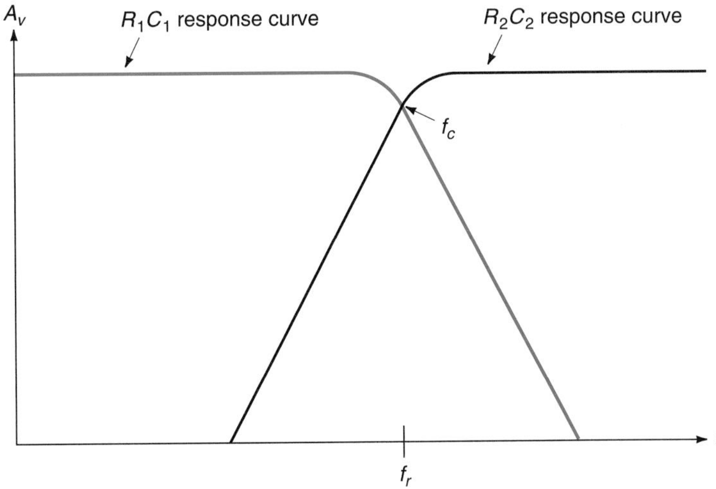

Figure 23.29 Frequency response of the positive feedback network.

For the circuit shown in Figure 23.28, $R_{in} = R_5$ and $R_f = (R_3 + R_4)$ when the diodes are off. Adjusting the value of R_4 has the effect of increasing or decreasing the A_{CL} of the circuit.

The diodes in feedback path prevent the circuit output from exceeding a predetermined value. If the oscillator output tries to exceed the magnitude of $(V_{R4} + V_{R5})$ by more than 0.7 V, one of the diodes turns on, shorting out R_3. This effectively reduces the closed-loop voltage gain. Essentially, the diodes are acting as *clippers* in this application.

Frequency Limits

As the operating frequency increases, the propagation delay of the op-amp can introduce a significant phase shift between its input and output terminals. As a result, the oscillating action of the circuit begins to lose its stability. Remember, the operation of the circuit depends heavily on the phase relationship between its input and its output. For most Wien-bridge oscillators, the upper frequency limit is below 1 MHz. The characteristics of the Wien-bridge oscillator are summarized in Figure 23.30.

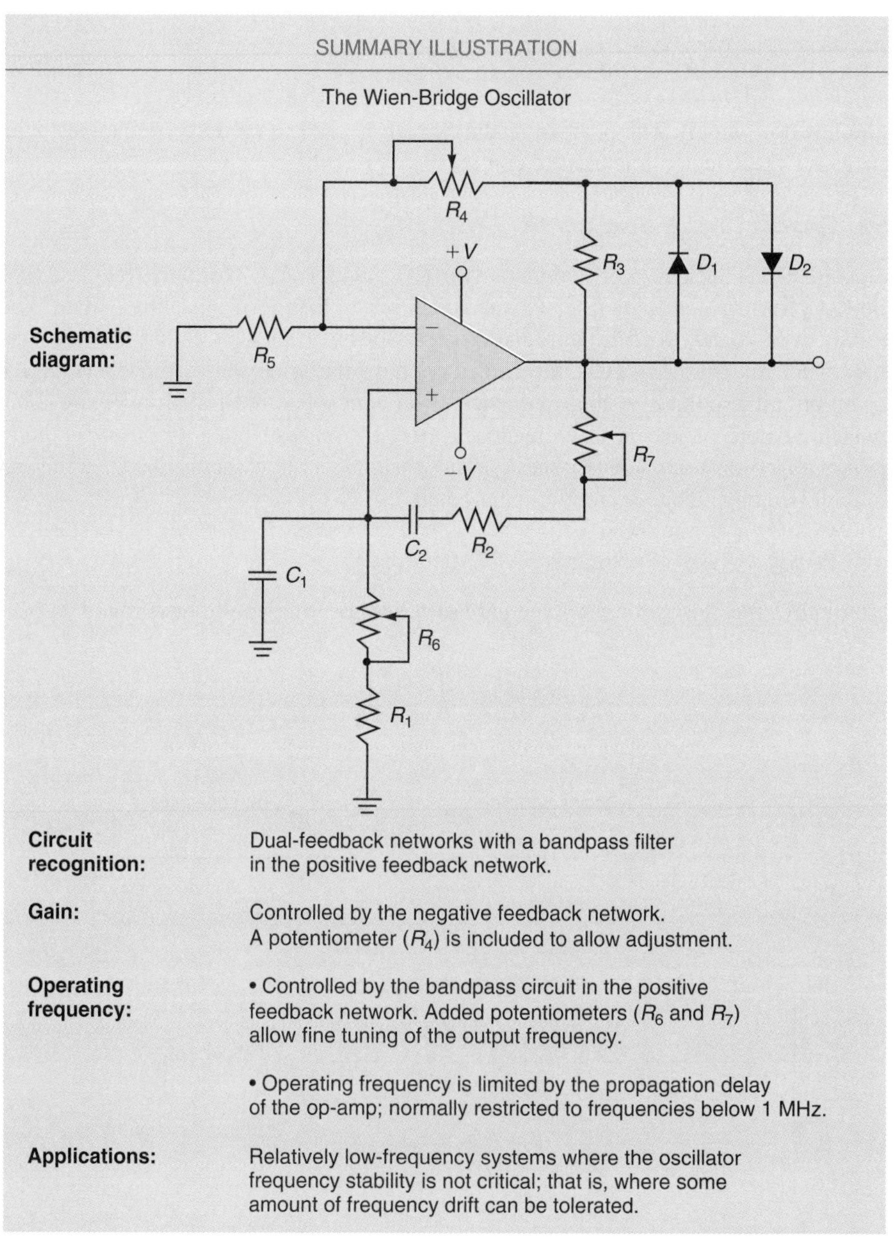

SUMMARY ILLUSTRATION
The Wien-Bridge Oscillator
Schematic diagram:
R4
+V
R3
D1
D2
R5
−V
R7
C2
R2
C1
R6
R1
Circuit recognition:
Dual-feedback networks with a bandpass filter in the positive feedback network.
Gain:
Controlled by the negative feedback network.
A potentiometer (R4) is included to allow adjustment.
Operating frequency:
• Controlled by the bandpass circuit in the positive feedback network. Added potentiometers (R6 and R7) allow fine tuning of the output frequency.
• Operating frequency is limited by the propagation delay of the op-amp; normally restricted to frequencies below 1 MHz.
Applications:
Relatively low-frequency systems where the oscillator frequency stability is not critical; that is, where some amount of frequency drift can be tolerated.

Figure 23.30

Section Review

1. Describe the construction of the Wien-bridge oscillator.
2. How does the *positive* feedback network of the Wien-bridge oscillator control its operating frequency?
3. Explain why potentiometers are normally included in the positive feedback network of a Wien-bridge oscillator.
4. Explain why a potentiometer is normally included in the *negative* feedback network of a Wien-bridge oscillator.
5. What op-amp characteristic limits the operating frequency of the Wien-bridge oscillator?

23.8 Discrete *LC* Oscillators: The Colpitts Oscillator

◄ OBJECTIVE 12

The **Colpitts oscillator** is *a discrete LC amplifier that uses a pair of tapped capacitors and an inductor to produce the regenerative feedback necessary for oscillations.* The resonant frequency of the *LC* tank circuit determines the operating frequency of the oscillator. A Colpitts oscillator is shown in Figure 23.31. Since this is a common-emitter (inverting) configuration, the feedback network must produce a 180° voltage phase shift for the feedback to be regenerative.

The Feedback Network

The key to understanding the Colpitts oscillator is in knowing how the feedback network produces a 180° voltage phase shift. The feedback network shown in Figure 23.31 is made of C_1, C_2, and L. The operation of the feedback network is based on the following key points:

- The amplifier output voltage is developed across C_1.
- The feedback voltage is developed across C_2.
- The voltage across C_2 is 180° out of phase with the voltage across C_1.

Therefore, the feedback voltage is 180° out of phase with the output voltage.

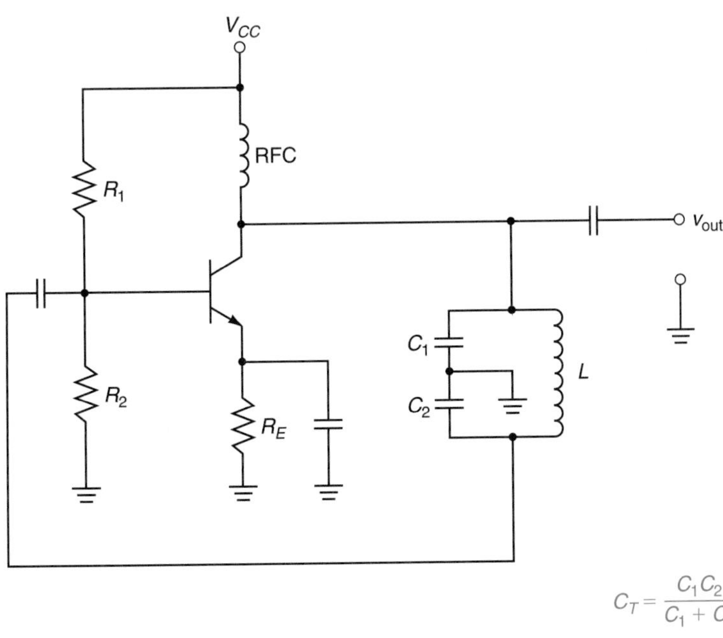

$$C_T = \frac{C_1 C_2}{C_1 + C_2}$$

$$f_r = \frac{1}{2\pi\sqrt{LC_T}}$$

Figure 23.31 A Colpitts oscillator.

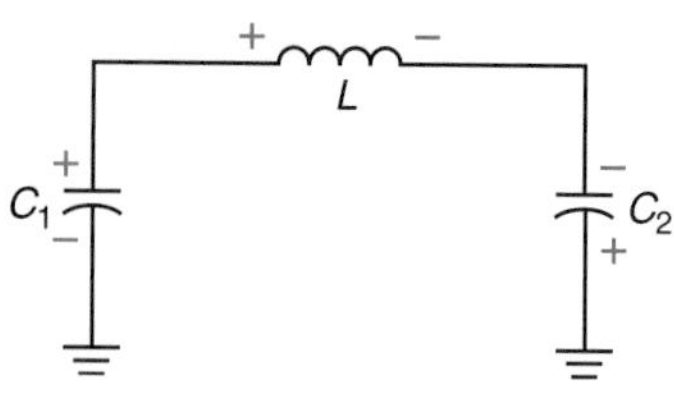

Figure 23.32

The first two points are fairly simple to see. C_1 is in parallel with the load, so any output voltage is developed across C_1. C_2 is in parallel with R_2, so any input voltage is developed across C_2.

The phase relationship between the capacitor voltages is a little tougher to visualize. If we redraw the feedback network as shown in Figure 23.32, it may be easier to understand. Let's assume for a moment that L has a voltage across it with the polarities shown in the figure. Since C_1 is connected from its *positive* side to ground, it has the polarities shown. Since C_2 is connected from the inductor's *negative* side to ground, it has the opposite polarity. As you can see, these voltages are 180° out of phase. This is how the necessary 180° phase shift in the feedback network is produced.

The value of the feedback voltage in the Colpitts oscillator depends on the α_v of the circuit. For this oscillator, α_v is the ratio of X_{C2} to X_{C1}. Since reactance is inversely proportional to capacitance, α_v is also equal to the ratio of C_1 to C_2. By formula,

$$\alpha_v = \frac{C_1}{C_2} \quad (23.15)$$

Circuit Gain

As with any other oscillator, $\alpha_v A_v$ must be slightly greater than one. As described earlier, v_{out} is developed across C_2, and v_f is developed across C_1. Since $v_{in} = v_f$, the value of A_v for the Colpitts amplifier can be found as

$$A_v = \frac{v_{out}}{v_{in}} \cong \frac{C_2}{C_1} \quad (23.16)$$

Operating Frequency

The frequency at which the Colpitts oscillator operates is the *resonant frequency* of the LC tank circuit. Figure 23.32 showed that C_1 is in *series* with C_2. Therefore, the operating frequency is found as

$$f_O = \frac{1}{2\pi\sqrt{LC_T}} \quad (23.17)$$

where

$$C_T = \frac{C_1 C_2}{C_1 + C_2}$$

Amplifier Coupling

As with any parallel resonant circuit, the feedback network loses some efficiency when loaded. To reduce the loading effects of R_L, Colpitts oscillators are commonly transformer-coupled to the load, as shown in Figure 23.33. For this circuit, the primary winding of T_1 is the feedback network inductance. The transformer reduces circuit loading because of the effect of the turns ratio on the reflected load impedance.

We can use capacitive coupling for the Colpitts oscillator provided that the following relationship is fulfilled:

$$C_C << C_T$$

where

C_T = the total series capacitance of the feedback network

By fulfilling this relationship, you ensure that the reactance of the coupling capacitor is much greater than the series combination of C_1 and C_2. This prevents circuit loading almost as well as transformer coupling does. Unfortunately, the relatively high reactance of the coupling capacitor causes a significant reduction in load voltage.

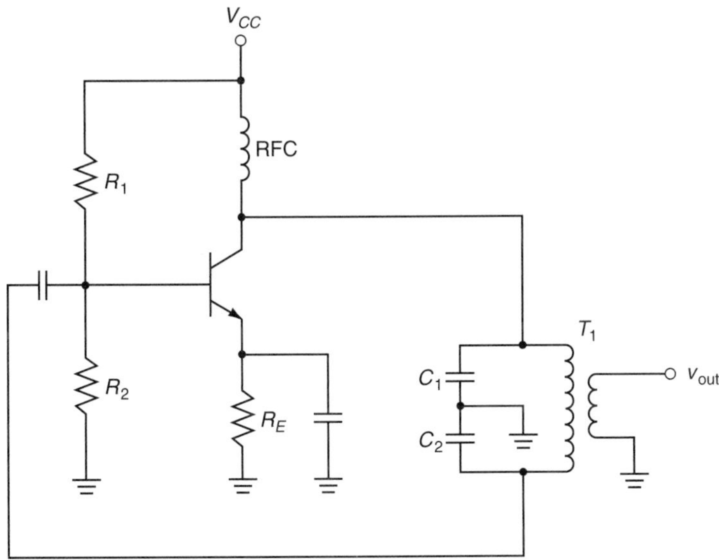

Figure 23.33 Transformer-coupled Colpitts oscillator.

Summary

The Colpitts oscillator uses a pair of tapped capacitors and an inductor (all in its feedback network) to produce the 180° feedback phase shift required for oscillations. The frequency of operation for the circuit is approximately equal to the resonant frequency of the feedback circuit.

Section Review

1. Describe the construction of the *Colpitts oscillator.*
2. Explain how the feedback network in the Colpitts oscillator produces a 180° voltage phase shift.
3. List the key points to remember about the operation of the Colpitts oscillator.
4. How do you calculate the value of A_v for a Colpitts oscillator?
5. Why is *transformer coupling* often used in Colpitts oscillators?
6. Is the following statement true or false? The higher the Q of the feedback network shown in Figure 23.31, the more stable the oscillator output frequency. Critical Thinking

23.9 Other *LC* Oscillators

Although the Colpitts is the most commonly used *LC* oscillator, several others are worth mentioning. In this section, we will take a very brief look at some of these other *LC* oscillators.

Hartley Oscillators

The **Hartley oscillator** is almost identical to the Colpitts oscillator. The primary difference is that the feedback network uses tapped inductors and a single capacitor. A Hartley oscillator is shown in Figure 23.34. ◀ *OBJECTIVE 13*

For the oscillator shown in Figure 23.34, the output voltage is developed across L_1, and the feedback voltage is developed across L_2. The attenuation caused by the feedback network (α_v) is found as

$$\alpha_v = \frac{L_2}{L_1} \tag{23.18}$$

A Practical Consideration: C_3 in Figure 23.34 is called a *blocking capacitor.* Its purpose is to prevent V_{CC} from being shorted to ground through the series combination of the RF choke (*RFC*) and L_1. The value of C_3 is too high (by design) to consider in any circuit frequency calculations.

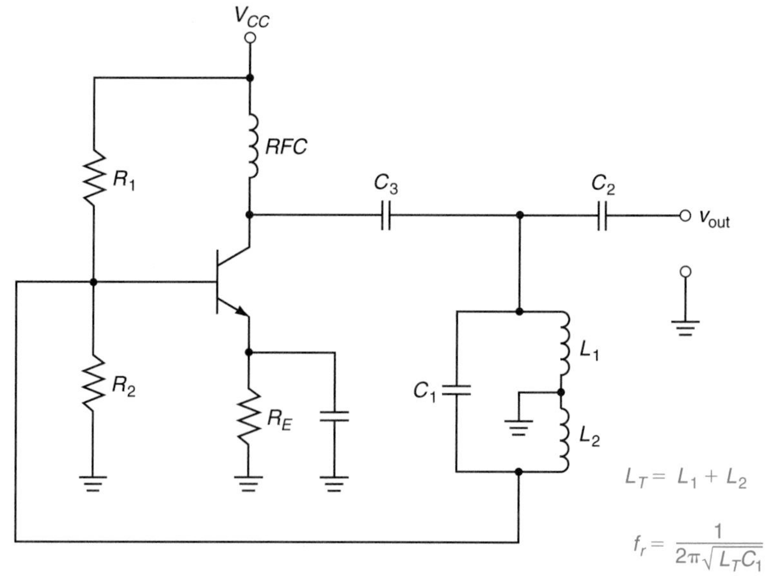

Figure 23.34 A Hartley oscillator.

Since the inductors in Figure 23.34 are in series, the total inductance is found as

$$L_T = L_1 + L_2$$

This value of L_T is used in the calculation of the circuit's resonant frequency.

The Clapp Oscillator

The **Clapp oscillator** is simply a Colpitts oscillator with an added capacitor in its feedback network. As shown in Figure 23.35, the added capacitor (C_3) is placed in series with inductor. This capacitor is normally much lower in value than C_1 or C_2. Because its value is so low, C_3 becomes the dominant component in all frequency calculations for the circuit. For the Clapp oscillator,

$$f_r \cong \frac{1}{2\pi\sqrt{LC_3}} \tag{23.19}$$

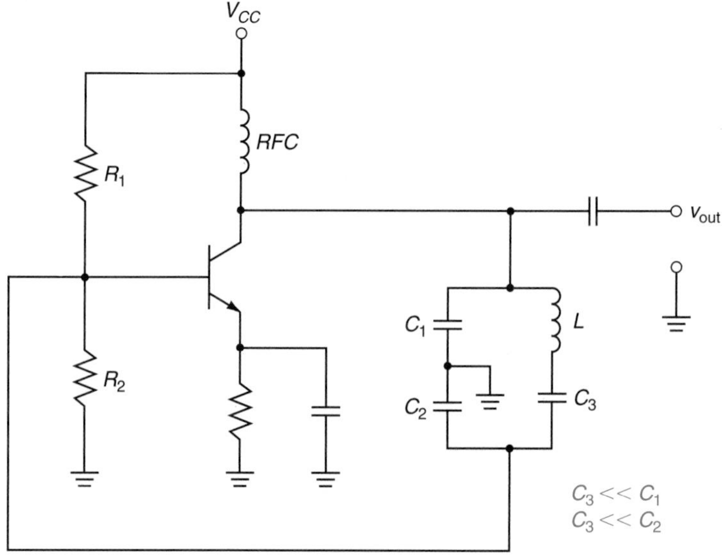

Figure 23.35 The Clapp oscillator.

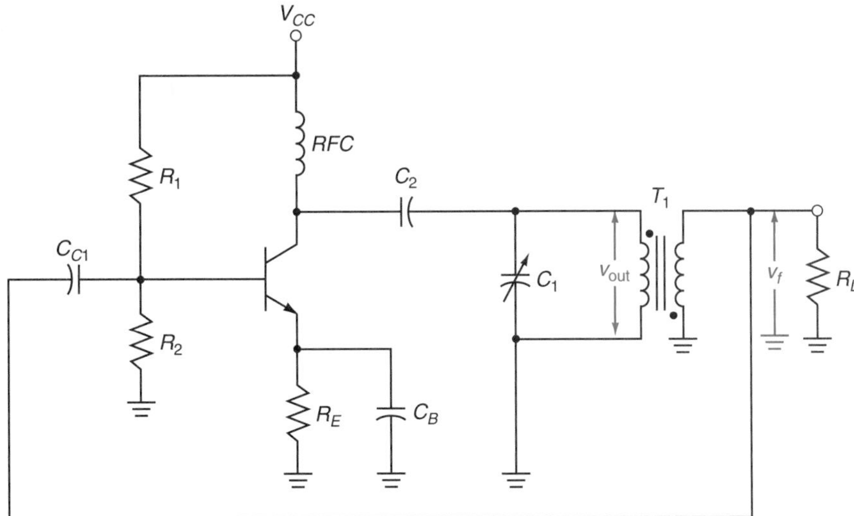

Figure 23.36 The Armstrong oscillator.

The Armstrong Oscillator

The **Armstrong oscillator** uses a transformer to achieve the 180° voltage phase shift required for oscillations. As shown in Figure 23.36, the output from the transistor is applied to the primary of the transformer (T_1), and the feedback signal is taken from the secondary. The polarity dots on the transformer symbol indicate a 180° voltage phase shift from primary to secondary. The magnitude of v_f depends on v_{out} and the turns ratio of the transformer. As shown in Figure 23.36, the load voltage equals v_f.

The capacitor in the output circuit (C_1) is included to provide the oscillator tuning. The resonant frequency of the circuit is determined by the adjusted value of C_1 and the inductance of the transformer primary. Note that C_2 is a blocking capacitor used to prevent the *RFC* and the primary of T_1 from shorting out the dc supply.

One Final Note

You have been introduced to several BJT-based *LC* oscillators. Any of the *LC* oscillators we have discussed can be made using either an FET or an op-amp. The key to dealing with the various oscillators is to learn the circuit-recognition features of the feedback networks. The recognition features of the *LC* oscillators described in this section are as follows:

Oscillator Type	*Recognition Feature(s)*
Hartley	Tapped inductors or a tapped transformer primary with a single parallel capacitor
Clapp	Looks like a Colpitts, with a capacitor added in series with the inductor
Armstrong	A transformer with a single parallel capacitor

Section Review

1. Where are the feedback and output voltages measured in the feedback network of a *Hartley* oscillator?
2. How does the construction of the *Clapp* oscillator differ from that of the *Colpitts* oscillator? What purpose does this difference serve?
3. How is the required 180° voltage phase shift accomplished in the feedback network of an *Armstrong* oscillator?
4. List the circuit-recognition features of the *LC* oscillators covered in this section.

23.10 Crystal-Controlled Oscillators

Most communications and digital applications require oscillators with extremely stable outputs. This can pose a problem for "conventional" oscillators. Various factors, such as component tolerances and thermal variations, can cause their operating frequency or amplitude to change over time. In any system where oscillator stability is critical, a *crystal-controlled oscillator* is normally used.

Crystals

OBJECTIVE 14 ➤

The key to the operation of a crystal is the **piezoelectric effect.** This means that *the crystal vibrates at a constant rate when it is exposed to an electric field.* When a crystal is placed between two metal plates, it vibrates at its resonant frequency as long as a voltage is applied. The frequency of the vibrations depends on the physical dimensions of the crystal. Crystals can also be made to vibrate by applying a signal to them at a given frequency. Several types of crystals can be used in oscillators, but the quartz crystal is the most common.

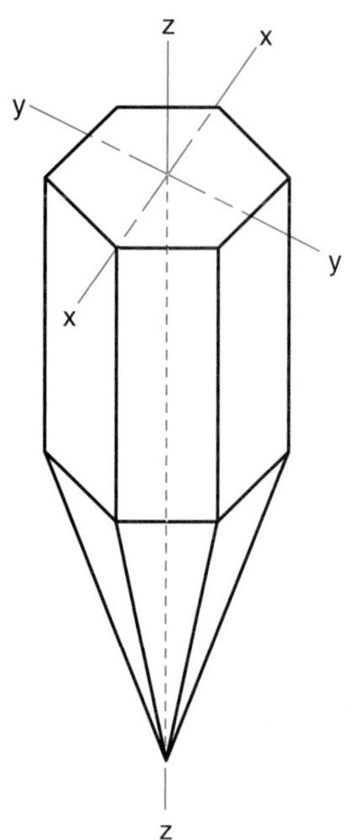

Figure 23.37
Quartz crystal.

Quartz Crystals

Quartz crystals are made of silicon dioxide (SiO_2). They develop as six-sided crystals, as shown in Figure 23.37. When used in an electronic component, a thin slice of crystal is placed between two conducting plates, like those of a capacitor. As stated earlier, the physical dimensions of the crystal determine its operating frequency.

The electrical operation of the crystal is based on its mechanical properties. However, we can still represent the crystal with an equivalent circuit. Figure 23.38a shows the schematic symbol for the crystal. Figure 23.38b shows the equivalent circuit for the device. The components in the equivalent circuit represent specific characteristics of the crystal, as follows:

C_C = the capacitance of the crystal itself

C_M = the mounting capacitance between the crystal and the mounting plates that hold it in place

L = the inductance of the crystal

R = the resistance of the crystal

The graph shown in Figure 23.28c represents the frequency characteristics of the crystal. At the low-frequency end of the curve, the impedance of the crystal is a function of

A Practical Consideration: There is only a slight difference between the values of f_S and f_p for a given crystal. The space left between these frequencies in Figure 23.38 is for illustrative purposes.

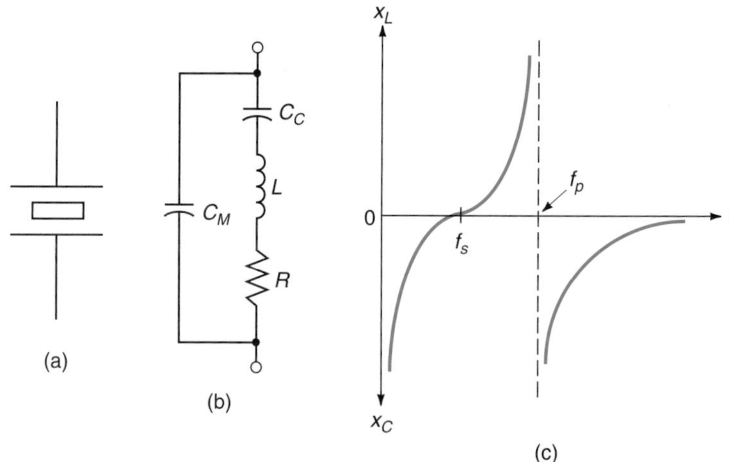

Figure 23.38 Crystal symbol, equivalent circuit, and frequency response.

the extremely high reactances of C_M and C_C. As the frequency increases, the series combination of C_C and L approaches resonance. At the *series resonant frequency* of the crystal (f_s), the net crystal reactance drops to near zero, and the total crystal impedance equals R (which is relatively low).

When the frequency increases above f_s, the net series reactance of the component increases rapidly. At the same time, the parallel combination of C_M and L approaches resonance. At the *parallel resonant frequency* of the crystal (f_p), its impedance approaches infinity. At frequencies above f_p, the decreasing reactance of C_M causes the net crystal impedance to drop off. The bottom line of all this is as follows:

- At f_s, the crystal has the characteristics of a *series* resonant circuit.
- At f_p, the crystal has the characteristics of a *parallel* resonant circuit.

This means that we can use a crystal in place of either a series LC circuit or a parallel LC circuit. If we use it in place of a series LC circuit, the oscillator operates at f_s. If we use it in place of a parallel LC circuit, the oscillator operates at f_p.

Overtone Mode

A given crystal produces outputs at its resonant frequency and at harmonics of that frequency. The resonant frequency is sometimes referred to as the *fundamental frequency,* and the harmonics are sometimes referred to as **overtones.** Crystals are limited by their physical dimensions to fundamental outputs of 10 MHz and less. However, by tuning circuits to the overtones, we can produce usable signals at much higher frequencies. This is referred to as operating in **overtone mode.**

CCO Circuits

A Colpitts oscillator can be modified into a **crystal-controlled oscillator (CCO),** as shown in Figure 23.39. As you can see, a crystal (Y_1) has been added in series between the LC circuit and the amplifier input. The crystal acts as a series-resonant circuit (at f_s). When operated at f_s, the impedance of the crystal drops to nearly zero, and the LC circuit is coupled directly to the amplifier input.

The frequency stability of the CCO comes from the extremely high Q of the crystal, which is typically in the thousands. Because of this high Q, the component acts as a series-resonant circuit over an extremely limited band of frequencies. This restricts the operation of the oscillator to the same limited band of frequencies. Placing a crystal in the same relative position in a Hartley or Clapp oscillator has the same result.

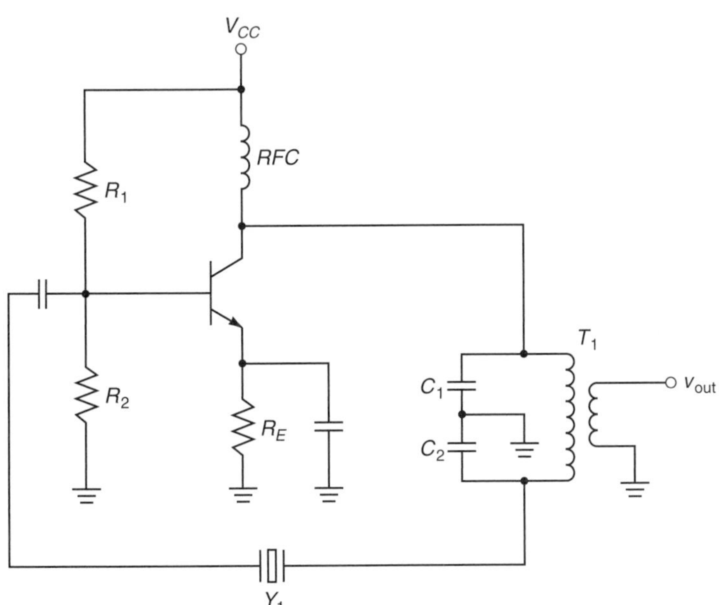

Figure 23.39 Crystal-controlled Colpitts oscillator.

Section Review

1. What is a crystal-controlled oscillator?
2. What is the *piezoelectric effect*?
3. Explain how a crystal will act as a *series-resonant circuit* at one frequency (f_s) and as a *parallel-resonant component* at another (f_p).
4. Explain how the addition of a *crystal* to an *LC* oscillator stabilizes the operating frequency of the circuit.

Critical Thinking

5. What prevents the circuit in Figure 23.39 from operating at the parallel-resonant frequency (f_p) of the crystal?

KeyTerms

The following terms were introduced in this chapter on the pages indicated:

Active filter, 772
Armstrong oscillator, 801
Attenuation factor (α_v), 791
Barkhausen criterion, 791
Bessel filter, 776
Butterworth filter, 775
Chebyshev filter, 775
Clapp oscillator, 800
Colpitts oscillator, 797
Crystal-controlled oscillator (CCO), 803
Damping, 792
Feedback, 789
Fidelity, 776
Hartley oscillator, 799
Midband gain ($A_{v(\text{mid})}$), 772
Multiple-feedback bandpass filter, 783
Negative feedback, 789
Order, 775
Oscillator, 772
Overtone, 803
Overtone mode, 803
Pass band, 772
Phase-shift oscillator, 793
Piezoelectric effect, 802
Pole, 775
Positive feedback, 789
Regenerative feedback, 790
Ripple width, 776
Sallen-Key filter, 779
Stability, 794
Tuned circuit, 772
VCVS, 779
Wien-bridge oscillator, 794

Practice Problems

1. An amplifier has values of f_0 = 14 kHz and BW = 2 kHz. Calculate the value of Q for the circuit.
2. An amplifier has values of f_0 = 1200 kHz and BW = 300 kHz. Calculate the value of Q for the circuit.
3. An amplifier with a geometric center frequency of 800 kHz has a Q of 6.2. Calculate the bandwidth of the circuit.
4. An amplifier with a geometric center frequency of 1100 kHz has a Q of 25. Calculate the bandwidth of the circuit.
5. Calculate the cutoff frequency and closed-loop voltage gain for the filter shown in Figure 23.40.

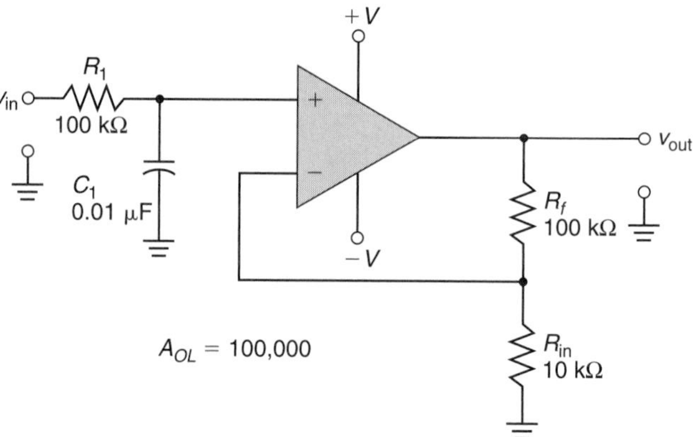

Figure 23.40

6. A filter like the one shown in Figure 23.40 has values of $R_1 = 82\ \text{k}\Omega$, $C_1 = 0.015\ \mu\text{F}$, $R_f = 150\ \text{k}\Omega$, and $R_{in} = 20\ \text{k}\Omega$. Calculate the cutoff frequency and closed-loop voltage gain of the circuit.

7. Calculate the cutoff frequency and closed-loop voltage gain of the filter shown in Figure 23.41.

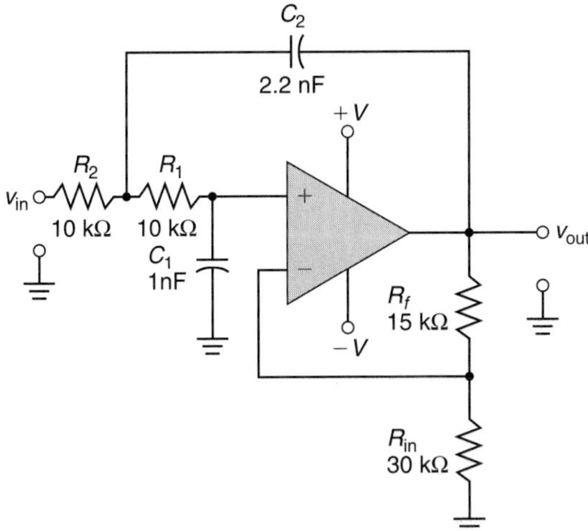

Figure 23.41

8. A filter like the one shown in Figure 23.10a has values of $R_1 = R_2 = 33\ \text{k}\Omega$, $C_1 = 100\ \text{pF}$, $C_2 = 200\ \text{pF}$, and $R_f = 220\ \text{k}\Omega$. Calculate the cutoff frequency and closed-loop voltage gain of the circuit.

9. Calculate the cutoff frequency and closed-loop voltage gain of the filter shown in Figure 23.42.

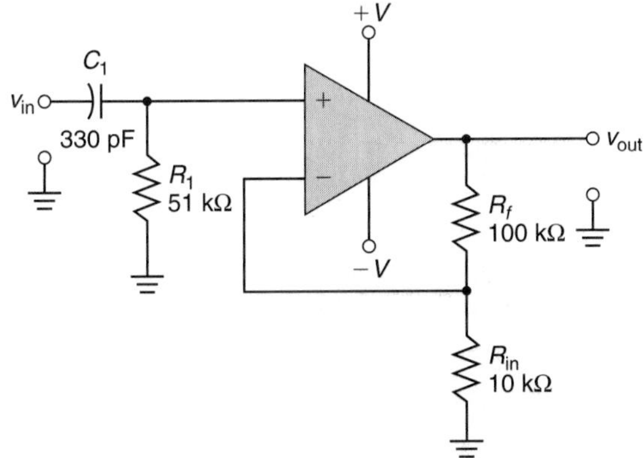

Figure 23.42

10. Calculate the cutoff frequency and closed-loop voltage gain of the filter shown in Figure 23.43.

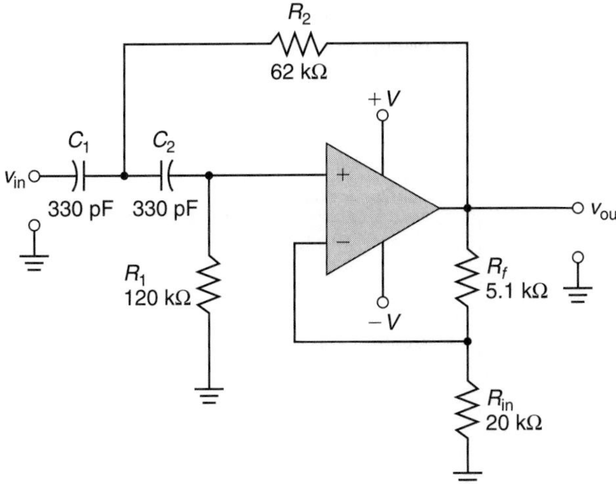

Figure 23.43

11. Calculate the values of f_{C1}, f_{C2}, bandwidth, geometric center frequency, and Q for the bandpass filter shown in Figure 23.44.

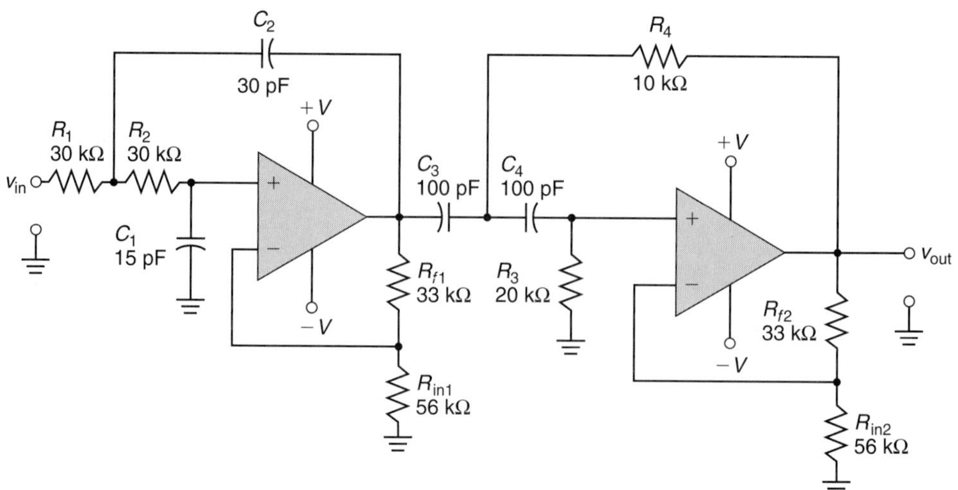

Figure 23.44

12. Calculate the values of f_{C1}, f_{C2}, bandwidth, geometric center frequency, and Q for the bandpass filter shown in Figure 23.45.

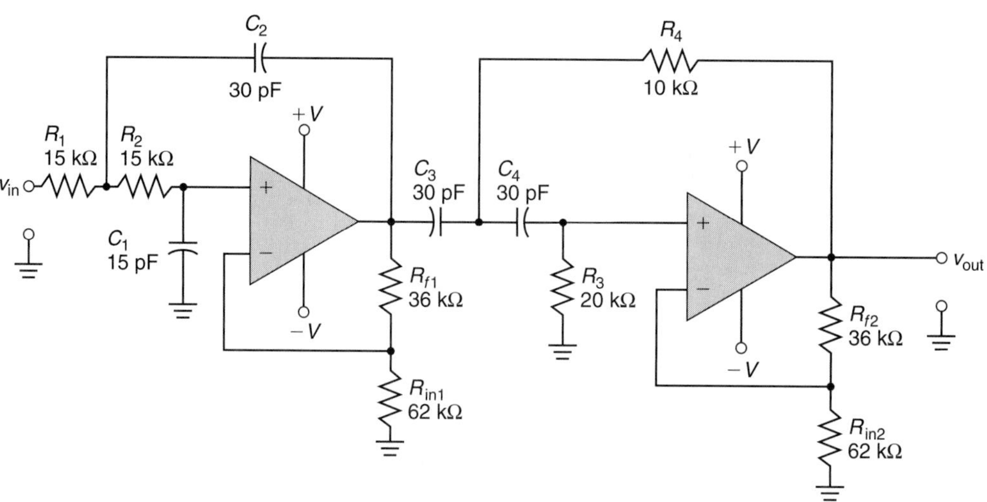

Figure 23.45

13. Calculate the values of f_{C1}, f_{C2}, bandwidth, geometric center frequency, and Q for the bandpass filter shown in Figure 23.46.

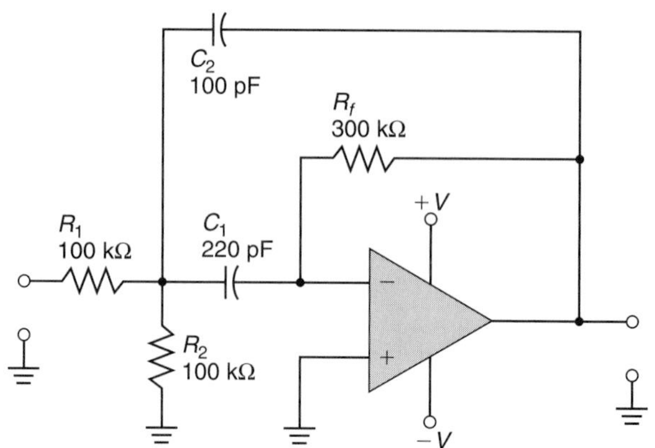

Figure 23.46

14. Calculate the values of f_{C1}, f_{C2}, bandwidth, geometric center frequency, and Q for the bandpass filter shown in Figure 23.47.

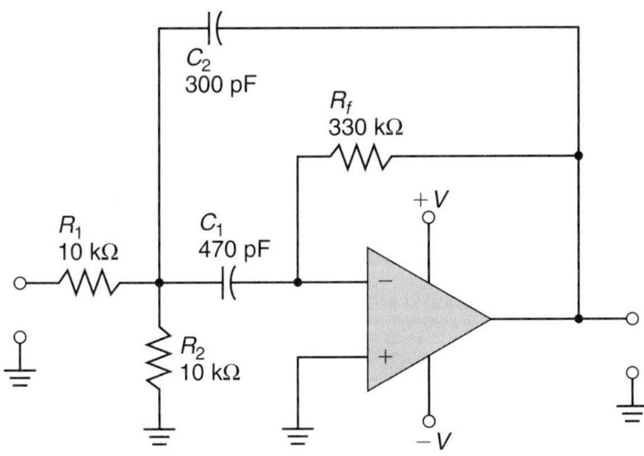

Figure 23.47

15. A multiple-feedback filter has values of $f_0 = 62$ kHz and BW = 20 kHz. Approximate the values of f_{C1} and f_{C2} for the circuit.

16. A multiple-feedback filter has values of $f_0 = 482$ kHz and BW = 200 kHz. Approximate the values of f_{C1} and f_{C2} for the circuit.

17. Calculate the value of A_{CL} for the filter shown in Figure 23.46.

18. Calculate the value of A_{CL} for the filter shown in Figure 23.47.

19. Calculate the values of f_{C1}, f_{C2}, bandwidth, geometric center frequency, and Q for the notch filter shown in Figure 23.48.

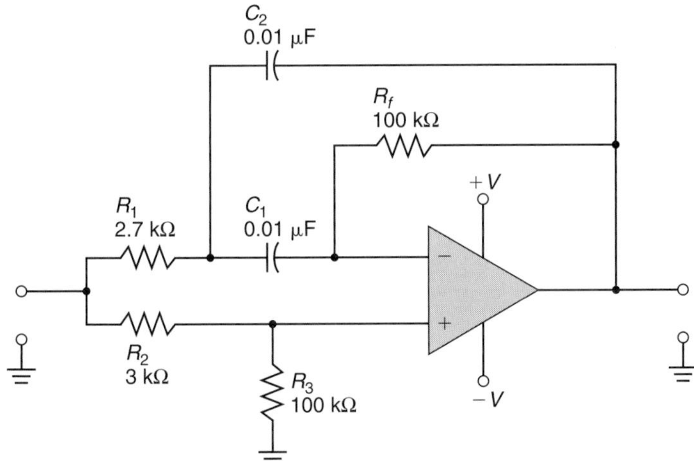

Figure 23.48

20. Calculate the values of f_{C1}, f_{C2}, bandwidth, geometric center frequency, and Q for the notch filter shown in Figure 23.49.

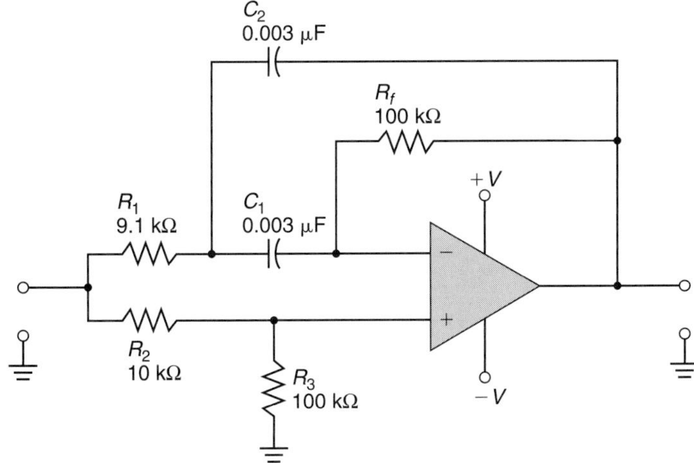

Figure 23.49

21. Calculate the operating frequency of the circuit shown in Figure 23.50.

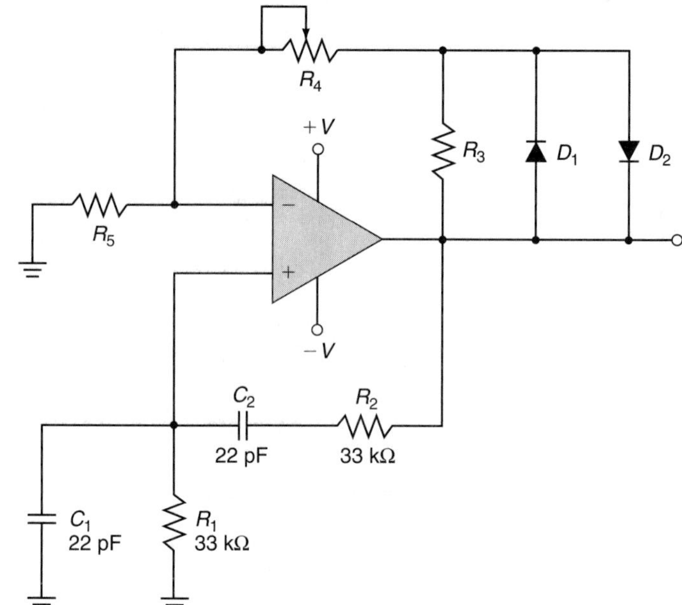

Figure 23.50

22. Calculate the operating frequency of the circuit shown in Figure 23.51.

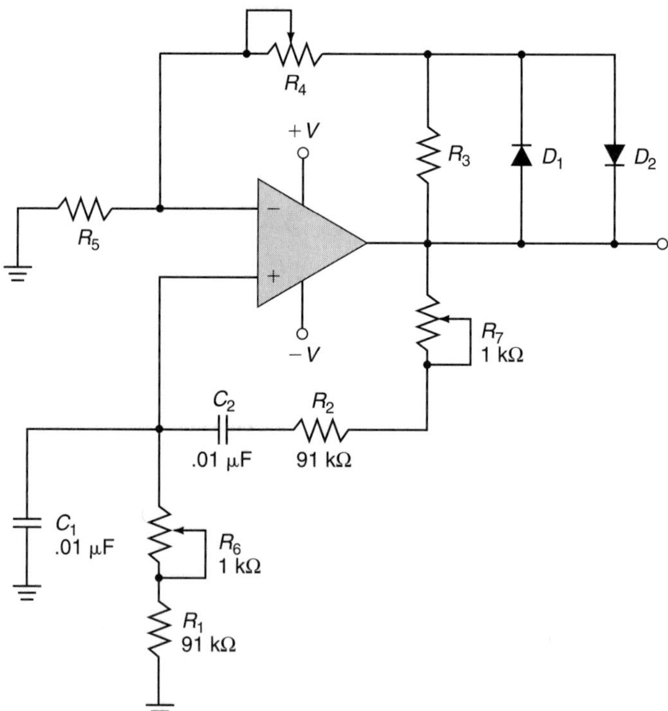

Figure 23.51

23. Calculate the operating frequency for the circuit shown in Figure 23.52.

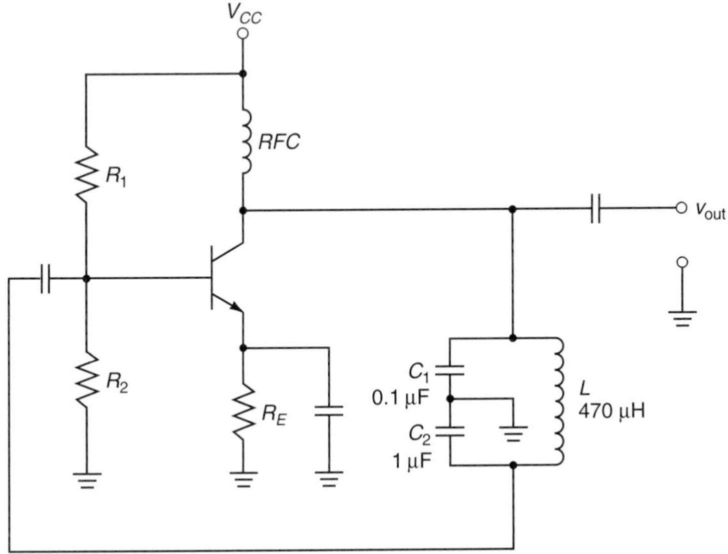

Figure 23.52

24. Calculate the value of α_v for the circuit shown in Figure 23.53.

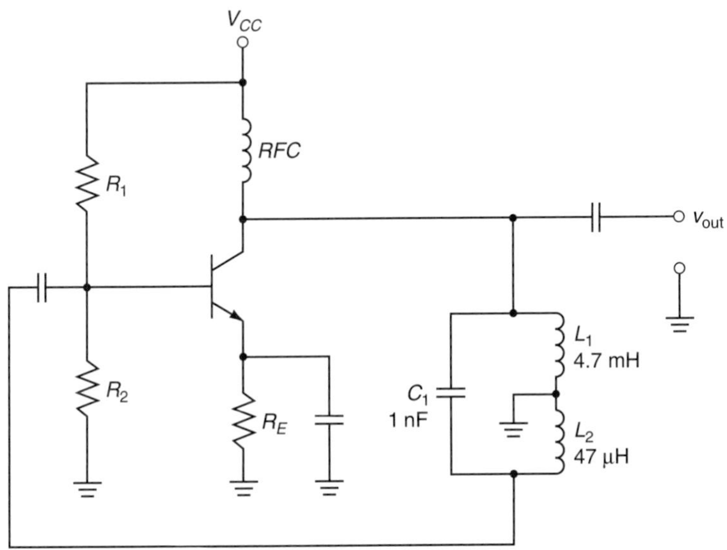

Figure 23.53

25. A circuit like the one shown in Figure 23.52 has values of $C_1 = 10$ nF, $C_2 = 1$ μF, and $L_1 = 3.3$ mH. Calculate the operating frequency, α_v, and A_v for the circuit.

26. A circuit like the one shown in Figure 23.52 has values of $C_1 = 1$ μF, $C_2 = 10$ μF, and $L_1 = 3.3$ mH. Calculate the operating frequency, α_v, and A_v for the circuit.

Looking Back

These problems relate to material presented in earlier chapters. The chapters are identified in brackets.

27. A parallel *LC* circuit has values of $L = 10$ μH, $R_W = 5\ \Omega$, and $C = 51$ nF. Calculate the equivalent parallel resistance (R_P) of the circuit when operated at resonance. [Chapter 14]

28. A sine wave has values of $V_{pk} = 12$ V and $f = 120$ Hz. Calculate the instantaneous value of the waveform at $t = 4$ ms. [Chapter 9]

29. An ac source with values of $V_S = 10$ V and $f = 60$ Hz is connected to a 0.01 μF capacitor. Calculate the value of the circuit current. [Chapter 12]

30. A half-wave rectifier is connected to a 910 Ω load. Calculate the average load voltage and current values. [Chapter 18]

31. A resistor is color-coded Red-Red-Black-Gold. Calculate the minimum and maximum values for the component. [Chapter 2]

32. A series circuit has values of $R = 100\ \Omega$ and $C = 1\ \mu F$. Calculate the time required for the capacitor to fully charge when a $+5\ V_{dc}$ input is applied to the circuit. [Chapter 16]

Pushing the Envelope

33. Figure 23.54 shows a notch filter that consists of a multiple-feedback bandpass filter and a summing amplifier. Using your knowledge of these circuits, describe the response of the filter to frequencies below, within, and above its bandwidth.

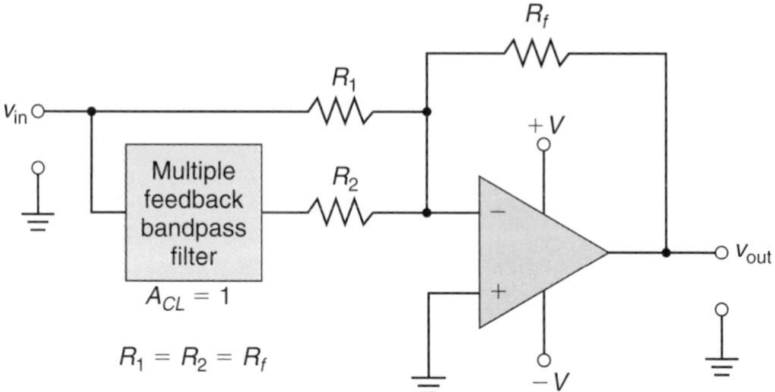

Figure 23.54

34. Despite its appearance, you've seen the dc biasing circuit in Figure 23.55 before. Determine the type of biasing used, and perform a complete dc analysis of the oscillator.

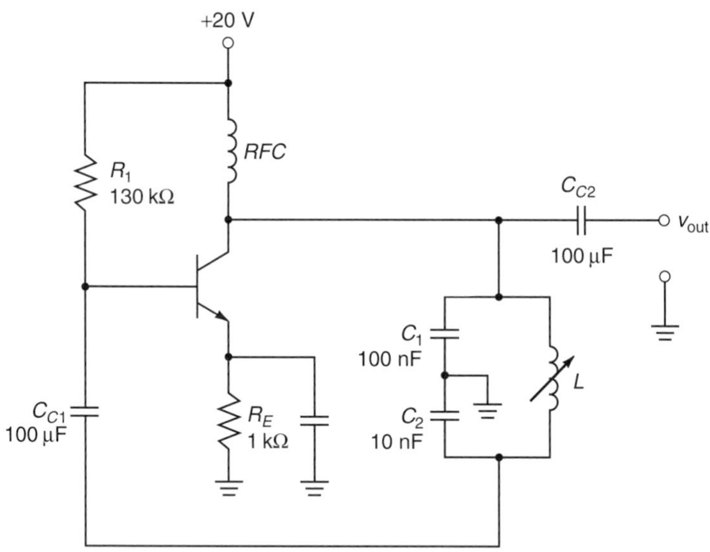

Figure 23.55

35. The feedback inductor illustrated in Figure 23.55 is shown to be a variable inductor. Determine the adjusted value of L that would be required for an operating frequency of 22 kHz.

Answers to the Example Practice Problems

23.1 60

23.2 $f_{C2} = 102.6$ Hz, $A_{CL} = 2$

23.3 $f_{C1} = 570$ Hz, $f_{C2} = 938$ Hz, BW = 368 Hz, $f_0 = 731$ Hz, $Q = 1.99$

23.4 33.6 Hz

23.5 $Q = 1.06$, BW = 31.7 Hz

Multisim File References

Figure	File Reference
23.8	ETF23_8.msm
23.10	ETF23_10.msm
23.18	ETF23_18.msm
23.28	ETF23_28.msm
23.39	ETF23_39.msm

chapter 24

Switching Circuits

Objectives

After studying the material in this chapter, you should be able to:

1. Describe and analyze the operation of the basic BJT, JFET, and MOSFET switches.
2. List the common *time* and *frequency* characteristics of rectangular waveforms, and describe how each is measured.
3. List and describe the factors that affect BJT and FET switching time.
4. Discuss the methods used to improve the switching times of basic BJT and FET circuits.
5. Compare and contrast the time-measurement techniques for *inverters* and *buffers.*
6. Describe and analyze the operation of *inverting* and *noninverting Schmitt triggers.*
7. List and describe the output characteristics of the three types of *multivibrators.*
8. Describe the internal construction and operation of the *555 timer.*
9. Describe and analyze 555 timer *astable* and *monostable multivibrators.*
10. Describe the operation of the 555 timer *voltage-controlled oscillator (VCO).*

Throughout our discussion of solid-state devices and circuits, we have concentrated almost entirely on their *linear* operating characteristics. Another important aspect of solid-state electronics is the way in which devices and circuits respond to *nonlinear* waveforms, such as *square waves. Circuits designed to respond to (or generate) nonlinear waveforms* are referred to as **switching circuits**. In this chapter, we will discuss the most basic switching circuits. As you will see, these circuits operate quite differently from linear circuits.

24.1 Introductory Concepts

To understand some of the more complex switching circuits, you must first be comfortable with the idea of using a BJT, FET, or MOSFET as a switch. In this section, we will discuss some of the switching characteristics of discrete solid-state devices. ◀ *OBJECTIVE 1*

The BJT as a Switch

A BJT can be used as a switch simply by driving the component back and forth between saturation and cutoff. A basic transistor switching circuit is illustrated in Figure 24.1a. As shown, a rectangular input produces a rectangular output. This relationship is easy to understand if you refer to the dc load line shown in Figure 24.1b.

When the input to the transistor is at $-V_{in}$, the emitter-base junction of the transistor is biased off, and the following (ideal) conditions exist:

$$V_{CE} = V_{CC} \qquad \text{and} \qquad I_C = 0 \text{ A}$$

As shown in Figure 24.2a, these conditions can be represented using an open switch.

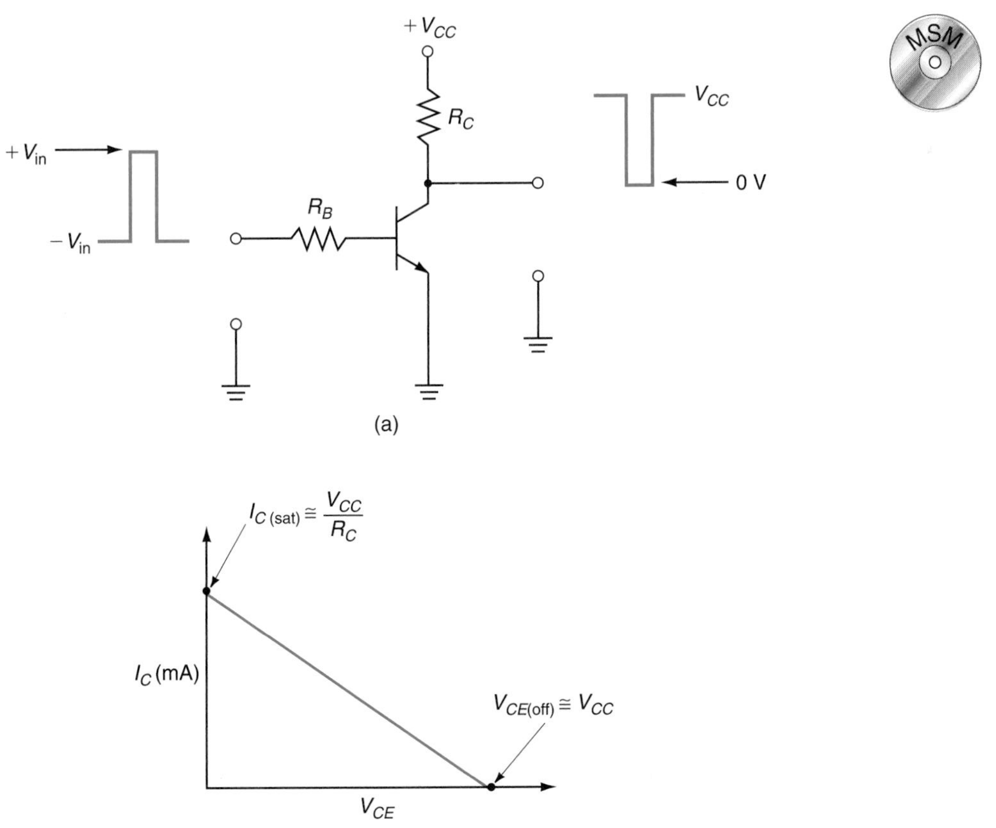

MSM

Figure 24.1 A basic transistor switch.

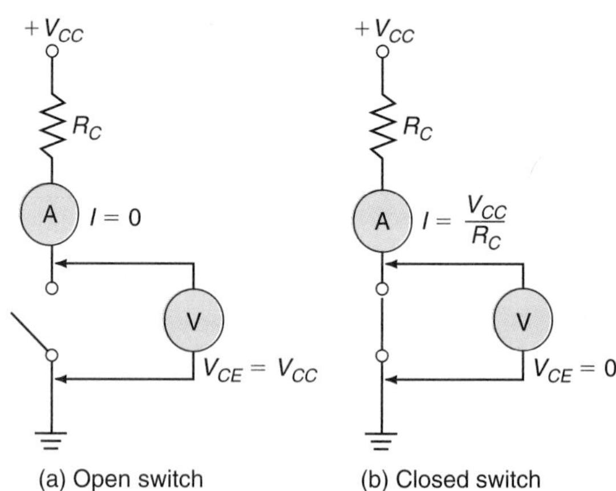

Figure 24.2 The "open" and "closed" transistor switch.

When the input goes to $+V_{\text{in}}$, the transistor is driven into saturation, and the following (ideal) conditions exist:

$$V_{CE} = 0 \text{ V} \qquad \text{and} \qquad I_C = \frac{V_{CC}}{R_C}$$

As shown in Figure 24.2b, these conditions can be represented using a closed switch.

To ensure that the transistor is driven into cutoff, $-V_{\text{in}}$ must be *at least* as negative as the emitter supply voltage (in this case, 0 V). To ensure that $+V_{\text{in}}$ drives the transistor into saturation, the circuit is usually designed so that $+V_{\text{in}} = V_{CC}$. Further, R_B must be low enough so that $I_B > \frac{I_{C(\text{sat})}}{h_{FE}}$. Example 24.1 uses several relationships from Chapter 19 to show how meeting these conditions ensure that a BJT is saturated when $+V_{\text{in}}$ is applied to its base.

Example 24.1

Determine the minimum high input voltage ($+V_{\text{in}}$) required to saturate the transistor switch shown in Figure 24.3.

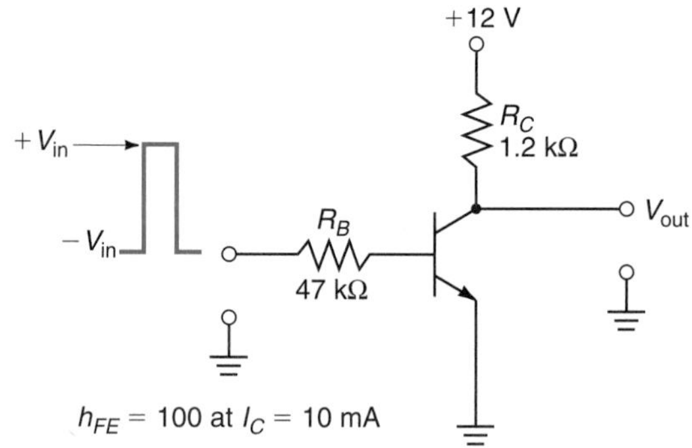

Figure 24.3

Solution: Assuming that the transistor is ideal, $V_{CE(\text{sat})} = 0$ V, and the value of $I_{C(\text{sat})}$ is found as

$$I_{C(\text{sat})} = \frac{V_{CC}}{R_C} = \frac{12 \text{ V}}{1.2 \text{ k}\Omega} = 10 \text{ mA}$$

Since $h_{FE} = 100$ when $I_C = 10$ mA, the base current required to saturate the transistor is found as

$$I_B = \frac{I_{C(\text{sat})}}{h_{FE}} = \frac{10 \text{ mA}}{100} = 100 \ \mu\text{A}$$

A Practical Consideration:
A circuit like the one shown in Figure 24.3 must be designed so that the collector current never meets or exceeds the $I_{C(\max)}$ rating of the BJT. As long as

$$\frac{V_{CC}}{R_C} < I_{C(\max)}$$

the BJT will be able to handle an input of $+V_{\text{in}} = V_{CC}$.

The base current generated in Figure 24.3 is found as

$$I_B = \frac{+V_{\text{in}} - V_{BE}}{R_B}$$

Using $I_B = 100\ \mu\text{A}$ and solving for $+V_{\text{in}}$, we get

$$+V_{\text{in}} = I_B R_B + V_{BE} = (100\ \mu\text{A})(47\ \text{k}\Omega) + 0.7\ \text{V} = 5.4\ \text{V}$$

Since $V_{CC} > 5.4$ V, using a value of $+V_{\text{in}} = V_{CC}$ is more than enough to saturate the transistor.

Practice Problem 24.1

A BJT switch like the one shown in Figure 24.3 has the following values: $V_{CC} = +10$ V, $R_C = 1\ \text{k}\Omega$, $R_B = 51\ \text{k}\Omega$, and $h_{FE} = 70$ at $I_C = 10$ mA. Determine the minimum value of $+V_{\text{in}}$ required to saturate the transistor.

So far, we have assumed that the transistor is an ideal component. That is, we have assumed that $V_{CE} = V_{CC}$ when the transistor is in cutoff, and that $V_{CE} = 0$ V when the transistor is saturated. In practice, the outputs from a practical transistor switch typically fall within the following ranges:

Condition	*Output Voltage* (V_{CE}) *Range*
Cutoff	Within 1 V of V_{CC}
Saturation	Between 0.2 and 0.4 V

The JFET as a Switch

The JFET switch differs from the BJT switch in two ways:

- A typical JFET switch has much higher input impedance.
- A *negative* input pulse, to an *n*-channel JFET, is used to produce a *positive* output pulse.

Both of these characteristics are illustrated in Figure 24.4. The input impedance of this switching circuit equals the value of R_G. Since R_G is usually in the megohm range, the circuit input impedance is very high.

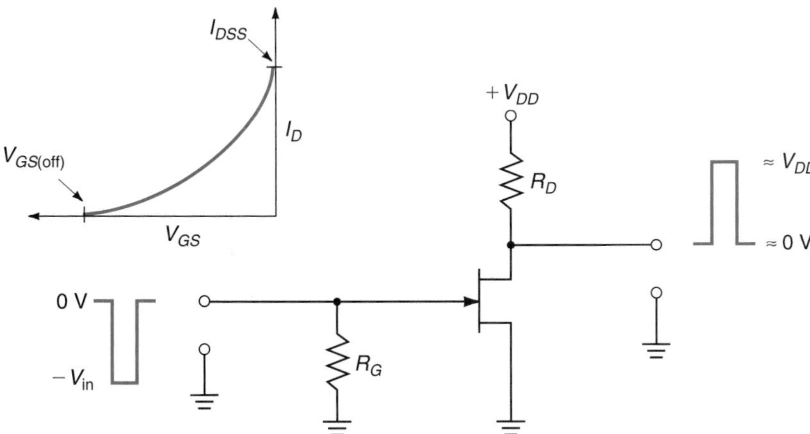

Figure 24.4 A JFET switch.

To understand the input/output voltage relationship of the circuit, we need to refer to the transconductance curve shown in Figure 24.4. When $V_{in} = 0$ V, $V_{GS} = 0$ V, and $I_D = I_{DSS}$. The output from the switching circuit is found as

$$V_{out} = V_{DD} - I_{DSS}R_D$$

Assuming that R_D has been selected properly, V_{out} is approximately 0 V when the input is 0 V. If $-V_{in}$ is more negative than (or equal to) $V_{GS(off)}$, the drain current drops to near 0 A, and

$$V_{out} \cong V_{DD}$$

Example 24.2 demonstrates the analysis of a basic JFET switch.

Example 24.2

Determine the high and low output voltage values for the circuit shown in Figure 24.5.

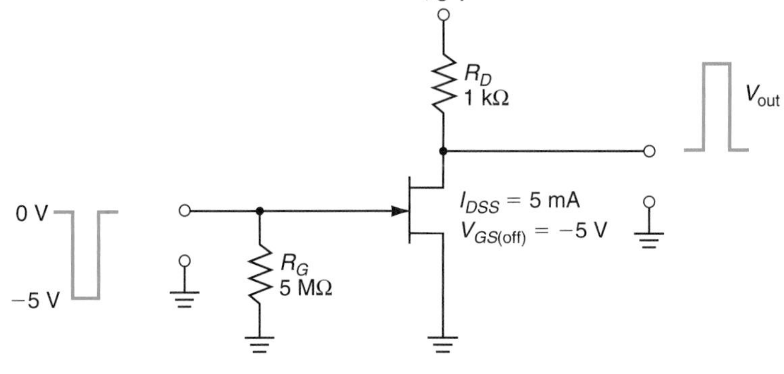

Figure 24.5

Solution: When $V_{in} = 0$ V, $I_D = I_{DSS}$, and

$$V_{out} = V_{DD} - I_{DSS}R_D = 5\text{ V} - (5\text{ mA})(1\text{ k}\Omega) = 5\text{ V} - 5\text{ V} = 0\text{ V}$$

When $V_{in} = V_{GS(off)} = -5$ V, $I_D = 0$ A, and V_{out} is found as

$$V_{out} = V_{DD} - I_DR_D = +5\text{ V} - 0\text{ V} = +5\text{ V}$$

A Practical Consideration: This example assumes that the JFET has only one value of I_{DSS} and $V_{GS(off)}$. In practice, JFETs typically have minimum and maximum values of I_{DSS} and $V_{GS(off)}$. When this is the case, you must use both values to determine the output voltage *ranges*.

Practice Problem 24.2

A JFET switch like the one shown in Figure 24.5 has the following values: $V_{DD} = +10$ V, $R_D = 1.8$ kΩ, $I_{DSS} = 5$ mA, and $V_{GS(off)} = -3$ V. The input pulse varies between 0 V and −4 V. Determine the high and low output voltage values for the circuit.

In Example 24.2, we once again assumed that the active component was ideal. In practice, the output values vary slightly from the ideal values of 0 V and V_{DD}.

The MOSFET as a Switch

A MOSFET switch has the input impedance advantage of a JFET switch and the input/output polarity relationship of a BJT switch. A MOSFET switch is shown in Figure 24.6. When the input to the circuit is at 0 V, $I_D = I_{DSS}$, and

$$V_{out} = V_{DD} - I_{DSS}R_D$$

As the above equation illustrates, a lower value of R_D produces an output that is closer to the value of V_{DD}. This point is illustrated in Example 24.3.

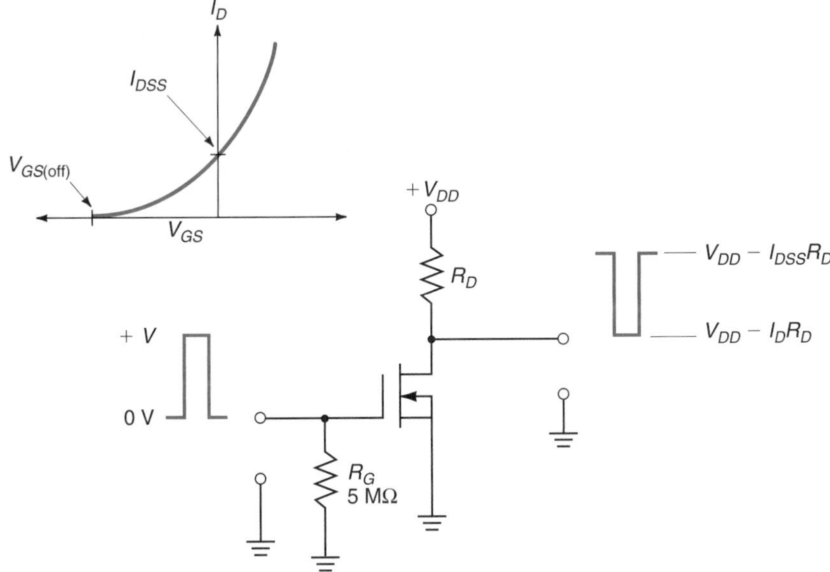

Figure 24.6 A MOSFET switch.

Example 24.3

Determine the output voltage for the circuit shown in Figure 24.7 when the input is at 0 V. Repeat the procedure for the circuit using $R_D = 100\ \Omega$.

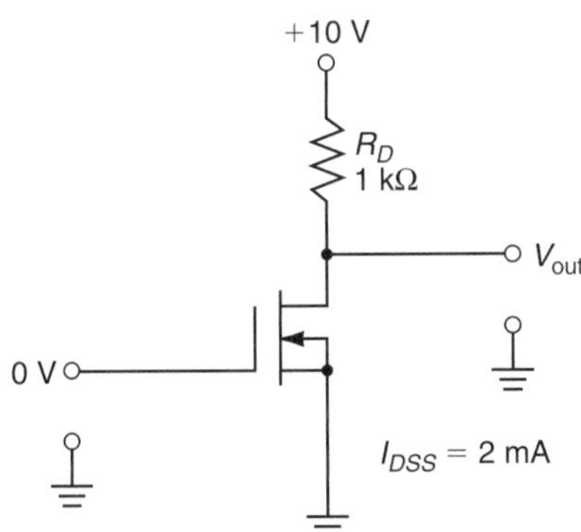

Figure 24.7

Solution: When the input is at 0 V, $I_D = I_{DSS}$, and

$$V_{out} = V_{DD} - I_{DSS}R_D = 10\text{ V} - 2\text{ V} = 8\text{ V}$$

If the drain resistor is replaced with a 100 Ω resistor, the output changes to

$$V_{out} = V_{DD} - I_{DSS}R_D = 10\text{ V} - 200\text{ mV} = 9.8\text{ V}$$

Practice Problem 24.3

A circuit like the one shown in Figure 24.7 has the following values: $V_{DD} = 18$ V, $R_D = 2$ kΩ, and $I_{DSS} = 500$ μA. Determine the value of V_{out} for the circuit when $V_{in} = 0$ V.

As demonstrated in Example 24.3, a lower value of R_D results in an output voltage that is much closer to V_{DD}. However, with a lower-value drain resistor, I_DR_D may not be sufficient to drop the output to 0 V when $I_D > I_{DSS}$. The CMOS switch shown in Figure 24.8 eliminates the trade-off problem that can be caused by the value of R_D. You

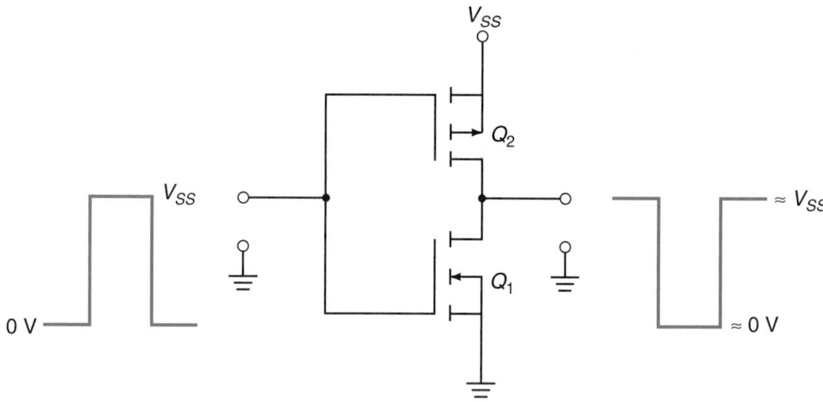

Figure 24.8 A CMOS switch.

may remember from earlier discussions that the two MOSFETs are always in opposite operating states. The following table is based on the relationship between the operating states of Q_1 and Q_2:

Q_1 State	*Q_2 State*	*Q_1 Resistance*	*Q_2 Resistance*
On	Off	Low	High
Off	On	High	Low

When Q_1 is on and Q_2 is off, the device resistances combine to produce an output that is very close to 0 V. When Q_2 is on and Q_1 is off, the device resistances combine to produce an output that is very close to $+V_{SS}$. As a result, the tradeoff caused by R_D in the basic MOSFET inverter is eliminated.

A Switching Application: The LED Driver

The LED driver shown in Figure 24.9 is used to couple a low-current output to a relatively high-current device. In this case, the output from circuit A is insufficient to drive the LED. The **driver** circuit (R_B and Q_1) solves the problem.

How does the circuit work? When the output from circuit A is 0 V, Q_1 is in cutoff (acting as an open switch), and the LED doesn't light. When the output from circuit A goes to +5 V, Q_1 saturates (acts as a closed switch), and I_C rises to a value that is sufficient to light the LED. Even though this driver circuit uses a BJT, drivers are commonly made with any of the active devices covered in this section.

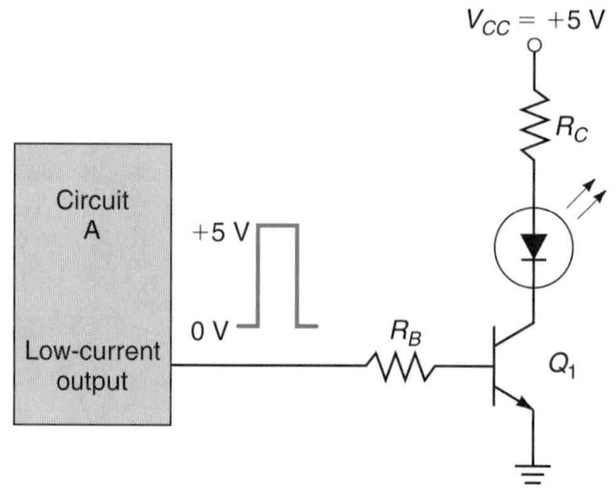

Figure 24.9 A typical application for a BJT switch.

Section Review

1. What is a *switching* circuit?
2. What is the primary difference between linear circuits and switching circuits?
3. Describe the operation of the BJT switch.
4. Describe the operation of the JFET switch.
5. Describe the operation of the MOSFET switch.
6. What is a *driver*?
7. When does the transistor shown in Figure 24.1 dissipate the greatest amount of power? Explain your answer. Critical Thinking
8. Using only the components shown, how could the circuit shown in Figure 24.9 be altered to light the LED when the output from circuit A is 0 V?

24.2 Basic Switching Circuits: Practical Considerations

In Chapter 10, you were introduced to rectangular waveforms and some related measurements. We will begin this section with a review of the rectangular waveform characteristics illustrated in Figure 24.10.

◄ *OBJECTIVE 2*

The time that a rectangular waveform spends in a high state is called the *pulse width* (PW). The time that it spends in a low state is called the *space width* (SW). The sum of the pulse width and the space width equals the *cycle time* (T_C) of the waveform. A *square wave* is a special-case rectangular wave that has equal values of pulse width and space width.

To this point, rectangular waveforms have been drawn using *ideal* (vertical) transitions from each dc level to the other. In reality, it takes some finite amount of time for any waveform to make the transitions between levels. These transitions are shown in the *practical* waveform in Figure 24.10. Note that the transitions are *not* vertical in this waveform. Since the transition times are not instantaneous, a standard must be set as to where time measurements are made on these waveforms. To provide a standard, the values of PW, SW, and T_C are always measured at the halfway point between the high and low values (as shown in the figure).

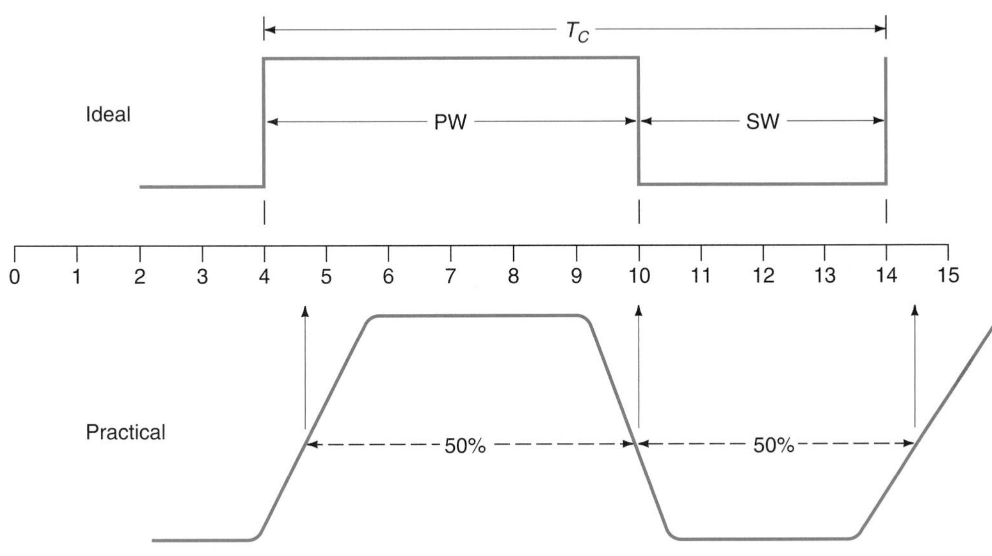

Figure 24.10 Rectangular waveform characteristics.

Another time-related measurement is referred to as *duty cycle.* The **duty cycle** of a waveform is *the ratio of pulse width to total cycle time, measured as a percentage.* By formula,

$$\text{duty cycle } (\%) = \frac{\text{PW}}{T_C} \times 100 \tag{24.1}$$

BJT Switching Time

OBJECTIVE 3 ➤

The output from an ideal BJT switch would change at the exact instant that the circuit input changes. In practice, it more closely resembles the collector waveforms shown in Figure 24.11. The V_B (or I_B) waveform represents an ideal input that changes states instantly at T_0 and T_3. The I_C and V_C waveforms differ from their ideal characteristics in two respects:

- There is a delay between each input transition and the start of the output transitions.
- The output transitions are not vertical, implying that they require some amount of time to occur.

The overall delay between the input and output transitions, as measured at the 50% points on the two waveforms, is referred to as **propagation delay.** There are four components of BJT propagation delay:

- **Delay time (t_d)** is *the time required for the BJT to come out of cutoff,* as measured between T_0 and T_1 in Figure 24.11. Delay time is the time required for I_C to increase to 10% of its maximum value, measured from the time that the input waveform makes its transition.
- **Rise time (t_r)** is *the time required for the transition from cutoff to saturation,* as measured between T_1 and T_2 in Figure 24.11. As shown in Figure 24.11, rise time is the time required for I_C to rise from 10% to 90% of its maximum value.
- **Storage time (t_s)** is *the time required for the BJT to come out of saturation,* as measured between T_3 and T_4 in Figure 24.11. Storage time is the time required for I_C to drop to 90% of its maximum value, measured from the time that the input waveform makes its transition. (Note that storage time is shown in the figure to begin at the point when the input signal changes.)
- **Fall time (t_f)** is *the time required to make the transition from saturation to cutoff,* as measured between T_4 and T_5 in Figure 24.11. As shown in Figure 24.11, fall time is the time required for I_C to drop from 90% to 10% of its maximum values.

Even though these times are defined it terms of I_C, it is important that you know the accompanying changes in V_C shown in Figure 24.11. Oscilloscopes make *voltage*-versus-time measurements, so you need know how each time relates to V_C.

The values of t_d, t_r, t_s, and t_f are all provided on the specification sheet of any given BJT. For example, the 2N3904 spec sheet lists the following maximum values under *switching characteristics:*

$$t_d = 35 \text{ ns} \qquad t_r = 35 \text{ ns} \qquad t_s = 200 \text{ ns} \qquad t_f = 50 \text{ ns}$$

If we add the values of delay time and rise time, we get the time required for the 2N3904 to make the transition from a high- to a low-output voltage: 70 ns. If we add the values of storage time and fall time, we get the time required for the 2N3904 to make the transition from a low- to a high-output voltage: 250 ns. If we add all four, we get a theoretical full-cycle switching time of 320 ns. It would seem that this total switching time could be used to calculate a *maximum* switching frequency, as follows:

A Practical Consideration: In switching circuit applications, frequency is often referred to as *pulse repetition rate* (PRR) or *pulse repetition frequency* (PRF).

$$f_{\max} = \frac{1}{t_d + t_r + t_s + t_f} = \frac{1}{320 \text{ ns}} = 3.125 \text{ MHz}$$

In practice, 3.125 MHz is far beyond the capability of a 2N3904-based switching circuit. The upper cutoff frequency for a BJT passing a square wave can be found using

$$f_{C2} = \frac{0.35}{t_r} \tag{24.2}$$

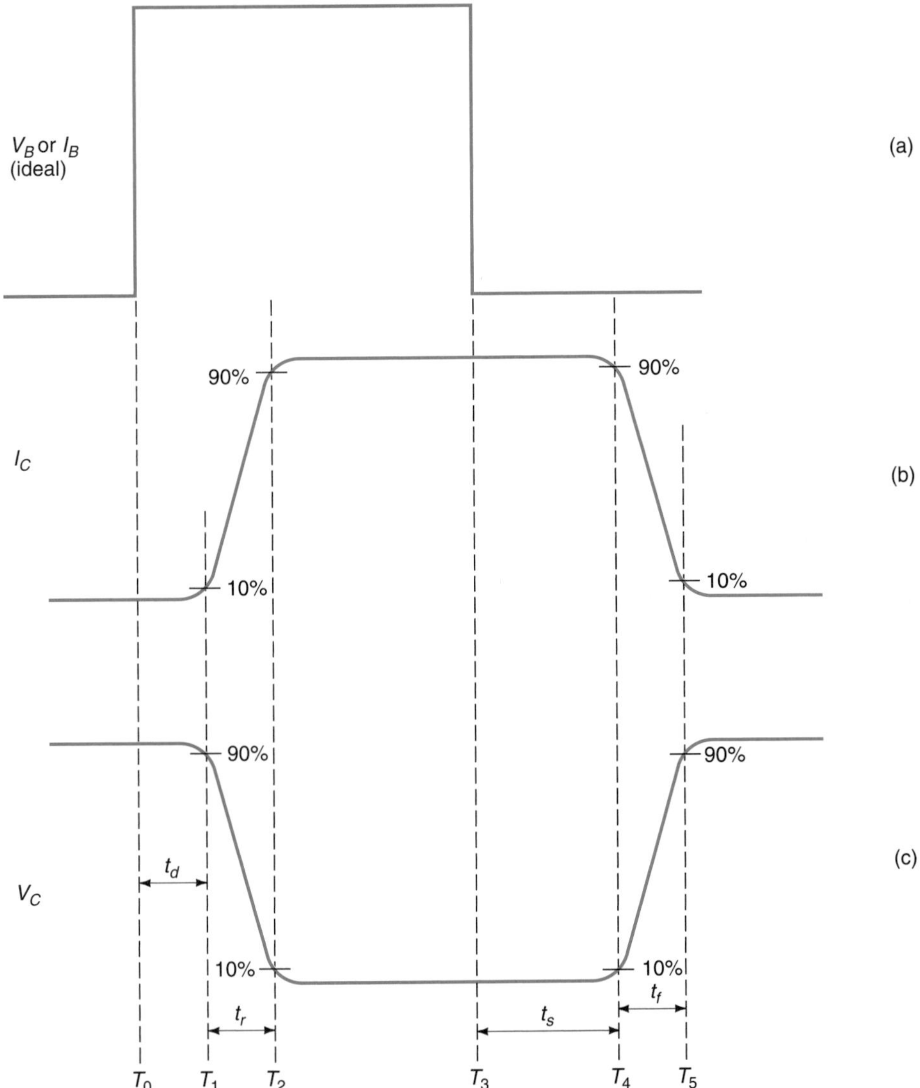

Figure 24.11 Sources of BJT propagation delay.

where

t_r = the rise time of the active device

To pass a square wave with minimal distortion, the upper cutoff frequency of a transistor needs to be at least 100 times the frequency of the input square wave. The following equation is used to calculate the *practical* upper frequency limit for a discrete switching circuit:

$$f_{max} = \frac{0.35}{100t_r} \tag{24.3}$$

Note: The derivation of equation (24.3) can be found in Appendix D.

Using this equation, the practical maximum switching frequency for a 2N3904 switching circuit is found as

$$f_{max} = \frac{0.35}{(100)(35 \text{ ns})} = 100 \text{ kHz}$$

As shown in Figure 24.12, if the input frequency increases beyond this practical limit:

- The leading and trailing edges of the circuit output become more and more rounded.

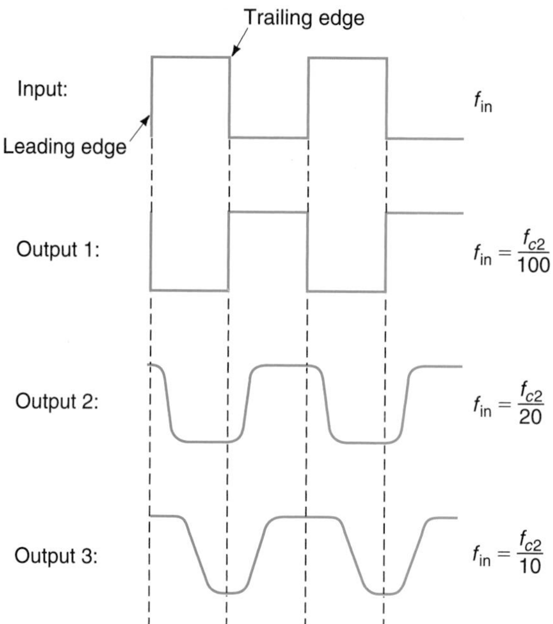

Figure 24.12

- The delays between the input and output transitions increase.
- The pulse width and space width become asymmetrical (PW ≠ SW), because $(t_d + t_r) < (t_s + t_f)$.

Improving BJT Switching Time

OBJECTIVE 4 ➤ Before you can understand the methods used to improve overall BJT switching time, we need to look at the sources of its various components. Both rise time and fall time are purely a function of the physical characteristics of the BJT. As such, there is nothing we can do to reduce these times short of replacing the BJT with one that has better t_r and t_f ratings. Although delay time and storage time also relate to the physical structure of the BJT, there are some things we can do to decrease both of these values.

Delay time occurs when the transistor is coming out of cutoff. We can decrease the time it takes for the transistor to come out of cutoff by:

- Applying a high initial value of I_B.
- Limiting the reverse bias to the minimum value required to hold the BJT in cutoff.

Storage time occurs when the transistor is coming out of saturation. When a transistor is in saturation, the base region is flooded with electrons. Storage time is the time required for these charge carriers to leave the base region, allowing a new depletion layer to begin developing. We can decrease the storage time of a BJT by:

- Limiting I_B so that I_C is slightly less than $I_{C(sat)}$.
- Applying a high initial reverse bias.

Combining these circuit requirements:

- I_B should be very high initially (to reduce delay time) and then settle down to some level just below that required for saturation (to reduce storage time).
- The reverse bias should be very high initially (to reduce storage time) and then decrease to the minimum value required to keep the transistor in cutoff (to reduce delay time).

A Practical Consideration: We can reduce t_d and t_s as described in this section. However, since the practical frequency limit is determined by rise time, improving these times does not increase the operating frequency limit of the switch.

The desired I_B and reverse-bias characteristics are both achieved by using a **speed-up capacitor.** A BJT switch with an added speed-up capacitor is shown in Figure 24.13. The means by which it accomplishes these functions can be explained using the waveforms shown in Figure 24.14.

The speed-up capacitor shown in Figure 24.13 effectively short circuits the resistor (R_B) during the time that the input is making its transitions. As indicated in Figure 24.14b, a

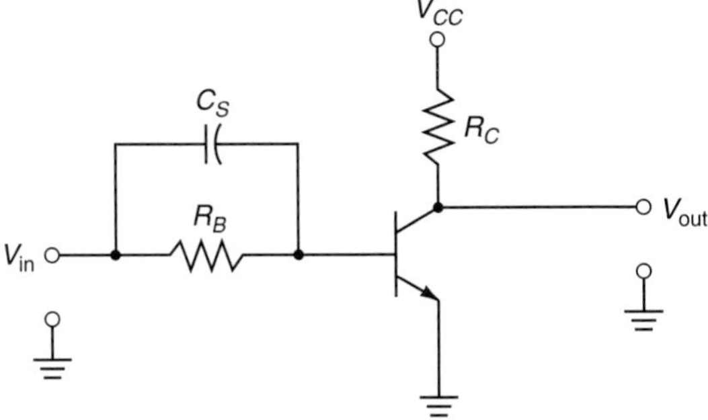

Figure 24.13 A speed-up capacitor (C_S).

positive spike is coupled to the base of the transistor when V_{in} makes its transition from 0 V to +5 V. This spike generates a high initial value of I_B. Then, as the capacitor charges, the spike returns to the 0 V level, and I_B decreases to some value slightly less than that required to fully saturate the transistor. When the input signal returns to 0 V, the charge on C_S drives the output of the RC circuit to −5 V. This is the high initial reverse bias that

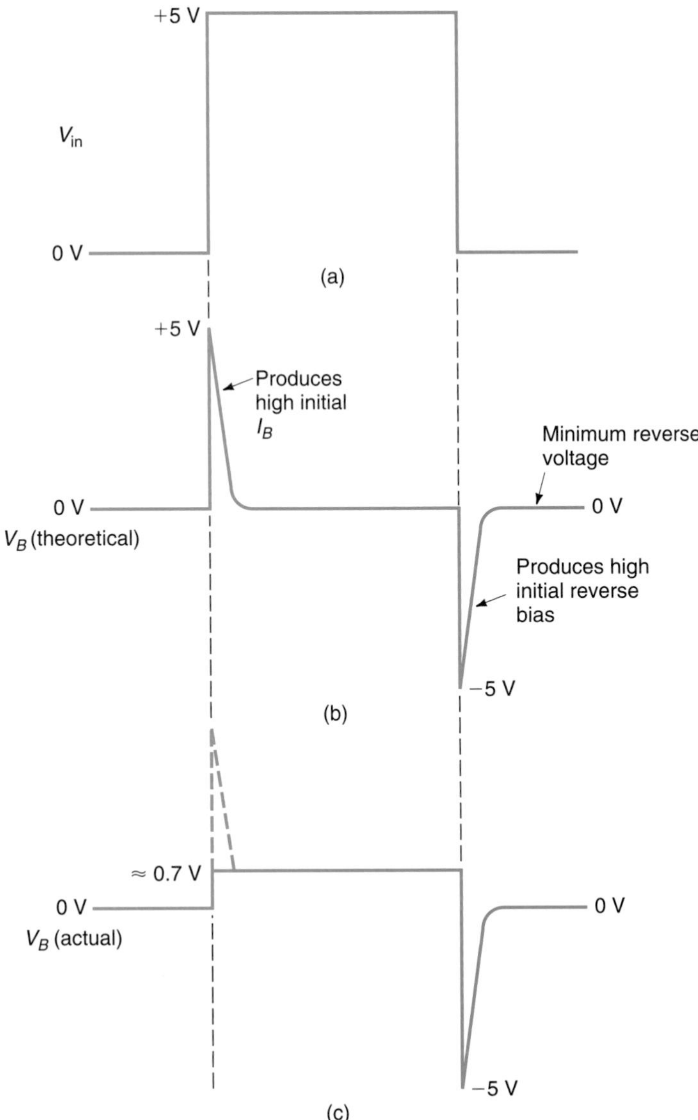

Figure 24.14 Waveforms that result from using a speed-up capacitor.

is needed to reduce storage time. Then, as the capacitor discharges, the reverse bias decreases to the minimum value required to keep the transistor in cutoff.

For the speed-up capacitor to be effective, its value must be chosen so that the *RC* time constant is very short compared to the minimum pulse width of the input. The value of C_S that fulfills this requirement is found using

$$C_S < \frac{1}{20R_B f_{max}} \tag{24.4}$$

Note: The value of f_{max} used in equation (24.4) is the practical value of f_{max} for the circuit, found as

$$f_{max} = \frac{0.35}{100t_r}$$

In Figure 24.14b, we simplified things by ignoring the effects of the emitter-base junction of the transistor. The actual waveform produced at the base of the transistor is shown in Figure 24.14c. As you can see, the positive input spike is clipped at the value of V_{BE} (approximately +0.7 V).

JFET Switching Time

JFET spec sheets usually list values of t_d, t_r, t_s, and t_f. However, some JFET spec sheets list only *turn-on time* and *turn-off time*. **Turn-on time (t_{on})** is *the sum of delay time and rise time.* **Turn-off time (t_{off})** is *the sum of storage time and fall time.* When only t_{on} is given, use this value in place of t_r in equation (24.3).

Using a speed-up capacitor reduces the turn-off time by initially supplying a gate-source voltage that is greater than $V_{GS(off)}$. However, the capacitor does little to improve the turn-on time of the component, because a JFET is a voltage-controlled device (i.e., there is no gate current).

Switching Devices

Groups of BJTs and JFETs, called **switching transistors,** are designed to have extremely low values of t_d, t_r, t_s, and t_f. For example, the 2N2369 BJT has the following maximum values listed on its data sheet:

$$t_d = 5 \text{ ns} \qquad t_r = 18 \text{ ns} \qquad t_s = 13 \text{ ns} \qquad t_f = 15 \text{ ns}$$

With the given rise time of 18 ns, the practical limit on f_{in} for the device would be found as

$$f_{max} = \frac{0.35}{100t_r} = \frac{0.35}{(100)(18 \text{ ns})} \cong 194 \text{ kHz}$$

This is nearly twice the acceptable f_{max} limit of the 2N3904.

Buffer Circuits

OBJECTIVE 5 ➤

So far, all of the switching circuits that we have looked at have been *inverters.* An **inverter** is *a switching circuit that produces a high output when its input goes low, and vice versa.* A **buffer** is *a switching circuit that produces an output that is in phase with its input.* Any emitter or source follower can be used as a buffer. Figure 24.15 compares the switching times of both inverters and buffers.

Section Review

1. What is a rectangular waveform?
2. Define each of the following terms: *pulse width, space width,* and *cycle time.*
3. What is a square wave?
4. Where are each of the values in Question 2 measured on a *practical* rectangular waveform?
5. Why do we need a standard for rectangular waveform time measurements?

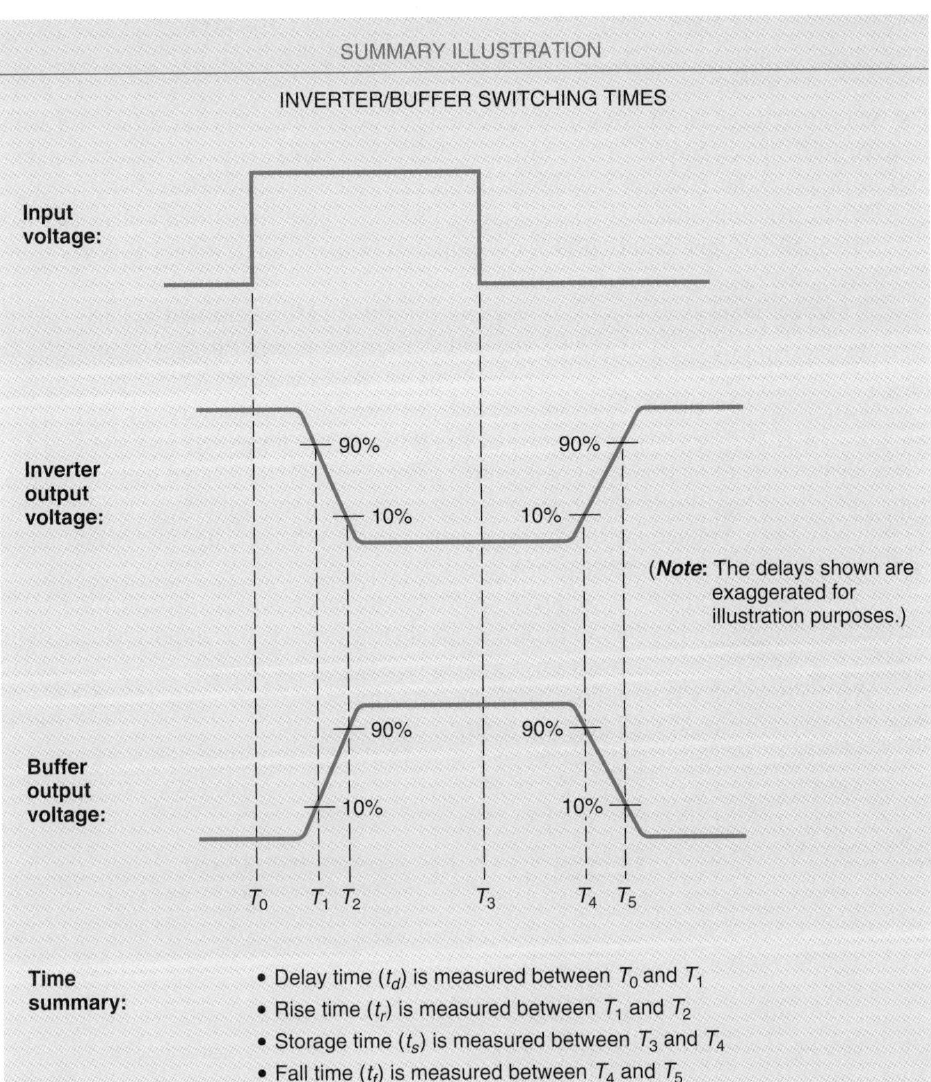
SUMMARY ILLUSTRATION
INVERTER/BUFFER SWITCHING TIMES
Input voltage:
Inverter output voltage:
90%
90%
10%
10%
(Note: The delays shown are exaggerated for illustration purposes.)
Buffer output voltage:
90%
90%
10%
10%
T_0 T_1 T_2 T_3 T_4 T_5
Time summary:
• Delay time (t_d) is measured between T_0 and T_1
• Rise time (t_r) is measured between T_1 and T_2
• Storage time (t_s) is measured between T_3 and T_4
• Fall time (t_f) is measured between T_4 and T_5

Figure 24.15

6. What is a *duty cycle*?
7. What is *propagation delay*?
8. What is *delay time*? Describe delay time in terms of V_C.
9. What is *rise time*? Describe rise time in terms of V_C.
10. What is *storage time*? Describe storage time in terms of V_C.
11. What is *fall time*? Describe fall time in terms of V_C.
12. Which BJT switching time determines the *practical limit* on f_{in} for the device?
13. How does the use of a *speed-up capacitor* reduce t_d and t_s?
14. What effect does a speed-up capacitor have on the *practical* frequency limit of a BJT? Explain your answer.
15. What is the primary difference between the *inverter* and the *buffer*?
16. List the points where t_d, t_r, t_s, and t_f are measured on the output voltage of an inverter.
17. List the points where t_d, t_r, t_s, and t_f are measured on the output voltage of a buffer.

24.3 Schmitt Triggers

OBJECTIVE 6 ➤ The **Schmitt trigger** is *a voltage-level detector that is similar to a comparator* but with some key differences. Figure 24.16 shows a typical input/output combination for a Schmitt trigger. Here's an overview of the circuit's response to the input waveform:

- When the input makes a *positive-going transition* past a specified voltage, called the **upper trigger point (UTP),** the output from the Schmitt trigger changes from $-V_{out}$ to $+V_{out}$.
- When the input makes a *negative-going transition* past a specified voltage, called the **lower trigger point (LTP),** the output from the Schmitt trigger changes from $+V_{out}$ to $-V_{out}$.

The trigger points may or may not be equal, but the UTP must always be more *positive* than the LTP. The concept of unequal trigger points is illustrated in Figure 24.17. When the positive-going transition of the input passes the UTP (at v_1), the output changes from $-V_{out}$ to $+V_{out}$. When the input begins its negative-going transition, it passes the UTP again (at v_2), without any change in output voltage. When the negative-going input passes the LTP, the output returns to $-V_{out}$.

The concept that the circuit output is unaffected by input voltages between the UTP and the LTP is illustrated further in Figure 24.18. As you can see, the output changes only when:

- The UTP is reached by a positive-going transition.
- The LTP is reached by a negative-going transition.

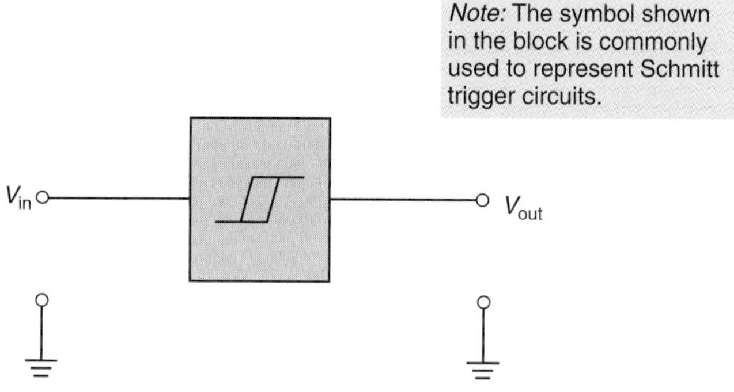

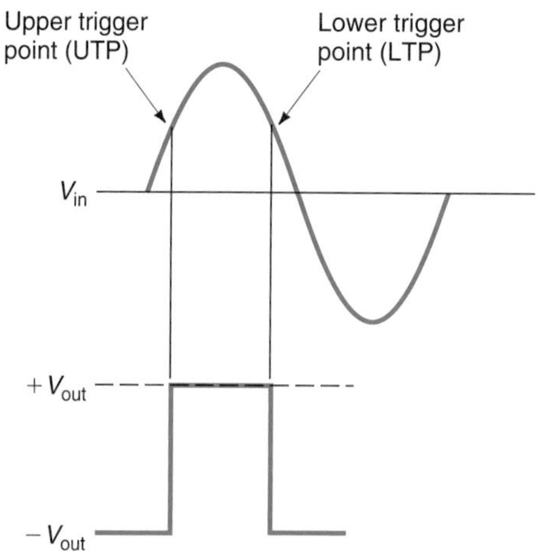

Figure 24.16 Schmitt trigger input and output signals.

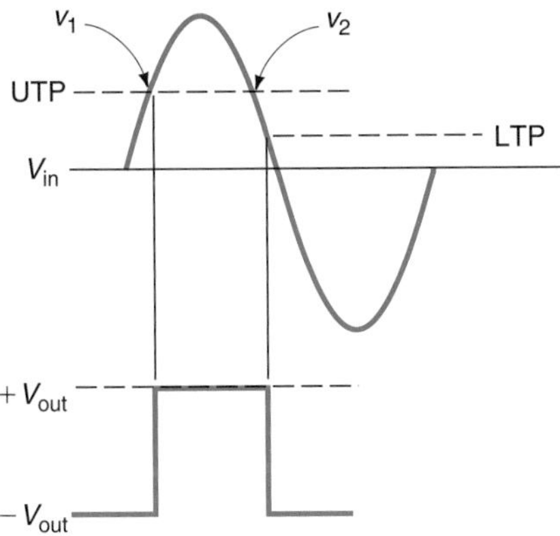

Figure 24.17 Unequal UTP and LTP values.

The range of voltages between the UTP and the LTP is referred to as **hysteresis.** If Figure 24.18 has values of UTP = +10 V and LTP = +2 V, the hysteresis of the circuit is the voltage range of $+2\text{ V} < V_{in} < +10\text{ V}$. No voltage in this range can cause the output of the circuit to change states. The fact that the Schmitt trigger can have different UTP and LTP values is one of the primary differences between this circuit and a common comparator.

Noninverting Schmitt Triggers

A noninverting Schmitt trigger uses a simple feedback resistor, which is connected as shown in Figure 24.19a. Note that $R_{in} = R_f = 10\text{ k}\Omega$. As you know, the high input impedance of the op-amp causes the current through the two resistors to be equal. Therefore, the voltages across the components are also equal. This is a key point in understanding the operation of this circuit.

We will start our analysis by assuming that $V_{out} = -4\text{ V}$ (its negative limit). When V_{in} reaches +4 V (at T_0), there are 4 V across each resistor (R_{in} and R_f). This means that the noninverting input is at 0 V. As the input goes more positive than +4 V, the voltage at the noninverting input becomes more positive than 0 V (the potential at the inverting input), and the output goes to +4 V.

The same principle applies to the negative alternation of the input cycle; the only difference is that the polarities are reversed. With the output of the circuit at +4 V, the input must pass −4 V before the noninverting input goes to a value that is more negative

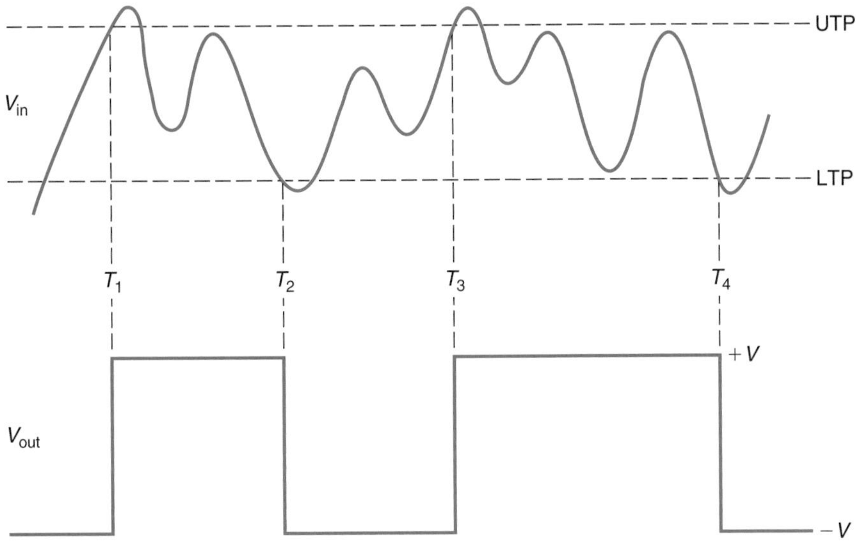

Figure 24.18 Schmitt trigger hysteresis.

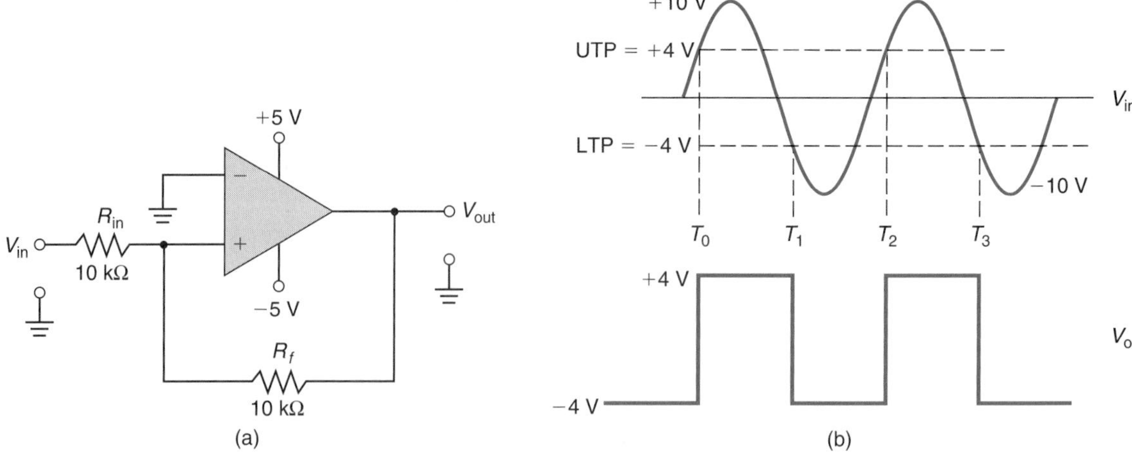

Figure 24.19 Noninverting Schmitt trigger operation.

than 0 V (the potential at the inverting input). When this happens, the output returns to −4 V, and the cycle repeats itself.

The Schmitt trigger shown in Figure 24.19a has UTP and LTP values of ±4 V. By changing the ratio of R_f to R_{in}, we can set the circuit for other trigger point values. Generally, the UTP and LTP values can be determined using the following relationships:

Remember: The value of V_{out} depends on the load resistance and the supply voltages. As long as the load resistance is greater than 2 kΩ, we can assume that $+V_{out} = +V - 1$ V and $-V_{out} = -V + 1$ V.

$$\text{UTP} = -\frac{R_{in}}{R_f}(-V_{out}) \tag{24.5}$$

and

$$\text{LTP} = -\frac{R_{in}}{R_f}(+V_{out}) \tag{24.6}$$

The following example demonstrates the use of these two equations.

Example 24.4

Determine the UTP and LTP values for the circuit shown in Figure 24.20.

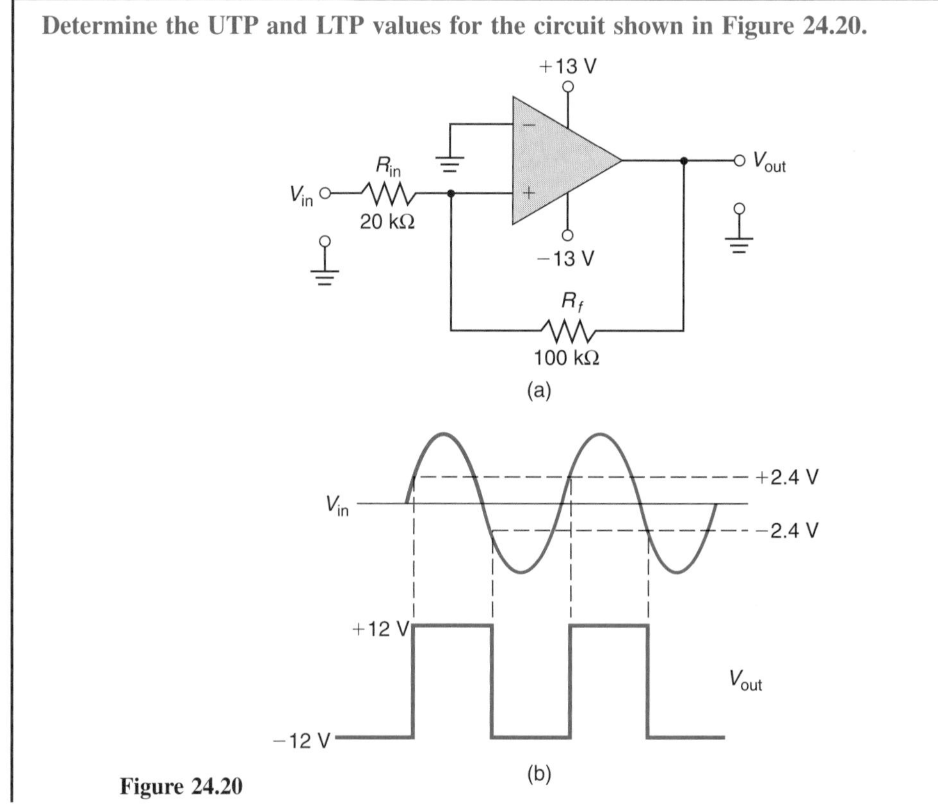

Figure 24.20

Solution: Using equation (24.5),

$$\text{UTP} = -\frac{R_{\text{in}}}{R_f}(-V_{\text{out}}) = (-0.2)(-12\text{ V}) = 2.4\text{ V}$$

Using equation (24.6),

$$\text{LTP} = -\frac{R_{\text{in}}}{R_f}(+V_{\text{out}}) = (-0.2)(+12\text{ V}) = -2.4\text{ V}$$

Practice Problem 24.4

A noninverting Schmitt trigger like the one shown in Figure 24.20 has values of $R_f = 33\text{ k}\Omega$ and $R_{\text{in}} = 11\text{ k}\Omega$. Determine the UTP and LTP values for the circuit.

The noninverting Schmitt trigger shown in Figure 24.20 is limited, because the *magnitudes* of the UTP and the LTP must be equal. However, the noninverting Schmitt trigger can be modified as shown in Figure 24.21 so that the magnitudes of the UTP and the LTP are not equal.

When the output from the Schmitt trigger in Figure 24.21 is *negative,* D_1 conducts and D_2 is off. As a result, R_{f2} is effectively removed from the circuit, and

$$\text{UTP} = -\frac{R_{\text{in}}}{R_{f1}}(-V_{\text{out}} + 0.7\text{ V}) \tag{24.7}$$

Remember: The value of V_{out} depends on the load resistance and the supply voltages. As long as the load resistance is greater than 2 kΩ, we can assume that $+V_{\text{out}} = +V - 1\text{ V}$ and $-V_{\text{out}} = -V + 1\text{ V}$.

When the output is *positive,* D_1 is off and D_2 conducts. As a result, R_{f1} is effectively removed from the circuit, and

$$\text{LTP} = -\frac{R_{\text{in}}}{R_{f2}}(+V_{\text{out}} - 0.7\text{ V}) \tag{24.8}$$

Since the UTP and LTP are determined by separate feedback resistors, these values are completely independent of each other.

Inverting Schmitt Triggers

An inverting Schmitt trigger is shown in Figure 24.22. In this circuit, the input (V_{in}) is applied to the inverting terminal. A feedback signal is applied to the noninverting terminal from the junction of R_1 and R_2.

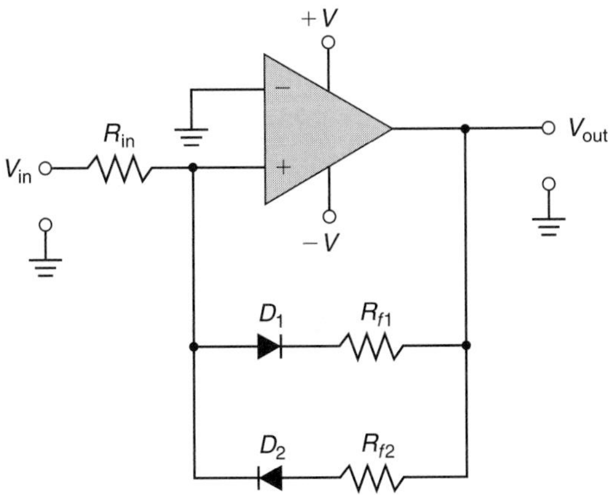

Figure 24.21 A noninverting Schmitt trigger with unequal UTP and LTP values.

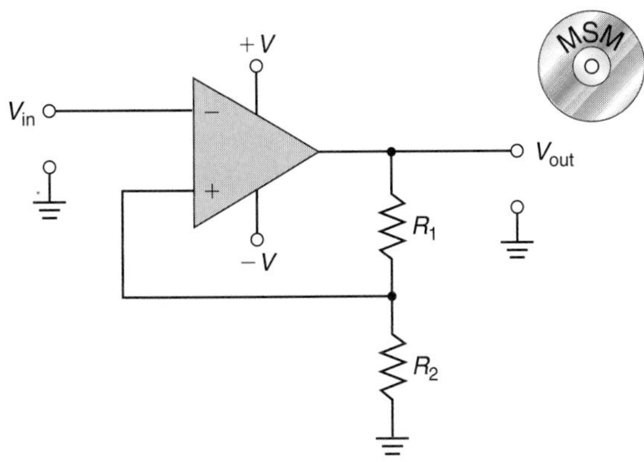

Figure 24.22 An inverting Schmitt trigger.

To understand the operation of this circuit, we must first establish the equations for the UTP and the LTP. The UTP for the circuit shown in Figure 24.22 is found as

Remember: The value of V_{out} depends on the load resistance and the supply voltages. As long as the load resistance is greater than 2 kΩ, we can assume that $+V_{out} = +V - 1$ V and $-V_{out} = -V + 1$ V.

$$\text{UTP} = \frac{R_2}{R_1 + R_2}(+V_{out}) \tag{24.9}$$

and the LTP is found as

$$\text{LTP} = \frac{R_2}{R_1 + R_2}(-V_{out}) \tag{24.10}$$

Example 24.5 demonstrates the use of these equations.

Example 24.5

Determine the UTP and LTP values for the circuit shown in Figure 24.23.

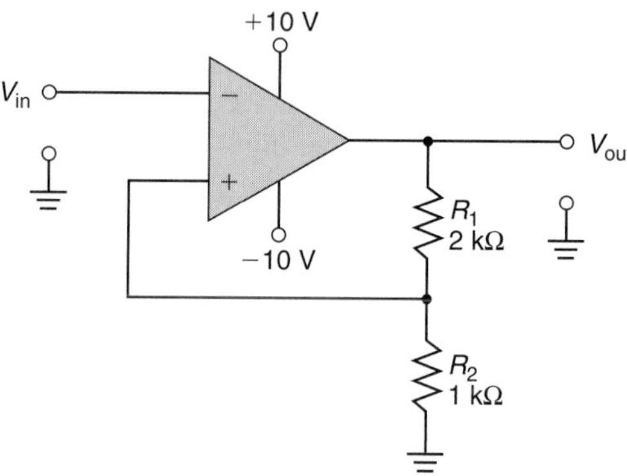

Figure 24.23

Solution: The UTP is found as

$$\text{UTP} = \frac{R_2}{R_1 + R_2}(+V_{out}) = \frac{1\text{ k}\Omega}{3\text{ k}\Omega}(+9\text{ V}) = +3\text{ V}$$

and the LTP is found as

$$\text{LTP} = \frac{R_2}{R_1 + R_2}(-V_{out}) = \frac{1\text{ k}\Omega}{3\text{ k}\Omega}(-9\text{ V}) = -3\text{ V}$$

Practice Problem 24.5

An inverting Schmitt trigger has ±10 V supply voltages and values of $R_1 = 3$ kΩ and $R_2 = 1$ kΩ. Determine the UTP and LTP values for the circuit. (Assume that $R_L > 10$ kΩ.)

We will use the values obtained in Example 24.5 to explain the operation of the inverting Schmitt trigger. Figure 24.24 shows the input and output waveforms for the circuit illustrated in Figure 24.23. As you can see, the input and output waveforms are out of phase by 180°.

When the output of the circuit is at +9 V, the voltage across R_2 is +3 V. As the input passes +3 V (at T_0), the output switches to −9 V. Now the voltage across R_2 is −3 V, so the output does not switch again until the input passes −3 V (at T_1). Note that the basic inverting Schmitt trigger (see Figure 24.22) always has UTP and LTP values that are equal in magnitude.

Remember: The value of V_{out} depends on the load resistance and the supply voltages. As long as the load resistance is greater than 2 kΩ, we can assume that $+V_{out} = +V - 1$ V and $-V_{out} = -V + 1$ V.

The inverting Schmitt trigger can be modified as shown in Figure 24.25 to provide UTP and LTP values that are *not* equal in magnitude. When the op-amp output is positive, D_1 is forward biased, and the UTP voltage is found as

$$\text{UTP} = \frac{R_{in}}{R_{f1} + R_{in}}(+V_{out} - 0.7\text{ V}) \tag{24.11}$$

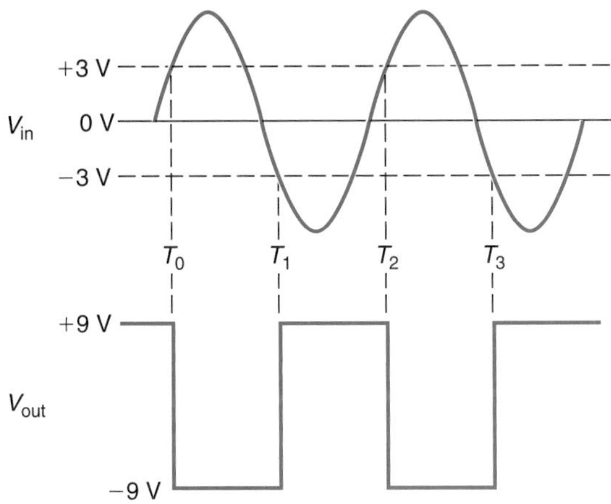

Figure 24.24 Input and output waveforms for an inverting Schmitt trigger.

When the op-amp output is negative, D_2 is forward biased, and the LTP voltage is found as

$$\text{LTP} = \frac{R_{in}}{R_{f2} + R_{in}}(-V_{out} + 0.7\text{ V}) \tag{24.12}$$

By using different feedback resistor values, the two trigger points can be set for *asymmetrical* (unequal) magnitudes.

Section Review

1. What is a *Schmitt trigger*?
2. What is the *upper trigger point* (*UTP*)?
3. What is the *lower trigger point* (*LTP*)?
4. What is *hysteresis*?
5. Explain the operation of the *noninverting* Schmitt trigger shown in Figure 24.19.
6. Explain the operation of the *noninverting* Schmitt trigger shown in Figure 24.21.
7. Explain the operation of the *inverting* Schmitt trigger shown in Figure 24.22.
8. Explain the operation of the *inverting* Schmitt trigger shown in Figure 24.25.
9. Compare and contrast the *Schmitt trigger* and the *comparator.*

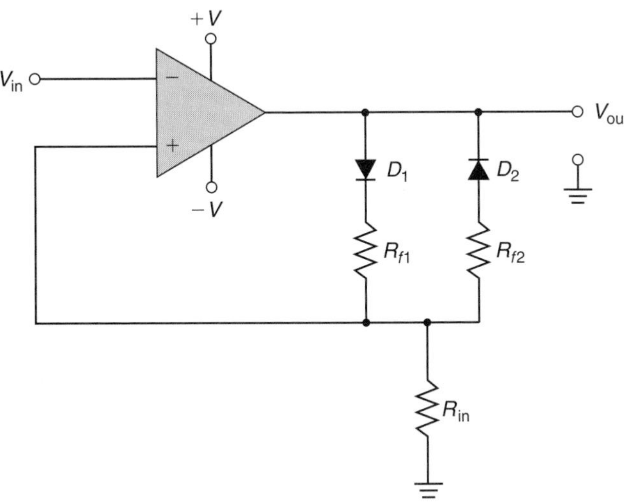

Figure 24.25 An inverting Schmitt trigger with asymmetrical UPT and LTP values.

OBJECTIVE 7 ➤ A **multivibrator** is *a switching circuit designed to have zero, one, or two stable output states.* The concept of stable output states is illustrated in Figure 24.26.

The **astable multivibrator** is *a circuit that switches back and forth constantly between two dc levels as long as it has a dc supply voltage input.* This circuit acts as a rectangular-wave oscillator, and is often referred to as a **free-running multivibrator.** The output from an astable multivibrator is shown in Figure 24.26a.

The **monostable multivibrator** is *a circuit that has one stable output state.* For the circuit represented in Figure 24.26b, $-V_{out}$ is the stable output state. When the circuit receives an input pulse, or *trigger,* it switches to the $+V_{out}$ state, then automatically reverts back to its stable output state ($-V_{out}$). The duration of the $+V_{out}$ state is determined by component values in the circuit. Since the monostable multivibrator produces a single output pulse for each input trigger, it is generally referred to as a **one-shot.**

The **bistable multivibrator** is *a circuit that has two output states,* as illustrated in Figure 24.26c. When the proper input trigger is received, it switches from one output state to the other. The circuit stays in that output state until another trigger is received, when it returns to its original state. The bistable multivibrator is generally referred to as a **flip-flop.** Flip-flops are covered extensively in any digital electronics course, so they are not covered in depth here.

As you have probably figured out by now, the term *astable* means not stable, the term *monostable* means *one stable state,* and the term *bistable* means *two stable states.* Though once constructed using discrete components, today multivibrators are almost always constructed using ICs. One of these ICs is the 555 timer, an 8-pin IC that can be used for a variety of switching applications.

The 555 Timer

OBJECTIVE 8 ➤ The **555 timer** is *a switching circuit contained in an 8-pin IC.* Its internal circuitry is shown in Figure 24.27. Inside the chip are two comparators (C_A and C_B), a flip-flop (FF), an inverter (I), two transistors, and a voltage divider (R_1 through R_3). The voltage divider is used to set the comparator reference voltages. Since $R_1 = R_2 = R_3$, the reference

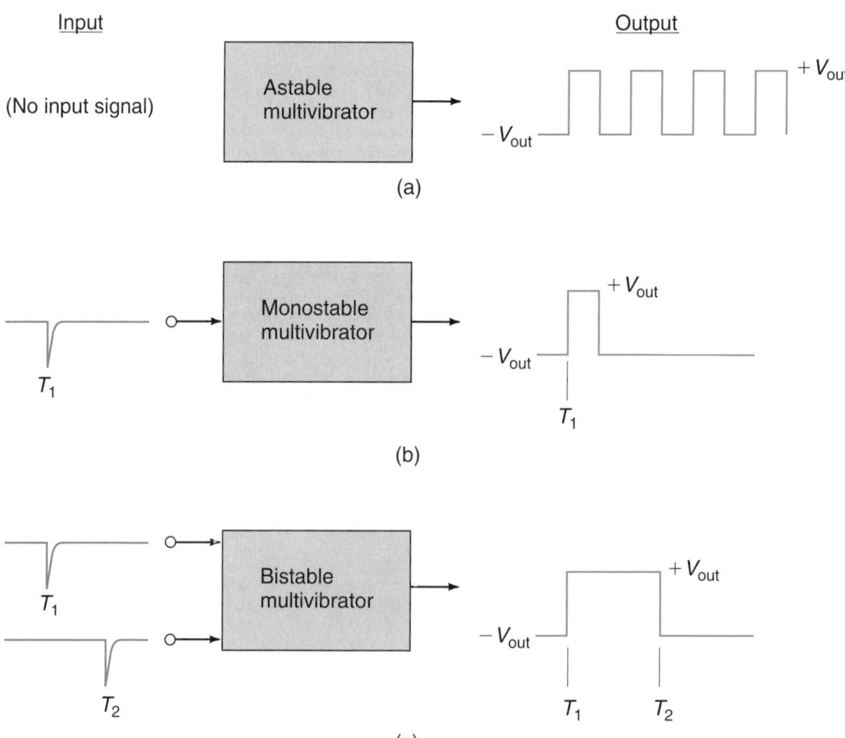

Figure 24.26 Multivibrator input/output relationships.

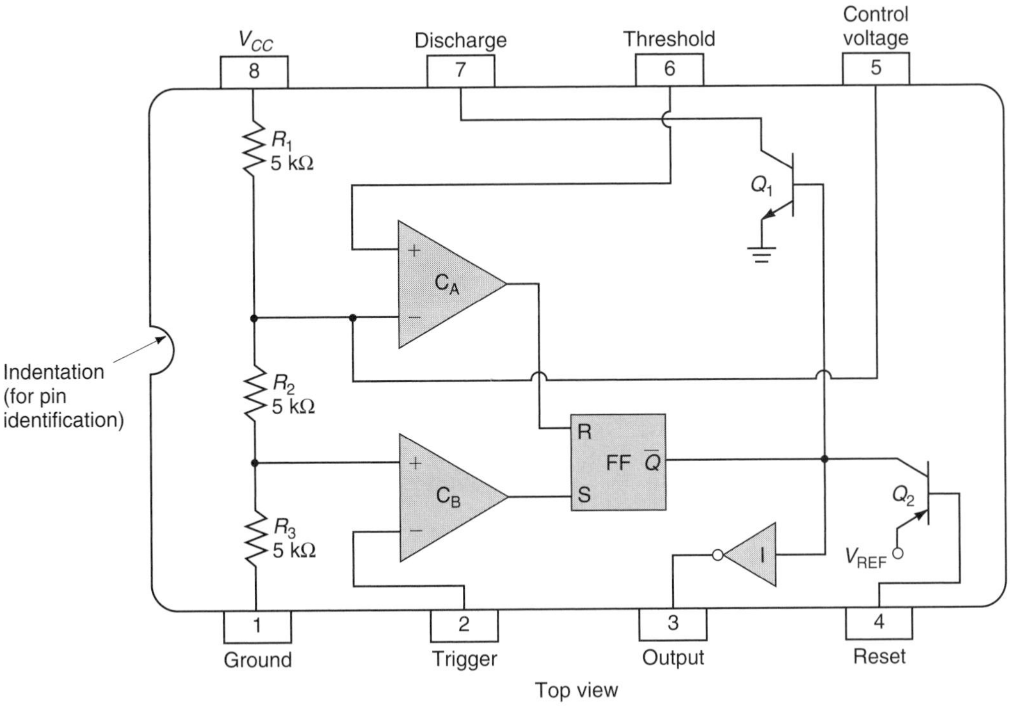

Figure 24.27
The 555 timer.

voltages for C_A and C_B are normally equal to $\frac{2}{3}V_{CC}$ and $\frac{1}{3}V_{CC}$, respectively. The only exception to this condition occurs when the control voltage input to the timer (pin 5) is used, as demonstrated later in this section.

Under normal circumstances, the comparators operate in one of three input/output combinations, as shown in Figure 24.28. Each of the output combinations depends on

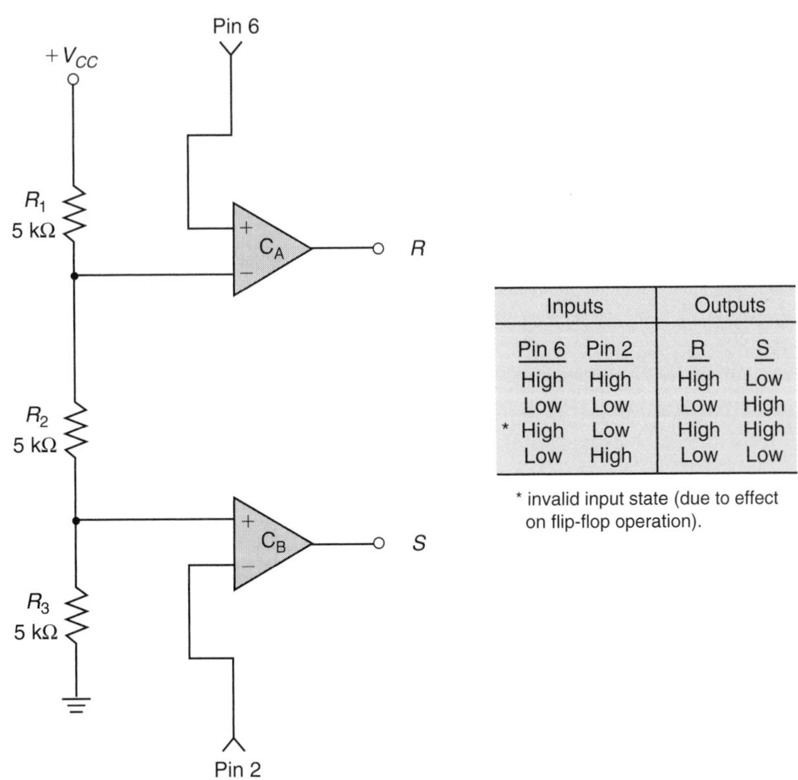

Inputs		Outputs	
Pin 6	Pin 2	R	S
High	High	High	Low
Low	Low	Low	High
* High	Low	High	High
Low	High	Low	Low

* invalid input state (due to effect on flip-flop operation).

Note: For each comparator, the inputs are designated as *high* or *low* with respect to the reference voltage.

Figure 24.28
Comparator input/output combinations.

Note: When the flip-flop inputs are not equal, the output follows the input at *R*.

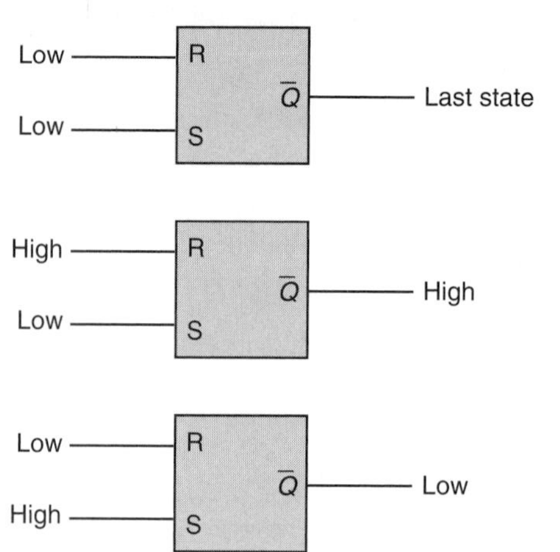

S	R	$\overline{Q}$
Low	Low	L.S.*
Low	High	High
High	Low	Low
High	High	Invalid

* Last state. The $\overline{Q}$ output remains in the state it was in before the input combination occurred.

Figure 24.29 Flip-flop input/output combinations.

the inputs at pins 2 and 6 of the timer. Note that it is possible to drive the outputs from both comparators high at the same time, but this is avoided because of the effect it will have on the flip-flop.

The flip-flop responds to its inputs as shown in Figure 24.29. As you can see, having two high inputs is considered to be "invalid," because there is no way to predict how the flip-flop will respond. Thus, the combination of pin 6 = high and pin 2 = low is never allowed for the 555 timer.

We need to clear up only a few more points—and then we will be ready to look at some circuits. First, the flip-flop output is tied to the discharge output (pin 7) via the *npn* transistor. When the output from the flip-flop is high, pin 7 is grounded through the transistor. When the output from the flip-flop is low, pin 7 appears as an open to any external connections. The inverter (pin 3) and the transistor (pin 7) both invert the flip-flop's output. The reset input (pin 4) is used to disable the 555 timer. When the reset input is low, the *pnp* transistor is biased on, and the output from the flip-flop is shorted. In all our circuits, pin 4 will be tied to V_{CC}, thus disabling the reset circuit.

The Monostable Multivibrator

OBJECTIVE 9 ➤ The 555 timer operates as a one-shot when it is wired as shown in Figure 24.30. Before we get into the operation of this circuit, a few observations should be made:

- The reset input to the 555 timer (pin 4) is connected to V_{CC}, which disables the reset circuit.
- Pins 8 and 1 are tied to V_{CC} and ground, respectively, as they must be for the IC to operate.
- The control voltage input (pin 5) is *not connected* (N.C.).
- Pins 6 and 7 are tied together, so the capacitor voltage (V_C) is applied to both pins.

Now, let's look at the operation of the circuit. The appropriate waveforms are shown in Figure 24.31. To start, we will assume that something caused the output of the flip-flop to be high before T_0. (You will see what caused this condition in a moment.) The inactive voltages shown between T_0 and T_2 in Figure 24.31 are explained as follows:

- By design, the trigger input (T) is approximately equal to V_{CC} when in its inactive state. As a result, the output from C_B is low when the trigger input is inactive.
- The high output from the flip-flop biases Q_1 *on,* which keeps the external capacitor (C) discharged. As a result, $V_C = 0$ V when the circuit is inactive.
- The high output from the flip-flop is applied to the *inverter.* A high input to the inverter results in a *low* (0 V) output from the circuit. Therefore, $V_{out} = 0$ V when the circuit is inactive.

Assuming that $\overline{Q}$ is high, we know that pins 3 and 7 are both *low.* The low output at pin 7 is applied to C_A via pin 6. With a low input to C_A, the output of C_A is low. Since

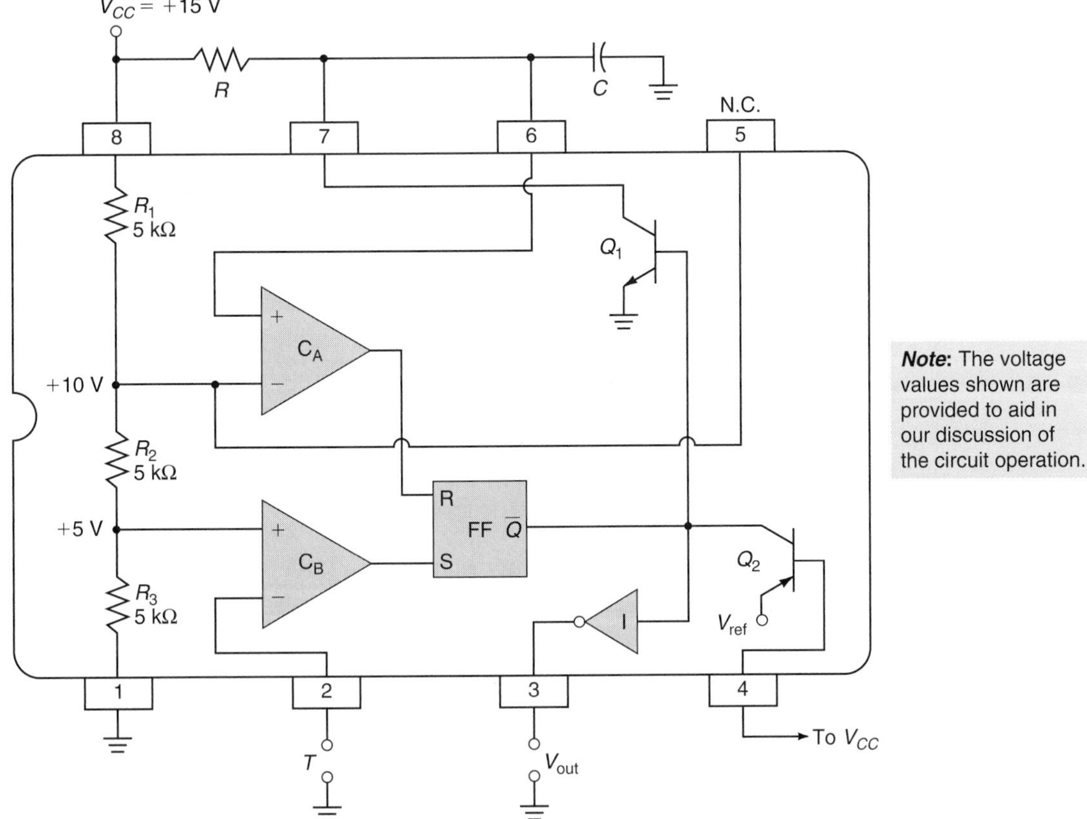

Note: The voltage values shown are provided to aid in our discussion of the circuit operation.

Figure 24.30 The 555 timer one-shot (monostable multivibrator).

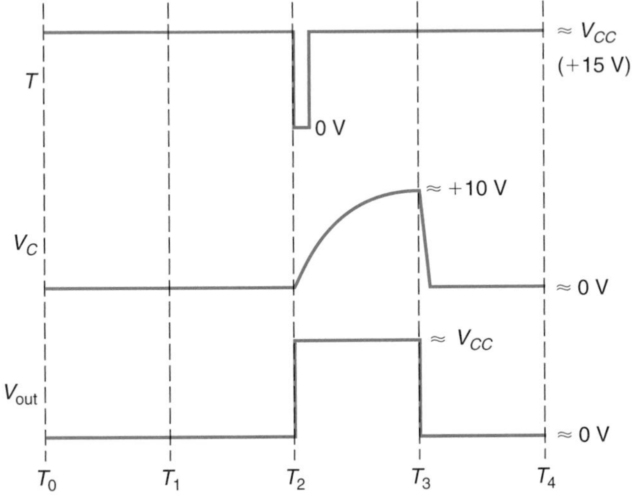

Figure 24.31 The 555 timer one-shot waveforms.

both inputs to the flip-flop are low, $\overline{Q}$ and the rest of the circuit voltages remain as shown until something happens to change them.

At T_2, the trigger (T) input is driven low, and the following sequence of events takes place:

- The low input to C_B drives the output of the comparator high.
- The inputs to the flip-flop are now unequal, and $\overline{Q}$ goes to the level of the R input, which is low.
- The low flip-flop output is inverted, and pin 3 goes high (as shown in Figure 24.31).

- The low flip-flop output biases the *npn* transistor off. Thus, the capacitor starts to charge toward V_{CC}. The charge time of the capacitor depends on the values of R and C.
- When $V_C > +10$ V, the input to pin 6 is high (as compared to the voltage at the inverting C_A input).
- The combined inputs to C_A cause the comparator's output to go high. Since the trigger signal is no longer present, the output from C_B has returned to a low.
- Since the inputs to the flip-flop are unequal again, $\overline{Q}$ switches to the level of R, which is high.
- The high output from the flip-flop causes pin 3 to go low again (shown at T_3).
- The high flip-flop output turns the *npn* transistor on, which discharges the capacitor. This returns all circuit potentials to their original values (shown at T_4).

As you can see, our initial assumption of $\overline{Q}$ = high is based on the fact that the flip-flop is always set to this value at the end of a trigger cycle.

The output pulse width (PW) of a 555 timer one-shot is determined by the values of R and C used in the circuit. The output PW of a 555 timer circuit can be found as

$$\text{PW} = 1.1RC \tag{24.13}$$

Example 24.10 demonstrates the use of this equation.

Example 24.6

The one-shot illustrated in Figure 24.32 has values of $R = 1.2\ \text{k}\Omega$ and $C = 0.1\ \mu\text{F}$. Determine the pulse width of the output.

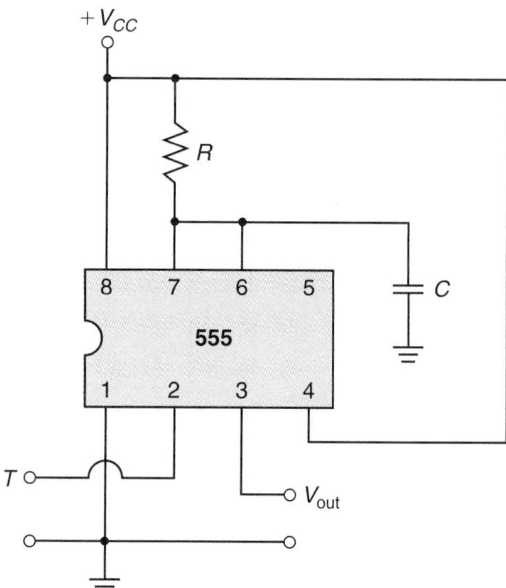

Figure 24.32

Solution: The output pulse width is found as

$$\text{PW} = 1.1RC = (1.1)(1.2\ \text{k}\Omega)(0.1\ \mu\text{F}) = 132\ \mu\text{s}$$

Practice Problem 24.6

A one-shot like that shown in Figure 24.32 has values of $R = 18\ \text{k}\Omega$ and $C = 1.5\ \mu\text{F}$. Determine the pulse width of the timer output signal.

Determining a Valid Trigger Signal

For a low pulse at pin 2 to be a valid trigger signal, it must be low enough to cause C_B to have a high output. When the control voltage input is not used, a valid trigger signal is one that fulfills the following relationship:

$$V_T < \frac{1}{3}V_{CC} \tag{24.14}$$

This equation is based on the fact that the (+) input of C_B is at $\frac{1}{3}V_{CC}$ when the control voltage input is not used. Therefore, the (T) input must go below this voltage level to cause the output of C_B to go high.

When a control voltage is applied to pin 5, it is dropped equally across R_2 and R_3 in the resistive ladder. Therefore, the voltage at the (+) input of C_B equals one-half the control voltage. In this case, a valid input trigger is one that fulfills the following relationship:

$$V_T < \frac{1}{2}V_{con} \tag{24.15}$$

where

V_{con} = the voltage at the control voltage input

One advantage of using the control voltage input to the 555 timer is that the circuit can be designed to be triggered by a lower amplitude pulse.

The Astable Multivibrator

The 555 timer operates as a free-running multivibrator when wired as shown in Figure 24.33. Note that the circuit contains two resistors and one capacitor, and that it does not have an input trigger from any other circuit. The lack of a trigger signal from an external source is the circuit-recognition feature of the free-running multivibrator.

The key to the operation of this circuit is the capacitor connection to both the trigger input (pin 2) and the threshold input (pin 6). Thus, if the capacitor is charged and discharged between the threshold voltage ($\frac{2}{3}V_{CC}$) and the trigger voltage ($\frac{1}{3}V_{CC}$), the timer produces a steady train of pulses, as illustrated in Figure 24.34. Since pins 6 and 2 are

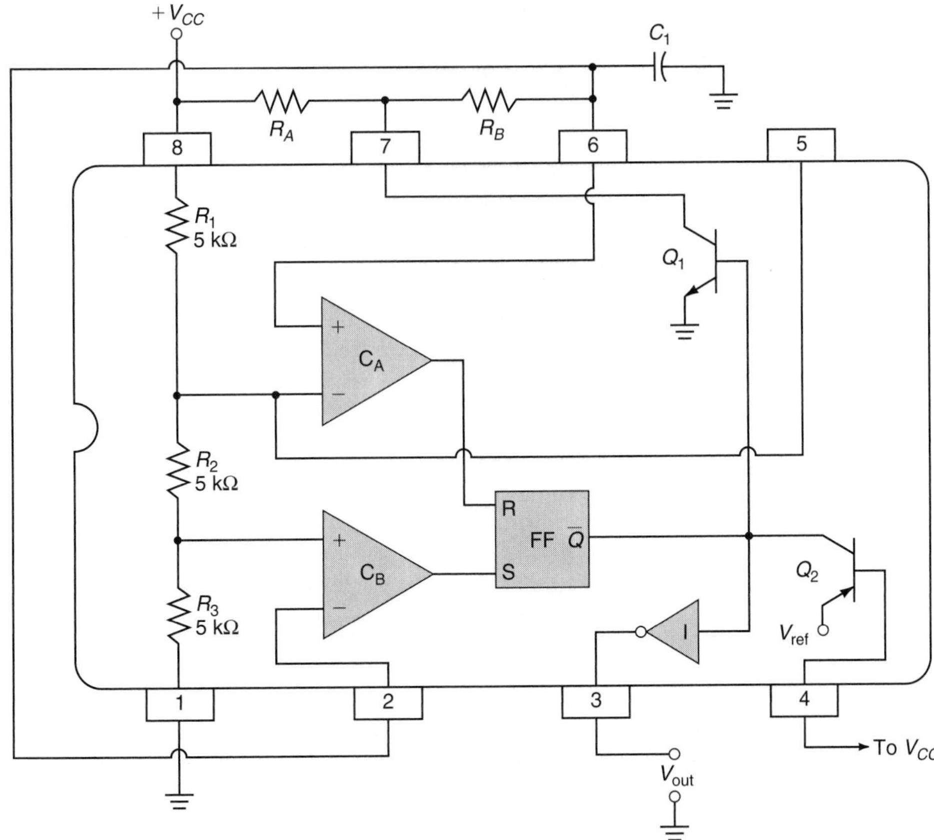

Figure 24.33 The 555 timer astable (free-running) multivibrator.

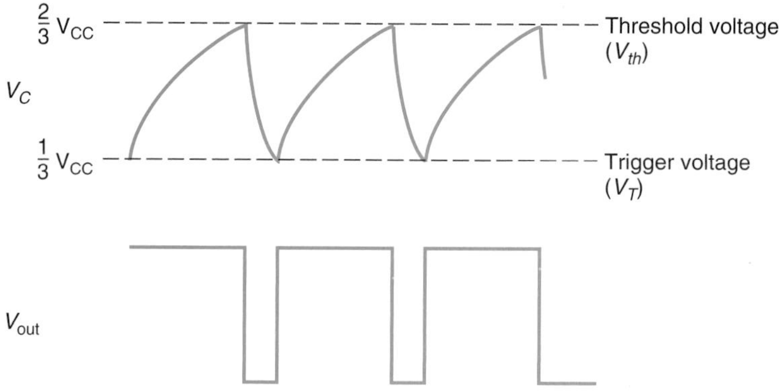

Figure 24.34 Free-running multivibrator waveforms.

tied together, their voltages are equal. When V_C charges to $\frac{2}{3}V_{CC}$, both inputs are high, and the output from the timer goes low. When V_C discharges to $\frac{1}{3}V_{CC}$, both inputs are low and the output from the timer goes high.

The charge–discharge action of C_1 is caused by the 555 timer internal circuitry. To see how this is accomplished, we will start by making a few assumptions:

- The capacitor is just starting to charge toward the threshold voltage (V_{th}).
- The output from the flip-flop is low, causing pin 3 to be high and pin 7 to be open.

With pin 7 open, the capacitor is charged by V_{CC} via R_A and R_B. As the capacitor charges, V_C eventually reaches V_{th}. At this point, both inputs to the 555 timer are high, which causes the output from the flip-flop to go high. The high output from the flip-flop causes the output (pin 3) to go low and also biases the *npn* transistor on. Thus, pin 7 is now shorted to ground (via the transistor), providing a discharge path for the capacitor (via R_B). When the capacitor discharges to the point at which $V_C = V_T$, both comparator inputs are low, which causes the output from the flip-flop to go low. When the flip-flop output changes, the timer output goes high, pin 7 returns to its open condition, and the cycle begins again.

The output frequency for the free-running multivibrator is found as

$$f_0 = \frac{1.44}{(R_A + 2R_B)C_1} \tag{24.16}$$

The duty cycle of the free-running multivibrator is the ratio of the pulse width to the cycle time. If you look at the waveforms shown in Figure 24.34, you'll see that the pulse width and cycle time of the circuit are determined by the charge and discharge times of C_1. The ouput pulse width of the circuit is found as

$$\text{PW} = 0.693(R_A + R_B)C_1 \tag{24.17}$$

and the duty cycle of the astable multivibrator can be found using

$$\text{duty cycle (\%)} = \frac{R_A + R_B}{R_A + 2R_B} \times 100 \tag{24.18}$$

Voltage-Controlled Oscillators

OBJECTIVE 10 ➤ A **voltage-controlled oscillator (VCO)** is *a free-running multivibrator whose output frequency is determined by a dc input control voltage (V_{con}).* A 555 timer operates as a VCO when wired as shown in Figure 24.35. When the dc control voltage is applied to pin 5 of the timer:

- The reference voltage for C_A equals V_{con}.
- The reference voltage for C_B equals $\frac{1}{2}V_{con}$ (because of the internal voltage divider).

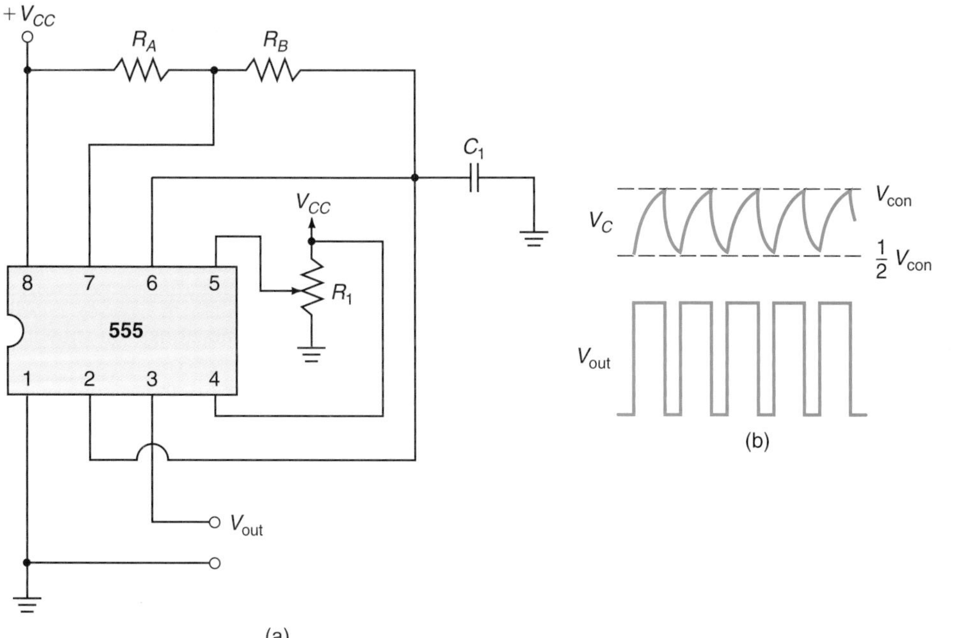

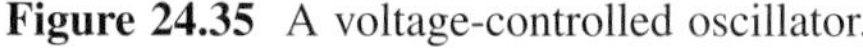
Figure 24.35 A voltage-controlled oscillator.

As the control voltage is increased (or decreased), the difference between the comparator reference voltages also increases (or decreases). As a result, the cycle time and operating frequency of the circuit also change.

One Final Note

In this chapter, we have barely scratched the surface of the subject of switching circuits. The study of digital electronics provides more complete coverage of the subject. However, this chapter has provided you with the fundamental principles of switching circuits to aid in your study of digital electronics.

Section Review

1. What is a *multivibrator*?
2. Describe the output characteristics of each of the following circuits:
   a. *Astable multivibrator* (*free-running multivibrator*).
   b. *Monostable multivibrator* (*one-shot*).
   c. *Bistable multivibrator* (*flip-flop*).
3. What is the *555 timer*?
4. List the input/output relationships for the 555 timer.
5. Briefly describe the operation of the monostable multivibrator (one-shot).
6. Briefly describe the operation of the astable (free-running) multivibrator.
7. Describe the operation of the *voltage-controlled oscillator* (*VCO*).

Key Terms

The following terms were introduced in this chapter on the pages indicated:

555 Timer, 832	Buffer, 824	Duty cycle, 820
Astable multivibrator, 832	Delay time (t_d), 820	Fall time (t_f), 820
Bistable multivibrator, 832	Driver, 818	Flip-flop, 832

Free-running multivibrator, 832
Hysteresis, 827
Inverter, 824
Lower trigger point (LTP), 826
Monostable multivibrator, 832
Multivibrator, 832
One-shot, 832
Propagation delay, 820
Rise time (t_r), 820
Schmitt trigger, 826
Speed-up capacitor, 822
Storage time (t_s), 820
Switching circuits, 813
Switching transistors, 824
Turn-off time (t_{off}), 824
Turn-on time (t_{on}), 824
Upper trigger point (UTP), 826
Voltage-controlled oscillator (VCO), 838

Practice Problems

1. Determine the minimum high voltage required to saturate the transistor shown in Figure 24.36.
2. A switch like the one shown in Figure 24.36 has the following values: $V_{CC} = +12$ V, $h_{FE} = 150$, $R_C = 1.1$ kΩ, and $R_B = 47$ kΩ. Determine the minimum high input voltage required to saturate the transistor.

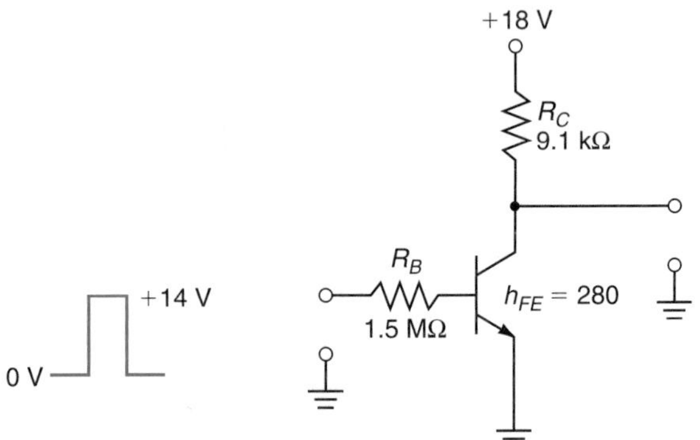

Figure 24.36

3. Determine the minimum high input voltage required to saturate the transistor shown in Figure 24.37.

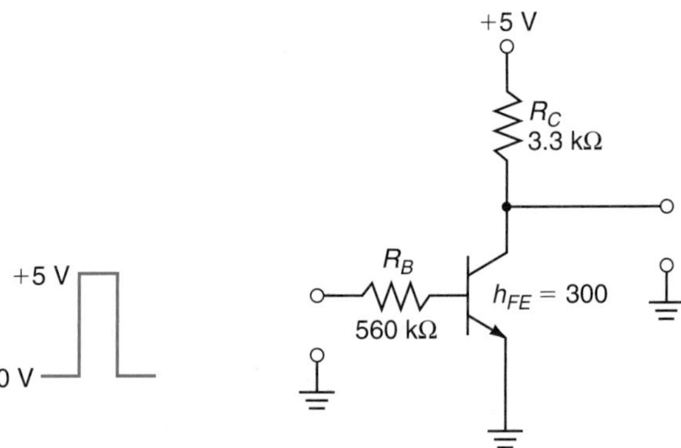

Figure 24.37

4. Determine the minimum high input voltage required to saturate the transistor shown in Figure 24.38.

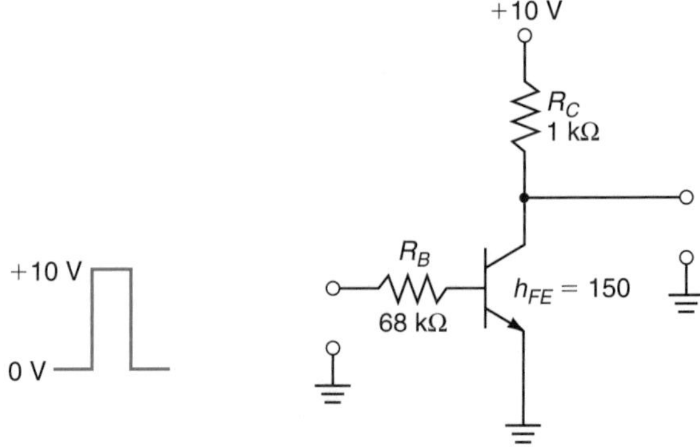

Figure 24.38

5. Determine the high and low output voltages for the circuit shown in Figure 24.39.

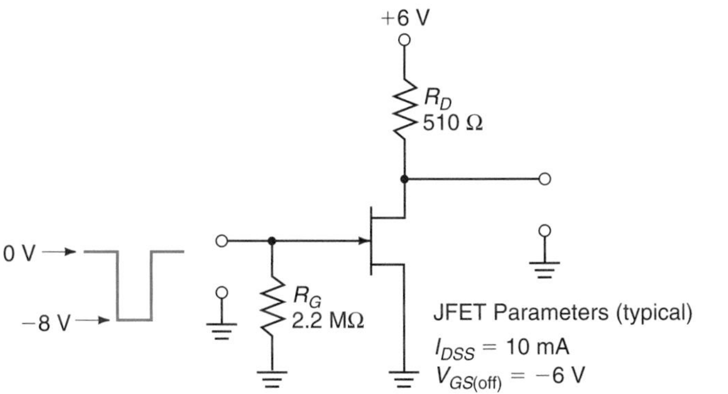

Figure 24.39

6. Determine the high and low output voltages for the circuit shown in Figure 24.40.

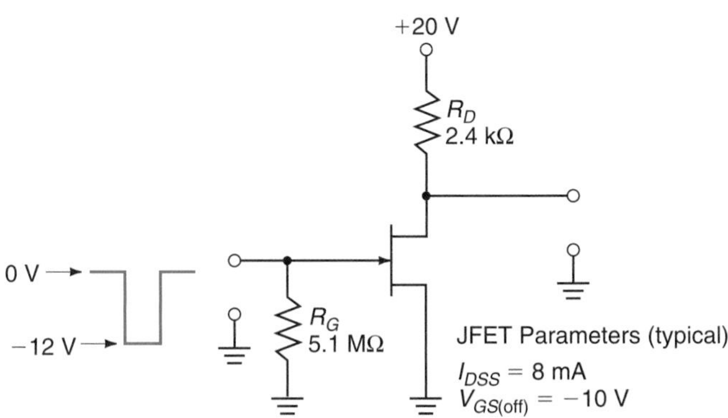

Figure 24.40

7. Determine the values of PW and T_C for the waveform shown shown in Figure 24.41.
8. Determine the values of PW and T_C for the waveform shown shown in Figure 24.42.
9. Calculate the duty cycle of the waveform shown in Figure 24.41.
10. Calculate the duty cycle of the waveform shown in Figure 24.42.

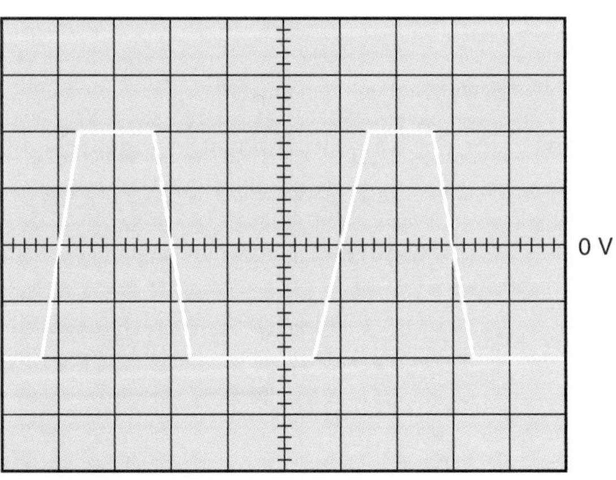

Time base: 10 μs/div

Figure 24.41

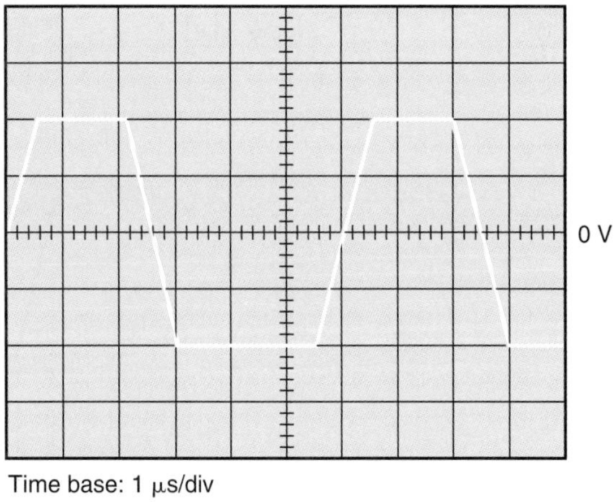

0 V

Time base: 1 μs/div

Figure 24.42

11. The 2N4264 BJT has the following parameters: $t_d = 8$ ns, $t_r = 15$ ns, $t_s = 20$ ns, and $t_f = 15$ ns. Determine the value of f_{C2} and the practical limit on f_{in} for the device.
12. The 2N4400 BJT has the following parameters: $t_d = 15$ ns, $t_r = 20$ ns, $t_s = 225$ ns, and $t_f = 30$ ns. Determine the value of f_{C2} and the practical limit on f_{in} for the device.

13. The BCX587 BJT has the following parameters: t_d = 16 ns, t_r = 29 ns, t_s = 475 ns, and t_f = 40 ns. Determine the value of f_{C2} and the practical limit on f_{in} for the device.

14. The MPS2222 BJT has the following parameters: t_d = 10 ns, t_r = 25 ns, t_s = 225 ns, and t_f = 60 ns. Determine the value of f_{C2} and the practical limit on f_{in} for the device.

15. The 2N4093 JFET has the following parameters: t_d = 20 ns, t_r = 40 ns, and t_{off} = 80 ns. Determine the value of f_{C2} and the practical limit on f_{in} for the device.

16. The 2N3971 JFET has the following parameters: t_d = 15 ns, t_r = 15 ns, and t_{off} = 60 ns. Determine the value of f_{C2} and the practical limit on f_{in} for the device.

17. The 2N4351 MOSFET has the following parameters: t_d = 45 ns, t_r = 65 ns, t_s = 60 ns, and t_f = 100 ns. Determine the value of f_{C2} and the practical limit on f_{in} for the device.

18. The 2N4393 JFET has the following parameters: t_{on} = 15 ns, and t_{off} = 50 ns. Determine the value of f_{C2} and the practical limit on f_{in} for the device.

19. Determine the values of t_d, t_r, t_s, and t_f for the waveform shown in Figure 24.43.

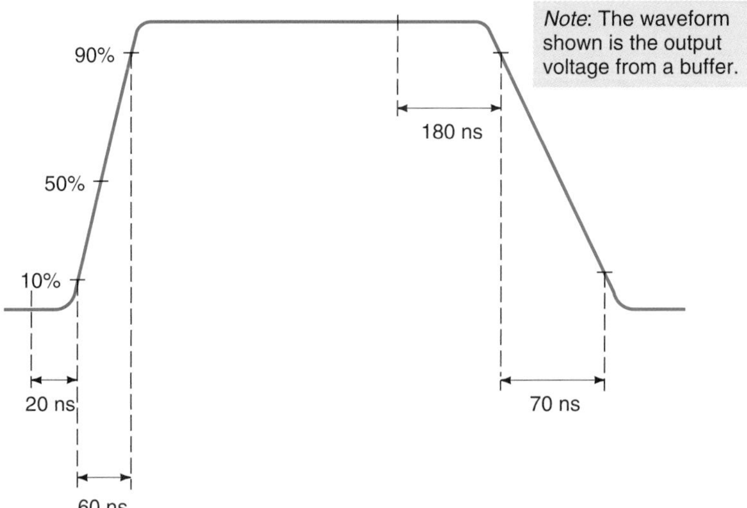

Figure 24.43

20. Determine the values of t_d, t_r, t_s, and t_f for the waveform shown in Figure 24.44.

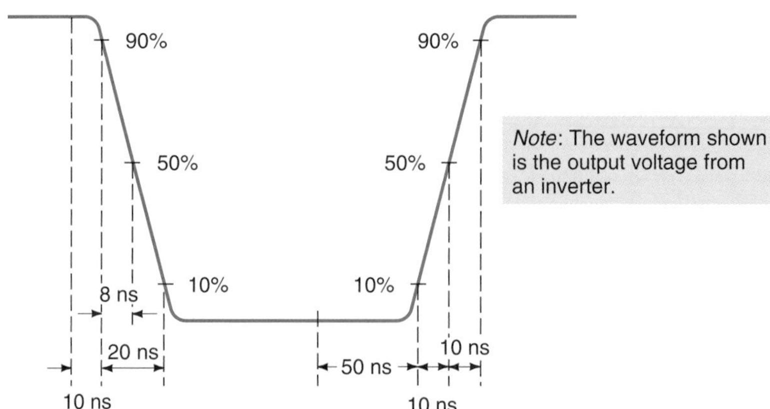

Figure 24.44

21. Determine the UTP and LTP values for the Schmitt trigger shown in Figure 24.45.

22. Determine the UTP and LTP values for the Schmitt trigger shown in Figure 24.46.

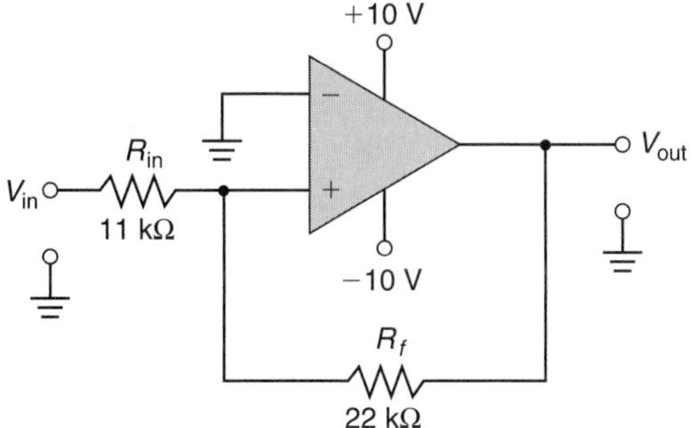

Figure 24.45

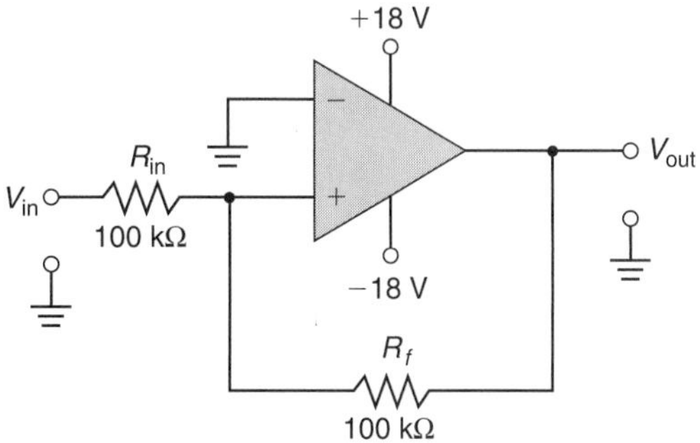

Figure 24.46

23. Determine the UTP and LTP values for the Schmitt trigger shown in Figure 24.47.
24. Determine the UTP and LTP values for the Schmitt trigger shown in Figure 24.48.

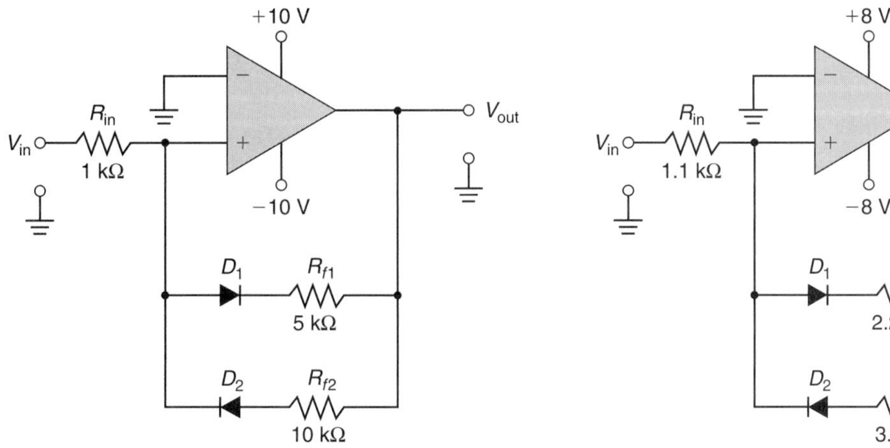

Figure 24.47

Figure 24.48

25. Determine the UTP and LTP values for the Schmitt trigger shown in Figure 24.49.
26. Determine the UTP and LTP values for the Schmitt trigger shown in Figure 24.50.

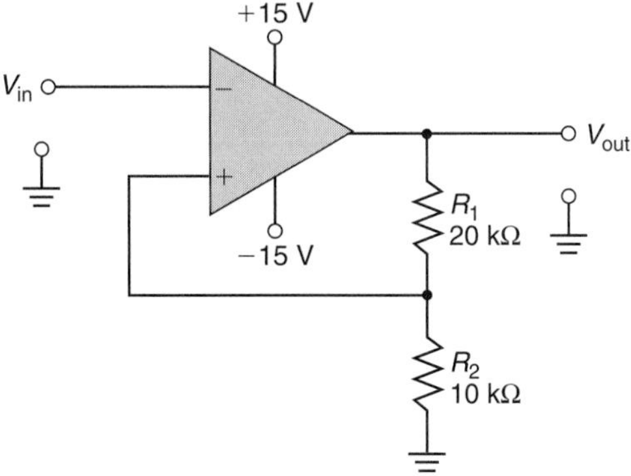

Figure 24.49

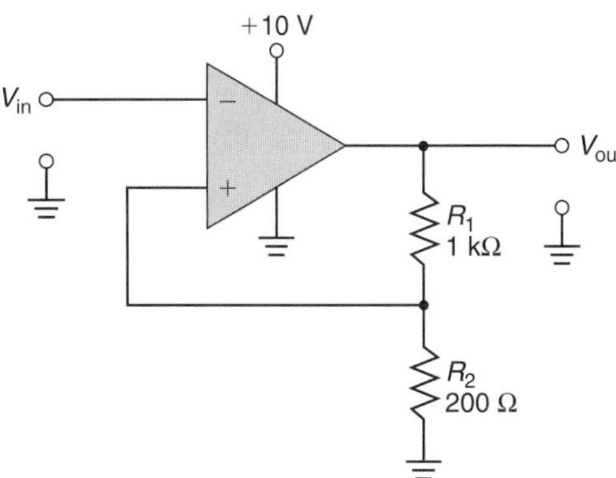

Figure 24.50

27. Determine the UTP and LTP values for the Schmitt trigger shown in Figure 24.51.
28. Determine the UTP and LTP values for the Schmitt trigger shown in Figure 24.52.

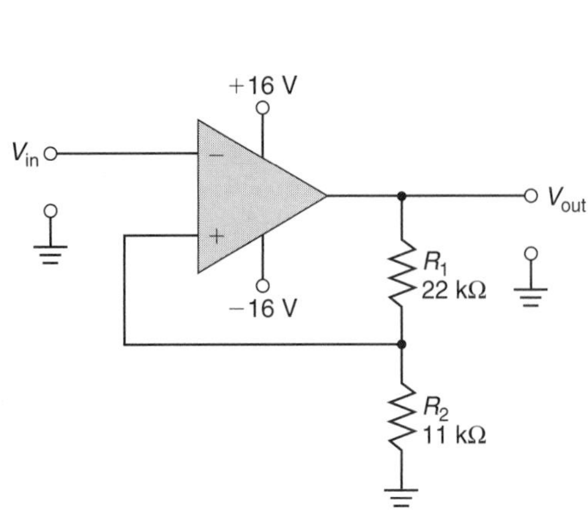

Figure 24.51

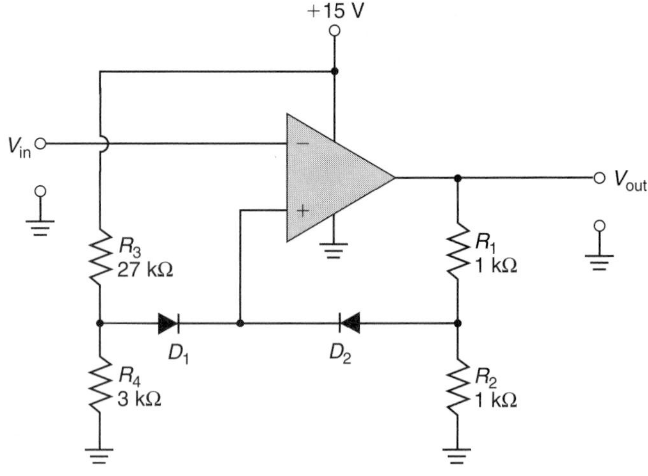

Figure 24.52

29. Calculate the output pulse width of the one-shot shown in Figure 24.53.

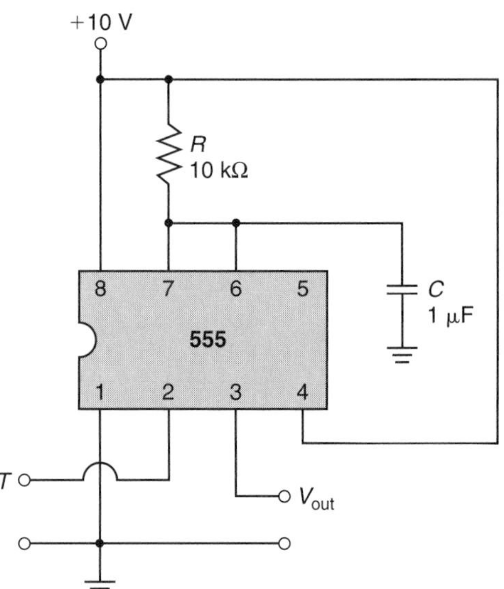

Figure 24.53

30. Calculate the output pulse width of the one-shot shown in Figure 24.54.

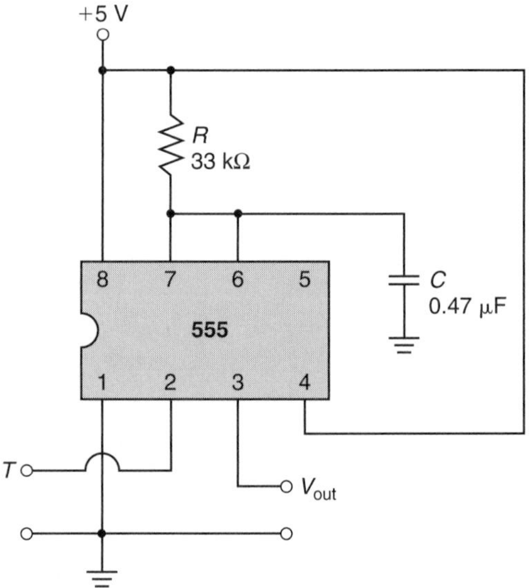

Figure 24.54

31. Calculate the value of $V_{T(\max)}$ for the switch shown in Figure 24.55.

32. Assume that the resistor values shown in Figure 24.55 are changed to $R_1 = 33$ kΩ and $R_2 = 11$ kΩ. Recalculate the value of $V_{T(\max)}$ for the circuit.

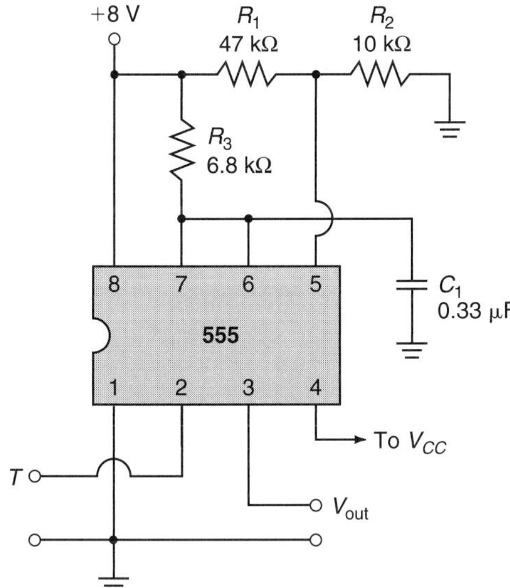

Figure 24.55

33. Calculate the operating frequency, duty cycle, and pulse width for the free-running multivibrator shown in Figure 24.56.

34. Calculate the operating frequency, duty cycle, and pulse width for the free-running multivibrator shown in Figure 24.57.

35. Assume that the value of R_B shown in Figure 24.56 is increased to 10 kΩ. Recalculate the operating frequency, duty cycle, and pulse width for the circuit.

36. Assume that the value of C_1 shown in Figure 24.57 is increased to 0.033 μF. Recalculate the operating frequency and pulse width for the circuit.

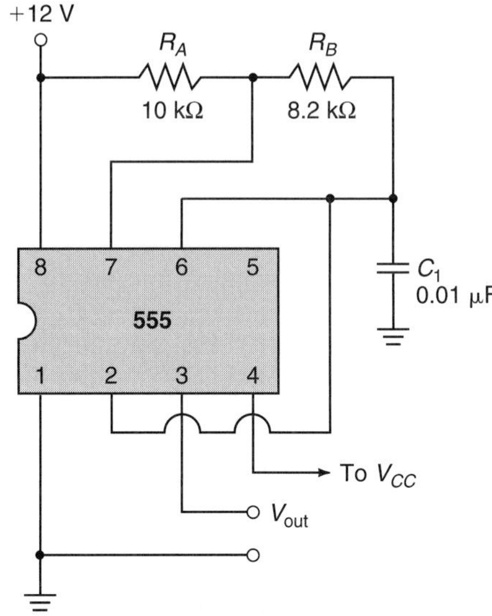

Figure 24.56

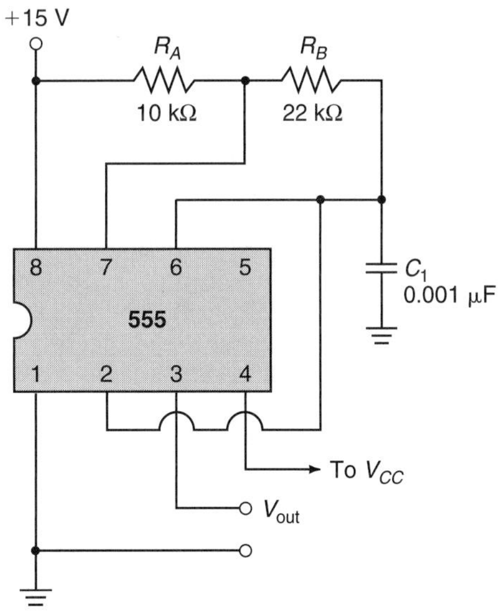

Figure 24.57

Looking Back

These problems relate to material presented in earlier chapters. The chapters are identified in brackets.

37. A series-resistive circuit has a source with values of $V_S = 8$ V and $R_S = 30\ \Omega$. The load resistance can be varied between 10 Ω and 100 Ω. Determine its maximum value of load power. [Chapter 4]

38. Determine the minimum and maximum values of current for the circuit described in Problem 37. [Chapter 4]

39. A collector-feedback bias circuit has the following values: $V_{CC} = 12$ V, $R_C = 10\ \text{k}\Omega$, $h_{FE} = 90$, and $R_B = 910\ \text{k}\Omega$. Determine I_{CQ} and V_{CEQ} for the circuit. [Chapter 19]

40. Calculate the reactance of a 33 μH inductor at 100 Hz. [Chapter 10]

Pushing the Envelope

41. The BJT shown in Figure 24.58a is faulty. Determine which (if any) of the transistors listed in the selector guide can be used as a substitute component. (Be careful; this one is more difficult than it appears!)

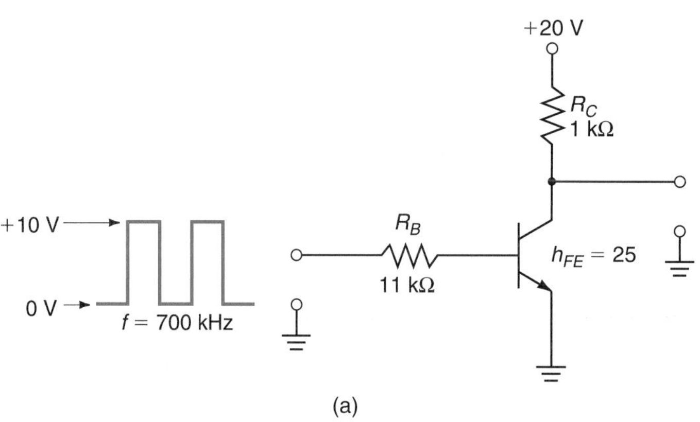

(a)

Pinout: 1-Base, 2-Emitter, 3-Collector
Devices are listed in order of descending f_T.

Device	Marking	Switching Time (ns) t_{on}	Switching Time (ns) t_{off}	$V_{(BR)CEO}$	h_{FE} Min	h_{FE} Max	h_{FE} @I_C (mA)	f_T Min (MHz)
NPN								
MMBT2369	1J	12	18	15	20	—	100	—
BSV52	B2	12	18	12	40	120	10	400
MMBT2222	1B	35	385	30	30	—	500	250
MMBT2222A	1P	35	385	40	40	—	500	200
MMBT4401	2X	35	255	40	40	—	500	250
MMBT3903	1Y	70	225	40	15	—	100	250
MMBT3904	1A	70	250	40	30	—	100	200
PNP								
MMBT3638A	BN	75	170	25	20	—	300	—
MMBT3638	AM	75	170	25	20	—	300	—
MMBT3640	2J	25	35	12	20	—	50	500
MMBT4403	2T	35	225	40	90	180	1	150
MMBT2907	2B	45	100	40	30	—	500	200
MMBT2907A	2F	45	100	60	50	—	500	200
MMBT3906	2A	70	300	40	100	300	10	250

(b)

Figure 24.58

42. Using a 555 timer, design a free-running multivibrator that produces a 50 kHz output with a 60% duty cycle. The supply voltage for the timer is +10 V, and it should have an inactive reset input pin.

43. Refer to Figure 24.59. The circuit was intended (by design) to fulfill the following requirements:
    a. The LED should light when a positive-going input signal passes a UTP value of 11.7 V.
    b. The LED should turn off when a negative-going input signal passes an LTP of 21.7 V.

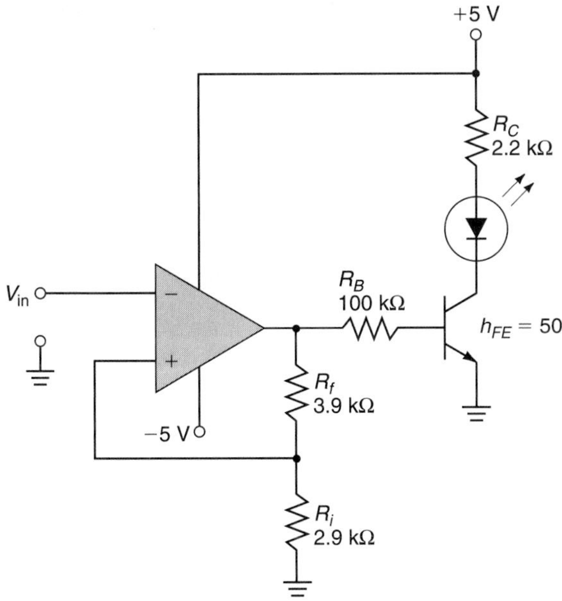

Figure 24.59

Unfortunately, there is a flaw in the circuit design. Find the flaw, and suggest a correction that will make the circuit work properly.

Answers to the Example Practice Problems

24.1 7.99 V

24.2 When $V_{GS} = 0$ V, $V_{out} = 1$ V. When $V_{GS} = -4$ V, $V_{out} = 10$ V.

24.3 17 V

24.4 UTP = +4 V, LTP = −4 V

24.5 UTP = +2.25 V, LTP = −2.25 V

24.6 29.7 ms

Multisim File References

Figure	File Reference
24.1	ETF24_1.msm
24.22	ETF24_22.msm
24.35	ETF24_35.msm

chapter 25

Discrete and Integrated Voltage Regulators

Objectives

After studying the material in this chapter, you should be able to:

1. List the purposes served by a *voltage regulator.*
2. Define *line regulation* and its commonly used units of measure.
3. Define *load regulation* and its commonly used units of measure.
4. Describe and analyze the operation of the *pass-transistor regulator.*
5. Describe the means by which short-circuit protection is provided for a pass-transistor regulator.
6. Describe and analyze the operation of the *shunt-feedback regulator.*
7. Discuss the need for overvoltage protection in a shunt-feedback regulator and the means by which it is provided.
8. List the reasons why the series regulator is preferred over the shunt regulator.
9. List and describe the various types of *linear IC voltage regulators.*
10. List and describe the common linear IC regulator parameters and ratings.
11. Compare and contrast the operation of *linear* and *switching regulators.*
12. List the four parts of the basic switching regulator and describe the function of each.
13. Describe the response of a switching regulator to a change in load resistance.
14. Describe the common methods for controlling power switch conduction.
15. List the circuit-recognition features for each of the common switching regulator configurations.
16. Describe the functions performed by a typical switching regulator IC.
17. In terms of their advantages, disadvantages, and applications, compare and contrast linear and switching regulators.

In Chapter 19, you were told that a power supply contains the three primary circuits represented in Figure 25.1:

- The *rectifier* converts the ac output from the transformer into pulsating dc.
- The *filter* reduces the variations in the rectifier output.
- The *voltage regulator* reduces the ripple in the filtered dc output and maintains a constant output voltage despite predictable variations in load demand.

◀ *OBJECTIVE 1*

In this chapter, we will look at a variety of practical voltage regulators.

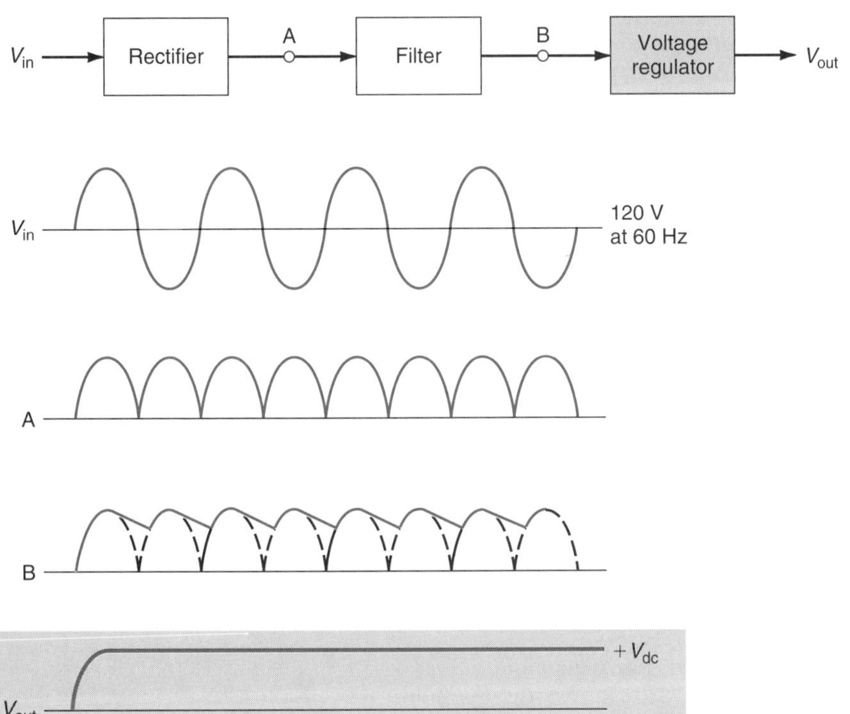

Figure 25.1 Basic power supply block diagram and waveforms.

25.1 Voltage Regulation: An Overview

The *ideal* voltage regulator maintains a constant dc output voltage, regardless of changes in its input voltage or its load current demand. Consider the ideal +10 V regulator shown in Figure 25.2a. A change in input voltage (ΔV_{in}) does not affect the dc output of the regulator ($\Delta V_{out} = 0$ V). This assumes that V_{in} does not decrease below the value required

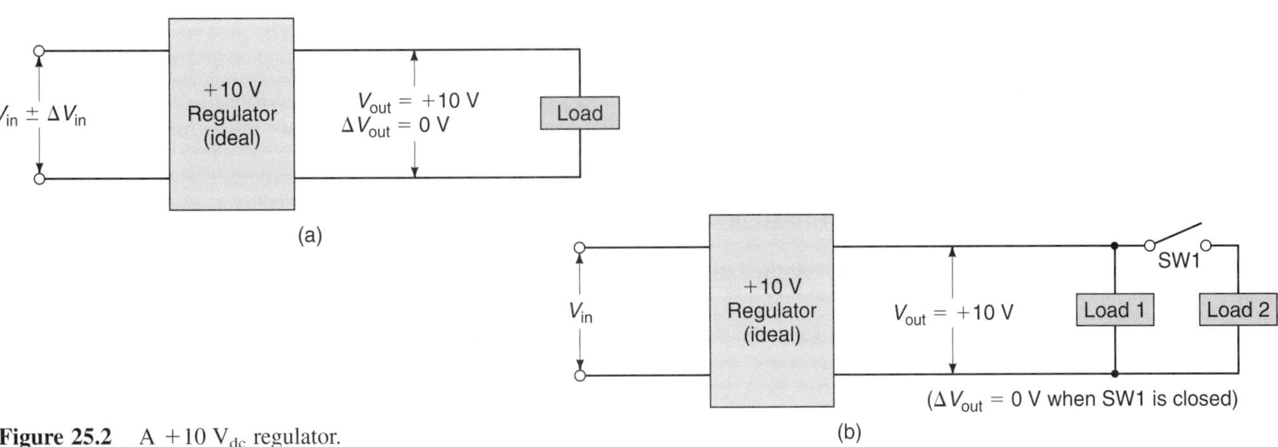

Figure 25.2 A +10 V_{dc} regulator.

to maintain the operation of the regulator. (A linear regulator cannot have a 10 V output if the input is less than 10 V.)

Figure 25.2b shows how the ideal regulator responds to a change in load current demand. Assuming that the two loads are equal, closing SW1 causes the current demand on the regulator to double. The ideal voltage regulator maintains a constant output voltage ($\Delta V_{out} = 0$ V) despite the change in load current demand.

Line Regulation

OBJECTIVE 2 ➤ The **line regulation** rating of a voltage regulator indicates *the change in output voltage that occurs per unit change in input voltage.* The line regulation of a voltage regulator is found as

$$\text{line regulation} = \frac{\Delta V_{out}}{\Delta V_{in}} \tag{25.1}$$

where

ΔV_{out} = the change in output voltage (usually in microvolts or millivolts)

ΔV_{in} = the change in input voltage (usually in volts)

Example 25.1 illustrates the concept of line regulation.

Example 25.1

A voltage regulator experiences a 10 μV change in output voltage when its input voltage changes by 5 V. Determine the value of line regulation for the circuit.

Solution: The line regulation of the circuit is found as

$$\text{line regulation} = \frac{\Delta V_{out}}{\Delta V_{in}} = \frac{10\ \mu\text{V}}{5\ \text{V}} = 2\ \mu\text{V/V}$$

This result indicates that the output voltage changes by 2 μV for every 1 V change in the regulator's input voltage.

Practice Problem 25.1

The change in output voltage for a voltage regulator is measured at 100 μV when the input voltage changes by 4 V. Calculate the line regulation rating for the regulator.

As stated earlier, the ideal regulator has a ΔV_{out} of 0 V when the input voltage changes. Thus, for the ideal voltage regulator,

$$\text{line regulation} = \frac{\Delta V_{out}}{\Delta V_{in}} = \frac{0\ \text{V}}{\Delta V_{in}} = 0$$

Based on the fact that the ideal line regulation rating is 0, we can say that *the lower the line regulation rating of a voltage regulator, the higher the quality of the circuit.*

Line regulation is commonly expressed in a variety of units. The commonly used line regulation units and their meanings are listed in Table 25.1.

Load Regulation

OBJECTIVE 3 ➤ The **load regulation** rating of a practical voltage regulator indicates *the change in regulator output voltage per unit change in load current.* The load regulation of a voltage regulator is found as

$$\text{load regulation} = \frac{V_{NL} - V_{FL}}{\Delta I_L} \tag{25.2}$$

TABLE 25.1 Commonly Used Line Regulation Units

Unit	Meaning
μV/V	The change in output voltage (in microvolts) per 1 V change in input voltage.
ppm/V	Parts-per-million per volt; another way of saying microvolts per volt.
%/V	The percentage change in output voltage per 1 V change in the input voltage.
%	The total percentage change in output voltage over the rated range of input voltages.
mV (or μV)	The actual change in output voltage over the rated range of input voltages.

A Practical Consideration: Technically, line regulation should have no unit of measure since it is a ratio of one voltage to another. However, the units listed in Table 25.1 are commonly used on spec sheets, so you need to know what they are.

where

V_{NL} = the no-load output voltage (i.e., the output voltage when the load is open)

V_{FL} = the full-load output voltage (i.e., the load current demand at its maximum value)

ΔI_L = the change in load current demand

Another way of expressing this is

$$\text{load regulation} = \frac{\Delta V_{\text{out}}}{\Delta I_L}$$

Example 25.2 illustrates the concept of load regulation.

Example 25.2

A voltage regulator is rated for an output current of I_L = 0 to 20 mA. Under no-load conditions, the output voltage is 5 V. Under full-load conditions, the output voltage from the circuit is 4.9998 V. Determine the load regulation for this circuit.

Solution: The load regulation of the circuit is found as

$$\text{load regulation} = \frac{V_{NL} - V_{FL}}{\Delta I_L} = \frac{5\text{ V} - 4.9998\text{ V}}{20\text{ mA}} = 10\ \mu\text{V/mA}$$

This rating indicates that the output changes by 10 μV for each 1 mA change in load current.

Practice Problem 25.2

A voltage regulator is rated for an output current of I_L = 0 to 40 mA. Under no-load conditions, the output voltage from the circuit is 8 V. Under full-load conditions, the output voltage from the circuit is 7.996 V. Determine the load regulation for this circuit.

Like the line regulation rating, the load regulation rating is commonly expressed in a variety of units. Table 25.2 lists some of the commonly used load regulation units and their meanings.

For the ideal regulator, $V_{NL} = V_{FL}$, and load regulation equals zero. Therefore, we can state that *the lower the load regulation rating of a voltage regulator, the higher the quality of the circuit.*

Section Review

1. What two purposes are served by a *voltage regulator*?
2. How would the *ideal* voltage regulator respond to a change in input voltage?
3. How would the *ideal* voltage regulator respond to a change in load current demand?

4. What is *line regulation*?
5. What is the line regulation value of an ideal voltage regulator?
6. What is the relationship between the quality of a voltage regulator and its line regulation rating?
7. List and define the commonly used line regulation ratings.
8. What is *load regulation*?
9. List and define the commonly used load regulation ratings.
10. What is the load regulation value of an ideal voltage regulator?

TABLE 25.2 Commonly Used Load Regulation Units

Unit	*Meaning*
μV/mA	The change in output voltage (in microvolts) per 1 mA change in load current.
%/mA	The percentage change in output voltage per 1 mA change in load current.
%	The total percentage change in output voltage over the rated range of load current values.
mV (or μV)	The actual change in output voltage over the rated range of load current values.
Ω	V/mA, expressed as a resistance value; this rating multiplied by the value of ΔI_L gives you the corresponding value of ΔV_L for the regulator.

25.2 Series Voltage Regulators

Series regulators are *circuits that have one or more devices placed in series with the load.* In this section, we will take a look at several common series voltage regulators.

Pass-Transistor Regulator

OBJECTIVE 4 ➤ The **pass-transistor regulator** is *a circuit that uses a series transistor to regulate load voltage.* The term *pass-transistor* indicates that the load current passes through the series transistor (Q_1), as shown in Figure 25.3.

As you know, a zener diode is designed to maintain a constant cathode-to-anode voltage as long as the device is properly biased. The key to the operation of the pass-transistor regulator is that the transistor base voltage is held relatively constant by the zener diode. Since Q_1 is an *npn* transistor, V_L is found as

$$V_L = V_Z - V_{BE} \tag{25.3}$$

What purpose is served by R_S? In many cases, the zener diode in Figure 25.3 requires more current (to maintain regulation) than the transistor base can allow. R_S provides an alternate path for this additional zener current.

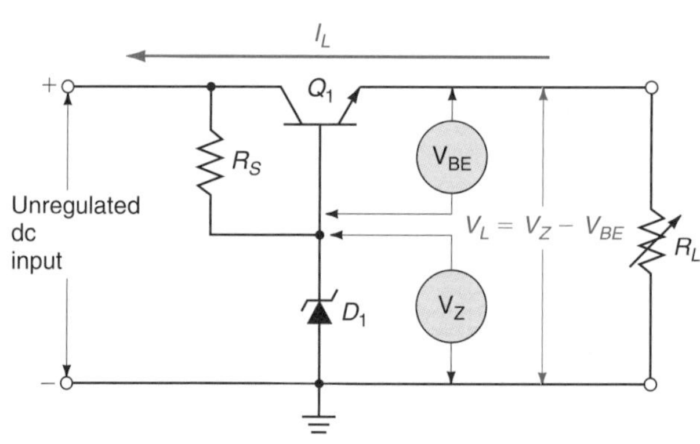

Figure 25.3 Pass-transistor regulator.

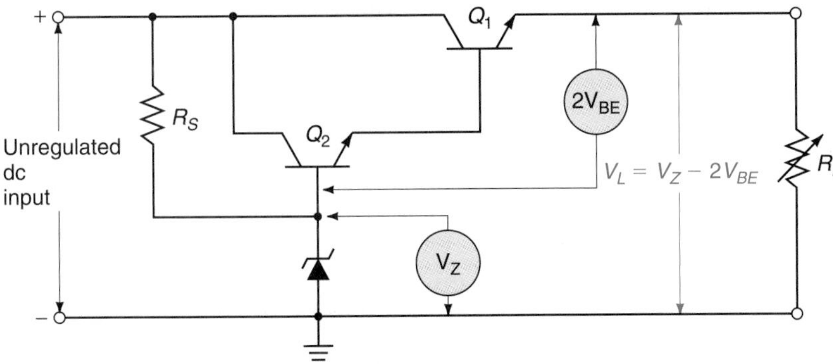

Figure 25.4 Darlington pass-transistor regulator.

If we rearrange this equation, we get

$$V_{BE} = V_Z - V_L \tag{25.4}$$

If the load resistance increases, the load voltage also increases. Since the zener voltage is constant, the increase in V_L causes V_{BE} to decrease. This reduces conduction through the pass-transistor. In turn, the load current decreases, and a relatively constant load voltage is maintained.

In a similar fashion, a decrease in load resistance causes V_L to decrease, which causes V_{BE} to increase. This increases conduction through the pass-transistor. In turn, the load current increases, and again, a relatively constant load voltage is maintained.

The problem with the pass-transistor regulator is that when the input voltage or load current values increase, the zener diode must dissipate a relatively high amount of power. This problem can be reduced by using a *Darlington pass-transistor regulator.*

Darlington Pass-Transistor Regulator

The **Darlington pass-transistor regulator** is *a circuit that uses a Darlington pair in place of a single pass-transistor,* as shown in Figure 25.4. The load voltage for the Darlington circuit is found as

$$V_L = V_Z - 2V_{BE} \tag{25.5}$$

Due to the high current gain of the Darlington pair, any increase in load current causes only a slight increase in zener current. Therefore, this circuit is not subject to the same power concerns mentioned earlier. However, zener conduction is affected by temperature, and the Darlington pair amplifies any increase in zener current. Therefore, the Darlington pass-transistor regulator must be operated at a relatively constant temperature.

Series Feedback Regulator

The **series feedback regulator** is *a series regulator that uses an error detection circuit to provide improved line and load regulation characteristics.* The block diagram for the series feedback regulator is shown in Figure 25.5.

The *error detector* receives two inputs: a *reference voltage* from the unregulated dc input voltage, and a *sample voltage* from the regulated output voltage. The error detector compares these two voltages and provides an output voltage that is proportional to the difference between them. This output voltage is amplified and used to drive the pass-transistor regulator.

The series feedback regulator is capable of responding very quickly to differences between its sample and reference input voltages. This gives the circuit much better line and load regulation characteristics than the other circuits we have discussed in this section.

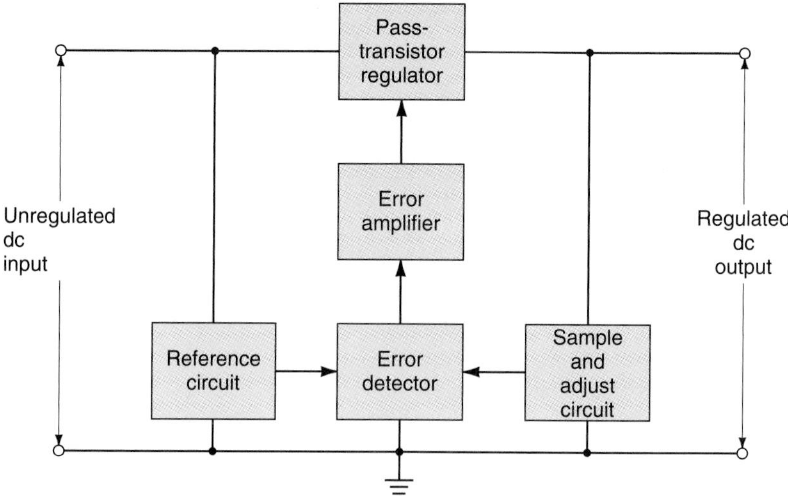

Figure 25.5 Block diagram for the series feedback regulator.

The schematic diagram for a series feedback regulator is shown in Figure 25.6. The sample-and-adjust circuit is the voltage divider that consists of R_3, R_4, and R_5. The reference voltage is set by the zener diode. Q_2 detects and amplifies the difference between the reference and the sample voltages and then adjusts the conduction of the pass-transistor accordingly.

The operation of this circuit is best illustrated by its response to a change in load resistance. Any increase in load resistance causes V_L to begin to increase. The voltage across the *sample-and-adjust* circuit equals the load voltage, so $V_{B(Q2)}$ increases. The increase in $V_{B(Q2)}$ causes $I_{C(Q2)}$ to increase, which causes $V_{C(Q2)}$ and $V_{B(Q1)}$ to *decrease.* The decrease in $V_{B(Q1)}$ reduces Q_1 conduction, causing the load current to *decrease.* The decrease in load current offsets the increase in load resistance, maintaining a constant load voltage. If the load resistance *decreases,* an opposite chain of events occurs, resulting in a decrease in pass-transistor conduction and load current. Again, this offsets the change in load resistance that initiated the sequence of circuit responses.

Short-Circuit Protection

OBJECTIVE 5 ➤ One weakness of the standard series regulator is the possibility of the pass transistor being destroyed by excessive current if the load is shorted. Adding a *current-limiting* circuit

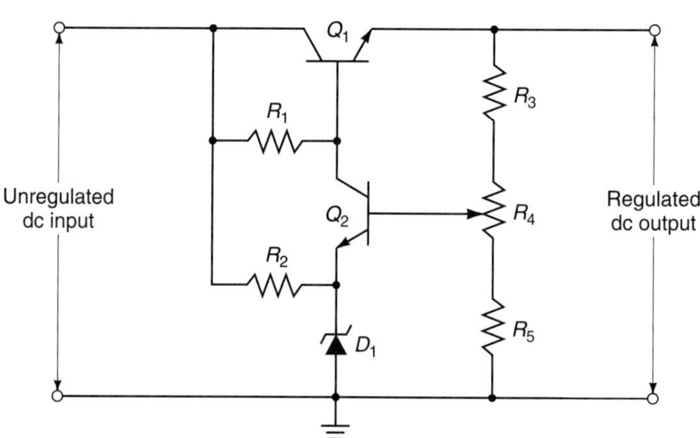

Figure 25.6 Schematic diagram for a basic series feedback regulator.

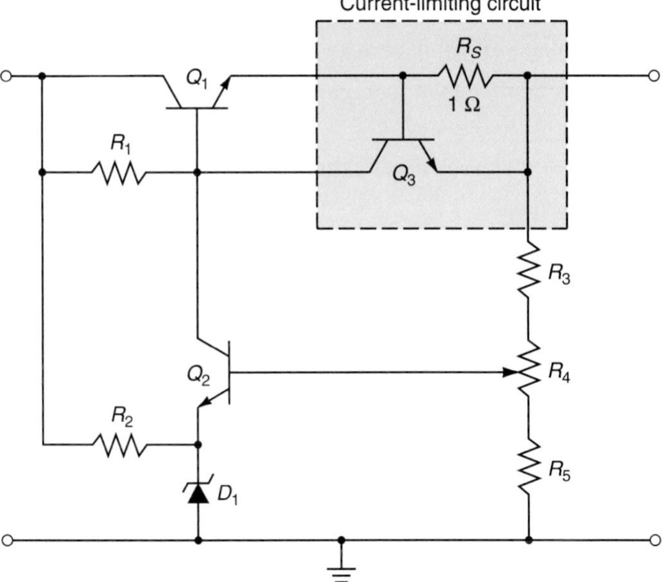

Figure 25.7 Series regulator current-limiting circuit.

like the one shown in Figure 25.7 can prevent this potential problem. For Q_3 to conduct, the voltage across R_S must reach approximately 0.7 V. This happens when

$$I_E = \frac{V_{BE(Q3)}}{R_S} = \frac{0.7\ \text{V}}{1\ \Omega} = 700\ \text{mA}$$

where I_E = the series current through the pass transistor. When I_E is less than 700 mA, Q_3 is in cutoff, and the circuit operates like any other series feedback regulator. If a shorted load causes I_E to increase above approximately 700 mA, Q_3 conducts, decreasing $V_{B(Q1)}$. This limits the conduction through the pass transistor, preventing any damage to the component. Note that the maximum current that can be drawn through the pass-transistor can be found as

$$I_{E(\max)} \cong \frac{V_{BE(Q3)}}{R_S} \tag{25.6}$$

Thus, the maximum allowable current through the pass transistor is set by using the appropriate value of R_S.

One Final Note

Another drawback to the series regulator is the fact that I_L passes through the pass transistor, which dissipates a significant amount of power. Also, the voltage drop across the pass transistor reduces the maximum output voltage. Even so, for reasons that you will see in the next section, the series regulator is still used more commonly than the shunt regulator.

Section Review

1. Describe the basic *pass-transistor regulator.*
2. Describe how the pass-transistor regulator shown in Figure 25.4 responds to a change in load resistance.
3. Explain how the *Darlington pass-transistor regulator* reduces the problem of excessive zener diode power dissipation.
4. Describe the response of the *series feedback regulator* to a change in load resistance.

5. Describe the operation of the current-limiting circuit shown in Figure 25.7.

6. List the disadvantages of using *series voltage regulators.*

Critical Thinking

7. How would the circuit shown in Figure 25.3 respond to slight variations in *line voltage*?

8. How would the circuit shown in Figure 25.6 respond to slight variations in *line voltage*?

25.3 Shunt Voltage Regulators

A **shunt regulator** is *a circuit that has a regulating transistor in parallel with the load.* In this section, we will look at the operation of the most common type of shunt regulator, the *shunt feedback regulator.*

Shunt Feedback Regulator

OBJECTIVE 6 ➤ The **shunt feedback regulator** is *a circuit that uses an error detection circuit to control the conduction through a shunt regulator transistor.* As shown in Figure 25.8, the shunt regulator transistor (Q_1) is connected to the two sides of the load. The sample and adjust circuit is (once again) a simple voltage divider. The reference circuit is made up of D_1 and R_1. The sample and reference voltages are applied to the base and emitter terminals of the error detector (Q_2), which controls the conduction through the shunt transistor.

The operation of the shunt feedback regulator is easiest to understand if we view the regulator transistor (Q_1) as being a variable resistor, R_{ce}. The value of R_{ce} varies *inversely* with the conduction of Q_1. By design, the value of R_{ce} is approximately midway between its extremes. If R_L *decreases,* V_L begins to decrease. The decrease in V_L causes V_{R3} to decrease, which decreases the conduction through both Q_2 and Q_1. The decreased conduction through Q_1 increases the value of R_{ce}, which increases the voltage at the collector of Q_1. Since $V_L = V_{C(Q1)}$, the increase in $V_{C(Q1)}$ offsets the initial decrease in R_L and V_L.

If R_L *increases,* V_L begins to increase, as does the conduction through Q_2 and Q_1. This reduces the value of R_{ce}, which decreases the values of $V_{C(Q1)}$ and V_L. Again, the circuit has offset the initial changes in R_L and V_L.

Overvoltage Protection

OBJECTIVE 7 ➤ Just as the series regulator must be protected against a shorted load, the shunt regulator must be protected from *input overvoltage conditions.* As the dc input voltage increases,

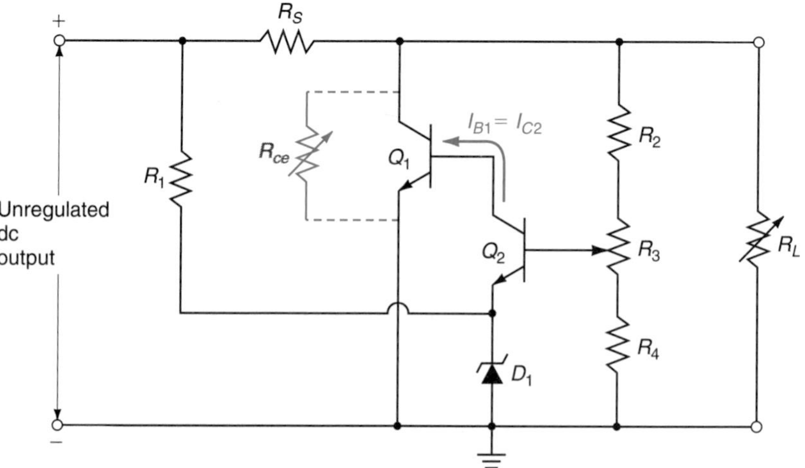

Figure 25.8 A shunt feedback regulator.

so does the conduction of Q_1. If the conduction through Q_1 increases too much, the power-handling capability of the device may be exceeded.

To ensure that any increase in unregulated dc input voltage will not damage the shunt transistor, the transistor used must have a $P_{D(max)}$ rating far greater than the maximum power dissipation than would ever be expected. Of course, there are always circumstances which can cause the unregulated dc input voltage to exceed its maximum rated value. To protect the circuit from such a circumstance, a *crowbar* circuit can be added to the input of the regulator. A **crowbar** is *a fail-safe circuit that is designed to protect a voltage-sensitive load (or circuit) from excessive input voltages.* (This type of circuit is discussed in Chapter 26.)

Another practical consideration involves the potentiometer, R_3. This potentiometer is included in the circuit to adjust for the dc output voltage. Adjusting R_3 varies the conduction of Q_1 and, thus, the value of R_{ce}. Because R_{ce} affects the value of V_L, varying R_3 sets the value of the regulated dc output voltage.

One Final Note

◄ *OBJECTIVE 8*

The primary problem with the shunt regulator is that it is a rather inefficient design. Both R_S and Q_1 dissipate a lot of power that is essentially wasted. Further, an input overvoltage condition is far more likely to occur than a shorted-load condition. In other words, the fault that can damage the shunt regulator is more likely to occur. This is why the series regulator is preferred over the shunt regulator.

Section Review

1. Describe the response of the shunt feedback regulator to a *decrease* in load resistance.
2. Describe the response of the shunt feedback regulator to an *increase* in load resistance.
3. Explain how an input overvoltage condition can destroy the regulator transistor in a shunt feedback regulator.
4. Discuss how a shunt regulator can be protected from an input overvoltage problem.
5. Why are *series regulators* preferred over *shunt regulators*?
6. Which of the two transistors shown in Figure 25.8 requires a higher power dissipation rating? Why? *Critical Thinking*

25.4 Linear IC Voltage Regulators

A **linear IC voltage regulator** is *a device that is used to hold the output voltage from a dc power supply relatively constant over a specified range of line and load variations.* Most IC voltage regulators are three-terminal devices, though some have more than three terminals. The schematic symbol for a three-terminal regulator is shown in Figure 25.9.

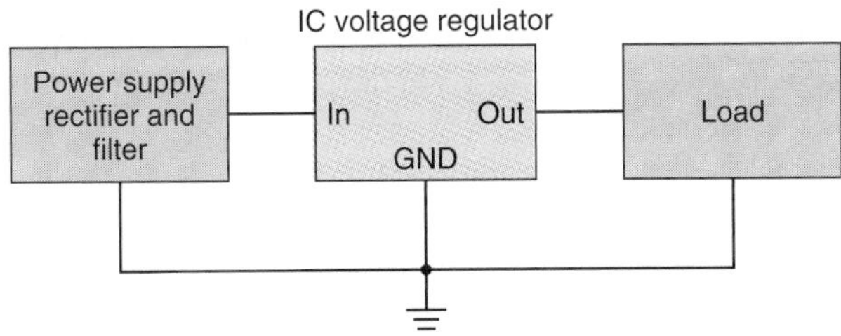

Figure 25.9 Schematic symbol for a three-terminal regulator.

OBJECTIVE 9 ➤ There are basically four types of IC voltage regulators: *fixed positive, fixed negative, adjustable,* and *dual tracking.* The **fixed-positive** and **fixed-negative regulators** provide specific output voltages. The **adjustable regulator** can be adjusted to provide any dc output voltage within its two specified limits. Both positive and negative variable regulators are available. The **dual-tracking regulator** provides equal positive and negative output voltages. Adjustable dual-tracking regulators are also available.

Regardless of the type of regulator used, the regulator input polarity must match the device's rated output polarity. In other words, positive regulators must have a positive input, and negative regulators must have negative input. Dual-tracking regulators require both positive and negative input voltages.

IC Regulator Specifications

OBJECTIVE 10 ➤ For our discussion on IC regulator specifications, we'll use the spec sheet for the LM317 voltage regulator. This device is an adjustable regulator whose output can be varied between +1.2 V and +32 V. The spec sheet for the LM317 is shown in Figure 25.10.

The **input/output voltage differential** rating indicates *the maximum difference between* V_{in} *and* V_{out} *that can occur without damaging the device.* For the LM317, this rating is 40 V. The differential voltage rating can be used to determine the maximum allowable value of V_{in} as follows:

$$V_{in(max)} = V_{out(adj)} + V_d \tag{25.7}$$

where

$V_{in(max)}$ = the maximum allowable unregulated dc input voltage
$V_{out(adj)}$ = the adjusted output voltage of the regulator
V_d = the input/output voltage differential rating of the regulator

This relationship is illustrated in Example 25.3.

Example 25.3

The LM317 is adjusted to provide a +8 V output. Determine its maximum allowable input voltage.

Solution: With a V_d rating of 40 V, the maximum allowable value of V_{in} is found as

$$V_{in(max)} = V_{out(adj)} + V_d = 8\text{ V} + 40\text{ V} = 48\text{ V}$$

Practice Problem 25.3

An adjustable IC voltage regulator has a V_d rating of 32 V. Determine its maximum allowable input voltage when adjusted for a +6 V output.

The line regulation rating of the LM317 is 0.04%/V (maximum). This rating is measured under the following conditions: $T_A = 25°\text{C}$, and $3\text{ V} \leq V_{in} - V_{out} \leq 40\text{ V}$. This indicates that the difference between the input and output voltages can be no less than 3 V and no greater than 40 V.

The load regulation rating for the LM317 depends on the value of V_{out}. If $V_{out} <$ +5 V, the output will not vary by more than 70 mV when the load current varies between 10 mA to approximately 1.5 A. If $V_{out} \geq +5$ V, the output voltage will change by no more than 1.5% under the same conditions.

The **minimum load current** rating indicates *the minimum allowable load current.* If the load current drops below a specified value, regulation of the output voltage is lost. For the LM317, this rating is 12 mA (maximum). The **ripple rejection ratio** is *the ability of the regulator to block any ripple voltage at its input.* For the LM317, input ripple is reduced by 60 dB (typical) at the output (a factor of approximately 1000).

Absolute Maximum Ratings

Power Dissipation	Internally limited
Input—Output Voltage Differential	40V
Operating Junction Temperature Range	
LM117	−55°C to +150°C
LM217	−25°C to +150°C
LM317	0°C to +125°C
Storage Temperature	−65°C to +150°C
Lead Temperature (Soldering, 10 seconds)	300°C

Preconditioning

Burn-In in Thermal Limit	**100% All Devices**

Electrical Characteristics (Note 1)

PARAMETER	CONDITIONS	LM117/217 MIN	LM117/217 TYP	LM117/217 MAX	LM317 MIN	LM317 TYP	LM317 MAX	UNITS
Line Regulation	$T_A = 25°C$, $3V \le V_{IN} - V_{OUT} \le 40V$ (Note 2)		0.01	0.02		0.01	0.04	%/V
Load Regulation	$T_A = 25°C$, $10\text{ mA} \le I_{OUT} \le I_{MAX}$							
	$V_{OUT} < 5V$, (Note 2)		5	15		5	25	mV
	$V_{OUT} \ge 5V$, (Note 2)		0.1	0.3		0.1	0.5	%
Thermal Regulation	$T_Z = 25°C$, 20 ms Pulse		0.03	0.07		0.04	0.07	%/W
Adjustment Pin Current			50	100		50	100	μA
Adjustment Pin Current Change	$10\text{ mA} \le I_L \le I_{MAX}$ $3V \le (V_{IN} - V_{OUT}) \le 40V$		0.2	5		0.2	5	μA
Reference Voltage	$3V \le (V_{IN} - V_{OUT}) \le 40V$, (Note 3) $10\text{ mA} \le I_{OUT} \le I_{MAX}$, $P \le P_{MAX}$	1.20	1.25	1.30	1.20	1.25	1.30	V
Line Regulation	$3V \le V_{IN} - V_{OUT} \le 40V$, (Note 2)		0.02	0.05		0.02	0.07	%/V
Load Regulation	$10\text{ mA} \le I_{OUT} \le I_{MAX}$, (Note 2)							
	$V_{OUT} \le 5V$		20	50		20	70	mV
	$V_{OUT} \ge 5V$		0.3	1		0.3	1.5	%
Temperature Stability	$T_{MIN} \le T_j \le T_{MAX}$		1			1		%
Minimum Load Current	$V_{IN} - V_{OUT} = 40V$		3.5	5		3.5	10	mA
Current Limit	$V_{IN} - V_{OUT} \le 15V$							
	K and T Package	1.5	2.2		1.5	2.2		A
	H and P Package	0.5	0.8		0.5	0.8		A
	$V_{IN} - V_{OUT} = 40V$, $T_j = +25°C$							
	K and T Package	0.03	0.4		0.15	0.4		A
	H and P Package	0.15	0.07		0.075	0.07		A
RMS Output Noise, % of V_{OUT}	$T_A = 25°C$, $10\text{ Hz} \le f \le 10\text{ kHz}$		0.003			0.003		%
Ripple Rejection Ratio	$V_{OUT} = 10V$, $f = 120$ kHz		65			65		dB
	$C_{ADJ} = 10\mu F$	66	80		66	80		dB
Long-Term Stability	$T_A = 125°C$		0.3	1		0.3	1	%
Thermal Resistance, Junction to Case	H Package		12	15		12	15	°C/W
	K Package		2.3	3		2.3	3	°C/W
	T Package					4		°C/W
	P Package					12		°C/W

Note 1: Unless otherwise specified, these specifications apply $-55°C \le T_j \le +150°C$ for the LM117, $-25°C \le T_j \le +150°C$ for the LM217, and $0°C \le T_j \le +125°C$ for the LM317; $V_{IN} - V_{OUT} = 5V$; and $I_{OUT} = 0.1A$ for the TO-39 and TO-202 packages and $I_{OUT} = 0.5A$ for the TO-3 and TO-220 packages. Although power dissipation is internally limited, these specifications are applicable for power dissipations of 2W for the TO-39 and TO-202, and 20W for the TO-3 and TO-220. I_{MAX} is 1.5A for the TO-3 and TO-220 packages and 0.5A for the TO-39 and TO-202 packages.

Note 2: Regulation is measured at constant junction temperature, using pulse testing with a low duty cycle. Changes in output voltage due to heating effects are covered under the specification for thermal regulation.

Note 3: Selected devices with tightened tolerance reference voltage available.

LIFE SUPPORT POLICY

FAIRCHILD'S PRODUCTS ARE NOT AUTHORIZED FOR USE AS CRITICAL COMPONENTS IN LIFE SUPPORT DEVICES OR SYSTEMS WITHOUT THE EXPRESS WRITTEN APPROVAL OF THE PRESIDENT OF FAIRCHILD SEMICONDUCTOR INTERNATIONAL. As used herein:

1. Life support devices or systems are devices or systems which, (a) are intended for surgical implant into the body, or (b) support or sustain life, and (c) whose failure to perform when properly used in accordance with instructions, for use provided in the labeling, can be reasonably expected to result in a significant injury of the

2. A critical component in any component of a life support device or system whose failure to perform can be reasonably expected to cause the failure of the life support device or system, or to affect its safety or effectiveness.

Figure 25.10 LM317 spec sheet. (Courtesy of Fairchild Semiconductor.)

Output Voltage Adjustment

The output voltage of the LM317 is adjusted using a voltage divider, as shown in Figure 25.11. The *complete* LM317 spec sheet (not shown here) contains the following equation for determining the regulated dc output voltage for the circuit:

$$V_{dc} \cong 1.25\left(\frac{R_2}{R_1} + 1\right) \tag{25.8}$$

The use of this equation is demonstrated in Example 25.4.

Example 25.4

Note: R_2 (in Figure 25.11) is used to set the regulator output voltage. The pin labeled ADJ (for *adjust*) is the same as the pin labeled GND in Figure 25.9.

R_2 in Figure 25.11 is adjusted to 2.4 kΩ. Determine the regulated dc output voltage for the circuit.

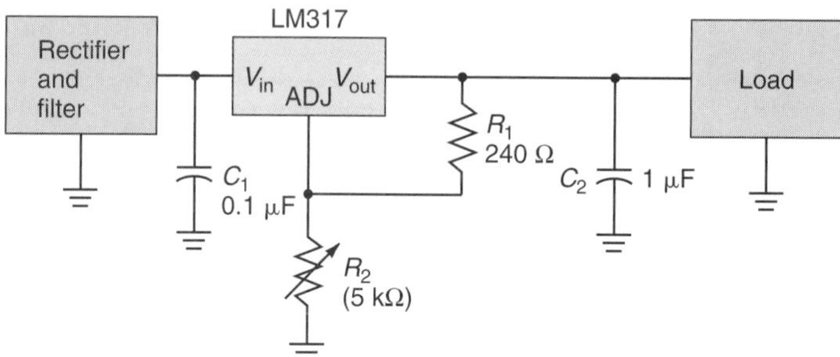

Figure 25.11

Solution: The regulated dc output voltage is found as

$$V_{dc} = 1.25\left(\frac{R_2}{R_1} + 1\right) = 1.25\left(\frac{2.4\text{ k}\Omega}{240\ \Omega} + 1\right) = 13.75\text{ V}$$

Practice Problem 25.4

R_2 in Figure 25.11 is adjusted to 1.68 kΩ. Determine the regulated dc output voltage for the LM317.

Note that the output voltage equation for an adjustable regulator is always provided on its spec sheet.

Figure 25.11 shows shunt capacitors connected to the regulator input and output pins. The input capacitor is used to prevent the input ripple from driving the regulator into self-oscillations. The output capacitor is used to improve the ripple reduction of the regulator.

Linear IC Regulator Applications: A Complete Dual-Polarity Power Supply

Figure 25.12 shows a complete dual-polarity power supply. The circuit uses matched fixed-positive and fixed-negative regulators to provide equal $+V_{dc}$ and $-V_{dc}$ outputs. As shown, the bridge rectifier is connected to a center-tapped transformer. The rectifier is wired in this fashion so that each regulator can respond independently to unequal load demands.

Capacitors C_1 and C_2 are the filter capacitors, typically in the mid- to high-microfarad range. C_3 and C_4 are the regulator input shunt capacitors, typically less than 1 μF in value. C_5 and C_6 are output ripple reduction capacitors and have values in the neighborhood of 1 μF.

One Final Note

Linear IC voltage regulators are extremely common. While we cannot possibly hope to cover the operation of every type, you should be able to deal with the common IC

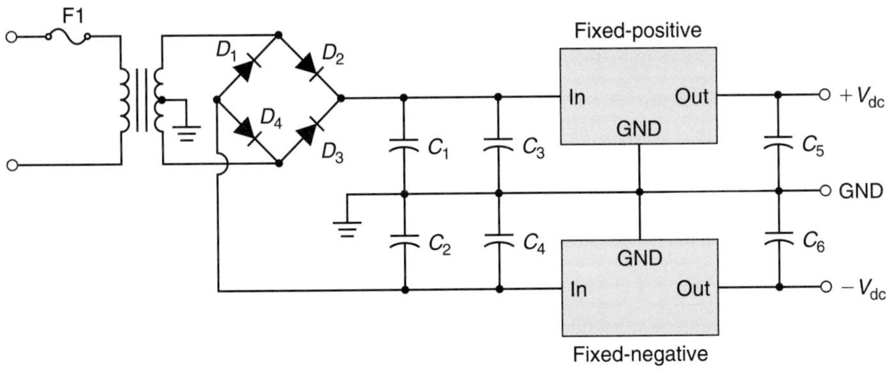

Figure 25.12 A complete dual-polarity power supply.

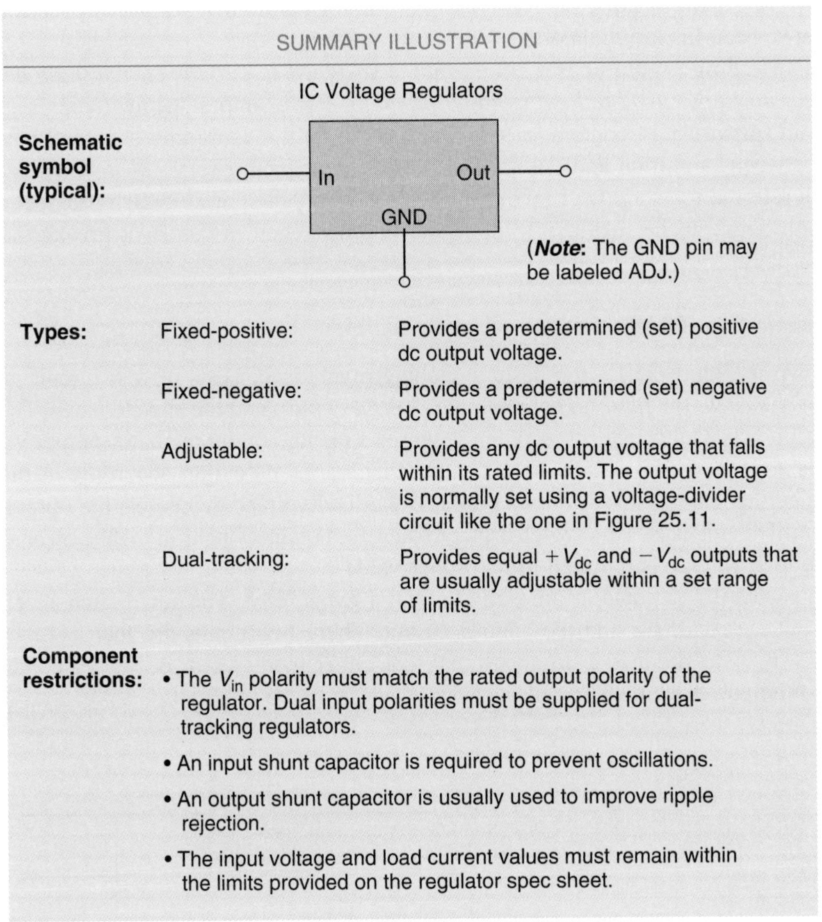

SUMMARY ILLUSTRATION

IC Voltage Regulators

Schematic symbol (typical):

(***Note***: The GND pin may be labeled ADJ.)

Types:		
	Fixed-positive:	Provides a predetermined (set) positive dc output voltage.
	Fixed-negative:	Provides a predetermined (set) negative dc output voltage.
	Adjustable:	Provides any dc output voltage that falls within its rated limits. The output voltage is normally set using a voltage-divider circuit like the one in Figure 25.11.
	Dual-tracking:	Provides equal $+V_{dc}$ and $-V_{dc}$ outputs that are usually adjustable within a set range of limits.

Component restrictions:

- The V_{in} polarity must match the rated output polarity of the regulator. Dual input polarities must be supplied for dual-tracking regulators.
- An input shunt capacitor is required to prevent oscillations.
- An output shunt capacitor is usually used to improve ripple rejection.
- The input voltage and load current values must remain within the limits provided on the regulator spec sheet.

Figure 25.13

voltage regulators. The characteristics of these regulators are summarized for you in Figure 25.13.

Section Review

1. What is an IC voltage regulator?
2. List and describe the four types of IC voltage regulators.
3. What *input polarity* (or polarities) is (are) required for each type of voltage regulator?
4. What is the *input/output differential* rating?

5. What is the *minimum load current* rating?
6. What is the *ripple rejection ratio* rating?
7. What type of circuit is normally used to provide the adjustment to the dc output voltage from an *adjustable regulator*?
8. Where can you find the V_{dc} equation for a given adjustable regulator?

25.5 Switching Regulators

OBJECTIVE 11 ➤ There are two fundamental types of voltage regulators: *linear regulators* and *switching regulators.* The **linear regulator** is *a circuit that provides a continuous path for current between the regulator input and the load.* All the regulators that we have covered to this point have been linear regulators.

The **switching regulator** is *a circuit that is designed so that the current path between the regulator input and the load is not continuous.* For example, the switching regulator shown in Figure 25.14 contains a pass transistor (Q_1) that is rapidly switched back and forth between saturation and cutoff. *The pass transistor in a series switching regulator* is often referred to as a **power switch.** When saturated, the power switch provides a current path between the regulator's input and output. When in cutoff, it breaks the conduction path.

Switching Regulator Operation

OBJECTIVE 12 ➤ As shown in Figure 25.15, the basic switching regulator consists of four circuit groups: the *power switch,* the *filter* and *clipper,* the *control circuit,* and the *switch driver.* The control circuit acts as a *comparator.* When it senses a change in the regulator output voltage, it sends a signal to the switch driver, which then varies its output according to the type of change that has occurred. For example, if the load voltage decreases, the output from the control circuit causes the switch driver to increase conduction through the power switch, and the load voltage is returned to its proper value.

OBJECTIVE 13 ➤ The switch driver contains an oscillator and a *gated latch.* The gated latch accepts inputs from the oscillator and the control circuit. These two inputs are then combined in such a way as to produce a driving signal that either increases or decreases conduction through the power switch.

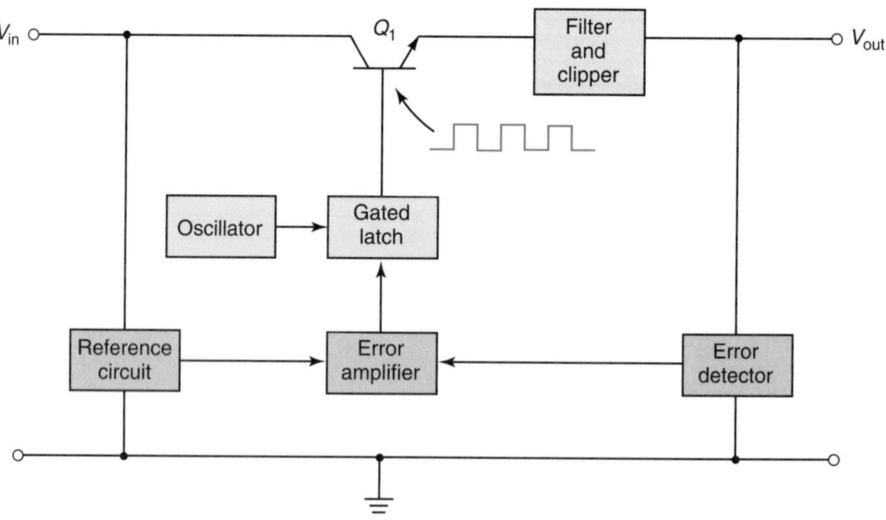

Figure 25.14 A typical switching regulator.

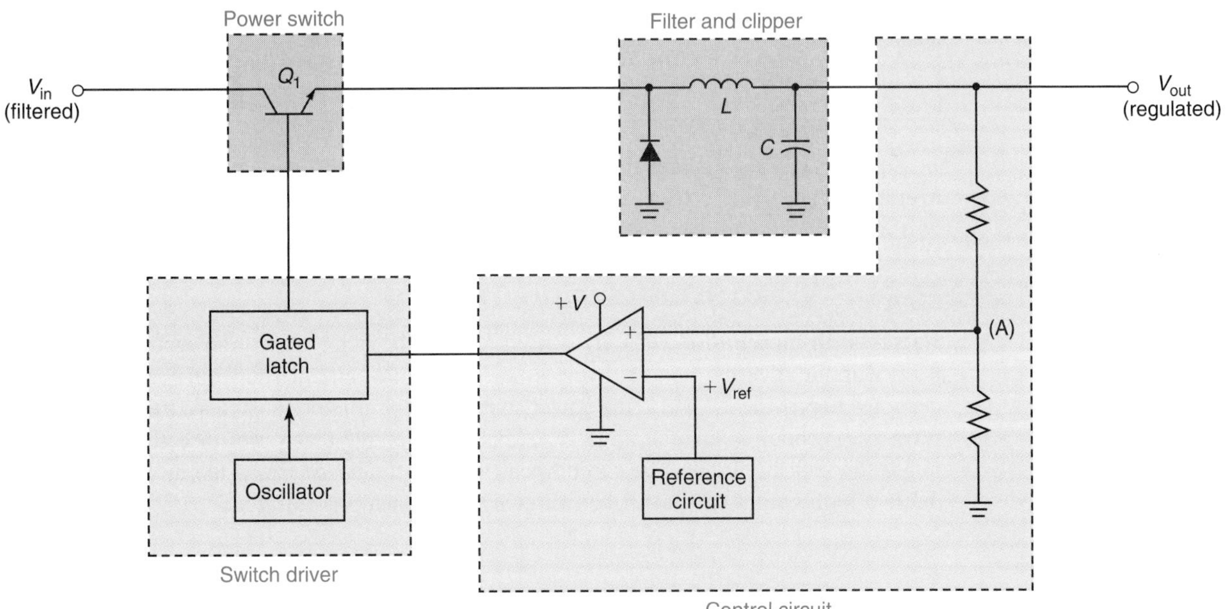

Figure 25.15 A basic switching regulator.

The output of the power switch is (more or less) a rectangular waveform. The filter-and-clipper circuit is designed to respond to this waveform as follows:

- The capacitor opposes any change in voltage, keeping the load voltage relatively constant.
- The inductor opposes any change in current, keeping the load current relatively constant.
- The diode clips the counter emf produced by the *LC* circuit, providing transient protection for the power switch (Q_1).

Now, let's take a look at the overall response of the regulator to a change in load resistance. If load resistance *decreases,* V_{out} begins to decrease, as does the voltage at point (A) in the control circuit. Since point (A) is connected to the comparator, the output from the comparator decreases, which signals the switch driver to increase conduction through the power switch. The increase in power switch conduction offsets the initial decrease in load resistance, returning V_{out} to its normal value. The circuit response to an *increase* in load resistance is exactly the opposite of the sequence described above.

Controlling Power Switch Conduction

As you know, the power switch is constantly driven back and forth between saturation and cutoff. The average (dc) value of the waveform produced at the emitter of the power switch can be found as

$$V_{ave} = V_{in}\left(\frac{T_{on}}{T_{on} + T_{off}}\right) \tag{25.9}$$

Note: The fraction in equation (25.9) is the *duty cycle* of the waveform (written as a ratio rather than a percentage). Therefore, V_{ave} can be described as the product of input voltage and duty cycle.

Example 25.5 demonstrates the use of equation (25.9).

Example 25.5

The regulator shown in Figure 25.14 has the following values: $V_{in} = 24$ V, $T_{on} = 5$ μs, and $T_{off} = 10$ μs. Calculate the dc average of the load voltage.

Solution: Using the values given, the average (dc) load voltage can be found as

$$V_{ave} = V_{in}\left(\frac{T_{on}}{T_{on} + T_{off}}\right) = (24\text{ V})\left(\frac{5\ \mu\text{s}}{15\ \mu\text{s}}\right) = 8\text{ V}$$

> ***Practice Problem 25.5***
>
> A regulator like the one shown in Figure 25.14 has the following values: V_{in} = 36 V, T_{on} = 6 μs, and T_{off} = 9 μs. Calculate the value of V_{ave} for the circuit.

Equation (25.9) indicates that varying the conduction of the power switch controls the average output voltage from a switching regulator. At this point, we will take a look at two methods that are used to control the conduction of the power switch.

Pulse-Width Modulation (PWM)

OBJECTIVE 14 ➤ **Pulse-width modulation (PWM)** is used to vary the pulse width of a rectangular waveform without affecting its total cycle time. The control circuit shown in Figure 25.16a is designed to provide PWM for the power switch. The oscillator generates a triangular waveform (V_o), while the error voltage (V_{error}) is a dc voltage that varies directly with the input to the control circuit. The gated latch (which actually performs the modulation) produces a high output whenever the following relationship is fulfilled:

$$V_o \geq V_{error}$$

For example, take a look at the first set of waveforms shown in Figure 25.16b. As you can see, the control voltage (V_C) goes high as V_o increases beyond the value of V_{error}. Then, as V_o drops below the value of V_{error}, the control voltage drops back to 0 V. Note that the control voltage is nearly a square wave.

Now take a look at the second set of waveforms shown in Figure 25.16b. If you compare the voltages shown to those in the first set of waveforms, you'll see that:

- The error voltage (V_{error}) has increased in value.

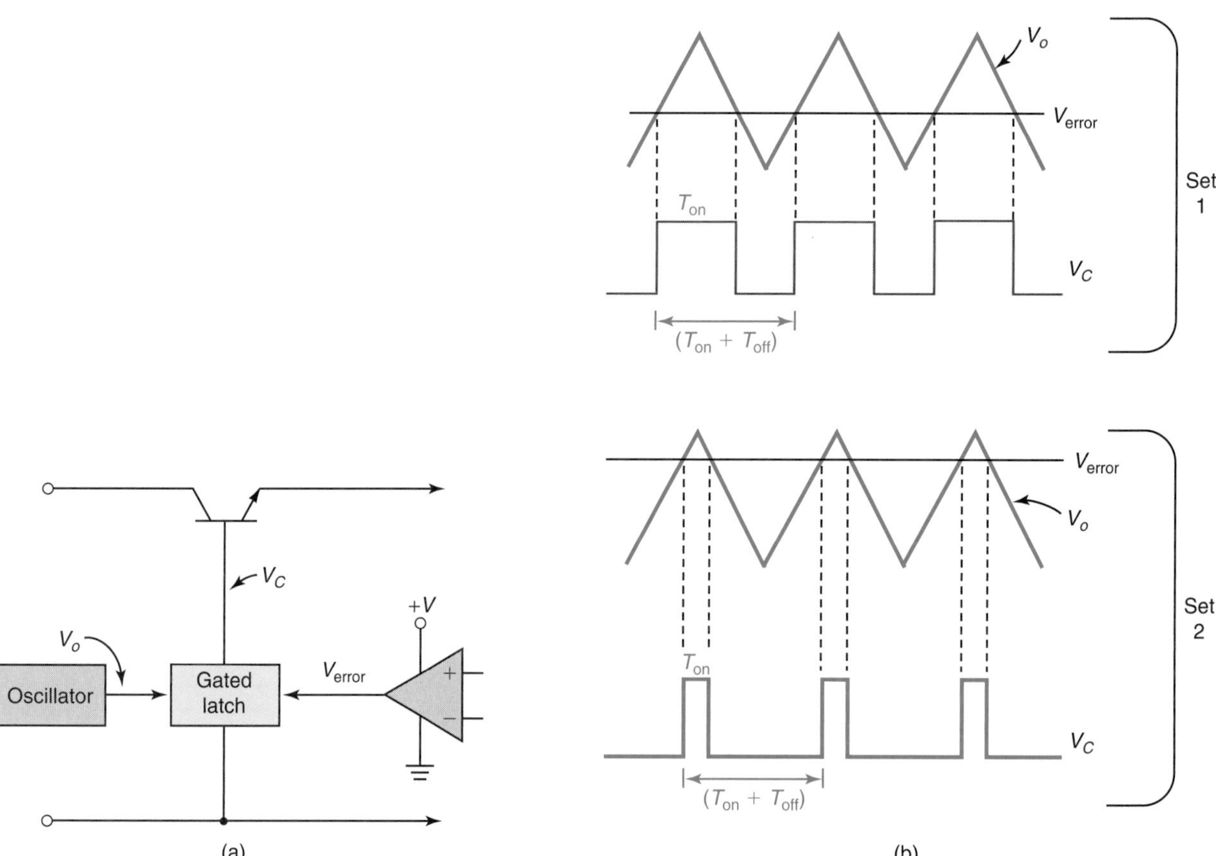

Figure 25.16 Pulse-width modulation.

- The pulse width (T_{on}) of the control voltage has decreased significantly.
- The total cycle time ($T_{on} + T_{off}$) has not changed.

Since the value of T_{on} has decreased, the average value of V_C has also decreased. Likewise, a decrease in V_{error} causes an increase in the average value of V_C. This is how PWM (as it applies to switching regulators) works.

Variable Off-Time Modulation

In Figure 25.17, the pulse width of the control voltage is fixed and the total cycle time is variable. This is known as **variable off-time modulation.** The gated latch accepts inputs from the error detector and an astable multivibrator. The square-wave output from the oscillator is gated to the power switch when the V_{error} is high, and it is blocked when V_{error} is low.

Figure 25.17b shows that the pulse width (T_{on}) does not change; however, the total cycle time ($T_{on} + T_{off}$) is altered by the V_{error} input. As the total cycle time varies, so does the average output from the power switch. Again, the average output voltage from the regulator is controlled.

Switching Regulator Configurations

So far, we have discussed only the **step-down regulator,** which produces a dc load voltage that is less than (or equal to) its rectified input voltage. The **step-up regulator** is *a switching regulator that provides a dc load voltage that is greater than its rectified input voltage.* As shown in Figure 25.18a, the power switch in a step-up regulator is a shunt component, and an inductor is placed in series with the input. During the on-time of the

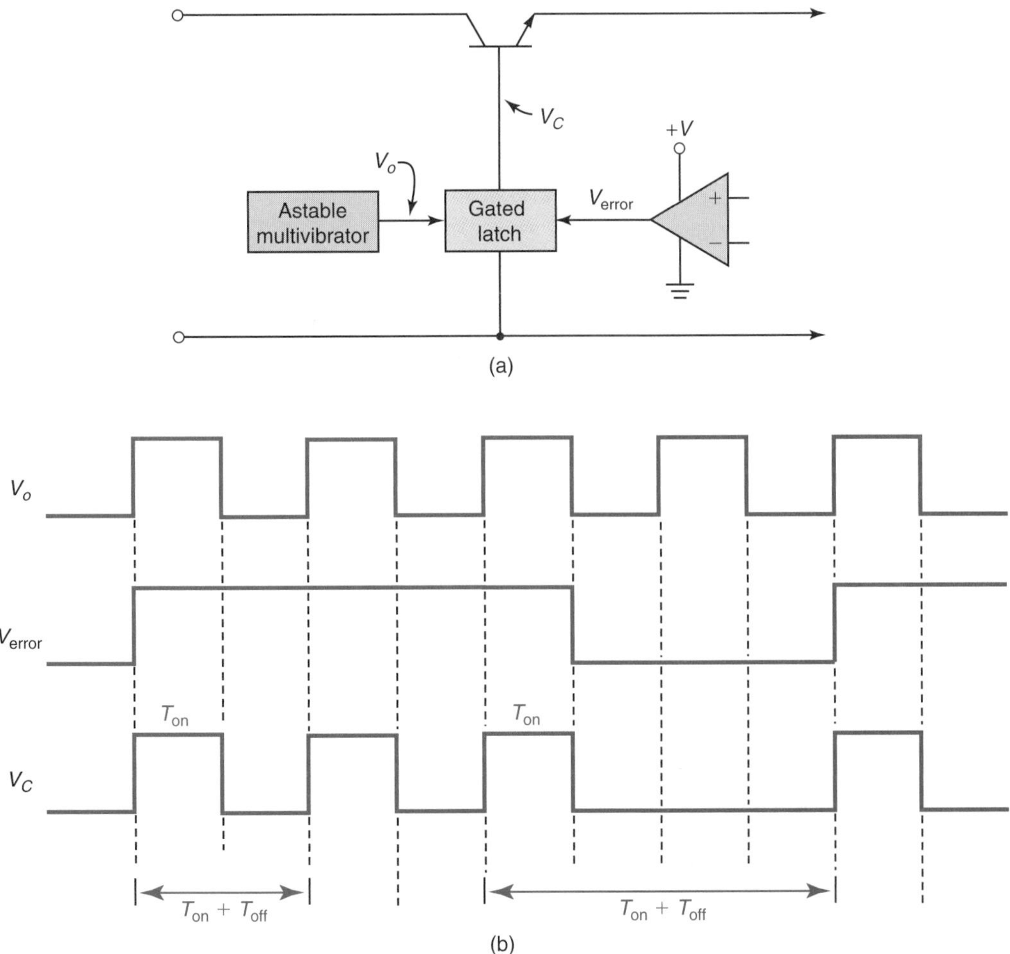

Figure 25.17 Variable off-time modulation.

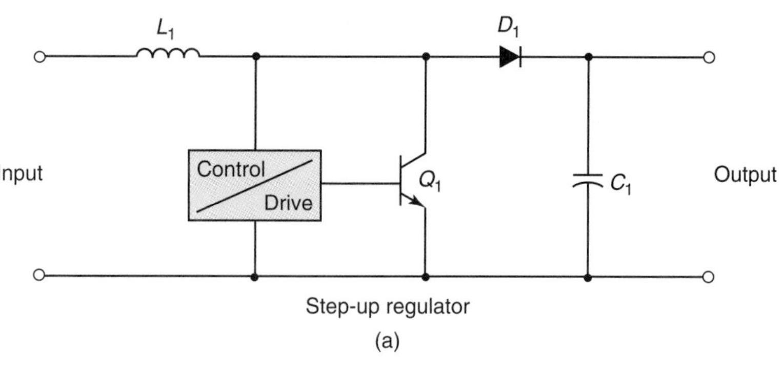

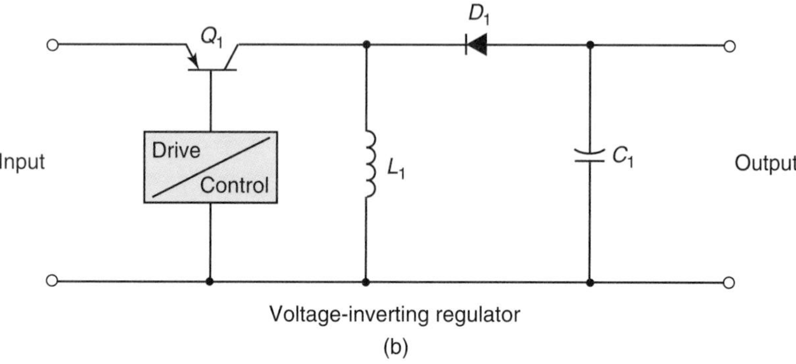

Figure 25.18 Basic switching regulator configurations.

transistor, the current through L_1 induces a voltage across its terminals. This voltage adds to the value of V_{in}, making it possible for the output voltage to be greater than the input voltage.

The **voltage-inverting regulator** shown in Figure 25.18b reverses the polarity of its rectified input voltage. Note the series power switch and shunt inductor connections characteristic of this configuration.

OBJECTIVE 15 ➤ Several points should be made regarding all these configurations. The control and switch driver circuits may vary from one switching regulator to another, but you should be able to determine which type of regulator you're dealing with by noting the position of the power switch and inductor, as follows:

- If the power switch is in series with the input and the inductor, it is a step-down regulator.
- If the power switch is a shunt component placed after the inductor, it is a step-up regulator.
- If the power switch is in series with the input and the inductor is a shunt component, it is a voltage-inverting regulator.

Another important point is that no "cookbook" equations are available to help with the analysis of these circuits. Switching regulators are complex circuits that can take as long as six worker-months to design, build, and debug. The detailed analysis of these circuits can be challenging.

IC Switching Regulators

OBJECTIVE 16 ➤ In many cases, the control functions and power switching of a switching regulator are handled by a single IC. A step-down regulator circuit using the Motorola MC34063 is shown in Figure 25.19. Note that the transistor labeled Q_1 in the IC is the power switch.

This particular circuit was designed to produce a 12 V/750 mA output with a 36 V rectified input. Note the positioning of the power switch and inductor, which is characteristic of a step-down regulator. Also, note the excellent efficiency rating of this circuit.

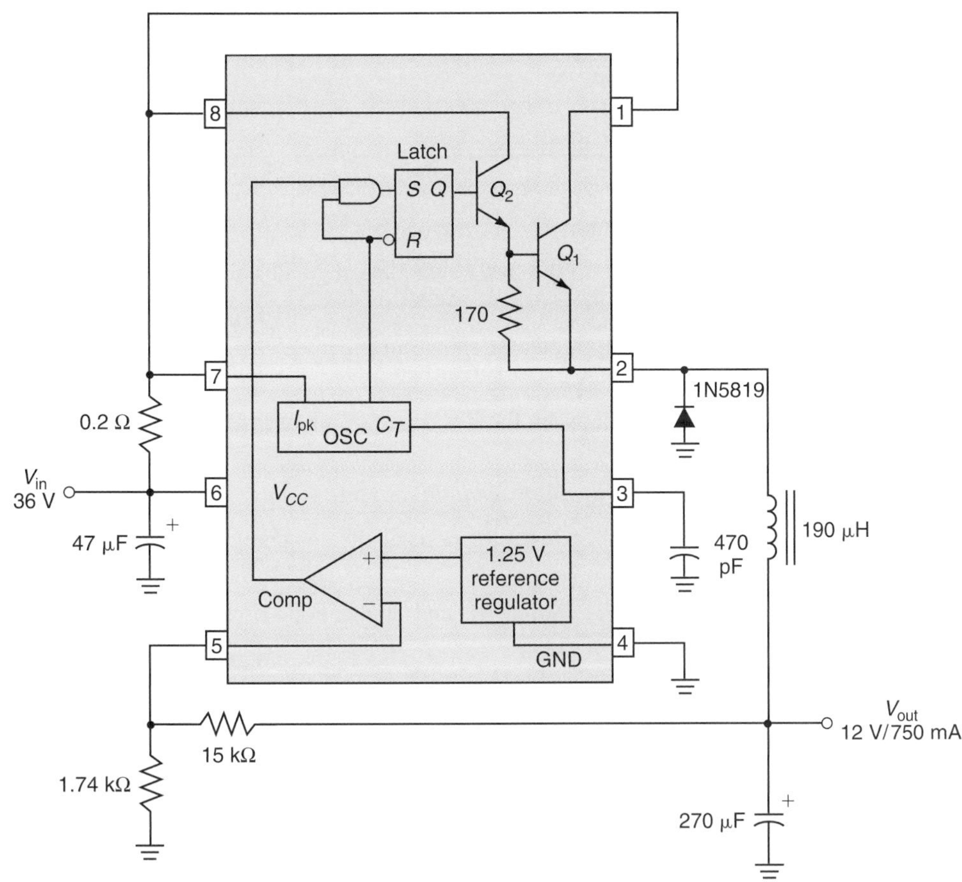

Test	Conditions	Results
Line regulation	V_{in} = 20 to 40 V, I_{out} = 750 mA	Δ = 15 mV or ± 0.063%
Load regulation	V_{in} = 36 V, I_{out} = 100 to 750 mA	Δ = 40 mV or ± 0.17%
Output ripple	V_{in} = 36 V, I_{out} = 750 mA	60 $mV_{p\text{-}p}$
Short circuit current	V_{in} = 36 V, R_L = 0.1 Ω	1.6 A
Efficiency	V_{in} = 36 V, I_{out} = 750 mA	89.5%

A maximum power transfer of 9.0 watts is possible from an 8-pin dual-in-line package with V_{in} = 36 V and V_{out} = 12 V.

Figure 25.19 The MC34063 switching regulator. (Copyright of Semiconductor Components Industries, LLC. Used by permission.)

Switching Regulators: Advantages and Disadvantages

One of the primary advantages that switching regulators have over linear regulators is their higher efficiency. Linear regulators are generally limited to efficiency ratings of less than 60%, but switching regulators can easily achieve ratings of 90%. Since the power switch dissipates relatively little power, the maximum power rating of the switch can be much lower than the rated output of the regulator. ◀ *OBJECTIVE 17*

Switching regulators can be built in a variety of configurations (as you were shown earlier in this section). In contrast, linear regulators can be designed only as *step-down* circuits. Even so, switching regulators do have some distinct disadvantages. The power switch generates a significant amount of noise, so unshielded switching regulators cannot be used in low-noise applications. Further, switching regulators are slower to respond to a change in load demand, because of the time required for the control and drive circuits to respond to a change in the error-detection circuitry.

Finally, the design of a switching regulator is far more complex and time consuming than that of a similar linear regulator. This adds to the production cost of the switching regulator.

Section Review

1. In terms of conduction characteristics, what is the primary difference between *linear* and *switching* regulators?
2. List the circuits that make up the basic switching regulator and describe the function performed by each.
3. Describe how the switching regulator shown in Figure 25.15 responds to a change in load demand.
4. Describe the relationship between the *duty cycle* of a power switch and its *average output voltage.*
5. Compare and contrast *pulse-width modulation* with *variable off-time modulation.*
6. List the circuit-recognition features for each of the switching regulators shown in Figure 25.18.
7. Discuss the advantages, disadvantages, and applications of linear and switching regulators.

Key Terms

The following terms were introduced in this chapter on the pages indicated:

Adjustable regulator, 858
Crowbar, 857
Darlington pass-transistor regulator, 853
Dual-tracking regulator, 858
Fixed-negative regulator, 858
Fixed-positive regulator, 858
Input/output voltage differential, 858
Line regulation, 850
Linear IC voltage regulator, 857
Linear regulator, 862
Load regulation, 850
Minimum load current, 858
Pass-transistor regulator, 852
Power switch, 862
Pulse-width modulation, 864
Ripple rejection ratio, 858
Series feedback regulator, 853
Series regulator, 852
Shunt feedback regulator, 856
Shunt regulator, 856
Step-down regulator, 865
Step-up regulator, 865
Switching regulator, 862
Variable off-time modulation, 865
Voltage-inverting regulator, 866

Practice Problems

1. A voltage regulator experiences a 20 μV change in its output voltage when its input voltage changes by 4 V. Determine the line regulation rating of the circuit.
2. A voltage regulator experiences a 14 μV change in output voltage when its input voltage changes by 10 V. Determine the line regulation rating of the circuit.
3. A voltage regulator experiences a 15 μV change in its output voltage when its input voltage changes by 5 V. Determine the line regulation rating of the circuit.
4. A voltage regulator experiences a 12 mV change in output voltage when its input voltage changes by 12 V. Determine the line regulation rating of the circuit.
5. A voltage regulator is rated for an output current of 0 to 150 mA. Under no-load conditions, the output voltage of the circuit is 6 V. Under full-load conditions, the output from the circuit is 5.98 V. Determine the load regulation rating of the circuit.
6. A voltage regulator experiences a 20 mV change in output voltage when the load current increases from 0 to 50 mA. Determine the load regulation rating of the circuit.

7. A voltage regulator experiences a 1.5 mV change in output voltage when the load current increases from 0 to 20 mA. Determine the load regulation rating of the circuit.

8. A voltage regulator experiences a 14 mV change in output voltage when the load current increases from 0 to 100 mA. Determine the load regulation rating of the circuit.

9. Determine the approximate value of V_L for the circuit shown in Figure 25.20.

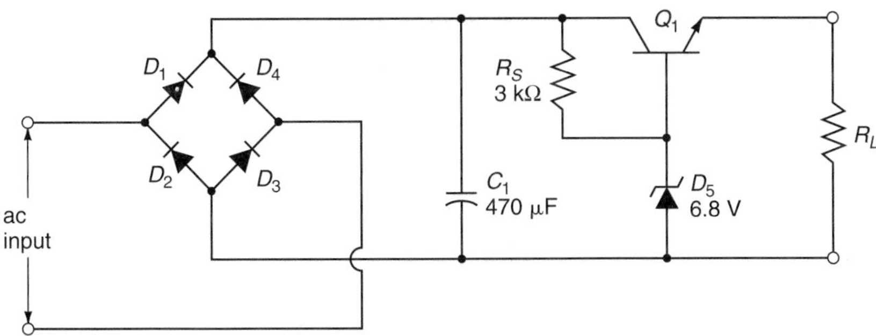

Figure 25.20

10. Determine the approximate value of V_L for the circuit shown in Figure 25.21.

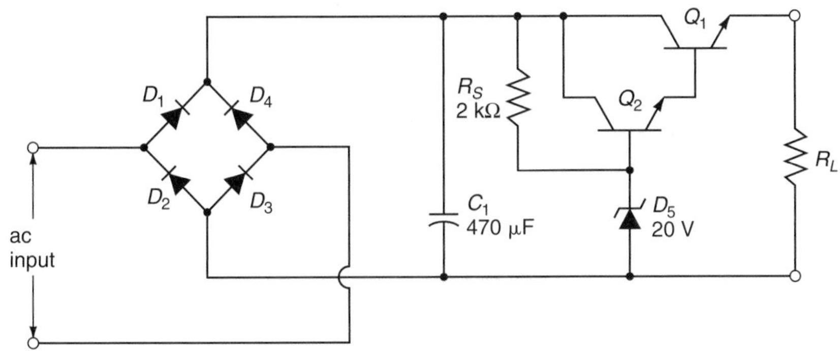

Figure 25.21

11. Determine the limit on the pass-transistor current for the circuit shown in Figure 25.22.

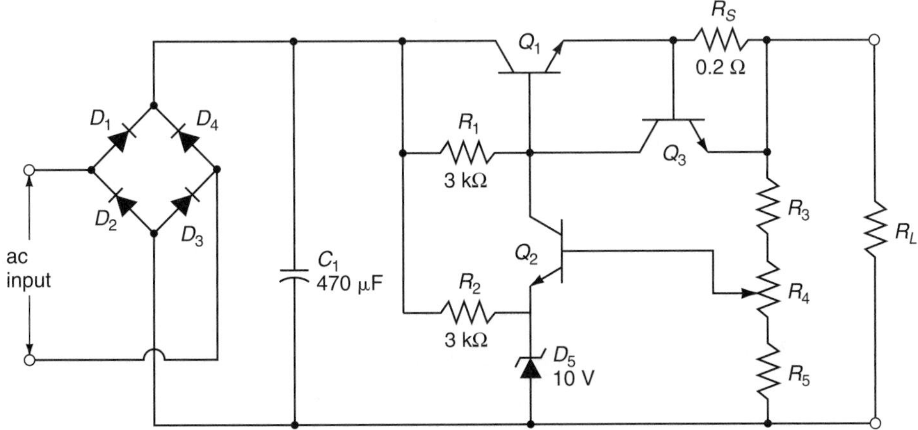

Figure 25.22

12. R_S in Figure 25.22 is changed to 1.2 Ω. Determine the limit on the pass-transistor current for the new circuit.

13. An adjustable IC voltage regulator is set for a +3 V output. The V_d rating for the device is 32 V. Determine the maximum allowable input voltage for the device.

14. An adjustable IC voltage regulator is set for a +6 V output. The V_d rating for the device is 24 V. Determine the maximum allowable input voltage for the device.

15. The LM317 is used in a circuit with adjustment resistance values of $R_1 = 330\ \Omega$ and $R_2 = 2.848\ \text{k}\Omega$ (adjusted potentiometer value). Determine the output voltage for the circuit.

16. The LM317 is used in a circuit with adjustment resistance values of $R_1 = 510\ \Omega$ and $R_2 = 6.834\ \text{k}\Omega$ (adjusted potentiometer value). Determine the output voltage for the circuit.

17. A power switch like the one shown in Figure 25.20 has the following values: $V_{in} =$ 36 V, $T_{on} = 12\ \mu\text{s}$, and $T_{off} = 48\ \mu\text{s}$. Determine the average output voltage from the power switch.

18. A power switch like the one shown in Figure 25.20 has the following values: $V_{in} =$ 24 V, $T_{on} = 10\ \mu\text{s}$, and $T_{off} = 40\ \mu\text{s}$. Determine the average output voltage from the power switch.

Looking Back

These problems relate to material presented in earlier chapters. The chapters are identified in brackets.

19. Calculate the wavelength of a 300 MHz sine wave. [Chapter 10]

20. A parallel LC circuit has values of $L = 100$ mH and $C = 10$ nF. Assuming that the winding resistance of the coil is 3 Ω, calculate R_P for the circuit at resonance. [Chapter 16]

21. A common-emitter amplifier has values of $I_E = 1$ mA, $R_C = 1.5\ \text{k}\Omega$, and $R_L =$ 7.5 kΩ. Calculate the value of A_v for the circuit. [Chapter 20]

22. Calculate the value of I_{ZM} for a 5.6 V, 2 W zener diode. [Chapter 19]

Pushing the Envelope

23. A +12 V regulator has a line regulation rating of 420 ppm/V. Determine the output voltage for this circuit when V_{in} increases by 10 V.

24. A voltage regulator has the following measured values: $V_{out} = 12.002$ V when $V_{in} =$ +20 V (rated maximum allowable input), and $V_{out} = 12$ V when $V_{in} = +10$ V (rated minimum allowable input). Determine the line regulation rating of the device in mV, %, and %/V.

25. A +5 V regulator has a load regulation rating of 2 Ω over a range of $I_L = 0$ to 100 mA (maximum). Express the load regulation of this circuit in V/mA, %, and %/mA.

26. A +15 V regulator has a 0.02 %/mA load regulation rating for a range of $I_L = 0$ to 50 mA (maximum). Assuming that the load current stays within its rated limits, determine the maximum load power that can be delivered by the regulator. (Assume that V_{out} increases as I_L increases.)

Answers to the Example Practice Problems

25.1 25 μV/V

25.2 100 μV/mA

25.3 38 V

25.4 10 V

25.5 14.4 V

Multisim File Reference

Figure	File Reference
25.11	ETF25_11.msm

chapter 26

Thyristors and Other Devices

Objectives

After studying the material in this chapter, you should be able to:

1. Describe the operation of the *silicon unilateral switch (SUS)*.
2. List and describe the methods commonly used to drive an SUS into cutoff.
3. Describe the construction of the *silicon-controlled rectifier (SCR)* and the methods used to trigger the device into conduction.
4. Discuss the problem of SCR *false triggering* and the methods commonly used to prevent it.
5. Discuss the use of the SCR as a *crowbar circuit* and a *phase controller.*
6. Discuss the similarities and differences between *diacs* and *triacs.*
7. List and describe the circuits commonly used to control triac triggering.
8. Describe the construction and operation of *unijunction transistors (UJTs).*
9. Compare and contrast *light emitters* and *light detectors.*
10. Discuss *wavelength* and *light intensity.*
11. Describe the operation and critical parameters of the common discrete photodetectors.
12. Discuss the operation and applications of *optoisolators* and *optointerrupters.*
13. State the purpose served by a *varactor.*
14. Describe the relationship between varactor *bias* and *junction capacitance.*
15. Describe the method by which a varactor can be used as the tuning component in a parallel (or series) *LC* circuit.
16. Describe a *surge* and list the common sources of power supply surges.
17. List the required characteristics of every surge-protection circuit.
18. Describe the forward characteristics of the *tunnel diode.*

In this chapter, we will look at three independent groups of electronic devices. First, we will discuss a group of components called *thyristors.* **Thyristors** are *devices designed specifically for high-power switching applications.* These components differ from other switching devices (such as BJTs and FETs) in that they are never used in linear applications (such as amplifiers). The second group of components we will discuss are known as **optoelectronic devices.** These are *devices that are controlled by or emit (generate) light.* Finally, we will take a brief look at several special applications diodes.

It should be noted that no central theme ties these groups of devices together. They are covered in the same chapter simply because it is convenient to do so.

26.1 Introduction to Thyristors: The Silicon Unilateral Switch

The simplest of the thyristors is the **silicon unilateral switch (SUS),** which is *a two-terminal, four-layer device that can be triggered into conduction by applying a specified forward voltage across its terminals.* As shown in Figure 26.1, the device contains two *n*-type and two *p*-type regions.

The SUS is sometimes referred to as a *Schockley diode* or a *current latch.* Note that the term *diode* is considered to be appropriate, because the device has two terminals: an *n*-type cathode (K), and a *p*-type anode (A).

SUS Operation

OBJECTIVE 1 ➤ If we split the SUS as shown in Figure 26.2a, we see that the device can be viewed as two transistors: one *pnp* (Q_1) and one *npn* (Q_2), with the collector of each connected to the base of the other.

A Practical Consideration: Thyristors are commonly classified as **breakover devices,** because each *can be triggered into conduction by a voltage that exceeds one (or more) breakover voltage rating(s).*

Figure 26.3a illustrates the response of an off (nonconducting) SUS to an increase in supply voltage. Assuming that both transistors are in cutoff, the total forward current through the device (I_F) is approximately equal to zero. Since the circuit current equals I_F, $V_{RS} = 0$ V, and $V_{AK} \cong V_S$. Note that these circuit conditions exist as long as V_{AK} is less than the **forward breakover voltage ($V_{BR(F)}$)** rating of the SUS, which is *the value of forward voltage that will force the device into conduction.*

When V_{AK} reaches the forward breakover voltage, two things happen:

- The device current (I_F) rapidly increases as the device is driven into saturation.
- V_{AK} rapidly decreases because of the low resistance of the saturated device.

The relatively low value of V_{AK} is identified as V_F in the operating curve.

The fact that the SUS drives itself into saturation is important. You may recall that the current through a saturated device is controlled by the values of resistance and voltage

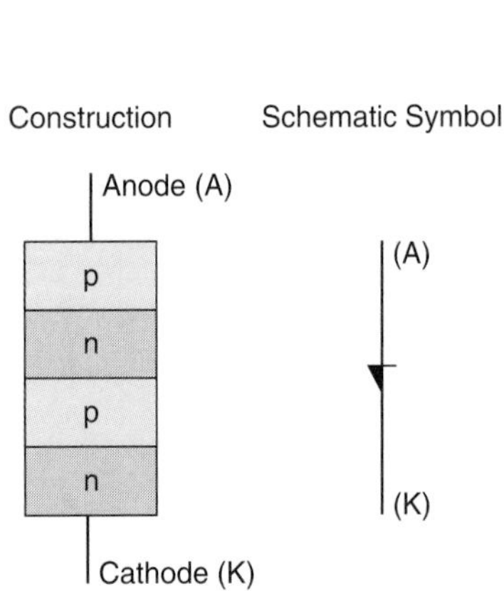

Figure 26.1 SUS construction and schematic symbol.

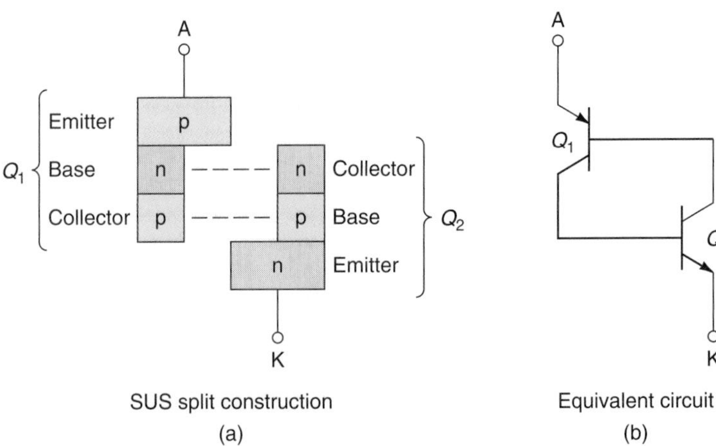

Figure 26.2 SUS split construction and equivalent circuit.

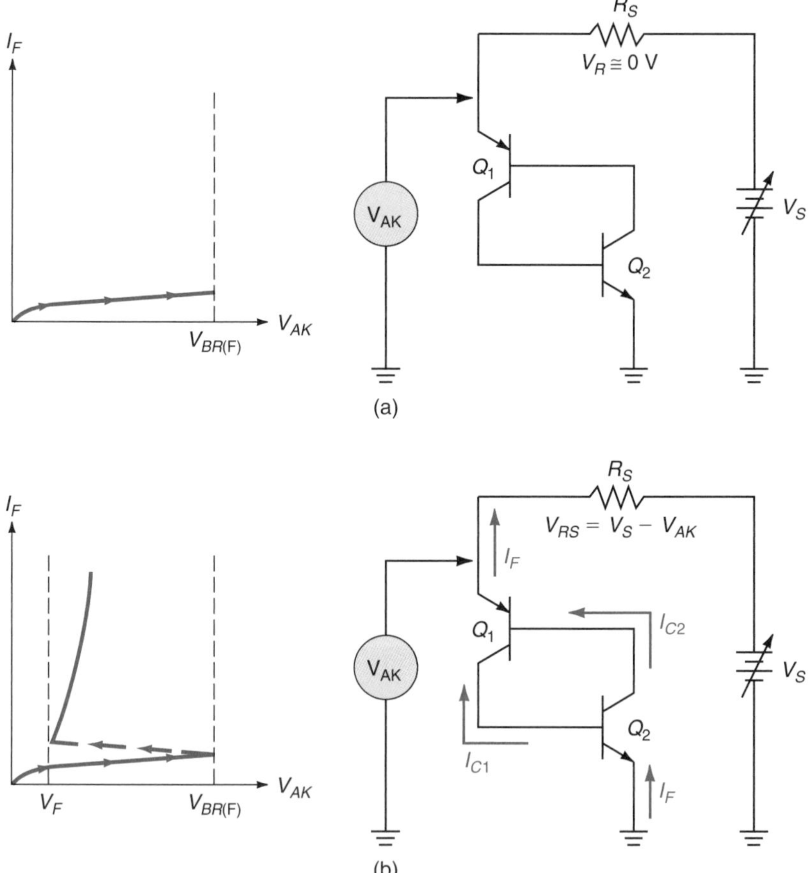

Figure 26.3 The response of an off (nonconducting) SUS to an increase in supply voltage.

that are external to the component. Thus, the forward current through the saturated SUS represented in Figure 26.3b is found as

$$I_F = \frac{V_S - V_F}{R_S} \tag{26.1}$$

where

V_F = the forward voltage drop across the device.

Equation (26.1) shows that the SUS requires a series current-limiting resistor (R_S). Without one, the SUS may be destroyed by excessive current.

Driving the SUS into Cutoff

◄ *OBJECTIVE 2*

Once an SUS is driven into conduction, it continues to conduct as long as I_F is greater than the **holding current (I_H).** The holding current rating of an SUS indicates *the minimum forward current required to maintain conduction.* Once I_F drops below I_H, the device is driven back into cutoff, and it remains there until V_{AK} reaches $V_{BR(F)}$ again.

Several methods are used to drive I_F below the value of I_H. The first, which is called **anode current interruption,** is illustrated in Figure 26.4. In Figure 26.4a, a series switch is opened to interrupt the path for I_F. In Figure 26.4b, a shunt switch is closed to divert the circuit current (I_T) around the SUS, again causing the device current to drop to zero.

In practice, *series interruption* is generally caused by the *opening of a fuse.* For example, thyristors are often used to protect voltage-sensitive loads from overvoltage conditions. When such a circuit is activated, the thyristor triggers and protects its load just

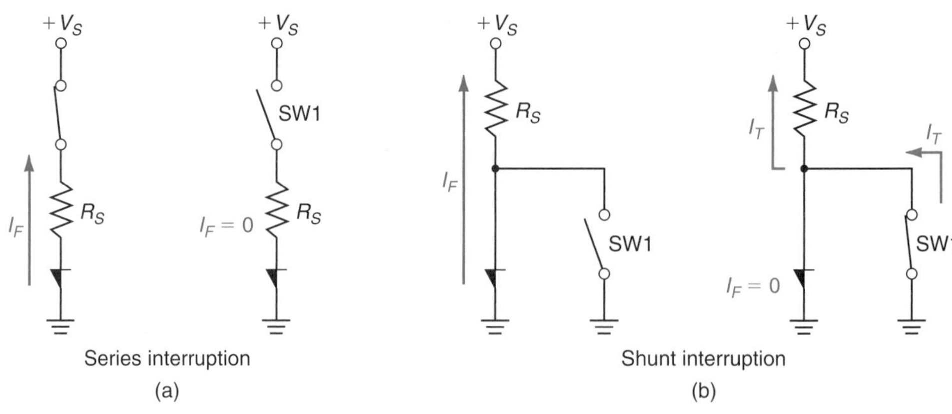

Figure 26.4 Anode current interruption.

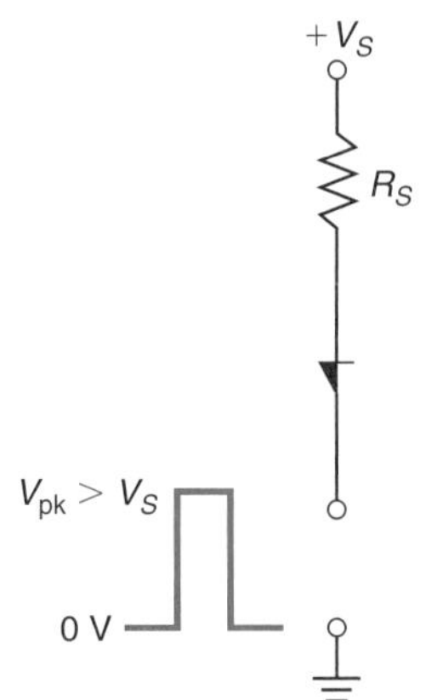

Figure 26.5
Forced commutation.

long enough for a fuse to blow. The switch in the *shunt interruption* circuit is normally a reset switch that is pressed when the overvoltage situation has passed. By closing the reset switch, the SUS is driven into cutoff, and normal circuit operation is restored.

Another method for driving an SUS into cutoff is called **forced commutation.** In this case, a reverse voltage is applied to the device, as shown in Figure 26.5. The reverse voltage causes the device current to drop below the value of I_H, and the SUS turns off.

SUS Specifications

Figure 26.6a shows the operating curve of the SUS. The part of the curve that falls between 0 V and the **reverse breakdown voltage ($V_{BR(R)}$)** is called the **reverse blocking region.** When the SUS is operating in this region, it has the same characteristics as a reverse-biased *pn*-junction diode.

Forward operation is divided into two regions, as shown in Figure 26.6b. The **forward blocking region** (off-state) is defined by two parameters: $V_{BR(F)}$, and $I_{BR(F)}$. As you know, $V_{BR(F)}$ is the value of forward voltage that causes the SUS to conduct. The **forward breakover current ($I_{BR(F)}$)** rating of the SUS is *the value of I_F at the point where breakover occurs.*

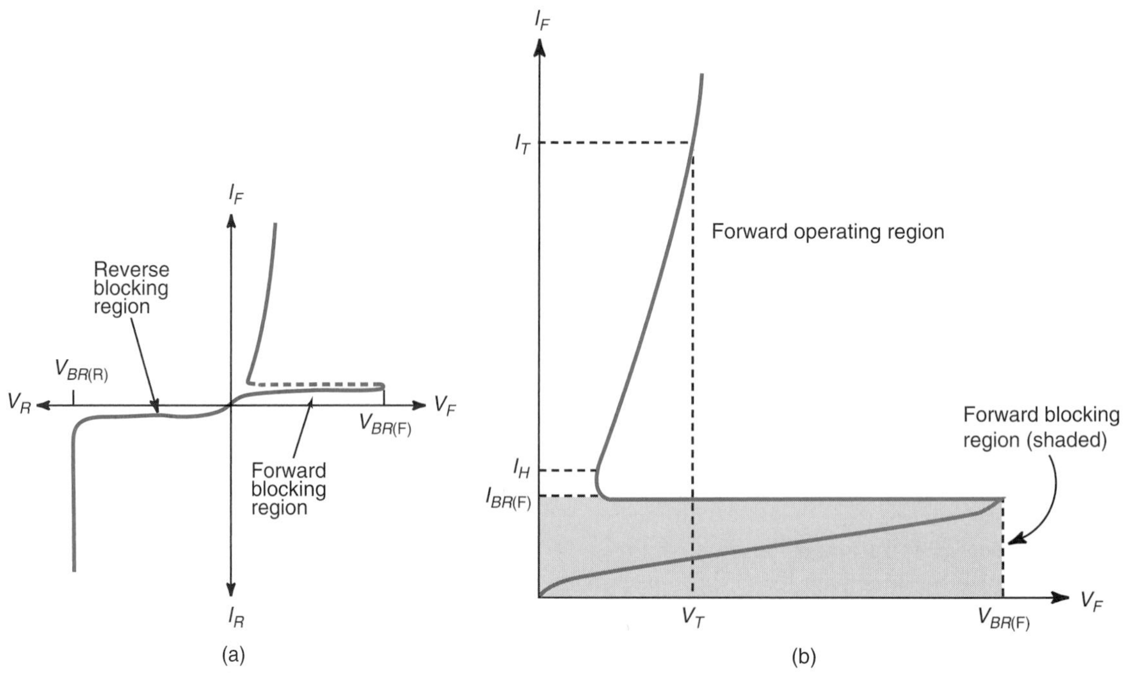

Figure 26.6 The SUS operating curve.

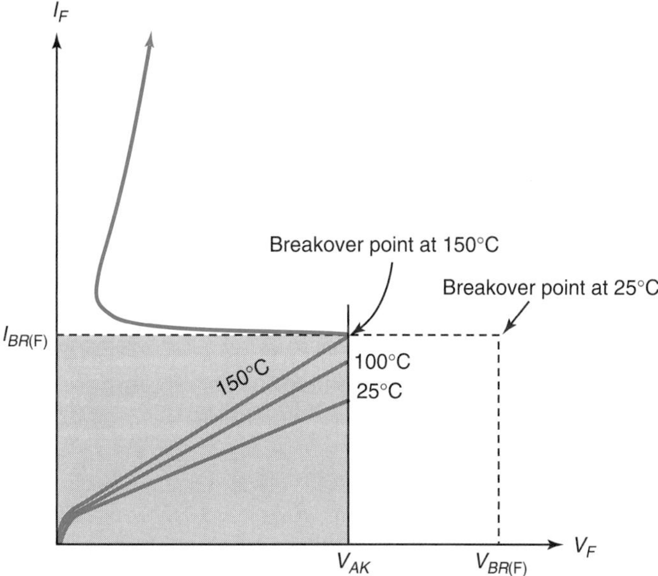

Figure 26.7 The effect of temperature on SUS forward conduction.

The importance of the $I_{BR(F)}$ rating is illustrated in Figure 26.7. As temperature increases, so does leakage current. At 150°C, the leakage current reaches the value of $I_{BR(F)}$, and the device is triggered into its on-state despite the fact that $V_{AK} < V_{BR(F)}$. Thus, an SUS can be triggered into conduction by a significant increase in operating temperature.

Two other SUS ratings are identified in Figure 26.6b. The **average on-state current (I_T)** is *the maximum average forward current for the SUS.* The series current-limiting resistor for an SUS is selected to ensure that I_F is held below this value. The **average on-state voltage (V_T)** rating is the value of V_F when $I_F = I_T$.

One Final Note

The SUS is rarely used in modern circuit design. However, it does serve as a valuable learning tool, because many commonly used thyristors are nothing more than variations on the basic SUS.

Section Review

1. What are *thyristors*?
2. What distinguishes thyristors from other switching devices, such as BJTs and FETs?
3. What is the *silicon unilateral switch* (*SUS*)?
4. Describe the construction of the SUS.
5. What is *forward breakover voltage*?
6. Describe what happens when the voltage applied to an SUS reaches the $V_{BR(F)}$ rating of the device.
7. Describe the two methods of driving an SUS into cutoff.
8. Explain how *temperature* can drive an SUS into its forward operating region.

26.2 Silicon-Controlled Rectifiers

◄ OBJECTIVE 3

The **silicon-controlled rectifier (SCR)** is *a three-terminal device that is very similar in construction and operation to the SUS.* As shown in Figure 26.8, the construction of the SCR is identical to that of the SUS, except for the addition of a *third terminal* called the **gate.**

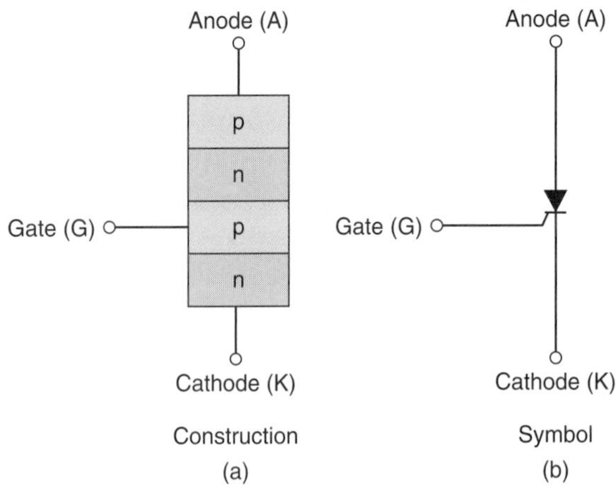

Figure 26.8 SCR construction and schematic symbol.

SCR Triggering

Figure 26.9a shows the two-transistor equivalent circuit for the SCR. Like the SUS, it can be triggered into conduction by applying an anode-to-cathode voltage (V_{AK}) that is equal to the $V_{BR(F)}$ rating of the device. The gate terminal provides an additional means of triggering the device, as shown in Figure 26.9b. When the gate pulse goes positive (T_1), Q_2 is forced into conduction. At that point, the device drives itself into saturation, just like the SUS. When the gate pulse is removed (T_2), forward conduction continues until I_F drops below the holding current (I_H) rating of the device. Note that the SCR is driven into the off-state using the same two methods we use for the SUS: anode current interruption, and forced commutation.

SCR Operating Curve

Figure 26.10a shows the operating curve of the SCR. Figure 26.10b shows the effects of gate current (I_G) on forward conduction. As the value of I_G increases, the device breaks over into its on-state at decreasing values of V_F. For example, let's assume an SCR has the value of V_{AK} shown in Figure 26.10b. At $I_G = 0$ A, the device is in the off-state. However, at $I_G = I_{G1}$, the value of V_{AK} is more than sufficient to drive the device

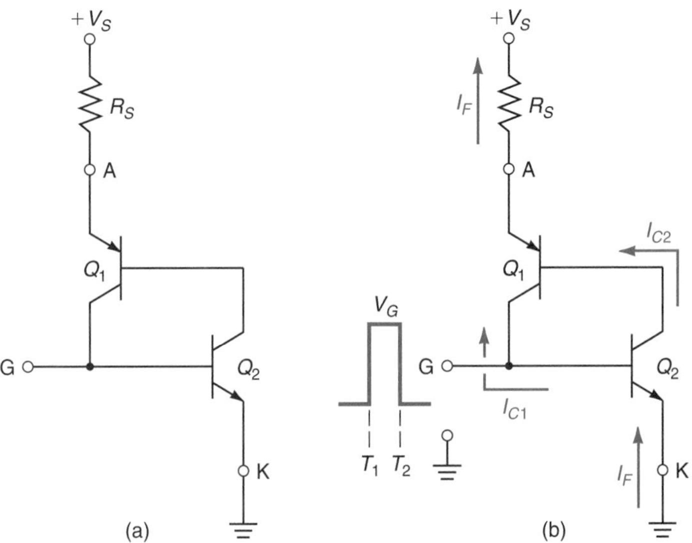

Figure 26.9 SCR gate triggering.

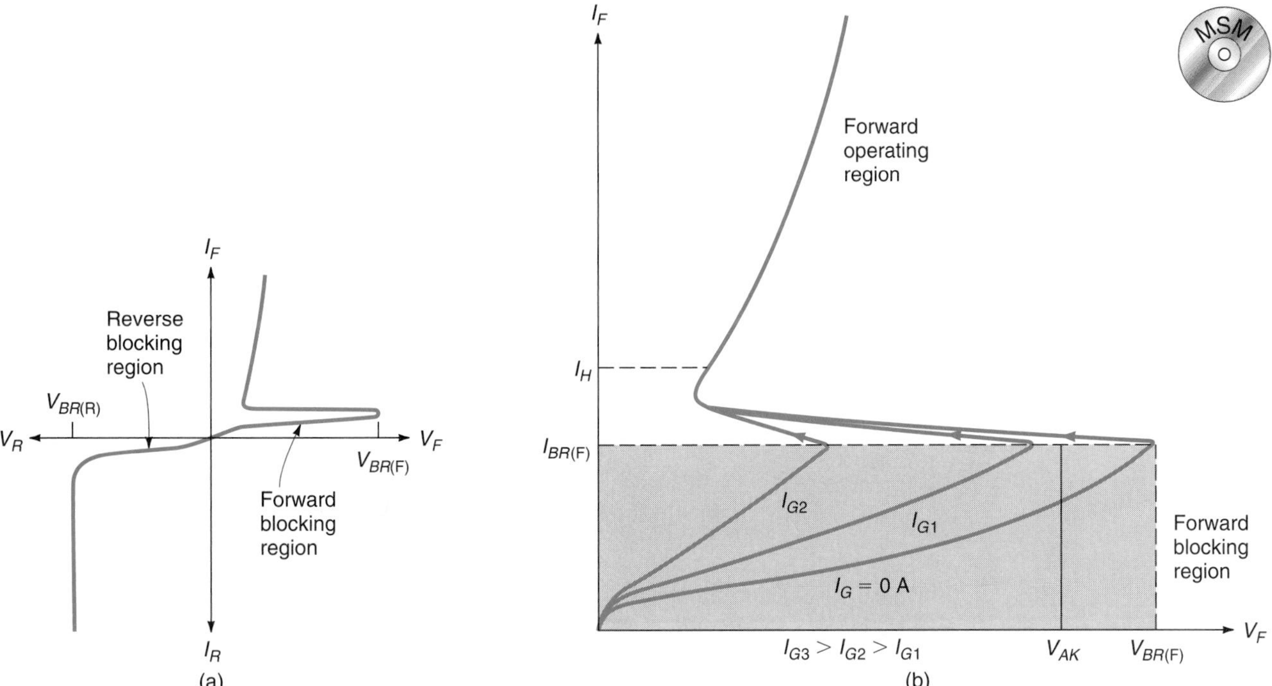

Figure 26.10 The SCR operating curve.

TABLE 26.1 SCR ratings

Specification	*Symbol*	*Definition*
Peak repetitive off-state blocking voltage	V_{RRM} or V_{DRM}	The maximum forward or reverse blocking voltage; the same thing as $V_{BR(F)}$.
Peak forward or reverse blocking current	I_{DRM} or I_{RRM}	The maximum forward or reverse blocking current; the same thing as $I_{BR(F)}$ and $I_{BR(R)}$.
Holding current	I_H	The minimum forward current required to maintain conduction.

Note: V_{RRM} is listed under the *maximum ratings* heading on the spec sheet. I_{RRM} and I_H are listed under the *electrical characteristics* heading.

into the on-state. Again, once the device turns on, the forward current must be driven below the component's I_H rating to return it to the off-state.

SCR Specifications

Many of the specifications for the SCR are identical to those for the SUS. Table 26.1 lists the specifications that have already been introduced. Note that the forward and reverse blocking voltages and currents are equal in magnitude (as they are for the SUS).

Several other SCR parameters warrant discussion. The **circuit fusing (I^2t)** rating indicates *the maximum forward surge current capability of the device, measured in ampere-square-seconds* (A^2s). For example, the circuit fusing rating of the 2N682–92 series SCRs is 93 A^2s. If the product of the surge current (squared) times the duration (in seconds) of the surge exceeds 93 A^2s, the device may be destroyed by excessive power dissipation. Example 26.1 demonstrates the determination of I^2t for a given circuit.

Example 26.1

The 2N682 is subjected to a 50 A current surge that lasts for 10 ms. Determine whether or not this surge will destroy the device.

Solution: The I^2t value in this case is found as

$$I^2t = (50\text{ A})^2(10\text{ ms}) = (2500\text{ A}^2)(10\text{ ms}) = 25\text{ A}^2\text{s}$$

Since this result is well below the maximum rating of 93 A^2s, the device can easily handle the surge.

Practice Problem 26.1

The 2N6394 has a circuit fusing rating of 40 A^2s. Determine whether or not the device will survive a 60 A surge that lasts for 15 ms.

If we know the circuit fusing rating for an SCR and the magnitude of the surge, the maximum safe duration of the surge can be found as

$$t_{max} = \frac{I^2t \text{ (rated)}}{I_S^2} \tag{26.2}$$

where

I_S = the known value of surge current.

We can also determine the maximum surge current for specific period of time by using a variation of equation (26.2):

$$I_{S(max)} = \sqrt{\frac{I^2t \text{ (rated)}}{t_S}} \tag{26.3}$$

where

t_S = the time duration of the current surge

The **gate nontrigger voltage (V_{GD})** rating for an SCR indicates *the maximum gate voltage that can be applied without triggering the SCR into conduction.* If V_G exceeds this rating, the SCR may be triggered into the on-state. The V_{GD} rating is important, because it points out one of the potential causes of *false triggering.*

OBJECTIVE 4 ➤ **False triggering** is *a situation where the SCR is accidentally triggered into conduction, usually by some type of noise.* For example, the 2N682–92 series of SCRs has a V_{GD} rating of 250 mV. If a noise signal with a peak value greater than 250 mV appears at the gate, the device may be triggered into conduction.

Another common cause of false triggering is a rise in anode voltage that exceeds the **critical rise (*dv/dt*)** rating of the SCR. The *dv/dt* rating of the 2N682–92 series SCRs is 30 V/μs. Obviously, 30 V of noise would never occur under normal circumstances. However, even a relatively small amount of noise can have a sufficient *rate of increase* to cause false triggering. To determine the rate of increase of a given noise signal in V/μs, the following conversion formula can be used:

$$\Delta V = \frac{dv}{dt}\Delta t \tag{26.4}$$

where

$\frac{dv}{dt}$ = the critical rise rating of the component

Δt = the rise time of the increase in V_{AK}

ΔV = the amount of change in V_{AK}

As Example 26.2 demonstrates, even a small amount of noise in V_{AK} can cause false triggering if the rise time of the noise signal is short enough.

A 2N682 SCR is used in a circuit where 2 ns (rise time) noise signals occur randomly. Determine the noise amplitude required to cause false triggering.

Example 26.2

Solution: With a critical rise rating of 30 V/μs, the amplitude that may cause false triggering is found as

$$\Delta V = \frac{dv}{dt}\Delta t = (30\ \text{V}/\mu\text{s})(2\ \text{ns}) = 60\ \text{mV}$$

As you can see, with a 2 ns rise time, only 60 mV of noise is required to cause false triggering.

Practice Problem 26.2

The C35D SCR has a critical rise rating of 25 V/μs. Determine the amplitude of noise needed to cause false triggering when the noise signal rise time is 100 ps.

Preventing False Triggering

False triggering is usually caused by noise in the gate circuit or a short rise-time noise signal in V_{AK}. Gate noise is generally reduced using one of the following two methods:

- The gate is connected (via a gate resistor) to a negative dc supply. This means that any noise signal must be greater than the negative gate voltage, plus V_{GD}, to trigger the SCR.
- A bypass capacitor is connected between the gate and ground to short out any noise signal. This is the most commonly used method, because it is very cost-effective.

Noise in V_{AK} is normally prevented by the use of a *snubber network*. A **snubber network** is *an RC circuit connected between the anode and cathode terminals of the SCR.* A snubber network is shown in Figure 26.11. The capacitor (C_S) effectively shorts any noise in V_{AK} to ground. The resistor (R_S) is included to limit the capacitor discharge current when the SCR turns on. Snubbers are used in most SCR circuits, because critical rise (dv/dt) problems are the most common cause of false triggering.

Note: False triggering can also be caused by temperature, as is the case with the SUS.

An SCR Application: The SCR Crowbar

◀ *OBJECTIVE 5*

A **crowbar** is *a circuit that is used to protect a voltage-sensitive load from excessive dc power supply output voltages.* A crowbar circuit is shown in Figure 26.12.

In the crowbar circuit, the SCR and the zener diode are normally off. If an overvoltage situation occurs, the zener starts to conduct, and the resulting current develops a voltage across R_G. This voltage is applied to the SCR gate, forcing the SCR into conduction. When the SCR conducts, it shorts the output from the power supply, protecting the load. Note that:

- The crowbar circuit is activated only by some major fault within the dc power supply. Once activated, the SCR continues to conduct until the power supply fuse opens.

Note: The SCR in a crowbar is turned off by a blown fuse in the dc power supply. This is a practical example of *anode current interruption.*

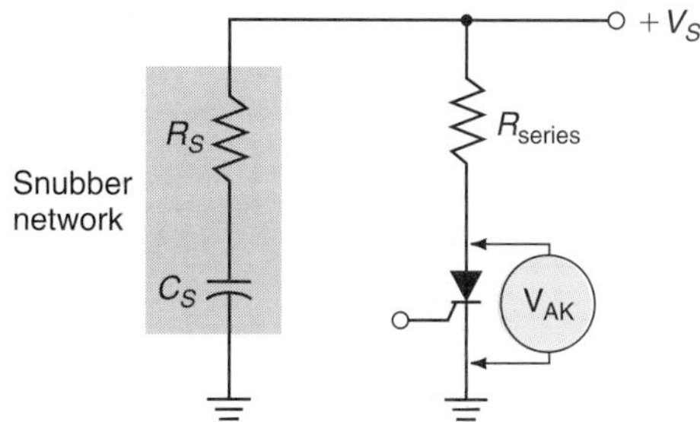

Figure 26.11 An *RC* snubber network.

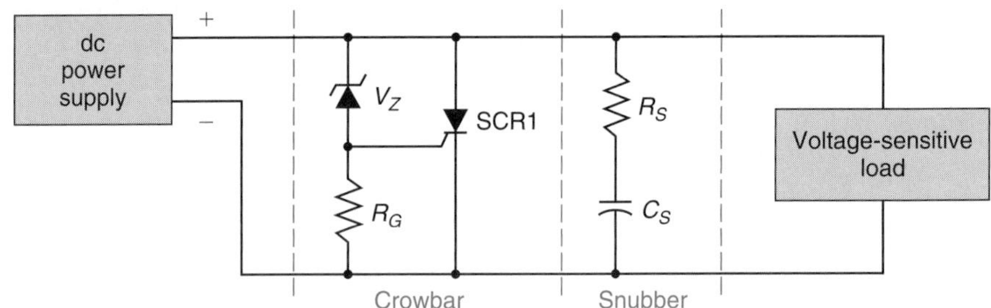

Figure 26.12 An SCR crowbar.

- The circuit fusing (I^2t) rating of the SCR must be high enough to ensure that the SCR isn't destroyed before the power supply fuse has time to blow.
- The SCR in the crowbar is considered to be expendable, which is why there is no series current-limiting resistor in the circuit. The object of the circuit is to protect the load, not the SCR.
- Many crowbar circuits are self-resetting after the overvoltage ceases. If it is not self-resetting, the SCR in a crowbar should be replaced after activation as a precaution.

An SCR Application: The SCR Phase Controller

A **phase controller** is *a circuit used to control the conduction angle through a load and, thus, the average load voltage.* A basic SCR phase controller is shown in Figure 26.13a. The **conduction angle** of a phase controller indicates *the portion of the input waveform (in degrees) that is coupled to the load.* Figure 26.13b shows three conduction angles as shaded areas. When the conduction angle (θ) is 180°, the entire alternation is coupled to the load. When the conduction angle is 90°, one-half the alternation is coupled to the load, and so on. By varying the setting of the potentiometer (R_2) in the phase controller (Figure 26.13a), we can vary the conduction angle (θ). Note that:

- The bypass capacitor between the SCR gate and cathode terminals (C_1) is included to prevent false triggering.
- SCRs have relatively low gate breakdown voltage ratings, so D_1 is included to prevent the negative alternation of V_{in} from being applied to the SCR gate-cathode junction.
- The load is in series with the SCR and acts as the series current-limiting resistor.

Note: Turning an SCR off by reversing the anode voltage polarity (as shown in Figure 26.13) is a practical example of *forced commutation.*

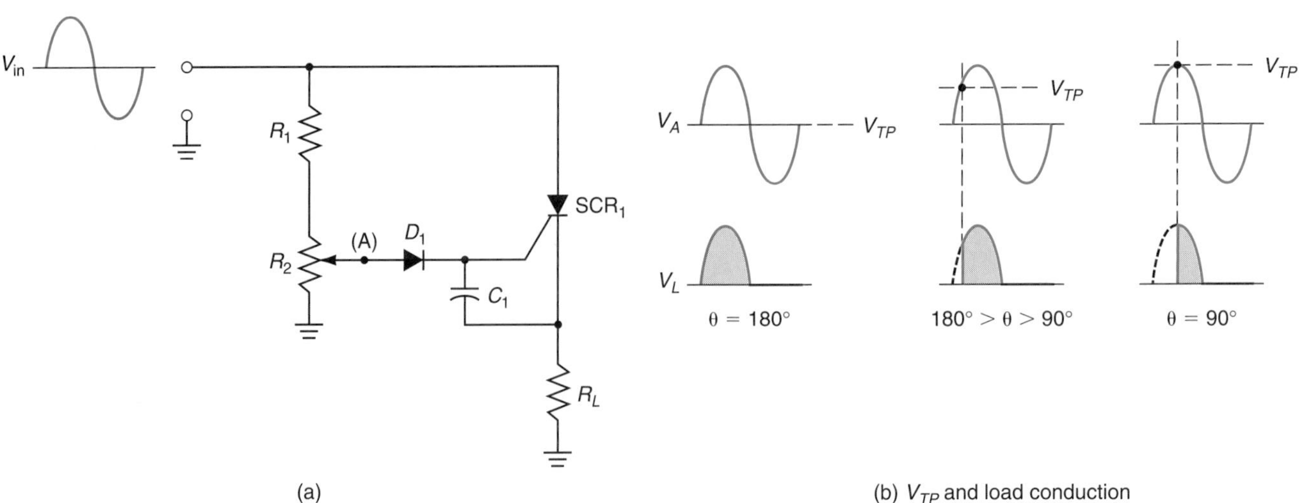

Figure 26.13 SCR phase controller.

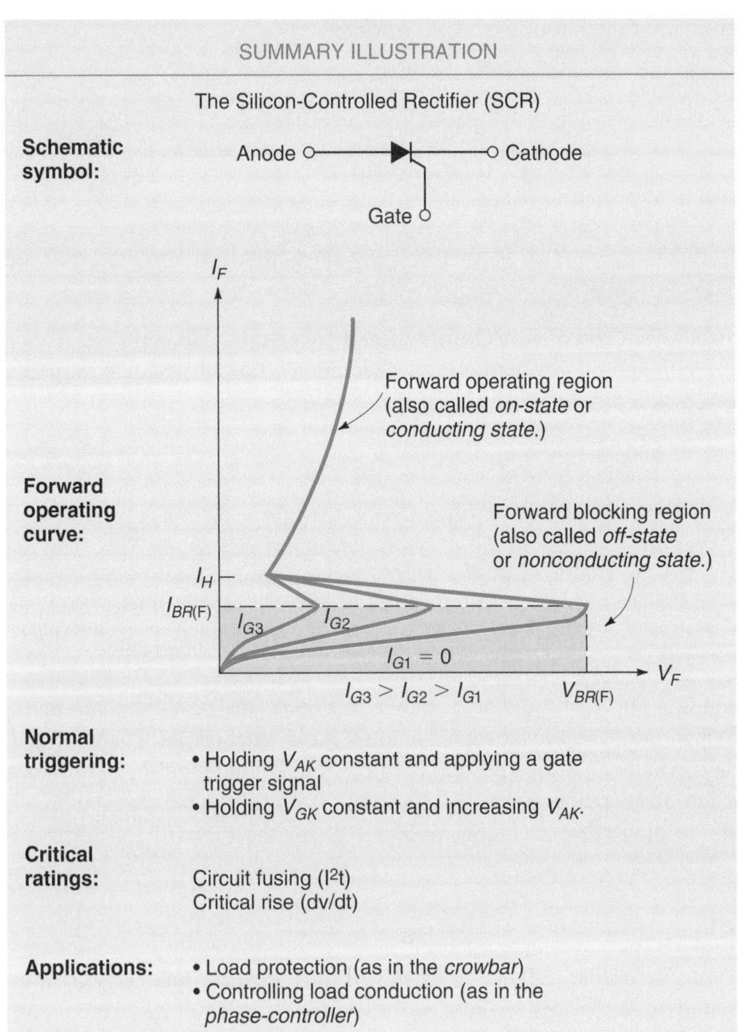

Figure 26.14

Summary

The SCR acts as a *gated* SUS. The gate terminal provides an additional means of triggering the device into conduction. Otherwise, the component has the same operating characteristics as the SUS. The characteristics of the SCR are summarized in Figure 26.14.

Section Review

1. Describe the construction of the SCR.
2. List the methods used to trigger an SCR.
3. List the methods used to force an SCR into cutoff.
4. Using the operating curve shown in Figure 26.10, describe the forward operation of the SCR.
5. What is the *circuit fusing (I^2t) rating* of an SCR? Why is it important?
6. What is *false triggering*?
7. Which SCR ratings are tied directly to false triggering problems?
8. How does a gate bypass capacitor prevent false triggering?
9. What is a *snubber network*? What is it used for?

10. What is the function of the SCR *crowbar*?
11. Describe the operation of the crowbar.
12. What is the function of the SCR *phase controller*?

26.3 Diacs and Triacs

Diacs and *triacs* are classified as **bidirectional thyristors.** The term *bidirectional* means that *they are capable of conducting in two directions.* For all practical purposes, the diac can be viewed as a bidirectional SUS and the triac as a bidirectional SCR.

Diacs

OBJECTIVE 6 ➤ The **diac** is *a two-terminal, three-layer device with forward and reverse operating characteristics that are identical to the forward characteristics of the SUS.* The diac symbol and construction are shown in Figure 26.15. The construction of the diac appears similar to that of a bipolar junction transistor (BJT), but the diac has no base terminal. Further, the size and doping levels of the diac regions are nearly identical (unlike the BJT, which has a small, lightly doped base region).

Note: Diacs are often referred to as *bidirectional diodes.*

The operating curve of the diac is shown in Figure 26.16. As you can see, the device acts as an open until the breakover voltage (V_{BR}) is reached. At that time, the device is triggered and conducts in the appropriate direction. Conduction then continues until the device current drops below its holding current (I_H) rating. Note that the breakover voltage and the holding current values are identical for the forward and reverse regions of operation.

Triacs

The **triac** is *a three-terminal, five-layer device with forward and reverse characteristics that are similar to the forward characteristics of the SCR.* The triac symbol and construction are shown in Figure 26.17. The primary conducting terminals are identified as *main terminal 1* (MT1) and *main terminal 2* (MT2). Note that MT1 is the terminal on the same side of the device as the gate. If we split the triac as shown in Figure 26.17c, you can see that we essentially have complementary SCRs that are connected in reverse-parallel.

Note: Triacs are also referred to as *triodes* and *bidirectional triode thyristors* (the JEDEC designated name).

Because of its unique structure, the operating curve of the triac is a bit more complicated than that of the diac. The triac operating curve, as shown in Figure 26.18, is divided into four quadrants, which are labeled I, II, III, and IV. Each quadrant has the MT2 and G (gate) polarities shown. For example, quadrant III indicates the operation of the device when MT2 is negative (with respect to MT1) and a negative potential is applied to the gate. The triac operating curve is normally plotted as shown to keep it as simple

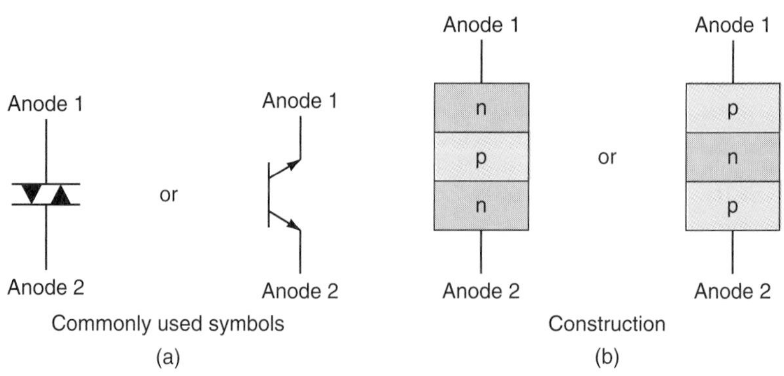

Figure 26.15 Diac construction and schematic symbols.

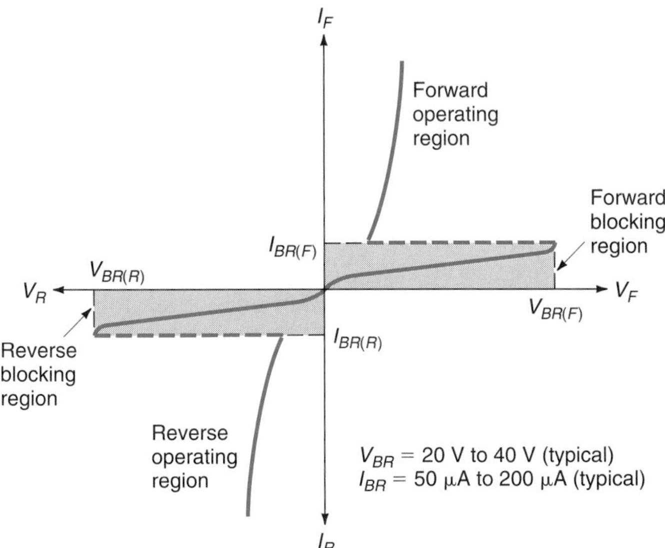

Figure 26.16 The diac operating curve.

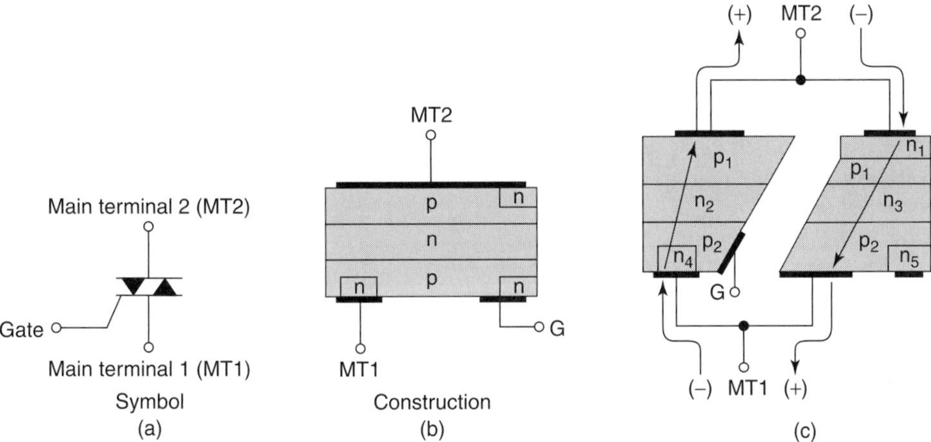

Figure 26.17 Triac symbol, construction, and currents.

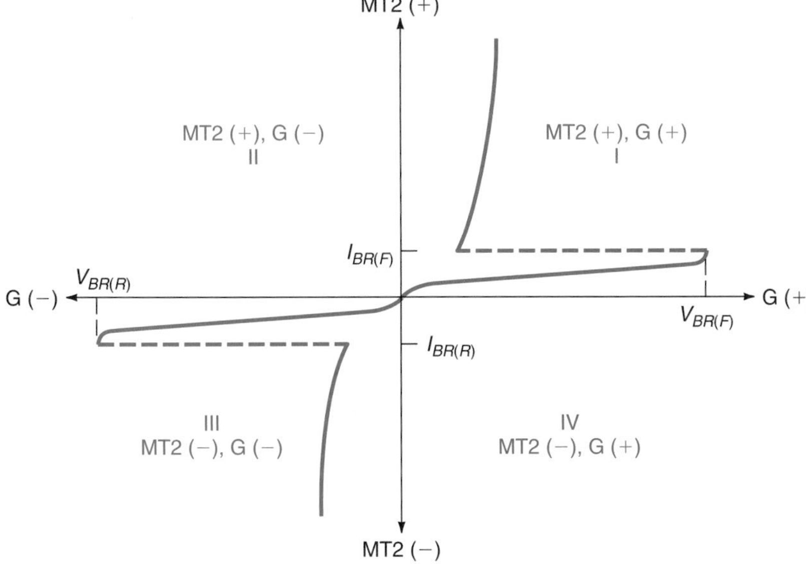

Figure 26.18 The triac operating curve.

as possible. The curve for quadrants II and IV is simply a mirror image of the one shown for quadrants I and III.

The operation in quadrant I is identical to the operation of the SCR. The device remains in the off-state until $V_{BR(F)}$ is reached or a positive **gate trigger voltage (V_{GT})** generates a gate current (I_G) sufficient to trigger the device. Note that when operated in quadrant I, the current through the triac is in the direction shown on the left side of Figure 26.17c.

The operation in quadrant III is simply a reflection of that in quadrant I. When MT2 is negative (with respect to MT1) and a negative gate trigger voltage is applied, it is triggered into reverse conduction. Quadrants II and IV indicate the unique triggering combinations of the triac, which can be triggered into conduction by either of the following two conditions:

- Applying a negative gate trigger while MT2 is positive (quadrant II).
- Applying a positive gate trigger while MT2 is negative (quadrant IV).

When quadrant II triggering occurs, the current through the device is in the direction indicated on the left side of Figure 26.17c. When quadrant IV triggering occurs, the current through the device is in the direction shown on the right side of Figure 26.17c.

Two points should be made at this time:

- Regardless of the triggering method used, a triac is forced into cutoff in the same fashion as the SCR.
- Except for the quadrant triggering considerations, the spec sheet for a triac is nearly identical to that of an SCR.

Controlling Triac Triggering

OBJECTIVE 7 ➤ The low gate voltages required to trigger the triac can cause problems in many circuit applications. For example, consider the 2N6157 triac shown in Figure 26.19a. Here, we see the approximate voltages required for quadrants I and II triggering. Three questions may arise about gate triggering voltages:

- What if we want the 2N6157 to be triggered only by a positive gate voltage?
- What if we want the 2N6157 to be triggered only by a negative gate voltage?
- What if we want positive and negative triggering but at higher gate voltage values?

The circuit shown in Figure 26.19b illustrates the method used to provide single-polarity triggering. Because the diode (D_1) is connected in series with the gate, the triac can be triggered only by a positive gate signal. The potential required to trigger the device equals the sum of the diode forward voltage and V_{GT}. By reversing the direction of the diode, we limit the triggering of the triac to negative gate voltages.

In many applications, the dual-polarity triggering characteristic of the triac is desirable, but the rated values of V_{GT} for the device are too low. In this case, using a diac in the gate circuit, as shown in Figure 26.19c, can raise the triggering levels. With the diac present, the triac can still be triggered by both positive and negative gate voltages. However, the voltage at (A) must overcome the values of V_{BR} (for the diac) and V_{GT}. Note that thyristor triggering control is the primary application for the diac.

Summary

The characteristics of the diac and triac are summarized in Figure 26.20.

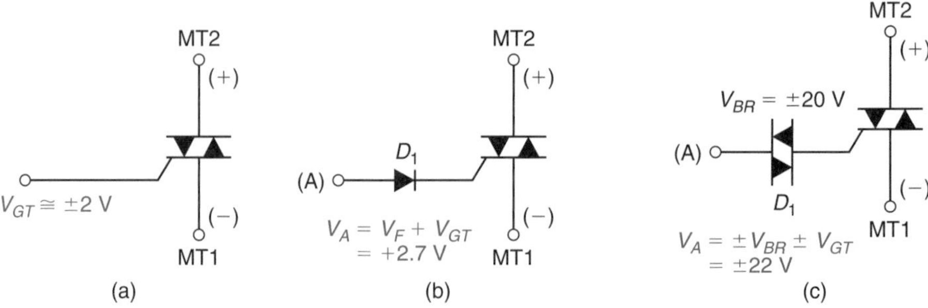

Figure 26.19 Controlling triac triggering.

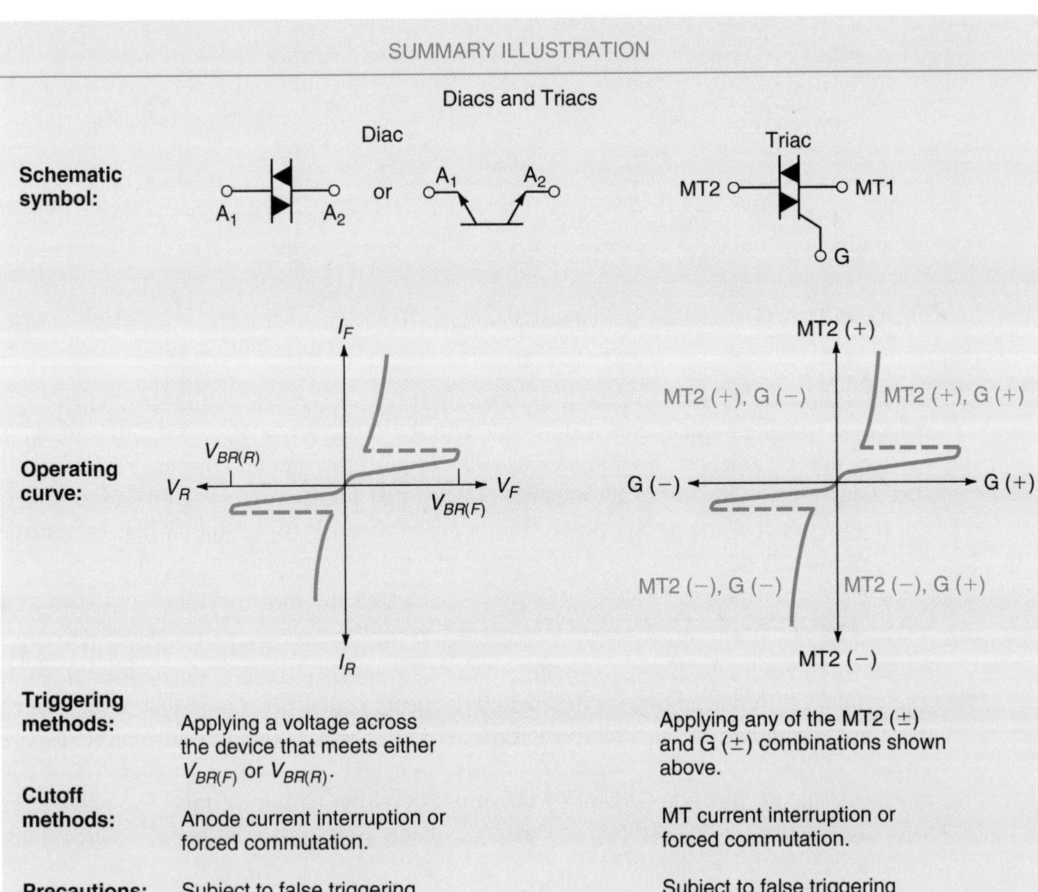

Figure 26.20

Section Review

1. What is a *bidirectional thyristor*?
2. What is a *diac*?
3. How is a diac triggered into conduction?
4. How is a diac driven into cutoff?
5. What is the most common diac application?
6. What is a *triac*?
7. Describe the four-quadrant triggering characteristics of the triac.
8. What methods are used to drive a triac into cutoff?
9. Describe the operation of each of the trigger control circuits shown in Figure 26.19.

26.4 Unijunction Transistors

OBJECTIVE 8

The **unijunction transistor (UJT)** is *a three-terminal switching device whose trigger voltage is proportional to its applied biasing voltage.* The schematic symbol for the UJT is shown in Figure 26.21a. The terminals of the UJT are called the *emitter, base* 1 (B1), and *base* 2 (B2). The switching action of the UJT is explained with the help of the circuit shown in Figure 26.21b. A biasing voltage (V_{BB}) is applied across the two base terminals.

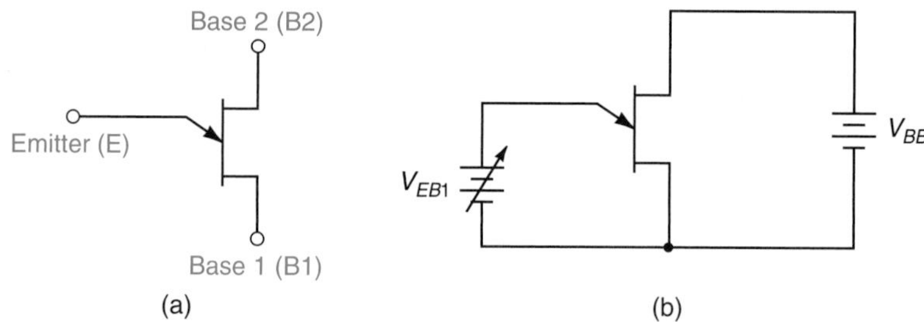

Figure 26.21 UJT schematic symbol and biasing.

The emitter–base 1 junction acts as an open until the voltage across the terminals (V_{EB1}) reaches a specified value, called the **peak voltage (V_P).** When $V_{EB1} = V_P$, the junction triggers into conduction. It continues to conduct until the emitter current drops below a specified value, called the **peak current (I_P).** At that point, it returns to its nonconducting state. Note that the peak voltage (V_P) is proportional to V_{BB}. Thus, when we change V_{BB}, we also change the value of V_P needed to trigger the device into conduction.

UJT Construction and Operation

As shown in Figure 26.22a, the structure of the UJT is very similar to the *n*-channel JFET. The primary difference between the two components is that the *p*-type (gate) material of the JFET surrounds the *n*-type (channel) material. The UJT equivalent circuit (Figure 26.22b) represents the component as a diode connected to a voltage divider. The diode represents the *pn*-junction formed by the emitter and the base materials. The voltage divider (R_{B1} and R_{B2}) represents the resistance of the *n*-type material between the emitter junction and each base terminal.

For the emitter–base diode to conduct, its anode voltage must be approximately 0.7 V more positive than its cathode. The cathode potential of the diode (V_K) is determined by the combination of V_{BB}, R_{B1}, and R_{B2}. Using the standard voltage-divider equation, the value of V_K is found as

$$V_K = V_{BB}\frac{R_{B1}}{R_{B1} + R_{B2}} \tag{26.5}$$

The resistance ratio in equation (26.5) is called the **intrinsic standoff ratio (η)** of the UJT. Adding 0.7 V to the value found in equation (26.5), and substituting η for the resistance ratio, we obtain an equation for the triggering voltage (V_P) for the UJT, as follows:

$$V_P = \eta V_{BB} + 0.7\text{ V} \tag{26.6}$$

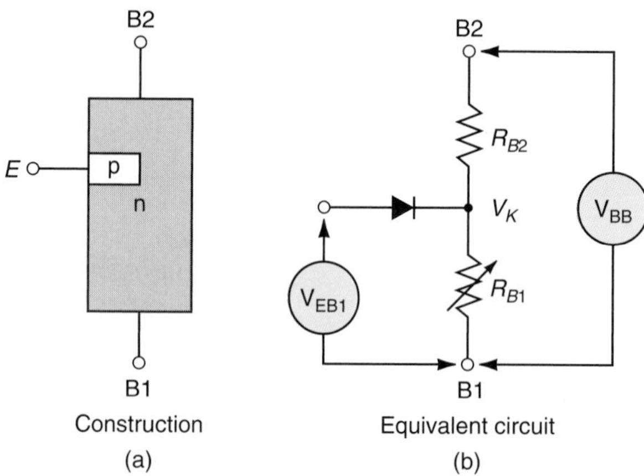

Figure 26.22 UJT construction and equivalent circuit.

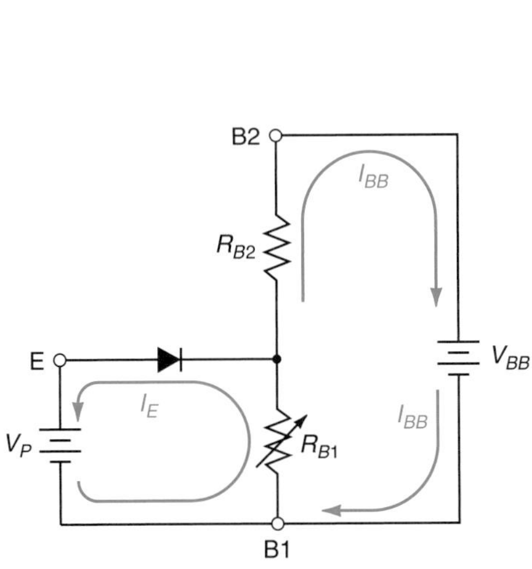

Figure 26.23 UJT currents.

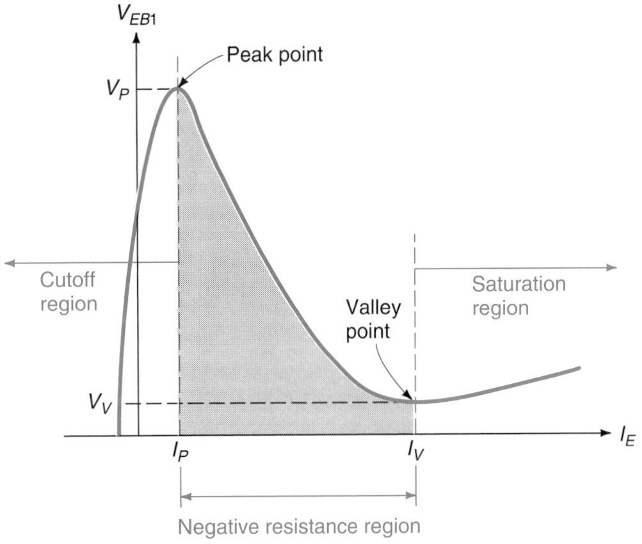

Figure 26.24 UJT operating curves.

The intrinsic standoff ratio for a given UJT is listed on its spec sheet. For example, the spec sheet for a 2N5431 UJT lists an intrinsic standoff ratio of $\eta = 0.8$ (maximum). If this UJT is connected to a +18 V biasing voltage, the maximum peak voltage for the component has a value of

$$V_P = \eta V_{BB} + 0.7 \text{ V} = (0.8)(+18 \text{ V}) + 0.7 \text{ V} = 15.1 \text{ V}$$

The UJT peak voltage is applied across the emitter–base 1 terminals. Because of this, almost all of I_E is drawn through R_{B1} when the device is triggered, as shown in Figure 26.23. Since I_E is drawn through R_{B1}, this resistance value drops off as I_E increases. The value of R_{B2} is not affected significantly by the level of I_E.

As the UJT characteristic curve illustrated in Figure 26.24 shows, V_{EB1} can be increased until the peak point is reached. When the peak point is reached, $V_{EB1} = V_P$, and the emitter diode begins to conduct. The initial value of I_E is referred to as the *peak emitter current* (I_P), but this term can be confusing. Peak current (I_P) is a *minimum* rating, because it is associated with the first point of conduction. Note that when the UJT is in its cutoff region of operation $I_E < I_P$.

When I_E increases above the value of I_P, the value of R_{B1} drops drastically. As a result, V_{EB1} actually decreases as I_E increases. The decrease in V_{EB1} continues until I_E reaches the **valley current (I_V)** rating of the UJT. When I_E increases beyond I_V, the device goes into saturation.

As the characteristic curve illustrated in Figure 26.24 shows, V_{EB1} and I_E are *inversely* related between the peak point and the valley point of the curve. The region of operation between these two points is called the **negative resistance region.** Note that the term **negative resistance** is commonly used to describe any device with current and voltage values that are inversely related.

Once the UJT is triggered, the device continues to conduct as long as I_E is greater than I_P. When I_E drops below I_P, the device returns to the cutoff region of operation and remains there until another trigger is received.

One Final Note

The UJT is technically classified as a thyristor-triggering device rather than a thyristor. Since triggering thyristors (such as SCRs and triacs) is the only common application for the UJT, the device is normally covered along with thyristors. Also, you will normally find the spec sheets for UJTs in the same section of a parts manual as the thyristors.

Section Review

1. What is the *unijunction transistor* (*UJT*)?
2. What is the *peak voltage* (V_P) value of a UJT?
3. What is the *peak current* (I_P) value of a UJT?
4. What is the *intrinsic standoff ratio* (η) rating of a UJT?
5. How does the value of η for a UJT relate to the value of V_P?
6. What is meant by *negative resistance*?
7. What current and voltage values define the limits of the negative resistance region of UJT operation?
8. What is the relationship between I_E and V_{EB1} in the negative resistance region of UJT operation?

26.5 Discrete Photodetectors

OBJECTIVE 9 ➤ **Light emitters** (like LEDs) are *components that generate light.* **Light detectors** are *components with electrical output characteristics that are controlled by the amount of light that they receive.* Before we discuss photodetectors, we need to consider a variety of parameters related to light.

Characteristics of Light

Light is electromagnetic energy that falls within a specific range of frequencies, as illustrated in Figure 26.25. The entire light spectrum falls within the range of 30 THz to 3 PHz. Light can be further classified as ultraviolet, infrared, and visible light, as shown.

OBJECTIVE 10 ➤ Two characteristics are commonly used to describe light. The first is *wavelength* (λ). In Chapter 9, you learned that **wavelength** is *the physical length of one cycle of a transmitted signal* (electromagnetic wave), found as

$$\lambda = \frac{c}{f} \qquad (26.7)$$

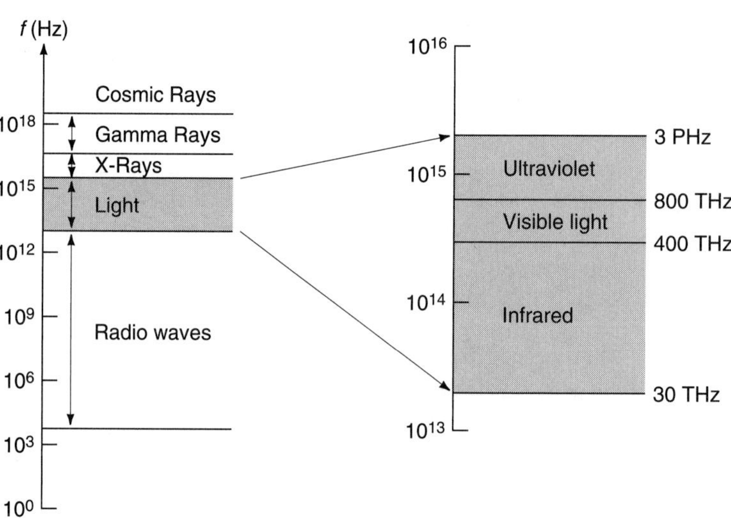

Figure 26.25 The light spectrum.

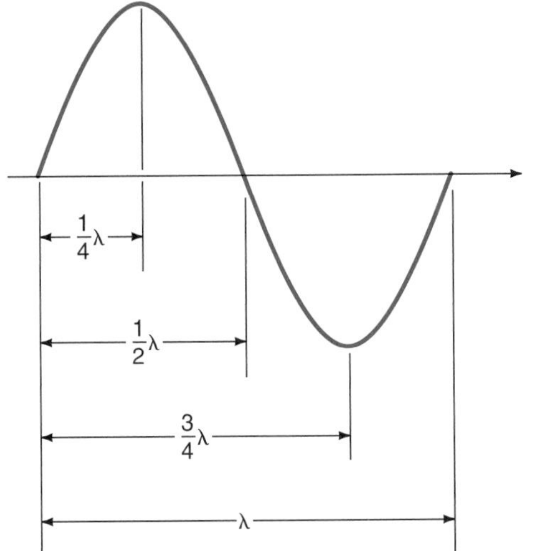

Figure 26.26 Wavelength.

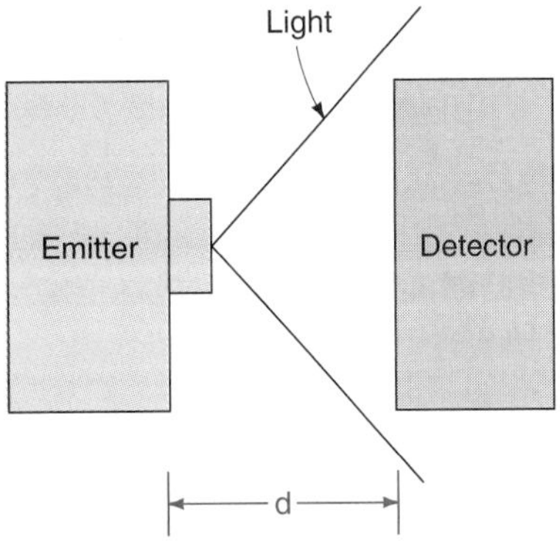

Figure 26.27 Emitter and detector.

where

λ = the wavelength of the signal, in nanometers

c = the speed of light, given as 3×10^{17} nm/s

f = the frequency of the transmitted signal

Note: Most optoelectronic devices are rated for wavelengths in nanometers (nm). This is why the values in equation (26.7) are expressed in nanometers.

As a review, the concept of wavelength is illustrated in Figure 26.26.

Wavelength is important because photoemitters and detectors are rated for specific wavelengths. As shown in Figure 26.27, light emitters are commonly used to drive light detectors. For a specific emitter to be used with a specific detector, both devices must be rated for the same approximate wavelength.

The second important parameter is *intensity.* **Light intensity** is *the amount of light per unit area that is received by a given photodetector.* Note that light intensity decreases as the distance (d) between the emitter and the detector increases. As you will see, the output characteristics of most photodetectors are controlled by the light intensity at their inputs (as well as the input wavelength).

The Photodiode

◄ *OBJECTIVE 11*

The **photodiode** is *a diode whose reverse conduction is controlled by light intensity.* When the light intensity at the optical input to a photodiode increases, the *reverse current* through the device also increases. This point is illustrated in Figure 26.28.

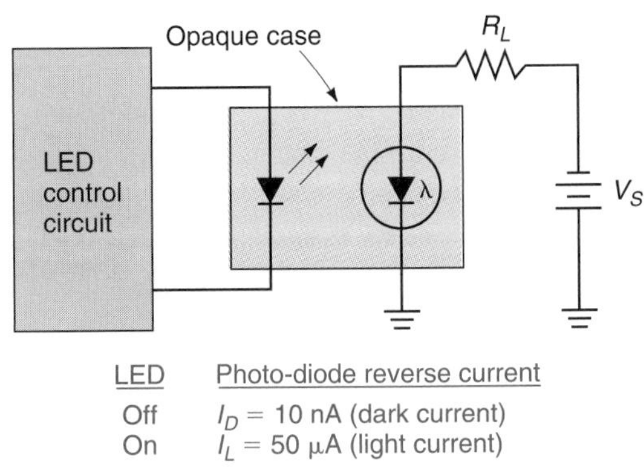

Figure 26.28

In Figure 26.28, a photodiode is enclosed with an LED in an *opaque* case. The term **opaque** is used to describe *anything that blocks light.* The photodiode shown receives a light input only when the LED is activated by its control circuit. When the LED is off, the photodiode reverse current is shown to be 10 nA. When the LED lights, reverse current increases to 50 μA. Note that a ratio of **light current (I_L)** to **dark current (I_D)** in the thousands is not uncommon.

The spec sheet for a photodiode lists its values of I_D and I_L. Several other photodiode ratings are also important. The first is the **wavelength of peak spectral response (λ_S).** This rating indicates *the wavelength that causes the strongest response in the photodiode.* For example, the MRD500 photodiode has an optimum input wavelength of 0.8 mm (800 nm). As Example 26.3 shows, this value can be used to determine the optimum input frequency for the device.

Example 26.3

Determine the optimum input frequency for the MRD500 photodiode.

Solution: Using the rated optimum wavelength of 800 nm, the optimum input frequency is found as

$$f = \frac{c}{\lambda} = \frac{3 \times 10^{17}\ \text{nm/s}}{800\ \text{nm}} = 375\ \text{THz}$$

Since 375 THz is in the infrared frequency spectrum, the MRD500 is classified as an infrared detector.

Practice Problem 26.3

The Motorola MRD821 photodiode has an optimum wavelength of 940 nm. Determine the optimum input frequency for the device.

The **sensitivity** rating of the photodiode is *a measure of its response to a specified light intensity.* For the MRD500, the sensitivity rating is 6.6 mA/mW/cm^2. This means that the reverse current of the device increases by 6.6 mA for every 1 mW/cm^2 of light applied to the device.

The **spectral response** of a photodiode is *a measure of its response to a change in input wavelength.* The spectral response of the MRD500 peaks at 800 nm. If the input wavelength goes as low as 520 nm or as high as 950 nm, the sensitivity of the device drops to 50% of its rated value.

Figure 26.29 summarizes the characteristics of the photodiode. Note that a second common schematic symbol for the photodiode is also shown.

The Phototransistor

The **phototransistor** is *a three-terminal photodetector whose collector current is controlled by the intensity of the light at its optical input* (base). Figure 26.30 shows one possible configuration for a phototransistor amplifier.

The phototransistor is enclosed with an LED in an opaque casing. When the light from the LED is varied by its control circuit, the change in light causes a proportional change in base (and therefore, in the collector and emitter) current through the phototransistor. The resulting change in voltage across R_E is applied to the inverting amplifier, which responds accordingly.

The phototransistor is identical to the standard BJT in every respect, except that its collector current is (ultimately) controlled by light intensity. Thus, the phototransistor shown in Figure 26.30 could be a linear amplifier or a switch, depending on the output of the LED control circuit. The biasing circuit (R_1 and R_2) is used to adjust the quiescent operation of the phototransistor. By adjusting R_1, the input voltage to the op-amp is varied, allowing the output to be set to 0 V. For this reason, R_1 in this circuit is called a **zero adjust.**

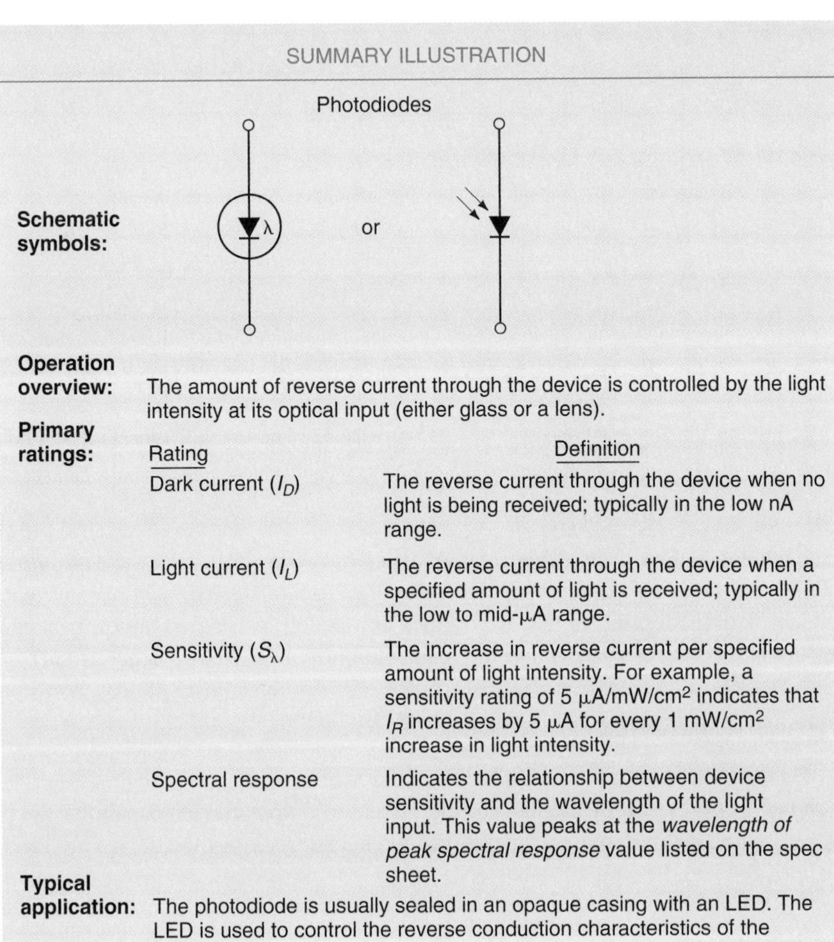

SUMMARY ILLUSTRATION

Photodiodes

Schematic symbols: [two photodiode symbols] or

Operation overview: The amount of reverse current through the device is controlled by the light intensity at its optical input (either glass or a lens).

Primary ratings:

Rating	Definition
Dark current (I_D)	The reverse current through the device when no light is being received; typically in the low nA range.
Light current (I_L)	The reverse current through the device when a specified amount of light is received; typically in the low to mid-μA range.
Sensitivity (S_λ)	The increase in reverse current per specified amount of light intensity. For example, a sensitivity rating of 5 μA/mW/cm^2 indicates that I_R increases by 5 μA for every 1 mW/cm^2 increase in light intensity.
Spectral response	Indicates the relationship between device sensitivity and the wavelength of the light input. This value peaks at the *wavelength of peak spectral response* value listed on the spec sheet.

Typical application: The photodiode is usually sealed in an opaque casing with an LED. The LED is used to control the reverse conduction characteristics of the photodiode.

Figure 26.29

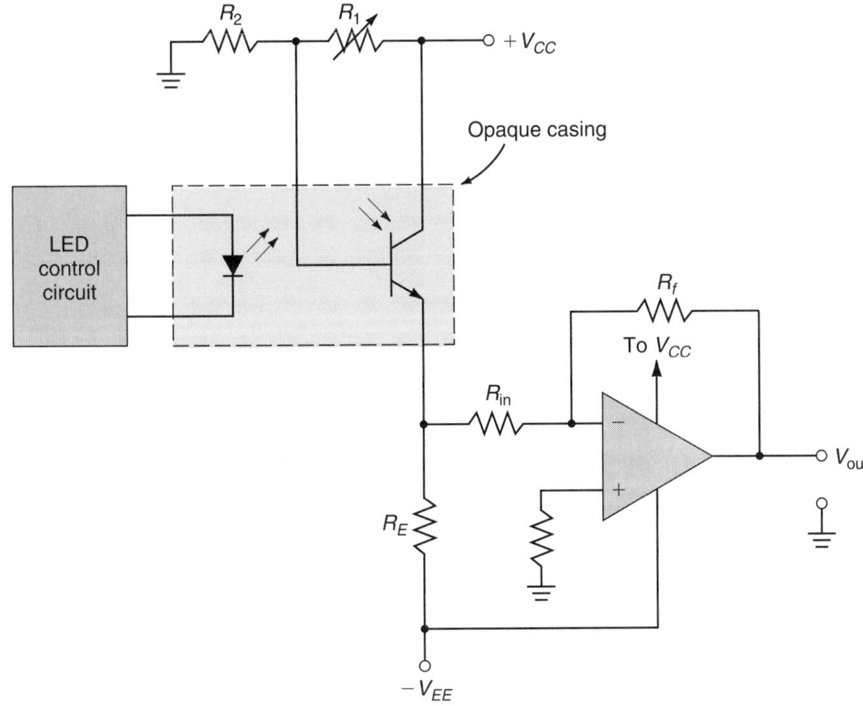

Figure 26.30 Optocoupling.

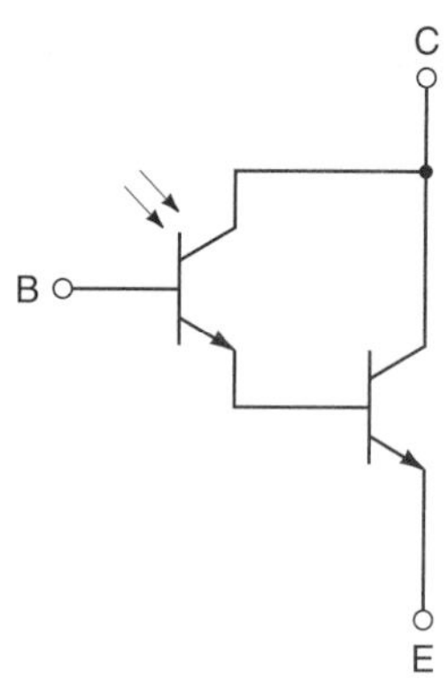

Figure 26.31 The photo-Darlington schematic symbol.

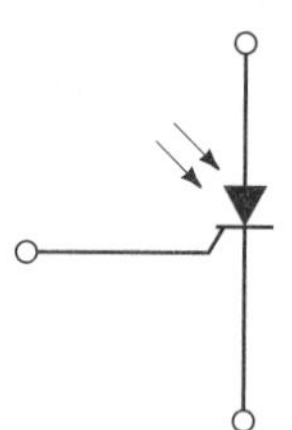

Figure 26.32 The LASCR schematic symbol.

The arrangement in Figure 26.30 is referred to as **optocoupling,** because *the output from the LED control circuit is coupled via light to the phototransistor/op-amp circuit.* The advantage of optocoupling is that it provides complete electrical isolation between the source (the LED control circuit) and the load (the phototransistor/op-amp circuit).

The Photo-Darlington

The **photo-Darlington** is *a phototransistor that is constructed in a Darlington configuration,* as shown in Figure 26.31. Except for the increased output current capability (as is always the case with Darlingtons), the operation and characteristics of the device are identical to those of the standard phototransistor.

The Light-Activated SCR

The **light-activated SCR (LASCR),** or *photo-SCR,* is *a three-terminal, light-activated SCR.* The schematic symbol for the LASCR is shown in Figure 26.32. Again, all the properties of the photodiode can be found in the LASCR, but in this case, the SCR is latched into its conducting state by the light input to the device. Thus, the LASCR can be used in an optically coupled phase-control circuit.

Section Review

1. What is the difference between a *light emitter* and a *light detector*?
2. What is *wavelength*?
3. Why is the concept of wavelength important when working with optoelectronic devices?
4. What is *light intensity*?
5. What is a *photodiode*?
6. In terms of the photodiode, what is *dark current* (I_D)? What is *light current* (I_L)?
7. What is the relationship between photodiode light current and dark current?
8. What is the *wavelength of peak spectral response* for a photodiode?
9. What is *sensitivity*?
10. Explain the *mA/mW/cm*^{2} unit of measure for sensitivity.
11. What is *spectral response*? Which photodiode rating is affected by spectral response?
12. What is a *phototransistor*?
13. What is the relationship between input light and collector current for a phototransistor?

14. What is the difference between the optical ratings of the phototransistor and those of the photodiode?
15. What is the difference between the *photo-Darlington* and the standard phototransistor?
16. What is the *LASCR*? How does it differ from the standard SCR?

26.6 Optoisolators and Optointerrupters

Optoisolators and *optointerrupters* are ICs that contain both a photoemitter and a photodetector. While the photoemitter is always an LED, the photodetector can be any one of the discrete photodetectors that we discussed in the last section. ◄ *OBJECTIVE 12*

Optoisolators

The **optoisolator** is an *optocoupler,* that is, *a device that uses light to couple a signal from its input device (a photoemitter) to its output device (a photodetector).* For example, an optoisolator with a transistor output contains the LED/transistor portion of the circuitry shown in Figure 26.30.

The typical optoisolator comes in a 6-pin DIP. Most optoisolator parameters and ratings are the standard LED and phototransistor specifications. However, several device specific ratings warrant discussion.

The **isolation source voltage (V_{ISO})** rating indicates *the voltage that will destroy the device if it is applied across the input and output pins.* For example, the 4N35–7 series optoisolators are rated at $V_{ISO} = 7500\ V_{pk}$. This rating indicates that the device can withstand up to 7500 V_{pk} across its input/output pins.

The **isolation current (I_{ISO})** rating indicates *the amount of current that can be forced between the input and output at the rated voltage.* For example, if 3550 V_{pk} is applied across the 4N35, it will generate 100 μA (maximum) between the input and output pins of the device.

The **isolation resistance (R_{ISO})** rating indicates *the total resistance between the input and output pins of the device.* For the 4N35–7 series, this rating is $10^{11}\ \Omega$ (100 GΩ).

Except for the ratings listed here, all the parameters and ratings for the optoisolator are fairly standard. However, two of the ratings we discussed earlier are not required for optoisolators. Since the photoemitter and detector are in the same chip, the *wavelength* and *sensitivity* ratings are not needed.

Optointerrupters

The **optointerrupter,** or *optical switch,* is *an IC optocoupler designed to allow an external object to block the light path between the photoemitter and the photodetector.* Two of the many common optointerrupter configurations are shown in Figure 26.33. Note that the photoemitter is on one side of the slot and the photodetector is on the other side.

The open gap allows the optointerrupter to be activated by some external object, such as a piece of paper. For example, let's say that an optointerrupter is positioned in a photocopier machine so that the edge of each sheet of paper passes through the slot as the paper moves through the machine. When no paper is in the gap, the light from the emitter reaches the phototransistor, causing it to saturate, and the collector voltage is low. When a piece of paper passes through the gap, the light is blocked, causing the phototransistor to go into cutoff. This causes its collector voltage to go high. The low and high output voltages are used to tell the machine whether or not a piece of paper is present at a specific location.

MAXIMUM RATINGS

Rating	Symbol	Value	Unit
INPUT LED			
Reverse Voltage	V_R	6	Volts
Forward Current — Continuous	I_F	60	mA
Input LED Power Dissipation @ T_A = 25°C Derate above 25°C	P_D	150 2	mW mW/°C
OUTPUT TRANSISTOR			
Collector-Emitter Voltage	V_{CEO}	30	Volts
Output Current — Continuous	I_C	100	mA
Output Transistor Power Dissipation @ T_A = 25°C Derate above 25°C	P_D	150 2	mW mW/°C
TOTAL DEVICE			
Ambient Operating Temperature Range	T_A	−55 to +100	°C
Storage Temperature	T_{stg}	−55 to +100	°C
Lead Soldering Temperature (5 seconds max)	—	260	°C
Total Device Power Dissipation @ T_A = 25°C Derate above 25°C	P_D	300 4	mW mW/°C

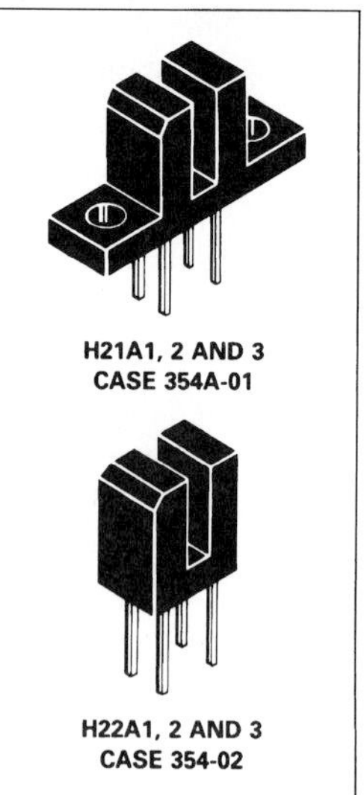

Figure 26.33 Optointerruptor cases. (Copyright of Semiconductor Components Industries, LLC. Used by permission.)

Section Review

1. What is an *optoisolator*?
2. List and define the common *isolation* parameters and ratings of the optoisolator.
3. Contrast the *optointerrupter* with the *optoisolator*.

26.7 Varactor Diodes

OBJECTIVE 13 ➤ The **varactor** is *a type of pn-junction diode with a relatively high junction capacitance when reverse biased.* The capacitance of the junction is controlled by the amount of reverse voltage that is applied to the device. This makes the component very useful as a *voltage-controlled capacitor.* Two commonly used schematic symbols for varactors are shown in Figure 26.34.

The ability of a varactor to act as a voltage-controlled capacitor is easy to understand when you consider the reverse-bias characteristics of the device. When a *pn*-junction is reverse biased, the depletion layer acts as an insulator between the *p*-type and the *n*-type materials. If we view the *p*-type and *n*-type materials as the plates and the depletion layer as the dielectric, it is easy to view the reverse-biased component as a capacitor.

OBJECTIVE 14 ➤ The value of **junction capacitance (C_t)** varies inversely with the width of the depletion layer. Since the width of the depletion layer increases with the amount of reverse bias, we can say that *junction capacitance increases as reverse bias decreases, and vice versa.* This relationship between V_R and C_t is further illustrated in Figure 26.35.

Varactor Ratings

The maximum ratings for the varactor are the same as those for the *pn*-junction diode and the zener diode. Since you are already familiar with these ratings, we will not discuss them here. The two ratings we will discuss are *diode capacitance* and the *capacitance ratio.*

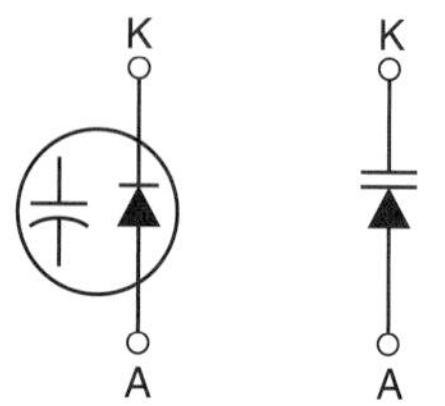

Figure 26.34
Varactor schematic symbols.

Diode Capacitance (C_t)

The **diode capacitance (C_t)** rating is self-explanatory. For the MV209 series varactors, the C_t rating is 26 to 32 pF when V_R is 3 V_{dc}. There is also a nominal value listing of

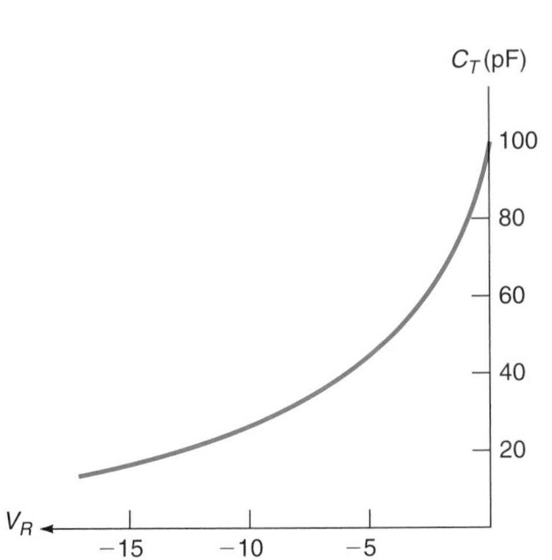

Figure 26.35 Varactor bias versus capacitance curve.

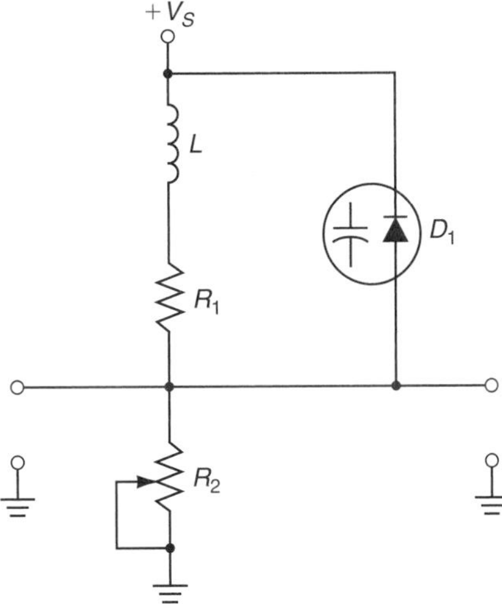

Figure 26.36 A varactor-tuned LC circuit.

29 pF. In other words, the MV209 would be rated as a 29 pF varactor diode, even though the actual value could fall anywhere within the specified range at $V_R = 3\ \text{V}_{\text{dc}}$.

Capacitance Ratio (C_R)

The **capacitance ratio (C_R)** rating tells you *how much the junction capacitance varies over the given range of voltages.* For the MV209 series, the C_R rating ranges from 5.0 to 6.5 for $V_R = 3$ to $25\ \text{V}_{\text{dc}}$. This means that the capacitance at $V_R = 3\ \text{V}_{\text{dc}}$ will be 5.0 to 6.5 times as high as it is at $V_R = 25\ \text{V}_{\text{dc}}$.

The higher the C_R rating, the wider the range of capacitance values. For example, let's say that we have two varactors, each one having a C_t rating of 51 pF when $V_R = 5\ \text{V}_{\text{dc}}$. Let's also assume that D_1 has a rating of $C_R = 1.5$, and that D_2 has a rating of $C_R = 4$ for $V_R = 5$ to $10\ \text{V}_{\text{dc}}$. To find the capacitance of each diode at $V_R = 10\ \text{V}_{\text{dc}}$, we would divide the value of C_t by C_R, as follows:

$$C_{D1} = \frac{51\text{ pF}}{1.5} = 34\text{ pF} \qquad \text{and} \qquad C_{D1} = \frac{51\text{ pF}}{4} = 12.75\text{ pF}$$

Thus, D_1 would have a range of values from 34 to 51 pF, and D_2 would have a range from 12.75 to 51 pF.

Whle the C_R rating of a varactor can be used to determine its capacitance *range,* it cannot be used to determine the diode capacitance at a specific value of V_R. Instead, we would use the component's capacitance versus voltage curve. For example, the diode represented by the curve in Figure 26.35 has a value of approximately 45 pF at $V_R = 5\ \text{V}_{\text{dc}}$.

A Varactor Application

◄ OBJECTIVE 15

Varactors are used almost exclusively in tuned circuits. A tuned LC tank circuit that uses a varactor as the capacitive component is shown in Figure 26.36.

There are two points that you should note. First, because the varactor is pointing toward $+V_S$, we know that it is reverse biased. Therefore, it is acting as a voltage-variable capacitor. Second, R_1 and R_2 form a voltage divider that determines the reverse bias across D_1, and therefore, its capacitance. By adjusting the setting of R_2, we can vary C_t, and therefore, the resonant frequency of the LC circuit.

Section Review

1. A *varactor* acts as what type of capacitance?
2. What is the relationship between the amount of reverse bias applied to a varactor and its capacitance?

3. What is the *capacitance ratio* (C_R) of a varactor?
4. Why is the C_R rating of a varactor important?
5. What is the relationship between varactor reverse bias and the *resonant frequency* of its tuned circuit?

26.8 Transient Suppressors

OBJECTIVE 16 ➤ **Transient suppressors** are *zener diodes with extremely high surge-handling capabilities.* These diodes are commonly used to protect voltage-sensitive circuits from surges, which can occur under a variety of circumstances. You may recall that a *transient* (or *surge*) is an abrupt, high-voltage (or current) condition that lasts for a very brief time. Most often, they are generated by electric motors, air-conditioning and heating units, arcing switches, and lightning.

The transient suppressor uses a configuration similar to that of a shunt clipper to protect the input of a power supply from any ac line surges. A basic surge-protection circuit is shown in Figure 26.37a. This circuit protects the power supply from surges that occur between the hot and the neutral lines. A more complete surge-protection circuit is shown in Figure 26.37b. As you can see, a transient suppressor is placed between each pair of incoming lines.

Transient suppressors have many other uses besides conditioning the ac voltage at the input to a power supply. They are used in many telecommunication, automotive, and consumer devices. Because of their wide range of applications, transient suppressors are available in a wide range of voltages.

OBJECTIVE 17 ➤ For a surge-protection circuit to operate properly, it must have several characteristics:

- The diodes used must have extremely high power dissipation ratings.
- The diodes must be able to turn on very rapidly.

For example, the Motorola 1N5908 through 1N6389 series of transient suppressors can dissipate as much as 1.5 kW for a period of slightly less than 10 ms. This amount of power would destroy any standard zener diode.

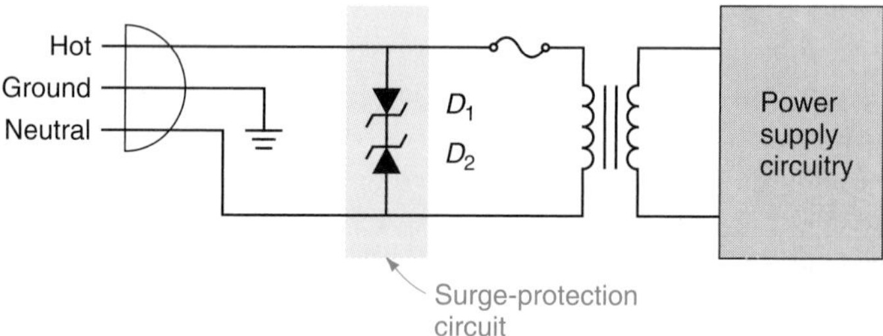

(a) Surge protection between hot and neutral lines

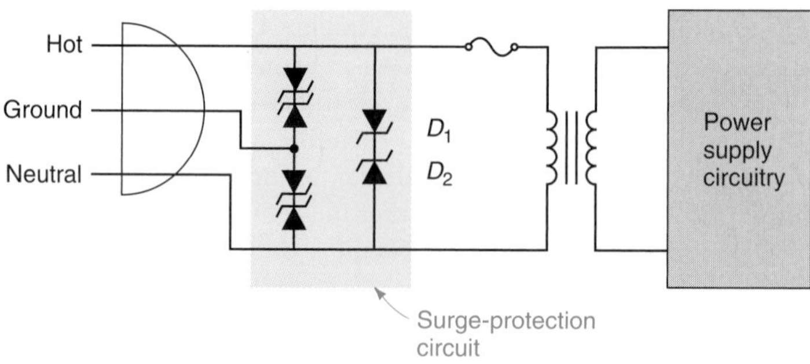

(b) Full surge protection to ground

Figure 26.37

MAXIMUM RATINGS

Rating	Symbol	Value	Units
Peak Power Dissipation (1) @ $T_L < 25°C$	P_{PK}	1500	Watts
Steady State Power Dissipation @ $T_L < 75°C$. Lead Length = 3/8″ Derated above $T_L = 75°C$	P_D	5.0 50	Watts mW/°C
Forward Surge Current (2) @ $T_A = 25°C$	I_{FSM}	200	Amps
Operating and Storage Temperature Range	T_J, T_{stg}	−65 to +175	°C

Lead Temperature not less than 1/16″ from the case for 10 seconds. 230°C

MECHANICAL CHARACTERISTICS

CASE: Void-free, transfer-molded, thermosetting plastic

FINISH: All external surfaces are corrosion resistant and leads are readily solderable and weldable

POLARITY: Cathode indicated by polarity band. When operated in zener mode, will be positive with respect to anode

MOUNTING POSITION: Any

ELECTRICAL CHARACTERISTIC ($T_A = 25°C$ unless otherwise noted) V_F* = 3.5V max, I_F** = 100A) (C suffix denotes standard back to back versions Test both polarities)

JEDEC Device	Device	Breakdown Voltage V_{BR} Volts Min	Breakdown Voltage @ I_T (mA)	Maximum Reverse Stand-Off Voltage V_{RWM}*** (Volts)	Maximum Reverse Leakage @ V_{RWM} I_R (μA)	Maximum Reverse Surge Current I_{RSM}† (Amps)	Maximum Reverse Voltage @ I_{RSM}† (Clamping Voltage) V_{RSM}(Volts)	Clamping Voltage: Peak Pulse Current @ I_{pp1}† = 1.0A V_{C1} (Volts max)	Clamping Voltage: Peak Pulse Current @ I_{pp2}† = 10A V_{C2} (Volts max)
1N6373	ICTE-5/MPTE-5	6.0	1.0	5.0	300	160	9.4	7.1	7.5
—	ICTE-5C/MPTE-5C	6.0	1.0	5.0	300	160	9.4	8.1	8.3
1N6374	ICTE-8/MPTE-8	9.4	1.0	8.0	25	100	15.0	11.3	11.5
1N6382	ICTE-8C/MPTE-8C	9.4	1.0	8.0	25	100	15.0	11.4	11.6
1N6375	ICTE-10/MPTE-10	11.7	1.0	10	2.0	90	16.7	13.7	14.1
1N6383	ICTE-10C/MPTE-10C	11.7	1.0	10	2.0	90	16.7	14.1	14.5
1N6376	ICTE-12/MPTE-12	14.1	1.0	12	2.0	70	21.2	16.1	16.5
1N6384	ICTE-12C/MPTE-12C	14.1	1.0	12	2.0	70	21.2	16.7	17.1

Figure 26.38 1N5908–6389 series transient suppressor ratings. (Copyright of Semiconductor Components Industries, LLC. Used by permission.)

Transient Suppressor Specifications

The maximum ratings and electrical characteristics shown in Figure 26.38 are for the 1N5908 through 1N6389 series transient suppressors. We will use the values shown for our discussion on typical suppressor specifications.

The maximum ratings are fairly standard, except for the **peak power dissipation rating (P_{PK}).** This rating indicates *the surge-handling capability of the suppressors.* Several notes (not shown) on the spec sheet indicate that this rating has some conditions:

- The operating temperature is 25°C.
- The surge has a duration of approximately 8 ms.
- The maximum number of surges is four per minute.

We will look briefly at each of these dependent conditions.

The temperature dependency of the P_{PK} rating is illustrated in Figure 26.39a. As you can see, the P_{PK} rating is derated (by percentage) at temperatures above 25°C. You may be wondering why a temperature curve is used when the spec sheet for the 1N5908 series suppressors lists a power derating value. The derating value listed in the maximum ratings table applies to the power dissipation (P_D) rating of the components, not to the P_{PK} rating. The time dependency of the P_{PK} rating is illustrated in Figure 26.39b. As shown, P_{PK} and surge duration (pulse width) vary inversely. As the duration of the surge increases, the power-dissipating capability of the suppressors decreases, and vice versa.

The **maximum reverse stand-off voltage (V_{RWM})** rating is extremely important. This rating indicates the maximum allowable peak reverse voltage for any circuit using the

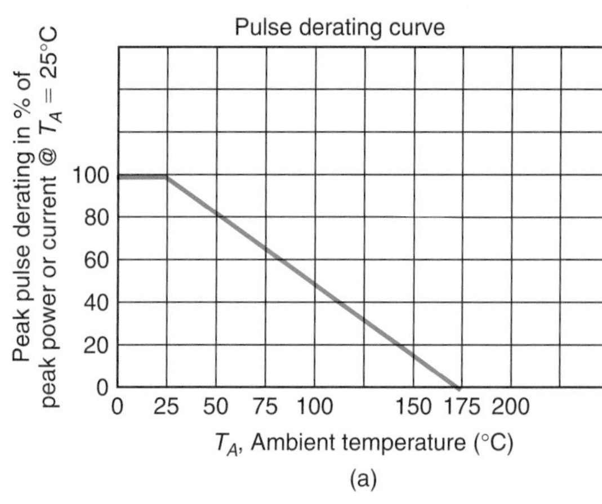

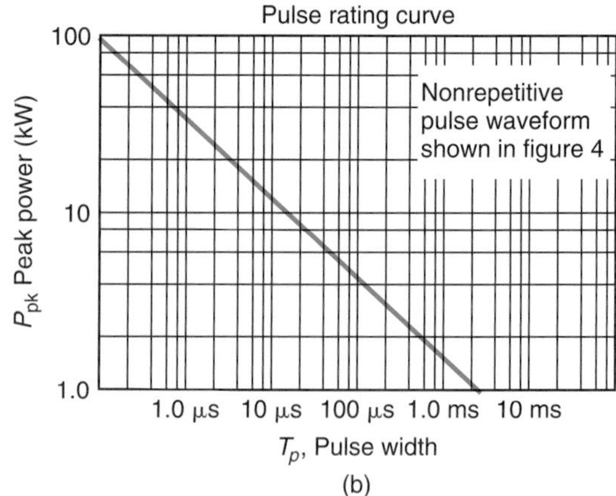

Figure 26.39 1N5908–6389 series transient suppressor derating curves. (Copyright of Semiconductor Components Industries, LLC. Used by permission.)

component. Refer back to Figure 26.37. Assuming that the circuit is driven by a 120 V_{ac} line (170 V_{pk}), the diodes must have V_{RWM} ratings that are greater than 170 V. Otherwise, they may turn on during normal circuit operation. Note that V_{RWM} is a primary consideration when substituting one suppressor for another.

Back-to-Back Suppressors

If you look closely at the ratings for the 1N6374 and the 1N6382, you'll see that their ratings are identical. You may be wondering, then, what the difference is between these two suppressors.

Figure 26.40
Back-to-back suppressor.

The 1N6382 is a **back-to-back suppressor.** This type of suppressor, whose schematic symbol is shown in Figure 26.40, is *a device that contains two transient suppressors that are connected internally,* like the two diodes shown in Figure 26.37. Note that back-to-back suppressors have no V_F rating, because they are designed to break down at the rated value of V_{BR} in both directions.

Section Review

1. What is a *transient suppressor*?
2. What is a *surge*? How are ac power line surges commonly generated?
3. What are the required characteristics for a surge-protection circuit?
4. How does a transient suppressor differ from a conventional zener diode?
5. Define each of the following ratings:
   a. *Peak power dissipation* (P_{PK}).
   b. *Maximum reverse stand-off voltage* (V_{RWM}).
6. Why don't back-to-back suppressors have V_F ratings?

26.9 Tunnel Diodes

OBJECTIVE 18 ➤

Tunnel diodes are used in the ultrahigh frequency (UHF) and microwave frequency range. Applications include amplifiers, oscillators, modulators, and demodulators. The schematic symbol and operating curve for the tunnel diode are shown in Figure 26.41a. The operating curve is a result of heavy doping of the tunnel diode, which is approximately 1000 times as high as the doping levels of *pn*-junction diodes. Figure 26.41b compares the curves of a *pn*-junction and a tunnel diode.

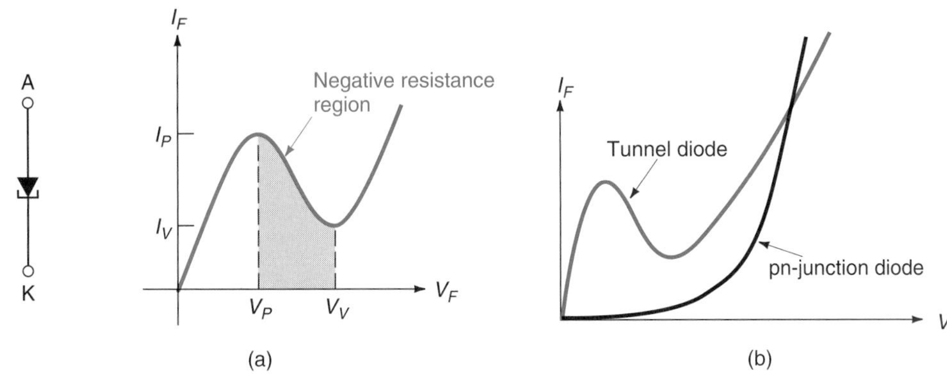

Figure 26.41 Tunnel diode symbol and characteristic curve.

In the forward operating region of the tunnel diode, we are interested in the area between the *peak voltage* (V_P) and the **valley voltage (V_V).** The *peak current* (I_P), which occurs when $V_F = V_P$, is the maximum value that I_F will reach under normal circumstances. As V_F increases to the value of V_V, I_F decreases to its minimum value, called the *valley current* (I_V). As you can see, the forward voltage and current vary inversely when the diode is operated between the values of V_P and V_V.

The region of operation between the peak and the valley voltage is referred to as the *negative resistance region.* The term *negative resistance* describes the dynamic resistance of the component over the range of V_P to V_V, found as

$$R_d = \frac{\Delta V}{\Delta I} = \frac{V_V - V_P}{I_V - I_P} \quad (26.8)$$

Since the dynamic resistance of the component is defined using inversely changing values of voltage and current, the calculated value of R_d is negative. This point is illustrated in Example 26.4.

Example 26.4

A tunnel diode has the following values: $I_P = 2$ mA at $V_P = 150$ mV, and $I_V = 100$ μA at $V_V = 500$ mV. Calculate the dynamic resistance of the device.

Solution: Using the values given, the dynamic resistance of the component is found as

$$R_d = \frac{\Delta V}{\Delta I} = \frac{V_V - V_P}{I_V - I_P} = \frac{500\text{ mV} - 150\text{ mV}}{100\ \mu\text{A} - 2\text{ mA}} = -184.2\ \Omega$$

Note that negative resistance is a mathematical concept describing the change that takes place over a range of values. At any point on the curve, the ratio of voltage to current still yields a positive value. Tunnel diodes are operated almost exclusively in the negative resistance region.

Tunnel Diode Oscillator

A basic tunnel diode oscillator, or **negative resistance oscillator,** is shown in Figure 26.42. The output frequency is approximately equal to the resonant frequency of the *LC* tank circuit. The key to sustaining oscillation is to provide the tank with additional current at the peak of each cycle. The tunnel diode acts like a switch, providing the necessary additional current.

The voltage divider (R_1 and R_2) provides the biasing voltage for the tunnel diode. When the tank circuit waveform is at its positive peak, the difference between this voltage and the biasing voltage equals the V_P rating of the diode. Thus, diode current is at its maximum value when the tank circuit waveform is at its positive peak. As the tank

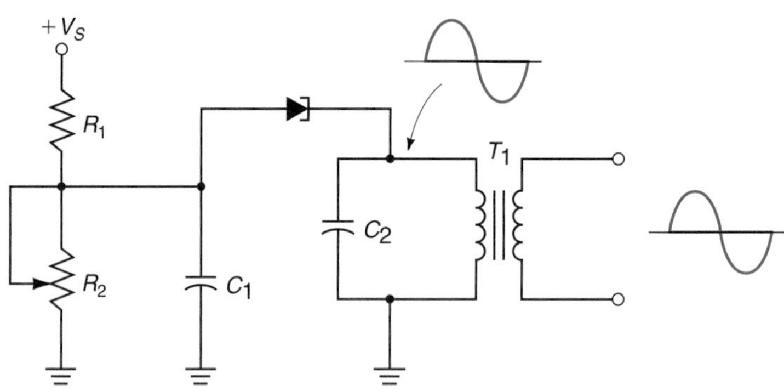

Figure 26.42 Tunnel diode oscillator.

circuit waveform decreases toward zero, the difference between this voltage and the biasing voltage approaches the V_V rating of the diode. Thus, diode current decreases as the sine wave voltage decreases.

Section Review

1. What is a *tunnel diode*?
2. What is the relationship between tunnel diode *forward current* and *voltage* when the component is operated in the *negative resistance region*?
3. Refer to the curves shown in Figure 26.41b. What tunnel diode characteristic accounts for the difference in the operating curves?
4. What is meant by *negative resistance*?

Key Terms

The following terms were introduced in this chapter on the pages indicated:

Anode current interruption, 873
Average on-state current (I_T), 875
Average on-state voltage (V_T), 875
Back-to-back suppressor, 898
Bidirectional thyristor, 882
Capacitance ratio (C_R), 895
Circuit fusing rating (I^2t), 877
Conduction angle, 879
Critical rise (dv/dt), 878
Crowbar, 879
Dark current (I_D), 890
Diac, 882
Diode capacitance (C_t), 894
False triggering, 878
Forced commutation, 874
Forward blocking region, 874
Forward breakover current ($I_{BR(F)}$), 874
Forward breakover voltage ($V_{BR(F)}$), 872
Gate, 875
Gate nontrigger voltage (V_{GD}), 878
Gate trigger voltage (V_{GT}), 884
Holding current (I_H), 873
Intrinsic standoff ratio (η), 886
Isolation current (I_{ISO}), 893
Isolation resistance (R_{ISO}), 893
Isolation source voltage (V_{ISO}), 893
Junction capacitance (C_t), 894
Light-activated SCR (LASCR), 892
Light current (I_L), 890
Light detectors, 888
Light emitters, 885
Light intensity, 889
Maximum reverse standoff voltage (V_{RSM}), 897
Negative resistance, 887
Negative resistance oscillator, 899
Negative resistance region, 887
Opaque, 890
Optocoupling, 892
Optoelectronic devices, 872
Optointerruptor, 893
Optoisolator, 893
Peak current (I_P), 886
Peak power dissipation rating (P_{PK}), 897
Peak voltage (V_P), 886
Phase controller, 880
Photo-Darlington, 892
Photodiode, 889
Phototransistor, 890

Reverse blocking region, 874
Reverse breakdown voltage ($V_{BR(V)}$), 874
Sensitivity, 890
Silicon-controlled rectifier (SCR), 875
Silicon unilateral switch (SUS), 872
Snubber network, 879
Spectral response, 890
Thyristors, 872
Transient suppressor, 896
Triac, 882
Tunnel diode, 898
Unijunction transistor (UJT), 885
Valley current (I_V), 887
Valley voltage (V_V), 899
Varactor, 894
Wavelength, 888
Wavelength of peak spectral response (λ_S), 890
Zero adjust, 890

Practice Problems

1. The 2N6237 SCR has a circuit fusing rating of 2.6 A^2s. Determine whether or not the component can withstand a 25 A surge that lasts for 120 ms.
2. The 2N6342 SCR has a circuit fusing rating of 40 A^2s. Determine whether or not the device can withstand an 80 A surge that lasts for 8 ms.
3. The 2N6237 SCR has a circuit fusing rating of 2.6 A^2s. Determine the maximum allowable duration of a 50 A surge through the device.
4. The 2N6342 has a circuit fusing rating of 40 A^2s. Determine the maximum allowable duration of a 95 A surge through the device.
5. The 2N6237 has a circuit fusing rating of 2.6 A^2s. Determine the maximum surge current that the device can withstand for 50 ms.
6. The 2N6342 has a circuit fusing rating of 40 A^2s. Determine the maximum surge current that the device can withstand for 100 ms.
7. The 2N6237 has a critical rise rating of $dv/dt = 10$ V/μs. Determine the anode noise amplitude at $t_r = 10$ ns that is required to cause false triggering.
8. The 2N6342 has a critical rise rating of $dv/dt = 5$ V/μs. Determine the anode noise amplitude at $t_r = 25$ ns that is required to cause false triggering.
9. The 2N2646 has a range of $\eta = 0.56$ to 0.75. Determine the range of V_P values for the device when $V_{BB} = +14$ V.
10. The 2N4871 UJT has a range of $\eta = 0.7$ to 0.85. Determine the range of V_P values for the device when $V_{BB} = +12$ V.
11. The 2N4948 UJT has a range of $\eta = 0.55$ to 0.82. Determine the range of V_P values for the device when $V_{BB} = +16$ V.
12. The 2N4949 UJT has a range of $\eta = 0.74$ to 0.86. Determine the range of V_P values for the device when $V_{BB} = +16$ V.
13. Determine the wavelength (in nm) for a 650 THz signal.
14. Determine the wavelength (in nm) for a 220 THz signal.
15. Determine the wavelength (in nm) for a 180 THz signal.
16. The visible light spectrum includes all frequencies between approximately 400 and 800 THz. Determine the range of wavelength values for this band of frequencies.
17. A photodetector has a value of $\lambda_S = 0.94$ nm. Determine the optimum operating frequency for the device.
18. A photodetector has a rating of $\lambda_S = 0.84$ nm. Determine the optimum operating frequency for the device.
19. The varactor shown in Figure 26.43 has the following values: $C_t = 48$ pF at $V_R = 3\ V_{dc}$, and $C_R = 4.8$ for $V_R = 3$ to $12\ V_{dc}$. Determine the resonant frequency for this circuit at $V_R = 3\ V_{dc}$ and $V_R = 12\ V_{dc}$.

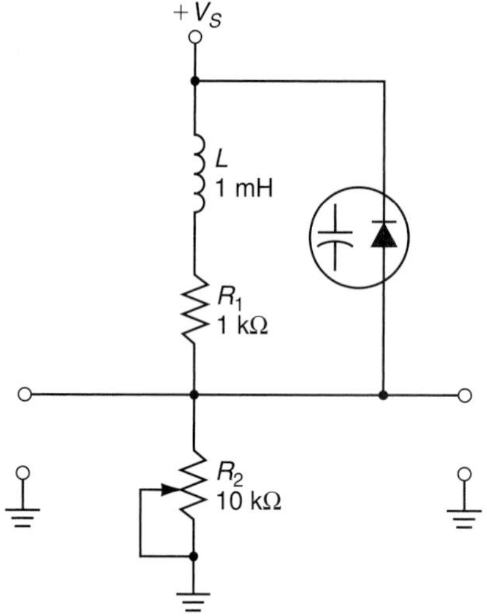

Figure 26.43

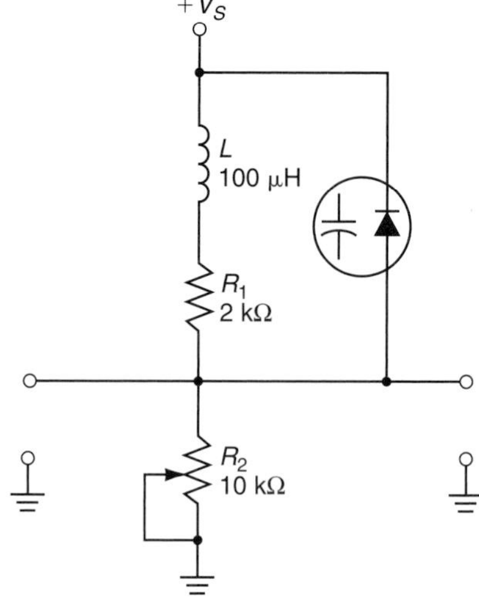

Figure 26.44

20. The varactor shown in Figure 26.44 has the following values: $C_t = 68$ pF at $V_R = 4\ \text{V}_{\text{dc}}$, and $C_R = 1.12$ for $V_R = 4$ to $10\ \text{V}_{\text{dc}}$. Determine the resonant frequency for this circuit at $V_R = 4\ \text{V}_{\text{dc}}$ and $V_R = 10\ \text{V}_{\text{dc}}$.

Looking Back

These problems relate to material presented in earlier chapters. The chapters are identified in brackets.

21. An emitter bias circuit has the following values: $V_{CC} = +12$ V, $V_{EE} = 12$ V, $R_E = 3$ kΩ, and $R_C = 1.5$ kΩ. Calculate $I_{C(\text{sat})}$ and $V_{CE(\text{off})}$ for the circuit. [Chapter 19]
22. A JFET has the following maximum values: $V_{GS(\text{off})} = -8$ V, and $I_{DSS} = 12$ mA. Calculate the maximum value of I_D when $V_{GS} = -3$ V. [Chapter 21]
23. A BJT has the following values: $t_d = 25$ ns, $t_r = 40$ ns, $t_s = 80$ ns, and $t_f = 50$ ns. Calculate the maximum practical switching frequency for the device. [Chapter 24]
24. The op-amp in a voltage follower has the following values: $Z_{\text{in}} = 1.5$ MΩ, $A_{OL} = 100{,}000$, and $Z_{\text{out}} = 80$ Ω. Calculate the values of A_{CL} and CMRR for the circuit. [Chapter 22]
25. A first-order, high-pass filter has values of $R = 33$ Ω and $C = 0.33$ μF. Draw the frequency response curve for the circuit. [Chapter 23]

Answers to the Example Practice Problems

26.1 For the application, $I^2t = 54\ \text{A}^2\text{s}$. The device cannot handle the surge.

26.2 2.5 mV

26.3 319.1 THz

Multisim File References

Figure	File Reference
26.10	ETF26_10.msm
26.43	ETF26_43.msm

appendix A

Conversions and Units

Conversions

Multiply	By	To Get
ampere-turns	1.257	gilberts
ampere-turns per cm	2.540	ampere-turns per inch
ampere-turns per inch	0.3937	ampere-turns per cm
ampere-turns per inch	0.4950	gilberts per cm
British thermal units	1054	joules
British thermal units	2.928×10^{-4}	kilowatt-hours
centimeters	0.3937	inches
centimeters	0.01	meters
centimeters	393.7	mils
centimeters	10	millimeters
circular mils	5.067×10^{-6}	square centimeters
circular mils	7.854×10^{-7}	square inches
circular mils	0.7854	square mils
cubic centimeters	3.531×10^{-5}	cubic feet
cubic centimeters	6.102×10^{-2}	cubic inches
cubic centimeters	10^{-6}	cubic meters
cubic centimeters	2.642×10^{-4}	gallons
cubic centimeters	10^{-3}	liters
cubic feet	2.832×10^{4}	cubic centimeters
cubic feet	1728	cubic inches
cubic feet	0.02832	cubic meters
cubic feet	7.481	gallons
cubic feet	28.32	liters
cubic inches	16.39	cubic centimeters
cubic inches	5.787×10^{-4}	cubic feet
cubic inches	1.639×10^{-5}	cubic meters
cubic inches	2.143×10^{-5}	cubic yards
cubic inches	4.329×10^{-3}	gallons
cubic inches	1.639×10^{-2}	liters
cubic meters	10^{6}	cubic centimeters

Multiply	*By*	*To Get*
cubic meters	35.31	cubic feet
cubic meters	6.1023×10^4	cubic inches
cubic meters	264.2	gallons
cubic meters	10^9	liters
degrees (∠)	60	minutes
degrees (∠)	1.745×10^{-2}	radians
degrees (∠)	3600	seconds
degrees per second	1.745×10^{-2}	radians per second
degrees per second	0.1667	revolutions per minute
degrees per second	2.778×10^{-3}	revolutions per second
dynes	1.02×10^{-3}	grams
dynes	2.248×10^{-6}	pounds
feet	30.48	centimeters
feet	12	inches
feet	0.3048	meters
feet	0.333	yards
gauss	6.452	lines per square inch
gilberts	0.7958	ampere-turns
gilberts per centimeter	2.021	ampere-turns per inch
horsepower	42.44	BTU per minute
horsepower	3.3×10^4	foot-pounds per minute
horsepower	550	foot-pounds per second
horsepower	1.014	metric horsepower
horsepower	10.7	kilogram-calories per minute
horsepower	0.7457	kilowatts
horsepower	745.7	watts
inches	2.54	centimeters
inches	10^3	mils
joules	9.486×10^{-4}	British thermal units
joules	10^7	ergs
joules	0.7376	foot-pounds
joules	2.39×10^{-4}	kilogram-calories
joules	0.102	kilogram-meters
joules	2.778×10^{-4}	watt-hours
kilometers	10^5	centimeters
kilometers	3281	feet
kilometers	10^3	meters
kilometers	0.6214	miles
kilometers	1093.6	yards
kilowatts	56.92	BTU per minute
kilowatts	4.425×10^4	foot-pounds per minute
kilowatts	737.6	foot-pounds per second
kilowatts	1.341	horsepower
kilowatts	14.34	kilogram-calories per minute
kilowatts	10^3	watts
kilowatt-hours	3415	British thermal units
kilowatt-hours	1.341	horsepower-hours
kilowatt-hours	2.665×10^6	joules
kilowatt-hours	860.5	kilogram-calories
kilowatt-hours	3.671×10^5	kilogram-meters
lines per cm^2	1	gauss
lines per in^2	0.155	gauss
ln (N)	0.4343	$\log_{10} N$

Multiply	*By*	*To Get*
$\log_{10} N$	2.303	ln (N)
lumens per ft^2	1	foot-candles
Maxwells	10^{-2}	microwebers (μWb)
meters	100	centimeters
meters	3.2808	feet
meters	39.37	inches
meters	10^{-3}	kilometers
meters	10^3	millimeters
meters	1.0936	yards
microfarads	10^{-6}	farads
microns	10^{-6}	meters
microwebers	100	Maxwells
miles	1.609×10^5	centimeters
miles	5280	feet
miles	1.6093	kilometers
miles	1760	yards
millihenries	10^{-3}	henries
millimeters	0.1	centimeters
millimeters	0.03937	inches
millimeters	39.37	mils
mils	2.54×10^{-3}	centimeters
mils	10^{-3}	inches
minutes ($\angle$)	2.909×10^{-4}	radians
minutes ($\angle$)	60	seconds ($\angle$)
quadrants ($\angle$)	90	degrees
quadrants ($\angle$)	5400	minutes
quadrants ($\angle$)	1.571	radians
radians	57.3	degrees
radians	3438	minutes
radians	0.637	quadrants
radians per second	57.3	degrees per second
radians per second	0.1592	revolutions per second
radians per second	9.549	revolutions per minute
revolutions	360	degrees
revolutions	4	quadrants
revolutions	6.283	radians
seconds ($\angle$)	4.848×10^{-4}	radians
square centimeters	1.973×10^5	circular mils
square centimeters	1.076×10^{-3}	square feet
square centimeters	0.155	square inches
square centimeters	10^{-4}	square meters
square centimeters	100	square millimeters
square feet	929	square centimeters
square feet	144	square inches
square feet	9.29×10^{-2}	square meters
square feet	3.587×10^{-3}	square miles
square feet	0.1111	square yards
square inches	1.273×10^6	circular mils
square inches	6.452	square centimeters
square inches	6.944×10^{-3}	square feet
square inches	645.2	square millimeters
square kilometers	10.76×10^6	square feet
square kilometers	10^6	square meters
square kilometers	0.3861	square miles
square meters	10.764	square feet
square meters	3.861×10^{-7}	square miles
square miles	2.788×10^7	square feet

Multiply	*By*	*To Get*
square miles	2.59	square kilometers
square millimeters	1.973×10^{3}	circular mils
square millimeters	0.01	square centimeters
square millimeters	1.55×10^{-3}	square inches
watts	5.692×10^{-2}	BTU per minute
watts	10^{7}	ergs per second
watts	44.62	foot-pounds per minute
watts	0.7376	foot-pounds per second
watts	1.341×10^{-3}	horsepower
watts	1.434×10^{-2}	kilogram-calories per minute
watt-hours	3.415	British thermal units
watt-hours	2655	foot-pounds
watt-hours	0.8605	kilogram-calories
watt-hours	367.1	kilogram-meters
watt-hours	10^{-3}	kilowatt-hours

Most of the values in this list were obtained from Appendix J of *Modern Electronics* by David Bruce (Reston Publishing Co., 1984).

Temperature Scales and Conversions

Scales

Scale	*Unit of Measure*
Fahrenheit (F)	°F
Rankin (R)[a]	°F
Celsius (C)	°C
Kelvin (K)[b]	K

[a]Sometimes referred to as *absolute Fahrenheit.*

[b]Sometimes referred to as *absolute temperature.*

Conversion Equations

$$T_C = \frac{5}{9}(T_F - 32) \qquad T_F = 1.8T_C + 32$$

$$T_C = T_K - 273.15 \qquad T_K = T_C + 273.15$$

$$T_F = T_R - 480 \qquad T_R = T_F + 480$$

International System of Units (SI)

Quantity	*Unit of Measure*	*Symbol for Unit*
acceleration	meters per second squared	m/s^2
angular acceleration	radians per second squared	rad/s^2
angular velocity	radians per second	rad/s
area	square meters	m^2
density	kilograms per cubic meter	kg/m^3
electric capacitance	farad	F
electric charge	coulomb	C
electric field strength	volts per meter	V/m
electric resistance	Ohms	Ω
flux of light	lumens	lm
force	newtons	N
frequency	hertz	Hz
illumination	lux	lx
inductance	henry	H
luminance	candela per square meter	cd/m^2
magnetic field strength	ampere per meter	A/m
magnetic flux	weber	Wb
magnetic flux density	tesla	T
magnetomotive force	ampere	A
power	watt	W
pressure	newtons per meter squared	N/m^2
velocity	meters per second	m/s
voltage, potential difference, electromotive force	volt	V
volume	cubic meters	m^3
work, energy, quantity of heat	joule	J

The material in this list was obtained from Appendix I in *Modern Electronics* by David Bruce (Reston Publishing Co., 1984).

appendix B

Resistor Standard Values and Color Codes

Precision Resistors

Precision resistors are commonly available in 0.1%, 0.25%, 0.5%, and 1% tolerances. The values listed here are the standard digit combinations for the resistors in this tolerance range. Resistors with a 1% tolerance are available only in the magnitudes shown in **bold**.

100	101	**102**	104	**105**	106	**107**	109	**110**	111	**113**	114
115	117	**118**	120	**121**	126	**127**	129	**130**	132	**133**	135
137	138	**140**	142	**143**	145	**147**	149	**150**	152	**154**	156
158	160	**162**	164	**165**	167	**169**	172	**174**	176	**178**	180
182	184	**187**	189	**191**	193	**196**	198				
200	203	**205**	208	**210**	213	**215**	218	**221**	223	**226**	229
232	234	**237**	240	**243**	246	**249**	252	**255**	258	**261**	264
267	271	**274**	277	**280**	284	**287**	291	**294**	298		
301	305	**309**	312	**316**	320	**324**	328	**332**	336	**340**	344
348	352	**357**	361	**365**	370	**374**	379	**383**	388	**392**	397
402	407	**412**	417	**422**	427	**432**	437	**442**	448	**453**	459
464	470	**475**	481	**487**	493	**499**					
505	**511**	517	**523**	530	**536**	542	**549**	562	**566**	569	**576**
583	**590**	599									
604	612	**619**	626	**634**	642	**649**	657	**665**	673	**681**	690
698											
706	**715**	723	**732**	741	**750**	759	**768**	777	**787**	796	
806	816	**825**	835	**845**	856	**866**	876	**887**	898		
909	920	**931**	942	**953**	965	**976**	988				

Precision Resistor Color Code

Precision resistors have *three* significant digits (instead of two), so they require a modified color code. This modified color code is represented in Figure B.1. In many cases, alphanumeric codes are used to indicate the values of lower-tolerance resistors. However, you will also see resistors that use the color code shown in the figure. Here are some example interpretations of the five-band code:

Red, Black, Gray, Brown, Blue = $208 \times 10^1 = 2080\ \Omega = 2.08\ \text{k}\Omega$ (0.25%)
Violet, Brown, Green, Red, Brown = $715 \times 10^2 = 71{,}500 = 71.5\ \text{k}\Omega$ (1%)
Green, Orange, Blue, Black, Green = $536 \times 10^0 = 536\ \Omega$ (0.5%)

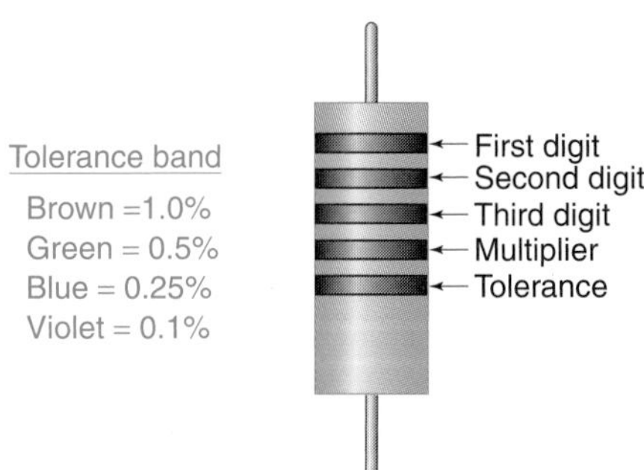

Figure B.1 The precision resistor color code.

appendix C

Specification Sheets

2N3905
2N3906

CASE 29-04, STYLE 1
TO-92 (TO-226AA)

Collector 3
2 Base
1 Emitter

GENERAL PURPOSE TRANSISTOR

NPN SILICON

MAXIMUM RATINGS

Rating	Symbol	Value	Unit
Collector-Emitter Voltage	V_{CEO}	40	Vdc
Collector-Base Voltage	V_{CBO}	40	Vdc
Emitter-Base Voltage	V_{EBO}	5.0	Vdc
Collector Current — Continuous	I_C	200	mAdc
Total Device Dissipation @ $T_A = 25°C$ Derate above 25°C	P_D	625 5.0	mW mW/°C
Total Power Dissipation @ $T_A = 60°C$	PD	250	mW
Total Device Dissipation @ $T_C = 25°C$ Derate above 25°C	P_D	1.5 12	Watts mW/°C
Operating and Storage Junction Temperature Range	T_J, T_{stg}	– 550 to + 150	°C

THERMAL CHARACTERISTICS

Characteristic	Symbol	Max	Unit
Thermal Resistance, Junction to Case	$R_{\theta JC}$	83.3	°C/W
Thermal Resistance, Junction to Ambient	$R_{\theta JA}$	200	°C/W

ELECTRICAL CHARACTERISTICS ($T_A = 25°C$ unless otherwise noted.)

Characteristic	Symbol	Min	Max	Unit
OFF CHARACTERISTICS				
Collector-Emitter Breakdown Voltage(1) (I_C1.0 mAdc, $I_B = 0$)	$V_{(BR)CEO}$	40	—	Vdc
Collector-Base Breakdown Voltage (I_C10 μAdc, $I_E = 0$)	$V_{(BR)CBO}$	40	—	Vdc
Emitter-Base Breakdown Voltage ($I_E = 10$ μAdc, $I_C = 0$)	$V_{(BR)EBO}$	5.0	—	Vdc
Base Cutoff Current ($V_{CE} = 30$ Vdc, $V_{BE} = 3.0$ Vdc)	I_{BL}	—	50	nAdc
Collector Cutoff Current ($V_{CE} = 30$ Vdc, $V_{BE} = 3.0$ Vdc)	I_{CEX}	—	50	nAdc

Figure C.1 The 2N3905–3906 specification sheet. (Copyright of Semiconductor Components Industries, LLC. Used by permission.)

ELECTRICAL CHARACTERISTICS (continued) ($T_A = 25°C$ unless otherwise noted.)

Characteristic		Symbol	Min	Max	Unit
ON CHARACTERISTICS(1)					
DC Current Gain		h_{FE}			—
($I_C = 0.1$ mAdc, $V_{CE} = 1.0$ Vdc)	2N3905		30	—	
	2N3906		60	—	
($I_C = 1.0$ mAdc, $V_{CE} = 1.0$ Vdc)	2N3905		40	—	
	2N3906		80	—	
($I_C = 10$ mAdc, $V_{CE} = 1.0$ Vdc	2N3905		50	150	
	2N3906		100	300	
($I_C = 50$ mAdc, $V_{CE} = 1.0$ Vdc)	2N3905		30	—	
	2N3906		60	—	
($I_C = 1]]$ mAdc, $V_{CE} = 1.0$ Vdc)	2N3905		15	—	
	2N3906		30	—	
Collector-Emitter Saturation Voltage		$V_{CE(sat)}$			Vdc
($I_C = 10$ mAdc, $I_B = 1.0$ mAdc)			—	0.25	
($I_C = 50$ mAdc, $I_B = 5.0$ mAdc)			—	0.4	
Base-Emitter Saturation Voltage		$V_{BE(sat)}$			Vdc
($I_C = 10$ mAdc, $I_B = 1.0$ mAdc)			0.65	0.85	
($I_C = 50$ mAdc, $I_B = 5.0$ mAdc)			—	0.95	
SMALL-SIGNAL CHARACTERISTICS					
Current-Gain — Bandwidth Product		f_T			MHz
($I_C = 10$ mAdc, $V_{CE} = 20$ Vdc, f = 100 MHz)	2N3905		200	—	
	2N3906		250	—	
Output Capacitance ($V_{CB} = 5.0$ Vdc, $I_E = 0$, f = 100 kHz)		C_{obo}	—	4.5	pF
Input Capacitance ($V_{CB} = 0.5$ Vdc, $I_C = 0$, f = 100 kHz)		C_{ibo}	—	10.0	pF
Input Impedance		h_{ie}			k ohms
($I_C = 1.0$ mAdc, $V_{CE} = 10$ Vdc, f = 1.0 kHz)	2N3905		0.5	8.0	
	2N3906		2.0	12	
Voltage Feedback Ratio		h_{re}			X 10^{-4}
($I_C = 1.0$ mAdc, $V_{CE} = 10$ Vdc, f = 1.0 kHz)	2N3905		0.1	5.0	
	2N3906		0.1	10	
Small-Signal Current Gain		h_{fe}			—
($I_C = 1.0$ mAdc, $V_{CE} = 10$ Vdc, f = 1.0 kHz)	2N3905		50	200	
	2N3906		100	400	
Output Admittance		h_{oe}			μmhos
($I_C = 1.0$ mAdc, $V_{CE} = 10$ Vdc, f = 1.0 kHz)	2N3905		1.0	40	
	2N3906		3.0	60	
Noise Figure		NF			dB
($I_C = 100$ μAdc, $V_{CE} = 5.0$ Vdc, $R_S = 1.0$ k ohms,	2N3905		—	5.0	
f = 10 Hz to 15.7 kHz)	2N3906		—	4.0	
SWITCHING CHARACTERISTICS					
Delay Time — ($V_{CC} = 3.0$ Vdc, $V_{BE} = 0.5$ Vdc, $I_C = 10$ mAdc, $I_{B1} = 1.0$ mAdc)		t_d	—	35	ns
Rise Time — ($V_{CC} = 3.0$ Vdc, $V_{BE} = 0.5$ Vdc, $I_C = 10$ mAdc, $I_{B1} = 1.0$ mAdc)		t_r	—	35	ns
Storage Time	2N3905	t_s	—	200	ns
	2N3906		—	225	
Fall Time — ($V_{CC} = 3.0$ Vdc, $I_C = 10$ mAdc, $I_{B1} = I_{B2} = 1.0$ mAdc)	2N3905	t_f	—	60	ns
	2N3904		—	75	

(1) Pulse Test: Pulse Width ≤ 300 μs, Duty Cycle ≤ 2.0%.

FIGURE 1 — DELAY AND RISE TIME EQUIVALENT TEST CIRCUIT

FIGURE 2 — STORAGE AND FALL TIME EQUIVALENT TEST CIRCUIT

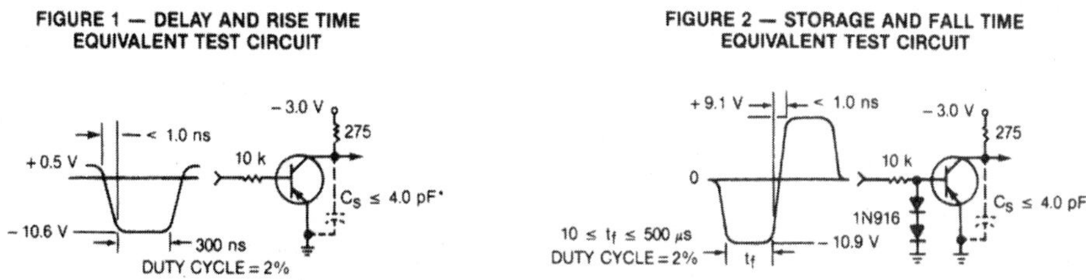

*Total shunt capacitance of test jig and connectors

Figure C.1 (Continued)

2N4013
2N4014

CASE 22-03, STYLE 1
TO-18 (TO-206AA)

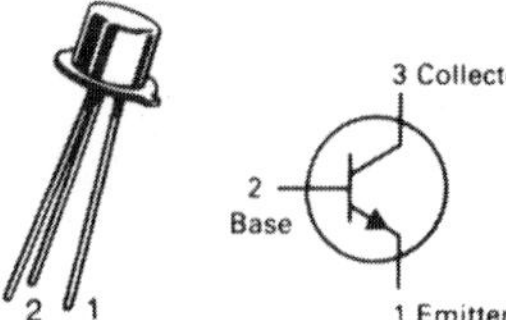

SWITCHING TRANSISTORS

NPN SILICON

MAXIMUM RATINGS

Rating	Symbol	2N4013	2N4014	Unit
Collector-Emitter Voltage	V_{CEO}	30	50	Vdc
Collector-Base Voltage	V_{CBO}	50	80	Vdc
Emitter-Base Voltage	V_{EBO}	6.0		Vdc
Collector Current — Continuous — Peak	I_C	1.0 2.0		Adc
Total Device Dissipation @ $T_A = 25°C$ Derate above 25°C	P_D	0.5 28.6		Watt mW/°C
Total Device Dissipation @ $T_C = 25°C$ Derate above 25°C	P_D	1.4 6.8		Watts mW/°C
Operating and Storage Junction Temperature Range	T_J, T_{stg}	−65 to +200		°C

ELECTRICAL CHARACTERISTICS ($T_A = 25°C$ unless otherwise noted.)

Characteristic		Symbol	Min	Typ	Max	Unit
OFF CHARACTERISTICS						
Collector-Emitter Breakdown Voltage(1) ($I_C = 10$ mAdc, $I_B = 0$)	 2N4014 2N4013	$V_{(BR)CEO}$	 50 30	 — —	 — —	Vdc
Collector-Emitter Breakdown Voltage ($I_C = 10$ μAdc, $V_{BE} = 0$)	 2N4014 2N4013	$V_{(BR)CES}$	 80 50	 — —	 — —	Vdc
Collector-Base Breakdown Voltage ($I_C = 10$ μAdc, $I_E = 0$)	 2N4014 2N4013	$V_{(BR)CBO}$	 80 50	 — —	 — —	Vdc
Emitter-Base Breakdown Voltage ($I_E = 10$ μAdc, $I_C = 0$)		$V_{(BR)EBO}$	6.0	—	—	Vdc
Collector Cutoff Current ($V_{CB} = 60$ Vdc, $I_E = 0$) ($V_{CB} = 40$ Vdc, $I_E = 0$) ($V_{CB} = 60$ Vdc, $I_E = 0$, $T_A = 100°C$) ($V_{CB} = 40$ Vdc, $I_E = 0$, $T_A = 100°C$)	 2N4014 2N4013 2N4014 2N4013	I_{CBO}	 — — — —	 0.12 0.12 — —	 1.7 1.7 120 120	μAdc
Collector Cutoff Current ($V_{CE} = 80$ Vdc, $V_{EB} = 0$) ($V_{CE} = 50$ Vdc, $V_{EB} = 0$)	 2N4014 2N4013	I_{CES}	 — —	 0.15 0.15	 10 10	μAdc
ON CHARACTERISTICS(1)						
DC Current Gain ($I_C = 10$ mAdc, $V_{CE} = 1.0$ Vdc) ($I_C = 100$ mAdc, $V_{CE} = 1.0$ Vdc) ($I_C = 100$ mAdc, $V_{CE} = 1.0$ Vdc, $T_A = -55°C$) ($I_C = 300$ mAdc, $V_{CE} = 1.0$ Vdc) ($I_C = 500$ mAdc, $V_{CE} = 1.0$ Vdc) ($I_C = 500$ mAdc, $V_{CE} = 1.0$ Vdc, $T_A = -55°C$) ($I_C = 800$ mAdc, $V_{CE} = 2.0$ Vdc) ($I_C = 1.0$ Adc, $V_{CE} = 5.0$ Vdc)	 2N4014 2N4013 2N4014 2N4013	h_{FE}	 30 60 30 40 35 20 20 25 25 30	 — — — — — — — —	 — 150 — — — — — —	—

Figure C.2 The 2N4013–4014 switching transistor specification sheet. (Copyright of Semiconductor Components Industries, LLC. Used by permission.)

2N5457 thru 2N5459

CASE 29-04, STYLE 5
TO-92 (TO-226AA)

1 Drain
3 Gate
2 Source

JFETs
GENERAL PURPOSE

N-CHANNEL — DEPLETION

Refer to 2N4220 for graphs.

MAXIMUM RATINGS

Rating	Symbol	Value	Unit
Drain-Source Voltage	V_{DS}	25	Vdc
Drain-Gate Voltage	V_{DG}	25	Vdc
Reverse Gate-Source Voltage	V_{GSR}	−25	Vdc
Gate Current	I_G	10	mAdc
Total Device Dissipation @ T_A = 25°C Derate above 25°C	P_D	310 2.82	mW mW/°C
Junction Temperature Range	T_J	125	°C
Storage Channel Temperature Range	T_{stg}	−65 to +150	°C

ELECTRICAL CHARACTERISTICS (T_A = 25°C unless otherwise noted.)

Characteristic		Symbol	Min	Typ	Max	Unit
OFF CHARACTERISTICS						
Gate-Source Breakdown Voltage (I_G = −10 μAdc, V_{DS} = 0)		$V_{(BR)GSS}$	−25	—	—	Vdc
Gate Reverse Current (V_{GS} = −15 Vdc, V_{DS} = 0) (V_{GS} = −15 Vdc, V_{DS} = 0, T_A = 100°C)		I_{GSS}	— —	— —	−1.0 −200	nAdc
Gate Source Cutoff Voltage (V_{DS} = 15 Vdc, I_D = 10 nAdc)	2N5457 2N5458 2N5459	$V_{GS(off)}$	−0.5 −1.0 −2.0	— — —	−6.0 −7.0 −8.0	Vdc
Gate Source Voltage (V_{DS} = 15 Vdc, I_D = 100 μAdc) (V_{DS} = 15 Vdc, I_D = 200 μAdc) (V_{DS} = 15 Vdc, I_D = 400 μAdc)	2N5457 2N5458 2N5459	V_{GS}	— — —	−2.5 −3.5 −4.5	— — —	Vdc
ON CHARACTERISTICS						
Zero-Gate-Voltage Drain Current* (V_{DS} = 15 Vdc, V_{GS} = 0)	2N5457 2N5458 2N5459	I_{DSS}	1.0 2.0 4.0	3.0 6.0 9.0	5.0 9.0 16	mAdc
SMALL-SIGNAL CHARACTERISTICS						
Forward Transfer Admittance Common Source* (V_{DS} = 15 Vdc, V_{GS} = 0, f = 1.0 kHz)	2N5457 2N5458 2N5459	$\|y_{fs}\|$	1000 1500 2000	— — —	5000 5500 6000	μmhos
Output Admittance Common Source* (V_{DS} = 15 Vdc, V_{GS} = 0, f = 1.0 kHz)		$\|y_{os}\|$	—	10	50	μmhos
Input Capacitance (V_{DS} = 15 Vdc, V_{GS} = 0, f = 1.0 MHz)		C_{iss}	—	4.5	7.0	pF
Reverse Transfer Capacitance (V_{DS} = 15 Vdc, V_{GS} = 0, f = 1.0 MHz)		C_{rss}	—	1.5	3.0	pF

*Pulse Test: Pulse Width ≤ 630 ms; Duty Cycle ≤ 10%.

Figure C.3 The 2N5457–5459 series JFET specification sheet. (Copyright of Semiconductor Components Industries, LLC. Used by permission.)

2N6426
2N6427

CASE 29-04, STYLE 1
TO-92 (TO-226AA)

Collector 3
Base 2
Emitter 1
1 2 3

DARLINGTON TRANSISTORS

NPN SILICON

MAXIMUM RATINGS

Rating	Symbol	Value	Unit
Collector-Emitter Voltage	V_{CEO}	40	Vdc
Collector-Base Voltage	V_{CBO}	40	Vdc
Emitter-Base Voltage	V_{EBO}	12	Vdc
Collector Current — Continuous	I_C	500	mAdc
Total Device Dissipation @ T_A = 25°C Derate above 25°C	P_D	625 5.0	mW mW/°C
Total Device Dissipation @ T_C = 25°C Derate above 25°C	P_D	1.5 12	Watts mW/°C
Operating and Storage Junction Temperature Range	T_J, T_{stg}	−55 to +150	°C

THERMAL CHARACTERISTICS

Characteristic	Symbol	Max	Unit
Thermal Resistance, Junction to Case	$R_{\theta JC}$	83.3	°C/W
Thermal Resistance, Junction to Ambient	$R_{\theta JA}$(1)	200	°C/W

(1) $R_{\theta JA}$ is measured with the device soldered into a typical printed circuit board.

ELECTRICAL CHARACTERISTICS (T_A = 25°C unless otherwise noted.)

Characteristic		Symbol	Min	Typ	Max	Unit
OFF CHARACTERISTICS						
Collector-Emitter Breakdown Voltage(2) (I_C = 10 mAdc, V_{BE} = 0)		$V_{(BR)CES}$	40	—	—	Vdc
Collector-Base Breakdown Voltage (I_C = 100 μAdc, I_E = 0)		$V_{(BR)CBO}$	40	—	—	Vdc
Emitter-Base Breakdown Voltage (I_E = 10 μAdc, I_C = 0)		$V_{(BR)EBO}$	12	—	—	Vdc
Collector Cutoff Current (V_{CE} = 25 Vdc, I_B = 0)		I_{CEO}	—	—	1.0	μAdc
Collector Cutoff Current (V_{CB} = 30 Vdc, I_E = 0)		I_{CBO}	—	—	50	nAdc
Emitter Cutoff Current (V_{BE} = 10 Vdc, I_C = 0)		I_{EBO}	—	—	50	nAdc
ON CHARACTERISTICS						
DC Current Gain(2) (I_C = 10 mAdc, V_{CE} = 5.0 Vdc)	 2N6426 2N6427	h_{FE}	 20,000 10,000	 — —	 200,000 100,000	—
(I_C = 100 mAdc, V_{CE} = 5.0 Vdc)	2N6426 2N6427		30,000 20,000	— —	300,000 200,000	
(I_C = 500 mAdc, V_{CE} = 5.0 Vdc)	2N6426 2N6427		20,000 14,000	— —	200,000 140,000	
Collector-Emitter Saturation Voltage (I_C = 50 mAdc, I_B = 0.5 mAdc) (I_C = 500 mAdc, I_B = 0.5 mAdc)		$V_{CE(sat)}$	 — —	 0.71 0.9	 1.2 1.5	Vdc
Base-Emitter Saturation Voltage (I_C = 500 mAdc, I_B = 0.5 mAdc)		$V_{BE(sat)}$	—	1.52	2.0	Vdc
Base-Emitter On Voltage (I_C = 50 mAdc, V_{CE} = 5.0 Vdc)		$V_{BE(on)}$	—	1.24	1.75	Vdc
SMALL-SIGNAL CHARACTERISTICS						
Output Capacitance (V_{CB} = 10 Vdc, I_E = 0, f = 1.0 MHz)		C_{obo}	—	5.4	7.0	pF
Input Capacitance (V_{BE} = 1.0 Vdc, I_C = 0, f = 1.0 MHz)		C_{ibo}	—	10	15	pF

Figure C.4 The 2N6426–6427 Darlington transistor specification sheet. (Copyright of Semiconductor Components Industries, LLC. Used by permission.)

ELECTRICAL CHARACTERISTICS (continued) (T_A = 25°C unless otherwise noted.)

Characteristic		Symbol	Min	Typ	Max	Unit
Input Impedance (I_C = 10 mAdc, V_{CE} = 5.0 Vdc, f = 1.0 kHz)	2N6426 2N6427	h_{ie}	100 50	— —	2000 1000	k Ω
Small-Signal Current Gain (I_C = 10 mAdc, V_{CE} = 5.0 Vdc, f = 1.0 kHz)	2N6426 2N6427	h_{fe}	20,000 10,000	— —	— —	—
Current Gain — High Frequency (I_C = 10 mAdc, V_{CE} = 5.0 Vdc, f = 100 MHz)	2N6426 2N6427	$\|h_{fe}\|$	1.5 1.3	2.4 2.4	— —	—
Output Admittance (I_C = 10 mAdc, V_{CE} = 5.0 Vdc, f = 1.0 kHz)		h_{oe}	—	—	1000	μmhos
Noise Figure (I_C = 1.0 mAdc, V_{CE} = 5.0 Vdc, R_S = 100 kΩ, f = 1.0 kHz)		NF	—	3.0	10	dB

(2) Pulse Test: Pulse Width ≤ 300 μs, Duty Cycle ≤ 2.0%.

Figure C.4 (Continued)

www.fairchildsemi.com

KA741/KA741E

Single Operational Amplifier

Electrical Characteristics

(V_{CC} = 15V, V_{EE} = −15V. T_A = 25°C, unless otherwise specified)

Parameter		Symbol	Conditions		KA741E			KA741			Unit
					Min.	Typ.	Max.	Min.	Typ.	Max.	
Input Offset Voltage		V_{IO}	R_S≤10kΩ		–	–	–	–	2.0	6.0	mV
			R_S≤50Ω			0.8	3.0	–	–	–	
Input Offset Voltage Adjustment Range		$V_{IO(R)}$	V_{CC} = ±20V		±10	–	–	–	±15	–	mV
Input Offset Current		I_{IO}	–		–	3.0	30	–	20	200	nA
Input Bias Current		I_{BIAS}	–		–	30	80	–	80	500	nA
Input Resistance		R_I	V_{CC} = ±20V		1.0	6.0	–	0.3	2.0	–	MΩ
Input Voltage Range		$V_{I(R)}$	–		±12	±13	–	±12	±13	–	V
Large Signal Voltage Gain		G_V	R_L ≥ 2KΩ	V_{CC} = ±20V, $V_{O(P-P)}$ = ±15V	50	–	–	–	–	–	V/mV
				V_{CC} = ±15V, $V_{O(P-P)}$ = ±10V	–	–	–	20	200	–	
Output Short Circuit Current		I_{SC}	–		10	25	35	–	25	–	mA
Output Voltage Swing		$V_{O(P-P)}$	V_{CC} = ±20V	R_L ≥ 10KΩ	±16	–	–	–	–	–	V
				R_L ≥ 2KΩ	±15	–	–	–	–	–	
			V_{CC} = ±15V	R_L ≥ 10KΩ	–	–	–	±12	±14	–	
				R_L ≥ 2KΩ	–	–	–	±10	±13	–	
Common Mode Rejection Ratio		CMRR	R_S ≤10KΩ, V_{CM} = ±12V		–	–	–	70	90	–	dB
			R_S ≤50KΩ, V_{CM} = ±12V		80	95	–	–	–	–	
Power Supply Rejection Ratio		PSRR	V_{CC} = ±15V to V_{CC} = ±15V R_S≤50Ω		86	96	–	–	–	–	dB
			V_{CC} = ±15V to V_{CC} = ±15V R_S≤10KΩ		–	–	–	77	96	–	
Transient Response	Rise Time	t_R	Unity Gain		–	0.25	0.8	–	0.3	–	μs
	Overshoot	OS			–	6.0	20	–	10	–	%
Bandwidth		BW			0.43	1.5	–	–	–	–	MHz
Slew Rate		SR	Unity Gain		0.3	0.7	–	–	0.5	–	V/μs
Supply Current		I_{CC}	R_L = ∞Ω		–	–	–	–	1.5	2.8	mA
Power Consumption		P_C	V_{CC} = ±20V		–	80	150	–	–	–	mW
			V_{CC} = ±15V		–	–	–	–	50	85	

www.fairchildsemi.com

Figure C.5 The KA741 electrical characteristics and operating curves. (Courtesy of Fairchild Semiconductor. Used by permission.)

appendix D

Selected Equation Derivations

Equation (5.6)

The total resistance of a two-branch parallel resistive circuit can be found as

$$R_T = \frac{1}{\dfrac{1}{R_1} + \dfrac{1}{R_2}}$$

This relationship is a form of equation (5.5). The denominator of the fraction can be rewritten as

$$\frac{1}{R_1} + \frac{1}{R_2} = \left(\frac{1}{R_1} \times \frac{R_2}{R_2}\right) + \left(\frac{1}{R_2} \times \frac{R_1}{R_1}\right) = \frac{R_2}{R_1R_2} + \frac{R_1}{R_1R_2} = \frac{R_1 + R_2}{R_1R_2}$$

Note that each fraction has been multiplied by one (in the form of a fraction). If we substitute the final value back into equation (5.5), we get

$$R_T = \frac{1}{\dfrac{1}{R_1} + \dfrac{1}{R_2}} = \frac{1}{\dfrac{R_1 + R_2}{R_1R_2}}$$

or

$$R_T = \frac{R_1R_2}{R_1 + R_2}$$

which is equation (5.6).

Equations (6.7), (6.8), and (6.9)

For two circuits to be resistive equivalents, they must provide the same resistance readings between any pair of terminals. For example, consider the circuits shown in Figure D.1. Assuming that the Δ and Y circuits shown are resistive equivalents, then the two must provide the same resistance readings between any pair of terminals. If we were to measure the resistance between the (x) and (y) terminals for each circuit shown in Figure D.1, we would obtain

$$R_{(x \to y)} = R_1 + R_2 \qquad \text{(for the Y circuit)}$$

and

$$R_{(x \to y)} = R_A \| (R_B + R_C) \qquad \text{(for the } \Delta \text{ circuit)}$$

Because the two are resistive equivalents, the above equations indicate that

$$R_1 + R_2 = R_A \| (R_B + R_C) = \frac{R_A(R_B + R_C)}{R_A + R_B + R_C} = \frac{R_A R_B + R_A R_C}{R_A + R_B + R_C} \tag{D.1}$$

The same technique is used to establish the following relationships:

$$R_1 + R_3 = R_C \| (R_A + R_B) = \frac{R_C(R_A + R_B)}{R_A + R_B + R_C} = \frac{R_A R_C + R_B R_C}{R_A + R_B + R_C} \tag{D.2}$$

and

$$R_2 + R_3 = R_B \| (R_A + R_C) = \frac{R_B(R_A + R_C)}{R_A + R_B + R_C} = \frac{R_A R_B + R_B R_C}{R_A + R_B + R_C} \tag{D.3}$$

Because these three equations describe the same circuit, they are *simultaneous equations;* that is, the three are true and interdependent. When two (or more) equations are true, then any combination of the equations also holds true. For example, if we *subtract* equation (D.3) from equation (D.2), we get

$$(R_1 + R_3) - (R_2 + R_3) = \frac{R_A R_C + R_B R_C}{R_A + R_B + R_C} - \frac{R_A R_B + R_B R_C}{R_A + R_B + R_C} = \frac{R_A R_C - R_A R_B}{R_A + R_B + R_C}$$

or

$$R_1 - R_2 = \frac{R_A R_C - R_A R_B}{R_A + R_B + R_C} \tag{D.4}$$

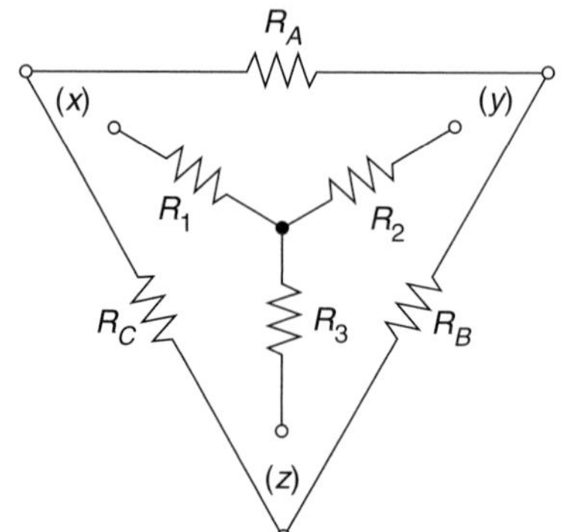

Figure D.1

Because equations (D.2) and (D.3) are true, equation (D.4) must also hold true. Now, if we add equation (D.4) to equation (D.1), we get

$$(R_1 + R_2) + (R_1 - R_2) = \frac{R_A R_B + R_A R_C}{R_A + R_B + R_C} + \frac{R_A R_C - R_A R_B}{R_A + R_B + R_C} - \frac{2R_A R_C}{R_A + R_B + R_C}$$

or

$$2R_1 = \frac{2R_A R_C}{R_A + R_B + R_C}$$

Dividing both sides of this equation by two gives us

$$R_1 = \frac{R_A R_C}{R_A + R_B + R_C}$$

which is equation (6.7). In a similar fashion:

1. Equation (6.8) is derived by subtracting equation (D.2) from (D.1) and adding the result to equation (D.3).
2. Equation (6.9) is derived by subtracting equation (D.1) from (D.3) and adding the result to equation (D.2).

Equations (6.10), (6.11), and (6.12)

The Y-to-Δ conversion equations are derived using the Δ-to-Y conversion equations. For convenience, the Δ-to-Y relationships are

$$R_1 = \frac{R_A R_C}{R_A + R_B + R_C} \tag{D.5}$$

$$R_2 = \frac{R_A R_B}{R_A + R_B + R_C} \tag{D.6}$$

and

$$R_3 = \frac{R_B R_C}{R_A + R_B + R_C} \tag{D.7}$$

Using equations (D.5) and (D.6), the ratio $R_1 : R_2$ can be found as

$$\frac{R_1}{R_2} = R_1\left(\frac{1}{R_2}\right) = \left(\frac{R_A R_C}{R_A + R_B + R_C}\right)\left(\frac{R_A + R_B + R_C}{R_A R_B}\right) = \frac{R_A R_C}{R_A R_B} = \frac{R_C}{R_B}$$

Based on this ratio, we can define R_C as

$$R_C = R_B \frac{R_1}{R_2} \tag{D.8}$$

Now we will use the same approach to define R_A. Using equations (D.7) and (D.5), the ratio $R_1 : R_3$ can be found as

$$\frac{R_1}{R_3} = R_1\left(\frac{1}{R_3}\right) = \left(\frac{R_A R_C}{R_A + R_B + R_C}\right)\left(\frac{R_A + R_B + R_C}{R_B R_C}\right) = \frac{R_A R_C}{R_B R_C} = \frac{R_A}{R_B}$$

Based on this ratio, we can define R_A as follows:

$$R_A = R_B \frac{R_1}{R_3} \tag{D.9}$$

Now we will substitute equations (D.8) and (D.9) for the defined variables in equation (D.5), as follows

$$R_1 = \frac{R_A R_C}{R_A + R_B + R_C} = \frac{\left(R_B \frac{R_1}{R_3}\right)\left(R_B \frac{R_1}{R_2}\right)}{\left(R_B \frac{R_1}{R_3}\right) + R_B + \left(R_B \frac{R_1}{R_2}\right)}$$

This equation probably looks a bit intimidating, but it has one very important characteristic. If you look at the compound fraction, you'll see that we have eliminated all the lettered variables other than R_B. To solve for R_B, we must start by simplifying the numerator of the compound fraction, as follows:

$$\left(R_B \frac{R_1}{R_3}\right)\left(R_B \frac{R_1}{R_2}\right) = \frac{(R_B R_1)(R_B R_1)}{R_2 R_3} = \frac{(R_B R_1)^2}{R_2 R_3}$$

Substituting this result into the numerator of the compound fraction, we get

$$R_1 = \frac{\frac{(R_B R_1)^2}{R_2 R_3}}{\frac{R_B R_1}{R_3} + R_B + \frac{R_B R_1}{R_2}} \tag{D.10}$$

At this point, we need to rewrite the equation so that all the terms have the same denominator. This can be accomplished by rewriting each term in the denominator of equation (D.10), as follows:

$$\frac{R_B R_1}{R_3} \times \frac{R_2}{R_2} = \frac{R_B R_1 R_2}{R_2 R_3}$$

$$R_B = R_B \frac{R_2 R_3}{R_2 R_3} = \frac{R_B R_2 R_3}{R_2 R_3}$$

$$\frac{R_B R_1}{R_2} \times \frac{R_3}{R_3} = \frac{R_B R_1 R_3}{R_2 R_3}$$

Now, using these values, equation (D.10) can be rewritten as

$$R_1 = \frac{\frac{(R_B R_1)^2}{R_2 R_3}}{\frac{R_B R_1 R_2}{R_2 R_3} + \frac{R_B R_2 R_3}{R_2 R_3} + \frac{R_B R_1 R_3}{R_2 R_3}}$$

Now the common denominator ($R_2 R_3$) can be dropped, leaving

$$R_1 = \frac{(R_B R_1)^2}{R_B R_1 R_2 + R_B R_2 R_3 + R_B R_1 R_3} = \frac{R_B R_1^2}{R_1 R_2 + R_2 R_3 + R_1 R_3}$$

Taking the reciprocal of the above relationship, we get

$$\frac{1}{R_1} = \frac{R_1 R_2 + R_2 R_3 + R_1 R_3}{R_B R_1^2}$$

Finally, multiplying each side of the equation by $R_B R_1$ gives us

$$R_B = \frac{R_1R_2 + R_2R_3 + R_1R_3}{R_1}$$

which is equation (6.11). Equations (6.10) and (6.12) are derived in the same fashion: by substituting equations (D.8) and (D.9) into equations (D.6) and (D.7).

Equation (9.5)

The average value of any curve equals the area under the curve divided by its length. For example, the average value of the sine-wave alternation shown in Figure D.2 can be found as

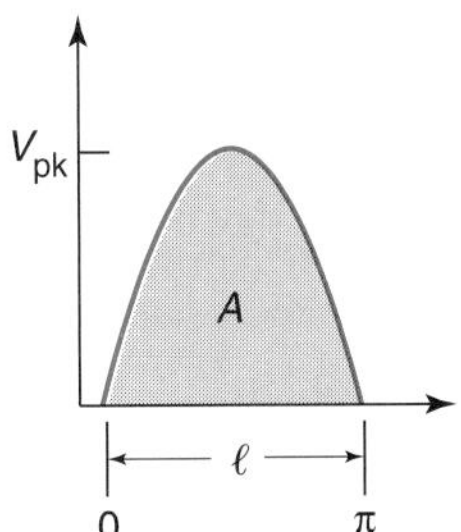

Figure D.2

$$V_{ave} = \frac{A}{\ell}$$

where

A = the value of the shaded area under the curve

ℓ = the difference between the zero points on the curve

To find the area under the curve, we start with the curve equation. The equation of the curve shown in Figure D.2 is given as

$$v = V_{pk} \sin \omega t \tag{D.11}$$

where

ωt = the phase angle, given as a product of angular velocity (ω), in radians, and time

Using calculus, the area under the curve is found as the integral of equation (D.11). By formula,

$$A = \int_0^{\pi} V_{pk} \sin \omega t \, d(\omega t) \tag{D.12}$$

Solving equation (D.12), we get

$$\begin{aligned} A &= \int_0^{\pi} V_{pk} \sin \omega t \, d(\omega t) \\ &= -V_{pk}[-\cos \omega t]_0^{\pi} \\ &= -V_{pk}[-\cos \pi - \cos 0] \\ &= -V_{pk}[-1 - 1] \\ &= 2V_{pk} \end{aligned}$$

Now we know that the area under the curve equals $2V_{pk}$. As shown in Figure D.2, the difference between the zero crossings of the waveform equals π (radians). Using these two values in equation (D.11),

$$V_{ave} = \frac{A}{\ell} = \frac{2V_{pk}}{\pi}$$

which is equation (9.5).

Equation (10.22)

Transformer secondary voltage (V_S) is found as

$$V_S = V_P \frac{N_S}{N_P}$$

This relationship was given as equation (10.17). Rewriting this equation for V_P, we get

$$V_P = V_S \frac{N_P}{N_S} \tag{D.13}$$

Transformer secondary current (I_S) is found as

$$I_S = I_P \frac{N_P}{N_S}$$

This relationship was given as equation (10.19). Rewriting this equation for I_P, we get

$$I_P = I_S \frac{N_S}{N_P} \tag{D.14}$$

Now, if we divide equation (D.13) by equation (D.14), we get

$$\frac{V_P}{I_P} = \frac{V_S \dfrac{N_P}{N_S}}{I_S \dfrac{N_S}{N_P}} = \frac{V_S}{I_S}\left(\frac{N_P}{N_S} \times \frac{N_P}{N_S}\right) = \frac{V_S}{I_S}\left(\frac{N_P}{N_S}\right)^2$$

Because $Z = \dfrac{V}{I}$, the above equation can be rewritten as

$$Z_P = Z_S\left(\frac{N_P}{N_S}\right)^2$$

which is equation (10.22).

Equations (15.46) and (15.47)

These derivations start with an assumption: *Every series resistive-reactive circuit has a parallel equivalent.* These two circuits are represented in Figure D.3.

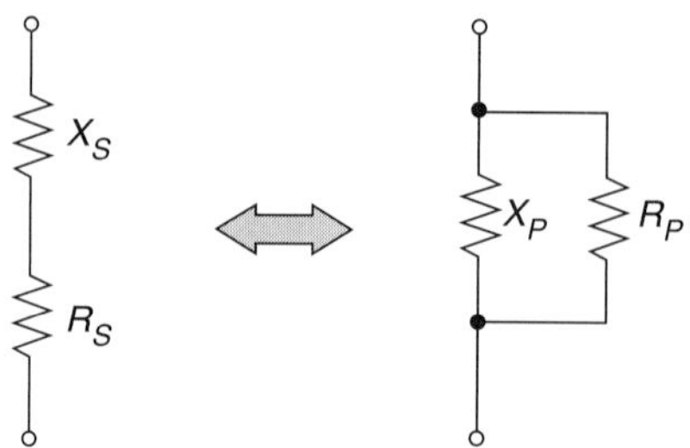

Figure D.3

The quality (Q) of series circuit shown is given as

$$Q = \frac{X_S}{R_S} \tag{D.15}$$

and the total impedance (in rectangular form) can be expressed as

$$Z_S = R_S \pm jX_S \tag{D.16}$$

For the parallel circuit shown in Figure D.3, the total impedance (Z_P) can be expressed as

$$Z_P = \frac{(R_P)(\pm jX_P)}{R_P + jX_P} \tag{D.17}$$

If we multiply the fraction in equation (D.17) by one in the form of

$$\frac{R_P \pm jX_P}{R_P \pm jX_P}$$

we get

$$Z_P = \frac{R_P X_P^2}{R_P^2 + X_P^2} \pm j \frac{R_P^2 X_P}{R_P^2 + X_P^2} \tag{D.18}$$

Our original assumption was that the two circuits shown in Figure D.3 are equivalent. Assuming that $Z_S = Z_P$, equation (D.18) can be rewritten as

$$Z_S = \frac{R_P X_P^2}{R_P^2 + X_P^2} \pm j \frac{R_P^2 X_P}{R_P^2 + X_P^2} \tag{D.19}$$

If we substitute equation (D.16) for the value of Z_S in equation (D.19), we get

$$R_S \pm jX_S = \frac{R_P X_P^2}{R_P^2 + X_P^2} \pm j \frac{R_P^2 X_P}{R_P^2 + X_P^2} \tag{D.20}$$

For the two sides of equation (D.20) to be equal, the real components must equal each other, and the imaginary components must equal each other as well. Therefore,

$$R_S = \frac{R_P X_P^2}{R_P^2 + X_P^2} \tag{D.21}$$

and

$$X_S = \frac{R_P^2 X_P}{R_P^2 + X_P^2} \tag{D.22}$$

Equations (D.21) and (D.22) can now be substituted into equation (D.15), as follows:

$$Q = \frac{X_S}{R_S} = X_S \frac{1}{R_S} = \frac{R_P^2 X_P}{R_P^2 + X_P^2} \times \frac{R_P^2 + X_P^2}{R_P X_P^2} = \frac{R_P^2 X_P}{R_P X_P^2}$$

or

$$Q = \frac{R_P}{X_P} \tag{D.23}$$

Now, using equation (D.23), we can rewrite equation (D.22) as

$$X_S = \frac{X_P}{1 + \left(\frac{1}{Q}\right)^2} \tag{D.24}$$

Assuming that $Q >> 2$ (which is normally the case with filters), then $\left(\frac{1}{Q}\right)^2 << 1$. Based on this assumption, equation (D.24) simplifies to

$$X_S \cong X_P \tag{D.25}$$

Based on equation (D.23), we know that $R_P = QX_P$. Or, based on equation (D.25),

$$R_P = QX_S \tag{D.26}$$

Because $X_S = QR_S$, as implied by equation (D.15),

$$R_P = QX_S = Q(QR_S)$$

or

$$R_P = Q^2 R_S$$

which is a form of equation (15.46).

Equation (15.47)

In a tuned *LC* circuit, X_L is the series reactance, and R_w is the series resistance. Replacing R_S in the R_P equation, we get

$$R_P = Q^2 R_w$$

Finally, replacing X_P with X_L in equation (D.24) gives us

$$Q = \frac{R_P}{X_L}$$

or

$$Q = \frac{R_P \parallel R_L}{X_L}$$

when a load is connected. This equation is a form of equation (15.47).

Equation (16.6)

The charge curve is a plot of the equation

$$\frac{I_t}{I_{pk}} = 1 - e^{-\frac{t}{\tau}}$$

Transposing the equation, we get

$$e^{-\frac{t}{\tau}} = 1 - \frac{I_t}{I_{pk}}$$

Taking the natural log (ln) of each side of the equation, we get

$$\ln\left(e^{-\frac{t}{\tau}}\right) = \ln\left(1 - \frac{I_t}{I_{pk}}\right)$$

or

$$-\frac{t}{\tau} = \ln\left(1 - \frac{I_t}{I_{pk}}\right)$$

Finally, multiplying each side of the equation by $(-\tau)$, we get

$$t = -\tau \ln\left(1 - \frac{I_t}{I_{pk}}\right) \tag{D.27}$$

Assuming that the inductor in a series *RL* switching circuit has no winding resistance, the total resistance in the circuit equals *R*. Therefore, the circuit current at any instant is found as

$$I_t = \frac{V_R}{R}$$

and the peak current is found as

$$I_{pk} = \frac{V_S}{R}$$

Using these relationships, the current ratio in equation (D.27) can be rewritten as

$$\frac{I_t}{I_{pk}} = \frac{\frac{V_R}{R}}{\frac{V_S}{R}} = \frac{V_R}{R} \times \frac{R}{V_S} = \frac{V_R}{V_S}$$

Using the ratio of V_R to V_S, equation (D.27) can be rewritten as

$$t = -\tau \ln\left(1 - \frac{V_R}{V_S}\right)$$

which is equation (16.6).

Equation (20.9)

Schockley's equation for the total current through a *pn* junction is

$$I_T = I_S(e^{Vq/kT} - 1) \tag{D.28}$$

where

I_S = the reverse saturation current through the diode
e = the *exponential constant* (approximately 2.71828)
V = the voltage across the depletion layer
q = the charge on an electron (approximately 1.6×10^{-19} V)
k = Boltzmann's constant (approximately 1.38×10^{-23} J/°K)
T = the temperature of the device, in degrees Kelvin (°K = °C + 273)

We can solve for q/kT at room temperature (approximately 21°C) as

$$q/kT = \frac{1.6 \times 10^{-19}\text{ J}}{(1.38 \times 10^{-23}\text{ J/°K})(294\text{°K})} = 40$$

Using this value, the equation for I_T is rewritten as

$$I_T = I_S(e^{40\text{ V}} - 1) \tag{D.29}$$

Differentiating the above equation give us

$$\frac{dI}{dV} = 40I_S e^{40\text{ V}} \tag{D.30}$$

If we rearrange equation (D.29) as

$$I_S e^{40\text{ V}} = (I_T + I_S)$$

we can use this equation to rewrite equation (D.30) as

$$\frac{dI}{dV} = 40(I_T + I_S) \tag{D.31}$$

Now, if we take the reciprocal of equation (D.31), we will have an equation for the ac resistance of the junction, r'_e as follows:

$$\frac{dV}{dI} = \frac{1\text{ V}}{40(I_T + I_S)} = \frac{1\text{ V}}{40}\,\frac{1}{I_T + I_S} = 25\text{ mV}\,\frac{1}{I_T + I_S} = \frac{25\text{ mV}}{I_T + I_S}$$

In this case, $(I_T + I_S)$ is the current through the emitter-base junction of the transistor, I_E. Therefore, we can rewrite the above equation as

$$r'_e = \frac{25\text{ mV}}{I_E}$$

Equation (20.16)

According to Ohm's law, the input current to a common-emitter amplifier can be found as

$$i_{\text{in}} = \frac{v_s}{Z_{\text{in}}}$$

As Figure D.4 shows, the input current to a common-emitter amplifier is divided between the resistors in the biasing network and the base of the transistor. Using the current-divider relationship, i_b can be found as

$$i_b = i_{\text{in}}\left(\frac{Z_{\text{in}}}{Z_{\text{in(base)}}}\right) \tag{D.32}$$

where

Z_{in} = the parallel combination of the biasing resistors and the base input impedance, $Z_{\text{in(base)}}$

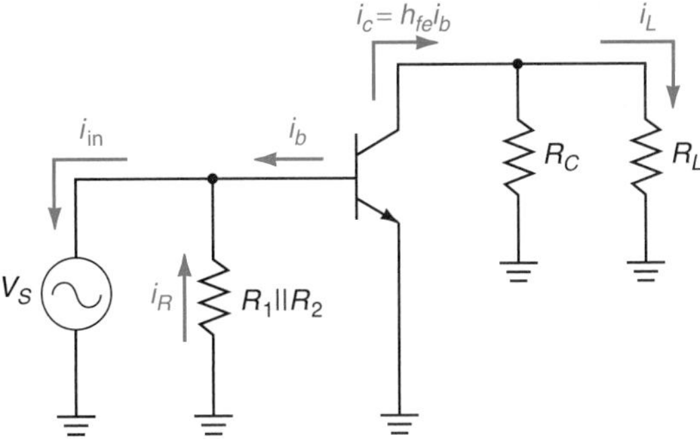

Figure D.4

Figure D.4 shows that the transistor output current (i_c) is equal to the product of the transistor ac current gain (h_{fe}) and the ac base current. By formula,

$$i_c = h_{fe} i_b$$

Substituting equation (D.32) for the value of i_b in the above equation, we get

$$i_c = h_{fe} i_{in} \left(\frac{Z_{in}}{Z_{in(base)}} \right) \tag{D.33}$$

As with the input circuitry, the output circuitry forms a current divider. The output current divider is made of the collector resistor (R_C) and the load. Once we know the value of i_c for a given common-emitter amplifier, the value of the ac load current can be found as

$$i_L = i_c \left(\frac{r_C}{R_L} \right) \tag{D.34}$$

where

r_C = the parallel combination of the collector resistor and the load resistance

The *effective* current gain of any amplifier equals the ratio of *ac load current* to *ac input current.* By equation,

$$A_i = \frac{i_L}{i_{in}}$$

Substituting the relationships we have established for the values of i_L and i_{in}, we get

$$A_i = \frac{i_c \dfrac{r_C}{R_L}}{\dfrac{v_s}{Z_{in}}}$$

or

$$A_i = h_{fe} i_b \left(\frac{r_C}{R_L} \right) \left(\frac{Z_{in}}{v_s} \right)$$

Because $\dfrac{Z_{in}}{v_s} = \dfrac{1}{i_{in}}$, the above equation can be rewritten as

$$A_i = h_{fe} \left(\frac{i_b}{i_{in}} \right) \left(\frac{r_C}{R_L} \right) \tag{D.35}$$

According to equation (D.32),

$$\frac{i_b}{i_{\text{in}}} = \frac{Z_{\text{in}}}{Z_{\text{in(base)}}}$$

Substituting this equation for the current ratio in equation (D.35), we get

$$A_i = h_{fe}\left(\frac{Z_{\text{in}} r_C}{Z_{\text{in(base)}} R_L}\right)$$

which is equation (20.16).

Equation (20.25)

The derivation of equation (20.25) is best understood by looking at the circuit shown in Figure D.5. Figure D.5a shows the ac equivalent of the emitter follower. Note that the input circuit consists of the source resistance (R_S) in parallel with the combination of R_1 and R_2. Thus, the total resistance in the base circuit (as seen from the base of the transistor) can be found as

$$R'_{\text{in}} = R_1 \| R_2 \| R_S$$

This resistance is shown as a single resistor in Figure D5b.

To determine Z_{out}, we start by writing the Kirchhoff's voltage equation for the circuit and solving that equation for i_e. The voltage equation is

$$v_{\text{in}} = i_b R'_{\text{in}} + i_e(r'_e + R_E)$$

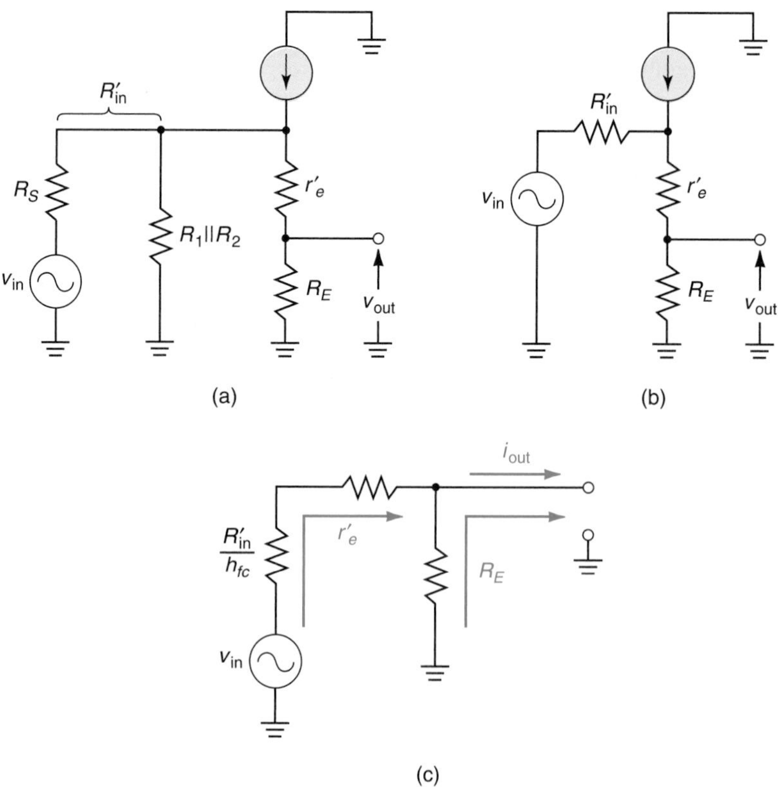

Figure D.5

Because $i_b = \dfrac{i_e}{h_{fc}}$, the voltage equation can be rewritten as

$$v_{\text{in}} = \frac{i_e R'_{\text{in}}}{h_{fc}} + i_e(r'_e + R_E)$$

Solving for i_e yields

$$i_e = \frac{v_{\text{in}}}{r'_e + R_E + \dfrac{R'_{\text{in}}}{h_{fc}}}$$

This equation indicates that the emitter "sees" a three-resistor circuit, as shown in Figure D.5c. If we thevenize the circuit, we find that the load sees R_E in parallel with the series combination of the other two resistors. Thus, Z_{out} is found as

$$Z_{\text{out}} = R_E \| \left(r'_e + \frac{R'_{\text{in}}}{h_{fc}} \right)$$

which is equation (20.25).

RC-Coupled Class A Efficiency

The *ideal* class A amplifier would have the following characteristics:

1. $V_{CEQ} = \dfrac{V_{CC}}{2}$
2. $V_{\text{PP}} = V_{CC}$
3. $I_{CQ} = \dfrac{I_{C(\text{sat})}}{2}$
4. $I_{\text{PP}} = I_{C(\text{sat})}$

These ideal characteristics will be used in our derivation of the maximum ideal efficiency rating of 25%.

Equation (20.38) gives us the following equation for load power:

$$P_L = \frac{V_{PP}^2}{8R_L}$$

Because $R_L = V_{\text{PP}}/I_{\text{PP}}$, we can rewrite the above equation as

$$P_L = \left(\frac{V_{\text{PP}}^2}{8}\right)\left(\frac{1}{R_L}\right) = \left(\frac{V_{\text{PP}}^2}{8}\right)\left(\frac{I_{\text{PP}}}{V_{\text{PP}}}\right) = \frac{V_{\text{PP}} I_{\text{PP}}}{8} \qquad \text{(D.36)}$$

Now we will use the characteristics above and equation (D.36) to determine the ideal maximum load power, as follows:

$$P_L = \frac{V_{PP} I_{PP}}{8} = \frac{V_{CC} I_{C(\text{sat})}}{8} = \frac{(2V_{CEQ})(2I_{CQ})}{8}$$

$$= \frac{4(V_{CEQ} I_{CQ})}{8} = \frac{V_{CEQ} I_{CQ}}{2} \qquad \text{(maximum, ideal)}$$

Now recall that the power drawn from the supply of an amplifier is found as

$$P_S = V_{CC} I_{CC}$$

Note: I_1 is being used to designate the current through the base biasing circuit.

where

$$I_{CC} = I_{CQ} + I_1$$

Because I_{CQ} is normally *much greater than* I_1, we can approximate the equation above to

$$P_S \cong V_{CC} I_{CQ}$$

Note that this equation is valid for most amplifier power analyses. We can also rewrite the above for the ideal amplifier, as follows:

$$P_S = 2V_{CEQ} I_{CQ}$$

We can now use the derived values of P_L and P_S to determine the *maximum ideal value* of η, as follows:

$$\eta = \frac{P_L}{P_S} \times 100 = P_L \frac{1}{P_S} \times 100 = \frac{V_{CEQ} I_{CQ}}{2} \frac{1}{2V_{CEQ} I_{CQ}} \times 100$$

$$= \frac{V_{CEQ} I_{CQ}}{4V_{CEQ} I_{CQ}} \times 100 = \frac{1}{4} \times 100 = 25\%$$

Class B Amplifier Efficiency

The value of $I_{C1(\text{ave})}$ is found as

$$I_{C1(\text{ave})} = \frac{V_{CC}}{2\pi R_L}$$

This equation is based on the relationship

$$I_{C1(\text{ave})} = \frac{I_{\text{pk}}}{\pi}$$

In the practical class B amplifier, $I_{C1(\text{ave})} >> I_1$. Therefore, we can assume that $I_{CC} = I_{C1(\text{ave})}$ in the *ideal* class B amplifier. Based on this point, the determination of the ideal maximum efficiency for the class B amplifier proceeds as follows:

$$\eta = \frac{P_L}{P_S} \times 100 = P_L\left(\frac{1}{P_S}\right) \times 100 = \left(\frac{V_{CC}^2}{8R_L}\right)\left(\frac{1}{V_{CC} I_{C1(\text{ave})}}\right) \times 100$$

$$= \left(\frac{V_{CC}^2}{8R_L}\right)\left(\frac{2\pi R_L}{V_{CC}^2}\right) \times 100 = \frac{2\pi}{8} \times 100 = 78.54\%$$

The final result can, of course, be assumed to be equal to 78.5%. Remember that this maximum efficiency rating is ideal. Any practical value of efficiency for a class B amplifier must be less than this value.

Equation (20.45)

The instantaneous power dissipation in a class B amplifier is found as

$$p = V_{CEQ}(1 - k \sin \theta) I_{C(\text{sat})}(k \sin \theta) \tag{D.37}$$

where

V_{CEQ} = the quiescent value of V_{CE} for the amplifier

$I_{C(\text{sat})}$ = the saturation current for the amplifier, as determined by the ac load line for the circuit

θ = the phase angle of the output at the instant that p is measured

k = a constant factor, representing the percentage of the load line that is actually being used by the circuit; k is represented as a decimal value between 0 and 1

The average power dissipation can be found by integrating equation (D.37) for one half-cycle. Note that a half-cycle is represented as being between 0 and π *radians* in the integration:

$$P_{\text{ave}} = \frac{1}{2\pi}\int_0^{\pi} p\,d\theta \tag{D.38}$$

Performing the integration yields

$$P_{\text{ave}} = \left(\frac{V_{CEQ}I_{C(\text{sat})}}{2\pi}\right)\left(2k - \frac{\pi k^2}{2}\right) \tag{D.39}$$

The next step is to find the maximum value of k in equation (D.39). The first step in finding this maximum value is to differentiate P_{ave} with respect to k, as follows:

$$\frac{dP_{\text{ave}}}{dk} = \frac{V_{CEQ}I_{C(\text{sat})}}{2}(2 - \pi k) \tag{D.40}$$

Next, we set the right-hand side of equation (D.40) equal to zero. This provides us with the following equations:

$$\frac{V_{CEQ}I_{C(\text{sat})}}{2} = 0 \qquad \text{and} \qquad 2 - \pi k = 0$$

Now we can solve the equation on the right to find the maximum value:

$$k = \frac{2}{\mu} \tag{D.41}$$

At this point, we want to take the value of k obtained in equation (D.41) and plug it into equation (D.14). Reducing this new equation gives us

$$P_{\text{ave}} = 0.1V_{CEQ}I_{C(\text{sat})} \tag{D.42}$$

Finally, we replace the values of V_{CEQ} and $I_{C(\text{sat})}$ in equation (D.42) using the following relationships:

$$V_{CEQ} = \frac{PP}{2} \qquad \text{and} \qquad I_{C(\text{sat})} = \frac{V_{CEQ}}{R_L}$$

This gives us

$$P_D = 0.1\left(\frac{V_{CEQ}PP}{2R_L}\right)$$

or

$$P_D = \left(\frac{V_{CEQ}PP}{20R_L}\right)$$

Now, because V_{CEQ} is equal to $\frac{PP}{2}$, we can rewrite the above equation as

$$P_D = V_{CEQ}\left(\frac{PP}{20R_L}\right) = \left(\frac{PP}{2}\right)\left(\frac{PP}{20R_L}\right)$$

or

$$P_D = \frac{PP^2}{40R_L}$$

which is another form of equation (20.45).

Equation (22.5)

The equation for the instananeous voltage at any point on a sine wave is given as

$$v = V_{pk} \sin \omega t$$

where

V_{pk} = the peak input voltage
$\omega = 2\pi f$
t = the time from the start of the cycle to the instant that v occurs

If we differentiate v with respect to t, we get

$$\frac{dv}{dt} = \omega V_{pk}(\cos \omega t)$$

Because the slew rate is a maximum rating of a given op-amp and the rate of change in a sine wave varies, we need to determine the point at which the rate of change in the sine wave is at its maximum value. This point occurs when the sine wave passes the reference voltage, or at $t = 0$. The value of dv/dt at $t = 0$ determines the maximum operating frequency of the op-amp, as follows:

$$\frac{dv}{dt}_{\max} = \omega_{\max} V_{pk}$$

or

$$\text{slew rate} = 2\pi f_{\max} V_{pk}$$

Finally, this equation is rearranged into the form of equation (22.5):

$$f_{\max} = \frac{\text{slew rate}}{2\pi V_{pk}}$$

The time required for a capacitor to charge to a given value is found as

$$t = RC\left[\ln \frac{V - V_0}{V - V_C}\right] \tag{D.43}$$

where

V = the charging voltage
V_0 = the initial charge on the capacitor
V_C = the capacitor charge of interest
ln = an operator, indicating that we are to take the *natural log* of the fraction

We can use equation (D.43) to find the time required for V_C to reach 0.1 V (10% of V), as follows:

$$t = RC\left[\ln \frac{V - 0\text{ V}}{V - 0.1\ V}\right] = RC\left[\ln \frac{V}{0.9\ V}\right] = RC[\ln (1.11)] = 0.1RC$$

Now we can use equation (D.43) to find the time required for V_C to reach $0.9V$ (90% of V) as follows:

$$t = RC\left[\ln \frac{V - 0\text{ V}}{V - 0.9\ V}\right] = RC\left[\ln \frac{V}{0.1\ V}\right] = RC[\ln (10)] = 2.3RC$$

Because rise time is defined as the time needed for V_C to fall from 90% of its maximum value to 10% of its maximum value, it can be found as

$$t_r = t_{(90\%)} - t_{(10\%)} = 2.3RC - 0.1RC = 2.2RC \tag{D.44}$$

Now the upper cutoff frequency of a given RC circuit is found as

$$f_{C2} = \frac{1}{2\pi RC}$$

Rewriting the f_{C2} equation, we get

$$RC = \frac{1}{2\pi f_{C2}}$$

Substituting this equation in equation (D.44), we get

$$t_r = 2.2RC = 2.2\left(\frac{1}{2\pi f_{C2}}\right) = \frac{2.2}{2\pi f_{C2}} = \frac{0.35}{f_{C2}}$$

or

$$f_{C2} = \frac{0.35}{t_r}$$

appendix E

h-Parameters

Hybrid parameters, or *h-parameters*, are transistor specifications that describe the component operating characteristics under specific circumstances. Each of the four h-parameters is measured under no-load or full-load conditions. The h-parameters are then used in circuit analysis applications.

The four h-parameters for a transistor in a common-emitter amplifier are:

h_{ie} = the base input impedance

h_{fe} = the base-to-collector current gain

h_{oe} = the output admittance

h_{re} = the reverse voltage feedback ratio

Note that all these values represent ac characteristics of the transistor, as measured under specific circumstances.

Before discussing the applications of h-parameters, let's take a look at the parameters and the method of measurement that is used for each.

Input Impedance (h_{ie})

The input impedance (h_{ie}) is measured with the output shorted. A shorted output is a full load, so h_{ie} represents the input impedance to the transistor under full-load conditions.

The measurement of h_{ie} is illustrated in Figure E.1a. As you can see, a capacitor is connected between the emitter and the collector terminals. This capacitor provides an ac short between the terminals when an ac signal is applied to the base-emitter junction of the transistor. With the input voltage applied, the base current is measured. Then, h_{ie} is determined as

$$h_{ie} = \frac{v_{\text{in}}}{i_b} \qquad \text{(output shorted)} \qquad \text{(E.1)}$$

Why short the output? You may recall that any resistance in the emitter circuit is reflected back to the base. This condition was described previously in the equation

$$Z_{\text{in(base)}} = h_{fe}(r'_e + r_E)$$

By shorting the collector and emitter terminals, the measured value of h_{ie} does not reflect any external resistance in the emitter circuit.

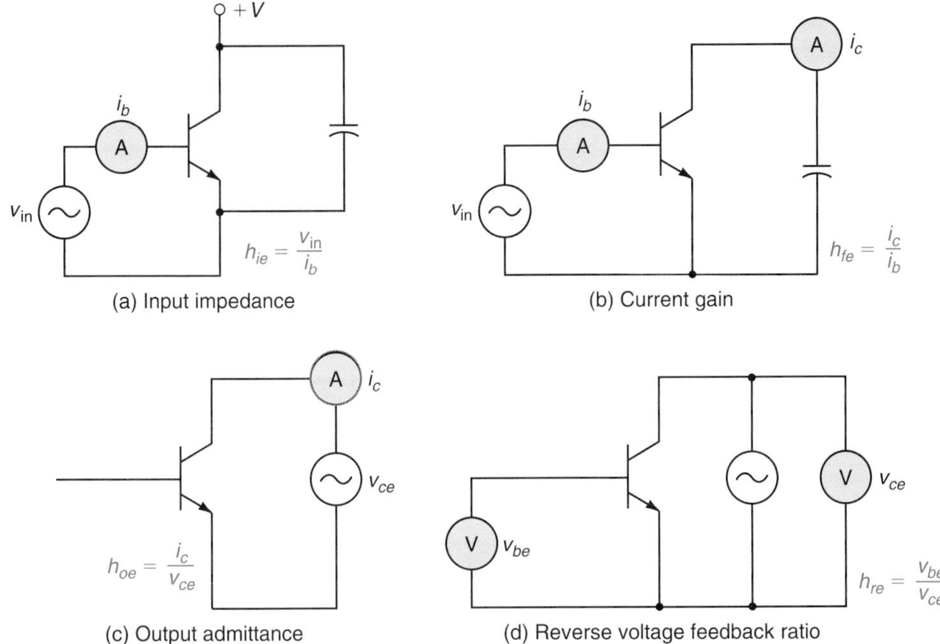

Figure E.1 The measurement of h-parameters.

Current Gain (h_{fe})

The base-to-collector current gain (h_{fe}) is also measured with an ac short connected across the emitter and the collector terminals. Again, this represents a full ac load, so h_{fe} represents the current gain of the transistor under full-load conditions.

The measurement of h_{fe} is illustrated in Figure E.1b. With a signal applied to the base and the output shorted, both the base and collector currents are measured. Then, h_{fe} is determined as

$$h_{fe} = \frac{i_c}{i_b} \quad \text{(output shorted)} \tag{E.2}$$

In this case, it is clear why the output is shorted; *If the output were left open, i_c would equal zero.* Shorting the output gives us a measurable value of i_c that can be reproduced in a practical test. In other words, anyone can achieve the same results simply by shorting the output terminals and applying the same signal to the transistor.

Output Admittance (h_{oe})

The output admittance (h_{oe}) is measured with the input open, as illustrated in Figure E.1c. As shown, a signal is applied across the collector-emitter terminals. Then, with this signal applied, i_c is measured. The value of h_{oe} is then determined as

$$h_{oe} = \frac{i_c}{v_{ce}} \quad \text{(input open)} \tag{E.3}$$

Measuring h_{oe} with the input open makes sense when you consider the effect of shorting the input. If the input to the transistor were shorted, there would be some base current (i_b). Because this current would come from the emitter, i_c would not be at its maximum value. In other words, by not allowing i_b to be generated, i_c is at its absolute maximum value. Therefore, h_{oe} is an accurate measurement of the maximum output admittance.

Reverse Voltage Feedback Ratio (h_{re})

The reverse voltage feedback ratio (h_{re}) indicates the amount of output voltage that is reflected back to the input. This value is measured with the input open.

The measurement of h_{re} is illustrated in Figure E.1d. A signal is applied to the collector-emitter terminals. Then, with the input open, the voltage that is fed back to the base is measured. The value of h_{re} is then determined as

$$h_{re} = \frac{v_{be}}{v_{ce}} \qquad \text{(input open)} \quad \text{(E.4)}$$

Because the voltage at the base terminal is always less than the voltage across the emitter-collector terminals, h_{re} is always less than one. By measuring h_{re} with the input open, you ensure that the voltage fed back to the base is always at its maximum possible value, because maximum voltage is always developed across an open circuit.

Circuit Calculations Involving h-Parameters

Circuit calculations involving h-parameters can be very simple or very complex, depending on the following:

- What you are trying to determine.
- How exact you want your calculations to be.

For our purposes, however, we are interested in only four h-parameter circuit equations:

$$A_i = h_{fe}\left(\frac{Z_{\text{in}}r_C}{h_{ie}R_L}\right) \qquad \text{(E.5)}$$

$$Z_{\text{in(base)}} = h_{ie} \qquad \text{(E.6)}$$

$$r'_e = \frac{h_{ie}}{h_{fe}} \qquad \text{(E.7)}$$

$$A_v = \frac{h_{fe}r_C}{h_{ie}} \qquad \text{(E.8)}$$

Equation (E.5) was introduced in Chapter 20. The only change in the equation as presented here is the substitution of h_{ie} for $Z_{\text{in(base)}}$, as given in equation (E.6). Equation (E.6) is relatively easy to understand if you refer back to the ac equivalent circuit for the common-emitter amplifier shown in Figure 20.15b. Looking at the ac equivalent circuit, it is easy to see that the only impedance between the base of the transistor and the emitter ground connection is the input impedance of the transistor, which is h_{ie}.

Equation (E.7) is derived using equations (20.15) and (E.6). You may recall that equation (20.15) defines $Z_{\text{in(base)}}$ as

$$Z_{\text{in(base)}} = h_{fe}r'_e$$

Substituting equation (E.6) for $Z_{\text{in(base)}}$, we obtain

$$h_{ie} = h_{fe}r'_e$$

Simply rearranging for r'_e gives us equation (E.7).

You may be wondering why we would go to such trouble to find r'_e with h-parameters when we can simply use

$$r'_e = \frac{25 \text{ mV}}{I_E}$$

The fact of the matter is, this equation provides only a rough approximation of r'_e. By using the h-parameter equation, we can obtain a much more accurate value of r'_e and, thus, calculate more closely the value of A_v for a given common-emitter amplifier.

Equation (E.8) is derived as follows:

$$A_v = \frac{r_C}{r'_e} = \frac{1}{r'_e} r_C = \frac{h_{fe}}{h_{ie}} r_C = \frac{h_{fe} r_C}{h_{ie}} \quad \text{(E.8)}$$

It should be noted that the entire subject of h-parameters and their derivations is far more complex than represented here.

Determining h-Parameter Values

The spec sheet for a given transistor will list the values of the device's h-parameters in the electrical characteristics portion of the sheet. This is illustrated in Figure E.2, which shows the electrical characteristics portion of the spec sheet for the Motorola 2N4400–4401 series transistors.

When the minimum and maximum h-parameter values are given, we must determine the geometric average of the two. For example, the geometric averages of h_{fe} and h_{ie} are found using

$$h_{fe(\text{ave})} = \sqrt{h_{fe(\text{min})} \times h_{fe(\text{max})}}$$

and

$$h_{ie(\text{ave})} = \sqrt{h_{ie(\text{min})} \times h_{ie(\text{max})}}$$

A Few More Points

In this section, we have concentrated on h_{ie} and h_{fe}, because these are the two parameters most commonly used for "everyday" circuit analysis. The other two h-parameters, h_{oe} and h_{re}, are used primarily in circuit development applications.

Whenever you need to determine the voltage gain of an amplifier, the required h-parameter values can be obtained easily from the spec sheet of the transistor. If a range of values is given, you can then follow either of two procedures:

- If you want an approximate value of A_v, use the geometric average of the h-parameter values listed.
- If you want the worst-case values of A_v, analyze the amplifier using both the minimum and maximum values of h_{fe} and h_{ie}. By doing this, you will obtain the minimum and maximum values of A_v, which represent the voltage gain limits for the amplifier.

2N4400, 2N4401

Characteristic		Symbol	Min	Max	Unit
Emitter-Base Capacitance (V_{BE} = 0.5 Vdc, I_C = 0, f = 100 kHz)		C_{eb}	—	30	pF
Input Impedance (I_C = 1.0 mAdc, V_{CE} = 10 Vdc, f = 1.0 kHz)	2N4400 2N4401	h_{ie}	0.5 1.0	7.5 15	k Ω
Voltage Feedback Ratio (I_C = 1.0 mAdc, V_{CE} = 10 Vdc, f = 1.0 kHz)		h_{re}	0.1	8.0	$\times 10^{-4}$
Small-Signal Current Gain (I_C = 1.0 mAdc, V_{CE} = 10 Vdc, f = 1.0 kHz)	2N4400 2N4401	h_{fe}	20 40	250 500	— —
Output Admittance (I_C = 1.0 mAdc, V_{CE} = 10 Vdc, f = 1.0 kHz)		h_{oe}	1.0	30	μmhos

Figure E.2 The 2N4400–4401 h-parameters. (Copyright of Semiconductor Components Industries, LLC. Used by permission.)

appendix F

Polar and Rectangular Notations

Vectors

Any value that has both magnitude and an angle (or direction) is referred to as a **vector.** Each of the instantaneous voltages shown in Figure F.1 is a vector. In contrast, *a value*

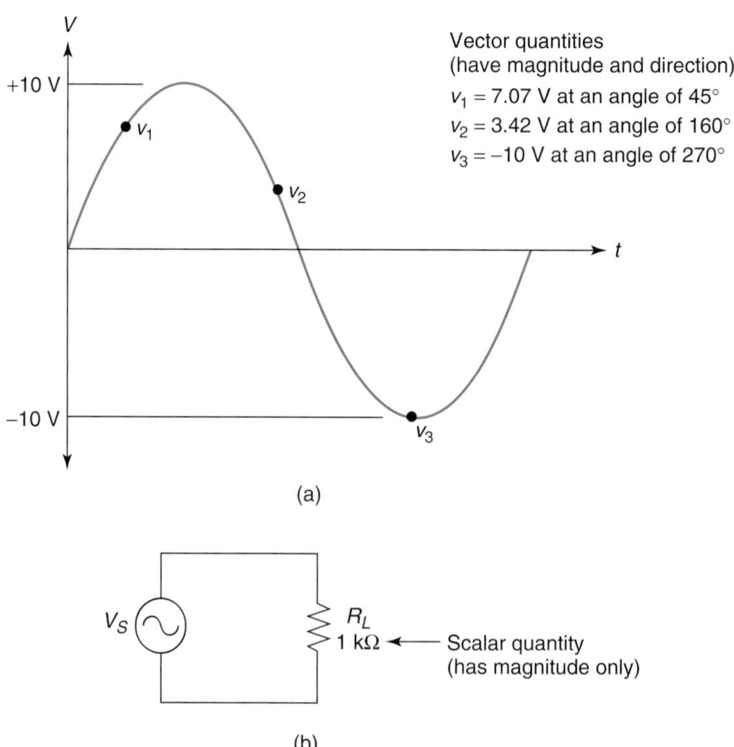

Figure F.1

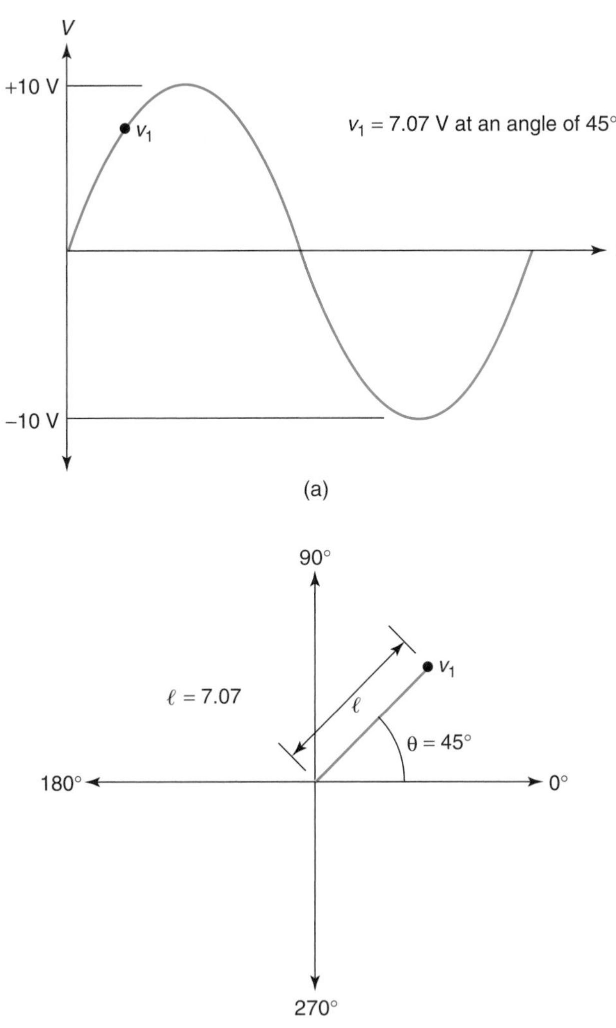

Figure F.2

that has magnitude only is referred to as a **scalar.** The value of a 1 kΩ resistor is a scalar, because resistance has no angle (or direction) associated with it.

Note that *both the magnitude and the direction of a vector must always be specified.* If the angle is omitted, the value is meaningless. (It would be like telling someone how far they need to go to get somewhere, without telling them which direction to take.)

Any vector can be plotted as a *line* on a graph. For example, a line representing the value of v_1 in Figure F.2a would be plotted as shown in Figure F.2b. As the figure demonstrates:

- *The length* (ℓ) *of the line* represents the *magnitude* of v_1.
- *The angle between the line and the positive horizontal* (θ) represents the *phase* of v_1.

This is the standard approach to graphing the values associated with any vector.

Polar Notation

In **polar notation,** the value of a vector is expressed as *a magnitude followed by an angle.* Figure F.3 shows several vectors; the polar notation for each vector appears with the graph.

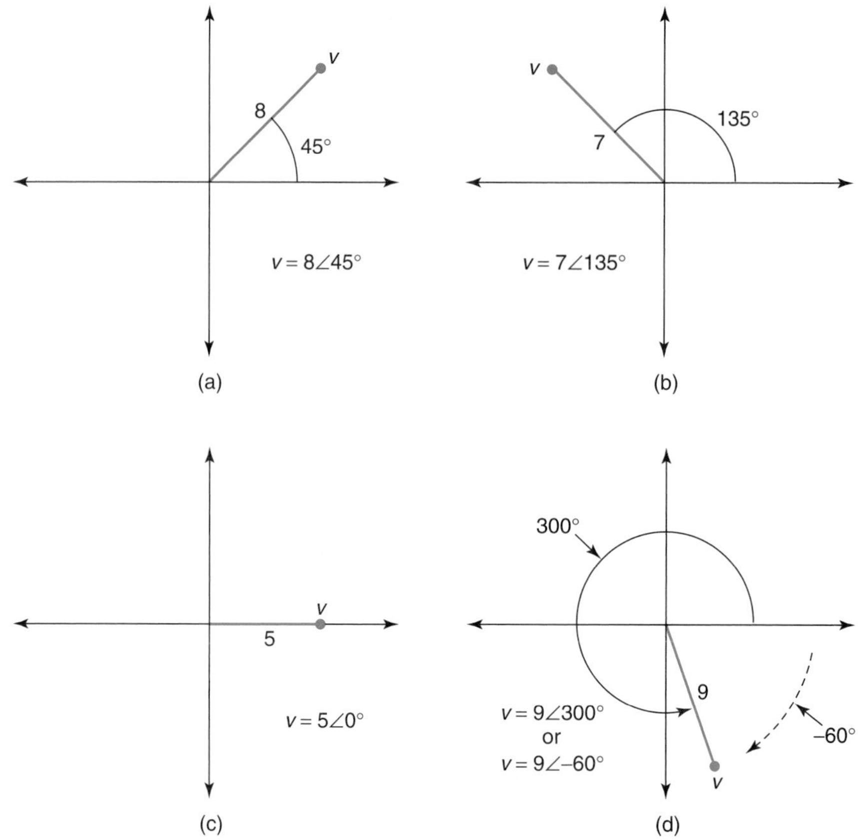

Figure F.3

Figure F.3a shows a vector that is eight units in length at a 45° angle from the positive horizontal (0° line). Therefore, its value is expressed as

$$v = 8\angle 45^\circ$$

(magnitude: 8; angle: 45°)

This is the form of any value written in polar notation.

A few points need to be made regarding the vectors shown in Figure F.3. In Figure F.3c, the vector is *positioned on the positive horizontal of the graph.* When this is the case, the vector is normally expressed as $v = 5\angle 0^\circ$, or $v = 5$. This demonstrates the only case where an angle is *not* specified in polar notation. *When this occurs, the value of the angle is automatically assumed to be 0°.*

Figure F.3d demonstrates that *there are two possible representations for any angle.* A *positive* angle indicates that *the vector angle is measured in a counter-clockwise direction from the positive horizontal.* For the vector shown, this angle is 300°. A *negative* angle indicates that *the vector angle is measured in a clockwise direction from the positive horizontal.* For the vector shown, this angle is −60°. Note that the difference between the angles is 360°, which is the value of one complete rotation.

Coordinate and Rectangular Notations

There are several ways to express the value of any point on a graph. For example, let's say that we want to express the value of the point shown in Figure F.4a. First, we can

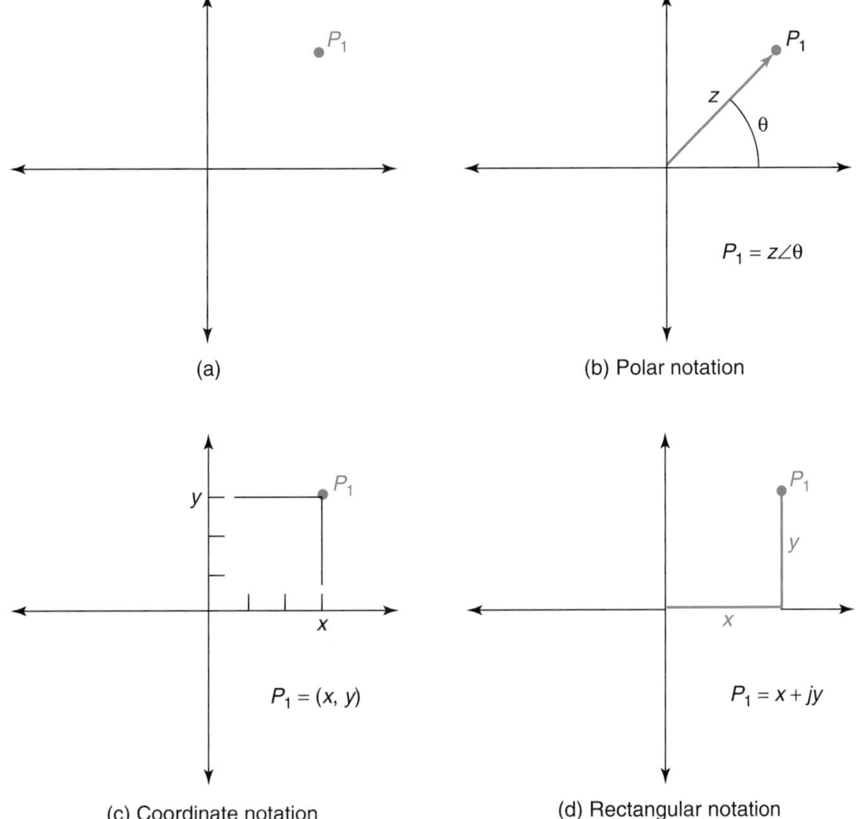

Figure F.4

express the value using *polar notation,* as shown in Figure F.4b. If we assume that Z represents the length of the vector, then the vector quantity is expressed as $Z\angle\theta$.

Another method to express the value of P_1, called **coordinate notation,** is illustrated in Figure F.4c. This method simply *identifies the point in terms of its x-axis and y-axis values,* or *coordinates.* (In essence, this is the approach used to identify points on a map.) If Figure F.4c has values of $x = 6$ and $y = 3$, then coordinate notation expresses the value of the point as

$$P_1 = (6, 3)$$

(x-coordinate: 6; y-coordinate: 3)

Note that every point on the graph has a *unique combination* of coordinates. That is, every point has its own (x, y) combination.

Rectangular notation can be viewed as a variation on coordinate notation. In this case, *the x and y coordinates are written in the form of an expression.* For example, look at the graph shown in Figure F4.d. The x and y values for the points are represented as vectors (rather than coordinates). These vectors are combined in the form of

$$x + jy$$

If we assume (again) that P_1 has values of $x = 6$ and $y = 3$, then rectangular notation expresses the point as $P_1 = 6 + j3$ (For now, we will assume that j in the expression merely identifies the y value. We will define j more accurately later in this section.)

The graph shown in Figure F.5 shows a series of points. Using the x and y values given, we can express each of the points in coordinate and in rectangular notation. The values

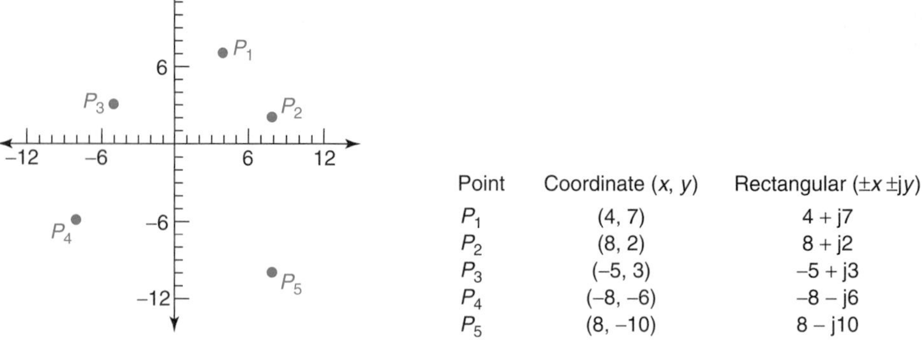

Point	Coordinate (x, y)	Rectangular ($\pm x \pm jy$)
P_1	(4, 7)	4 + j7
P_2	(8, 2)	8 + j2
P_3	(−5, 3)	−5 + j3
P_4	(−8, −6)	−8 − j6
P_5	(8, −10)	8 − j10

Figure F.5

of each point are listed in the figure. If you compare the P_4 and P_5 values with the first three, you'll see that $+x$ and $-y$ values are distinguished by *the sign preceding the j.* That is:

- A *positive y* value is expressed in the form of $+jy$.
- A *negative y* value is expressed in the form of $-jy$.

This is always the case with rectangular notation.

The third column heading in Figure F.5 indicates that every rectangular notation value can be expressed in the form of

$$\pm x \pm jy$$

The *j* that appears in this expression is actually referred to as the ***j*-operator.** As this name implies, *j* represents an operation (just like the signs $+$, $-$, $\times$ and $\div$).

The *j*-operator represents *a vector rotation of 90°.* A *positive j-operator* indicates that *the vector rotates in a positive direction.* A *negative j-operator* indicates that *the vector rotates in a negative direction.* Both of these operations are illustrated in Figure F.6.

When a value is expressed in rectangular notation, *it is being expressed as the sum of two vectors.* For example, consider the point shown in Figure F.7. P_1 is represented as the sum of:

- A vector four units in length along the *positive x*-axis of the graph.
- A vector six units in length at a 90° angle to the first. The direction of this vector is indicated by the $(+j)$ operator in the expression.

One more point: *In rectangular notation, every value is written as the sum of two vectors. Any number written as the sum of two vectors* is referred to as a **complex number.**

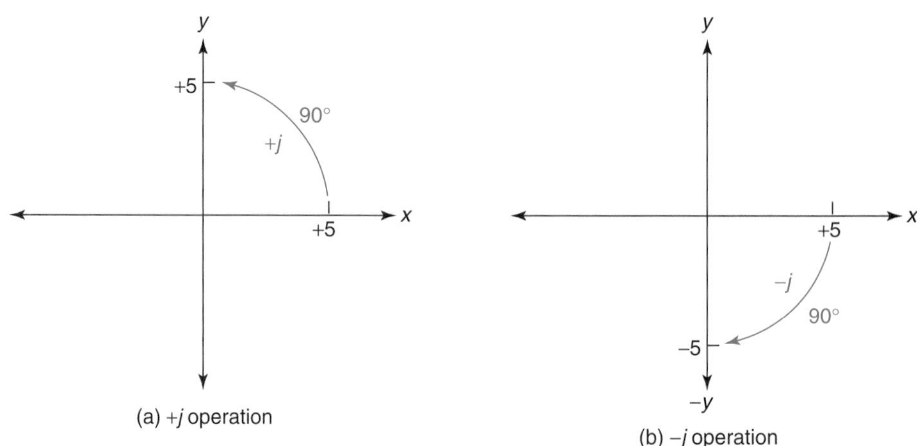

Figure F.6

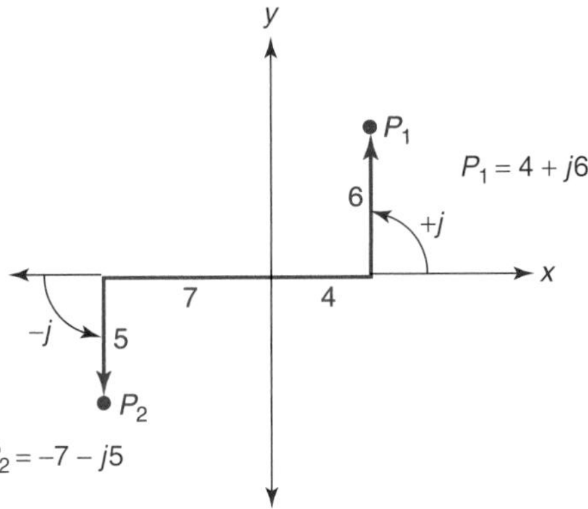

Figure F.7

Vector Notation Conversions

Every point on a graph has both a *polar* representation and a *rectangular* representation. For example, Figure F.8a shows a point (P_1) with both of its representations. Because the polar and rectangular representations shown describe the same point, we can say that

$$z\angle\theta = x + jy$$

for the point.

If we combine the polar and rectangular representations shown in Figure F.8a, we get the *right triangle* shown in Figure F.8b. Because the representations form a right triangle,

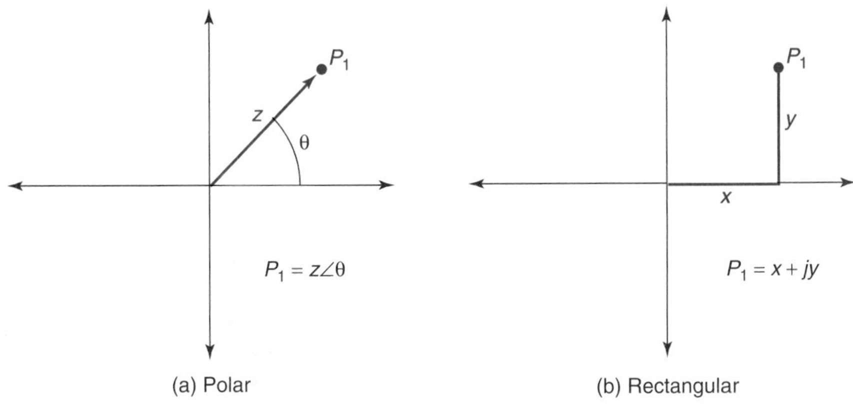

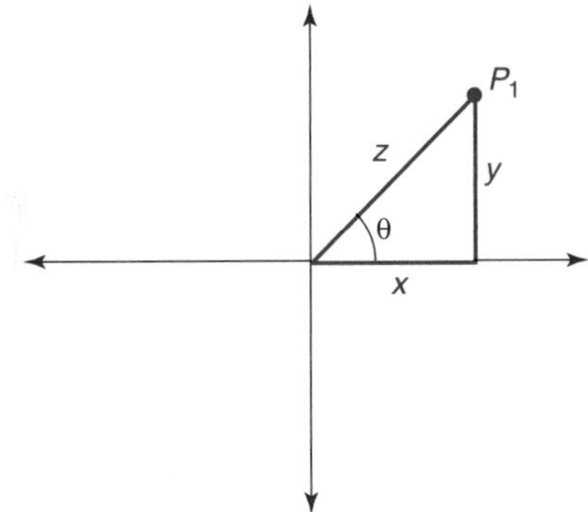

(c) Combined polar and rectangular representations

Figure F.8

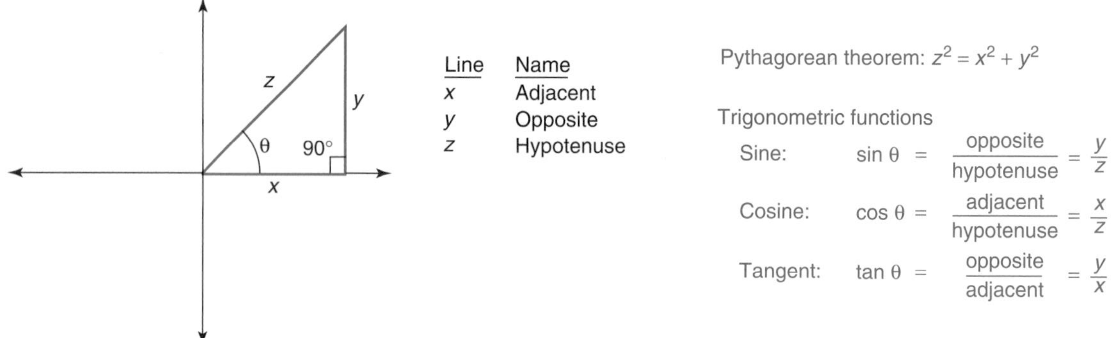

Figure F.9

the relationship between them must adhere to the principles of every right triangle. In other words, the relationships between $(z\angle\theta)$ and $(x + jy)$ are the same as those for any right triangle.

Right Triangle Relationships

Four fundamental relationships are used to describe every right triangle. These relationships are listed in Figure F.9.

The **Pythagorean theorem** describes the relationship between the line lengths in any right triangle. It states that for any right triangle, *the square of the hypotenuse is equal to the sum of the squares of the other two sides.* By formula,

$$z^2 = x^2 + y^2 \tag{F.1}$$

Each **trigonometric function** shown in Figure F.9 serves to *relate angle θ to two sides of the triangle.* If you look closely at the "trig" functions listed in the figure, you'll see that they contain every possible combination of two sides in the triangle. These trig functions are identified as follows:

$$\sin\theta = \frac{y}{z} \tag{F.2}$$

$$\cos\theta = \frac{x}{z} \tag{F.3}$$

and

$$\tan\theta = \frac{y}{x} \tag{F.4}$$

where

x = the side *adjacent* angle θ

y = the side *opposite* angle θ

z = the *hypotenuse* (the side opposite the angle)

Each trig function is nothing more than a *ratio.* For example, sin θ for a given right triangle equals *the ratio of the length of y to the length of z.* The reason that we are interested in the length ratio is demonstrated in Figure F.10. Here, we have two triangles with the same angle θ. For a given angle θ, the ratio of y to z is constant, even though the actual line lengths may change. For the two triangles shown in Figure F.10, $y_1 \neq y_2$, and $z_1 \neq z_2$. However,

$$\frac{y_1}{z_1} = \frac{y_2}{z_2}$$

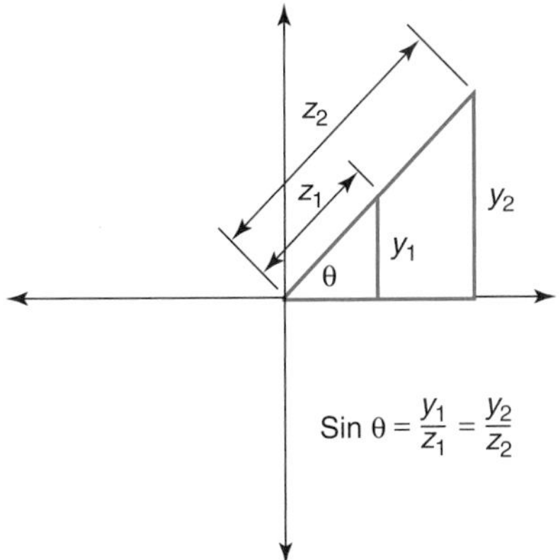

Figure F.10

Because the ratios are constant, *the sine function for any angle is constant,* regardless of the actual line lengths. For example,

$$\sin 30° = 0.5$$

This means that the ratio of y to z equals 0.5 when $\theta = 30°$, regardless of the actual lengths of y and z.

The constant-ratio characteristic of the sine function also holds true for the cosine and tangent functions.

Polar-to-Rectangular Conversion

By transposing equations (F.2) and (F.3), we obtain the relationships needed to convert a *polar* value to *rectangular* form. These relationships are as follows:

$$x = z \cos \theta \tag{F.5}$$

and

$$y = z \sin \theta \tag{F.6}$$

Example F.1 demonstrates the process for converting a vector value from polar to rectangular form.

Example F.1

A vector has a value of $V_1 = 15\angle 60°$. Convert this value to rectangular form.

Solution: The value of x can be found as

$$x = z \cos \theta = (15) \cos (60°) = (15)(0.5) = 7.5$$

and the value of y can be found as

$$y = z \sin \theta = (15) \sin (60°) = (15)(0.866) \cong 13$$

Therefore, $15\angle 60° = 7.5 + j13$.

Rectangular-to-Polar Conversion

Rectangular-to-polar conversion is performed using modified versions of the *Pythagorean theorem* and the *tangent function.* Given a complex number in the form of $(x + jy)$, the length of the vector can be found as

$$z = \sqrt{x^2 + y^2} \tag{F.7}$$

The value found using this equation is the length of $z\angle\theta$. To determine $\angle\theta$, we can use the *inverse-tangent function.* This function can be expressed as

$$\theta = \tan^{-1}\frac{y}{x} \tag{F.8}$$

where

x, y = the components of the complex number
$\tan^{-1}$ = the inverse-tangent function

The inverse-tangent basically reverses the tangent operation. (The tangent of the angle gives us the ratio, so the inverse-tangent of the ratio gives us the angle.) Example F.2 shows how equations (F.7) and (F.8) are used to convert a complex number to polar form.

Example F.2

Convert the result from Example F.1 back to polar form.

Solution: In Example F.1, we calculated a value of $7.5 + j13$. Converting this value back to polar form,

$$z = \sqrt{x^2 + y^2} = \sqrt{(7.5)^2 + (13)^2} = \sqrt{225.25} \cong 15$$

and $\angle\theta$ can be found as

$$\theta = \tan^{-1}\frac{y}{x} = \tan^{-1}\frac{13}{7.5} = \tan^{-1}(1.733) \cong 60°$$

Putting It All Together

Throughout the text, you will be confronted with calculations involving both polar and complex numbers. When working these types of problems, you will find that notation conversions are often used to simplify the process. The relationships used to convert each vector notation to the other are summarized in Figure F.11.

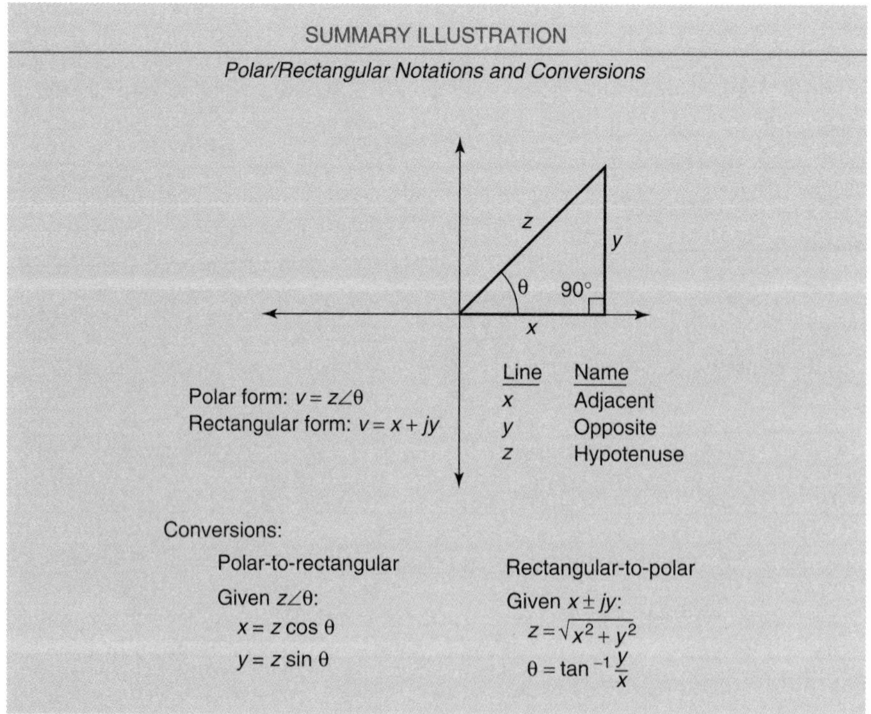

Figure F.11

appendix G

Glossary

3 dB frequency A term commonly used to describe a circuit cutoff frequency, based on the fact that gain drops by 3 dB at that frequency.
3 dB point Another name for the 3 dB frequency.
555 Timer An 8-pin IC designed for use in a variety of switching applications.

AC Beta (β_{ac}) The ratio of ac collector current to ac base current; listed on specification sheets as *small-signal current gain* (h_{fe}).
AC Emitter resistance (r'_e) The dynamic resistance of the transistor emitter; used in voltage gain and impedance calculations.
ac Equivalent circuit The representation of a circuit that shows how the circuit appears to an ac source.
Acceptor atoms Another name for *trivalent* atoms.
Active components Components with values that are controlled (in part) by an external dc power source.
Active filter A tuned op-amp circuit.
Adjustable regulator A regulator whose dc output voltage can be set to any values within set limits.
Admittance (Y) The reciprocal of impedance, measured in Siemens.
Air gap The area of free space (if any) between two magnetic poles.
Alternating current (ac) A term used to describe current that is *bidirectional;* that is, the flow of charge changes direction periodically.
Alternation A term used to describe either the positive or the negative transition of a waveform; also referred to as a *half-cycle*.
American wire gauge (AWG) A system that uses numbers to identify standard wire sizes (cross-sectional areas).
Ampacity From *amp*ere cap*acity*, the rated limit on current for a given wire gauge.
Ampere (A) The basic unit of current; a rate of charge flow equal to one coulomb per second.
Ampere-hours (Ah) A capacity rating for a battery, equal to the product of current (in amperes) and time (in hours). This rating, which is a constant for a given battery, provides a means of determining how long a battery will last at a given load demand.
Ampere-turns (At) The product of the coil current and the number of turns in a coil. Usually represented as NI in equations (N = number of turns, I = coil current).
Amplification The process of strengthening an ac signal.
Amplifier A circuit that can be used to increase the power level of a sine wave or other waveform.
Amplifier input impedance The impedance of an amplifier as seen by its source.
Amplifier output impedance The impedance of an amplifier as seen by its load.
Amplitude The maximum value of a changing voltage or current.
Angular velocity (ω) The speed of rotation for the rotor in a sine-wave generator; the rate at which the phase of a sine wave changes, and therefore, the rate at which its instantaneous values change.
Anode The *p*-type material of a diode.
Anode current interruption A method of driving an SUS or SCR into cutoff by breaking its current path or shorting current around the component.
Apparent power (P_{APP}) The product of alternating voltage and current in a reactive circuit, measured in volt-amperes (VA). The term derives from the fact that only a portion of the value obtained is actually dissipated by the reactive components.
Armstrong oscillator An oscillator that uses a transformer in its feedback network to achieve the required 180° voltage phase shift.
Astable multivibrator A switching circuit that has no stable output state; a square-wave oscillator. Also called a *free-running multivibrator*.

Atom The smallest particle of matter that retains the physical characteristics of an element. In its simplest model, the atom is made of *protons* and *neutrons* (which form the nucleus) and orbiting *electrons*.

Atomic number The number of protons in the nucleus of an atom. Elements are commonly identified by their atomic number.

Attenuation The signal loss caused by the frequency response of a component or a circuit.

Attenuation factor (α_v) The ratio of feedback voltage to output voltage. The value of α_v is always less than one.

Atto- (a) The engineering notation prefix with a value of 10^{-18}.

Autotransformer A special-case transformer made of a single coil with three terminal connections; in essence, a *tapped inductor*.

Average ac power The average power generated over the course of each complete cycle of an ac waveform; on a power graph, the value that falls halfway between 0 W and peak power. Often referred to as *equivalent dc power*.

Average breakdown voltage The voltage that will force an insulator to conduct; rated in kV/cm.

Average forward current (I_0) The maximum allowable value of dc forward current for a diode.

Average frequency (f_{ave}) The frequency that lies halfway between the cutoff frequencies of a bandpass or notch filter.

Averaging amplifier A summing amplifier that provides an output proportional to the average of its input voltage levels.

Back-to-back suppressor A single package containing two transient suppressors that are connected cathode-to-cathode or anode-to-anode.

Balanced A term used to describe the bridge operating state in which the bridge current is 0 A. A bridge is balanced when the voltages on the two sides of the bridge are equal.

Band Another name for an electron orbital shell (in atomic theory) or a range of frequencies (in frequency analysis).

Band-stop filter A filter designed to reject all frequencies between two cutoff frequencies. Also referred to as a *notch filter*.

Bandpass filter A filter designed to pass all frequencies between two cutoff frequencies.

Bandwidth (BW) The frequency spread between the cutoff frequencies of a component or circuit, equal to the difference between those frequencies.

Barkhausen criterion The relationship between the circuit feedback factor (α_v) and voltage gain (A_v) required for proper oscillator operation.

Barrier potential The natural difference of potential that exists across a *pn* junction.

Base One of three bipolar junction transistor (BJT) terminals. The other two are the *collector* and the *emitter*.

Base bias A BJT biasing circuit that consists of a single base resistor connected between the base terminal and V_{CC} and no emitter resistor. Also known as *fixed bias*.

Base cutoff current (I_{BO}) The maximum amount of reverse current through a reverse-biased base-emitter junction.

Base-emitter junction One of two *pn* junctions that make up the bipolar junction transistor (BJT). The other is the *collector-base junction*.

Base-emitter saturation voltage ($V_{BE(sat)}$) The maximum value of V_{BE} for a transistor in saturation.

Battery A component that produces a difference of potential (emf). Batteries convert chemical, thermal, or nuclear energy into electrical energy.

Bel (B) A ratio, expressed as a common (base 10) logarithm; in terms of power gain, the ratio of output power to input power, expressed as the common log of the ratio.

Beta (β) A Greek letter often used to represent the base-to-collector current gain of a bipolar junction transistor (BJT).

Beta curve A curve showing the relationship among beta, collector current, and operating temperature.

Beta-dependent circuit A BJT biasing circuit whose *Q*-point values are affected by changes in the dc beta (h_{FE}) of the transistor.

Beta-independent circuit A BJT biasing circuit whose *Q*-point values are not affected significantly by changes in the dc beta (h_{FE}) of the transistor.

Bias A potential applied to a device to obtain a desired mode of operation.

Biased clamper A clamper that allows a waveform to be shifted above or below a dc reference other than 0 V.

Biased clipper A shunt clipper that uses a dc voltage source to bias the clipping diode.

Bidirectional thyristor A thyristor capable of conducting in two directions.

Biomedical systems Systems designed for use in diagnosing, monitoring, and treating medical problems.

Bipolar junction transistor (BJT) A three-terminal, solid-state device whose output current, voltage, and/or power are normally controlled by its input current.

Bistable multivibrator A switching circuit with two stable output states. Also called a *flip-flop*.

Bode plot A normalized graph that represents frequency response as a change in gain versus operating frequency.

Branch A term used to identify each current path in a parallel circuit.

Branch current A term used to describe the current generated through each branch of a parallel circuit.

Breadboard A physical base used for the construction of experimental circuits.

Breakdown voltage (V_{BR}) The voltage that may force an insulator (such as a capacitor dielectric or reverse-biased *pn* junction) to break down and conduct.

Bridge rectifier A four-diode rectifier used to control the direction of conduction through a load; the most commonly used of the various power-supply rectifier circuits.

Buffer Any circuit used for impedance matching; a switching circuit that does not produce a voltage phase shift.

Bulk resistance The natural resistance of a forward-biased diode.

Butterworth filter An active filter characterized by flat passband response and a 20 dB/decade/pole roll-off rate.

Bypass capacitor A capacitor used to establish an ac ground at a specific point in a circuit.

Capacitance (C) The ability of a component to store energy in the form of a charge, measured in farads (F); the characteristic of a component that opposes any change in voltage.

Capacitance ratio (C_R) The factor by which the capacitance of a varactor varies over a rated range of reverse voltage (V_R) values.

Capacitive filter A power-supply filter that uses one or more capacitors to oppose any variations in load voltage.

Capacitive reactance (X_C) The opposition provided by a capacitor to any sinusoidal current.

Capacitor A component designed to provide a specific measure of capacitance.

Capacity A measure of how long a battery will last at a given output current, measured in ampere-hours (Ah); the amount of charge that a capacitor can store, measured in coulombs per volt, or farads (F).

Carbon-composition resistor A component that uses carbon to provide a desired value of resistance.

Cascade A group of series connected filters (or other circuits). Each circuit in a cascade is referred to as a *stage*.

Cascode amplifier A low-C_{in} amplifier used in high-frequency applications.

Cathode The *n*-type terminal of a diode.

Cathode-ray tube (CRT) A vacuum tube that is used as the video display for many TVs, video games, etc.

Cell A single unit designed to produce electrical energy through *thermal* (chemical) or *optical* (light) means.

Center frequency (f_0) The frequency that equals the geometric average of the cutoff frequencies of a bandpass or notch filter. More accurately referred to as the *geometric center frequency*.

Center-tapped transformer A transformer that has a lead connected to the center of the secondary winding.

Channel The material that connects the source and drain terminals of an FET.

Charge (Q) A force that causes two particles to be attracted to or repelled from each other.

Chebyshev filter An active filter characterized by an irregular pass-band response and high initial roll-off rates.

Choke A low-resistance inductor designed to provide a high reactance within a designated frequency range.

Circuit A group of components that performs a specific function.

Circuit fusing (I^2t) rating The maximum forward surge current capability of an SCR.

Circular-mil (CM) A unit of area found by squaring the diameter (in mils) of a conductor.

Clamper Circuits used to change the dc reference of an ac signal. Also known as *dc restorers*.

Clapp oscillator A Colpitts oscillator with an added capacitor (in series with the feedback inductor) used to reduce the effects of stray capacitance.

Class A amplifier An amplifier with a transistor or FET that conducts throughout the entire input cycle.

Class AB amplifier An amplifier with two transistors, each of which conducts for slightly more than 180° of the input signal cycle.

Class B amplifier An amplifier with two transistors, each of which conducts for 180° of the input signal cycle.

Clipper A circuit used to eliminate a portion of a waveform or limit the input to a circuit to some predetermined value. Also known as a *limiter*.

Closed-loop voltage gain (A_{CL}) The gain of an op-amp with a feedback path; always lower than the open-loop voltage gain (A_{OL}).

Coefficient of coupling (k) A measure of the degree of coupling that takes place between two (or more) coils.

Coercive force The magnetomotive force required to return the flux density within a material to zero.

Collector One of three terminals of a bipolar junction transistor (BJT). The other two are the *base* and the *emitter*.

Collector-base junction One of two *pn* junctions that make up the bipolar junction transistor (BJT). The other is the *base-emitter junction*.

Collector biasing voltage (V_{CC}) A dc supply voltage connected to the collector of a BJT as part of its biasing circuit.

Collector curve A curve that relates the values of I_C, I_B, and V_{CE} for a given transistor.

Collector cutoff current (I_{CO}) The maximum reverse current through the collector of a transistor in cutoff.

Collector-emitter saturation voltage ($V_{CE\,(sat)}$) The maximum V_{CE} for a given transistor when the device is in saturation.

Collector-feedback bias A bias circuit constructed so that V_C has a direct effect on V_B.

Colpitts oscillator An oscillator that uses a pair of tapped capacitors and an inductor to produce a 180° voltage phase shift in the feedback network.

Combination circuit Another name for a series-parallel circuit.

Common-base amplifier A BJT amplifier that provides voltage and power gain.

Common-collector amplifier A BJT amplifier that provides current and power gain. Also called an *emitter follower*.

Common-drain amplifier The FET counterpart of a common-collector amplifier. Also referred to as a *source follower*.

Common-emitter amplifier The only BJT amplifier with a 180° voltage phase shift from input to output; exhibits current, voltage, and power gain.

Common-gate amplifier The FET counterpart of a common-base amplifier.

Common-mode gain (A_{CM}) The gain that a differential amplifier provides to common-mode signals; typically less than one.

Common-mode rejection ratio (CMRR) The ratio of differential gain to common-mode gain for a differential amplifier.

Common-mode signals Identical signals that appear simultaneously at the two inputs of an op-amp.

Common-source amplifier The FET counterpart of a common-emitter amplifier.

Common terminal The terminal of a dc power supply common to two (or more) voltage sources; used as the *reference* point.

Communications systems Systems designed to transmit and/or receive information.

Comparator A circuit used to compare two voltages; typically used to compare a changing voltage to a reference voltage.

Compensating resistor A resistor connected to the noninverting input to an op-amp to compensate for differences in the input currents.

Complementary MOS (CMOS) A logic family made up of MOSFETs.

Complementary-symmetry amplifier A class B circuit configuration using complementary transistors. Also called a *push-pull emitter follower*.

Complementary transistors Two BJTs, one *npn* and one *pnp*, that have nearly identical electrical characteristics and ratings.

Compliance The maximum undistorted peak-to-peak output voltage from an amplifier.

Computer systems Systems designed to store and process information.

Conductance (G) A measure of the ability of a circuit to conduct (pass current), measured in Siemens.

Conduction angle For a phase controller, the portion of the input (in degrees) that is coupled to the load.

Conduction band The energy band outside of the valence band (or shell).

Copper loss Power dissipated by any current-carrying conductor, generally applied to the winding resistance of a coil.

Coulomb (C) The basic unit of charge; equal to the total charge of approximately 6.25×10^{18} electrons.

Coulomb's law A law stating that the magnitude of force between two charges is directly proportional to the product of the charges and inversely proportional to the square of the distance between them.

Counter emf A term used to describe the voltage induced across an inductor by its own changing magnetic field. This induced voltage opposes the change in current that originally generated the changing magnetic field. Also referred to as *kick emf.*

Coupling When energy is transferred between two or more components (or circuits).

Coupling capacitor A capacitor connected between amplifier stages to provide dc isolation between the stages while allowing an ac signal to pass without distortion.

Covalent bonding One means by which atoms are held together. In a covalent bond, each atom shares electrons with surrounding atoms.

Critical rise (*dv/dt*) rating The maximum rate of increase in V_{AK} for an SCR that will not cause false triggering.

Crowbar An SCR circuit used to protect a voltage-sensitive load from excessive dc power supply output voltages.

Crystal-controlled oscillator An oscillator that uses a quartz crystal to produce an extremely stable output frequency.

Current-controlled resistance Any resistance whose value is determined by the magnitude of a controlling current.

Current gain (A_i) The ratio of circuit output current to input current.

Current source A source designed to provide an output current that remains relatively constant over a range of load resistance values.

Current (*I*) The directed flow of charge through a circuit, measured in amperes (A).

Cutoff A BJT operating state in which I_C is nearly zero.

Cutoff frequency (f_C) Any frequency at which the power ratio of a circuit drops to 50% of its maximum value.

Cycle The complete transition through one positive alternation and one negative alternation of a waveform.

Cycle time For a sine wave, the time required for one 360° cycle; for a rectangular wave, the sum of pulse width and space width. Also referred to as *period.*

Damping The fading and loss of oscillations that occurs when $\alpha_v A_v < 1$. See *Barkhausen criterion.*

Dark current The reverse current through a photodiode with no active light input.

Darlington pair Two BJTs connected to provide high current gain and input impedance. The emitter of the input transistor is connected to the base of the output transistor, and the collectors of the two are tied together.

dBm reference The ratio of a power value to 1 mW, expressed as 10 times the common log of the ratio.

dB power gain The ratio of output power to input power, expressed as 10 times the common log of the ratio.

dB voltage gain The ratio of output voltage to input voltage, expressed as 20 times the common log of the ratio.

DC Alpha (α) The ratio of dc collector current to dc emitter current.

DC Amplifier Any amplifier that provides midband gain at 0 Hz.

DC Beta (β_{dc} or h_{FE}) The ratio of dc collector current to dc base current.

DC Load line A graph of all possible combinations of I_C and V_{CE} for a given BJT biasing circuit.

DC offset A term used to describe a dc value (other than zero) that acts as the reference for a sine wave (or other waveform).

DC power supply A circuit that converts ac line voltage to one or more dc operating voltages for a given electronic system; a piece of equipment with dc outputs that can be adjusted to any value within its design limits.

DC reference voltage A dc voltage used to determine the input signal level that will trigger a circuit such as a comparator or Schmitt trigger.

DC restorer A circuit used to change the dc reference of an ac signal. Also called a *clamper.*

Decade A frequency multiplier equal to 10.

Decade box A device that contains a series of potentiometers, each with different resistance ratings. Adjusting the individual pots allows the user to set the overall (i.e., total) resistance to specific values over a wide range.

Decade scale A scale where each increment is 10 times the value of the previous increment.

Decibel (dB) A logarithmic representation of a number; in terms of power, a ratio of output power to input power, expressed as 10 times the common logarithm of the ratio.

Delay time (t_d) The time required for a BJT to come out of cutoff when provided with an input pulse; the time required for I_C to increase to 10% of its maximum value after the input pulse is received.

Depletion layer The area around a *pn* junction that is depleted of free carriers. Also called the *depletion region.*

Depletion-mode operation Using an input voltage to reduce the channel size of an FET.

Depletion MOSFET (D-MOSFET) A MOSFET that can be operated in both the depletion mode and the enhancement mode.

Designator code An IC code that indicates circuit type and operating temperature range.

Diac A two-terminal, three-layer breakover device with forward and reverse characteristics identical to the forward characteristics of the silicon unilateral switch (SUS).

Dielectric The insulating layer between the plates of a capacitor.

Dielectric absorption The tendency of a dielectric to absorb charge.

Differential amplifier A circuit that amplifies the difference between two input voltages; the input circuit of the op-amp, driven by the inverting and noninverting inputs.

Differentiator A circuit whose output is proportional to the rate of change in its input signal.

Diffusion current The current that "wanders" across a *pn* junction as it increases or decreases in size.

Digital circuits Circuits designed to respond to specific alternating dc voltage levels.

Digital multimeter (DMM) A meter that measures voltage, current, and resistance. Many DMMs are capable of making additional measurements, depending on the model.

Diode A two-electrode, solid-state component designed to control the direction of current.

Diode bias A class AB biasing circuit; a variation on the class B amplifier that uses two diodes in place of the resistor(s) between the bases of the two transistors.

Direct current (dc) A term used to describe current that is unidirectional; that is, charge flows in one direction only.

Discrete A term used to describe electronic devices packaged in individual casings (as opposed to integrated circuits).

Distortion Any unwanted change in the shape of a waveform.

Domain theory One possible explanation for the source of magnetism. This theory states that atoms with like magnetic fields join together to form magnetic "domains," with each acting like a magnet.

Donor atoms Another name for *pentavalent atoms*.

Doping The process of adding impurity elements to intrinsic (i.e., pure) semiconductors to improve their conductivity.

Drain The FET counterpart of the transistor (i.e., BJT) collector.

Drain-feedback bias The MOSFET counterpart of collector-feedback bias.

Driver A circuit used to couple a low-current output to a relatively high-current device.

Dual-gate MOSFET A MOSFET constructed with two gates to reduce gate input capacitance.

Dual in-line package (DIP) switches SPST switches grouped in a single case.

Dual-polarity power supply A dc supply that provides both positive and negative dc output voltages.

Dual-trace oscilloscope An oscilloscope with two independent traces that can be used to compare two waveforms. Each trace has its own set of vertical controls.

Dual-tracking regulator A voltage regulator that provides equal +dc and −dc output voltages.

Duty cycle The ratio of pulse width to cycle time, usually given as a percentage.

Dynamic values Values that change as a direct result of normal circuit operation. Dynamic values are measured under specified conditions.

Eddy current A circular flow of charge generated in the core of a transformer by a changing magnetic field. The term *eddy* is normally used to describe anything that flows in a circular motion (such as air, water, charge, etc.).

Effective value Another name for an *rms value*. It is derived from the fact that an rms value provides the same heating *effect* as its dc equivalent.

Efficiency (η) The ratio of output power to input power, typically given as a percentage; for an amplifier, the ratio of ac load power to amplifier dc input power.

Electrodes The terminals of a battery.

Electrolyte The chemical in a battery that interacts with the terminals, producing a difference of potential between them; a chemical used to produce the relatively high capacitance of an electrolytic capacitor.

Electromagnetic wave A waveform that consists of perpendicular electric and magnetic fields.

Electron-hole pair A free electron and its matching valence band hole.

Electronics The field dealing with the design, manufacturing, installation, maintenance, and repair of electronic equipment.

Electronic tuning A term used to describe the use of a varactor to control the tuning of a circuit.

Electron-volt (eV) The energy absorbed by an electron when subjected to a 1 V difference of potential.

Element A substance that cannot be broken down into a combination of simpler substances.

Emitter One of the three terminals of a bipolar junction transistor (BJT). The other two are the *base* and the *collector*.

Emitter bias A bias circuit that consists of a split power supply and a grounded base resistor.

Emitter-feedback bias A BJT biasing circuit designed so that the emitter voltage (V_E) directly affects the value of the base voltage (V_B).

Emitter follower See *common-collector amplifier*.

Energy gap The difference in the energy levels of two electron orbital shells.

Enhancement-mode operation The mode of operation where an input voltage increases the channel size of a MOSFET.

Enhancement MOSFET (E-MOSFET) A MOSFET that is capable of enhancement-mode operation only. See also *depletion MOSFET*.

Engineering notation A form of scientific notation that uses standard prefixes to designate specific ranges of values.

Epicap Another name for a *varactor*.

Exa- (E) The engineering notation prefix with a value of 10^{18}.

Exponential curve A curve whose rate of change is a function of a variable exponent.

Extrinsic Another word for *impure*.

Fall time (t_f) The time required for a BJT to make the transition from saturation to cutoff; the time required for I_C to decrease from 90% to 10% of its maximum value.

False triggering When a noise signal triggers an SCR (or other thyristor) into conduction.

Farad (F) The unit of measure of capacity. One farad is a capacity of one coulomb per volt.

Feedback bias A term used to describe a circuit that "feeds" a portion of the output voltage or current back to the input to control the circuit operation.

Feedback path A signal path from the output of an amplifier back to its input.

Fempto- (f) The engineering notation prefix with a value of 10^{-15}.

Field-effect transistor (FET) A three-terminal, voltage-controlled device used in amplification and switching applications; a component whose output characteristics are controlled by its input voltage.

Filter A circuit designed to pass a specific range of frequencies while rejecting others; in power supplies, a circuit that reduces the variations in the output from the rectifier.

Fixed bias See *base bias*.

Fixed-negative regulator A voltage regulator that provides a set negative dc output voltage.

Fixed-positive regulator A voltage regulator that provides a set positive dc output voltage.

Flip-flop A bistable multivibrator; a circuit with two stable output states.

Flux density The amount of flux per unit area; an indicator of magnetic field strength.

Forced commutation Driving an SUS into cutoff by applying a reverse voltage to the device.

Forward bias A potential used to reduce the resistance of a *pn* junction.

Forward blocking region The off-state (i.e., nonconducting) region of operation for a thyristor.

Forward breakover current ($I_{BR(F)}$) The value of I_F at the point where breakover occurs for a given thyristor.

Forward breakover voltage ($V_{BR(F)}$) The forward voltage that forces an SUS into conduction; the value of V_F at the point where breakover occurs.

Forward operating region The on-state (i.e., conducting) region of device operation.

Forward voltage (V_F) The voltage across a forward-biased *pn* junction.

Free-running multivibrator See *astable multivibrator*.

Frequency (f) A measure of the rate at which the cycles of a waveform repeat themselves, measured in cycles per second. See *Hertz*.

Frequency counter A piece of test equipment that measures the number of cycles per second of a given waveform.

Frequency response Any changes that occur in a circuit as a result of a change in operating frequency.

Frequency response curve A curve showing the relationship between operating frequency and gain.

Full load A load that draws maximum current from the source. A full load occurs when load resistance is at its *minimum* value.

Full-cycle average The average of all the instantaneous values of a waveform voltage or current throughout one complete cycle. Sometimes referred to as *full-wave average*.

Full-wave average Another name for the *full-cycle average* of a waveform.

Full-wave rectifier A circuit that converts negative alternations to positive alternations (or vice versa) to provide a single-polarity (pulsating dc) output. A full-wave rectifier provides two output pulses for each full-cycle input.

Full-wave voltage doubler A circuit that produces a dc output that is twice the peak value of a sinusoidal input. The term *full-wave* indicates that the entire input cycle is used to produce the dc output.

Fundamental frequency The lowest frequency in a harmonic series. The lowest frequency generated by an oscillator or other signal source.

Fuse A component designed to open automatically if its current exceeds a specified value.

Gain A multiplier that exists between the input and the output of an amplifier (or filter).

Gain-stabilized amplifier An amplifier that uses a partially bypassed emitter (or source) resistance to stabilize its voltage gain against variations in device characteristics. Also referred to as a *swamped amplifier*.

Galvanometer A current meter that indicates both the magnitude and direction of a low-value current.

Gate The FET counterpart of the BJT base.

Gate bias The FET counterpart of BJT *base bias*.

Gate nontrigger voltage (V_{GD}) The maximum gate voltage that can be applied to an SCR without triggering the device into conduction.

Gate reverse current (I_{GSS}) The maximum amount of gate current generated when an FET is reverse biased.

Gate-source cutoff voltage ($V_{GS(\text{off})}$) The value of V_{GS} that reduces JFET drain current (I_D) to zero.

Gate trigger voltage (V_{GT}) The value of gate-cathode voltage (V_{GK}) that triggers an SCR into conduction.

General amplifier model An amplifier model that represents the circuit as one having gain, input impedance, and output impedance; used to predict the output voltage for any combination of input voltages.

Germanium One of the semiconductor materials used to produce solid-state components. Another is *silicon*.

Giga- (G) The engineering notation prefix with a value of 10^9.

Ground-loop path The conducting path between the ground connections throughout a circuit.

Half-cycle A term used to describe a negative or positive alternation of a waveform.

Half-cycle average The average of all the instantaneous values of a waveform voltage or current throughout one alternation of a cycle. Sometimes referred to as *half-wave average*.

Half-wave average Another name for the *half-cycle average* of a waveform.

Half-wave rectifier A circuit that eliminates the negative or positive alternations of a waveform to provide a single-polarity (pulsating dc) output. The half-wave rectifier produces one output pulse for each full-cycle input.

Half-wave voltage doubler A circuit that produces a dc output approximately equal to twice its peak input voltage.

Harmonic A whole-number multiple of a given frequency.

Harmonic series A group of frequencies that are harmonics of the same fundamental frequency.

Hartley oscillator An oscillator that uses a tapped inductor (pair) and a parallel capacitor in its feedback network to produce the 180° voltage phase shift required for oscillations.

Heat sink A large metallic object that helps to cool components by increasing their effective surface area.

Henry (H) The unit of measure of inductance. One henry is the amount of inductance that produces a 1 V difference of potential when current changes at a rate of 1 A/s.

Hertz (Hz) The unit of measure for frequency, equal to one cycle per second.

High-pass filter A filter designed to pass all frequencies above its cutoff frequency.

Holding current (I_H) The minimum forward current (I_F) required to maintain conduction through a thyristor.

Hole A gap in a covalent bond.

h-parameters See *hybrid parameters*.

Hybrid parameters (h-parameters) Transistor specifications used to describe and analyze the operation of small-signal circuits, measured under full-load or no-load conditions.

Hysteresis The time lag between the removal of a magnetizing force and the drop in flux density.

Hysteresis curve A curve representing flux density (B) as a function of the field intensity (H) of a magnetizing force.

Hysteresis loss A loss of energy that occurs whenever a magnetic field of one polarity must overcome any residual magnetism from the previous polarity.

Ideal voltage source A voltage source that maintains a constant output voltage regardless of the resistance of its load(s). A voltage souce with 0 Ω internal resistance.

Impedance (Z) The total opposition to current in an ac circuit, consisting of resistance and/or reactance; the geometric sum of resistance and reactance, expressed in ohms.

Impedance mismatch A term used to describe a situation where source impedance does not equal load impedance.

Inductance (L) The ability of a component with a changing current to induce a voltage across itself or a nearby circuit by generating a changing magnetic field, measured in henries (H). The ability of a component to store energy in an electromagnetic field. The property of a component that opposes any change in current.

Induction The process of producing an artificial magnet.

Inductive filter A power supply filter that uses one or more inductors to oppose any variations in load current. See also *capacitive filter*.

Inductive reactance (X_L) The opposition that an inductor presents to a changing current, measured in ohms.

Inductor A component designed to provide a specific amount of inductance; a component that opposes a change in *current*.

Industrial systems Systems designed for use in manufacturing environments.

Input bias current The average quiescent value of dc biasing current drawn by the signal inputs of an op-amp.

Input offset current Any difference in the noninverting and inverting input currents to a differential amplifier caused by differences in the input transistor beta ratings.

Input offset voltage Any difference between the base-emitter voltages in a differential amplifier that produces an output offset voltage when the signal inputs are grounded.

Input/output voltage differential The maximum allowable difference between V_{in} and V_{out} for a voltage regulator.

Instantaneous value The magnitude of a voltage or current at a specified point in time.

Integrated circuit (IC) A semiconductor that contains groups of components or circuits housed in a single package.

Integrated rectifier A power supply rectifier constructed on a single piece of semiconductor material and housed in a single package.

Integrated resistors Microminiature resistors made using semiconductor materials other than carbon. Because they are so small, many can be housed in a single casing.

Integrator A circuit whose output is proportional to the area of the input waveform.

Interbase resistance The resistance between the base terminals of a UJT, measured with the emitter-base junction reverse biased.

Intrinsic Another word for *pure*.

Intrinsic standoff ratio (η) The ratio of emitter–base 1 resistance to the total interbase resistance of a UJT.

Inverter A basic switching circuit that produces a 180° voltage phase shift; a logic-level converter.

Inverting amplifier A basic op-amp circuit that produces a 180° voltage phase shift from input to output. The op-amp equivalent of the common-emitter and common-source amplifiers.

Inverting input The op-amp input that shifts any input signal by 180°. When a signal is applied to this input, the input and output signals are 180° out of phase.

Isolation resistance (R_{ISO}) The total resistance between the input and output pins of an optoisolator.

Isolation source voltage (V_{ISO}) The voltage that, if applied across the input and output pins of an optoisolator, will destroy the device.

Isolation transformer A transformer with an equal number of primary and secondary windings; one designed for equal primary and secondary voltages; used to provide dc isolation between two electronic circuits or systems.

***j*-operator** A mathematical operation resulting in a vector rotation of 90°.

Junction capacitance The capacitance produced by a reverse biased *pn* junction. This capacitance decreases when the amount of reverse bias increases, and vice versa.

Kick emf See *counter emf.*

Kilo- (k) The engineering notation prefix with a value of 10^3.

Kilowatt-hour (kWh) A practical unit for measuring energy, equal to the amount of energy used by a 1000 W (1 kW) device that is run for one hour.

Kirchhoff's current law A law stating that the algebraic sum of the currents entering and leaving a point must equal zero.

Kirchhoff's voltage law A law stating that the algebraic sum of the voltages around a closed loop must equal zero.

Knee voltage (V_K) The point on a current-versus-voltage graph where current suddenly increases or decreases.

LDMOS A high-power MOSFET that uses a narrow channel and a heavily doped *n*-type region to obtain high drain current and low drain resistance.

Leakage current A low-value current that may be generated through a capacitor (or reversed-biased *pn* junction) indicating that the resistance of the dielectric (or junction) is not infinite.

Left-hand rule A memory aid that helps you to remember the relationship between the direction of *electron flow* and the direction of the resulting magnetic field.

Lenz's law A law stating that the voltage induced across an inductor always opposes its source (in keeping with the law of conservation of energy).

Level detector Another name for a comparator used to compare an input voltage to a fixed dc reference voltage.

Lifetime The time between electron-hole pair generation and recombination.

Light Electromagnetic radiation that falls within a specified range of frequencies.

Light-activated SCR (LASCR) An SCR that is triggered into conduction by light at its optical input (gate).

Light current The reverse current through a photodiode with an active light input.

Light detector Any optoelectronic device that responds to light.

Light emitter Any optoelectronic device that generates light.

Light-emitting diode (LED) A diode designed to produce one or more colors of light under specific circumstances.

Light intensity The amount of light per unit area received by a given photodetector. Also called *irradiance*.

Limiter See *Clipper*.

Linear A term used to describe any value that changes at a constant rate.

Linear IC voltage regulator A device used to hold the output voltage from a dc power supply relatively constant over a specified range of line voltages and load current demands.

Linear power supply A power supply whose output is a linear function of its input; any power supply whose output is controlled by a linear voltage regulator.

Line regulation The ability of a voltage regulator to maintain a stable dc output, despite variations in line voltage; a rating that indicates the change in regulator output voltage that will occur per unit change in line voltage.

Load analysis A method of predicting the effect that a change in load has on the output from a given circuit.

Load The part of a circuit that absorbs (i.e., uses) power.

Loaded-*Q* The quality (*Q*) of a filter when a load is connected to the circuit.

Load regulation The ability of a regulator to maintain a constant dc output despite changes in load current demand; a rating that indicates the change in regulator output voltage that will occur per unit change in load current.

Logarithmic scale A scale with frequency intervals that increase from each increment to the next at a geometric rate.

Logic family A group of digital circuits with nearly identical characteristics.

Logic levels A term used to describe the two dc levels used to represent information in a digital circuit or system.

Lower trigger point (LTP) A Schmitt trigger reference voltage. When a negative-going input reaches the LTP, the Schmitt trigger output changes state.

Low-pass filter A filter designed to pass all frequencies below its cutoff frequency.

Magnetic field The area of space surrounding a magnet that contains magnetic flux.

Magnetic flux (ϕ) A term used to describe the lines of force produced by a given magnet.

Magnetic force The force that a magnet exerts on the objects around it.

Magnetic induction Using an external magnetic force to align the magnetic domains within a given material, thereby magnetizing the material.

Magnetic shielding Insulating an instrument (or material) from magnetic flux by diverting the lines of force around that instrument (or material).

Magnetomotive force (mmf) The force that produces magnetic flux; the magnetic equivalent of electromotive force.

Majority carrier In a given semiconductor material, the charge carrier that is introduced by doping; the electrons in a pentavalent material; the holes in a trivalent material. See also *Minority carrier.*

Matched transistors Transistors that have the same operating characteristics.

Matter Anything that has weight and occupies space.

Maximum power transfer theorem A theorem stating that maximum power transfer from a voltage source to a variable load occurs when the load and source resistances are equal.

Maximum ratings A spec sheet table which lists device parameters that must not be exceeded under any circumstances.

Maximum reverse standoff voltage (V_{RWM}) The maximum allowable peak or dc reverse voltage that will not turn on a transient suppressor.

Maximum zener current (I_{ZM}) A zener diode rating indicating the maximum allowable value of device reverse current.

Mega- (M) The engineering notation prefix with a value of 10^6.

Metal-film resistor A precision resistor with a low temperature coefficient.

Metal-oxide-semiconductor FET (MOSFET) An FET designed to operate in the enhancement mode. A D-MOSFET can operate in the depletion and enhancement modes, while an E-MOSFET is limited to enhancement mode operation.

Micro- (μ) The engineering notation prefix with a value of 10^{-6}.

Microweber (μWb) A practical unit of measure of magnetic flux, equal to 100 Maxwells (i.e., lines of force).

Midpoint bias Having a Q-point centered on the dc load line; a biasing condition in which amplifier output current and voltage are half their maximum values.

Mil A unit of length equal to one-thousandth of an inch.

Milli- (m) The engineering notation prefix with a value of 10^{-3}.

Minimum load current For a voltage regulator, the value of I_L below which regulation is lost.

Minority carrier In a given semiconductor material, the charge carrier that is not introduced by doping; the holes in a pentavalent material; the electrons in a trivalent material. See also *Majority carrier.*

Model A representation of a component or circuit.

Momentary switch A switch that makes or breaks the connection between its terminals only while the contact button is pressed.

Monostable multivibrator A switching circuit with one stable output state. Also called a *one-shot.*

MOSFET An FET that can be operated in the enhancement mode. See also *Metal-oxide-semiconductor FET.*

Multicolor LED An LED that emits different colors when the polarity of its biasing voltage changes.

Multiple-feedback filter A bandpass (or band-stop) filter that contains a single op-amp with two feedback paths; one resistive, and one capacitive.

Multisource circuit A circuit with more than one voltage and/or current source.

Multivibrators Switching circuits designed to have zero, one, or two stable output states.

Mutual inductance (L_M) The process by which the magnetic field generated by one inductor induces a voltage across another inductor in close proximity; the inductance generated between two or more inductors in close proximity.

Nano- (n) The engineering notation prefix with a value of 10^{-9}.

Negative clamper A circuit that shifts an entire input signal below a dc reference voltage.

Negative feedback A type of feedback in which the feedback signal is 180° out of phase with the amplifier input signal.

Negative resistance Any device whose current and voltage vary inversely.

Negative resistance oscillator Any oscillator whose operation is based on a negative resistance device, such as a tunnel diode or UJT.

Negative resistance region A term used to describe any region of device operation where current and voltage vary inversely.

Negative temperature coefficient An indicator of the effect of temperature on resistance for a given material. A negative temperature coefficient indicates that resistance varies inversely with temperature.

Neutron In the simplest model of the atom, the particle in the nucleus that provides mass but no charge.

Noise Any unwanted interference with a transmitted signal.

No-load output voltage (V_{NL}) The output from a voltage source when its load terminals are open; the maximum possible output from a voltage source.

Node A term used to describe any point that connects two or more current paths.

Nominal zener voltage The rated value of V_Z for a given zener diode.

Noninverting amplifier An op-amp circuit that does not produce a voltage phase shift from input to output.

Nonlinear distortion A type of distortion caused by driving the base-emitter junction of a transistor into its nonlinear operating region.

***n*-Order harmonics** The frequencies (other than the fundamental) in a harmonic series.

Normally closed (NC) switch A push-button switch that normally *makes* a connection between its terminals. Activating the switch *breaks* the connection.

Normally open (NO) switch A push-button switch that normally has a *broken* path between its terminals. Activating the switch *makes* the connection.

Norton's theorem A theorem stating that any resistive circuit or network, no matter how complex, can be represented as a current source in parallel with a source resistance.

Notch filter Another name for a *band-stop filter.*

***npn* Transistor** A BJT with *n*-type emitter and collector materials and a *p*-type base.

***n*-Stage filter** A term commonly used to refer to a cascaded circuit, where *n* is the number of stages.

***n*-Type material** A semiconductor that has added pentavalent impurities.

Nucleus The core of an atom. In its simplest model, the nucleus of an atom consists of protons and neutrons. (Hydrogen has no neutrons.)

Octave A frequency multiplier equal to two.

Octave scale A scale where the value of each increment is two times the value of the previous increment.

Offset null Input pins on an op-amp that are used to eliminate output offset voltage.

Ohm (Ω) The unit of measure for resistance; the amount of resistance that limits current to 1 A when 1 V is applied.

Ohm's law A law which states that current is directly proportional to voltage and inversely proportional to resistance.

One-shot Another name for the *monostable multivibrator*.

Opaque A term used to describe anything that blocks light.

Open circuit A physical break in a conduction path.

Open-loop voltage gain (A_{OL}) The gain of an op-amp with no feedback path.

Operational amplifier (op-amp) A high-gain dc amplifier that has high input impedance and low output impedance.

Optocoupler A device that uses light to couple a signal from its input (i.e., a photoemitter) to its output (i.e., a photodetector). Also called an *optoisolator*.

Optoelectronic devices Devices that are controlled by and/or emit light.

Optointerruptor An IC optocoupler designed to allow an external object to block the light path between its photoemitter and its photodetector.

Optoisolator See *Optocoupler*.

Oscillator An ac signal generator; a dc-to-ac converter.

Oscillator stability A measure of an oscillator's ability to maintain constant output amplitude and frequency.

Oscilloscope A piece of test equipment that provides a visual representation of a voltage waveform for various amplitude and time measurements.

Output current rating A transformer (or device) rating that indicates the maximum allowable value of secondary (or device output) current.

Output offset voltage A voltage that may appear at the output of an op-amp, caused by an imbalance in the differential amplifier.

Output short-circuit current The maximum output current for an op-amp, measured under shorted-load conditions.

Output voltage rating A transformer rating that indicates the ac output voltage from a transformer when provided with a 120 V_{ac} input.

Overtone A whole-number multiple of a given frequency. Another word for *harmonic*.

Overtone mode A term used to describe the operation of a filter (or other circuit) that is tuned to a harmonic of a given frequency.

Parallel circuit A circuit that provides more than one current path between any two points.

Parallel *RC* circuit A circuit that contains one or more resistors in parallel with one or more capacitors. Each branch in a parallel *RC* circuit contains only one component (either resistive or capacitive).

Parallel *RL* circuit A circuit that contains one or more resistors in parallel with one or more inductors. Each branch in a parallel *RL* circuit contains only one component (either resistive or inductive).

Parallel-equivalent circuit The equivalent of a series-parallel circuit, consisting entirely of components connected in parallel.

Parameter A limit.

Pass band The range of frequencies passed by a tuned circuit.

Passive components A term generally used to describe components with values that are not controlled by an external dc power source.

Pass-transistor regulator A regulator that uses a series transistor to maintain a constant load voltage.

Peak current (I_P) One of two current values that define the limits of a negative resistance region of operation for a device such as a tunnel diode or UJT. The other is the *valley current* (I_V).

Peak inverse voltage (PIV) The maximum reverse voltage that a diode is exposed to in a given circuit.

Peak power dissipation (P_{PK}) rating A rating that indicates the maximum power a transient suppressor can dissipate.

Peak repetitive reverse voltage (V_{RRM}) The maximum amount of reverse bias that a *pn* junction can safely handle.

Peak value The maximum value reached by either alternation of a waveform.

Peak voltage (V_P) One of two voltage values that define the limits of a negative resistance region of operation for a device such as a tunnel diode or UJT. The other is the *valley voltage* (V_V).

Peak-to-peak value The difference between the peak values of a waveform.

Pentavalent element An element with five valence electrons.

Permeability (μ) A measure of the ease with which lines of magnetic force are established within a given material.

Permittivity A measure of the ease with which lines of electrical force are established within a given material.

Peta- (P) The engineering notation prefix with a value of 10^{15}.

Phase The position of a given point on a waveform relative to the start of the waveform, usually expressed in degrees.

Phase angle (θ) The phase difference between two or more waveforms.

Phase controller A circuit used to control the conduction phase angle through a load and, thus, the average load voltage.

Phase-shift oscillator An oscillator that uses three *RC* circuits in its feedback network to produce a 180° phase shift.

Phasor A vector with a phase angle that can change under certain circumstances; a vector that can have a rotational value.

Photo-Darlington A phototransistor with a Darlington-pair output.

Photodiode A diode whose reverse conduction is light-intensity controlled.

Phototransistor A three-terminal photodetector whose collector current is controlled by the intensity of the light at its optical input (base).

Pico- (p) The engineering notation prefix with a value of 10^{-12}.

Piezoelectric effect The tendency of a crystal to vibrate at a fixed frequency when subjected to an electric field.

Pinch-off voltage (V_P) With V_{GS} held constant, the value of V_{DS} that generates maximum drain current (I_D).

Plates The conductive surfaces of a capacitor.

***pnp* Transistor** A BJT with *p*-type emitter and collector terminals and an *n*-type base.

Polar notation A method of expressing the value of a vector as a magnitude followed by an angle.

Polarity dots Dots placed in the symbol of a transformer that indicate the phase relationship between the transformer input and output voltages.

Pole The moving contact in a mechanical switch; each *RC* circuit in an active filter.

Poles The points where magnetic lines of force leave (and return to) a magnet, referred to as the *north-seeking pole* (*N*) and the *south-seeking pole* (*S*).

Positive clamper A circuit that shifts an entire input signal above a dc reference voltage.

Positive feedback A type of feedback in which the feedback signal is in phase with the circuit input signal; the type of feedback used to produce oscillations. Also referred to as *regenerative feedback.*

Positive temperature coefficient An indicator of the effect of temperature on resistance for a given material. A positive temperature coefficient indicates that resistance varies directly with temperature.

Potentiometer A three-terminal resistor whose value can be adjusted (within limits) by the user; a variable resistor.

Power (*P*) The amount of energy used per unit time, measured in watts. One watt of power is equal to a rate of *one joule per second.*

Power factor (PF) The ratio of resistive power to apparent power. The power factor of a sine wave is equal to cos θ.

Power gain (A_p) The ratio of output power to input power for a component or circuit.

Power rating A measure of a component's ability to dissipate heat, measured in watts.

Power supply A group of circuits that combine to convert ac to dc.

Power supply rejection ratio (PSRR) The ratio of a change in op-amp output voltage to a change in supply voltage.

Power switch The power transistor in a switching regulator that controls conduction through the load.

Primary The transformer coil that serves as the component input.

Primary cell A cell that cannot be recharged. Often called *dry cells,* because they contain dry electrolytes.

Primary impedance The total opposition to current in the primary of a transformer.

Primary resistance The winding resistance of a transformer's primary coil.

Propagation delay The time required for a signal to get from the input of a component or circuit to the output, measured at the 50% points on the two waveforms.

Proton In the simplest model of the atom, the particle in the nucleus that is the source of positive charge.

Prototype The first working model of a circuit or system.

***p*-Type material** Silicon or germanium that has trivalent impurities.

Pulse width (PW) A term generally used to describe the positive alternation of a rectangular waveform.

Pulse-width modulation Modulating a rectangular waveform so that its pulse width varies while its overall cycle time remains unchanged. Essentially, it varies the duty cycle of the modulated waveform.

Push-pull emitter follower See *Complementary-symmetry amplifier.*

***Q* (tank circuit)** A measure of the quality of a tank circuit; the ratio of energy stored in the circuit to energy lost per cycle.

***Q*-point** A point on a load line that indicates the values of the output current and voltage for an amplifier without an active input signal.

***Q*-point shift** A condition in which a change in operating temperature (or device substitution) indirectly causes a change in the *Q*-point values of output current and voltage.

Quality (*Q*) A numeric value that indicates how close an inductor (or capacitor) comes to having the power characteristics of its *ideal* model. Sometimes referred to as the *figure of merit* of the component. The ratio of filter center frequency to bandwidth.

Quiescent At rest.

Radian The angle formed at the center of a circle by two radii and an arc of equal length. (*Note: Radii* is the plural form of *radius.*)

Ramp Another name for a sawtooth waveform; also used to describe the actual transition from one peak to the other.

***RC* time constant (τ)** The time required for the capacitor voltage in an *RC* switching circuit to increase (or decrease) by 63.2% of its maximum possible ΔV.

Recombination A term used to describe the process of a free electron giving up its energy and returning to a valence band orbit.

Rectangular notation A variation on coordinate notation where the *x* and *y* coordinates are written in the form of $x + jy$.

Rectangular waves Waveforms that alternate between two dc levels.

Rectifier A circuit that converts ac to pulsating dc.

Regenerative feedback Another name for *positive feedback.*

Regulation A rating that indicates the maximum change in regulator output voltage that can occur when input voltage and load current vary over their entire rated ranges.

Relative conductivity The conductivity of a material as compared to the conductivity of copper, expressed as a percentage.

Relative permeability (μ_r) The ratio of a material's permeability to that of free space.

Relative permittivity The ratio of a material's permittivity to that of a vacuum.

Reluctance ($\mathcal{R}$) The opposition that a material presents to magnetic lines of force; the magnetic equivalent of *resistance.*

Residual flux The magnetism that remains in a material after a magnetizing force has been removed.

Residual flux density The flux density that remains in a given material after any magnetizing force is removed.

Resistance (R) A static opposition to current, measured in ohms.

Resistive-capacitive (*RC*) circuit A circuit that contains any combination of resistors and capacitors.

Resistive-inductive (*RL*) circuit A circuit that contains any combination of resistors and inductors.

Resistivity The resistance of a specified volume of an element or compound.

Resistor A component designed to provide a specific amount of resistance; a component used to limit current in a circuit.

Resolution For a potentiometer, the change in resistance per degree of control shaft rotation. High-resolution pots (i.e., those with low ΔR per degree ratings) allow more exact control over the adjusted value of the potentiometer.

Retentivity The ability of a material to retain its magnetic characteristics after a magnetizing force has been removed.

Reverse bias A potential that causes a *pn* junction to have maximum resistance.

Reverse blocking region The thyristor off-state (i.e., nonconducting) region of operation between 0 V and its reverse breakdown (or breakover) voltage rating.

Reverse breakdown voltage (V_{BR}) The minimum reverse voltage that causes a device to break down and conduct in the reverse direction.

Reverse current (I_R) The current through a reverse-biased *pn* junction.

Reverse voltage (V_R) The voltage across a reverse-biased *pn* junction.

RF amplifier Radio frequency amplifier; the input circuit of a communications receiver.

Right-hand rule A memory aid that helps you to remember the relationship between the direction of *conventional current* and the direction of the resulting magnetic field.

Ring magnet A magnet that forms a closed loop and, therefore, has no identifiable poles. The center of a ring magnet has no magnetic flux.

Ripple rejection ratio The ratio of regulator input ripple to its output ripple.

Ripple voltage The remaining variation in the output from a filter or voltage multiplier.

Rise time (t_r) The time required for a BJT to go from cutoff to saturation; the time required for I_C to increase from 10% to 90% of its maximum value.

Roll-off A term used to describe the decrease in gain that occurs over a range of frequencies when a filter is operated outside of its pass band.

Roll-off rate The rate at which gain rolls off, normally given in dB per octave or dB per decade.

Root-mean-square (rms) value The value of voltage or current that, when used in the appropriate power equation, gives you the average ac power (equivalent dc power) of the waveform.

Rotary switch A switch with one or more poles and a series of throws.

Rotor The rotating part in a motor or generator.

Rowland's law A law stating that magnetic flux is directly proportional to magnetomotive force (mmf) and inversely proportional to reluctance.

Saturation The BJT operating region in which I_C reaches its maximum value.

Sawtooth A term used to describe a waveform that constantly changes at a linear rate.

Scalar A value that has only magnitude (as opposed to one that has magnitude and direction).

Schmitt trigger A voltage-level detection circuit.

Scientific notation A system that represents a given number as the product of a number and a whole-number power of 10. The number is always written so that the most significant digit falls in the units position.

Secondary cell A cell that can be recharged. Often called *wet cells*, because they contain liquid electrolytes.

Secondary impedance The total opposition to current in the secondary circuit of a transformer; generally assumed to equal the load resistance.

Secondary The transformer coil that serves as the component output.

Selector guide A publication that groups components according to their critical ratings and/or parameters.

Self-bias An FET biasing circuit that uses a source resistor to establish a negative V_{GS}.

Self-inductance The ability of an inductor with a changing current to generate a changing magnetic field that, in turn, induces a voltage across the component.

Semiconductors A group of components that began to replace vacuum tubes during the early 1950s; elements that are neither insulators nor conductors. Semiconductors are smaller, cheaper, more efficient, and more rugged than their vacuum-tube counterparts.

Sensitivity A rating that indicates the response of a photodetector to a specified light intensity.

Series-aiding inductors A two-inductor series connection where the magnetic poles are such that the flux produced by the coils adds together. This results in a total inductance that is greater than the sum of the individual component values.

Series-aiding voltage sources Sources connected so that the current supplied by the individual sources are in the same direction.

Series circuit A circuit that contains only one current path.

Series clipper A clipper that is in series with its load.

Series-equivalent circuit The equivalent of a series-parallel circuit, consisting entirely of components connected in series.

Series feedback regulator A series regulator that uses an error-detection circuit to improve the line and load regulation characteristics of other pass-transistor regulators.

Series-opposing voltage sources Sources connected so that the current supplied by the individual sources are in opposition to each other.

Series-opposing inductors A two-inductor series connection where the flux generated by each coil opposes the flux generated by the other. This results in a total inductance that is less than the sum of the individual component values.

Series-parallel circuit A circuit that contains both series and parallel elements. Also referred to as a *combination circuit*.

Series regulator A voltage regulator that is in series with its load.

Shell A term used to describe the orbital paths of electrons.

Short circuit An extremely low-resistance path between two points that normally does not exist.

Shorted-gate drain current (I_{DSS}) The maximum possible value of I_D for a given JFET. Also called *zero-gate-voltage drain current*.

Shunt A term used to describe a component or circuit connected between a signal path and ground.

Shunt clipper A clipper that is in parallel with its load. The circuit provides a waveform output when the diode is reverse biased (i.e., not conducting).

Shunt feedback regulator A circuit that uses an error detector to control the conduction of a shunt transistor.

Shunt regulator A voltage regulator that is in series with its load.

Siemens (S) The unit of measure for conductance. One siemen equals one volt per ampere.

Signal A term that is commonly used in reference to a waveform.

Silicon One of the semiconductor materials commonly used to produce solid-state devices. Another is *germanium*.

Silicon-controlled rectifier (SCR) A three-terminal device very similar in construction and operation to a silicon unilateral switch (SUS). A third terminal, called the *gate*, provides an additional means of triggering the component.

Silicon unilateral switch (SUS) A two-terminal, four-layer device that can be triggered into conduction by applying a specified forward voltage across its terminals.

Simultaneous equations Two or more equations used to describe relationships that occur at the same time; any group of concurrent equations with a common solution.

Slew rate The maximum rate at which op-amp output voltage can change, usually measured in V/μs.

Slope The rate at which a value changes, measured as a change per unit time.

Snubber network An *RC* circuit that is connected between an SCR anode and cathode to eliminate false triggering.

Solder A high-conductivity compound that has a relatively low melting point; used to affix components to each other and/or PC boards.

Source The part of a circuit that supplies power. The FET counterpart of the BJT emitter.

Source follower The common-drain amplifier; the FET counterpart of the BJT emitter follower.

Space width (SW) A term generally used to describe the negative alternation of a rectangular waveform.

Specification sheet A listing of all the important parameters and operating characteristics of a device, circuit, or system.

Spectral response A measure of a photodetector's response to a change in input wavelength. Spectral response is measured in terms of device sensitivity.

Speed-up capacitor A capacitor used in the base circuit of a BJT to help the device switch more rapidly between saturation and cutoff.

Square wave A special-case rectangular wave with equal pulse width and space width. The duty cycle of any square wave is 50%.

Stage A term used to describe each filter (or amplifier) in a cascade.

Static values Values that are constant. Values that do not change as a result of normal circuit operation.

Step-down regulator A voltage regulator whose output voltage is less than its input voltage.

Step-down transformer A transformer with fewer secondary windings than primary windings, resulting in a secondary voltage that is less than the primary voltage.

Step-up regulator A switching regulator whose dc output voltage is greater than its dc input voltage.

Step-up transformer A transformer with fewer primary windings than secondary windings, resulting in a secondary voltage that is greater than the primary voltage.

Stop band The range of frequencies outside of an amplifier's pass band.

Storage time (t_s) The time required for a BJT to come out of saturation; the time required for I_C to fall to 90% of its maximum value after an input has been received.

Substrate The foundation material of a MOSFET.

Subtractor A summing amplifier that provides an output proportional to the difference between two inputs.

Summing amplifier A circuit that produces an output voltage proportional to the sum of its inputs.

Superposition theorem A theorem stating that the response of a circuit to more than one source can be determined by analyzing the circuit's response to each source (alone) and combining the results.

Surface-mount component (SMC) A term used to describe a type of IC package designed to be mounted on the surface of a PC board. SMCs are smaller than their DIP counterparts.

Surge current The high initial current in a power supply; any current caused by a nonrepetitive, high-voltage condition.

Susceptance (*B*) The reciprocal of reactance, measured in Siemens.

Swamped amplifier See *Gain-stabilized amplifier*.

Switch A device that allows you to make or break the connection between two or more points in a circuit.

Switching circuit A term generally used to describe any circuit with a square (or rectangular) wave input.

Switching power supply A power supply that contains a switching regulator.

Switching voltage regulator A voltage regulator that alternates between being fully on and fully off, resulting in very high efficiency.

Symmetrical waveform A waveform with cycles that are made up of identical halves. Waveforms can be symmetrical in time and/or amplitude.

Taper A measure of the rate at which the resistance of a potentiometer changes as the control shaft is rotated between its extremes. Tapers are designated as either *linear* or *nonlinear*.

Telecommunications The area that deals with the transmission of data between two or more locations.

Tera- (T) The engineering notation prefix with a value of 10^{12}.

Terminals Another name for the *leads* on a component.

Theorem A relationship between a group of variables that has never been disproved.

Thermal contact Placing two or more components in physical contact with each other (or a common surface) so that their operating temperatures are equal.

Thevenin resistance (R_{TH}) The resistance measured across the output terminals of a circuit with the load removed. To measure R_{th}, the voltage source must be removed and replaced with a wire.

Thevenin voltage (V_{th}) The voltage measured across the output terminals of a circuit with the load removed.

Threshold voltage ($V_{GS(th)}$) The value of V_{GS} that generates a channel through an E-MOSFET, beginning device conduction.

Throw The nonmoving contact in a mechanical switch.

Thyristors Devices designed specifically for high-power switching applications. Also called *breakover devices*.

Time base The oscilloscope control used to set the period of time represented by the space between adjacent major divisions along the horizontal axis.

Time constant (τ) A time interval on any universal rise (or decay) curve that is constant and independent of magnitude and operating frequency; the time required for the current (or voltage) in a switching circuit to increase (or decrease) by 63.2% of its maximum possible ΔI (or ΔV).

Tolerance The range of values for a resistor, given as a percentage of its nominal (i.e., rated) value.

Toroid An inductor with a doughnut-shaped core.

Traces The conductors that connect components on a printed circuit board.

Transconductance (g_m) A ratio of a change in drain current (I_D) to a change in V_{GS}, measured in microsiemens (μS) or micromhos (μmhos).

Transconductance curve A plot of all possible combinations of V_{GS} and I_D for a field-effect transistor.

Transducer Any device that converts energy from one form to another.

Transformer A two-coil component that uses electromagnetic energy to pass an ac signal from its input to its output while providing dc isolation between the two.

Transient An abrupt current or voltage spike; a surge.